Selected Radioactive Isotopes

Naturally occurring radioactive isotopes are designated by a mass number in blue (although some are also manufactured). Letter m indicates an isomer of another isotope of the same mass number. Half-lives follow in parentheses, where s, min, h, d, and y stand respectively for seconds, minutes, hours, days, and years. The table includes mainly the longer-lived radioactive isotopes; many others have been prepared. Isotopes known to be radioactive but with half-lives exceeding 10^{12} y have not been included. Symbols describing the principal mode (or modes) of decay are as follows (these processes are generally accompanied by gamma radiation):

- α alpha particle emission
- β^- beta particle (electron) emission
- β^+ positron emission
- EC orbital electron capture
- IT isomeric transition from upper to lower isomeric state
- SF spontaneous fission

Group	Atomic No.	Symbol	Name	Atomic weight	Oxidation states	Col. 1	Col. 2	Col. 3	Electron configuration
18/VIII	2	He	Helium	4.00260		4.216	0.95 at 26 atm	0.1785†	$1s^2$
13/IIIB	5	B	Boron	10.811	3	4275	2365	2.34	$1s^2 2s^2 p^1$
14/IVB	6	C	Carbon	12.011	±4,2	5100*	3825*	2.26	$1s^2 2s^2 p^2$
15/VB	7	N	Nitrogen	14.0067	±3,5,4,2	77.344	63.15	1.251†	$1s^2 2s^2 p^3$
16/VIB	8	O	Oxygen	15.9994	-2	90.188	54.8	1.429†	$1s^2 2s^2 p^4$
17/VIIB	9	F	Fluorine	18.99840	-1	85.0	53.55	1.696†	$1s^2 2s^2 p^5$
18/VIII	10	Ne	Neon	20.1797		27.10	24.55	0.900†	$1s^2 2s^2 p^6$
13/IIIB	13	Al	Aluminum	26.98154	3	2740	933.5	2.70	$[Ne]3s^2 p^1$
14/IVB	14	Si	Silicon	28.0855	4,2	2630	1683	2.33	$[Ne]3s^2 p^2$
15/VB	15	P	Phosphorus	30.97376	±3,5,4	553	317.3	1.82	$[Ne]3s^2 p^3$
16/VIB	16	S	Sulfur	32.066	±2,4,6	717.82	392.2	2.07	$[Ne]3s^2 p^4$
17/VIIB	17	Cl	Chlorine	35.4527	±1,3,5,7	239.18	172.17	3.214†	$[Ne]3s^2 p^5$
18/VIII	18	Ar	Argon	39.948		87.45	83.95	1.784†	$[Ne]3s^2 p^6$
10	28	Ni	Nickel	58.6934	2,3	3005	1726	8.90	$[Ar]3d^8 4s^2$
11/IB	29	Cu	Copper	63.546	2,1	2840	1356.6	8.96	$[Ar]3d^{10} 4s^1$
12/IIB	30	Zn	Zinc	65.39	2	1180	692.73	7.13	$[Ar]3d^{10} 4s^2$
13/IIIB	31	Ga	Gallium	69.723	3	2478	302.92	5.91	$[Ar]3d^{10} 4s^2 p^1$
14/IVB	32	Ge	Germanium	72.61	4	3107	1211.5	5.32	$[Ar]3d^{10} 4s^2 p^2$
15/VB	33	As	Arsenic	74.9216	±3,5	876 (subl.)	1090 (28 atm.)	5.78	$[Ar]3d^{10} 4s^2 p^3$
16/VIB	34	Se	Selenium	78.96	-2,4,6	958	494	4.79	$[Ar]3d^{10} 4s^2 p^4$
17/VIIB	35	Br	Bromine	79.904	±1,5,7	331.85	265.95	3.12	$[Ar]3d^{10} 4s^2 p^5$
18/VIII	36	Kr	Krypton	83.80	0,2	120.85	116	3.75†	$[Ar]3d^{10} 4s^2 p^6$
10	46	Pd	Palladium	106.42	2,4	3240	1825	12.0	$[Kr]4d^{10}$
11/IB	47	Ag	Silver	107.868	1	2436	1235.08	10.50	$[Kr]4d^{10} 5s^1$
12/IIB	48	Cd	Cadmium	112.41	2	1040	594.26	8.65	$[Kr]4d^{10} 5s^2$
13/IIIB	49	In	Indium	114.82	3	2350	429.78	7.31	$[Kr]4d^{10} 5s^2 p^1$
14/IVB	50	Sn	Tin	118.710	4,2	2876	505.12	7.31	$[Kr]4d^{10} 5s^2 p^2$
15/VB	51	Sb	Antimony	121.760	±3,5	1860	903.91	6.69	$[Kr]4d^{10} 5s^2 p^3$
16/VIB	52	Te	Tellurium	127.60	-2,4,6	1261	722.72	6.24	$[Kr]4d^{10} 5s^2 p^4$
17/VIIB	53	I	Iodine	126.9045	±1,5,7	457.5	386.7	4.93	$[Kr]4d^{10} 5s^2 p^5$
18/VIII	54	Xe	Xenon	131.29	0,2,4,6	165.1	161.39	5.90†	$[Kr]4d^{10} 5s^2 p^6$
10	78	Pt	Platinum	195.08	2,4	4100	2042.1	21.45	$[Xe]4f^{14} 5d^9 6s^1$
11/IB	79	Au	Gold	196.9665	3,1	3130	1337.58	19.3	$[Xe]4f^{14} 5d^{10} 6s^1$
12/IIB	80	Hg	Mercury	200.59	2,1	629.88	234.31	13.55	$[Xe]4f^{14} 5d^{10} 6s^2$
13/IIIB	81	Tl	Thallium	204.383	3,1	1746	577	11.85	$[Xe]4f^{14} 5d^{10} 6s^2 p^1$
14/IVB	82	Pb	Lead	207.2	4,2	2023	600.65	11.35	$[Xe]4f^{14} 5d^{10} 6s^2 p^2$
15/VB	83	Bi	Bismuth	208.9804	3,5	1837	544.59	9.75	$[Xe]4f^{14} 5d^{10} 6s^2 p^3$
16/VIB	84	Po	Polonium	(209)	4,2,6	—	527	9.3	$[Xe]4f^{14} 5d^{10} 6s^2 p^4$
17/VIIB	85	At	Astatine	(210)	±1,3,5,7	610*	575	—	$[Xe]4f^{14} 5d^{10} 6s^2 p^5$
18/VIII	86	Rn	Radon	(222)	0,2	211.4	202	9.73	$[Xe]4f^{14} 5d^{10} 6s^2 p^6$
10	110	Uun	(Ununnilium)	(269)		—	—	—	$[Rn]5f^{14} 6d^8 7s^{1*}$
11/IB	111	Uuu	(Unununium)	(272)		—	—	—	$[Rn]5f^{14} 6d^9 7s^{1*}$
12/IIB	112	Uub	(Ununbium)	(277)		—	—	—	$[Rn]5f^{14} 6d^{10} 7s^{2*}$
13/IIIB	113	Uut	(Ununtrium)						
14/IVB	114	Uuq	(Ununquadium)	(285)		—	—	—	$[Rn]5f^{14} 6d^{10} 7s^2 7p^{2*}$
15/VB	115	Uup	(Ununpentium)						
16/VIB	116	Uuh	(Ununhexium)	(289)		—	—	—	$[Rn]5f^{14} 6d^{10} 7s^2 7p^{4*}$
17/VIIB	117	Uus	(Ununseptium)						
18/VIII	118	Uuo	(Ununoctium)	(293)		—	—	—	$[Rn]5f^{14} 6d^{10} 7s^2 7p^{6*}$
	64	Gd	Gadolinium	157.25		3545	1585	7.90	$[Xe]4f^7 5d^1 6s^2$
	65	Tb	Terbium	158.9253		3500	1629	8.23	$[Xe]4f^9 6s^2$
	66	Dy	Dysprosium	162.50		2840	1685	8.55	$[Xe]4f^{10} 6s^2$
	67	Ho	Holmium	164.9303		2968	1747	8.80	$[Xe]4f^{11} 6s^2$
	68	Er	Erbium	167.26		3140	1802	9.07	$[Xe]4f^{12} 6s^2$
	69	Tm	Thulium	168.9342		2223	1818	9.32	$[Xe]4f^{13} 6s^2$
	70	Yb	Ytterbium	173.04		1469	1092	6.97	$[Xe]4f^{14} 6s^2$
	71	Lu	Lutetium	174.967		3668	1936	9.84	$[Xe]4f^{14} 5d^1 6s^2$
	96	Cm	Curium	(247)		—	1620	13.5	$[Rn]5f^7 6d^1 7s^2$
	97	Bk	Berkelium	(247)		—	—	14*	$[Rn]5f^9 7s^2$
	98	Cf	Californium	(251)		—	1170*	—	$[Rn]5f^{10} 7s^2$
	99	Es	Einsteinium	(252)		—	1130*	—	$[Rn]5f^{11} 7s^2$
	100	Fm	Fermium	(257)		—	1800*	—	$[Rn]5f^{12} 7s^2$
	101	Md	Mendelevium	(258)		—	1100*	—	$[Rn]5f^{13} 7s^2$
	102	No	Nobelium	(259)		—	1100*	—	$[Rn]5f^{14} 7s^2$
	103	Lr	Lawrencium	(262)		—	1900*	—	$[Rn]5f^{14} 7s^2 7p^*$

The A& B subgroup designations, applicable to elements in rows 4, 5, 6 and 7, are those recommended by the International Union of Pure and Applied Chemistry.

Side 1

Catalog Number WLS-18806

P.O. Box 5229, Buffalo Grove, IL 60089-5229
1-800-727-4368 FAX 1-800-676-2540

ENGINEERING MATERIALS TECHNOLOGY
Structures, Processing, Properties, and Selection

Fifth Edition

JAMES A. JACOBS
Professor of Technology
Norfolk State University

THOMAS F. KILDUFF
Professor Emeritus
Thomas Nelson Community College

Upper Saddle River, New Jersey
Columbus, Ohio

Library of Congress Cataloging-in-Publication Data
Jacobs, James A.
Engineering materials technology : structures, processing, properties, and selection / James A. Jacobs, Thomas F. Kilduff.—5th ed.
p. cm.
Includes bibliographical references and index.
ISBN 0-13-048185-8
1. Materials. I. Kilduff, Thomas F. II. Title.

TA403.J26 2005
620.1'1—dc22

2003064977

Editor in Chief: Stephen Helba
Executive Editor: Debbie Yarnell
Associate Editor: Kimberly Yehle
Production Editor: Louise N. Sette
Production Supervision: Carlisle Publishers Services
Copyeditor: Pamela Rockwell
Design Coordinator: Diane Ernsberger
Cover Designer: Bryan Huber
Production Manager: Matt Ottenweller
Marketing Manager: Jimmy Stephens

This book was set in Times Roman and Univers by Carlisle Communications, Ltd. It was printed and bound by Courier-Kendallville. The cover was printed by Coral Graphics.

Pearson Education Ltd.
Pearson Education Singapore Pte. Ltd.
Pearson Education Canada, Ltd.
Pearson Education—Japan
Pearson Education Australia Pty. Limited
Pearson Education North Asia Ltd.
Pearson Educación de Mexico, S.A. de C.V.
Pearson Education Malaysia Pte. Ltd.

10 9 8 7 6 5 4 3 2 1
ISBN 0–13–048185–8

As authors we are indebted to the many people whose constructive suggestions and cooperation helped bring this book to completion. However, most of all, we dedicate this book to our wives, Martha and Virginia, and daughters, Sherri, Tammi, Jeanene, Liz, and Suzy, to whom we owe a major debt for their understanding, unfailing support, optimism, and tangible support as we toiled for many years.

Contents

Preface

This fifth edition is based on input from students, faculty, and specialists in the field whose valuable feedback, for those of you who used earlier editions, resulted in considerable format changes—namely breaking larger modules into smaller modules and grouping them into units. We know the book is used in a variety of courses and hope the redesign improves flexibility.

This edition continues to cover basic concepts in materials science, engineering, and technology dealing with traditional as well as advanced materials. We added coverage of surface science and engineering to reflect new technologies resulting from microtechnology, improved characterization and analysis instruments, and the need for improved quality and reliability. By beginning with the materials cycle for life-cycle analysis, readers gain a perspective on the connectivity of the many topics that comprise materials science and engineering.

Awareness of new technologies like nanotechnology, biologically inspired and morphing materials has been creeping into popular culture, as seen in movies like *Spiderman* and *Terminator*. This edition includes introductory concepts for nanotechnology, morphing, and smart materials because they have become areas of considerable research, with some applications already showing up in consumer products. Materials are so pervasive that they impact any new technology, ranging from fuel-cell vehicles to computers, and become central to current events such as the collapse of the World Trade Center and the tragedy of NASA's shuttle *Columbia*. Alex King's statement in the *MRS Bulletin* captures this perspective: "Materials science and engineering forms the exclusive knowledge pathway through which fundamental discoveries are turned into useful products." This book uses news events, case studies, and end-of-module assessments to promote comprehension and retention. The featured internet addresses also facilitate links to websites for current developments, further explanations, and illustrations.

A major challenge in writing a revision centers on adding appropriate new information and retaining key content while keeping the book length reasonable. We moved some assessment items to the Instructor's Manual *and* provided websites addresses for sources of the latest data on material properties, applications, and comparisons.

Our first edition was written for students in engineering and industrial technology programs. Since then, we have learned that many other people found our text refreshing and useful to them, including professors, secondary school teachers, and industrial trainers with the responsibility of teaching materials science and technology subjects. We continually encourage readers to become "materials observers" and point out the relevance of this important subject in everyday products and systems. Another effective teaching tool for learning, in evidence since the original text, is the use of integrated common themes: (1) relationships among structure, properties, processing, and applications of materials, (2) the importance of considering the total materials cycle in materials synthesis, selection, processing, and economics, (3) the need for "green" design and manufacturing of systems and products for a sustainable environment, (4) new and/or improved technologies that influence many aspects of engineering materials technology such as micro/nanotechnology, recycling, surface engineering, smart materials, intelligent structures, and biomimetics, (5) encouraging the reader to explore the many resources and databases outside of this book, including the Internet, and (6) the role of standard practices in all aspects of design, testing, processing, manufacturing, selection, and applications of materials.

To present the fundamentals of materials science on a level appropriate to readers who have not completed formal courses in chemistry, physics, or mathematics beyond algebra is uppermost in our minds. As a consequence, terminology is defined and the maximum number of graphics is used, such as equations, graphs, tables, and micrographs. Illustrative problems requiring calculations are numerous whenever a law or principle can be expressed as an equation. Emphasis is consistently placed on the proper use of units in solving equations using both the English and International System (SI) of units. Substitution of units as well as numbers in equations is stressed as essential for arriving at a correct answer. Problems in materials science, in general, require a degree of creative analysis of concepts coupled with research into databases in order to solve them.

ORGANIZATION

The reorganization of this edition aims for improved flexibility and involves grouping modules into units that open with Pause & Ponder sections to present case studies or applications of materials developments. These are intended to spark the reader's interest and provide a reason for learning the content. The self-contained modules consist of

- Module Outcomes that present what the reader should be able to do after studying the module
- Solved problems and numerous illustrations and examples representing a full range of products, systems, and technologies that support concepts covered within the module
- Website addresses to provide further information, illustrations, and animations
- Applications & Alternatives sections that reinforce the concepts presented
- Self-Assessments with open-ended essay questions, problems to solve, and objective questions
- References & Related Readings sections that provide source information, plus a listing of useful handbooks, journals, periodicals, and similar data
- Listings of experiments and updates from past National Educators' Workshops, which can be obtained by using information in the Instructor's Manual

The experiments and demonstrations have been tested and thoroughly peer-reviewed and offer some excellent activities to supplement the text. The answers to all the Self-Assessment problems and questions are available in the Instructor's Manual, which offers a test bank from the Self-Assessments and other useful supplements.

The book's organization—including numerous tables, extensive appendices, and key terms highlighted in bold type—makes it comprehensive enough to serve as a ready-reference book on materials for the immediate future and later. Please become familiar with the table of contents and the index. Using the table of contents, you can locate many useful aids in the appendix, such as Greek symbols, SI/U.S. customary conventions and conversions, tables of properties, ASTM abbreviations, hardness-to-strength conversions, trade names, materials selection guides, and more. Within modules, you will find useful visual aids such as our hardness scale comparison figure. All of these features are cross-referenced in the index.

UPDATES AND OTHER RESOURCES

Since 1986, Jim Jacobs, as creator of the National Educators' Workshop series, has codirected these annual workshops, which focus on emerging concepts in engineering materials, science, and technology with emphasis on laboratory experiments. Among the major sponsors of these workshops and related research were the National Aeronautics and Space Administration

(NASA); the Materials Science and Engineering Laboratory at National Institute of Standards and Technology (NIST); Air Force Office of Scientific Research; the Federal Laboratories with the Department of Energy (DOE); automakers Ford, General Motors, and Daimler-Chrysler; the National Composite Center; the Boeing Company, Seagate Technologies, IBM, and several host universities. Faculty from community colleges, four-year colleges, and major universities from across the United States, Canada, and other nations gathered at these sessions to share ideas on more effective materials education. Many leading corporate and governmental agencies with materials science and engineering laboratories provided updates at these workshops on the full range of materials developments. The seminars presented at the workshops provided us with an opportunity to learn of the latest improvements. Much of the fine work of all the contributors to National Educators' Workshop: Updates has been compiled into a useful CD-ROM, *EMSET 2* (*Experiments in Materials Science and Engineering 2*) to complement this book as well as those used in other courses. This CD-ROM has over 350 experiments and demonstrations as well as numerous color images, web links, and other instructional aids that should serve as a valuable supplement to a variety of courses.

ACKNOWLEDGMENTS

Until you write a textbook, especially one as extensive as *EMT*, you cannot fully appreciate the work involved. We have been most fortunate to receive so much input from so many sources. From those hundreds of educators, researchers, engineers, scientists, and students who participated in the National Educator's Workshops series, we gained valuable insights and resources that helped us to keep our material current and appropriate.

We thank our colleagues in community colleges and four-year institutions who provided their suggestions. The thorough review by William K. Dalton, Purdue University; Jihad M. Albayyari, Eastern Michigan University; and Edward V. Clancy, California State Polytechnic University, Pomona, were instrumental to our revisions.

We welcome feedback from readers of our work and invite you to contact us through Prentice Hall.

Illustrating a text of this nature is a very difficult job. We appreciate the assistance of the numerous colleagues, technical societies, companies, and governmental laboratories, most of whom are listed in each figure's caption, for providing the valuable information and illustrations. We extend a special "thank you" to Luther Gammon of Boeing Materials Technology, who provided a wealth of photomicrographs for this book and *EMSET 2*. He also provided an entire short course on "Microscopy of Fiber-Reinforced Polymer Composites" for *EMSET 2*. As with previous editions, editors and contributors to *Advanced Materials & Processes* were very helpful in supplying the photographs and illustrations that improve readability. Uncredited photos were taken by Jim Jacobs, and many other uncredited illustrations were developed by Jim Jacobs and Tom Kilduff.

We expressed our sincere appreciation to Martha Jacobs and Jennifer Winnett Taylor for their tireless assistance in reformatting the manuscript and working with illustrations. Our detailed index resulted from the thorough work done by Danae Thompson. The editors and staff listed on the copyright page gave us needed support and understanding as we worked to meet deadlines and ensure a quality revision.

We especially wish to acknowledge invaluable assistance from the following people for their various input over several editions: Mitchel Lifland, Eastern Kentucky University; Patricia Olesak, Purdue University; Dale Wilson and his students from Tennessee Technological University; students at Norfolk State University, Tidewater Community College, and Norfolk Ford Assembly Plant; Ed Widener, Purdue University; Clint Bertrand, Texas A&M University; Jim Nagy, Erie Community College; Michael Wehrein, Lethbridge CC, Alberta, Canada; Rhonda Housley, Spartanburg Technical College; Carl Metzloff, Erie Community College; Larry Helsel, Eastern Illinois University; Sarah Joy Nichols, Cape Fear Community College; Richard Cowan, Ohio Northern University; Seth Bates and Richard Chung, San Jose State

University; James A. Clum, SUNY–Binghamton; Joseph Neville, Wentworth Institute of Technology; Kathleen Kitto, Western Washington University; David A. Smith, Lehigh University; James DeLaura, Central Connecticut State University; Wayne Elban, Loyola College; Steve Piippo, Richland High School; Roy Bunnell, Mike Schweiger, Irene Hays, Denis Stracher, and Eugene Eschbach, Battelle, Pacific Northwest National Laboratories; James E. Gardner and Charles E. Harris, NASA Langley Research Center; Laurie A. George, American Dental Association, Health Foundation; Robert Berrettini, Editor of *JME*; Hugh Baker, ASM International; Irwin Poston, General Motors Technical Center; Thomas C. Holka and Bennie S. Bailey, Ford Motor Company; Leonardo E. Boulden, Howmet Corporation; Douglas Craig, Director of Metals and Ceramics at Oak Ridge National Labs; Jonice Adams and Anna Fraker, Materials Science and Engineering Laboratories, NIST; Louis Nenninger, II, Ceramic engineer for Owens-Illinois; Sankar Sastri, New York Technical College; Peter Beardmore, Director of Ford Motors' Materials Research Staff; George G. Marra, Deputy Director, U.S. Forest Products Laboratory, Madison, Wisconsin; Charles V. White, GMI Engineering and Management Institute; Cindy Neilson, Ceramic engineer, IBM; Walter E. Thomas, Western Carolina University; Frank J. Rubino, Middlesex County College; Howard Hull, New York City Technical College; Charles Flanders, Texas A & M University; Mario J. Restive, Mohawk Valley Community College; Richard I. Phillips, Southwest Missouri State University; Peter Route, Wentworth Institute of Technology; Lynda L. Anderson, SCCC; Pravin Raut, Savannah State College; Alden P. Gandreau, Hudson Valley Community College; William Wait, SUNY–Oswego; and many others.

UNIT 1

Nature & Structure

PAUSE & PONDER

Pause to examine the photo, and ponder the excitement of a cold morning, December 17, 1903, as the Wright brothers successfully flew the first motorized airplane on the windswept sand dunes of Kill Devil Hill, North Carolina. The "first flight" capped centuries of dreams and efforts by humans to soar like the birds. When the two bicycle makers from Ohio launched their flying machine, they demonstrated application of innovative design to commonly available materials of their time—wood, cotton fabric, and steel.

While December 17, 2003 marked the Centennial of Flight, it also celebrated a half-century of space travel. Just as the Wright brothers and their fellow pioneers of flight made unique uses of materials to achieve flight, 50 years of space travel also saw ever-improving uses of conventional materials such as aluminum and glass in order to build superior, aerodynamically designed craft. This 50-year period, known as the Space Age, also saw the introduction of space-age materials. These "designed materials" signified a departure from early practices of using available bulk materials. *Designed materials* involved the new materials such as syn-

The 1903 Flyer used conventional materials in innovative ways. (Library of Congress Digital Images)

thetic fibers, high-temperature plastics, toughened ceramics, and other innovations that achieved the result of reducing weight while adding strength and improving heat resistance.

What can we expect from the next century of aerospace technology?

According to predictions, within 25 years the future of aerospace will advance at an even faster pace as new materials and processes technology emerge to complement other technologies. For example, nanoscale science and engineering will allow the precise. construction of structures at the molecular level—atom by atom. Smart materials will be able to *morph*, or change their shape, on electronically operated flight surfaces. Evolving *biologically inspired materials* that mimic plants and animals will also provide unique properties. Nanoscale processing will allow ceramic and polymers to be shaped without machining. For travel at higher speeds, new nanocoatings will provide thermal and wear-resistant surfaces for superior TPS (thermal protection systems). As envisioned by NASA, design advances in nanotechnology may translate into the future generation of spacecraft.

Achieving these innovations will require the same type of determination and application of new technology that the Wright brothers employed. That aspect of progress does not change. Innovators must grasp fundamentals. That is as true today as it was a century ago, even as new principles of science and technology evolve.

This book and related instructional resources offer you the opportunity to learn fundamentals in materials science and technology that will help you understand and participate in our exciting future!

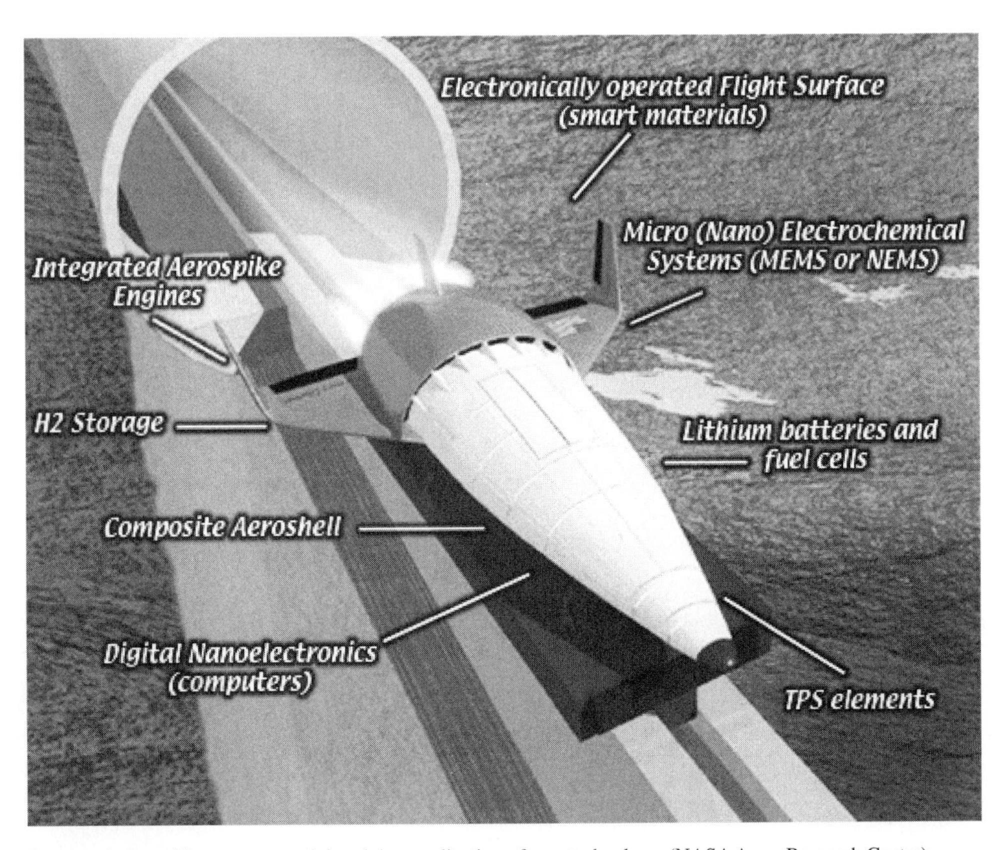

Concept design of future spacecraft involving application of nanotechnology. (NASA Ames Research Center)

Module

1

Engineering Materials Technology

After studying this module, you should be able to do the following:

1–1. Describe the broad field of engineering materials technology and the contributions of materials science and engineering to materials developments, processing, and manufacturing.

1–2. Explain the significance and consequences of the five stages in the materials cycle and recount how life cycle analysis promotes green design and manufacturing for a sustainable environment.

1–3. Describe the characteristics of the ideal engineering material, and discuss obstacles to changing from one material to another.

1–4. Cite examples of recent developments in materials science and engineering that have affected technologies in such fields as information, transportation, production, medicine, and entertainment.

1–5. Synthesize the personal and societal values gained from the study of engineering materials technology in terms of career preparation, consumerism, and citizenship.

1–6. Use the website addresses denoted by the icon to gain insight into current developments and evolving content in materials science and technology.

A World of Stuff

Our world is full of "stuff" ranging from cellular phones to roller blades. The *stuff* of that stuff is materials such as wood, metal, plastic, or glass. You cannot have your stuff without *materials stuff*. The materials stuff gets cooler by the day because of the rapid advances in materials science and technology (MST). This section aims to stimulate your interest in the science and technology that make all our stuff possible. Some of the terms used may not be familiar to you, but they will be defined later in the book. This section and the rest of this book should make you a smart materials observer. As a result, you will become a better consumer of stuff and a wiser user of materials.

As you will read in Module 2, we live in a period of time that may be labeled the "Materials Age." Much of the explosion in new technology is made possible largely due to new and improved materials and processes. For example, telephones and watches (Figure 1–1) have evolved due to advances in semiconductor and liquid crystal display (LCD) technology. The phone itself is but one component of a complex telecommunications system involving satellites and transmission lines made of metals, glasses, and plastics.

Materials science and engineering (MSE, defined later) is a relatively new field of study compared to such disciplines as physics or civil engineering. However, the multidisciplinary nature of MSE is fostering improvements in all our stuff, from artificial skin for burn victims

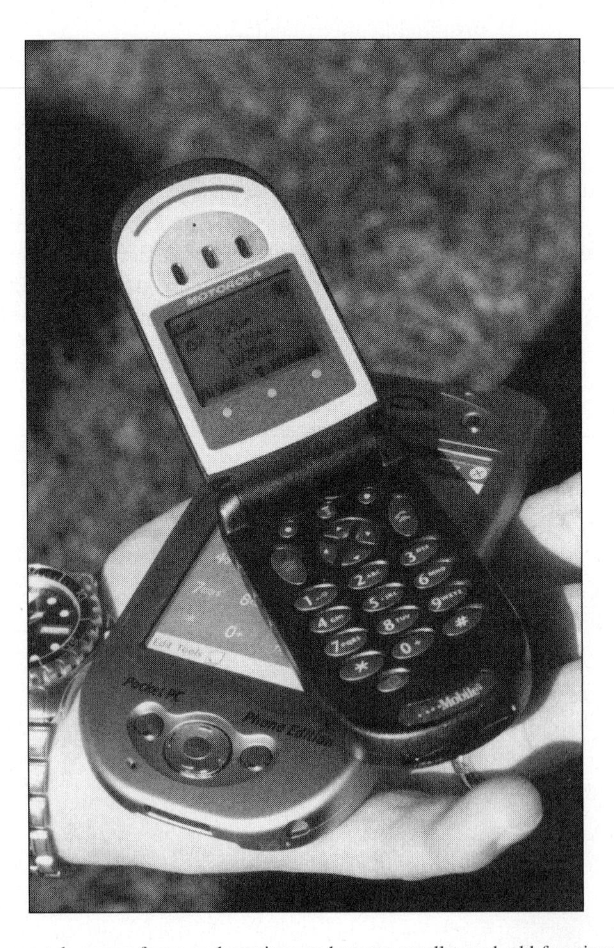

Figure 1-1 Cell phones take many forms and continue to become smaller and add functions. Here we see a cell phone resting on top of a combination cellphone-PDA. The combination provides all the PDA functions, such as phone book, scheduling, music (MP3) playback, as well as Internet access and computer interface while offering cell phone multifunctions such as color display, speakerphone, voice record, voice-activated dialing, multilanguage support, time and date display, a 250-entry phonebook, Quickstore phone numbers, three-way calling, caller ID, pager, and two-way radio. Materials science and technology, including nanotechnology, will continue to reduce sizes further, perhaps putting all these functions into ballpoint-pen-size devices.

to frames for bicycles. So MST and MSE provide new materials and, of equal importance, better means of producing (manufacturing) both the raw materials and the finished products. The cell phones shown in Figure 1–1 continue to add features and functions. New phones owe their smaller sizes and expanded capabilities to MSE progress in improving microelectronic methods for doping silicon with metal atoms at scales down to 0.3 micrometers (0.3×10^{-6} m). Also, advances in materials and processes technology have led to smaller, lighter, and even flexible batteries that last longer, thus allowing expanded functions in cell phones and watches as well as in computers and other devices. Much of the credit for miniature and microscale devices is owed to our NASA-led space missions.

Designers of our stuff use new and improved materials to make changes in existing products, such as dual-wheel skates, or create totally new designs, such as the in-line wheel skates seen in Figure 1–2a. Polyurethane plastic wheels on skates have the advantages of improved traction, greater shock absorption, and lighter weight over the older steel and rubber wheels. The in-line designs are more maneuverable than the dual-wheel design. Other features of the skates that add to comfort and use for roller hockey are the polymer memory foam in the ankle area; 70-millimeter, super-high-rebound hockey profile wheels; and precision bearings. Notice the protective helmets worn by the skaters in Figure 1–2a. They are examples of how plastics have been able to make skating, biking, and other sports safer. A combination of a tough, ABS plastic outer shell and a lightweight urethane, dual-density foam inner cushion both protects the skaters' heads and provides comfort.

When you purchase recreation equipment or other products, do you understand the descriptions of materials and design provided with the products, such as the data just given for in-line skates? Your study of materials science and technology will help you to do so.

We often take for granted the stuff we eat without giving much thought to how we keep our food safe and fresh for consumption. Canning with glass (a ceramic material) and "tin cans" has been done for more than a century. Figure 1–3a shows some of the early food containers from the beginning of the twentieth century. Recent decades witnessed dramatic improvements in most forms of container materials (Figure 1–3b). Many "tin" cans (actually made of tin-free steel) are now quite thin and have only two pieces (top and body) unlike the older, three-piece soldered or welded can. Aluminum, although more costly than steel, is used for most beverage cans. Can you imagine why? Many forms of polymers, including paper and plastics, often laminated with aluminum foil, are also used as packaging to keep beverages, such as juice and milk, fresh without refrigeration. You will learn more about these types of composite containers later in the book.

(a)

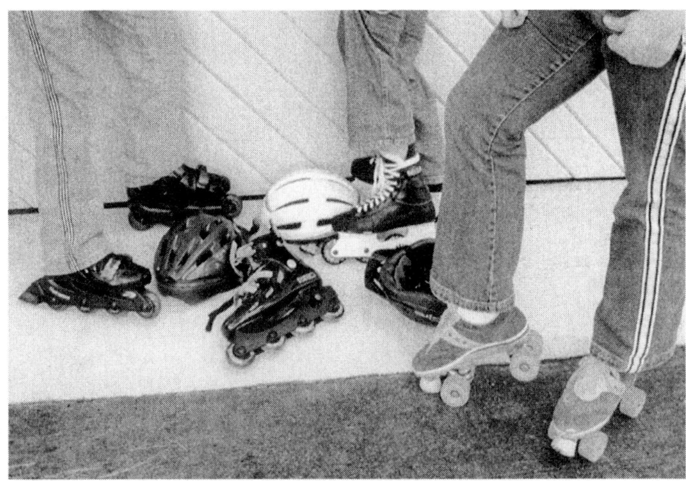

(b)

Figure 1-2 (a) Natural polymers such as cotton in jeans and synthetic polymers such as polyester in the girls' shirts and acrylic lenses in sunglasses are the stuff of our world. Synthetic polymer in the polyurethane plastic wheels on skates have improved traction and shock absorption and lighter weight than the older steel wheels. (b) Protective helmets worn by Amanda, Elizabeth, and Jennifer are also made of a tough synthetic polymer, ABS plastic, which forms the outer shell, while a lightweight urethane, dual-density foam acts as an inner cushion.

(a)

(b)

Figure 1-3 (a) Early food containers of glass and steel used the common ceramic and metal materials of the late nineteenth and early twentieth centuries. (b) Advances in materials technology brought new containers made of two-piece aluminum and steel cans and "barrier" packages made of laminated composites of plastics, aluminum, and paper. These new containers are lightweight, will not break, and offer long shelf life for their contents.

STOP!.. *Look at various kinds of cans to observe their construction and materials. Then look at the construction of other containers, such as your toothpaste tube. Tear it apart to see its laminated composite construction of polyethylene, paper, and aluminum.*

As you learn basic and advanced concepts about engineering materials technology throughout this book, you will also read about connections to applications in your everyday life as well as trends in new technology that may affect you. Your newfound knowledge of materials technology may lead you into a career path that will involve selecting and applying materials improvements.

Your study of materials science and technology should help you to (1) develop an awareness of the importance of engineering materials in everyday life; (2) recognize society's dependence on materials; and (3) appreciate the value of knowledge of engineering materials technology for yourself as a consumer, citizen, and member of the technological workforce.

Many forward-thinking businesses seize upon advances in materials and science technology to gain an advantage over their competitors. Their engineers can select from ***designed materials*** rather than settling for ***off-the-shelf materials*** (materials with limited structure and properties). Throughout this module and the text, we will cite many examples of unique applications of materials and processing technology that set the pace for superior products and systems in such fields as transportation, construction, communication, recreation, health, and entertainment. The examples will help you to develop important competencies in engineering materials technology that will serve you well.

We have designed this book to provide you with an understanding of how materials science and engineering fit into the many aspects of technology and society. Most industries, such as the automotive, aircraft, sports, computer, and entertainment industries, rely heavily on materials technology. The illustrations, case studies,—including Pause & Ponders and Applications & Alternatives—module questions, and other topics in the book were selected to give you a feel for the significance of both engineering materials and the technical problem solving required in modern industry and society. Enjoy!

Materials Science and Technology—*http://MST-Online.nsu.edu**

1.1 ENGINEERING MATERIALS TECHNOLOGY

Materials are the matter of the universe. These substances have properties that make them useful in structures, machines, devices, products, and systems. The term ***properties*** describes the behavior of materials when subjected to some external force or condition. For example, the tensile strength of a metal is a measure of the material's resistance to a pulling force. The ***family of materials*** (which will be explained in more detail in Module 2) consists of four main groups: ***metals*** (e.g., steel), ***polymers*** (e.g., plastics), ***ceramics*** (e.g., porcelain), and ***composites*** (e.g., glass-reinforced plastics). The materials in each group have similar properties and/or structures, as will be described later.

Engineering materials is a term often loosely used to define most materials that go into ***products*** and ***systems***. A telephone is a product that is part of a telephone system composed of many telephones, wires, fiber optics, switches, computers, and so on. *Engineering materials* can also more specifically refer to materials whose structure has been designed to develop specific properties for a given application. For example, *engineering plastics* are

> . . . those plastics and polymeric compositions for which well-defined properties are available such that engineering rather than empirical [trial and error] methods can be used for the design and manufacture of products that require definite and predictable performance in structural applications over a substantial temperature range. (ASTM, 1990, p. 167)

In other words, engineering plastics such as polycarbonates and acetals could replace more ***conventional engineering materials*** such as steel and wood because their properties are competitive for structural components such as piping, cams, and gears. On the other hand, general-purpose plastics, such as polystyrene and vinyls, do not possess the properties to carry heavy loads but serve as packaging, upholstery, and so on.

The field of ***materials engineering*** "deals with the synthesis** and use of knowledge [structure, properties, processing, and behavior] in order to develop, prepare, modify, and

*The symbol found throughout this book will link you to Internet sites related to topics being covered. The dynamic nature of the Web brings frequent changes, so some of these sites, while available at the time of writing, may not be up now.

**Several terms, such as *synthesis* and *process*, introduced in this module will be discussed later in the book. The index allows easy cross-reference to most terms.

apply materials to specific needs" (National Research Council, 1989, p. 20). Materials engineers have become very much in demand as we seek to improve the efficiency of products.

Materials science and engineering (MSE) has become a major field of study, one critical to many other fields. As defined by a National Academy of Sciences study, MSE involves the generation and application of knowledge relating the *composition, structure*, and *processing* of materials to their *properties* and *uses*. The "science" focuses on discovering the nature of materials, which in turn leads to theories or descriptions that explain how structure relates to composition, properties, and behavior. The "engineering," on the other hand, deals with the use of science in order to develop, prepare, modify, and apply materials to meet specific needs. The field is often considered an engineering science because of its applied nature. **MSE** is interdisciplinary or multidisciplinary, embracing such areas as metallurgy, ceramics, solid-state physics, and polymer chemistry.

Engineering materials technology (EMT) covers fields of applied science related to materials, materials processing, and the many engineering specialties dealing with materials, such as research and development, design, manufacturing, construction, and maintenance. Many new processes have evolved for the manufacture of engineering materials. Through these processes, coupled with the design of engineering materials, we now enjoy the benefits of superior engineering materials. **Materials science and technology (MST)** is another label for this field of study.

1.2 THE MATERIALS CYCLE—LIFE-CYCLE ANALYSIS

To better understand engineering materials technology, it is useful to view the ***materials cycle***, which can be broken into phases, as shown in Figure 1–4. The following explanation of each of the stages will provide you with insight into the importance of materials, how they affect

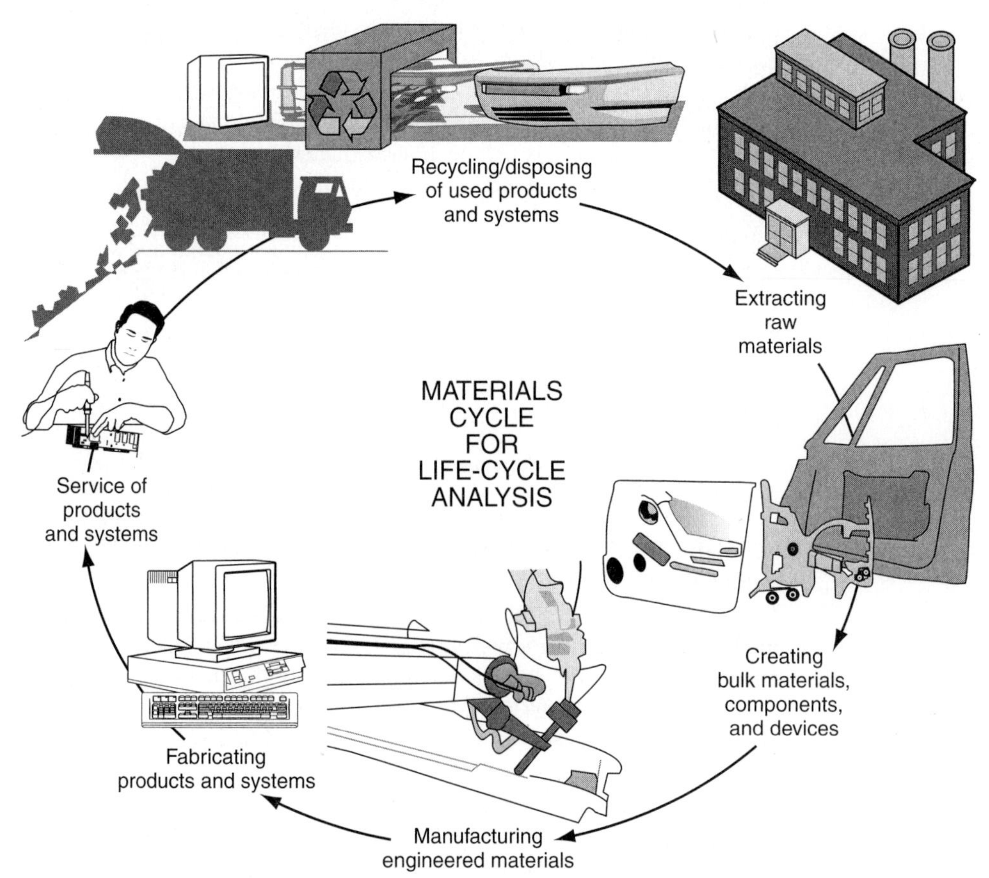

Figure 1-4 Materials cycle—life-cycle analysis

our lives, and their total impact on society. As explained later in this module and discussed throughout the book, the materials cycle allows for ***life cycle analysis* (LCA)** of products to determine their impact on the global environment. Each phase of the cycle presents a challenge for improving efficiencies in dealing with the materials and processes involved in products in terms of cost and environmental impacts.

1.2.1 Extracting Raw Materials

The earth has provided us with the basic ingredients for producing an unlimited variety of materials. The basic building blocks of these materials are the 111 or so known chemical elements that comprise the *periodic table of elements* (shown on the inside front cover and in Table A-1 of the appendix). Some of these elements, such as gold, are readily available for use in the form of *solid materials*. Pure carbon in the crystalline form of diamond, the hardest known material, also takes little processing. But other elements are locked inside materials in the earth, ocean, or atmosphere. For example, it is often necessary to mine ore to obtain raw materials. Aluminum (Al), for example, is the most abundant metal; it comprises more than 15% of the earth's crust. However, when it is extracted from the earth, Al is combined with oxygen (O) in the form of alumina (Al_2O_3). It takes large amounts of electricity to *extract* Al from bauxite (Al_2O_3) ore. It takes 95% less energy to produce aluminum by *recycling* scrap aluminum products.

Synthesis involves transforming the atoms and molecules of gases, liquids, and solid elements by chemical and physical means to form other, solid materials. The synthesis of polymers from ***raw materials***, including coal, petroleum, water, and air, yields substances for producing *bulk materials*, such as bulk nylon. Synthesis requires constant research by scientists in order to improve the techniques.

Although it also deals with atoms and molecules, ***materials processing*** "includes control of structure at higher levels of aggregation and may sometimes have an engineering aspect" (National Research Council, 1989, p. 224). Materials processing yields bulk materials, like nylon tubing; individual components, like ceramic jet nozzles; devices, like semiconductors for computers; structures, like automobile frames; and systems, like fiber-optic communications systems.

To describe the complex intertwining of science with the technological applications involved in materials processing and manufacturing, Lyle H. Schwartz (1996) coined the term ***scitech***. Historically, materials processing has been considered more technology than science for, generally speaking, trial and error was used to make advances in technology rather than scientific inquiry and theoretical analysis. More recently, however, applications developed in such fields as applied chemistry, solid-state physics, and fracture mechanics have brought about a merging of science with engineering and the development of the discipline of materials science and engineering. The syntheses phases of materials technology have traditionally been considered to involve more basic science, requiring as they do the work of scientists such as metallurgists, polymer chemists, and nuclear physicists. However, the development of the of advanced ceramics and advanced composites as well as of electronic materials has caused a melding of disciplines. Many universities now offer degrees in MSE because of the need to ensure that the synthesis phase incorporates considerations for the processing of raw materials. These degrees might be considered *scitech* degrees. The label *materials science and technology* reflects this new approach to MSE.

One example of emerging materials technology is ***shape-limited synthesis***, a new method developed to produce materials that blend synthesis with processing by beginning with one of the chemical agents already in the form of its final shape, as discussed in Module 20. This technology contrasts with the conventional procedure of first synthesizing liquids, particles, powders, or pellets, which are then processed by casting or molding into the final shape. The new techniques are used in the production of advanced ceramics and composites.

During the raw materials extraction phase, there must be a concern for the by-products of refining and synthesis. Are gases released to the atmosphere that may have adverse effects on people, animals, vegetation, and even on other materials? Is the water used in separation processes returned safely to rivers and streams or into the ground? Does the waste contain heavy metals or toxic chemicals that would be dangerous to the animal food chain (insects, birds, fish, cattle, and then humans)? The techniques of life cycle analysis should promote reducing the impact of materials on the environment.

1.2.2 Creating Bulk Materials, Components, and Devices

Bulk materials are the products of synthesis, materials extraction, refinement, and processing. There are many bulk materials with which you are familiar, such as fir plywood, sheet steel, acrylic tubing, window glass, copper wire, and concrete. Bulk materials are usually made in large quantities through continuous processing and then supplied to manufacturers of components and devices.

Components include gears, electrical wires, screws, nuts, jet engine turbine blades, brackets, levers, and the thousands of constituent parts that go into many products and systems. ***Devices***, which include microprocessors, resistors, switches, and heating elements, are usually more complex than components and are designed to serve a specific purpose. ***Products*** are individual units, such as roller-blade skates, television sets, chairs, and telephones. ***Systems*** are an aggregate of products, components, and devices. For example, a telephone system is made up of millions of products (e.g., telephones, microwave transmitters, and computers), components (e.g., optical fiber, copper wire, and lasers), and devices (e.g., switches, relays, and microprocessors).

The source of bulk materials varies by material type. Some producers take materials through all stages, from extraction of raw materials to production of finished parts, but this situation is an exception. Normally, specialists are involved in the various stages. For example, production of bulk polymers (plastic, rubber, paint, adhesives) is usually done by materials makers known as *polymer manufacturers*. These manufacturers are often located near oil refineries because most plastics are petroleum based (see Figure 14–5). The flow diagram for polymer production shows that the manufacturer begins with the raw materials of crude oil, natural gas, trees, or cotton to make chemical compounds such as ethane, trichloromethane, hydrogen fluoride, and ethylene chloride. From these raw materials, monomers (single molecules) are produced, including ethylene, methylmethacrylate, and vinyl chloride. Through heat, pressure, and the addition of chemicals (catalysts), the monomers are polymerized to form long-chain polymers. The final polymer (resin) is a bulk "virgin material" ready to go to the next stage of production, which is normally handled by ***fabricators***. Polymer manufacturers may sell monomers to ***processors***, who polymerize the monomers into bulk plastics such as acrylic sheet or PVC [poly(vinyl chloride)] tubing. The processors then sell their bulk plastics to fabricators, who then produce engineered materials such as acrylic lenses for automobile brakelights.

Figure 1–5 depicts how advances in engineering materials technology have helped to improve properties of engineering materials. As the graph shows, traditional materials like stone and cast iron had very low strength-to-density ratios (*specific strength*, defined in Module 15). In the latter half of the twentieth century, specific strength in engineered materials has increased to over 10×10^6 in. for advanced composites, compared to less than 0.5×10^6 in. for traditional, "off-the-shelf" engineering materials of wood, stone, and metals (National Research Council, 1989, p. 20). Figure 1–6 shows some applications of advanced ceramics and advanced composites in new materials for engines.

Before the industrial revolution, production was done manually and individual components were cast or cut and *wrought* (shaped with heat and force) from bulk materials. Next, the bulk materials were turned into finished "customized" products; this required considerable skill and labor. Mass production, which evolved throughout the industrial revolution, made it possible to produce large numbers of parts using machines that reduced the amount of human labor required. More jobs were created as demand for affordable products grew in developed countries. As metallurgy began to evolve as a science aimed at better understanding the nature, properties, and processability of metals and their alloys, the materials science and engineering (MSE) field was in a formative stage. Then, during the middle of the twentieth century, MSE began to grow in importance as society began to demand more from technology. Conventional materials could no longer meet the need for higher operating temperatures, greater strength-to-weight ratios, and directional properties. It became necessary to engineer materials as well as products and systems.

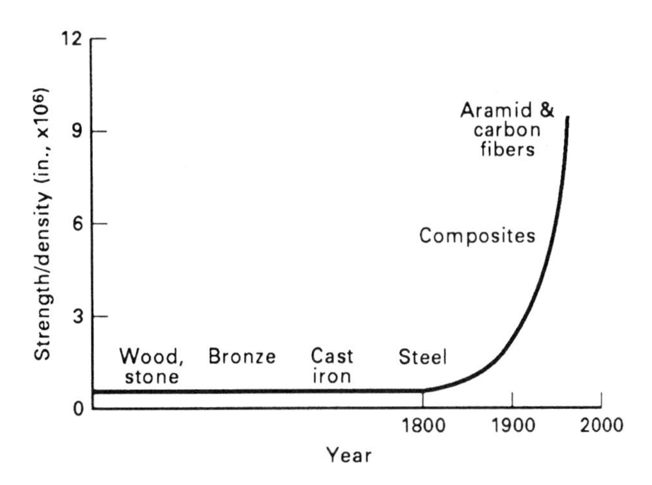

Figure 1-5 Creating bulk materials and components using innovative techniques yields large gains. Modern materials have 50 times higher strength-to-weight ratios than cast iron and other early engineering materials. (Oak Ridge National Laboratory)

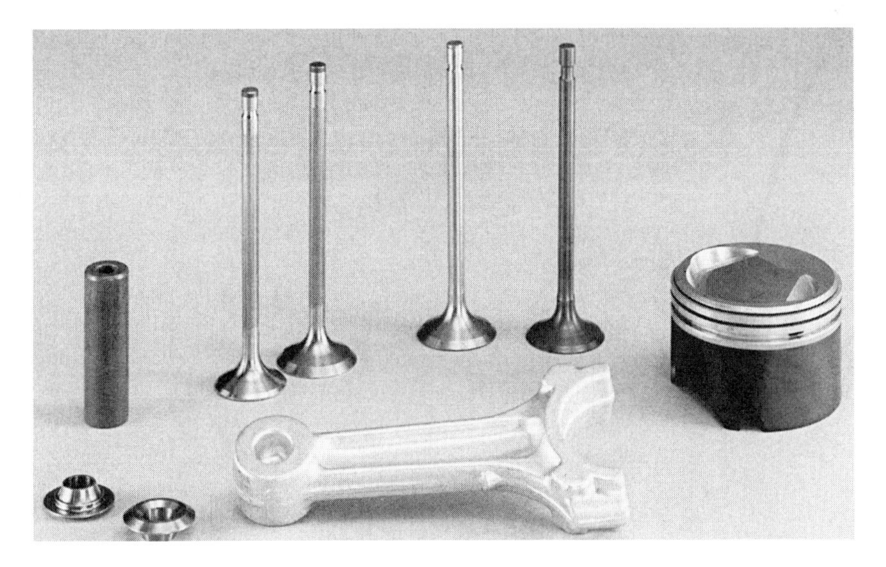

Figure 1-6 More efficient engine components made of *advanced ceramics* and *advanced composites* replace *traditional materials*. Clockwise from upper left: silicon-nitride piston pin, two titanium valves, rapid solidification (RS) aluminum valve, silicon-nitride valve, metal-matrix composite (MMC) piston, RS aluminum connecting rod, and two titanium valve spring retainers. (*Advanced Materials & Processes*)

1.2.3 Manufacturing Engineered Materials

The manufacturing phase may involve bulk materials like plastic sheets or may create a bulk material, component, or product. The new-generation engineering materials are often ***designed materials***; they have been *engineered* to provide designated properties. In other words, instead of selecting from a list of available materials, designers may specify the desired properties for their needs and then rely on materials engineers and technologists to create materials to suit the need. ***Advanced composites*** are examples of engineered materials because engineers and technicians determine how reinforcing fibers should be aligned to withstand the stresses that a product will encounter under service conditions. These composites may be made of plastic resins that can withstand higher temperatures than general-purpose plastics. ***Advanced ceramics***, one class of engineered materials, may be tougher than ceramics that are normally brittle. When manufacturing an engineered ceramic, chemical vapors of silicon carbide form a matrix as they are infiltrated into reinforcing fibers of silicon carbide for net-shaped advanced ceramic composite parts.

Another group of engineered materials is ***smart materials***, a term referring to a variety of liquids and solids that have the ability under predetermined conditions to alter their properties in response to stresses. For example, smart glass will darken when an electrical current is passed through a laminated grid in the glass. Electrical current can also thicken certain fluids. Optical fibers, metal fibers, and electrorheological (ER) fluids inserted into solids such as plastic composites, aluminum sheets, or concrete can sense stresses and cracking to provide early warning of probable failure. These fibers might be used in engineering applications such as aircraft wings and highway bridges to provide early warning of failure. The fibers could also be used to change the stiffness of automotive springs, helicopter blades, or golf clubs. Smart products using piezoelectric devices include smart skis and bicycle shock absorbers.

Smart skis—*http://www.acx.com/cool_smartski.html*
Smart bicycle shocks—*http://www.acx.com/cool_shock.html*

Biomimicking has been employed in the search for smart materials to study and attempt to mimic nature's wonders. Examples of such research include efforts to reproduce the way that spiders produce very strong fibers and that mollusks build their shells, which may lead to better techniques for fabricating integrated circuits for computers and microprocessors. ***Biologically inspired materials*** is a new term used to label materials resulting from mimicking or imitating processes and structures of plants and animals.

Figures 1–7 and 1–8 depict how the system of manufacturing technology for engineered materials works. The progress of composite science and technology involves an integrated program, as seen in Figure 1–7, in which materials science combines with the studies of engineering mechanics (study of forces in a materials system), durability, and engineering design, which then converge to address issues of product life cycle, fundamental laws (e.g., physics and chemistry), the interaction of materials components quality and reliability, and finally, the system's effect on cost. The tennis racquet in Figure 1–8 is an example of an advanced composite product that has resulted from the systems approach to manufacturing. The pointer on

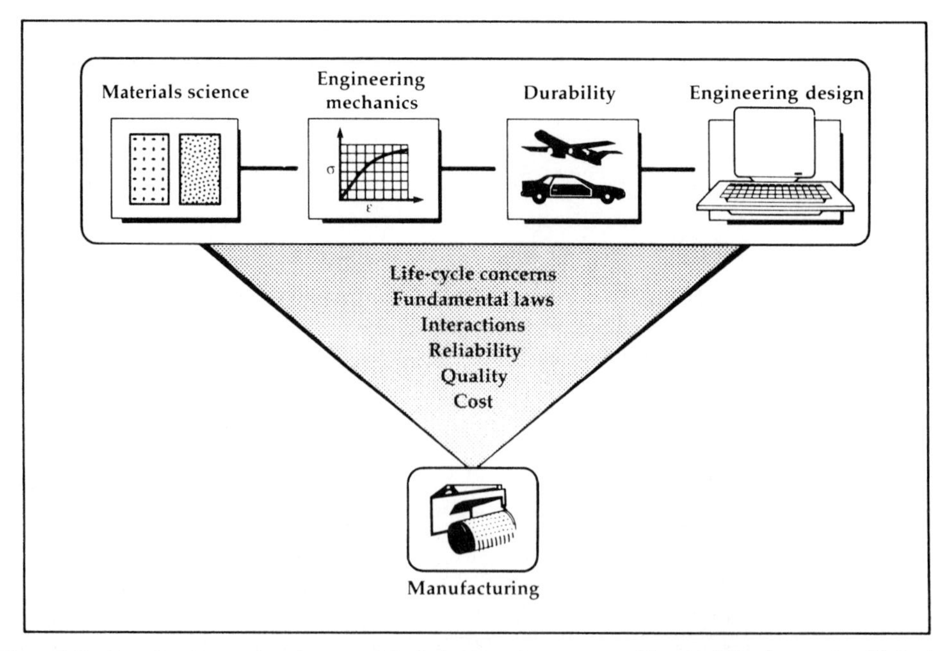

Figure 1-7 Manufacturing engineering materials. An integrated program combines traditional strengths with the development of manufacturing processes to yield an expanding science base, new engineering methodologies and practice, and a unified life-cycle approach to manufacturing. (*Advanced Materials & Processes*)

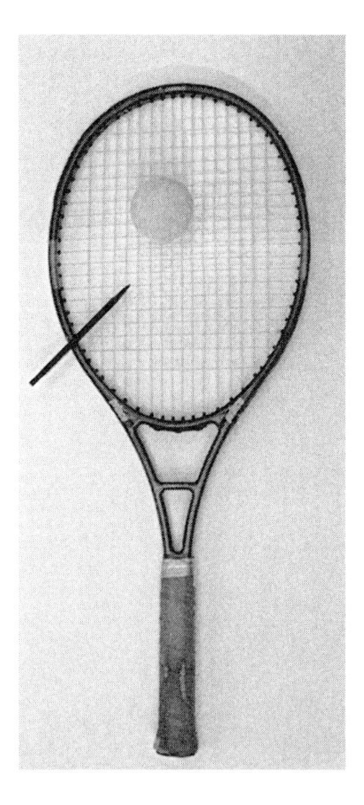

Figure 1-8 Left: Engineered graphite epoxy-framed tennis racquet provides directional properties for a large sweet spot (at arrow) and considerable power. Right: Force (at arrow) improperly exerted on the frame, instead of the strings, caused the racquet to fail catastrophically.

the left indicates the "sweet spot" of the racquet. The graphite fibers within the racquet's epoxy matrix frame were "engineered" to provide the directional properties that could withstand impact from a tennis ball on the sweet-spot region. An equivalent amount of force improperly applied to the racquet in another region, indicated by the pointer at the upper right, results in racquet failure. The directional properties of the materials in the frame were engineered only for normal impact from proper use of the racquet.

Many new processes have evolved for the manufacture of engineering materials, as shown in Figure 1–9. At each stage of the materials cycle, we must look at various issues, ranging from materials and production costs, to environmental impacts, to consumer acceptance. Note from Figure 1–9 how producers of engineering materials strive to address issues such as lowering cost, reducing processing steps, recycling materials, and improving air quality (reducing VOCs, i.e., volatile organic compounds, to improve air quality).

1.2.4 Fabricating Products and Systems

Once engineering materials have been manufactured, they are assembled into many useful products and systems. New fabricating techniques are evolving as a result of improvements in manufacturing engineering. ***Manufacturing engineering*** is the study of the techniques of turning bulk materials into finished products and systems. Once a field of empirical methods in which processes were handed down from generation to generation, with improvements coming gradually as a result of observation and analysis, manufacturing has recently become a field for detailed study and application of theoretical principles. There are now college programs leading to degrees ranging from associate to doctorate in manufacturing technology and manufacturing engineering.

A clear understanding of the total materials cycle is now required of the entire technological team of crafters, technicians, technologists, engineers, and scientists. Computer-integrated manufacturing (CIM), a ***systems approach*** to the manufacture of products, places the materials engineers and technologists together with the design engineers, technicians, manufacturing engineers, plus the environmental engineers and even marketing personnel. New laws regarding the environment and natural resources are intended to ensure that this

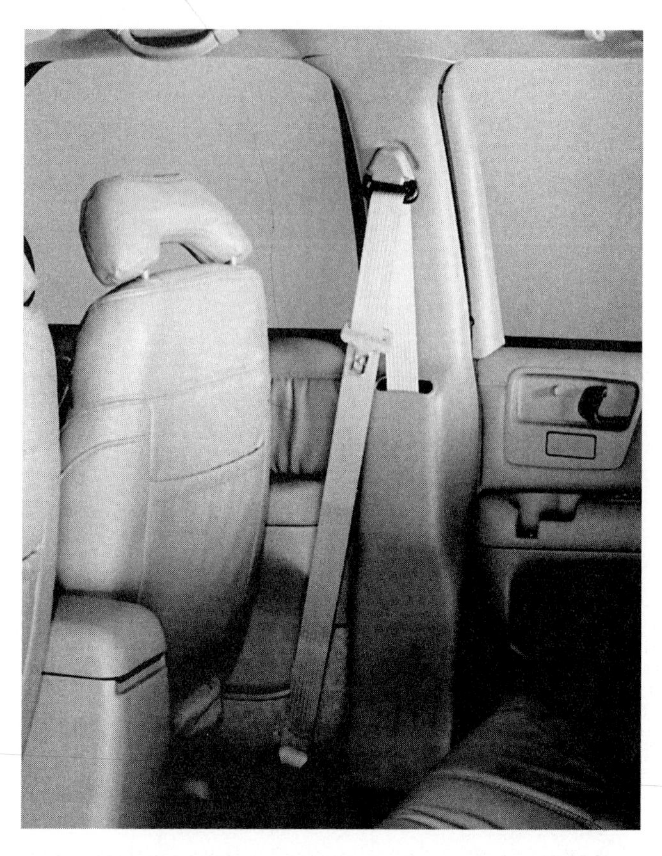

Figure 1-9 Manufacturing engineering materials with improvements. An engineering thermoplastic ABS resin, used in these interior door panels, elevates this grade of Cycolac® to be a competitor with the popular polypropylene plastic. Advantages of the ABS include reduced glare (for customer acceptance) and molded-in color, which is preferred by processors because of the potential to eliminate painting, lower component cost, reduce emission of VOCs (volatile organic compounds), and enhance recycling. (G.E. Plastics)

technological team uses life cycle analysis (refer to Figure 1–4) at all stages of the materials/product cycle so that appropriate safeguards are implemented to protect the environment as well as present and future human generations.

The changes in the way products are fabricated have led to numerous changes in the workplace. The number of manufacturing jobs for crafters and production workers (skilled, semiskilled, and unskilled labor) have decreased drastically. Jobs now call for educated people who can design products and manufacturing systems; set up, monitor, troubleshoot, and service manufacturing equipment; and serve as liaison (go-betweens) among various stages of the total materials/product cycle to ensure that quality is always improving and customer needs are met.

 Manufacturing—*http://www.manufacturingiscool.com/*

1.2.5 Service of Products and Systems

The shifts in manufacturing have resulted in a larger service workforce and a smaller manufacturing workforce; as well, the complexity of products makes it harder for consumers to make repairs on their own. Today, special diagnostic equipment is required to troubleshoot everything from automobile systems to robots to appliances. This situation has led to a demand for better quality in products and systems and to improved, long-term warranties. Manufacturers are very interested in analyzing materials that fail so that they can improve materials engineering and product design.

1.2.6 Recycling and Disposal

Proposed laws, and some laws already on the books, involve life cycle analysis by placing responsibility for disposal of materials and products on the manufacturer once the product service life is over. ***Design for assembly*** emphasizes easy product assembly by robots and other automated equipment; in contrast, ***design for disassembly*** is a concept that places recycling at the beginning or design stage of the materials cycle to ensure that waste going into municipal landfills will be minimized. Thus, the last stage of the materials cycle can become the first stage of the cycle by the resurrection of material through recycling. Most materials can be recycled; it is, however, very difficult for manufacturers to develop a full materials cycle that will ensure recycling. Examples of industries with successful recycling programs are steel and aluminum. These programs have been successful because of the tremendous savings realized through using recycled materials in the production of bulk materials. In other cases, it has taken legislation to force recycling, particularly of glass and paper and even of aluminum, by requiring deposits on beverage containers as a financial incentive for people to return cans and bottles. In addition, federal, state, and local laws have restricted the amount of municipal solid waste that can be placed in landfills. Clean air and water regulations have also restricted the amount and type of waste that can be incinerated or dumped into the ocean.

But much remains to be accomplished to develop the proper attitudes and habits among our citizens if we are to make the total materials cycle efficient and thus protect the environment and natural resources for future generations. Data from life cycle analysis will promote wiser recycling and disposal decisions. One example of the problems involved concerns materials for packaging. Manufacturers of polystyrene packaging (often improperly called Styrofoam) worked with McDonald's and other fast-food restaurants to head off complaints based on the fact that most plastics are not biodegradable. Citizen groups had called for elimination of the polystyrene clamshells used to package hot dogs, hamburgers, and other sandwiches as well as for drink cups. Manufacturers set up a recycling system to reclaim the polystyrene, but because of the pressure and the fear of losing business, most fast-food chains moved back to paper packaging before the recycling system could be fully implemented. But was this the best action to take? Consider that paper will *not* biodegrade in the modern landfill because landfills are designed to stop materials from breaking down and leaching into the groundwater. So packaging foods in paper might actually create more solid waste because this type of paper is difficult to recycle.

Complex issues are involved in decisions regarding the use of plastics versus paper or glass. Most of our local, state, and federal legislators do not have a technical background and so are not qualified to decide some of these issues. It is therefore vital that citizens prepare themselves with knowledge of materials technology in order to provide input. Recycling and the disposal of materials will continue to be major issues for the foreseeable future.

1.2.7 Sustainable Environmental Technologies

In order to maintain life on earth, we must ensure that our technologies do not result in harmful impacts on the quality of air, water, and soil. ***Sustainable environmental technologies*** are practices that ensure we do not pollute and we make the best use of our natural resources. ***Green manufacturing***, a major global effort, seeks to make all phases of the materials cycle environmentally friendly, thereby placing the least amount of stress on the delicate balance of nature. Materials issues are a key to green manufacturing.

Achieving the goals of green manufacturing requires a beginning-to-end view of a product in the manner we depicted by the materials cycle in Figure 1–4. Life cycle analysis (LCA) looks at all the factors involved in a product by making an inventory of the inputs (e.g., raw materials and energy) and outputs (e.g., products and by-products). A ***life cycle inventory (LCI)*** is a listing of raw materials, bulk materials, gas emissions, product waste, recyclables, and so on. LCA results in a life cycle inventory. Using the LCI to evaluate possible impacts, such as depletion of old-growth trees and water pollution, should lead to a thorough interpretation of the factors and minimize harmful results. LCA thereby provides a perspective on the impact of manufacturing.

The need to comply with laws that regulate manufacturing's impact on the environment is evident; manufacturers also recognize the need to demonstrate to the public that they are environmentally responsible. The push for green manufacturing came from a variety of forces, but today most major manufacturers and their industries seek to be seen as environmentally friendly. The auto steel industry has built up a good recycling infrastructure, to the extent that most automobile steel goes through salvage yards and is returned to make new steel. In fact, the system is so well established that modern steelmaking actually requires scrap steel (see Figure 10–1, A flowline of steelmaking). Aluminum, newsprint paper, and bottle glass are other examples of successful recycling. But we have a long way to go to reduce the waste that crowds landfills with many of our engineering materials. As we saw from Figure 1–4, recycled materials can be great substitutes for raw materials, but it takes a major effort to create cost-effective systems.

Plastics are a particular challenge to our need to reduce the amount of products buried as wasted materials instead of being recycled. Many plastics lack an efficient recycling infrastructure. But as will be discuss in Module 15, some success with recycling has been achieved with certain packaging plastics like polyethylene terephthalate and high-density polyethylene. Nylon makers have begun a system to keep nylon carpet out of landfills. Recycling of the carpet yields a basic raw material, caprolactam, used in making nylon 6. The recycling of nylon carpet saves not only the raw material (oil) but also the energy used in processing. This is often the result of recycling. Recycling aluminum, as we have mentioned, can save 95% of the electricity required to process bauxite ore into aluminum ingots.

Mistakes costly to the environment and society have been made in mining, well drilling, and tree harvesting. Mine tailings left behind after mining ore are washed into streams, rivers, and lakes by rain, leaching the concentrated minerals that might be harmful to plants and animals. Oil drilling can result in harmful spillage. Clear-cutting of trees can result in severe soil erosion and can upset wildlife. On the other hand, if we did not extract these raw materials from the earth, civilization would not benefit from these valuable resources, people would not have employment in related fields, and we would not have the thriving commerce that makes possible a good quality of life. In other words, we have made trade-offs—some conscious, some not. For example, the plastics industry, a comparatively young industry that evolved when oil and gas seemed limitless, offered us sanitary packaging, great strength-to-weight ratios in materials, improved design flexibility, lower-cost products, and other advantages over the traditional metal, wood, paper, and ceramics. There was not much knowledge or concern then about pollution, global warming, the ozone layer, or the scarcity of landfill space for municipal solid waste.

We now live with the mistakes and bad environmental impacts that resulted from our ignorance of the new plastics materials class. Have we learned how to safeguard the environment from new materials? Environmental concerns about the just-emerging nanomaterials have already cropped up. Will nanoparticles interact with organisms and create new, deadly bacteria? What are the pollution implications of nanomaterials? Engineers and scientist must work with environmentalists to ensure safe practices as nanotechnology matures. Now that we recognize the problems of limited resources, energy waste, harmful by-products of materials processing, and materials disposal, the expectation is that everyone will apply knowledge of materials to ensure that our natural resources are best utilized and that engineering materials technology safeguards the environment.

1.3 MATERIALS SELECTION

We are surrounded by materials but we rarely think about how these materials are selected. Why was your desk made of solid wood, plywood, or plastic-laminated particleboard? Why have so many plastics replaced steel and zinc in automobiles? Why is using foamed polystyrene plastic to package fast food controversial? Although you might take for granted the materials that make up your products, you can be sure that the designers did not. People who design homes, cars, aircraft, clothing, furniture, and other products or systems devote a lot of attention to the selection of the materials they use. Materials selection might make or break a company. But how do the designers decide what the best material is?

There are hundreds of different types of plastics. They range from very soft to quite hard, inexpensive to very expensive, and transparent to opaque. Wood, too, is available in numerous varieties, ranging from the very soft, lightweight balsa wood used for model airplanes to a group of birch and beech trees that are so heavy and hard that they are sometimes called ironwood. *Lignum vitae*, one of the hardest and heaviest woods, has been used for bearings and pulley sheaves on ships since ancient times. Metals in combination with other metals or nonmetal elements known as ***alloys*** include many varieties of steels (iron and carbon), aluminum alloy, brass (copper and zinc), and hundreds of other alloys. Steel, the most common production metal, is found in the bodies of cars, railroad-train wheels, and piano wire. Some glasses, such as most drinking glasses, are very delicate and cannot stand much temperature change, but thermally hard glass is used for windows in spacecraft and in furnaces (see Figure 23-11). Cement is one type of ceramic, as is glass. Clays are ceramic raw material; they are used for making dishes, toilets, and spark plug insulators, among other things.

Faced with the nearly limitless range of materials available, how do designers and architects make a materials selection for products and buildings? What selection criteria are most important?

1.3.1 The Ideal Material

The engineer, technologist, technician, or architect searches for the ideal materials to suit the designated need. What is an ***ideal material***? Among other characteristics, we can list the following:

1. Endless and readily available source of supply
2. Cheap to refine and produce
3. Energy efficient
4. Strong, stiff, and dimensionally stable at all temperatures
5. Lightweight
6. Corrosion resistant
7. Without harmful effects on the environment or people
8. Biodegradable
9. Numerous secondary uses

Finding the ideal material for a specific product is a very complex process. Even products that have been made for centuries pose such problems.

Selecting materials for musical instruments has always been a challenge; the piano, for example, is made up of a wide variety of materials. The modern, conventional piano in Figure 1–10 is made of such metals as cast iron for the soundboard, brass for the foot pedals and hinges, and various steel alloys for the piano wire and turning pins. The piano also uses several natural polymers in its walnut cabinet, cotton-felt hammer heads, and plastic keys (formerly made of the natural material ivory, which is now illegal).

With over 100,000 engineering materials from which to choose, selecting the best material for a given application or part involves an understanding of the many relationships that exist between a selection and a few important variables: (1) design, (2) performance characteristics of materials (including properties and structures of materials), and (3) manufacturing processes. A cost variable is implicit in all of these choices.

1.3.2 Selection and Compromise

Compromise is the rule, not the exception, in materials selection. For example, ***space-age materials*** like graphite/epoxy composites and Kevlar (aramid fiber) epoxy composites possess strength-to-weight ratios 3 to 5 times greater than those of steel. So why don't automakers use graphite epoxy and aramid epoxy composites to make car bodies, axles, and driveshafts?

Figure 1-10 Materials of entertainment. The piano is an example of materials selection involving such properties as appearance (beautiful wood colors and grain pattern), durability (repeated striking of hammers on strings), and acoustics (musical sound quality).

Wouldn't these composites allow for much more fuel-efficient vehicles? The space-age materials *would* be much more fuel efficient—but because these materials can cost 15 to 40 times as much as steel, a $21,000 Ford might cost $300,000. Who could afford to buy it?

1.3.2.1 Properties of materials. With so many materials to chose from, the designer needs to use many aids in materials selection. Periodicals, websites, and handbooks can provide current data and performance criteria for structural (load-bearing) materials:

1. Strength (tensile, compressive, flexural, shear, and torsional)
2. Resistance to elevated temperatures
3. Fatigue resistance (repeated loading and unloading)
4. Toughness (resistance to impact)
5. Wear resistance (hardness)
6. Corrosion resistance

Technical society and manufacturer's references include numerical data about the performance criteria (properties) of metals, polymers, and ceramics and have updates on newer materials such as aramid fibers, zinc aluminum alloys, and superalloys. Various periodicals, such as *Advanced Materials Processes, Machine Design*, and *Modern Plastics*, publish annual materials selectors that provide general information on properties of a long list of materials. There are also many handbooks, such as the multiple-volume *Metals Handbook* and the series of *Engineered Materials Handbooks,* that cover nonmetals; both are published by ASM International. General databases from handbooks provide much detail, but the final selection often requires consulting the materials manufacturers' own properties database for their product lines.

Properties of materials receive extensive treatment throughout this book, and appendix tables present the properties of representative materials. Throughout we provide numerous examples of alternative designs, materials, and processes from many fields of technology. These examples help to illustrate the complexity of materials selection. As pointed out earlier, materials selection is not simply a matter of making decisions about cost and properties.

NASA photos and animation—*http://oea.larc.nasa.gov/Photo/photo.html*

1.3.3 Obstacles to Change

Switching from traditional materials like steel and concrete to newer materials like plastic-based composites seems a simple, straightforward approach for the contemporary designer. The newer materials are often superior. But sometimes there are complications. Often, lack of experience with a new material causes designers to hesitate. Departures from tried-and-true materials may also be costly, as was the case with plumbing pipes. Materials selection is a problem that requires an algorithm for its solution.

Change is often hampered by obstacles, but sometimes changes occur by chance or *serendipity*. In the history of materials development there are some notable examples of a successful new material resulting from accident or an apparent failed development. The Post-it Notes system from the 3M Company is an example of a very successful product, now available in many forms (Figure 1–11), that came about from a failed effort to develop a strong adhesive; that instead, a weak adhesive was the result. The adhesive that did not stick became "temporarily permanent" markers that had just the right amount of adhesion to stick to paper and other surfaces without damaging them.

3M Company innovations—*http://3m.com/*

1.3.4 Materials Systems

Materials rarely exist in isolation; they interact with other materials and are selected to complement one another. In a successful ***materials system,*** each component is compatible with the others while contributing its distinctive properties to the overall characteristics of the system of which it is a part. A state-of-the-art cell phone is a good example (Figure 1–1). The casing might be a tough ABS plastic, which houses a microchip (a solid-state ceramic device) that provides memory and sound-transmission capabilities. Copper leads join the circuitry together. There might be a battery and a ceramic light-emitting diode to show when the battery is low. The acid in the battery must be isolated to prevent corrosion, and the copper leads must be insulated so that they do not short out. Each component is made of materials that meet the

Figure 1-11 Failure turned to success in the form of Post-it note pads. An alert 3M Company researcher realized that a weak adhesive had value for bookmarkers and other temporary notes as seen in the wide variety of Post-it products.

demands of the physical and chemical environment normally encountered when using the system.

1.3.5 Materials Selection in This Book

Many of the illustrations and tables in this book provide examples of materials selection. The Self-Assessment at the end of each chapter will provide practice in materials selection as well as test other knowledge gained from the module text.

Stop!.. *You are also encouraged to constantly observe materials applications in your everyday life. Keep a materials journal in which you write materials applications as "observations." Next to the observations section, place a section entitled "analysis"; record there the positive or negative aspects of the observed use of materials. For some observations, state how you could make materials substitutions to improve on the designer's choice. Discuss these improvements with others. As you progress through the book, your observations and analyses will become more sophisticated. This valuable knowledge will serve you well as a consumer, as an intelligent citizen, and in your career.*

1.4 TECHNOLOGICAL LITERACY

The National Academy of Engineering report *Technically Speaking: Why All Americans Need to Know About Technology* provides a definition of technology:

> In the broadest sense, ***technology*** is the process by which humans modify nature to meet their needs and wants. However, most people think of technology only in terms of its artifacts: computers and software, aircraft, pesticides, water-treatment plants, birth-control pills, and microwave ovens, to name a few. But technology is more than its tangible products. An equally important aspect of *technology is the knowledge and processes necessary to create and operate products*, such as engineering know-how and design, manufacturing expertise, various technical skills, and so on. *Technology also includes all of the infrastructure necessary for the design, manufacture, operation, and repair of technological artifacts*, from corporate headquarters and engineering schools to manufacturing plants and maintenance facilities. (pp. 2, 3)

The report lays out the benefits of technological literacy and spells out the characteristics of a technologically literate citizen:

> Society continues to become more complex. Corporate "downsizing," company restructuring, and similar trends that aim at fewer workers for greater profits place people in jeopardy of losing their jobs. To avoid being a victim of technological advancements, we must be prepared to change jobs several times during our working lifetime. To be agile in the job market, one must become ***technologically literate***, that is, understand the language and concepts of technology to understand new technological advances. Technological literacy is not only important from a career standpoint, it is also required to function in our ever-changing technological society.

1.4.1 The Materials Consumer

Each year, an increasing number of newly developed materials are substituted for more familiar materials that possess limited properties. Today, 40% or more of a manufactured item's price represents materials cost. To be an informed and intelligent consumer requires a basic understanding of materials. The selection of a product can be improved by a greater knowledge of the nature and properties of the materials used in the product. After a product has been purchased, the problem of failure, sometimes caused by the *abuse* or *misuse* of the product, can be lessened by such knowledge. Learning about the structure of materials, hence how materials behave, should permit an intelligent analysis of a failure and possibly pinpoint its source and cause. A knowledgeable consumer stands a much better chance of success in demanding remedial action from both manufacturers and retailers of faulty products than does one with less knowledge of the behavior of materials and a poor technical vocabulary with which to explain such behavior.

Wood is our oldest building material and possesses unique structure, warmth, and beauty; perhaps more significant is its ability to renew itself, making it equal in importance to the newer materials (Figure 1–12). The complex structures (Figure 1–13a) and wide varieties of wood, a natural composite, give it many desirable properties. Figure 1–12b illustrates a wall system to replace older wood "stick" construction; various forms and types of plastics replace wooden two-by-fours. Steel studs and plastic/concrete wall systems are also gaining wider use in residential and commercial construction. Still, materials technology always looks for improvements and wood often finds stiff competition, in for example, a new design for drumsticks (Figure 1–13b). In 1850, about 90% of U.S. energy came from wood, in both the raw state and refined into charcoal for fuel in glass and iron making. By 1900, wood accounted for only 20% of the nation's fuel and continued to decline. The use of wood as a fuel is now on the rise in both home and industry. The lumber and paper industries, for example, make extensive use of wood residues, including waste chemical by-products and bark. Wood does not figure as a major fuel for meeting the void to be left by depleted fossil fuels, but it holds promise as a substitute for more energy-intensive materials.

In construction materials for similar phases of building, a comparison of the fuel required for the preparation of building products shows that wood requires the least energy and plastic requires nearly six times and steel about eight times the amount of fuel needed for wood products. The increased use of concrete, glass, and other ceramics also shows great promise for building and manufacture because the raw materials are as numerous as the sands of a desert. The scarcity of old-growth trees and competition for high-grade wood to make lumber, furniture, and so on has also resulted in the development of new composites that may be superior to wood for many construction and nonbuilding applications. Consumers can make choices among products made of new composites as well as natural and synthetic polymers. Knowledge of the properties of materials will aid in decision making (Figure 1–13b and c).

1.4.2 The Intelligent Citizen

The technology of materials also provides us with the knowledge necessary to make decisions based on personal values relating to political, social, and ecological issues. *A better-informed citizen is a better citizen*, who is much needed in today's changing technological society in which the great issues over energy and materials resources are being debated and voted on. Citizens who comprehend key concepts of materials science and technology, such as the materials cycle and life cycle analysis, will be able to guide good legislation and participate in green manufacturing practices like reduced resources consumption and recycling. Long-range industrial research and development of new alloys and nonmetals, intensification of programs to conserve and reclaim metals, enlargement of the search for and development of new domestic sources, and the utilization of ocean resources are all affected by issues that find a source in the political, social, economic, or ecological spheres within our society. The fact that the United States is dependent on foreign sources for most metals (with the exception of iron and copper), plus the negative effects of our high rate of consumption of the world's known reserves, places greater emphasis on a good working knowledge of the technology of materials so that citizens can guide their government representatives in making the correct decisions on matters that will have a lasting effect on our lives and living standards.

(a)

(b)

Figure 1-12 (a) Most contemporary homes in the United States use the "stick-framed" system for walls. This system is hundreds of years old and does not take advantage of newer approaches to engineering walls. (b) A developmental wall system combines an experimental corrugated strand product with low-density polyphenylene oxide foam and reclaimed hamburger packages. The panel is covered with a glass-reinforced plastic that is an excellent fire retardant. (General Electric Plastics)

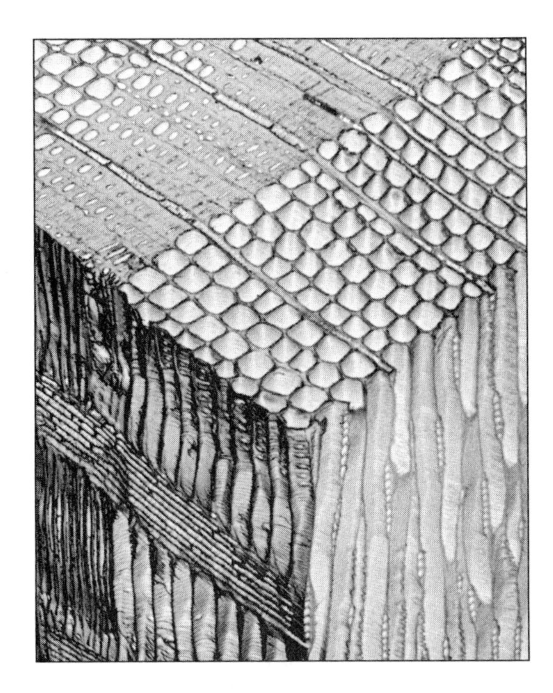

Figure 1-13 Wood's complex structure. (a) Electron microscope reveals microstructure. (Dr. Wilfred A. Côté, SUNY College of Environmental Science and Forestry)

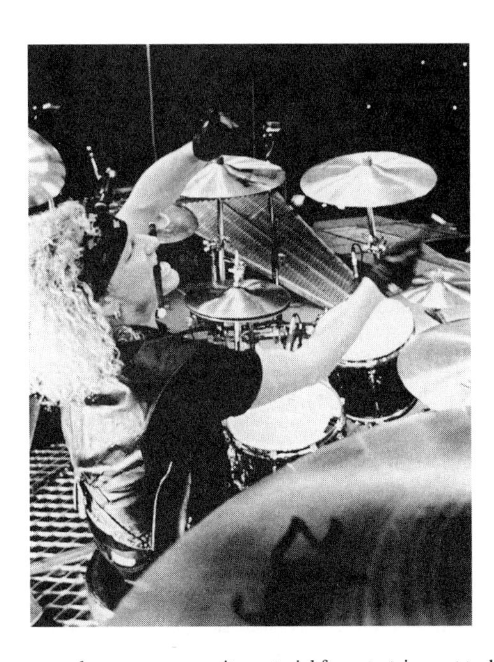

Figure 1-13 (b) Improving on wood—a new composite material for entertainment technology. The drumstick combines the strength and vibration absorption efficiency of an aluminum core with the strong sound-making ability of the cover molded of a thermoplastic polyurethane resin. The resin is hard enough to produce strong sound and not wear out, but soft enough not to damage the drum sets. When the covers wear out, they can be replaced for less than half the cost of wooden drumsticks. The first major endorsement of the new advanced high-efficiency alloy drumstick (AHEAD) was by a heavy metal rock star, drummer Matt Sorum of the band Guns N' Roses. (*Advanced Materials & Processes*)

Figure 1-13 (c) A wide range of natural and synthetic polymers. There are natural polymers, including the teak-laminated cabinet, leather boots with rubber soles, cotton jeans, and skin and hair of Reid and Connor. The chair is made of such synthetic polymers as polyurethane and nylon fabric. The carpet of natural fiber wool Berber covers a plywood (composite wood) floor. Teak, used on the cabinet, is a beautiful wood, but quite expensive. Laminating 1 millimeter of teak over particleboard (made of wood chips) makes beautiful furniture that is available at a more economical price.

APPLICATIONS & ALTERNATIVES

As consumers demand better products, the technological team of technicians, scientists, engineers, crafters, and technologists will improve materials and processes to achieve better design and higher quality. One example of applying new materials and processing technology is seen in the optimum compaction low void (OCLV) bicycle in Figure 1–14.

To produce a world-class competition bike, the U.S. company Trek Bicycle Corporation set out to design an agile, lighter, and more durable frame. The uniquely constructed one-piece main frame consists of 65% carbon fiber strands that have been compacted in an epoxy resin matrix by a process that yields less than 1% void (air pockets) in the finished composite frame. Trek advertises the OCLV carbon frames as the world's lightest, stiffest, and strongest. Of their frames made of ZX aluminum alloy, Trek states that the alloy achieves desirable properties by combining aluminum, with its low density, with magnesium, for tensile strength, and zinc, to improve corrosion resistance.

Attached to the rear of the OCLV frame is a 6061 T aluminum alloy rear triangle that is welded by tungsten inert gas (TIG) for a light and stiff structure. Trek maintains that human crafters provide the precision and passion for bike assembly that robots do not possess. It is easy to accept the passion of humans, but many will question whether humans perform with greater precision than robots.

As you see in Figure 1–14, the OCLV model bike uses various materials, including parts made of rubber, aluminum, steel, and plastic, that complement the carbon composite frame. As with many newer product designs, adhesive bonding plays a role in joining dissimilar materials. Advanced bonding technology (ABT) relies on epoxy resin to glue components into alignment to achieve the designed geometry without the need for postfabrication adjustments, as required in other fabrication techniques. Teflon-impregnated composite bearings are also bonded into the rear aluminum triangle and main frame to eliminate maintenance and lubrication.

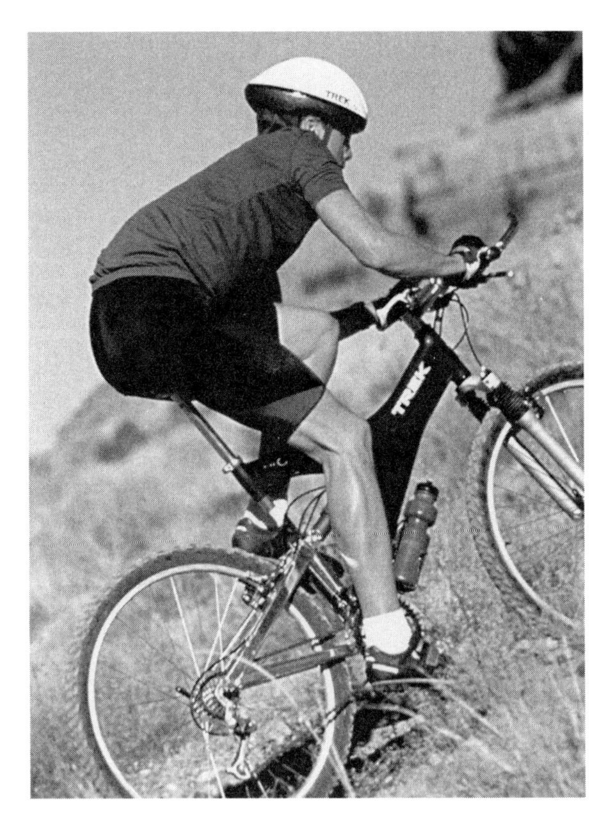

Figure 1-14 Advances in engineering materials technology allow for improvements in many products. Advanced composites developed by the aerospace industry are now found in many types of sporting equipment. The OCLV frame for the performance competition racer, seen here, uses carbon fiber composite technology similar to that used for stealth fighters and the space shuttle. (Trek Bicycle Corporation)

The Trek bicycle and its innovations illustrate the aggressive way in which forward-thinking businesses seize on advances in materials technology to propel themselves past their competitors. Their engineers selected from ***designed materials*** rather than settled for **off-the-shelf materials** (those readily available materials with limited structure and properties that you would find in a warehouse or hardware store).

What materials systems can you recall? Watch for uses of materials systems as you read newspapers and magazines or observe products because often an idea used in one field can be modified and employed in another. Can you see applications in other fields for materials systems that use the principles discussed here?

SELF-ASSESSMENT

1-1. Recall the Pause & Ponder and Applications & Alternatives in light of what you have read in this module. Which statement is MOST TRUE about materials technology?

- **a.** Materials science and engineering usually follows the lead of developments in other technologies.
- **b.** Industry and the consumer are quick to accept new materials and processes innovations.
- **c.** Developments in materials science and engineering have allowed for innovations in most other technologies.
- **d.** Existing specifications and laws make it easy to introduce new materials in fields such as building construction and commercial aircraft.

Questions 1–2 through 1–5: Choose the BEST TERM to match the definition:

a. Engineering plastics
b. Engineering materials
c. Engineering materials technology
d. Materials
e. Materials science
f. Materials engineering
g. Conventional materials
h. Materials science and engineering

1-2. A term that loosely defines most materials in products and systems.

1-3. A field that deals with developing, preparing, modifying, and applying materials to a specific need.

1-4. A broad term that covers fields of applied science related to all aspects of materials, materials processing, and the many engineering specialties dealing with materials such as those used in manufacturing and construction.

1-5. A field that involves the generation and application of knowledge relating the composition, structure, and processing of materials to their properties and uses.

1-6. Which concept of the materials cycle has been LEAST AFFECTED by recent trends?

a. Concern for the entire cycle that includes effects of harmful by-products and unrecycled old products
b. Ability to design materials to meet ever-increasing demands of technology rather than use "off-the-shelf" materials
c. Need for a more highly educated workforce to support new manufacturing techniques
d. Desire to achieve the maximum properties from any given material

1-7. Transforming gaseous, liquid, and solid elements by chemical and physical means, where atoms or molecules are combined to form solid materials is called

a. Wroughting
b. Synthesis
c. Analysis
d. Processing

1-8. Control of structure at higher levels of aggregation to yield bulk materials is called

a. Wroughting
b. Synthesis
c. Analysis
d. Processing

1-9. Composites produced with innovative processes that yield high strength and stiffness plus weight reduction are

a. Advanced composites
b. Fiberglass
c. Reinforced concrete
d. Hydrocarbons

1-10. Manufacturing engineering has evolved into a sophisticated field of study. Which two aspects of fabricating products and systems contribute most to the new approaches of this field?

a. Systems approach
b. Chemistry
c. Automation
d. Mathematics

1-11. What approach to design reflects a new concern for disposing of products after their useful life?

a. JIT
b. Design for assembly
c. Design for disassembly
d. CIM

1-12. Which is *not* an obstacle to acceptance of new engineering materials and processes?

- **a.** Lack of experience by designers and fabricators
- **b.** Concern for the safety of new techniques
- **c.** Need for compromise between favorable properties and cost
- **d.** Concern for keeping up with one's competition

1-13. What factor in materials selection usually dominates the final choice?

- **a.** Cost
- **b.** Processibility
- **c.** Recyclability
- **d.** Weight

1-14. Life cycle analysis aids in ensuring

- **a.** High profits
- **b.** Sustainable environment
- **c.** Minimal materials
- **d.** Improved cycles

1-15. Which of the following best fits the category of a materials system?

- **a.** Gold
- **b.** Battery
- **c.** Plastic spoon
- **d.** Titanium

1-16. Name the method used during design and manufacturing to determine the impact of a product or system on the environment.

- **a.** Analysis
- **b.** TQM
- **c.** Life cycle analysis
- **d.** Just in time

REFERENCES & RELATED READINGS

AMATO, IVAN. *Stuff: The Materials the World Is Made Of.* New York: Basic Books (HarperCollins), 1997.

AMERICAN SOCIETY FOR TESTING AND MATERIALS. *Compilation of ASTM Standard Definitions*, 7th ed. Philadelphia, PA: ASTM, 1990.

ASHBY, MICHAEL F. "Materials in Mechanical Design," *Journal of Materials Education*. Vol. 15, pp. 143–166, 1993.

ASM INTERNATIONAL. *ASM Engineered Materials Handbook Desk Edition*, 1995.

ASM INTERNATIONAL. *ASM Materials Engineering Dictionary*, 1992.

ASM INTERNATIONAL. "Green Manufacturing," *Advanced Materials and Processes*. January 1997, pp. 25–28.

ASM INTERNATIONAL WEB NEWS–http://asminternational.org, February 23, 2003.

BALL, PHILIP. *Made to Measure: New Materials for the 21st Century*. Ewing, NJ: Princeton University Press, 1997.

CALLISTER, WILLIAM D., Jr. *Materials Science and Engineering: An Introduction*, 6th ed. New York: John Wiley & Sons, Inc., 2003.

CANBY, THOMAS Y. "Advanced Materials: Reshaping Our Lives," *National Geographic*, December 1989, pp. 746–781.

HUMMEL, ROLF E. *Understanding Materials Science: History-Properties-Applications*. New York: Springer-Verlag, 1998.

MOTOROLA WEBSITE– http://www.motorola.com/, January 23, 2003.

MULHALL, DOUGLAS. "Ban what we don't yet understand . . . to be green." *Smalltimes*, Sept./Oct. 2002, pp. 33–35.

NATIONAL ACADEMY OF ENGINEERING. *Technically Speaking: Why All Americans Need to Know About Technology*. Washington, D.C.: National Academy Press, 2002.

NATIONAL RESEARCH COUNCIL. *Materials Science and Engineering for the 1990s: Maintaining Competitiveness in the Materials Age*. Washington, D.C.: NRC, 1989.

ROGERS, CRAIG A. "Intelligent Materials Systems: The Dawn of a New Materials Age," Blacksburg, VA: Center for Intelligent Materials and Structures, Virginia Polytechnic Institute and State University, 1992.

SCHWARTZ, LYLE H. "Industry, Government Cooperation: New in a Changing World," *Advanced Materials & Processing*, February 1996, pp. 110–116.

SME. "New Nylon 6 Grades Environmentally Friendly," *Molding Systems*, October 1999, pp. 8, 9.

3M COMPANY WEBSITE. http://3m.com/profile/looking/1980.html, October 21, 1999.

Periodicals

Advanced Materials and Processes
Automotive Engineering
Machine Design
Manufacturing Engineering
MIT's Technology Review
Popular Science

Module 2

Nature & the Family of Materials

After studying this module, you should be able to do the following:

2–1. Using the materials and civilization time line, give an example of how materials technology developments sparked new technology in the Scientific Age.

2–2. Use the periodic table of elements and illustrations to determine and/or calculate the atomic structure of elements, ionization potential, and electronegativity.

2–3. Apply the SI rules for prefixes in expressing multiples of quantities.

2–4. Use sketches and explanations to describe the microstructure and macrostructure of solid-state materials plus bonding forces within materials.

2–5. Discuss the significance of fullerenes and the polymorphs of carbon.

2–6. Name the five main groups that make up the family of materials; cite examples of materials within each group; and sketch the typical microstructure for ceramics, metals, polymers, and composites.

2–7. Describe intelligent materials systems (smart materials) and discuss some goals of materials science and engineering (MSE) for this class of materials.

2–8. Use the website addresses denoted by the icon* to further explore the nature of materials.

*The symbol found throughout the book will link you to Internet sites related to topics being covered. The dynamic nature of the Web brings frequent changes, so some of these sites, while available at the time of writing, may not be up now.

A study of history shows that materials technology had major influences on societies. Periods of civilization have been named for the materials that dominated each era, from the earliest time of primitive people in the Stone Age through the Bronze Age and the Iron Age. The era we live in has been given many names, including the Scientific Age, the Information Age, the Space Age, and even the Materials Age. Figure 2-1(a) provides a historical perspective of the concurrent development of civilization and materials technology and a graphical comparison of the relative importance of materials through the ages. Note that it divides materials into four groups: metals, polymers, composites, and ceramics. In this module, we will provide an in-depth look at these groupings as we study the family of materials.

The Stone Age, which includes prehistory, was so named because it was then that civilization began to make great evolutionary strides through the development of stone implements. Naturally occurring flint permitted shaping, and people discovered that flint and other stones could be joined with sticks to produce tools and weapons. Clay dug from the earth was formed and dried into pottery. People of this period (which might more accurately be called the Ceramic, Hide, and Wood Age) thus used naturally occurring "engineering materials." Note in Figure 2-1(b) and (c) that Stone Age practices persisted into the sixteenth century in North American cultures and into the nineteenth century in some primitive cultures located in remote jungles.

The Bronze Age represented another major leap forward in "materials technology" as the first metal alloy, the melting together of copper and tin, to produce bronze, was developed. Bronze was a superior metal useful for shields, knives, ornaments, cups, and urns. Note in Figure 2-1(c) that ceramics, still widely used, were also improving during the Bronze Age, which saw the development of glass and composite bricks (clay and straw). The numerous materials-related quotations in ancient sources, like Deuteronomy 8 in the Old Testament, in which Moses describes the Promised Land as "a land whose stones are iron, and out of whose hills you can dig for copper" is evidence that a large proportion of the people must have known and possessed these materials in those early times. The discovery of iron, around 1100 B.C., ushered in the Iron Age. Iron became of great importance to society, for its abundance and fine engineering properties made it superior to bronze. Eventually, the alloying of iron with carbon and other elements to produce steel made it the superior engineering material. To this day, steel has been the dominant engineering material, although often challenged. From the Iron Age and on into the industrial revolution and the Scientific Age, thousands of iron and steel alloys have been developed by metallurgists to meet the demands of society.

Throughout most of this history, humans have also initiated chemical reactions to produce new substances from natural materials, as in the production of pottery from clay, but many of these reactions were discovered accidentally. ***Chemical reactions*** are made up of two main ingredients or substances. One group of substances, called ***reactants,*** may consist of one or more existing substances. The second category of substances, called ***products,*** are the new substances produced by the reaction. Stated another way, a chemical reaction is a process in which atoms in the reactant molecules are rearranged to form the molecules of the products. The reaction can be represented in chemical shorthand by a ***chemical equation,*** which uses the symbols for the elements found in the periodic table. These equations tell us more than just what substances react together and what substances are produced; they show us how many atoms or molecules of each reactant take part in the reaction as well as how many atoms or molecules of each product are formed. In other words, a chemical equation is a good bookkeeping device that keeps track of the atoms, molecules, and elements involved in the reaction. Ordinary chemical equations are symbolic statements that show both qualitatively and quantitatively what has taken place. (*Qualitative* denotes a description of something, whereas *quantitative* refers to a measurement of how much of something is present). These equations primarily involve the outer shell electrons (valence electrons) of the atoms. The various chemical reactions discussed throughout this book, such as corrosion of materials (Module 5), most forms of which are electrochemical in nature—that is, they involve the flow of electrons—can be explained by writing chemical equations. Chemical equations facilitate understanding of the reactions of the environment upon materials.

Pure aluminum, the third most abundant element in the earth's crust, is a product of the discovery of a chemical reaction. Aluminum remained a rare and expensive metal until 1886,

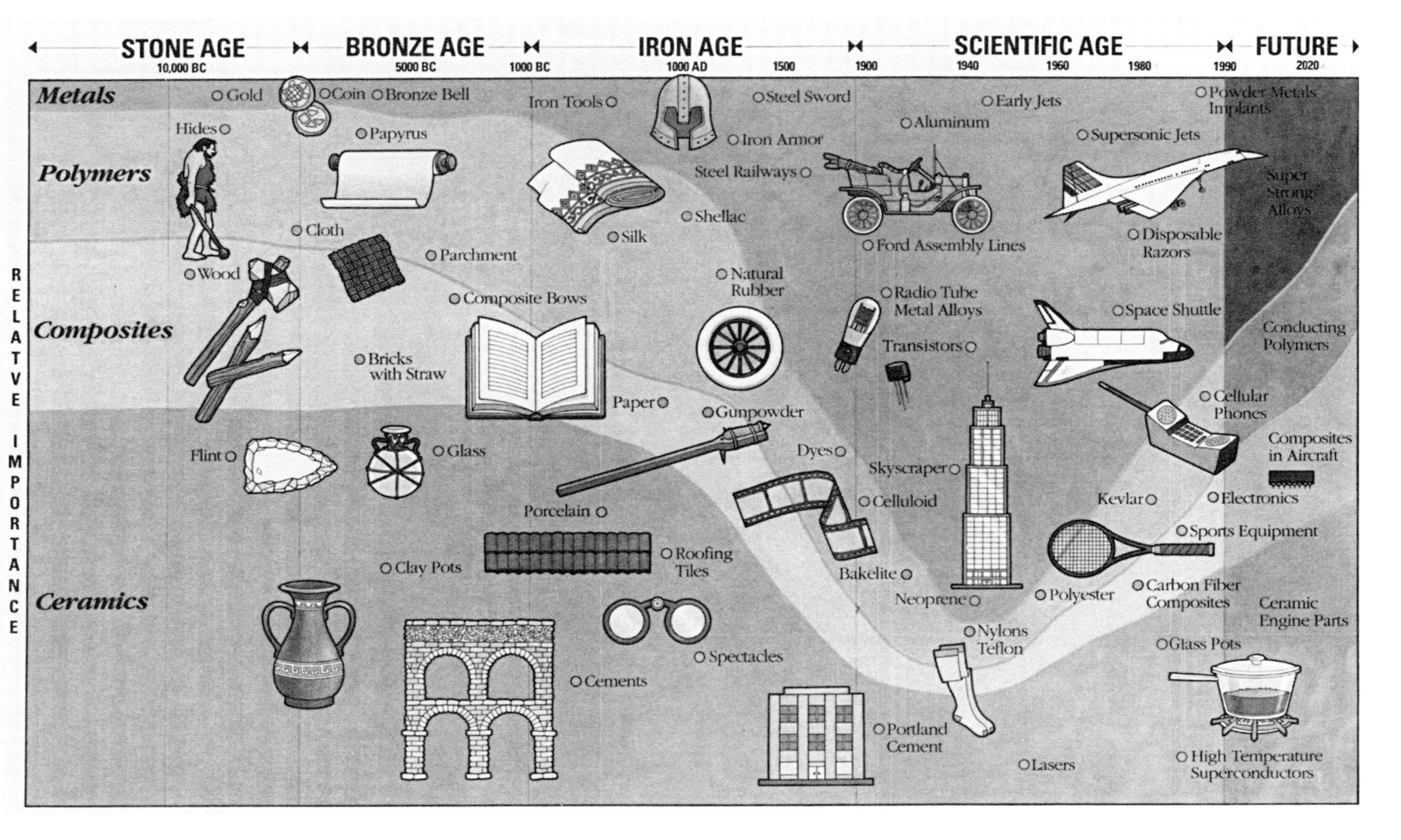

Figure 2-1 (a) Materials and civilization time line. (From Future Materials exhibit, Franklin Institute Science Museum)

(b)

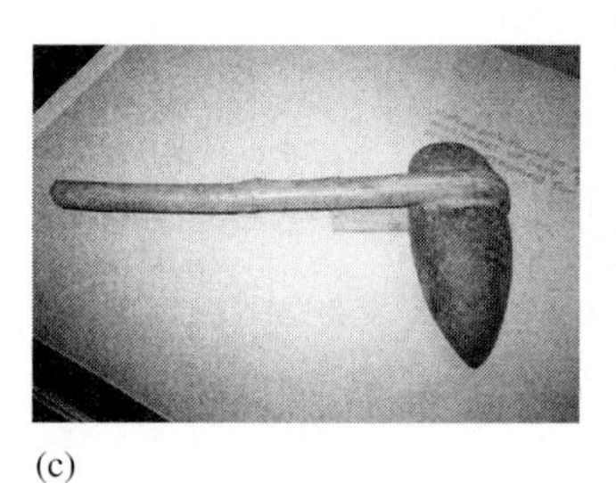

(c)

(d)

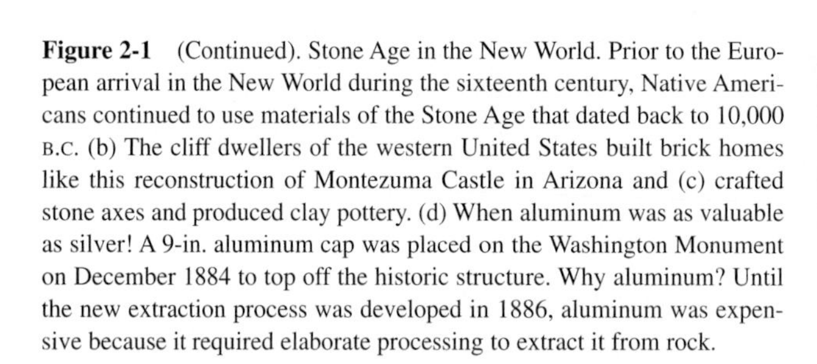

Figure 2-1 (Continued). Stone Age in the New World. Prior to the European arrival in the New World during the sixteenth century, Native Americans continued to use materials of the Stone Age that dated back to 10,000 B.C. (b) The cliff dwellers of the western United States built brick homes like this reconstruction of Montezuma Castle in Arizona and (c) crafted stone axes and produced clay pottery. (d) When aluminum was as valuable as silver! A 9-in. aluminum cap was placed on the Washington Monument on December 1884 to top off the historic structure. Why aluminum? Until the new extraction process was developed in 1886, aluminum was expensive because it required elaborate processing to extract it from rock.

when an American, Charles Martin Hall, and a Frenchman, Paul Heroult, independently devised a practical process for electrolytic production of aluminum. According to his sister's comprehensive laboratory reports, Hall discovered the aluminum process about a month before Heroult. Aluminum oxide has a melting point of 2060°C, which is too high for convenient and economical electrolysis in the molten state; however, the addition of energy in the form of heat by passing an electric current over the aluminum mineral reactants (bauxite and cryolite) produces pure aluminum. The basic process can be represented by a simple chemical equation, $2Al_2O_3 \rightarrow 4Al + 3O_2$, which shows the aluminum oxide on the left as the reactant and the new substances (products) produced, namely the pure aluminum (Al) plus oxygen (O) on the right of the arrow. Aluminum was as valuable as gold prior to development of the Hall–Heroult process; in fact, Napoleon commissioned a full set of "priceless" aluminum dinnerware. In 1884, just prior to the new aluminum extraction process, the Washington monument was capped with a valuable 9-inch aluminum pyramid (Figure 2-1d). Still used today, the Hall–Heroult process is the largest consumer of electricity in the United States. Approximately ten times as much energy is required to produce a ton of aluminum as is needed to produce a ton of steel, making the recycling of aluminum products most worthwhile. With the discovery of the new process for producing aluminum, however, iron and steel had competition from a lightweight, corrosion-resistant metal.

Chemistry also was responsible for the development of plastics and rubbers, including Bakelite, Celluloid, Neoprene, nylon, and Teflon (Figure 2-1), which were economical substitutes for natural materials such as ivory, tortoiseshell, wood, and silk. Experimentation also continued with new combinations of metals (alloys) to meet the needs of modern communications, transportation, and construction. New types of steel made possible the building of tall skyscrapers, which supported the growth of large cities. Soon the radio vacuum tube was supplanted by transistors, which were eventually replaced by the microchip as transistors were reduced in size and combined with other circuit components to produce integrated circuits made with semiconductor materials. Today, it is exciting to follow the application of new electronics materials to computer technology in such areas as artificial intelligence (AI), which mimics human reasoning, and virtual reality (VR), which expands the dimensions of visualization using computer graphics.

As we move into the future, the results of research and development of the Scientific Age or Age of Materials are moving technology away from the long-standing dominance of iron and steel. Notice in Figure 2-1 that the curves indicating the relative importance of metals, polymers, composites, and ceramics are altering. Advanced materials like the fiber Kevlar used in aircraft and spacecraft, high-performance ceramics for engine parts, and high-temperature ceramic superconductors are causing the zones of relative importance for ceramics and composites to widen. Continued use of powder metals, super-strong alloys, and conducting polymers will maintain the importance of the polymer and metal technologies, but they will not retain the dominance of recent eras.

The new products and systems to be developed in the twenty-first century will call for improved materials—the advanced materials and the smart materials. These materials will be produced through scitech as materials engineers work with manufacturing engineers and materials scientists to invent new materials processing. As you study the nature and family of materials, keep in mind the history of materials and recognize that technology can cause many shifts in their relative significance. It is important for you to understand the similarities and differences among the main groups of materials and to keep abreast of new developments.

Franklin Institute Science Museum—*Http://sln.fi.edu*
Smithsonian Institution—*Http://www.si.edu*

2.1 NATURE

Through an understanding of the nature and structure of materials, one can predict how materials should behave when exposed to certain forces and environments. The nature of materials results from their composition and structure, which determine their properties. In this module, we will present basic concepts of atomic and molecular structure that lay a foundation for the discussion throughout the book on the nature, properties, and uses of materials. The ability to grasp these fundamental materials technology concepts not only allows you to deal with today's materials and processes but also prepares you for new developments in materials.

2.1.1 Internal Structure

The structure of materials, like the structure of a building, develops through the joining of smaller units. Particles of earth form bricks, bricks are stacked together to form walls, walls form rooms, and rooms make a building. ***Matter*** is anything that has mass and occupies volume. It exists in one of four physical states: solid, liquid, gas, or plasma. ***Plasma*** exists only under drastic conditions. At very high temperatures, matter is shredded into positively charged particles. These charged particles move about in a cloud of negatively charged electrons. On the surface of the sun, hydrogen and helium gases are changed into plasma by the heat of the sun (6000°C). A ***state*** of matter is the condition of a substance, plasma, gas, liquid, or solid at a particular temperature. A ***solid*** is a sample of matter that has a fixed volume and a fixed shape. A ***liquid*** is a sample of matter that has a fixed volume but assumes the shape of the container it occupies. A ***gas*** is a sample of matter that has neither fixed volume nor fixed shape. Any substance may assume any of the four states of matter under varying conditions, in particular, as the temperature changes. This last statement describes the convertibility of these states of matter, which is a property of most substances. Remember that in all four states we are dealing with the same substance (see Figure 3-39 for a graphic picture of three states of water).

The terms *mass* and *weight* are often used interchangeably. ***Mass*** is the amount of matter in an object. Mass is a scalar quantity (no direction or sense), which signifies that it possesses only a magnitude. It is the term used to express an object's resistance to change in its state of motion or at rest. The greater that resistance, the greater the mass. On earth, the force that brings about a change in motion in an object is the force of gravitational attraction between the earth and the object. This gravitational force is called ***weight,*** a vector quantity whose magnitude and direction must be defined. The magnitude and direction of the weight depends on the mass of the object, the mass of the earth, and the distance of the object from the center of the earth. The SI unit for measuring the mass of an object is the kilogram (kg); the SI unit for measuring force is the newton (N). Remember, if an object such as the lunar lander is moved to the moon, its mass stays the same as what it was on earth, but its weight would change because it would be subjected to a different force of gravitational attraction.

When two or more ***atoms*** are joined by covalent bonds, the unit of matter formed is called a ***molecule.*** An ***element*** is a substance that cannot be broken down any further by chemical reaction. Its atoms all have the same atomic number (A, to be defined later). Examples of elements are carbon, oxygen, or sulfur. Elements occur in nature as mixtures of their isotopes. Some elements, such as phosphorus, have only a single naturally occurring isotope. Elements combine to form compounds. An example of a compound is table salt, NaCl. A ***compound*** is a substance that can be broken down into elements by a chemical reaction. Molecules combine through chemical reactions to form polymeric substances known as ***polymers.*** Elements, molecules, and compounds can all exist in gas, liquid, or solid form.

2.1.1.1 Atomic structure.

An ***atom*** is the smallest particle of an element that possesses the physical and chemical properties of that element, and it is the smallest particle of an element that can enter into a chemical change. Atoms are the basic building blocks of matter. The average diameter of an atom is only about 10^{-10} meter. It would take more than 10^6 atoms aligned edge to edge to make up the thickness of this page.

What the atom really looks like or how the charged particles distribute themselves and move within the atom is not known exactly. Consequently, scientists resort to various models of an atom to help explain its characteristics. One model, the planetary model, appears to be the most popular in explaining the nature of an atom (see Figure 2-2); however, it is far from being the most accurate. In the model the atom is pictured as a sphere with a very dense cen-

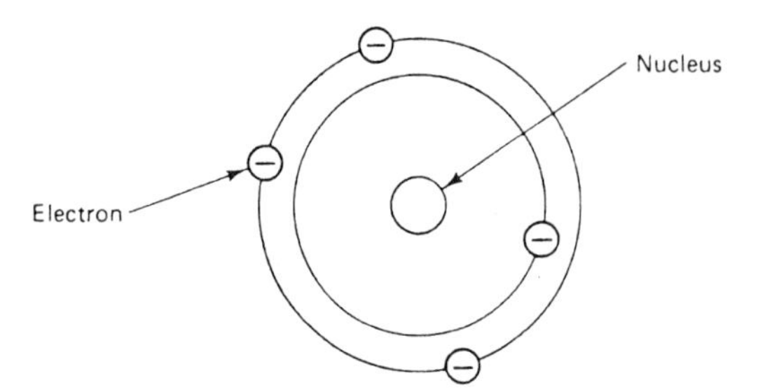

Figure 2-2 Planetary model of an atom.

ter, the nucleus, containing the protons and neutrons and around which negatively charged particles (the electrons) are moving in circular or elliptical orbits, much like the motion of the planets around the sun. The diameter of the nucleus is about 1/10,000 the size of an atom, but it contains more than 90% of the mass of the atom. If one represented the nucleus as a golf ball, the electrons would revolve in a sphere with a radius of about $1\frac{1}{2}$ miles from the nucleus.

For most atoms and their isotopes the nucleus is spherical and stable. Long ago, in studying ***radioactivity*** (i.e., the spontaneous disintegration of a nucleus), it was determined that the nuclei of all the isotopes of elements with atomic numbers greater than 83 were "unstable"; that is, they decompose into nuclei of other elements by giving off alpha particles, electrons, or gamma rays. However, atoms with atomic numbers between 51 and 67—and more specifically europium (Eu) and holmium (Ho), which contain 63 and 67 protons, respectively—are unstable and have "highly deformed" nuclei. Europium and holmium are part of a group of elements called the ***lanthanides,*** or rare earth elements, that belong in Period 6 but are separately listed below the main table. After studying europium and holmium, Argonne National Laboratory determined in 1998 that they engage in a rare form of radioactivity called ***proton emission.*** The rate of ejection depends on the shape of the nucleus, which was determined to be in the shape of a football. (For further information on radioactivity see Module 7.)

Another model of the atom, sometimes sketched in three dimensions, attempts to show the spaces within which the electrons are confined as fuzzy volumes. The atom can be thought of as a sphere of empty space with 90% of its mass located in its center or nucleus. For our purposes, the nucleus consists of protons and neutrons. A ***proton*** is a particle of matter that carries a positive electrical charge equivalent to the negative charge on an electron. ***Neutrons*** are uncharged particles in the nucleus with a mass nearly equal to the proton's mass. The ***atomic number*** Z of an element is equal to the number of protons in the nucleus, which equals the number of electrons in the atom in its balanced, neutral, or equilibrium state. The periodic table (see Table A-1A) arranges the elements in order according to their atomic numbers (Z). An element may be defined as a substance whose atoms all have the same atomic number (**Z**).

All atoms of an element contain the same number of protons, but they can contain different numbers of neutrons. In other words, these atoms are different, but their positive and negative electric charges are identical. In fact, atoms of most elements are not identical. Later we will see that the number and arrangement of electrons in an atom, not the number of neutrons, determines the chemical properties of an element. Atoms of an element with different numbers of neutrons are called ***isotopes.*** Isotopes of the same element share the same chemical properties because they differ from each other in mass only. Another way of describing an isotope is to say that atoms of an element that have different masses are isotopes. For example, hydrogen has three isotopes (see Figure 2-3). Tin has ten naturally occurring isotopes.

Illustrative Problem

Given the following isotopes

(a) ${}^{14}_{7}X$ (b) ${}^{11}_{5}X$ (c) ${}^{12}_{6}X$ (d) ${}^{13}_{6}X$ (e) ${}^{15}_{7}X$

Find:

(a) Which of the above have the same mass numbers?
(b) Which have the same number of neutrons?
(c) Which are isotopes of the same element?

Solution

(a) a and d (b) b and c (c) a and e are isotopes of N and c and d are isotopes of C

Note:

${}^{14}_{7}N$ and ${}^{15}_{7}N$ Isotopes of nitrogen N
${}^{12}_{6}C$ and ${}^{13}_{6}C$ Isotopes of carbon C
${}^{11}_{5}B$ One of two isotopes of boron B

Complete each of the following symbols and list the number of protons, neutrons, and electrons respectively in each atom.

(a) Al (b) C (c) Cl

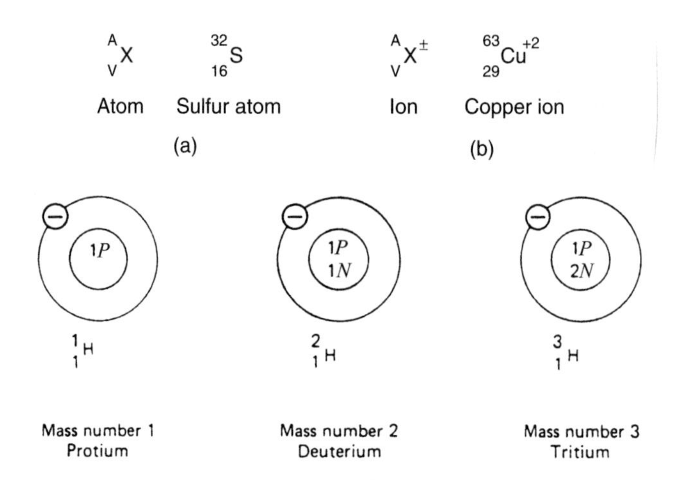

Figure 2-3 (a) Standard notation for describing the composition of the nucleus of an atom. A, the superscript, is the symbol for the mass number, V, the subscript, is the atomic number, and X represents the symbol for the element. (b) An atom that has either lost or gained electrons, an ion, carries a net electric charge. This charge is equal to the number of protons minus the number of electrons. This net charge is indicated by a + or − number to the right and above the Cu, the symbol for the element Cu. This copper ion has a mass number A of 63. Its atomic number Z indicates that it has 29 protons. If it had 29 electrons, it would be a balanced atom with no electric charge. However the +2 charge indicates that it only has 27 electrons, which leaves a positive electric charge of +2. (c) Isotopes of hydrogen. 99.985% of hydrogen is in the form of protium. Deuterium represents only 0.015% of hydrogen. Tritium is an unstable isotope of hydrogen.

The total number of protons and neutrons in the nucleus is the ***mass number of the nucleus* (A)**. Protons and neutrons are referred to as ***nucleons.*** These nucleons are composed of still smaller particles called ***quarks,*** which might be considered the fundamental building blocks of all materials. The mass of an atom is concentrated in the nucleus while the electrons account for most of the ***volume.*** Using the new terminology introduced in this paragraph we can define an isotope of a given element as an atom with the same atomic number Z but different mass number A. ***Mass,*** or simply the quantity of matter of a substance, is expressed in the SI system of units using kilograms (kg). The mass of a hydrogen atom is 1.6735×10^{-27} kg or, using grams, 1.6735×10^{-24} g. Such small numbers are inconvenient to use; therefore, atomic mass is expressed in several other ways, including ***atomic mass units.*** An atomic mass unit (u) is defined as being equal to $\frac{1}{12}$ the mass of a single carbon-12 atom; the mass of one C atom is defined as exactly 12 u. Another way to display or convey the meaning of the average mass of an atom is to express the mass relative to $\frac{1}{12}$ the mass of a C atom. This ***relative average atomic mass*** (or ***average atomic mass***) has no units. For example, the average mass of an oxygen atom (with three isotopes) is 15.9994 times $\frac{1}{12}$ the mass of a C atom, so the relative average mass of an O atom is 15.9994. This atomic mass is often called ***atomic weight,*** but this is a misnomer because weight is different from mass, as previously explained. (Refer to the key for the elements in Table A-1A for use of the term *atomic weight.*)

The standard notation for expressing the composition of the nucleus is to write the atomic number (A) at the lower left (a subscript) of the symbol of the element and the mass number (Z) of its nucleus as a superscript at the upper left. For example, a carbon atom is expressed as ${}^{12}_{6}C$, indicating that the C atom has 12 nucleons in the nucleus of which 6 are protons. From that we know that the carbon atom has $12 - 6 = 6$ neutrons. All atoms have zero charge overall. The nuclear charge (+Z) must be balanced by the total number of the surrounding electrons, each of which has a negative charge of −1. This nuclear charge (Z) differentiates the atoms of an element from those of another. An element may also be defined as a substance whose atoms all have the same *atomic number* (A). Figure 2-3 shows this notation below the subshells for the isotopes of hydrogen. Hydrogen exists in three isotopic forms of which tritium is the unstable form; that is, it decomposes spontaneously into nuclei of other elements. All but two of the elements with atomic numbers less than 83 have one or more stable isotopes. Elements with atomic numbers 43 and 61 have no stable isotopes.

Illustrative Problem

Given the following atom $^{32}_{16}S$

Find:

(a) Mass number A
(b) Number of protons
(c) Number of electrons
(d) Number of neutrons
(e) Atomic number
(f) Number of nucleons

Solution

(a) A = 32 (b) Z = 16 (c) Z = 16 (d) Z = 16 (e) 32

Note: (1) Protons and neutrons are both nucleons. In an atom both the number of electrons and the number of protons represent the atomic number of the atom.

(2) Sometimes the atomic number is omitted since the symbol implies it. An isotope of oxygen is often written as ^{16}O (read oxygen-16) with the atomic number Z of 8 left out.

Electrons. In summary, an atom (see Figure 2-4) of an element is made up of a dense core, the nucleus, consisting of protons and neutrons surrounded by even smaller particles called ***electrons.*** The electrons carry a negative (−) electrical charge much like the charge at the negative terminal of a battery. Those electrons that occupy the outermost ring or shell from the nucleus are ***valence electrons. Valence*** is the capacity of an element to combine with other elements. For example, one oxygen atom can combine with two hydrogen atoms to produce the water molecule H_2O; oxygen is said to have a valence of 2. All elements in the same group in the periodic table have the same primary or principal valence. For example, sodium, potassium, and the other elements in Group 1 have a valence of 1. For Groups V to VIII, the principal valence is equal to $8 - n$, where n is the group number. Chlorine, in Group VII, has a valence of $8 - 7 = 1$. Some elements have more than one valence or valence number. If the atom is in a balanced or equilibrium state, it possesses the same number of protons as electrons and the electrical charge carried by the atom is zero (see Figure 2-4). The negative (−2) charge of the valence electrons of the oxygen atom in Figure 2-4 is balanced by the positive charge of the protons in the nucleus.

The process of pulling away (removing) or adding valence electrons from a balanced or neutral atom is called ***ionization.*** The atoms with unbalanced electrical charge are called ***ions*** (see Figures 2-5 and 2-6). The tendency of atoms to lose their valence electrons and the energy required to remove these electrons will be discussed in the treatment of the periodic table and its underlying concepts.

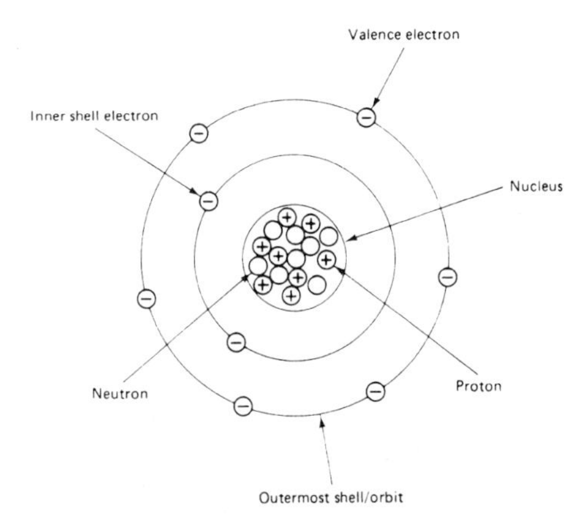

Figure 2-4 Subatomic parts of the oxygen atom, A.N. 8. This atom is shown in an equilibrium, balanced, or electrically neutral state.

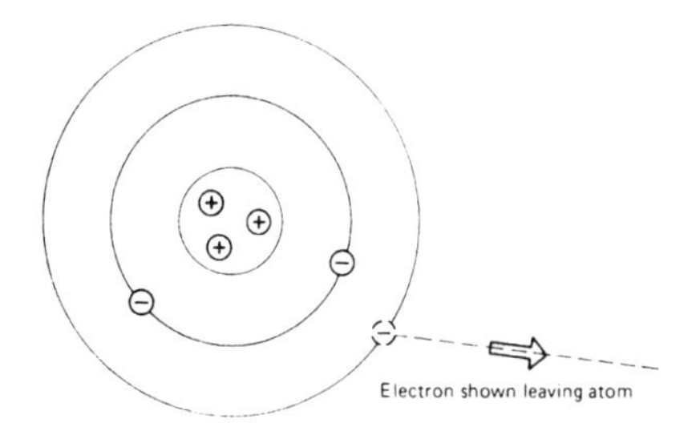

Figure 2-5 Positive ion or cation.

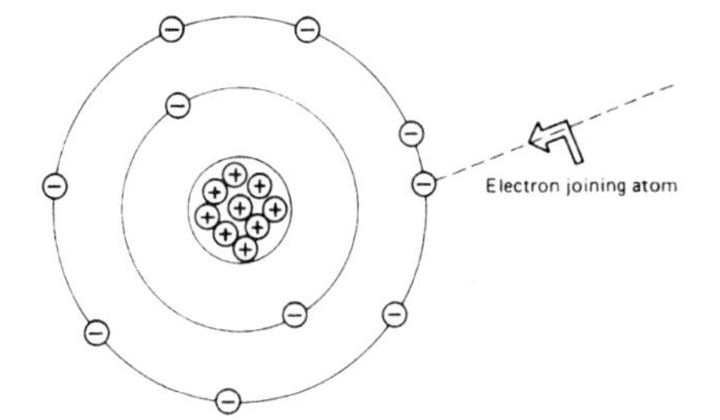

Figure 2-6 Negative ion or anion.

Electronic Structure. Many new theories developed between 1900 and 1930 laid the basis for a new branch of physics called ***quantum mechanics***. One theory was the idea that light is quantized—that it is made up of discrete amounts of energy. In a similar manner, we consider our present-day model of an atom with its electrons as having both wavelike and particlelike characteristics. The position of electrons in relation to the nucleus of an atom, according to this new theory, must be described in terms of a probability distribution (electron cloud) rather than by the precisely determined values of classical theory. Four parameters, called ***quantum numbers,*** characterize an electron as to its size, shape, and spatial orientation. Shells or principal energy levels are specified by a principal quantum number, n, where $n = 1$ through 7 or are by the labels K, L, M, N, O, P, and Q (see Figure 2-7a). The second quantum number, l, signifies the sublevel, subshell, or orbital (s, p, d, f). The number of energy states for each subshell is determined by the third quantum number, m_l. For the s subshell, there is a single energy state; for p, there are three; for d, there are five; and for f, there are seven. The fourth quantum number, m_s, relates to the electron spin moment, for which there are two values possible ($+\frac{1}{2}$ and $-\frac{1}{2}$), one for each of the spin orientations. Spin momentum controls the manner in which electrons fill their subshells and also the magnetic properties of materials (see Module 6). Table 2-1 contains an example of sublevels for selected elements, which are given lowercase-letter names.

Three basic rules define the order in which electrons fill their subshells or establish their electron configurations. One says that the electrons occupy their subshells or orbitals in such a way as to minimize the energy of the atom. A second rule states that only two electrons may exist in the same orbital, and these electrons must have opposite spins. The last rule says that when orbitals of equal energy are being filled, the electrons initially occupy these orbitals singly. The result of this rule is that electrons try to get as far apart from each other as possible. To accomplish this, they favor occupying empty orbitals of equal energy as opposed to pairing up with other electrons in half-filled orbitals. As a result, atoms tend to have as many unpaired electrons as possible. Iron (Fe), atomic number 26, has six electrons in the 3d orbital, of which four are not paired together. The sixth electron fills in the 3d orbital pairs with one of the previous five. In Fe, the unpaired inner electrons are responsible, in part, for the ferromagnetism that Fe exhibits. ***Paired electrons*** in the inner shells produce no resultant magnetic dipole moment. (For further discussion of ferromagnetism and magnetic domains see Module 6.)

The orbital diagram in Figure 2-7 demonstrates how electrons fill orbitals. Figure 2-7c shows three ways to depict the electron configuration of the element carbon (C). As in Figures 2-7a and b, each subshell is broken down into individual orbitals of equal energy. Electrons are shown as arrows. An arrow pointing up corresponds to one type of spin ($\frac{1}{2}$) and an arrow pointing down to the other ($-\frac{1}{2}$). Electrons in the same orbital with opposing (opposite)

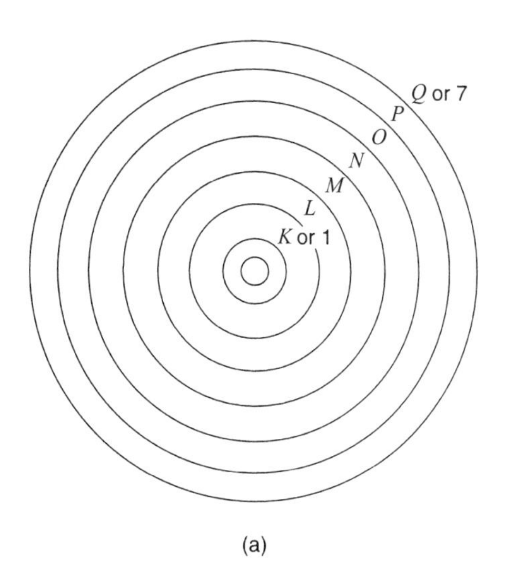

(a)

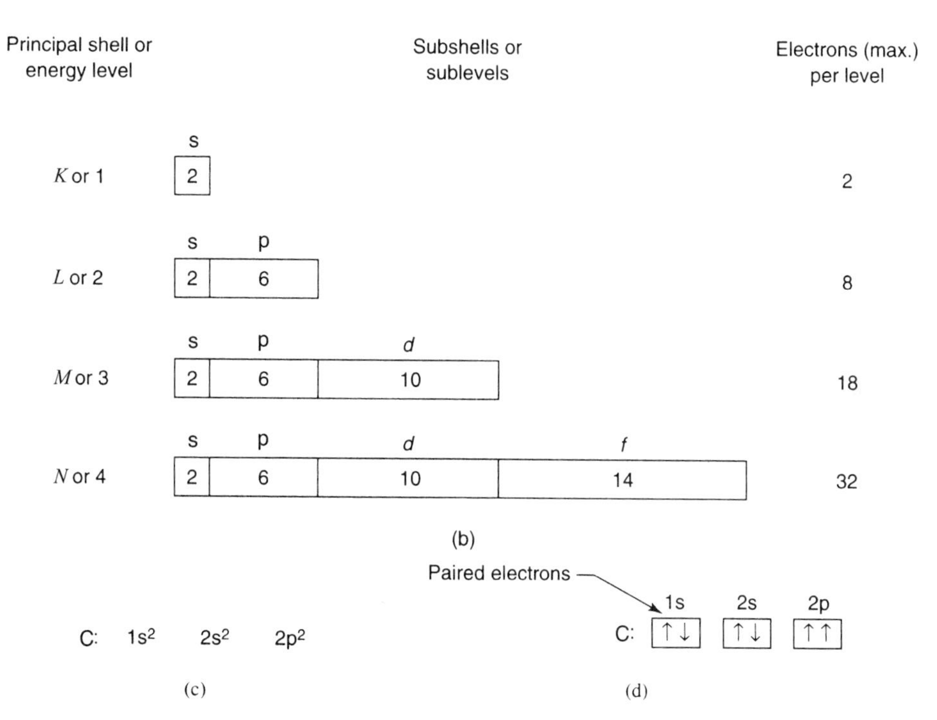

Figure 2-7 (a) Electron principal energy levels or orbits. (b) Block diagram of the first four energy levels showing their respective sublevels or orbitals along with the maximum number of electrons per energy level. Orbitals are used to describe the locations of electrons within the energy levels. They refer to volumes of space around the nucleus of an atom where an electron could probably be found. An orbital can be occupied by one electron or by two paired electrons only. An s orbital is spherical in shape; p orbitals have identical dumbbell shapes oriented around the *x*-, *y*-, and *z*-axes; and d orbitals take the shape of a four-leaf clover. (c) Electron configuration for the element carbon (C), atomic number A.N. 6 in Period 2 of the periodic table, shows three different ways of representing the location of its six electrons. The arrows indicate the directions of the spinning electrons.

spins (↑↓) are said to be paired. The electrons in the 1s and 2s orbitals, for example, are paired. Electrons in different, singly occupied orbitals of the same subshell have parallel spins (arrows pointing in the same direction). The orbital diagram represents the carbon atom in its most stable, or ground, state. The order in which all electrons fill subshells and thus minimize the energy of the atom, as established by experiment, is (with but a few exceptions) as follows:

1s, 2s, 2p, 3s, 3p, 4s, 3d, 4p, 5s, 4d, 5p, 6s, 4f, 5d, 6p, 7s, 5f, 6d, 7p

TABLE 2-1 ELECTRON CONFIGURATIONS OF SELECTED ELEMENTS

Element	Atomic number	K 1 s	L 2 s p	M 3 s p d	N 4 } orbits s p d f orbitals
Oxygen (O)	8	2	2 4		
Sodium (Na)	11	2	2 6	1	
Aluminum (Al)	13	2	2 6	2 1	
Silicon (Si)	14	2	2 6	2	
Titanium (Ti)	22	2	2 6	2 6 2	2
Iron (Fe)	26	2	2 6	2 6 6	2
Arsenic (As)	33	2	2 6	2 6 10	2 3

There are other techniques or methods used to determine the order of filling, but they are not discussed in this text.

The energy levels in atoms have been given letter names starting with K, the first and lowest energy level (closest to the nucleus), and proceeding outward to Q, the seventh level, where the electrons have the highest energy. Each level can hold only a certain number of electrons at any one time. To determine the maximum number of electrons at any energy level, the relationship $2n^2$ can be used. The number of the energy level is substituted for n. For example, the third or M energy level can contain a maximum of 2×3^2, or 18, electrons. Figure 2-7a is a simple sketch of all the energy levels drawn concentrically to the nucleus of an atom. Figure 2-7b lists some energy levels along with the maximum number of electrons permitted at each level.

The manner in which the electrons of an atom distribute themselves in the ground, or lowest energy, state is the ***electron configuration*** of the atom. Learning the electron configuration of atoms permits us to predict what the atoms will do in the presence of other atoms. In other words, the number of electrons in an atom's outermost energy level (the valence electrons) is the key that controls the chemical properties of an element. As an example, hydrogen, with an atomic number of 1 in the ground state, has its one electron in the K level. Carbon has six electrons; two fill the K level and the remaining four are in the L level. Atoms that have a completed energy level are chemically inert; that is, their atoms do not react with other atoms. They do not attract other electrons, nor do they readily lose their outer electrons. Hence, these atoms possess atomic stability. Helium with its two electrons has a complete K energy level. Some atoms have the tendency to build up the number of electrons in their levels to eight. (This tendency is known as the ***octet rule*** or the ***rule of eight.***) Once they arrive at eight, the next electron goes into the next higher level. Those atoms that have only one electron in their outermost energy level are extremely reactive; they tend to lose their electron. Examples of such elements are lithium and sodium. The K level has only one sublevel, the s sublevel. Using a block diagram, the energy levels could be represented as shown in Figure 2-7b. The electrons fill these sublevels in a particular way. The first 18 elements of the periodic table build up their permitted number in a routine manner. Once the K level reaches its maximum of two electrons, the electrons begin filling up the L level until a maximum of eight electrons is reached, as with neon (Ne). Then the third energy level from the nucleus (M level) with its three sublevels (s, p, d) begins filling to its maximum of 18. This point is reached with the element argon (Ar, atomic number 18). Note that both neon (Ne) and argon (Ar) have a maximum possible number of electrons in their outermost energy levels. These elements are classified as chemically ***inert gases;*** that is, they are very stable elements and do not react with other elements. Neon (Ne), with its ten electrons, has an electron configuration of Ne: $1s^2\ 2s^2\ 2p^6$. Note that it has a full L, or 2, level of eight electrons (octet rule). This neon core of electrons can be represented as [Ne]. Argon, atomic number 18, can be represented in a similar way, as [Ar]. Again, note that its outermost level (M, or 3) has eight electrons. To represent the electron configuration of titanium (Ti, atomic number 22) showing the order in which the electrons fill the orbitals and using the argon core symbol, we write Ti: [Ar] $3d^2\ 4s^2$. (The longer version of the Ti electron configuration follows in the next paragraph.) Table 2-1 contains the electron configurations of other elements.

Using a shorthand notation, the electronic configuration of oxygen is ^{8}O: $1s^2$, $2s^2$, $2p^4$. Note that the L energy level has only six electrons, but the p sublevel could contain a maximum of eight electrons. As a result, oxygen tends to accept two additional electrons to form a stable outer energy level. By definition, oxygen is an acceptor of electrons and hence is classified as a ***nonmetal.*** A ***metal*** such as aluminum, with its electronic configuration ^{13}Al: $1s^2$, $2s^2$, $2p^6$, $3s^2$, $3p^1$, has a stable L energy level below the farthest energy level, the third or M shell. The three valence electrons in the M level are close to those in the L level and possess similar bonding energies, which results in these three electrons having a tendency to stay with their atom. In addition, to achieve stability, aluminum would need the addition of five more electrons, which is unlikely to occur. Hence, aluminum is classified as a metal. One final example is titanium (Ti), with the electronic notation ^{22}Ti: $1s^2$, $2s^2$, $2p^6$, $3s^2$, $3p^6$, $3d^2$, $4s^2$. This element has been chosen to show that the filling of electron levels is somewhat irregular. Note that two electrons have filled the 4s sublevel, whereas the 3d sublevel is far from being filled, with only two electrons (see Figure 2-7b). For an example of the use of this notation, refer to Section 3.2.9.

Illustrative Problem

Write the electronic configuration for the element sulfur.

Solution

Step 1. Determine the chemical symbol and atomic number for sulfur from the periodic table.

S, atomic number 16

Step 2. Determine the total number of electrons in the sulfur atom.

16

Step 3. Write the order in which the orbitals are filled (see subsection 2.1.1.1).

1s, 2s, 2p, 3s, 3p, 4s, 3d, etc.

Step 4. Just below the orbitals in step 3 write the number of electrons per orbital for each orbit and level (refer to Figure 2-7).

2 2 6 2 6 2 10

Step 5. Write the configuration:

$$S = 1s^2\ 2s^2\ 2p^6\ 3s^2\ 3p^4$$

Step 6. Check by adding the number of superscripts, which represent the number of electrons in each orbit, to arrive at the sum of 16.

In contrast, some elements have only one electron in their outermost energy level. Lithium (Li), atomic number 3, with two electrons in the K level and only one electron in the L level, is extremely reactive. The same can be said of sodium (Na) and potassium (K). For elements with atomic numbers greater than 18, starting with potassium (K), the routine followed by the next 18 elements in filling their electron levels is somewhat different (see the previous discussion of order of filling).

Through an understanding of the electronic structure of atoms, the properties of elements can be ascertained and the bonding of atoms of elements together to form molecules, compounds, or alloys can be explained. Additionally, electrons determine the size of atoms, the degree of electrical and thermal conductivity of materials, and the optical characteristics of materials. When atoms take part in chemical reactions, there is an energy change or transfer, usually resulting in the release of energy in the form of heat. This energy change brings about a rearrangement of the electrons, and the end result is an electron configuration with lower energy than any of the reactant elements. Only the outermost (valence) electrons are rearranged. The electrons in the inner energy levels (or shells) are too stable, or are shielded from this activity. More will be said about this phenomenon when chemical reaction and bonding of atoms are discussed later in this module.

Periodic Table. Elements are the basic building blocks of all macroscopic matter. Just as musicians experiment with musical notes to produce new music, materials scientists experiment with new combinations and arrangements of these elements to produce new and exciting materials that may become part of a technological device. Each has its own chemical symbol, a shorthand abbreviation of its chemical name, and *each has its own specific properties.* It is important to reiterate that elements with the same number of valence electrons have similar properties. Scientists have long recognized that some elements had similar properties and others, such as sodium and chlorine, had very different properties. All versions of the modern periodic table arrange the elements in order of increasing atomic number, which usually matches the order of increasing atomic mass.

Refer to the periodic table on the inside front cover (or Table A-1A) for the following discussion. The horizontal rows of the table, called **periods** are arranged in order of increasing atomic number. The periods may or may not be numbered from 1 to 7, but the period in which an element appears tells us how many electron energy levels the atoms of this element has. In other words, each period corresponds to the filling, or at least the partial filling, of the energy levels. The first period of the table consists of the elements hydrogen (H) and helium (He). The sixth period is quite long, with 32 member elements starting with cesium (Cs). To accommodate this period in a table with 18 columns (groups), 14 members are extracted and placed at the bottom of the table. This series of elements that follows lanthanum (La) are called *lanthanides,* or rare earth elements. Nd-Fe-B magnets, about 70,000 tons of which are used worldwide in a year, are one example of their use. The last or seventh period is incomplete (some members have not been discovered). A 14-member set of elements is also extracted and placed at the bottom of the table. The elements in this block follow the element actinium (Ac) and are called *actinides.*

The vertical columns are called *groups* or *families*. There are several methods for naming the vertical groups in the periodic table, two of which are used in this edition. For example, the element neon (Ne) is in Group 18 or in Group VIII and bromine (Be) is in Group VII or 17. Group VII or 17 are called *halogens*: F and Cl are gases, Br is a liquid, and I is a solid at room temperature. Because the halogens are very reactive, none is found free in nature. Halogen compounds, such as HCL and HF, are used in the production of coatings using the chemical vapor deposition process discussed later in this text.

Looking at the electron configurations of all the elements in Group VII, you will note that they all have seven outer shell or valence electrons. The significance of this is that the properties of an element are largely determined by the electron configuration of the outermost or valence electrons. Hence, all the elements in a group have similar properties. If you look at two elements alongside each other in the periodic table (both in different groups, such as phosphorus, P, and sulfur, S, you see that their electron configurations are quite different, and therefore they should and do have different properties. Group VIII or 18 consists of elements known as inert gases or noble gases. Their energy levels are completely full of valence electrons. As a consequence, they are mostly inactive; they do not get involved in making chemical compounds when they combine with other elements. The noble gas argon (Ar) is used in gas welding to provide a shield against the intrusion of oxygen into the weld metal. Xenon (Xe), atomic number 54, plays a role, in the form of short-lived radioisotopes, in an automatic sensory device that is strategically placed in some 80 different locations throughout the world to help police the Comprehensive Nuclear Test Ban Treaty by detecting small underground nuclear explosions. The sensor automatically samples large volumes of the atmosphere for the presence of any noble gases, principally xenon, that might escape from an underground explosion because they are inert and do not react with soil.

Periodic Tables of the Elements: Kiwi Web—*http://www.chemistry.co.nz/periodic_tables.htm*
Periodic Table of Comic Books—*http://www.uky.edu/~holler/periodic*
Periodic Table—*http://www.periodictable.com*

The elements can also be classified broadly into three basic groupings (metals, metalloids, and nonmetals) due to certain common properties. ***Metals*** have certain characteristic properties: They are solids at room temperature (with the exception of mercury), they are malleable and ductile, and they conduct electricity. ***Nonmetals*** (a total of 17 elements) are poor conductors of heat, they are mostly insulators, and they tend to be brittle and fracture easily. ***Metalloids,*** or ***semimetals,*** have some properties like the metals (they conduct electricity) but they share other properties with nonmetals. The seven elements adjacent to the zigzag border between the metals and the nonmetals (B, Si, Ge, As, Sb, Te, and At) are brittle and are poor conductors of heat and electricity. Silicon and germanium, for example, are called *semiconductors* and they play a major role in our technological world (see Module 22).

The atomic radius of an atom (size of the atom) decreases as one moves from left to right across a period. As one moves down a particular group of elements, the size of the atoms increases. In general, ***ionization potential***—the energy required to cause the loss of a valence electron—follows the periodic relationship and decreases as the atomic numbers of the elements increase in a given periodic group. The first ionization potential is a measure of the energy needed to remove the first electron from the atom in its ground or equilibrium state. The lower the ionization potential, the greater the metallic character. For all elements, the ionization potential is usually expressed in the periodic table in electron volts (eV) or other energy terms, such as calories per mole. For example, it takes 13.6 eV of energy (313 kcal/mol) (shown as 13.598 in Table A-1B) to free the electron from the hydrogen atom. A detailed discussion of the ionic bonding between sodium and chlorine to produce a molecule of sodium chloride (NaCl) can be found in the subsection on ionic bonding (electron swapping) on p. 51.

This binding energy increases as the size of the atom decreases. In other words, the farther away an electron is from its nucleus, the smaller the force of attraction between it and its positively charged nucleus. Metallic elements on the far left of the periodic table readily ionize, whereas atoms of nonmetals located on the right of the table more readily share or accept additional electrons and become negative ions. The observation to be made is that the tendency of atoms to ionize depends on their relative positions in the periodic table. ***Electronegativity*** is the degree to which an atom attracts electrons. It is measured on a scale from 0 to 4.1, as shown in Table A-1B. The nonmetallics, which accept electrons to produce negative ions (anions), are the most electronegative elements, and Groups VIB and VIIB contain the most electronegative elements, with a value of 3.98 for fluorine (F) and 0.7 for francium (Fr), the lowest value.

Electropositive elements are metallic in nature and give up electrons in chemical reactions to produce positive ions (cations). The most electropositive elements are found in Group 1A. Another term that describes this phenomenon is ***electron affinity,*** which is a measure of the attraction between the electron and the nucleus. Figure 2-8 is a graph based on the Pauling electronegativity scale showing the percentage of ionic bonding (character) in a compound versus the difference in electronegativity between two elements in the compound. If the difference between the electronegativity of two elements is greater than 1.7, the bond is ionic. The higher the electronegativity value for an element, the greater is the ability of that element to attract electrons to itself. The greater the difference in electronegativities between two atoms forming a bond, the more polar is the bond (see the subsection on polar covalent bonding or covalent electron sharing, p. 52).

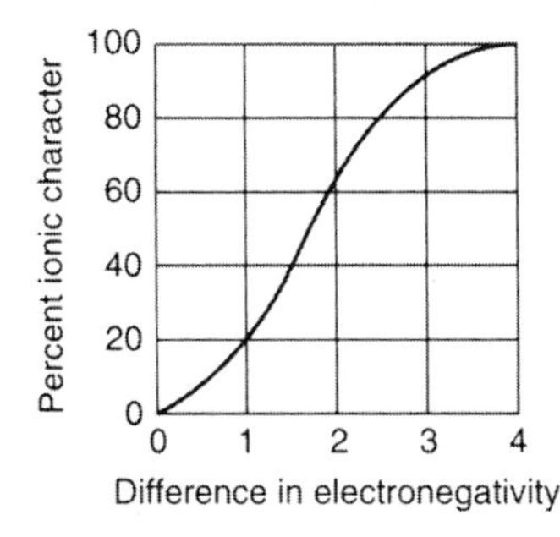

Figure 2-8 Percent ionic character of a single bond plotted as a function of the difference in electronegativities of the two bonded atoms.

Illustrative Problem

Determine the type of bonding in the compound MgO.

Solution Using Table A-1B, the electronegativity for oxygen is found to be 3.44, and for magnesium it is 1.31. Entering a difference of 2.13 in the graph shown in Figure 2-8, the percent ionic bonding is about 68%, leaving 32% of the bonding covalent. A more accurate solution is obtainable using data in the small table at the top of Table A-1B.

2.1.1.2 Molecular structure and bonding. We mentioned previously that atoms join together to form molecules and that molecules combine to form various compounds. A ***compound*** is a substance containing two or more elements combined in fixed proportions. For example, sodium, Na, combines with chlorine, Cl, to make the compound sodium chloride (NaCl), as described in the next section. Compounds lose the characteristics of their elements. In some molecules, the atoms are of the same element. Fluorine (F) exists as a diatomic molecule under normal conditions because two fluorine atoms combine to form one fluorine molecule. All the Group 17 VIIB elements form diatomic molecules. In this section, we will gain further understanding about how the 100 or so elements combine or chemically bond to form not only more than 3 million organic compounds, but the various inorganic and bulk metals such as iron and copper.

Molecules and compounds are formed by chemical bonding. ***Chemical bonding*** can be explained simply as the end product of the interaction of the electrical forces of attraction and repulsion between oppositely charged or similarly charged particles of matter. Chemical bonds are formed by the electrons in the atom's outermost energy levels or regions. In our study of these electron energy levels, we have learned that, generally, atoms that have eight electrons in their outermost orbit are very stable (in equilibrium state). This tendency to start filling the next-higher level once an atom has eight electrons in an energy level is known as the *octet rule*, described previously. Atoms with fewer than eight outermost electrons attempt to seek this stable condition, the electron configuration of the nearest noble gas (Group VIII) in the periodic table, by bonding with each other so that each atom can attain the stable configuration of eight electrons (valence electrons) in its outermost energy level. One way that atoms can achieve this stability is by sharing electrons with other atoms that also need eight electrons in their outermost energy levels. For purposes of study, we can divide the various types of bonding into two groups, strong or ***primary bonding*** and weak or ***secondary bonding.*** In the first group, we include covalent, ionic, and metallic bonding. The second group includes weak atomic bonds, electric dipoles, polar molecules, and the hydrogen bond. The forces that produce secondary bonds are known as ***van der Waals forces.*** We emphasize that atoms of one material do not bond solely by one particular bonding mechanism. One type of bond can predominate in a material, but in many materials the various bonding types are represented and mixed-bond types of materials are produced.

Polar Covalent Bonding, Covalent Electron Sharing, or Electron-Pair Bonding. The sharing of electrons between two or more atoms is known as ***covalent*** or ***shared electron-pair bonding.*** As an example, fluorine (F) atoms have a total of nine electrons, two in the inner energy level and seven in the outermost level. Two fluorine atoms can share one of the electrons with the other to form a single covalent bond, depicted with a single dash representing the bond. Each (—bond) is a pair of shared electrons (see Figure 2-9a). Bonding of the elements is indicated with the small dash (–), each indicating a covalent bond. Covalent bonding gets its name from the fact that a pair of valence electrons is part of the electron structure of each atom bonding the two atoms together. In general, organic compounds (complex compounds containing carbon and/or hydrogen) are formed by covalent bonding. A ***triple covalent*** bond is one in which two atoms *share three* of their electrons with each other. Figure 2-9b illustrates this with nitrogen (N). Nitrogen needs eight electrons to fill the outer energy level, but it has only five. Hence, it shares three with another nitrogen atom, forming a triple bond.

Organic chemistry is the chemistry of carbon compounds. Of all the millions of known chemical compounds, more than 95% are compounds of carbon. The simplest organic compounds are the ***hydrocarbons (HCs),*** which contain only hydrogen and carbon. ***Alkanes,*** also known as *saturated* HCs, are a series of related compounds that contain only single bonds. Figure 2-10 shows the first three compounds in this series: methane, ethane, and propane—three out of

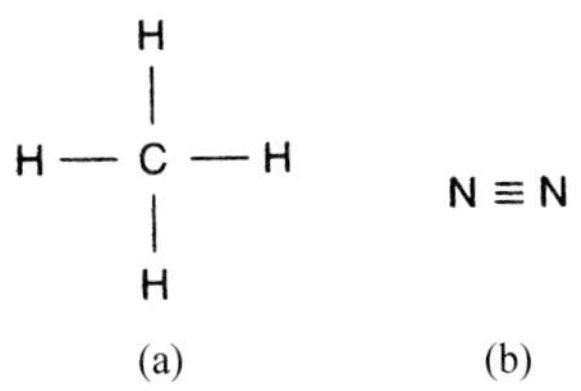

H
|
H — C — H
|
H

(a)

N ≡ N

(b)

Figure 2-9 (a) Single bond. (b) Triple covalent bond.

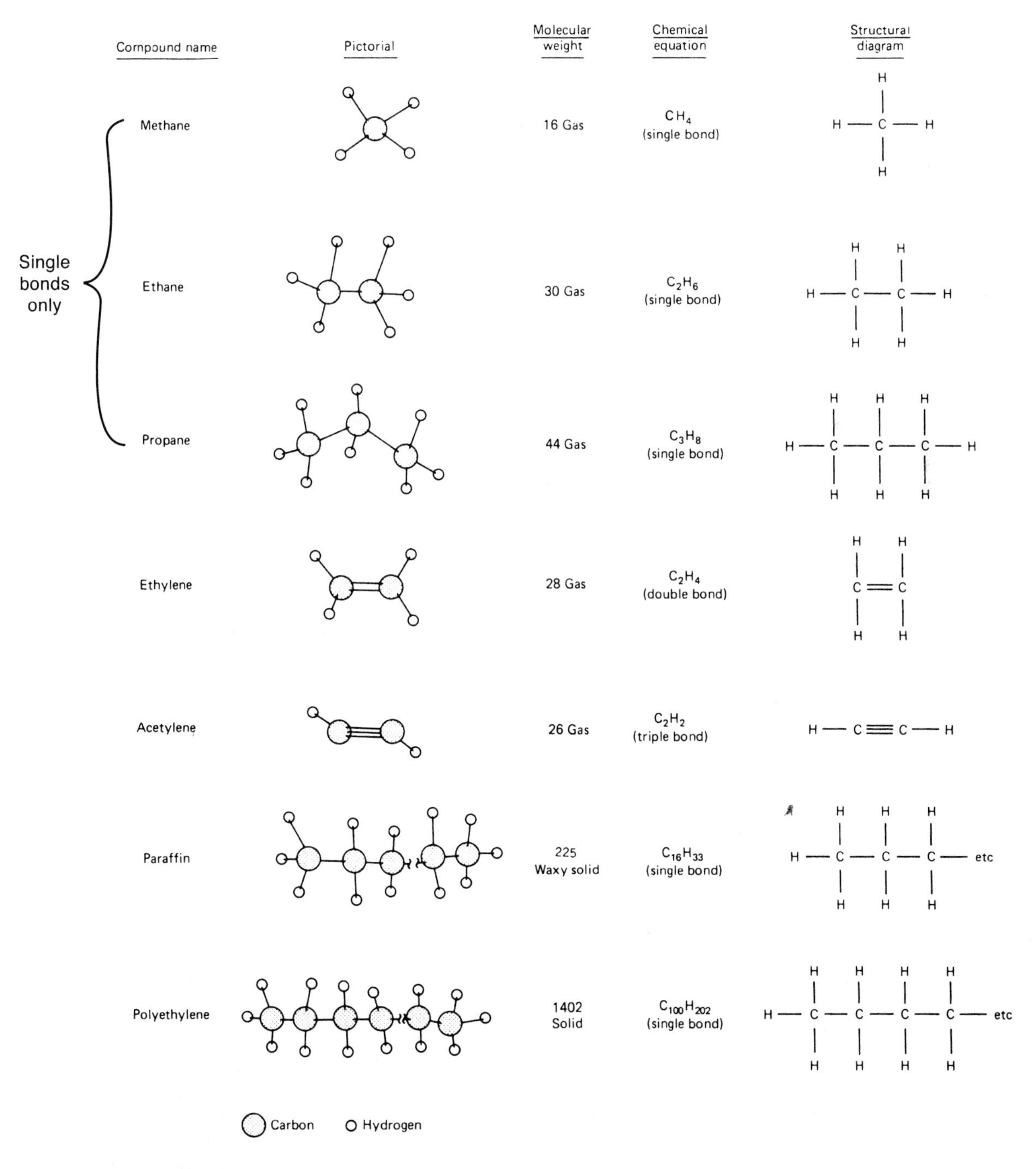

Figure 2-10 Hydrocarbons.

H
H — C — H
H

CH_4 molecule

Figure 2-11 Covalent bonding of methane.

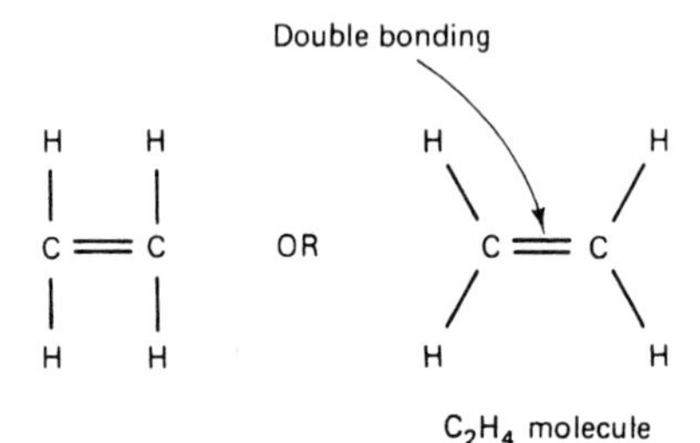

Figure 2-12 Covalent bonding of ethylene.

many that have single bonds. They all have a continuous, open-ended chain of carbon atoms and possess similar chemical properties. Each dash in Figure 2-11 represents the single bond of a hydrogen atom with a carbon atom to form the compound methane ethylene. (The C in these figures represents the carbon atom and the H represents the hydrogen atom.) Ethylene is the first member of a family of HCs called ***alkenes.*** The structure of ethylene as shown in Figures 2-10 and 2-12 is the basic compound for producing polyethylene. More than half of the production of ethylene goes into the manufacture of polyethylene. In the form of ethylene glycol, it is also the major component of most brands of antifreeze. Notice in Figure 2-10 that as the molecular weight increases, the compounds change from a gaseous state to a liquid and finally to a solid as more and more molecules bond to their chains of molecules. Each alkene contains a carbon-to-carbon double bond. Acetylene is the first member of the ***alkyne*** family that contains a triple bond. Collectively, alkenes and alkynes are called *unsaturated* HCs because they can add more hydrogen atoms to form saturated HCs or alkanes. One unusual feature of unsaturated HCs is that they can also add to each other to form very large molecules called *polymers* (discussed in Module 14).

Benzene and similar compounds such as toluene are referred to as ***aromatic hydrocarbons*** because some of the first benzenelike compounds had strong aromas. Today, any compound that contains a benzene ring or has similar properties to those of benzene is called *aromatic*. Aromatic HCs are all liquids that are lighter than water. They are used mainly as solvents.

Figure 2-13 shows many ways of representing benzene. All possible structures have double bonds and some have triple bonds, but benzene does not react that way chemically. Because the three pairs of electrons that would form double bonds are not localized (they seem to lie above and below the plane of the carbon atoms), the popular way of representing benzene is by a circle within a hexagon. The circle represents the six unassigned electrons and the hexagon represents the ring of six carbon atoms. This ring of electrons resists being disrupted, providing the benzene molecule with a stable structure.

The chemical formula (Figure 2-13) may be considered as a one-dimensional description of a chemical compound or molecule, the structural description as a two-dimensional view, and the pictorial as a three-dimensional view. Covalent bonding occurs between atoms of nonmetallic elements such as hydrogen, fluorine, chlorine, and carbon and in molecular compounds such as water, carbon dioxide, ammonia, and ethylene. By sharing electrons, the atoms achieve stability through the creation of interatomic electrical forces, which bond the atoms together to form molecules and compounds. Research has shown that double or triple bonds are stronger than single bonds between the same two atoms. Carbon forms four single covalent bonds.

The many forms (***polymorphism***) of carbon serve to illustrate the array of properties achieved by variations in bonding and the location of atoms. Carbon exists as diamond, the hard-

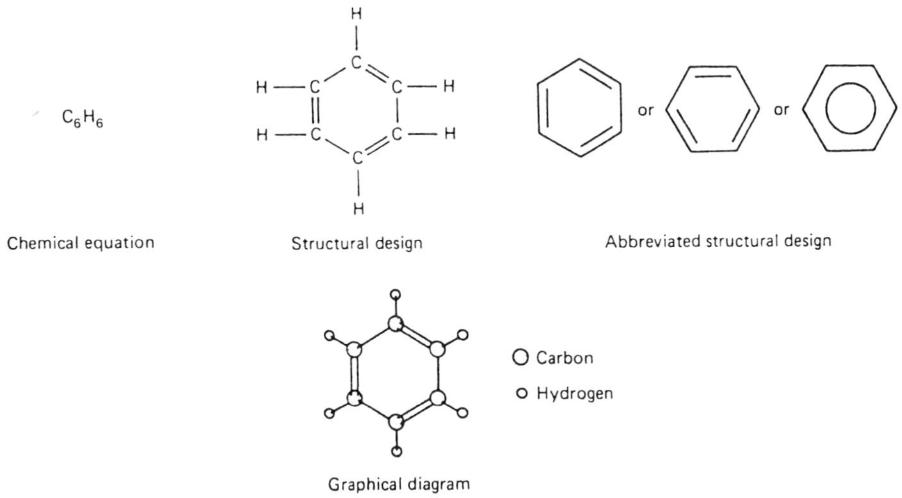

C_6H_6

Chemical equation

Structural design

or or

Abbreviated structural design

Carbon

Hydrogen

Graphical diagram

Figure 2-13 Benzene or benzene (aromatic) ring.

est natural material known, and graphite, an effective lubricant. Graphite is the stable form of carbon at temperatures up to 3000°C and pressures of 10^4 atmospheres (1 atmosphere = 14.7 pounds per square inch) and higher. Diamond is the stable form of carbon at very high pressures (see Section 20.3.4 for discussion of natural/synthetic diamond). ***Diamond*** has a cubic symmetry (Figure 2-14a) with perfect covalent bonding that joins all atoms into a single molecule. ***Graphite*** develops into sheets with good covalent bonding, but between sheets van der Waals bonds are weak and shear easily, thus allowing the sheets to slide past each other (Figure 2-14b). Figure 2-14c depicts the recently discovered carbon form ^{60}C, which draws its name, ***buckminsterfullerene,*** from Buckminster Fuller, the inventor of geodesic domes. Two common names for this molecule are ***fullerene*** and ***buckyball,*** because it has a soccer-ball shape, composed of 20 hexagons and 12 pentagons. Fullerene fibers, also known as ***nanotubes,*** are carbon fibers with tubular structures. They are potentially 30 to 100 times stronger than steel but only one-sixth its weight. The fibers measure about 1 nanometer in diameter. Single-wall carbon nanotubes conduct electricity better than copper and conduct heat better than diamond. They have 100 times the tensile strength of steel and are used in radar absorption devices, batteries, superconductors, coatings and fibers.

The discovery of fullerenes (^{60}C molecules, Figure 2-14c), announced by Richard E. Smalley in 1985, was confirmed by the visual proof provided by the scanning tunneling microscope, discussed in Module 3. This new polymorph of carbon holds much promise as materials scientists and engineers develop the techniques for processing it. Generated at about 2500°C, the sooty substances can be produced as helixes, much like DNA, the basic building block of all living cells. The helix form of fullerene may make possible the creation of resistance-free electrical superconductors. On the other hand, adding sodium atoms to ^{60}C spheres will yield electrical insulators.

Among other applications of fullerenes are superstrong fibers for construction materials or thin diamond films for hard-coat cutting tools and electronic circuitry. When joined with ammonia atoms, buckyballs become magnetic, and when the ^{60}C tubular molecules are alloyed in steel—an alloy of iron, carbon, and other elements—greater strength is achieved while using conventional steelmaking techniques.

Fullerenes can be expected to play a key role in many materials applications because this unique substance provides alternatives in numerous areas of materials science and engineering. The development of ^{60}C will mirror that of many other new materials and be hastened by the availability of powerful new tools, such as the scanning tunneling microscope, to aid in unlocking the secret of structure. Buckyballs, which are much like charm bracelets that allow one to attach many charms of different types, permit the combination of numerous elements and

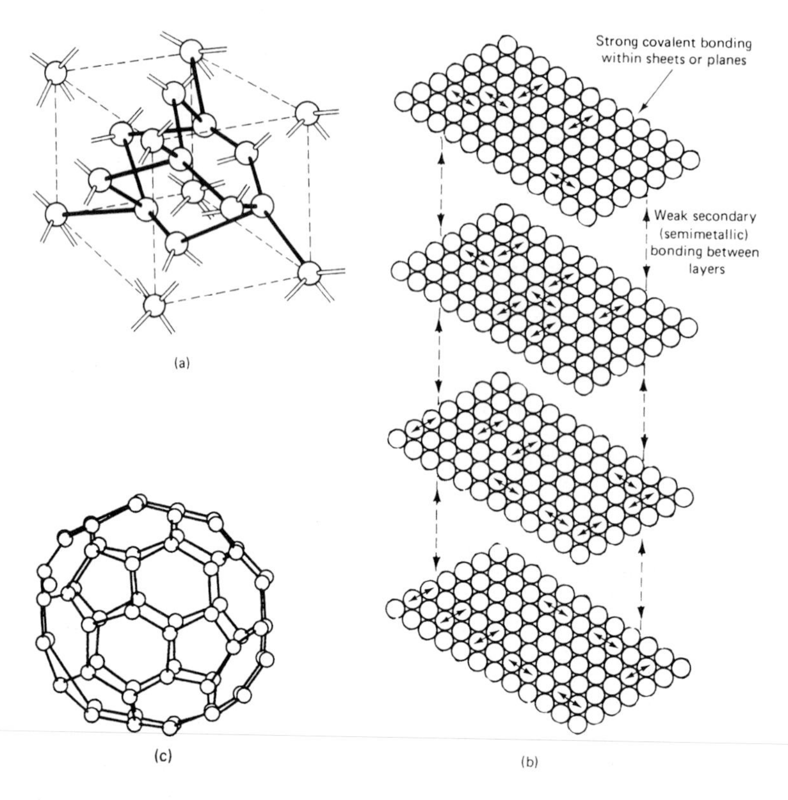

Figure 2-14 Polymorphism of carbon. (a) Diamond's cubic structure: Each carbon atom forms strong covalent bonds with four other carbon atoms to develop a tetrahedron in the same manner as silicon and germanium. Diamond cutters use a sharp instrument and a sharp blow to split the crystal along cleavage planes (111) to produce perfect smaller jewels. (b) Layered structure in graphite: van der Waals bonding between layers allows easy cleavage into sheets, thus providing good lubrication properties. (c) C^{60}, or fullerene, molecules have a spherical shape, like a soccer ball, composed of 60 carbon atoms covalently bonded together.

molecules. Chemists and materials scientists will be at work for many decades producing new materials.

Ionic Bonding (Electron Swapping). Instead of sharing electrons, some elements actually swap or ***transfer*** electrons to other elements, a type of bonding called ***ionic.*** In general, inorganic compounds bond ionically. Elements having outer electron levels that are almost full or almost empty tend to gain or lose electrons to complete their outer energy levels. A classic example of ionic bonding of two elements is sodium (Na) and chlorine (Cl). From the periodic table, we learn that neutral sodium has 11 electrons, one of which is in the outer energy level. This single valence electron is weakly attracted to the positive nucleus. Chlorine (Cl) has 17 electrons, leaving 7 electrons in the outer or third energy level. As you recall, chlorine needs 8 electrons in its third energy level to attain stability. Since each atom attempts to attain equilibrium (lowest energy state) by completing its required number of electrons for each energy level, the sodium atom gives up its single electron through the action of a strong driving force of the chlorine atom to form a stable outer shell (Figure 2-15). This results in a temporary unstable condition that is soon changed by the sodium atom, which has now become a positive ion (***cation***) due to the loss of one of its electrons. The positively charged sodium atom (cation) interacts with the chlorine atom, which is now a negative ion (***anion***) due to the gain of the electron from the sodium atom. This interaction is a strong electrical (electrostatic) force between oppositely charged ions forming the ionic bond, which produces the molecule of sodium chloride (NaCl), common salt.

Each of these elements has its own characteristic properties. Sodium is an active metal and chlorine is a poisonous gas. Yet combined as an ionic molecule, they have the well-known properties of common salt. Sodium and chloride ions are also different from their respective

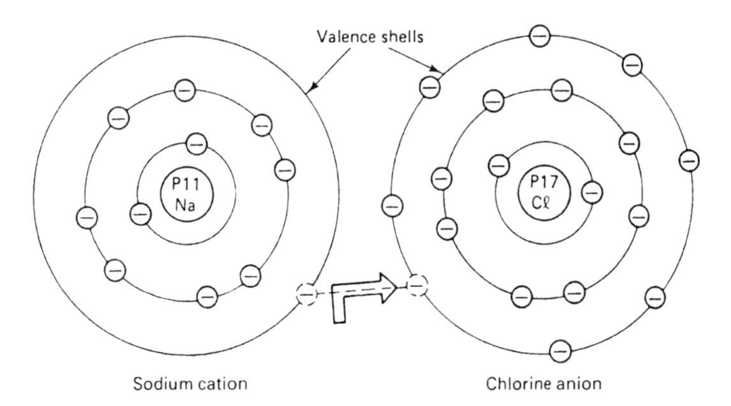

Figure 2-15 Ionic bonding (electron swapping) of a sodium chloride molecule. Upon bonding, both atoms have the equivalent of eight valence electrons in their outer shells. The orbitals within the shells and the subshells are not shown in this figure.

neutral substances (atoms). The greenish-yellow element chlorine, whose atoms have seven valence electrons, is very active. Chloride ions, in the formation of NaCl, have satisfied the octet rule with eight electrons in their outermost shells, and thus are stable. Otherwise, Cl^{-1} ions are extremely active, reacting with metals to produce corrosion. All these bonds are electrostatic in origin, so that the chief distinction among the various ways in which atoms bond to form molecules is in the distribution of the electrons around the atoms and molecules. Metals can be characterized by the tendency of their atoms to lose their valence electrons. An ***active metal*** is one that loses its electrons quite readily. Similarly, an ***active nonmetal*** is one whose atoms readily accept electrons. We also know that nonmetallic elements have a larger number of valence electrons than do metallic elements.

Metallic Bonding (Electron Swarming). Metals contain one, two, or three valence electrons. These are shielded from the strong attractive forces of the positive nucleus by the inner electrons and thus they bond to the nucleus relatively weakly. Consequently, when in the company of other metal atoms, these metal atoms will lose their weakly held valence electrons, which, in turn, enter a common free-electron cloud, swarm, sea, or gas. These free or delocalized electrons can then move in three dimensions. Once these electrons leave their atoms, the atoms become positive ions (cations). An electrostatic balance is maintained between the cation and the ***electron cloud,*** which results in the cations arranging themselves in a three-dimensional pattern, as shown in Figure 2-16. This electrostatic balance is the glue that bonds the metallic structure. The ever-moving free-electron cloud acts like a matrix, surrounding the positive metallic ions and providing rigidity. If these ions were represented as spheres, their space arrangement in bulk metal could be likened to a box full of table tennis balls.

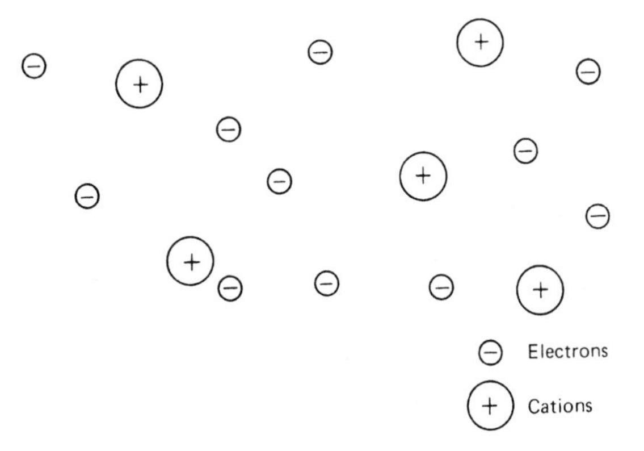

Figure 2-16 Metallic bonding or electron swarming.

Secondary Bonding. The van der Waals bonds, which are much weaker forces of attraction, produce bonding between atoms and between molecules. Group VIIIA elements (inert gases) have a full complement of valence electrons in their outer levels. Thus, they have no inclination to lose, gain, or share electrons. At ordinary temperatures, they remain as single atoms (monatomic). Only at very low temperatures will these gases condense, and it is the presence of van der Waals forces that permits this condensation. A similar situation occurs with covalently bonded molecules that have achieved equilibrium by sharing electrons. The van der Waals forces also permit these molecules to condense at low temperatures.

All molecules consist of distributions of electrical charge. When two molecules are in close proximity, these electrical charge distributions interact and create intermolecular forces. The van der Waals forces of attraction between molecules have their origin in the forces of attraction of the nucleus (positive) of one molecule for the electron charge (negative) of a neighboring molecule, which form fluctuating dipoles in the two molecules. Such forces increase with an increase in the number of electrons per molecule because the electrons in the large molecule are more readily polarized by these forces. In a ***polar molecule*** or ***electric dipole,*** the charges are polarized; that is, both the positive and negative charges are localized within the molecule. Figure 2-17 illustrates these two situations. A hydrogen molecule is a simple example of a ***nonpolar molecule.*** This diatomic molecule, in which both atoms are of the same element and are covalently bonded together, is sketched in Figure 2-18. Each atom shares the bonding electrons equally, producing an electrical charge distribution that is symmetrical about a line joining the two nuclei.

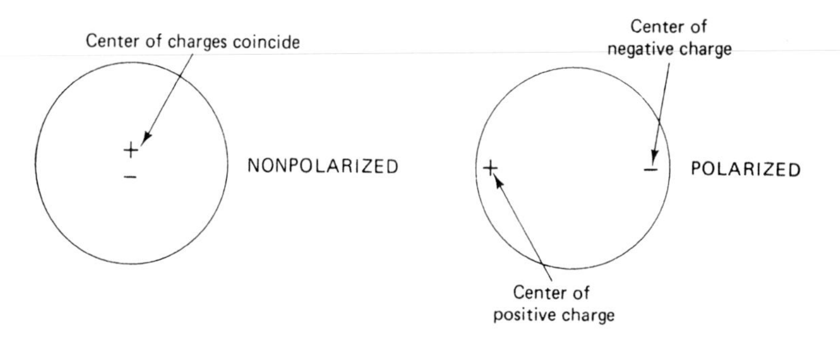

Figure 2-17 Schematic of a nonpolarized and a polarized molecule.

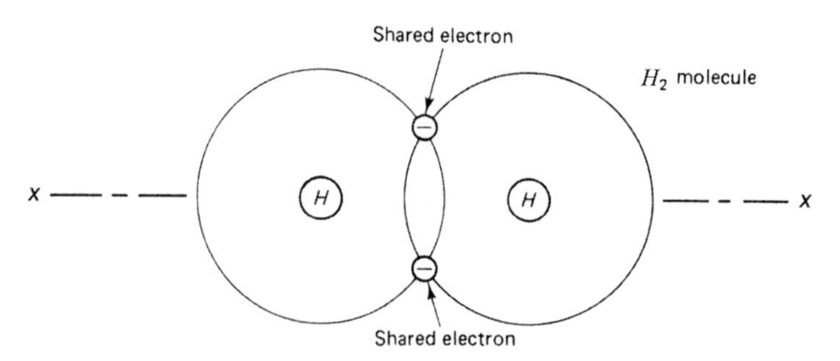

(a) Both atoms share two electrons equally forming a single covalent bond to complete each atom's outer energy level (K level).

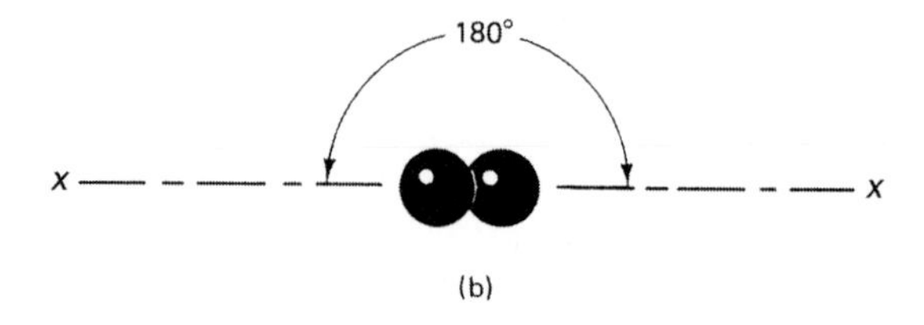

(b)

Figure 2-18 (a) A nonpolar diatomic hydrogen molecule. (b) A linear hydrogen molecule.

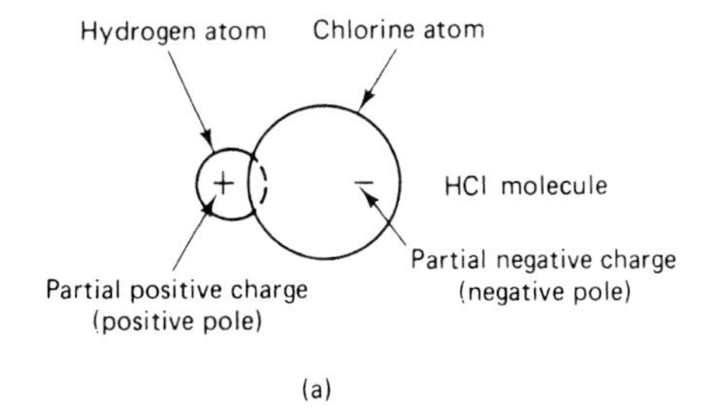

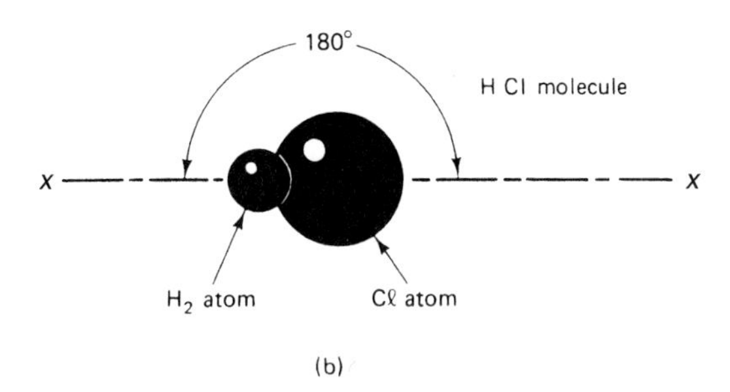

Figure 2-19 (a) A polar diatomic molecule of hydrogen chloride (HCl). (b) A linear hydrogen chloride molecule.

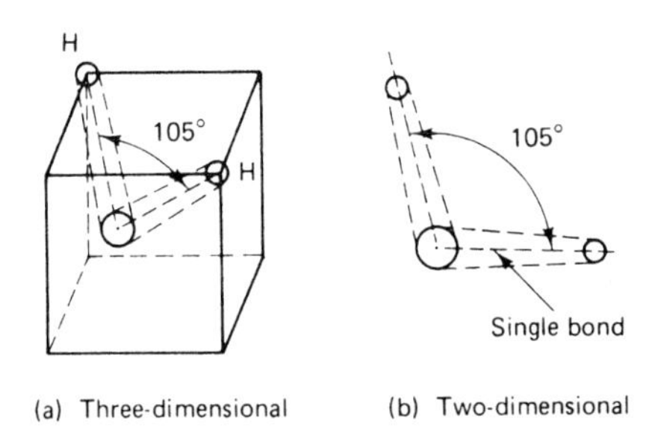

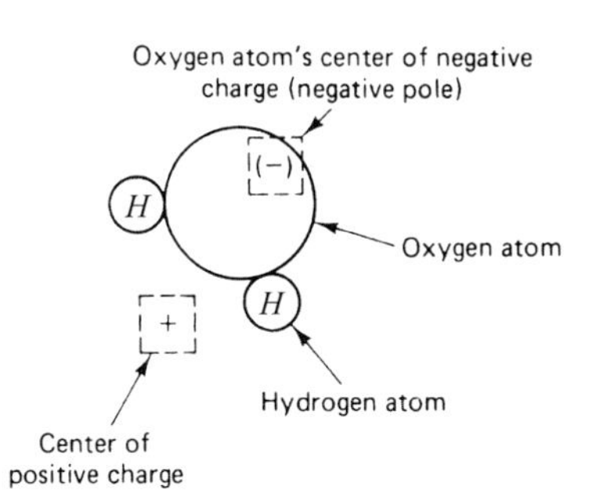

Figure 2-20 Various sketches of a water molecule (H_2O).

Any diatomic molecule, such as hydrogen chloride (HCl), in which the atoms are different from each other is polar (Figure 2-19a). The negative pole is located at the atom that is more electronegative (attracts electrons more strongly). In a polar molecule, the electrons are not shared equally. The chlorine atom, being more electronegative, attracts more electrons than the hydrogen atom does, which causes the shared electrons to be associated more closely with the chlorine nucleus. Figure 2-19b is a three-dimensional sketch of this molecule showing that, like the hydrogen molecule, it has an electrical charge distribution with a symmetry around the line x–x.

A good example of a polar molecule containing more than two atoms is water (H_2O), sketched in Figure 2-20. Oxygen has a greater attraction for electrons than does hydrogen. As a result, the four electrons being shared in the two covalent bonds remain closer to the oxygen atom. This unequal sharing of electrons has the effect of creating a partial positive charge between the hydrogen atoms and a partial negative charge on the oxygen, which produces a *polar covalent bond*. The three-dimensional sketch (Figure 2-20a) of the water molecule, with the oxygen atom in the center of a cube and two hydrogen atoms occupying two of the eight corners, shows a bent molecule (not symmetrical), with an angle between the two hydrogen atoms of about 105° (Figure 2-20b). Most covalent bonds are polar. The asymmetrical charge distribution on the water molecule results in the formation of regions of positive (+) and negative (−) charges located at a maximum distance from each other (Figure 2-20c). If the charge distribution were symmetrical, the positive (+) and negative (−) charges would cancel each other, as both centers would be at the center of the molecule. Thus, each center of charge can exert an attractive force on an adjacent charge of opposite sign. This statement explains, in part, how water molecules bond together to form liquid water. As pictured, the hydrogen atoms bond on one side of the oxygen, leaving the protons of the hydrogen atom farther from the nucleus of the oxygen atom.

The covalent bonds between the hydrogen and oxygen atoms in a water molecule, being polar, result in an asymmetrical charge distribution. The oxygen atom, having a slight negative charge, attracts a positive hydrogen atom belonging to an adjacent water molecule and forms a ***hydrogen bond,*** sometimes called a ***hydrogen bridge.*** This bond is the strongest of the secondary bonds. Figure 2-21 shows one schematic for representing the hydrogen bond between two water molecules in the liquid phase. The relatively high boiling temperature as well as other extraordinary properties of water are attributable to the bonding action of the hydrogen bridging.

As do all bonds, the hydrogen bond affects the properties and behavior of materials, particularly polymeric materials. In thermoplastics, the hydrogen bond joins long, chainlike molecules to each other. These relatively weak bonds can be easily loosened or broken by heating, permitting flow to take place. This explains why thermoplastics can be converted by heating to a soft, flexible state and then back to rigid, solid plastic material upon cooling.

Our discussion of chemical bonding is well summarized in Table 2-2, which describes one secondary bond and the three primary bonds. Further discussion of these chemical bonding mechanisms comes up in dealing with the various groups within the family of materials listed in Table 2-3.

2.1.2 Solid State

A ***solid*** is a sample of matter that has a fixed volume or size and a fixed shape that it retains indefinitely without any need for external confinement (see Figure 3-4). Picturing the atom as a sphere of microscopic size, atoms that comprise a solid are packed closer together than the same atoms that comprise a liquid or gas. We can say that solids do not flow like liquids nor do they expand like gases. The atoms of a solid do not possess the kinetic energy or the motion they would possess if they were in a liquid or gas form. The atoms are quite subdued and do not act as independent units, but move with their neighboring atoms. As we shall see, solids can be grouped into several categories, such as metals, plastics, ceramics, and composites. The attractive forces between atoms in a solid are much greater than those between atoms in a liquid or gas. These forces may position the atoms in some orderly geometric arrangement to form a ***crystalline structure*** (as in most metals) in which the atoms vibrate, rotate, and oscillate around fixed locations, maintaining a minimum dynamic equilibrium between adjacent atoms.

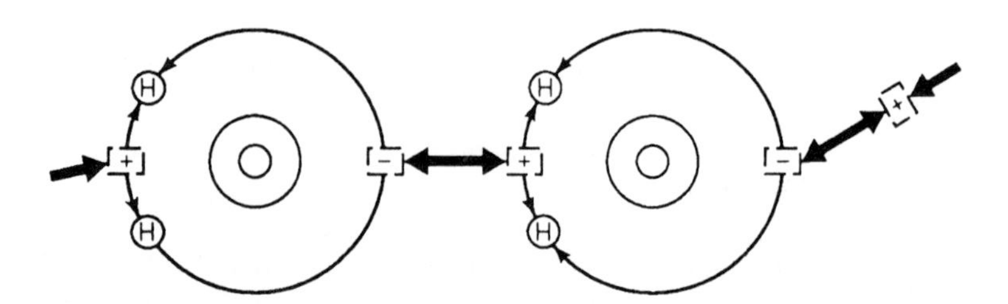

Figure 2-21 Hydrogen bond (bridge) in liquid water.

TABLE 2-2 SUMMARY OF ATOMIC BONDING

Type of bond	Number of electrons "shared"	Types of atoms involved	Remarks
van der Waals (molecular)	0	Same	Weak electrostatic attraction due to asymmetrical electrical charges in electrically neutral (as a whole) atoms or molecules
Ionic	1 (or more transferred)	Different	Strong electrostatic attraction
Covalent		Same or different	Electron pair "revolves" in common orbit about both nuclei, one atom supplying one electron
Normal	2		
Coordinate	2	Different	Electron pair "revolves" in common orbit about both nuclei, one atom supplying both electrons, for other atom supplying none (quite rare)
Metallic	∞	Same	General attraction of a very large number of positive (metallic) ions for a dispersed cloud of electrons

Source: C. O. Smith, *The Science of Engineering Materials*, 2nd ed., Englewood Cliffs, N.J.: Prentice-Hall, 1977.

From this, we can conclude that the geometric arrangement of atoms is related to the energy content. If the arrangement is orderly, this implies a minimum energy content. On the other hand, a disorderly, random distribution of particles (producing an ***amorphous structure,*** as in many plastics) is an indicator of motion and, hence, unbalanced forces acting on the atoms. Thus, the system of particles needs to increase its internal energy to sustain such motion, from the small energy demand required by vibration to the larger energy demanded by translational motion. Another indicator of energy is temperature. The higher the temperature of the system of particles, the higher the disorder of the particles will be and the more energy will be possessed by the system. With solids, we know that most of the motion of the atoms is confined to a vibrational type.

2.1.3 Processing and Structure

Throughout this book, you will learn how internal structure is altered as a result of processing. If materials are allowed to cool slowly as they transform from the liquid to the solid form, they will achieve equilibrium, which results in their natural internal structure and a given set of properties. However, if the cooling process is accelerated or interrupted, an entirely different structure and set of properties often results. For example, in casting molten metals to produce a desired shape, the rate of cooling is normally around 1 kelvin (K) per second. But with use of a rapid-solidification processing technique, the molten metal may be cooled at a rate of 10^8 kelvin per second, which results in disorderly (amorphous) structure and unique magnetic properties not available in the normally cooled metal. Module 3 covers the rapid solidification process.

Structure can also be changed when materials are in the solid state. If the proper ingredients have been mixed together during material formulation, it becomes possible to reheat the solid, then vary the rate of cooling to obtain a variation in properties such as hardness or electrical conductivity. The process of bending or compressing a solid material will alter the internal structure in a manner that will possibly increase the hardness and improve strength.

2.2 FAMILY OF MATERIALS

There is nearly a limitless variety of materials. With so many materials, how can one be expected to understand them? The method used in this book is to consider all materials as members of a big family. Materials that possess common characteristics are then placed into their

TABLE 2-3a PRIMARY MATERIALS

Metals
Polymers
Ceramics

own group within the family. Even though overlaps exist in the grouping system, it is easier to understand materials when relationships are identified.

Table 2-3a shows the primary materials, consisting of (1) metals, (2) polymers, and (3) ceramics, that form the family of materials. Table 2-3b shows the full family, comprised of four main groups: (1) *metallics,* (2) *polymers,* (3) *ceramics,* and (4) *composites.* A fifth group of *other materials* is used for materials that do not fit well into the four main groups. Each group then divides into subgroups. Some examples of each subgroup are listed to illustrate common materials in the subgroup. Within any system, such as transportation, communications, or construction, numerous *materials systems* exist. These systems must use materials from the groups within the material's family to meet their special needs. Figure 2-22 illustrates the variety of materials used in the telecommunications system. The wise materials user recognizes the many options available due to the compatibility between groups and then proceeds to develop a materials system that meets a specific need.

2.2.1 Other Systems of Grouping

The grouping system used in this book follows closely the grouping systems common to engineering. Slight differences do exist among grouping systems. For example, one system might divide materials into metals, polymers, ceramics, glass, wood, and concrete. Some systems divide all materials into three groups—organics, metals, and ceramics—and then distinguish between natural and synthetic materials. Others characterize materials as crystalline or amorphous. Regardless of the system, some inconsistencies develop because of the nearly limitless variety of materials. Still, the advantages of grouping outweigh the limitations. In this section, we will present the system of grouping for this book and study briefly the general nature of each group and subgroup. Detailed study of each group follows in subsequent modules.

2.2.2 Metallics

Metallics, or metallic materials, include metal alloys. In a strict definition, *metal* refers only to an element such as iron, gold, aluminum, and lead. (See the discussion of metals and nonmetals in Section 2.1.1.1.)

2.2.2.1 *Metallic* defined. The definition used for a metal will differ depending on the field of study. Chemists might use a definition for metals different from that used by physicists.

- ***Metals*** are elements that can be defined by their properties, such as ductility, toughness, malleability, electrical and heat conductivity, and thermal expansion.
- ***Metals*** are also large aggregations (collections of millions of crystals composed of different types of atoms held together by metallic bonds; see Figure 2-16). Metals usually have fewer than four valence electrons, as opposed to nonmetals, which generally have four to seven. The metal atom is generally much larger than the atom of the nonmetal.
- ***Alloys*** consist of metal elements combined with other elements. Steel is an iron alloy made by combining iron, carbon, and some other elements. Aluminum–lithium alloys provide a 10% saving in weight over a conventional aluminum alloy.

2.2.2.2 Types of metallics. Although metals comprise about three-fourths of the elements that we use (see metal groups in the periodic table), few find service in their pure form. There are several reasons for not using pure metals—they may be too hard or too soft, or they

TABLE 2-3b FAMILY OF MATERIALS

Group	Subgroup	Examples
Metallics (metals and alloys)	Ferrous	Iron
		Steel
		Cast iron
	Nonferrous	Aluminum
		Tin
		Zinc
		Magnesium
		Copper
		Gold
	Powdered metal	Sintered steel
		Sintered brass
	Metallic nanotubes	
Polymerics	Human-made	Plastics
		Elastomers
		Adhesives
		Paper
	Natural	Wood
		Rubber
	Animal	Bone
		Skin
Ceramics	Crystalline compounds	Porcelain
		Structural clay
		Abrasives
	Glass	Glassware
		Annealed glass
Composites	Polymer based	Plywood
		Laminated timber
		Impregnated wood
		Fiberglass
		Graphite epoxy
		Plastic laminates
	Metallic based	Boron aluminum
		Primex
	Ceramic based	Reinforced concrete
		CFCC
	Cermets	Tungsten carbide
		Chromium alumina
	Other	Reinforced glass
		MEMS
Others	Electronic materials	Semiconductors
		Superconductors
	Lubricants	Graphite
	Fuels	Coal
		Oil
	Protective coatings	Anodized aluminum
	Biomaterials	Carbon implants
		OLEDS
	Smart materials	Shape-memory alloys
		Shape-memory polymers
	Nanomaterials	Carbon nanotubes
	Biologically inspired materials	Self-healing fabrics

may be too costly because of their scarcity—but the key factor normally is that the desired property sought in engineering requires a blending of metals and elements. Thus, the combination forms (alloys) find greatest use. Therefore, *metals* and *metallics* become interchangeable terms. Metallics are broken into subgroups of ferrous and nonferrous metals.

Ferrous. ***Ferrous*** is a Latin-based word meaning "iron." Ferrous metals include iron and alloys of at least 50% iron, such as cast iron, wrought iron, steel, and stainless steel. Each

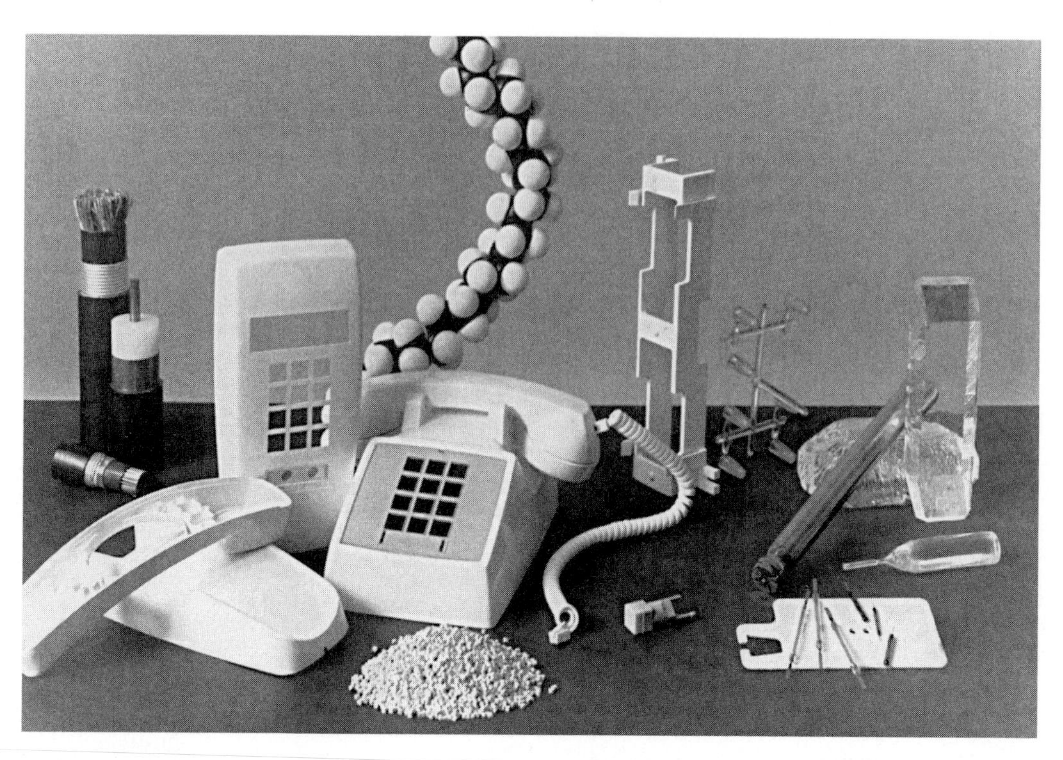

Figure 2-22 Family of materials in communication. One of the major factors influencing performance of telecommunications equipment is the performance of materials used in its construction. Plastics (polymers) for outdoor, undersea, and underground cables; magnetic alloys (metals); fiber optics; synthetic single crystals (ceramics); and fiber-reinforced rubber (composites) are five principal areas in which materials research and development discoveries greatly influence telecommunications. (Bell Laboratories)

of these alloys is highly dependent on possessing the element carbon. Steel is our most widely used alloy. Sheet steel forms car bodies, desk bodies, cabinets for refrigerators, stoves, and washing machines; it is used in doors, "tin" cans, shelving, and thousands of other products. Heavier steel products, such as plate, I-beams, angle iron, pipe, and bar, form the structural frames of buildings, bridges, ships, automobiles, roadways, and many other structures.

Nonferrous. Metal elements other than iron are called ***nonferrous*** metals. The nonferrous subgroup includes common lightweight metals such as titanium and beryllium and common heavier metals such as copper, lead, tin, and zinc. Among the heavier metals is a group of white metals, including tin, lead, and cadmium; they have lower melting points, about 230° to 330°C. Among the high-temperature (refractory) nonferrous metals are molybdenum, niobium, tantalum, and tungsten. Tungsten has the highest melting point of all metals: 3400°C. Metal alloys other than iron are called *nonferrous* alloys. The possible combinations of nonferrous alloys are practically limitless.

Powdered metals. Alloying of metals involves melting the main ingredients together so that on cooling, the metal alloy is generally a nonporous solid. Powdered metal is often used instead of solid metal because it is undesirable or impractical to join the solid elements through alloying or to produce parts by casting or other forming processes. Powdered metal is sometimes called ***sintered*** metal. As shown in Figure 2-23, sintering consists of producing small particles, compacting them, and *sintering* (applying heat below the melting point of the main component). The squeezing pressure along with the added heat bonds the metal powder into a strong (sometimes porous) solid. Powdered metals can be ferrous, nonferrous, or a combination of ferrous and nonferrous elements with nonmetallic elements. Many powdered metal parts are specifically designed to be porous, for example, bearings, cam followers, and medical implants. The ability to control porosity (density) also allows for control of the modulus of elasticity.

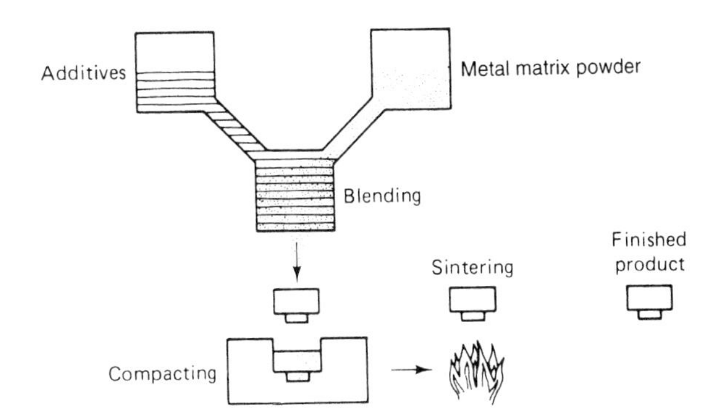

Figure 2-23 Powdered metal.

2.2.3 Polymerics

Polymeric materials basically are materials that contain many parts. *Poly* means "many" and *mer* comes from *meros,* which means "parts." A ***polymer*** is a chainlike molecule made up of smaller molecular units (monomers). The ***monomers,*** made up of atoms, bond together covalently to form a polymer that usually has a carbon backbone.

Figure 2-10 showed a simple sketch of an ethylene monomer. The atoms of carbon (C) and hydrogen (H) form covalent bonds. Many of the ethylene monomers in Figure 2-10 have joined into a polyethylene polymer. Thousands of polyethylene polymers join together to form polyethylene plastic. The same polymerization process is responsible for the formation of human-made rubbers and plastics; natural fibers; wood; rubber; animal bone; skin; and the tissues of humans, animals, and insects.

2.2.3.1 Plastics. The term ***plastic*** is used to define human-made polymer resins containing carbon atoms covalently bonded with other elements, along with organic and inorganic substances. The word ***plastic*** also means "moldable" or "workable," as in dough or wet clay. Plastic materials are either liquid or moldable during the processing state, after which they turn to a solid. After processing, some plastics cannot be returned to the plastic or moldable state; they are ***thermosetting plastics*** or ***thermosets.*** *Thermo* means "heat," and *set* means "permanent." Common thermosetting plastics include epoxy, phenolic, and polyurethane. Other plastics can be repeatedly reheated to return to the plastic state; they are ***thermoplastics.*** Examples of thermoplastic plastics are acrylics (e.g., Plexiglas or Lucite), nylon, and polyethylene. Although at present most plastics are produced from oil, they can also be made from other organic (carbon) materials, such as coal or agricultural crops, including wood and soybean. As our limited supply of oil is depleted, the major sources of polymerics will change.

2.2.3.2 Wood. Of all the materials used in industry, ***wood*** is the most familiar and most used. Wood is a natural polymer. In the same manner that polymers of ethylene are joined to form polyethylene, ***glucose*** monomers polymerize in wood to form ***cellulose*** polymers ($C_6H_{10}O_5$). Glucose is a sugar made up of carbon (C), hydrogen (H), and oxygen (O). Cellulose polymers join in layers with the gluelike substance ***lignin,*** which is another polymer.

2.2.3.3 Elastomers. Prior to World War II, most rubbers (*elastomers*) consisted of natural rubbers. Today, the use of synthetic or human-made rubbers far exceeds that of natural rubbers. An ***elastomer*** is defined as any polymeric material that can be stretched at room temperature to at least twice its original length and return to its original length after the stretching force has been removed. Some stretch to over ten times their original length. Elastomers are able to store energy, so they can return to their original length and/or shape repeatedly.

Elastomers have a molecular, amorphous structure similar to that of other polymeric materials. This *amorphous*, or shapeless, structure consists of long, coiled-up chains of giant molecules

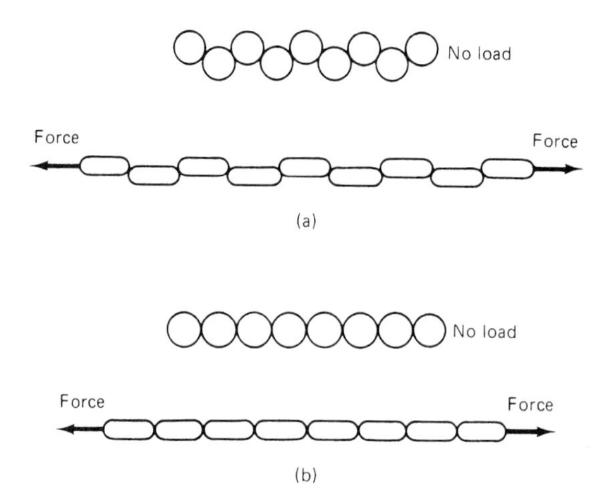

Figure 2-24 Elastomer structure: (a) Bond straightening, (b) bond lengthening.

(polymers) that are entangled with each other. Adjacent polymers are not strongly bonded together. As Figure 2-24a shows, when a tensile force (pulling force) is applied, these coils straighten out (***bond straightening***) and snap back like springs to their original coiled condition on removal of the force. When this same type of force is strong enough, not only do the polymers increase in length as a result of the bonds lengthening between individual atoms, they also lengthen (***bond lengthening,*** Figure 2-24b) due to the unwinding coils, resulting in a temporary structure that approaches crystallinity, or a more orderly structure. Remember that with crystallinity comes strength.

To further increase the strength of the elastomers, the process of ***vulcanization*** is used to form the necessary cross-links (strong bonds) between adjacent polymers (to be discussed later in Section 14.2.3). Vulcanization is a chemical process (invented by Charles Goodyear in 1839) that produces covalent bonding between adjacent polymers with the help of a small amount of sulfur. Cross-links tie the polymers together to produce a tough, strong, and hard rubber for many uses in industry, such as in automobile tires.

2.2.3.4 Other natural polymers. A most amazing natural polymer is human skin, which has as yet no substitute equal to it. Animal skin or hide in the form of fur and leather has limited industrial use because synthetic materials have been developed that offer greater advantages to the designer than those of the natural polymers. Medical science continues to study such natural polymers as bones, nails, and tissues of human beings and animals in order to synthesize these materials for replacement when they are damaged due to injury or illness. Bioengineering and biomechanics are newer fields that integrate engineering and medicine to solve material problems in the treatment of humans.

2.2.4 Ceramics

Ceramics are crystalline compounds that combine metallic and nonmetallic elements. ***Glass*** is grouped with ceramics because it has similar properties, but most glass is amorphous. Included in ceramics are porcelain, such as pottery; abrasives, such as emery used on sandpaper; refractories (materials with good resistance to heat), such as tantalum carbide, with a melting temperature of about 3870°C; and structural clay, such as brick. Ceramics, including glass, are hard, brittle (no internal slip), stiff, and have high melting points. Ceramics primarily have ionic bonds, but covalent bonding is also present. Silica is a basic unit in many ceramics. The internal structure of silica has a pyramid (tetrahedron) shape, as shown in Figure 2-25a, which includes both the graphical diagram and the chemical equation. Silicate tetrahedrons join into chains, as shown in Figure 2-25b. In the figure, note how the larger oxygen (O) atoms surround the small silicon (Si) atom. The silicon atom occupies the space between

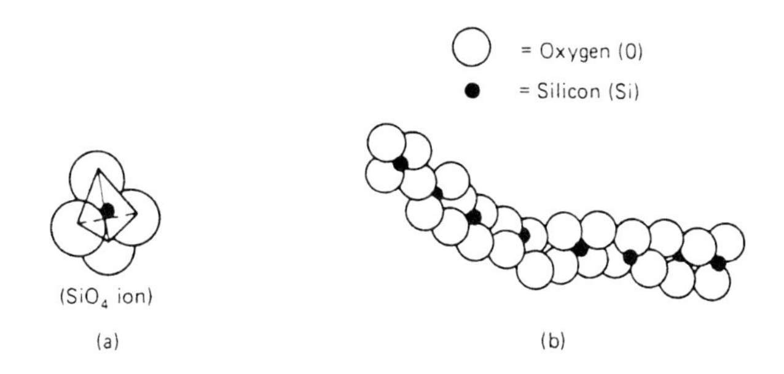

Figure 2-25 Ceramic structure: (a) Silicate tetrahedron, (b) single chain of silicate.

(interstice) the oxygen atoms and shares four valence electrons with the four oxygen atoms. Silica chains are extremely long and join together in three dimensions. They are held together by ionic bonds, whereas individual silica tetrahedrals bond together covalently. Silica is combined with metals such as aluminum, magnesium, and other elements to form a wide variety of ceramic materials.

2.2.5 Composites

By strict definition, a ***composite*** is a material containing two or more integrated materials (constituents or phases), with each material keeping its own identity. Normally, the combining of the materials serves to rectify weaknesses possessed by each constituent when it exists alone. By this strict definition, many natural materials exist as composites; for example, wood is a combination of cellulose and lignin. But wood and natural materials are not classified as composites. Some familiar composites include plywood, laminated dimes and quarters, and shoe soles.

Although most of the groups of the family of materials could be classed as composites because of the way they are placed in service (such as painted steel or case-hardened steel), the composite classification commonly refers to materials developed to meet the demands of the building, electronic, aerospace, and auto industries, where they find an ever-increasing use because they can be designed to be stronger, lighter, stiffer, and more heat resistant than natural materials or to possess properties required by technology that are not available in a single material. Composites allow the designer to select the right combination of materials that will perform safely at the lowest cost.

The subgroups of composites shown in Table 2-3 include polymer based, metallic based, ceramic based, cermets, and others. It is also possible to classify composites by their structure. Composite structures include layers, fibers, particles, and any combination of the three. ***Layered composites,*** shown in Figure 2-26, consist of ***laminations,*** like a sandwich. The laminations are usually bonded together by adhesives, but can be bonded by other means, such as welding. Fibers and particles are integrated into composites by suspending them in a ***matrix*** or by the use of *cohesive* forces. The matrix is the material component, such as plastic, epoxy cement, rubber, or metal, that surrounds the fibers or particles. ***Cohesive*** forces involve the molecular attraction of one constituent to the other.

Examples of layered composites (Figure 2-26a) include plywood, laminated boards such as particleboard (wood chips covered with plastic sheets), thermocouples (different metals welded together), safety glass (plastic sandwiched between sheets of glass), cardboard, and Alclad (made of a copper alloy core that gains corrosion resistance from external layers of aluminum). ***Particle composites*** include concrete (gravel, rock, and sand in a Portland cement matrix) and particleboard made of wood chips in a resin matrix (Figure 2-26b). ***Fiber composites*** (Figure 2-26c), such as reinforced plastics (FRPs), include fiberglass (glass fibers in a polyester plastic matrix), hardboard (vegetable fibers in an adhesive matrix), B Al composite (boron fibers in an aluminum matrix), reinforced glass (wire mesh in a glass matrix), and graphite epoxy (graphite fibers in an epoxy plastic matrix). The reinforcing phase is usually

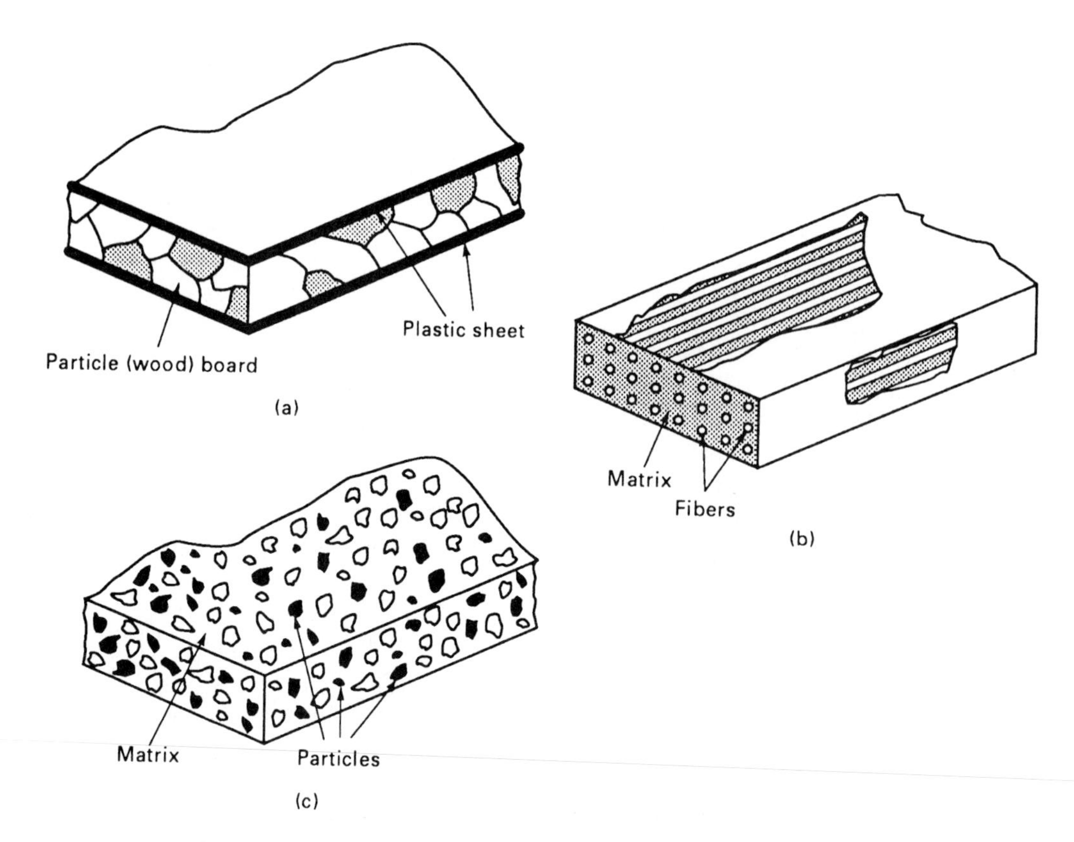

Figure 2-26 Typical composites: (a) Layered composite, (b) particle composite, (c) fiber composite.

stiffer and stronger than the matrix phase and serves as the load-carrying constituent. In general for fiber composites, one phase acts as a reinforcement for the other phase.

2.2.6 Other Materials

Another set of materials includes those that do not group well with the others. Comparing the materials in the "other" group with the rest of the family of materials in Table 2-3 reveals that most come from the metallics, polymers, ceramics, and composite groups, but their applications justify a separate study. Some are discussed in Module 9.

The new materials systems described in Module 1 lead to new possibilities for grouping materials. Each major group of the family of materials has some materials from a newly evolving subgroup known as intelligent material systems. ***Intelligent material systems, smart materials,*** or ***smart structures*** are materials and materials systems designed to mimic biological organisms. They offer the ultimate materials system, one that can insert control and feedback into a material structure. They take their cue from biological elements like muscles, nerves, and bodily control systems that adapt to environmental changes. Just as alchemists long yearned to turn lead into gold, so too, materials scientists and engineers seek to endow materials with the abilities to (1) make controlled adaptations to changes in stress or heat, (2) do self-repairs, and (3) provide feedback information on conditions that may have caused a material failure. As in the movie *Terminator II,* science fiction has long predicted "morphing" (the ability to change shape and structure) and ***biomimetics*** (materials mimicking living tissue). And as foretold in science fiction, materials designers today continue to develop materials structures that do allow for adaptation to unpredictable environments and information feedback to designers and users throughout the life of the material or structure. Monitoring systems using piezoelectric devices (explained in Module 27) are embedded into or adhesively bonded onto structures such as bridges, aircraft skin, building walls, and similar load-bearing structures. Connected to recording instruments, these intelligent structures provide information to help predict material/structural failures caused by aging, earthquakes, storms, or other stresses.

Nitinol, an alloy of nickel and titanium, illustrates how "smart materials" can be tailored to provide unique properties. The alloy is a ***shape memory alloy (SMA).*** SMAs, through heat treating, can be trained to hold a shape that they will always "remember," even if they are drastically bent out of shape. An eyeglass frame temple made of an SMA can be wrapped around one's fingers and then snapped back into its original shape. Through training, it is possible to set the shape memory so that it will change from a new shape back to its designed shape through the application of a desired temperature with 61% accuracy. Nitinol is available in a variety of forms, including wire, sheet, ribbon, and springs and has a high operating temperature, abrasion and corrosion resistance, and high tensile strength. The nitinol alloy could be used in a variety of ways, ranging from mobile robotic fingers activated by electrical currents to stress/strain sensors embedded in bridges and other structures.

We have begun to learn from the evolutionary traits of biological structures and have begun to imitate nature in the design of synthetic materials. ***Biologically inspired materials*** that result from analysis of nature should contribute to the array of smart materials. Some of the design goals for smart materials and intelligent structures are

1. Cost-efficient, durable structures whose performance matches the demands made on the structure
2. Ability to change properties, color, shape, and means of handling external physical loads to repair damage or make repairs from damage
3. Possession of the five senses of smell, taste, hearing, sight, and touch
4. Ability to learn, grow, survive, and age with grace and simplicity
5. Transmission of information back to designers and user
6. Adaptive features and the intelligence to reduce mass and energy needs
7. Flexibility to allow for specification of materials and structural requirements for designs that are both affordable and that fulfill design objectives*

The few examples of intelligent materials systems cited here and in other parts of this book reveal "the dawn of a new materials age," as explained by Rogers (1992). Be alert to this evolution of new materials. The concepts learned here will allow you to comprehend the innovations.

Biologically inspired materials represent a new class of evolving materials that spring from research at the "nanoscale" that achieve structure and properties similar to plants and animals. These materials have the ability to: 1) self-assemble, 2) self-heal, 3) morph or change shape in response to stress, 4) change color, etc. For example, major NASA research at NASA Langley focuses on air and space craft that would change configuration much like a bird uses its feathers for efficient flight. The NASA funded Biologically Inspired Materials Institute (BIMat) states

> Materials with orders-of-magnitude improvements in performance and reliability are a crucial element of these new space systems. *Our objective is to leap over current barriers in performance-to-weight ratios and to achieve reliabilities beyond any achieved so far.* However, such improvements cannot use old approaches. New concepts and approaches are prerequisite to the design of materials systems optimized for multifunctional uses that range from strengthening (stiffness and toughness) to sensing/actuating and self-healing.**

*List adapted from Craig Rogers, 1992.

**From Ilhan A. Aksay lecture, "Materials World: the Developments during Last 12,000 Years & its Future" NASA Langley, Hampton, VA June 3, 2003.

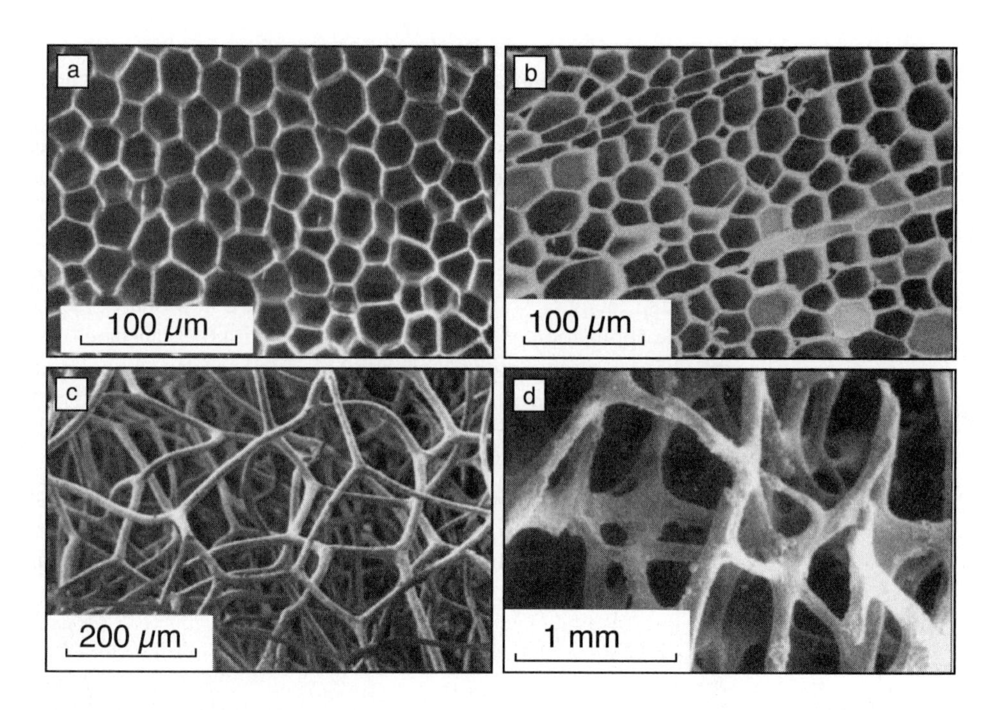

Figure 2-27 Examples of cellular solids in natural materials: (a) cork, (b) balsa wood, (c) sponge, and (d) trabecular bone. (L. J. Gibson and M. F. Ashby *Cellular Solids: Structure and Properties,* 2nd ed., Cambridge, U.K.: Cambridge University Press, 1997)

BIMat depicts cellular biomimetic material, as seen in Figure 2-29 on p. 66.

 BIMat NASA URETI—*http://princeton.edu/~cm/*

2.2.6.1 Cellular solids. Many solid materials may have large internal voids of various forms, such as prismatic or polyhedral cells. The cellular solids exist naturally or can be processed in order to produce cellular structure. Figure 2-27 shows some examples of cellular solids in natural materials: cork, balsa wood, sponge, and trabecular bone. Figure 2-28 shows some examples of engineered honeycombs. In later modules, we will explore how cellular structure offers unique properties in such applications as cores of structural sandwich panels used for snow skis or lightweight doors and building panels. Cellular solids can absorb the kinetic energy of an impact, as found with football helmet pads and auto bumpers. "Styrofoam" cups and thermal insulation building foams are examples of closed-cell foam. Metal foams with open cells can have high strength-to-weight ratios and also allow fluids to flow, as in heat-exchange devices. Foams are also now made from polymers, metals, ceramics, and glasses.

APPLICATIONS & ALTERNATIVES

This module introduced the family of materials grouping system as a means to help organize your thinking about the infinite variety of engineering materials. Applications of materials are seen everywhere, including aircraft. New breeds of commercial airplanes use the latest in the applica-

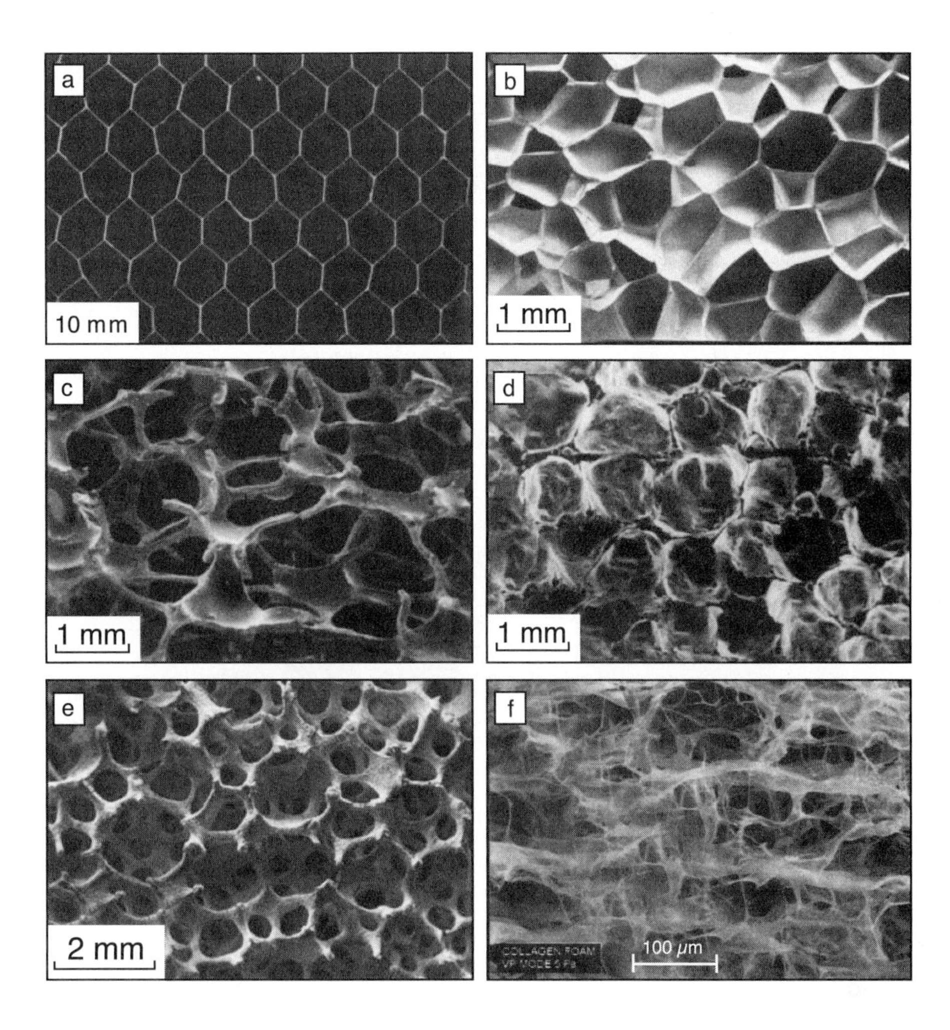

Figure 2-28 Engineered cellular materials: (a) aluminum honeycomb, (b) closed-cell polyethylene foam, (c) open-cell nickel foam, (d) closed-cell glass foam, (e) open-cell zirconia foam, and (f) a collagen-based porous scaffold used in tissue engineering. (L. J. Gibson and M. F. Ashby *Cellular Solids: Structure and Properties*, 2nd ed., Cambridge, U.K.: Cambridge University Press, 1997)

tion of engineering materials technology, as in the Boeing 7E7 (Figure 7-30), that succeeded the Boeing 777. Several illustrated discussions throughout this book will focus on the 777 and the innovative application of design, materials, and manufacturing techniques used on this new breed of airplane. With an estimated development cost of $5 billion and an average per plane cost of $100 million, builders of the 777 relied on newer approaches to product design, development, and manufacture. One such approach to design/manufacturing, ***concurrent engineering,*** was a highly coordinated team approach that commenced when the project began. All parties (designers, manufacturers, suppliers, and customers) provided input throughout the project. Another useful tool was ***simulation,*** which saved time and money by using computers and testing equipment to replace prototypes, live test beds, and other traditional approaches to development. Another approach used in the project involved ***extensive component testing*** to ensure the reliability of each component, such as engines or landing gears, and is often done prior to assembly of the aircraft.

Following the design and construction of the 777, aerospace companies investigated ways to rethink the use of advanced composites and aluminum alloys for aircraft. The study concluded that composites offer such advantages as better durability, reduced maintenance requirements, and opportunities for continued materials improvements. On the other hand, the

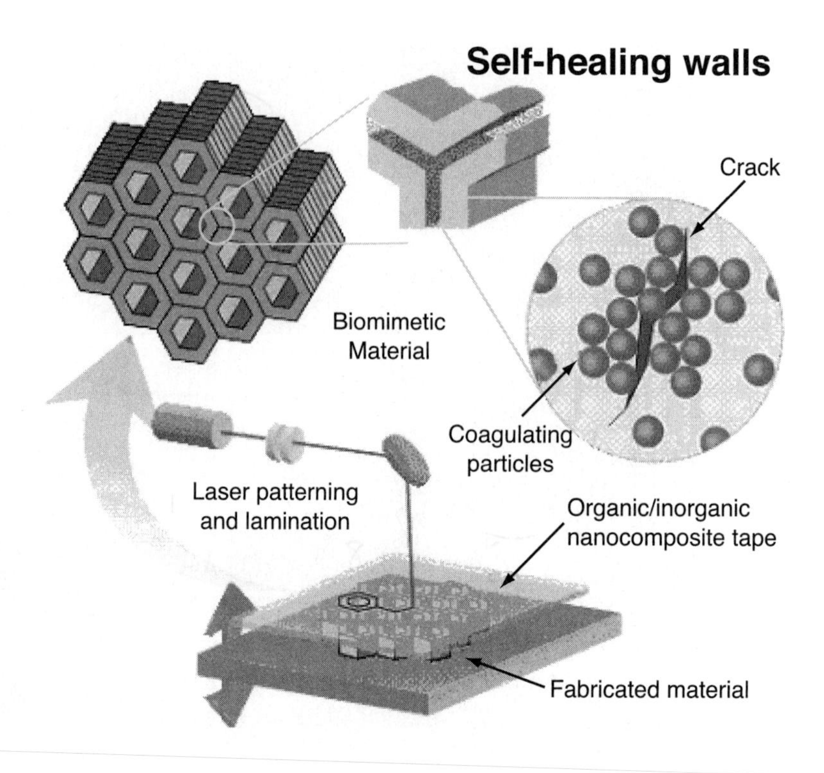

Figure 2-29 Cellular biomimetic material from BIMat website.

study found that the weight of aluminum alloys is not significantly more than advanced composites and that alloys do not bring significant cost savings.

The 7E7 design employs graphite combined with toughened epoxy resin for the primary material for wings and fuselage. Titanium matrix composites reinforced with graphite fibers are also used in the wings.

Airbus, (Boeing's only competitor for large jet liners) introduced its jumbo A380 that will transport 555 passengers. The Boeing 7E7 is a smaller (200–250 passenger) but more efficient aircraft in terms of fuel usage and maintenance. Lighter materials and improved jet engines result in lower operating costs, reduced takeoff noise, and reduced airport space, allowing it to use smaller airports with no major modifications. Reduced maintenance will result from the application of sensors throughout the aircraft made of some of the smart materials and intelligent structures discussed in this book.

The estimated development cost of the 7E7 is $10 billion, twice that of its predecessor, the 777. The initial two design versions of the 7E7 (basic and stretch) offer flying ranges of 7200 to 8000 nautical miles (13,300 to 14,800 kilometers).

Following the 9–11 terrorist attacks on the World Trade Center and the Pentagon, commercial air travel has been in a tailspin. Recovery of the industry (both aircraft builders and airline companies) has been slow. As Airbus and Boeing compete for new business, the stakes are high. Some feel the fate of Boeing's entire commercial jet business rests in the hands of its 7E7. You may wish to keep abreast of developments in the news at the following website:

 *www.boeing.com/commercial/7e7*

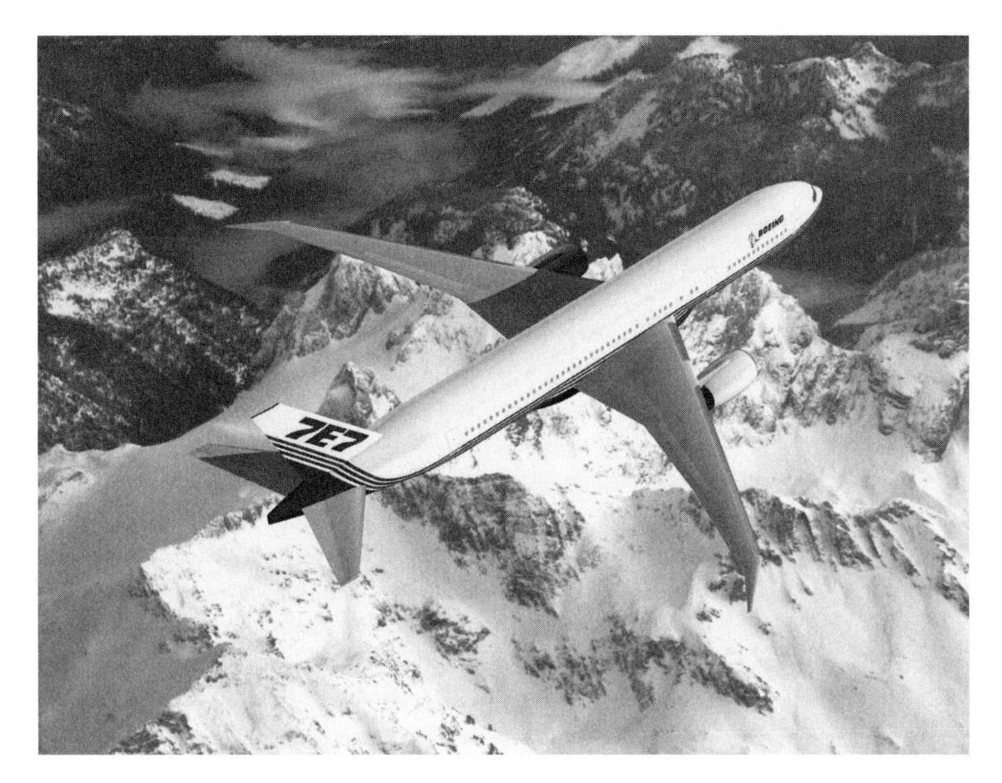

Figure 2-30 Boeing 7E7 jet aircraft relies on advanced design concepts and advanced materials for a more efficient airliner. (The Boeing Company)

SELF-ASSESSMENT

2-1. The largest consumer of electricity in the United States is the refining of the metal ______.

a. Steel **b.** Copper **c.** Aluminum

2-2. A ______ is a substance that cannot be broken down any further by a chemical reaction.

a. Molecule **b.** Compound **c.** Element **d.** Atom

2-3. A compound is a substance that can be broken down by chemical reactions into ______.

a. Elements **b.** Molecules **c.** Atoms

2-4. Protons and neutrons are known as ______.

a. Quarks **b.** Electrons **c.** Neutrinos **d.** Nucleons

2-5. Another name for an electron shell is ______.

a. Quantum **b.** Principal energy level **c.** Energy state **d.** Ion

2-6. Two electrons can exist in the same orbital if they have ______.

a. The same energy **b.** The same polarity **c.** Opposite spins

2-7. The maximum number of electrons in the L, or second, energy level is ______.

a. 2 **b.** 18 **c.** 8 **d.** 0

2-8. Inert gases can also be called ______.

a. Stable **b.** Group 5 elements **c.** Noble gases **d.** Active

2-9. The core electrons in neon (Ne) number ______.

a. 2 **b.** 10 **c.** 18 **d.** 6

2-10. Horizontal rows in the periodic table are called ______.

a. Groups **b.** Transition elements **c.** Periods **d.** Metals

2-11. Metalloids consist of how many elements?

a. 6 **b.** 10 **c.** 7

2-12. The electronegativity of iron (Fe) is __________.

a. 1.83 **b.** 3.98 **c.** 0.93

2-13. What is the principal valence of Group VIII elements?

a. 16 **b.** 6 **c.** 8 **d.** 12

2-14. ${}^{16}_{8}O$, ${}^{17}_{8}O$, and ${}^{18}_{8}O$, are the isotopes of oxygen. What is the atomic number of oxygen? How does the number vary among the three isotopes?

2-15. The number of which nucleons never changes when an atom becomes an ion?

2-16. Consulting the periodic table name the elements represented by X in each of the following:

a. ${}^{26}_{12}X$ **b.** ${}^{48}_{22}X$ **c.** ${}^{184}_{74}X$

2-17. What other information does the atomic number convey in addition to the number of protons in the nucleus of an atom?

2-18. How many protons are there in the nucleus of one atom of the following elements?

a. hydrogen **b.** oxygen **c.** boron

2-19. List the number of protons, neutrons, and electrons in each of the following atoms:

a. ${}^{40}Ca$ **b.** ${}^{2}H$ **c.** ${}^{19}F$

2-20. What are the chemical symbols and the atomic numbers of one atom of the following:

a. magnesium **b.** calcium **c.** zinc

2-21. The key word in describing ionic bonding is *transfer*. What key word can be used to describe covalent bonding?

2-22. What is the name of the recently discovered carbon structure that resembles a soccer ball?

2-23. Give examples of ionic, covalent, and metallic bonding.

2-24. Express the average interatomic distance in units of nanometers and inches.

2-25. What is the relationship between processing and structure?

2-26. In which family group belong the materials with large chainlike molecules that consist of smaller molecules or monomers? Are they generally amorphous or crystalline?

2-27. When a metallic contains at least 50% iron, what is its classification?

2-28. What is the name for a metallic consisting of metal elements and other elements? Give three examples.

2-29. Define *composites*. List three composite structures.

2-30. In the ceramic grouping, most materials are crystalline, but which one has an amorphous structure?

2-31. List three examples of lightweight nonferrous metals or alloys, three examples of heavier nonferrous metals, and one example of a refractory metal.

2-32. What is the subgroup of metallics in which metal particles are joined by pressure and sintering heat?

2-33. What subgroup covers human-made polymers containing carbon atoms covalently bonded together with other elements? List the two divisions of this polymeric material and give an example of each.

2-34. Name three natural polymeric materials.

2-35. How do elastomers differ from other polymeric materials? How can the strength of rubber be improved?

2-36. How does the energy shortage threaten the supply of plastics and rubber?

2-37. What is the prime purpose of selecting a composite material over material from the other family groups?

2-38. List three materials that were grouped in the "other" family group of materials.

2-39. Define cellular solids and cite one example. Sketch an example of a cellular solid.

2-40. Describe how biologically inspired materials may act as smart materials.

2-41. A distinction that sets intelligent structures apart from other advanced materials is

a. Cost per unit volume, which is often higher than traditional materials

b. Ability to adapt to changes in environment

c. More exotic processing

d. Benefits from superior properties, which offset materials cost

REFERENCES & RELATED READINGS

AMATO, IVAN. *Stuff: The Materials the World Is Made Of.* New York: Harper Collins, 1997; Avon Books (Bard), 1998.

ASHBY, M. F. "Technology of the 1990s: Advanced Materials and Predictive Design," *Philosophical Transactions of the Royal Society of London*, Vol. A322, 1987, pp. 393–407.

ASIMOV, ISAAC. *Building Blocks of the Universe.* New York: Abelard, 1957.

BELLMEYER, FRED W. *Synthetic Polymers.* Garden City, NY: Doubleday, 1972.

BETTS, JOHN E. *Physics for Technology.* Reston, VA: Reston, 1964.

BROWN, THEODORE L., AND H. EUGENE LEMAY, JR. *Chemistry: The Central Science,* 2nd ed. Englewood Cliffs, NJ: Prentice-Hall, 1981.

CHRISTIANSEN, G. S., AND PAUL H. GARRETT. *Structure and Change: An Introduction to the Science of Matter.* San Francisco: W.H. Freeman, 1960.

COTTERILL, RODNEY. *The Cambridge Guide to the Material World.* Cambridge, U.K.: Cambridge University Press, 1989.

CRAIG, DOUGLAS F. *"Structural Ceramics," National Educators' Workshop, NEW: Update 91*, Oak Ridge National Laboratory, Oak Ridge, TN, November 12, 1991.

EASTERLING, KEN. *Tomorrow's Materials.* London: The Institute of Metals, 1988.

EWEN, DALE, AND LERAY HEATON. *Physics for Technology Education.* Englewood Cliffs, NJ: Prentice-Hall, 1981.

GAMOW, GEORGE A. *Mr. Tomkins Explores the Atom.* New York: Macmillan, 1945.

GIANCOLI, DOUGLAS C. *Physics.* Englewood Cliffs, NJ: Prentice-Hall, 1980.

HUMMEL, ROLF E. *Understanding Materials Science: History, Properties, Applications.* New York: Springer-Verlag, 1998.

Module 3

Synthesizing, Processing, & Structure of Solid Materials

After studying this module, you should be able to do the following:

3–1. Discuss the relationships between processing and structure of materials.
3–2. Recall the synthesis and processing capabilities of contemporary materials technology at the various scales (macro, micro, nano, and atomic).
3–3. Explain how nanotechnology will enhance materials science and engineering in terms of synthesis, processes, structure, and applications.
3–4. Use sketches, unit cell structures, and standard notations to explain and solve problems related to crystallography. Determine the unit cell structure of given elements from a periodic table of elements for metallurgists.
3–5. Explain the alloying and doping process of materials in terms of solutions, alloying elements, packing factors, crystal imperfections, and impurities and use appropriate sketches. Cite specific practical applications for alloying and doping.
3–6. Using cooling curves as an aid, recall the effects of solidification in pure metals, alloys, and nonmetals.
3–7. Recognize Greek symbols and determine their meaning and proper pronunciation.
3–8. Apply SI rules of usage, make conversions of U.S. customary units to SI (and vice versa), and use both for computations.
3–9. Use the website addresses denoted by the * icon to seek current developments and emerging technology related to processing and structure.

*The symbol found throughout the book will link you to Internet sites related to topics being covered. The dynamic nature of the Web brings frequent changes, so some of these sites, while available at the time of writing, may not be up now.

3.1 SYNTHESIZING AND PROCESSING

Solid materials are products of the conversion of raw materials into bulk materials. With many polymers, synthesis is a part of raw materials production. Newer materials, such as artificially structured materials, are also synthesized. Raw materials processing or synthesis by conventional means provides us with familiar structures, such as polycrystalline metals, amorphous plastics, and crystalline ceramics. ***Artificially structured materials***, which are synthesized on micro-, nano-, and atomic scales, have unique engineered structures that may be ultrapure and/or possess electronic circuitry and optical properties based on a special design. In the modules on metals, polymers, ceramics, and composites, you will learn that a newer process, rapid solidification of metal, provides an amorphous structure rather than the natural polycrystalline structure. Chemical vapor deposition (CVD) and molecular beam epitaxy (MBE) are recent synthesis techniques that allow for the production of "designed materials." Many ***"beam" processing*** and synthesis techniques that use laser, ion, and electron beams to produce micro- and nano-scale structures, are discussed later.

Manufacturing processes are traditionally used to convert bulk synthesized or processed materials into structural shapes and final subassemblies of products. Manufacturing processes will usually change the structure of the bulk material, thereby requiring a variety of "secondary processes" to ensure that the final product has the necessary structure. ***Near-net-shape*** and ***net-shape processing*** aim to achieve a final product or specified structure through the initial synthesis or raw materials processing, thereby avoiding secondary processes. Aluminum is an ideal example. Although aluminum is the most plentiful metal in our earth's crust, it requires enormous amounts of electrical and chemical energy to extract it from bauxite ore. Great savings are realized by recycling, which makes it practical and desirable for all concerned. Often, however, bulk aluminum must go through many ***secondary processes***, such as machine cutting, welding, grinding, anodizing, and heat treating, to achieve a finished product. Using near-net-shape processing to go from the raw aluminum to the final product, while achieving the desired structure and properties, would save even more energy. In Module 20, we will explain how pressureless infiltration processes produce near-net-shape metal-matrix-composite (MMC) parts.

Table A-6 lists the prefixes used in the worldwide measurement system SI, (taken from the French words "*Système Internationale*" as explained in Table A-1). Note that *micro* and *nano* are prefixes that stand for 10^{-6} (one millionth) and 10^{-9} (one billionth), respectively. The following statements may give you some appreciation of the "smallness" of the "invisible to the human eye" atoms and molecules that form human beings and materials. One thousand (10^3) particles of some material with a diameter of 10 μm (the prefix μ stands for 10^{-6} and m is the symbol for meter) could fit inside the period at the end of this sentence, as would several hundred thousand nanometers (nm). A 6-foot-5-inch basketball player is about 2 billion nanometers (nm) tall. One billionth of a meter, 1 nm, is about four times wider than an atom or about $\frac{1}{10,000}$ the width of a human hair. We saw in Figure 2–1 the time line of materials development up to the present. The immediate future of materials technology can be classified as the *nanocomposite era* (defined later in this module).

The winners of the race to place products in the marketplace will be those who develop improved instrumentation with which to explore and manipulate matter at the nano and atomic scales. All developed countries and many developing countries are investing heavily to improve or maintain their competitiveness with the advanced materials that are used in products like high-definition television (HDTV), superior computer technology (neural networks, multimedia, artificial intelligence, virtual reality), improved transportation (magnetic-levitation trains and surface seacraft), better communications and entertainment technology, and smart homes and smart highways. With instruments such as the scanning tunneling microscope (described later), materials scientists may view and manipulate the nano and atomic levels of materials for better understanding and control of their structure and improved synthesis and processing. Working at these scales, it is becoming possible for them to produce new structures such as ***nanocomposites***—materials with atoms and molecules arranged in unique phases that provide designed properties.

National Institute of Standards & Technology—*www.nist.gov*
MicroScapes—"Too Small for the Human Eye," a series of photographs taken through high-power microscopes—*http://www.lucent.com/microscapes/*

3.1.1 Nanotechnology

This section will expose you to a few basic concepts of nanotechnology. Module 28 also covers concepts of nanotechnology and microelectromechanical systems (MEMS); since it offers revolutionary opportunities to generate new classes of materials, nanotechnology deserves coverage throughout the book. Nanotechnology provides the potential to design and manipulate material structures at the atomic and molecular levels, as we have said. The transistors, capacitors, and diodes that emerge from the nation's laboratories have dimensions of less than 0.25 micrometers. More useful as a unit of measure is the ***nanometer (nm)***, the next smallest unit, defined as 1 billionth of a meter. The diameter of most atoms is 0.1 to 0.4 nm; the diameter of human hair ranges from 40 to 120 μm, which to express in nanometers would require multiplying by 1000. (See Figure 8-21 for a comparison of synthetic ceramic fibers with human hair.)

Table 3-1 lists the building blocks of our world of materials, beginning with (1) nature's elements of atoms, molecules, and cells; next, (2) the nanoscale components that have become possible through the new nanotechnology that allows the processing of nanoparticles to form nanosheets or nanofilms, which then roll up into nanotubes; (3)***microfabrication***, which includes the modern techniques of lithography and thin films that permit fabrication of integrated circuits, surface engineering, and MEMS devices; (4)finally, at the macroscale, fabrication from bulk materials, which includes both ancient and newer processes, such as casting, machining, and forming to produce engine blocks, gears, soda cans, and the like.

Nanotechnology has spurred global efforts to enhance the manufacturing of classes of materials that have perfect precision. These manufacturing processes are called ***molecular manufacturing, nanofabrication manufacturing, or molecular nanotechnology. Nanotechnology*** is defined as the processing of materials into microelectromechanical structures (MEMS) and devices whose sizes vary from hundreds to hundredths of a micrometer (mm). Nanotechnology is also the science of producing and tailoring devices. The National Science Foundation (NSF) defines it as involving novel materials in the size range of 1 to 100 nm that are made in novel ways, for example, nanotubes and DNA conductors. Present-day applications of ***nanoelectromechanical systems (NEMS)*** can be found in air bags triggers, digital light projection in theaters, and the controls for the splitting of colors in an inkjet printer. ***Microelectromechanical systems (MEMS)***, made with the same silicon fabrication methods used

TABLE 3-1 THE BUILDING BLOCKS OF THE WORLD OF MATERIALS

1. Nature's elements
   - Atoms
   - Molecules
   - Cells
2. Nanoscale components
   - Nanoparticles
   - Nanotubes
3. Microfabrication
   - Lithography
   - Thin films
4. Fabrication from bulk materials

Source: Adapted from L. C. Frechette. "MEMS and Nanotechnology in 21st Century." Presented at NEW: Update 2001, October 14–17, 2001, University of Maryland.

TABLE 3-2 APPLICATIONS OF MEMS TECHNOLOGY

- MEMS: low cost, small, multifunctional integrated systems:
  - Ex: Sensors and actuators
- MEMS as bridge between nanoscale elements and the macroworld
  - Ex: Lab-on-a-chip and nanotube probe
- MEMS as tools for nanoscale research
  - Ex: Probing in cellular biology
- Nanotechnology can enable:
  - Nanostructured materials by design
  - Nanoelectronics, optoelectronics, magnetics
  - Bionanosensors and drug delivery

Source: Adapted from L. C. Frechette. "MEMS and Nanotechnology in 21st Century." Presented at NEW: Update 2001, October 14–17, 2001, University of Maryland.

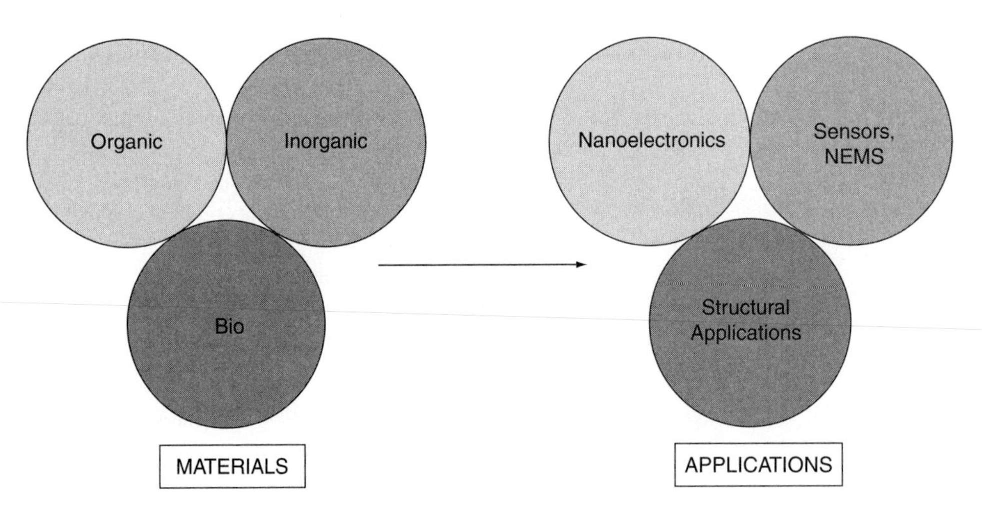

Figure 3-1 Research and development in nanotechnology involves producing materials at the nanoscale with organic, inorganic, and biological atoms and molecules for a wide array of applications. (NASA Ames)

to produce computer chips, are gradually giving way to the next step in machine miniaturization, devices built on the nanometer scale. Table 3-2 lists some application of MEMS. Nanotechnology research and development (R&D) now focuses on producing materials at the nanoscale from organic, inorganic, and biological materials for a wide array of applications, as seen in Figure 3-1. Figure 3-2 illustrates how sheets of nanoscale grapheme can be rolled into three forms of nanotubes, and Figure 3-3 shows an image of actual rows of carbon nanotubes (CNT).

 NASA Ames —*http://www.ipt.arc.nasa.gov*

3.2 STRUCTURE OF SOLID MATERIALS

In simplified form, structure is the "cause" that produces the properties ("effect") possessed by a material. In fact, there are many levels of structure, from the atomic to the macroscopic (visible to the human eye), each with its own chemical composition and distribution of its structural components (i.e., atoms, molecules, crystals, and phases, or what can be called the basic building blocks of materials). All these material structures give rise to an infinite array of materials, each possessing its own unique set of properties. All who work with materials or use them must constantly be aware that subjecting a material to external forces and/or conditions (impact, changes in temperature, corrosive conditions, and radiation) will, in the majority of instances, have an effect on its structure with attendant changes in its

CNT is a tubular form of carbon with diameter as small as 1 nm.
Length: few nm to microns.

CNT is configurationally equivalent to a two dimensional graphene sheet rolled into a tube.

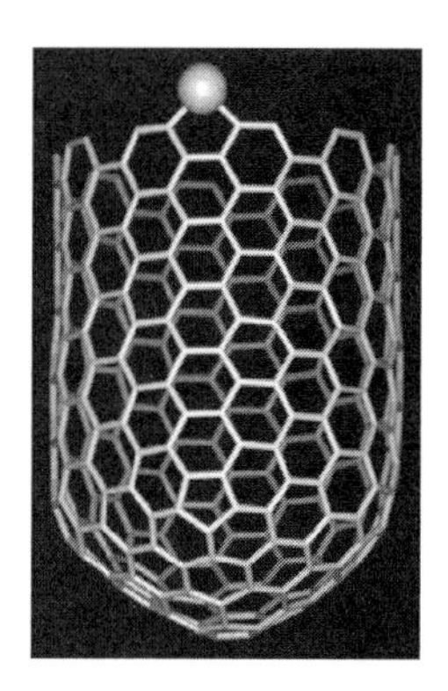

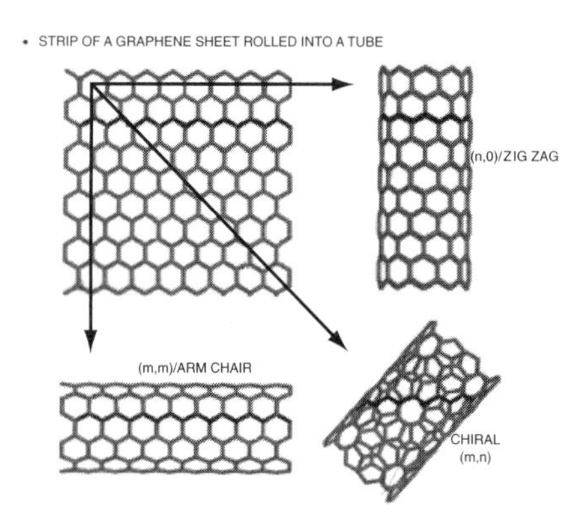

CNT exhibits extraordinary mechanical properties: Young's modulus over 1 Tera Pascal, as stiff as diamond, and tensile strength~200 GPa.

CNT can be metallic or semiconducting, depending on chirality.

Figure 3-2 Producing carbon nanotubes (CNT) begins with sheets of nanoscale grapheme. Depending on the three orientations, tubes can be rolled into three forms of nanotubes. (NASA Ames Research Center)

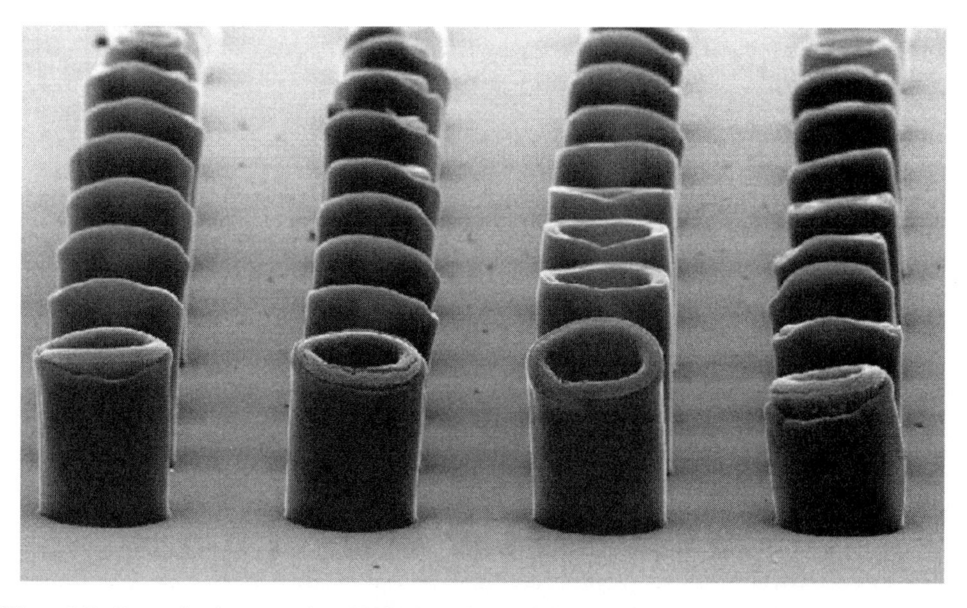

Figure 3-3 Rows of carbon nanotubes. (NASA Ames Research Center)

properties. Some of these changes can be handled quite satisfactorily by the material if the design process was correct in every respect. A final observation on the awareness of the many structures in a material is that it helps us understand the role of materials scientists, who are constantly working to develop materials with different structures or with different combinations of structures with the right mix of properties to help solve our technological problems.

Definitions of solids most always contain the phrase "definite volume and shape," together with the fact that solids maintain their shape to a varying extent when subjected to external mechanical forces. Solids occur in two basic forms: crystalline and amorphous. A true ***crystalline solid*** possesses an ordered, three-dimensional, geometric arrangement that repeats itself. A metal (though not perfectly crystalline) would be representative of a crystalline solid. An ***amorphous solid***, on the contrary, contains no repetitious pattern of atom locations to any extent. The classic example of an amorphous solid, glass, is sometimes referred to as a supercooled liquid because of the random nature of its atomic arrangement. Figure 3-4 illustrates the different atomic arrangements of three states of matter, emphasizing the different degrees of disorder and closeness of the atoms' positions in relation to each other.

In our treatment of ***crystallography*** (the study of crystalline structures), we will limit ourselves to the orderly arrangement of the atoms in their microscopic world. In so doing, we represent atoms, ions, or molecules essentially as spheres of varying sizes occupying points at various distances from each other in space—hence, the need for an ***axis system***. Such a system is shown in Figure 3-5. The arrowhead on each axis points in the positive direction. The negative directions, therefore, are opposite to the arrowheads. The x-, y-, and z-axes are, in this

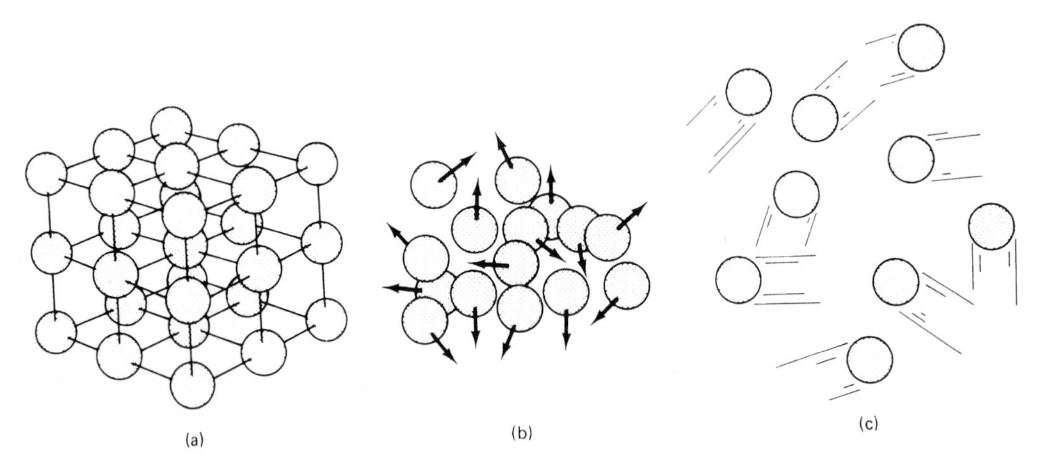

Figure 3-4 How atoms are arranged in (a) a crystalline solid, (b) a liquid, (c) a gas. (D. C. Giancoli, *Physics*, Prentice-Hall, Inc., Englewood Cliffs, NJ, 1980)

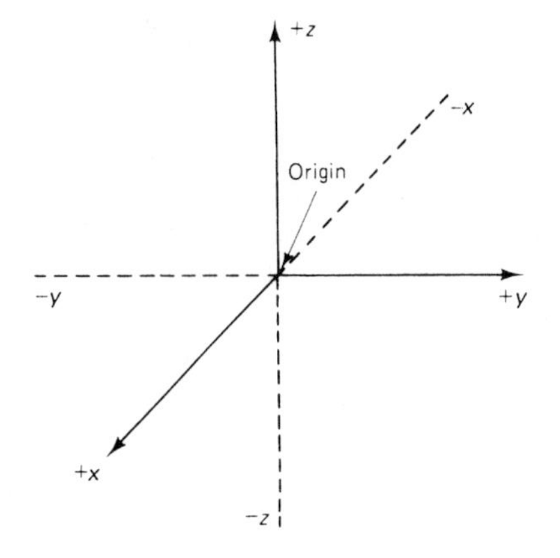

Figure 3-5 A standard axis system.

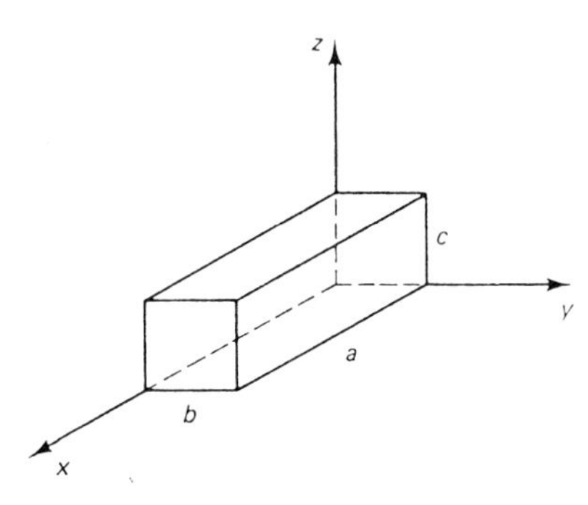

Figure 3-6 A box aligned on an axis system.

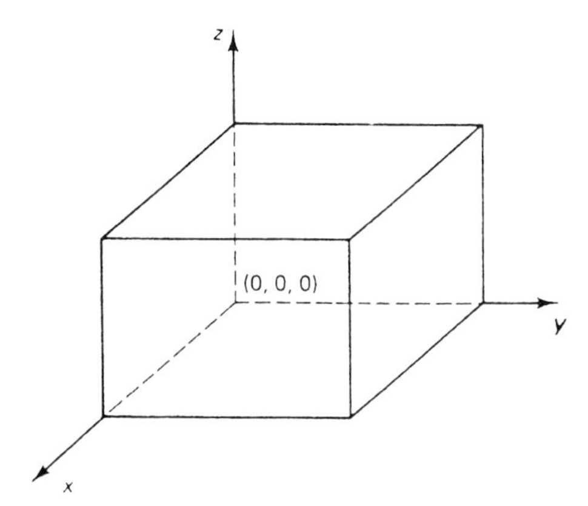

Figure 3-7 A three-dimensional box aligned on an axis system with the origin at the lower-left rear corner.

case, perpendicular to each other, with their origins coinciding at any convenient point. A box using this axis system with dimensions *a*, *b*, and *c* is sketched in Figure 3-6 If this solid figure were a cube, these dimensions would be of the same magnitude. However, in one important crystal system they are not equal in magnitude. Another interpretation of these letters (or *lattice parameters*) is that they represent distances from the origin to points on the axes. The letter *b* would, for example, represent the perpendicular distance from the origin to a point on the *y*-axis.

The three axes in Figure 3-7, which are infinite in length, form three mutually perpendicular planes that are also infinite in size. A box is drawn using these axes as sides to help illustrate some basics concerning crystal systems. The *x–y* horizontal plane forms the base of the box. The top of the box is contained in a plane parallel to the *x–y* plane. A portion of the *y–z* plane contains the rear end of the box (as looked at by the viewer), and the front end would be contained in a plane parallel to the *y–z* plane. There are countless planes parallel to these three primary (*x–y*, *y–z*, and *x–z*) planes, which can be referred to using a standard technique called *Miller indices* (see Section 3.2.3.1).

3.2.1 Unit Cells and Space Lattices

In our study of crystals, the axis system just described (called a ***simple cubic crystal system***) is modified slightly for the sake of uniformity. The standardized axis system shown in Figure 3-8 retains the *x*-, *y*-, and *z*-axes. The box is now called a *unit cell* (defined later). The angles between the principal planes are named α (Greek letter alpha), β (Greek letter beta), and γ (Greek letter gamma). For example, in Figure 3-9, α is the angle measured in degrees between the *x–y* and *x–z* planes; angle β, between the *x–y* and *y–z* planes; and angle γ, between the *x–z* and *y–z* planes. The sides of the box (unit cell), labeled *a*, *b*, and *c*, are the lattice parameters in the *x*, *y*, and *z* directions, respectively. These distances are also known as ***intercepts***. To describe a particular axis or crystal system adequately, all six of the preceding dimensions are needed.

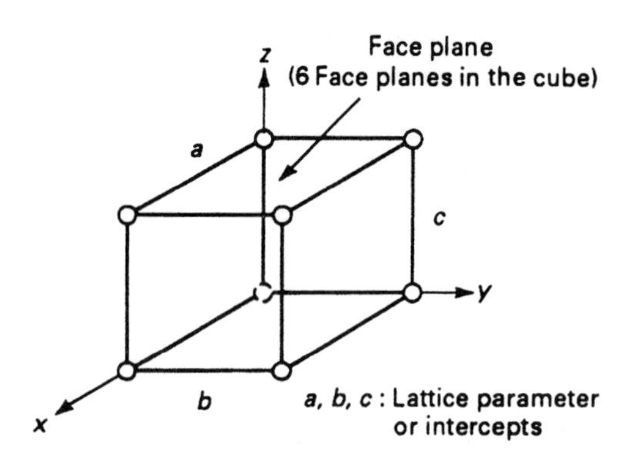

Figure 3-8 A simple crystal lattice unit cell with equal intercepts a, b, and c.

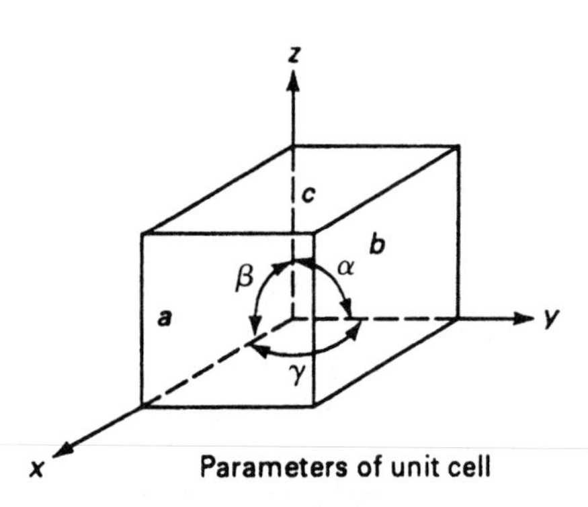

Figure 3-9 Angles between planes and intercepts in a unit cell.

The term ***unit cell*** is used to describe the basic building block or basic geometric arrangement of atoms in a crystal. You can compare a unit cell to a single brick in a brick wall. Knowing that atoms are located at each corner of the single brick, it is easy to picture the atomic structure of a crystal, with such a unit cell repeating itself in three-dimensional space. If you repeat the unit cell in all three dimensions, you create a crystalline structure with a definite pattern. This larger pattern of atoms in a single crystal is known as a ***space lattice or crystal lattice***. A space lattice is three sets of straight lines at angles to each other, constructed to divide space into small volumes of equal size, with atoms (ions or molecules) located at the intersections of these lines or between the various lines. We must remember that the lines and points in a space lattice are only imaginary. The lattice concept is used to show the positions of atoms, molecules, or ions in relation to each other. Atoms may be represented by circles, spheres, table tennis balls, or tennis balls located either at the intersection of these lines or between these lines. A part of a space lattice is sketched in Figure 3-10, with small spheres representing atoms. We must also remember that the actual atoms in solids are located as close to each other as possible, thus attaining the lowest possible energy level. Two atoms closest to each other would be represented by two spheres touching each other. The closer the atoms are, the denser the solid.

 Material Science & Technology Website —*http://mst-online.nsu.edu/*

3.2.2 Crystal Systems

Most pure substances form crystals in the solid state. In this section, we emphasize crystal systems composed of atoms and ions, but the structural particles of crystalline solids can be atoms, ions, or molecules. Solid methane, CH_4, a molecular solid, has a face-centered cubic

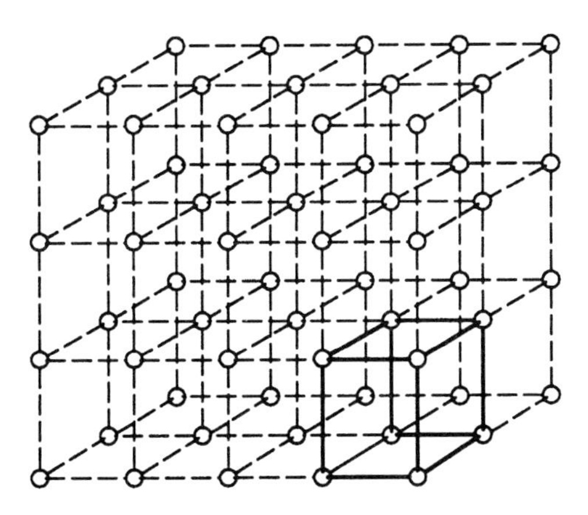

Figure 3-10 A space or crystal lattice with a simple cubic unit cell outlined.

(fcc) structure, which means that there is a CH_4 molecule at each corner and at the center of each face in its unit cell. The forces acting among these structural particles may be metallic bonds, interionic attractions, van der Waals forces, or covalent bonds. Most metals, a good many ceramic materials, and certain polymers crystallize when they solidify. Most metals have a cubic close-packed, hexagonal close-packed, or body-centered cubic lattice structure. There are some exceptions. Tin (Sn) and indium (Im) form body-centered tetragonal structures. (Figure 3-14 shows a bcc tetragonal unit cell.) Table 3-3 indicates this tetragonal structure as XX, showing, among other things, that this structure is not a common one for pure elements in the periodic table. Zirconia (ZrO_2), most important in improving toughness (see Module 19), forms a tetragonal unit cell structure. Martensite, a form of tough steel, gains its hardness and strength from its hard, brittle, tetragonal phase that is supersaturated with carbon.

The cubic system's basic structure has been depicted in Figure 3-5 to Figure 3-9 assuming that the sides of the boxes described are all equal in length ($a = b = c$). It is possible for atoms to form three different patterns or arrangements within this cube. These patterns, of which there are a total of 14, are known as ***Brevais lattices*** and will be discussed later. Each pattern can be described in detail by using a unit cell as a reference. The *cubic axis system*, with its mutually perpendicular axes and equal lattice parameters, has already been described. The cubic coordinate axis system allows us to form three basic unit cells by placing atoms, ions, or molecules at various lattice positions.

The simple ***cubic unit cell*** (Figure 3-11) consists of eight atoms located at each corner of the cube. This structure does not exist for any of the elements in the periodic table because the space in the center would be filled by other atoms. The point to be made is that this representation of atoms in a solid shows only the location of atoms. It must be remembered that if you represented these eight atoms with hard rubber balls and arranged them in accordance with this simple cubic unit cell, all eight atoms would be touching each other.

Another unit cell (Figure 3-12) is known as the ***body-centered cubic (bcc)***. It is similar to the simple cubic unit cell but contains an additional atom located in the center of the cube. The third type of unit cell formed from the cubic axis system is the ***face-centered cubic (fcc)***. One atom at each corner and one in the center of each of the cube faces make up the complement of atoms. There is no atom at the center of the cube (Figure 3-13).

For both the bcc and fcc structures, the same type of atoms must fill both the center/face-centered locations, as do the cell corners. If, in fact, a different type of atom fills the center position, from those at the cell corners in a body-centered cubic structure, the structure is correctly called a simple cubic (sc) structure. An alloy made up of about 50% copper and 50% zinc forms such a structure.

The ***tetragonal crystal system*** has similar unit cells to the cubic, but the sides are not equal. As an example, the ***body-centered tetragonal (bct)*** crystal lattice unit cell is shown in

TABLE 3-3 PERIODIC SYSTEM OF THE ELEMENTS FOR FERROUS METALLURGISTS

PERIODIC SYSTEM OF THE ELEMENTS

Adapted Primarily for Ferrous Metallurgists

Atomic size factors (in parentheses) are % smaller (−) or larger (+) than gamma (FCC) iron at 75 F. Lattice environment (Coordination No.) is taken into account; CN is 12 except 6 for interstitials H, B, C, N & O. Groups VI, VIb, VII & VIIb form ionic compounds with the metals. Atomic size is based largely on work of W. Hume-Rothery and associates and L. Pauling (Some values, such as those for H & O, are approximate). Alloying valences are those of Pauling.

H-1 ▲(−58) ⊗ XX ◀

0	I	II	III	IV	V	VI	VII
He-2 FCC (Others)	Li-3 ⊗(+23) BCC* HCP†	Be-4 ●(−11) HCP* BCC ◗	B-5 △(−29) ⊗ XX ▮	C-6 ◭(−34) ⊗ XX ◆	N-7 ◭(−36) ⊗ XX ◆	O-8 ◭(−33) ⊗ XX	F-9
Ne-10 FCC	Na-11 ⊗(+50) BCC* HCP	Mg-12 ⊗(+27) HCP	Al-13 ◐(+14) FCC ◗	Si-14 ●(+7) XX ◗	P-15 ●(+2) XX ◗	S-16 ●(+1) XX ▮	Cl-17 XX

0	Ia	IIa	IIIa	IVa	Va	VIa	VIIa	VIII			Ib	IIb	IIIb	IVb	Vb	VIb	VIIb
Ar-18 FCC	K-19 ⊗(+86) BCC	Ca-20 ⊗(+56) FCC* BCC	Sc-21 ⊗(+29) HCP* BCC	Ti-22 ◐(+16) HCP* BCC ◗	V-23 ●(+6) BCC ◗	Cr-24 ●(+1) BCC ◗	Mn-25 ●(+1) XX* FCC† ◀	Fe-26 ● (0) BCC* FCC	Co-27 ●(−1) HCP* FCC ◀	Ni-28 ●(−1) FCC ◀	Cu-29 ●(+1) FCC ◆	Zn-30 ●(+6) HCP ◗	Ga-31 ●(+12) XX	Ge-32 ●(+9) XX ◗	As-33 ●(+11) XX ◗	Se-34 ●(+11) XX	Br-35 XX
Kr-36 FCC	Rb-37 ⊗(+97) BCC	Sr-38 ⊗(+71) FCC* HCP‡	Y-39 ⊗(+42) HCP* BCC	Zr-40 ⊗(+27) HCP* BCC ▮	Cb-41 ◐(+15) BCC ▮	Mo-42 ●(+10) BCC ◗	Tc-43 ●(+8) HCP	Ru-44 ●(+6) HCP ◆	Rh-45 ●(+6) FCC ◀	Pd-46 ●(+9) FCC ◀	Ag-47 ◐(+14) FCC	Cd-48 ⊗(+20) HCP	In-49 ⊗(+25) XX	Sn-50 ⊗(+23) XX ◗	Sb-51 ⊗(+27) XX ◗	Te-52 ⊗(+27) XX	I-53 XX
Xe-54 FCC	Cs-55 ⊗(+112) BCC	Ba-56 ⊗(+76) BCC	La-57 ⊗(+48) HCP* FCC‡	Hf-72 ⊗(+26) HCP* BCC ▮	Ta-73 ◐(+16) BCC ▮	W-74 ●(+11) BCC ◗	Re-75 ●(+9) HCP ◆	Os-76 ●(+7) HCP ◀	Ir-77 ●(+8) FCC ◀	Pt-78 ●(+10) FCC ◀	Au-79 ◐(+14) FCC ◆	Hg-80 ⊗(+25) XX	Tl-81 ⊗(+36) HCP* BCC	Pb-82 ⊗(+39) FCC	Bi-83 ⊗(+35) XX	Po-84 ⊗(+40) XX	At-85
Rn-86	Fr-87	Ra-88	Ac-89 ⊗(+49) FCC														
Alloying Valence	1	2	3	4	5	6	6	6	6	6	5.56	4.56	3.56	2.56 Note 2	1.56 Note 2	(2) Note 3	(1) Note 3

Note 1: The rare-earth (lanthanide, 58-71) and actinide (90-103) series are omitted.
Note 2: Valence is 4 for C; 3 for N and P.
Note 3: (1) and (2) are not alloying valences.

SUBSTITUTIONAL SOLID SOLUTIONS

- ● FAVORABLE SIZE FACTOR: 0 TO ± 13%
- ◐ BORDERLINE SIZE FACTOR: ± 14 TO ± 16%
- ⊗ UNFAVORABLE SIZE FACTOR: > ± 16%

INTERSTITIAL SOLID SOLUTIONS

- ▲ FAVORABLE SIZE FACTOR: > (−40%)
- ◭ BORDERLINE SIZE FACTOR: (−30) TO (−40) %
- △ UNFAVORABLE SIZE FACTOR: < (−30%)

STRUCTURE

- BCC - BODY CENTERED CUBIC
- FCC - FACE CENTERED CUBIC
- HCP - HEXAGONAL CLOSE PACKED
- XX - NOT BCC, FCC OR HCP USUALLY MORE COMPLEX
- * - STRUCTURE AT 75 F
- † - ALSO FCC ‡ - ALSO BCC

TYPE OF GAMMA IRON (FCC) FIELD IF ALLOYED WITH IRON

- ◗ GAMMA LOOP, LIKE Cr
- ▮ LIMITED GAMMA LOOP, LIKE B
- ◀ OPEN GAMMA REGION, LIKE Ni
- ◆ LIMITED GAMMA REGION, LIKE C

Prepared by O. O. Miller

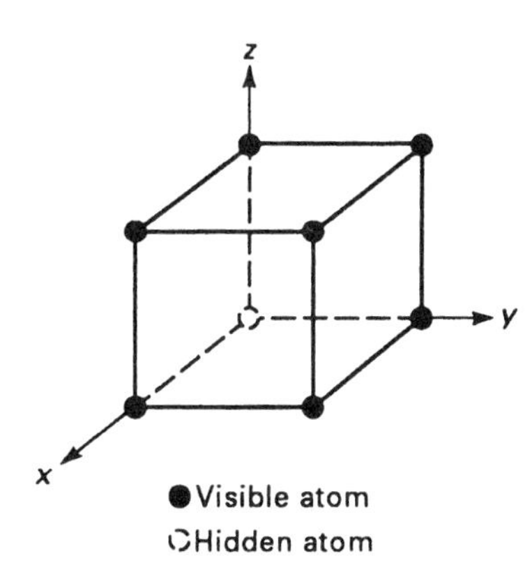

Figure 3-11 A simple cubic unit cell showing atoms only at the corners of the cube.

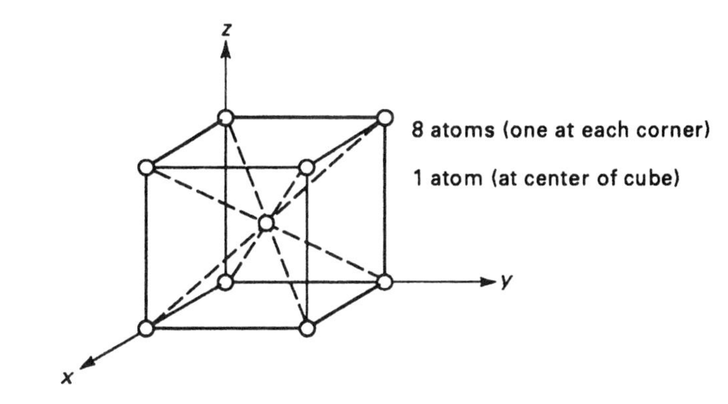

Figure 3-12 A body-centered cubic unit cell.

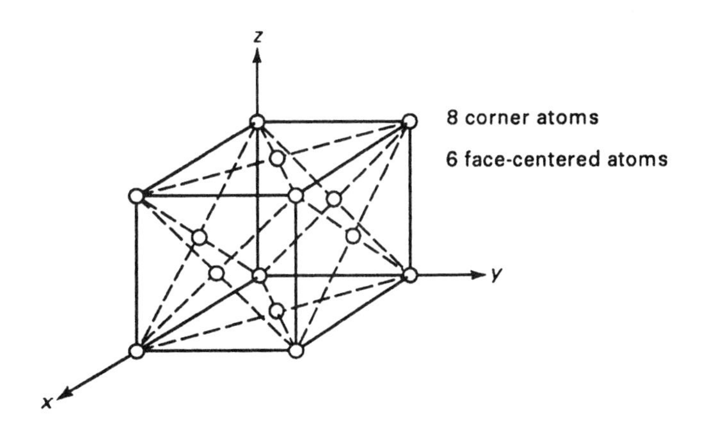

Figure 3-13 A face-centered cubic unit cell.

Figure 3-14. Tin forms a tetragonal unit cell. The tetragonal is similar because the axes are all normal to each other. The difference lies in the length of the intercepts. The x- and y-intercepts have the same magnitude; the z-intercept is larger than the x- or y-intercept. Martensite, a combination of iron and carbon that is contained in a hard steel, has its atoms of iron and carbon in a tetragonal lattice structure (Module 9). The tetragonal crystal lattice unit cell for barium titanate ($BaTiO_3$), with the central Ti ion not in the same plane as the four O ions in the side faces, illustrates this more complex crystal structure (**perovskite** structure) for materials that have three different types of ions. This shift in the relative positions (displacement) of these ions results in the formation of a local, permanent, ionic dipole for each unit cell. The strength of the dipole is related to the magnitude of the atomic displacement, which can be altered by either an external force or an electric field. Such a crystal can be used as

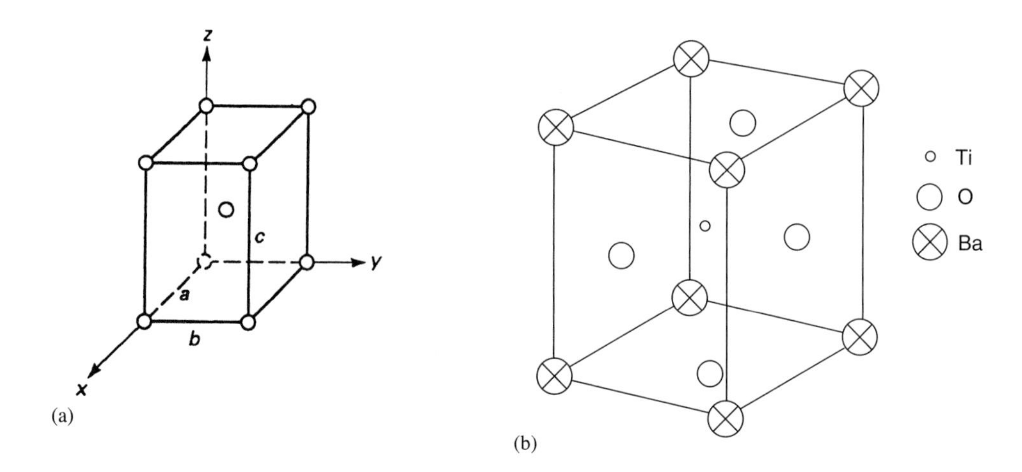

Figure 3-14 (a) A body-centered tetragonal crystal lattice unit cell. (b) The tetragonal crystal lattice unit cell (perovskite structure) for barium titanate ($BaTiO_3$) with the central Ti ion not in the same plane as the four oxygen atoms in the side faces. This shift in the relative positions (or displacement) of these ions results in the formation of a local electric dipole. The strength of the dipole is related to the magnitude of the atomic displacement, which can be altered by either an external force or an electric field. Many important electrical/magnetic ceramics, including ionic materials such as magnetite (lodestone), Fe_3O_4, have this structure. A common application is found with ignitors in barbecue grills.

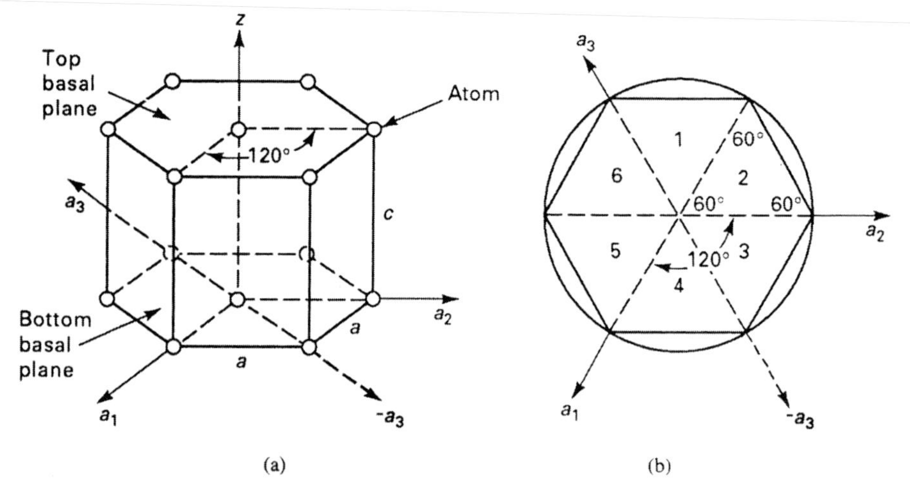

Figure 3-15 (a) A hexagonal crystal lattice unit cell. (b) Top plane of unit cell showing six equilateral triangles.

a transducer to convert mechanical energy into electric voltage, and vice versa. A new automotive emission control technology cuts the use of precious metals by as much as 70% while removing more pollutants than standard catalytic converters. This new technology uses a mixture of metal oxides in crystal form (perovskites) to filter engine exhaust while using less palladium or platinum. See Module 27 for further discussion of dielectric properties, ferroelectrics, dipole moments, and piezoelectric materials.

The ***hexagonal crystal system*** (Figure 3-15a) can best be described using three axes (a_1, a_2, and a_3) in the x–y (horizontal) plane 120° apart and a fourth axis (z) at 90° to the x–y plane. The intercepts along the three axes in the horizontal plane are equal in length ($a = a = a$), but the fourth intercept, labeled c, is of a different length. This unit cell is basically made up of two parallel planes (top and bottom basal) separated by a distance equal to the dimension c. The atoms shown in the figure trace out a right hexagonal prism. Each of these two planes can be divided into six equilateral triangles, with each side equal to the intercept a (Figure 3-15b).

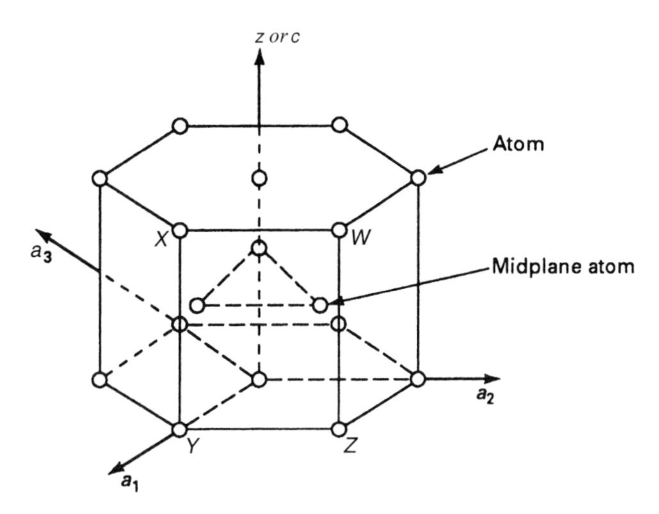

Figure 3-16 A close-packed hexagonal crystal lattice unit cell.

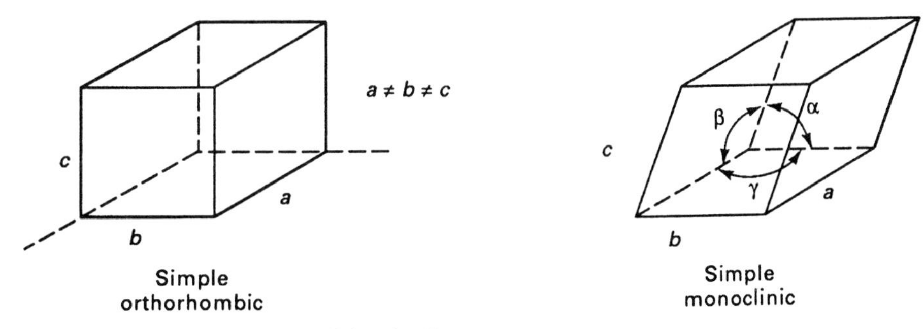

Figure 3-17 Orthorhombic and monoclinic unit cells.

TABLE 3-4 CRYSTAL SYSTEMS SUMMARY

System	Intercepts	Angles between axes
Cubic	$a = b = c$	$\alpha = \beta = \gamma = 90°$
Tetragonal	$a = b \neq c$	$\alpha = \beta = 90°$
Hexagonal	$a = a \neq c$	$\alpha = ß = 90°; \gamma = 120°$
Monoclinic	$a \neq b \neq c$	$\alpha = \gamma = 90° \pm \beta$
Orthorhombic	$a \neq b \neq c$	$\alpha = \gamma = 90°$

Atoms of solid materials do not form the purely hexagonal unit cell as in Figure 3-15 because they cannot satisfy equilibrium conditions by being so far apart. In other words, they would be unstable. Consequently, they form the hexagonal unit cell, called ***close-packed hexagonal (cph)***, as shown in Figure 3-16, with its three midplane atoms at a distance of c/2. Zinc, titanium, and magnesium form cph unit cells.

There are a total of seven crystal systems, of which three have been mentioned. Two examples of some of the unit cells from the orthorhombic and monoclinic crystal systems are shown in Figure 3-17. Not discussed are the triclinic and rhombohedral systems. Using these seven axis systems, atoms can form 14 patterns in space (Brevais lattices). As an example, using unit cells for the cubic crystal system, atoms can form three different patterns: simple cubic (Figure 3-11), bcc cubic (Figure 3-12), and fcc cubic (Figure 3-13). Table 3-4 lists the intercepts and the angles between axes for five of seven crystal systems with which the earth's elements form their particular atomic structures.

3.2.3 Crystallographic Planes, Miller Indices, and Crystal Directions

The amazing thing about crystallographic structures is that any variation in the arrangement of atoms or the stacking of the planes formed by the atoms has a pronounced effect on the properties of the solid material. Atoms within solids move with each other. Solids tend to deform in the directions along the planes that are the most closely packed because it is easier for an atom (or ion) to move in these close-packed planes. Planes of atoms move, or slip, in parallel directions much like two adjacent cards in a deck of playing cards. This movement of atoms in a solid results from the application of some type of external force, such as a compressive or tensile load. Less force is needed to move the atoms from one position to the next adjacent one in close-packed planes. The degree of movement of planes of atoms gives an indication of the ductility and strength of the particular material. Therefore, it is important to know something about how the principal crystal structures differ in the location of the atoms in their space lattices/unit cells and something about the planes in these structures. The bcc crystal structure is not a close-packed structure. Of the volume of the bcc unit cell, 68% is occupied by atoms and the remainder is empty space. In contrast, the atomic packing factor (APF) (see Section 3.2.6) for a fcc unit, cell is 0.74, indicating that the atoms are packed as close together as possible.

First, let us consider how atoms (or ions) move in a plane. A simple experiment can show this phenomenon. Line a number of pennies up in two rows, all touching, with an equal number in each row. If you push one row with your finger in one direction parallel to the row and do the same with the other row, but in the other direction, each row will move easily. Next, if you line up the objects in less than a perfect arrangement with maybe only one of the objects projecting into the other row and then push the rows as before, the result is very little movement in the direction of the pushing force and considerable movement in many directions. Later you will learn that metals and metal alloys that arrange themselves with fcc unit cells exhibit the most ductility mainly because they possess the greatest number of close-packed planes in the greatest number of directions. Also, if not parallel, these planes will intersect and movement of atoms may be hindered or stopped completely at some stage.

The number of atoms in any atom plane depends on the stacking sequence of the planes. The atom planes in these crystal lattices that have the largest number of atoms are the ***close-packed planes.*** Representing atoms (or ions) as spheres, the closest packing of atoms occurs when each atom has 12 other atoms of equal size around it. Of all the crystalline materials found in nature, the most important fall into only three stacking sequences. The stacking of four atom planes in a particular order creates the fcc arrangement of atoms. Of course, in a solid material there would be an infinite number of planes stacked in this particular way. The fcc metals such as Al, Cu, and Si therefore have four sets of nonparallel, close-packed planes in their unit cells, as discussed later. The bcc unit cells for elements such as Cr, Li, or Mo contain no close-packed planes. A different sequence of stacking forms cph unit cells. As a result, they have two hexagonal bases (basal planes), as shown in Figure 3-15. Numerous exercises can be developed to better visualize these structures. In so doing, remember that no adjacent planes can be identical. Any deviation from the stacking sequences mentioned are considered ***stacking faults***.

In the *bcc crystal system,* we note in observing the bcc unit cell in Figure 3-12 and Figure 3-18 that, by use of a cutting plane through this cell, the plane containing the most atoms would be the one that passed through four of the eight corner atoms and the one in the center of the cube. All figures are expanded views; actually, the atoms are much closer. This diagonal plane could be a (110) plane, as sketched in Figure 3-18. It should be noted that, although these planes contain the most atoms, the atoms are not as closely packed as they could be. A view of this diagonal plane from a position at right angles to it, as in Figure 3-19, shows that the spaces are not occupied by atoms, the diagonal has a length of $4r$ (r is the radius of each atom), and the atoms touch each other along the bcc unit cell body diagonal.

In the *fcc unit cell*, the planes with the most atoms can be represented by plane (111), as sketched in Figure 3-20, or by seven other planes with the same number of atoms. Observe that the plane contains three face-centered atoms and three corner atoms. Figure 3-21 shows

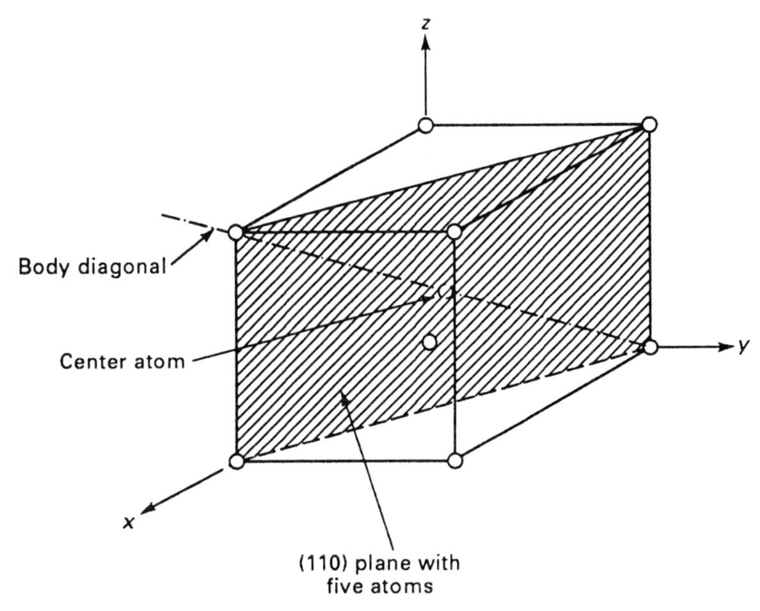

Figure 3-18 A bcc unit cell showing (110) plane.

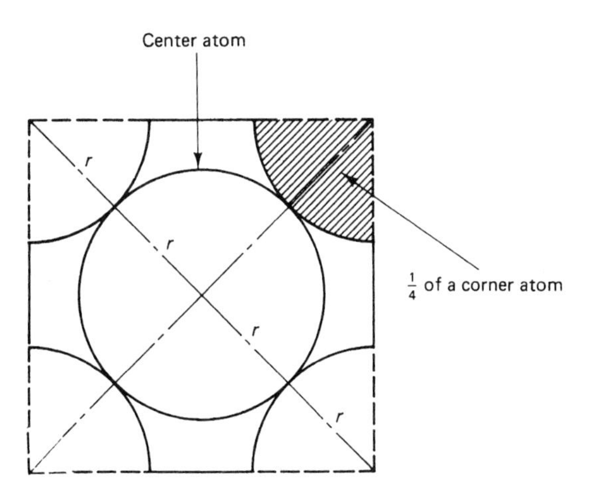

Figure 3-19 A bcc unit cell showing (110) plane.

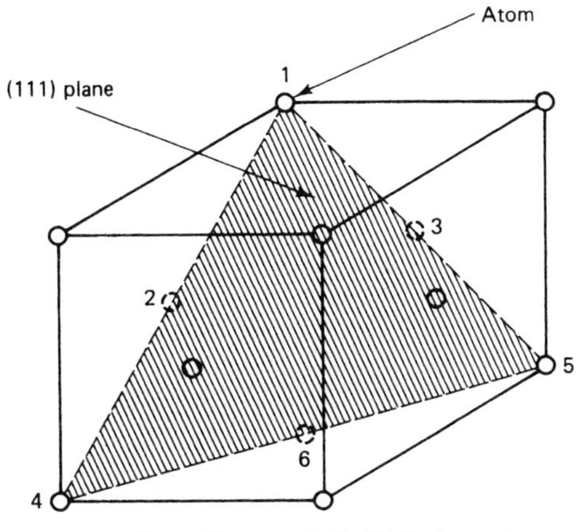

Figure 3-20 Normal view of diagonal plane (111) in fcc unit cell showing six numbered atoms.

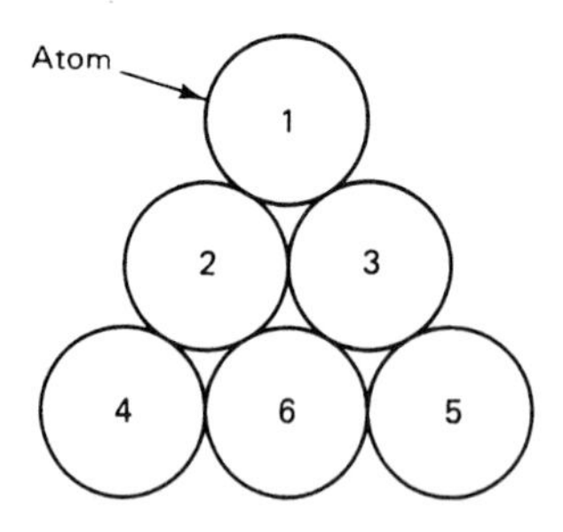

Figure 3-21 (111) plane in fcc unit cell.

these atoms all in contact with each other with a minimum of unoccupied space between them. With atoms of the same size, six atoms are touching each atom in such a plane.

The *cph unit cell* readily tells us the most closely packed planes of atoms are the basal planes and all planes parallel to them. This means that a metal with a cph crystal lattice can slip in only one direction. In other words, this metal would be brittle. The bcc or fcc metals, having a greater number of slip planes, can slip in many directions and are thereby classified as ductile metals. Recall that, at this stage, we are talking about the atomic structure of millions of atoms in one crystal. The vast majority of solid materials are polycrystalline (i.e., composed of millions and millions of individual crystals whose axes are oriented in random directions). This random orientation could result in a blocking effect; that is, a crystal could stop the slippage of atoms in any adjacent crystal. More details of the movement of atoms will be discussed later.

3.2.3.1 Miller indices for atomic planes and directions. Because atoms move, it is useful to be able to describe the location of atoms in a unit cell as well as the direction of their movement. The sites or locations of atoms and/or points in a unit cell are described by their axial coefficients expressed in unit-cell dimensions. These coefficients (also called *coordinates* or *intercepts*) are the respective distances along the x-, y-, and z-axes from the origin of the axis system. By convention, the origin is located at the lower-left rear corner of the unit cell (Figure 3-7). However, it may be moved temporarily to another position if it is so indicated. In Figure 3-7, for example, the location of the atom shown at the origin of the axis system would have the coordinates 0,0,0, as indicated. The atom above the origin on the z-axis in Figure 3-8 at a distance of 1 unit length would have the coefficients 0, 0, c or 0, 0, 1. In Figure 3-13, the face-centered atom in the x–y plane is located by the coefficients 1/2, 1/2, 0. Note that the distances making up the coordinates are separated by commas. The planes in a unit cell are designated by their **Miller indices.** The procedure for determining the Miller indices of a plane is illustrated in the following problems.

Illustrative Problem

In Figure 3-22, two planes, A and B, are shown with their intercepts. The four-step procedure for determining the Miller indices of plane A is as follows:

Step 1. Record the intercepts in order of the x-, y-, and z-axes.

x-axis	intercept 1
y-axis	intercept $\frac{1}{2}$
z-axis	intercept $\frac{1}{3}$

Step 2. Clear any fractions by taking the reciprocals of the intercepts.

x-axis	$1/1 = 1$
y-axis	$1/\frac{1}{2} = 2$
z-axis	$1/\frac{1}{3} = 3$

Step 3. Clear any fractions arrived at in step 2 by multiplying by a common multiplier.

No fractions

Step 4. Record the numbers (indices) in parentheses with no commas separating the numbers.

(123)

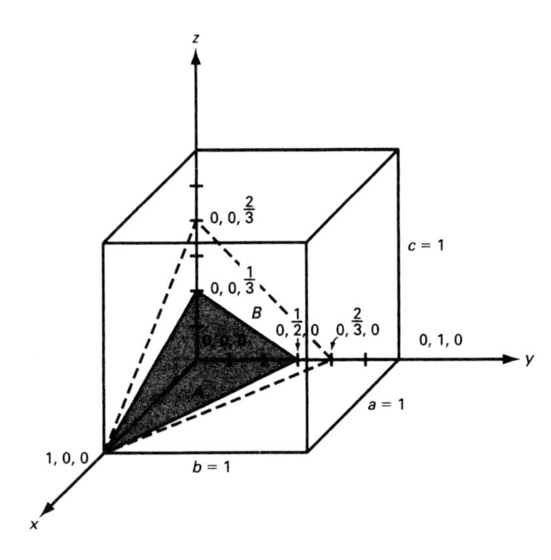

Figure 3-22 Miller indices for atomic planes for a cubic unit cell. Plane A is shaded; plane B is drawn with dashed lines.

Illustrative Problem

Determine the Miller indices for plane B.

Step 1. The intercepts are:

x-axis	1
y-axis	$\frac{2}{3}$
z-axis	$\frac{2}{3}$

Step 2. Taking reciprocals:

x-axis	$\frac{1}{1} = 1$
y-axis	$1/\frac{2}{3} = \frac{3}{2}$
z-axis	$1/\frac{2}{3} = \frac{3}{2}$

Step 3. Clearing fractions:

Common multiplier = 2

$1 \times 2 = 2 \quad \frac{3}{2} \times 2 = 3 \quad \frac{3}{2} \times 2 = 3$

Step 4. Record numbers (indices):

(233)

Illustrative Problem

Determine the Miller indices for the face plane in Figure 3-8 that parallels the x–z plane at a distance $b = 1$.

Step 1. Intercepts:

x-axis	∞ (plane intercepts x-axis at infinity)
y-axis	1
z-axis	∞ (plane intercepts z-axis at infinity)

Step 2. Reciprocals:

x-axis	$1/\infty = 0$
y-axis	$1/1 = 1$
z-axis	$1/\infty = 0$

Step 3. No fractions.

Step 4. (010) indices.

Figure 3-18 shows the (110) plane in a bcc unit cell, and Figure 3-20 shows a sketch of the (111) in a fcc unit cell.

3.2.3.2 Crystal directions. Many materials are anisotropic*; consequently, their properties are not the same in all directions. Miller indices are used to specify directions in a crystal structure, As with Miller indices for specifying points and planes in a crystal lattice system, there is a procedure for determining the Miller indices for directions. The first step is to determine the coordinates of two points that lie in the particular line of direction. The first point, sometimes called the "head point," is farthest from the origin. Using the origin as the second point simplifies the procedure. The second step is to subtract the second point from the first point. The third step calls for the clearing of any fractions to obtain indices with the lowest integer values. The fourth step is the writing of the indices in square brackets without commas. Negative integer values are indicated by the use of a bar placed over the integer. The procedure is illustrated in the following problem (refer to Figure 3-13).

Illustrative Problem

In Figure 3-23, find the direction for a line passing through point A with coordinates ½, 1, 0 and the origin of the unit cell.

Step 1. Axis system, defined in Figure 3–23.

Step 2. Head point coordinates: $\frac{1}{2}$, 1, 0

Tail point coordinates: $-$ 0, 0, 0

$\frac{1}{2}$, 1, 0

Step 3. Clear fractions (multiply by 2) and reduce to lowest integers: 1 2 0.

Step 4. Enclose in brackets without commas: [120]

In more complex crystal systems, the determination of directions requires temporary relocation of the origin to another point in the unit cell to simplify the procedure. If the properties of a crystal measured along two different directions are identical, the two directions are termed *equivalent*. Examples of equivalent directions are the face diagonals or the body diagonals in a cubic crystal. Equivalent directions are referred to as a *family of directions*. For example, the directions for the face diagonals (a family of directions) are specified by using spread or angle brackets: ⟨1, 1, 0⟩. This family consists of 12 directions, which can be arrived at by taking all the various ways (permutations) of expressing the three digits, 1, $\bar{1}$, and 0. The negative indices are represented by a bar over the appropriate index.

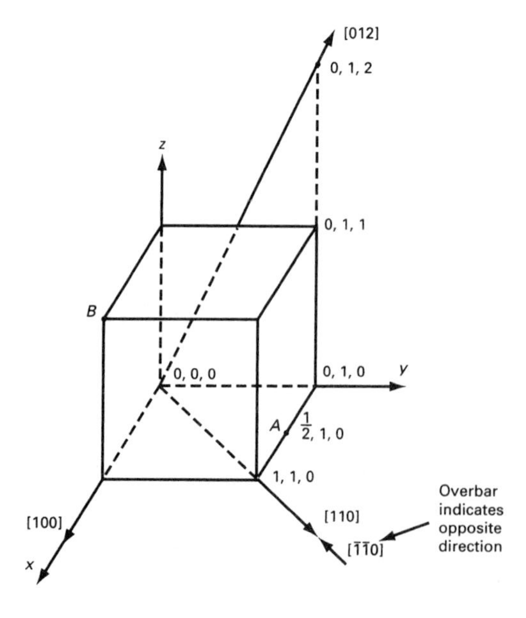

Figure 3-23 A cubic unit cell with some intercepts and crystal directions indicated. Note the use of square brackets and the omission of commas to express Miller indices of direction.

*An anisotropic material exhibits different values of a property in different directions with respect to a fixed axis system. Isotropic and anisotropic materials will be discussed further in Module 4, in the section on ductility.

Illustrative Problem

The plane *XYZW* is shown in Figure 3-16. Find (1) the Miller indices for the plane *XYZW* and (2) the Miller indices for the direction whose head point is *Z* with the first point being the origin.

Solution Miller indices for the plane

Step 1. Intercepts are:

$$\begin{aligned} a_1 \text{ axis intercepts} &= 1 \\ a_2 \text{ axis intercepts} &= \infty \\ a_3 \text{ axis intercepts} &= -1 \\ z \text{ or } c - \text{ axis intercepts} &= \infty \end{aligned}$$

Step 2. Taking reciprocals:

$$\begin{aligned} 1/1 &= 1 \\ 1/\infty &= 0 \\ 1/-1 &= -1 \\ 1/\infty &= 0 \end{aligned}$$

Step 3. Clearing fraction:

No fractions involved.

Step 4. Record numbers:

$$\begin{array}{r} 1\,0\,\bar{1}\,0 \\ -\underline{0\,0\,0\,0} \\ 1\,0\,\bar{1}\,0 \end{array}$$

Solution Miller indices for the direction

Step 1. Coordinates of the head point:

$$1\,0\,\bar{1}\,0$$

Step 2. Subtracting the coordinates of the origin (all are zero) from the coordinates of the head point does not change the numbers.

Step 3. Clearing fractions: No fractions involved

Step 4. Write indices in square brackets without commas:

$$[1\,0\,\bar{1}\,0]$$

Note:

1. $1/\infty = 0$, which is the reciprocal of ∞.
2. $1/0 = 0$.
3. In the indices given for the plane, the second number and the fourth number are zero, which indicates that this plane is parallel to those axes (a_2 and z or c).
4. The plane and the direction in this problem both have the same numbers, which indicates they are perpendicular to each other.
5. The Miller indices for hexagonal crystal lattices have four numbers as opposed to the cubic crystal lattices, which have three numbers.

3.2.4 Coordination Number

To describe how many atoms are touching each other in a group of coordinated atoms, the term ***coordination number*** (CN) is used. The CN is the number of neighboring atoms immediately surrounding each atom. Note in Figure 3-16 that each upper and lower basal plane of a cph (close-packed hexagonal) unit cell contains an atom at its center. Each atom touches six atoms in its own plane, plus three atoms above and below in adjacent planes. Consequently, the CN for these atoms would be 12. The number of nearest atoms is dependent on two factors: (1) the type of bonding and (2) the relative size of the atoms or ions involved. In our discussion of

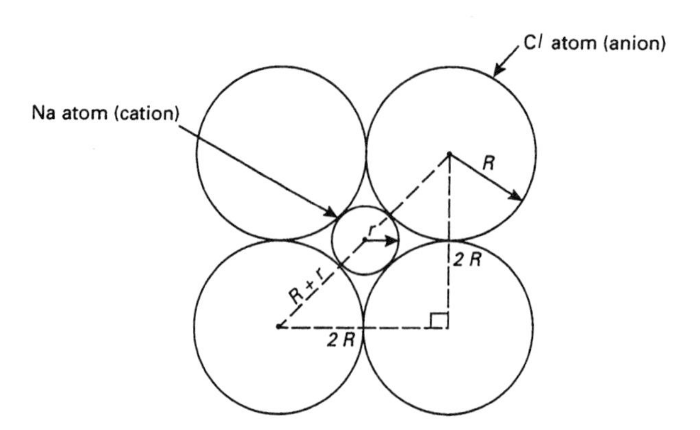

Figure 3-24 A face plane in a NaCl crystal.
Note: For clarity, two Cl anions located along the axes perpendicular to the page are not shown.

bonding, for example, we learned that valence electrons determine the type of bonding as well as the number of bonds an atom or ion can have. Carbon (C) in Group IV has four covalent bonds and therefore a CN of 4. The Group VII elements, such as chlorine (Cl), form only one bond (CN is 1). Ionic bonding involves ions of different charges, hence, different sizes. Atoms change in size during ionization. Cations lose electrons, making them smaller; anions gain electrons, making them larger. The limiting factor in this case is the ratio of the size (radii) of the combined atoms of ions. The minimum ratios of atomic (ionic) radii produce various CNs.

Figure 3-24 represents the five ions occupying one of the six faces of the fcc unit cell for NaCl. The Na ion is just the right size to fit between the Cl ions at the corners of the unit cell. Thus, the ions are closely packed, with each cation separated from other cations by a layer of anions. Each cation and each anion are shared equally by six oppositely charged ions. Therefore, a CN of 6 describes this geometric arrangement. The center-to-center distance between the Na ion (radius $=r$) and its six neighboring Cl ions (radius $=R$) is determined to be the hypotenuse of an isosceles right triangle (two legs or sides of equal length). This triangle is outlined in Figure 3-24 with its sides labeled $2R$. Using Pythagorean Theorem $x^2 + y^2 = z^2$ produces

$$(2R)^2 + (2R)^2 = (2r + 2R)^2$$

Taking the square root of both sides gives the expression

$$\sqrt{2}(2R) = (2r + 2R)$$

Solving for the radius (r) of the smaller ion in terms of the larger ions:

$$r = (\sqrt{2} - 1)R = 0.414R$$

These data show that the smallest ion that will just fill the spaces between six larger ions or that just touches six ions must have a radius that is at the minimum 0.414 times the size of the larger ions. This same result can be obtained by using a proportion between the isosceles triangle in Figure 3-24 and the standard 45–45–90 degree right triangle.

$$\frac{1}{2R} = \frac{\sqrt{2}}{(2r + 2R)}$$

Cross multiplying:

$$r + R = \sqrt{2}\,R$$

TABLE 3-5 MINIMUM RADII RATIOS FOR CNS

CN	$\frac{r}{R}$
3	≥ 0.155
4	≥ 0.225
6	≥ 0.414
8 (bcc)	≥ 0.732
12 (cph or fcc)	1.0

r, radius of smaller atom; R, radius of larger atom; $\geq$, greater than or equal to.

Solving for *r:*

$$r = \sqrt{2}\,R - R$$
$$r = R(\sqrt{2} - 1)$$
$$r = R(0.414)$$

Table 3-5 reveals that to have a CN of 6, the ratio of r/R must be at least 0.414.

In the case of sodium chloride (NaCl), r/R is $0.95/1.81 = 0.520$. This being greater than 0.414, NaCl qualifies for a CN of 6. Note that as the difference between r and R decreases, higher CNs are possible. A CN of 12 is the maximum, which occurs when the atoms (ions) have the same radius and the ratio becomes 1. In other words, as the r gets smaller than R (radius of surrounding atoms), the fewer neighboring atoms can make contact or touch the smaller atoms. With a CN of 12, each atom has contact with 12 other atoms. Each atom in a fcc or cph unit cell meets this description, provided that their radii are of similar size. Table 3-5 lists the minimum radii ratios for some common CNs.

3.2.5 Allotropy/Polymorphism

It is of interest to list some common materials and show their crystal structure and melting point to illustrate the concept of the movement of atomic planes within a solid. Table 3-6 lists several elements that exist in more than one crystal structure, depending primarily on the temperature. This phenomenon is known as ***allotropy*** or ***polymorphism*** (*poly*, "many"; *morph*, "shape"). Actually, an allotropic material can, after changing to one structure, reverse the phenomenon and return to its previous structure; a polymorphic material does not possess this reverse phenomenon. Over one-fourth of the elements are allotropic. Steel owes its extensive use to this property. Because of allotropy, steel can be produced from iron. When iron is heated to above 910°C, its structure changes from bcc to fcc, allowing for a much greater absorption of

TABLE 3-6 CHARACTERISTICS OF SELECTED METALS

Element (symbol)	Crystal structure (20°C)	Atomic radius (10^{-10} m)	Melting point (K)
Beryllium (Be)	cph	1.14	1562
Aluminum (Al)	fcc	1.43	933
Titanium (Ti)	cph	1.46	1945
	bcc 877°C		
Iron (Fe)	bcc	1.24	1810
	fcc 910°–1399°C	1.27	
	bcc 1399°–1538°C		
Cesium (Cs)	bcc	2.62	300
Tungsten (W)	bcc	1.37	3680

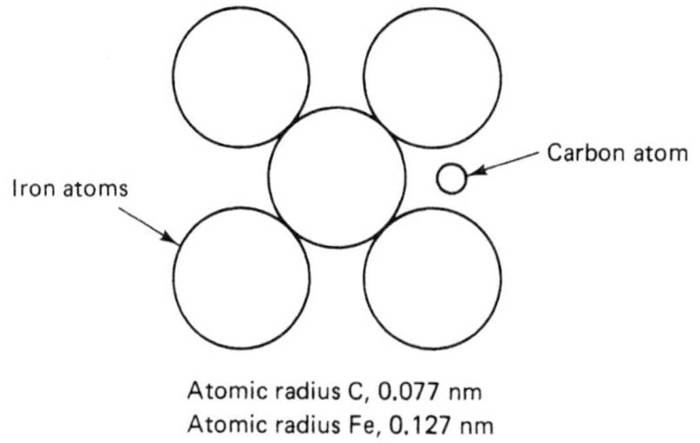

Carbon atom

Iron atoms

Atomic radius C, 0.077 nm
Atomic radius Fe, 0.127 nm

Figure 3-25 A face plane of a fcc unit cell for iron showing a carbon atom located in an interstice.

carbon atoms (2% maximum). Figure 3-25 shows one carbon atom in an fcc unit cell of iron. Given its allotropic property and its relatively cheap price when compared with the many advanced materials now available in today's marketplace, recent advances made in the research and development of lightweight, high-strength steels, have made steel a great competitor, especially in the transportation industry.

In general, we now know that any allotropic change in an element produces different properties. All the properties of steel change almost instantaneously at 910°C. The best example of a change in properties with a change in crystal structure is that of carbon (C). One polymorph of pure carbon is graphite—a black, greasy, low-strength material. A second polymorph is diamond, the hardest naturally occurring substance. A third naturally occurring polymorph is the recently discovered fullerenes (see Figure 2-14). A final example is zirconia (ZrO_2), a ceramic used to toughen alumina (Al_2O_3) and described in detail in Module 19 (see Section 19.3). ZrO_2 is polymorphic with a fcc structure at high temperature. As it cools, it changes first to a simple tetragonal structure, followed by a further change to a monoclinic polymorph. Figure 3-26 is a sketch of the change in volume of ZrO_2 as it goes through these transformations on being cooled. These changes require the consumption of energy that would otherwise help a crack propagate through the ceramic alumina. The end result is alumina toughened by this addition of zirconia.

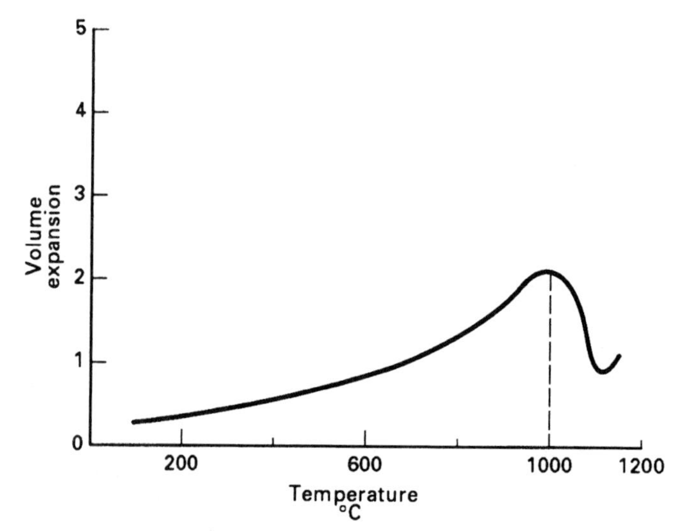

Figure 3-26 Volume expansion curve for ZrO_2 emphasizing the effects of the polymorphic transformation on the rates of expansion and the abrupt changes in volume.

3.2.6 Volume Changes and Packing Factor

In discussing crystal structure changes, we mentioned that every change in atomic structure brings changes in the properties of the solid. One of these changes is volume. When steel transforms from a bcc to a fcc structure, it decreases in volume. The explanation for this phenomenon involves the density of atoms in the various unit cells. ***Density*** is the ratio of the mass to the volume of a substance, which stays constant provided it is nonallotropic. With mass measured in kilograms (kg) and volume in cubic meters (m^3), density has the units of kilograms per meter cubed (kg/m^3). Other typical units are g/cm^3.

The ***atomic packing factor (APF),*** or *packing factor (PF),* is the ratio of the volume of atoms present in a crystal (unit cell) to the volume of the unit cell. In calculating the volume of an atom, we assume the atom is spherical. The difference between the PF and unity (1) is known as the *void fraction,* that is, the fraction of void (unoccupied or empty) space in the unit cell. Using the *simple cubic crystal unit cell* sketched in Figure 3-27, with atoms of equal radius (R) located at each corner, the volume of the cell occupied by the eight atoms is equivalent to one atom. Each corner atom contributes one-eighth of its volume. Therefore, the volume of one atom is $4/3\pi R^3$. The (1 0 0) face plane of this simple cubic unit cell shows four corner atoms touching each other (see Figure 3-28). The relationship of the radius R of an atom to the lattice parameter or edge length a of the unit cell is $2R$. The volume of the unit cell is a^3. The volume of atoms is $(\pi/6)a^3$. Thus,

$$PF = \frac{\text{volume of atoms}}{\text{volume of unit calls}} = \frac{(\pi/6)a^3}{a^3} = \frac{\pi}{6} \approx 0.52$$

Solving for the PF in terms of the radius R produces the same results. The void factor is therefore $1 - 0.52 = 0.48$. What the calculation tells us is that only about half (52%) of the space in the simple cubic unit cell is occupied by the atoms. This is too inefficient, so atoms of metals do not crystallize in this structure. Remember, the closer the atoms come to each other, the less energy they have and the more stable is their structure.

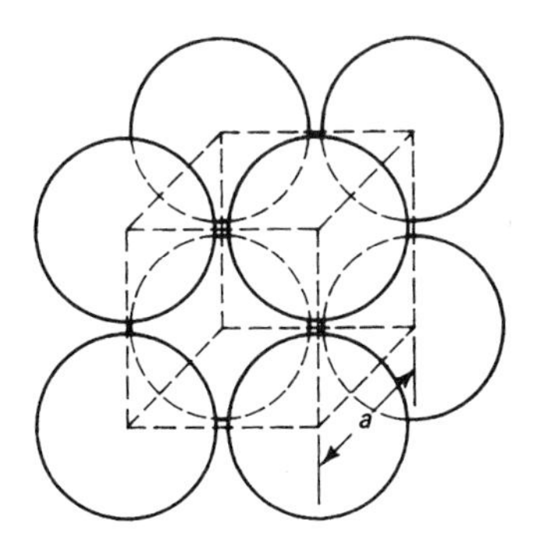

Figure 3-27 A simple cubic unit cell.

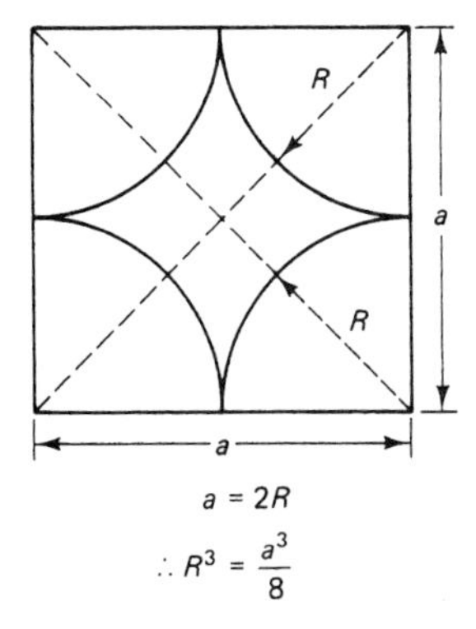

Figure 3-28 (100) face plane of simple cubic unit cell.

The bcc unit cell, as you recall, is quite similar to the simple cubic unit cell with the addition of one atom in the very center of the unit cell. Therefore, the bcc unit cell contains the equivalent of two atoms. For the fcc unit cell, there are four net atoms. Each of the eight corner atoms contributes one-eighth of an atom. Each of the six face atoms contributes one-half. The total is 1 + 3, or 4, atoms. Notice that the atomic radius (R), or a as used in the formula, cancels out in all these calculations, which tells us that the PF is not dependent on the radius of the spheres being packed if all the atoms are of the same size. The fcc structure has the maximum PF for a pure metal. The cph structure also has a PF of 0.74. Finally, we should note that the coordination number varies directly with the PF. As an example, the $CN_{bcc} = 8$ and $PF_{bcc} = 0.68$; the $CN_{fcc} = 12$ and $PF_{fcc} = 0.74$.

In our study of metals, we will see that pure iron will change its structure on heating from bcc to fcc at 910°C. Knowing the PF for both these structures will lead us to the conclusion that iron will contract in volume as it is heated above 910°C. This change in structure forms the basis for the production of steel as well as for the heat treatment of steel.

3.2.7 Crystal Imperfections

So far in our study of crystal structures we have dealt with atom arrangements that are perfect in every way. These structures have ***long-range order*** because the orderly arrangement of atoms extends throughout the entire material, forming a regular gridlike lattice or pattern. To grow a perfect crystal requires laboratory conditions, and even such a controlled environment achieves limited success. Of course, this lack of perfection in the microstructure of materials used by people is far from being all bad. If it were not for imperfections of many kinds in solid materials, these solids would not possess the properties that we desire them to have. An example would be the heat-treating process used with high-carbon steel to change the properties of the steel to suit certain conditions demanded by our present technological age. Without imperfections in the crystalline arrangement of atoms, these processes would be severely limited in, if not incapable of, changing the structure and hence the properties of steel. The whole semiconductor industry owes much of its existence to the imperfections in bonding arrangements of the atoms' outer shell electrons. Therefore, it is now essential to delve into the several imperfections the atomic structure of a solid so that we can learn how to take advantage of such disorder.

It is important to point out here that we are still talking about crystalline materials and not amorphous (noncrystalline) materials. We know that *amorphous* materials have no regular atomic structure. Known as ***short-range-order*** materials, their order is limited to an atom's nearest neighboring atoms. For example, water has short-range order due to its covalent bonding between the hydrogen and oxygen atoms. Even with its secondary bonding (hydrogen bridging), the water molecules are randomly joined. Glass and polymers also have short-range order. This is not to say that amorphous materials are of little use. Many materials produced by processes such as powder metallurgy and rapid solidification are amorphous in structure and possess remarkable properties needed for some specific application.

Crystal imperfections fall into two categories: those involving impurity atoms and those in which there is some disorder in the atomic structure brought about by something other than impurity atoms.

3.2.8 Crystal Impurities

The word ***impurity*** derives from its use in referring to the small percentage of copper in sterling silver that distinguishes sterling from pure silver. One also talks about the impurities in copper that reduce its conductivity. In many instances, these impurities are purposely added to improve a material's properties and/or reduce its cost. Such a solid, called an *alloy,* is a material composed of two or more elements at least one of which is an element that possesses metallic properties. ***Wrought alloys*** are those that can be mechanically deformed. Cast alloys are those alloys whose brittleness is such that they cannot be cold-worked. Brass is an alloy consisting of copper to which has been added some zinc. The addition of zinc has a great effect on the hardness, strength, ductility, and conductivity of the pure copper. Our objective in the following discussion is to explain in simple terms why adding these impurities (the zinc atoms) to the copper atom produces such differences in the properties between the pure metal and the alloy.

Before proceeding, definitions of the words *solution, solvent, solute, mixture, alloy,* and *diffusion* are required. A ***solution*** is a homogeneous mixture of chemically distinct substances that forms a phase. A ***phase*** is defined briefly as a physically distinct material that has its own structure, composition, or both. Uses of phases in solids are discussed in this module. The atoms or molecules of one substance are uniformly distributed throughout the other on a random basis. The substance present in the greatest proportion is the ***solvent.*** The other substance or substances present are the ***solutes.***

A ***mixture***, on the other hand, is a material that has no fixed composition and contains more than one phase. The components (substances) can be identified and separated by physical means. Thus, a mixture of sugar and salt crystals is not homogeneous nor is it random. One can readily see the two distinct crystalline phases and, with some patience, segregate one crystal phase from the other. However, if you dissolve sugar and salt in water in a dilute concentration, they form a liquid solution. In this situation, the salt and sugar lose their individual identities by dissolving in water.

Air is an example of a solution of many gases dissolved in another gas. A similar situation occurs in solids, producing solid solutions. A ***solid solution*** is simply a solution in the solid state that consists of two kinds of atoms combined in one type of space lattice. In other words, when two elements are soluble in each other in the solid state, they produce a solid solution. ***Alloys***, then, are a combination of a metal and one or more other elements forming either a mixture or a solid solution. Steel is a mixture of iron with a bcc structure and cementite. Brass is a solid solution because its single-phase structure is all fcc; it is a solid solution of copper (solvent) to which some zinc (solute) has been purposely added. The zinc atoms are diffused into the atomic structure of the copper on a random, uniform basis. ***Plastic alloys*** are blends of polymers or copolymers with other polymers or elastomers; they are also called *polymer blends.* Styrene-acrylonitrile is a plastic alloy. To describe the amount of material in a mixture or an alloy, an expression is used that contains "wt" (for weight) and "%" ("for each hundred"). As an example, in describing monel metal later in this module, the expression 70 wt % Ni states that 70 grams of nickel (Ni) are present in 100 grams of this alloy of nickel and copper.

Diffusion comes from the Latin verb meaning "to pour out" or "to spread out." In materials science, it means the intermingling in solid materials of atoms (in metals), ions (in ceramics), or molecules (in polymers). This active movement of particles is fairly well understood through our experiences with gases and liquids. A bottle of perfume uncapped in one part of a room will soon disclose its presence to people in a distant corner of the room by the diffusing of perfume atoms through the air in the room. Salt or sugar dissolved in water will diffuse throughout the water and form a solution in which the salt or sugar atoms will be evenly distributed throughout the water.

Our main preoccupation in our present study of materials is with solids. Diffusion takes place in solids, too. We know that in a gaseous or liquid state the individual atoms and molecules are relatively far apart, offering little opposition to other atoms migrating through them. In solids, the atoms are held tightly and close together. If a metallic crystalline solid were formed with a perfect crystalline structure, atoms would find it impossible to move about. But in our study of imperfect crystal structures, we will learn that point defects are the rule and not the exception. These defects are one main reason why atoms of a solid can actively move about within the atomic structure of the solid. Combine the presence of vacancies (the absence of an atom or atoms in a lattice site) with the fact that each atom possesses sufficient energy to cause it to vibrate about its position in the lattice structure and it is fairly easy to visualize why certain atoms possessing higher average energy in the crystal structure can break their bonds and "jump" from one lattice site to one that is not occupied (a vacancy) in the lattice structure. Once the atom moves to a new site, it leaves behind another vacancy. In other words, the atom exchanges positions with a vacancy.

The phenomenon of diffusion in solid materials is especially important in understanding the manufacturing and functioning of semiconductor materials; the carburizing of steel in surface or case hardening; the production of metal alloys including steel, the primary alloy of iron; and the heat treating of aluminum alloys, called *precipitation hardening*. ***Precipitation*** is a change in the solid state (change in phase) brought about by diffusion. How this precipitation is controlled to produce strong aluminum alloys is discussed in Module 5. These metallurgical processes are discussed elsewhere in this book.

In polymeric materials, diffusion is aided by defects in the molecular structure similar to those in metals. The diffusion of a penetrant into a solid polymer is of great importance. If the penetrant is a gas, the gas may permeate through the solid (even glass). Plastic materials can absorb liquids, gases, or vapors. Such permeation frequently causes swelling, a change in the chemical structure, or environmental stress cracking. Polypropylene (PPO) stress cracks when exposed to mineral oil. This diffusion of a fluid through a porous body is described as ***permeation***. To permeate is to pass into, pass through, penetrate, or spread through something, such as ink passing into and spreading throughout a blotter. The term is used to describe the action of atoms diffusing through a crystal structure. Polymer films are designed to prevent the diffusion of gases and water vapor into foods that have been wrapped in the film. Liquid solvents may permeate a polymer and produce a softening of the polymer, color changes, and odors.

The ***doping*** (alloying by adding a very small amount of alloy) of silicon with phosphorus in the alloying of silicon to increase its electrical conductivity is another diffusion process that depends on point defects in the silicon structure. Many sintering processes, such as powder metallurgy and the large number of welding and brazing processes for the joining of two metals by local ***coalescence*** ("to grow", "to unite into a single body"), depend on the transport of atoms by diffusion. Most surface modification treatments depend on atomic diffusion (see Module 13).

Because the internal energy possessed by atoms is related to temperature, an increase in temperature will increase the rate of diffusion. Of course, if there are no vacancies in the structure, very little, if any, diffusion will occur. Another factor that affects the diffusion of atoms, molecules, or ions is the type of bonding of the matrix atoms (i.e., strong bonding requires more energy to break the bonds). An example is high-melting-point solids. Smaller permeating atoms stand a better chance of diffusion through a structure of larger atoms, as in the case of carbon atoms diffusing through a structure of iron atoms. Our study of the microstructure of solids tells us that a lattice structure that contains loosely packed atoms (less dense) will offer less resistance to diffusing atoms than one whose structure contains tightly packed atoms. Finally, diffusion depends on time. This translates usually into allowing sufficient time for diffusion to take place at some higher temperature, as in an oven or furnace.

In summary, diffusion is the process depended on by materials scientists to change the microstructure of solids and thus vary the properties of the many solid-state materials in use by our society. Having completed a brief background discussion of solvents, solutes, solid solutions, alloys, and the very important phenomenon of diffusion, we will return to the subject of crystal imperfections. The first of two categories of such imperfections in a crystal structure involves, as mentioned, impurity atoms, which produce two types of solid solutions, substitutional and interstitial.

3.2.9 Substitutional Solid Solutions

In a substitutional solid solution (Figure 3-29), the solute atoms replace some of the solvent atoms in a crystal structure of the solvent. Using brass as our solid solution, up to about 40% of the copper atoms can be replaced by zinc atoms. This is possible because the atoms of copper and zinc are much alike. They are about the same size (the atomic radius of the copper atom is 1.278×10^{-10} m and of zinc, it is 1.39×10^{-10} m) and their electron configurations are comparable:

$^{29}Cu: 1s^2, 2s^2, 2p^6, 3s^2, 3p^6, 4s^1, 3d^{10}$
$^{30}Zn: 1s^2, 2s^2, 2p^6, 3s^2, 3p^6, 4s^2, 3d^{10}$

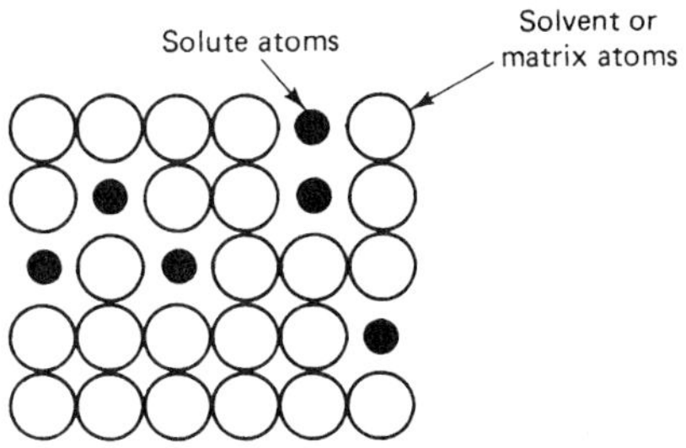

Solute atoms

Solvent or matrix atoms

Figure 3-29 Substitutional solid solution.

Their crystalline structures are both fcc with a CN of 12. Another good example of a solid solution is monel. Monel is a solution of copper in nickel (about 70% Ni and 30% Cu). The range of solubility goes from practically no nickel to almost 100% nickel. Again, these two elements have a common crystalline structure (fcc), and the atomic radius of nickel is 0.1246 nm. Therefore, monel is a substitutional solid solution.

Another interesting fact about this type of solid solution is that atoms may fill only one type of site in the lattice structure of the solvent atoms. For example, in the alloy of copper and gold, the majority of copper atoms occupy the face-centered sites, and the gold atoms, the corner sites of the face-centered cubic unit cell. An ***ordered substitutional solid*** solution is formed. As a rule, two distinct elements may form a substitutional solid solution if the sizes of their atoms do not differ by more than 15%. Of course, there are further restrictions on the degree of solubility brought about by any differences in their crystal or electron structure. The solute or impurity atoms, although of similar size, may be larger or smaller than the solvent atom, which will produce only a slight distortion in the lattice structure. Copper and zinc are very close in size, which limits solubility (about 40%). A Cu–Zn phase diagram shows the beta (β) phase, with Cu and Zn atoms randomly dispersed at lattice points. Upon cooling, the phase changes to an ordered structure (β'), which is still bcc but the Cu atoms are located at the corners and the Zn atoms are at the center of the unit cells. In most cases, the addition of solid solution elements tends to decrease ductility as strength increases. However, the addition of Zn (up to about 30 wt %) to Cu increases ductility along with strength (46,000 psi).

3.2.10 Interstitial Solid Solutions

The second type of solid solution formed by impurity atoms is the ***interstitial*** (Figure 3-30). If the impurity atoms take up sites in the lattice structure that are normally unfilled or unoccupied by the pure (solvent) atoms, they form an interstitial solid solution. These normally unfilled voids or vacant spaces are called ***interstices***. In the fcc unit cell, we know there is a relatively large interstice in the center and smaller interstices near each corner atom. It is worthwhile to point out that steelmaking is made possible because of the formation of an interstitial solid solution. First, we know that iron is allotropic. At temperatures below 910°C, iron is in the bcc form. Above that temperature, the bcc structure changes to fcc to accommodate a higher energy level of the atoms. In the fcc structures, carbon atoms can form in the interstices of the iron unit cell. At temperatures below 910°C, the bcc structure contains no room for the carbon

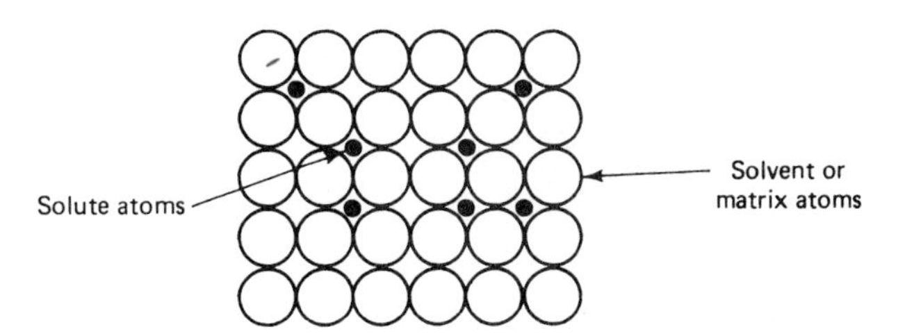

Figure 3-30 Interstitial solid solution.

atoms to fit between the iron atoms. This fact forms the basis for many of the heat-treating procedures used to produce a multitude of steels with the many different properties required by our technological society. We will discuss heat treating in detail later in the book.

3.2.11 Crystal Defects

The second category of crystal imperfections or lattice defects, a disorder of the crystal structure, is brought about by some mechanism such as thermal agitation of the crystal during its formation, the effects of gravity, or the result of high-energy radiation. Such deviations from the perfect crystal are classified for purposes of explanation as *point defects, line defects,* and *area defects*. Actually, defects occur in all combinations.

Point defects (one or two atom positions), the simplest and best understood, affect only the small volume of the crystal surrounding a single lattice site. One such point defect is a ***vacancy***, that is, the absence of an atom at a lattice site in the otherwise regular crystal (see Figure 3-31). As a result, the electronic bonding of the adjacent atoms is disrupted, which changes the effective radii of these atoms. This weakens the crystal. If sufficient vacancies were produced by heating of a crystalline solid, the crystal structure would lose its long-range symmetry and order, resulting ultimately in porosity or a change to a fluid. It is important to note that these local imperfections in the crystal structure produce a disequilibrium that has a great effect on the important properties of crystalline solids, such as density, mechanical strength, diffusion, and electrical conductivity. By themselves, point defects do not affect strength as much as they affect diffusion—the migration of atoms.

Another point defect is called an ***interstitial defect*** or ***interstitialcy***. This is produced by the presence of an extra atom in a void, the space between normal lattice positions. This interstitial atom may be added specifically as an alloying element or it might be an impurity atom indigenous to the solid. Other impurity atoms may deposit themselves in a lattice position reserved for atoms of the solid. Regardless of how they are formed, these point defects produce local aberrations in the atomic arrangement in the crystal, which produce varying degrees of local disorder in the bond structure and energy distributions in the solid. It is worthwhile to mention that disorder can occur below the atomic level (i.e., at the subatomic level). Imperfections in the electronic structure of atoms exist also. This fact has been capitalized on by the semiconductor industry, which produces materials with varying electrical properties. Further treatment of point defects in the electronic structure of atoms will be included under the topic of electronic bond theory of nonmetallic crystalline solids in Section 27.1.

It should be evident by now that imperfections in the atomic, ionic, and electronic structure of solids do not all have a negative effect on a solid's properties. Some imperfections can improve certain properties; others may degrade some properties. All imperfections do not affect the strength of solids.

So far we have discussed point defects in terms of metallic crystals, with the exception of the brief mention of nonmetallic crystals with "imperfections" in their electronic structures. As we know, all metals crystallize when they solidify, provided that they are given enough time to form their inherent structures. So do most ceramics and some polymeric (plastic) materials. In the case of ceramic materials, the crystal structures are more complex than those of metals, and the bonding is ionic, covalent, or a combination of metallic and covalent. With

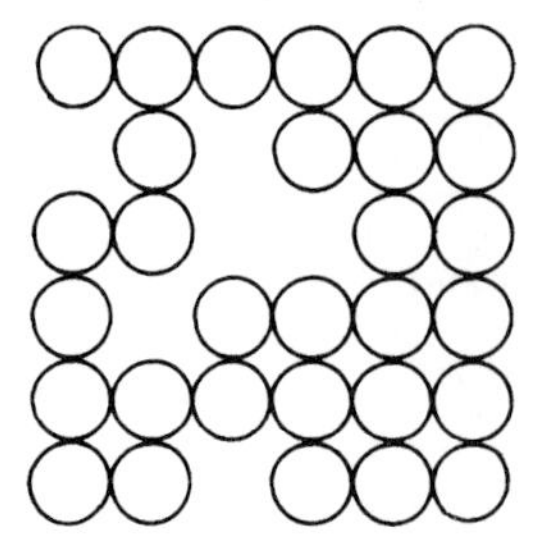

Figure 3-31 Point defects (vacancies).

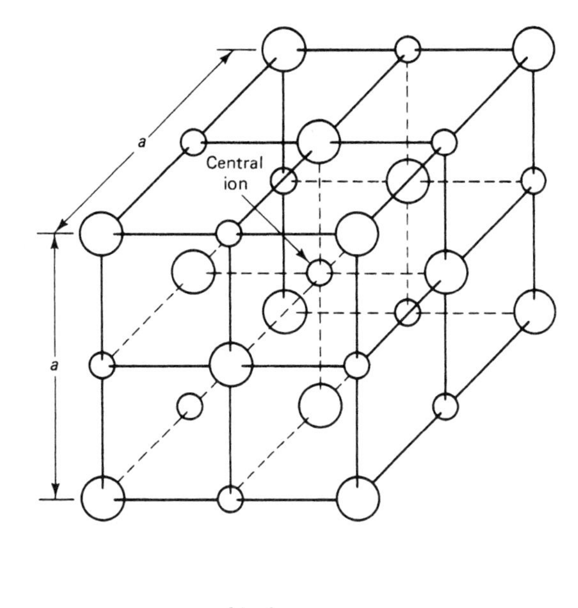

Figure 3-32 A unit cell of NaCl.

ionic bonding, the atoms, of course, become ions. In the ionic solid sodium chloride (NaCl), each Na^+ cation finds itself surrounded by six equally spaced Cl^- anions, and each Cl^- anion is surrounded by six Na^+ cations (C= 6). Figure 3-32 is a sketch of a model of the unit cell for NaCl upon which the crystalline structure is built. Each unit cell has 4 sodium ions and 4 chlorine ions associated with it. In this model the very center of the cube is occupied by a sodium ion. Each of the other 12 sodium ions is shared by three adjacent unit cells, so each contributes one-fourth of an ion to the unit cell sketched. The total number of sodium ions, therefore, is $1 + 1/4(12) = 4$. The 14 chlorine ions distribute themselves in a similar fashion and contribute a total of 4 chlorine ions to the unit cell under discussion. Each set of ions forms an fcc structure with equal numbers of sodium and chlorine ions. The radii ratio of sodium to chlorine is 0.54, with the Na ion being almost half the size of the Cl ion. This difference in size of the ions places restrictions on other ions that could replace them. Furthermore, the replacement ions must have the same number of exterior (valence) electrons. With these restrictions in mind, point defects in ionic or ceramic crystals are the rule rather than the exception. A vacancy may consist of pairs of ions of opposite charge. An interstitialcy would consist of a displaced ion located at an interstitial site in the lattice structure.

The second type of imperfection, the (one-dimensional) ***line defect,*** is also known as a dislocation (see Figure 3-33). A ***dislocation*** is a linear array of atoms along which there is some imperfection in the bonding of the atoms. An undeformed crystal lattice is represented by

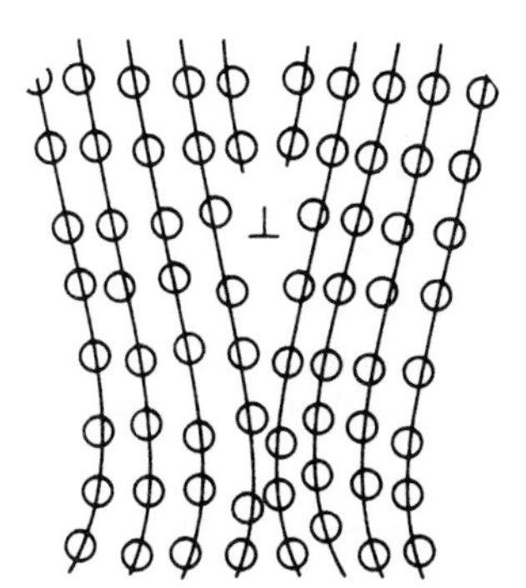

Figure 3-33 Line defects (dislocation).

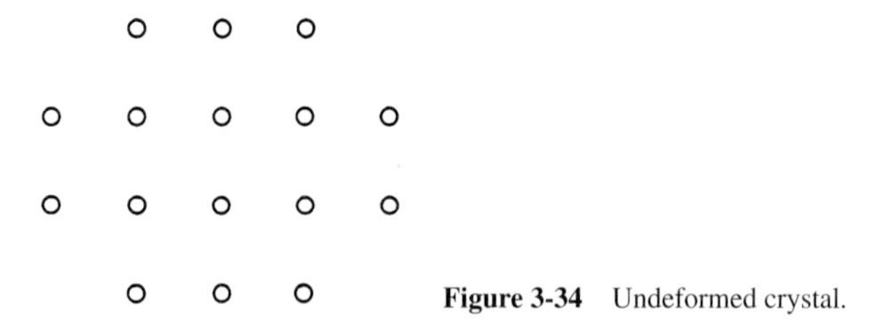

Figure 3-34 Undeformed crystal.

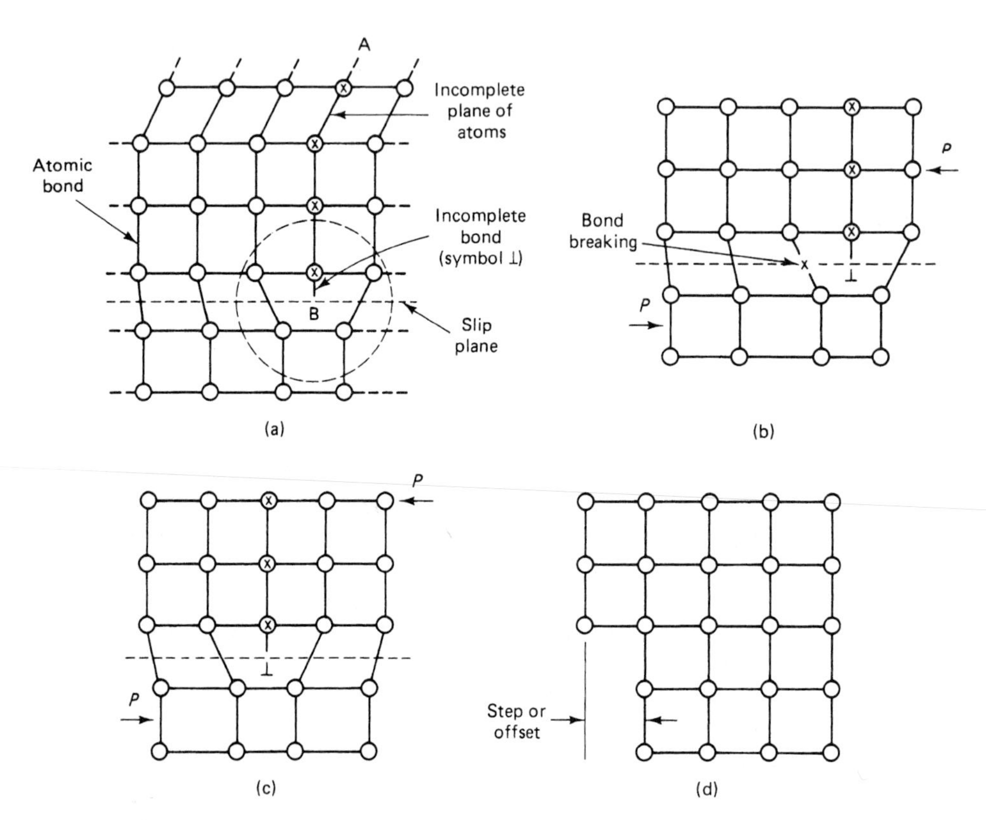

Figure 3-35 Edge dislocation movement (slip). (a) Dislocation—incomplete row of atoms above slip plane. (b) Shear force (P), causing dislocation to move. (c) Dislocation moved one row to the left. (d) Dislocation reaches surface of crystal, producing plastic strain (deformation).

Figure 3-34. Figure 3-35a and Figure 3-35b show what appears as an extra or incomplete plane of atoms (A–B) that causes distortion of the crystal structure. The two-dimensional representation also has atoms behind each atom, shown as open dots. The atoms in the area circled by a dashed line represent a center or core of poorly bonded atoms that extend back into the material along a line normal to the paper. This line defect is known as an ***edge dislocation***, whose symbol is (⊥). A force P, acting as shown, would cause the rows of atoms to move in the direction of the force, each row moving one at a time much like dominoes striking each other (Figure 3-35c). As each row of atoms moves, the next follows in turn until all planes of atoms in the area have been displaced sufficiently to make all planes continuous, as shown in Figure 3-35d.

The line of local disturbance represented by the five atoms circled in Figure 3-35a, which may extend to the boundaries of the crystal, is a region of higher energy. As stated previously, this region contains the line defect or dislocation. The dislocation plane contains this line. Above the line the atoms are under a compressive stress, while those atoms below the line are experiencing a tensile stress. These bonding forces are not as strong as in a perfect crystal lattice, and thus a relatively small shear force can break the bonds, allowing the dislocation to move. The bonds re-form after the dislocation passes.

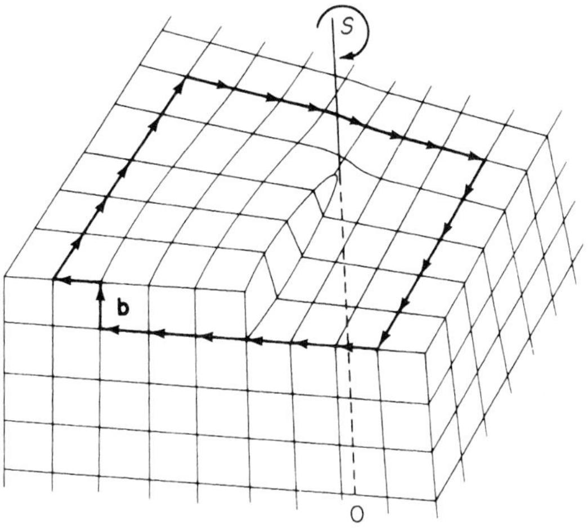

Figure 3-36 A screw dislocation. (Arthur L. Ruoff, *Materials Science* © 1973. Reprinted by permission of Prentice-Hall, Inc., Englewood Cliffs, NJ)

The successive passage or slipping of planes of atoms has been likened to the sliding of a large, heavy rug. To move the rug requires a large force; however, if you make a wrinkle in the rug and push the wrinkle a little at a time, the rug can be moved. The small movement of the rug to make the wrinkle can be thought of as the slipped portion of the rug; the other portion of the rug, the unslipped region. The wrinkle is the dislocation that separates these two regions. The displacement of atoms (***slip***) is in a direction perpendicular to the dislocation line and/or plane. Where the direction of slip is parallel to the dislocation line, the line defect is called a ***screw dislocation***, denoted by the symbol , depicted in Figure 3-36.

As with point defects, many line defects are actually combinations of edge and screw dislocations, producing curved dislocation lines or loops that start and end within the crystal. Dislocations originate during crystallization or plastic (inelastic) deformation. More will be said about dislocations when the topic of plastic deformation is treated later in this book. However, before leaving dislocations, we make the following observations. The ideal crystal structure would contain no deformations. Experience with human-made, near-perfect crystals, or whiskers, has indicated that such whiskers contain great strength. The reason a relatively little force is able to deform a crystal structure with strongly bonded atoms or ions is that only a few atomic bonds need to be broken and re-formed when dislocations are present in the crystal structure. In view of our present inability to produce near-perfect crystalline solids, the problem resolves into one of determining how to control this movement of dislocations by hindering the movement (strengthening the solid) or facilitating it (temporarily weakening the solid for some purpose).

In summary, line defects have a great deal to do with the strength of a solid. An abundance of them will cause a mutual interference in their movement through a crystal, preventing the planes of atoms from slipping, thereby strengthening the material. The presence of a few dislocations increases the ductility of a crystalline solid.

Area defects (interfacial or two-dimensional) are the third type of imperfection and exist in the form of grain boundaries (Figure 3-37). As each crystal grows, it establishes its own axis system, on which the atoms/ions orient themselves. Adjacent crystals with their differently oriented lattice structures close in on each other. The last atoms to take up position in a crystal find it more difficult to occupy normal lattice sites. Consequently, a transition zone is formed that is not aligned with any of the adjoining crystals. The atoms making up the grain boundary possess greater disorder, and hence greater energy, than their counterparts within the crystals themselves. Furthermore, the atoms are less efficiently packed together. These factors signify that the atoms in the grain boundaries are ready to act as sources of new crystal formation (nucleation sites) once the right conditions are met. Second, they assist in the diffusion of atoms through the solid. Third, they offer resistance to the movement of

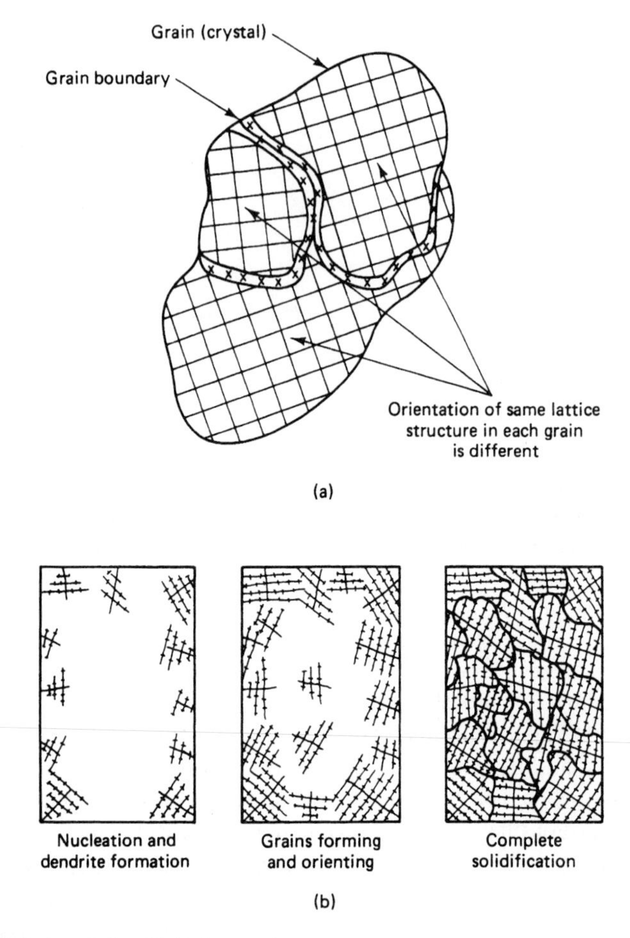

Figure 3-37 (a) Grain boundaries. (b) Grain growth.

dislocations and therefore modify the strength and the ability of materials to plastically deform. Fourth, they act as sinks for vacancies. A solid with a large number of individual crystals (also called ***grains***) has more grain boundaries than a solid with a lesser number of grains. Fine-grained material will, at normal temperatures, be stronger than coarse-grained material. Figure 3-37 shows grain boundaries with different orientations of grains and grain growth in two dimensions. Remember that a solid contains crystals, each having the same lattice structure. What is different between the individual grains is the orientation of that structure within each grain. A technique for revealing the details of the grain structure of solids uses reflected light through a metallographic microscope (Figure 3-38).

3.3 PHASES/PHASE DIAGRAMS

3.3.1 Phases

A ***phase*** is a homogeneous part or aggregation of material that differs from another part due to difference in structure, composition, or both. The difference in structures forms an interface between adjacent or surrounding phases. Some solid materials have the capability of changing their crystal structure with varying conditions of pressure and temperature, which cause these materials to change phase.

All living things maintain an internal pressure equal to the external, atmospheric pressure. Due to its construction an automobile tire can maintain pressure greater than its external pressure. This internal pressure plus the atmospheric pressure acting on both the gauge and the tire is a measure of the total pressure. For example, if the tire gauge registers 220 kPa (gauge pressure), adding the atmospheric pressure of 101 kPa would give a total pressure of 321 kPa.

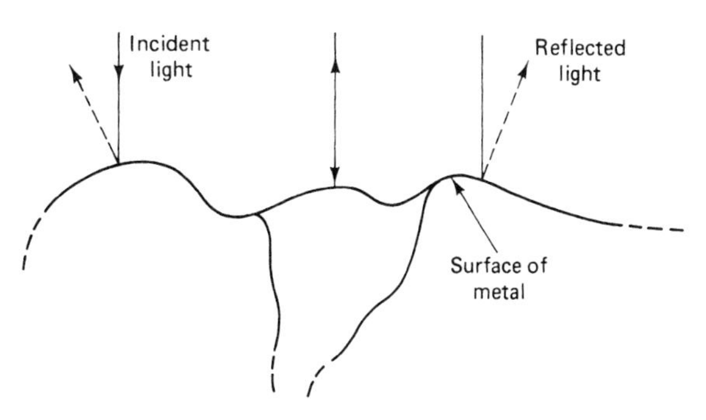

Figure 3-38 Identification of individual grains by reflected light.

One ***standard atmosphere*** (atm) is defined as the pressure supporting a column of mercury exactly 76 cm in height at sea level and 0°C. One atmosphere = 1013.25 mb (mb is the abbreviation for millibar). Another way of expressing pressure is using millimeters of mercury (mmHg): 1 atm = 760 mmHg = 29.92 in. = 14.7 psi (psi is the abbreviation for pounds per square inch). The pressure unit ***torr*** is defined exactly as 1/760 of a standard atmosphere. Thus, 1 atm = 760 torr; torr and mmHg are therefore interchangeable. The SI unit of pressure, the pascal (Pa), is related to an atmosphere by the expression 1 atm = 101 kPa, and 1 psi = 6.895 $\times 10^3$ Pa. Another unit of pressure is the bar used on weather maps: 1 ***bar*** = 1×10^5 N/m^2 or 100 kPa. Therefore, 1 atm is slightly more than 1 bar.

Water can exist as a gas, liquid, or solid. Note that each of these general phases contains the same *components* (i.e., the same basic chemical elements or chemical compounds). The single component present in water is the compound H_2O, made up of hydrogen and oxygen. Water in the general, solid phase can have different phases because it forms different crystal structures under different conditions of temperature and pressure. At the triple point of water (see Figure 3-39a) all three distinct phases can coexist with identical compositions, each having unique atomic structures and properties. Being allotropic, like many metallic elements, iron can also have different crystal structures and hence different phases (see Figure 3-40). At room temperature iron has a bcc structure. Heated to above 910°C (1670°F), iron's structure transforms to fcc. (These allotropic or polymorphic forms are also referred to by Greek letters: α (alpha), γ (gamma), and δ (delta).) Then, on heating above 1394°C (2541°F), iron's structure changes back to bcc.

It is worthwhile to pause briefly to recall that *phases*, being physically distinct with their own characteristic crystal structure, must of necessity possess different properties, for no elements found in nature are 100% pure. During the processing of metals, for example, additional impurities are introduced unintentionally, due partly to the high costs involved in trying to eliminate the impurities.

Phase change materials change phases depending on the surrounding temperature. This thermal cycling can continue indefinitely. As this material absorbs heat, it begins to shift from a solid to a liquid. After absorbing this excess energy, the material is completely liquefied. If the ambient temperature drops, the phase change material begins to shift from a liquid to a solid, releasing the stored energy as heat. The clothing industry has taken advantage of this unique material by encapsulating it into microscopic balls of heat-resistant plastic similar to that used in a dishwasher-safe dinnerware. This plastic material, in turn, is used to coat fabric that is subsequently made into clothing which responds to human body ambient temperatures. Phase change material is designed to maintain the midpoint of a narrow temperature range, which can be changed for different situations by changing the length of the hydrocarbon molecules that make up the materials. Changing the proportions of these molecules will also change the temperature range over which the material will react. When wearing such clothing, the wearer essentially is surrounded by air at a constant temperature. As he or she exercises and generates heat, the material absorbs the excess energy and changes to a liquid. When the temperature drops too much, the phase change material refreezes, shifting from a liquid to a solid and releasing stored energy in the form of heat.

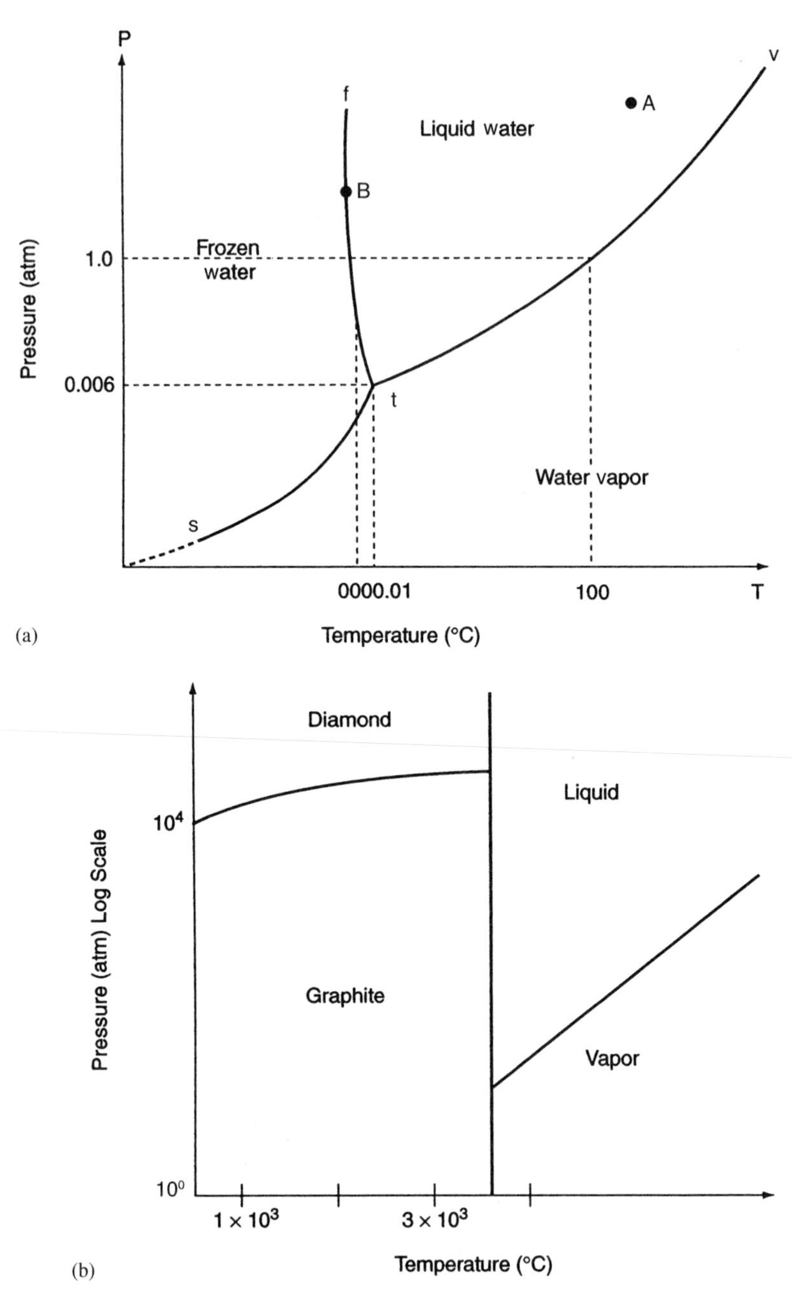

Figure 3-39 (a) Phase (P–T) diagram for water—unary phase. The SI temperature at the triple point is 273.16 K (note that scales are not linear). (b) Phase diagram for carbon. Graphite is the stable form of carbon at temperatures up to 3000°C and pressures of 10^4 atm and higher. Diamond is the stable form of carbon at very high pressures. See Figure 2–14 for the polymorphs of carbon.

3.3.2 Phase Diagrams

In our previous discussion we learned that atoms possess higher energies at elevated temperatures than at some lower temperature. This higher energy permits the atoms to diffuse more quickly. At lower temperatures, this kinetic energy is lessened. One can see that this reduced atomic mobility, coupled with the disequilibrium energy of a crystal structure at positions of high-energy content, presents a broad spectrum of possibilities for the diffusion (precipitation) of atoms, producing a phase transformation in the solid state of a system.

Phase or equilibrium diagrams serve as maps for finding one's way through the many solid-state reactions that occur in materials. Equilibrium conditions may be defined as the slow heating and/or cooling of materials that permits any phase change to occur. ***Equilibrium***, in

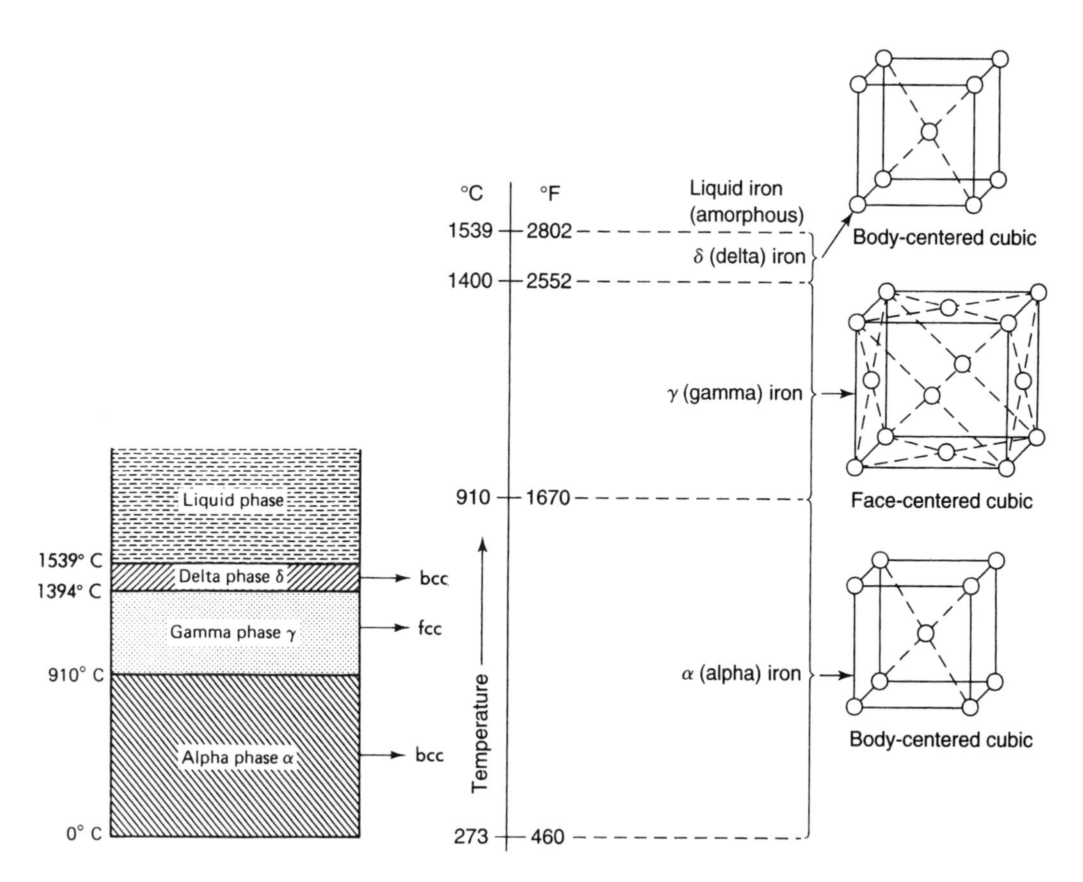

Figure 3-40 Allotropic forms of iron (three phases: bcc, fcc, bcc).

terms of a material, has also been defined as a condition of minimum energy. It should be noted that in pure terms equilibrium is a condition neither reachable nor attainable, but for practical purposes we can treat processes as if they can reach equilibrium. The condition of equilibrium might be compared to a metal spring that is neither stretched nor compressed. Once the spring is stretched or compressed, a force is set up inside the spring that tends to return the spring to its free-length position. The spring stores up the work done to stretch or compress it in the form of potential energy. Anywhere disorder occurs in the crystalline structure of a solid (grain boundaries, imperfections, etc.) there is a place of high energy similar to that of the spring under the action of some external force. The magnitude of the energy difference between the low- and high-energy states is proportional to the magnitude of the disorder.

To describe completely the conditions of equilibrium for a particular system, the three externally controlled variables of temperature, pressure, and composition must be specified. Normally, phase diagrams record the data when the pressure is held constant under normal atmospheric conditions. Consequently, phase diagrams especially are graphical representations of a material system under varying conditions of temperature and composition. To determine the data necessary to plot these many phase diagrams, specialized equipment and techniques are used. Cooling curves for a pure metal, a pure iron, a metal alloy, and a nonmetal are sketched in the following figures to illustrate the phase changes that occur, if any. Note that for pure iron, three of the four phase changes occur while the iron is in the solid state. The α-iron stage existing at the higher temperatures is nonmagnetic.

A ***cooling curve*** is a graphical plot of the structure of a material (usually a pure metal or metal alloy) over the entire temperature range through which it cools under equilibrium conditions. Figure 3-41b and Figure 3-41c show the structures that the particular metal or alloy is in at any particular temperature and time. Three structures are indicated for a pure metal or alloy: (1) liquid phase (melt), (2) a combination structure consisting of both a liquid and a solid phase, and (3) a final solid phase near or at room temperature. For the temperature to remain constant, between points *b* and *c*, as in Figure 3-41a, there must be a balance between the heat

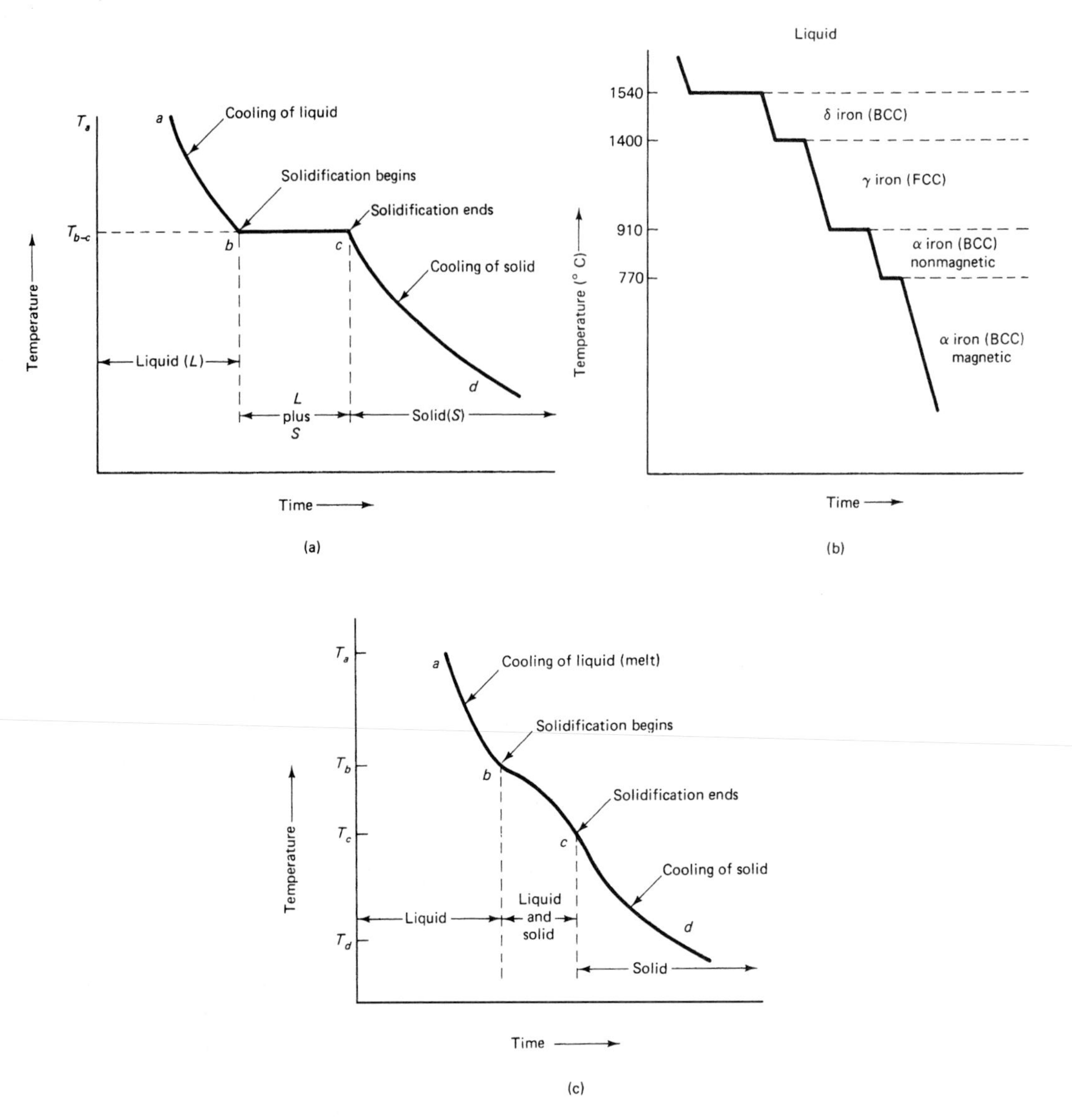

Figure 3-41 (a) Cooling curve for pure metal: (1) Heat pure metal to point T_a (liquid phase or melt); (2) cooling of liquid metal $a - b$; (3) at point *b*, pure metal starts to precipitate out of solution; (4) point *c*, *precipitation* (pure metal completely solid); curve from *b* to *c* straight horizontal line showing constant temperature $T_b - T_c$ because thermal energy absorbed in change from liquid to solid; (5) more cooling of solid pure metal from *c* to *d* and temperature begins to fall again; change in volume follows same pattern. (b) Cooling curve for pure iron. (c) Cooling curve for a metal alloy: (1) Two metals heated to point *a* (liquid phase or melt, with both metals soluble in each other); (2) cooling of alloy $(A+B)$ in liquid phase; (3) point *b*, solidification begins; (4) point *c*, solidification complete (both metals in solid form); sloped *b–c* due to changing from liquid to solid over range of temperature $T_b - T_c$ because metals *A* and *B* have different melting/cooling temperatures; (5) further cooling from *c* to *d* of solid-state metal alloy.

withdrawn from the metal and the heat supplied. In this instance, the metal is giving up heat (latent heat of fusion), which balances the heat removed. The result is a net change of zero, as measured by a thermometer. As a metal cools from a higher to a lower temperature, its electrons require less thermal energy as a result of the relatively more ordered, more dense, and more bonded positions occupied by the atoms. As the electrons surrender their excess thermal energy (in the form of heat), they move closer to their nuclei and slow their motions to match the energy level of their new positions (equilibrium position). Cooling curves for nonmetals such as glass or plastics (Figure 3-42) show no clearly defined melting points.

To capture the structure of an alloy at a particular temperature, samples are taken at various temperatures and quickly cooled. Metallurgical microscopes using visible light

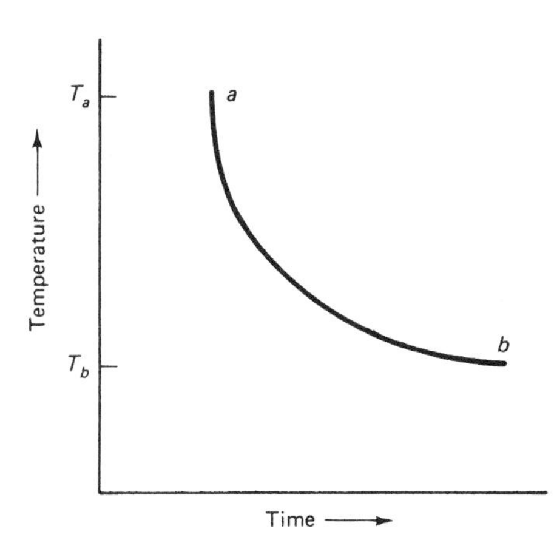

Figure 3-42 Cooling curve for a nonmetal. No phase transformation occurs in a nonmetallic heated to temperature T_a and allowed to cool to room temperature T_b. With no phase changes, a continuous plot of temperature and time produces a smooth curve.

study the grain structure, including the size, shape, and distribution of grains, with a maximum magnification of about 2000X. X rays, gamma rays, and electron beams are also used to probe the crystal structure with great resolution and magnification, and electron microscopy permits the detailed chemical analysis of selected areas within the solid material, as well as the character of an individual dislocation. The data collected are used to construct various representations of the alloy, particularly of the structure under varying conditions of temperature, pressure, and composition. Most, if not all, of these graphical representations indicate the structure of the system under *equilibrium conditions*. Therefore, sufficient time must be allowed for atoms to diffuse and thus establish equilibrium conditions at any particular pressure, temperature, and composition. In practice, equilibrium conditions are normally not achieved, either because of the length of time involved (and attendant cost) or because producing a solid phase that is in a ***nonequilibrium*** condition is desired.

The phase diagram for water makes a good departure point for an understanding of phase diagrams for systems comprising more than one component as we have said, water can exist as a gas, liquid, or solid, and each of these general phases contains the same components. The single-component phase of water is the compound H_2O, so at the triple point of water the system under study would contain a definite amount of H_2O. The one-component diagram for a definite amount of water shown in Figure 3-39 is a graphical plot of pressure along the vertical axis versus temperature along the horizontal axis, expressed in SI units, and represents the physical changes that water (H_2O) undergoes. Note that regardless of the ***physical changes*** (i.e., liquid to solid, water to ice), there is no alteration of the chemical composition of the water; only its structure is changing. On the other hand, ***chemical change*** always alters the composition of a material. Observe in Figure 3-43 how a change in temperature chemically changes the composition of the metal alloy plotted along the horizontal axis.

The pressure–temperature fixed composition diagram shown in Figure 3-39 actually contains three curves. The fusion curve is the line that separates the solid from the liquid phase. The curved line separating the liquid from the vapor phases, called the ***vaporization curve***, is a plot of the various pressures and temperature combinations at which the liquid and vapor phases coexist in equilibrium with each other. The same can be said of the sublimation curve separating the solid from the vapor phases. ***Sublimation*** is the process by which a solid changes directly to its vapor state without passing through its liquid phase. Each curve forms a boundary between two different phases.

Curve *T–F*, the fusion curve, represents where the liquid and solid phases are in equilibrium. In other words, it is a graph of the freezing point versus pressure. At 1 atm, the freezing point of water is 0°C. At this pressure, water is a liquid if its temperature is between 0°C and 100°C, a solid if below 0°C, and a vapor if above 100°C.

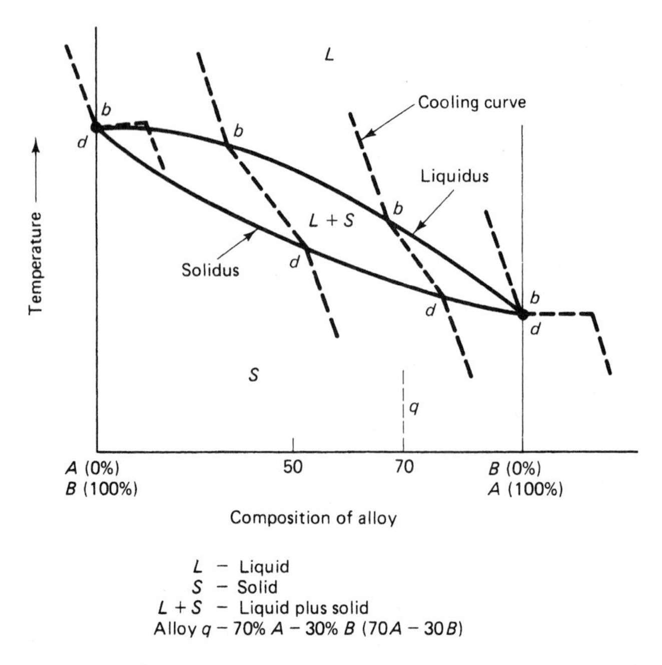

Figure 3-43 A set of hypothetical cooling curves for an alloy system phase diagram. If cooling curves for several combinations of two metals (metal alloy) are plotted as shown here, a phase or equilibrium diagram forms. Points *b*, in cooling curves where solidification begins, are connected to form a smooth line (*liquidus line*). Points *d*, where solidification is completed for each alloy, are also connected by a smooth line (*solidus line*). More discussion of phase diagrams for metals and ceramics follows in the text.

Curve *S–T*, the sublimation curve, represents where the solid and vapor phases are in equilibrium. It is a graphic plot of the sublimation point versus the pressure. Carbon dioxide, for example, in the solid phase (called dry ice) sublimates at atmospheric pressure.

Curve *T–V*, the vaporization curve, is a plot of those points where the liquid and vapor phases are in equilibrium. It is therefore a graphic of the boiling point versus pressure.

Curve *T–F* is unusual in that it has a negative slope (slopes upward to the left). For substances that normally contract upon freezing this curve would have a positive slope (sloping upward to the right). Water expands upon freezing, so a lower temperature at a higher pressure is required to cause the liquid to freeze. When water freezes, the molecules form a hexagonal lattice structure. This open structure accounts for water's unique property of expanding on freezing and being less dense as a solid than as a liquid. At 4°C this structure begins to form and at 0°C it is complete. The volume of a given amount of water decreases as it is cooled from room temperature until its temperature reaches 4°C. Below 4°C the volume increases and therefore the density decreases.

Special note should be made of the point where all three curves intersect. This point, known as the ***triple point***, is where all three phases coexist. For water, the triple point occurs at a temperature of 273.16 K (~0°C) and a pressure of 610.47 N/m2.

In our study of metal alloys, not only do we have more than one component but we want to study the solid-state reactions while varying the amounts of these components. To accomplish this, several approaches are used. Fortunately, in practice, most metallurgical processes are conducted at normal atmospheric conditions, which simplifies matters considerably. If the cooling curves for many combinations of two metals (metal alloy) or ceramics were plotted as in Figure 3-43, a phase, or equilibrium diagram, would be formed. Points *b* in the cooling curves, at which solidification began, can be connected to form a smooth line, or *liquidus.*

Points *d*, at which solidification is complete for each alloy, are also connected by a smooth line, a *solidus*. There are various types of phase diagrams of metal alloys, depending on whether the two components are soluble, insoluble, or partially soluble in each other in the liquid or solid phase.

The ***liquidus*** is the name of the curve that represents the temperatures at which solidification begins. It marks the lower boundary of the liquid phase, and thus it is a solubility curve that shows the limit of solubility of the components *A* and *B* in the liquid. The ***solidus*** is the curve that passes through all points at which solidification is completed. It marks the upper boundary of the solid solution phase, and therefore it is the solubility limit of both components A and B in the solid solution. Between these curves, where a two-phase, slushy region of liquid plus solid exists, the process of solidification takes place. It is evident that the points where the liquidus and solidus curves meet represent compositions of 100% A or 100% B (i.e., pure metals). Between these two points are an indefinite number of varying compositions of the two pure metals A and B. Metal A could represent Cu and metal B, Ni.

3.3.3 Phase Diagrams (Equilibrium Diagrams) for Metallic Systems

Building on our knowledge of phase diagrams presented in Section 3.3.2, we will describe "road maps" that guide and direct those involved in the development, fabrication, heat treatment, and design of alloys, using some simple diagrams to illustrate their main features. Before beginning this task, note that various rules must be observed to ensure correctness of the diagrams. These rules are derived from Gibbs' formulations developed over 100 years ago, the most important of which is the ***Gibbs phase rule.***

The Gibbs phase rule is expressed in equation form as

$$P + F = C + 2$$

where P is the number of phases in equilibrium; F is the variance or number of degrees of freedom, or the number of variables such as pressure, temperature, or composition that can be varied without affecting the number of phases in equilibrium; and C is the number of components (elements, compounds, or solutions) in a particular system. The number 2 in the equation stands for temperature and pressure, the two variables that can be allowed to change.

In the phase diagram for water in Figure 3-39 the pressure is plotted as ordinate and the temperature as abscissa. This is a ***unary phase diagram***, meaning that it consists of one component—in this instance, water (H_2O). The series of curves in the diagram show the division between the three phases in which water can exist depending on temperature and pressure conditions. Point A in the liquid region is defined by a certain combination of pressure and temperature. The number of components (C) at this point is 1, a unary diagram. The number of phases (P) is 1 because only liquid water exists in this region. The phase rule tells us that the number of degrees of freedom is

$$\begin{aligned} F &= C - P + 2 \\ &= 1 - 1 + 2 \\ &= 2 \end{aligned}$$

The 2 means that, within the limits of the liquid phase, the pressure, temperature, or both can be changed and the phase would still be liquid. No phase changes occur and equilibrium is maintained.

Point B is on the fusion curve (T–F), the boundary between the liquid and solid phases of water. The number of components has not changed and therefore $C=1$. The number of phases is $P = 2$, because anywhere along this curve, liquid and solid phases of water coexist in equilibrium. Using the phase rule to find the number of variables that can be changed without changing any phases in equilibrium, we have

$$\begin{aligned} F &= C - P + 2 \\ &= 1 - 2 + 2 \\ &= 1 \end{aligned}$$

This single degree of freedom means that if the pressure is changed, the temperature must also be changed to stay anywhere on that boundary line. We can no longer change either the pressure or the temperature independently and stay on the boundary line where the liquid and solid phases of water coexist in equilibrium.

Point *T*, at the intersection of the three curves, the triple point of water, is where the three phases of water coexist. The phase rule tells us that the number of degrees of freedom for this position on the phase diagram is

$$\begin{aligned} F &= C - P + 2 \\ &= 1 - 3 + 2 \\ &= 0 \end{aligned}$$

Zero degrees of freedom means that we cannot change pressure or temperature. If we did, the three phases would no longer coexist in equilibrium. It also means that the pressure and temperature are fixed for this condition to exist. The word *invariant* is also used to describe this particular point.

Point *V*, which is known as a ***critical point***, means that no matter how high the temperature and pressure rise beyond this point, the vapor phase will never change to a liquid.

Materials science deals primarily with the solid and liquid phases of materials. Second, the pressure variable has only a small effect on materials. Third, metallic phase diagrams are usually limited to showing the interactions between only two components and thus are known as ***binary diagrams***. For example, the iron–iron carbide diagram and the phase diagram for brass have only two components. Consequently, phase diagrams in materials science deal with temperature changes versus changes in the composition of materials. This means that the phase rule can be simplified as follows:

$$P + F = C + 1$$

Centre for Materials Science—*http://www.cms.fy.chalmers.se/*

3.3.4 Solid Solution Strengthening

What are the effects of adding other substances to a pure element? The results produced can be many and varied. One result of the addition of impurities might be the formation of three different solid phases. On solid solutions may result. (A solid solution is simply a solution in the solid state that consists of two kinds of atoms combined in one type of space lattice.) A third possibility is that one or more compounds could form. Finally, two or more of the preceding results could coexist, depending on the prevailing conditions of pressure, temperature, and degree of concentration of the components of the system. For example, an fcc structure of solvent Cu atoms will substitute Ni atoms in a solid solution of copper and nickel, known as a ***substitutional solid solution***. The substitution of Zn atoms for Cu atoms in a fcc structure forms brass, a disordered or random substitutional solid solution (see Figure 3-44a). If the ordering were 100%, a compound would be formed. The alloy Cu–Au (Figure 3-44b) has Cu atoms occupying the face-centered sites and Au atoms occupying the corner sites of the fcc unit cell.

Ordered solid solutions form generally at lower temperatures, or they come into existence when a disordered solid solution becomes unstable at a lower temperature. The fundamental lattice structure may or may not change during this particular transformation. It is

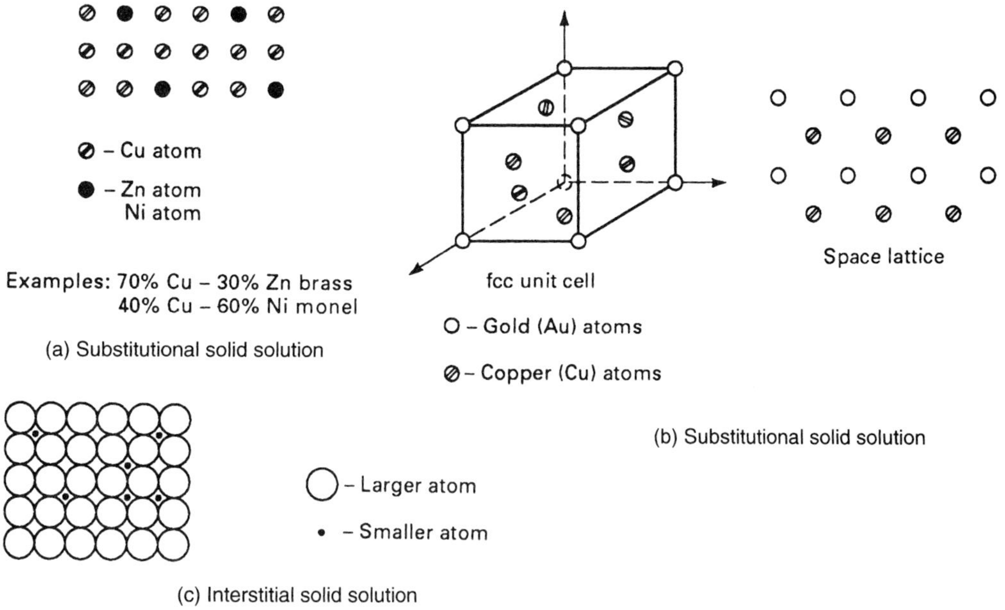

⊘ - Cu atom
● - Zn atom
Ni atom

Examples: 70% Cu – 30% Zn brass
40% Cu – 60% Ni monel

(a) Substitutional solid solution

fcc unit cell

O - Gold (Au) atoms

⊘ - Copper (Cu) atoms

Space lattice

(b) Substitutional solid solution

O - Larger atom
• - Smaller atom

(c) Interstitial solid solution

Figure 3-44 (a) Disordered (random) substitutional solid solution. Cu and Zn atoms occupy lattice sites at random. (b) Ordered substitutional solid solution in which Au and Cu atoms occupy restricted sites in the space lattice.

important to point out again that this modification in the arrangement of the atoms brings on an alteration in physical properties. In some instances, the ordered arrangement is harder and has greater electrical conductivity than the disordered arrangement.

Another type of crystal structure is formed when some atoms take up positions between the regular solvent atoms, that is, in the vacant spaces between lattice points. These solute atoms and the solid solutions formed are interstitial. The carbon atoms in ferrite (alpha iron or alpha ferrite) occupy the interstices (spaces) in this bcc allotropic form of iron. For solute atoms to form interstitial solid solutions (Figure 3-44c), they must be small enough to fit into the interstices between the normal lattice points.

It is time to ask, is it possible to determine ahead of time whether an element will form a separate phase or a solid solution when added to another element? In other words, can we predict the degree of solubility occurring in certain systems? Certainly, this question lies at the heart of metals technology in developing alloys (solid solutions) that possess the ductility for subsequent forming operations or two-phase materials with greater strength and hardness than those of the pure substances alone. A set of general rules known as the Hume–Rothery solubility rules provides this guidance. By following the ***Hume–Rothery solubility rules***, we can produce solid solution alloys that result in solid solution strengthening. As with all rules, however, they are not foolproof. For example, Cu and Ag meet all the requirements set forth by the rules but they have limited solid solubility. Solid solution alloys have greater strength than the pure metals. The difference in atomic size between the solvent and the solute atoms produces greater strength because there is a greater disruption of the initial crystal or lattice structure, making slip more difficult. The larger the amount of the alloying or solute atoms, the greater the strengthening. (See section 25.2.4 for a discussion of dispersion strengthening of composite materials when the limit of solubility may be exceeded.) The large size difference between the atoms also produces a disruption of the initial lattice.

Rule 1 The difference in atomic diameters between the solvent atom and the solute atom should be less than 15%. ***Atoms must be of similar size.*** Note that smaller atoms generally produce greater strengthening than larger atoms. For example, Be in Cu produces greater strength properties than Al when added to Cu.

Rule 2 The metals must have the same crystal or lattice structure.

Rule 3 The atoms of the metals must have the same valence. Such elements do not form compounds. Near neighbors in the periodic table such as Fe and Co tend to form solutions in one another. If the valence difference is significant, then the metals may form compounds, not solutions.

Rule 4 The atoms of the metals must have approximately the same electronegativity. With different electronegativities the tendency is to form compounds (note the case of Na and Cl to form NaCl).

A pause to discuss these rules with regard to other previous definitions of solid solutions is worthwhile at this time. Assume two metals are soluble in each other to some extent. If the size difference in their atoms is less than 8% and the other conditions are satisfactory, there is almost complete solid solubility. In other words, the two metals are soluble in each other in all possible proportions. Examples of this are Monel (Ni–Cu) and brass (Cu–Zn) (Figure 3-45). Referring to Table 10–1B, copper's atomic radius is 1.57 Å (1 Å = 10^{-10} m) and nickel's is about 1.62 Å. * Expressed in nanometers, these radii are 0.157 nm and 0.162 nm, respectively. Both have a fcc structure. Note their position in the periodic table, where they are elements 29 and 28, respectively. The fcc Cu–Ni alloys can range from near 0% Ni to almost 100% Ni. There are some 25 alloys used in industry that are based on this mutual solubility of copper and nickel in all proportions. These nickel alloys, known as ***Monel alloys***, contain about 30% copper and have combinations of high strength and good corrosion resistance. Some high-strength Monels are, in addition to being nonmagnetic, equivalent to heat-treated steels having tensile strengths approaching 200,000 psi. In contrast to pure Cu, which has relatively low strength, the alloys of Cu and Ni permit much higher strength because of the interaction between the different atoms in the solid solution, producing increased resistance to slip.

Another common alloy, ***brass***, is a solid solution of Cu and Zn (Figure 3-45b). Zinc's atomic radius is about 0.153 nm; it is element 30 in the periodic table. Cu and Zn atoms differ in size by less than 15%. In this system, Zn atoms can replace Cu atoms up to a maximum of 50% and produce disordered substitutional solid solutions. We can add tin (Sn) to Cu; Sn, element 50 in the periodic table, has an atomic radius of about 0.146 nm, which indicates that we are approaching the 15% limit. In fact, only about a maximum of 10% of the Cu atoms can be substituted by Sn atoms to produce the single-phase alloys known as ***bronzes***. The term *bronze* was originally used to describe copper–tin alloys. Now it is used for any copper alloy that contains up to approximately 12% of a principal alloying element. It is incorrectly applied to some alloys that are special brasses. Architectural bronze, a trade designation for an alloy that contains 57% Cu, 40% Zn, and 3% Pb, is actually a brass.

In summary, the effects of solid solution strengthening are realized in the properties of the material, which is consistent with the basic premise that a change in the structure of a material results in a change in its properties. Some of the effects that are produced in solid solution alloys as opposed to the pure metals are as follows:

1. Tensile strength and yield strength are greater than in the pure metals.
2. Hardness is also increased.
3. Ductility is usually decreased, as well as electrical conductivity.
4. High-strength capabilities are increased, as well as resistance to creep at high temperatures.

For an interstitial solid solution to form, the added atoms must be sufficiently small to fit into the interstices between the solvent atoms. From our study of atoms and the periodic table, we know that the atom's radius is measured from the center of the atom to the outermost electron. We also know that this diameter decreases as we move from left to right in a period of the table (Table 3-3). Taking these facts into consideration, the five elements

* An angstrom (Å) is a unit of length sometimes used when referring to the wavelengths of electromagnetic energy.

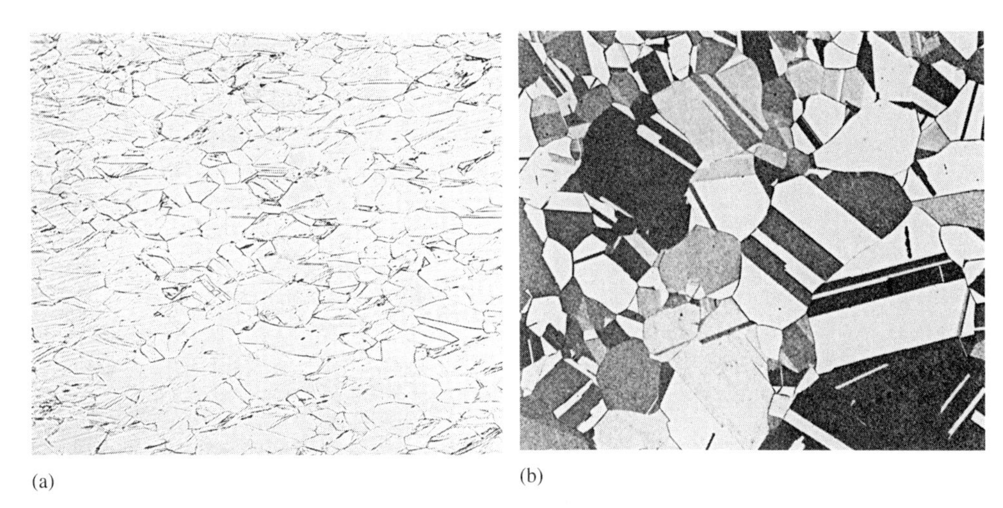

(a) (b)

Figure 3-45 Solid solutions. (a) Monel (Ni, 67.5%; Cu, 30.18%). (b) Brass (Zn, 70%; Cu, 30%). (Buehler Ltd.)

with radii less than 0.1 nm are hydrogen (H), carbon (C), boron (B), nitrogen (N), and helium (He). A classic example is the interstitial solid solution formed by C in Fe in the production of steel. Pure Fe has a bcc structure at room temperature. This phase is called α (alpha) iron. In this form the Fe is relatively weak and incapable of being shaped for commercial use. Above 912°C, iron, being allotropic, changes to an fcc structure known as γ (gamma) iron. From our study of unit cells of the various lattice structures, we know that the interstices between atoms in a bcc structure are quite small. In the fcc unit cell, a relatively large interstice exists in the center. Carbon added to α–Fe allows only 0.025% of the C atoms to take up positions in the iron bcc structure. If, instead, the γ-Fe is heated to above 912°C and then the C atoms are added, the γ-Fe will accommodate up to 2% of the C atoms. This interstitial solid solution is given the name ***austenite***. The formation of steel and cast iron will be treated extensively later when phase equilibrium diagrams and heat treating are discussed.

It may be fair to conclude from a study of the formation of compounds in liquid or gaseous states that most chemical compounds are nonmetallic, and second, that they involve the exchange or sharing of electrons. Third, one of the elements involved must have a positive valence and the other a negative valence (oxidation states). Fourth, the algebraic sum of these oxidation states must be zero in the compound formed. Experience indicates that two metals show no inclination to join together in a chemical way to produce compounds. The latter statement is true for metals in the liquid and gaseous states. Here we are discussing solid-state reactions. In solids, two metals can combine to form ***intermediate alloy phases*** in which their compositions are intermediate between the two pure metals, with crystal lattices differing generally from those of the pure metals. The intermediate phase can occur alone or accompanied by the pure metal or their solid solution phases. These later phases are then called *terminal* phases. Many intermediate phases have great technological importance in the strengthening of solid materials. They will eventually be called on, as examples, in a discussion of a heat-treating process to precipitate out at some lower temperature to impede the slippage of planes of atoms over one another, thus producing an increase in the strength and hardness of steel. (Intermediate phases are discussed in more detail in Section 3.3.5.)

The degree of solid solution strengthening is controlled primarily by two conditions. First, the larger the difference in the size of the original solvent atoms, the greater the strengthening effect. Both the yield strength and the tensile strength as well as hardness increase over that of the original pure metal due to the amount of disruption of the atomic lattice of the original atoms. The more distortion, the more difficulty atoms have trying to move (slip). Resistance to creep and retention of strength at high temperatures increase. Ductility and electrical conductivity suffer in most cases. Second, the greater the amount of alloying (solute) atoms added, the greater the increase in strength of the solvent (original) atoms. Again, this addition must be within the limits of solubility mentioned previously.

The liquid (or molten) state of a metal may consist of a single pure metal. It may comprise a solution of two or more metals that are completely soluble in each other. A third possibility is that the liquid state represents two immiscible metals. Finally, it could contain two insoluble liquid solutions. Note that the word *insoluble* is synonymous with *immiscible.* Substances that are not soluble (that are **insoluble**) are **immiscible.** Water and oil have no solubility; that is, they are insoluble, or immiscible. Very few metals are insoluble in each other in the liquid state. Aluminum and lead come close to being immiscible in both the liquid and solid states. In the solid state they solidify into two separate layers with a clear line of contact showing no appreciable degree of diffusion. But most metals are soluble in each other to some degree in their liquid phases. When these metals are cooled under equilibrium conditions from their liquid phases, they solidify and produce various solid phases. Assuming that they are completely soluble in the liquid phase, on cooling they may be

1. Completely miscible in the solid state
2. Insoluble in the solid state
3. Partly soluble in the solid state

Other products may also be formed, such as intermediate phases. Also, if the liquid phases are not completely soluble, other results may be produced in the solid state. To understand why and how these various results occur requires a good knowledge of the construction and use of phase diagrams.

The process of solidification begins with the formation of seed crystals or embryos from the liquid. Seed crystals, embryos, or nuclei (do not confuse these with the nucleus of an atom) subsequently grow into full crystals. For these nuclei to form, ***supercooling*** or undercooling of the liquid phase is necessary; that is, a temperature decrease below the equilibrium temperature is needed to make the phase transformation proceed at a measurable rate. This supercooling is an indication of the removal of the latent heat of solidification. Again, we can call on our example of the spring that, once deformed, stores up energy within itself. Supercooling is comparable to the amount the spring is deformed. For a nucleus to form—a process called ***nucleation***—energy is needed to create the nucleus by overcoming surface tension effects. Nucleation can be aided by impurity atoms, which are almost always present. As each of these nuclei grows larger, it forms dendritic, or treelike, crystals with a random distribution of crystallographic planes relative to one another (see Figure 3-46). These crystals, called ***grains,*** are surrounded by regions of high energy known as ***grain boundaries.*** If many nucleation sites are formed, a fine grain size in the solid state is the result. To obtain this fine grain, a minute amount of some element is purposely added. The particles of the element added act as impurities, on which the nuclei can form. In other words, the average grain size is proportional to the degree of supercooling.

Phase, equilibrium, or constitutional diagrams, regardless of their name, are a convenient way to display the phase microstructures of a particular alloy system, the temperatures and compositions where these phases change, and the quantities of the phases at equilibrium. In so doing, phase diagrams aid in understanding the structures, hence the properties, of alloys. With this information one can learn how to modify the structures by heating or cooling treatments to attain a desired set of properties. Phase changes can appear or disappear, but normally these changes occur when the temperature is lowered. In so doing the resulting microstructures may have obtained equilibrium (become stable) or nonequilibrium (nonstable or metastable) characteristics. Metastable phases may persist over time, with significant consequences. For example, a metastable phase may possess a greater strength while in this condition, which might be the original goal of the heat treater who is controlling the heating or cooling of a metal alloy. Phase diagrams play an important role in predicting a phase transformation, its resultant microstructure, and its characteristics.

We may now wish to know what the *actual composition of the two-phase region* (Figure 3-47) is for a particular alloy at a specific temperature. Figure 3-47a shows a typical phase diagram for two metallic elements, and Figure 3-47b shows ceramic materials that are completely

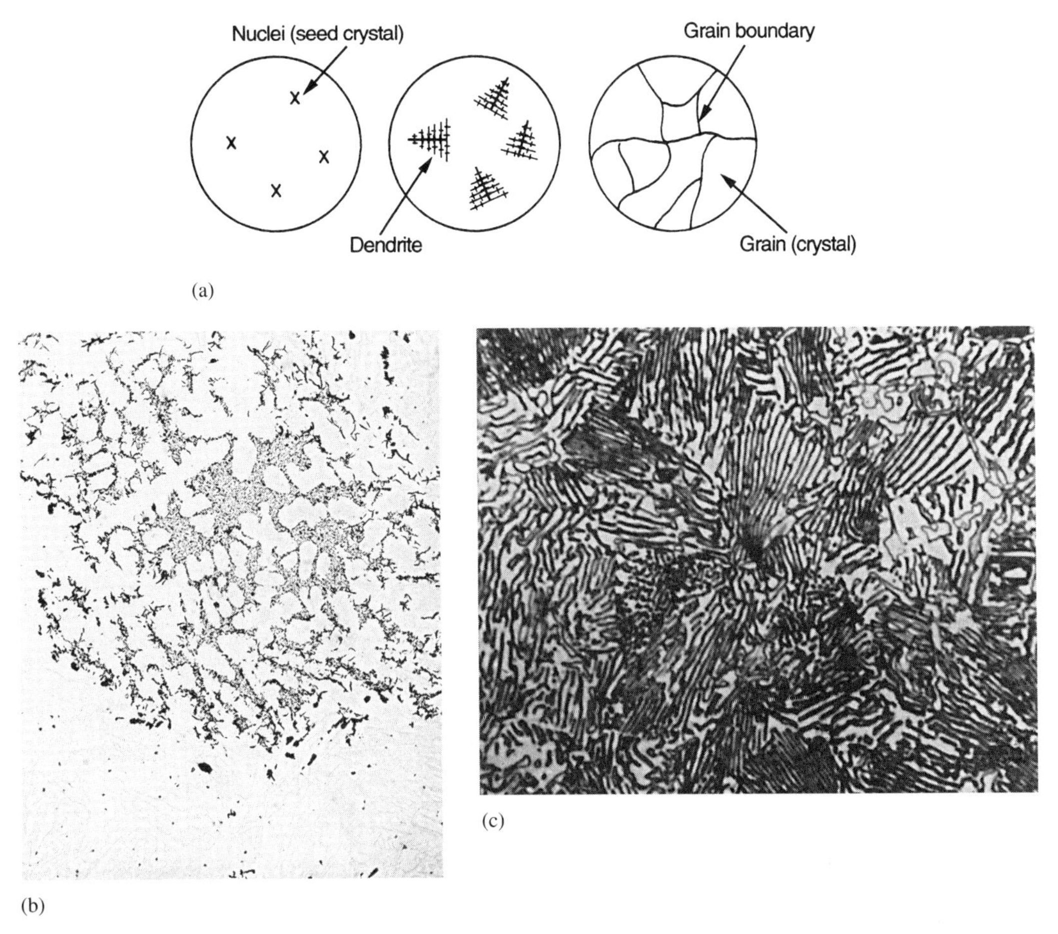

Figure 3-46 (a) Grain formation. (b) Dendrites of "tough pitch" copper, dendrites outlined by copper oxides (200X). (Buehler Ltd.) (c) Grains of "high-carbon" steel. (American Iron and Steel Institute)

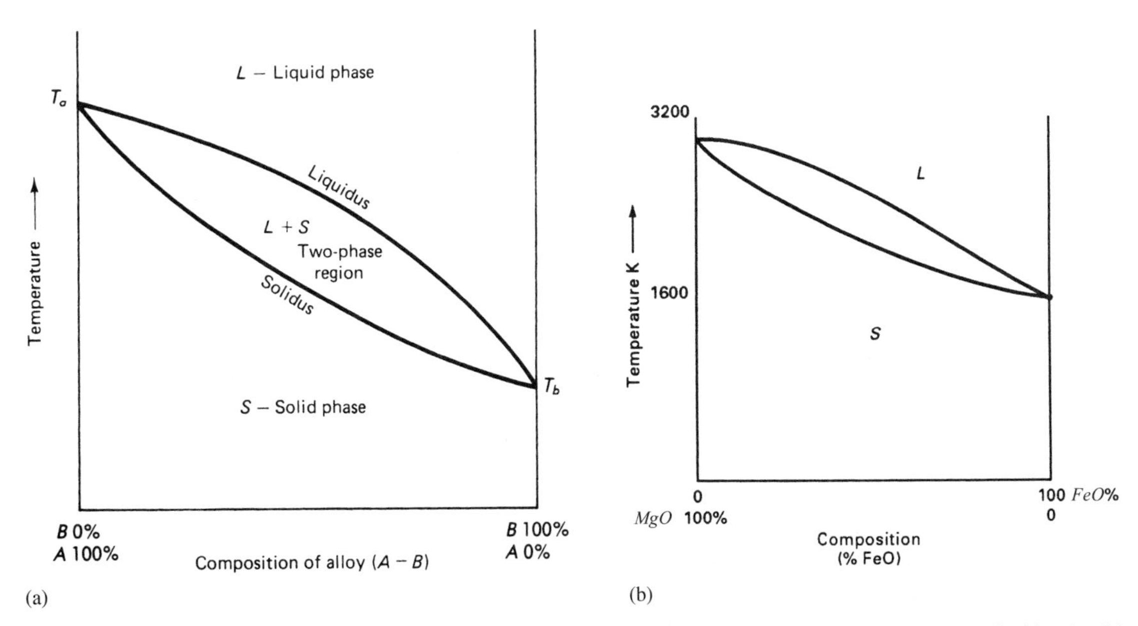

Figure 3-47 Phase diagrams of metals and ceramics. (a) Metals *A* and *B* completely soluble in each other in both liquid and solid phases. (b) Phase diagram representing two ceramic materials completely soluble in each other in the liquid and solid states.

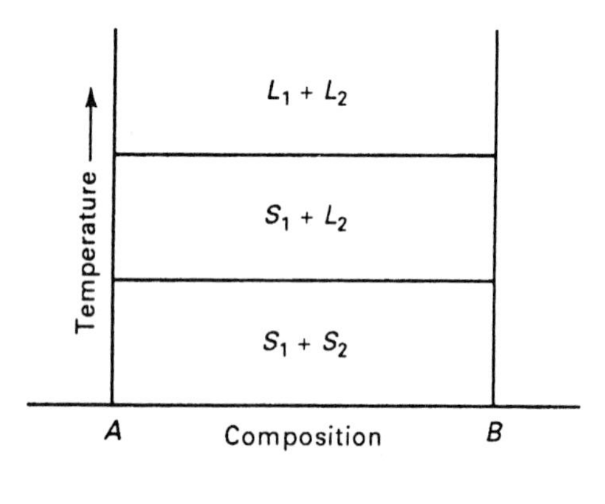

Figure 3-48 Phase diagram for metals *A* and *B* completely insoluble in each other in both liquid and solid phases (hypothetical).

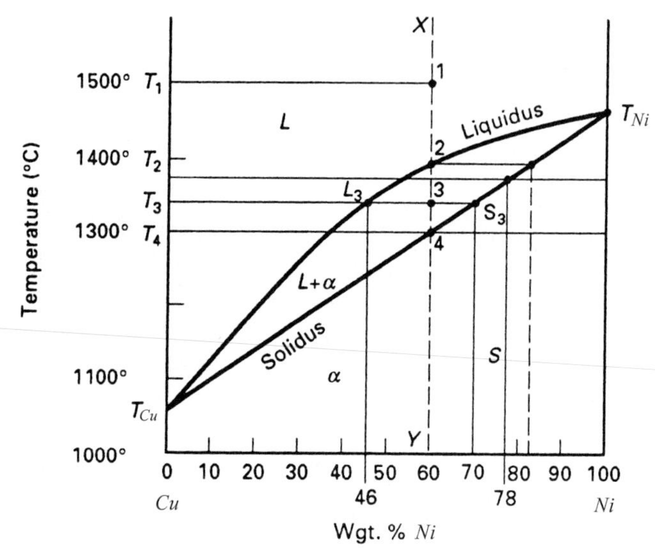

Figure 3-49 Cu–Ni phase diagram. Alloy *X–Y* is 60% Ni–40% Cu.

soluble in each other in both liquid and solid phases. The vertical axis at the left represents a pure metal *A* with a melting point of T_a. The vertical axis at the right represents metal *B* that has a melting temperature less than metal *A*. The upper curved line of the cigar-shaped $L + S$ region is the liquidus, and the lower line is the solidus. The pure metals and any composition of the two are in the liquid (melt) phase upon heating above the liquidus. Upon cooling to the solidus, they solidify. When an alloy is at a temperature above the solidus but below the liquidus, it exists as part liquid and part solid in a two-phase region labeled $L + S$. Figure 3-48, for comparison purposes, is a hypothetical phase diagram for two metals that are completely insoluble in each other, both in the liquid and solid phases.

Using an alloy with a composition of 60% Ni–40% Cu, we will plot this composition in Figure 3-49 to determine first its chemical composition at any given temperature. Vertical line *X–Y* represents our alloy. At point 1 the alloy is at temperature T_1 in the liquid solution phase. Upon slow cooling, point 2 is reached at a temperature of T_2. Our alloy is now entering the two-phase region. Part of our alloy is liquid and part has formed a solid solution. To determine the composition of these two phases at any point, such as at 3, a horizontal temperature line, called a ***tie-line***, is drawn through point 3. This tie-line will cross both the liquidus and solidus curves at points labeled L_3 and S_3, respectively. From both these points, drop vertical lines to the abscissa (horizontal axis) to determine the composition of these two phases at the temperature T_3. The liquid phase formed so far has a composition of 54% Cu and 46% Ni, determined by proceeding from point 3 along the tie-line to the liquidus L_3, then vertically down to the horizontal axis at the point labeled 46% Ni. Similarly, the composition of the solid phase is found to be 70% Ni–30% Cu. This time we proceeded along the tie-line to the solidus at S_3 and vertically downward to the horizontal axis to read 70%. Note that the

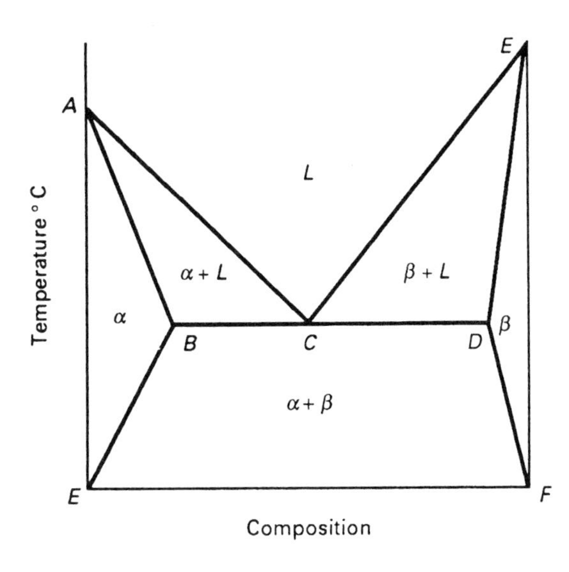

Figure 3-50 Two-phase regions of a phase diagram showing partial solid solubility.
ACE—liquidus
ABCDE—solidus
BE and *DF*—solvus lines
C—eutectic point

tie-line drawn across the two-phase region ties in or connects the adjacent phases. In this instance, the solid phase is on the right and the liquid phase is on the left of the two-phase region. This is an excellent way of identifying any two-phase region. As the alloy continues to cool to temperature T_4, observe, using the same procedure, that both phases continue to increase in the percentage of copper. Upon reaching the solidus at temperature T_4, the last liquid, very rich in Cu, solidifies at the grain boundaries. Through diffusion of the atoms, all the solid solution will be at an overall composition of 60% Ni–40% Cu. The solid solution is designated α solid solution.

The tie-line procedure plays an important role in identifying two-phase regions in a phase diagram. Figure 3-50, a phase diagram that roughly resembles the Al–Si phase diagram, shows the two-phase regions labeled. Greek letter symbols are commonly used to identify those solid phases first to form. By drawing a tie-line across these regions and knowing what phases are present at the extremities of the tie-line, the two-phase region can readily be labeled. Note the complete solidus line, showing that any phase below it is a solid phase. Lines *BE* and *DF* are solvus lines, or ***lines of maximum solubility*** of the terminal or first-formed solid solutions. Note the reduction in the limits of solubility of these solutions to absorb other phases as the temperature decreases.

The next problem to be tackled is to determine the *relative amounts of liquid and solid existing at a specific temperature in a two-phase region.* So far we know that at temperature T_3 (see Figure 3-49) two phases exist: a solid phase with a composition of 70% Ni–30% Cu and a liquid phase with a composition of 54% Cu–46% Ni. How much of the liquid has solidified upon being slowly cooled to the temperature T_3? To determine this percentage, we use what is known as the ***lever rule*** (also known as the ***inverse-ratio law***, as we shall soon demonstrate). The law gets its name from treating the tie-line as though it were a lever (fulcrum) supported at the point where the vertical line representing our alloy *X–Y* crosses the tie-line. Figure 3-51 is an enlarged sketch of the tie-line in Figure 3-49.

The complete tie-line from L_3 to S_3 represents the total weight (100%) of the two phases present at temperature T_3. The vertical line (*X–Y*) representing our alloy cuts the tie-line into two parts or lever arms; one part forms L_3 to 3, and the second forms 3 to S_3. The length of the first lever arm, *L*3 to 3, represents the amount of solid phase present; the right lever arm, 3 to S_3, represents the amount of liquid phase present. Note that the left arm or part contains L_3 on the liquidus line, yet it represents the amount of solid phase. The right lever arm contains a point S_3 on the solidus, but this arm represents the amount of liquid phase present. Thus, the lever rule states that these two lengths are inversely proportional to the amount of the phase present in the two-phase region. Observe that point 3 is closer to the solid-phase region, and therefore the quantity of the liquid phase is greater than the quantity of the solid phase at that temperature. The lengths of these two parts or lever arms of the tie-line are measured in units of composition.

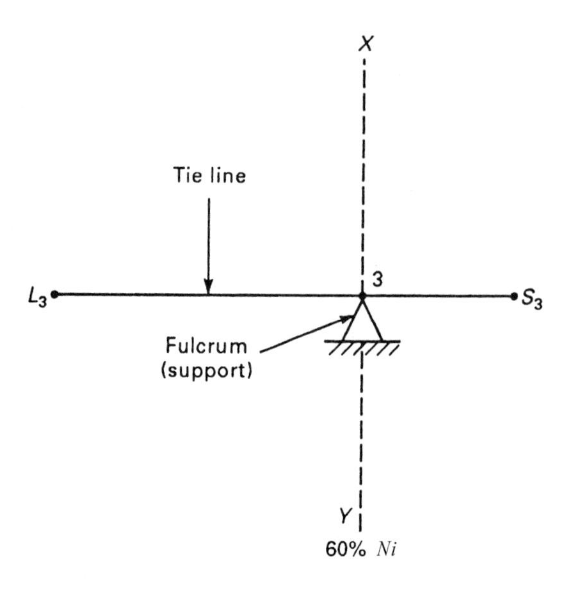

Figure 3-51 Tie-line from Figure 3–49.

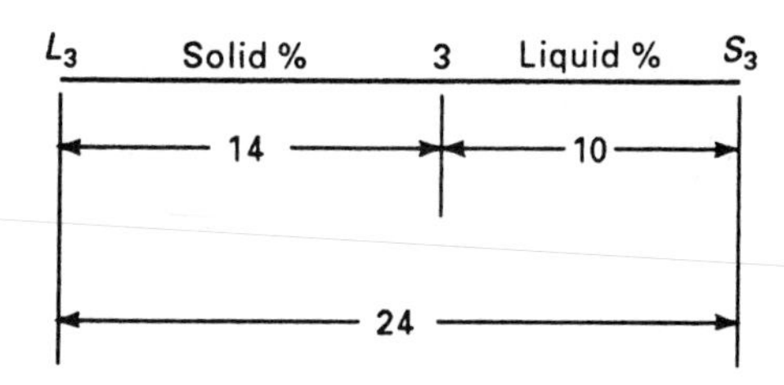

Figure 3-52 Tie-line from Figure 3–51.

Now let's apply this lever rule to our alloy 60% Ni–40% Cu. Figure 3-52 is another sketch of the L_3-3-S_3 tie-line in Figure 3-49 (a repeat of Figure 3-51. The tie-line's length in terms of composition is 24, or $70 - 46 = 24$, leaving out the percent sign. The length of L_3 to 3 (left lever arm) is 14, leaving 10 for the length of the right arm. Next ask yourself: What percent of the entire line is the right lever arm? To find the answer, you would take the length of 3 to S_3 and divide it by the total length of the tie-line, L_3 to S_3, and multiply the result by 100%:

$$\frac{10}{24} \times 100\% = 42\% \text{ (liquid formed)}$$

Using the same procedure, we find that the left part (left arm) represents the percent of solid formed:

$$\frac{14}{24} \times 100\% = 58\% \text{ (solid formed)}$$

Let's take another look at Figure 3-49 to note another metallurgical process that is taking place: ***diffusion***. For our purposes, note that the solid formed at T_2 consists of a composition of about 82% Ni–18% Cu. Assuming slow equilibrium cooling, our alloy will ultimately solidify completely, with a composition of about 60% Ni–40% Cu. What this tells us is that sufficient time must elapse for the atoms to migrate through the solid phase to satisfy the principles of equilibrium; that is, given sufficient time, atoms will diffuse away from places of high concentration such that a uniform distribution of atoms in the crystal structure is once again achieved. The reheating of a solid metal to give atoms greater energy to diffuse or move toward equilibrium (hence, more uniform composition) is termed ***homogenization***. Homogenizing eliminates the segregation of phases, such as coring (see below), that form during solidification. Such segregation leads to nonuniform properties. Annealing, specifically normalizing (see Module 9), is an example of a homogenization process involved in the heat treating of metals.

Research into the diffusion process discloses that the rate of diffusion slows as the difference in concentration of different atoms is reduced. Second, diffusion in a solid slows with reduced temperature. Therefore, in practice it is extremely difficult to achieve equilibrium cooling. Also, slow cooling produces large grain size, which is usually undesirable. Faster cooling rates are the rule, not the exception, and they prevent complete diffusion. The end result is that the initial crystals formed are of one composition. As the crystal grows in a dendrite fashion by atoms attaching themselves to the original crystal, the composition of the total crystal changes. The original higher-melting-point central portion is surrounded by lower-melting-point solid solutions. This condition is known as ***coring*** or ***dendritic segregation***. Segregation is always present to some extent when metals are melted and subsequently solidified in the making of steel (steel ingots) and in the casting of metal parts. A metallurgical process that attacks the problem of cored structures will be discussed later.

Illustrative Problem

Determine the relative amounts of liquid and solid existing at temperature T_1. (See Figure 3-49.)

Solution Using the lever rule and the previously determined compositions of the two phases existing at that temperature, draw a horizontal tie-line at $T_1 = 1336°F$. The complete line (Figure 3-51) from L_3 to S_3 represents the total weight (100%) of the two phases present at this temperature. The vertical line (X–Y) representing the alloy cuts the tie-line into two parts or lever arms. Lever arm L_3 to 3 represents the amount of solid phase present. Lever arm L_3 to S_3 represents the amount of liquid phase present. Note that these two lengths are inversely proportional to the amounts of phase present. The length of arm L_3 to S_3 is a bit smaller than arm L_3 to 3, hence, one would expect the amount of solid to be a bit larger than the amount of liquid. Referring to Figure 3-52, the estimated lengths of the arms are determined to be L_3 to $3 = 60 - 46 = 14$, and L_3 to $S_3 = 70 - 60 = 10$.

$$\text{Solid phase: } \frac{14}{24} \times 100\% = 58\%$$

$$\text{Liquid phase: } \frac{10}{24} \times 100\% = 42\%$$

Illustrative Problem

Determine the composition of alloy X–Y (40% Cu–60% Ni) at point 2 in Figure 3-49.

Solution The tie-line through this isotherm (at T_2) intersects the solidus at about 60% Ni–40% Cu, giving the composition of the liquid phase. The solid phase at T_2 is 82% Ni–18% Cu.

Illustrative Problem

Determine the composition of the last remaining liquid phase of alloy X–Y in Figure 3-49 just prior to the alloy completely forming a solid solution.

Solution The tie-line constructed through point 4 (T_4–4) intersects the liquidus. Reading vertically downward, the percent composition of the alloy X–Y at this temperature immediately prior to forming a solid solution is about 62% Cu–38% Ni.

To continue our study of phase diagrams, the next step is understanding the *eutectic reaction*. The word ***eutectic*** is taken from the Greek and means "to melt well." Figure 3-53 labels the intersection of the liquidus and solidus as the eutectic point (E). The temperature that corresponds to it is labeled T_E, which is the lowest temperature at which a liquid solution will remain completely liquid. The alloy 57% A–43% B has the lowest melting point of any alloy in the system AB. Not only does it have the lowest melting temperature but this eutectic, a mechanical mixture, solidifies completely at this temperature rather than over a range of temperatures. Alloy 80%A–20%B begins to solidify at temperature T_1 and completes its solidification at temperature T_E. The two-phase region to the left of the eutectic point is composed of metal A and liquid. The two-phase region to the right of the eutectic point is made up of metal B and

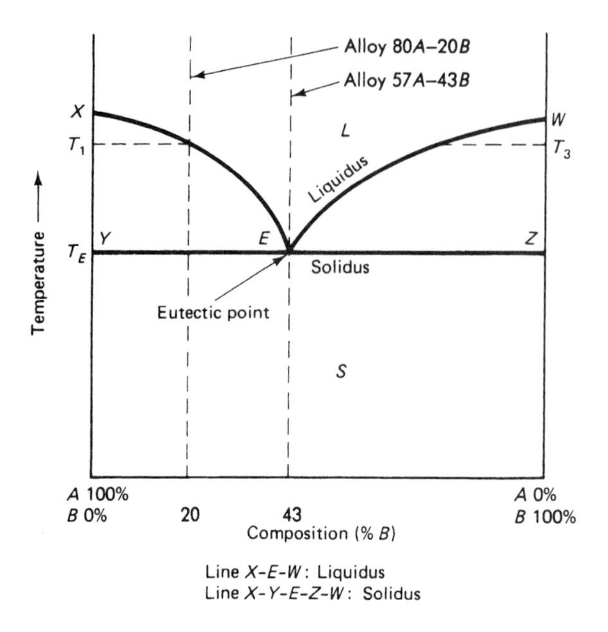

Figure 3-53 Metals *A* and *B* completely soluble in the liquid phase and completely insoluble in the solid phase (hypothetical).

liquid. Once they reach the solidus, all these alloys will solidify as separate phases. Thus, the region below the solidus consists of a mixture (not a solution) of two solid phases. This eutectic reaction for the eutectic alloy, in our example 57% *A*–43% *B*, can be written in equation form as follows:

$$L \rightleftharpoons S_1 + S_2$$

The double arrow indicates reversibility. Cooling the liquid phase produces two solid phases (S_1, S_2). Heating the two solid phases produces a liquid (melt) (L). For the lowest melting composition of the system *AB*, the eutectic composition 57% *A*–43% *B*, its two solid phases will be completely distinguishable from each other when viewed under a metallurgical microscope. The parallel wavy-line pattern formed, as viewed by the microscope, is known as a lamellar type (Figure 3-54). It can be compared to the stripes of a zebra.

Referring to Figure 3-53, alloys to the left of the eutectic mixture are known as ***hypoeutectic*** mixtures. Those mixtures to the right of the eutectic are known as ***hypereutectic***.

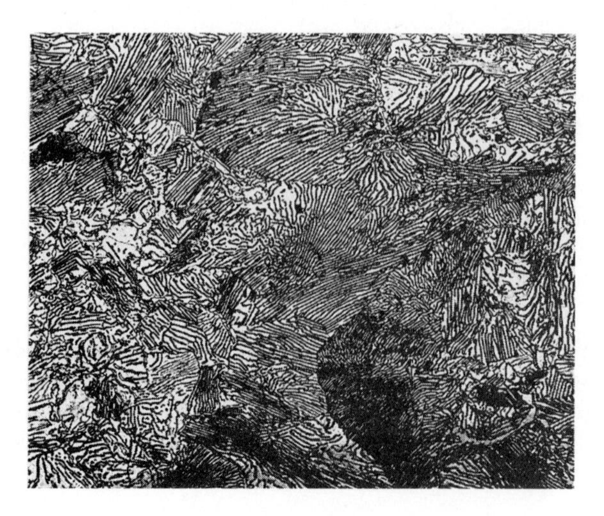

Figure 3-54 Two-phase lamellae microstructure. (Buehler Ltd.)

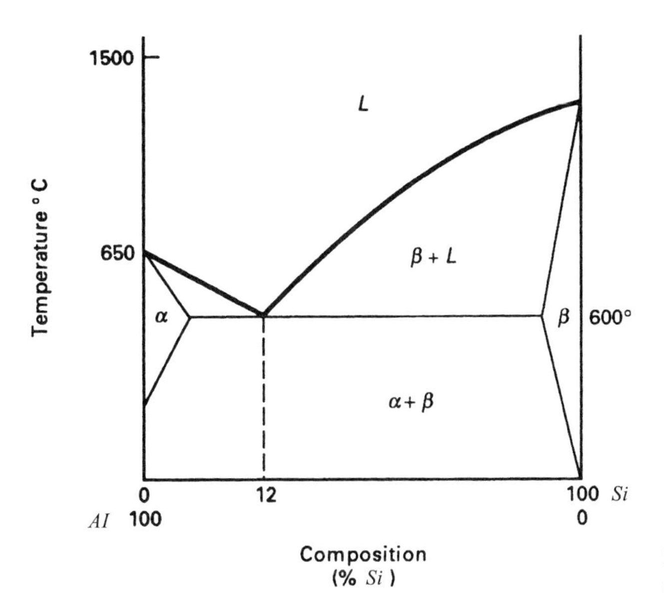

Figure 3-55 Aluminum–silicon phase diagram.

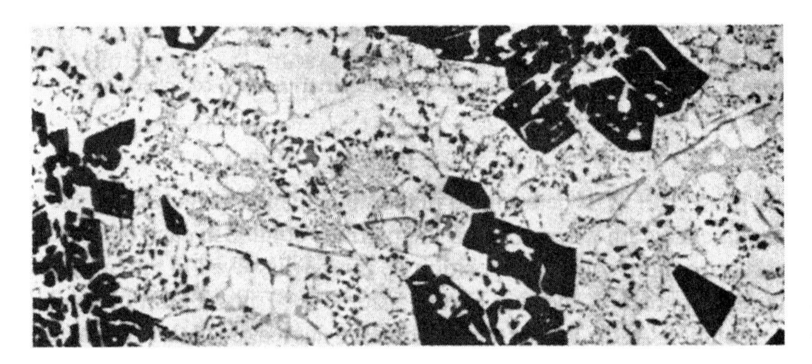

Figure 3-56 Microstructure of Al–Si alloy as cast X250, unetched. (Buehler Ltd.)

Those alloys closer to the eutectic composition will contain more eutectic mixture in the solidified alloy. Note that all phase diagrams depicting the complete insolubility of two metals in the solid state have been labeled as hypothetical (see Figure 3-48); most metals are soluble in each other to some degree. An alloy that comes close to being completely insoluble is aluminum silicon (AlSi) (see Figure 3-55), a family of alloys that forms the basis for a number of commercial alloys. Silicon is only slightly soluble in aluminum. Figure 3-55 is a sketch of the Al–Si phase diagram. Figure 3-56 is a photomicrograph of an Al–Si alloy.

Our interest now turns to the more realistic situation in which the metals are to some degree soluble in each other in the solid state. This situation is depicted in Figure 3-57. Notice that this diagram is basically the same as the hypothetical phase diagram, Figure 3-53, with the addition of points Y and Z and the lines extending from these points. Also note the similarities of Figure 3-55 and Figure 3-57. Lines Y–S and Z–T are *solubility curves,* also known as *solvus lines.* Before learning more about these lines, note that there are three areas of single-phase solutions, three regions of two-phase solid solutions, and one eutectic mixture. The solid alpha (α) phase is composed mostly of metal A with some metal B. The solid beta (β) phase is rich in metal B along with some metal A. The α phase, which is called a *terminal* phase, differs from pure metal A by the presence of some metal B. Using a tie-line, the two-phase region between α phase and the liquid (L) phase is labeled $\alpha + L$. The main solid phase, $\alpha + \beta$, is determined in the same manner. Figure 3-57 shows the typical shape of a phase diagram where A and B are partially soluble in each other. As the temperature increases, the more solid (composition B) can be dissolved in the liquid. Note that the solvus lines end at points Y and Z in

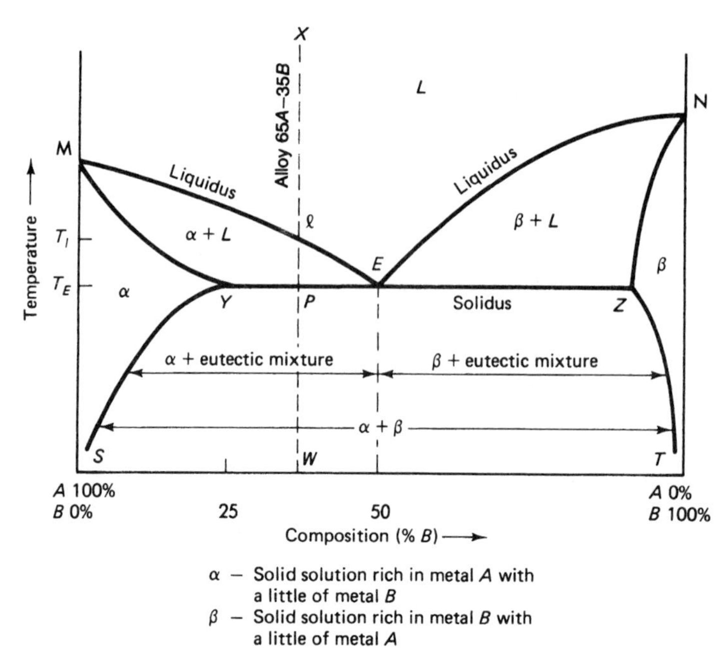

Figure 3-57 Typical eutectic-type phase diagram: metals *A* and *B* completely miscible in the liquid phase and partially miscible in the solid phase.

Figure 3-57. Point *Y* is the limit of solubility of metal *B* and solid solution. At the lowest temperature the degree of solubility is a minimum (98% A–2% B); that is, only 2% *B* is soluble in *A*, but as the temperature increases, this degree of solubility of *B* in *A* increases to a maximum limit (75% *A*–25% *B*).

In terms of a particular alloy (65% *A*–35% *B*), the phase transformations that take place as it is cooled from somewhere in the liquid phase can be briefly explained as follows; At point ℓ (Figure 3-57) on vertical line *X*–*W* and temperature T_l, the liquidus line is reached. Solid solution α is beginning to form. As the alloy cools, this solid solution becomes richer in metal β until the solidus is reached at point P and temperature T_E. The last liquid to solidify has a composition of 50% *A*–50% *B*. Assuming that diffusion has kept pace with nucleation and growth of crystals, the solid phase has a composition of 75% *A*–25% *B*. Applying the lever rule, we could determine the relative amounts of these two phases present. Note that the last liquid to solidify has the eutectic composition and therefore begins to solidify and form a eutectic mixture consisting of alternate layers of crystal of α phase and crystals of β phase. At this point we are just below temperature T_E. Here is where the solvus lines come into the action. If you were to draw in the tie-line for this temperature, it would be pretty close to the line *Y*–*Z*. As the temperature continues to lower, the tie-lines that you might draw get longer and longer. Add to this observation the fact that the solvus line *S*–*Y* shows the maximum solubility of metal *B* in metal *A* at various temperatures, and that this solubility decreases as the temperature is reduced to room temperature. Metal *B* is in solution in the α solid phase. Therefore, the excess β phase must precipitate out of solution. At room temperature our alloy will consist of α phase with some excess β phase precipitated within it, plus the eutectic mixture. This eutectic mixture is also made up of both α and β phases.

3.3.5 Intermediate Phases

The phase diagrams discussed up to this point are binary alloy phase diagrams, such as copper–nickel, or binary eutectic phase diagrams similar to Figure 3-57, which could easily represent a lead–tin alloy system. In both these types of phase diagrams, there are only two solid phases α and β, called *terminal solid solutions,* that exist over a range of compositions near the extremities of the diagram. Other alloy systems contain ***intermediate solid solutions*** *or* ***intermediate phases***. These phases are formed at other than the two extremes in composition. The copper–zinc system (not shown) has, in addition to its two terminal phases, four intermediate phases or intermediate solid solutions. None of these four phases extends to the extremities of the phase diagram.

In some solid systems discrete intermediate compounds are formed. They are composed of a metal and a nonmetal and have distinct chemical formulas. Intermediate compounds formed in metallic phase diagrams are known as ***intermetallic compounds*** *or* ***intermetallics***. They have a narrow range of composition (see mullite in Figure 7-12), and some have a fixed composition that is represented on a phase diagram as a single vertical line (see alloy *AB* in Figure 3-58 and Figure 3-59. Their atomic bonding can be either mixed ionic/metallic or mixed covalent/metallic. Their properties include brittleness coupled with good high-temperature resistance and a high strength-to-weight ratio. Consequently, intermetallics like Ni_3Al and Ti_3Al are finding applications in the aerospace industry. Aluminum casting alloys often contain intermetallic phases between grains, which are prone to form because of the low solubility of the many alloying elements in aluminum. Refer to the phase diagram for mullite (Figure 7-12) and section 24.2.3 for the use of intermetallics as matrix materials for metal matrix composites.

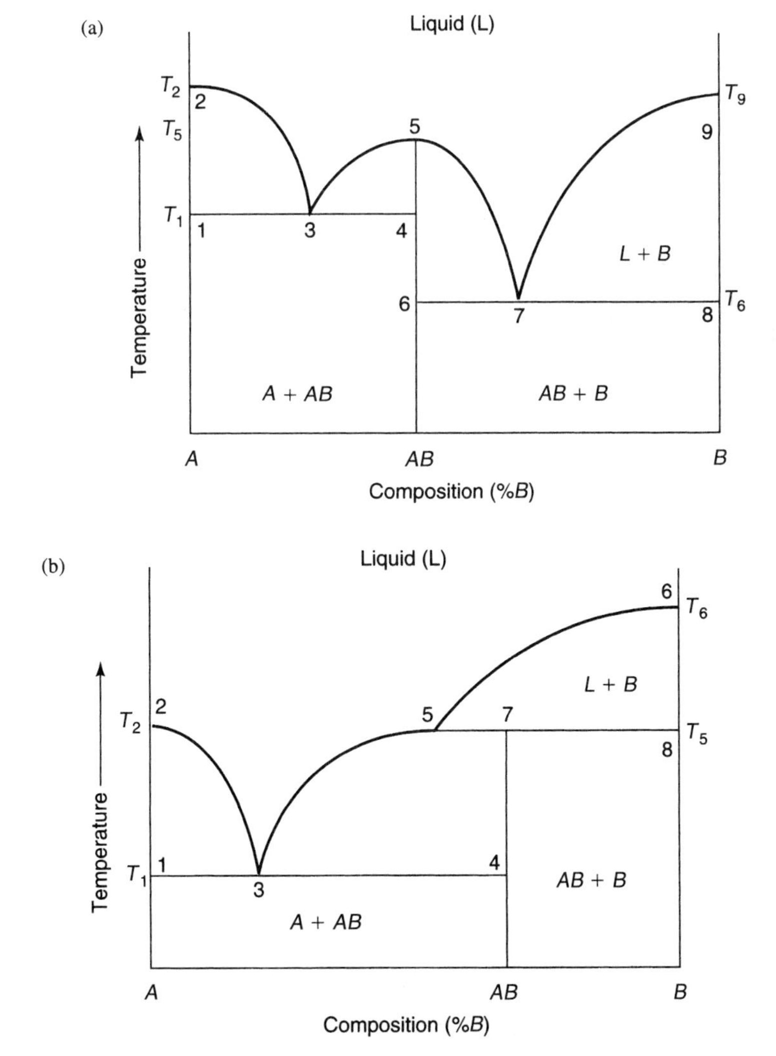

Figure 3-58 (a) Binary phase diagram with an intermediate, congruent intermetallic (*AB*). *Note*: Points 3 and 7 are eutectic points. (b) Binary phase diagram with an intermediate, incongruent intermetallic (*AB*). *Note*: *AB* never reaches the liquidus. Point 5 is a peritectic point.

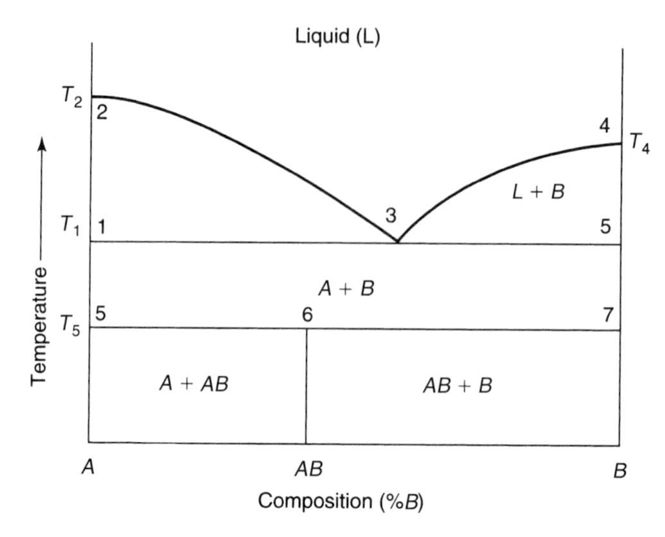

Figure 3-59 Binary phase diagram with an intermediate, dissociating intermetallic stable only over a limited temperature range. *Note*: T_5 represents the upper limit of stability upon heating.

Intermetallics are stoichiometric compounds with characteristic metal atom ratios, for example, CuZn, AlLi, Ni_3Al, Al_3V, AlSb, Mg_2S, and Ti_3Al.

These compounds follow the rules set forth in ***stoichiometry*** (pronounced "stoy-key-om-eh-tree"), the calculation of the quantities of chemical elements or compounds involved in chemical reactions. Chemical equations balance the starting materials (reactants) needed to produce a certain amount of product. In addition, it permits determining the cost of the final product from the cost of the starting materials by allowing one to calculate the amount of starting materials necessary to obtain the correct amount of product. In ceramic materials, stoichiometry is used to define the ionic state for ionic compounds, wherein there is an exact ratio of cations to anions as predicated by the chemical formula. A ceramic is nonstoichiometric if there is any deviation from this exact ratio. Using calcium chloride as an example, each calcium ion (cation) has a +2 charge (Ca^{2+}), and associated with each chlorine ion (anion) is a single negative charge Cl^{-1}. Thus, there must be twice as many Cl^{-1} anions as there are Ca^{2+} cations in the compound calcium chloride, which has the formula $CaCl_2$. The positive-negative attraction is what holds the ions together. In this compound a Ca atom gives up its two outermost electrons, transferring one electron to each chlorine atom. The calcium atoms become the positively charged Ca^{2+} ions (cations). Each chlorine atom has an extra electron, producing a negatively charged chlorine Cl^{-1} ion (anion). In other words, a stoichiometric ceramic is one in which all crystallographic lattice positions are filled according to the normal chemical formula. The chemical formula of a compound indicates the ratio of cations to anions, or the composition that results in the balancing of the charges. In general, intermetallics possess strong ionic/covalent bonding, which gives them nonmetallic properties such as poor ductility and poor electrical conductivity. An intermetallic might contain but a single component. Therefore it is represented, on a phase diagram, as a vertical line and labeled by a chemical formula such as NiAl or MgSn.

Congruent phase changes occur when one phase changes into another phase at constant temperature (isothermal) and without any change in chemical composition. Pure metals and some intermediate compounds change phase congruently. In Figure 3-58a, the congruently melting compound is a solid from room temperature up to its melting point (T_5). In other words, a congruent intermetallic melts immediately once it reaches the liquidus line. An ***incongruent*** melting compound (Figure 3-58b) would not melt directly but would decompose into a liquid plus one of the pure metals. In ceramics some compounds are not stable through the complete temperature range. In other words, the compound may form only over a limited temperature range (Figure 3-59). The word ***dissociating*** is used to describe these compounds. Both congruent and incongruent compounds are formed in ceramics.

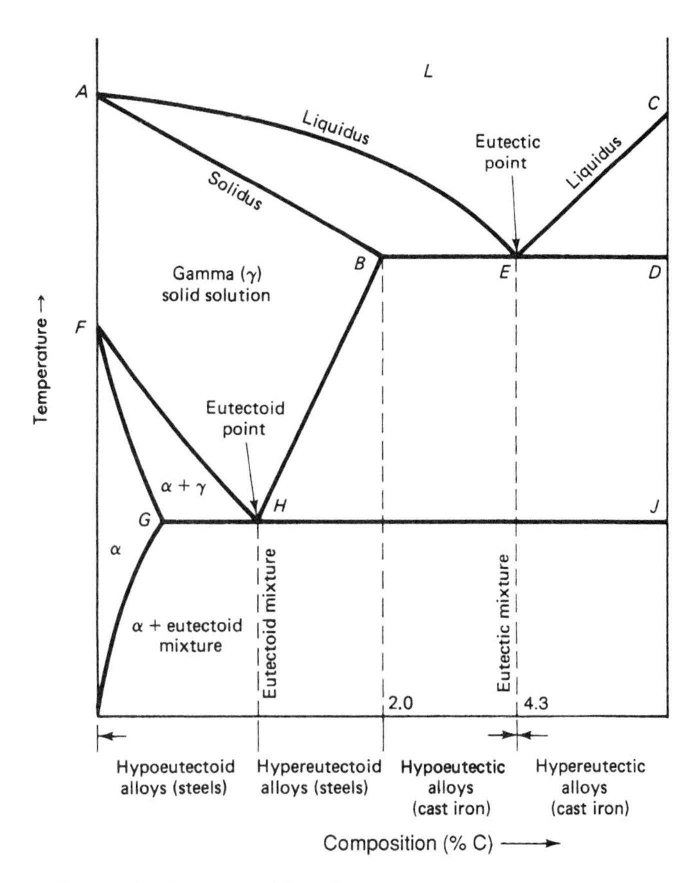

Figure 3-60 Phase diagram showing a eutectoid reaction.

One last solid-state reaction will be discussed, the ***eutectoid*** reaction, which takes place within the solid state. The suffix-***oid*** means "resembling" or "like." Therefore, the eutectoid reaction is like the eutectic reaction. As you recall, the eutectic reaction involves the transformation of a liquid phase into two solid phases ($L \rightleftharpoons S_1 + S_2$). The eutectoid reaction then is the transformation of a solid phase (S_1) into two new solid phases (S_2 and S_3). An appropriate equation describing this reaction is

$$S_1 \rightleftharpoons S_2 + S_3$$

Figure 3–60 shows the many similarities of this type of reaction to the eutectic reaction (see Figure 3-57). In industry the eutectoid reaction is most important because it forms the basis for heat treatment of many metallic materials. Point E is the eutectic point; point H is the eutectoid point located well below it in the solid region of the diagram. As shown, the gamma solid solution at point H transforms into the eutectoid mixture, made up of two solid phases. The prefixes ***hypo***- and ***hyper***- are attached to particular compositions of metals on either side of the eutectoid mixture to express their relation to the mixture in a fashion similar to the way they are related to the eutectic composition. This very crude diagram will be refined and studied in more detail in Module 9, where the diagram will be called the iron–iron carbide phase diagram—the heat treater's main road map to the processing of steel and cast iron.

3.3.6 Multiphase Metal Alloys

So far our discussion has involved solute and solvent atoms that have the same type of crystal structure and thus form relatively pure ***homogeneous*** mixtures or solutions. Now we advance

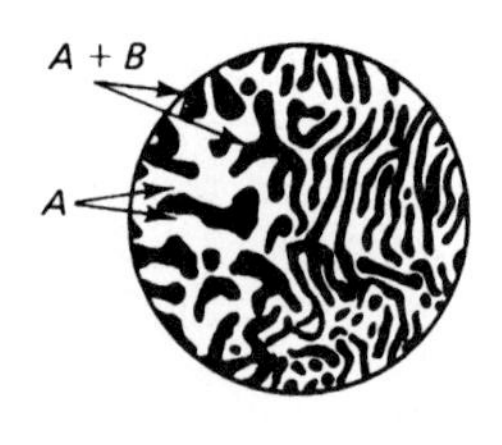

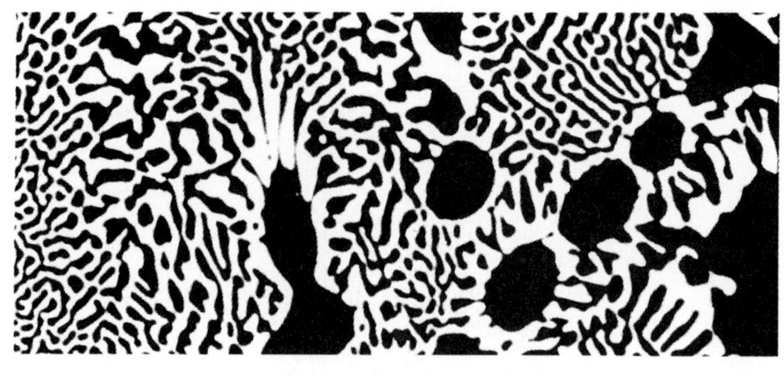

Figure 3-61 Microstructure of a multiphase metal alloy. Two different compositions (*A* and *A*+*B*) in the solid state. 65 Al–Cu alloy as cast X250, $Fe(NO_3)_3 \cdot 9H_2O$. (Buehler Ltd.)

to alloys classified as mixtures of more than one phase that differ from each other in their composition or structure. Under the microscope, the boundaries between phases can be seen in some metal alloys (Figure 3-61). Solder is an example of this type of multiphase metal alloy, with the metals tin and lead present in two separate phases. Steel is a more complex multiphase mixture of different phases, some of which are solid solutions. There are many more ***multiphase metal alloys*** than single-phase solid solution alloys, due primarily to the greater flexibility of their properties, which arises from the fact that they have more variables that can be controlled to produce differing properties. In other words, the properties of multiphase metal alloys such as steel depend not only on the phases present but on the structure, amount, shape, distribution, and orientation of these phases (Figure 3-61). The ability to control these phases allows us to produce desired properties in metals. For example, low-carbon ductile steel is used for auto bodies and harder and more brittle steel is used for metal files.

3.3.7 Phase or Structural Transformations

Just how can we bring about these transformations to produce a metal alloy with desirable properties? First, let's consider the atomic world once more. We know that a few metal atoms can change their structural arrangement as a result of a change in temperature. Remember the name of this phenomenon? *Allotropy*. We know that there is a definite limit of solubility of one solid material in another and that this limit also depends on temperature. If this limit is reached, we see atoms coming out of solution to form a multiphase metal alloy. Remember the name of this phenomenon? *Precipitation*. We know that atoms can move about in a solid just as cigarette smoke moves in a room, but certainly not as fast. Remember this phenomenon? *Diffusion*. And the rate at which the atoms move varies with the temperature. At higher temperatures the atoms move at faster rates because they have more energy. What energy? *Thermal energy*. Where does this energy come from? From an external source of heat. The more heat, the more the energy and the greater the movement. By now you should have a pretty good idea about the answer to the next question. In looking over the various ways that we can change atomic structure in a metal alloy, what is the one common quality or characteristic in all of them? Yes, it is thermal energy.

3.4 SINGLE-CRYSTAL CASTINGS

Ordinary metal castings produce polycrystalline structures. Because impurities gather at grain boundaries, these many crystals or grains of random orientation possess weak atomic bonds across the boundaries. Techniques to improve across-boundary strength have included improved mold material, alloying, and producing refined ***equiaxed*** castings. Then casting technology undertook to control the direction of grain boundaries and finally eliminated them for certain applications. ***Directional solidification*** (Figure 3-62) produces long grains growing continuously and parallel from one end of a casting to the other, with the longitudinal growth

Figure 3-62 Photograph of metal castings of turbine engine airfoils showing the progress in grain control from equiaxed (right), to directionally solidified (left), to monocrystal (center). (Howmet Turbine Components Corporation)

designed to be in the direction that will withstand greatest stress. The next technological development involved the selection of one grain and the control of its solidification into a ***single-crystal*** casting (***monocrystal***, Figure 3-62). The monocrystal eliminates the need for boundary strengtheners, which depress melting points and limit high-temperature use; allows high treatment temperatures; and improves high-temperature corrosion resistance. The monocrystal turbine blade seen in Figure 3-62 represents the growing acceptance of single-crystal castings for high-temperature use. Turbine blades for the main engines of the space shuttle were originally cast as directional solidified and then as single crystal.

3.5 STRUCTURES RESULTING FROM NONEQUILIBRIUM CONDITIONS

A new breed of material, Metglas, developed in the late 1970s by Allied Corporation, is an amorphous metal alloy. It is produced by bringing a molten metal alloy at about 1000°F into contact with rapidly moving and relatively cool substrate (or chill block) in a continuous casting process. The drastic quenching operation is similar to that used in conventional solid-state thermal processing of metals such as steel to transform austenite to martensite. It differs because the cooling is so rapid; for most metallic glasses the minimum rate is 10^5 kelvin/second (K/s). To ensure a uniform cooling rate at all points within the metal, one dimension of the liquid layer must be kept to a minimum. Thus, foil, wire, or powder are the forms currently produced. The magnitude of the cooling rate (about 1 million degrees per second) allows but 1 millisecond (ms) for the metal to solidify. The end result is that atomic diffusion is prevented, which in turn precludes any nucleation and growth of crystals. A glassy state consisting of one chemically homogeneous phase, with atoms packed in a random arrangement similar to that of glass or liquid metal, is the result. Alloy compositions produced by this process consist of transition elements such as Fe and Ni with small percentages of metalloids (B, C, Si, P). Also, mixtures of transition elements such as B and Ni are common. Initial applications of this material are in vacuum furnace brazing of engine parts for turbines, in which the older, less desirable brazing transfer tape is replaced by Metglas brazing foils. The ductility of Metglas foil is greater than that of transfer tape. It is capable of being folded back on itself without fracture and this permits it to be converted to various preforms as well as punched to exact shape to conform with various brazing-joint designs.

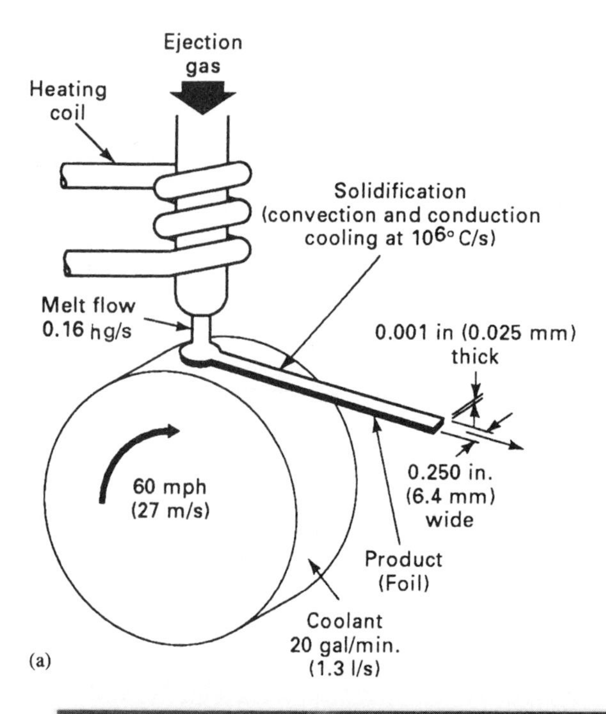

Figure 3-63 (a) A jet of molten metal is solidified rapidly when it strikes a liquid-cooled rotating drum. (b) This striking photo shows a melt-spinning device at the National Institute of Standards and Technology. Glowing metal, melted by coils at top right, hits a whirling wheel and flies off as rapidly solidified ribbon. Time photography produced multiple images. (Photo courtesy of NIST)

Many new techniques have been developed since the introduction of Metglas. Known by the name ***rapid-solidification processing***, these techniques employ large departures from local equilibrium in the formation of homogeneous structures. Powders for rapid solidification are produced mainly by *inert gas atomization* and *centrifugal atomization*, both of which produce spherical particles in the size range 20 to 100 μm in diameter by employing cooling rates of up to 10^6 K/s. Other processes are *melt spinning* and *self-quenching*, which produce ribbons/filaments that are pulverized into gritlike powders (Figure 3-63). Powder consolidation can be accomplished by *hot isostatic processing (HIP), hot extrusion, or dynamic compaction.* The highly alloyed powders are sintered and pressed into bulk form. In some processes the powders are pressed into "green" compacts and sintered without ever coming into contact with the air (see Sections 12.1 and 20.1).

The materials produced by rapid solidification are finding applications in many areas, including as reinforcing filaments in ceramic-matrix composites and as surface-modification materials (surface alloying) for hard facing of tools, dies, valve seats, and turbine blade tips. This new technology makes it possible to produce alloys that could never be produced with conventional ingot-casting techniques. By adjusting the cooling rate, a variety of new materials can be produced with highly desirable mixes of properties. Amorphous magnetic materials with atomic structures that impose no obstacles to the movement of magnetic domain boundaries, low hysteresis losses, and high permeability (see Section 6.3) have been developed for various electronic devices such as sensors and motors. New fine-grained homogeneous crystalline materials (Al and Mg alloys) for use as tough tool steels and nickel-based superalloys are now available. The ferrous amorphous metals produced by the new technology are "soft" magnetic materials, meaning that they possess no preferred orientation of crystals and hence no easy direction of magnetization. In practical terms, these new materials have a vast potential for use as metal cores in electrical transformers, in which billions of kilowatt-hours are lost in overcoming the resistance of the presently used silicon metal alloy cores to changes in the direction of magnetization. Other uses for such a family of materials are being developed by a multitude of industrial concerns. An amorphous metal alloy called Liquidmetal is a recent alloy said to be twice as strong as steel. It is composed of Ti, Cu, Zr, and Be. The atoms of these elements are of different sizes, so they do not readily form crystals. However, they can join to produce an alloy up to an inch thick that will not shrink when it solidifies and can reportably be cast with a precision down to 1 micrometer. High-elastic-modulus and corrosion-resistant cast alloy VIT-001 with a zirconium base has a yield strength of 1900 MPa compared to 860 MPa for cast 17–4 stainless steel.

Materials Science and Technology Gateway website with links to many other sites—*http://MST-Online.nsu.edu*>

3.6 LABORATORY ACTIVITIES AND CLASSROOM DEMONSTRATIONS WITH PROCESSING AND STRUCTURE

The *Experiment in Materials Science, Engineering and Technology (EMSET2) CD-ROM* provides numerous experiments and demonstrations related to processing and structure as a result of a cooperative effort by scores of educators and materials specialists from industry. See Prentice-Hall's website for ordering information on the latest edition of *Experiments in Materials Science (CD-ROM)*, ISBN 1–13–030534 at *www.prenhall.com*

APPLICATIONS & ALTERNATIVES

Understanding the natural structure of materials permits the manipulation of structures to achieve results different from those that would normally occur. Normally, metals must be heated to the liquid state before they can be formed into complex shapes. Less complex shapes can be achieved by heating metals to a point in which they become ***plastic***, or formable. These processes include extrusion, drawing, bending, and so on. Some softer metals and thinner, harder metals can be cold-formed at room temperature.

Thixotropic ("thicks-oh-tropic") ***metals*** exist in a semisolid state; they liquefy while under shearing stress and then solidify when left standing. Jello is semisolid but, of course, much softer

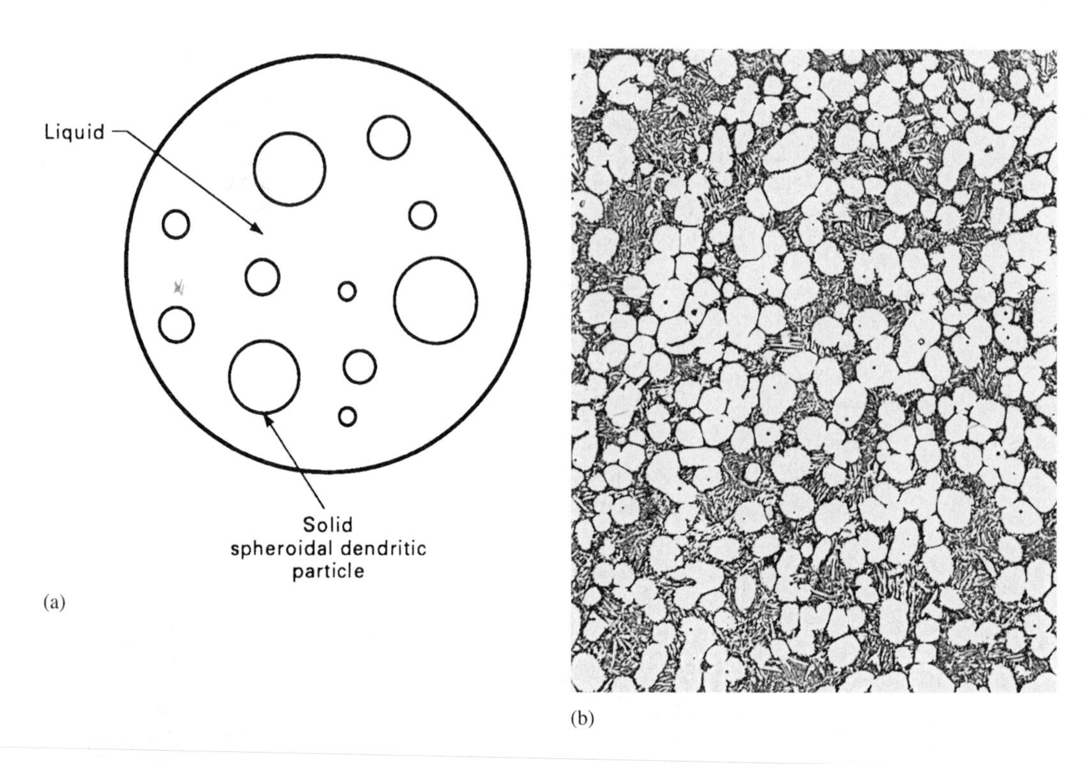

Figure 3-64 Semisolid, thixotropic alloys form nearly spheroidal dendrites. (a) Diagram of SSM structure. (b) Photomicrograph reveals how continuous stirring during solidification of semisolid material (aluminum casting alloy) results in a microstructure of large equiaxed solid particles within the formerly liquid eutectic mixture. The solid particles agglomerate during the SSM processing, affecting flow behavior. (ASM International)

than a thixotropic metal. It is possible for metal alloys to enter the thixotropic state by stirring metal vigorously as it cools from the liquid state en route to the solid state. When metal cools under ***equilibrium*** conditions (slow, natural change from liquid to solid), as shown in Figure 3-64, crystals form: first, by ***nucleating*** (seed crystals begin to form), then with further cooling growing to form ***dendrites*** (tree-shaped structures), and finally growing into ***grains*** or ***crystals***. If the metal is stirred vigorously while in a state between liquid and solid (much like the mush of melting snow, which is part ice crystals and part water), the dendrites will form spherelike shapes (Figure 3-65), rather than the normal treelike dendrites, to produce a ***semisolid material (SSM)***.

 Thixotropic injection molding processes with image gallery—*http://www.thixomat.com*>

Casting and ***molding*** are shaping operations in which the starting material is a heated liquid or semifluid. When referring to metals, the term ***casting*** is normally used, whereas ***molding*** is the common term used for plastics. Both processes are used in glassworking. Figure 11-18a shows the sequence of operations in the investment casting of titanium alloys. ***Injection molding***, shown in Figure 3-65, utilizes a screwing action to force thixotropic magnesium through a nozzle into a mold or die made of very hard tool steel or similar die stock. The mold has a cavity formed in the reverse shape (female) of the part to be produced. Although injection molding has been often used for molding plastic and ceramic parts, but not metals, Dow Chemical developed a process for the injection molding of magnesium parts made possible through the use of thixotropic alloys.

In Figure 3-65, you will notice that cold magnesium pellets (in a thixotropic state) are fed through a hopper into an argon atmosphere, then into the heated barrel of the injection molding machine. Argon is an ***inert gas*** and, as such, protects the magnesium from oxygen in the factory atmosphere. The shearing action of the screw, acting on the magnesium pellets, transforms them into a thixo-extrudate at 580°C. When enough of the extruded magnesium slurry (solid and liquid) collects in the heating zone chamber, it is injected by the force of the high-speed shot system, which pushes against the thrust bearing, causing the reciprocating screw to inject slurry through

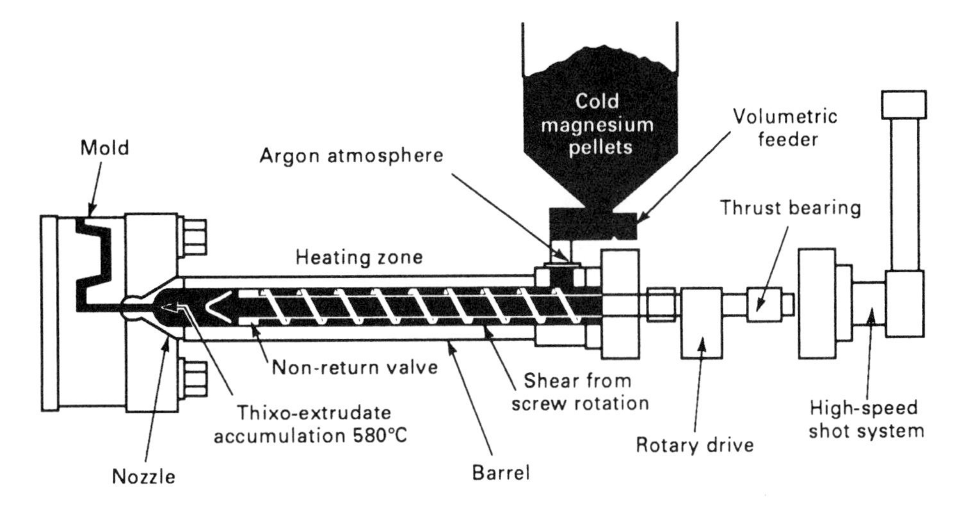

(a)

(b)

Figure 3-65 (a) Dow Chemical's thixotropic magnesium injection molding process. (b) SSM magnesium alloy parts formed by thixotropic injection molding. (ASM International and Thixomat Corp.)

the nonreturn valve past the nozzle into the mold cavity. The molded SSM part is equivalent to a part produced through melting and casting in the liquid state. The injection process offers advantages over die casting: improved mold cycle time, less downtime, and improved part quality.

These processing techniques illustrate the desire to apply technology to provide an end product of the highest quality with as few steps as possible. In this module, we have provided information on the basic structure of materials and how structure is altered by processes and synthesis. The quest for improved techniques that can form and treat materials to produce the ideal material for the perfect product or system is never-ending. It requires constant study to keep up with new developments. This module can help prepare you to deal with today's materials and processes and to adapt to coming changes.

SELF-ASSESSMENT

3-1. What type of dendrites form in thixotropic metals?

a. Treelike

b. Spheroidal

c. Bcc

d. Fcc

3-2. The ability of a material to exist in different space lattices is called

a. Allotropic
b. Crystalline
c. Solvent
d. Amorphous

3-3. Contemporary synthesizing/processing capabilities are best at which level of the materials technology scale?

a. Atomic (0.1 nm)
b. Macro (cm–m)

3-4. Amorphous metals develop their microstructure as a result of ___________.

a. Dendrites
b. Directional solidification
c. Slip
d. Extremely rapid cooling

3-5. In an alloy, the material that dissolves the alloying element is the ___________.

a. Solute
b. Solvent
c. Matrix
d. Allotrope

3-6. Applications & Alternatives focused on changing structure through processing. Which processes produce improved equiaxed grains?

a. Investment casting
b. Shell molding
c. Thixotropic molding
d. Injection molding

3-7. What advantage comes from equiaxed grains?

a. Lower processing costs
b. Uniform strength
c. Directional strength
d. Reduced density

3-8. What distinguishes a cooling curve for a metal from a cooling curve for a nonmetal?

a. Phase transformations occur with nonmetals.
b. No phase transformations occur with nonmetals.

3-9. What is the volume of a piece of iron having a mass of 864 g?

a. 10.8 cm^3
b. 109.78 cm^3
c. 20.8 cm^3
d. 200.6 cm^3

3-10. What is the coordination number (CN) for the fcc structure formed by ions of sodium and chlorine that in the chemical compound NaCl?

a. 6
b. 8
c. 14
d. 16

3-11. What pressure is normally used in constructing a phase diagram?

a. 100 psi
b. Depends on material
c. Ambient
d. Normal atmospheric pressure

3-12. What line on a binary diagram indicates the upper limit of the solid solution phase?

- **a.** Liquidus
- **b.** Eutectic
- **c.** Eutectoid
- **d.** Solidus

3-13. A ceramic material cannot be represented by a phase diagram.

- **a.** T (true)
- **b.** F (false)

3-14. What are the Miller indices for the basal plane (base of the box) in Figure 3–15?

- **a.** (0001)
- **b.** (001)
- **c.** (111)
- **d.** (101)

3-15. What is the crystal direction for a line from the origin passing through point A in Figure 3-23?

- **a.** [110]
- **b.** [222]
- **c.** [120]
- **d.** [001]

3-16. What holds the atoms (ions) together in a compound such as NaCl are electrostatic forces between ___________.

- **a.** Atom and ion
- **b.** Covalent bonds
- **c.** Electrons and nuclei
- **d.** Neutrons

3-17. What is the CN for a cph unit cell?

- **a.** 5
- **b.** 8
- **c.** 12
- **d.** 24

3-18. What is the magnitude of the highest atomic-packing factor when atoms are of the same size?

- **a.** 0.74
- **b.** 0.81
- **c.** 21
- **d.** 81

3-19. How many directions of slip are there in cph systems?

- **a.** 3
- **b.** 4
- **c.** 5
- **d.** 6

3-20. Single-phase materials have a single-crystal structure.

- **a.** T (true)
- **b.** F (false)

3-21. Single-crystal structured materials are single-phase materials.

- **a.** T (true)
- **b.** F (false)

3-22. Diffusion of atoms through a solid takes place by two main mechanisms. One is diffusion through vacancies in the atomic structure. Another method of diffusion is ___________.

- **a.** Cold
- **b.** APF
- **c.** Substitutional
- **d.** Interstitial

3-23. Two possible substances formed by alloying two or more metals are solid solution and ___________.

a. Substance
b. Interstitial
c. Mixture
d. Compound

3-24. Traditional ceramics have poor tensile strength due to what?

a. Ionic bonding within grains
b. Metallic bonding within grains
c. Poor bonding across grain boundaries
d. Good bonding across grain boundaries

3-25. Grain boundaries ___________ movement of dislocations through a solid.

a. Improve
b. Inhibit
c. Do not affect

3-26. Iron can be alloyed with carbon because it is ___________ .

a. Crystalline
b. Amorphous
c. A mixture
d. Allotropic

3-27. Metals can be cooled only to crystalline solids.

a. T (true)
b. F (false)

3-28. Solid crystals that do not form unit cells as do regular crystals are ___________.

a. Allotropes
b. Twins
c. Quasicrystals
d. Alloys

3-29. A single crystal of zinc, like many other single-metal crystals, is fairly ductile. There is nothing to prevent slippage of planes of atoms over one another. Explain why some of these metals, like zinc, become brittle when they are in the polycrystalline state, although they are ductile in single-crystal form.

3-30. The atomic radius of copper is 1.278 10^{-10}. Express this radius in terms of nanometers (nm) and in inches (in.).

3-31. How many degrees do the angles α, β, and γ measure in the cubic axes system.

3-32. Sketch an fcc unit cell.

3-33. Using Table 3–2, determine the crystal structure of lead.

3-34. Obtain a rubber band. Stretch it as far as it will go without breaking while holding it in contact with your moistened lips. Observe the temperature change. Hold the band in this stretched condition for about 30 seconds. Note any change in temperature. Release the band suddenly to its original, unstretched length and touch it to your lips. Note the temperature change. Rubber is an elastic polymer when stressed. The stressed molecules align themselves and local crystallization occurs. When the stress is released, the molecules return to their original arrangement. Determine if energy is absorbed or released when an amorphous material crystallizes. Record all your observations and comments.

3-35. Using Figure 3-24, sketch the isosceles right triangle described in the accompanying text, label its sides, and using the proportional method derive the equation $(r = R)^2 = 2R^2.r + R = R\sqrt{2}$.

3-36. Metals are classified as crystalline materials. Name one metal that is an amorphous solid and name at least one recent application in which its use is saving energy or providing greater strength and/or corrosion resistance.

REFERENCES & RELATED READINGS

ASM INTERNATIONAL, www.asminternational.org

CALLISTER, WILLIAM D., JR. *Materials Science and Engineering; An Introduction,* 6th ed. New York: John Wiley and Sons, Inc., 2003.

CZERWINSKI, FRANK. "Injection Molding Magnesium," *Advanced Materials and Processes*, November, 2002, pp. 31–33, 2002.

DAVIS, J. R., ED., *ASM Materials Engineering Dictionary.* Materials Park, OH: ASM International, 1993.

DREXLER, K. ERIC. *Nanosystems: Molecular Machinery, Manufacturing, and Computation.* New York: John Wiley & Sons, Inc., 1992.

FREDERICK, P. S., N. L. BRADLEY, AND S. C. ERICKSON. "Injection Molding Magnesium Alloys," *Advanced Materials and Processes*, November 1988, pp. 53–56.

MALLARDI, JOSEPH L. *From Teeth to Jet Engines.* Howmet Corporation, 1992.

MCLEAN, D. *Mechanical Properties of Metals.* Melbourne, FL: R. E. Krieger, 1977.

MEYYAPAN, MEYYA. "Nanotechnology: Opportunity and Challenges," National Educators' Workshop [NEW:Update 2002], San Jose, CA, October 13, 2002.

NANTOTECHNOLOGY WEBSITE—http://mrsec.wisc.edu/nano

O'BRIEN, STEPHEN P. "A Chemical Perspective of Strategy and Design of Strategy Nanoscle Materials: The Science Behind Nanotechnology," National Educators' Workshop [NEW:Update 2002], San Jose, CA, October 16, 2002.

SEMTC. *Navy Manufacturing Letter*, February 1995, p. 6.

Periodicals

Advanced Materials & Processes

Journal of Materials Education

UNIT 2

Properties of Materials

PAUSE & PONDER

This unit provides an introduction to properties of materials and to the environments these properties must counteract in order to meet their service needs. How does one tell the difference among the materials that are used to produce the many products we use in our daily lives? The kitchen table, the fishing rod, the automobile tire, or the roof of the house in which we live—each material has a list of identifying features by which it can be identified. These features are known as properties or characteristics. Producing a myriad of materials with the many different properties required by society demands constant application of new science and technology. Many, if not most, of the new materials that are engineered, as described in Module 1, take on their new properties as a result of the application of heat or cold, which, in turn, changes the material's atomic structure. Advances in technology provide new tools to materials engineering and manufacturing and improve the confidence with which the limits of failure or product reliability of products can be predicted. An inability to do so in this world economy means losing markets for products and consequent failure in the marketplace.

An example of an adoption of a new technology or the use of technology in new ways is cryogenics. ***Cryogenics*** is the science that deals with the production of very low temperatures and their effects on the properties of materials. Various objects are being treated cryogenically—golf balls, brass instruments, cutting tools, firearms, and electronic circuit boards. When subjected to a temperature of −320°F, which is the temperature of liquid nitrogen, the crystal (atomic) structure of these materials is changed such that its performance is enhanced. Cryogenics has been using materials science and technology with engineering materials for nearly 20 years in a limited way; its use with all types of materials is ongoing. For example, cryogenic surgery (cryosurgery) uses the same cold temperature of liquid nitrogen to selectively destroy human tissue by freezing it. Other uses of these cold temperatures are whole-body freezing and storage of human cells as a future source of new tissues or organs. Cryogenic processing has been applied to engine parts for race cars. As seen in the drawing, cryogenic treatments have resulted in reduced wear in bearings, brakes, camshafts, transmissions, and differentials. Cylinder blocks do not warp, piston rings seal better, piston rods exhibit less flexing, and brake rotors better resist cracking. The photo shows some of the wear-resistant and toughened parts.

Camshaft
Cryogenics: Reduce wear on lobes, bearings, reduce breakage.
Coatings: reduce wear on lobes, bearings.

Valve Train
Cryogenics: Valve Springs last longer, lose less spring force. Reduced rocker arm breakage, longer valve life, less seat erosion, less stem and insert wear. Reduce pushrod flex.
Coatings: Reduce temperature of valves and springs.

Cylinder Heads
Cryogenics: Reduce warping, increase resistance to detonation damage.
Coatings: Higher thermal efficiency, less heat loss.

Transmission-Differential
Cryogenics: Reduce wear and breakage on bearings, gears, shift forks, synchronizers.
Coatings: Reduce wear on gears.

Brakes
Cryogenics: Reduce wear on rotors and pads. Reduce rotor warping and cracking.
Coatings: Dissipate heat faster. Reduce heat absorption by calipers to keep brake fluid from boiling.

Engine Block/Crankshaft
Cryogenics: Reduce cylinder wear. Reduce block shift under racing loads. Reduce crank wear and breakage.
Coatings: Reduce wear on crank journals.

Controlled Thermal Processing, Inc.
Cryogenics 98 Coatings

Wheel Bearings
Cryogenics: Reduced bearing wear.

Connecting Rods
Cryogenics: Reduce rod breakage.

Pistons
Cryogenics; Less skirt wear, better ring sealing more resistance to detonation damage. Reduced wrist pin wear. Reduce distortion.
Coatings: Heat reflecting coating reduces crown temperature, reduces detonation. Lubrication coatings reduce piston skirt scuffing.

(a)

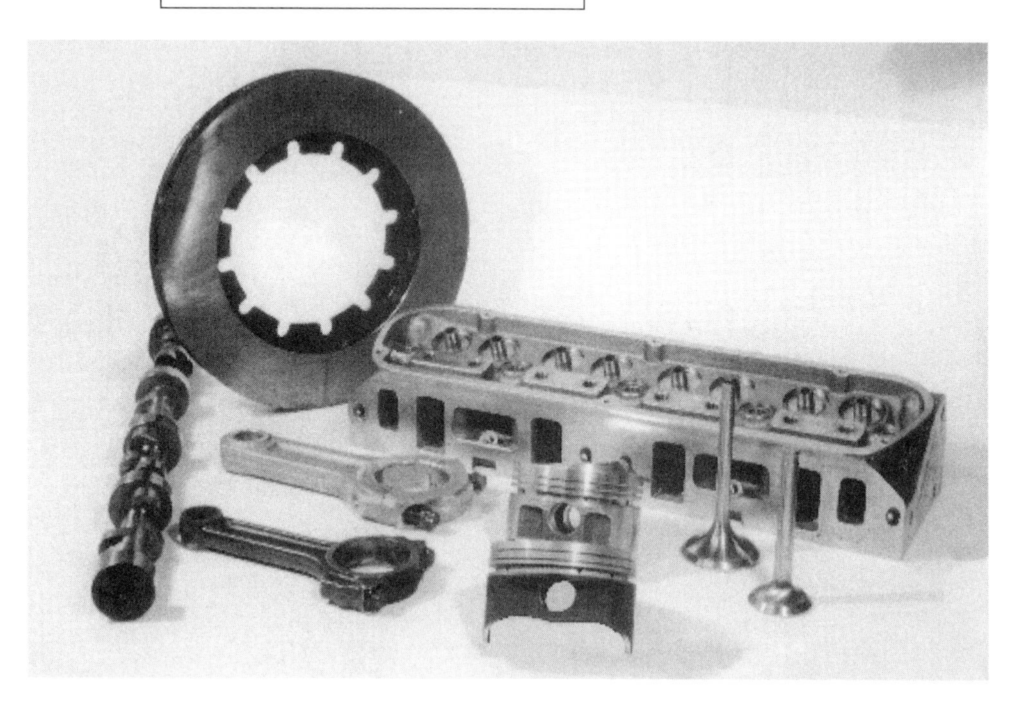

(b)

(a) Cryogenic processing coupled with coatings of engine parts in race cars has increased the life of high-performance race cars three to five times.
(b) Improved wear and tougher engine components allow them to endure increased stresses. (Courtesy Controlled Thermal Processing, Inc.)

Module 4

Mechanical Properties Defined

After studying this module, you should be able to do the following:

4–1. List and define the basic properties of materials. Differentiate among the many types of strengths that materials may possess.

4–2. Use diagrams, explanations, and calculations to determine physical and mechanical properties of materials.

4–3. Analyze service conditions (stresses and environments) to determine properties critical to the ability of parts or products to survive.

4–4. Define the relationship between stress and strength, ductility and stiffness, and forces and loads. Explain the differences between elasticity and plasticity and stress and strain.

4–5. Use diagrams, explanations, and calculations to determine shear, creep, and brittle failures.

4–6. Use the website addressed denoted by the * icon for the most current developments related to properties of materials.

American Society for Testing and Materials—*http://www.astm.org/*

*The symbol found throughout the book will link you to Internet sites related to topics being covered. The dynamic nature of the Web brings frequent changes, so some of these sites, while available at the time of writing, may not be up now.

4.1 PROPERTIES AND ENVIRONMENTAL VARIABLES

The properties of a material are those characteristics that help modify and distinguish one material from another. Taken as a whole, these qualities define a material. All properties are observable and most can be measured quantitatively. Properties are classified into two main groups, chemical and physical properties. ***Chemical properties*** are associated with the transformation of one material into another. Iron rusts when it combines with oxygen to produce an iron oxide through a chemical reaction. ***Physical properties*** involve no change in the composition of the material. Density, strength, and hardness are examples of such properties. Physical properties are, in turn, arbitrarily subdivided into many categories. These subdivisions bear names such as ***mechanical, metallurgical, fabrication, general, magnetic, electrical, thermal, optical, thermonuclear***, and ***electro-optical***. Regardless of the name, physical properties result from the response of the materials to some ***environmental variable***, such as a mechanical force, a temperature change, or an electromagnetic field (which includes all the radiation in the electromagnetic radiation spectrum). For purposes of this book, physical properties will be divided primarily into mechanical, thermal, electrical, magnetic, and optical.

4.2 MECHANICAL PROPERTIES

In selecting a material for a product, such as a piston in an internal combustion engine, a designer is interested in properties like strength, ductility, hardness, or fatigue strength. These are some mechanical properties of a material. ***Mechanical properties*** are defined as a measure of a material's ability to carry or resist mechanical forces or stresses. Additionally, mechanical properties of a material affect how the material can be worked. For example, very low hardness can be detrimental to a metal's machining performance. High ductility causes a metal to tear as the chip is formed, resulting in a poor finish. When selecting materials for many applications, mechanical properties rank highest in importance. Terms such as *durability*, *formability*, *rigidity*, *strength*, and *toughness* are often used to describe these properties. Three tests supply the most useful information for most applications—the tensile, impact, and hardness tests. Durability can be assessed in part by the hardness test, strength can be described by the results of tensile testing, and impact testing provides adequate data for determining the rigidity of a material.

When any matter is at rest, the atomic or molecular structure is in equilibrium. The bonding forces in this structure resist any attempt to disrupt this equilibrium. One such attempt may be an external force or load. Stress (see Section 4.2.1) results from ***forces*** such as tension, compression, or shear that pull, push, twist, cut, or in some way deform or change the shape of a piece of material. Often, this deformation is so minute that only delicate instruments can detect it. Figure 4-1 shows a universal testing machine used to apply loads to material specimens to detect minute deformation of materials under load. The standard tensile test, the measure of nearly a dozen important properties of engineering materials, produces experimental data needed to design structures and machines. The data are obtained by fracturing a specimen of material intended for a specific design project. ***Fracture*** or ***rupture*** occurs when a material or structure separates into two or more pieces. Most often this would be termed a ***failure*** of the material, defined as the inability of a material, structure, or component to perform its designed function. The failure criterion used in design is different for different materials. For concrete, the criterion of failure is crushing strength or compression. For elastomers, the failure criterion is the tear strength. For composite materials, it is tensile strength. For ductile metals in particular, the design criterion is yield strength. Many factors, such as the amount of cold working and strain hardening, affect metal strength. Work done on metals directly affects metal strength, producing a range of values for its yield strength. Metal in an annealed condition would exhibit a low value, while a strain-hardened metal would be approaching its tensile strength. Depending upon the design criteria, a failure may occur prior to fracture. Beams, such as floor beams, that are designed not to deflect beyond a standard amount under normal loads may be termed a failure if deflection exceeds this stated amount. In this case, no fracture

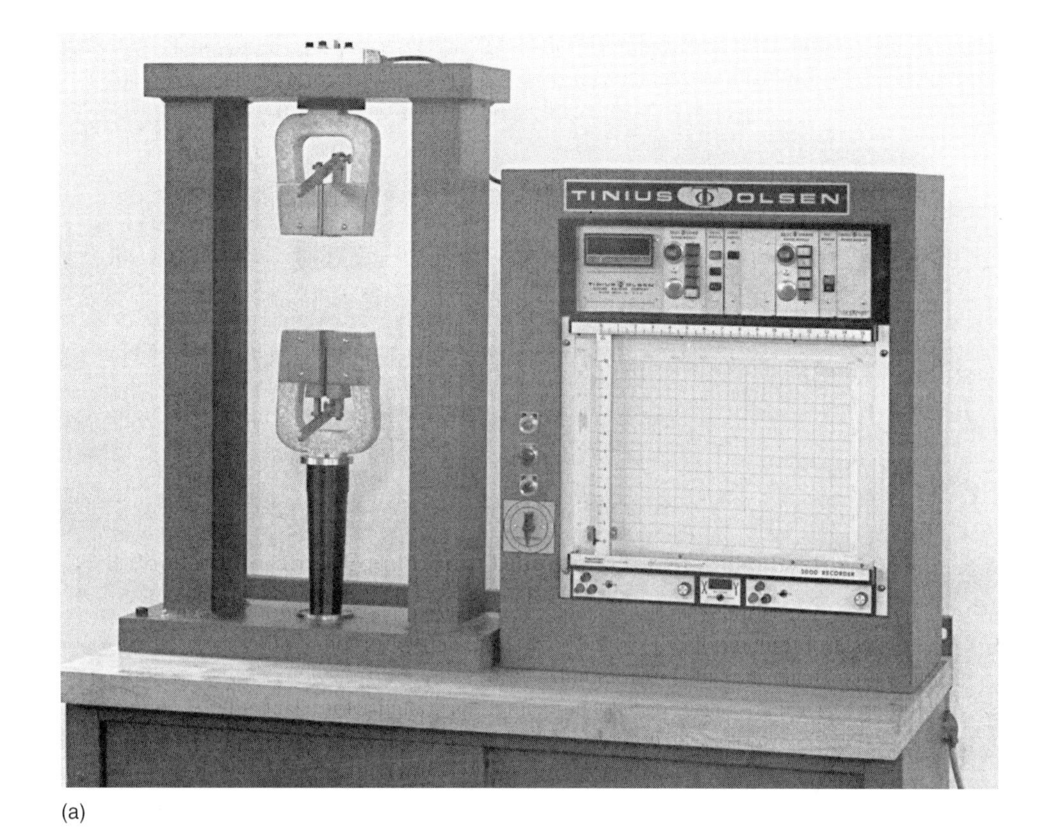

(a)

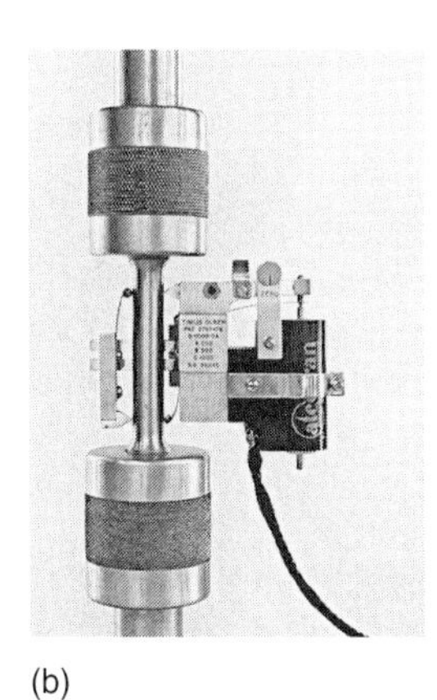

(b)

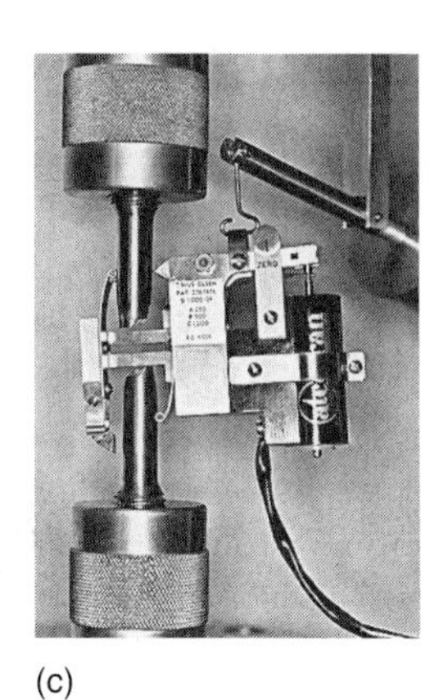

(c)

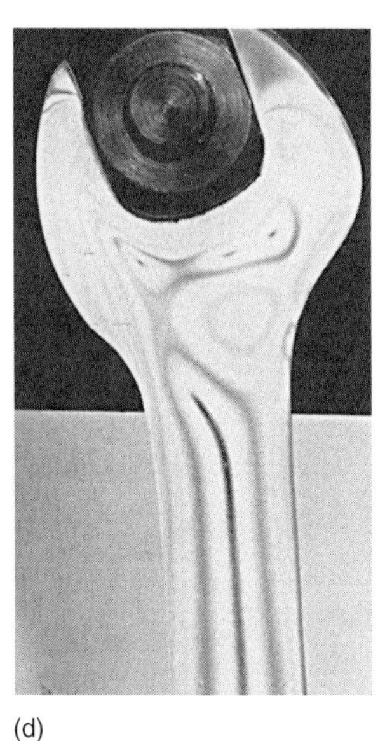

(d)

Figure 4-1 (a) Universal testing machine. The unloaded grippers would hold the specimen and strain indicators seen in (b) and (c). (b) Before stressed and (c) after breaking. (Tinius Olsen Testing Machine Co.) (d) Photoelastic model of wrench for analysis of stress. (Measurement Group Inc., Raleigh, NC)

has occurred. Similarly, an aluminum desk chair that collapses and buckles is a failure even though it has not fractured.

4.2.1 Stress

Stress, defined as the resistance offered by a material to external forces or loads, is measured in terms of the force exerted per area [pounds per square inch (psi)]. Normal stress is that applied perpendicular to a surface, that is, tension or compression. The corresponding SI units are newtons per square meter (N/m^2) or pascal (Pa). One pascal (1 Pa) equals 1 N/m^2. Another way of defining stress is to say that it is the amount of force (F) divided by the area (A) over which it acts. Using σ (the Greek letter sigma) as the universal symbol for normal stress, we say mathematically that

$$\sigma = \frac{F}{A}$$

The assumption is that the stress is the same on each particle of area making up the total area (A). If this is so, the stress is uniformly distributed. When a load or force is applied to an object, we are unable to measure directly the stress produced by this force in the material. What we do instead is measure the force, identify the area over which the force acts, and measure the area as well. These two quantities can then be used to calculate the stress produced in the material by the previous relationship. With the use of polarized light and models made of photoelastic plastic, it is possible to detect concentrations of stress, as seen in Figure 4-1d.

Illustrative Problem

Calculate the stress on a 10-ft circular rod with a diameter of 1 in. that is in tension due to a pulling force of 50 lb. Sketch the rod in two dimensions, showing the force and the area over which the stress is acting (to resist the external force). Express your answer in pounds per square inch (psi).

Solution

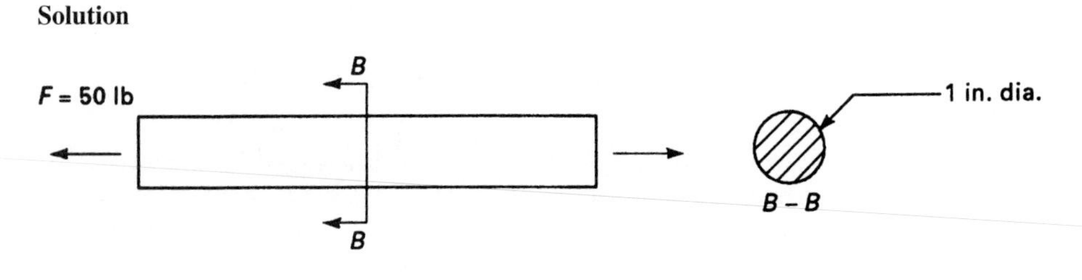

$$\sigma = \frac{F}{A} \qquad \text{area } A = \frac{\pi D^2}{4}$$

$$= \frac{50\text{ lb}}{0.785\text{ in}^2} \qquad = \frac{\pi(1)^2}{4}$$

$$= 63.7\text{ psi} \qquad = 0.785\text{ in}^2$$

4.2.2 Strain (Unit Deformation)

Strain, or ***unit deformation***, is defined as the unit change in the size or shape of material as a result of force on the material. Many times we assume that a solid body is rigid; that is, when the body is loaded with some force, the body keeps its same size and shape. This is far from correct. Regardless of how small the force, a body will alter its shape when subjected to a force. In other words, the body will change its dimensions. The change in a physical dimension is called ***deformation*** (δ, Greek letter delta). Figure 4-2a illustrates a rod of original length (l_0) and original diameter (d_0) placed under a tensile load (F) and elongated (stretched). Its change in length (Δl) equals the difference in the two dimensions ($l - l_0$) in the direction of the force. (In this example, Δl is exaggerated for illustrative purposes.) This is known as total deformation, δ (Greek lowercase letter delta).

The relationship between the total deformation and the unit deformation can be expressed mathematically. Using the symbols δ for total deformation (or Δl), l_0 for original length, and ϵ (Greek letter epsilon) for unit deformation, we can write the equation $\delta = \epsilon \times l_0$. Note that the original diameter has been reduced in size and produces a corresponding change in the lateral direction (at right angles to the direction of the load). Note too that in one instance one dimension increased and the other decreased. The change in the length is called a ***total axial*** or ***longitudinal deformation***. The change in the lateral dimension is known as a ***total lateral deformation***. The ratio of the total axial deformation to the original length is known as the ***unit axial*** or ***longitudinal strain***, ϵ. The linear units are not canceled and are kept as part of the term. Therefore, in mathematical terms,

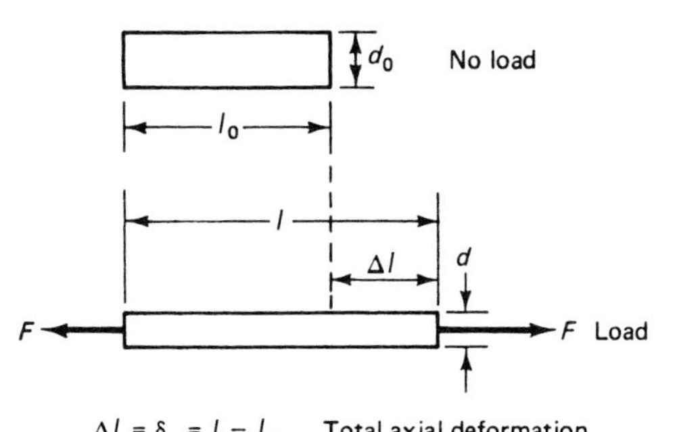

Note: Deformations exaggerated for illustrative purposes

(a)

W F
E E
G

E – gripped ends, may be threaded, plain, or with hole for gripping by machine
W – reduced width to insure specimen breaks in middle – round on round specimens and flat on flat specimens
G – marked gage length to precisely measure the change in length before, during and after test
F – fillet to reduce stress concentrations

(b)

Figure 4-2 (a) Rod under a tensile load. (b) Standard tensile test specimen or gauge.

$$\epsilon_{\text{long.}} = \frac{\Delta l \text{ mm}}{l_0 \text{ mm}} = \frac{\delta \text{ mm}}{l_0 \text{ mm}} \quad \text{(longitudinal unit deformation)}$$

$$\epsilon_{\text{lat.}} = \frac{\Delta d \text{ mm}}{d_0 \text{ mm}} = \frac{\delta \text{ mm}}{d_0 \text{ mm}} \quad \text{(lateral unit deformation)}$$

In summary, when a piece of material (a body) is subjected to a load, it will not only deform in the direction of the load (axial deformation) but it will also deform in a lateral direction (at right angles to the direction of the tensile or compression load). The ratio of the lateral unit deformation or strain ($\epsilon_{\text{lat.}}$) to the unit longitudinal deformation or strain ($\epsilon_{\text{long.}}$), given the symbol μ (Greek letter mu), is known as ***Poisson's ratio***:

$$\mu = \frac{\epsilon_{\text{lat.}}}{\epsilon_{\text{long.}}}$$

The unit longitudinal deformation is larger than the unit lateral deformation and therefore Poisson's ratio is less than 1. For steel, it is about 0.3.

We demonstrated our discussion of deformation by using a tensile load (Figure 4-2a). If a body was loaded in compression, the length of the body would have decreased and its width increased. One major reason for conducting tension and compression tests using standardized equipment and specimens or gauges (see Figure 4-2b) is to determine the data needed to plot stress–strain diagrams for the material under investigation so that other meaningful relationships can be determined.

4.2.3 Stress–Strain Diagrams and Hooke's Law

The stress–strain diagram is used to determine how a certain material will react under load. Figure 4-3 is a stress–strain diagram for a low-carbon (mild) steel. Strain values (mm/mm) are plotted along the horizontal axis (abscissa), and stress values (MPa, megapascal) are plotted along the vertical axis (ordinate). The straight-line portion of the diagram up almost to the yield point is known as the ***elastic region***. Within this range of stresses, the material will return to its original dimensions once the load, hence, the nominal stress, has been removed. (Nominal stress is the load divided by the original cross-sectional area. Actual stress can vary due to changes in area caused by lateral deformation.) In the elastic region, each increase in stress will produce a proportionate increase in strain, a phenomenon known as ***Hooke's law*** ($\sigma = E\epsilon$), where E is the constant of proportionality (the ***elastic modulus***—the word *modulus* means a ratio or constant of proportionality). The elastic modulus, also known as Young's modulus, can be obtained graphically by measuring the tangent of the slope angle in the elastic region of the stress–strain curve (see Figure 4-3). Figure 4-4 shows an ancient grape press made mostly of wood used in wine

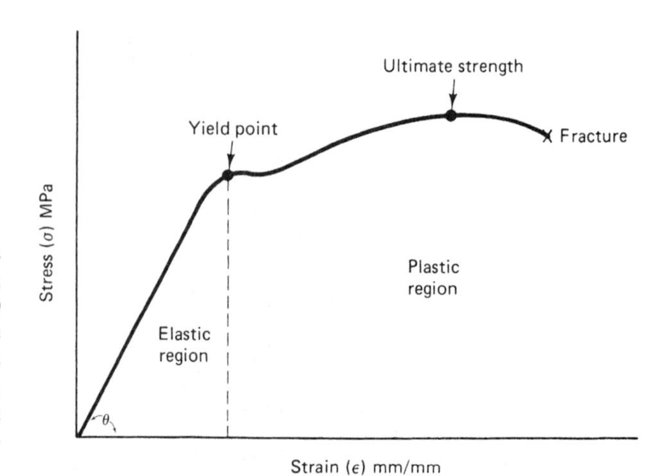

Figure 4-3 Engineering symbols for the sides of the triangle shown in the elastic region of the curve, θ (theta) is the slope angle, E is the modulus of elasticity or Young's modulus and is computed $E = \tan \theta$, and Δ is the symbol that means "a measurable change in" the stress–strain diagram.

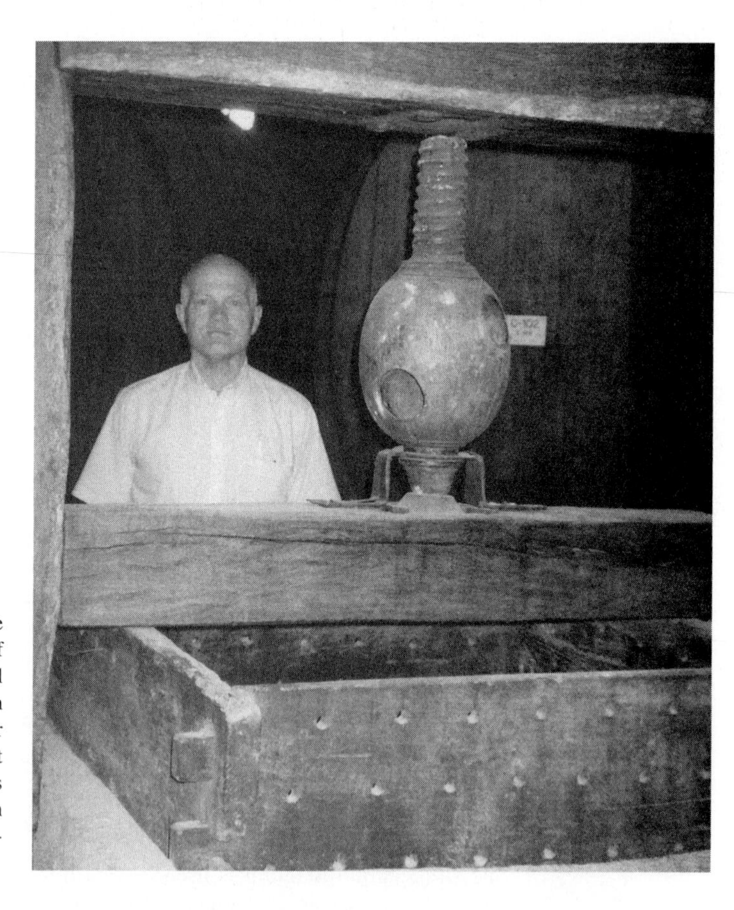

Figure 4-4 Ancient grape press exemplifies early use of wood for many structural components. With torsion force applied to the threader shaft with a handle (not shown) compressive forces transfer to the platen, which compresses the grapes to produce juice for wine making. Note the comparable size of the cross beams and side columns and the wooden turn screw. What type of stresses would the components undergo while the press compresses grapes?

Young's modulus or elastic modulus (E) is defined as the ratio of engineering stress (σ) to engineering strain (ϵ) in the linear or elastic region of the stress–strain diagram. It is also a measure of the interatomic bonding forces in a material. The higher the magnitude of these bonding forces, the stiffer the material or the higher the resistance of the material to being deformed. When a material is loaded with external forces in the elastic region, the material with the highest modulus value experiences the least amount of deformation or strain, which may be either elastic or plastic deformation. If the material reverts back to its normal size and shape upon removal of the load, it is ***elastic deformation***. If the applied force or load is removed and the material is permanently deformed (changed in shape), the material is said to have undergone ***plastic deformation***.

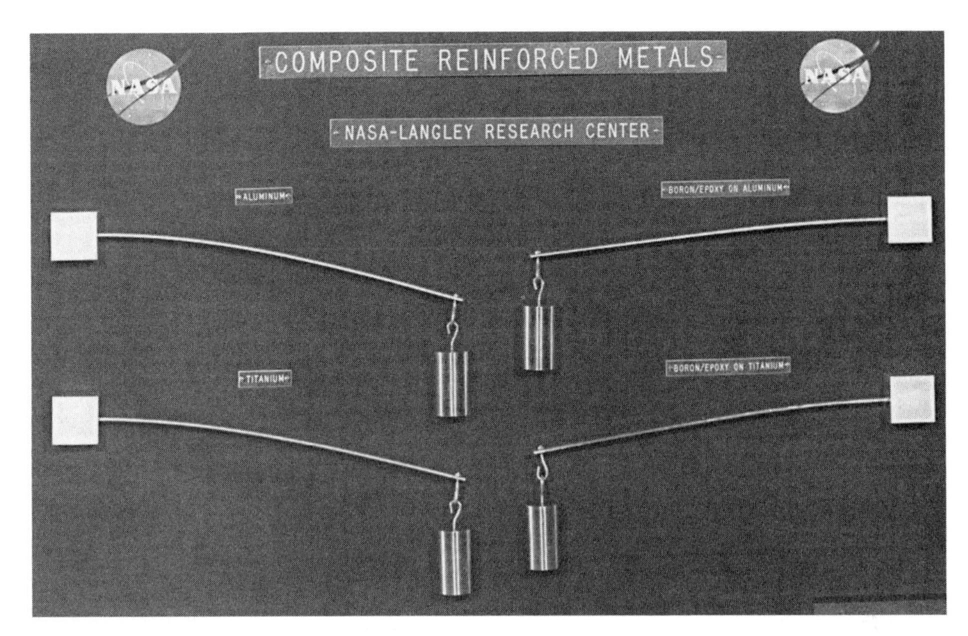

Figure 4-5 The stiffness of metals versus metal composites. (NASA)

Illustrative Problem

Using the same rod as in the Illustrative Problem in Section 4.2.1, determine how much the rod has stretched under the tensile force of 50 lb. Assume that the modulus of elasticity for the steel with which the rod is fabricated is 30×10^6 psi and the tensile force is below the yield strength.

Solution Using Hooke's law, $\sigma = E\epsilon$ and solving for ϵ, we find that

$$\epsilon = \frac{\sigma}{E} = \frac{63.7 \text{ psi}}{30 \times 10^6 \text{ psi}} = 2.12 \times 10^{-6} \text{ in./in.} \quad \text{and}$$

$$\delta = \epsilon \times l = 2.12 \times 10^{-6} \text{ in./in.} \times 10 \text{ ft.}$$

$$= 2.12 \times 10^{-5} \text{ in./in.} \times 1 \text{ ft} = 212 \times 10^{-5} \text{ ft}$$

Beyond the yield point, the material will continue to deform, but with less stress than before, because it has begun to yield. In this region, known as the ***plastic region***, plastic deformation takes place, and when the load is removed, the material will not return to its original dimensions. It now has a ***permanent set***. Note also that it takes less nominal stress to break the metal specimen than it does to reach the ultimate strength. Because the material has yielded, the original cross-sectional area of the specimen has been reduced in size so that less material is available to resist the load.

The ***yield point*** represents the dividing line or transition from the elastic to the plastic region of the curve. When the stress reaches the yield point, a large increase in strain occurs with little increase in stress. The ***modulus of elasticity, elastic modulus, tensile modulus, Young's modulus, modulus of elasticity in tension***, or ***coefficient of elasticity***, given the symbol E, is the ratio of the stress to the strain in the elastic region of the stress–strain diagram.* The ***tensile modulus*** is approximately equal to the ***compressive modulus of elasticity*** within the proportional limit (elastic limit of the diagram). Note that this ratio expresses the slope of the straight-line portion of the curve. Regardless of the name, this modulus is an indication of the stiffness of the material when subjected to a tensile load. The stiffness of a material is defined as the ratio of the load to the deformation produced. The higher the value of Young's modulus, the stiffer the material. As demonstrated in Figure 4-5, reinforced metal composites possess greater stiffness than nonreinforced metals.

Not all materials produce stress–strain diagrams on which there is a clear indication of the start of yielding as the load is increased (see Figure 4-6). Cast iron is an example.

*The SI unit for E is the pascal (Pa).

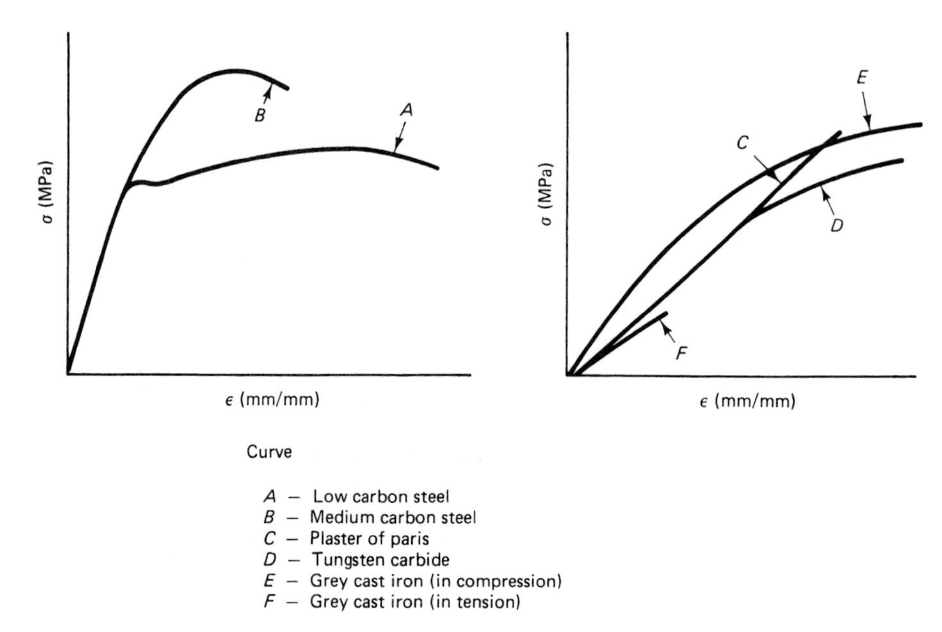

Figure 4-6 Typical stress–strain diagrams.

This situation should not be interpreted to mean that cast iron does not exhibit elastic properties under moderate loads. In other words, such materials are elastic, with strains returning to zero when moderate loads, hence stresses, are removed. The modulus of elasticity for these materials is sometimes taken as the slope of a tangent to the stress–strain curve at the origin.

4.2.4 Ultimate Strength or Tensile Strength

Ultimate strength or ***tensile strength*** is the maximum stress developed in a material during a tensile test. It is a good indicator of the presence of defects in the crystal structure of a metal material, but it is not used too much in design because considerable plastic deformation occurs in reaching this stress. For brittle materials, tensile strength is still a valid criterion. Most gray cast irons are specified by their tensile strengths. Plastic deformation is not all bad; however, in many applications the amount of plastic deformation must be limited to much smaller values than that accompanying the maximum stress. The ultimate shear strength (Section 4.2.7) is about 75% of the ultimate tensile strength.

4.2.5 Yield Strength

Yield strength is the strain corresponding to the elastic limit. It is the lowest stress at which plastic deformation occurs. For most design purposes, the yield strength is assumed to be the same in tension as in compression. Strain is measured at various points in the engineering stress–strain diagram. The strain corresponding to the elastic limit is called the ***yield point strain*** (ϵ_{yp}). The elastic limit replaces the yield point in those metals that do not show a yield point on the stress–strain diagram. Several face-centered cubic materials, for example, such as copper and aluminum, do not have a well-defined yield point. The stress at the yield point strain is the yield strength (F_{ys}). The strain corresponding to the tensile strength is called ***uniform strain*** (ϵ_u), because the strain to this point is uniformly distributed throughout the specimen or gauge cross section. The engineering strain at the fracture point (ϵ_f) may also be used to express ductility of the metal specimen or sample. Low-carbon steel is one of just a few materials that exhibit a point where the strain increases without an accompanying increase in stress, which poses a problem in deciding when plastic deformation begins for such materials. By agreement, a practical approximation of the elastic limit, called the ***offset yield strength***, is used. It is the stress at which a material exhibits a specified plastic strain. For most applications, a plastic strain of 0.002 in./in. can be tolerated, and the stress that produces this

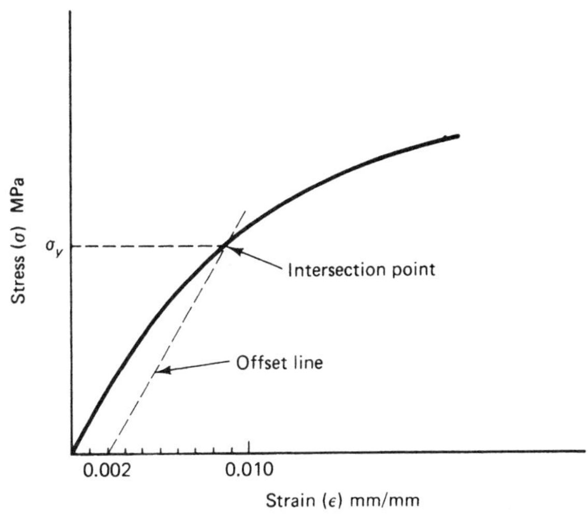

Figure 4-7 Determining offset yield strength using a stress–strain diagram.

strain is the yield strength, sometimes expressed as 0.2% strain. The yield strength is determined by drawing a straight line, called the *offset line*, from the 0.2% strain value on the horizontal axis parallel to the straight-line portion of the stress–strain curve. The stress at which this offset line intersects the stress–strain curve is designated as the yield strength of the material at 0.2% offset. In some cases the offset can be specified as 0.1% or even 0.5%.

Figure 4-7 shows a typical stress–strain curve for an aluminum alloy with no pronounced yield point. The offset, offset line, and point of intersection of the offset line with the stress–strain curve are shown. When reporting yield strength, care must be taken to include the amount of offset as well as the value of the stress at the intersection with the data. In general, the yield strength of metals is much higher than that of other materials. For brittle materials, yield strength differs very little, if at all, from tensile strength. As an example, for class 40 gray iron, both strengths are 40,000 psi. Ductile materials show a wide difference in these strengths. Figure 4-6 compares typical stress–strain curves for various materials. The yield strength in shear for a ductile material is determined to be about one-half (0.577) the yield strength in tension.

4.2.6 Resilience

The ***modulus of resilience*** (R), represented by the area under the straight-line portion (elastic region) of the stress–strain curve, is a measure of the energy per unit volume that the material can absorb without plastic deformation. If the unit of volume is 1 meter cubed, the SI units of resilience are MN · m/m^3. Figure 4-8 shows that this area is a right triangle.

4.2.7 Shear Stress

A second family of stresses is known as ***shear stress*** or ***shearing stress***. The symbol τ (Greek letter tau) represents a shear stress. A shearing force produces a shear stress in a material, which, in turn, results in a shearing deformation. Figure 4-9 shows a shearing load in action. In Figure 4-9a, the shear force, F, produces an angular deformation of a block of material, not a lengthening or shortening. (This deformation is exaggerated for illustrative purposes.) In Figure 4-9b, a block of material is subjected to a shear force that, if larger than the shear strength of the material, will shear a section out of the block. A hole in a metal plate can be produced by the action of a punch and hammer that delivers an impact blow, causing the metal in the plate to fracture by shear.

Shear strain, γ (Greek letter gamma), then, is the deformation (δ_s—read as "delta sub s") produced by the shear force F (see Figure 4-9a) divided by the dimension h, or

$$\gamma = \frac{\delta_s \text{ in.}}{h \text{ in.}}$$

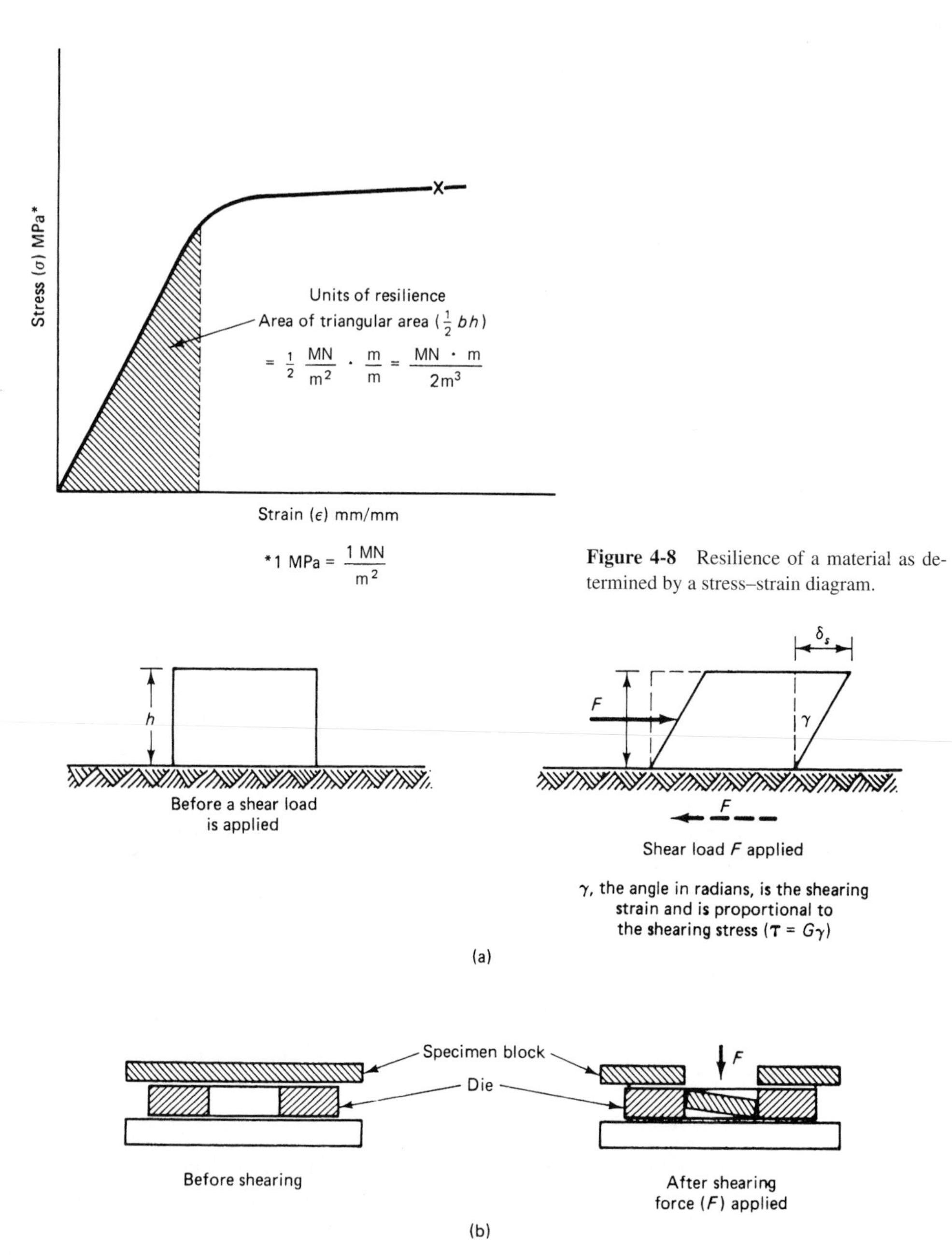

Figure 4-8 Resilience of a material as determined by a stress–strain diagram.

Figure 4-9 Shearing stress and strain. See Figure 4–10 for sheet metal shearing. (a) Before shear load. (b) After shear load.

Note that this ratio is also an expression for the tangent (ratio of the opposite side of an angle in a right triangle to the adjacent side of that angle) of the angle labeled γ in radians.

Illustrative Problem

A small pad subjected to a shearing force is deformed at the top of the pad 0.006 in. (see δ in (see Figure 4-9a). The height of the pad (h) is 2 in. What is the shearing strain on the pad?

Solution $\gamma = \delta/h = 0.006 \text{ in.}/2 \text{ in.} = 0.003 \text{ rad}$

Note: $\gamma =$ is an angular measurement in radians or degrees (π rad $= 180°$).

A shear force acts parallel to the area over which it is exerted, producing a shearing stress (τ) and a shear deformation (δ). A torque wrench used to tighten a nut on a bolt produces a shear force on the bolt that acts along a circular path. The atoms along this path resist a tendency to slide past one another. The shear stress, called ***torsion***, is discussed more thoroughly later in this module. (Figure 4–1d shows this stress in a model of a plastic wrench, Figure 4-26 shows a failure produced by torsion.)

A stress–strain diagram can be plotted using shear stress and shear strain. Such a diagram will show a definite straight-line portion (elastic region) in which the shearing stress is directly proportional to the shearing strain. Like the normal stress–strain ratio, the ratio of the two shear quantities, G, is known as the ***modulus of rigidity*** or ***shear modulus*** of elasticity. In mathematical terms,

$$G = \frac{\tau}{\gamma}$$

with units of psi or pascals. Finally, we can state, as we did with normal stress and normal strain, that the two quantities can be set equal to each other (Hooke's law), and the preceding equation can be written $\tau = G\gamma$, where G, the constant of proportionality, is the shear modulus. In addition to the simple loads we have been discussing, more complex forces are often imposed on materials, producing corresponding stresses such as a bending force (see Figure 4-18) and a twisting force (see Figure 4-26).

Illustrative Problem

A piece of structural steel with a modulus of rigidity of 12×10^6 psi is subjected to a shearing stress of 15,000 psi. Find the strain on the steel.

Solution Using Hooke's law and solving for shearing strain yields

$$\gamma = \frac{\tau}{G} = \frac{15{,}000 \text{ psi}}{12 \times 10^6 \text{psi}} = 0.00215 \text{ rad}$$

4.2.8 Ductility

Ductility is measured by the amount of strain (plastic deformation) at failure. Stated another way, ductility measures the amount of deformation a material can withstand without breaking. It is determined in a tensile test, as briefly described in Figure 4-2 and Figure 4-3. One should not confuse ductility with stiffness. Ductility involves a plastic deformation, whereas stiffness is related to elastic properties of a material. The modulus of elasticity *(E)* is a measure of stiffness, which is an indicator of a material's elastic tensile properties (the bonding forces that bond atoms in a material). A material that can undergo large plastic deformation without fracture is called a ***ductile material.*** A ***brittle material,*** on the other hand, shows an absence of ductility. Consequently, a brittle material shows little evidence of forthcoming fracture by yielding, as a ductile material would do. A brittle fracture is a sudden fracture. A ductile material, by yielding slightly, can relieve excess stress that would ultimately cause fracture. This yielding can be accomplished without any degradation of other strength properties. Figure 4-10a and Figure 4-10b show stress–strain curves for both a ductile and a brittle material. Note the difference in the amount of plastic deformation shown by each curve prior to fracture.

Figure 4-10c and Figure 4-10d show an application of the plastic deformation of a material undergoing a metal-forming operation called ***deep drawing.*** Flat sheets of metal (in this case, steel) are drawn and stretched by a punch and die set into the shape of a can. The metal is subjected to a variety of tensile and compressive forces that must be controlled to not exceed the strength of the metal. Other effects of this cold working are discussed in Module 5. If a failure of the metal occurs through tearing or excessive thinning of the can walls, the operation fails. Redrawing, resorted to when containers (shells) are too difficult to draw in one operation, produces longer sides in the container with reduced diameters. Greater can lengths can be produced by multiple drawing, which involves drawing, annealing, and then redrawing to a greater depth. The term annealing, to be explained further in Module 9, is a heat-treating operation that, among other things, restores ductility to metals after they have been cold-worked. Experience gained in deep drawing reveals that ***cold-rolled (cold-worked)*** sheets of metals with the desired grain orientation contain an ***anisotropy*** (material properties that differ in direction) conducive to successful drawing operations. In determining this anisotropy, tensile specimens are cut from cold-rolled sheets and subjected to an elongation of 15–20%, and strains are measured to determine the degree of anisotropic properties.

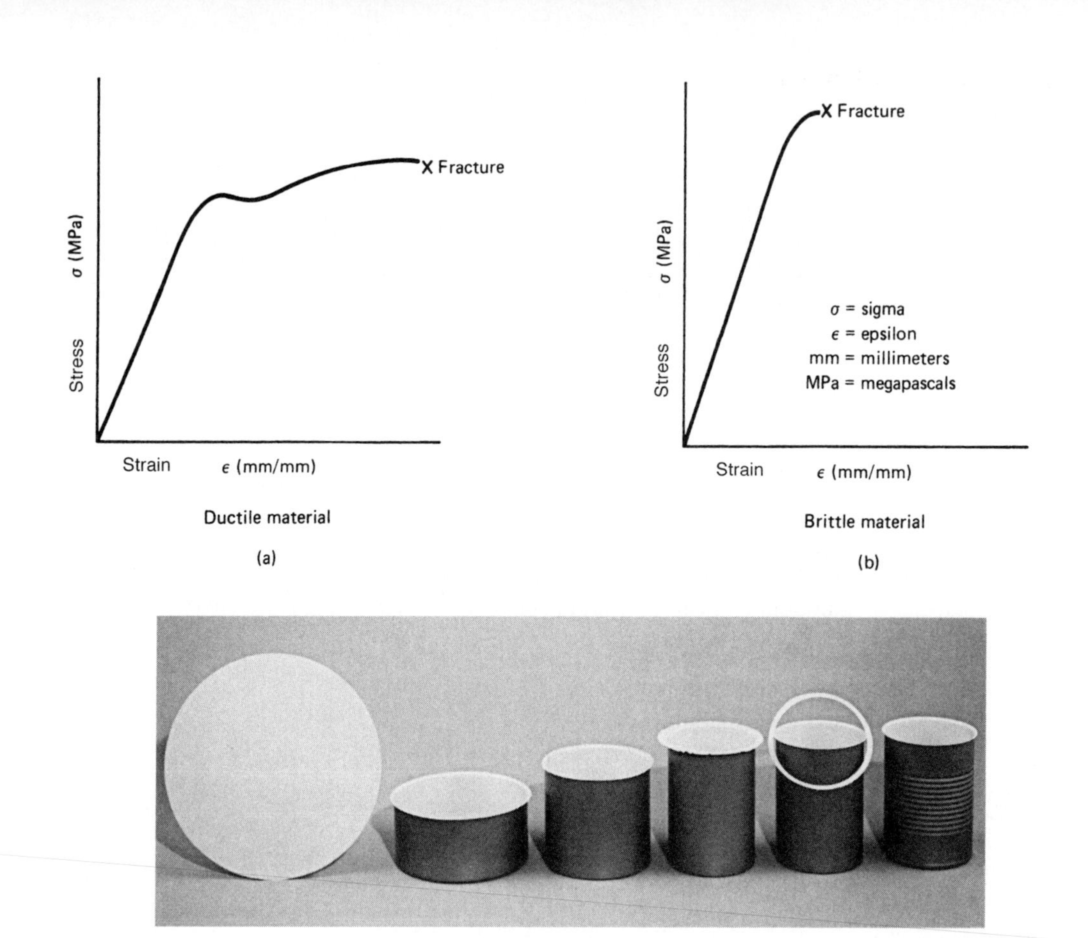

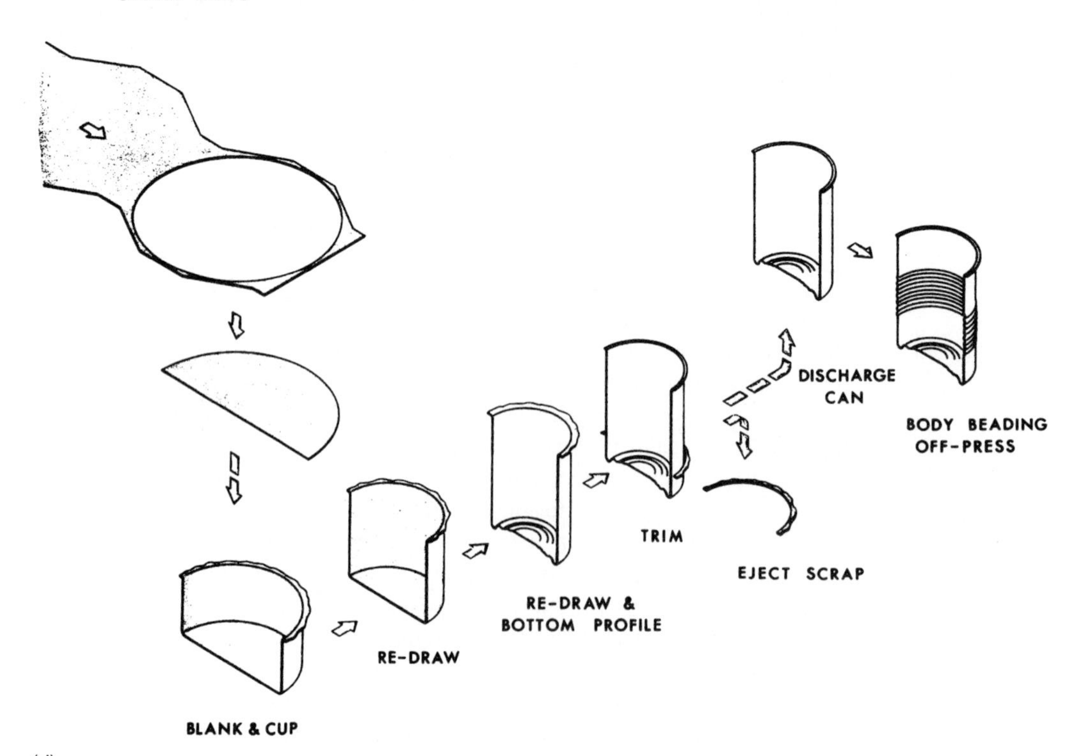

Figure 4-10 Stress–strain curves for (a) ductile and (b) brittle material. (c) Ductile and malleable metals allow complex forming. Photograph of the various stages in deep drawing a tin can. Beginning with a blank of 8.3 to 8.5 in. in diameter, a tin can is drawn with final dimensions of 3 3/16 in. diameter and 4 3/8 in. height. (American Can Co.) (d) Diagram showing the stages required to produce a two-piece can. (American Can Co.)

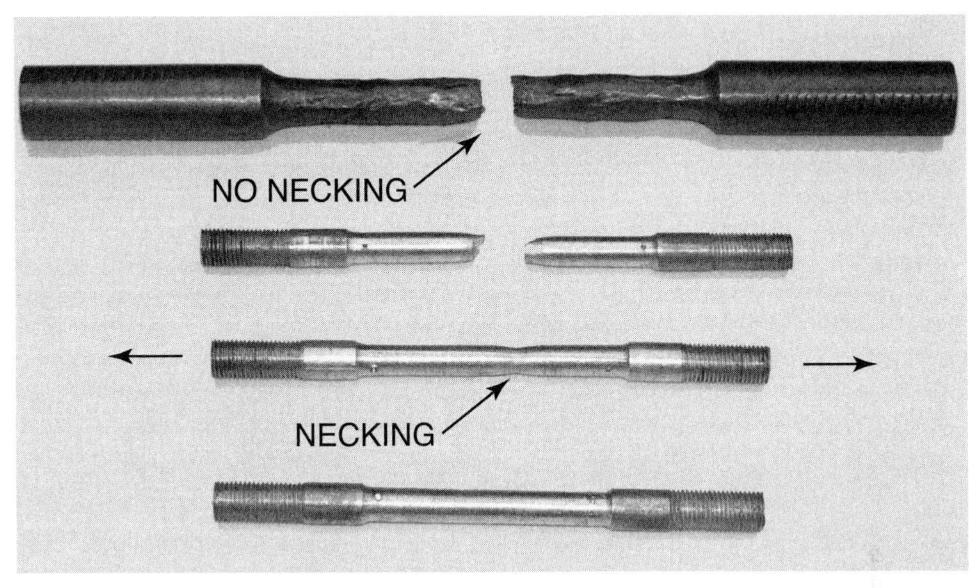

(e)

Figure 4-10 (continued) (e) Tensile specimens; 3 specimens of 6061-T6, ductile aluminum seen before test, necked during test, and broken from ultimate load, and one bronze alloy that was worked hardened (note hammer marks), making it brittle and causing it to break without necking.

Ductility is measured in either of two ways with a tensile test. In the first method, the ***percent elongation*** is the ratio of the change in length of a specimen from zero stress to failure compared to the original length; the quotient is then multiplied by 100%. In terms of a mathematical relationship, these factors can be written as

$$\% \text{ elongation} = \frac{l_F - l_O \times 100\%}{l_O}$$

where

l_F = length at fracture
l_O = original length

Illustrative Problem

A test specimen with an extensometer clamped to it is undergoing a tensile test in a universal testing machine (Figure 4-1a–c). Its original length is 2.0 in. and its final length is 2.8 in. What is its percent elongation?

Solution $\% \text{ elongation} = \dfrac{l_F - l_O}{l_O} \times 100\% = \dfrac{2.8 - 2.0}{2.0} \times 100\% = 40\%$

The second method, ***percent reduction in area***, measures the change in the cross-sectional area of a specimen, compares it to the original cross-sectional area, and multiplies the quotient obtained by 100%:

$$\% \text{ RA in area reduction} = \frac{A_O - A_F \times 100\%}{A_O}$$

where

A_O = original cross-sectional area
A_F = cross-sectional area at fracture

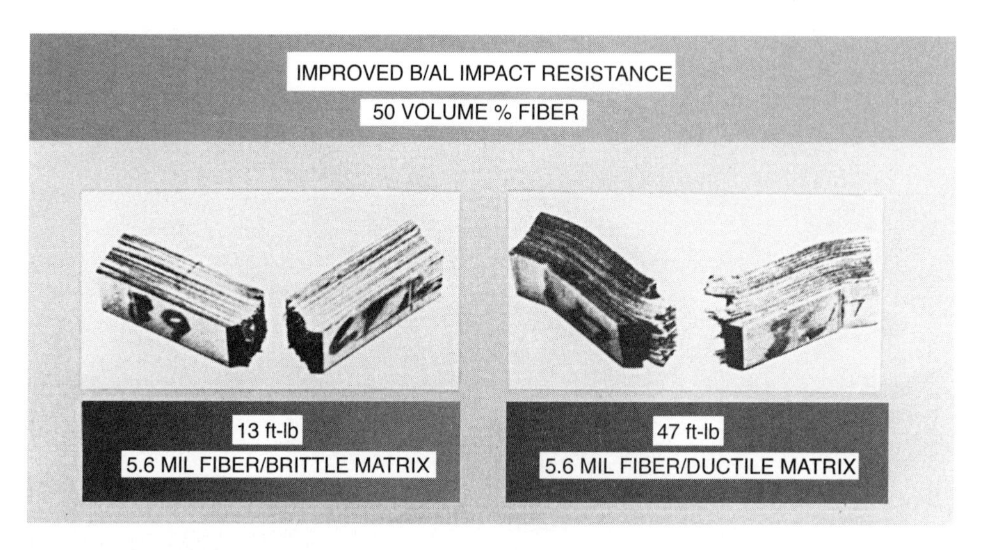

Figure 4-11 Photographs of impact specimens of boron–aluminum fiber composites. (NASA)

Both these measures of ductility do not take into account the nonuniform effect on elongation and area reduction due to necking. During a tensile test in which a specimen is being stretched, its lateral direction contracts (see Figures 4-1 and 4-2). Beyond a certain point, this lateral contraction becomes more pronounced. A drastic change in diameter occurs in a localized area or ***neck.*** Both necking and rupture (fracture) follow after the load exceeds the specimen's ultimate strength. This necking occurs in ductile materials. Figure 4-10e shows two different alloy, tensile specimens—ductile aluminum and brittle, worked-hardened bronze—that illustrate the necking of ductile alloys and lack of necking in brittle materials. See Figure 4-3 to locate the values on the stress-strain diagram.

It is customary to consider a specimen that has 5% or less elongation as a brittle material. Brittle materials should not be considered as having inferior strength, rather such materials lack the ability to plastically deform under load. Figure 4-11 is a pair of photographs showing the effects of increased ductility on the impact resistance of a specimen of boron–aluminum composite material. Pure metals can undergo elongations of 40–60% without rupture, whereas typical superalloys, because of their different-size atoms, have ductilities of less than 20%.

Illustrative Problem

A cylindrical metal specimen undergoes a tensile test. Its original diameter is 10 mm with a gauge length of 50 mm. At the end of the test the diameter in the necked region is reduced by 4 mm. Calculate the percent reduction in area and the true strain at fracture.

Solution

Using $A = \pi d^2/4$ original cross-sectional area $(A_0) = \pi d^2/4 = 78.54\ \text{mm}^2$

Final cross-sectional area $(A_F) = \pi(6)^2/4 = 28.27\ \text{mm}^2$

Using $\%\text{RA} = (A_0 - A_F)/A_0 \times 100\%$ $\%\text{RA} = 64\%$

Using $\varepsilon_f = \ln[100/(100 - \text{RA})] = \ln[100/23] = \ln 4.3478 = 1.469$

 Materials Science and Engineering Laboratory–NIST—*http://www.msel.nist.gov*

4.2.9 Toughness

The ability or capacity of a material to absorb energy during plastic deformation is known as ***toughness.*** The ***modulus of toughness*** (T), equal to the total area under the stress–strain curve up to the point of rupture (see Figure 4-12), represents the energy per unit volume of a material required to produce fracture under static conditions. Toughness can also be expressed in terms of the ease or difficulty in propagating a crack. It can be measured by the amount of energy absorbed by a material in creating a unit area of crack. A tough material would have no

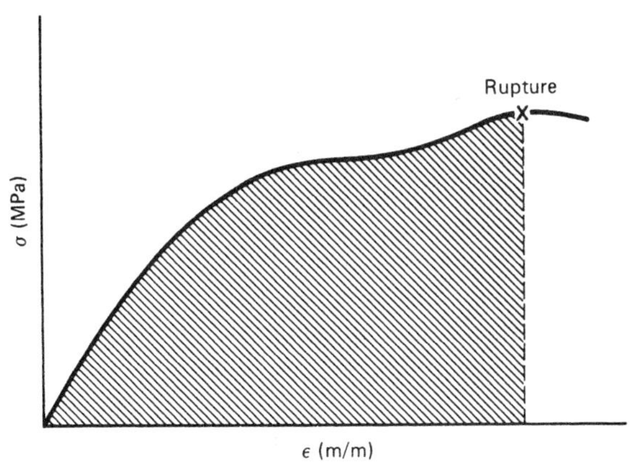

Figure 4-12 Modulus of toughness.

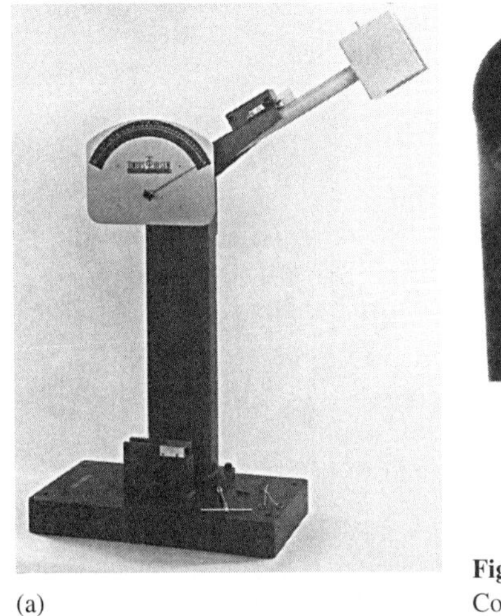

(a)

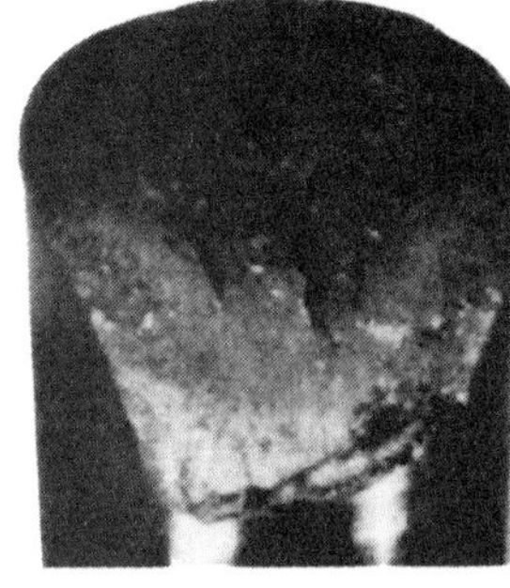

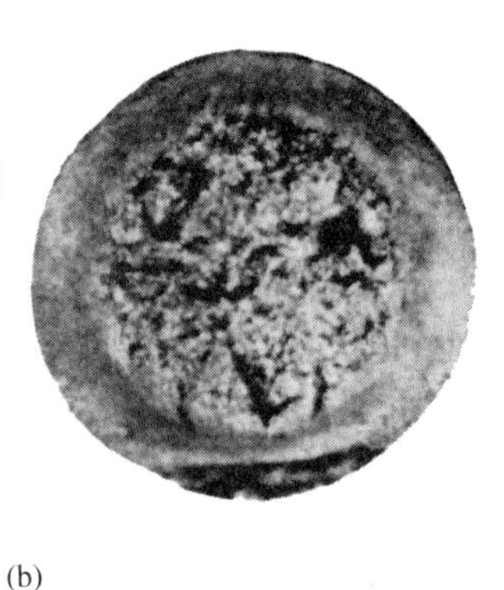

(b)

Figure 4-13 (a) Impact testing machine. (Tinius Olsen Testing Machine Co.) (b) Impact fractures. (John Deere Co.)

defects in its microstructure (not normally attained). Some crack-stopping mechanisms, such as inserting fibers in a metal matrix composite, are used to increase its toughness (see Figure 4-11). Having a good amount of ductility makes many materials tough because the plastic flow permitted by the ductility prevents a concentration of stress at the tip of a crack.

Impact is defined as a sudden application of a load confined to a localized area of a material. Exemplified by the striking of a material with a hammer, this relatively quick application of force, as opposed to a slow or static loading of a material, can cause considerable damage to a material that cannot adequately redistribute the stresses caused by the impact. Ductile materials usually survive impact due to their microstructure, which allows slip to take place. Most metals have good toughness and thus have good impact resistance. Due to their inherent nature as compounds of metals and nonmetals, ceramics do not possess the ability to redistribute stresses and plastically deform. Consequently, they have poor toughness, poor impact resistance, and poor fracture toughness.

Impact tests are also used to indicate the relative toughness of a material. Impact testing is considered ***dynamic testing,*** in contrast to the slower, ***static testing,*** such as tensile or compression testing. Notched-bar impact tests using either of the two standard notched specimens, the ***Charpy*** or the ***Izod,*** reveal the material's behavior in sustaining a shock load. The notched specimens of the material under test have either a keyhole or a V-notch cut to specifications. A heavy pendulum swinging through its arc strikes and breaks the specimen. By recording the initial and final elevations of the pendulum, the difference in potential energy can be calculated. This difference in potential energy is the impact energy absorbed by the specimen during failure. Figure 4-13a shows a universal impact tester for conducting Charpy, Izod, and tension impact tests on materials. Figure 4-11 and Figure 4-13b show impact failures. Customarily used for nonmetallic materials, the Izod test employs V-notched specimens to determine the resistance of the material to crack propagation. The impact energy absorbed by a specimen during failure is expressed in foot-pounds (ft · lb) or joules (J), both of which are used to describe the toughness of a material. The Charpy impact test, developed by a French scientist in 1905, played a key role in understanding fracture problems of merchant ships during World War II when it was used to determine which hull plates were subject to fracture. Today, it is used in many industries including bridge construction and the construction of pressure vessels. Impact

(a)

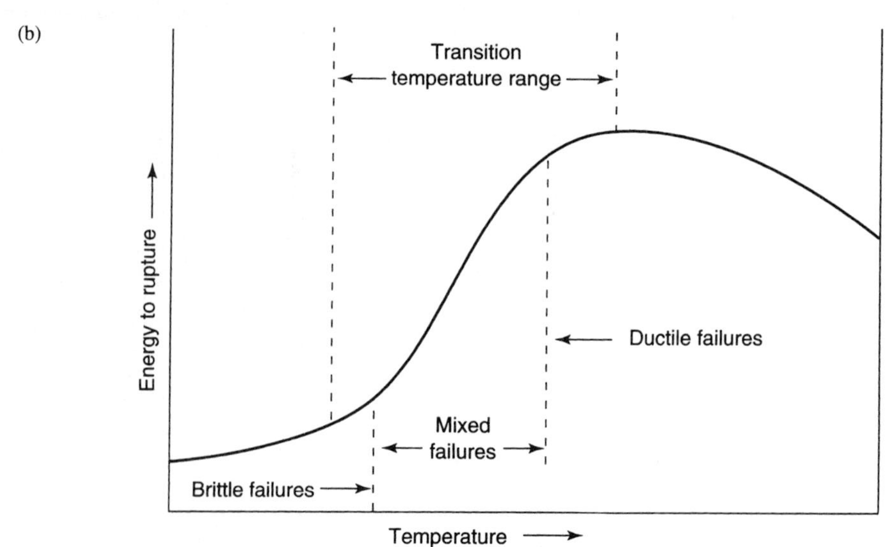

Figure 4-14 (a) Designers of the Hard Rock Café located at Universal Studios in Orlando, Florida, chose polycarbonate plastic sheet for the dramatic, covered arched walkway. To ensure that the plastic glazing would endure Florida's hurricanes, the complete window system underwent dynamic impact tests in which projectiles were fired from an air cannon. The plastic glazing passed local building codes for hurricane resistance and met Occupational Health and Safety Administration (OSHA) guidelines for fall-through protection for skylights and other overhead glazing applications. (G.E. Plastics) (b) Variation of temperature versus energy to rupture in impact testing of metals.

tests serve the building construction industry, which must be concerned with how storms will affect building materials (see Figure 4-14a).

In contrast to the relatively small effect of temperature on the strength and ductility of metals, temperature has a marked effect on the impact resistance of notched bars. Figure 4-14b shows, in general form, the effects of temperature over a wide range on the impact resistance of metals. Some materials are incapable of absorbing low-energy fractures at low temperatures. Below some critical temperature, a particular metal will experience a brittle failure with little absorption of energy. Cracks normally propagate (grow) most readily along specific crystallographic planes by ***cleavage.*** Another main path for cracking is an ***intergranular*** (along the grain boundaries) path, particularly when the grain boundaries are weakened by inclusions or segregation. A cleavage fracture is readily recognized by its crystalline or "rock candy" appearance. Another common fracture form produced is the ***chevron pattern,*** which consists of several crack fronts propagating at different levels in the material from a common origin. A pattern of ridges fans away from the origin. A chevron pattern and its point of origin are visible to the naked eye. These examples illustrate ***brittle fracture,*** which occurs in high-strength metals or metals that have poor toughness and ductility. In many cases the cause of this type of fracture is impact as opposed to overload.

Figure 4-15 is a photomicrograph of a cloverleaf deck socket that failed on a new class of U.S. Navy transport ships. These sockets are flame-cut from deck plating and then welded

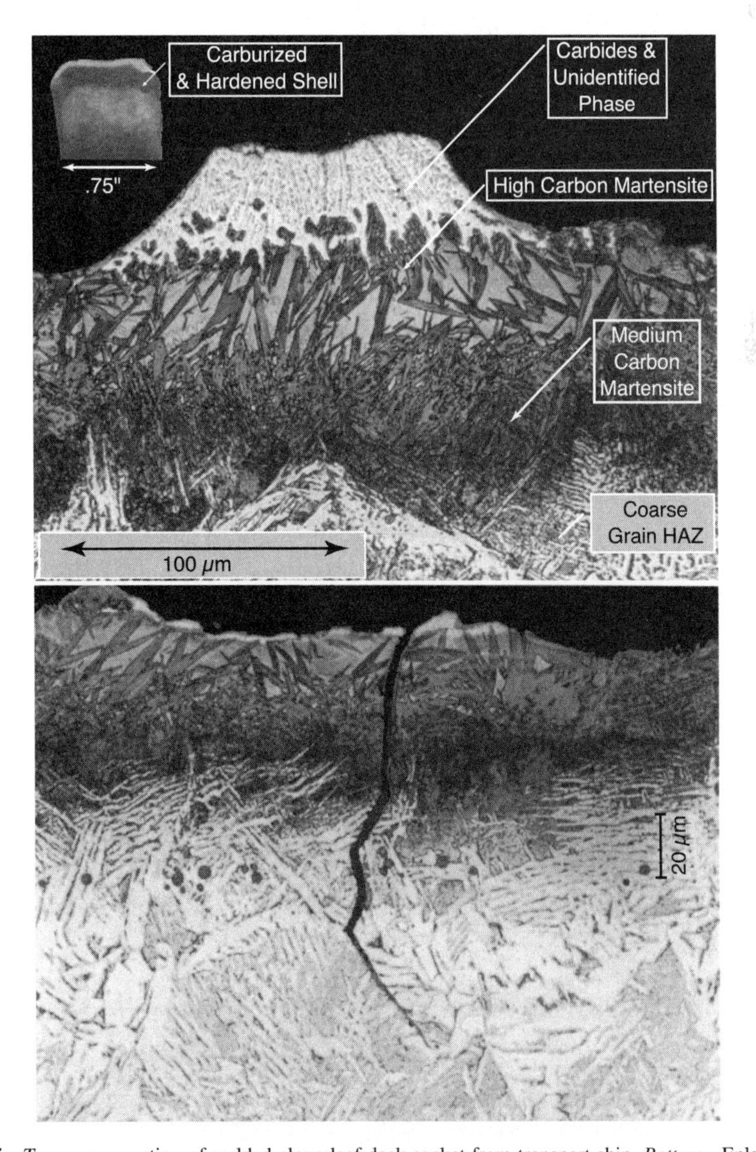

Figure 4-15 *Top*—cross section of welded cloverleaf deck socket from transport ship. *Bottom*—Enlargement reveals crack in failed HAZ or heat affected zone. (*Advanced Materials & Processes*)

to the decks. They are used to fasten vehicles being transported by the ships securely to the deck to permit these ships to sail under varying sea conditions. Various cracked specimens were analyzed using SEM stereo section fractography (SSF), which uses a scanning electron microscope. This technique is used to carefully evaluate the causes of cleavage initiation (beginning cracking) in the material through simultaneous observation of both the fracture surface and the underlying microstructure. In this case the SSF technique successfully identified the dominant local features that triggered cleavage initiation. The ultimate cause of the failure of the sockets was a combination of overload failure and the transition from ductile to brittle fracture produced by high brittleness in the carbon-rich transformed martensite layer introduced by the cutting flame used in the manufacture of the socket. In addition, high residual stresses were induced in the socket as a consequence of the manufacturing processes. Milling and cutting were substituted for a coining operation, and several steps were recommended to reduce the hard and brittle surface layers from the sockets. (Further information related to brittle failures can be found in Section 4.3.)

Above some critical temperature, failures in metals are ductile, with a much greater absorption of energy. Between these temperatures there is a zone in which types of failures are mixed (see Figure 4-14b). In some steels the ***transition temperature range*** may be abrupt, but with other steels it may extend over a considerable range. Coarse grain size, strain hardening, or the addition of certain elements tend to raise the transition range of temperature, whereas fine grain size, heat treatments, and the addition of certain elements (e.g., carbon) tend to enhance notch toughness even at rather low temperatures. Not all materials have a distinct transition temperature. Metals with an fcc crystal structure do not have such a characteristic, whereas bcc metals do. Aluminum and copper-based alloys (fcc crystal structure) retain their ductility at very low temperatures. Ceramic materials experience the ductile-to-brittle transition at temperatures exceeding 1000°C. Crystalline and amorphous polymers both are brittle at low temperatures, with corresponding low impact strength. Selecting these materials should require a very narrow transition temperature range occurring below room temperature.

Knowledge of the ductile-to-brittle transition was not possessed by materials engineers during World War II, when Liberty ships were built with steel that had a high ductility transition temperature; that is, this type of steel could not absorb low-energy fractures at low temperatures, with the result that cracks continued to propagate until the ship broke apart. Many of these ships, which sailed on the Murmansk supply route bringing supplies from the United States to the Soviet Union, have never been recovered after sinking somewhere in the North Atlantic. Brittle fractures have plagued the airline industry as well. An abnormal amount of cycles was produced, for example, by the number of short flights at high altitudes in the mountainous Hawaiian area.

4.2.10 Processing Properties

Malleability, workability, and ***formability*** are some terms related to ductility that describe, in a general way, the ability of materials to withstand plastic deformation without the occurrence of negative consequences (rupture, cracking, etc.) as a result of undergoing various mechanical processing techniques (see Figure 4-10). Terms such as ***weldability*** and ***machinability*** are mentioned here as additional examples of terms used to generally describe the reaction of materials to various manufacturing and/or fabricating processes in industry. In the case of copper alloys (brass, for example) an increase in lead content and hardness improves machinability but reduces ductility (cold workability). An improvement in machinability is always achieved at the expense of cold workability. On the other hand, increases in copper content improve workability but reduce an alloy's machinability. Weldability defines the capacity of a metal or combination of metals to be welded to form a designed structure with weld joints that will perform satisfactorily. Other characteristics included in the term are the ease with which the welding process is accomplished and the absence of weld defects. Sheet metal formability defines the ability of the metal to undergo the desired shape changes without failures, such as necking and tearing. In essence, it describes the relative ease with which a metal can be shaped through plastic deformation. Several techniques are used to test the formability of sheet metals.

4.2.11 Flexural or Bending Strength

A ***beam*** is a structural member that bends or flexes when subjected to forces perpendicular to its longitudinal axis. In beam strength determination this axis is called the beam's ***neutral axis (NA)*** because no normal stresses are assumed to exist along its length. The neutral axis is the edge view of a neutral plane. Figure 4-16a is a sketch of a ***simple-supported beam*** (a beam supported at its

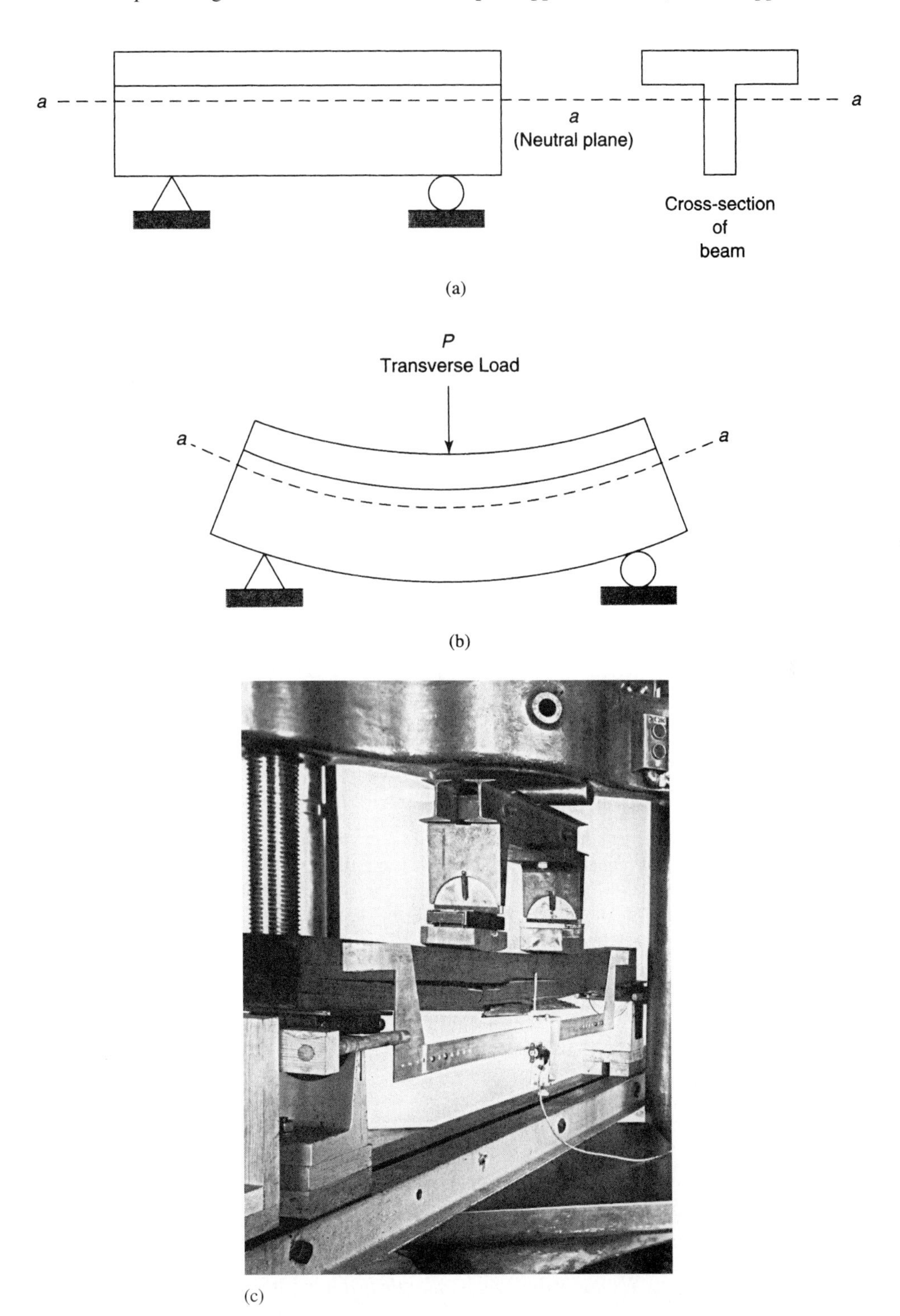

Figure 4-16 (a) Simple-supported beam with a transverse load. (b) Loaded beam showing deformation (deflection). (c) Wood specimen undergoing a static bending (flexural) test and showing evidence of failure due to tension. (Forest Products Laboratory, Forest Service, USDA)

ends with a pin and a roller). The transverse load, or force P, bends the beam (causes it to deform, i.e., deflect), thus producing both shear and normal stresses in the beam. The load in this example would cause the beam to deflect, as in Figure 4-16b, resulting in normal stresses (compressive) near the top surface and normal stresses (tensile) at the bottom. These bending stresses are normal to any cross-sectional area through the beam, which makes their direction parallel to the neutral axis of the beam. For ceramic materials, the maximum bending stress in tension at failure is called the ***bend strength*** and is often referred to as the ***modulus of rupture (MOR).*** The presence of flaws, their distribution in a ceramic material, and their attendant adverse effects on the strength of the material are not taken into consideration when using this indicator of strength.

Assuming that the beam material is ***homogeneous*** (same material nature throughout) and ***isotropic*** (same properties in all directions throughout the beam material), the normal stresses will be at a maximum near the top and bottom surfaces of the beam. These normal stresses (both compressive and tensile) are known as ***flexural, fiber,*** or ***bending stresses.*** The ***flexure formula,***

$$\sigma = \frac{Mc}{I}$$

relates these stresses (σ) to the ***bending moment*** (M), the maximum distance (c) from the beam's neutral axis (where bending stresses are zero) to the outer surfaces of the beam, and the rectangular ***moment of inertia*** of the cross-sectional area of the beam (I). A *flexure test* is performed using a simple supported beam loaded, as shown in Figure 4-16c. The maximum bending stress and deformation (deflection) are recorded for increments of load P, and these data are plotted to obtain a stress–strain diagram. The maximum bending stress developed at failure is known as the ***flexural strength.*** For those materials that do not crack, the maximum bending or flexural stress is called the ***flexural yield strength.*** A bend test used to determine the ductility of certain materials should not be confused with this flexure test.

A very recent U.S. Steel laminated product adequate for automobile sheet applications takes advantage of this distribution of bending stresses. Called **steel-plastic-steel (SPS),** this sandwich material consists of light-gauge steel skins (the top and bottom layers, or laminates) and a plastic core (center) . The polymer core (polypropylene) has sound-dampening characteristics in addition to having low cost and low density. SPS laminates with the same bending stiffness as solid sheet steel weigh considerably less, corresponding to the weight loss that would occur if aluminum were used instead of steel.

Illustrative Problem

Referring to the beam in Figure 4-16b, the bending moment (M) produced by the force (P) at the center of the beam is 3530 ft·lb.The beam is a T beam with a distance of 5 in. (c) from its centroid to the bottom surface of the beam. Calculate the stress at the bottom surface of the beam midway between the supports. The moment of inertia (I) for the beam's cross section is 136 in^4. Also indicate in your answer whether the stress is tensile or compressive.

Solution

$$\begin{aligned}\sigma &= \frac{Mc}{I}\\ &= \frac{3530 \text{ ft}\cdot\text{lb} \times 5 \text{ in.}}{136 \text{ in}^4}\left(\frac{12 \text{ in.}}{1 \text{ ft}}\right)\\ &= 1560 \text{ psi} \qquad \text{(tensile)}\end{aligned}$$

4.2.12 Fatigue (Endurance) Strength

Service conditions commonly involve many repetitions of applied stress or reversals of stress. We know that when a ***simple beam*** is subjected to a downward-acting, transverse load (see Figure 4-17), the material in the top half of the beam will be compressed (the stress s_c is

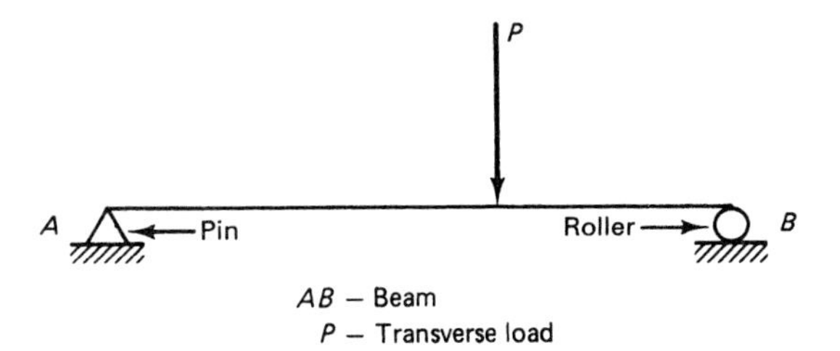

Figure 4-17 A simple beam with a concentrated, transverse load.

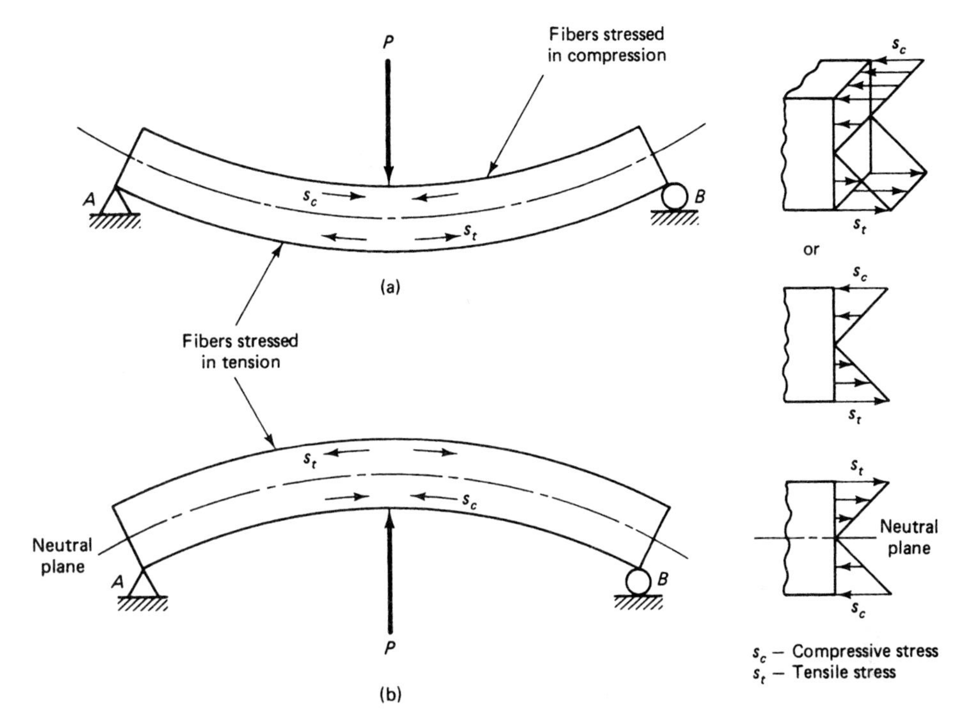

Figure 4-18 A simple beam deflected by a cyclic, transverse load.

compression) and the lower half of the beam will be subjected to a tensile stress (s_t),* as demonstrated in Figures 4-18 and 4-19. Roller and pin supports are termed ***simple supports*** to differentiate them from fixed supports, which can also resist rotational forces. A shaft with a pulley or gear can be compared to the horizontal beam referred to previously. The load or force, in this instance, is a pulley or gear force that causes the shaft to deflect (bend), particularly if sufficient bearing supports are not provided. Now, if a transverse load is applied and removed in some cyclic fashion, the material in the beam or shaft would go from a condition of zero stress to a stressed condition and back to a condition of zero stress. This type of cyclic loading is called *repetitious*. If the stress in the material changes due to the loading from compression to tensile, or vice versa, this is known as ***stress reversal.*** The latter condition is easier to visualize with a point on the surface of a rotating shaft turning at 1000 revolutions per minute whose bending stress changes from compression to tension 1000 times a minute. In practice, many failures have occurred under such conditions when the stresses developed were well below the ultimate stress and frequently below the yield strength. These failures are called ***fatigue*** (or ***endurance***) ***failures*** (see Figure 4-20). Fatigue causes over 75% of failures of mechanical parts and components in service. One example is the Boeing 737 operated by Aloha Airlines in 1988 that lost a canopy without warning in level flight over the Pacific

**Note*: θ is the universal symbol for stress. However, lowercase s is sometimes substituted for θ, as seen in Figures 4-18 and 4-19.

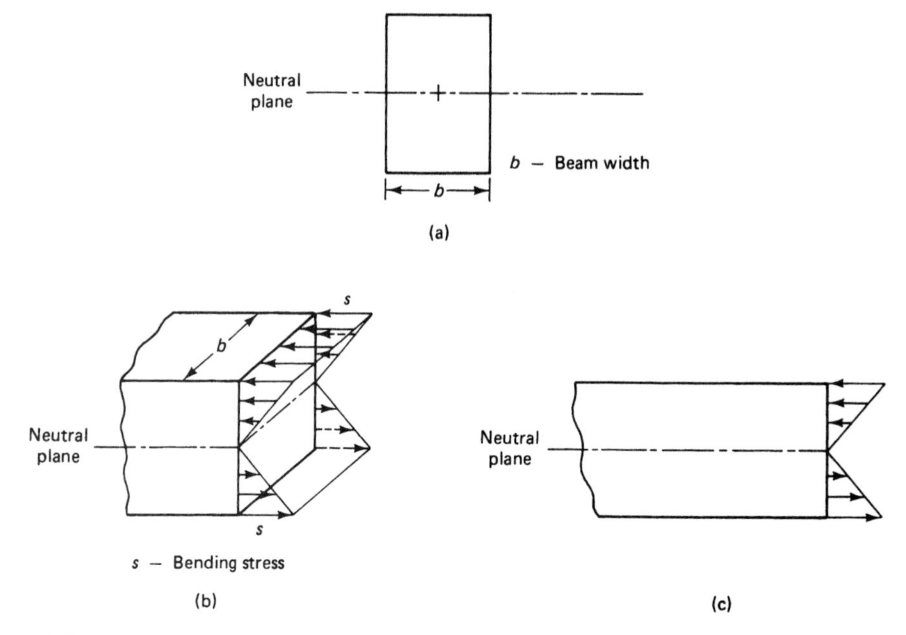

Figure 4-19 (a) End view of beam showing cross section. (b) Front view (three-dimensional) of a portion of beam showing distribution of tensile and compressive bending stresses acting on a transverse plane. (c) Front view (two-dimensional) showing same stress distribution as in (b).

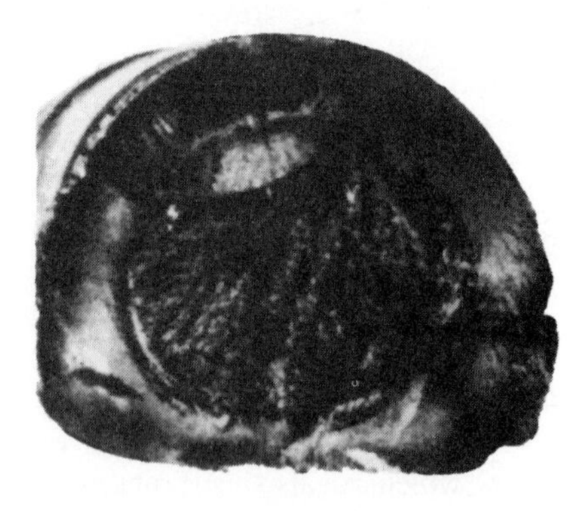

Figure 4-20 Combined fatigue fracture due to combined bending and torsional loads on an axle. The outer surface was hardened. A crack began at the top surface. The rough surface in the center indicates the last area to fracture instantly when the axle broke. (John Deere Co.)

Ocean. Subsequent investigation revealed that the canopy was weakened as a result of extreme corrosion and fatigue of the aluminum alloy used as a skin material. The fatigue analysis concluded that the low stresses developed due to excessive cycles of pressurizing and depressurizing of the cabin.

Many fatigue failures start as a tiny crack whose origin is often traced to an inspection stamp, tool mark, or other defect on the surface. The crack produces a stress concentration that assists in the growth of the crack until eventually the area of material remaining to withstand the stress is insufficient, which results in a sudden fracture (see Figure 4-20). Not all fatigue failures originate from discernible stress concentrations. Stresses less than the yield strength can cause microscopic movement on slip planes at the surface, which ultimately serve as the origin of the failure. Much empirical research is done using fatigue testing machines to determine the strength of materials under fatigue loading. Machines developed by the National Aeronautics and Space Administration use computers that simulate aircraft or spacecraft flights to test the material until it fails under varying conditions of load and ambient temperatures. The results of these many tests are recorded on semilog or log-log paper to produce *s* (stress)–*N* (cycles) diagrams. Typical ***s–N diagrams*** are shown in Figure 4-21 for a fiberglass

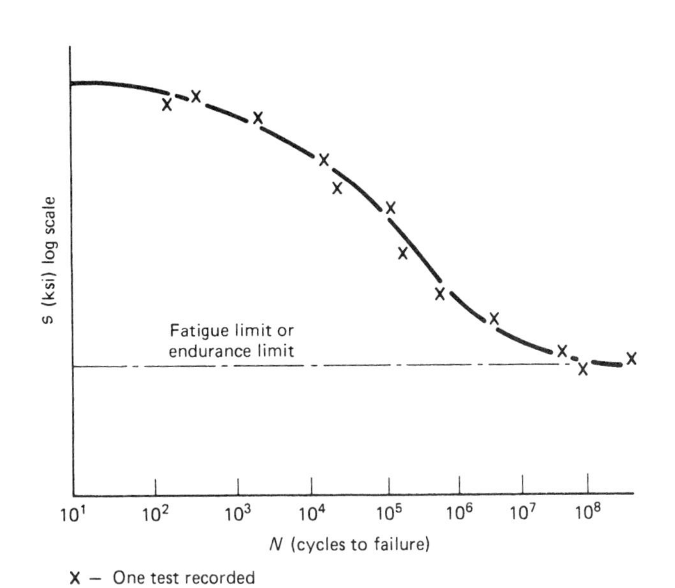

Figure 4-21 *s–N* diagram for a typical fiberglass composite material.

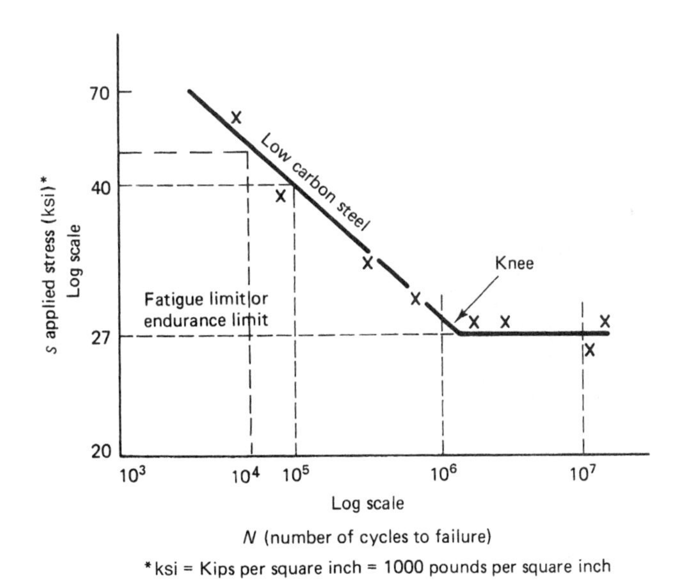

Figure 4-22 *s–N* diagram for a typical low-carbon steel.

composite, in Figure 4-22 for a typical low-carbon steel, and in Figure 4-23 for some nonferrous metals and some plastics. Figure 4-24a is a photograph of a typical fatigue fracture showing two very distinct surfaces. The *s–N* curve for a low-carbon steel (see Figure 4-22) shows an abrupt break, or *knee*, at which point the curve tends to approach a horizontal line. In other words, with any stress below the fatigue or endurance limit, a low-carbon steel can be cycled continuously without fracturing. Aluminum and other nonferrous materials fracture at relatively low stresses after many cycles. For many ferrous alloys, the endurance limit is about one-half the tensile strength of the metal. Furthermore, they exhibit no fatigue limit, which means that there is no stress below which they will not fracture. Therefore, when speaking of the fatigue of nonferrous materials, it is necessary to express both the stress and the number of cycles in describing the life of the material. See Secton 4.2.9 for an example of fatigue failure in an Hawaiian aircraft.

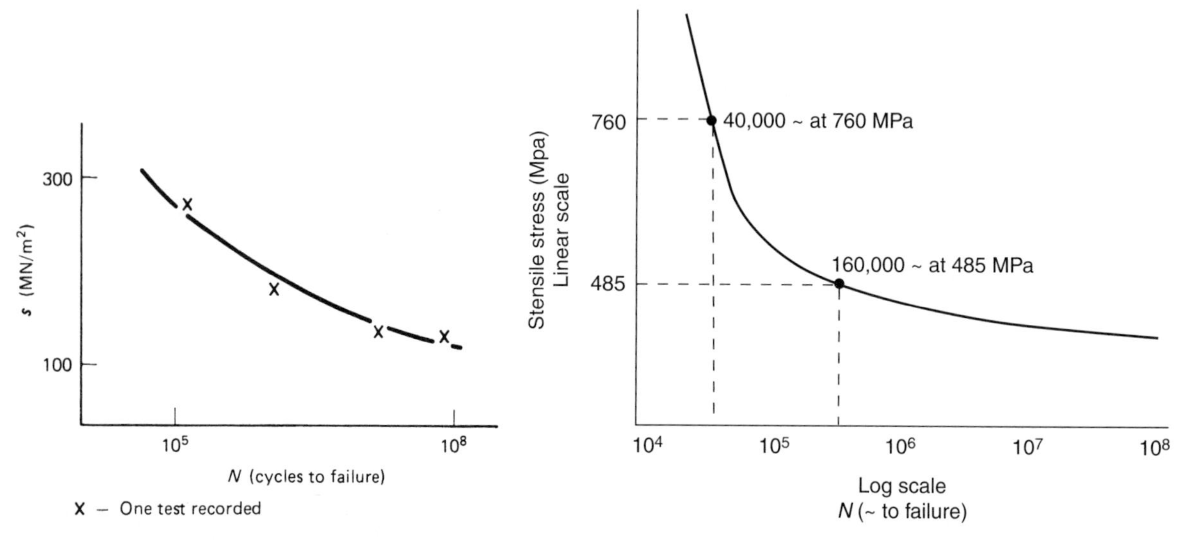

Figure 4-23 *s–N* diagram for nonferrous metal (e.g., aluminum and copper) and some plastics (e.g., nylon). Typical *s–N* curve for shot peened gears. Reducing high residual compressive stress at the surface of the tooth roots enhances their fatigue life. Shot peening proves very effective for parts subject to high-cycle fatigue loading (> 10^4 to 10^5 cycles).

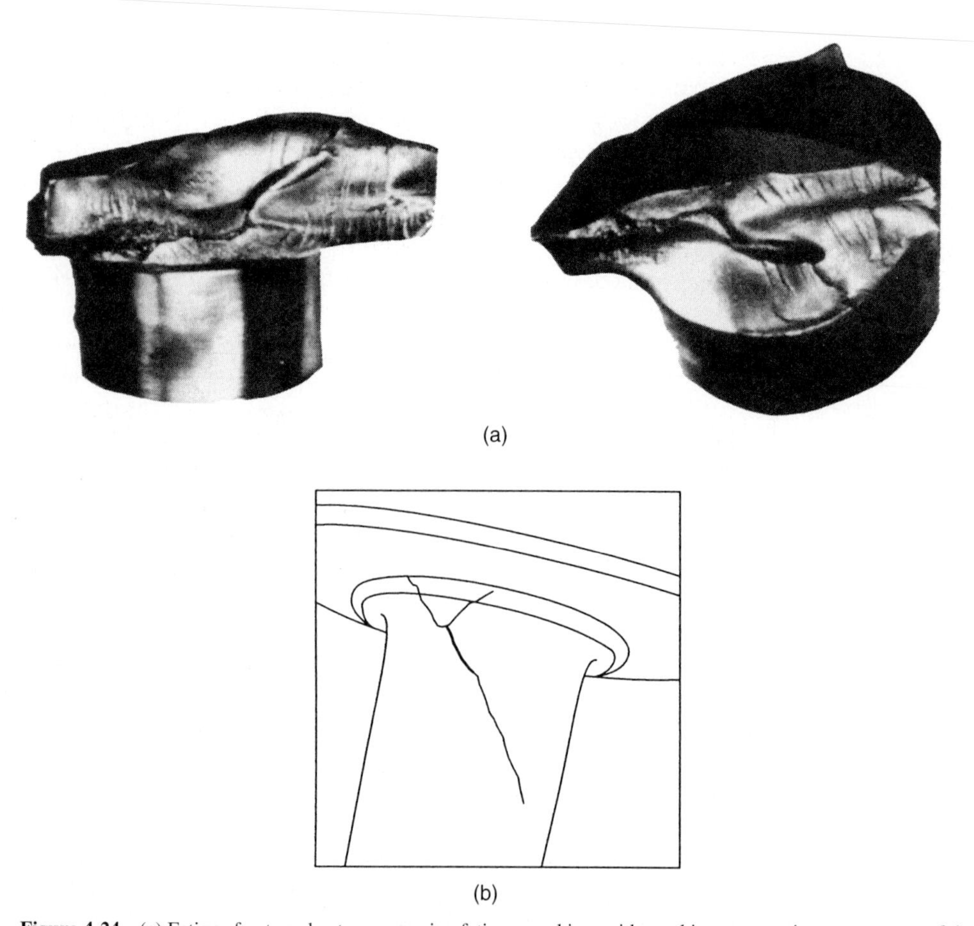

Figure 4-24 (a) Fatigue fracture due to progressive fatigue cracking, with cracking progressing across most of the section before the final overload fractures the remaining metal. (John Deere Co.) (b) Sketch illustrating torsional fatigue on a crankshaft. (Adapted from *Shot Peening Applications,* 7th ed., Metal Improvement Co. Inc., p. 43)

It is conservatively estimated that over 50% of the failures occurring under service conditions are fatigue failures. In view of the expense involved in conducting tests, particularly when attempting to duplicate service conditions in a laboratory, a ***fatigue*** or ***endurance ratio*** is sometimes used instead. This ratio, varying between 0.25 and 0.45 depending on the material, compares the fatigue limit to tensile strength. For many ferrous alloys the endurance limit is about one-half the tensile strength of the metal. The fatigue strength of a bolt after 10^7 cycles is only about 10–20% of the ultimate tensile strength. The shape of the metric threads plus the martensitic microstructure of a high-strength bolt means that fatigue is the limiting mechanism for fasteners. The ***fatigue*** or ***endurance limit*** is the maximum stress below which a material can presumably endure an infinite number of stress cycles (Figure 4-21 and Figure 4-22). ***Fatigue strength*** is the maximum stress that can be sustained for a specified number of cycles without fracture. In other words, fatigue strength can be any value on the ordinate of the *s–N* diagram. To make it meaningful, the corresponding *N* value must be reported also. In Figure 4-22 the fatigue strength of about 40 kpsi corresponds with an *N* of 10^5 cycles. The fatigue limit, as determined empirically, is generally below the yield strength. Most design stresses are lower than the fatigue or yield strength of a material primarily because of the adverse effects of surface conditions on the strength of materials.

Photos & Animation—LAVA (Langley Animation & Video Archives)—
http://oea.larc.nasa.gov/Photo/photo.html

4.2.13 Creep

Creep is a slow process of plastic deformation that takes place when a material is subjected to a constant condition of loading (stress) below its normal yield strength. Creep occurs at any temperature. However, at low temperatures, slip (movement of dislocations) is impeded by impurity atoms and grain boundaries. At high temperatures, the diffusion of atoms and vacancies permits the dislocations to move around impurity atoms and beyond grain boundaries, which results in much higher creep rates. The word *creep* implies, then, that a material plastically deforms or flows very slowly under load as a function of time. After a certain amount of time has elapsed under constant load, the ***creep strain*** (plastic deformation) will increase and some materials will rupture. This rupture, or fracture, is known as ***creep rupture.*** The time required for failure to occur is the rupture time. Higher stress or higher temperature reduces the rupture time and increases the creep rate. Aluminum alloys begin to creep at around 100°C. Aluminum aircraft wing panels are creep formed. During ***creep forming,*** a workpiece is loaded against a die or tool for a specified time. The load is normally applied using pressure or vacuum at temperatures around 160°C. The process results in predetermined shapes that require little or no machining. The amount of creep resulting from such operations must be predetermined through numerous creep tests. In addition, springback must be taken into consideration. When the forming or bending force is removed at the end of the deforming operation, elastic energy remains in the deformed part, causing it to recover partially to its original shape. This elastic recovery, called ***springback,*** is defined as the increase in the included angle of the bent part relative to the included angle of the forming tool after the tool is removed.

Although some materials will creep at low or room temperatures, this type of plastic deformation is usually associated with high temperatures. Polymeric materials creep at room temperature; this low-temperature creep is called ***cold flow.*** Both steam and gas turbines, used for propulsion of ships or the generation of electricity, operate, by necessity, at high temperatures over many years. The extremely close fits between turbine blades and their casings prevent the escape of steam or gas past the blades. If these turbine blades were allowed to change their dimensions at any time during their expected service life, during which they are under large centrifugal loads (rotating at extremely high speeds) at extremely high temperatures for long periods, possibly weeks, the failure would be catastrophic. Consequently, the material from which these turbine blades are fabricated must possess, among other necessary properties, high creep resistance. Continuing research over the years has produced materials for

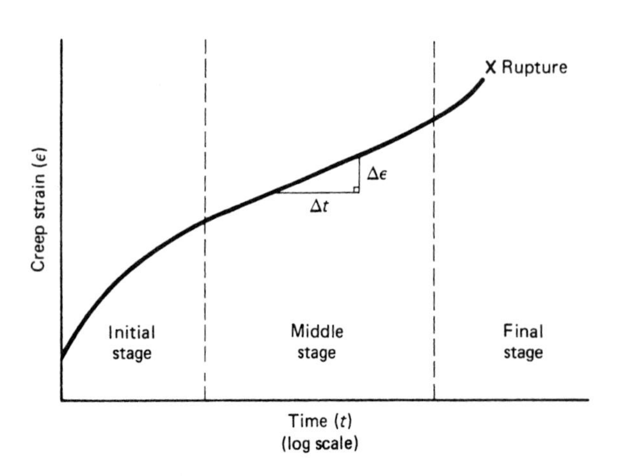

Figure 4-25 Typical creep curve.

medium- and high-temperature service such as the many titanium alloys and the "superalloys" composed of iron, nickel, and/or cobalt. Module 9 discusses investment casting of such alloys.

Tensile creep tests develop data at a constant level (stress) and at a constant temperature and determine the creep rate of a material. The amount of creep strain when plotted against time, as in Figure 4-25, indicates that the creep curve formed can be divided into several stages, with the last stage ending with a rupture of the material. The ***creep rate*** is determined, at any point, by the slope of the curve (drawing a tangent to the curve at the point). The creep rate, the slope of the middle portion of the curve, is nearly constant and represents the minimum rate. Once the creep enters the last stage, rupture soon follows. Creep rate is usually expressed in percent creep strain per hour. A typical rate might be $10^{-4}\%$ per hour.

Engineering materials must perform satisfactorily throughout their ***service lives.*** In many applications, service life can extend well beyond 20 years. A determination must be made as to the maximum allowable deformation that can be tolerated during the expected service life of a material before the material is chosen for the particular application. Research in polycrystalline materials with varying amounts of grain-boundary areas has shown that, at high temperatures, ***coarse-grained materials*** possess more creep resistance than do fine-grained crystalline materials. At low temperatures, ***fine-grained materials*** offer more creep resistance. These statements, including the phenomenon of creep itself, can be explained in terms of the movement and stoppage of mobile dislocations (linear arrangements of atoms making up imperfections in the atomic structure of solid materials) throughout the crystal structure. Creep, then, like other properties of materials, is dependent on the structure of materials.

4.2.14 Torsional Strength

Torsion refers to the process of twisting. A body such as a circular rod (e.g., a shaft for transmitting power), as shown in Figure 4-26, is under torsion as a result of a force acting to turn one end around the longitudinal axis of the rod while the other end is fixed or twisted in the opposite direction. The material resists this twisting action by generating a similar twist internally. The product of the force P and the radius r perpendicular to the line of action of the force is called a ***torque.*** The external torque produces both a stress and a deformation of the rod. The stress is classed as a shear stress, which causes the atoms to twist past each other. The ***shear deformation*** is measured in terms of the angle γ (Greek letter gamma) and the angle θ (Greek letter theta), as indicated in Figure 4-27a. As a result of the torque T, the point C on the surface of the rod moves to point C'. This deformation can be measured by angle γ on the surface of the rod or by angle θ shown on the cross-sectional area of the rod, known as the ***angle of twist.*** Both angles are exaggerated in the figure for illustrative purposes. Forces can combine to cause failures, as seen in Figure 4-20.

A torsion-test machine measures the torque applied to a specimen of material along with the corresponding angle of twist (θ). The results can be plotted as a *torque–twist diagram* (see

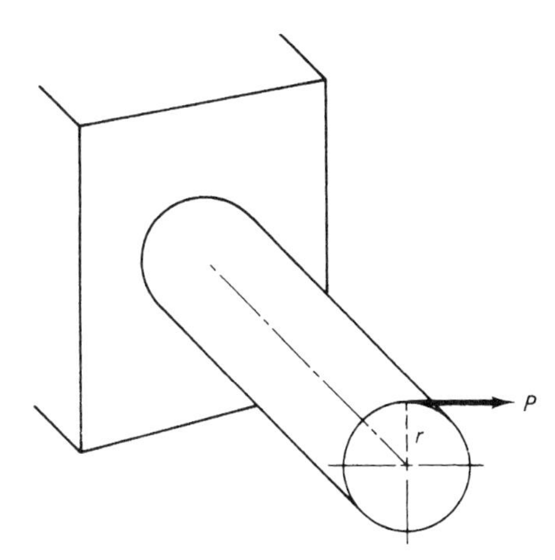

Figure 4-26 Circular rod under torsion.

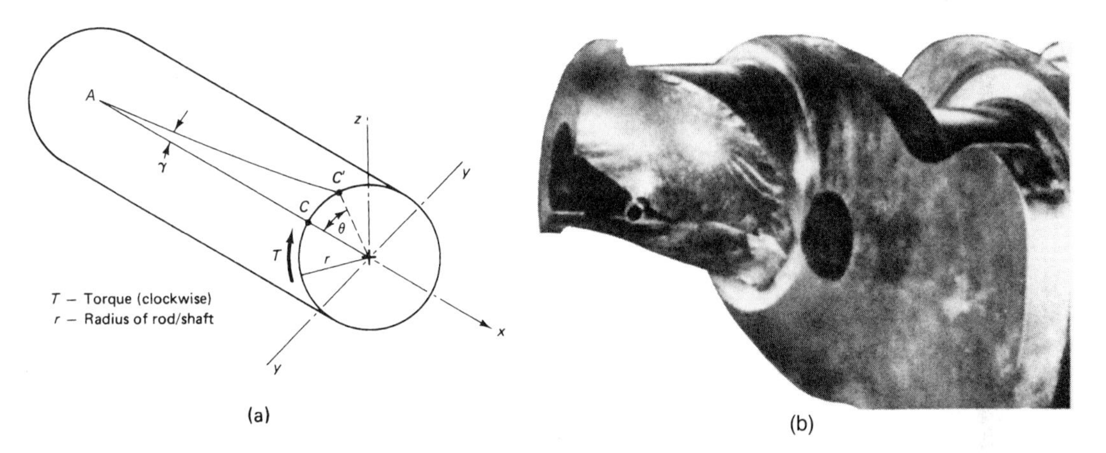

Figure 4-27 (a) Circular rod undergoing a torque, with resulting deformation. (b) Torsional fracture. Torsional (twisting) loads produce spiral types of failure. Note the curved line from (*A*) to (*C′*) is part of a helix.

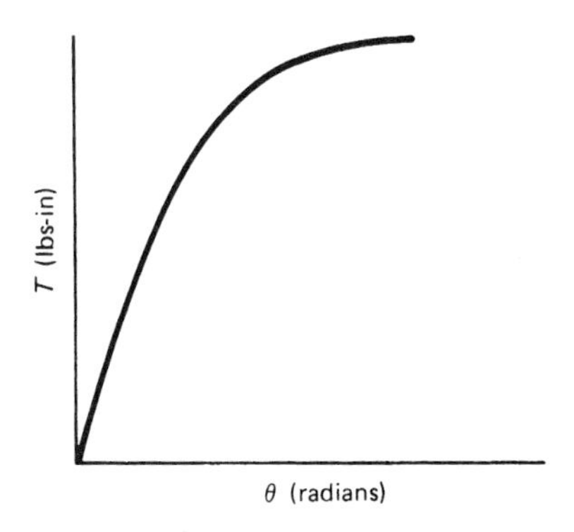

Figure 4-28 Typical torque–twist diagram.

Figure 4-28), which resembles an ordinary stress–strain diagram as obtained by the usual tensile-test procedure. The torsional stress is the shear stress (τ) produced in the material by the applied torque and is calculated using the torsion formula for circular shafts:

$$\tau_{\max} = \frac{T_r}{J}$$

where r is the radius of the cross-sectional area and J is the centroidal ***polar moment of inertia.*** The maximum torsional stress occurs when r is a maximum, as indicated in the preceding

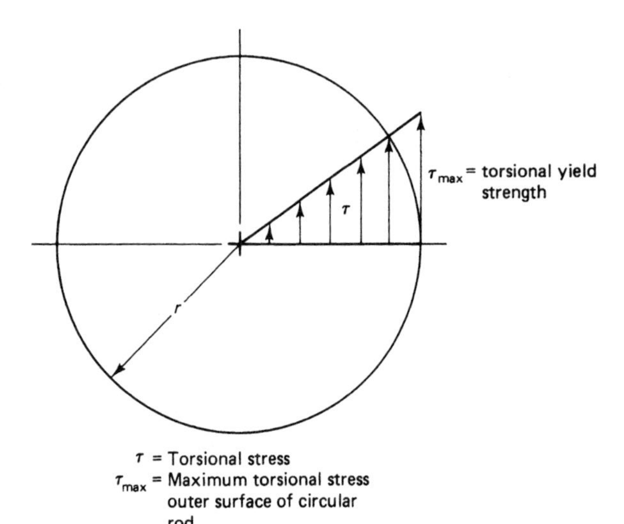

Figure 4-29 Distribution of torsional stress in a circular rod.

equation and illustrated in Figure 4-29. The maximum torsional stress occurring at the outer surfaces of the circular rod is called the ***torsional yield strength.*** It relates to that point on the torque–twist diagram where the curve begins to depart from a straight line. Torsional yield strength roughly corresponds to the yield strength in shear. The ***ultimate torsional strength*** or ***modulus of rupture*** expresses a measure of the ability of material to withstand a twisting load. It is roughly equivalent to the ultimate shear strength. The ***torsional modulus of elasticity*** as determined by the torque–twist diagram, known as the ***modulus of rigidity,*** is approximately equal to the shear modulus or the modulus of elasticity in shear (G).

Illustrative Problem

How much torque (T) is needed to produce a shear stress of 79.6 MPa on the surface of a 40-mm-diameter shaft with a polar moment of inertia (J) equal to 25.12×10^{-8} m^4?

Solution

P

T

$$T = Pr \qquad \tau = \frac{Tr}{J}$$

$$r = \frac{D}{2} = 0.02 \text{ m} \qquad T = \frac{\tau J}{r}$$

$$= \frac{79.6 \text{ MPa} \times 25.12 \times 10^{-8} \text{ m}^4}{0.02 \text{ m}}$$

$$= \frac{79.6 \times 10^6 \text{ N}/m^2 \times 25.12 \times 10^{-8} \text{ m}^{4^2}}{0.02}$$

$$= 999.8 \text{ N}\cdot\text{m} \cong 1 \text{ kN}\cdot\text{m}$$

Shafts are members that are subjected to the twisting moments commonly called *torques*. When a shaft is loaded down with a torque, it is undergoing torsion. A great many applications in engineering involve shafts of solid or hollow circular sections. The application of an external torque (the applied load) is resisted by an internal torque, which produces a shearing stress. An example of an external torque would be the force acting on the tooth of a gear or a belt tension on a pulley mounted on a shaft. For this situations the preceding ***torsion formula*** can be expressed in terms of the maximum radius (symbol c) and with the maximum torque (T), written as $\tau_{max} = Tc/J$. The moment of inertia of an area is computed with respect to an axis. For shafting, the area is the cross-sectional area and the axis is perpendicular to the plane of the area. This moment of inertia, the polar moment of inertia, in written with the symbol J. An area where the axis is in the plane of the area of the moment of inertia is called a *rectangular* or, simply, the *moment of inertia* and expressed with the symbol I. For circular solid shafts, the polar moment of inertia is expressed in terms of the radius r, or the diameter d:

$$J = \pi r^4/2 = \pi d^4/32$$

Using Hooke's law for shear stress (see Section 4.2.3), $\tau = G\gamma$ and the torsional formula $\tau_{max} = Tr/J$, the angle of twist (θ) can be expressed as

$$\theta = TL/JG$$

Illustrative Problem

A mild steel solid shaft 2 meters (m) long and 50 millimeters (mm) in diameter withstands a torque of 2000 N · m. The shear modulus for this steel is 84 *G* Pa. Determine the maximum shear stress in the shaft and the angle of twist of the shaft.

Solution

$$J = \pi d^4/32 = 6.14 \times 10^{-7}\ \text{m}^4$$

$$\tau_{max} = Tc/J = 81.5 \times 10^6\ \text{N/m}^2$$

$$\theta = TL/JG = 0.078\ \text{rad, or } 4.47^\circ \qquad \text{(radians or degrees)}$$

An analysis of the angle of twist equation shows that a low shear modulus results in higher values of angular displacement. Materials scientists are developing wire products that are composite materials, with core materials that differ from the outer materials. Certain low-modulus materials can now be combined with high-strength core materials to provide a composite material with unique properties. Even more interesting are composites that combine relatively soft outer materials and a correspondingly harder core. These composites can provide exotic properties or bolster the strength of another material in ways that a monolithic material would be incapable of doing (see Drawn Filled Tube [DFT] in Module 25).

4.2.15 Hardness

Hardness is a measure of a material's resistance to penetration (local plastic deformation) or scratching. One of the oldest and most common hardness tests, based on measuring the degree of penetration of a material as an indication of hardness, is the Brinell test. ***Brinell hardness numbers*** (HB) are a measure of the size of the penetration made by a 10-mm steel or tungsten carbide sphere with different loads, depending on the material under test. The indentation size is measured using a macroscope containing an ocular scale. ***Vickers hardness numbers*** (HV) employ a diamond pyramid indentor. Otherwise, the two tests are basically similar. ***Rockwell hardness testers***, using a variety of indentors and loads with corresponding scales, are direct-reading instruments (i.e., the hardness is read directly from a dial). The hardness number, for example, 65 HRC, indicates the reading came from the C scale using a diamond cone indentor and a 150-kg load. It is therefore important in reporting Rockwell hardness readings to include the scale letter so that the person wishing to use the information knows the type of indentor as well as the size of load used in the test. (See Table 4–1 and Figure 4-30 .) The National Institute of Standards and Technology (NIST) and ASTM committees are working on a revision to ASTM E18, *Standard Test Methods for Rockwell Hardness and Rockwell Superficial Hardness of Metallic Materials*. The goal of the revision is to arrive at consistent Rockwell hardness (HR) standard values for the makers of hardness testers, test blocks, and users of the machines and blocks worldwide.

Few, if any, tools are subjected to such an elaborate heat treatment as nail hammers, an item that weighs no more than 20 oz. The striking face and entire claw section including the eye (or crotch) are first hardened to about 62 HRC by induction (see the section on surface hardening in Module 13). Because of variations in section thickness of the claw, two different induction frequencies are used. The striking face and the rim around the striking face are tempered by induction using a coil arrangement that results in 50 to 58 HRC on the striking face and somewhat lower hardness values around the face (called ***ring tempering***). Claws are also

TABLE 4-1 HARDNESS SCALES COMPARISONS

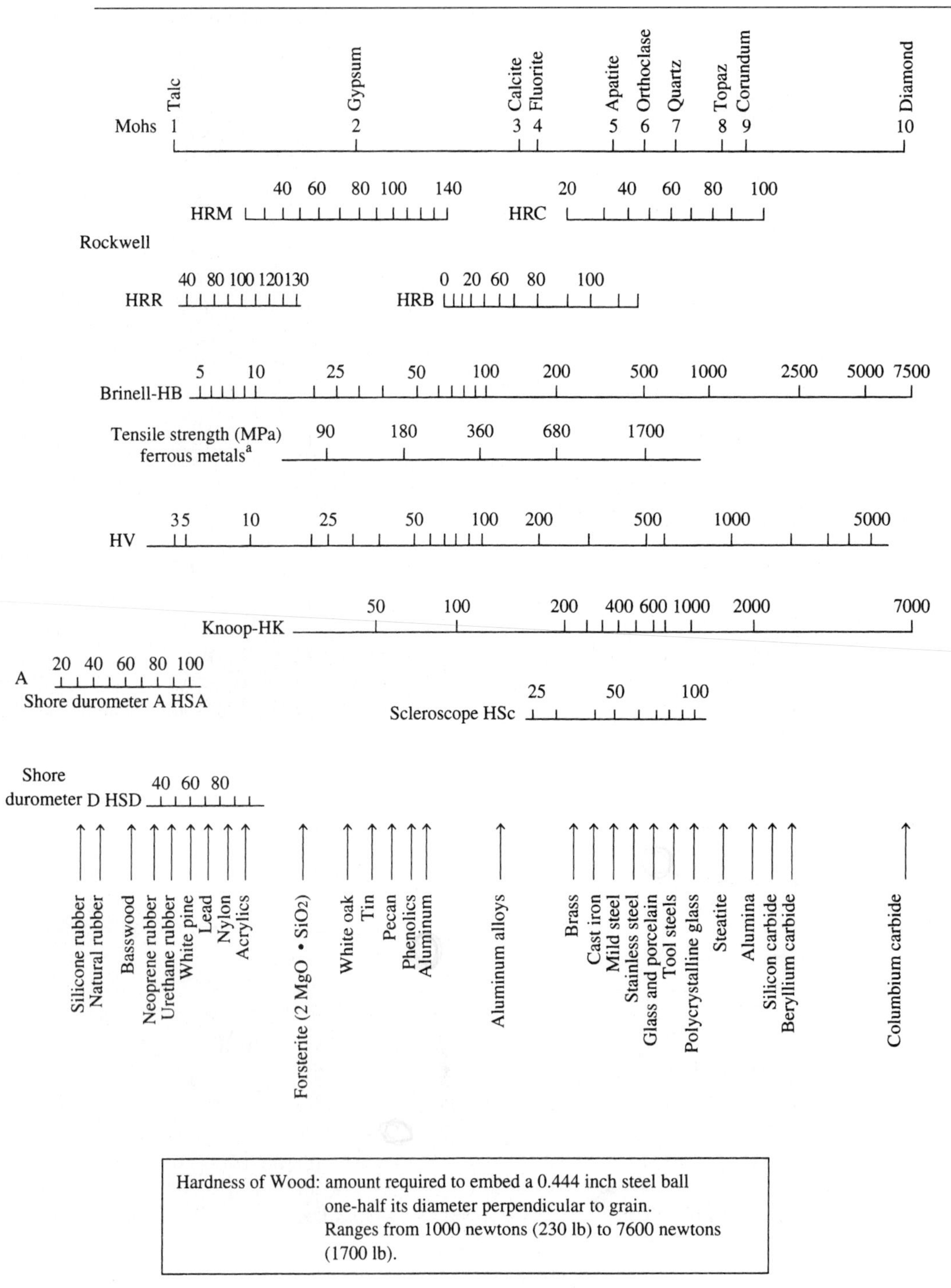

[a]HB × 3.6 for HB up to 175
HB × 3.4 for HB over 175

[a]HB × 3.6 for HB up to 175
HB × 3.4 for HB over 175

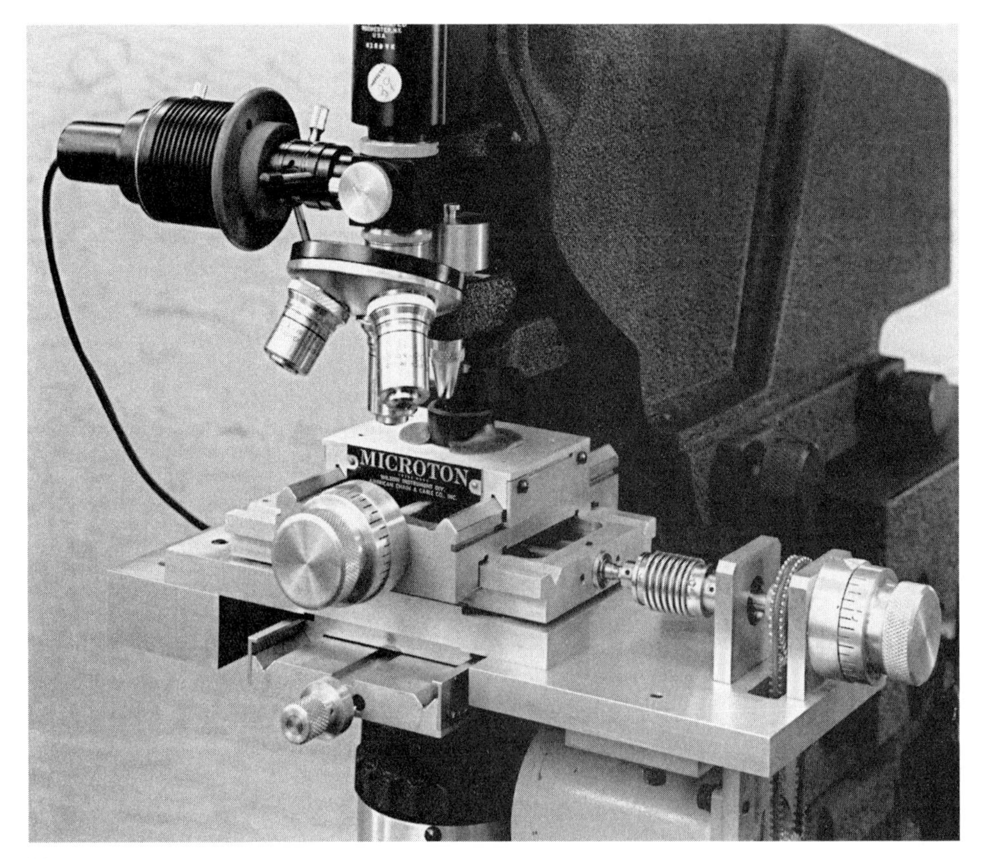

(a)

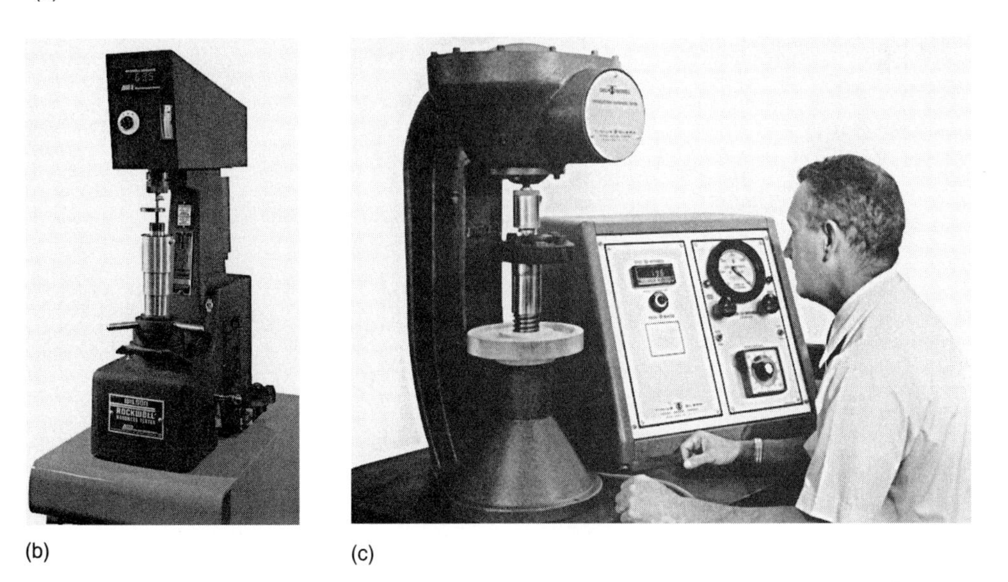

(b) (c)

Figure 4-30 (a) Photograph of Tukon microhardness tester. (Measurement Systems Div., Page Wilson Corp.) (b) Photograph of Rockwell hardness tester. (Measurement Systems Div., Page Wilson Corp.) (c) Photograph of Air-O-Brinell metal hardness tester with digital readout of Brinell values. (Tinius Olsen Testing Machine Co.)

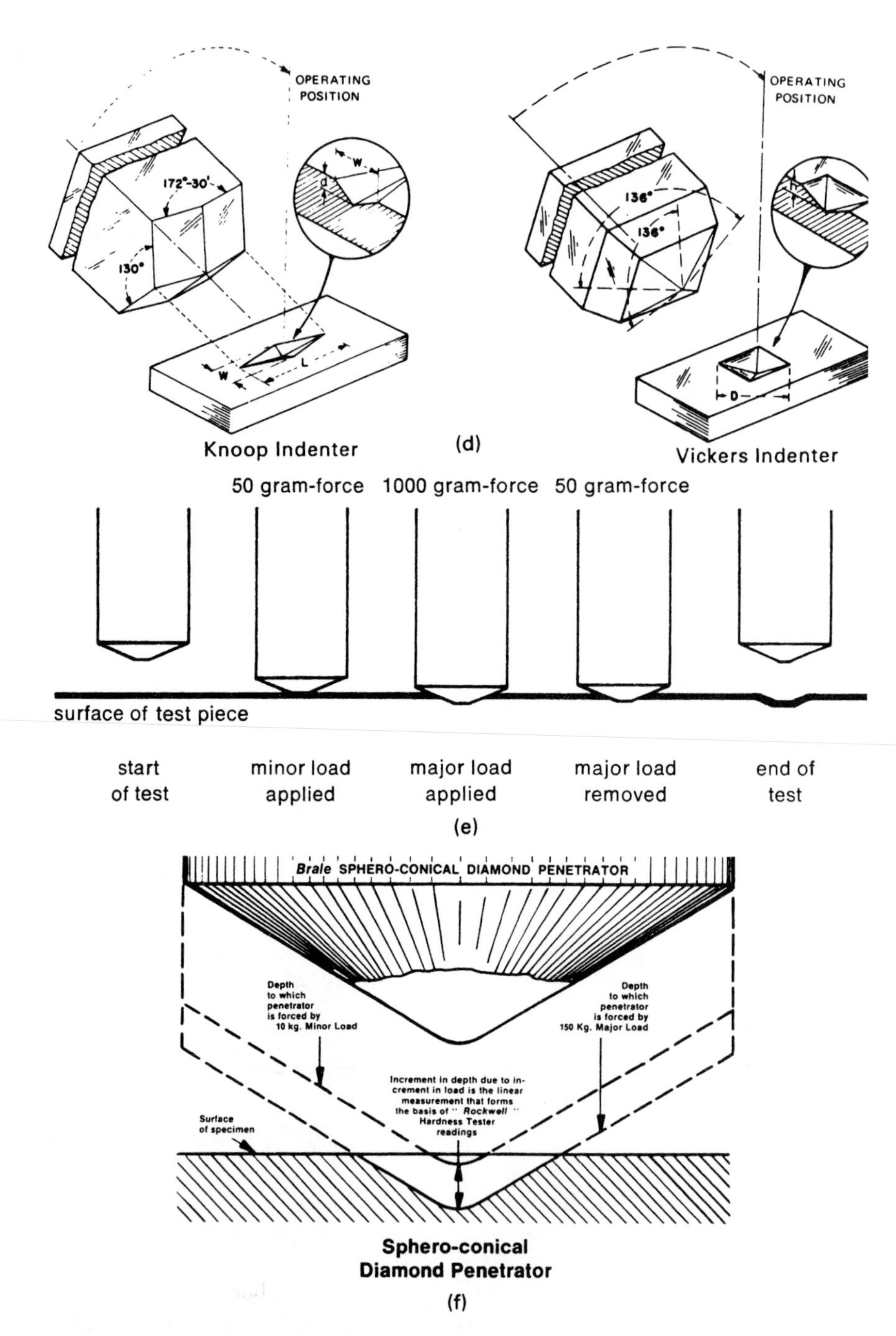

Figure 4-30 continued (d) Microhardness penetrator (Knoop and Vickers) indentations. (Wilson Instrument Division of ACCO) (e) Various standard loads for the Rockwell hardness tester. (Wilson Instrument Division of ACCO) (f) Brale sphero-conical diamond penetrator. (Wilson Instrument Division of ACCO)

induction tempered. Specified hardnesses of the crotch are 44 to 52 HRC; the claw tips, 47 to 55 HRC. The heads are then bulk tempered in a furnace at 600°F.

Barcol hardness is a surface hardness value obtained by measuring the penetration resistance of a given material to a sharp steel point under a spring load. The Barcol Impressor is an instrument that measures hardness on a 0–100 scale. Other tests used to report hardness include a file test (resistance to scratching), a ***Scleroscope*** HSc test (measures the rebound of a small weight bounced off the surface of the material) and a comparison test that also uses scratch resistance. The latter test compares a material's hardness to some ten known minerals arranged in order of hardness. Mostly used by mineralogists, the ***Mohs scale*** classes hardness of all materials between 1 (the hardness of talc) and 10 (hardness of diamond). The scale is based on the ability of a hard material to scratch a softer material (Table 4-1). Table A-15 is a graphical picture showing the relationships of several hardness scales to the approximate tensile strength of carbon and alloy steels.

Very thin materials (e.g., coatings, foils, plated surfaces), very brittle materials (e.g., glass or silicon), and very small parts (e.g., gears in a wristwatch) require special care in hardness testing due primarily to their thinness and/or size. Furthermore, laboratory research in such materials necessitates hardness testing on a microscale in determining the differences in hardness over the minute area of a single grain of metal or between the middle of a crystal and the grain boundary area. For such purposes a microhardness tester (Figure 4-30a) with loads and indentations that are so small that indentations require microscopic viewing with appropriate scales for accurate measurement is used.

Table 4-1 provides a comparison of the approximate hardness of a variety of materials using 11 different hardness scales. These readings may be compared to the tensile strengths of steel given in Table A-11. Some correlation exists between hardness and tensile strength, but it is only approximate. For example, the tensile strength of steel (but not other materials) is about 500 times the HB as listed in the table. As a general rule, the tensile strength of a given ductile metal can be estimated from the hardness reading only within an error of less than 10%.

Illustrative Problem

Given the formula for determining the Brinell hardness number (HB), where P is the applied load (kg), D is the diameter of the steel ball (mm), and d is the diameter of the indentation (mm),

$$HB = \frac{P}{(\pi D/2)(D - \sqrt{D^2 - d^2})}$$

what is the Brinell hardness number if a 10-mm ball with a load of 3000 kg produces an indentation with a diameter of 2.88 mm?

Solution

$$HB = \frac{3000}{\pi(10/2)[10 - \sqrt{100 - (2.88)^2}]}$$

$$= \frac{600}{\pi(10 - 9.576)} = \frac{600}{1.333} = 450$$

 Materials Science and Technology Gateway website—*http://MST-Online.nsu.edu*

4.3 NANOTECHNOLOGY

Applying nanotechnolgy to materials development offers means to greatly enhance physical and mechanical properties. Buckyballs and carbon nanotubes (CNT) allow modifying structure at the molecular level. Figure 4-31 lists some of the structural and mechanical applications possible with the improved properties of CNT. As Figure 4-32 illustrates, CNTs offer the

- High-strength composites
- Cables, tethers, beams
- Multifunctional materials
- Functionalize and use as polymer backbone plastics with enhanced properties like "blow molded steel"
- Heat exchangers, radiators, thermal barriers, cryotanks
- Radiation shielding
- Filter membranes, supports
- Body armor, space suits

Figure 4-31 Mechanical and structural applications with CNT.

- The strongest and most flexible molecular material because of C–C covalent bonding and seamless hexagonal network architecture.
- Young's modulus of over 1 TPa vs 70 GPa for aluminum, 700 GPa for C-fiber
- Strength-to-weight ratio 500 times > for Al; similar improvements over steel and titanium; one order of magnitude improvement over graphite/epoxy
- Maximum strain approx. equal to 10%, much higher than any material

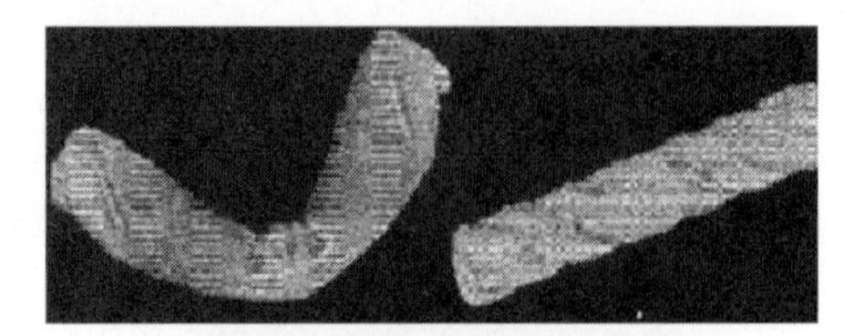

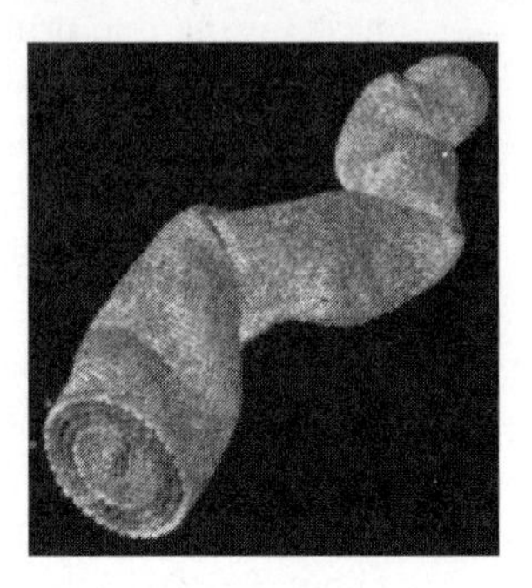

Figure 4-32 Carbon nanotube properties.

strongest and most flexible molecular material due to carbon-to-carbon covalent bonding and architecture with a seamless hexagonal network. As the figure shows, theses structures dramatically increase strength-to-weight ratios and increase maximum strain.

SELF-ASSESSMENT

4-1. Cite one property of a material that cannot be observed.

a. Conductivity **b.** Hardness **c.** None can be observed **d.** All can be observed

4-2. Give two examples of a mechanical property.

a. Thermal resistance **b.** Wear resistance **c.** Hardness **d.** Strength

4-3. Scissors used in the home cut material by concentrating forces that ultimately produce a certain type of stress within the material. Identify this stress.

a. Bearing stress **b.** Shearing stress **c.** Compressive stress

4-4. Given m for a material as 0.25 and the strain in the lateral direction as 1.50×10^{26} mm/mm, what is the axial deformation expressed in inches?

a. 6×10^{26} in./in. **b.** 0.6×10^{25} in./in. **c.** Both (a) and (b)

4-5. An aluminum rod 1 in. in diameter ($E = 10.4 \times 10^6$ psi) experiences an elastic tensile strain of 0.0048 in./in. Calculate the stress in the rod.

a. 49,920 ksi **b.** 49,920 psi **c.** 49,920 msi

4-6. Express the stress in Question 4–5 in SI units with an approved prefix.

a. 0.3442 GPa **b.** 3.442×10^8 Pa **c.** 34.42 MPa

4-7. A 6-ft steel bar is deformed 0.01 in. in an axial direction. What is the unit deformation?

a. 1.389×10^{24} **b.** 1.389×10^{24} in./in. **c.** 1389×10^{27} in./in.

4-8. A shaft supported by thrust bearings placed 4 ft apart will fail if the total deformation exceeds 0.00025 in. What is the maximum allowed strain?

a. 3.7×10^{26} in./in. **b.** 5.2×10^{26} in./in. **c.** 15.6×10^{25} in./in. **d.** None of the above

4-9. The minimum yield stress for a material is 48,000 psi. What is the factor of safety if the allowed working stress is 24,600 psi?

a. 0.51 **b.** 1.95 **c.** 3.90

4-10. Express the yield stress units of Question 4–9 in megapascals, the recommended units in the SI system. Use appropriate abbreviations.

a. 331 MPa **b.** 331 GPa **c.** 331 MPA **d.** All of the above

4-11. A 1-in.-diameter steel circular rod is subject to a tensile load that reduces its cross-sectional area to 0.64 in^2. Express the rod's ductility using a standard unit of measure.

a. 18.5% **b.** 1.85% **c.** 18.5 **d.** (a) and (c)

4-12. If a force (push or pull) of 40 lb is used to turn a wrench such that the 40-lb force is applied at right angles to the wrench at a distance of 16 in. from the center of its jaw, what torque is produced?

a. 640 in.-lb **b.** 640 lb-in. **c.** 640 #-in. **d.** (b) and (c)

4-13. An angle of twist of 0.48° can be expressed as how many radians?

a. 8.4×10^{23} **b.** 84 rad **c.** 0.0084 rad **d.** (a) and (c)

4-14. A standard-weight pipe with a nominal diameter of 1 in. has an outside diameter (OD) of 1.315 in. and a thickness of 0.133 in. Calculate its polar moment of inertia.

a. 0.247 in^3 **b.** 0.247 in^4 **c.** 2.47×10^{-3} in^4 **d.** (a)–(c)

4-15. Determine the bending stress on the top surface of a rectangular beam whose height is 11 1/2 in. if its moment of inertia is 951 in^4 and its bending moment is 82,856 lb-in.

a. 499.34 psi **b.** 499.34 lb/sq. in. **c.** 499.34 #/in^2 **d.** (a)–(c)

4-16. What term is used to describe the low-temperature creep of polymerics?

a. Springback **b.** Creep rupture **c.** Cold flow **d.** Creep forming

4-17. What two factors can increase the creep rate of a material?

a. Pressure **b.** Temperature **c.** Stress **d.** Humidity

4-18. What hardness test expresses hardness on a scale of 0 – 100?

a. Rockwell **b.** Shore **c.** Barcol **d.** Brinell

4-19. A hardenable carbon steel has a Rockwell C hardness of 50. What is its approximate tensile strength using Table A-15?

a. 258 psi **b.** 255,000 psi **c.** 1700 MPa **d.** 255 MPa

4-20. A 1-in.-diameter steel rod 3 ft long is used between two immovable (rigid) walls. The rod is welded to the two walls at each of its two ends. The ambient temperature increases by 100°F. The modulus of elasticity for the steel is 30×10^6 psi and the CTE is 6.5×10^{-6}/°F.

a. Make a sketch of the problem using appropriate welding symbols. Note on the sketch that the walls do not move or deform. This sketch will show the rod welded at both ends to the walls.

b. Describe what is happening to the rod. What is acting on the rod from the outside? What is happening to the atoms in the steel rod?

c. If the rod were free to move, describe its movement and calculate how much it would move and in what direction.

d. Make another sketch with the walls replaced by force vectors representing the walls acting on the rod.

e. Calculate the stress acting inside the rod.

f. Calculate the force on the rod that would be necessary to produce the same stress as calculated in part (e).

4-21. A 3000-kg load using a 10-mm indentor produces a 3.2-mm impression on a ferrous metal specimen that yields a Brinell hardness of 364 (HB = 364). Using the relationship of HB to the tensile strength (TS), determine the approximate TS value for the material. Also, knowing the relationship between TS and the endurance limit (EL), determine the EL value for the material.

4-22. Refer to Table 4–1 and the accompanying text. What prefix would be added to a Rockwell hardness reading taken from a major load of 150 kg with a diamond penetrator?

REFERENCES & RELATED READINGS

American Society for Testing and Materials. *1996 Annual Book of ASTM Standards.* Philadelphia, PA: ASTM, 1996.

American Society for Testing and Materials. *UNSearch, Version 2.0,* software. Philadelphia, PA: ASTM, 1992.

ASM International. *ASM Handbook,* VOL. 1, *Properties and Selection: Irons, Steels and High Performance Alloys;* VOL. 2, *Properties and Selection: Nonferrous Alloys and Special Purpose Materials*; VOL. 8, *Mechanical Testing*, Materials Park, OH: ASM International, 2001, 1989.

ASM International. "MATERIALS SELECTION," *Advanced Materials and Processes,* Special Issue, June 1999.

Cubberly, W. H., and Ramon Bakerjian, eds. *Tool and Manufacturing Engineers' Handbook.* Dearborn, MI: Society of Manufacturing Engineers, 1989.

Ferrara, Michael. "STRESS RELIEF META-LAX EXTENDS LIFE OF ENGINE COMPONENTS," *Turbo & Hi-Tech Performance,* May 1995, pp. 62–65.

Johnston, Norman J. "INTRODUCTION TO HIGH PERFORMANCE COMPOSITES," *National Educators' Workshop, NEW: Update 92,* Oak Ridge National Laboratory, Oak Ridge, TN, November 13, 1992.

McGhee, Douglas B. "IMPACTS OF NEW HARDNESS STANDARD," *Advanced Materials and Processes,* October 1999, pp. 165–167.

Oberg, E., et al. *Machinery's Handbook.* New York: Industrial Press, 2003.

Provan, James W. "AN INTRODUCTION TO FATIGUE," *Journal of Materials Education,* VOL. 11, NO. 1/2, 1989, pp. 1–105.

Periodicals

Advanced Materials and Processes
ASTM Standardization News
Geotechnical Testing Journal
Journal of Composites Technology and Research
Journal of Materials Education
Journal of Materials Engineering and Performance
Journal of Testing and Evaluation
Materials Performance

Materials World Modules—*http://mwm.ms.nwu.edu/*

Module

5

Chemical, Thermal, and Tribological Properties & Corrosion

After studying this module, you should be able to do the following:

5–1. Use diagrams with explanations to describe electrochemical reactions in materials.

5–2. Explain the causes of corrosion using sketches for clarity.

5–3. Know methods of corrosion protection.

5–4. List thermal properties and calculate thermal effects on materials.

5–5. State how materials degrade due to water absorption and biological attachments and give some methods to prevent such degradation.

5.1 ELECTROCHEMISTRY—ELECTROLYSIS, CORROSION, AND ITS PREVENTION

Electrochemistry is the study of chemical reactions that produce electricity and of how electricity can be used to cause chemical reactions. Electricity drives chemical reactions, as in electrolysis and the refinement of aluminum. Oxidation and reduction (redox) reactions (discussed later) involve the transfer of electrons and, by definition, when electrons move there exists an electrical current. In this technological world the many electronic devices that permeate our existence, like handheld computers, portable radios, watches, and automobiles, depend on electricity produced by chemical reactions. An electrochemical cell is a device that uses oxidation and reduction reactions to produce an electrical current. Flashlights, pocket calculators, portable computers, and watches are just some of many devices that depend on batteries that use oxidation-reduction (redox) reactions. Fuel cells depend on oxidation and reduction reactions that convert chemical energy directly into electrical energy. Many industrial processes use electrical energy to generate chemical reactions that, in turn, produce a particular product. An understanding of the working of an electrochemical cell (see Figure 5-1), which will be discussed in Section 5.1.3, can be put to good use not only in designing materials and structures that can resist corrosion but also in the proper selection of materials that will be subjected to some adverse environmental condition. Although they are beyond the scope of this book, studying the laws of thermodynamics and the concepts of enthalpy, entrophy, and free energy will provide a greater understanding of the chemical reactions that can be obtained. ***Thermodynamics*** is a branch of physics that describes in terms of ***macroscopic*** variables (detectable by our senses) the reversible transformations of heat into other forms of energy and the laws that govern such conversions.

5.1.1 Chemical Properties and Reactions

Chemical properties are those that determine how a material interacts with gaseous, liquid, or solid environments. The most familiar chemical property is ***corrosion resistance,*** the ability

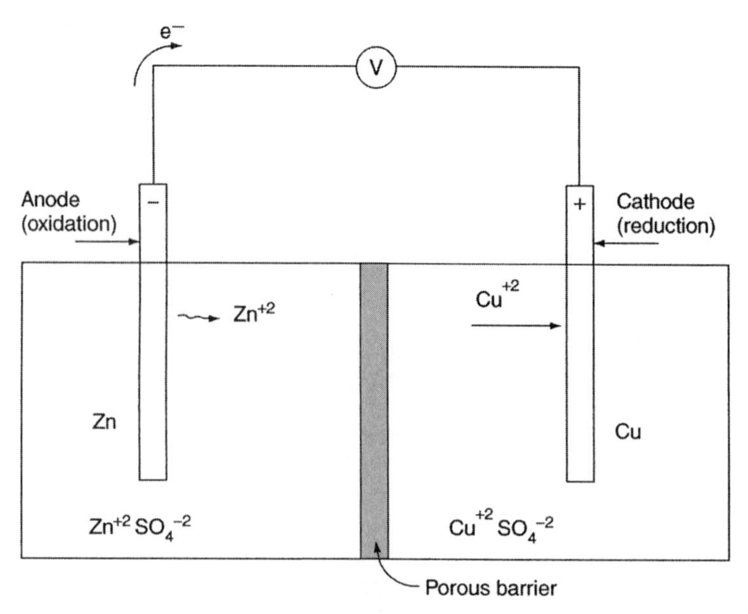

Figure 5-1 Electrochemical cell. This device uses a spontaneous chemical reaction to produce an electric current as in flashlights, pocket calculators, portable radios, and many other devices that rely on batteries. The porous partition allows the movement of ions through the cell but prevents the electrolyte solution in each half-cell from mixing. The electrons lost by oxidation of the zinc atoms leave the Zn electrode and flow through the external wire to the copper electrode, where they reduce the Cu ions and, thus, the Cu atoms coat the surface of the copper electrode.

to resist oxidation. Common examples include the ability of aluminum to resist rusting when exposed to air and moisture and the ability of rubber to withstand sunlight (ultraviolet rays) without drying and cracking. Many conditions or environments threaten materials. Polluted air, for example, is filled with elements emitted from gasoline engines, home furnaces, and industrial plants; these elements combine with materials and cause damage that shortens their service life. The American Lung Association has a Healthy Home program that certifies new homes for good indoor air quality. The program's stringent requirements include a sealing out of moisture and pollutants with a tight, well-insulated building shell and foundation and filtering all air supplied to the house. Air pollution is also damaging some of our ancient architectural treasures. The Parthenon (an ancient Greek temple) and copper and stone statues are rapidly deteriorating from corrosive, polluted air. Corrosion in automobiles is promoted by salty water from ocean spray or road salt used on icy roads. In addition, some materials react in a hazardous way to high temperatures and fire. Many plastics emit poisonous gases when burned. But sometimes it is desirable for materials to deteriorate or be biodegradable so that the natural environment can break down this material, thus reducing solid waste.

5.1.2 Electrolysis

Electrolysis is the decomposition of a material by the passage of electricity through it. The electrolysis of water is an excellent example of this chemical reaction, brought about by the flow of electrons in which water is broken down into the two gases of oxygen (O_2) and hydrogen (H_2). Figure 5-2 is a schematic of the apparatus needed to decompose water by electrolysis. A few drops of hydrosulfuric acid (H_2SO_4) are added to water to make it more conductive. The acid is an excellent electrolyte, which allows for the migration of ions through the water from one metal electrode to the other. The external circuit provides the surface of the cathode with electrons, which transfers them to the electrolyte containing the H ions. Hydrogen ions are attracted to the cathode, where they pick up or gain electrons to become H atoms. *Reduction*, the gaining of electrons, thus takes place and atoms of H are formed. As atoms, they combine with other H atoms to produce molecules of hydrogen gas (H_2). At the anode, the water molecules (H_2O) that have lost electrons decompose, releasing O_2. This process is called *oxidation*. The oxidizing reaction that occurs at the anode results in the metal atoms of the electrode being ionized; these then enter the electrolyte, and because the metal ions leave the anode, it corrodes. When electrons are lost from the metal of the anode in the ionizing process and leave through the electrical connection, they are first transferred from the electrolyte to the surface of the anode, where oxidation takes place.

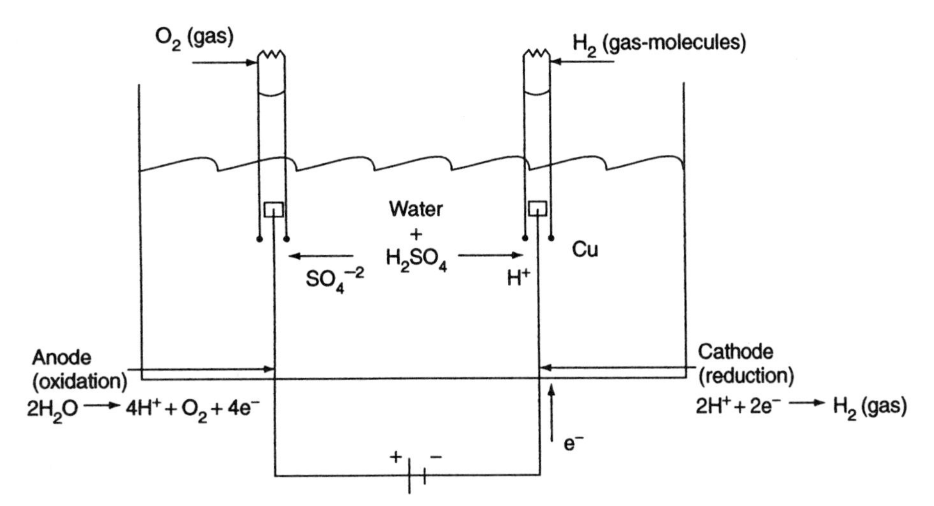

Figure 5-2 Electrolysis of water. H^+ ions gain electrons at the cathode and form H atoms. H atoms then form H_2 molecules of gas. At the anode, the H_2O decomposes to replace the H^+ ions lost at the cathode and release O_2.

The chemical changes that take place (oxidation and reduction) can be added together and the net effect is represented by the following chemical equation, which represents both the oxidation and reduction reactions.

$$2H_2O \rightarrow 2H_2 + O_2$$

The electric charge is carried through the electrolyte not by the free electrons, but by the movement of positive ions toward the negative electrode (cathode) and the movement of negative ions toward the positive electrode (anode). This migration of ions, referred to by some as the *internal circuit*, is known as ***ionic conduction.*** The external circuit connecting the electrodes with a source of electrical energy contains charge carriers that are the free electrons. The flow of electrons in this circuit is known as ***metallic*** or ***electronic conduction.***

Some examples of electrolysis used in industry are the refining of aluminum from bauxite, the production of chlorine from salt, the production of sodium metal (Na) and chlorine gas (Cl) (shown schematically in Figure 5-3), and electroplating of metals such as chromium and silver. In this last process, the object to be plated is made to be the cathode and the plating material, the anode. In ***electroplating*** (see Figure 5-4) the combining of metal ions and electrons (purposely augmented by an electrical potential) produces a metal that plates out and covers the cathode surface. Electricity can produce chemical change by the process of electrolysis. The synthesis reaction discussed in Section 5.1.3 can be reversed as in the case of molten salt. Molten salt can be changed into sodium, a metal and chlorine (a gas), by passing an electric current through it.

5.1.3 The Electrochemical Cell

It is not possible to directly observe the transfer of electrons, the formation of ions, or the formation of compounds in redox reactions, but by separating these two dependent reactions, the electron transfer can occur through an external circuit and be observed by instruments. Such a device is called an ***electrochemical cell,*** which uses a spontaneous chemical reaction to produce an electric current. In simpler terms, an electrochemical cell contains two pieces of different metals in contact with each other that are placed in a liquid medium capable of conducting ions, an electrolyte. Known as a ***voltaic*** or ***galvanic cell,*** an electrochemical cell consists of two half-cells with the anode (negative electrode) and the cathode (positive electrode) both connected by a wire to provide the external circuit (Figure 5-4). Both electrodes are dipped into their respective electrolytic solutions, which permit the passage of ions from

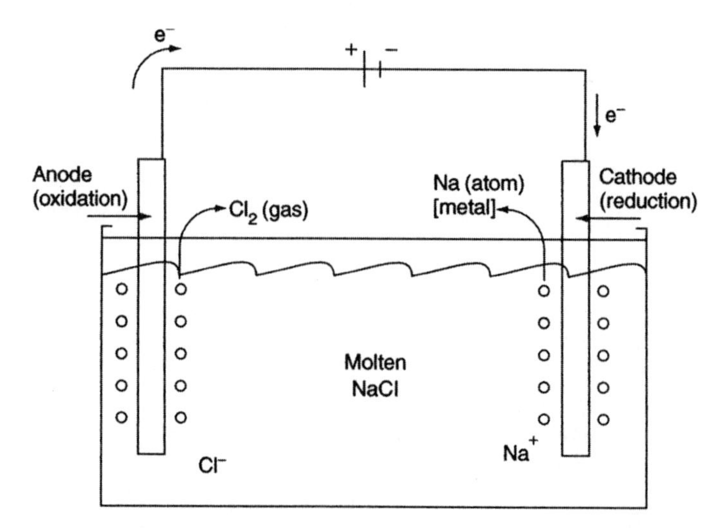

Figure 5-3 Electrolysis of NaCl. At the anode, Cl^- ions give up electrons to the electrode to form Cl atoms, which combine to form Cl_2 molecules (gas). At the cathode, Na^+ ions accept e^- from the cathode to produce Na atoms, which form liquid Na metal.

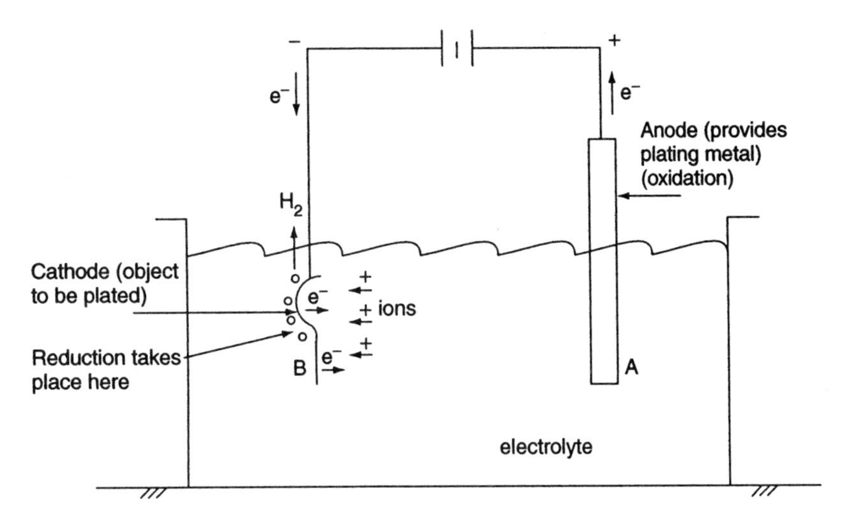

Figure 5-4 Schematic of electroplating (electrochemical plating). The anode acts as a sacrificial electrode supplying the plated metal. Tin plating of steel is done mostly by this method. Metal ions from the anode migrate to the cathode and are reduced to atoms by gaining electrons from the cathode. Chromium provides the hardest plating metal for applications requiring wear resistance.

TABLE 5-1 ORDER IN WHICH DISSIMILAR METALS PRODUCE ELECTROMOTIVE-FORCE (GALVANIC) SERIES

	Electromotive series (corrosion potential)	
↑ Greater oxidizing agent	Magnesium 2+[a]	Active or Anodic (greatest corrosion) more active ↑
	Beryllium	
	Aluminum 3+	
	Zinc 2+	
	Chromium 3+	
	Iron (steel) 2+, 3+	
	Cadmium 2+	
	Tin 2+	
	Nickel 2+	
	Cobalt	
	Lead 2+	
	Copper 1+	
	Silver 1+	
	Titanium 1+	
	Platinum 2+	
Greater reducing agent ↓	Gold 3+	Noble or Cathodic (least corrosion) more passive ↓

[a] Oxidation numbers.

Anodic coatings are more active in the galvanic series than the substrate upon which the coatings are applied. Noble coatings refer to metal surface coatings that are less active in the *galvanic* series than the substrate upon which the coatings are applied.

one electrode to the other (the internal circuit), thus completing the entire electrical circuit. The electrons produced by an oxidation reaction at the anode travel through the external circuit to the cathode, where they cause the reduction reaction to occur. Anions given off at the cathode move toward the anode, and cations move in the opposite direction, thereby completing the circuit inside the cell. When a metal ion in solution gains electrons at or near the electrodes, the porous barrier keeps the two solutions from mixing without impeding the movement of the ions between the electrodes. The flow of electrons pushed out by the anode and flowing through the external circuit produces a potential difference, which is measured by a voltmeter. By using this technology with one hydrogen electrode, standard potentials for the various metals and alloys can be measured and used to construct tables such as the activity series of metals and Table 5-1. An electrochemical cell is capable of defining either electrochemical

ELECTRON CONDUCTOR

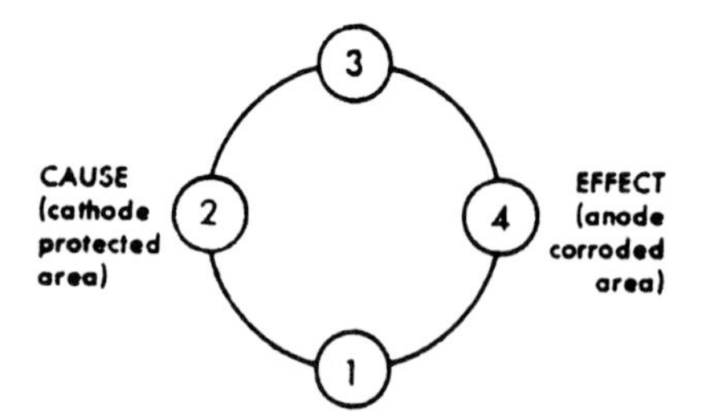

CONTINUOUS LIQUID PATH (electrolyte)

1. Electrolyte (continuous liquid path, usually water in the form of condensate, salt spray, etc.)
2. Cause (cathode—the cause of corrosion—the area through which electricity flows.)
3. Electron conductor (in a structure, usually a metal-to-metal contact, e.g. rivets, bolts, spot welds, etc.)
4. Effect (anode—the surface or object which corrodes.)

Figure 5-5 The corrosion circle, a graphic picture of the four essential elements needed for corrosion to occur.

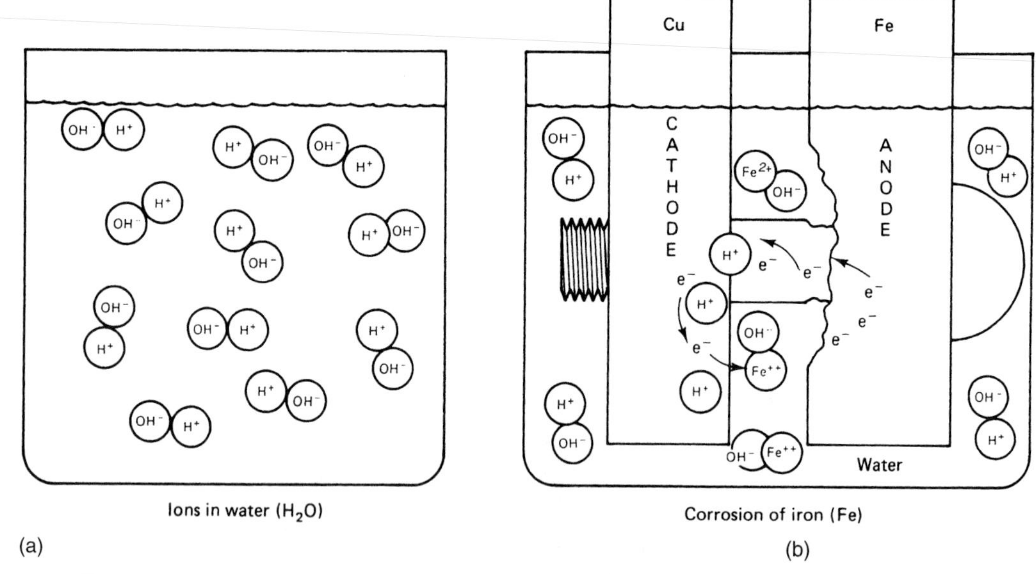

Figure 5-6 (a) Electrochemical corrosion depicting ionized water, and (b) the corrosion of iron (Fe).

corrosion or electroplating. These processes differ in their purpose, in the source of the electric potential by which a current is caused to flow, and by what type of chemical reaction occurs at the cathode.

The electrochemical cell illustrated in Figure 5-5 is also helpful in explaining electrochemical corrosion. The cell consists of four basic elements: physically connected electrodes, an anode, a cathode, and electrolytes (electrolytic solution) such as water that contains ions. The anode is the electrode that loses or releases electrons as a result of oxidation occurring at that electrode. The cathode is the electrode that gains or receives electrons as a result of reduction occurring at that electrode. Figures 5-6 and 5-7 graphically represent these conditions. When corrosion occurs, two chemical reactions occur simultaneously—oxidation and reduction. These two simultaneous processes are known as a ***redox reaction*** or process. Oxidation occurs at the anode while reduction takes place at the cathode. To prevent or control

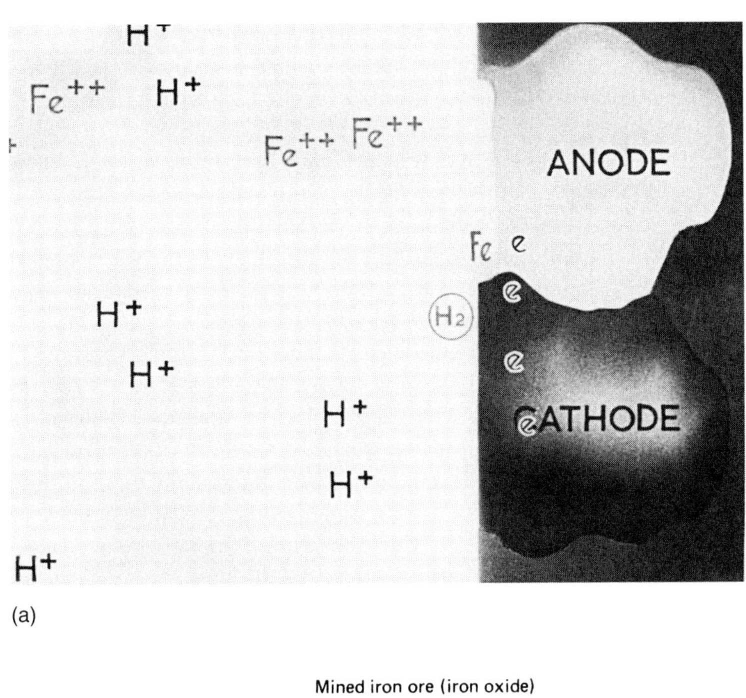

(a)

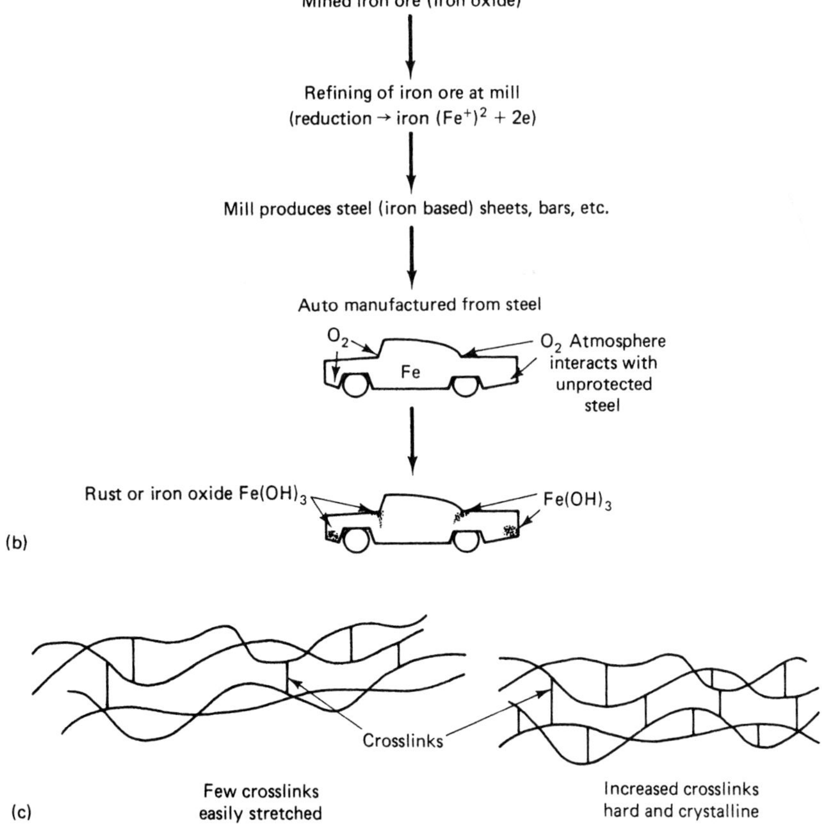

(b)

(c)

Figure 5-7 (a) Stress cells. (b) Intergranular corrosion. (c) Stress cells (anodic regions or anodes) in a bent nail produced by cold working will corrode faster than other areas. A cold-worked steel bar that is bent produces a similar condition.

electrochemical corrosion, it is only necessary to eliminate one of the four basic elements. The flow of electrons is from the anode (negative terminal or electrode) through the external wire connection to the cathode (positive electrode or terminal). At the same time the electrical current composed of ions leaves the anode, enters the electrolyte, and proceeds to the cathode, where it combines with the electrons that have flown through the external circuit to form some solid metallic material such as rust. Note that the anode is being corroded because of its loss of electrons and the formation of metallic ions. In actual corrosion circumstances, corrosion occurs because the anodic and cathodic areas on the surface of a material are connected by moisture or some other ionic solution that acts as an electrolyte, thus providing all four elements needed to complete an electrochemical cell.

Electrolytes are substances that give conducting solutions when mixed in water. Most electrolytes are dissolved in water by dissociation; that is, the molecules break down into ions (see Figure 5-6). Water is a molecular substance that has only a slight electrical conductivity, but when an ionic solid is dissolved in it, the ions from the solid separate from each other in the solution and can move about carrying their charge with them. Molten salt is an electrolyte. Figure 5-6 shows polyionic ions in an aqueous electrolyte. In ***dry cell batteries*** (see Figure 11-21, which describes a nickel–metal hydride, a primary cell battery), the electrolyte is a moist paste that fills the space between the electrodes. The alkaline dry cell battery (primary cell) uses potassium hydroxide (KOH) in the form of a gel as an electrolyte. Dry cell batteries are classified as ***primary cells*** because once the reaction ends (reaches equilibrium), the battery is dead and cannot be recharged. The ***lead–acid battery*** or ***secondary cell*** can be recharged by reversing these reactions by passing an electric current through them that has a voltage of opposite sign and is larger than the cell voltage. A dilute sulfuric acid solution is the electrolyte. Figure 11-21 includes a description of how a typical power cell (secondary cell battery) works. A 12-V battery has six single lead cells connected in series.

Chemical reactions occur when the composition of a material is changed and the material acquires different physical and chemical properties (rusted metal, a boiled egg, or tarnished silver) from the original material. We learned in Module 1 that physical changes do not change the composition of the materials, as in the case of freezing water or dissolving sugar in a glass of water. A chemical reaction is a process in which some chemical bonds are broken, others are formed, and the chemical energy changes as a result of the reaction. These reactions can produce electricity and electricity can cause chemical reactions. Chemical reactions can be divided into two broad classes: synthesis and decomposition or dissociation. A ***synthesis reaction*** has two elements combining to produce a compound. Metals react with nonmetals to produce a compound that is generally ionic.

Metals may be considered ***donor substances*** because they donate or lose one or more electrons in an electrochemical process; nonmetals are sometimes called ***acceptor substances*** because they accept or gain one or more electrons. For example,

$$2Na + Cl_2 \rightarrow NaCl$$

The product in this reaction is table salt; the two separate metals, the reactants in the form of elements, are, however, dangerous to humans. The ionic bonding of this compound was discussed in Module 2 along with the terms *ionization, ions, anions*, and *cations*. ***Polyatomic ions*** are charged particles containing two or more covalently bonded atoms, for example, OH^{-1} (hydroxide ion) and SO_4^{+2} (sulfate ion). Note that the atoms of hydrogen (H_2) and oxygen (O_2) in the hydroxide ion are covalently bonded together, plus there is an "extra" electron, which gives this molecule a negative charge (see Figure 5-6 showing the hydroxide ion in an aqueous electrolyte). The electric charge on an ion is referred to as its ***oxidation state***. This concept will be explained further when discussing galvanic corrosion. If an electric circuit is passed through the water, it can be decomposed into hydrogen and oxygen. This ***dissociation reaction*** can be shown as

$$2H_2O \rightarrow 2H_2 + O_2$$

The products formed are two gases.

5.1.3.1 Oxidation-reduction (redox) reactions and electron transfer reactions.

As we have said, the two previous chemical equations are examples of oxidation-reduction (redox) reactions. In the past, *oxidation* was defined as the addition of oxygen to an element or compound and *reduction* as the removal of oxygen from an oxide. This was so because many elements combine with oxygen and oxygen is very abundant on the earth's surface in the form of metal oxides. To free these metals from their oxides requires the chemical process of reduction. Iron is separated from its oxide using this process:

$$Fe_2O_3 + 3CO \rightarrow 2Fe + 3CO_2$$

Carbon monoxide (CO) removes the oxygen from the iron oxide and is thereby converted to carbon dioxide (CO_2). The iron oxide is said to be reduced to iron. At the same time, the CO is oxidized to CO_2. Figures 5-8a and b depict the refining of iron and the rusting of iron; Figure 5-8c shows the oxidation of polymeric materials such as old erasers, rubber bands, and automobile tires by the addition of cross-links in the resulting network structure. Because many other substances can be used to reduce oxides to their elements, the name for oxidation and reduction reactions has been broadened to ***electron-transfer reactions***, to reflect the fact that electrons are lost or gained in the process. Electrons are transferred from an element that is oxidized to an element that is reduced. Substances that can oxidize other substances are said to be oxidizing agents. Any substance from which electrons can be readily removed, such as a metal, may reduce other substances and is therefore a reducing agent. In sum, the reaction between a metal and a nonmetal is an electron-transfer reaction in which the metal behaves as a reducing agent and is oxidized and the nonmetal behaves as an oxidizing agent and is reduced.

For greater understanding, these two processes, oxidation and reduction, can be separated into two parts, called ***half-reactions***. A separate equation represents each process. The two half-reaction equations can then be added together to construct an equation that represents the entire reaction. Using the synthesis equation for forming the compound of table salt from sodium and chlorine can demonstrate this procedure:

$$2Na + Cl_2 \rightarrow 2NaCl \qquad \text{(the overall equation)}$$

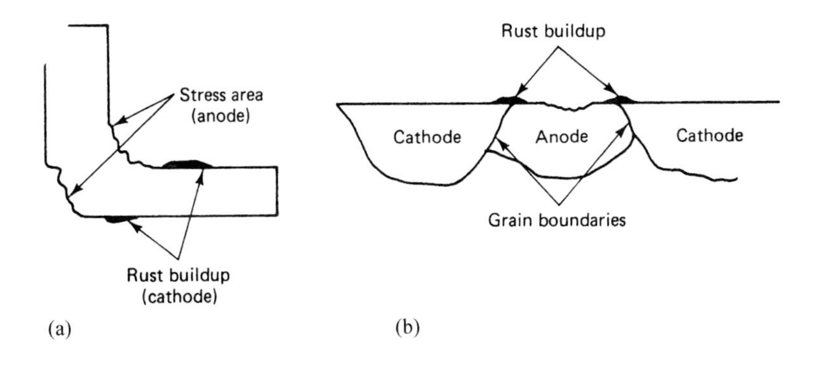

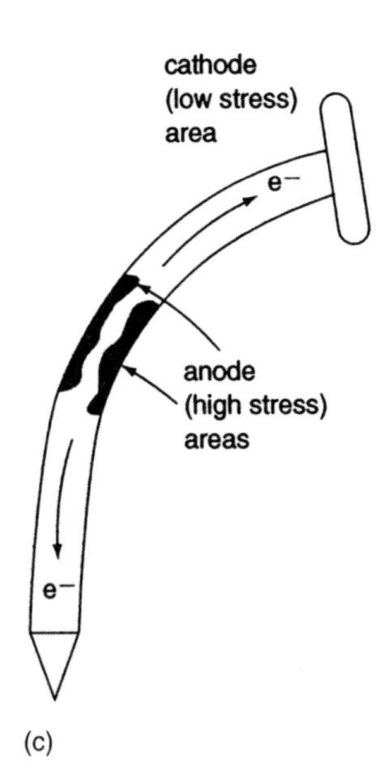

Figure 5-8 (a) Oxidation of iron. Iron ions form at the anode and hydrogen at the cathode in local cell action. (INCO.) (b) Cycle of iron and steel seeking natural equilibrium. (c) Typical thermosetting polymer, increasing the number of cross-links produces a network structure.

The half-reaction representing oxidation is

$$2Na \rightarrow 2Na^+ + 2e^-$$

The half-reaction representing reduction is

$$Cl_2 + 2e^- \rightarrow 2Cl^-$$

Note in both half-reactions that ions are formed. When combined in the final equation to form the compound NaCl, both Na and Cl are each in a balanced electron state. In our discussion of electrolysis, the two elements were separated from their compound form to produce the metal sodium and the gas chlorine (see Figure 5-3).

5.1.4 Corrosion of Metals

A recent study done with the cooperation of the National Association of Corrosion Engineers (NACE) showed that the total cost of metallic corrosion to the U.S. economy is $279 billion, or approximately 3.2 % of the nation's Gross Domestic Product. Among the many recommendations of the study were achieving better corrosion management by using preventive strategies at every level of involvement (owner, operator, user, government, and general public), increasing awareness of large corrosion costs, changing the misconception that nothing can be done about corrosion, improving education and training of staff, using advanced design practices for better corrosion management, and advancing corrosion technology through research, development, and implementation. The poem "Rust's a Must" pretty well summarizes the need to recognize that the corrosion of metals merits constant attention.

Rust's a Must

Mighty ships upon the ocean suffer from severe corrosion,
Even those moored at dockside are rapidly becoming oxide.
Alas, that piling in the sea is mostly Fe_2O_3.
And where the ocean meets the shore, you'll find there's Fe_3O_4.
Cause when the wind is salt and gusty, things are getting awful rusty.
We can measure, we can test it, we can halt it or arrest it;

We can gather it and weigh it; we can coat it, we can spray it;
We examine it and dissect it; we cathodically protect it.
We can pick it up or drop it, but heaven knows, we'll never stop it.
So here's to rust, no doubt about it: most of us would starve without it.

T. R. B. Watson, 1916–1996, author and poet, President of National Association of Corrosion Engineers International, 1964–65. *Advanced Materials and Processes*, March 2002, with permission.

Without any protection, most metals will oxidize (corrode). According to the Battelle Institute in Columbus, Ohio, the corrosion of metals amounts to a loss of $300 billion annually. More than 200,000 bridges in the United States are in need of major repairs because of it. Designed to last for 45 years, bridges are failing in 12 to 18 years. There are two aspects to corrosion loss, in that as products and equipment corrode they must be replaced by expending limited material resources, plus the corroded materials end up in a landfill if not recycled. The production of new replacement equipment also means the expenditure of energy resources that may also produce pollutants harmful to the world's atmosphere, as in the case of burning fossil fuels. Corrosion, an electrochemical process is also, unfortunately, a natural one. Most metals are found in nature in the form of compounds such as a metallic oxide, gold being the one metal that is found in its metallic form. Once metals are freed from their combined form, they are driven to return to their natural states, that is, to reach their equilibrium state or state of lowest possible energy. This tendency is the root cause of the electrode potential possessed by metals that has been described briefly in this module.

Ceramics do not possess this tendency, because they occur in nature as compounds and remain so while in use. Therefore, they do not possess the force that would result in their corroding. Their natural form also accounts for poor conductivity of ceramics, because their electrons are all locked up in bonding their atoms. Glasses that contain Na or K ions are, however, subject to leaching, in which these ions diffuse from the surface of the glass into the surrounding medium, which is many times a liquid such as water. These ions belong to the alkali metals (Group 1A) in the periodic table, which we have described in previous modules as being highly active with only one valence electron per atom. This fact results in rather weak bonding strength. Polymer materials will also corrode, but they do so only because of some chemical reaction between their environment and the molecules that comprise these materials.

Corrosion may be defined as the deterioration of a material by interaction between the material and its environment. It is not considered an inherent property of a material because corrosion is the result of the material's reaction with its environment. Remove the environment and no corrosion will occur. The root cause of corrosion is the instability of materials, particularly metals, which leads them to revert to their natural states (to seek equilibrium at a lower state of energy). Materials differ in the amount of this tendency, or electric potential. Iron rusts because it has a large driving force to revert to equilibrium. To determine if a particular metal will corrode in the presence of another metal (dissimilar metals), its potential (redox potential) is measured and tabulated along with that of other metals. Many environments include atmospheric air containing moisture that provides for ionic conductivity. Metals in aqueous (containing water) environments always disturb the electrochemical equilibrium, with the result that electrochemical reactions whereby metals are transformed into the ionic state and the products of corrosion, including the loss of material and the formation of nonmetallic scale occur. This change from a metallic into a nonmetallic state results in the lowering of the energy of the system as the metal converts to a lower-energy form. Corrosion of steel, for example, produces rust, a nonmetallic corrosion by-product. This combining of the ions and electrons is the solid material rust, $Fe(OH)_2$. Corrosion may also be described as the conversion (oxidation) of a metallic element from a zero or lower valence state to a positive or higher valence state.

Some metals lose their chemical reactivity and become inert or passive. This condition, known as ***passivity,*** possessed by such metals and their alloys as Cr, Fe, or Ti, results in the formation of a highly adherent but very thin oxide film on the metal surface. This barrier coating prevents further corrosion by breaking the electrical circuit or by stopping the diffusion of ions, as in the case of polarization. Stainless steel and aluminum exemplify passivity. ***Polarization*** describes the condition of many metals when the rate of corrosion has decreased with time, a condition caused by the diffusion of ions within the electrolyte. A decrease in ion concentration at the cathode or an increase in metal ion concentration at the anode will decrease the corrosion rate, which depends on the cell potential and is changed, in turn, by the ion concentration. Metals will experience higher corrosion rates if the electrolyte is a moving fluid, which increases the ion removal rate, thus reducing the polarization.

Corrosion is affected by increases in temperature of the ambient environment (air, liquid, or solid) as well as increases in the velocity of the surrounding medium. Such increases in the fluid velocity produce still further corrosion of the material called *erosion* (to be discussed later). In addition to these processes, metals will corrode more easily if they or portions of them are under strain imposed either by some industrial process or by some unexpected application of excessive force. Strained metals are more active (more anodic) than unstrained metals. A bent nail (see Figure 5-8) or dented steel automobile fender will readily show this phenomenon in a graphic way.

5.1.5 Types of Corrosion

A listing of the different forms of corrosion tends to impress the reader with the scope of the problem and to bring home the importance of attending to it early on in the design process. A brief description of these varied forms of corrosion follows; further reference to them can be found throughout the text.

Uniform corrosion includes general rusting of ferrous metals, which leaves a scale or deposit over an entire exposed surface. ***Pitting corrosion*** results from localized surface defects. One or two examples might illustrate this type of corrosion. A marine organism attached to a metal surface of a boat hull will cause a pit to form, which can result in deep, isolated holes under the surface of the metal. Pitting of a chromed automobile bumper occurs when the chromium barrier coating is broken, exposing the steel underneath. This type of corrosion is more devastating than uniform corrosion because of the localization of the attack. If a steel tank suffers this type of corrosion, it could lead to a hole causing a leak of the contents. Stainless steels and aluminum are not recommended for use with saltwater because they are subject to pitting; Monel and copper alloys are more suitable. ***Cavitation corrosion*** is the result of high-velocity motion between a metal and a liquid. If the pressure accompanying a high-velocity motion drops to the vapor pressure, the liquid will vaporize and form a vapor cavity at the metal surface. At a slight increase in pressure, this bubble will collapse, causing a concentrated liquid impact, causing erosion (metal particles are removed from the metal's surface) and pitting, depressions, and pockmarks to appear on the metal surface. ***Crevice corrosion*** involves the junction of two metals exposed to a corrosive environment. Its basic causes have been identified as (1) a difference in oxygen concentration and (2) a deficiency in metal-ion concentrations. A concentrated solution is one that has a relatively large amount of dissolved solute. For example, a drop of water on a metal surface may have a greater concentration of oxygen near the edge of the drop than in the middle, which would aid the corrosion of the metal by the forming of local anodic and cathodic regions. ***Fretting corrosion*** is produced by vibrations that result in the striking/rubbing at the interface of close-fitting, highly stressed metal surfaces. Splines, keyways, and bearings are examples of where this takes place. ***Intergranular corrosion*** occurs along grain boundaries (see Figure 5-8b). Grain boundaries found in heavily stressed or cold-worked areas are anodic regions because they have more crystal defects, which generate higher energy levels than defect-free crystal regions, such as those in the interior of a crystal or grain. When heated to temperatures between 500° and 925°C for sustained periods, stainless steels become sensitized to this form of attack. The high temperatures cause the precipitation of chromium carbides along the grain boundaries. A chemical reaction occurs between the chromium and carbon atoms, which diffuse to the grain boundaries in the form of particles of chromium carbide. In so doing, a chromium-depleted zone is created adjacent to the grain boundaries, which is highly susceptible to corrosion because it has lost its former alloying composition. Sensitizing can be brought about by welding these steels. Through proper heat treating, or by lowering the carbon content, this condition can be rectified.

Glasses are degraded by a process called ***weathering,*** which involves the ***leaching*** ("to extract a soluble substance from some materials", "to dissolve and be washed away") of ions under high humidity conditions. Degradation of plastics is a chemical interaction with the ultraviolet rays of sunlight. Ceramic materials such as bricks degrade by ***spalling*** (flaking or chipping) due to high thermal stress. ***Selective leaching,*** or ***preferential corrosion,*** is found in solid solution alloys and results in the preferential removal of one element or component as a consequence of corrosion. The most common example is the ***dezincification*** of brass containing at least 15% Zn during which both Zn and Cu are attacked and the Cu ions are replated onto the brass while the zinc is selectively leached, leaving only a porous mass of copper in the dezincified region of the metal. Al, Fe, Co, Cr, and other elements are vulnerable to this preferential corrosion. ***Erosion*** is a form of stress-assisted corrosion, for example, the friction and wear caused by fluid impingement upon the interior surface of metal pipes at locations such as elbows.

When metals (especially dissimilar metals) are placed in an electrolytic solution, one metal becomes the anode and the other becomes the cathode, just as in a battery. This is ***galvanic corrosion.*** Cations of one metal (the anode) enter the ionized water to join anions. Left behind are electrons, which travel through the metal conductor to the cathode. Figure 5-6b illustrates the electrochemical corrosion of an iron bolt that is connected to a piece of copper, possibly a washer or nut. The positive-charged ions (Fe) enter the water to form bonds with the negative hydroxide ions (OH^{-1}). Metallic bonds are broken to form ionic bonds. Electrons (e^{-1}) left behind in the iron travel through the iron bolt to the copper, where they meet positive hydrogen ions (H^{+1}) and become neutral hydrogen atoms (H). The anode (the iron bolt) is pitted as it loses atoms. Galvanic corrosion involves the same redox reactions that occur in voltaic cells (batteries), the difference being that they are unwanted. Redox reactions occur between some metals and not others. For example, Cu and Ag produce oxidation-reduction reactions but the combination of Cu and Zn does not. This difference in behavior involves a study of electrode potential, electromotive force, electric charge density, and other concepts beyond the scope of this book.

It is possible to determine how active a metal will be by referring to the electromotive-force series in Table 5-1, which uses seawater as an electrolytic solution. The metals on the top are stronger oxidizers and thus are active anodes. The bottom metals are less active and become reducing agents (cathodes). Greater galvanic action results from joining metals that are more widely separated in the table. In Table 5-2 the different elements and alloys are ranked according to their tendencies to lose their electrons in a particular environment. Developed for particular environments, such as seawater, freshwater, or industrial atmospheres, these series are based upon the electromotive-force series, which compares the tendencies of a metal to give up its electrons. If a metal in the form of an electrode connected to a standard hydrogen or platinum electrode along with a standard electrolyte has a greater tendency to give up its electrons than does the hydrogen or platinum, then its potential is termed "negative" and the metal is anodic with respect to the hydrogen or platinum electrode. Metals are listed in tables according to their tendency to oxidize, to be active, or to be anodic, as in Table 5-1. The metals are so arranged that one can determine which of two dissimilar metals will oxidize in the presence of another, assuming there is a medium (electrolyte) in contact with both metals

TABLE 5-2 POSSIBLE CORROSION RATES FOR METALS IN THREE GENERAL ATMOSPHERES

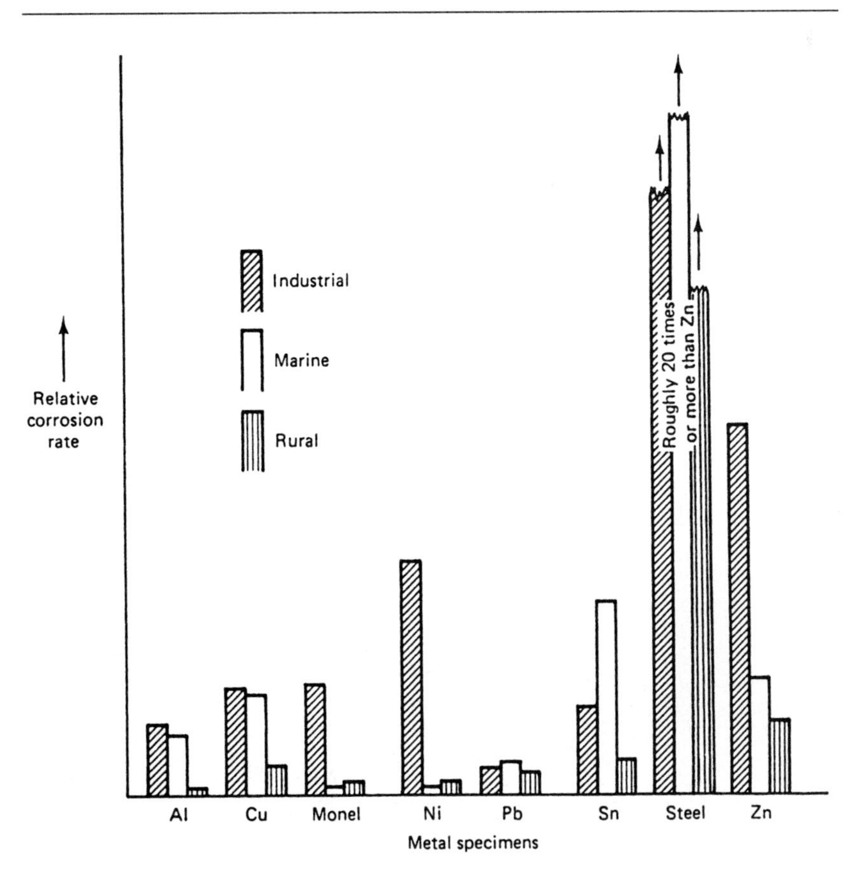

that is capable of transmitting ions (ion exchange). Usually, an aqueous solution (containing water) is the most common medium for corrosion. In Table 5-1 Zn is located high on the list of metals, indicating it is more active or anodic, and more capable of being corroded, than those metals listed below it. Fe is below Zn in the listing, which indicates why Fe many times is coated (galvanized) with Zn to protect the Fe from corrosion.

It is important to realize that not all galvanic series are equal. Each material has a tendency to behave differently in different electrolytes. Note that the metals listed in Table 5-1 have a superscript, identified by the asterisk as an oxidation number. ***Oxidation number*** or ***oxidation state (OS)*** designates the number of electrons that an atom loses, gains, or otherwise uses in joining other atoms in compounds. In the oxidation process, the OS of elements increases, and in reduction the OS of an element decreases. The OS of an atom, by itself, is zero. In the equation for the oxidation of Zn,

$$Zn \rightarrow Zn^{+2} + 2e^{-2}$$

The OS of the Zn atom on the left in its free state is 0 and it increases to +2 on the right because it has lost $2e^{-1}$. Redox reactions must always occur together. The total number of electrons associated with oxidation must equal the total number associated with reduction.

5.1.6 Corrosion Prevention and Protection

The large variety of environments to which a material can be subjected makes corrosion protection quite complex. On the other hand, the polymers that have been developed are so resistant to our natural environment that they are nearly indestructible and have become solid waste pollutants. Some common methods of corrosion protection for metals include protective coating, materials selection and design, cathodic protection, stress relieving, insulation, and alloying. Figure 5-9 shows that materials are tested and then selected because of their corrosion resistance. Many rules of good design are intended to reduce the opportunity for corrosion to occur.

Corrosion prevention requires large expenditures of capital to prevent the incipient loss of the world's limited material wealth. Protection from corrosion takes many forms. To prevent corrosion requires a good knowledge of the design of materials, the environments in which materials spend their service lives, and the ability to translate these factors into the proper selection of the best materials for the intended application. ***Design*** also involves a knowledge of the many types of corrosion and their causes, including the techniques required to reduce or eliminate corrosion. If the distance between the anode and the cathode is designed to be as large as possible, then corrosion is lessened. Eliminating the difference in compositions in a material will prevent the setting up of localized cells. Alloying of elements such as aluminum or chromium with steel produces oxide layers that make the metal passive. Conversely, alloying of certain dissimilar metals produces galvanic cells within the metal that enhances corrosion. (See Module 8 for a further discussion of design.). ***Two-phase alloys*** have more potential to corrode than a single-phase metal. In steel, ferrite is anodic to cementite, which can cause galvanic corrosion. Grain boundaries are regions of higher stress in relation to the interior of the grains. Fine grains are anodic to coarse grains in the same metal. High ***applied stressed areas*** accelerate corrosion and reduce the time required to produce failure. Such areas can be produced by cold working (see Figure 5-8c).

Corrosion produces cracks on the surface of metal parts that grow with time and produce fractures. Some fractures are catastrophic. One parameter for highly loaded fasteners is the possibility of environmental-assisted cracking (EAC), specifically from ***hydrogen embrittlement.*** Hydrogen embrittlement or **hydrogen stress cracking** is a type of failure resulting from applied or residual tensile stresses. Hydrogen in its atomic form (H) as opposed to its molecular form (H_2) diffuses through the crystal structure, leading to crack formation. Supposedly this dissolved hydrogen interferes with dislocation movement. As cracks grow and rapidly propagate, brittle fracture occurs catastrophically. The sources of atomic hydrogen are many, including water vapor under elevated temperatures, pickling of steels in hydrogen sulphate (H_2SO_4), electroplating, and heat treatments. High-strength steels, especially martensitic, are

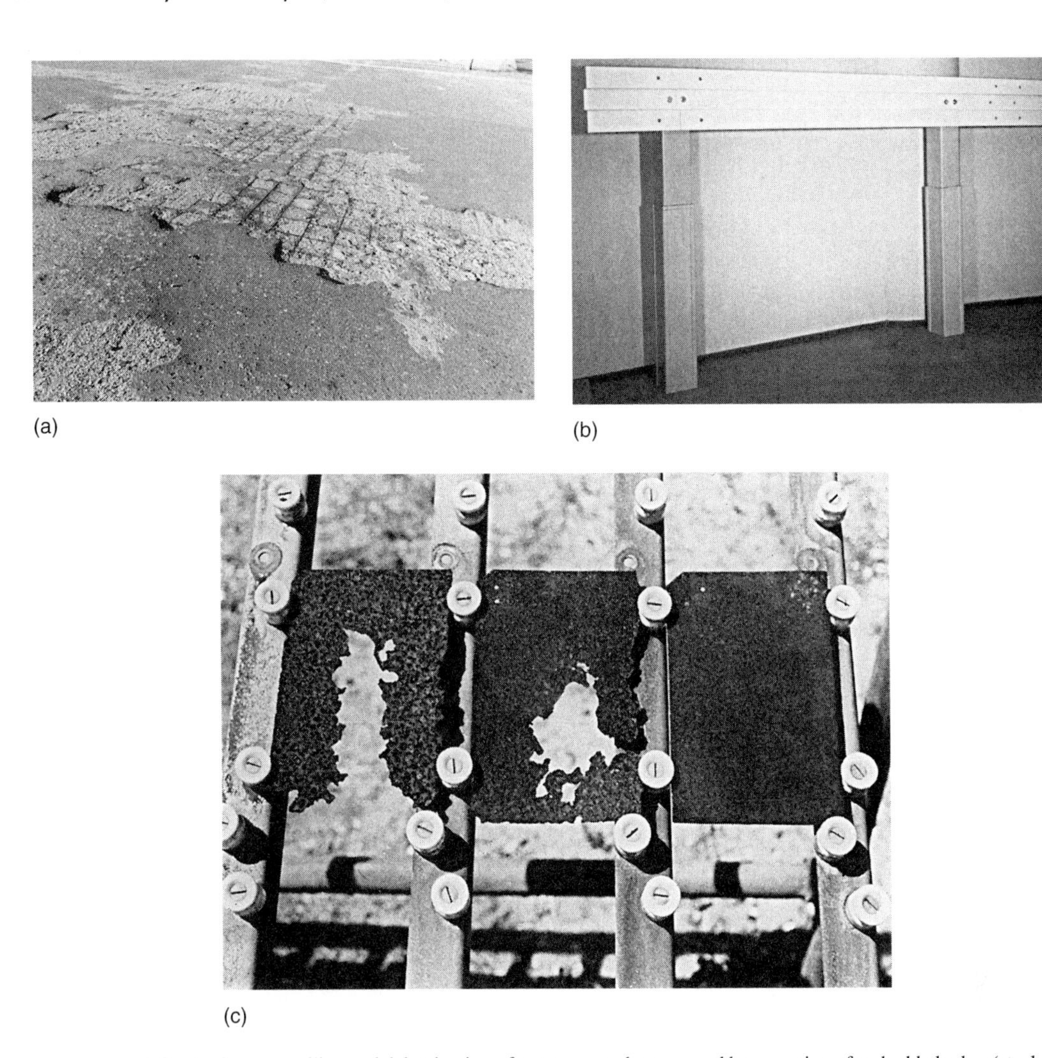

Figure 5-9 (a) Severe spalling and delaminating of concrete roadway caused by corrosion of embedded rebar (steel reinforcing) initiated by deicing salts. (W. R. Grace & Co.) (b) Plastic–glass composite highway guardrail made of pultruded glass-fiber-reinforced polyester is noncorrosive, maintenance-free, and stronger than similar structures of steel and steel-reinforced concrete; they are also comparable in cost. (*Advanced Materials & Processes*) (c) Corrosion of steels in a marine atmosphere. *Left*: Low-copper steel. *Center*: Ordinary steel. *Right*: Nickel-copper-chromium steel. (The International Nickel Co.)

susceptible, as are steels with increased hardness and high strength. FCC alloys are less affected. Reducing the hardness and tensile strength of steels and baking the steels at elevated temperature to drive out hydrogen are some techniques of reducing the incidence of hydrogen embrittlement.

The corrosion protection methods for steel fasteners must consider acidic environments, which permit hydrogen to be absorbed into the bulk material and may lead to delayed fractures. For this reason it is important in the manufacture of fasteners to include an additional post-heat treatment to remove residual hydrogen. Ford replaced steel lugs with titanium wheel lugs in the Partnership for a New Generation Vehicle (PNGV). With its high strength and low density, titanium does not need the expensive protective coatings required by ferrous steels. Also, a 9-gram titanium nut replacing a steel nut weighing 16 grams represents a 44% weight reduction. In addition to cutting weight, titanium lugs prevent galvanic corrosion with magnesium wheels. If two metals are to be connected in some aqueous atmosphere, corrosion could be slowed by choosing metals that are close to each other (see Table 5-1). To control or prevent galvanic corrosion, it is necessary to eliminate only one of the four essential elements previously mentioned: anode, cathode, electrolyte, or physical contact between the anode and the cathode.

Stress relieving through heat treatment, structural changes, and design considerations provide more homogeneous metals with nearly equal energy levels. In addition to heat treating, there are alternative techniques for producing stress relief. For example, the ***Meta-Lax*** process is used by auto and marine engine builders for subresonance stress relieving. It is effective in reducing the resonant frequency of a material and thus relieving the internal stress buildup as a consequence of a manufacturing process. Magnetic Processing Systems, Inc. has developed a pulsed magnetic treatment that relieves stress and improves fatigue properties of metal parts having diameters up to 300 mm.

Minimizing ***surface roughness*** is helpful. Roughness decreases corrosion resistance because the rougher the surface, the more the chance of collecting entrapped corrosive media. A correct choice of filler wires in ***welding*** that are the same or slightly more noble in composition than the base metal can reduce corrosion. The use of insulating washers, bushings, and gaskets in ***bolted connections*** is a good design practice. Neutralizing or "buffering" the electrolyte by adding inhibitors, as is done with antifreeze coolants in automobile radiators, has proven to be a successful practice. These inhibitors slow the anode–cathode reactions, thereby protecting the cast-iron engines and the brass–copper or aluminum radiators. The avoiding of surfaces that will collect liquids and debris by incorporating drainage or sealing is an example of the many good rules of design that are intended to reduce the opportunity for corrosion to occur.

The atmosphere provides excellent electrochemical mechanisms to promote corrosion. Oxygen and ozone (a highly oxidizing allotrope of oxygen) contribute to smog at the earth's surface (but provide a protective layer in the upper atmosphere); they also cause the degrading of rubber products. Sulfur dioxide (SO_2) from burning fuel oil and coal, ammonia, hydrogen and hydrogen sulfide, dusts, and salts exist in varying amounts depending on location (marine, industrial, or rural). These active chemicals produce different rates of corrosion for the materials exposed to them. Because of these atmospheric conditions, the U.S. Air Force stores its excess aircraft in open storage areas in Arizona, where the effects of atmospheric corrosion are nil. As Table 5-2 indicates, materials must be tested and then selected based on their corrosion resistance in the many different ***environmental conditions*** under which structures and devices need to function during their designed operating lives, including those new influences and atmospheres that appear almost daily. The large variety of environments to which a material can be subjected makes corrosion protection quite complex, making the study of such conditions another category of specialized knowledge that needs to be acquired. It is vital to provide steel with some form of corrosion resistance, such as adding elements (e.g., chromium) or painting or coating the surfaces.

5.1.6.1 Protective coatings. Protective coatings are the most familiar methods for preventing corrosion. Paint, varnish, oil, and a variety of polymeric and ceramic coatings prevent oxygen and moisture from reaching the metal to be protected. Automobile bodies are dipped into protective coating solutions. Zinc-coated steel (galvanized steel) is an example of both protective coating and cathodic protection. Engelhard Corporation uses electrophoresis to coat automobile radiators. Some protective coatings occur "naturally." For example, although aluminum oxidizes quickly, in doing so it forms an oxide film on the surface that prevents oxygen from reaching the metal below. Iron and steel do not oxidize as rapidly as aluminum, but the scale on iron is soft and porous, which allows oxygen to penetrate farther into the metal until it corrodes away to nothing. A low-alloy steel known as Cor-Ten uses a small percentage (less than 0.5%) of copper to produce a compact oxide; therefore, this oxide coating can also serve as a barrier and reduces corrosion to less than one-fourth that of regular steel. Two common ***coatings*** for high-strength ferrous fasteners are zinc dichromate and a three-layer phosphate-zinc-organic (PZO) coating. Zinc dichromate consists of an electroplated zinc layer with a chromium conversion top layer. The topcoat is an electrical insulator that resists the flow of electrical current. The zinc layer is anodic, or ***sacrificial,*** and corrodes preferentially when exposed to a corrosive environment, thus providing ***cathodic protection*** by serving as the ***sacrificial anode.*** If a Zn coating on steel is broken, the Fe is still protected and the corrosion product (oxides of Zn) protects the Zn from further corrosion. When tested in a salt spray test, such fasteners (e.g., ASTM B117) should survive up to 400 hours before red rust appears. The three-layer PZO coating consists of an initial layer of zinc phosphate, which provides a small amount of corrosion resistance; its main function, however, is to promote the adhesion of other coatings. The second layer is a zinc-rich inorganic resin made of mixed metal oxides. The zinc provides anodic protection to the underlying iron. The final layer is an aluminum-impregnated epoxy resin that protects iron; the aluminum makes the coating compatible with aluminum materials.

If you coat iron with Cu or Sn, the iron is protected, provided the coating is not disrupted by cracks or damaged by impact. Once the underlying iron is exposed, it will corrode (oxidize) because iron is more active (more anodic) than either Cu or Sn (refer to Table 5-1 to confirm this statement). If one is in the process of designing a structure that calls for connecting two dissimilar metals, find their relative position in a table like Table 5-1 to identify the metal on the anodic end of the table that will corrode. Keep in mind that alloy additions strongly affect the corrosion potential of metals. Stainless steel, for example, is more cathodic than low-carbon steel. Copper and its alloys rarely corrode galvanically because copper is cathodic to such metals as iron, aluminum, magnesium, tin, or lead. The further apart two metals are in Table 5-1, the greater the potential for corrosion when they are connected in the presence of an electrolyte. Suppose instead of Cu or Sn you choose Zn to coat the iron and the coating of Zn is damaged or cracked. The iron substrate is still protected from corroding because the Zn is more active (more anodic) than iron so the Zn will be attacked (oxidized), producing Zn atoms that protect the Zn from further corrosion. Fastening Zn anode plates on boat hulls (steel) or Zn collars on propeller shafts (bronze) prevents galvanic attack between more noble metals.

Protective coatings and ***surface engineering*** (see Module 13) receive considerable attention throughout this book, but it often must be pointed out that the preparation of materials for coatings and application of coatings, plus the disposal of gases, liquids, and solids used in coatings, cause harmful effects on the environment. Heavy metals, chlorofluorocarbons (CFCs), and volatile organic compounds (VOCs) have been targeted as polluters. Some new techniques that reduce this hazard include low-temperature surface engineering fluidized bed technology, which is based upon diffusion at low temperatures of metallic elements such as chromium, nitrogen, and carbon into the base metal. These treatments are used to improve the wear resistance of metal parts. The parts to be treated are immersed in a specially designed fluidized bed and typically heated to 560°C, at which point the reactive gases containing the basic elements needed for the formation of the required layer are introduced into the bed. The bed itself consists of inert particles such as aluminum oxide coated with the element to be diffused into the surface layer. A new coating known as PremAir[a], developed by Engelhard Corporation and Ford Motor Company, offers the potential for a new system of "smog-eating radiators" (see Figure 5-10). This catalyst system may turn car radiators into smog-eating, air-filtration systems to complement the catalytic converters now used in exhaust systems and may have the effect of cleaning all sources of smog from the atmosphere as air flows over vehicle radiators. Not only will this help animal life but it will also reduce corrosion. The developer's testing indicates the system will destroy 97% of VOCs, 90% of nitrogen oxides (NO_x), and 96% of carbon monoxide (CO) from the air it contacts.

Most present-day ***lubricants*** for threaded fasteners are based upon metal powders of lead, nickel, and other heavy metals that are a source of health or environmental risks. These lubricants are also the main ingredients in the production of galvanic corrosion. Dow-Corning Corporation has developed ultrapure, high-temperature, metal-free lubricants made of high chromium, nickel, or molybdenum alloys to overcome the problems of these older lubricants. Galvanic corrosion came to the forefront when automakers began to substitute

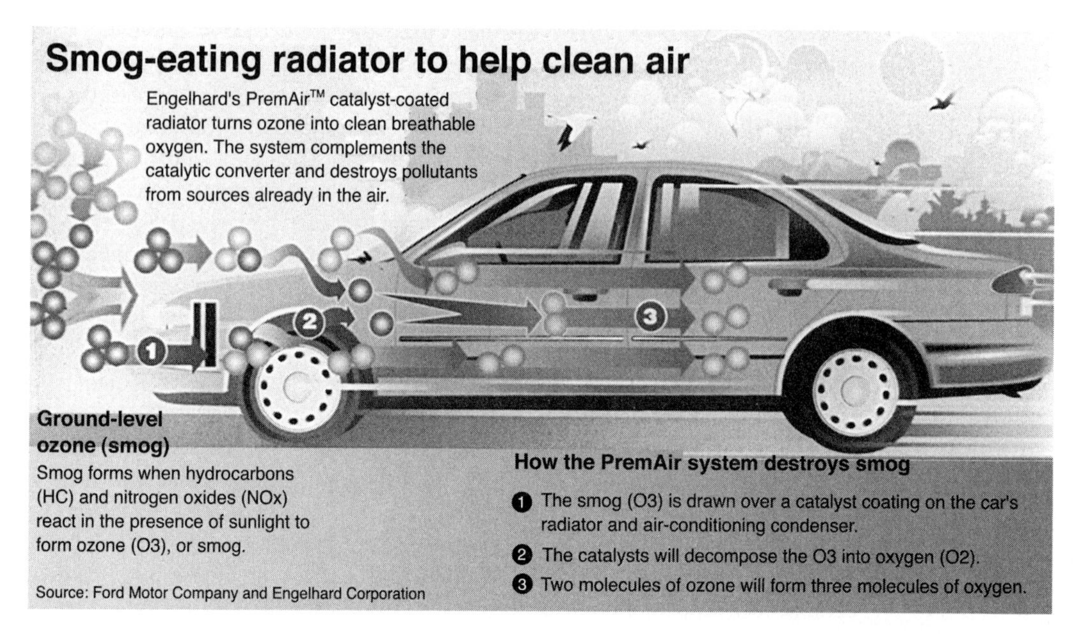

Figure 5-10 Polluted air not only harms animal life but also accelerates corrosion. Smog-eating radiators use PremAir[a] coatings to convert ozone (O_3)—the main component of smog—into breathable oxygen (O_2). (Ford Motor Company)

lighter-weight metals in their campaign to produce more efficient automobiles. Bolt-on applications for hoods and trunk lids where aluminum can be substituted for steel presented little problem. The different metals can be kept from making contact with each other by using polymeric materials. When greater use of such metals was attempted, as in the case of using aluminum for welded-in panels for roofs and floor pans, then a greater problem had to be solved: how to join an aluminum roof to a steel body frame. Water, air, and salts can go between the metals and form an electrolyte, causing the two dissimilar metals to act as a battery, resulting in destructive corrosion. Texas Instruments developed a transition metal made of two metals (aluminum and steel) joined under such high pressure that they bond at the molecular level. This type of bonding eliminates the galvanic reaction. Such sandwich material is cut into strips and placed between the dissimilar metals to prevent them from making direct contact. The parts can then be welded in the conventional way. This new technique is undergoing testing so that it can be integrated into the assembly line without reducing assembly-line speed.

5.1.6.2 Materials selection. The subject of ***materials selection,*** which this text stresses, is in the forefront of the battle to save our limited natural resources from the ravages of corrosion. Materials selection must involve the placing of materials into the proper environment to prevent corrosion. The heating of stainless steels to high temperatures when undergoing heat treating or welding causes a depletion of chromium and a partial loss of corrosion protection. The depletion appears as chromium carbide particulates that precipitate along the grain boundaries, which results in the grain boundary region being susceptible to ***intergranular corrosion.*** The technique of alloying steel with titanium (Ti) or niobium (Nb), called ***scavenger elements,*** which have a greater tendency to form carbides with carbon than does chromium, keeps chromium in the solid solution. Measures to prevent ***selective leaching*** involve redissolving the chromium and reducing the amount of carbon below 0.03 wt % C to reduce the carbide formation.

Illustrative Problem

A structure is being designed that has solid metal plates joined with nickel bolts. Electrochemical corrosion is anticipated due to humid environmental conditions (at least 50% humidity). The choice of materials for the solid metal plates, which require considerable tensile strength, is limited to copper and plain carbon steel. Select the best materials for the plates, and explain your reasoning.

Solution Consulting Table 5-1, the best material is plain carbon steel, as steel is more anodic (active) than nickel. Steel is closer to nickel than is copper, thus reducing the potential for galvanic action between the two metals.

5.1.6.3 Cathodic protection. Protecting metals from local soil conditions that are unusually corrosive (i.e., with high concentrations of sulfides, chlorides, or hydrogen ions) may necessitate neutralizing the soil by adding limestone, protective coatings or wrappings, or some form of cathodic or anodic protection. Large structures made of steel or iron that are in contact with soils or water, such as pipelines or ship hulls, can be protected by the use of a sacrificial anode, as mentioned in Section 5.1.6.1 (see Figure 5-11). This anode is made from an active metal (anodic to the metal being protected) and connected directly or through an electrical wire to the metal object being protected. Oxidation occurs at the active metal, which slowly dissolves in use. The iron or steel object acts as the cathode, whose surface acquires electrons from the anodic reaction. Magnesium is one of these metals that finds universal use in cathodic protection. Magnesium drain plugs provide cathodic protection in auto engines; the plug is the anode and the engine block is the cathode. Magnesium anodes in domestic hot water heaters are sacrificial anodes that protect the steel tank. Cathodic protection of a buried steel pipe or structure is accomplished by connecting the pipe or structure (the cathode) to a sacrificial anode of magnesium by a wire conductor and passing a direct current through the soil (the electrolyte in this case) to the pipe or structure, completing the circuit through the wire conductor (see Figure 5-11). By impressing a direct current on a buried steel pipe (the cathode) through the use of a buried graphite anode (the sacrificial anode), the steel pipe is protected by the current passing through the electrolyte (the soil) from the anode to the pipe (cathode).

By keeping all surfaces of the materials dry and the humidity below 50%, the effectiveness of an electrolytic substance can be reduced. If this is not possible, consideration should be given to electrically insulating one from the other. The relative size of the anode and cathode also plays a role in preventing or slowing corrosion due to galvanic action, a rule known as the ***anode–cathode area ratio.*** If you make the anode surface large compared with the cathode, corrosion will be slowed or stopped, as in the case of a steel bolt joining two aluminum plates. The anode in this case is the two plates and the cathode is the bolt. If you reverse this situation,

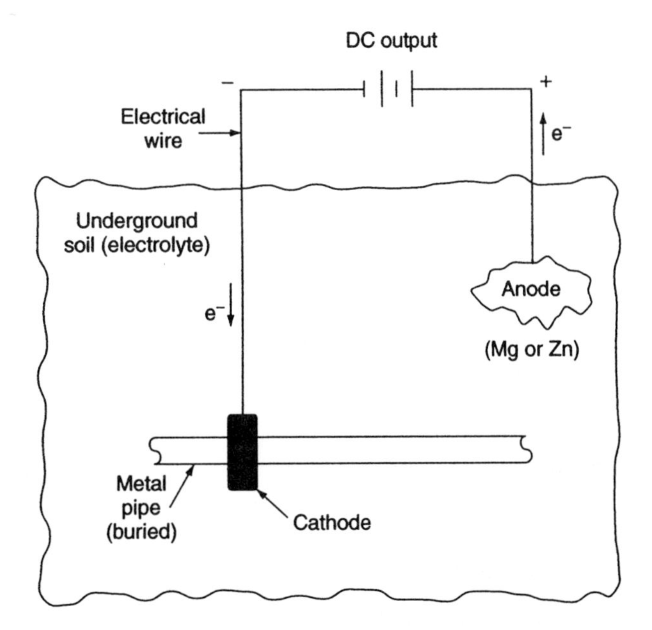

Figure 5-11 Cathodic protection of a buried pipe using a sacrificial anode and a source of electrical current.

with the steel plate as the anode, the steel bolt would be corroded in a rapid fashion. Chromium (Cr)-plated steel provides a hard coating especially where wear resistance is desired, but if the chrome plating is scratched, the steel substrate will corrode. The small exposed area of steel becomes anodic. The chromium is cathodic to the steel, causing the steel to corrode. This situation violates the rule rendering a small ratio: The higher the ratio of the area of the anode to the area of the cathode, the less severe the corrosion.

5.1.6.4 Shot Peening ***Shot peening*** is a cold-work process in which the surface of a metal is bombarded with small media called *shot*. The shot should be as hard or harder than the parts being peened. Media typically used for shot peening are small spheres with no sharp edges or broken particles made from cast steel. Ceramic or glass beads are used where iron contamination on the peened surface is a concern. Laser shock peening (LSP) produces a much deeper compressive layer and with less cold working. Each piece of shot acts as a tiny peening hammer that produces a small indentation or dimple on the surface. To produce a dimple, the surface must be subjected to a tension stress that causes the necessary deformation. At the same time, the material below the dimple, being subjected to a compressive force, attempts to seek an equilibrium condition and restore the surface to its original shape. The overall effect of the number and overlapping of dimples is the formation of a narrow layer of residual compressive stress at and below the surface. The maximum compressive stress formed is at least as great as half the yield stress of the metal being peened. Surface hardness also increases as a result of the cold working.

Compressive stresses are beneficial in increasing resistance to surface fatigue failures (pitting), corrosion fatigue, stress corrosion, cracking, galling, and fretting. The accompanying cold working results in work hardening, reduction in porosity, and greater intergranular corrosion resistance. Low-plasticity burnishing (LPB) is similar to roller burnishing or surface rolling, in which the surface of a component is cold-worked by a hard, polished roller to improve surface finish by eliminating scratches, tool marks, and pits, surface imperfections that are sources of corrosion. Because fatigue is the major cause of gear failure, gear teeth are shot peened or burnished primarily to improve their fatigue life. LPB substantially improves high cycle fatigue performance, primarily due to the ability to retain surface compressive stresses at elevated temperatures. Figure 4-23b (p. 162) shows a typical *s–N* curve with a linear, vertical scale *s* plotting the tensile stress in MPa and a logarithmic, horizontal scale of the number of load cycles, *N*. The graph illustrates that a reduction in stress produces an exponential rise in the fatigue life. As illustrated, a 35% reduction in stress improves the fatigue life by some 400%.

Peen forming is used to produce a compound, convex curvature on the peened side in aircraft wing skins to increase resistance to flexural bending fatigue. Hard anodizing of high-strength aluminum improves the fatigue resistance of coated materials when shot peening is used prior to anodizing. A follow-on process that requires shot peening, first developed by Metal Improvement Company, Inc. called ***Chemically Assisted Surface Engineering*** or ***C.A.S.E.,*** extends the surface fatigue life (pitting of the contact faces) beyond that of shot peening alone. First applied to racing gears, it has been extended to passenger car transmission gears and boat propellers. Using chemicals, this process produces a polished surface (superfinishing) that removes the surface asperities left over by shot peening and provides smooth contact surfaces with no peaks where microwelding can occur and pitting can initiate. Only low microindentations remain, which act as reservoirs for lubricant retention and provide locations for surface micromovement to occur under sliding and rolling pressures.

5.2 THERMAL PROPERTIES

A knowledge of thermal properties is vital to those who work with materials to make parts that will serve at temperatures other than as fabricated or that perform some heat-transfer function. For example, when heated deliberately or by hot working, a metal rod exhibits **three thermal effects:** (1) The rod absorbs heat, (2) it expands, and (3) it transmits heat. Each of these situations has a property that measures and describes the degree to which each effect is present. ***Absorption*** of heat is characterized by the property's heat capacity (c_p). ***Expansion*** is usually described by the *coefficient of thermal expansion* (α, Greek letter alpha). ***Heat transmission*** is identified by *thermal conductivity* (k).

5.2.1 Heat Capacity—Specific Heat

The efficiency of a material in absorbing heat is known as heat capacity. ***Heat capacity (c)*** is the amount of heat required to raise the temperature of a material a given amount. The larger the heat capacity, the greater the amount of heat needed to produce a temperature change. Its value depends on both the size of the material and its composition. Using kilocalories (kcal), or 1000 cal, it takes 1 kcal of heat to raise 1 kilogram (kg) of water by 1°C. The mechanical equivalent of heat or the amount of work equivalent to 1 cal of heat is 4.186 joules (J). Thus, 4.186 J = 1 cal or 4.186×10^3 J = 1 kcal. The ***specific heat*** of a material (c_p) is the amount of heat necessary to raise the temperature of exactly 1 gram of a material by exactly 1 degree (1°C or 1 K). Therefore, the amount of heat necessary to raise the temperature of a given material is the specific heat times the mass of the material times the change in the temperature. Water requires a large expenditure of energy to produce a change in temperature. The specific heat of liquid water, 4.186 J/kg · °C, is much higher than most other materials. Substances with high specific heat values do not change their temperature appreciably. The amount of heat required to change the temperature of a substance is expressed by the relationship:

$$Q = c_p m \Delta t$$

where Q is the amount of heat expressed in cal or kcal that flows into or out of a substance assuming the pressure on the material stays constant, m is the mass of the substance, and Δt is the temperature change. Solving for c_p gives the specific heat of the substance. Using SI units, the units of c_p are kilojoules per kilogram per °C (kJ/kg · °C). Table 5-3 shows the specific heat for some representative materials under constant pressure conditions.

Illustrative Problem

The specific heat of iron is 0.11 kcal/kg · °C. How much heat is required to raise 50 kg of iron from 20° to 100°C?

Solution Using the preceding expression,

$$Q = mc_p\Delta t = 50\text{ kg} \times 0.11\text{ kcal/kg}\cdot{}^\circ\text{C} \times 80^\circ\text{C} = 440\text{ kcal}$$

Illustrative Problem

Express the specific heat of aluminum 900 J/kg · °C in units of kcal/kg · °C.

Solution

$$900\text{ J/kg}\cdot{}^\circ\text{C}\ (1\text{ cal}/4.186\text{ J})\ (1\text{ kcal}/1000\text{ cal}) = 0.215\text{ kcal/kg}^\circ\text{C}$$

TABLE 5-3 SPECIFIC HEAT (C_p) FOR SOME SELECTED MATERIALS AT CONSTANT PRESSURE (UNITS OF J/kg · K)

Metals	C_p	Ceramics	C_p	Polymers	C_p
Ag	230	Al_2O_3	775	PE	2100
Al	900	BeO	1050	PP	1800
Cu	386	Silica	750	PS	1350
Fe	450			PTFE	1050
Ge	322			Nylon	1675
Mg	1020				
Si	700				
Ti	523	Water	4186		
Steel	500	Wood	1700		
Kovar	460	Human body	3470		

5.2.2 Thermal Expansion

Nearly all solid materials expand when heated and contract when cooled. Isotropic materials expand or contract equally in all directions. This temperature deformation results from the changes in the distances between adjacent atoms due to changes in thermal energy. The ***coefficient of linear thermal expansion*** (α) describes this effect in the following equations for unit thermal strain (ϵ_t), or unit deformation due to temperature change:

$$\varepsilon_t - \frac{\Delta l}{l_0} = \alpha \Delta t$$

where Δl is the change in length of some dimension, l_0 is the original length of that dimension, Δt is the change in temperature, and α, the constant of proportionality, is the coefficient of linear thermal expansion. In expressing the coefficient of thermal expansion, the term α $(C°)^{-1}$ is read "the coefficient is expressed per degree Celsius." Another notation, $°F^{-1}$, is read in a similar manner. The expression $\Delta l/l_0$ is another way of indicating unit deformation (ϵ, strain) brought about by temperature changes. To determine the total linear deformation (δ_t) due to a temperature change:

$$\delta_t = \varepsilon_t \times l_0$$

The coefficient for thermal expansion (CTE) for lamellar composite materials must match closely the CTE for constituent materials, otherwise the composite material could delaminate at the interfaces. The CTEs for some selected materials are shown in Table 5-4.

In metals there is a close correlation between the coefficient (α) and the melting point (T_m). The lower the coefficient, the higher the T_m value. Knowing the T_m value gives an indication of the strength of the bonding forces. A bimetallic strip (see Figure 5-12) exemplifies the use of the coefficient of thermal exp[illegible]on in temperature-control relays. With a temperature change, two metals bonded togeth[illegible]d having different coefficients will bend, making electrical contact with a switch. If the two metals A and B are thin, the induced stresses created by the difference in expansion will cause the metallic strip to bend. The degree of bending can be magnified by the use of a bimetallic coil made of the two materials. The expansion of the coil can be used to break or make an electrical circuit when the desired level of the variable being controlled, such as temperature, is reached.

E materials, designed and developed by Brush Wellman, Inc., are a family of metal matrix composites made up principally of beryllium (Br) and beryllium oxide platelets. Three grades—E20, E40, and E60—are now produced by blending the beryllium and beryllium oxide powders into a homogenous mixture to create isotropy of physical and thermal properties. Used principally in electronic packaging heat sinks, each alloy can be tailored to contain specific properties by varying the volume percentage ratio of beryllium and beryllium oxide. One

TABLE 5-4 COEFFICIENT OF (LINEAR) THERMAL EXPANSION ($\propto$) FOR SOME REPRESENTATIVE MATERIALS

Units of $-10^{-6} \times °C^{-1}$					
Metals	CTE ($\propto$)	Ceramics	CTE ($\propto$)	Polymers	CTE ($\propto$)
Ag	20	Al_2O_3	7	PE	100–200
Al	25	BeO	9	PP	58–100
Cr	17	SiC	5	PS	60–90
Fe	12	Si	3	PTFE	100
Mg	25	Si_3N_4	2.7	Nylon	6/6 80
Ti	9	Silica 96%	0.8	Epoxy	45–100
Steels	11–17				
Brass	18				
Kovar	5				
Invar	0.9				

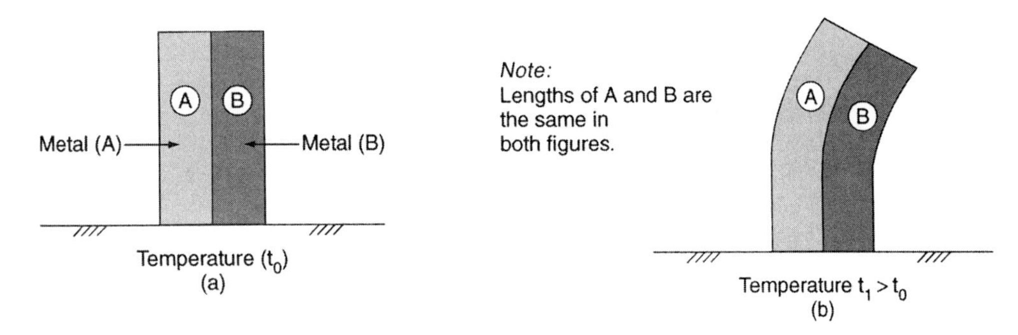

Figure 5-12 A bimetallic strip. Note that temperature t_1 is greater than t_0. After the passage of a discernible amount of time at t_1, bending is produced.

property that can be controlled by the proper mixing of constituents is the coefficient of thermal expansion (CTE or COT). The CTE determines the longevity of solder joints on circuit boards. The better the COT match is between joined materials, the less strain is imposed on the joining materials such as solder or adhesive bonds. Less strain translates into better fatigue life for the joints and longer life for electronic circuit boards. The CTE for E60 alloy is 6.1 per degree Celsius. The mix is hot isostatically pressed (HIP) into fully dense blocks for further processing into finished blanks and subsequent machining into components. Applications are found in ***avionics*** (the branch of electronics dealing with the development, manufacture, and use of electronic devices and equipment in aviation and astronautics) to improve the failure rate of electronic components by reducing the vibrations experienced in flight and launch by lessening the dynamic stresses exerted on the solder or adhesive-bonded joints of electronic packaged devices. Materials with high elastic modulus (high specific stiffness) have a positive effect on the transmissibility of vibrations. The elastic modulus for the E family of materials is about 320 GPa, with an average density of about 2.30 g/cc.

In the design of materials systems, deformation due to temperature must be taken into account. Changes in temperature are indications of heat transfer. An example of such a case is the use in a composite structure of two different materials that might have excessive differences in their coefficients of expansion. What is a possible outcome? What force or condition is recognized by ship and road builders who install expansion joints in their structures? What is the cause of the repetitious clinking sound produced by metal wheels running on steel railway tracks? In this discussion so far, all examples indicate that there is no resistance to the expansion of materials due to temperature changes. What happens to the materials if the expansion or contraction is resisted by an external force?

Illustrative Problem

A steel rod 2 m long is fastened at one end to a fixed support and is free to expand or contract. If the ambient temperature rises by 25°C, how much will the rod expand or contract?

Solution

$$\alpha_{steel} = 11.7 \times 10^{-6} {}^\circ C^{-1} \qquad \text{(see Table 5–4)}$$

$$l_0 = 2 \text{ m } \Delta t = 25^\circ C$$

$$\delta_t = \Delta l = l_0 \alpha \Delta t$$

$$\delta_t = 2 \text{ m}(11.7 \times 10^{-6} {}^\circ C^{-1})(25^\circ C)$$

$$\delta_t = 0.585 \text{ mm or } 5.85 \times 10^{-4} \text{ m} \qquad \text{(total deformation expansion)}$$

Polymers have small bonding forces (van der Waals bonds) and consequently larger expansion coefficients than metals. In summary, the covalent and ionic bonded materials have the lowest coefficients (strongest bonding), polymers the highest coefficients (weakest bonding), with metals somewhere between these limits.

5.2.3 Thermal Conductivity

Thermal ***conductivity,*** the ability to transmit heat, and electrical conductivity are both related to a material's atomic structure. Thermal energy can be conducted through a material by two mechanisms: the movement of free electrons and the movement of phonons (lattice vibrations). The ability of a nonmetallic solid to transmit heat depends primarily on ***phonons*** (quanta of energy). Phonons or "particles" behave somewhat like gas atoms and transfer the solid's lattice vibrations from a region of higher energy (high temperature) to a region of lower energy. As the temperature increases, they move with the speed of sound through the material and encounter more collisions, which results in a lowering of the thermal conductivity. Using electron band theory, conduction through conductor metals, semiconductors, and insulators is discussed in Module 27.

Thermal conductivity of materials has a varied relationship to temperature. In some materials thermal conductivity initially decreases with increasing temperature up to a certain point and then increases. Iron and graphite are examples of this type of behavior. With other materials the conductivity increases continuously with increased temperature. Glasses, nylon, and platinum are some examples. In general, metals have comparatively high thermal conductivities and polymers have rather low values. Oriented polymers can have much greater conductivities than unoriented polymers as in the case of filaments of polyethylene. Amorphous ceramics tend to have lower thermal conductivities than more crystalline ceramics.

In metals, the free, or conduction, electrons also serve as carriers of thermal energy received from the phonons. Elements added to metals reduce thermal conductivity, much as porosity and other crystal imperfections do in all materials. Polymers, lacking free electrons and having a less crystalline structure (orderly atom arrangement), have poor thermal conductivities. Such a characteristic makes polymers good thermal insulators. A ***coal-based carbon foam,*** recently developed by West Virginia University researchers and now being marketed for commercial use, is made from bituminous (high-sulfur) coal; it absorbs shock and is lightweight, strong, and noncombustible. It could be used to insulate Navy ships against fire and protect military vehicles from explosions. Other potential applications are in high-strength parts for the aerospace industry.

Thermal conductivity (k) is a measure of heat flow through a material. It relates heat flow (the flow of heat energy per unit area, per unit time) to the temperature gradient. The term ***temperature gradient*** (in symbols, $\Delta t/\Delta d$) describes a temperature difference per unit distance. Using symbols, the heat flow can be expressed as q/A, where q represents the heat energy in units such as watts, Btu/hr, or calories per second (cal/s); A is the area through which the heat passes (in square units, such as square meters or square feet). The change in temperature (Δt) can be expressed in °C, °F, or K. Using these symbols, thermal conductivity can be written

$$\frac{q}{A} = k \times \frac{\Delta t}{\Delta d}$$

This relationship shows that thermal conductivity (k) is a constant of proportionality. Solving for k gives

$$k = \frac{q}{A} \cdot \frac{\Delta d}{\Delta t}$$

Substituting SI units into this equation, the units of k are

$$k = \mathrm{W/(m \cdot K)} \quad \text{or} \quad k = \mathrm{W \cdot m^{-1} \cdot K^{-1}}$$

Note that the superscript, -1, is not an exponent but rather indicates that the base quantity is to be divided as shown in the preceding equation.

TABLE 5-5 THERMAL CONDUCTIVITIES (k) FOR SOME REPRESENTATIVE MATERIALS

[unit W/m · K]

Metals		Ceramics		Polymers	
Ag	450	Al_2O_3	35	PE	0.4
Al	300	BeO	220	PP	0.12
Cu	483	SiC	95	PS	0.12
Fe	134	Silica glass	2	PTFE	0.24
Ge	60	SiO_2	1.5	Nylon	0.24
Mg	170			Wood	0.16
Ti	30				
Steel (carbon)	50				
Stainless steel	15				

Illustrative Problem

An insulating material placed inside the walls of a home is 6 in. thick. The heat flow is calculated to be 4.2 Btu/hr when the inside temperature is 70°F and the outside temperature is 0°F. Determine the thermal conductivity of the insulating material as the heat flow passes through 1 ft^2 of the wall. In addition to expressing k in units of Btu/(hr · ft · °F), convert your units to SI units using watts, meters, and Kelvin temperature. Using Table 5-5 identify what insulating material was used.

Solution

$$k = \frac{q}{A} \cdot \frac{\Delta d}{\Delta t}$$

$$= \frac{4.2 \text{ Btu}}{1 \text{ ft}^2 \cdot \text{hr}} \cdot \frac{6 \text{ in}}{70^\circ\text{F}} \left(\frac{1 \text{ ft}}{12 \text{ in.}} \right)$$

$$= 0.03 \text{ Btu}/(\text{hr} \cdot \text{ft} \cdot {}^\circ\text{F})$$

Using conversion Table A-7, we have

$$1 \text{ Btu}/\text{ft} \cdot \text{hr} \cdot {}^\circ\text{F} = 1.729 \text{ W}/\text{m} \cdot \text{K}$$

$$k = \frac{0.03 \text{ Btu}}{\text{hr} \cdot \text{ft} \cdot {}^\circ\text{F}} \left(\frac{\text{ft} \cdot \text{hr} \cdot {}^\circ\text{F}}{1 \text{ Btu}} \cdot \frac{1.729 \text{ W}}{\text{m} \cdot \text{K}} \right)$$

$$= 0.052 \text{ W}/\text{m} \cdot \text{K} \qquad \text{(polymer material)}$$

5.2.4 Thermal Resistance

As we have said, the thermal conductivity of most materials is temperature dependent. In other words, the value changes with a change in temperature. Refractory materials, however, have thermal conductivities that have a minimum dependence on temperature. Insulating materials have low thermal conductivities and can retard the transfer of heat. Porous materials such as textiles, rock, wood, cork, foamed plastics, and human-made insulating tiles are good insulators partly because of their ability to trap air, which itself has a low thermal conductivity. Furthermore, the entrapped air is free from the circulating currents that aid in the transfer of heat. Numerous applications for insulating materials exist in today's society, particularly in the area of heat (energy) conservation. Wood is a good insulating material, but its flammability is a large drawback. Even though it remains strong as it burns, with charring action a positive factor in its resistance to burning (the char acts as an insulator material), once wood loses its strength, it fails catastrophically. Although organic materials like wood and polymers are flammable, certain materials (additives) such as chlorine, phosphorus, or metallic salts can reduce the chances that combustion will occur or spread. In building construction wood is often impregnated with fire-retardant chemicals. (Further information can be found in Module 14 dealing with ***flame retardants,*** charring, and ablation.) Steel does not burn but it loses strength at high temperatures, causing collapse by drooping or otherwise giving way (see Module 18).

Insulating materials are therefore used to protect buildings from fire. Mineral fiber, made from spun molten iron slag, and gypsum-based products in the form of blankets or boxes are used to encase building elements. These materials remain inert to fire, do not burn, and do not conduct heat readily. Gypsum is a hydrated sulfate of calcium with the chemical formula $CaSO_4 \cdot 2H_2O$ that occurs in sedimentary rocks. When heated to above 90°C, the chemical bonds in gypsum break down and release water, which cools a substrate. Cementious-sprayed fire-resistant materials are cement- or ceramic-fiber-based materials, commonly referred to as "flung dung," that are sprayed on in liquid form and then left to harden to protect steel beams and columns. Intumescent materials are sprayed, painted, applied as putty, or wrapped in sheets onto a surface to be protected. (***Intumescent*** means the material swells under heat.) When this type of material is exposed to high heat, it foams up or "blooms." Once foamed, the material hardens and remains inert, exploiting the foam's air bubbles as an insulating cushion. Also, as the materials expand, they can stop the spread of fire by sealing gaps and cracks. Endothermic foaming materials developed to protect electrical cables in nuclear power plants are formed into blankets to wrap steel structures and electrical wires. (***Endothermic*** is defined as designating or produced by a chemical change in which there is an absorption of heat from the environment.) When exposed to intense heat these materials chemically absorb heat energy, keeping the substrate materials cool. They also release chemically bound water to further cool the surface of the substrate, and so combine intumescent and heat-absorbing properties. A new filler material by Dow Corning, trademarked 'Instill,' is enclosed within a thin barrier material of metal foil and is evacuated to a high vacuum, producing a vacuum-insulating panel with an R-value six times as high as traditional glass fiber insulation. One inch of this panel is equivalent to 6 in. of glass fiber. This new fiber material is made of open-cell polystyrene foam. The open cells allow all the gas or air within to escape, thus enhancing the heat-blocking effect of the vacuum. Styrofoam is an example of closed-cell construction. The Dow's new material will be used in refrigerated trucks, vending machines, and shipping packages; all of these applications will produce considerable energy savings.

Figure 5-13 shows two vital applications that exemplify several thermal properties mentioned in this brief discussion. Module 19 discusses thermal shock and thermal shock resistance.

5.3 WATER ABSORPTION AND BIOLOGICAL RESISTANCE

Natural and synthetic polymers are subject to biological attacks and the absorption of water. When left untreated, dried wood will absorb moisture and then serve as a good environment for the growth of fungi (small plants) and insects which feed on cellulose and lignin and cause deterioration. Unprotected dry wood is also a host for termites and other insects. Figure 5-14 shows two examples of the biological degradation of wood. In Figure 5-14a mildew is forming on stained T111 siding (thin cedar laminate on exterior plywood). The mildew formed due to high moisture exposure and lack of sunlight since this portion of the house is mostly in the shade. Without treatment, the mildew will cause the wood to rot. In Figure 5-14b the rotting of fir fascia board resulted from moisture entering the ends of the board, which caused rotting below the painted surface. The rotting will spread to other boards in contact as well as attract swarming termites since this corner of the house sits next to many trees and heavy underbrush.

Some synthetic polymers swell through water absorption, which causes deterioration and provides a good environment for damaging microorganisms. Organisms, such as barnacles, living on the metals can cause corrosion through oxidation. ***Biofilms*** are communities of bacteria that collect in sticky layers and form an adhesive bond to many substrates, from human teeth to boat hulls. As the biofilms grow through cell multiplication, they form pillars and mushroom-shaped structures, which can clog sewer pipes, foul machinery, and contaminate drinking water systems. In drinking water pipes the films cause corrosion and disease. To counter biofilms, industry has developed enzymes to dissolve the films and blocking agents to stop their formation.

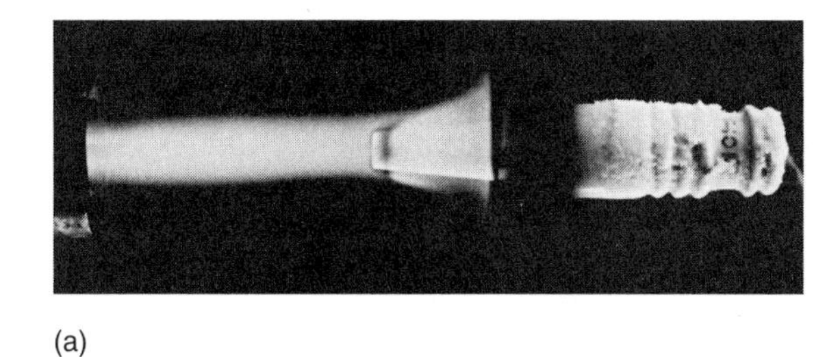

(a)

(b)

Figure 5-13 (a) Thermal resistance. The high alumina (Al_2O_3) content of spark-plug insulators can withstand thermal shocks from below zero (−73°C) to white hot heat (over 1000°C). Note the frost still on the top of the insulator. (American Ceramics Society) (b) Various thermal insulating materials systems protect space vehicles from the high heat of reentry into the earth's atmosphere. (Corning Glassworks)

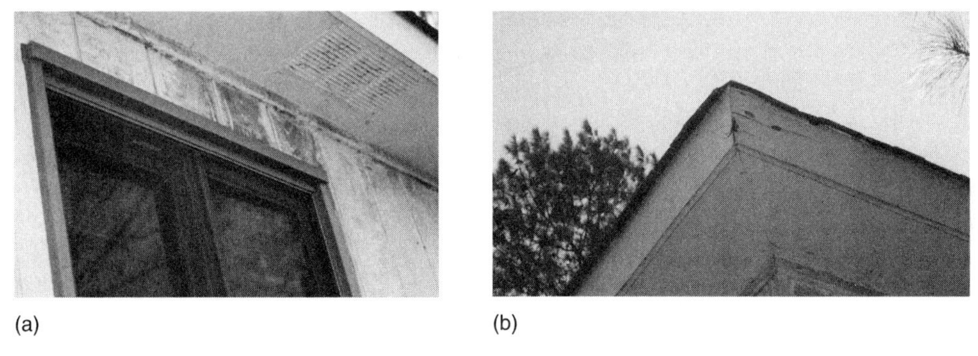

(a) (b)

Figure 5-14 (a) Mildew on stained plywood will eventually cause the wood to rot. (b) Intersecting corner of fascia board and lack of seal from rain allows moisture absorption and biological attack resulting in rotting boards.

SELF-ASSESSMENT

5-1. Define a chemical property.

5-2. When selecting a material that will be subjected to abrasion, what property would you expect the material to possess to a great degree?

5-3. How does the chemical industry protect its tanks and piping from corrosion due to the action of highly corrosive liquids?

5-4. What would your explanation be for the development of surface cracks on the surface of four-year-old automobile tires?

5-5. Name one method that large automobile manufacturers use to inhibit corrosion of the steel chassis parts of trucks and passenger vehicles.

5-6. Wood and other polymers swell and rot or degrade due to

a. Water absorption
b. Ozone depletion
c. Infrared light
d. UV reflection

5-7. Unprotected wood serves as host for

a. X rays
b. Covalent bonds
c. Solvents
d. Termites

5-8. Communities of bacteria that form sticky layers that can harm teeth and boat hulls are

a. Termites
b. Solvents
c. Biofilms
d. Biogens

5-9. Thermal resistivity is as important as thermal conductivity. Illustrate both properties by citing an example of each from your own experience or knowledge.

5-10. Explain why water is chosen to cool automobile engines. What is the effect of using more than the recommended percentage of antifreeze in the automobile's cooling system?

5-11. Even though heat and work are different forms of the same quality (energy), they have customarily been expressed using different units. Name two customary units used to measure heat energy.

5-12. The coefficient of thermal linear expansion may be expressed in units such as m/m/°C. What must you do to convert these units to °F? Would you do the same thing if you were given units of m/m/K?

5-13. Is the coefficient of linear expansion the same for iron and iron alloys (steels)? If not, how do you explain the difference, and what is the implication for designing steel structures?

5-14. Both the coefficient of thermal expansion and the thermal conductivity of most materials are temperature sensitive. Explain this statement in terms of any changes in their respective values. Can you find a material with a coefficient of thermal expansion that is nearly zero?

5-15. Under the heading Properties, develop a list of at least two thermal properties. Next to each, write a service condition in which the property is important from your own experience.

5-16. The specific heat (specific heat capacity) of water is 1.0 kcal $\times$ kg^{-1} $\times$ K^{-1}. Write this value using a different set of SI units.

a. 4184 J/kg K
b. 4184 J kg^{-1} K^{-1}
c. 4.184 $\times$ 10^3 J/kg K
d. All of the above

5-17. The specific heat of Alclad 2024 is listed in Table A-11. Express the specific heat in SI units of kcal/kg.

5-18. Which smart material might work a robot finger control or an antenna for a cellular phone because of its ability to remember its original shape after being deformed?

a. MMC
b. Piezoelectric
c. Titanium
d. Shape memory alloy

5-19. Biomimetics can help intelligent material systems technology by

a. Providing feedback and control
b. Ensuring osseointegration
c. Aiding improved hybrids
d. Prompting biotransfer

5-20. What is a main concern when selecting a material for bioengineering?

- **a.** Ability to resist heat
- **b.** Ability to provide feedback and control
- **c.** Compatibility with human and animal systems
- **d.** Good formability

REFERENCES & RELATED READINGS

AMERICAN SOCIETY FOR TESTING AND MATERIALS. *1996 Annual Book of ASTM Standards*. Philadelphia, PA: ASTM, 1996.

AMERICAN SOCIETY FOR TESTING AND MATERIALS. *Unsearch, Version 2.0*, Software. Philadelphia, PA: ASTM, 1992.

ASM INTERNATIONAL. *ASM Handbook*, Vol. 13, *Corrosion*. Materials Park, OH: ASM International, 2000.

ASM INTERNATIONAL. "MATERIALS SELECTION," *Advanced Materials and Processes*, Special Issue, June 1999.

BERKE, NEAL S., AND LAWRENCE R. ROBERTS. "Reinforced Concrete Durability and ASTM," *ASTM Standardization News*, January 1992, pp. 46–51.

BOSICH, JOSEPH F. *Corrosion Prevention for Practicing Engineers*. New York: Barnes & Noble, Inc., 1970.

DEXTER, S. C., ED. *Biologically Induced Corrosion*. Houston, TX: NACE, 1986.

OBERG, E., ET AL. *Machinery's Handbook*, New York: Industrial Press, 2003.

Periodicals

Advanced Materials and Processes
ASTM Standardization News
Corrosion and Performance
Materials Performance and Research
Journal of Materials Education
Journal of Materials Engineering
Journal of Composites Technology
Surface Engineering

Module 6
Electrical, Magnetic, & Optical Properties

After studying this module, you should be able to do the following:

6–1. Use diagrams and explanations to determine and define electrical, magnetic and optical properties of materials.

6–2. Solve problems related to electrical, magnetic, and optical properties of materials.

6–3. Use tables and references to determine values for electrical, magnetic, and optical properties for making calculations.

6–4. Relate electrical, magnetic, and optical properties in materials selection criteria.

6–5. Use the website addresses denoted by the icon* to find Internet sites with the most current developments related to properties and materials selection.

*The symbol found throughout the book will link you to Internet sites related to topics being covered. The dynamic nature of the Web brings frequent changes, so some of these sites, while available at the time of writing, may not be up now.

6.1 ELECTRICAL PROPERTIES

Before beginning our discussion of electrical properties, a review of some common terms used to describe electrical behavior of materials is essential. First, it is important to realize that electrical behavior is affected not only by the structure of the material and how it was processed but also by the material's environment.

Using a simple conducting wire circuit to which a source of electrical potential such as a battery or generator is applied (see Figure 6-1b), Ohm's law relates mathematically the electrical properties of current flow, resistance to current flow, and the voltage drop or potential difference across a resistance. Stated mathematically, $I = E/R$. Ohm's law states that the current flow (I), measured in amperes (A), is directly proportional to the applied voltage (E), measured in volts (V), and is inversely proportional to the resistance (R) to the current flow, measured in ohms (Ω) (the Greek capital letter omega, see Appendix Table A-3). If the potential drop across a resistance is 1 volt when the current in it is 1 ampere, then the resistance must have unit value. This unit is called the *ohm*. The ***ohm*** is a resistance across which is a potential drop of 1 volt when the current in it is 1 ampere. Ohm's law applies only to metal conductors.

Figure 6-1a is a plot of current versus voltage for a direct current (dc) circuit (Figure 6-1b). It shows that in this type of electrical circuit the resistance (R) is a constant, but we also know that resistance is affected by temperature changes, increasing as the temperature increases. This resistance might be considered as the macroscopic definition of ***electrical resistance*** because it is related to the size, shape, and properties of the materials that comprise the electrical circuit. For example, a copper wire 1000 ft in length with a diameter of 0.1 in. has the same resistance of 1 Ω as a steel rod 1000 m long with a diameter of 1 cm. The unit of current is the ampere, which is equal to a rate of flow of electric charge of 1 coulomb per second.

Illustrative Problem

The current in a circuit is 0.25 A when a potential difference (voltage) of 120 V is applied across its terminals. Find the value of the resistance.

Solution

$$I = \frac{E}{R} \quad \text{and} \quad R = \frac{E}{I}$$

$$= \frac{120\text{ V}}{0.25\text{ A}} = 480\ \Omega$$

Note that in an electrical circuit the electrical current is capable of producing several effects. The principal ones are heat, magnetism, and electrolysis. A current in a lamp heats the filament to incandescence and produces illumination. A current in the electromagnet of a telegraph key magnetizes the iron core and causes a pivoted piece of iron to be attracted to it, thereby producing a click. An electrical current in water causes the liberation of hydrogen and oxygen by electrolysis. Electric current is sometimes confused with ***electric charge.*** Using the analogy of an electrical generator or battery circulating charges in an electrical circuit to a pump circulating water in a pipeline, the difference between electrical current and electric charge can be distinguished. The quantity of water being circulated would be expressed in gallons, and the electric charge being circulated would be measured in ***coulombs.*** The flow rate of water past any point in the pipeline would be expressed in gallons per second; the flow of electricity in the electrical circuit would be in coulombs per second, or ***amperes.*** The flow of electric charges can be demonstrated by rubbing an amber rod (a resinous material whose electrons are only loosely bound by the atoms) with a piece of cloth, producing a charge of electricity on the rod, whose electrons leave the rod by conduction and deposit themselves on the cloth. The rod now has an excess of positive charge (each proton in the nucleus can be considered a positive charge) and the cloth is now negatively charged, so both materials have acquired a charge. To demonstrate that these charges can flow, connect the amber rod to the earth

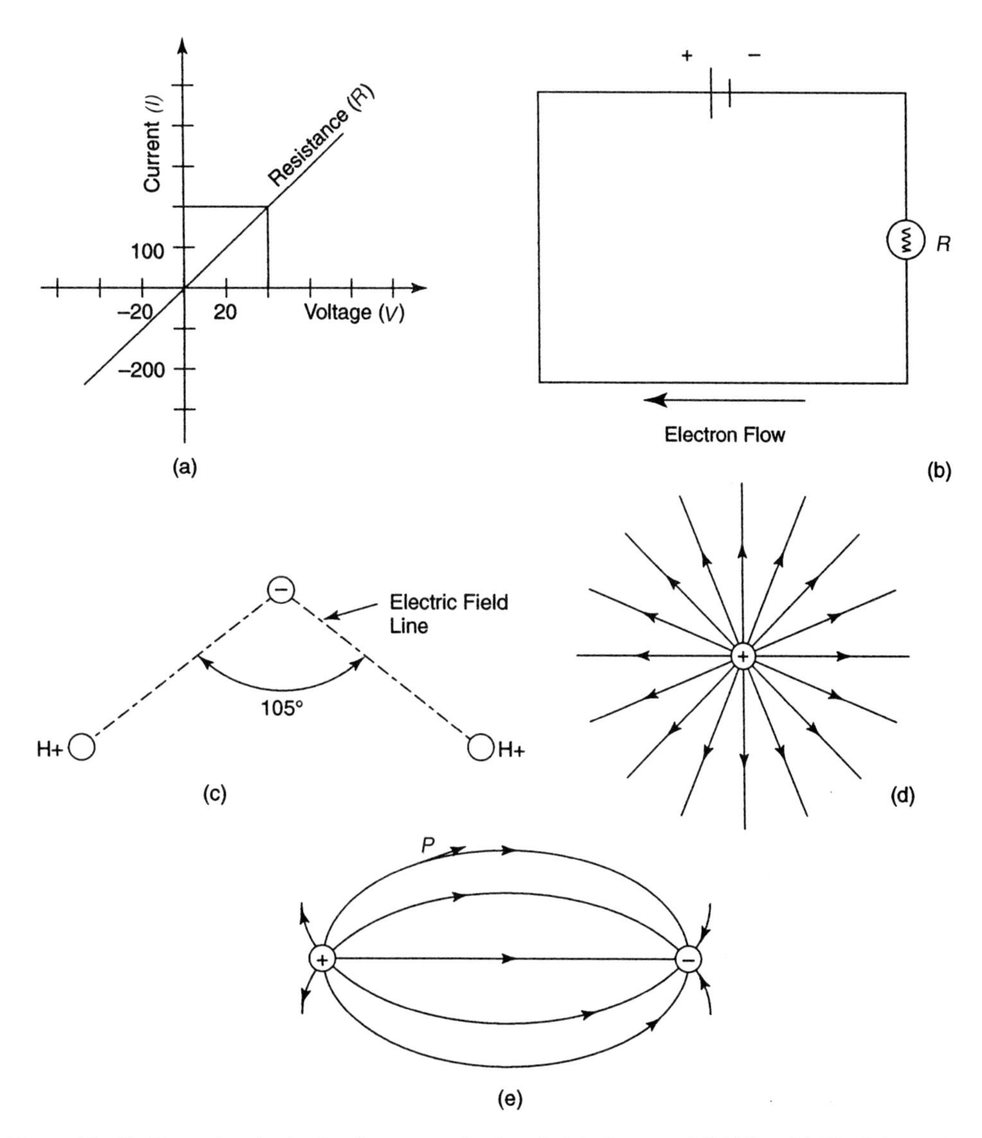

Figure 6-1 Resistance in a dc circuit-polar water molecule—electric charges and field lines (a). Ohm's law states that resistance in metal conductors is a constant in (b) direct current circuits. The ratio of V/I is called the resistance (R). A material is ***ohmic*** whenever the resistance is constant over a range of potential differences (voltages). (c) A water molecule is a polar molecule due to the separation of charges between the hydrogen and oxygen atoms. (d) A schematic of a single positive electric charge showing the electric field lines starting at a positive charge. The greater the number, the greater the charge; and the denser the lines near the charge, the stronger the field. (e) The electric field around two charges of opposite sign. The direction of the lines is from the positive to the negative charge. At point P, the direction of the field is tangent to the lines. These lines are continuous (closed curves) and they always point from the north toward the south pole of a magnet.

(ground the rod). The rod will lose its positive charge, because electrons flow from the earth and neutralize it. An excellent demonstration of this experiment can be done using a Teflon rod and a piece of fur. The flow of electric charge is further discussed in Section 6.2, on dielectrics.

Just as two material bodies (mass m and mass M) exert forces on each other, so too do two charged bodies (charge q and charge Q). The ***law of universal gravitation*** deals with two bodies possessing a mass. It states that there is an attractive force existing between the two masses that is directly proportional to the product of their two masses and inversely proportional to the distance (r) between them ($F = mM/r^2$), whereas ***Coulomb's law*** concerns itself with charged bodies. Using the symbol q to represent a charge on a body, Coulomb's law states that the electric forces exerted on one another by two charged bodies are directionally proportional to the product of their charges (qQ) when the bodies are at rest and inversely proportional to the distance (r) between them, $F = qQ/r$. This force is now referred to as the ***electromagnetic force.*** One major distinction between the two laws is that Coulomb's law deals with both attractive and repulsive forces. These opposite forces can be represented by the use of + or − signs, as in $+q$ and $-q$. From prior knowledge we remember the statement that like charges repel and unlike charges attract each other. The precise magnitude of the electric charge between an electron and a proton has been experimentally determined. It is the smallest found in nature. Using the SI unit of charge, the coulomb (C), the magnitude of the charge on the electron is $e = -1.602 \times 10^{-19}$ C. The mass of an electron, noted here for informational purposes only, is 9.11×10^{-31} kg.

Charges give rise to an ***electric field,*** the space around a charged body. *Field* is a construct that helps us understand how these separate concepts of gravity and electric charge can exert forces on objects at some distance. The earth exerts a force (your weight) on your body, which can be calculated. Electric fields can be visualized by using ***electric field lines.*** Figure 6-1d is a sketch of a single positive point charge showing the electric field lines (or lines of force) emanating radially out of the charge. These field lines extend outward into space. The closer the lines are to each other, the stronger the force is in that region. A negative point charge would have its electric field pointing inward. If another point charge was placed near this charge, it would feel a force because of the electric field from the first charge. The interaction between the two charges produces a force within this field. Figure 6-1e shows the electric field lines surrounding two charges of opposite sign. The direction of the field at any point is shown by the arrow at point P drawn tangent to the line. The field lines are directed from the positive charge to the negative charge.

Any charge located at a distance from another charge will experience a force either of attraction or repulsion. Just as the earth is influenced by the presence of some other object with mass, one charge can impose a force on another even if they are separated by large distances. Electric fields spread through space with the speed of light and they carry energy. They can be calculated by measuring the magnitude and direction of the electric force acting on a small test charge or probe placed in the field. Note that this force is a vector quantity, as is the electric field. The SI units for electric field are newtons per coulomb, or N/C. A map of the field lines about charged bodies can be made experimentally by sprinkling tiny bits of straw or hair on an insulating plate held between the bodies.

6.1.1 Resistivity

Resistivity (or ***volume resistivity***) is the term used to describe the relationship between electric current and the applied electric field. It is a measure of the resistance to the flow of current at a microscopic level, that is, as explained in terms of atoms, the basic building blocks of all solid materials. The resistivity, ρ (Greek letter rho), depends on the behavior and number of free, or conduction, electrons and not on the shape of the conductor, as does resistance. Like density, this inherent property of a material will change as the structure changes, as in the alloying of metals or the doping of semiconductor materials (Section 27.6.2). It depends on the movement of charge carriers-electrons in metallic conductors or ions in ionic materials. In fact, electrical resistivity is the reciprocal of electrical conductivity (σ, the Greek letter sigma) ($\rho = 1/\sigma$). This electrical property is discussed later in this module.

Microstructure plays a large role in resistivity. Any imperfections in the crystalline structure—atoms out of their normal positions, dislocations, or grain boundaries, to mention but a few—increase the collisions between electrons. This, in turn, prevents the transfer of energy in the form of electron flow to some intended user. Figures 6-2a and b are sketches of the effects of an increase in temperature on metallic conductors and semiconductors, respectively. With metallic conductors, the increased vibrational energies of the atoms as a result of an increase in energy in the form of heat make the passage of free electrons through the structure even more difficult. The ***mean free path,*** the average distance an electron can travel as a wave without hitting or deflecting off a positive-ion core (atom) in the lattice structure, is decreased. Consequently, the mobility of the electrons decreases, which pro-

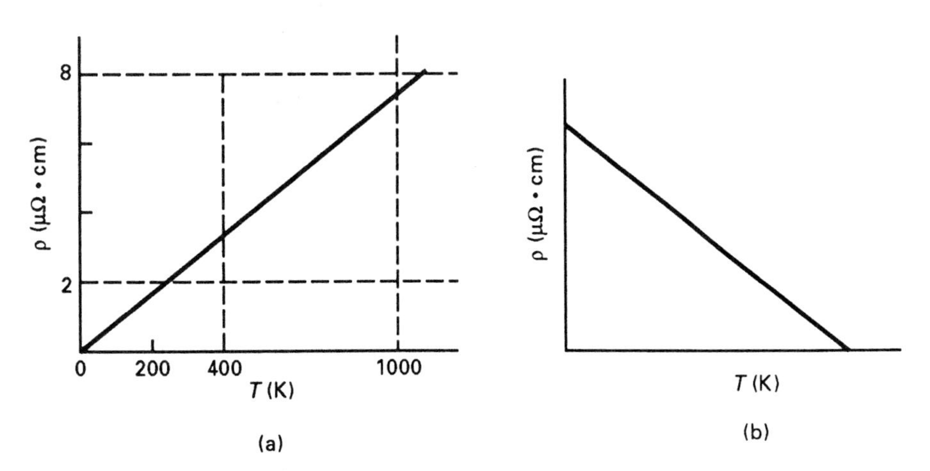

Figure 6-2 (a) Resistivity (ρ) versus temperature (*T*) for a metallic conductor. (b) Resistivity versus temperature for a semiconductor material.

duces an increase in the resistivity. For semiconductor materials, the resistivity decreases (conductivity increases) with an increase in temperature because more charge carriers become available to act as conduction electrons. Carbon's resistivity decreases with an increase in temperature.

An interesting illustration of the requirements imposed on an insulating material is the silicon-based ***transistor chip.*** Minuscule transistors form the basis of all modern computing, switching on and off billions of times a second. Some available IC circuits are only 5 μm thick but, on average, present-day ICs are about 400 μm thick. The chip's insulating layer of silicon dioxide, known as the gate oxide, is its smallest feature; typically, it is 24 atoms thick. The gate-oxide lies between the gate electrode, which turns the current flow on and off, and the channel through which the current flows and acts as an insulator to prevent a short circuit. The thickness of the gate oxide and the length of the electrode are two variables that control the transistor's switching speed. Changes in these variables have resulted in a doubling of a transistor's switching speed every 18 to 24 months. (Bell Labs recently reported the construction of an ultrathin, five-atom-thick layer of silicon dioxide.) Each layer must be uniform in structure and smooth (i.e., flawless) to prevent a short circuit.

As transistors are made smaller, electrical current leaks from the microscopic components, which means that more power is needed for them to operate and therefore more heat is released. The result, if unchecked, would be short battery lives and computers that would be too hot to be useable. One solution to this problem has been to build transistors in a thin layer of silicon on top of which is embedded a layer of insulating material. The result is a "depleted substrate" transistor with 100 times lower leakage. Another solution involves the use of a new dielectric material (high k gate dielectric) in place of silicon dioxide between the gate and the active area of a transistor. This new material reduces leakage by more than 10,000 times. The doping of semiconductor materials also lowers the resistivity by increasing the number of charge carriers. Additional information on this subject appears in Section 27.6.

In the discussion of resistance, we stated that resistance varies directly with the length l and indirectly with the uniform cross-sectional area A of a conductor. To write this as a mathematical statement, we need a constant to make the units agree on both sides of the equation. Thus,

$$R = \rho\frac{l}{A}$$

where ρ is the proportionality constant or resistivity in ohm-centimeters, assuming that l and A are expressed in centimeters. Note that a conductor 1 cm in length with an area of 1 cm^2 has a resistance R equal to the resistivity ρ.

TABLE 6-1 DIELECTRIC PROPERTIES AND RESISTIVITY FOR SOME COMMON MATERIALS

Material	Resistivity ($\Omega \cdot$m at 20°C)	Dielectric constant, κ (10^6 Hz)	Dielectric strength (10^6 V/m)	Dielectric-loss factor, $\kappa \tan \delta$ (10^6 Hz)
Mica	10^{11}	7	79	
Nylon 6/6	10^{13}	3.5	17	0.04
$BaTiO_3$	10^9	1600.0		
Al_2O_3	10^{12}	9	98	0.001
Steatite	10^{12}	6	8	
Phenolics-transparent	10^{10}	8	6	0.05
Paraffin	10^{15}	2.3	10	
Water	10^{12}	78		
Polyethylene	10^{15}	2.3	20	0.0001
Soda-lime-silica glass	10^{13}	7	10	0.01
Silver	16×10^{-9}			
Copper	17×10^{-9}			
Tungsten	56×10^{-9}			
Aluminum	29×10^{-9}			
Steel (301)	7×10^{-7}			
Graphite	10^{-5}			
SiC	0.1			

Illustrative Problem

(a) A metallic wire 100 cm long has a diameter of 0.05 cm. If it has a resistance of 0.08 Ω when 10 A of current are running through the circuit at standard temperature, find the wire's resistivity in $\Omega \cdot$cm. (b) Using Table 6-1, identify the material from which the wire was fabricated.

Solution

(a)
$$R = \rho(l/A) \qquad \text{therefore, } \rho = R/(A/l)$$

$$A = \frac{\pi d^2}{4} = 0.7854 \times (0.05)^2 = 1.96 \times 10^{-3}\ \text{cm}^2$$

$$\rho = 0.08\ \Omega \times 1.96 \times 10^{-3}\ \text{cm}^2/10^2 = 1.7 \times 10^{-6}\ \Omega \cdot \text{cm}$$

(b) Either silver or copper would be adequate answers.

In selecting conducting materials for the express purpose of generating heat from the flow of electricity, the material must have a carefully selected variety of properties, such as a moderate resistivity, excellent resistance to oxidation, and a capability to operate effectively at high temperatures without failure. Table 6-1 lists the electrical resistivities of some selected solids from three families of materials. Figure 6-3 illustrates an industrial use for a human-made nonconductive material. Note in our discussion that where heat is involved in the flow of electric charge, efficiency is reduced. The incandescent light bulb is a good example. This bulb gets hot when used; a fluorescent bulb remains cool. Consequently, the fluorescent bulb is more efficient in the use of electricity (energy). As a matter of fact, the incandescent light is only about 3% efficient; that is, 97% of the energy needed to produce light is lost mainly to heat. Compare the lighting characteristics of light-emitting diodes (LEDs), discussed in Section 27.7.

Summarizing, the resistance R of an electrical circuit is a function of the shape, size, and nature of the solid material in the circuit. Just like specific heat or density, resistivity, ρ, is a function of the intrinsic nature of the material itself. Instead of thinking of resistivity, one can think in terms of conductivity, σ. Low resistivity and high conductivity refer to a similar situation, but in different terms. To further reinforce the understanding of resistivity, remember that 1 lb of aluminum has the same resistivity as 1 g of aluminum, whereas the resistance of 1 lb of aluminum is very different from that offered by 1 g.

Figure 6-3 Pultruded fiberglass boom is superior to older metal booms in terms of mechanical, weathering, and electrical properties. The boom exhibits less than 100 microamp leakage over any 2-foot section after a 48-hour full immersion soak while a 100-KV (dc) test voltage is applied. (Morrison Molded Fiber Glass Co.)

6.1.2 Electrical Conductivity

As we have mentioned, the reciprocal of resistivity is known as ***conductivity.*** It is a measure of how readily electrons flow through a material. Conductivity, represented by the symbol σ (Greek letter sigma), is determined for a particular material by measuring the amount of electric charge that passes through a unit cube of the material per unit of time. Conductivities of common metals range from 10,000 to 550,000 $\Omega^{-1}\ \text{cm}^{-1}$. (See Section 24.3 for information on increasing conductivities of graphite fibers.) Consequently, σ is dependent on three factors:

1. n, the number of charge carriers in a cubic centimeter (cm^3) of material
2. q, the charge per carrier (coulombs/carrier)
3. μ, the mobility of each carrier, or the ease of movement of the charge carriers (e.g., electrons in a metal)

$$\mu = \frac{\text{velocity of the carriers (cm/s)}}{\text{voltage gradient (V/cm)}}$$

The product of these three factors, with units of $1/(\Omega \cdot \text{cm})$ or $(\Omega \cdot \text{cm})^{-1}$, is the conductivity. Expressed in mathematical terms,

$$\sigma = nq\mu$$

This calculation agrees with units obtained for conductivity if the basic equation for resistivity is used ($\rho = RA/l$).

Illustrative Problem

A tungsten wire is used in an electrical circuit at a temperature of 20°C. Find its electrical conductivity.

Solution Conductivity is the reciprocal of resistivity. From Table 6-1, tungsten's resistivity is $56 \times 10^{-9}\ \Omega \cdot \text{m}$, so the conductivity is $1/56 \times 10^{-9}\ \Omega \cdot \text{m} = 17.9 \times 10^{6}\ (\Omega \cdot \text{m})^{-1}$, or $\Omega^{-1} \cdot \text{m}^{-1}$.

Illustrative Problem

A rectangular block of aluminum has the dimensions 25 mm × 25 mm × 50 mm.

(a) Make a three-dimensional sketch of the block and place dimensions on it.

(b) What is the resistivity of the block, in $\mu\Omega\cdot$cm, measured between the two squared ends, at 20°C?

(c) Determine the electrical conductivity for aluminum at 20°C.

Solution

(a)

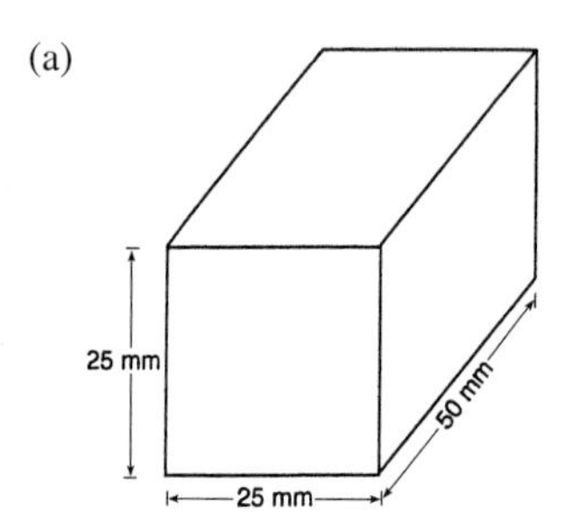

(b) The cross-sectional area is 25 mm^2 = 6.25 cm^2.

$$R = \rho\frac{l}{A} = \frac{2.9 \times 10^{-6}\ \Omega\cdot\text{cm} \times 5\ \text{cm}}{6.25\ \text{cm}^2}$$

$$R = 2.32\ \mu\Omega$$

(c) $\sigma = 1/\rho = 1/2.9\ \mu\Omega\cdot\text{cm} = 3.45 \times 10^5\ (\Omega\cdot\text{cm})^{-1}$

6.2 DIELECTRIC PROPERTIES

6.2.1 Dielectrics

Dielectric properties describe the response of dielectric materials (poor electrical conductors) to electric fields. The term ***dielectric*** refers to the polarization of nonconductive materials caused by an electric field. In other words, dielectric-properties are dependent on the polarization of the material. Most dielectric materials are ceramics or polymers. A material or medium such as a gas or vacuum is called dielectric because it permits the passage of the lines of force of an electrostatic field but does not conduct the current. Subjecting these materials to a magnetic field also produces similar effects, although the sources of the force are different. A dielectric has two functions to perform: One is to act as an insulator and the second is to increase the capacitance of a condenser (capacitor) beyond that which can be derived from an air gap or vacuum between the plates of the capacitor. As a covering for an electrical wire, plastic coating is acting as an insulator. Ceramics are also good electrical insulators, that is, dielectric materials.

Even though these materials do not conduct electrical current, they are affected by the applied electric field. This effect, known as ***polarization,*** results in a shift in the distribution of charge within the material, forming an electric dipole. The magnitude of the polarization (P) can be expressed in the units C/m^2. ***Dipoles*** are atoms/ions or groups of atoms/ions that have an unbalanced or asymmetrical charge. They have a positive pole and a negative pole; in other words, in a dipole the centers of positive and negative charges do not coincide. Figure 6-4a is a sketch of a single dipole. The centers of positive and negative charges are indicated by the letter q, both charges being equal in magnitude. They are separated by a distance d. The vector labeled p is the ***dipole moment,*** defined as the product of one charge multiplied by the distance d, or $p = qd$. The units of dipole moment are coulomb-meters (C-m). The water molecule is a permanent dipole and has a dipole moment of about 6×10^{-30} C-m. When an external electric field is applied to a material containing dipoles, a force (or torque) will come to

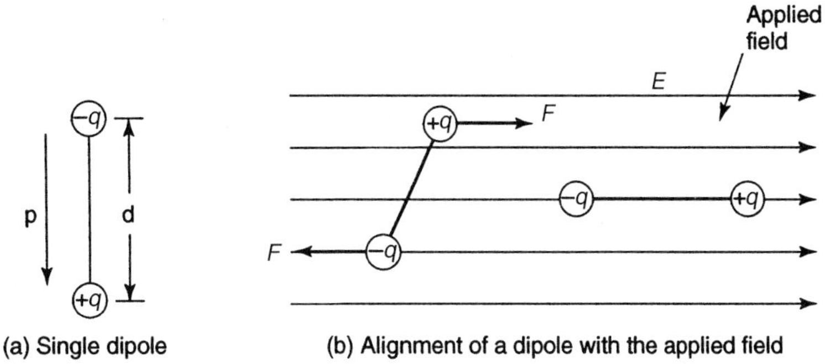

Figure 6-4 (a) A single dipole; (b) a single dipole undergoing a force from an applied electric field that will orient it with the applied field.

bear on a dipole to orient it with the applied field (see Figure 6-4b). This process of dipole alignment, termed *polarization,* is affected by many structural features, such as the presence of permanent dipoles, ease of molecule movement, moisture, cracks, and grain boundary imperfections. It should be remembered that atoms and ions move (i.e., they translate as well as rotate within their structures).

6.2.1.1 Electronic polarization. When discussing the bonding of atoms and the polarity of some molecules in Module 2, we used such expressions as "attraction of electrons," "regions of positive charge," or "charge distribution." The unequal positive charge on one side of the water molecule sketched in Figure 6-1c has a strong covalent bond that holds the molecule together. The two hydrogen atoms in the molecule are represented as positive point charges in relation to the negative point charge of the oxygen atom. The two hydrogen atoms are separated by an angle of about 105°. There is a net positive charge on each hydrogen atom because the electrons spend only part of their time there. The shared electrons are more likely to be found around the oxygen atom. When an external electric field (E, in volts) is applied to a dielectric material, the electron clouds around the atoms and molecules shift off the center of the positive nuclear charge and concentrate on one side of the nucleus near the positive end of the field. The atoms then act as a temporary or induced dipole. A temporary dipole can induce another dipole in an adjacent atom. The two dipoles then experience a force of attraction. This effect can occur in molecules as well. When the field is removed, the polarization disappears (see Figure 6-5a).

6.2.1.2 Ionic polarization. The applied field causes a separation of charges. The cations are displaced in one direction and the anions in the opposite direction. The cations are attracted toward the negative electrode or the negative side of the field, and the anions move toward the positive electrode or the positive end of the field. Polarization may change the overall dimensions of a material (see Figure 6-5b).

6.2.1.3 Molecular, dipole, or orientation polarization. Molecular polarization is associated with materials that have natural dipoles. Some examples are water molecules, some glass, polar polymers, and crystalline ceramics such as barium titanate ($BaTiO_3$). When the external field is removed, the polarization can be retained because of the presence of permanent electric dipoles in the material (see Figure 6-5c).

The extent of polarization depends on the following factors: the type of atoms, the type of bonding, the frequency of the applied field, and the mass of the material involved. Each of the polarization mechanisms occurs in a limited range of frequencies. Electronic polarization can respond to radiation in the visible portion of the electromagnetic spectrum. Ionic polarization will be experienced in the infrared (IR) region. Molecular polarization responds to frequencies in the subinfrared range. The larger the mass involved, the more sluggish is the

Electronic field
E = 0

E ≠ 0

Electron

Nucleus

Nonpolarized
centers of + and – charge coincide

Polarized
center of negative charge displaced

(a) Electronic polarization

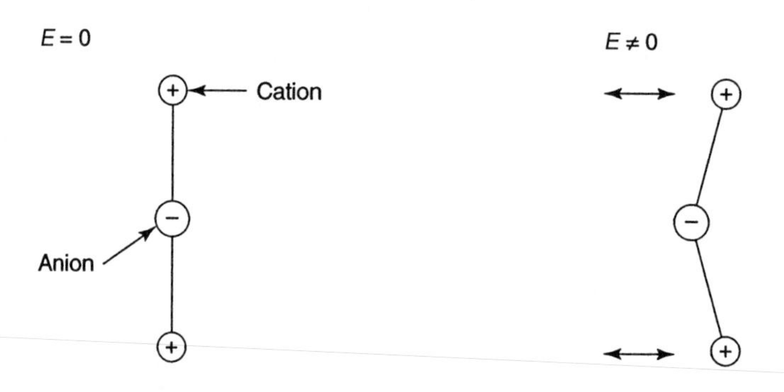

(b) Ionic polarization

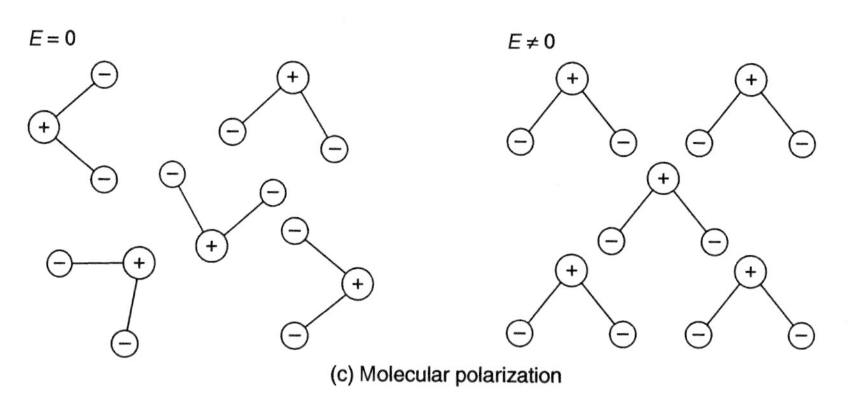

(c) Molecular polarization

Figure 6-5 (a) Electronic polarization: a nonpolarized atom with its centers of charge coinciding at the nucleus when there is no electric field ($E = 0$). When an external electric field is applied ($E \cong 0$), the charge centers separate producing a polarized atom. (b) Ionic polarization showing a single ion with distorted bonds when subjected to the applied field. (c) Molecular polarization showing the natural dipoles disoriented when no electric field is applied, and aligned with the field when the field is imposed.

response to the external field. Electrons, being less heavy than protons, can respond faster to fields as they are applied and removed.

Dielectrics and their properties are important in the design and fabrication of both ceramic and polymer composites. ***Dielectrometry*** uses electrical techniques to measure the changes in dielectric loss factor and capacitance during the cure of a resin in a laminate. Ceramics are widely used as insulators and as substrates in metal matrix composites, where the need is for high resistivities and dielectric strength. High thermal conductivity might also be needed if heat must be dissipated.

6.2.2 Dielectric Constant

The ***capacitance (C)*** of a condenser or capacitor is a measure of its ability to store electricity. It is determined by measuring the amount of charge (Q) that must be placed on a conductor to raise its potential by 1 V. Mathematically, $C = Q/V$, where C is expressed in farads (coulombs per second), Q in coulombs, and V in volts. Solving the expression for $Q = CV$, it can be seen that C is a constant of proportionality with no units. A farad (F), the unit of electric capacitance, is defined as the capacitance of a capacitor, between the plates of which there appears a potential (E) of 1 V when it is charged by a quantity of 1 coulomb of electricity. A farad is a fairly large unit, and most capacitors have small values of capacitance, So prefixes such as *pico-* or *micro-* are used to express smaller units of the farad (1 pF $= 10^{-12}$ F).

Two conductors separated from each other by some insulating material (including air) form a condenser. The conductors make up the plates of the condenser, and the insulation between the plates is the dielectric. The condenser is charged by connecting its plates to a battery or a source of electrical energy. Electrons will flow and collect on one plate until equilibrium is reached and the potential difference between the two plates equals the electromotive force (emf) of the battery. The ratio of the condenser charge (Q) to the potential difference (E) defines the capacitance of the condenser. The capacitance of a parallel-plate condenser with free space as the dielectric is proportional to the area of the plates and inversely proportional to the distance between the plates. Expressed mathematically, we have

$$C = \epsilon_0 \frac{A}{d}$$

where A is the area of the plates, d is the thickness of the dielectric or the plate separation, and ε_o (Greek letter epsilon) is the ***permittivity constant of free space*** with a value of 8.85×10^{-12} F/m. Note that C depends on the geometry of the arrangement.

Illustrative Problem

A condenser is made of two thin sheets of copper with a total area of 1 m^2. The sheets are separated at a distance of 0.05 mm by air. What is the capacitance of the condenser?

Solution

$$C = \epsilon \frac{A}{d} = \frac{8.85 \times 10^{-12}\ \text{F/m} \times 1\ \text{m}^2}{5 \times 10^{-5}\ \text{m}}$$

$$= 0.177\ \mu\text{F}$$

If a dielectric such as mica is placed between the plates of the same parallel-plate condenser using the same potential difference E, the charge Q on the plates will be greater and will produce a corresponding increase in the capacitance C of the condenser. The dielectric permits a higher potential difference to be applied than when air is used. This result can be expressed:

$$C = \kappa \epsilon_0 \frac{A}{d}$$

where κ (Greek letter kappa) is the ***dielectric constant*** or ***relative permittivity*** of the dielectric, that is, a measure of its ability to achieve a desired capacitance when the material is placed between two conducting plates. It is a dimensionless number (a ratio of similar quantities, having no units) that indicates the reduction of the field due to the dielectric. (See Table 6-1 for dielectric constants for some common materials.)

Illustrative Problem

Using the preceding problem with only air between the plates of the condenser ($\kappa = 1$), insert paper with a dielectric constant of 3.5 between the plates. Find the capacitance.

Solution

$$0.05 \text{ mm} = 5 \times 10^{-5} \text{ m}$$

$$C = \kappa \varepsilon_0 \frac{A}{d}$$

$$C = 3.5 \times 8.85 \times 10^{-12} \left(\frac{\text{F}}{\text{m}} \times \frac{1 \text{ m}^2}{5 \times 10^{-5} \text{ m}} \right)$$

$$C = 6.19 \times 10^{-7} \text{ F} = 0.619 \text{ } \mu\text{F}$$

Note: Adding paper increased the capacitance almost fourfold.

When a dielectric material such as a polymer, glass, or ceramic is placed between the plates of a condenser, an induced charge Q appears on the plates without any change in the voltage. This naturally increases the capacitance of the condenser. The increased charge is due to the polarization of the entire volume of the dielectric, in turn due to the applied electric field.

Dielectrics are used in alternating current (ac) circuits with a wide range of frequencies. The higher the frequency, the faster the dipoles must switch directions to allow the device to perform satisfactorily. If the dipoles have difficulty moving, as is the situation with complex organic molecules, dipole friction results, with attendant energy losses. These effects show up in the value of the dielectric constant. Temperatures also affect the dielectric constant. In general, the dielectric constant increases with an increase in temperature.

High values of the dielectric constant are needed for capacitors, whereas materials having low values are used as insulating materials. ***Capacitors*** perform many different functions, including filtering (separating) electrical signals with different frequencies, blocking direct current, permitting alternating current to pass, and serving as energy storage devices. Much research is devoted to the miniaturization of electrical circuits, including capacitors, which means that a continual search is underway to find materials, such as ceramics, that not only have high dielectric constants but have such values at higher and higher frequencies and temperatures for use in, for example, dielectric heating applications. Using electrically conductive carbide and nitride ceramic powders and nanomaterials technology, ***ultracapacitors*** have been developed that have an attractive combination of low cost and high performance. They have numerous potential applications in electric propulsion systems, improved radios, and higher-power radar. Combined with batteries, ultracapacitors will optimize performance and extend battery life in such applications as cellular phones.

Ferroelectrics are a group of dielectric materials that exhibit spontaneous polarization, i.e., polarization in the absence of an electric field. Such materials must contain permanent electric dipoles. This ferroelectric behavior is temperature dependent. Above the Curie temperature ferroelectric behavior is lost. Some ferroelectric material such as barium titanate ($BaTiO_3$) at its Curie temperature (120°F) changes its structure from a distorted tetragonal structure to a normal cubic perovskite structure (Section 3.2, Crystal Systems). Ferroelectrics are analogous to ferromagnetic materials that display permanent magnetic behavior (Section 6.3.2). These materials have extremely high dielectric constants at relatively low applied field frequencies. Barium titanate ($BaTiO_3$) is an example of such a material when it exists in its tetragonal crystal structure. In this structure it is also piezoelectric (see Section 27.4 for further information on piezoelectric materials). Table 6-1 lists some additional properties of barium titanate. Capacitors made from ferroelectric materials can be significantly small as a consequence of their high dielectric constants. There is also a need for materials that have low (≤ 3) dielectric constants for use in coatings, films, matrix resins, and fibers, particularly in the electronics and aerospace industries. A low dielectric constant polyimide fiber with high thermal stability and good tensile properties similar to standard textile fibers has been developed for possible use in printed circuit boards and in aircraft composites.

6.2.3 Dielectric Strength

Dielectric strength is the voltage gradient (voltage per unit thickness, such as volts per mil), that produces electrical breakdown through the dielectric. It is an expression of the maximum voltage that a dielectric can withstand before electrical discharge occurs through the material. This insulating strength is dependent not only on the usual items, such as bond type and crystalline structure, but on moisture absorption and the nature of the applied electrical energy. The source of electrical energy might be a direct current (dc) or an alternating current (ac), or a combination of the two. If the magnitude of the dielectric field is sufficient, it will overcome the attraction of the electrons to their positive-charged nuclei and produce a ***leakage*** and eventual rupture of the material. The breakdown voltage has then been reached. With an ac source, the continual reorientation of the material's atoms and electrons, in addition to the deformation of the paths of the electrons, results in a hysteresis loss (explained later) that produces heat in the dielectric material. Therefore, the breakdown voltage will vary inversely with the frequency of the source voltage. Some typical ranges of values of dielectric strength for some nonmetallic materials, expressed in 10^6 volts per meter (10^6 V/m), are listed in Table 6-1.

6.2.4 Dissipation Factor/Dielectric-Loss Factor

When alternating current provides the voltage (sinusoidal) needed to maintain the charge on a capacitor or any dielectric, the current leads the voltage by $90° - \delta$, where the angle δ is called the *dielectric-loss angle*. If the dielectric were perfect, the angle δ would be zero and no loss would occur. In this situation, the current would lead the voltage by 90°. Most of the time, however, perfection is not attained and the current and voltage are therefore out-of-phase, resulting in a loss of electrical energy. This loss of energy, hence power, usually takes the form of heat. The tangent of the angle δ ($\tan \delta$) is known as the ***dissipation factor***. Multiplying $\tan \delta$ by the dielectric constant (κ) expresses the ***dielectric-loss factor, dielectric heat-loss factor, dielectric coefficient,*** or ***loss tangent.*** Dielectric heating is the heating of dielectric materials by subjecting them to a high-frequency, alternating electric field. This heating assists in the bonding or drying of materials (see Figure 6-6). Dielectric loss is affected by frequency and temperature changes. The loss usually increases at low frequencies and as the temperature increases. Table 6-1 lists the dielectric-loss factor for a few nonmetallic materials.

6.3 MAGNETIC PROPERTIES

6.3.1 Magnets, Magnetic Field, Electromagnetic Force, Magnetic Domains

A ***magnet*** is a body that has the property of attracting iron and steel and that, if suspended freely, will turn so as to point in a certain direction. Any material that a magnet attracts is known as a magnetic material. A rod of steel can be made into a magnet simply by rubbing it with another magnet or by placing it within a coil of wire carrying an electrical current. If the rod is made of hard steel, it will retain its magnetism for a long time after the electrical current has ceased. It then becomes a ***permanent magnet.*** Soft iron can be magnetized in the same manner, but upon interruption of the electric current it loses most of its magnetism. What magnetism remains is called ***residual magnetism.*** As long as the current continues to flow through the wire, the soft iron will be magnetized and act as an ***electromagnet.*** Magnetism and electricity are therefore closely related, a relationship not discovered until the nineteenth century. Permanent magnets made of iron can remain magnetized for long periods of time; however, if

(a)

Microstructure of Si_3N_4-6% Y_2O_3-2% Al_2O_3 After Annealing at 1200° C for 20 Hours

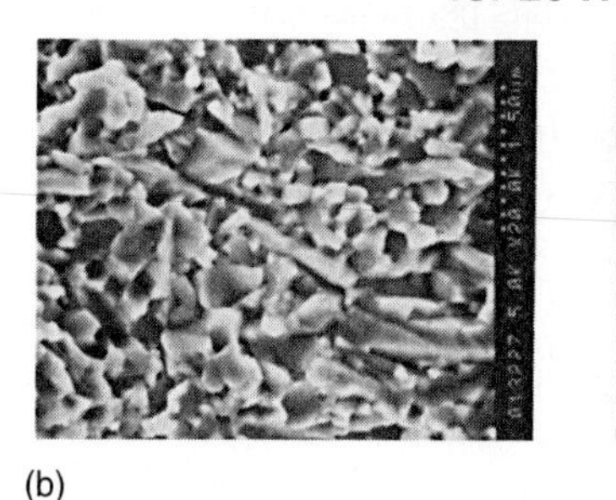

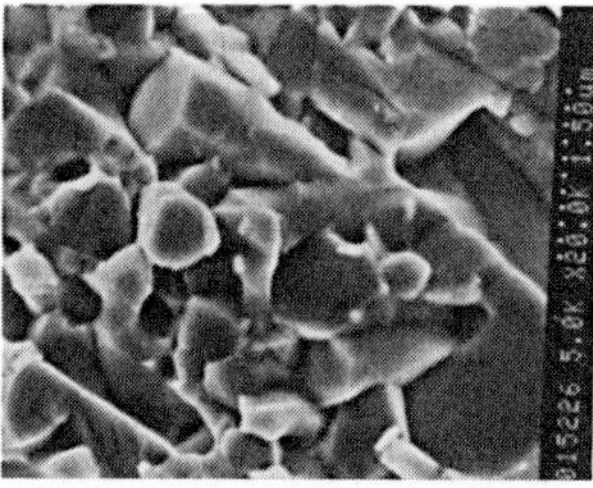

(b)

Figure 6-6 Dielectric heating using microwaves assists in the drying/sintering/curing of materials as well as the bonding of composites using adhesives that have different dielectric-loss factors than the materials being joined. (a) Microwave ovens are used in ceramic processing to achieve lower temperatures, reducing the possibility of inflicting damage on materials due to high-temperature processing. (b) Contrasting ceramic microstructures from conventional and microwave heating. (ORNL)

you drop a magnet, you may cause it to become disoriented and thus lose some or all of its magnetism. Heating a magnet may cause the same effect. Above a characteristic temperature, known as the ***Curie temperature*** (T_c) (1043 K for iron), magnets cannot be made.

A magnet (named after its place of discovery) will attract iron objects such as pins or paper clips. If a magnet is suspended from a thread, its ***north pole*** will point toward the earth's geographic north. This is the principle of the magnetic compass. The other pole pointing toward the south is called the ***south pole***. Magnetism shows its greatest effects at the ends of a magnet, the **poles**, and practically none at its center. If you break a magnet in half, you produce two new magnets, each with a north and a south pole. There is no such thing as a magnet with only one pole. A suspended magnet will turn to a general north-south orientation with one pole always pointing in the same direction. If two magnets are brought near one another, each exerts a force on the other. When a north pole is brought near the north pole of another magnet, the force is repulsive and can be felt before the magnets touch each other. An attractive force is felt when two unlike magnetic poles are brought near one another. Besides iron, other elements such as cobalt, nickel, and gadolinium show strong magnetic effects. These materials are classified as ***ferromagnetics*** (see Section 6.3.3). All other materials only show extremely small magnetic effects.

Why can a piece of iron be turned into a bar magnet (see Figure 6-7a) whereas a piece of aluminum cannot? Why do lodestones (made from the mineral magnetite) have magnetic fields? The understanding of the magnetic properties of bulk materials, that is, the magnetic

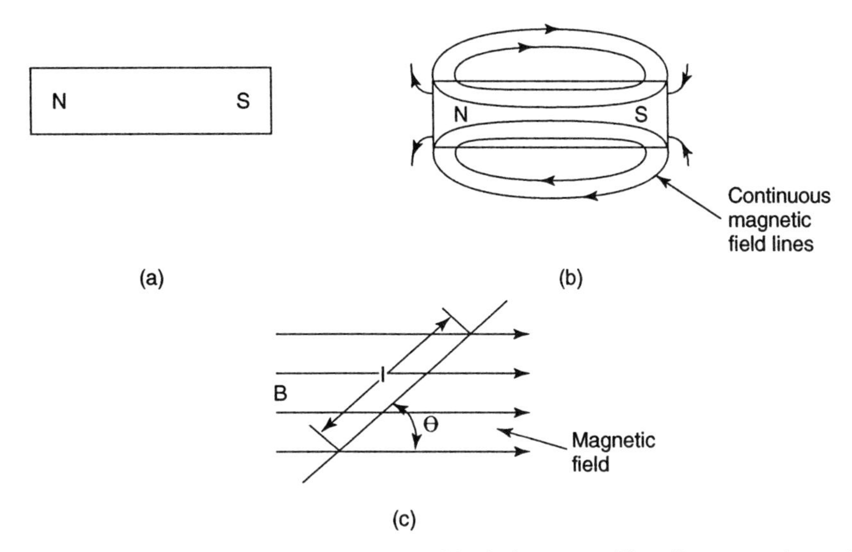

Figure 6-7 (a) Bar magnet. (b) Magnetic field lines outside of a bar magnet. These lines are continuous (closed curves) and they always point from the north toward the south pole of a magnet. (c) A magnetic field *B* with a length of wire (*l*) and crossing it at an angle (θ) to the horizontal.

behavior exhibited by these materials, is related to the atoms that make them up. A magnet creates a magnetic field that acts for a considerable distance, like gravitational and electric forces. We can visualize the magnetic field by constructing magnetic field lines. It is this field that interacts with the magnetic field of an adjacent magnet or an electric field (see Figure 6-7). *A magnetic field exerts a force on an electric current and electric currents produce magnetic fields.* Magnetic and electric forces combined are called ***electromagnetic force,*** one of the four fundamental forces in nature. The lines of a magnetic field, due to a current in a straight wire, form circles around the wire and the field exerts a force on magnets placed adjacent to it. Figure 6-7b shows magnetic field lines that are continuous (closed curves), whereas electric field lines begin and end on the electric charges. The magnetic field is represented by a vector with the symbol ***B,*** whose magnitude is defined as the torque exerted on a compass needle when it makes a certain angle within the magnetic field. The torque applied to the needle orients the needle so that it is parallel to the magnetic field line at that point. B is sometimes referred to as ***magnetic flux density*** or ***magnetic induction.*** The force on a current-carrying wire caused by a magnetic field is determined by

$$F = I \times l \times \mathrm{B} \times \sin\theta$$

where θ is the angle the wire makes with the magnetic field as measured from the horizontal (illustrated in Figure 6-7c), *I* is the value of the current in the wire, and *l* is the length of wire in contact with the magnetic field B. The direction of the magnetic field can be defined as the direction that a north pole of a compass needle would point when placed at that point. As we have said, an electric current produces a magnetic field. Placing a compass needle near an electric wire carrying a current will cause the compass needle to deflect. Conversely, if a magnet is placed near a wire carrying a current, the magnet exerts a force on the wire. The SI unit for this force B is the tesla (T). 1 T = 1 N/(A − m). The older name for the tesla was the weber per meter squared (1 Wb/m^2 = 1 T). The earth's magnetic field at its surface is about 0.5 T. A refrigerator magnet used to hold paper notes has a magnetic field of about 10^{-2} T. A strong electromagnet can produce a magnetic field strength of 3 T. Strong, superconducting magnets have magnetic fields over 10 T. One application of electromagnetic force is the ***electromagnet*** made up of a ***solenoid,*** a long coil of wire consisting of many loops, and a piece of iron placed inside it. When carrying an electric current, each coil of wire produces a magnetic field. The magnetic field of the electromagnet consists of the magnetic field produced by the current plus

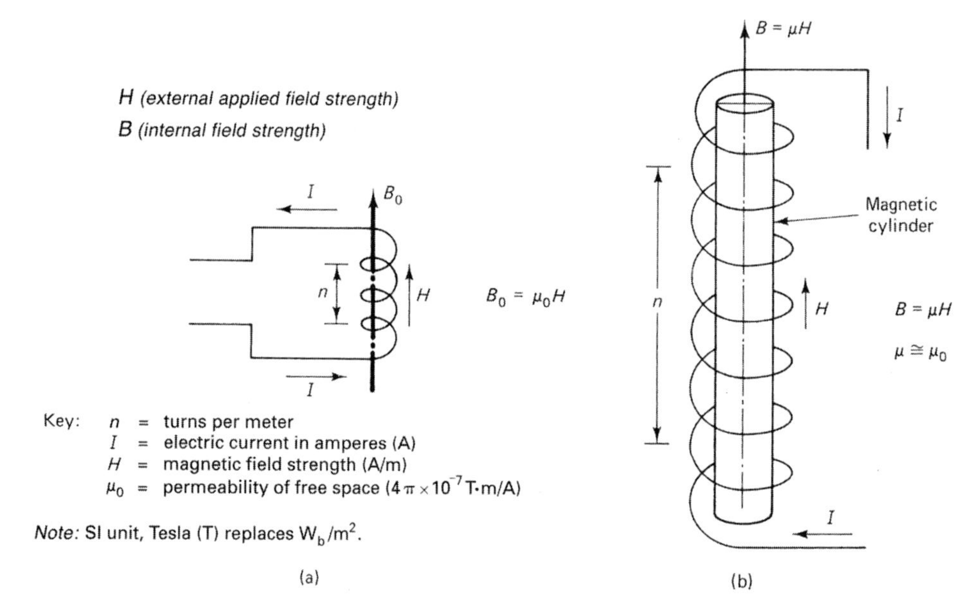

Figure 6-8 (a) Magnetic helical coil with a vacuum core. (b) Magnetic helical coil with a solid material core.

the magnetic field produced by the alignment of the atomic domains in the soft iron core. Figure 6-8 shows a solenoid and an electromagnet; the densely packed continuous magnetic field lines that would be formed inside the coil in (b) are not shown.

Illustrative Problem

An electric wire carries a 30-A current. It is oriented in a magnetic field whose width is 12 cm at an angle of 40°. The magnetic field (B) is 0.90 T. Find the force acting on the wire. See Figure 6-7c.

Solution The sketch shows that if the angle $\theta = 90°$, then the sine of 90° is 1. This makes the wire perpendicular to the field and the force on the wire a maximum. If the angle is 0°, then the wire would be parallel to the field and sine 0° would be 0. For the case where the angle is 40°, and 0.9 T or 0.9 $\text{N}(\text{A} - \text{m})^{-1}$, where N is neutrons of force and A is amperes of current flow,

$$F = I \times l \times \text{B} \times \sin\theta$$

$$F = 30 \text{ A} \times 0.12 \text{ m} \times 0.90 \text{ T} \times 0.64 = 0.1 \text{ N}$$

6.3.2 Permeability Classes of Magnetic Materials, Magnetic Domains

In a magnetic circuit, the ***permeability*** (μ) of a material, a measure of the ease with which a magnetic field can be set up through a material, can be compared to the conductivity in an electrical circuit. The magnetic field, represented by the symbol (H) with units of amperes per meter (A/m), corresponds to the voltage or potential drop. The magnetic flux density, magnetic induction, or magnetic field (B), expressed in units of webers per square meter (Wb/m^2), corresponds with the electric current. (For the SI unit, see Fig 6-8a). The ratio of B to H (B/H) is known as *permeability*, μ (Greek letter mu), with units of webers per ampere meter (Wb/A·m). An important distinction between electrical and magnetic circuits is that permeability is not constant. As H changes, so does μ. Figure 6-8 illustrates the permeability of a vacuum or free space in the core of an electrical coil through which an electrical current is passing. When a core of material replaces the vacuum in the coil, a change in the flux density, B, is detected. This change is reflected in a new value of μ. If the material is made of iron, cobalt, or nickel, an exceedingly large rise in B is noted.

Relative permeability (μ_r) is the ratio of two permeabilities and therefore has no units. The permeability of a material, μ, is thus compared mathematically with the permeability of free space

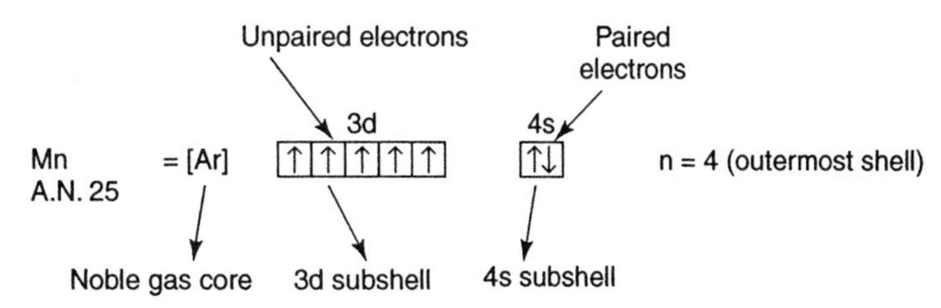

Figure 6-9 Orbital filling diagram for manganese (atomic number 25). The noble gas core consists of a shorthand notation for the orbital diagram for argon (Ar, atomic number 18). The electron configuration is written as Mm:[AR]$4s^2$ $3d^5$.

(vacuum), $\mu_r = \mu / \mu_0$. This ratio may be used to classify magnetic materials into three groupings depending on the interaction of the electrons (internal dipoles) and an external magnetic field:

1. Diamagnetic material if $\mu_r < \mu_0$.
2. Paramagnetic material if $\mu_r > \mu_0$.
3. Ferromagnetic material if $\mu_r >> \mu_0$.*

Diamagnetic materials are solids composed of atoms with completely filled electron shells, so the interaction is nil. Some examples are copper (Cu), silver (Ag), gold (Au), and organic polymers. ***Paramagnetic*** materials contain unpaired electrons in their inner shells that form atomic dipoles that are free to rotate with an external field, producing some interaction and rendering a relative permeability μ slightly greater than 1. Like diamagnetic materials, paramagnetic materials only exhibit magnetism when in the presence of an external magnetic field. Some examples of paramagnetic materials are the metals aluminum (Al), chromium (Cr), and titanium (Ti). Figure 6-9 is a sketch of the electron configuration of manganese (Mn) showing the unpaired electrons in its third inner shell. ***Ferromagnetism*** is present in some metallic materials that possess a permanent magnetic moment. Iron (Fe), cobalt (Co), nickel (Ni), and some metal alloys exhibit ferromagnetism. The magnetic behavior can be increased in magnitude by strong magnetic fields. The explanation of this type of behavior uses the word ***domain,*** which is a tiny region about 1 mm in length or width. Each domain acts like a tiny magnet with a north pole and a south pole and is the result of the electronic structure of individual atoms. These regions are visible using powerful electron microscopes. If the magnetic material is demagnetized ($B = 0$), as in heating, the domains are arranged in a random fashion, as seen in Figure 6-10. This random arrangement cancels out any magnetic effects of the domains. A magnet's domains are preferentially aligned in a single direction. These domains are capable of rotating as well as changing size. They will rotate so as to be more parallel to an external field and grow in size at the expense of their neighboring domains, depending on the strength of the external field. If a ferromagnetic material is heated, the increased movement of the atoms leads to randomization of the domains and their magnetic moments. All ferromagnetic materials begin to lose their magnetic properties as their temperatures increase. The thermal oscillations produced tend to destroy the magnetic domains. At the ***Curie temperature*** (T_c), the randomization is complete and the material is no longer ferromagnetic. The value of T_c varies with the material. For iron, $T_c = 1045$ K. As the temperature decreases below T_c, ferromagnetism returns gradually in a manner similar to a lake freezing over in the winter. Steel, even if demagnetized, has little packets of atoms that are aligned roughly in the same direction. Using domains, one can readily understand why an iron nail placed in the vicinity of a magnet will become magnetized in a direction dictated by the location of the north-seeking end of the magnet used to magnetize the nail. Proof of the existence of domains is readily obtained by using a colloidal solution of iron-oxide particles spread over the polished surface of a piece of magnetic material. When viewed under a microscope, the particles of iron oxide will be concentrated around the domain boundaries.

*The symbol >> means "much greater than."

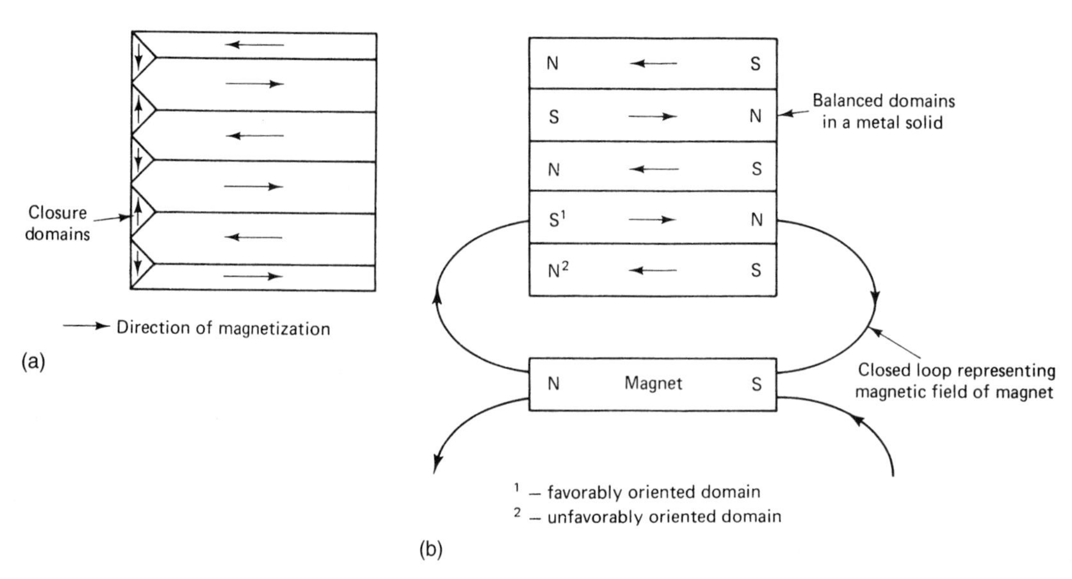

Figure 6-10 (a) Schematic of magnetic domains in a crystal at $B = 0$. (b) Balanced domains in an external magnetic field.

6.3.3 *B-H* Hysteresis (Magnetization) Curves for Ferromagnetic Materials

In our earlier discussion of permeability, we used the terms *magnetic flux density (B)* and *magnetic field (H)*. The symbol B can be accurately described in several ways, such as magnetic flux density, internal magnetic field, induced field, magnetic induction, and magnetization with a symbol M. The symbol H also has several ways to describe external magnetic field strength, applied field, or field intensity. If we plot these two quantities, we arrive at a *B-H* or magnetization curve for a ferromagnetic material. Such curves are depicted in Figures 6-11 and 6-12 for soft- and hard-magnetic materials (discussed in Sections 6.3.4 and 6.3.5). Referring to Figure 6-11, an "unmagnetized" material at point a begins to receive an increasing H as the external magnetic field is built up, possibly by use of an electrical coil winding as shown in Figure 6-8. Inside the material the domains are growing and they reach point b, where the maximum B_s (saturization) or $B_{(max)}$ is reached. At this point, the material is essentially one large domain. ***Remanence*** or ***remanent flux density*** with a symbol Br is a quantity of B, the magnification remaining in the material after the external applied field H is removed. Point c in Figure 6-11 represents the remanence for this curve. Upon decreasing the external magnetic field (H), the domain tends to return to a balanced domain arrangement; but because of the many structural imperfections, such as crystal defects, grain boundaries, or inclusions, the balanced domain state is not reached. Such a situation can be compared to the plastic region of a stress–strain curve. Point c represents the remanent or remanent magnetization with $H = 0$. As the magnetic field increases in the opposite direction, that is, as H increases in the negative direction ($2H$), point d is reached and $B = 0$. The corresponding value of H at point d is known as the ***coercive field*** or ***force.*** This value of H is needed to demagnetize the material (balanced domains). Most domains can be turned around at this point. Continuing to increase the field in the negative direction produces a similar domain growth, but in the opposite direction ($-B_s$). Reversing the direction of the field a second time produces a curve that will plot out a ***hysteresis loop.*** Notice that the curve does not pass through the origin, point a in this cycle. The fact that these curves do not retrace themselves on the same path is called ***hysteresis.*** In such a cycle, much energy is transformed into thermal energy (friction) due to the reordering of the domains. This energy loss is proportional to the area of the hysteresis loop.

6.3.4 Soft-Magnetic Material

Using Figure 6-11 again and treating it as a representative hysteresis curve for a soft-magnetic material, note that for small values of H, a relatively large value of B is produced. Second,

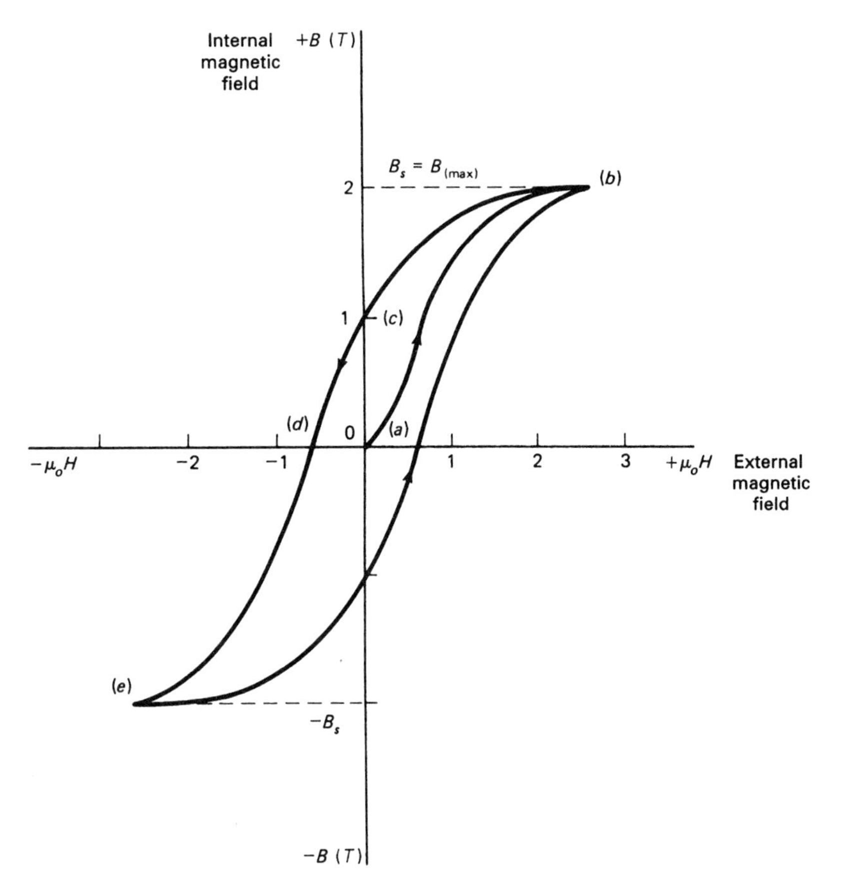

Figure 6-11 Representative *B-H*, or hysteresis, curve for soft-magnetic materials.

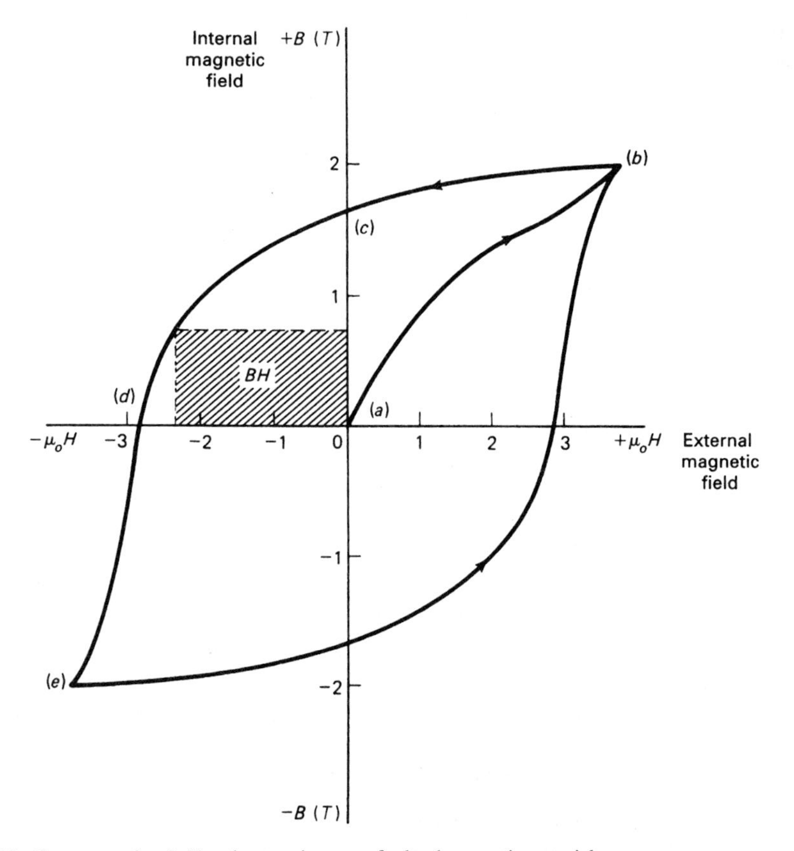

Figure 6-12 Representative *B-H*, or hysteresis, curve for hard-magnetic materials.

note that relatively little remanent magnetism needs to be canceled. Third, the area enclosed by the loop is relatively small. Such a material would find use as a transformer core. A ferromagnetic material can be demagnetized by reversing the magnetic field repeatedly while decreasing its magnitude. Iron (Fe), nickel (Ni), cobalt (Co), and certain alloys are ferromagnetic at room temperature (RT). Some alloys are ferromagnetic only at low temperatures. Soft-magnetic material is used for electromagnets because the field can be switched off, reversing the field with little loss of energy. Tape recorder heads are demagnetized in this way.

6.3.5 Hard-Magnetic Material

Figure 6-12 is a typical hysteresis curve for a hard- or permanent-magnetic material. Comparing this curve with the similar curve for a soft-magnetic material in Figure 6-11 and assuming that the scales for the axes are the same for both curves, several major differences can be observed. Hard-magnetic materials gain their magnetization in manufacture during which the domains are preferentially aligned in one direction. Their essential property is their resistance to demagnetization. In other words, they must retain their magnetization when the external magnetic field is removed. A relatively large coercive field is, then, characteristic of this property, as well as a large hysteresis loop. Note the differences in the values of points c and d in both figures. The area of the largest rectangle that can be inscribed in the second quadrant of the hysteresis loop in Figure 6-12, called the ***B-H product,*** is a measure of the power of a permanent magnet. Materials that make permanent magnets are said to have high ***retentivity***. Table 6-2 is a typical listing of magnetic properties for some nickel alloy steels.

As elsewhere in this text, time and space limit discussion of past and ongoing research and development efforts in the field of new materials (metals, ceramics, and composites), including those being developed for permanent magnets, the market for which is currently over $3 billion and growing. The market for polymer-blended neodymium-iron-boron (NdFeB) permanent magnet is growing at a 30% annual rate. NdFeB magnets, also known as ***rare earth iron magnets,*** are produced by a sintering process called ***rubber isostatic pressing (RIP).*** These high-performance magnets are being used in both the electronics and power industries. When used in large electric motors, they contribute to the world's efforts to reduce global warming because they improve the efficiency of these motors substantially. Due to their effect on the earth's environment, these magnets could be classed as "***green magnets.***"

TABLE 6-2 MAGNETIC PROPERTIES OF NICKEL ALLOY STEEL*[a]

Steel[b]		Temperature						
Type	Condition	°C	°F	Initial permeability, μ_i	Maximum permeability, μ_{max}	Remanence, B (gauss)	Coercivity, H (oersteds)	Hysteresis loss (ergs/cm^3/cycle)
1015 (0.10 Ni)	NT	20	68	190	1590 (H = 4.7[c])	11,400	4.0	2.7×10^4
		93	200	260	1680 (H = 4.2)	11,100	4.0	2.3×10^4
		165	392	260	1850 (H = 3.9)	10,250	3.8	1.9×10^4
		300	572	290	1900 (H = 3.3)	8,500	2.9	1.5×10^4
4340 (1.88 Ni)	NT	25	77	43	330 (H = 30)	13,700	22.5	11.4×10^4
		114	237	34	440 (H = 25)	12,700	13.9	7.6×10^4
		167	333	40	420 (H = 28)	12,000	18.0	7.9×10^4
		302	576	85	460 (H = 23)	10,250	11.0	4.9×10^4

Source: Courtesy of Inco.

**Note*: units not SI.

[a]These data are based on the initial magnetization curve and the magnetic hysteresis loop for applied field strengths (magnetizing forces) up to about 90 oersteds.

[b]N, normalized; NN, double normalized; Q, quenched; T, tempered.

[c]Oersted.

6.3.6 Smart Electronic Materials: Electrorheological and Magnetorheological Materials

The word *rheology* literally means "a knowledge of flow." Today, rheology is an interdisciplinary science that, in the main, deals with all aspects of deformation of materials under the influence of an external force (load, electrical or magnetic field, etc.). In fact, recent attention in the field concentrates on materials that can be classified as neither solid nor liquid. This intermediate state of matter between the solid and fluid states is represented by many rheological materials, such as biological fluids, polymer melts, solutions, emulsions, and pastes. Rheology uses terms such as *elasticity, viscosity, solvents, solutes, gels*, and *suspensions*, many of which have been described previously. Two examples of such "smart" materials and their potential for future applications are ***electrorheological*** and ***magnetorheological*** materials, which are introduced in Module 15.

Electrorheological fluids change their flow characteristics when subjected to an electric field. The response, almost instantaneous, to such a field is a progressive gelling of the fluid that is proportional to the field strength. The fluid, composed of two components—-a carrier fluid and the suspended particles in the fluid—does not conduct electricity. The suspended particles—typically, polymers, minerals, and ceramics, which may take the form of suspended or dissolved particles or molecules—polarize and link themselves together when an electrical field is applied across the fluid (See Sections 2.1.1.2 and 6.2.1 for discussions of polarized molecules and polarization.) This linking produces a gelling of the carrier fluid, inhibiting its flow. Once the applied field is removed, the polarization of the particles ceases and the fluid can then flow freely (see Section 15.4). Many future applications are envisioned in the development of automatic transmissions, brakes, clutches, hydraulic valves, and actuator devices.

A second example of smart electronic materials is ***magnetorheological materials.*** Similar in many respects to electrorheological materials, these materials use magnetic particles suspended in a nonmagnetic fluid (the host fluid). The host fluid could be a light machine oil or vegetable oil. A simple system, such as iron filings suspended in a vegetable oil and a horseshoe magnet, could amply demonstrate the principles of magnetorheological behavior. When a magnetic field is applied, the magnetically susceptible particles in the nonmagnetic fluid align themselves and remain stationary, thus restricting the movement of the suspension fluid. The viscosity of these materials is thus a function of the magnetic field in which they are immersed. The greater the field, the more solidified is the slurry. Applications for such materials are as varied and numerous as they are for electrorheological materials, especially for controlling fluid flow, as in hydraulic braking. A recent application is in controlling the suspension of truck seats by automatically adjusting the seat to changing driving conditions. During routine driving, a height sensor notes the downward movement of the seat and sends a signal to a microprocessor, which activates an electromagnet. The electromagnet quickly stiffens the fluid in a liquid state in a shock absorber to a jellylike near-solid. The fluid is silicon-based hydraulic oil plus an additive that permits a greater suspension of iron filings. Delphi Automotive System's new Magnaride shock absorber liquid is a combination of microscopic iron particles and conventional fluid. It changes viscosity when exposed to a magnetic field from the shock absorber piston. The change is rapid—ten times as fast as a conventional shock—and the suspension damping is attered without the need for mechanical valves, which allows a wide variation in shock action.

6.4 OPTICAL PROPERTIES

6.4.1 Characteristics of Light

Optical properties describe a solid material's response to exposure to electromagnetic radiation and specifically to visible light, a narrow portion of this spectrum (see Figure 6-13). This response has been described in terms of the interaction that occurs between this incident radiation and the material's atoms, ions, or electrons. These interactions and specialized terms are discussed briefly throughout the text under the topics of absorption, reflection, refraction,

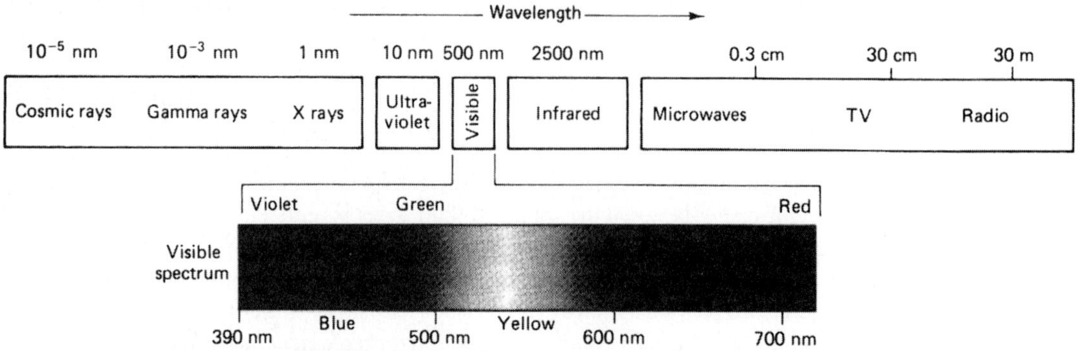

Figure 6-13 Wavelengths of electromagnetic radiation characteristic of various regions of the electromagnetic spectrum. (Theodore L. Brown and H. Eugene LeMay Jr., *Chemistry*, 2nd ed., Englewood Cliffs, NJ: Prentice-Hall, Inc., 1981)

color, phosphorescence, electroceramics, polarization, electromagnetic spectrum, electron band theory, reflection of light, refraction, transmission, luminescence, solar cells, and lasers.

Light waves are often compared to water waves. The study of wave motion usually begins using examples such as wave motion in a body of water made by a boat, waves produced by vibrating a rope or string, or those sound waves produced by some vibrating object. Such waves are considered ***mechanical waves.*** All require some medium in order to travel. For example, without a medium, sound waves could not be heard. If one observes leaves floating on a lake when waves pass by, the leaves simply oscillate up and down and are not carried forward with the waves. A wave consists of oscillations that move without carrying matter with them. Rather, they carry energy from one place to another. Visualize ocean waves hitting a beach during a hurricane to understand the significance of this statement.

Wavelength (λ) is the distance between any two successive identical points on a wave. The frequency of a wave (f) is the number of complete cycles per second or the number of crests that pass a given point per unit time. ***Frequency*** is usually specified in hertz (Hz), where 1 Hz = 1 cycle per second. The period (T) is $1/f$, or the time required for one complete cycle. For example, if a wheel rotates at a frequency of three revolutions per second, then each revolution takes 1/3 s. The wave velocity (v) is the speed at which wave crests appear to move. A wave crest that travels a distance of one wavelength in the time period T has a velocity equal to l/T or $v = l/f$.

Illustrative Problem

If a wave with a wavelength of 14 cm is traveling at a velocity of 34 m/s, what is the wave's frequency?

Solution Convert the units, 14 cm = 0.14 m. Using the previous formula $v = l/f$ and solving for f, $f = v/l$. Substituting values, $f = 34$ m/s/0.14 m = 243 Hz.

Electromagnetic (EM) waves, or EM radiation, resemble mechanical waves. They are produced by electric charges that are oscillatory. Generally, EM waves are produced by accelerating electrons and other charged particles, and they carry energy from one region of space to another. The different forms of EM radiation are represented by the EM spectrum (see Figure 6-13). They all travel in free space at the same velocity—the velocity of light—at 3.00 $\times 10^8$ m/s, represented by the letter c. Visible light can be considered an electromagnetic wave, with wavelengths between about 400 nm and 750 nm.

Illustrative Problem

Calculate the wavelength of a 60-Hz EM wave.

Solution Using the previous formula and substituting c for v, this new relationship is $c = \lambda f$. Solving for λ and substituting values:

$$\lambda = c/f = 3.00 \times 10^8 \text{ m/s}/60 \times 10^{-1\text{s}} = 5.0 \times 10^6 \text{ m}$$

(Note the size of this wavelength.)

When light travels from one medium into a second medium where its speed is greater, the ray bends away from the common normal. When the ray enters a medium where the speed of light is less, the ray bends toward the normal. In other words, the ***index of refraction*** depends on the wavelength. The ratio of the speed of light in vacuum to the speed in a given material (v) is the ***index of refraction (n)*** of that material. Because the index will vary with the wavelength of the light, a particular wavelength is specified when dealing with materials. As wavelength changes with the index, the frequency of a wave remains the same as the wave passes from one medium to another. The refractive index (n) is determined by the chemical composition and the crystallographic properties in a material. Changes in it can be brought about by compositional changes, crystallographic changes, or intermingling of phases. Usually these changes cause a material's physical properties to change. Microscopic analysis of the ***morphology*** (shape and structure) of fiber-reinforced polymeric materials (composites) uses the refractive index of polymeric materials since most are semitransparent to light.

Illustrative Problem

Using the index of refraction for diamond of 2.42 and using yellow light with a wavelength of 589 nm (see Figure 6-13), find the speed of light in diamond.

Solution

1. $n = c/v$.

2. Substituting and solving for v,

$$v = c/n = 3 \times 10^8 \text{ m/s}/2.42 = 1.24 \times 10^8 \text{ m/s}$$

The frequency of light does not change when it enters another material, but the wavelength of light in a material differs from the wavelength of light in a vacuum. We can conclude that the index gives us a measure of the speed of light in a transparent material, that is, it is a measure of the material's ***optical density.*** The greater the index, the greater is the optical density of the material and the smaller the speed of light in that material. Light can be pictured as traveling in rays (light perpendicular to a wave front). When a light ray strikes a smooth surface it is reflected, making an angle θ (Greek letter theta) with a line normal to the surface, at the point of reflection. The angle θ_i that the reflected ray makes with the normal obeys the equality known as the **law of reflection,** namely $\theta = \theta_i$. See Figure 6-14a.

Light is energy in the form of radiation. It is regarded as being composed of either ***photons*** (a quantum of electromagnetic radiation) in the energy range 2.5 to 5.6×10^{-19} J or electromagnetic waves in the wavelength range of 800 nm down to 380 nm (see Figure 6-13). Photons can be viewed as being packets of energy. Light, being a form of electromagnetic radiation, interacts with the electronic structure of the atoms of a material. The initial interaction is one of absorption; that is, the electrons of atoms on the surface of a material will absorb the energy of the colliding photons of light and move to the higher energy states. The degree of absorption depends, among other things, on the number of free electrons capable of receiving this photon energy. The electrons can then do several things. They can have sufficient energy to jump into higher energy states, which permits them to be accelerated within an electric field and thereby conduct electricity (photoconductive effect, see Sections 6.4.4 and 27.3.1 for further discussion of photoconductivity), they can collide with atoms and release their excess energy in some form of electromagnetic radiation, or they can convert their excess energies to atoms in the form of thermal energy.

When light proceeds from one medium into another (a metal, semiconductor materials, or electrically insulating materials), it may be transmitted through the materials, it may be reflected at the interface between the two media, or it may be absorbed. Figure 6-14 shows the light ***intensity (I)*** with units of watts per square meter represented by I_0, the incident light ray; I_r, the reflected light; and I_t, the transmitted light. ***Reflectivity*** is a ratio of I_r/I_0 (see Figure 6-14). ***Transmissivity*** is the ratio of I_t/I_0, and ***absorptivity*** is the ratio of absorbed light to the incident light. Light radiation is absorbed within solid materials by interactions between the light, a form of electromagnetic radiation (Figure 6-13), and the material's electrons. The results may be electronic polarization (see Figure 6-5) and/or, as we have said, electron transfers from one energy state to another (see electron band structure in Section 27.1).

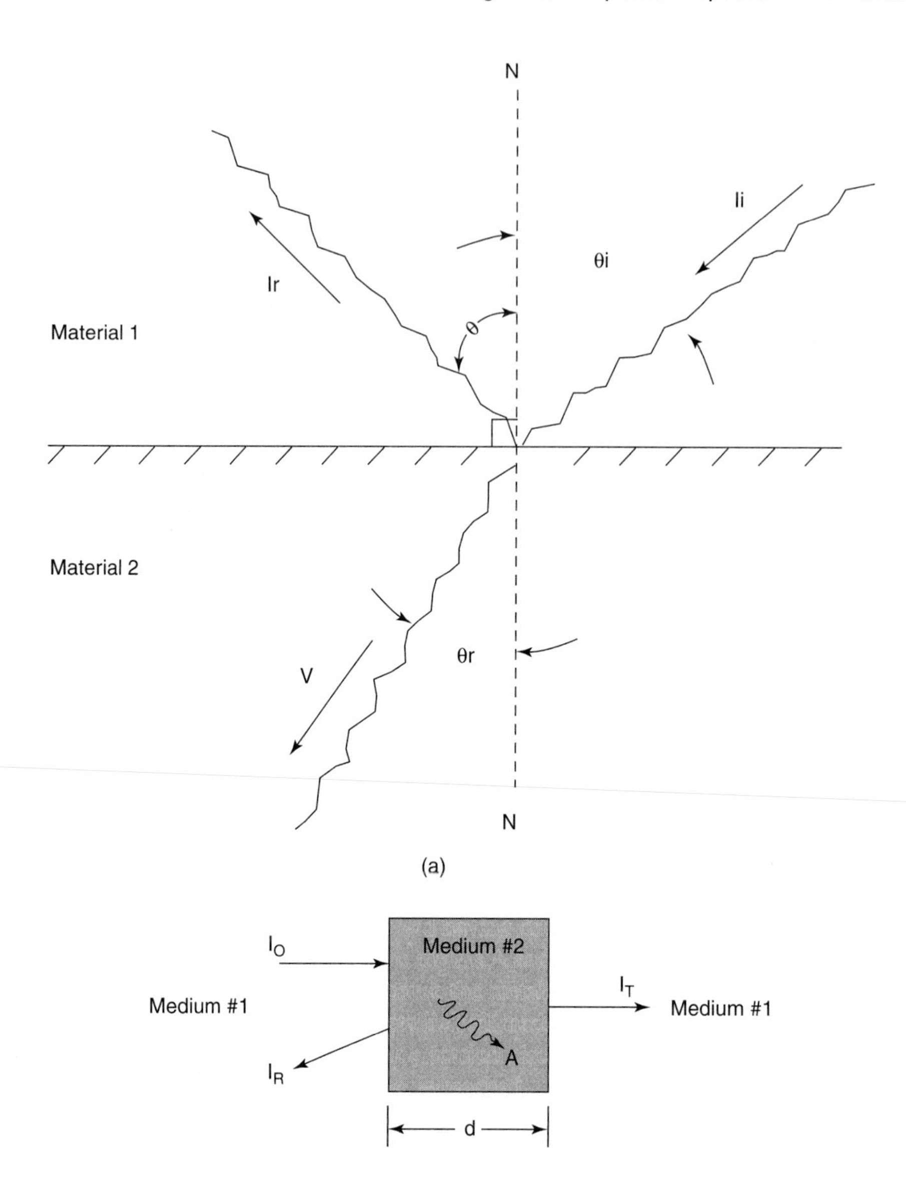

Figure 6-14 (a) Refraction of light, showing the angle of incidence (θ_i) and angle of refraction (θ_r), both measured from a common normal *N*. (b) The transmission of light through a transparent medium of thickness *d* is shown. I_0 is the incident light, I_r is the reflected light, and I_t represents the transmitted light leaving the material. The absorbed light in medium #2 is shown by a wiggly arrow labeled A. Both versions of medium #1 are the same material. Refraction is generally accompanied by reflection.

The light from a sodium lamp is an example of the effect of passing a current through a gaseous medium. The high-pressure sodium (HPS) lamp, invented in the 1960s, can last 6.5 years if operated for 10 hours per day (24,000 hr). It puts out 140 lumens per watt, versus 15 lumens per watt for an incandescent lamp. The light is produced by passing an electric arc through vapors of sodium and mercury, resulting in a temperature of greater than 2192°F in the thin-walled "arc tube," which is made of translucent, polycrystalline Al_2O_3. The arc tube, about 5/16 in. diameter × 4 in. long, is enclosed within a glass bulb. High-intensity lighting gives truer colors. Metal halide vapor emits white light, which is more expensive and shorter-lived.

Light is composed of low-energy photons (E_p is less than 10 eV) represented by infrared (IR), visible light, and ultraviolet (UV) in the electromagnetic spectrum (see Figure 6-13). UV light is being tested for curing solvent-free urethane paint on autobodies as a technique to eliminate the release of VOCs into the atmosphere from the spraying and baking process newly painted autobodies must undergo. As light enters a nonmetallic material, it is absorbed (Figure

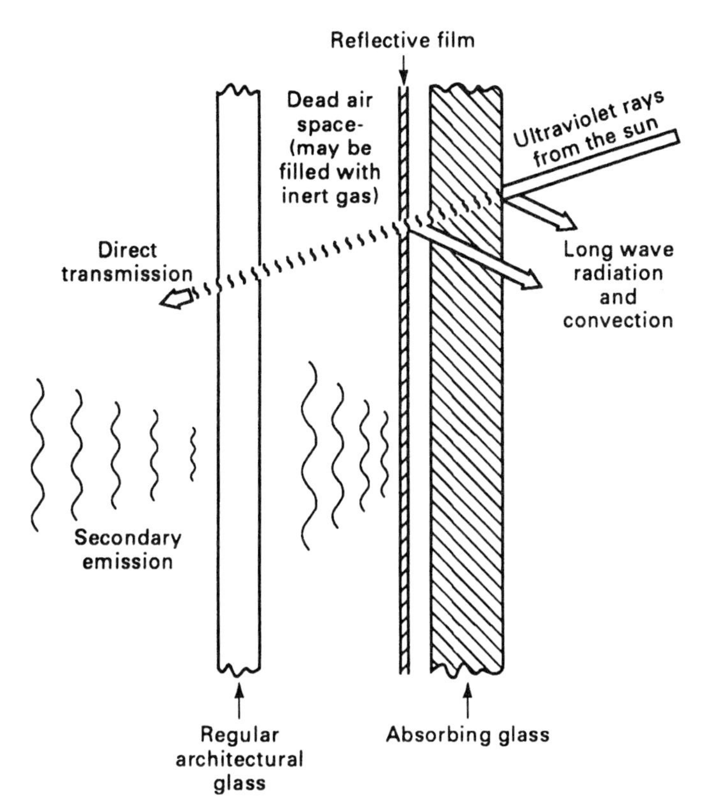

Figure 6-15 Energy-efficient, low-emissive, double-pane glazing with a reflective film.

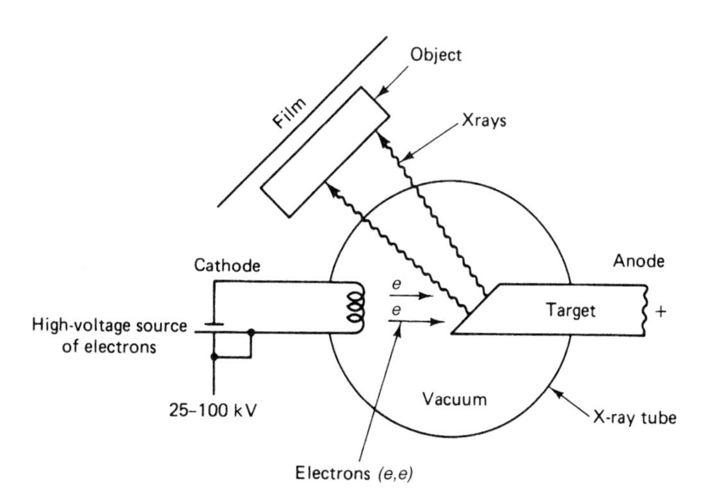

Figure 6-16 Schematic of an X-ray tube.

6-15) by the action of several mechanisms, such as electronic polarization (Figure 6-5) and electron transitions. The electron transitions depend on the electron energy band structure of the material. The interaction process characteristic of photons depends on their energy (E_p). Low-energy photons interact principally by ionization or excitation of the outer orbitals in the atoms of a solid. High-energy protons (E_p greater than 10^4 eV) are produced by X rays and gamma rays (Figure 6-16). The minimum energy (E_p) required to excite and/or ionize the component atoms of a solid is called the *absorption edge* or *threshold.*

When crystalline materials are exposed to radiation, the change in their properties is due to a displacement of atoms from their normal locations in the crystal structure, that is, the creation of crystal defects. The total energy of the incident radiation is measured in terms of dose,

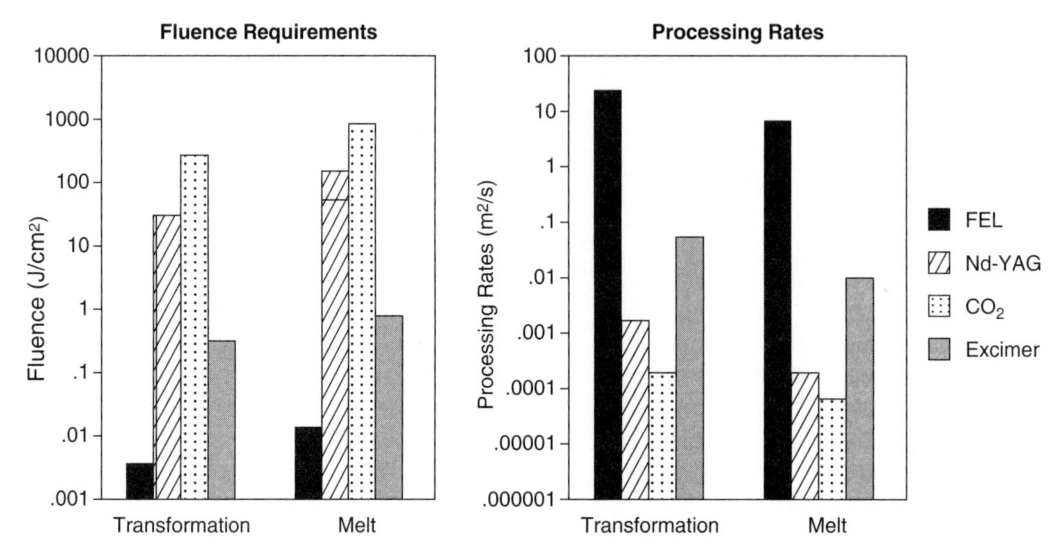

Figure 6-17 Calculated laser fluence and processing rates for metal surface processing with conventional lasers and the UV demonstration FEL. Large-scale exploitation of UV surface processing will require sources of at least a few tens of kilowatts, light that costs under a cent per kilojoule, and a full wavelength tunability. (TJNAF, Newport News, VA)

the amount of radiation affecting the material, or the dose rate, which is the rate of energy deposition. Irradiation conditions are often given in terms of the total number of particles incident on the sample, that is, the ***fluence*** (fluence equals the summation of flux, or the particles per unit time deposited, over irradiation time; see Figure 6-17). A greater understanding of this topic requires a study of the mechanisms by which energy is transferred from the radiation (in this case, light) to the atoms of a solid. In the case of metals, only the energy transferred to the nuclei by the beam of energetic, charged particles such as electrons, photons, and fission fragments is effective in creating lattice defects. The crystalline structure can also be converted to an amorphous one by sufficient exposure to radiation (see Table 6-4). In metals, the band gap is zero and photons of all energies can be absorbed to increase the kinetic energy of the conducting electrons and holes. In polymers (covalently bonded materials), photons of suitable energy are capable of exciting vibrations directly in both molecules and crystalline structures. The chemical bonding can be altered by electronic excitation, which can also create cross-links and break primary bonds.

When light is transmitted into the interior of a transparent solid, it experiences a decrease in its velocity. The result is that light bends as shown in Figure 6-14b. As we have said, this phenomenon is known as refraction and the index of refraction (n) of a solid material depends on the wavelength of the incident light. For air, the index is nearly equal to 1, whereas water has a higher index of refraction, 1.33. Other indices of refraction for some materials are as follows: silica glass, 1.46; corundum, 1.76; polyethylene, 1.51; and polypropylene, 1.49. Lucite (Plexiglas) has an index of 1.51 and diamond's is 2.42. When light passes from one medium to another having a different index of refraction, some of the light is scattered at the interface between the two media. Both of these media could be transparent materials.

6.4.2 Photoelectric Effect, Photons, and Lasers

Certain metallic materials will become positively charged when exposed to electromagnetic radiation. Because such radiation can carry no electron charge, the conclusion is that the radiation must interact with the electrons attached to the metallic atoms and cause the ejection of electrons from the surface of the metal. Such a phenomenon is known as the ***photoelectric effect,*** and the ejected electrons are called ***photoelectrons***. Not all electromagnetic radiation produces this effect; it occurs only when a certain frequency of radiation is reached (***threshold frequency.***) The threshold frequencies differ for different metallic elements. As an example, the metal sodium will produce this effect once the frequency of the incident radiation reaches 5.6×10^{14} Hz. The corresponding wavelength of 5.4×10^{-7} m places such radiation in the visible-light portion of the electromagnetic spectrum. Zinc requires a threshold frequency of 8.0×10^{14} Hz, which is in the ultraviolet region of the spectrum. The intensity of the radiation plays no part in producing the photoelectric effect. To explain this phenomenon adequately, Einstein proposed in 1905 that this radiation be treated not as a series of waves (wave phenomenon), but as a series of particles (light photons) that interact with single electrons. X-ray films and the electron microscope reveal some examples of radiation that exhibit wavelike properties. The pattern of electrons that projects itself onto the X-ray film and the highly magnified images of the internal structure of a material as viewed under an electron microscope are diffraction patterns of waves typical of those exhibited by the diffraction of light through a grating. Most of the questions dealing with light radiation confine themselves to large-scale bodies such as glass lenses. When we begin to probe into the land of the atom with its subatomic electrons, we use the particle nature of radiation to explain the results of such interactions of radiation, not with a whole surface but with individual electrons.

Present-day understanding of the fundamental structure of matter reveals four fundamental forces in nature, of which gravity and the electromagnetic (the electric) force are two. These forces act over large distances and so are called *long-range* forces. Electromagnetic force also binds atoms and molecules into ordinary matter. In the nucleus of an atom there are strong and a weak electromagnetic forces. The strong force is responsible for holding the nucleus together. It acts between all nucleons. The protons in the nucleus attract each other via the strong force at the same time they repel each other by the weak force. Neutrons, being electrically neutral, can only attract other neutrons or protons. The strong force operates only at short distances. If two nucleons are less than about 10^{-15} m apart, the force is very strong; it drops to zero at distances greater than that amount.

Quarks and electrons are the basic building blocks of the world's structure. A theory developed in the early 1980s, **quantum chromodynamics (QCD),** uses an analogy to explain quarks. An ordinary atom consists of electrons bound to the nucleus by electromagnetism, the force generated by particles of light. Similarly, the proton and neutron form clusters of quarks bound together by the strong force—the "gluing" force— generated by particles "playfully" called *gluons*. Now, protons and neutrons can be thought of as "atoms" made up of quarks. Two quarks, called *up* and *down*, are found in everyday matter. By using high-energy accelerators, four other quarks have been found fleetingly under these extreme conditions—*strange, charm, top*, and *bottom*. The quantum chromodynamics theory forms the basis for the experiments with the continuous electron beam accelerator conducted at the Jefferson Laboratory in which experimenters seek to build understanding of everyday matter by examining nuclei in terms of quarks and their interactions via the strong force.

A photon of light (a quantum of electromagnetic radiation) exchanges its energy with a single valence or conduction electron. We have learned that such electrons possess the highest kinetic energies and are the outermost electrons in an atom. When a photon of light with sufficient energy ($E = hc/\lambda$) strikes a valence electron, it causes the excited electron to leave a solid metal. Note that the photon reacts with a single electron, not with an entire atom or a solid metal electrode. The kinetic energy (KE) of the ejected electron (now called a *photoelectron*) can be expressed in two terms. The first term, the work function (W), is that part of a photon's energy (work) expended in digging out or removing the valence electron from the surface of the metal. The second term represents the remainder of the photon's energy, appearing in the form of the kinetic energy of the escaping electron. The photon disappears, having done its job of transferring its energy, and a photoelectron is emitted from the surface. The electrons in a metal that are eligible to receive energy from colliding photons may be in various energy levels of an atom. Those with the highest energies will require the least energy expenditure to break free from the surface; hence, they will have the greatest kinetic energy. Using electron volts (eV) as units, this energy transfer can be expressed mathematically as

$$\text{KE} = E - W$$

Suppose that violet light is incident on a piece of cesium (Cs). Referring to reference books to find the photoelectric properties of metallic elements, we find the threshold frequency for Cs to be 4.6×10^{14} Hz, the work function, W, is 1.9 eV, and the wavelength (λ) of violet light is 4×10^{-7} m. To find the frequency, f, of violet light, we use

$$f = \frac{c}{\lambda} = \frac{3 \times 10^8 \text{ m/s}}{4 \times 10^{-7} \text{ m}} = 7.5 \times 10^{14} \text{ Hz}$$

Note that this frequency exceeds the threshold frequency for Cs. The photon's energy, E, is next calculated using

$$E = \hbar f$$

where $\hbar$ is Planck's constant (6.625×10^{-34} J·s) and 1 eV = 1.60×10^{-19} J. Thus,

$$E = \frac{(6.625 \times 10^{-34} \text{ J·s})\,(7.5 \times 10^{14} \text{ Hz})}{1.6 \times 10^{-19} \text{ J}} = 3.1 \text{ eV}$$

The kinetic energy (KE) of the photoelectron emitted can then be determined:

$$\text{KE} = E - W = 3.1 - 1.9 = 1.2 \text{ eV}$$

Luminescence is the reemission of photon energy at wavelengths in the visible spectrum as the result of the absorption of electromagnetic radiation from some outside source. If re-emission occurs for times much less than 1 second while the material is absorbing the radiation, the phenomenon is called ***fluorescence.*** In a fluorescent lamp, electrons emitted by the incandescent cathodes collide with electrons of the mercury atoms that fill the tube. The collisions cause the emission of radiation in the invisible ultraviolet range. This radiation, in turn, strikes the fluorescent or phosphor material coating the inner side of the tube and causes this material to emit radiation in the visible range of the spectrum. Should the light emitted continue after the radiation producing it has been removed, the light is called ***phosphorescence.*** Color TV picture tubes use phosphors that phosphoresce. Phosphorescence is an important optical property of ceramics. Ceramic phosphors are used in many applications such as TV screens and fluorescent lights. Phosphorescent ceramic coatings are used in photocopy lamps. Various dopants are added to produce the emission of certain colors.

When a surface is exposed to radiation, it can absorb part or all of the incident radiation (energy). The fraction of the energy absorbed is the emissivity (e) of the surface. For a black body, the perfect absorber, $e = 1$; for the perfect reflector, a white body, $e = 0$. ***Photoconductivity*** describes the increase in conductivity experienced by some semiconductive materials when illuminated by a light source. This increase in conductivity occurs when light is absorbed by photon-induced electron transistors, which produce additional charge carriers. Cadmium sulfide, a photoconductive material used in light meters, generates when struck by incident light radiation a photon-induced current that is a measure of the rate at which photons of light strike the photoconductive material. Gallium arsenide (GaAs) converts light directly into electricity. It is used in light-emitting diodes (LEDs) and transistors. (See Section 6.27.3 for further discussion of photoconductive devices.)

By synchronizing the light emission from many electrons, one can produce light that is monochromatic (one frequency) and coherent (photons are in phase with each other). More and more materials are being found whose electromagnetic radiation can be controlled by technicians. These include solids, liquids, and gases. Semiconductor materials can produce laser action that emits photons with wavelengths in the range 7-8 $\times 10^{-7}$ m. A ***laser*** (*l*ight *a*mplification by *s*timulated *e*mission of *r*adiation) is a special example of the emission phenomenon we have been discussing. The light produced is further amplified through stimulating the emission. To defeat lung cancer, two newly developed technologies use laser light to detect and destroy cancer cells. One such technology uses certain wavelengths of laser light to illuminate the organ; cancerous tissues fluoresce in reddish hues. Another technology called ***photodynamic therapy*** involves injecting the patient with a photosensitive drug that concentrates in cancerous tissues; when activated by the red laser light, the drug destroys these cells. A ***maser*** (*m*icrowave *a*mplification by *s*timulated *e*mission of *r*adiation) differs from lasers mainly in the wavelength of the radiation produced.

The Laser Processing Consortium (LPC), a collaboration involving nine U.S. corporations and companies, seven research institutions, and the Thomas Jefferson National Accelerator Facility (TJNAF) in Newport News, Virginia, has completed its first major demonstration of high-power infrared (IR) light, the ***free-electron laser (FEL).*** The FEL is a unique, high-power, high-efficiency, tunable light source for basic research, industrial processing, and military applications. "Free electrons" are electrons that are free from their respective nuclei in the atomic structure of materials and that travel at the speed of light. Conventional lasers suffer from limitations in cost, power, and choice of wavelength. Figure 6-17 compares the fluence (a measure of the irradiation on a surface) and the processing rates of commercial lasers with the FEL. Commercially available excimer lasers are limited to tenths of a kilowatt, cost tens of cents per kilojoule, and have a few isolated specific wavelengths. Table 6-3 compares the characteristics of some commercial lasers with those demonstrated by the FEL. The FEL can be tuned over a much wider range, which allows it to lase a variety of materials with different absorbencies with a single-laser system (see ***absorbtivity*** in this section).

The cost-effectiveness of the FEL is increased by its ability to recover almost all energy back into the accelerating electron beam. This recycling of the electron beam extracts power that was not radiated as light. Its short pulse length, measured in picoseconds (ps), also allows for more efficient coupling of light with the target materials. A pulse rate between 18 and 74 megahertz (MHz) can drastically increase the rate at which a material is lased; thus, speed deposition times can be reduced in laser-based deposition systems. These characteristics favor future application both in the military and for industrial applications calling for processing of large volumes of materials (***light-based manufacturing***).

Currently, the FEL operates in the IR region of the electromagnetic spectrum, but the ultraviolet (UV) region is favored by industry because UV light has higher absorbency in most relevant materials. A UV FEL could replace wet chemical methods and provide new capabilities for surface modification and other valuable procedures. Thus, FEL manufacturing would become a method of "green manufacturing," helping to reduce the pollution of the earth in which we live. Most surface modification treatments of over 40 billion pounds of fiber produced in the United States use wet chemical processes. Experiments with the surface modification of nylon and polyester to enhance certain properties such as surface friction, filtration, and visual characteristics have already begun (see Figures 6-18a and b and 16-19). A NASA project involves a "*Solar Sail*," a kilometer-wide or larger sail that spacecraft could unfold in interplanetary space to harness solar winds (the winds are made up of protons). These sails could consist of a solid polyimide base material coated with aluminum. The FEL would be

TABLE 6-3 PERFORMANCE COMPARISON OF SELECTED LASERS

Laser	Pulse width	Repetition rate	Output energy (per pulse)
CO_2	100 ns	50 kHz	0.1 J
Excimer	10-30 ns	~100 Hz	0.3 J
IR-DEMO FEL	~0.5-2 ps	18.7 or 37.4 MHz	25 μJ
$Ti:Al_2O_3$	100 fs	80 MHz	10 nJ
Nd:YAG (Q-switched)	3-20 ns	10 Hz-50 kHz	1.5 J-80 μJ
Nd:YAG (ultra-fast)	30 ps	10 Hz	50 mJ
Alexandrite	30-100 ns	20 Hz	0.5 J
HF	400 ns	100 Hz	50 mJ

Source: TJNAF, Newport News, VA.

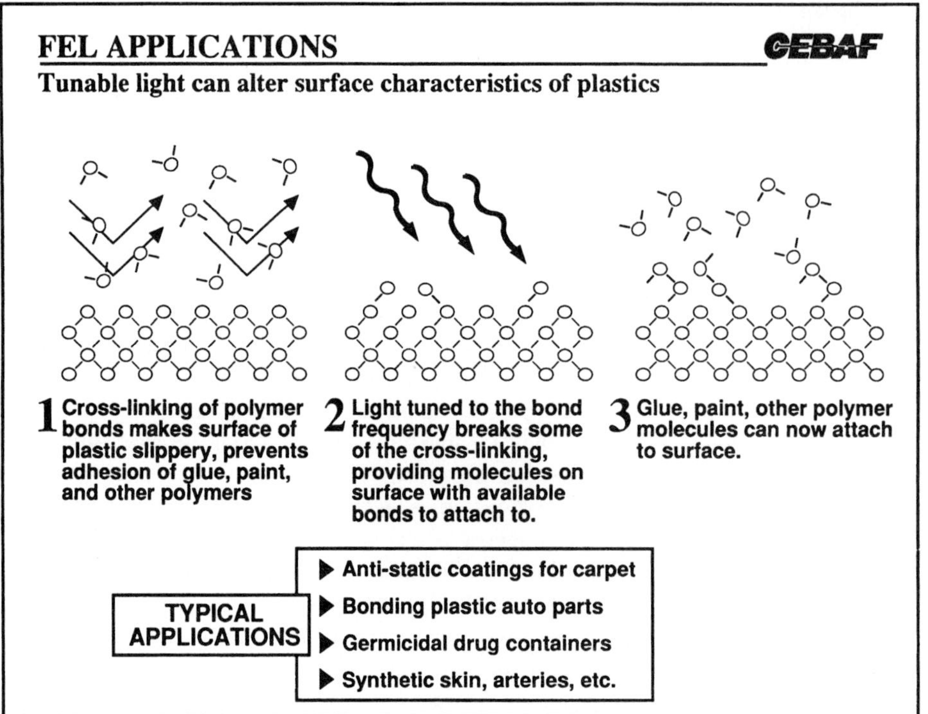

(a)

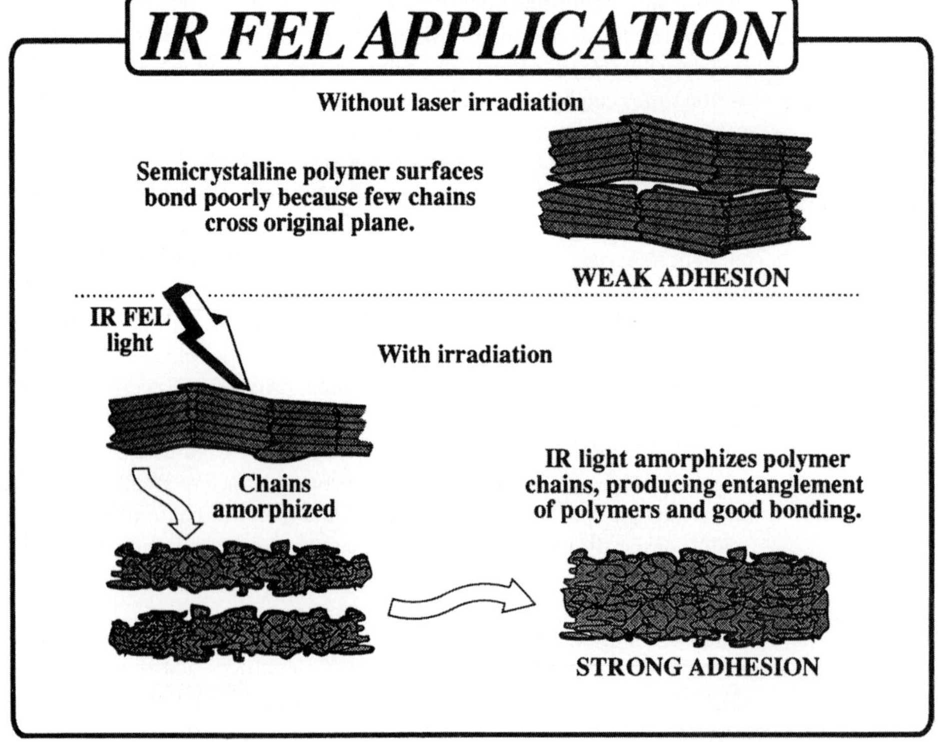

(b)

Figure 6-18 (a) Light-based manufacturing involving modifying materials' surfaces by using the UV portion of the electromagnetic spectrum. Laser processing treatments can give a polymer or fiber product new friction, filtration, wetting, or visual characteristics. (TJNAF, Newport News, VA) (b) Surface modification by light can add value to existing products, serve as the value basis for new products, or be incorporated into special interfaces of packaging materials or structural composites. In this application, surface amorphization (carefully tailored light energy absorption in the transformed region in the surface layers of a polymeric material) produces greater adhesion, with the additional benefit of eliminating the environmental costs of processing with wet chemistry. (TJNAF, Newport News, VA)

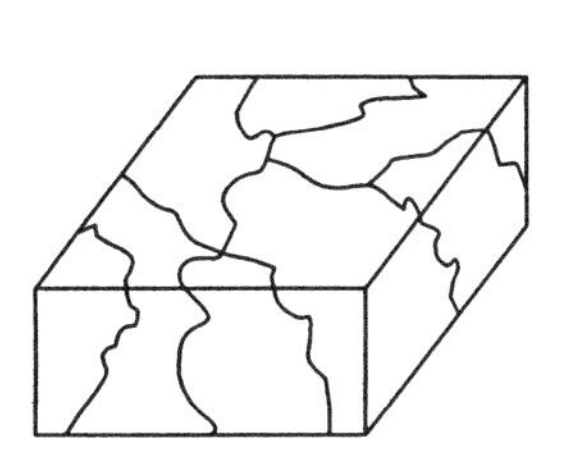

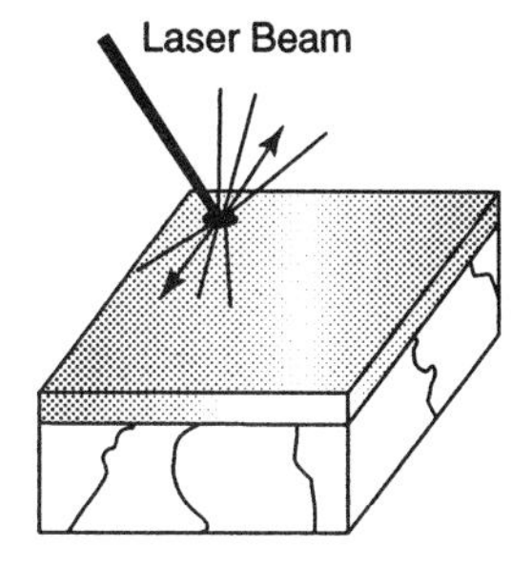

Figure 6-19 FEL application to surface processing of metals.

used to machine away most of the polyimide used as ribbing to strengthen the aluminum. Preliminary tests showed that the laser could ablate the polyimide without affecting the aluminum. The steel industry is interested in using the FEL to remove from carbon steels oxide layers that build up during the milling process. These layers are normally removed by acid etching in pickling vats, a process that uses hazardous materials, calling for unusual care and raising environmental concerns.

Another process earmarked for the FEL laser is the forming of amorphous layers on metals to produce harder, more corrosion-resistant surfaces. Figure 6-19 shows a laser forming a thin surface layer on a metal surface to eliminate grain boundaries. Other future applications for FEL are the production of particulate-free films for use in semiconductor microelectronics and optical materials by providing atom-by-atom vaporization of substrate materials. Laser-assisted chemical vapor deposition (CVD) would provide greater control over chemistry and eliminate the need to heat substrates to high temperatures. Scientists at the TJNAF have developed a procedure to use the FEL to bond chemicals to the surface of fabrics by creating chemical reactions that produce protective layers that make the material germ-free. Using materials such as nylon and the bacterium *E. coli*, the technique is promising. One highly important application would be to develop this laser process to work against anthrax. Table 6-4 sets forth the status (level of maturity) of some representative industrial applications of FELs, using the categories of commercializable, developable, and prospective.

The FEL represents another development in surface modification technology. ***Surface modification*** techniques are mentioned throughout this book, particularly in reference to heat-treatment surface techniques such as case hardening and nitriding, electroplating, diamond films, titanium coatings, plating, layered composites, chemical vapor deposition (CVD), sputtering, chemical vapor infiltration (CVI), toughening, surface engineering, and the like. A recent laser technique reported by QQC, Inc. converts a surface layer into a superhard coating not only on metals but also on ceramics and plastics. The process uses three high-powered lasers—excimer, yttrium–aluminum–garnet, and carbon dioxide—each producing a different wavelength of light to overlap the surface. The lasers vaporize a thin layer of material, creating an electrically charged ***superheated plasma*** of iron atoms. After the beam is removed, the atoms fall to the surface and bond together. Elements can be added to the surface of a material by sending a stream of particles across it. The lasers break apart the particles, adding their constituent atoms to the plasma. One application is the hardening of a hot forging punch with a

TABLE 6-4 REPRESENTATIVE INDUSTRIAL APPLICATIONS OF FELS

Level of maturity	Polymer surface processing	Micromachining	Electronic materials processing	Metal surface processing
Commercializable				
Demonstrated major market	Surface texturing Surface conductivity	Fuel injectors Spinnerets	Flat panel displays Large-area photovoltaics	Laser glazing and annealing Surface carburizing and nitriding
Cost/capacity barrier	Surface amorphization	Slitting coated films	Large-area diamond coating	
Developable				
Demonstrated	Antimicrobial nylon	Microtexturing	Pulsed laser deposition	Adhesive bond pretreatment
Development needed	Pulsed laser deposition	Subthreshold ablation		Solvent-free cleaning Removal of corrosives
Prospective				
Preliminary results Research development needed	Surface activation of carbon fibers for composites	Nanometer-scale surface contouring for chemical catalysis	High-density CD-ROMs Embedded layer silicon processing	Thermal barrier coatings (TBCs) Functionally gradient coatings Metglas coatings
Enabling facilities needed				Shape memory alloys (SMAs)

Source: TJNAF, Newport News, VA.

layer of high-strength chromium–cobalt alloy, which increases its wear life from 1 hour to 12 days. Specific locations on a material can also be hardened with nanometer precision. Advances in this new technology will propel many improvements in materials, manufacturing processes, and product quality. Surface modification is discussed further in Module 13.

Many uses for these well-controlled sources of energy will continue to be discovered. Boring holes, welding delicate parts of the human eye, detecting and measuring pollutants in the atmosphere, and carrying human communications through space with little or no interference are some of the possibilities. Present applications of the optical behavior of materials, including glass and plastics that produce light through the interaction of electromagnetic radiation with the material's electrons, are numerous. Some common examples are worth mentioning. The coating of TV screens with zinc sulfide (ZnS), a semiconductor material on which electrons act to produce light, is one. A second is the phototube that "reads" the sound track on motion-picture film, ultimately producing an electrical signal that can be amplified and broadcast to an audience. Solar batteries that use semiconductor materials and the exposure meters carried by photographers for properly setting their cameras are additional examples of the many applications of such devices.

6.4.3 X Rays

X rays, discovered by W. Roentgen in 1895, can be produced in two ways. The first process uses a high-energy electron capable of knocking an electron from an inner energy level of an atom completely out of the atom. This removal of an electron from, say, the K level, immediately creates a hole into which an L-shell electron can fall. As an L-shell electron falls back to the lower-energy K level, the sudden decrease in potential energy is emitted as a photon of electromagnetic energy characteristic of X rays (atomic radiation). The hole produced in the L shell calls for the transition of another electron from an outer level to the L shell, which produces an additional X ray. Several X rays are emitted as electrons cascade down to fill the lower-energy-level vacancies, or holes, until eventually the atom captures an electron from the surrounding region and changes from an ion to a neutral atom in its lowest equilibrium condition. The wavelength of the X ray depends on the particular energy levels involved, as do the colors in visible light emitted from a given electron drop from one energy level to another. X rays lie in the wavelength range between 1×10^{-12} and 1×10^{-8} m. Another process of generating X rays (also called ***cathode rays***) is through the sudden braking (deceleration) of high-energy electrons by aiming them at a metal target, which results in a conversion of part of the electron's kinetic energy into a quantum of electromagnetic radiation, or X ray. Such a technique is shown schematically in Figure 6-16.

6.4.4 Thermionic Emission

Another way to exchange energy with the free electrons in a solid metal is to heat the metal to such an extent that the thermal energy of the electrons is sufficient for them to be emitted from the surface. This is known as ***thermionic emission***, the type of emission that occurs in a typical low-power vacuum diode, whose cathode consisting of a nickel tube coated with barium oxide is heated by an insulated filament contained within the tube.

6.4.5 Transmission

Optical devices possess many desirable properties, but the most important is their ability to transmit light. One measure of this ability is ***transmittance***, defined as the percentage of an incident light ray remaining after passing through 10 mm of a material. It can also be expressed as a ratio of transmitted-light intensity to incident-light intensity (***intensity*** is the measure of the strength of a light source). Some materials, such as metals, are ***opaque*** (will not let light pass through). When white light is incident on a metal surface, most of the light is reflected and some is refracted into the surface of the metal where the photons strike the electrons, thus transferring discrete amounts of their energy to the electrons and raising them to higher energy levels. (See Figure 6-14). In copper, the 3d electrons absorb through collisions the energy of the photons corresponding to blue wavelengths and reflect the others, which give a reddish color. The energy absorbed causes the atoms concerned to increase their energy levels, which allows them to increase their vibrations, which, in turn, results in an increase in temperature, an indicator of heat.

In covalent and ionic solids, the electrons are bound to the atoms by the very nature of the bonds. It takes more energy to break such bonds and free the electrons with sufficient energies to move them to higher energy levels. Therefore, visible light passes through such solids without reacting with the atom's electrons. Such solids are termed ***transparent***. If impurities are present, the photons of the visible light could react with the electrons of the impurity atoms, which are not involved in covalent or ionic bonding. In a ruby, the chromium atoms absorb the photons of blue and green light but allow the remaining photons to pass through, which are mostly of wavelengths that correspond to red light. Many polymer materials are transparent. Others are ***translucent***; that is, these materials, because of their structure, allow some light to pass through, but most of the light is scattered due to the reflection and refraction of the light as it transmits across phase boundaries and interacts with pores in the structures. Research into new ceramic materials has allowed the production of single-phase, pore-free ceramics that are transparent.

The bending of light rays is due to the fact that the speed of light within a dense medium is less than in air. Many of us have had the experience of trying to pick up an object in a shallow pool of water, only to find that the object's actual location is displaced from its apparent location. As discussed earlier in Section 6.4.1, the ratio of the speed of light in air (c) to its speed in a medium (v_m) is called the ***index of refraction*** of the medium (n).

$$n = c/v_m$$

Illustrative Problem

Given the index of refraction in a medium of 1.5, how fast would ordinary light travel in this medium? What material has an index with this value?

Solution Solving the preceding equation for v_m,

$$v_m = c/n = (3 \times 10^8 \text{ m/s})/1.5 = 2 \times 10^8 \text{ m/s} \qquad \text{(polyethylene)}$$

The index of refraction is related to polarization in a material by the expression $n = \kappa^{1/2}$, where κ is the dielectric constant. The index may vary with the change in density of materials as well as in certain crystal directions (i.e., n is anisotropic). The index also varies with frequency. Using a wavelength of 589 nm, some selected indices are: diamond, 2.4; polyethylene, 1.5; water, 1.33; and air, 1.0. Another way in which the index may be defined is in terms of the angles of incidence (θ_i) and refraction (θ_r). Figure 6-14 shows how these angles are measured from a common normal.

$$n = \frac{\sin\theta_i}{\sin\theta_r}$$

Illustrative Problem

Visible light traveling in air enters a medium with an index of refraction of 1.5 at an angle of incidence of 60°. Find the angle of refraction.

Solution

$$n = \frac{\sin\theta_i}{\sin\theta_r} \qquad \sin\theta_r = \frac{\sin\theta_i}{n}$$

$$\sin\theta_r = \frac{0.866}{1.5}$$

$$= 0.577$$

$$\theta_r = 35.3^\circ$$

The refractive index n applies to the medium into which the incident light penetrates. Solids with dense atomic packing generally have higher indexes of refraction. In crystals, the index of refraction is different for different directions, which can result in the incident rays being split into two polarized rays.

In explaining the phenomenon of polarized light, we use the concept of the electromagnetic wave. All electromagnetic radiation of whatever frequency consists of two fields, one electrical and the other magnetic. Figure 6-20 shows these two fields, designated by the vectors **E** and **B**, separated by 90°; that is, for each E there is a corresponding B set up in a plane at right angles to each other. For simplicity, only two fields are shown; it must be remembered that any radiation is composed of many, many waves with **E** and **B** vectors pointing in all directions. Using just the **E** field and Figure 6-21, if you position yourself as standing out on the x-axis and looking back at the origin of our x-y-z axis system, the many **E** fields could be represented by the sketch. These **E** vectors could also be represented by their respective components in the y and z directions, thus simplifying the representation. A schematic of unpolarized

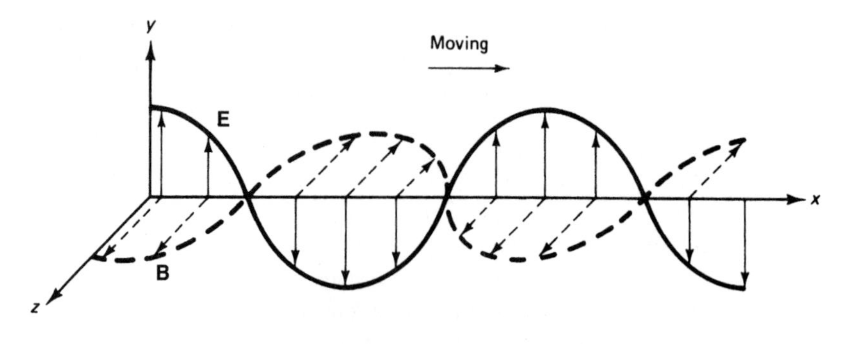

Figure 6-20 Electromagnetic fields in a light wave.

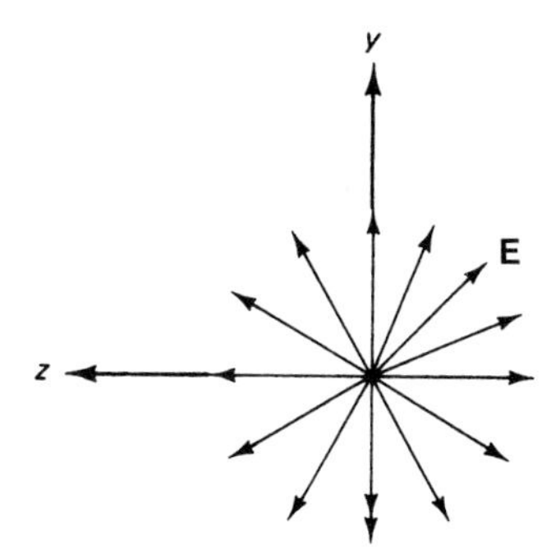

Figure 6-21 A point view of electric field vector **E** in an electromagnetic radiation of unpolarized light.

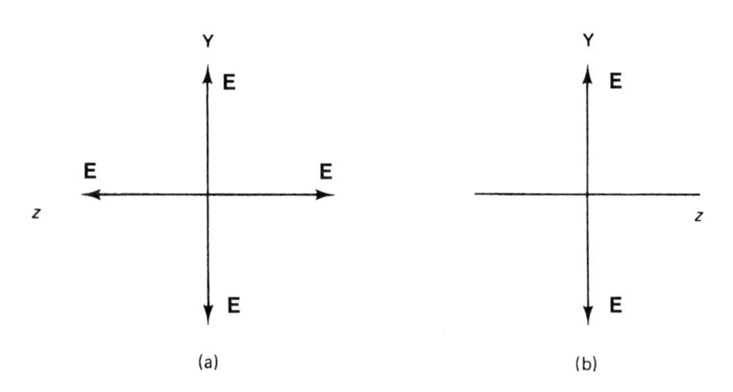

Figure 6-22 (a) Unpolarized light with **E** vector resolved into *y*- and *z*-axes. (b) Polarized light in *y* direction.

light would then look like the sketch in Figure 6-22a. This view of ***unpolarized*** light represents **E** vectors that oscillate back and forth along the *y*- and *z*-axes. If the radiation consists only of **E** vectors along only one axis, it is known as *polarized*. Polarization can be accomplished by the use of a filter that absorbs the electric field oscillations in one particular direction. A representative sketch of polarized light is shown in Figure 6-22b. The light in this example is polarized in the *y* direction.

Unpolarized and polarized light can be further understood by comparing them to vibrating a rope. If you fix a rope at one end and vibrate the other end up and down in all directions, it would be a model for ordinary, or unpolarized, light. If you vibrate the rope in just one direction, say, the vertical direction (up and down), this would be representative of polarized light. Furthermore, if the same rope were pulled through the slots in a picket fence, the rope could be vibrated only in this up-and-down manner. The picket fence then could serve as a model of an optical filter (a polarizer) that permits light vibrations in only one direction and absorbs (prevents) other vibrations from being transmitted. When light is transmitted through material, the light can become polarized due to the differences in the material structures in different directions. A material whose structure varies with the orientation of the material is not isotropic (that is, it is anisotropic). As you recall, isotropy is a quality of a material that has identical properties in all directions. The term ***optical anisotropy*** refers to a material having different optical properties in different directions.

Being insulators, most polymeric materials have no free electrons. In addition, their internal structure contains boundaries and pores that interact with light and cause it to be transmitted in different directions at different speeds, hence at different refraction indexes. The polarization of light is used in many applications, such as liquid crystal displays (LCDs), Polaroid sunglasses, and photoelasticity, where polarized light is used to reveal the amount of anisotropy in a transparent-polymer (polystyrene) model of a loaded structure (see Figure 6-23). When subjected to stresses ($\theta = P/A$), certain transparent materials become ***birefringent***, or double refracting. These substances will divide an incident ray of light into two beams that travel through the material at different speeds. In particular, these two beams are polarized at

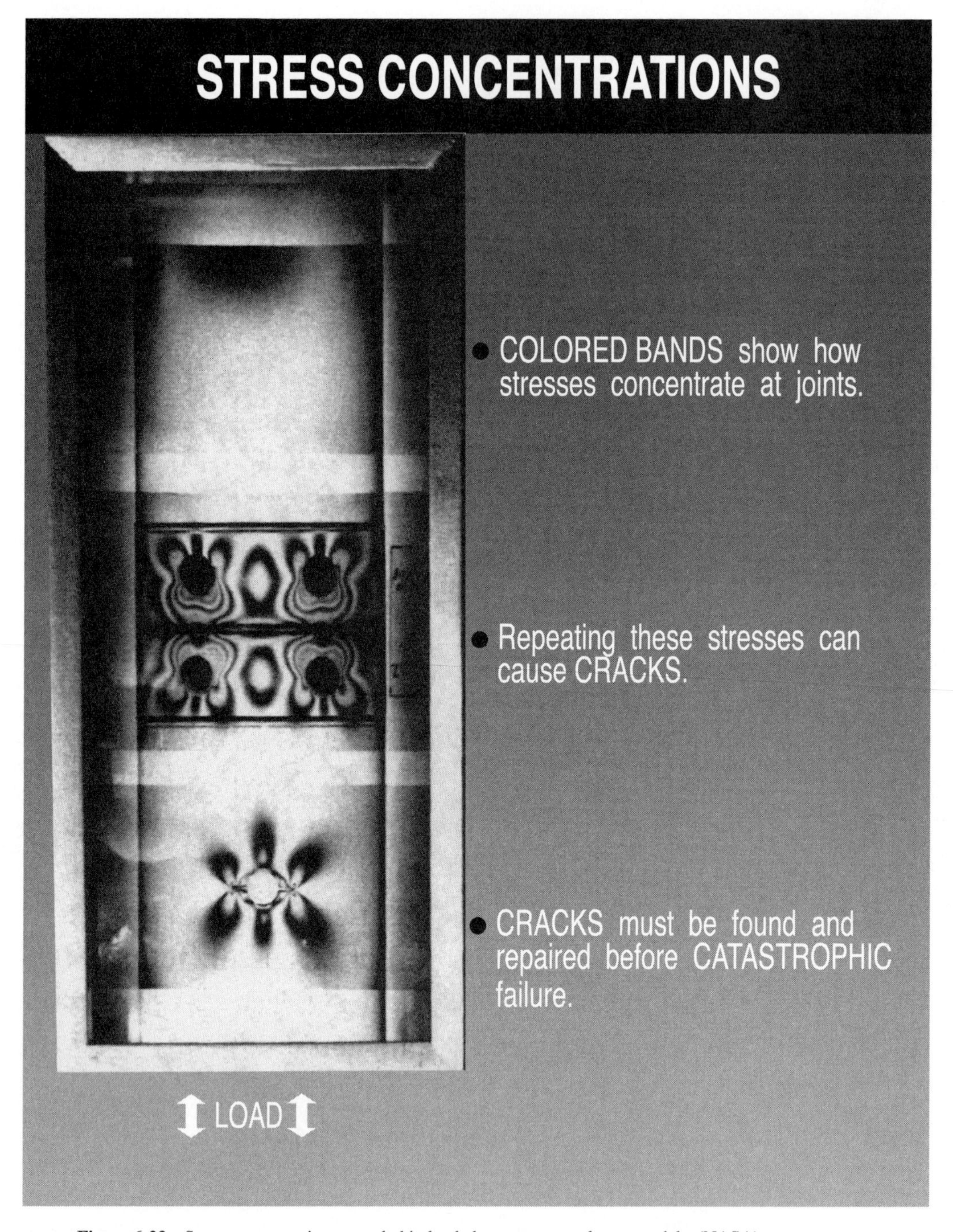

Figure 6-23 Stress concentrations revealed in loaded transparent-polymer models. (NASA)

right angles to each other (see Figure 6-22). This material property of birefringence is taken advantage of in the construction of structural models—of buildings, for example—using transparent materials having double-refraction properties. The models are subjected to forces (strains), and the internal stresses produced are analyzed by translating the stress fringes or isochromatics into stress magnitudes. An ***isochromatic*** appearing as a dark band (see Figure 6-23) represents a region or locus of constant difference in principal stress. A pattern of alternate bright and dark lines represents the variation in refraction produced in the model. A variation of this technique is to coat an actual structure with a photoelastic coating such as Photostress. When the part is subjected to loads, the strains are transmitted to the coating, which then becomes birefringent.

6.4.6 Photoelasticity

Some materials will transmit light only in a particular direction. If light waves such as those in Figure 6-21 were directed into such material, the resulting emerging beam would have vibrations (oscillations) in one plane only, as shown in Figure 6-22. Other materials rotate the plane of polarized light. If two pieces of such material are arranged so that their polarized planes are perpendicular, no light would emerge. Such polymeric materials experience this rotation as a result of the realignment of their macromolecules due to internal stress, an effect known as ***photoelasticity***.

6.5 NANOTECHNOLOGY

Research at the NASA Ames Research Center and across the globe is moving toward dramatically improving the electrical and optical capabilities of materials. Figures 6-24 and 6-25 depict a carbon nanaotube (CNT) with electrical conductivity 6 orders of magnitude greater than copper. These CNTs can have either metallic or semiconductor characteristics depending on ***chirality*** (the ability of objects to have "handedness"), that is, the existence of left/right (mirror image) opposition. "A chiral molecule is one that is not superimposable on its mirror image." CNTs permit tailoring electronic properties through application of external magnetic fields or mechanical forces such as bending, as seen in Figure 6-24.

- CNT quantum wire interconnects
- Diodes and transistors for computing
- Capacitors
- Data Storage
- Field emitters for instrumentation
- Flat panel displays
- THz oscillators

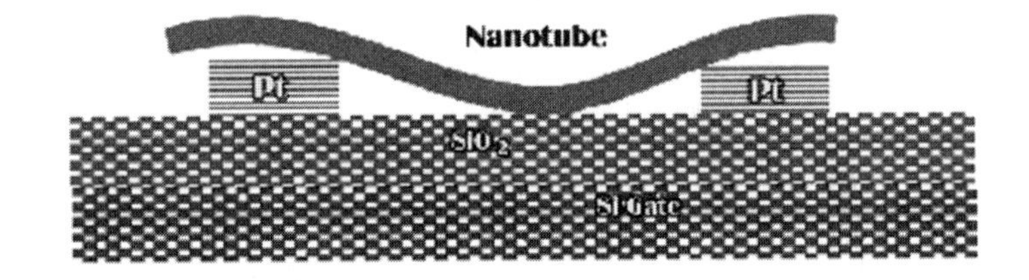

Challenges
- Control of diameter, chirality
- Doping, contacts
- Novel architectures (not CMOS-based)
- Development of inexpensive manufacturing processes

Figure 6-24 CNT electronic application: can be metallic or semiconducting, depending on chirality, with a "tunable" bandgap and properties that can be tailored through application of external magnetic field or mechanical deformation; very high current-carrying capacity; excellent field emitter with high aspect ratio and small tip radius of curvature that are ideal for field emission; can be functionalized. (NASA Ames)

- CNT based microscopy: AFM, STM...
- Nanotube sensors: force, pressure, chemical...
- Biosensors
- Molecular gears, motors, actuators
- Batteries, Fuel Cells: H_2, Li storage
- Nanoscale reactors, ion channels
- Biomedical
  - in vivo real-time crew health monitoring
  - Lab on a chip
  - Drug delivery
  - DNA sequencing
  - Artificial muscles, bone replacement, bionic eye, ear...

Challenges
- Controlled growth
- Functionalization with probe molecules, robustness
- Integration, signal processing
- Fabrication techniques

Figure 6-25 CNT applications for sensors, NEMS, bio: CNT quantum wire interconnects; diodes and transistors for computing; capacitors; data storage; field emitters for instrumentation; flat-panel displays, THz oscillators.

 Fun with Chirality website-—*http://www.rod.beavon.clara.net/chiralit.htm*

Chirality-—*http://chirality.ouvaton.org/homepage.htm*

SELF-ASSESSMENT

6-1. The dielectric constant for water is

a. 7×10^6 Hz **b.** 12×10^6 Hz **c.** 78×10^6 Hz

6-2. A conductivity in Ω^{-1} cm^{-1} for a common metal would be

a. 4000 **b.** 100,000 **c.** 700,000

6-3. A dielectric is a nonconductive material whose polarization is caused by

a. A magnetic field **b.** An electric field **c.** IR radiation

6-4. Dielectric constants are in demand by the

a. Machine tool industry **b.** Electronics industry **c.** Oil and gas industry

6-5. If the Mn atom loses two electrons, its paramagnetic properties

a. Stay the same **b.** Increase **c.** Decrease

6-6. The Na atom has an atomic number of 11. It is

a. Diamagnetic **b.** Paramagnetic **c.** Neither

6-7. The atomic number of Cl is 17, which tells us that Cl is

a. Diamagnetic **b.** Paramagnetic **c.** Ferromagnetic

6-8. A pair of electrons with opposing spins has

a. A large magnetic field **b.** A small magnetic field **c.** No magnetic field

6-9. An example of a transition element is

a. B **b.** C **c.** Sc

6-10. The wavelength of microwaves is about

a. 10^{-5} nm **b.** 30 m **c.** 0.3 cm

6-11. The hysteresis loss (ergs/cm^3/cycle) for 4340 steel at 25°C is

a. 2.7×10^4 **b.** 1.5×10^4 **c.** 11.4×10^4

6-12. High-energy protons are produced by

a. Alpha rays **b.** X rays **c.** Beta rays

6-13. Electronic polarization deals with the polarity of

a. Atoms **b.** Ions **c.** Protons

6-14. The bending of light is known as

a. Reflection **b.** Refraction **c.** Incidence

6-15. The index of refraction for diamond is

a. 0.67 **b.** 1.36 **c.** 2.42

6-16. Optical anisotropy refers to a material with __________ optical properties.

a. Good **b.** Poor **c.** Different

6-17. The size of the forbidden gap in conductors is

a. Large **b.** Zero **c.** Small

6-18. An electric current of 1 coulomb of charge passing a point in 1 second is known as a(n)

a. Micron **b.** Ampere **c.** Angstrom

6-19. The SI prefix μ (mu) stands for

a. 10^{-6} in. **b.** 10^{-6} **c.** 10^{-6} m

6-20. High-T_c superconductors have an advantage over normal conductors because they offer no __________ to the flow of current.

a. Strain **b.** Resistance **c.** Dielectric **d.** Permeability

6-21. Using Table 6-1 and a room temperature of 68°F, determine the resistance of a copper wire with a length of 100 m and radius of 1 mm.

6-22. Would a material that has an electrical resistivity of $10^{12}\ \Omega \cdot \text{cm}$ make a good conductor or a good insulator?

6-23. Express a dielectric strength of 2000 V per mil in units of (a) V/cm; (b) V/m.

REFERENCES & RELATED READINGS

AMERICAN SOCIETY FOR TESTING AND MATERIALS. *2003 Annual Book of ASTM Standards*. Philadelphia, PA: ASTM, 2003.

AMERICAN SOCIETY FOR TESTING AND MATERIALS. *UNSearch, Version 2.0*, Software. Philadelphia, PA: ASTM, 1992.

ASM INTERNATIONAL. *ASM Handbook*, Vol. 13, *Corrosion*. Materials Park, OH: ASM International, 2000.

ASM INTERNATIONAL. "*Materials Selection*," *Advanced Materials and Processes*, Special Issue, June 2003.

OBERG, E., et al. *Machinery's Handbook*. New York: Industrial Press, 2003.

Periodicals

Advanced Materials and Processes

ASTM Standardization News

Materials Performance and Research

Journal of Materials Education

Journal of Materials Engineering

UNIT 3

Evaluation, Standards, Design, & Selection

PAUSE & PONDER

This unit builds on your study of properties and will provide background for modules in this unit and beyond. The following case study provides insight into failure analysis while relating the investigation to evaluation techniques used on material components as a part of failure analysis. Also, you will read how engineers, technicians, and technologists investigate product failures that may cause revisions in design and standards for certifying products safe for various types of applications.

November 12, 2001. American Airlines Flight 587, an Airbus A-300 (see figure) crashed into the New York borough of Queens, killing 265 passengers and crew. Initial analysis of the flight voice and data recorders caused members of the National Transportation Safety Board's (NTSB) investigating team to believe the Airbus was hit twice by powerful wake turbulence from the plane immediately ahead. Wake turbulences from other aircraft create extremely strong forces on trailing aircraft and in the past have caused fatal crashes. The vertical stabilizer (tail fin) and both engines broke off the Airbus prior to crashing (see figure).

A year later, the NTSB reported findings from extensive failure analysis by engineers at NASA's Langley Research Center. The analysis seems to rule out a structural flaw in the plane's all-composite tail as the cause of the crash. Both manufacturer (Airbus) and the LaRC team conducted extensive loads calculations, which revealed that the aerodynamic and internal loads that the vertical stabilizer experienced were significantly above the ultimate loads specified by the FAA and French certification standards.

LaRC conducted a series of nondestructive examinations (NDE) to define the areas of damaged and undamaged structure, which included an instrumented tap test, Lamb wave test, thermography, ultrasound mapping, and X-ray radiography. In the briefing the NTSB explained they would go into a destruction phase by performing more intrusive examinations where components would be disassembled. The text of the briefing, conducted by an NTSB member, continued:

> The major attachment points between the vertical stabilizer and the fuselage are six lugs on the fin, made of composite materials that fit into six metallic fittings in the fuselage, with each lug and fitting set held together by a metal pin. All six pins remained in place in the fittings on the fuselage during the accident. Major portions of three of the lugs were found still attached to the aircraft at the wreckage site, while the other three lugs broke through the hole for the pin, leaving the larger piece of these lugs attached to the tail fin that was recovered from the water. All of these lug pieces will be sent to other facilities for CT-Scans that will permit higher-resolution examination than available here.

American Airlines Airbus A-300 ascending.

Vertical stabilizer.

Vertical stabilizer (tail fin) attachment point.

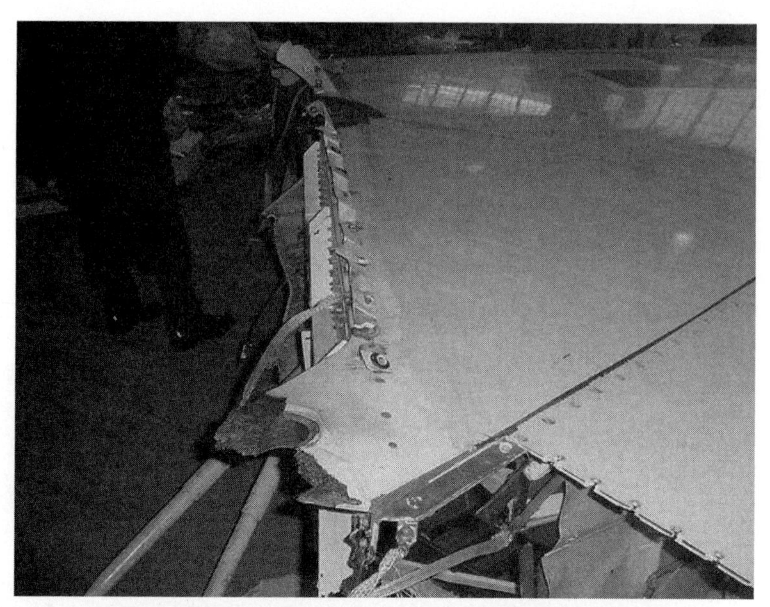

Right side forward and center attachment points.

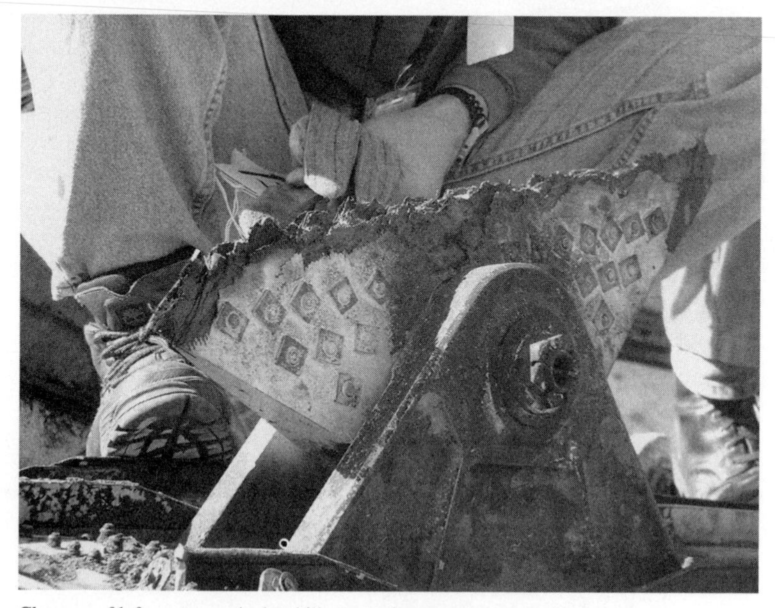

Close-up of left center vertical stabilizer attachment point, at crash site.

We are continuing with a highly meticulous exercise—determining whether the stabilizer was built to its design specifications and whether the design specifications and certification standards were adequate. We have cut four small sections—called coupons—from undamaged areas of the stabilizer. These will be tested—here at NASA Langley—to generate basic information on the material used in the manufacture of the fin, to see if the material was manufactured to desired specifications. You will see the rectangular holes when you view the wreckage.

As you may know, these composites are a series of layers of carbon fiber fabric or tape, with each layer laid down in a desired orientation. In this examination, we will go layer by layer to check the number, stacking sequence and orientation of each.

We will be moving into the damaged areas of the stabilizer, examining areas where the layers have separated to determine if they were caused during the fin's departure from the aircraft or if they were pre-existing.

In short, we have many months of work ahead of us here at NASA Langley.

But that isn't the whole story. Our work will not only document the damage and the physical specifications of the composite structure, it will also involve development and validation of internal and external loads, failure analysis and subsequent structural testing. That work has been going on here, as well as in France, Germany and Washington, D.C.

Our engineers have estimated the aerodynamic loads on the tail fin. After those load estimates are verified, a finite element model analysis of the vertical stabilizer will allow us to determine the reactions at the attachment points and the resulting internal loads experienced by the tail fin structure. Airbus has provided a finite element model (FEM) of the fin. NASA engineers are evaluating that model and will provide key analysis and expertise to us as the investigation continues.

All of this is aimed at determining, first, what loads were experienced by this aircraft and, second, how this particular tail fin performed with reference to its design specifications.

In a related matter, we are reviewing all design and certification documentation to make sure Airbus met the requirements during the certification process in the 1980s, and that we understand the basis for the design specifications.

I want to describe one other area of this investigation in which NASA Langley is providing assistance. As you may know, flight 587 is believed to have twice encountered wake turbulence produced by a Boeing 747 that had taken off directly ahead of it. We have provided NASA the FDR information from the accident aircraft and the 747 that gives temperature, wind speed and direction, and acceleration data. NASA will then provide a model of the wake vortices produced by the 747. The model will help us confirm external winds and to calculate loads on the vertical fin produced by the vortices. This will assist us in determining what role wake turbulence played in the accident sequence.

In addition, we are considering the need to duplicate the flight loads experienced in the cockpit through the use of motion-based simulation. We would like to study the effect of those motion-induced loads on the flight crew's ability to control the airplane. One promising method is the use of NASA's Vertical Motion Simulator at NASA Ames in California. It has a very large range of motion as compared to the typical simulator's range.

Getting back to composites, we have made arrangements to acquire exemplar components for possible work in the future. We have purchased a used A300–600 rudder that has undergone NDE at Sandia National Laboratory in New Mexico. We also may use the stabilizer and rudder that have been removed from an airliner that experienced an upset event five years ago. We have not yet determined how best to employ these exemplars, but possibilities include cutting into them for examination or testing them to failure.

In the examination bay, you will see the vertical stabilizer [see figure] basically intact, with the coupons and some of the lugs removed. You will see a section of the empennage to which the stabilizer was attached. And you will see the rudder, broken in several pieces. I believe we are making good progress in determining the sequence of the break-up and separation of the rudder, although we won't have conclusive results for some time.

. . . Was all of this damage the result of the separation (of tail fin) from the aircraft, or could some of it have been pre-existing? Could the damage have been induced during recovery? We are sorting that out, one piece at a time. Beware of those who will approach you with theories on how this accident occurred based on the photos you provide. If it were that easy, we'd have solved this a long time ago.

> We are fortunate to be able to tap the expertise that is available here at NASA Langley. We expect to have work going on here throughout the length of this investigation. I don't know how long that will be, but we intend to find out what caused this tragedy, the second deadliest airline accident in U.S. history.*

At the time of writing this case study much work lay before the investigation team. We encourage you to go to the website listed below to gain more information.

NSTB—*http://www.ntsb.gov/*

* Briefing—http://www.ntsb.gov/Speeches/blakey/mcb020503.htm
Photos—http://www.ntsb.gov/Events/2001/AA587/tailcomp.htm

Module

7

Testing, Failure Analysis, Inspection, & Standards

After studying this module, you should be able to do the following:

7–1. Explain the types and uses of industrial standards in design and manufacturing.

7–2. Describe the role of technical societies, government agencies, and private industry in developing and using standards for engineering materials.

7–3. Explain reasons for the degradation and failure of materials and the role of failure analysis, factors of safety, postprocess thermal treatments, and corrosion protection.

7–4. Recall purposes and methods for nondestructive testing, inspection, quality control, and quality assurance.

7–5. Relate properties of materials and the effect of their cost on materials selection criteria.

7.1 TESTING, STANDARDS, AND INSPECTION

7.1.1 Testing

Materials are tested to determine their basic properties. In particular, the testing of materials has as its objective the determination of numerical values for properties. As stated previously, most properties of engineering materials can be classified into the major categories of physical, mechanical, chemical, thermal, electromagnetic, and optical. An example of the innovative testing techniques used are those developed to investigate the high-temperature properties of advanced materials. Current metallic alloys are capable of withstanding temperatures to only about 700°C (1300°F). These high-temperature techniques test materials at temperatures approaching 2200°C (4000°F) and fall into two categories: macromechanical testing, such as static/cyclic biaxial loading of a hollow specimen, and micromechanical testing, such as a stereo image processing system. ***Macromechanical systems*** provide data on the strength, fatigue, and tribological properties of materials, mainly for engineering design of components; ***micromechanical systems*** permit detailed experiments for determining and describing specific mechanisms of failure. The terms *testing* and *inspection* are often used interchangeably, but they should not be.

Table 7-1 provides a summary of the general types of tests used to determine the properties discussed in this module. Modern society relies on materials testing of (1) raw materials, (2) product subassemblies or components, (3) products, and (4) systems. Materials testing is widespread throughout industry and government. Who does the testing? Everyone! Read the caption in Figure 7-1. Think about the liability of Arrow Dynamics, the company who designed and built the Drachen Fire, and Busch Gardens, the company who operates this thrill ride that moves through spiral rolls and travels 60 miles per hour! Every aspect of this amusement must meet rigid quality standards to ensure the safety of riders while protecting the designers, builders, and operators of the ride from product liability. This includes the steel structural pipes and other structural members that make up the track. Also involved are the wheels, bearings, and suspension system that connects the cars to the track. Quality standards also extend to the restraint system that holds the "brave souls" in the cars and include the composite body that encloses the cars and makes the seats. But who gets involved in the many tests that ensure this quality?

7.1.2 Nondestructive Testing (NDT), Nondestructive Inspection (NDI), and Nondestructive Evaluation/Examination (NDE)

Inspection differs from testing in its objectives. The ***objectives of inspection*** are to examine parts of materials for the presence of ***discontinuities*** and ***defects***. In other words, inspection estimates the degree to which a product conforms with design specifications. Materials characterization, as discussed in Module 8, may be part of the inspection procedure. The American

TABLE 7-1 GENERAL TYPES OF MATERIALS TESTING

Mechanical Tests Used to evaluate mechanical properties of materials where there is a measure of the material's ability to carry or resist mechanical forces such as tension, compression, shear, torsion, and impact. Tests can be *static*, in which the stress is applied slowly, or ***dynamic***, in which the stress is applied suddenly. Testing usually results in permanent damage to the specimen.
Nondestructive Tests Used to examine materials and components in ways that do not harm the future usefulness and service. They detect, locate, measure, and evaluate discontinuities, defects, and other imperfections to assess integrity, properties, and composition, and to measure geometric characteristics.
Physical Tests Used to determine physical properties of materials when there is no change in composition of the material. They include tests to determine density and electrical, optical, thermal, and mechanical properties.
Chemical Tests Used to determine chemical properties of materials where there is transformation of one material into another. They include tests of the abilities of materials to resist oxidation, corrosion, and rust.

Figure 7-1 Taming the beast. Brave souls challenge the mighty Drachen Fire, the fearsome roller coaster at Busch Gardens in Williamsburg, Virginia. As one of the world's largest roller coasters, it stands 150 feet tall, reaches speeds of more than 60 mph, and features many first-of-a-kind thrill-ride elements. Hurtling along 3550 feet of electric-blue colored track, riders are turned upside down six times. (Courtesy Busch Gardens and Arrow Dynamics, Inc.)

Society for Nondestructive Testing (ASNT) defines a ***discontinuity*** as an interruption in the normal physical structure or configuration of a part, such as a crack or porosity. A discontinuity may or may not be detrimental to the usefulness of a part. A ***defect***, however, is a discontinuity whose size, shape, location, or properties adversely affect the usefulness of the part or exceed the design criteria for the part. ASNT has five major inspection systems for detecting such discontinuities: (1) radiographic testing (RT), (2) ultrasonic testing (UT) (see Figure 7-2), (3) eddy-current testing (ET), (4) magnetic-particle testing (MT), and (5) liquid-penetrant testing (PT). In addition to these five major systems, inspection techniques using holography, microwaves, liquid crystals, infrared radiation, and leak testing are available when circumstances dictate. ***Computer simulation*** of materials processing is also developing into a reliable means for improving quality processes. Sophisticated simulation allows process engineers to look at most variables prior to building molds for casting and other process components.

As is evident from the preceding information, inspection techniques involve tests, most of which are classed as ***nondestructive testing*** (***NDT***) and ***nondestructive evaluation*** (***NDE***). NDE was once relegated to the role of inspecting a finished product or a product already in service for defects. Today, NDE has been incorporated into the manufacturing cycle (Intelligent Manufacturing), where it provides vital information for feedback into the cycle at various points. This amounts to real-time improvements in the process and greater quality assurance, plus better product quality and reliability. A listing of NDE technologies involved with present-day materials includes ultrasonics, acoustics, holography, microwave, thermography, radiography, tomography, and spectroscopy. Several of these NDE methods or techniques will be described throughout the text.

Illustrative Problem

A client has recently purchased a new brass lamp. A neighbor asked if the lamp was made of solid brass or plated brass. How can the owner determine the answer to this question?

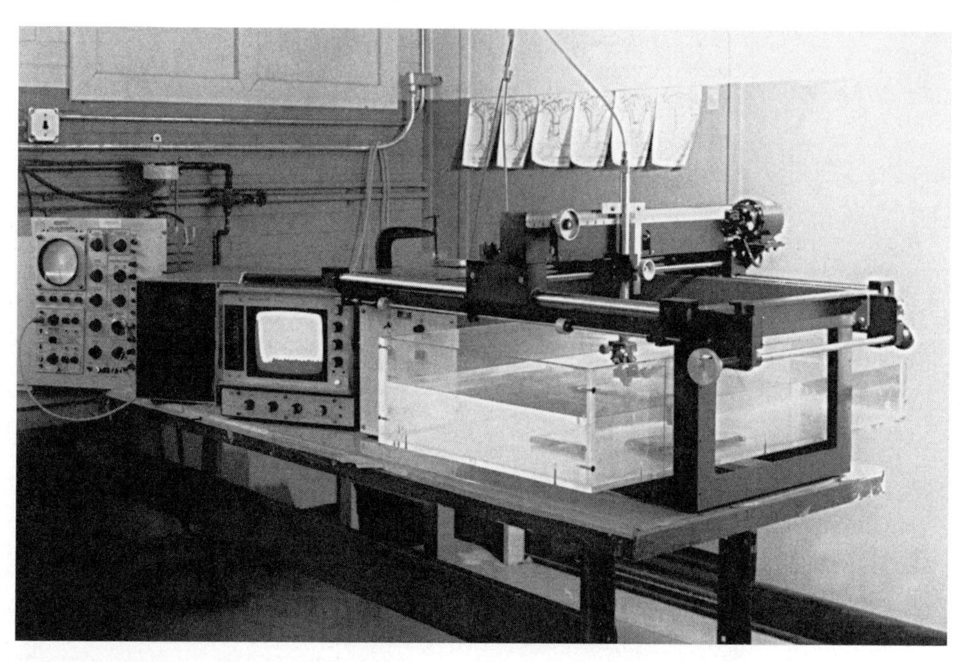

Figure 7-2 Ultrasonic-testing setup. (NASA)

Solution Using a small refrigerator-type magnet, the owner could place the magnet on the lamp. If the magnet is attracted to the lamp, then the lamp is made of a steel alloy that has been plated with brass. If the magnet fails to adhere to the lamp, then the lamp is made of brass.

Yahoo Materials Science Site—
http://dir.yahoo.com/science/engineering/material_science

7.2 WEAR, FRICTION, TRIBOLOGY, AND WEAR RESISTANCE

The degradation of engineering materials caused by the insidious attack of chemicals and/or wear amounts to a staggering financial loss and a large drain on the world's limited resources. We have previously described corrosion, that is, the chemical removal of atoms. ***Wear***, in the various forms of abrason, adhesion, and erosion, also takes its toll by removing part of the materials by the mechanical action of solids, liquids, or gases. Phenomena in the environment, like constantly changing temperatures, can cause wear from repeated expansion and contraction. Rates of wear are difficult to measure beforehand, even though there are many different types of tests for materials under various environmental conditions. ***Tribology***, the science and technology dealing with interfacing surfaces, focuses on design, friction, lubrication, and wear behavior to avoid problems with mating parts.

Every object moves in accordance with the forces acting on the object. A ***force*** may be defined as any influence that can cause a change in the state of motion of an object. The magnitude (size) of forces is expressed in units such as pounds (U.S. system) or newtons (SI system) and in terms of their direction expressed as up or down, left or right. Forces are vector quantities since they have both a magnitude (size) and a direction (N or S, left or right). If these forces (**F**) balance each other, there is no net force left over and the object does not accelerate (change its motion). If the net force is not zero, then this force will cause the object to accelerate (a), producing motion. The acceleration is expressed $a = F/m$, in which the mass (formerly called ***weight*** in the U.S. system of units) of the object is represented by m. Skydivers jumping out of an airplane from great heights fall at a constant, or terminal, speed. Constant speed indicates that divers reach a point in their fall in which they no longer accelerate; that is, their speed no longer changes. The terminal speed, in this case, is approximately 125 mph. What produces this result is the balancing of the forces acting on the diver. The force of gravity acts downward on the diver; the opposing force is friction, which is the result of the diver moving through the medium of air. The frictional force produces air resistance.

Friction, or frictional force, acts not only on objects falling in air but also between any object that moves over or through matter. An automobile tire moving over a highway, a fish swimming through water, a tennis racket contacting a ball, a metal piston moving within a cylindrical surface, and a tool insert in contact with a moving metal surface are examples of situations in which friction is present. Frictional motion is further defined as relative motion, since it refers to two bodies or objects either in contact with each other, such as a brake pad and a disk, or one body moving through a medium, such as air or water. The only time one can rule out the presence of friction is when the object under study is moving through a perfect vacuum, but this environment is difficult to achieve, even in outer space.

Practically all mechanical manufacturing processes involve friction, usually in a negative way, although friction can be considered as either a positive or a negative factor. For instance, it certainly is a positive factor when trying to move an automobile with an inoperable engine by pushing it. People who drive on ice-covered roads also recognize the significance of this statement. Even the simple act of walking would not be possible without it. On the other hand, some 20% of the power developed by an automobile engine is lost to internal friction, but lubrication of rubbing surfaces can reduce such friction in machinery. Friction is generally looked upon as a negative or undesirable feature that must be overcome by expending some form of energy and financial resources. As a result, much effort is spent on designing processes that reduce friction to the greatest extent possible.

Friction is a force that impedes sliding (see Figure 7-3). It originates at the surfaces of bodies in contact where irregularities of these surfaces can interlock. Bodies then adhere at the touching points of the irregular surfaces. Harder materials tend to "plow out" softer materials with which they are in contact. In Figure 7-3, friction (**F**), a vector quantity, is shown acting on a solid block, sketched in two dimensions, sitting on a flat horizontal surface. The block represents any object being subjected to friction. The block has mass, so it therefore has a gravitational force called weight (**W**) that acts in a vertical direction. The weight is opposed by the reaction force of the surface on the block called a normal force (*N*). Since the **W** force and the **N** force are equal in magnitude, there is no unbalanced force and thus no acceleration or movement in the vertical direction. If we now try to move the block by pushing it to the left with a pushing force (**P**), the block generates or reacts to this force with a frictional force (**F**). Note that the direction of the friction force is always opposite to the impending relative motion between the surfaces in contact. When the maximum force of friction is exerted just before the box starts to slide, the relationship between the friction force, the normal force, and the coefficient of friction (μ) is $F = \mu N$.

It has been determined experimentally that the coefficients of friction are nearly independent of the size of the contact area between metal surfaces. For polymer surfaces, including wood, this statement does not apply. Up to this point, the discussion has involved the case of ***static friction***, and the term *static* could have been used throughout. Static friction is the frictional force sufficient to prevent relative motion between surfaces. Another form of fric-

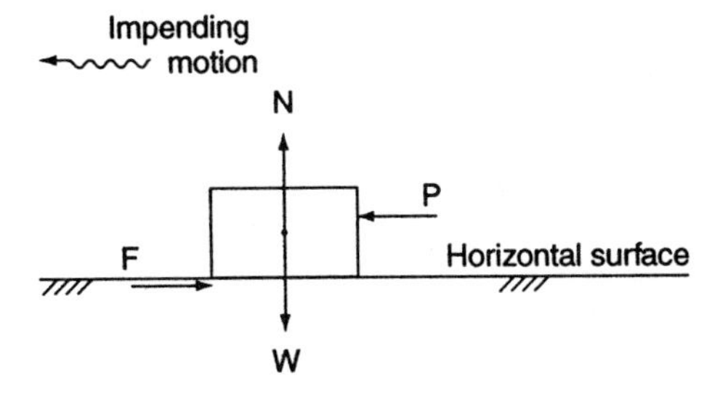

Figure 7-3 A friction force (**F**) generated by a pushing force (**P**) acting on a solid block resting on a horizontal surface.

TABLE 7-2 SOME REPRESENTATIVE COEFFICIENTS OF FRICTION

Materials	μ_s	μ_k
Glass on glass	0.94	0.35
Steel on aluminum	0.61	0.47
Steel on dry steel	1.20	0.85
Steel on lubricated steel	0.12	0.07
Teflon on Teflon	0.04	0.04

tion is ***kinetic friction***, or ***sliding friction***, where there is relative (sliding) motion at the interface of the surfaces in contact. Generally, the coefficient of kinetic friction is less than the coefficient of sliding friction for two surfaces (see Table 7-2 for some representative materials and their coefficients). To distinguish between the different coefficients, the subscripts s and k are used. For example, the coefficient of kinetic (sliding) friction (μ_k) for steel sliding on dry steel is 0.85. Note that this coefficient is a dimensionless quantity (no units). For glass sliding on glass, the coefficient of static friction (μ_s) is 0.94.

As we have said, one major effect produced by friction is wear, or the gradual removal of material from the surface of an object. Manufacturing operations contain numerous examples of wear—tool wear (saw blades, drilling bits, inserts), erosion in piping, molds caused by fluid flow or by particles contained in the flowing medium that impinge on the conducting surfaces, and general obsolescence of machinery as a result of the effects of friction over a long period of operation. Additionally, the mechanical properties of a material affect how well the material can be worked. For example, very low hardness can be detrimental to a metal's machining performance. High ductility causes a metal to tear as the chip is formed, resulting in a poor finish. When selecting materials for many applications, mechanical properties rank highest in importance. Terms such as ***durability***, ***formability***, ***rigidity***, ***strength***, and ***toughness*** are often used to describe these properties, but three tests supply the most useful information for most applications, namely, the tensile, impact, and hardness tests. Durability can be assessed in part by the hardness test, strength can be described by the results of tensile testing, and impact testing provides adequate data for determining the rigidity of a material. Figure 7-4 shows four common tests for determining the rate of wear of some common materials under attack by oxidation, acid corrosion, and erosion by mechanical testing. This figure points out that one factor in the struggle to control wear in materials is to protect their surfaces from these types of attacks. In this case, ceramic (TiC) is used to coat the material (graphite) with a hard surface impervious to chemical attack, wear, or their combination. Further discussion of some of these techniques to provide wear resistance is found in each of the succeeding modules (see Section 19.1.4).

According to developers at Argonne National Laboratory, a new ***coating*** called ***near-frictionless carbon (NFC)*** or, commercially, Slick 001 is much harder and slicker than Teflon. This carbon-based material has a coefficient of friction of only 0.001, which is one-twentieth that of the previous record holder, ***molybdenum disulfide***. For comparison, Teflon's coefficient of friction is approximately 0.04. Produced quickly and in large quantities, this nontoxic material adheres to a variety of surfaces including plastic. NFC is produced by using a proprietary gas-containing carbon that is converted into a plasma "soup" of atomic elements from which the carbon elements are allowed to drift onto a substrate, forming a coating. ***Wear resistance*** of the new material was tested by sliding a sapphire ball over the coating with a pressure equal to 145,000 psi (normal atmospheric pressure is 14.5 psi). After 5 million cycles, the new material lost just 1.0 μm (0.0004 in.) of its thickness. Promising applications are to be found in automobile turbocharger rotors and fuel-injector components, oil-less bearings, and spacecraft mechanisms.

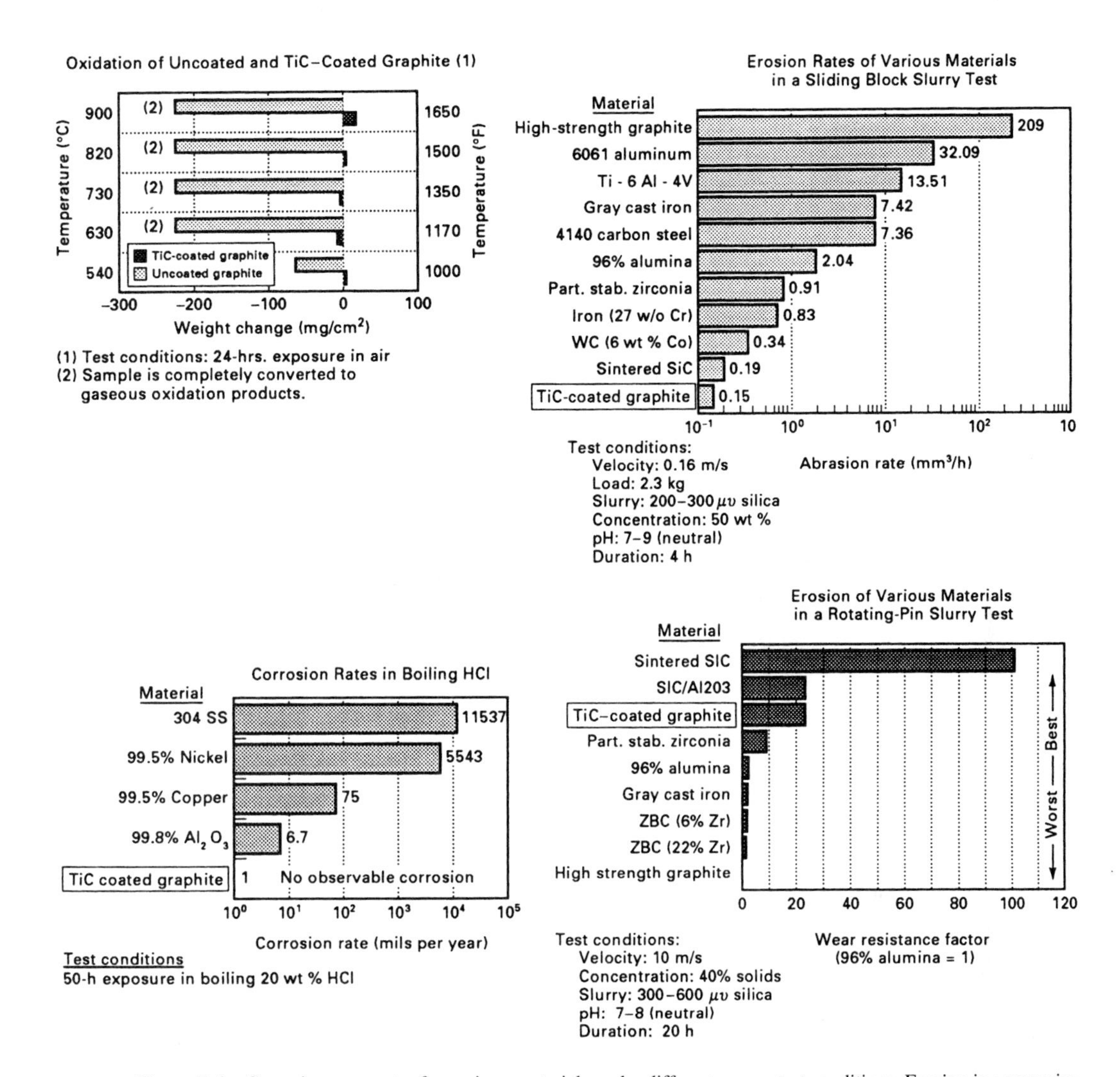

Figure 7-4 Corrosion wear rates for various materials under different source test conditions. Erosion is a corrosion mechanism.

7.3 FAILURE ANALYSIS

Up to this point in the module, we have discussed many of the physical properties involved in mechanical failures of materials. Other properties of materials also produce failures in materials, such as thermal or chemical properties. Failures of materials do not only cost vast sums of money annually, they contribute to loss of human life, although many failures in equipment, structures, and systems are not all attributable to material failures, but are the results of other factors not discussed in this book.

When a failure occurs, ***failure analysis*** is undertaken to help determine whether the failure was due to a design or material deficiency. An examination of the fracture tries to reconstruct the sequence of events and the cause of the fracture. Much information can be determined by analysis of the path a crack makes as it propagates through the material, the origin of the crack, and the cause of the crack initiation (i.e., material flow, oxidation/corrosion, tensile stress overload, thermal shock, etc.). Often, most of these data can be determined from visual observation by a trained investigator. If not, low-power microscopy, scanning electron microscopy, an electron probe, and other sophisticated instruments might be needed to analyze the chemical makeup of the fracture surfaces to aid in determining where responsibility for the

(a)

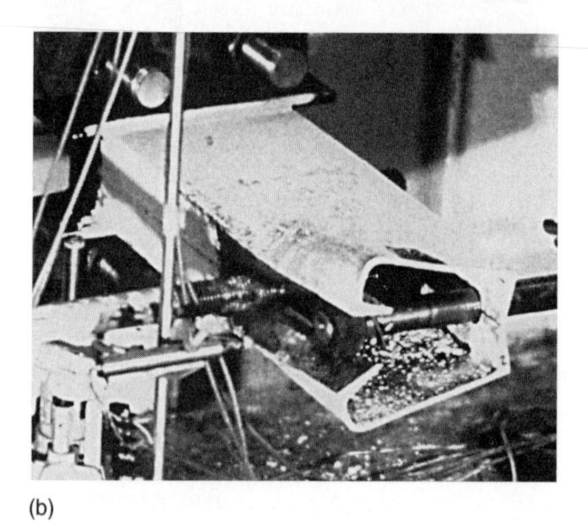

(b)

Figure 7-5 (a) The National Institute for Standards and Technology (NIST) investigated the Hyatt Regency Hotel disaster by using a mock-up and actual samples of box beams that supported the failed walkways. (b) Close-up shows joint pulling loose.

fracture lies. This highly technical area within the broad scope of failure analysis is known as ***fractography***. Fractography involves the recognition and interpretation of the various features of a fracture enabling the viewer to perform improved failure analysis and to better understand the relationships of the fracture mode to the microstructure. Atlases of fractographs are available containing over a thousand fractographs that are invaluable for a greater understanding of the causes and mechanisms of failure of engineering materials. A finding of responsibility not only is critical as an element in liability suits but is equally important in future design, specification writing, modifications, selection of materials, and the like. A classic example of failure analysis is the investigation of the Hyatt Regency Hotel disaster, in which 113 people were killed and many injured when concrete walkways collapsed in 1981 (Figure 7-5).

7.3.1 Factors of Safety

In the design of machine parts and structural members, it is the responsibility of the designer to provide a margin of safety, that is, to ensure that a machine part is safe for operation under normal operating conditions. ***Factor of safety*** (or design factor, N) is defined as the ratio of ultimate strength to allowable stress. ***Allowable stress***, also called ***design stress***, ***working stress***, or ***safe stress***, is the maximum level of stress that a part will be permitted to endure under operating conditions. In determining the allowable stress, considerations of the type of load on the part (static, impact, fluctuating, reversed, etc.) and the material from which a part is made are paramount. A material's ductility, as well as its strengths and stiffness, must be evaluated because the mode of failure of a ductile material varies dramatically from that of a brittle material such as grey cast iron, some plastics, or some heat-treated steels. For example, experience warrants that for fatigue loading (reversal of stress from tension to compression on a repetitive basis), a ductile material be selected with a percent elongation of 10% or higher to ensure against a brittle-type failure. The need for a factor of safety is most evident when knowledge is sketchy about (1) possible unexpected high loads on a structure, (2) the degree of residual stresses within a material due to manufacturing processes, (3) homogeneity of the internal structure of a material throughout its entirety, (4) loading history of the material since its fabrication, and (5) numerous but necessary assumptions used in the analysis and design procedures that can lead to appreciable errors. For ductile metals and plastics under static or steady tensile loading, the yield strength (σ_{ys}) is used as the basis for determining the allowed working stress. In most cases the factor of safety used is 2. Thus, the allowed working stress or design stress (σ_{des}) is calculated:

$$\sigma_{des} = \sigma_{ys}/2$$

For brittle metals such as cast iron, the ultimate tensile strength (σ_{uts}) with a factor of safety of 4 is used:

$$\sigma_{des} = \sigma_{uts}/4$$

The design shear stress or maximum allowed stress is calculated using the material's yield stress and a factor of 2:

$$\tau_{max} = \sigma_{ys}/2$$

In dealing with ceramics, a brittle material due to the presence of defects and cracks of varying sizes that will make the tensile strength vary, the design stress will, of necessity, employ probability design methods such as Weibull statistics. Even though ceramics and other brittle materials are quite capable of handling compressive loads, the concern with tensile strength is due to the fact that even ceramics behave as other materials do. When subjected to temperature changes that incur changes in microstructure and induce internal thermal stresses, the result is tension failures, with sometimes disastrous results.

Computing the required factor of safety can be quite involved. However, in some cases the factor of safety is specified in codes promulgated by organizations such as the American Institute of Steel Construction (AISC) and the American Society of Mechanical Engineers (ASME), as well as in building codes and company policies, as mentioned in Section 7–1.

7.4 STANDARDS AND TESTING ORGANIZATIONS

Over a two-week period in August 1999, four people died on three separate amusement rides. Additionally, several people suffered heat exhaustion after being trapped upside down for more than an hour on a thrill ride. These incidents took place in spite of good safety records, focused attention on detail, careful fabrication, proper maintenance, thorough inspection, and riders' observations of safety rules on these amusement rides. Attention to these same issues carries over to buildings, transportation vehicles, and most of our products and systems. The need to hold to rigid standards to achieve safety of life and property cannot be taken lightly.

TABLE 7-3 WHO SETS STANDARDS FOR MATERIALS TESTING AND CONDUCTS MATERIALS TESTING?

1. Technical societies
2. Governmental agencies and research centers
3. Private laboratories
4. Manufacturing companies
5. Wholesalers and retailers
6. Consumer groups
7. Individual consumers

Table 7-3 lists categories of agencies, societies, and organizations who (1) develop standards for materials testing and/or (2) conduct materials testing. ***Materials testing standards*** are rules and procedures established for the testing of materials to ensure objectivity and common practices for making judgments about materials. Just as the standard measurements for the layout of lines on a tennis court ensures that everyone is playing within standard boundaries, it is vital that everyone have clear, standard rules on how to test materials so the results have clear meaning. This sounds simple and straightforward, but development of materials testing standards is a complex topic that we will only touch on here. Many people make full careers of developing standards that the entire civilized world relies on for commerce and safety. CSSinfo, Ann Arbor, Michigan, is a distributor of engineering standards information and specializes in creating on-line access and distributions systems for industrial standards. CSSinfo has launched a new World Wide Web site devoted to the dissemination of standards information from over 200 worldwide standards-developing organizations, including ISO, IEC, ASTM, ASME, ANSI, EIA, SAE, API, BSI, CSA, ACI, IEEE, NSF, UL, AAMI, and EN.* On-line searching for documents and standards information is free of charge. See the websites listed at end of Section 7.1.

7.4.1 Technical Societies

Groups of people, made up of technical specialists from private industry, government, and universities plus individuals with common interests, come together in technical societies to agree on what standards are best for their particular industry. Some examples of technical societies are the Society of Plastic Engineers (SPE), Society of Automotive Engineers (SAE), Society of Manufacturing Engineers (SME), ASM International (ASMI), American Society for Testing and Materials (ASTM), American National Standards Institute (ANSI), and International Standards Organization (ISO). These societies often cooperate in standards development. For example, ASTM publishes the ***Annual Book of ASTM Standards***, which involves coordinating 140 main and 2034 subtechnical committees and can include representatives from ASMI, SAE, and ANSI. The *Annual Book* consists of approximately 60 volumes and thousands of standards that serve as guides to materials testing and other related concerns. The ASTM standards are used around the world to set legal criteria and objective means for analysis of materials. For example, ASTM devotes about 40 standards to impact testing. One of these is *ASTM E23, Izod Impact Test of Metallic Materials*.

7.4.2 Governmental Agencies and Research Centers

Federal, state, and local government agencies become involved in many forms of materials testing. Other testing and development of materials are conducted at regional research centers that are under contract to federal agencies, such as the Department of Energy's Oak Ridge National Laboratory (ORNL) and Pacific Northwest Laboratory (PNL). The National Institute of Standards and Technology (NIST), formerly the National Bureau of Standards (NBS), is a federal agency that falls under the jurisdiction of the U.S. Department of Commerce. NIST is

* —http://www.cssinfo.com

deeply involved in developing materials testing standards and conducts a broad range of tests of materials. For example, the bureau set up the test of the structural connection of the concrete walkway that collapsed in the Hyatt Regency hotel (Figure 7-5). NIST was involved in testing the walkway connection because its experts in materials testing could help explain the fault in the system and thus help avoid similar disasters and perhaps develop new standards. The National Aeronautics and Space Administration (NASA), another federal research agency, has research centers across the United States that perform materials testing on all sorts of materials relating to space and aeronautics and also many associated technologies, such as aircraft runways or satellites.

7.4.3 Private Laboratories

Many private laboratories test materials for both private businesses and government agencies. For example, Underwriter Laboratories (UL) tests and approves appliances, tools, motors, and similar products. In addition to large private labs such as UL, you can find small, private materials companies operating in many communities that test a wide range of materials and products. To assist the designer in selecting a plastic material that is to undergo long-term exposure to elevated temperatures while under load, the UL developed a temperature index (UL 746B) that lists a temperature for each polymeric material. The index temperature is determined by exposing samples to circulating air at various temperatures for 10,000 hours. The temperature that causes a sample to lose 50% of its strength or toughness is the index rating temperature.

Figure 7-6 illustrates the use of standards for building construction materials. The CoreGuard security wall system was designed for high-security areas such as hospital corridors, schools, computer centers, libraries, or anywhere high-impact materials are required. Note the UL label; the panel system was designed and tested according to UL 263 and ASTM E-119 standards for 1-hour and 2-hour fire-rated wall assemblies.

American National Standards Institute—*http:/web.ansi.org*
ASM International—*http:/www.aminternational.org*
Materials Research Society—*www.mrs.org*
Minerals Metals and Materials Society—*www.tms.org*

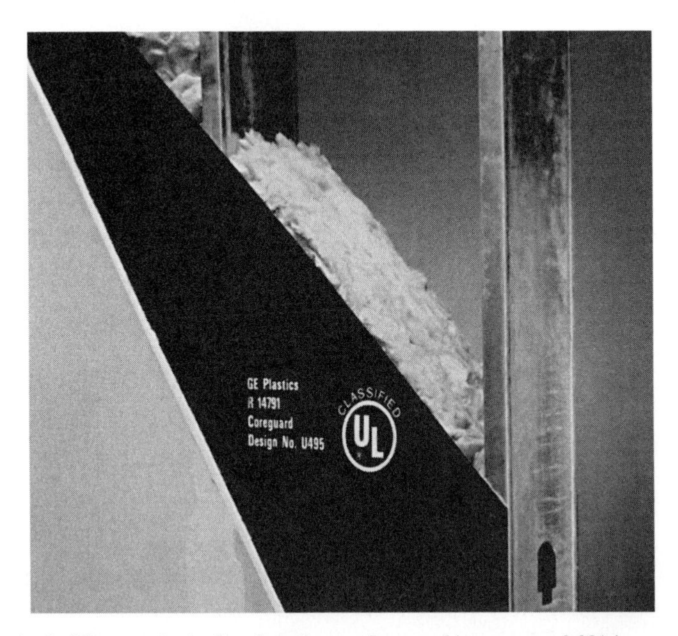

Figure 7-6 Meeting building standards. CoreGuard's one-piece panel incorporates 0.081-in. opaque recycled polycarbonate sheet and 1⅝-in. or 1¾-in. fire-rated gypsum wallboard. The panel provides at least 600 ft/lb of impact strength. (G.E. Plastics)

Underwriters Laboratory—*www.ul.com*
SAE (Society of Automotive Engineers)—*www.sae.org*
CSSinfo,Ann Arbor, Michigan [a disseminator of standards information]—
www.cssinfo.com

7.4.4 Manufacturing Companies

Large manufacturing companies rely on their materials testing labs to test all sorts of materials, ranging from raw materials and material stock from outside vendors (to ensure that they meet specifications) to the components and products they make themselves to be sure they will meet quality standards. For example, manufacturers may test nuts and bolts to ensure that they have sufficient strength, and they may crash-test a car to see how the materials and components respond to improved design. See Section 7.4.3. for information on UL 746B test for plastic materials. Smaller manufacturers may not be able to afford their own testing lab, so they will rely on private labs. Even large manufacturers use private labs such as UL to certify their products. Mecmesin, Inc. of Santa Rosa, California, has designed and manufactured a fully automated pneumatic pick-and-place machine (PCM VersaTest) for testing tennis balls in compression (see Figure 7-7). Pneumatic grips pick a ball from a carousel of 12 balls and place the ball between specially designed platens. A compression test cycle is carried out, the ball is twisted through 90°, and the test is repeated. Then the ball is automatically returned to the carousel. Two automatic procedures have been developed: one for the ball, and one for preconditioning each ball prior to testing. A report giving test data for each ball and statistics for

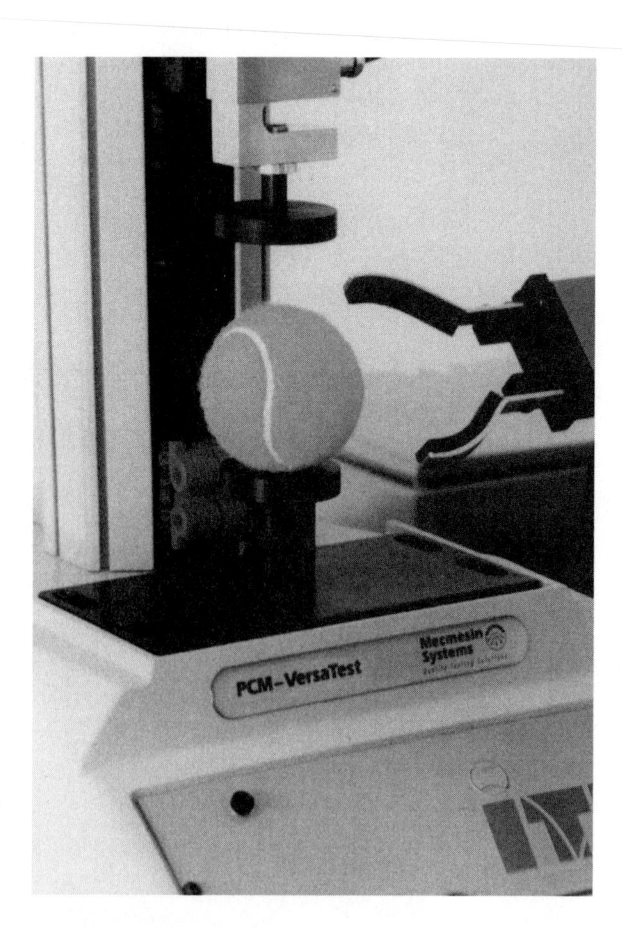

Figure 7-7 Fully automated tennis ball compression tester (PCM VersaTest) is computer controlled, featuring a pneumatic pick-and-place machine. (Mecmesin, Inc.)

each batch of balls is automatically printed. This testing system can be adapted for any compression-or tension-type testing of materials or components within a range of 0 to 550 lb.

7.4.5 Wholesalers and Retailers

To ensure product quality, wholesale products distributors and retail stores may have their own lab or use private labs to test samples of products that they will sell. This quality testing is important to ensure reliable products that will keep the confidence of customers. Can you think of some examples of retailers and the products they might test?

7.4.6 Consumer Groups

Many nonprofit consumer organizations also conduct tests of products. A few months before the Christmas season, you may see consumer groups talking on television about toys considered dangerous as a result of their testing. Consumers Union tests a wide variety of consumer products and publishes the results in various publications, such as *Consumer Reports*. This magazine is a useful resource to learn more about materials and product testing.

7.4.7 Individual Consumers

The ultimate testers of materials and products are consumers. Manufacturers try to stay in touch with their customers to ensure satisfaction. As newer materials, such as composites and advanced ceramics, appear on the market in new products, consumer acceptance is of great concern to producers. Even though extensive testing goes into these materials and products, the final and most important testing is by you, the consumer, whose feedback to the manufacturers helps them to improve products as well as to validate their testing procedures.

PAUSE & LOOK

Look around your room, lab, or office and locate the familiar UL label on an electrical cord or piece of equipment. Discuss the type of material and product testing that may have been done on it. Do some research into UL testing to determine how it helps you, the consumer.

Two terms used in conjunction with testing and inspection are ***quality control*** **(QC)** and ***quality assurance*** **(QA)**. The objective of QC is to determine *statistically* how much testing and inspection are required to ensure that products will meet design specifications and service life expectations. For example, in the production of an automobile engine, each part cannot, because of high cost, be given a complete inspection to see if it meets exact dimensional tolerances, if the metal used is free of all external and internal defects, and if each part has the necessary strength. Rather, a *statistical sample* is taken of a batch of parts as they are produced. Through statistical analysis, the sample is determined to be representative (probably the same) as all parts made in the process. Whole engines are also chosen statistically to be samples of the engine-making process. They are tested to failure (destructive test) as a means of providing quality assurance (QA) for the engine itself. QA, the goal of any QC program, refers to the total set of operations and procedures used in manufacturing a product whose goals are conformity of a product to design specifications.

7.5 MATERIALS SELECTION

In the section Pause & Ponder and throughout this book, new materials and processes are described. Often, 10 to 20 years elapse before new materials come into common use. It requires that much time for development of a reliable ***database*** to ensure that the potential (properties,

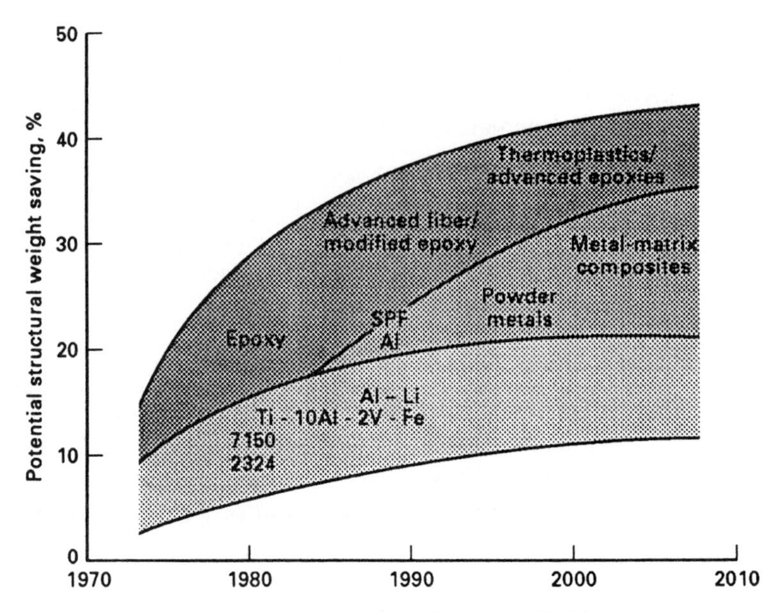

Figure 7-8 Potential weight savings trend for future structural materials. (NASA)

processing techniques, cost, etc.), as well as any risk assessment, is well known. The normally high cost of committing to production with new materials dictates an ***evolutionary*** rather than a ***revolutionary*** procedure for introducing new materials. Computer simulation coupled with rapid prototyping (see Module 15) is helping to shorten the time from concept to reliable materials technology.

Advanced composites serve to illustrate the point. In Unit 8, we discuss the many benefits of advanced composites, including the favorable properties studied in this unit. However, these materials have been accepted only slowly, even in the aerospace market where their benefits (especially strength-to-weight ratios) are so desirable, as seen in Figure 7-8. Among the reasons aerospace designers have moved cautiously is that material failure in a commercial passenger aircraft carries enormous consequences. Another reason is the need for refinement of the new design technology and automation of the fabrication of composites. Costs for advanced composites have declined. Graphite fibers that were around $35 a pound in the early 1980s dropped to around $20 per pound in the mid-1990s. Carbon fiber used as reinforcements ranges in cost from a low of $6 per pound to a high of $200 per pound for high-modulus varieties. However, fabrication costs will remain high until automated fabrication of composites, with the help of robotics, is developed further. The Applications & Alternatives section that follows demonstrates why the properties available in advanced composites make them the material of choice when cost is not an overriding consideration. Still, in applications where cost is a major concern, expect to see the trend toward composites grow (Figure 7-8) because their properties make them so attractive that databases, design parameters, and processing/manufacturing technology will quickly advance. In rebuilding the world's transportation infrastructure, for example, low-cost polymer matrices that are environmentally durable and safe will gain acceptance. As you will see in later modules, the fabrication of automobiles, trucks, trains, and oceancraft is also moving toward use of these composites because they offer the right properties to meet new demands related to energy efficiency and environmental impact.

A knowledge of how and why materials parts fail in service is essential for effective materials selection. The ASMI Center for Materials Solutions is developing a system to provide the user with an on-line source for materials information. One goal of this program is to manage *materials solution case histories* and match them with current materials manufacturing and in-service problems. The case histories will be categorized by materials, including alloy numbers or another suitable designation (aluminum, copper, glasses, etc.), processing (heat treating, surface treatment, casting, etc.), and in-service effects (wearing, failure, corrosion, etc.). The case histories will be a guide to solutions of materials problems. Some popular data banks—Alloy Finder, Alloy Digest, and MAPP–98— are also available. ***Alloy selection*** mandates that a matrix be developed, setting forth material requirements based on design criteria and the performance demands made on the material. Using the matrix, these various electronic data banks may be consulted to find the alloy and the thermomechanical conditions that best satisfy those requirements. As discussed in Module 3, low-energy materials and processes must also be considered, due to cost and environmental issues. Also, in developing countries, which may lack natural resources and sophisticated technology, wood and other polymers and cement and other ceramics must be favored.

Later you will learn how properties of polymers, metals, ceramics, and composites qualify these materials to be selected for critical parts in a variety of products and systems. With a knowledge of the properties of materials and practice in using this knowledge, wise material selection should result.

7.6 LABORATORY ACTIVITIES AND CLASSROOM DEMONSTRATIONS WITH PROPERTIES

The ***Experiment in Materials Science, Engineering and Technology (EMSET2) CD-ROM*** provides numerous experiments and demonstrations related to properties as result of a cooperative effort by scores of educators and materials specialists from industry. See Prentice-Hall's website, www.prenhall.com for ordering information on the latest edition of *Experiments in Materials Science (CD-ROM).*

APPLICATIONS & ALTERNATIVES

Advances in technology have provided new tools for materials engineering and manufacturing and have improved the confidence with which ***product reliability*** (the ability of the product to perform as specified by the designer) can be predicted. Companies may find it very difficult to compete in a global economy if they do not use the new technology and instead rely on only ***subjective*** (personal experience and feelings) rather than objective data (data that can be measured with instruments). Figure 7-9 depicts a high-tech, instrumented, materials-testing setup that you might find in a typical manufacturing plant. Consumers demand that quality be built into products, and the way to ensure built-in quality is to test every aspect of the product design and manufacture before releasing it to the public. Quality involves both performance and durability.

Sports and recreation are big businesses throughout the world. Both professional and amateur athletes continue the search for the winning edge. College courses offered in materials engineering for sports equipment focus on improving performance through engineered materials and new designs. Designers began to apply the new materials evolved for the aerospace industry to a wide variety of equipment—bicycles, snow skis, race cars, baseball bats, hang gliders, and tennis rackets, just a few of the products that have profited from such innovations.

A baseball bat should have the proper weight and balance, be able to transfer energy from the batter to the ball efficiently, and be able to withstand the energy transfer from the strongest batter without breaking. To determine the batting quality of aluminum bats, manufacturers evaluate them with a drop-weight impact test, as seen in Figure 7-9a. The falling tip inputs controlled energy to the bat, and a computerized data-acquisition system (wired into the testing machine) generates data curves showing the ratio of energy absorbed to energy reflected. The closer the impact to the "sweet spot," the less energy is absorbed by the bat. *Sweet spots,*

(a)

(b)

Figure 7-9 High-tech testing. (a) Testing aluminum bats to determine the sound produced, "ping," and to increase energy reflectivity or performance. The reflectivity of the bat depends on the properties of the selected aluminum alloy and the wall thickness of the hollow bat. The designer's goal is to increase the "sweet spot" area where the maximum force is transferred to the ball. (b) The curves generated by the impact tester provide a "sweet spot" response. This location (5 in. from the end of the bat) is where maximum energy is transferred from the bat to the ball. (Courtesy GRC Instruments)

as found on bats, tennis rackets, badminton rackets, and golf clubs, are the best zones for hitting a ball or birdy to achieve maximum control and distance. Less energy absorption by a bat means the ball should travel farther and with less strain on the batter. The curves shown on the monitor in Figure 7-9b indicate the falling tip has hit 12.5 cm from the fat end of the bat, which is the heart of the sweet spot. Therefore, the load from the impact tip is reflected away with minimum absorption.

The game of golf is going high tech (see Figure 7-10). Over 26.5 million Americans play golf and spend more than \$5 billion a year on clothing and equipment. It is estimated that the exploding population of golfers is around 50 million worldwide. The latest innovations in the game to make it less frustrating involve golf clubs, balls, and teaching techniques. In the early

Figure 7-10 Golf clubs reflect the manner in which sporting equipment makers quickly adopt new materials to benefit from unique properties. Athletes and hobbyists are looking for an edge to improve their performance. Seen on the right, an older steel shaft with persimmon-headed driver. The putter, iron, and 3 wood with aluminum head and titanium face all use graphite epoxy shafts. Sometimes the trend shifts back; persimmon shows up on expensive new woods.

1990s titanium was used to produce clubs, improving the size of the sweet spot (see Module 28). Today's high-tech clubs have sweet spots the size of a half dollar. Made of lightweight but strong titanium, the club heads are not as heavy, which allows the golf club shafts to be longer so as to improve the driving distance. Supposedly these new innovations in design may exceed the authorized standards imposed on their construction, because their illegal springlike effect when the club face flexes upon contact with the ball, gives the ball greater velocity and driving range. Many players fear that imposing tougher standards would limit further innovations in club design and the use of new materials. Visit a sporting goods shop and notice the wide variety of golf clubs now available. Read the literature and compare the prices. While in the shop, observe the other equipment to see how materials technology is meeting the desires of sports and recreational enthusiasts.

SELF-ASSESSMENT

7-1. Factors of safety are defined either in terms of the ultimate strength of a material or its yield strength. In other words, by the use of a suitable factor, the ultimate or yield strength is reduced in size to what is known as the design stress or safe working stress. Which factor of safety would be more appropriate for a material that will be subjected to repetitious, suddenly applied loads?

7-2. Product liability court cases have risen sharply in recent years because of poor procedures in selecting materials for particular applications. Assuming that a knowledge of a material's properties is a valid step in the selection process, cite two examples where such lack of knowledge could or did lead to failure or unsatisfactory performance.

7-3. Make a sketch and fully dimension an Izod impact test specimen.

7-4. At what level of yield stress do most fatigue failures occur?

7-5. From your experience, what typical automobile part requires complete fatigue testing prior to development?

7-6. Identify the origin of most fatigue failures and explain the significance of your answer in terms of handling a metal workpiece in the machining process.

7-7. Which agency publishes the *Annual Book* of standard test methods used worldwide for evaluation of materials?

a. NASA **b.** NIST **c.** ASTM **d.** SPE

7-8. What symbol provides assurance that an appliance or product meet standards for home and office usage?

a. UL **b.** CU **c.** AP **d.** UA

7-9. When selecting a material that will be subjected to abrasion, what property would you expect the material to possess to a great degree?

7-10. How does quality control differ from quality assurance?

7-11. A test that would find voids within a composite component is

a. UT **b.** LT **c.** AP **d.** PT

7-12. The objectives of inspections are to find

a. Quality
b. Discontinuities and defects
c. Overtime
d. Materials standards

REFERENCES & RELATED READINGS

AMERICAN SOCIETY FOR TESTING AND MATERIALS. 2003 *Annual Book of ASTM Standards*. Philadelphia, PA: ASTM, 2003.

ASM INTERNATIONAL. *Complete ASM Handbooks*. Materials Park, OH: ASM International, 2003.

ASM INTERNATIONAL. *Failure Analysis*, CD-ROM. Materials Park, OH: ASM International, 1999.

ASM INTERNATIONAL. *Practical Failure Analysis*. Materials Park, OH: ASM International, 2002.

BLAU, PETER, ED. *ASM Handbook*, VOL. 18, *Friction, Lubrication, and Wear Technology*. Materials Park, OH: ASM International, 1992.

DIXON, J. I. *Failure Analysis: Techniques and Applications*. Materials Park, OH: ASM International, 1992

ESAKUL, K. A. *Handbook of Case studies in Failure Analysis*. Materials Park, OH: ASM International, 1992

OBERG, E., et al. *Machinery's Handbook*. New York: Industrial Press, 2003.

PROVAN, JAMES W. "An Introduction to Fatigue," *Journal of Materials Education*, VOL. 11, No. 1/2, 1989, pp. 1–105.

RUMBLE, JOHN. "Computerizing ASTM Test Methods," *ASTM Standardization News*, July 1992, pp. 34–38.

Periodicals

Advanced Materials and Processes
ASTM Standardization News
Journal of Materials Education
Journal of Testing and Evaluation

Module

8

Materials Design, Selection, & Characterization

After studying this module, you should be able to do the following:

8–1. Discuss the relationships between materials under development and the future fabrication of these materials into new products for use by the consumer.

8–2. Explain the use of computers in the designing of new materials.

8–3. List at least six basic ingredients that could be used in a combinatorial approach to materials design.

8–4. Discuss compromise and trade-off in materials selection and other materials selection factors, including tools used for selection.

8–5. Define an algorithm and describe examples of using algorithms to solve an engineering problem.

8–6. Name techniques and instruments used to determine materials structures, including the capabilities of the optical, electron, and scanning tunneling microscope and the scanning SQUID microscope.

8–7. Characterize components of substrate and coating on a material specimen, including oxide inclusion, substrate roughness, void, pore, crack, cohesive strength between particles, particles, phases, and grains. Explain qualitative and quantitative materials characterization.

8–8. Explain the reliance of materials designers on the laws of nature as a source of inspiration in the development of new materials for components and products demanded by our advanced technological society.

8–9. Explain why there is an urgent need for engineered materials.

8–10. Use the website addresses denoted by the icon* to seek out current developments and emerging technology related to materials research and design.

*The icon found throughout this book will link you to Internet sites related to topics being covered. The dynamic nature of the Web brings frequent changes, so some of these sites, while available at the time of writing, may not be up now.

8.1 MATERIALS RESEARCH

Down through the ages the fascination aroused by watching a bird fly has stirred human attempts to duplicate this seemingly effortless feat imparted to a bird by the laws of nature. A present-day NASA Morphing Project was inspired by the phenomenon of bird flight (see Figure 8–1). Birds' wings never stay the same; they constantly move as a whole as do their many different feathers. One goal of NASA's project is to create an aircraft wing that could flex and respond to the environment. This "active aeroelastic wing" for improving aircraft maneuverability, also known as "wing warping," was tested by a NASA F/A fighter aircraft to demonstrate improved maneuverability through improved roll control for high-performance aircraft. (See Figure 8-2.)

Figure 8-1 Eagle in flight. By studying nature or using biomimetics (mimicking nature) aircraft designers, materials scientists, and engineers hope to be able to replicate highly efficient and maneuverability of insects and birds in the next generations of aircraft. (NASA Langley Research Center—http://avst.larc.nasa.gov/morphing.html)

Figure 8-2 This concept aircraft from the NASA Langley Research Center (LaRC) Morphing Project seeks to develop and test technologies and concepts for efficient, adaptable air vehicles. The word *morphing* in general use means "to visibly change form or structure," but in the context of this program, "morphing" can include small and large changes using structures and fluids for control. The major areas of research include smart materials, adaptive structures, microadaptive flow control, and biologically inspired flight systems. The long-term vision for the program is to provide technologies for aerospace vehicles that efficiently adapt to diverse, multivariable conditions in flight. The project's outcomes will apply to nearly all classes and sizes of air vehicles. (http://avst.larc.nasa.gov/morphing.html)

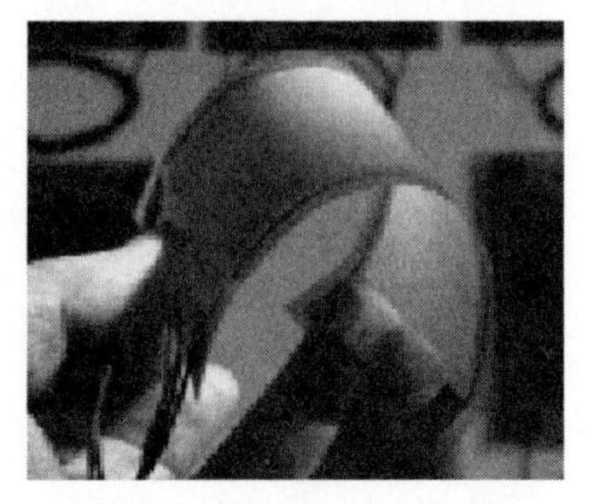

Figure 8-3 Morphing shapes. Thin, flexible film containing piezoelectric material that responds to the bend by producing a voltage that's detected by the electrodes seen at the bottom left of the image. Laugley Research Center is also developing customized variations of piezoelectric materials. These substances link electric voltage to motion. If you contort a piezoelectric material, a voltage is generated. Conversely, if you apply a voltage, the material will contort. Combined with microelectronics, these materials could lead to a radical advance in airplane design. (NASA Langley Research Center—http://avst.larc.nasa.gov/morphing.html)

These developing structural systems, known as "smart," "intelligent," or "adaptive" structures, have more recently, been called *adaptronic*. These structures employ thermally, electrically, or magnetically activated actuator systems made of piezoelectronic or electro- and magnetorheological fluids (see Figure 8-3). At the present time, this technology is spreading to applications in the automobile and railroad industry, among others.

One of many approaches to developing a new wing for aircraft is changing the air flow over the wing surface to gain better control during different conditions of flight, possibly by using air jets to create "virtual surfaces" that would limit drag and increase fuel efficiency or by developing a new wing material that would let the wings warp a few centimeters, thus changing the way air flows over them. Thinner wing skins, controlled by sensors in the wing, may allow the outer wing panels to twist up a few degrees. This would give a pilot better control during different modes of flight, such as takeoff or landing. Another approach is to perfect the existing shape of the wing to suit a particular flight environment by using shape-changing materials, such as piezoelectric materials. Shape-memory alloys are in use today in such devices as cell phone antennas, teeth braces, and eyeglass frames that spring back into shape after being distorted. Another avenue of research is developing materials to make self-healing skins or a plasticlike material that could be added to a wing surface to close punctures and other damage caused by space debris or bird impacts. Materials that reduce or neutralize vibrations in aircraft especially from high-speed flight are also a legitimate research goal.

All researchers dream about developing a material that possesses great properties. Many of these dreams can and do turn into reality. Sometime in the near future aircraft will sense the environment and adapt their shape to the existing flight conditions. Some time in the future a material might be found that could replace all the different components in a machine by doing all the work itself, a material that could, for example, replace a computer and all its components, such as hard drives, with a single block of material. Many of the "dreams" like these will be realized, and eventually those that are funded by federal agencies such as NASA will be spun off to the civilian economy. See the color section, C-30 Biologically Inspired Systems, for another illustration.

8.1.1 Combinatorial Materials Science

The continuing and accelerating developments in the design of materials and systems have improved the functionality of products demanded by our highly advanced technological society and the speed with which they enter the marketplace. Traditionally, industry used to take six years to design and deliver a new car. Today, the design cycle has been reduced to two years or less and customers can order and receive a new car custom assembled to their specifications in five days. Making new materials requires searching for the right combination of complex materials, that is, the right combination of metal alloys, polymers, composite materials, ceramics, chemical compounds, particles, elements, molecules, atoms . . . Where does it stop? Nanoparticles, nanotubes, protons, quarks?

The design of new materials has been assisted by the dramatic increase in computer resources, analytical instrumentation, and automation. Technologies using the ***combinatorial approaches*** permitted by the invention of sophisticated instruments have greatly increased the speed with which scientific undertakings can be brought to fruition. Figure 8-4 exemplifies this approach—the flow from the original design of materials at the micro and nano level to the design of a new material and its route via digital communication to its final synthesis as a new product coming off the factory floor. ***Combinatorial methods*** create large "libraries" of materials compositions and test those compositions systematically in parallel for specific properties. Past approaches that tested one composition at a time were time-consuming. Now it is possible for new materials discoveries to come faster, better, and cheaper. Lastly, this new approach facilitates

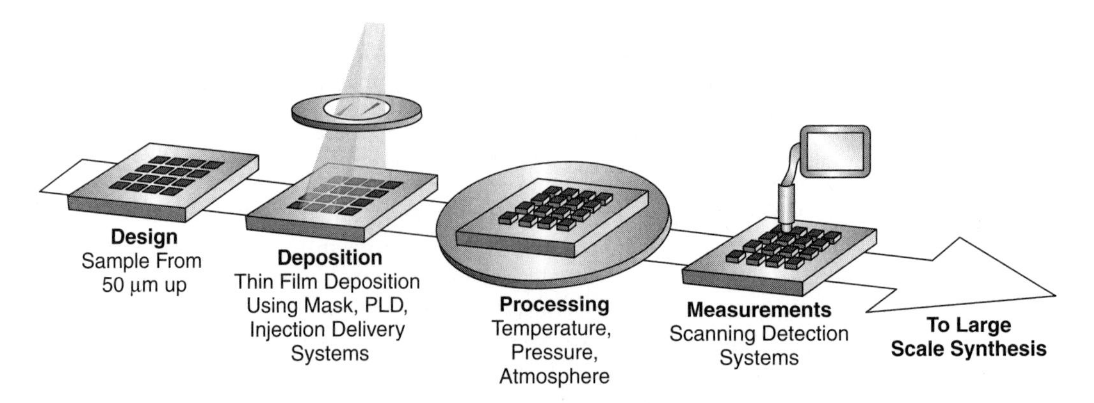

Figure 8-4 Flow of the combinatorial approach to the discovery of new materials. (Ichiro Takeuchi, "Combinatorial Approach to Materials Discoveries," *National Educators' Workshop:2001*. NASA/CP-2002-211735, June 2002, pp. 217–239)

a central scientific goal, namely, the fundamental understanding of the structure–property relationships of materials behavior. ***Combinatorial materials research*** is also an environmentally sound approach that leads naturally to a reduction in the resources used in research.

The automated system used in this type of research employs artificial intelligence and techniques that simultaneously test thousands of formulations. At the same time, the equipment systematically creates and tests thousands of samples. This ***combinatorial chemistry*** has had a profound impact on the pharmaceutical industry. Researchers have been able to discover new types of catalysts as well as many types of new materials and new drugs and genetic therapies that are the foundation for the biotechnical revolution now in progress. Designed by chemical engineers, these automated combinatorial chemistry systems use an infrared sensor technology to screen each sample material combination to evaluate its performance. Using artificial intelligence and both hybrid networks and genetic algorithms, the equipment mimics the thought processes of chemists and can screen many hundreds of precursors in the same amount of time that researchers previously used to evaluate one possible combination of materials. Interim results represent both material combinations that have failed and those that are successful, and the final results in this recycling operation that progressively creates better and better materials are entirely new combinations.

Material scientists are now beginning to adopt these techniques in their never-ending search for faster, cheaper, and better materials. From our studies we have learned that ***five main materials parameters*** have the greatest effect on properties of materials: (1) composition, (2) dopants, (3) processing, (4) defects, and (5) microstructure. The task of evaluating all these parameters would involve literally millions of experiments using one-at-a-time analysis. Moreover, materials are composed of not just one or two known ingredients but many, and some of these have not been seen before by scientists. Add in the vast array of polymeric materials, and the value of combinatorial methods is clear. The new combinatorial process therefore dramatically accelerates the experimental search for new materials.

The following descriptions merely touch the surface of this search for new materials. Much of the vocabulary may be unfamiliar due to the sophistication of the research that is taking place both in industry and in academia. We describe briefly some of the tools used by these researchers as a first step in attempting to understand the amazing processes being developed to quickly identify new combinations of substances with structures that will produce a set of properties needed to solve a problem to the benefit of mankind.

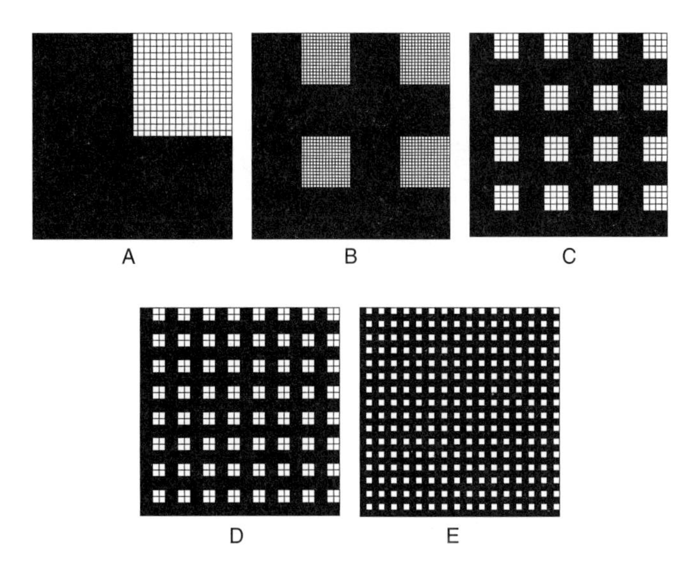

Figure 8-5 Quaternary mask patterns for fabricating libraries. Masks are applied successively from A to E. Each mask is used in four depositions, each time rotated by 90°. A total of 1024 (4^5) different combinations of precursors can be obtained as a result. ($4 \times n$ depositions and 4^n combinations, where $n = 5$, renders 1024 discrete combinations of precursors that can be created on a single chip). (Ichiro Takeuchi, "Combinatorial Approach to Materials Discoveries," *National Educators' Workshop:2001*. NASA/CP-2002-211735, June 2002, pp. 217–239)

8.1.2 Combinatorial Research: Terms and Techniques

First, we will review some basic terms. A ***catalyst*** is a substance that enters into a chemical reaction (or chemical change) in such a way as to make the reaction go faster without itself being consumed in the reaction. A ***chemical reaction,*** as defined previously, is a process in which one set of substances is converted to a new set of substances, called *products*. Two simple substances such as carbon monoxide (CO) and hydrogen (H_2) when heated and put under high pressure in the presence of a catalyst zinc oxide (ZnO) and chromium oxide (Cr_2O_3) will chemically react to produce methanol (CH_3OH). This chemical reaction describes the ***synthesis*** of methanol, which is the forming of a more complex substance from simpler substances.

Stoichiometry, defined briefly previously in Module 3, is the branch of chemistry that deals with calculation of mass implied in chemical equations. In other words, it refers to the calculation of the quantities of chemical elements or compounds involved in chemical reactions. This knowledge is vital in making new materials from simpler substances. ***Throughput*** is the amount of material put through a process in a given period, as by computer. ***Synthesis*** is the making or formation of the whole from its constituents, discussed in greater detail in Module 1.

Combinatorial research methods begin with the use of ***masking forms,*** shown in Figure 8-5, devices used in making thin-film libraries. There are three types, namely ***shadow masks, lithographic masks,*** and ***moveable shutter masks.*** The first two are used as a broad screen for elemental components; and the third is more useful for optimizing the composition of materials. Shadow masking allows for the rapid preparation of ***libraries*** (an array of materials of different compositions produced by various deposition techniques) containing 100–1000 compositions per square inch. On top of the primary mask, which separates different samples on the substrate underneath, a sequence of secondary masks can be overlaid, through which controlled quantities of various thin-film precursors can be deposited.

A series of precisely positioned **shadow masks** allow selective depositions of different combinations of precursors (e.g., rayon is one of the precursors for carbon fiber) at different

Figure 8-6 Size comparison of a of thin-film library with a U.S. penny. (Ichiro Takeuchi, "Combinatorial Approach to Materials Discoveries," *National Educators' Workshop:2001*. NASA/CP-2002-211735, June 2002, pp. 217–239)

positions on a substrate in order to generate diverse compositional variation across a library. Using a binary (composed of two elements) masking treatment, 1024 different material compositions can be generated. The key to the making of a thin-film library using such a vacuum deposition system is the masking strategy. With the correct masking strategy, the number of thin-film compositions that can be obtained increases exponentially with the number of deposition steps. Libraries containing from 100 to over 1000 discrete chemical compositions on a 1 in. × 1 in. square substrate have been made in thin-film or powder form using selective vapor deposition and liquid dispensing techniques, respectively. Figure 8-6 shows the size of a library to a U.S. penny. Each sample in the library is formed from a multiple-layered precursor.

After the deposition of the precursor layers, a thermal annealing technique enables the diffusion of the elemental layers before a high-temperature synthesis of the final phase. Following the deposition step, libraries undergo controlled thermal treatments to diffuse, mix, and crystallize compounds, if desired.

Figure 8-7 shows a combinatorial thin-film library used in screening for superconductor materials. This high-temperature superconductor had been the subject of long, laborious research for a material that will conduct electricity at temperatures approaching normal room

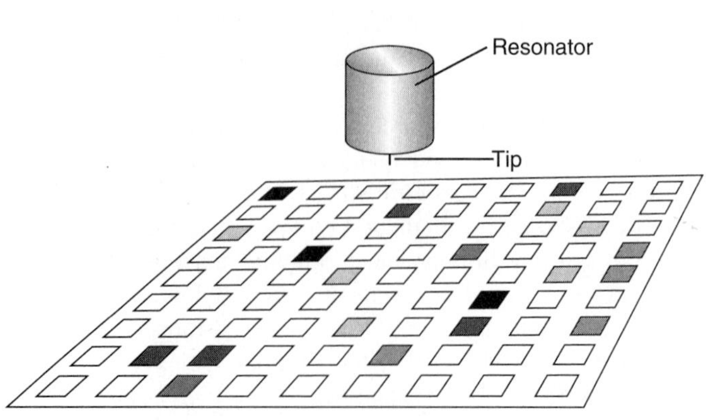

Figure 8-7 Thin-film library of superconductor materials. Combinatorial capacitor library used in the research for high-temperature superconductors materials. (Ichiro Takeuchi, "Combinatorial Approach to Materials Discoveries," *National Educators' Workshop:2001*. NASA/CP-2002-211735, June 2002, pp. 217–239)

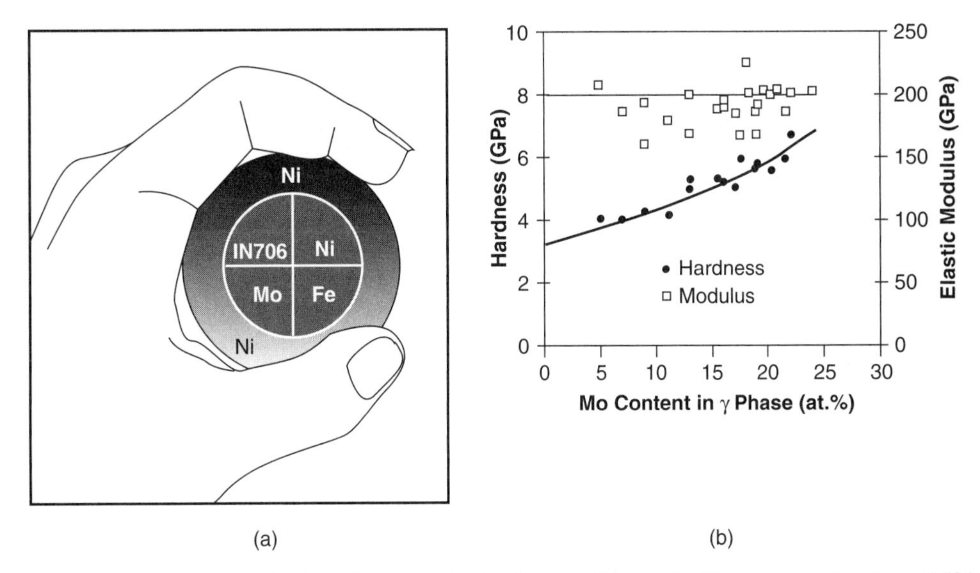

Figure 8-8 Diffusion multiple. (a) A diffusion multiple of Ni, Fe, Mo, and the Ni-based superalloy Inconel 706 (IN706). (b) Hardness variation with Mo content in one phase of the Ni-Mo-Fe trijunction where all three of these elements meet. (Adapted from Zhao Ji-Cheng, et al., "A Diffusion Multiple Approach for the Accelerated Design of Structural Materials," *MRS Bulletin*, April 2002, p. 325)

temperature. The library dramatically reduces the time scale for exploring the universe of elemental compositions for such advanced materials. These thin-film depositions, once generated, can now be examined by sophisticated scanning detection systems to observe the characteristics of these new superconductor candidates. These instruments permit the rapid screening of libraries for desired attributes.

A ***diffusion multiple*** is an assembly of three or more different metal blocks, in intimate interfacial contact, that is subjected to a high temperature to allow for thermal interdiffusion. Figure 8-8 is a diffusion multiple of Ni, Fe, Mo, and superalloy Iconel 706. Diffusion multiples are used to map critical data for structural materials design, such as phase diagrams and materials properties (hardness and elastic modulus). In this example, ten ternary-phase diagrams were mapped from this single sample. It would probably take more than a thousand alloys to map these same diagrams. This diffusion multiple was produced in order to explore the phase relationships and the properties of this alloy as well as the effects of these elements on modifying its properties. At the tri-junction where these three elements meet, extensive interdiffusion occurred. The diffusion process created compositional variations of all the single-phase regions, including all of the intermetallic compounds. Advanced microanalytical techniques—**electron probe microanalysis (EPMA)** and **electron back scatter diffraction (EBSD)** plus nanoindentation—permit assessment of the effects of composition variations on the alloys's properties. Diffusion multiples provide a quick way to generate critical experimental data to test theoretical predictions. They can be designed with different shapes and forms to achieve different purposes. In the case of high-temperature coating applications, diffusion barriers are needed to prevent desired elements in the coating from diffusing into the substrate, thus possibly degrading the coating's oxidation resistance.

8.2 MATERIALS DESIGN

Different materials provide different libraries of properties. These properties, in turn, determine which materials are suitable for an application and how they may be processed. Materials properties can be modified on purpose, and they are changed when they are processed into final form or shape. Intelligent use of materials requires a sound appreciation of the strong links among product design, manufacturing processes, and materials properties.

Product design is a cyclic, iterative process that starts with a market need and proceeds through conceptual design, detail design, and manufacturing to arrive at the selling of the product. "Iterative" describes the back-and-forth flow of information that links all the steps in the cycle. Module 19 describes three general design models used by engineers to design ceramics. The complex mathematical expressions used in these models are solved with the aid of powerful computer programs and analytical software, and the results of these analyses determine, as much as possible, whether or not a material would meet the requirements of a particular application.

In Module 1 we stated that conventional materials could no longer meet the need for certain material properties, and as a consequence it has become necessary to "engineer" materials. We further stated that this new generation of engineered materials are referred to as ***designed*** materials. Once upon a time, designers, after considerable deliberation, would select from a list of available materials one that possessed a majority of the properties required to manufacture a new product. Those days are gone forever. Now engineering designers rely on materials scientists and materials engineers to create materials that have the desired characteristics required by the new product or device.

Figure 8-9 represents a model of this process for developing new materials. In fact, the heading of this section could have read "***Materials by Design***" as opposed to "Materials Design." In this new ***original design*** process materials designers begin from scratch to create a new, engineered material offering new, unique combinations of properties that will allow a new product, structure, or system to fulfill new technological needs. A recently developed ***computer model,*** for example, will help engineers predict the properties of coatings that are mixtures of ceramics and metals applied to engine parts via plasma spray to enable jet engines to operate at higher temperatures for longer periods of time by simply inputting the percentages of the mixture. Reportedly more than 90% accurate, this model rules out mixtures that do not provide the needed properties. The model determines the properties of various mixtures of the components at a range of temperatures and other conditions.

The following two sections provide further examples of original design in the never-ending search for advanced materials never before seen in this highly technological world. The next step will be the transmittal of the new material's specifications via a digital network to the factory floor, where computer-controlled processing and fabricating machines and

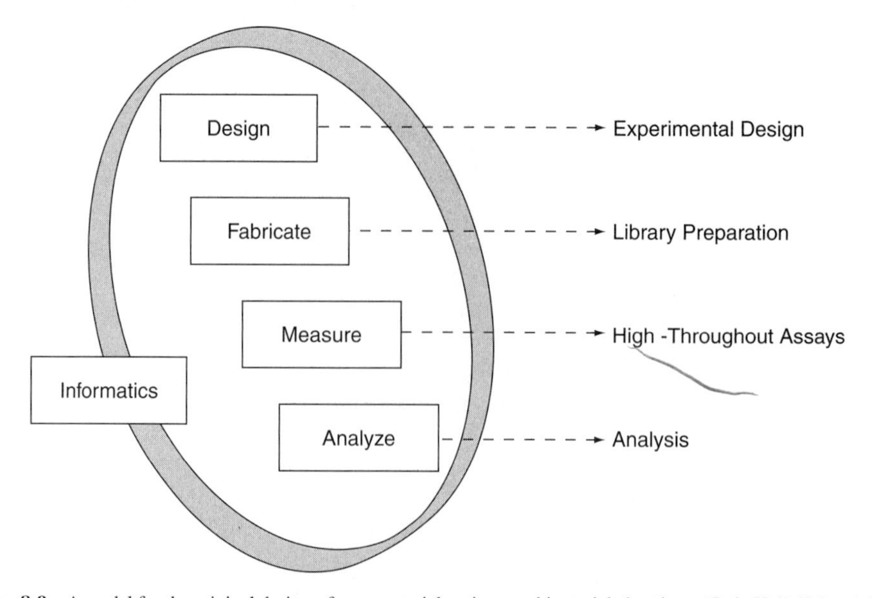

Figure 8-9 A model for the original design of new materials using combinatorial chemistry. (Dale Hall, "Materials Research at NIST—Century of Progress," *National Educators' Workshop:2001*. NASA/CP-2002-211735, June 2002, p. 25)

robots will transform them into the new material and then into the form, shape, size, and number of products required.

8.2.1 Modeling Technology

High-quality models can now be made quickly and conveniently available throughout the design process. These models reduce the time to market and cost as well as increase product quality. The leading product technologies in use today are ***stereolithography (SLA), selective laser sintering (SLS),*** and similar processes. For example, the making of a composite tool starts with the building of a pattern, usually of low weight, rigid, and with good heat transfer and a low thermal expansion coefficient (CTE). Using a range of innovative, seamless epoxy pastes, large mock-ups and dimensional models enable a composite tool builder to extrude material where needed. A new technique of moldmaking is the Custom Moldmaking Process Technology, which bridges the gap between rapid soft tooling of fiberglass or urethane prototype castings and conventionally machined all-metal tools. Laminated metal structures can be produced by ***ultrasonic consolidation.*** A continuous solid-state ultrasonic joining creates a metallurgical bond between layers of metal tape, and continuous fibers or meshes of varying types can be embedded in the laminate, or metal feedstocks containing embedded particulates or short fibers may be employed.

8.2.2 Materials Optimization

Composite structures are often designed to achieve maximum efficiency from the mechanical characteristics of the material. The nature of composite materials and processes, however, often results in parts with very poor repairability and maintainability. A materials optimization system in manufacturing uses a "best practice" process that encourages design engineers to optimize product attributes such as weight, thickness, and materials. This leads to higher ***confidence levels***, improved product quality, and a reduction in overall product weight. For any product there are hundreds of thousands of raw materials to pick from. Each material type, class, and grade has its own unique set of properties. Some are better for human contact, and others may be cheaper to process. Engineers tend to be conservative; as a result they apply a large ***safety margin*** to their analysis to reduce risk of failure. (See Section 7.3.1, "Factors of Safety," for further discussion of this topic.) This type of analysis leads to ***overdesigned parts*** that are many times heavier than required, resulting in a significant increase in overall cost of the product. Access to a ***central repository*** of materials property information leads to the sharing of expert knowledge, with attendant cost savings in raw materials and processing. Figure 8-10 is a graphic model of new methods for the design of new materials that apply the combinatorial approach to redefine the scientific method and use the newly developed tools discussed in this module.

8.3 MATERIALS SELECTION

Recall (from Module 1) that the Wright brothers' "1903 flyer" used a limited number of materials because those were the available materials at that time. Henry Ford and his designers at the Ford Motor Company as well began work in 1903 with limited materials for automobile design. To celebrate the Ford Centennial, designers rebuilt the Model "T" with contemporary engineering materials. (See the websites below for more information.) Today, designers face a different problem: The vast array of available materials makes selection difficult. We will explore some of the tools and techniques they use.

Henry Ford Museum & Greenfield Village—*www.hfmgv.org*
Ford Centennial—*http://www.ford.com/en/ourCompany/centennial/default.htm*

8.3.1 Algorithm for Materials Selection

Engineering requires clearly stated, unambiguous steps for problem solving. ***Algorithms*** are well-defined methods for solving specific problems. Computer programs are written after an algorithm has been developed to lay out clearly the steps that the program is to solve. For example, you could write a simple algorithm to calculate the strength required of a light pole to withstand the pushing forces (compression) from a light fixture. A much more complex algorithm would be required to select a piston-connecting rod for an internal combustion engine. The first problem requires only the selection of a material of suitable size/strength to hold up the light fixture, and almost any material would suffice as long as it was sufficiently strong and pleasing to the user. On the other hand, a connecting rod will undergo many types of mechanical stress, ranging from compressive to tensile to torsional to gravity forces, in addition to thermal stress from the combustion chamber. How does the designer match component requirements with available materials?

8.3.2 Selection Tools

To aid in the creation of materials selection algorithms, databases must be available to answer questions on material suitability. A materials ***database*** involves tables that list properties of materials, such as tensile strength, hardness, corrosion resistance, and the ability to withstand heat. Thousands of reference books are available with such data. Much of these data are computerized to allow easier access.

Certain graphical techniques aid the designer in materials selection. In Figure 8-10 a graphical plot of two important properties shows a relationship of stiffness (Young's modulus) to weight (density) of major groups of materials. For example, ellipses encircle the stiffness-to-density ratios of most engineering ceramics and engineering alloys. By encircling major materials groups, a designer can see that both groups of materials possess high stiffness, but at the expense of high density. On the other hand, engineering composites, as a group, possess stiffness that approaches or may equal that of engineering ceramics and alloys; however, com-

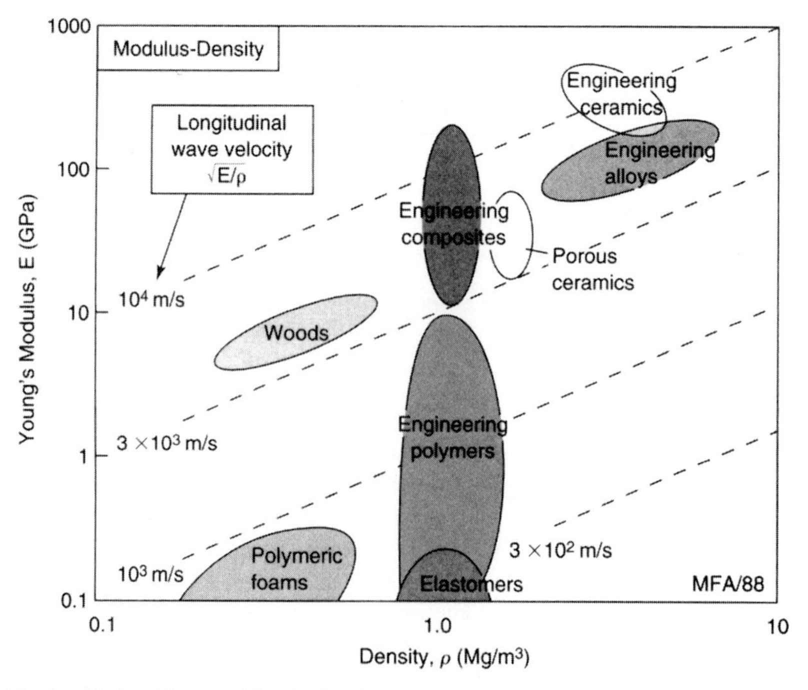

Figure 8-10 Graphical tool for materials selection. The materials selection chart depicts a plot of Young's modulus and material density. Boundaries encircle the clustering plots of these two properties to allow comparisons of groups of materials. (Michael F. Ashby, "Materials in Mechanical Design," *Journal of Materials Education*, Vol. 15, pp. 143–166, 1993)

posites have the advantage of much lower density. Similar graphical plots can be made for a variety of property ratios to narrow the choice of materials down to design requirements.

8.3.3 Properties of Materials

Periodicals, websites, and handbooks can provide current data and performance criteria for structural (load-bearing) materials in terms of

1. Strength (tensile, compressive, flexural, shear, and torsional)
2. Resistance to elevated temperatures
3. Fatigue resistance (repeated loading and unloading)
4. Toughness (resistance to impact)
5. Wear resistance (hardness)
6. Corrosion resistance

Technical society and manufacturers' references give values of the performance criteria (properties) of metals, polymers, and ceramics, with updates on newer materials such as aramid fibers, zinc aluminum alloys, and superalloys. Various periodicals, such as *Advanced Materials Processes, Machine Design*, and *Modern Plastics*, have annual materials selectors that provide general information on properties for a long list of materials. There are also many handbooks, such as the multiple-volume *Metals Handbook* and the series *Engineered Materials Handbook*, which cover nonmetals; both are published by ASM International.

The tables as found in the Appendix of this book provide comparative data on selected ceramics, metals, and nylon. The designer must use such data to determine if a material has appropriate physical, mechanical, and chemical properties to withstand the service conditions to which a part will be subjected. For example, from Table A-9 we learn that nylon is an engineering thermoplastic and is used for some bearing applications. By turning to Table 19-2, "Typical Properties of Selected Engineering and Technical Ceramics," which compares selected plastics, metals, and ceramics, you will note that nylon has a melting point of 215°C and a maximum service temperature of 422 K (149°C). On the other hand, alumina (used for spark plugs and computer modules) has a melting point of 2050°C and a maximum service temperature of 2222 K. Alumina is also much harder than nylon: 9 Mohs versus 2 Mohs. The modulus of elasticity (given in megapascal, MPa) is a comparison of stiffness; alumina is very stiff (modulus of 34.5×10^4 MPa) and brittle, whereas nylon is flexible (modulus of 0.33×10^4 MPa). From the comparison of hardness and temperature resistance, it is clear why ceramics are selected for furnaces, ovens, and other applications where high abrasion and high heat are present. On the other hand, nylon is selected for rope, tubing, and gears and bearings in electrical appliances.

The two tables just discussed provide general data on properties for simple comparisons. Selection of specific materials requires many more detailed specifications. General databases from handbooks will provide much detail, but the final selection often requires consulting the properties databases for the product lines of materials manufacturers.

8.3.4 Materials Systems

Although databases are imperative in the initial selection steps, there are other factors that complicate materials selection. Materials rarely exist in isolation; they interact with other materials. So a combination of materials must be selected that complement one another. In a successful ***materials system***, each component is compatible with the others while contributing its distinctive properties to the overall characteristics of the system of which it is a part. A state-of-the-art telephone is a good example (see Figure 1-1). The casing might be a tough ABS plastic that houses a microchip (a solid-state ceramic device), which provides memory and sound-transmission capabilities. Copper leads join the circuitry together. There might be a battery and a ceramic light-emitting diode to show when the battery is low. The acid in the battery must be isolated to prevent corrosion, and the copper leads must be insulated so that they do not short out. Each component is made of materials that meet the demands of the physical and chemical environment normally encountered when using the system.

8.3.5 Additional Selection Criteria

Existing specifications have a lot of influence on the choice of material. These specifications or *standards* are used when redesigning an improved model of the product. When the materials selection algorithm results in the selection of a new material, the material might not be covered by current specifications from such standardization agencies as the National Institute of Standards and Technology, Underwriters' Laboratory, fire departments, the American Society for Testing and Materials, or the U.S. Food and Drug Administration. The Occupational Safety and Health Act (OSHA) of 1970 sets forth conditions of safety that must be met by those involved in the manufacture or use of goods and services in the United States. It might take considerable time for these agencies to alter their specifications to include the new material, or they might not approve its use.

Availability is another concern of the designer. Will the material be easily available in the quantities and sizes required by production demands? Also, will it be available in the shapes required? Aluminum extrusions, for example, are available in many varieties of standard shapes, such as round, oval, and square. In the past, designers were limited to existing materials such as metal alloys, woods, or concrete. Now, however, it is possible to start from scratch at the synthesis stage to have materials engineers design a materials system to provide properties to meet the expected needs.

Processability, the ease with which raw materials can be transformed into a finished product, is of paramount concern. Much of the current focus is on low-energy processing. Companies may have difficulty processing the new material on existing equipment. Can they afford to invest in new equipment? Today, the reverse question is usually asked: Can we afford not to use the new material and process? If we do not, the competition might make the change and run us out of business with their superior product. Many new technologies are now available.

Near-net-shape production involves incorporating numerous separate parts into a single, integrated assembly, thus saving overall production costs. This new technology is receiving considerable attention from various industries. Plastics, ceramics, and composites have become increasingly competitive because they suit near-net-shape production. These products and parts are suitable for plastics like polystyrene, nylon, polypropylene, polycarbonate, and glass-filled epoxy. The metal industry has also brought out new processes and refined older ones to simplify and integrate parts for near-net shapes. Net-shape and near-net-shape processing aims to improve product reliability while reducing costs of materials and processing. One such example, the "Super Plug," a patented automobile door hardware module (see Figure 8-11), replaced the 61 separate, stamped, and formed steel, molded plastic, and rubber door parts with one plastic/glass composite system made of XENOY®, a polycarbonate/polyester blend with 30% glass fiber. The single system not only saves materials and processing costs but also provides manufacturing cost savings at assembly and yields overall weight reduction that results in fuel savings for the life of the vehicle. Another picture of the module is shown with the door in Figure 15-15.

The near-net-shape concept is a growing trend because it is consistent with the new thinking about design and manufacturing; that is, concept design, engineering analysis, materials selection, and processing are now orchestrated in a team approach. In traditional practice, design was handled by an engineering group. It sent the drawings to manufacturing, and that department was responsible for determining how to produce the part. Often, there was conflict in the demands that each group made on the other. The new CAD/CAM technologies force an integrated approach. Materials selection, central to the design process, has great bearing on manufacturing techniques. ***Expert systems*** directed by computer-based ***artificial intelligence*** are becoming available to help weigh the increasing number of variables and levels of optimization inherent in materials.

Quality and ***performance*** are two aspects that result in consumer satisfaction. The high cost of most durable goods and the competition for customer acceptance has resulted in extended

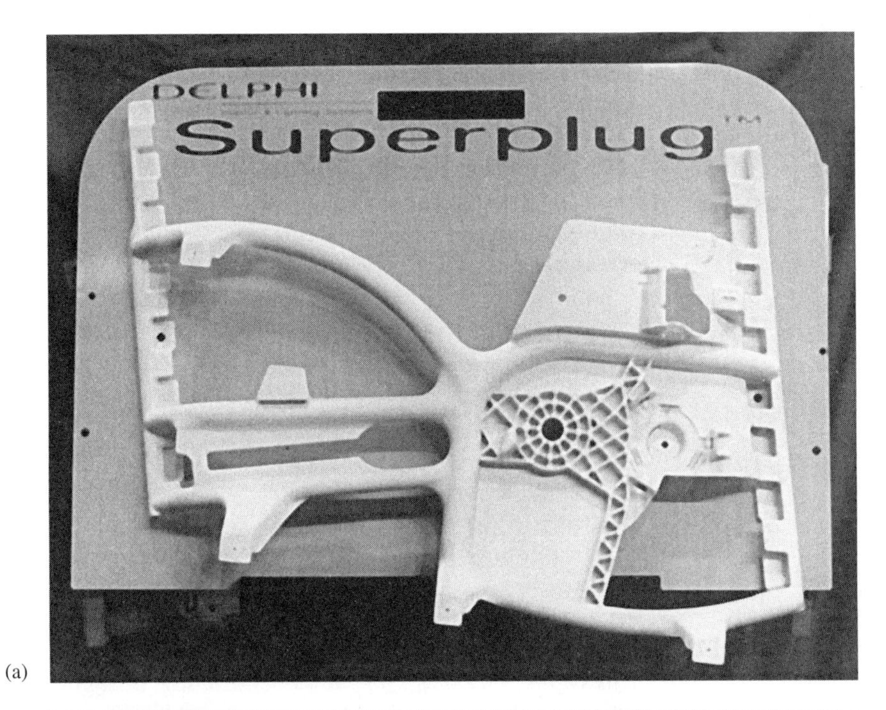

(a)

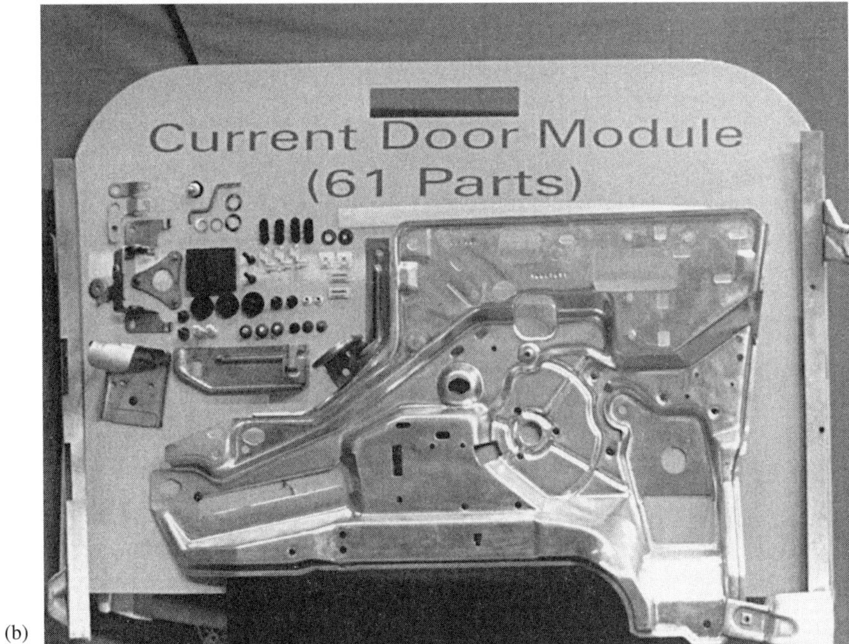

(b)

Figure 8-11 (a) G.E. Plastics Super Plug™ is an example of net-shape processing. The glass-reinforced plastic composite provides savings in both production and operating costs over the current door module (b), which is made of stamped steel and uses up to 61 parts. (G.E. Plastics)

warranties. Materials selection must therefore ensure that parts will not rust, break under repeated stress, or fail to perform in any other way for the predicted service life of the product.

Consumer acceptance includes many factors beyond excellent quality and high performance; there are also societal aspects. As a whole, society as well as governmental agencies are taking a closer look (life cycle analysis and inventory) at manufactured products. Any product has to be considered in terms of its total life cycle. What are the results of the processing methods? Are polluting gases being released into the environment, or are toxic metals and chemicals being flushed into our rivers and streams? During use, does the product safeguard our health? At the end of the product's useful life, how can it be disposed of safely? Municipal solid waste is a hidden product cost that we pay in the form of higher taxes and a poorer quality of life. In a previous example, fast-food restaurant chains moved away from polystyrene packages because the public believed that these plastic containers were more harmful to the environment than paper packaging. Soft-drink manufacturers are moving toward *reusable* plastic bottles. Currently, poly(ethylene terephthalate) (PET) is used for 2-liter soft-drink bottles; it is a good recycling plastic. But a more durable bottle of polycarbonate can be rewashed and refilled just as many glass bottles are now handled. Why hasn't polycarbonate been adopted?

Design for disassembly has become a theme in much of product design by major corporations. Europe, which has a higher degree of ecological concern, has led the way. In order to facilitate recycling, manufacturers of small appliances and durable goods are establishing procedures to ensure that products can be broken into components for easy sorting prior to recycling. Among the procedures are reducing the variety of plastics, adding labels to plastics for easy identification of plastic type, and eliminating screws and adhesives so that parts will disassemble easily. See Figure 15–26 for examples of labels.

Newer computer software programs offer design systems that make products easier to fix. The ***Design for Service*** program takes its place alongside the previous software programs ***Design for Assembly*** and ***Design for Manufacturability***. Design for manufacturing (DFM) is a process for integrating the benefits and capabilities of the manufacturing process into the design of the product so that the best product can be produced in the shortest possible time and with limited resources. The new programs help product designers consider repair issues early in the design stage. Objectives of the programs include making repairs less costly and extending the functioning life of products. Environmental issues like recycling are also directly addressed by this new computer software, which may have customers fixing products rather than tossing them out. In addition, this software augments previous software that addresses the need for disassembly of a product for whatever reason.

More often than not, ***cost*** is the primary selection criterion that will determine the final choice of materials. In other words, if several materials have the specified physical, mechanical, and chemical properties and are suitable for the processing technique selected, the lower-cost materials would be the logical choice. Determining cost is not as simple as it may seem. For example, a variety of plastics, including PET [poly(ethylene terephthalate)] and HDPE (high-density polyethylene), have replaced glass as containers for soft drinks, milk, and juices. Although the initial cost of plastic may be greater, the plastic bottles provide savings due to their toughness (less breakage) and the savings in shipping (PET and HDPE are much lighter than glass bottles).

The traditional "stick-framed" home (Figure 8-12a) uses 2×4 studs (sticks) and a wall material such as sheetrock for the inside and brick veneer and perhaps vinyl siding for the exterior. Recycled plastics provide an alternative to the very old and inefficient stick-framed house. General Electric has developed factory-built composite wall systems (Figure 8-12b) that combine a corrugated engineered-wood by-product with a low-density reinforced PPO [poly(phenylene oxide)] foam and recycled polystyrene, which is covered with a fire-retarding, glass-reinforced plastic. This type of wall panel can cut down on building costs and promote recycling while improving overall quality.

Product liability is the civil (as opposed to criminal) liability of the manufacturer to an ultimate user for injury resulting from a defective product. ***Caveat emptor*** (let the buyer beware) was once the rule. Today, numerous liability laws are in effect. For those involved in materials, particularly materials selection in the design process, the trend is for courts and juries in a product liability action to identify members of the design team as being responsible for some fault. Therefore, it is imperative to obtain and use the latest information about materials selection, particularly materials' long-term characteristics.

(a)

(b)

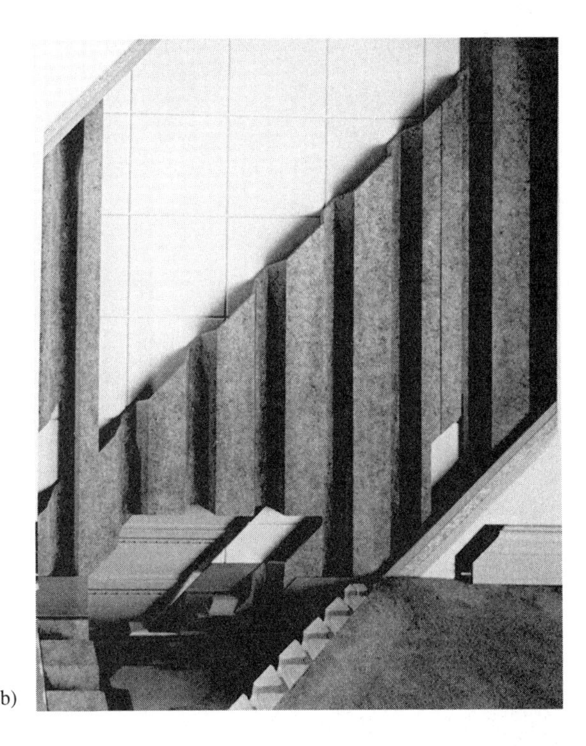

Figure 8-12 (a) Most contemporary homes in the United States use the "stick-framed" system for walls. This system is hundreds of years old and does not take advantage of newer approaches to engineering walls. (b) A developmental wall system combines an experimental corrugated strand product with low-density polyphenylene oxide foam and reclaimed hamburger packages. The panel is covered with a glass-reinforced plastic that is an excellent fire retardant. (G.E. Plastics)

8.4 DETERMINING MATERIAL STRUCTURE

The kinetic theory of matter describes atoms in a solid as being at their lowest energy state. In this state, their motions have slowed considerably. These atoms of solids are so close to their neighboring atoms that they can no longer act as independent particles. They vibrate about their equilibrium positions, depending on the degree of remaining kinetic energy (thermal energy) they may possess. But they can also move in conjunction with their neighboring atoms.

In the discussion about bonding in Module 2, we learned that two or more atoms can chemically join each other, depending on their electronic structures. We also learned that atoms, ions, and molecules are of different sizes.

The internal structure of solids determines to a great extent the properties of the solid or how a particular solid material will perform or behave in a given application. By understanding microstructure and its origins, one can develop a basis for achieving specific properties tailored to the application. ***Metallography*** is the study of the interrelationship between the structural characteristics and the physical and mechanical properties of metals and alloys. Photomicroscopy and microhardness testing are closely allied to this study. Structural interpretation of these materials is made possible by sectioning, mounting, grinding, polishing, and etching a specimen, which can be viewed microscopically. Evaluating images of these specimens is very important to the maintenance of close control over production processes. Further, the interpretation of these views is invaluable in materials research, much of which is devoted to the development of new materials as well as to determining the cause of present-day materials failures. When dealing with ceramic materials alone, the term ***ceramography*** is applied to the detailed instructions for how to saw, mount, grind, etch, examine, interpret, and measure ceramic microstructures. The study of the internal structure of ceramics is pursued in various ways, using X rays as well as electron, proton, or neutron beams to disclose crystal size, crystal structures, crystal imperfections, bonding types, spacing of adjacent atoms and adjacent planes, and the different atoms present. Some knowledge of the more important tools in the field of spectroscopy and microscopy that permit a still-deeper probe into the internal structure of materials is vital to a greater understanding of the effects of structure on the behavior of natural and human-made materials. Figure 8-13 shows photomicrographs of varying magnification produced with the aid of various instruments used in materials science.

IBM Research—*http://www.research.ibm.com/topics/serious/materials*
NIST Gallery—*http://www.nist.gov/public_affairs/gallery/gallery.html*

8.4.1 Spectroscopy

White light, when refracted (changing direction when passing from one medium to another), is broken up into a number of different colors. Each spectral color corresponds to a particular wavelength. Every chemical element (as well as its atoms) produces a characteristic ***spectrum,*** which may be detected and measured by a spectroscope. Spectroscopy studies substances by analyzing their spectra. In other words, ***spectroscopy*** permits the precise measurement of the wavelengths of the radiations of atoms, which are discrete and distinct for each atom.

When atoms are exposed to high energy from an outside source, such as infrared, atomic absorption, arc-spark, or glow discharge, they become excited and their electrons move to higher energy levels. The interchanges of energy between radiation and matter take place in discrete units called ***quanta*** (discussed in Modules 4 and 27). Under excitation, an atomic substance will reach a state of dynamic equilibrium in which some atoms are releasing energy at the same rate that it is being absorbed by others. An atom loses energy when its excited electrons become unstable and transit to a lower energy level, giving off a ***photon*** of light. The wavelength of this photon is related to the magnitude of the change of energy. It is these transitions that, when observed by the spectroscope, produce atomic spectra (Figure 8-14), whose discrete and distinguishing wavelengths correspond to the various electron transitions between energy levels.

These studies permit scientists to increase their knowledge of atoms, electronic structures, and various bonding mechanisms. ***Raman spectroscopy*** directs light of a specified frequency at a sample and measures the frequency of the reflected light. Differences in frequency are caused by the coupling of the light photons with bonds in the materials, and they indicate the vibration and the rotation of the molecules in the material. A later spectroscopy called **gradient-field Raman (GFR)**, is similar to Raman spectroscopy in that its resolution is in nanometers rather than micrometers.

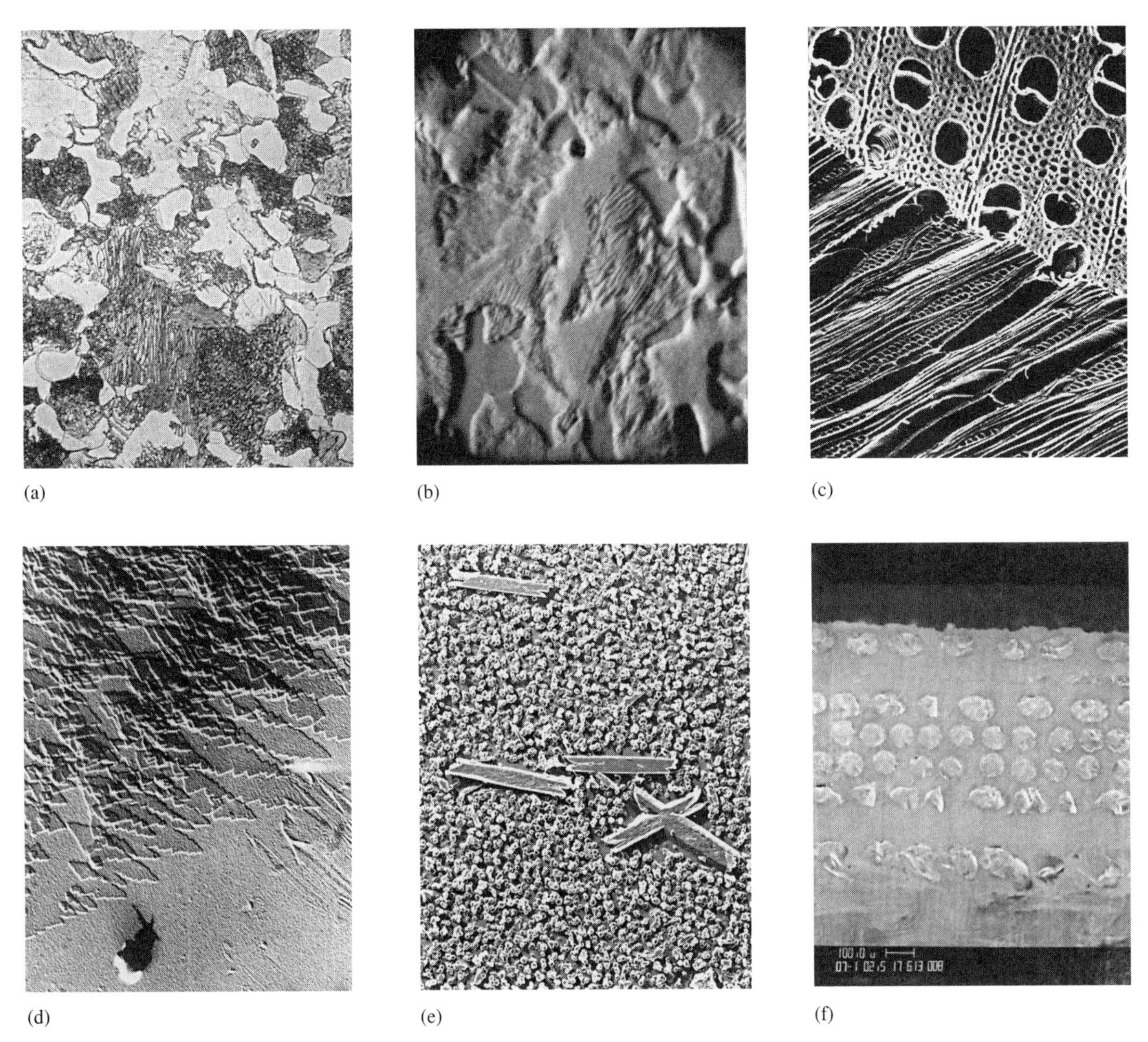

Figure 8-13 Photomicrographs of structural features of some representative materials. (a) Steel (SAE/AISI 1045) 1000×. (NASA) (b) Steel (SAE/AISI 1045) SEM 2550×. (NASA) (c) Wood (maple) 415×. (U.S. Forest Products Lab) (d) Plastic (polyethylene) dendrite crystals. (Dr. Philip H. Geil—University of Illinois) (e) Glass ceramic: crystals growing in amorphous glass, 17,900×. (Corning Glass Works) (f) Fiber metal composite. (NASA)

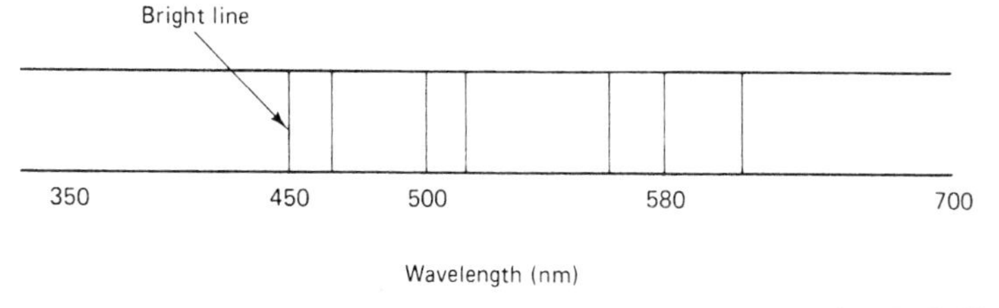

Figure 8-14 Atomic spectrum of sodium in the visible region. (Adapted from D. C. Giancoli, *Physics*, Prentice-Hall, Inc., Englewood Cliffs, NJ, 1980)

8.4.2 X-Ray Diffraction

One of the most useful tools in the study of crystal structures of solids is ***X-ray diffraction***. An X-ray diffractometer is used in this process. X rays have wavelengths about equal to the diameters of atoms (10^{-10} meters) or about the same length as the spacing between atoms (or ions) in solids. When these X rays are directed at a solid with a crystalline structure, the waves are refracted. The equation, known as *Bragg's law*, $n\lambda = 2d \sin \theta$, relates the wavelength, λ,

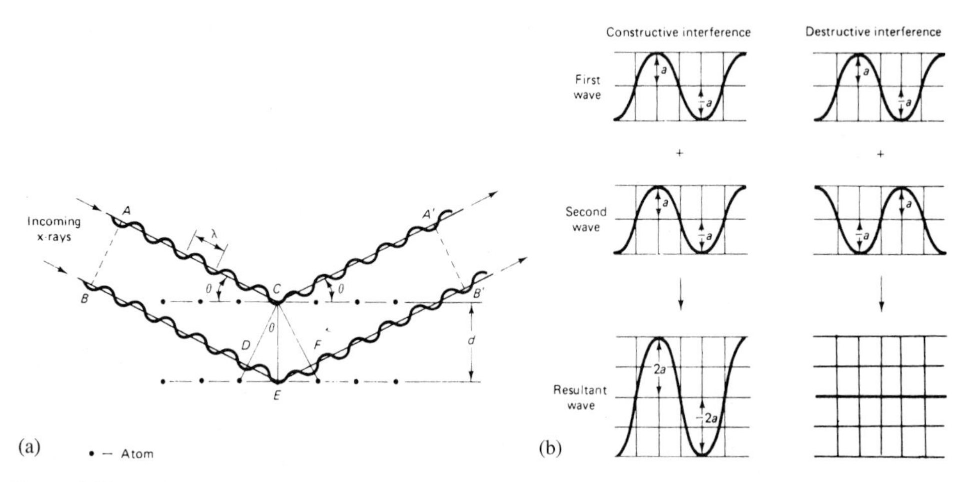

Figure 8-15 (a) Scattering of x-rays by atoms in parallel planes. (b) Constructive and destructive interference of waves. (Theodore L. Brown and H. Eugene LeMay, Jr., *Chemistry—The Central Science*, 2nd ed., Prentice-Hall, Inc., Englewood Cliffs, NJ, 1981)

to the distance between planes of atoms, *d*, and the glancing angle between the incident beam and the plane of atoms, *d*. Figure 8-15 shows a sketch of this relationship. If the incident beam strikes the planes of atoms at some arbitrary angle, the reflected beam may be nonexistent, because the refracted rays from the atomic planes will be out-of-phase and produce destructive interference (Figure 8-15b) and cancellation of the reflected beam. It has been determined that, at a particular angle θ, the reflected beam will be in-phase and produce constructive interference (Figure 8-15b). Mathematically, this can be expressed as $\lambda_1 = n\lambda_2$, where $n = 1, 2, 3, \ldots$; the distance traveled from the different parallel planes of atoms represents an integral number (n) of wavelengths. Referring to Figure 8-15b, this statement may be written as

$$ACA' = BEB' - n\lambda \qquad \text{where } n = 1, 2, 3, \ldots$$

Knowing the wavelength of the incident beam of X rays, the glancing angle can be measured experimentally, and solving Bragg's law for $d = n\lambda/2 \sin \theta$, the interplanar distance can be calculated. Figure 8-16 shows a photographic plate on which the constructive waves produce a series of dots, indicating that the X rays are scattered from crystals at only certain angles. All the various angles at which diffraction occurs are determined by measurements on the

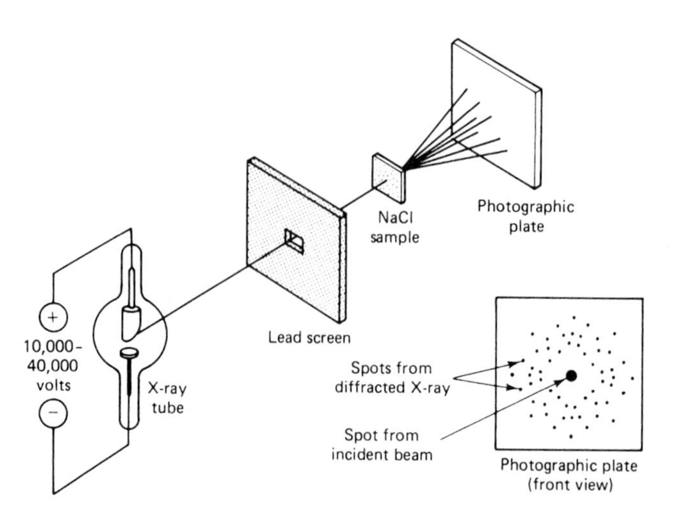

Figure 8-16 The X-ray diffraction pattern for NaCl and the experimental method by which it is obtained. (T. L. Brown and H. E. LeMay, Jr., *Chemistry—The Central Science*, 2nd ed., Prentice-Hall, Inc., Englewood Cliffs, NJ, 1981)

photographic film. By studying the directions of diffracted X-ray beams, as well as the intensities produced by this powerful tool, much can be learned about the crystal structure of solids.

8.4.3 Optical Microscopy

The human eye can distinguish between two points in space only if the two points are approximately 0.1 mm apart at a distance of 25 cm. Microscopic objects perhaps only 0.5 mm apart with the eye focused at the normal reading distance of 25 cm can be seen as an image magnified by a factor of 200 or more. The compound microscope fulfills this requirement by having two lenses, the objective and the eyepiece, to give two stages of magnification. The quality of the microscope image is determined by the magnification, degree of diffraction, distortion due to aberrations of the lenses, and contrast. ***Metallographs*** are optical, metallurgy microscopes capable of photographing the images of solid specimens (see Figures 8-17 and 8-18). Opaque specimens, such as metallurgical samples, are typically observed with a ***reflection microscope,*** in which light is reflected from the sample. Specimens often show very poor contrast between adjacent areas, even though these areas differ in chemical composition or morphology. As a result, most metallographic specimens are acid-etched to make specimen details visible. The ***etching*** reacts with the grain boundary areas at a different rate than the grains themselves. When incident light rays strike these areas with their different crystal orientations, as well as their difference in rate of etching, the reflected light travels back to the eye in different amounts (see Figure 8-19). If most of the incident light is reflected back to the eye, the grain will appear bright. Other crystals reflecting less light will appear darker. A ***photomicrograph*** of a specimen taken through a metallograph (Figure 8-20) shows the details of the microstructure, including the grain boundaries. Photomicrographs of various materials are shown throughout this book.

8.4.4 Electron Microscopy

The maximum magnification produced by optical microscopes is about 2000X. This limitation is imposed by the wavelengths of visible light, which limit the resolution of minute details in the specimen being observed. ***Resolving power*** is a term used to describe the ability of

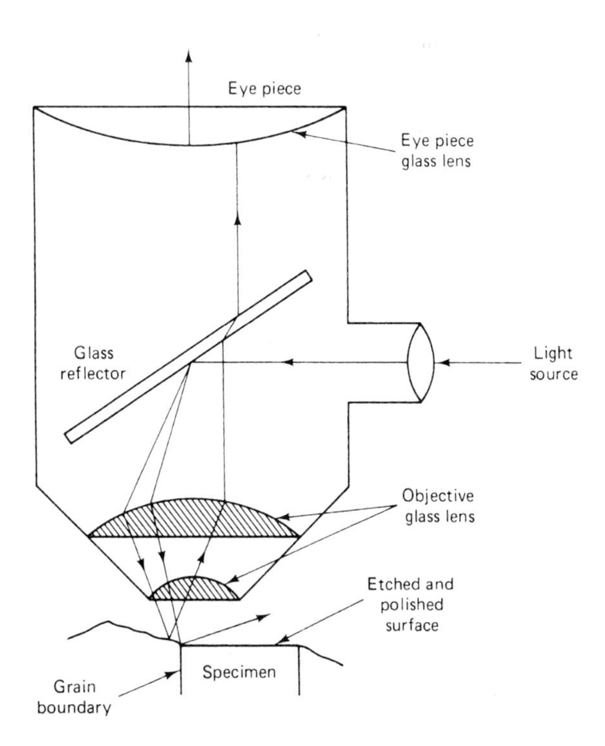

Figure 8-17 Schematic of an optical system of a metallographic microscope detecting a grain boundary in a crystal.

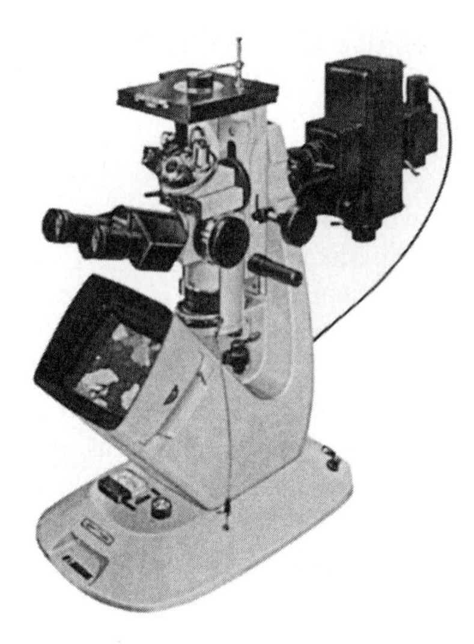

Figure 8-18 Photograph of a metallograph. (Buehler Ltd.)

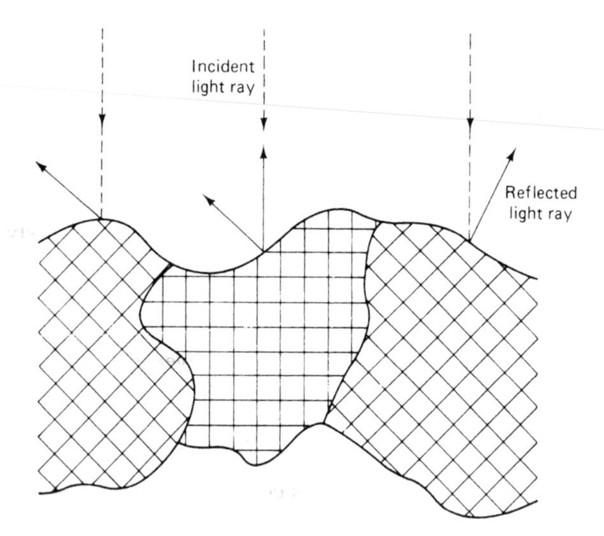

Figure 8-19 Sketch showing identification of individual crystals by reflected light rays.

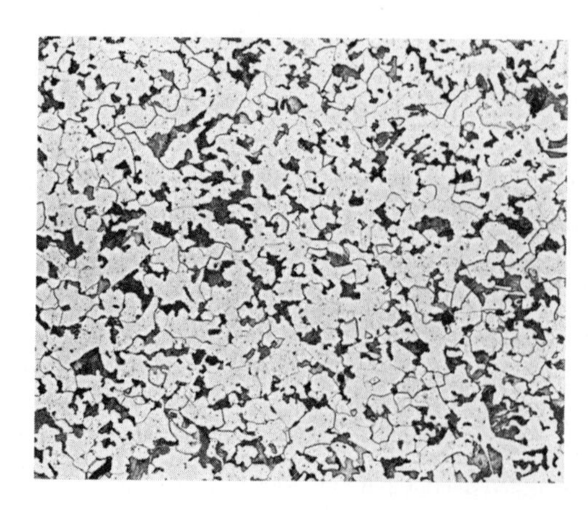

Figure 8-20 A photomicrograph of low carbon steel. (Buehler Ltd.)

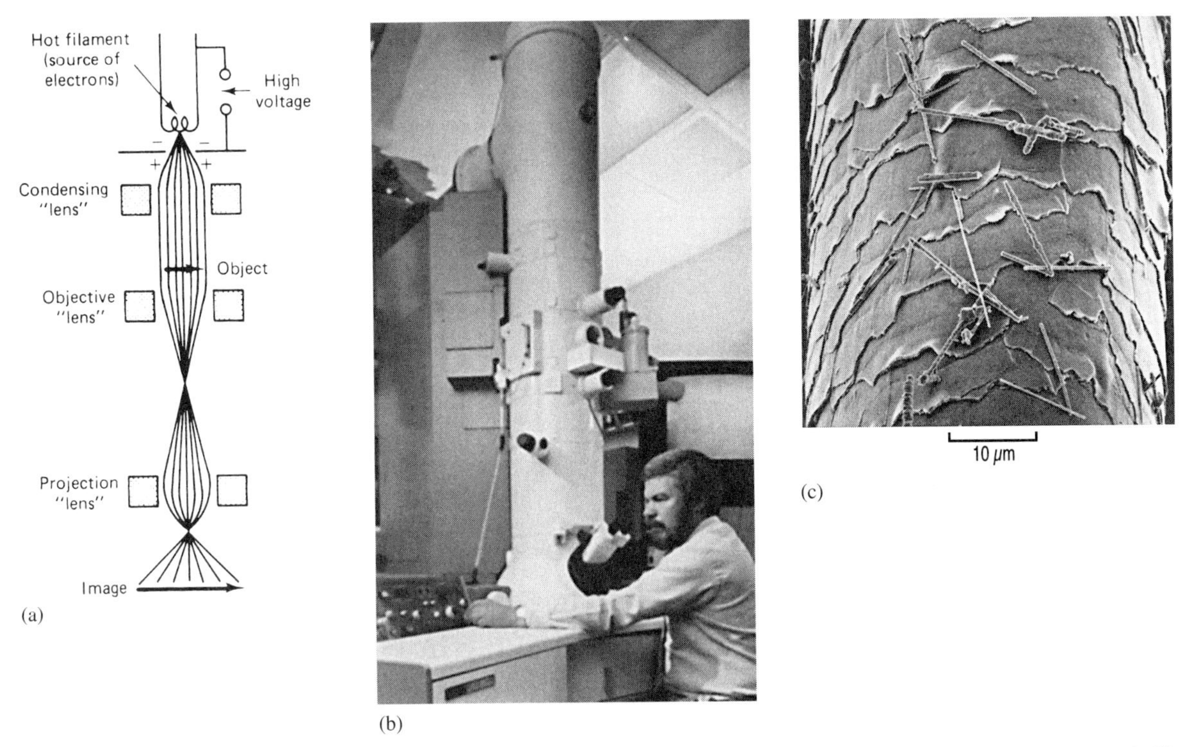

Figure 8-21 Transmission. (a) Electron microscope. The squares represent magnetic field coils for the "magnetic lenses." (Douglas C. Giancoli, *Physics—Principles with Applications*. © 1980, reprinted with permission of Prentice-Hall, Inc., Englewood Cliffs, NJ) (b) Photograph of a transmission electron microscope (TEM). (c) TEM micrograph comparing whiskers of ceramic (SiC) to a human hair (ORNL).

a lens to reveal detail in an image. Optical microscopes (discussed previously) do not have the ability to discern details smaller than 10^{-6} m. To observe objects in the 10^{-10}-m range, high-energy electrons with wavelengths around 4×10^{-11} are used. These electrons are similar to visible light in that they are both forms of electromagnetic radiation. By using magnetic lenses to focus the electrons (see Figure 8-21a), their resolving power can be increased greatly to give magnifications over 200,000X. Remember that metallic specimens are opaque to electrons (i.e., electrons are absorbed by the metallic atoms).

The ***transmission electron microscope (TEM)*** (see Figure 8-21) was a driving force in the development in electro-optics earlier in this century. In brief, TEM works by putting an unfocused electron beam incident to a very thin sample. By using the transmitted electrons, magnifying projection optics form an image of the sample on photographic film, with a resolution of a few billionths of a meter. Because the electron beam passes through the specimen, details of the internal microstructure are made observable. However, solid materials are highly absorptive of electrons, and specimens must be prepared in the form of a very thin foil. The transmitted beam is projected onto a fluorescent screen or photographic film so that the image may be viewed. ***Whiskers*** are short, crystalline fibers with outstanding strength that have an average diameter of about 0.05 μm. These whiskers are shown in Figure 8-21c compared to a human hair (the larger, dark object).

The ***scanning electron microscope (SEM)*** scans the surface of a specimen with an electron beam and the reflected (back-scattered) beam of electrons is collected and displayed at the same scanning rate on a cathode-ray tube. The image on the screen contains the surface features (with good depth of field) of the specimen, which may be photographed. The major requirement for the specimen is that it be electrically conductive. Accessory equipment allows analysis of the chemical composition of very localized areas of the sample.

An example of the use of electron microscopy can be found in the research and development of synthetic crystalline materials produced by ***molecular beam epitaxy (MBE)***. These materials, not found in nature, have extremely thin layers of atoms deposited alternately on a semiconducting substrate or base. By controlling the chemical composition and thickness of the layers, various crystalline structures are formed with varying electrical, mechanical, and optical properties. The MBE method (described in Module 27) would not be possible without the prior development of electron microscopes with small wavelengths (0.2 nm) and large magnifications, which permitted scientists to verify that their efforts were indeed producing a crystal with extremely thin atomic layers of alternating composition. Atomic layers as thin as one ten-millionth of a centimeter were determined to exist in these new human-made materials.

Knowing that (1) changes in the internal structure of solids are accompanied by a change in energy and (2) the release of this energy in the form of heat (***exothermic***) or the absorption of this heat (***endothermic***) can be detected and measured leads to another technique that increases our knowledge about the internal structure of solids. ***Differential thermal analysis (DTA)*** techniques produce temperature patterns that can be interpreted to obtain information about the various structural changes that solids undergo as a result of the application of different external forces. Finally, even the vibration of atoms in a molecule can be detected using infrared spectroscopy. Such techniques are used, for example, to identify unknown polymeric materials.

Micro Worlds with Tutorials—*http://www.ibl.gov/MicroWorlds/*
Oak Ridge National Laboratory-STM and AFM-Animated Image Gallery—
http://www.ornl.gov/img/afm.html

8.4.5 Digital Instrumentation

Scientists had been trying to make accurate microscopic 3-D images for more than 150 years, with no success up to now. They had to rely on artist's renderings of what the object should look like. Existing microscopes were unable to explore the internal details of samples when they were larger than the diameter of a human hair. The procedure for viewing an object was to embed it in wax, slice it into dozens of paper-thin sections, and view the dyed sections on slides under a microscope. But this procedure was still only two-dimensional, resulting in distorted and damaged images due to the cutting and reassembly process. The result was a grossly inaccurate picture.

A new microscope is revolutionizing the way scientists can obtain 3-D images of almost anything they care to look at. These new images can be rotated and viewed from any angle. Opened up, they reveal the sample's interior. Up to this time no one knew, for example, how the two parts of Velcro attach, or what the interface between layers in a composite material looked like. This new microscope uses ***digital volumetric imaging***, in which the sample to be viewed is stained with a fluorescent dye and embedded in a black liquid plastic. A laser shoots light through the sample and a digital camera snaps an image of the fluorescent surface. When the laser light hits the sample, it penetrates only the outermost layer, resulting in no visual overlap from one image to the next. A blade then slices off the outer layer, which is discarded, and the sample's newly exposed face is imaged. The process of slicing–imaging is repeated up to 1000 times until there is nothing left of the sample. A computer program compiles all the views into a simple 3-D image that represents the interior as well as the surface of the sample. The resulting data are transformed to a disk and, with special software and a computer workstation, can be viewed, analyzed, and manipulated. With this method scientists at Sandia National Laboratories working on micromechanical devices (see Figure 8-22) such as tiny robots have viewed for the very first time the tiny gears and screws just a few hundred micrometers long that can determine whether or not the dimensions of the manufactured parts match the computer drawings.

Digital imaging systems combine the high image quality of film and the real-time viewing of video, all at an affordable price and increased ease of use. Digital displays provide 256 levels per color. The following factors make a strong case for incurring the cost of upwards of

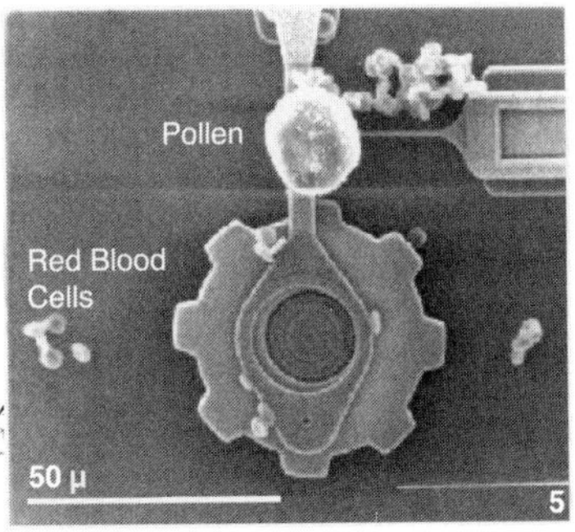

Figure 8-22 Size of micromechanical device compared to pollen and red blood cells. Note 50μ scale. (Sandia National Laboratory)

$8000 for a digital system: (1) best images obtained quickly, (2) delivery by e-mail, (3) unlimited duplication, and (4) ease of storage and retrieval.

Sandia National Lab Image Gallery—*http://mems.sandia.gov/scripts/index.asp*

The ***scanning tunneling microscope (STM)*** is a newer form of electron microscope, invented at IBM in the 1980s by two Swiss scientists, Binnig and Rohrer, that uses a very fine metal tip to scan across a sample at a height of a few atomic spacings. (See Figure 8-23.) At such spacings, a tunneling current occurs when only a few volts are applied between the tip and the sample. This tunneling current is sensitive to the tip-to-sample spacing and is used to measure and control the height of the tip. In this way, scanning micrographs of the surface with a height resolution of about 1 atom and lateral resolution of about 1 to 3 atomic spacings can be produced. Also, because the electron energy distribution of the tunneling current depends on the electronic properties of the sample, it is used to select or view different types of surface atoms. This instrument made it possible to see the new polymorph of carbon known as *fullerenes*. With the STM users don't actually *see* individual atoms. Instead, the STM scans the surface of an atom and a computer converts the data into a three-dimensional image.

Scanning probe microscopy (SPM) involves many related technologies, all of which operate by scanning a fine probe tip over a sample surface (Figure 8-24 shows a schematic of the probe and surface) to generate a high-magnification (10^9 X), three-dimensional image. Piezoelectric ceramics, discussed in Module 27, control the probe motions in the x, y, and z directions with nm resolution (see Figure 8-23c). SPM can be used under air, liquid, or vacuum conditions. What is noteworthy is that the sizes of these pieces of equipment range from palm- to desk-size. Their importance is so momentous because SPMs are the connecting bridge between our world and the world of the atom and the molecule. Scientists who are now crossing this bridge are engaged in developing a new technology (discussed in various sections of this book), which is called by various names: *nanotechnology*, *molecular manufacturing*, or *molecular nanotechnology*. Regardless of the name, this new technological revolution will someday bring us the capability to manufacture materials as nature does by assembling individual atoms (see Figure 8-23b) and/or molecules to construct whatever material is needed. A few examples of available techniques using SPM follow:

Atomic force microscopy (AFM)—a cantilever probe senses electrostatic interactions with the specimen (see Figure 8–23). It can work with samples in their natural state.

Lateral force microscopy (LFM)—frictional response between the cantilever probe and the surface is plotted.

Modulated force microscopy—maps the variation in sample compliance (hardness) due to changes in cantilever modulation.

(a)

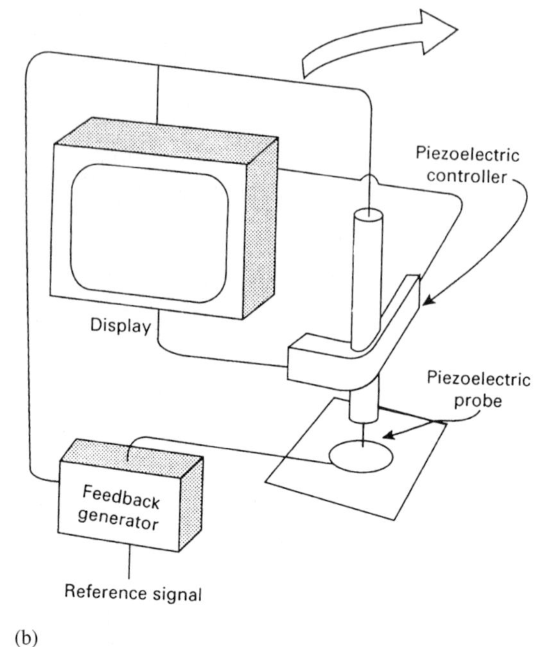

(b)

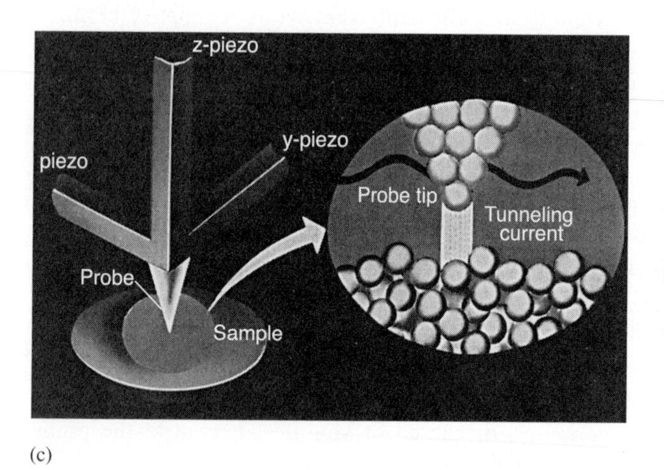

(c)

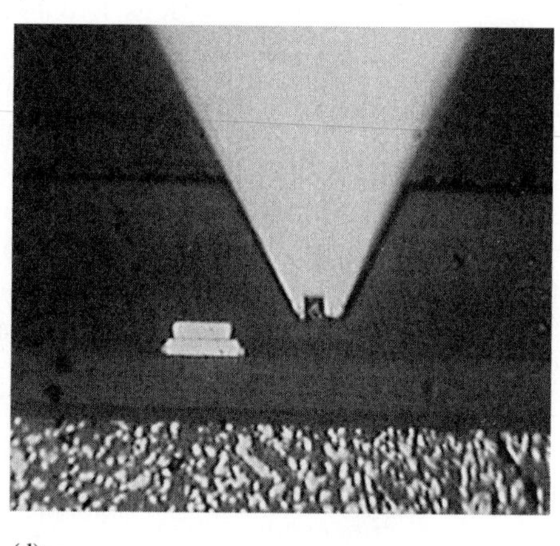

(d)

Figure 8-23 (a) Atomic force microscope has a resolution of 10×10^{-10} and a scan range of 100^{-6} m. (Park Scientific Instruments) (b) Scanning tunneling microscope (STM) schematic diagram. (c) Enlarged diagram of probe. (d) Enlarged photo shows tungsten probe tip 0.2 nanometers wide to probe conductor and semiconductor materials.

Electric force microscopy (EFM)—measures the electrostatic force of attraction or repulsion between the tip and the sample.

Magnetic force microscopy (MFM)—measures the magnetic force of attraction or repulsion between the tip and the sample.

Scanning thermal microscopy (SthM)—uses an electrically resistive probe to map thermal conductivity or temperature variations between the tip and the sample.

Multimode Variable Temperature Scanning Microwave Microscope and ***Scanning Superconducting Quantum Interference Device (SQUID)***, seen in Figures 8-25 and 8-26, are two innovative technologies for mapping compositions and physical properties of interest. These new technologies have lifted the constraints on the scope of materials system and properties that can be addressed using combinatorial materials science. Using such instruments with a substrate as small as 1 cm^2, thousands of different compositions

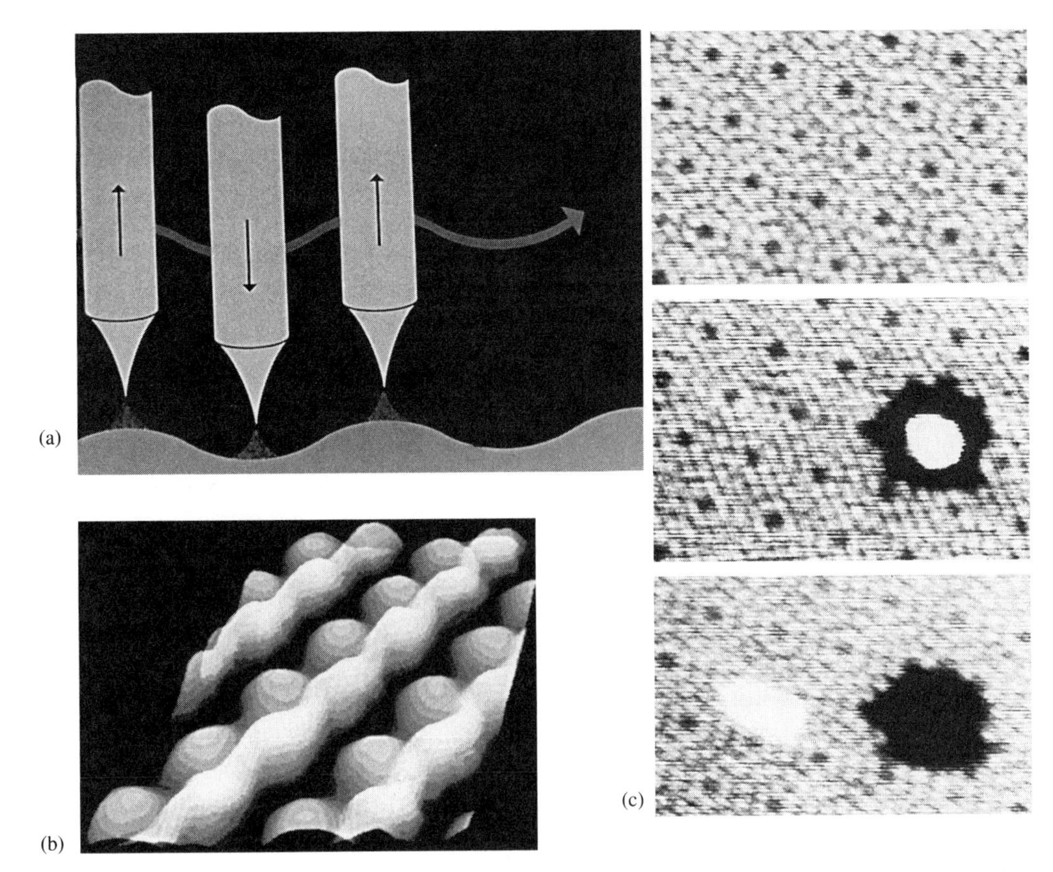

Figure 8-24 (a) Schematic showing how tungsten probe moves along and scans silicon surface. (b) Image from STM of gallium arsenide (GaAs) atoms. (c) Nanoscale composite mixing—top: STM image of silicon surface; middle: a voltage pulse of +3 volts applied between probe tip and surface led to formation of a mound of silicon atoms (large white spot) surrounded by a black matlike area; bottom: a second voltage pulse of +3 volts leads to removal of the mound, then with the probe in a different position, application of pulse −3 volts led to a cluster (white spot) being deposited on the surface, leaving a large empty area (large black spot). (IBM Research)

Figure 8-25 Multimode Variable Temperature Scanning Microwave Microscope. (Ichiro Takeuchi, "Combinatorial Approach to Materials Discoveries" National Educators' Workshop: 2001. NASA/CP - 2002–211735. June 2002, pp. 217–239)

Figure 8-26 Scanning SQUID microscope. (Ichiro Takeuchi, "Combinatorial Approach to Materials Discoveries" National Educators' Workshop: 2001. NASA/CP - 2002–211735. June 2002, pp. 217–239)

can be integrated, synthesized, and screened for desired properties in a single experiment. New techniques are continually being developed. An example is the SPM with an SEM. This technique combines into a powerful instrument the best features of both microscopes in the study and characterization of materials (see Section 8.5). The scanning ion-conductance microscope (SICM) and the laser force microscope (LFM) are helping to advance synthesis of high-temperature superconductors and to improve the understanding of all materials.

In 1951, the ***field-ion microscope*** was introduced. It can provide direct images of individual atoms in a solid surface. It achieves its high magnification from a radial projection of ions from the tip of a sharp needle to a fluorescent screen. The needle, called a *field emitter* or, more commonly, a tip, has a radius of curvature at its end of around 10 nm. Ions are generated above the tip surface in a process known as ***field ionization***. The tip is placed in a vacuum chamber and a low-pressure *imaging gas* is introduced. An electric field is also applied. The imaging gas atoms in the vicinity of the tip are ionized by an electron tunneling from the tip surface to the specimen surface. The positive field ions formed follow the electric field lines away from the tip surface to the fluorescent screen, where they produce image spots that are representations of individual atoms. From the symmetry of the pattern of spots, the Miller indices (Section 3.2.3) of the observed planes of atoms can be determined. The field-ion microscope is similar in design to the ***field-emission microscope*** invented in 1936.

Only a few of the many sophisticated instruments and techniques allow scientists today to measure the features of materials, define their properties, and manipulate their structure at the atomic and molecular levels to increase scientific knowledge, assist in the manufacture of products, and contribute to the process of analyzing applications. As stated earlier regarding synthesis and nanoscale mixing, nanoscale technology is used in processing/syntheses involving probe microscopes and beam processing (MBE, laser,

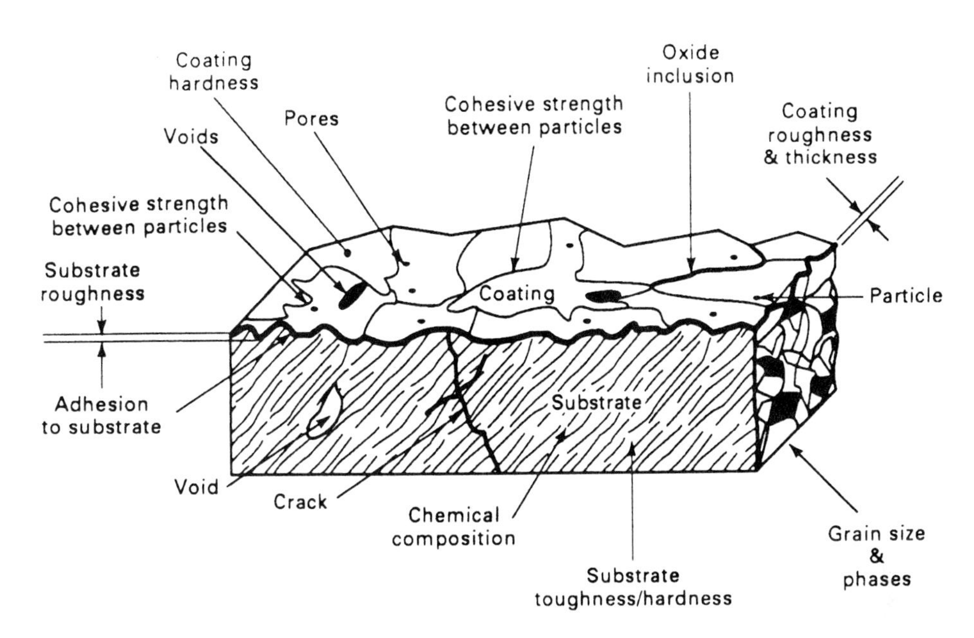

Figure 8-27 Characterization components of a substrate material and coating.

etc.) at scales of 50 nm or less. These techniques are employed on advanced metals, ceramics, and composites.

8.5 MATERIALS CHARACTERIZATION

Just as humans have fingerprints unique to each person, materials have unique characteristics. ***Characterization*** is the process of identification and analysis of unique properties of a material's microstructure, which are revealed as defects, crystal structure, and impurities, as well as macrostructural properties such as surface roughness. Characterization may use some of the same tools and techniques as those used for destructive or nondestructive testing and evaluation (see Module 7), but characterization focuses more on structure and composition.

The analysis can be qualitative or quantitative; it can be used to determine chemical or physical characteristics of the material. ***Qualitative characterization*** has been practiced since early times as crafters examined wood, metal, or stone to determine the suitability of that material for the tool or product they would craft. Figure 8-27 shows some of the characteristics derived from an analysis of a material's substrate and coating. ***Quantitative characterization*** results from the ability to link computers with instruments of analysis to capture large amounts of data with which to make a comparison against a standard. Quantitative characterization is a recent technology that has evolved as a result of the need for improvement over older evaluation techniques that could not provide the precision data required to construct advanced materials. Very hard materials or materials such as ceramic fibers in softer polymer composite matrices are often difficult to prepare for analysis by conventional metallographic techniques. Different means of preparing the same specimen for analysis can yield different results. For characterization techniques to be useful, they must provide true micro/macrostructural information. On the specimen shown in Figure 8-27, polishing with silicon carbide paper or fixed diamond platens seemed to indicate voids in the coating. However, examination of the same specimen after diamond lapping revealed that it was the polishing medium that caused pullout of the coating.

To overcome problems of inaccurate characterization, both improved specimen preparation and analysis may be required. For example, ***digital analysis*** is now possible for images obtained with microscopes and X rays or generated from acoustic (sound) devices. Digital analysis provides data on features such as porosity, grain size, and surface flaws. ***Artificial intelligence (AI)***, the use of computers to solve problems using methods similar to those used by humans, has been employed to make quantitative characterization and statistical analysis of structures that would be too tedious for humans. Advances in digital analysis have come on the heels of methods to enhance optical and other imaging techniques with technology such as machine vision, which feeds special data into image analyzers. Improved methods of spectrometry, such as flame atomic absorption, allow chemical characterization of trace elements in high-performance alloys down to a few parts per million (ppm). These analyses permit the use of chemically characterized alloyed parts to be qualified for use in very critical applications, such as aircraft gas turbines.

The new analysis tools discussed in this module, such as the atomic force microscope, allow characterization of coatings, adhesives, electronics, and biomaterials. As an example, a combinatorial analysis of polymer films might be done to study wetting and dewetting. (The discussion of adhesives in Module 17 covers *wetting*.) ***Dewetting*** results when a liquid is applied to a surface such as adhesive or solder, but the liquid recedes before fully covering the surface.

SELF-ASSESSMENT

8-1. An aircraft capable of changing the shape of its components to improve maneuverability would be

a. Warping **b.** Biomimicking **c.** Morphing **d.** Hawking

8-2. An old saying is "What mankind can dream, technology can achieve." Do you believe that to be true? If so, give one example from the efforts of materials engineers to create new materials.

8-3. The further development of which particular technology (communications, transportation, automobile, materials, medical, energy, or other) would, in your opinion, contribute the most to the enhanced well-being of mankind in the twenty-first century?

8-4. A person who delves into the nature of matter, explores processes that produce engineered materials, and searches for the reasons why things behave as they do is involved in what type of work?

a. Materials research **b.** MEMS **c.** Futurology **d.** Metallurgy

8-5. The design of a product critically depends on the properties of the materials from which it is made. Describe what two properties you would pick to meet certain criteria and explain why you chose them.

8-6. Cite two examples of new processes for fabricating products explained in this module.

8-7. Name one way high-quality models affect product quality.

8-8. List three factors (or parameters) that affect the properties of materials.

8-9. To be successful, a product must do more than meet its technical specifications. Other factors, which could be called external factors, must be taken into consideration in designing a product. Name one of these factors and explain its relationship to the designing of a product.

8-10. What are the methods used to determine unique properties by identifying and analyzing (both quantitatively and qualitatively) features such as voids, roughness, grain size, coating, and hardness?

a. Characterization **b.** Slip **c.** Diffusion **d.** Synthesis

8-11. Cite an example of man's observation of nature and its laws and the effects of that observation on the design of materials for aircraft.

8-12. What is the substance that accelerates a chemical reaction without being consumed itself in the reaction?

a. Reagent **b.** Solvent **c.** Catalyst **d.** Substrate

8-13. The science of measuring the proportions in which chemical elements combine in a chemical reaction is known by what term?

8-14. Discuss the "safety margin" and confidence levels in the design of products.

8-15. Estimate the number of materials available for use by materials designers in the creation of new materials.

8-16. Explain what role, if any, processing plays in the final selection of a material.

8-17. What is one final step a newly discovered material for a particular product application must pass before acceptance by the consumer ?

8-18. Name two constraints involved in the selection of materials.

8-19. What determines to a great extent the properties of a solid material or how a solid material will behave in a given application?

Select the *best* match from the following techniques and instruments for Questions 8–20 to 8–25.

a. Spectroscopy **b.** X-ray diffractometer **c.** Optical microscope

d. Electron microscope **e.** Scanning tunneling microscope **f.** Nanotechnology

8-20. Uses magnetic lenses to focus electron beams that provide over 200,000X magnification.

8-21. Provides magnification of about 2000X and is a relatively low-cost instrument.

8-22. A general term for techniques that utilize the differences in wavelength spectra produced by chemical elements.

8-23. A method to manipulate structures from hundreds to hundredths of a micrometer.

8-24. Used to measure the angles of electrons glancing off material specimens.

8-25. Uses very fine probes that move over the surface of a material to provide images of individual atoms.

8-26. What electron microscope made it possible to see the new polymorph of carbon known as fullerine?

8-27. The new technology that uses laser light, slicing–imaging, and computer programing to obtain internal 3-D views of solid materials is known by what name?

8-28. A characterization technique involving visual clues to a crafter or inspector is

a. Nanoscale **b.** Quantitative **c.** Qualitative **d.** Artificial

8-29. A characterization technique involving instruments and methods such as AI is

a. Nanoscale **b.** Quantitative **c.** Qualitative **d.** Artificial

REFERENCES & RELATED READINGS

ASM International. "Advances in Characterization and Testing of Engineering Materials," *Advanced Materials and Processes*, November 2002.

ASM International. "Characterizing Thermal Spray Coatings," Special Issue, *Advanced Materials and Processes*, April 1992.

Davis, J. R., ed. *ASM Materials Engineering Dictionary*. Materials Park, OH: ASM International, 1993.

Drexler, K. Eric. *Nanosystems: Molecular Machinery, Manufacturing, and Computation*. New York: John Wiley & Sons, Inc., 1992.

Hall, Cecil E. *Introduction to Electron Microscopy*, Melbourne, FL: R. E. Krieger, 1983.

Hall, Dale. "Materials Research at NIST—Century of Progress," *National Educators' Workshop:2001*. NASA/CP-2002-211735. June 2002. pp. 1–26.

INGERSOL RAND, *Compressed Air*, October/November 1994, p. 16.

LOVELAND, R. P. *Photomicrography: A Comprehensive Treatise*, Vols. 1, 2. Melbourne, FL: R. E. Krieger, 1981.

MCDONALD, ALAN F. "Scanning X Microscopy," *National Educators' Workshop, NEW: Update 89, Standard Experiments in Engineering Materials, Science, and Technology*. NASA Langley Research Center, Hampton, VA, October 19, 1989.

TAKEUCHI, ICHIRO. "Combinatorial Approach to Materials Discoveries," *National Educators' Workshop:2001*. NASA/CP-2002-211735, June 2002, pp. 217–239.

WICKRAMASINGHE, H. KUMAR. "Scanned-Probe Microscopes," *Scientific American*, October 1989, pp. 98–105.

ZHAO. JI-CHENG, ET AL. "A Diffusion Multiple Approach for the Accelerated Design of Structural Materials," *MRS Bulletin*, April 2002, p. 325.

Periodicals

Advanced Materials & Processes

Journal of Materials Education

UNIT 4

Metallics

PAUSE & PONDER

On September 11, 2001, terrorists flew two Boeing 767 airliners, at 120 mph and full of jet fuel, into the World Trade Center (WTC) in New York. Why did each majestic tower survive the initial impact of aircraft, but eventually collapse?

As seen in the aerial photo, the WTC design involved structural steel framing to provide an "open plan," with trusses placed in a manner to resist the lateral loading of 13,000 tons of wind pressure. This truss design works much like a soda can. Liquid and internal pressure in a soda can provide rigidity to an otherwise weak aluminum shell. In the same manner, each tower consisted of relatively thin structural steel columns closely spaced together to form a

The majestic World Trade Center stood as a tribute to advances in structural engineering and the unique use of structural steel as well. It was a center for cooperation in world trade until evil motives resulted in its collapse on 9/11/01. The tragedy claimed the lives of about 3000 innocent victims, who represented many cultures from around the world. (Corbis Digital Stock)

hollow exterior tube much like the soda can's thin sides. An inner core tube of structural steel columns provided housing for elevators, stairways, and service spaces. The floor trusses that spanned the opening between the inner core and exterior columns provided large, column-free spaces on concrete flooring for offices, restaurants, and similar purposes. The central core tube below where the aircraft severed them held the towers for 53 minutes after impact to the first tower and 33 minutes after impact to the second tower, thus allowing many to escape prior to collapse.

Due to the redundancy of columns provided in the design, the initial impact of the airliners into the external tubes did not cause their collapse. Redundant columns helped bear the load as it was transferred from the destroyed columns. Some structural engineers involved in the study of the disaster believe that upon impact the explosions blew away insulation that had been sprayed onto the steel floor trusses. Then, the subsequent heat of the fires from the jet fuel weakened, but did not melt, the structural steel. Jet fuel produced a much hotter fire than engineers calculated for a normal office fire. As the trusses distorted, they pulled away steel bolts and welded connectors from the central core tube and other shell tube. As a truss collapsed, with all the weight of concrete flooring, furniture, office equipment, and so on, it "pancaked" onto the next floor, adding the weight of each subsequent floor in a chain reaction. After the first tower was hit, it stood for nearly an hour. But as the structural steel softened from intense heat and the trusses collapsed, it took only 11 seconds for the tower to come down.

Structural steel provides great strength that allows for building majestic sky scrapers like the WTC. However, engineers know it weakens under fire. Figure 18-16 shows how wood, an ancient building material, often withstands fire better than structural steel due to its ablation properties. Had the WTC been "hardened, like the Pentagon which was also hit by terrorists'

One tower rises toward completion in the center of this aerial photo and the other tower rises in the lower right, revealing the open plan design for construction. A tube of closely spaced structural steel columns formed the exterior wall, and an inner core tube of structural steel columns provided for elevators, stairs and service equipment. (Getty Images Inc.—Hulton Archive Photos)

airliners on 9-11," which included superior fire insulation, is it possible it would not have collapsed? Even if Pentagon-type hardening might save buildings from such attacks, the prohibitive cost makes it impractical.

Following the initial study of the WTC collapse by a team of civil engineers, the Materials Science and Engineering Laboratory at NIST made further failure analyses of components. Among many aspects of the failed parts, this study looked at grain structure in an effort to determine the amount of heat and its effects on the integrity of the components.

This unit on metallics will introduce you to structure, processing, properties, and application of metals and alloys. In Module 7 you gained insights into failure analysis; this unit will provide you with concepts of metallics for a better understanding of how metals gain their properties as a result of processing and structure as well as how they might fail in events such as the WTC collapse.

World Trade Center New York Pictorial History including construction of the WTC—*http://iserve.wtca.org/History/*

Module

9

Nature & Thermal Processing of Metals

After studying this module, you should be able to do the following:

9–1. Describe basic alloying, forming, casting, and thermal processing used by industry to change metal microstructures and thus vary their properties.

9–2. Analyze phase diagrams and photomicrographs to determine thermal processes, basic structures, and properties of ferrous and nonferrous metals.

9–3. Explain the effect on grain size, and metal properties (tensile strength, hardness, and toughness) as a result of strain hardening and various types of thermal processing.

9–4. Describe how degrees of both hardness and toughness are achieved through thermal treatments.

9–5. Use tables in the text, other handbooks, and periodicals to compare alloy content, cost, properties, processability, and applications in order to select metallic materials.

9–6. Use the website addresses denoted by the icon* to seek out current developments and emerging technology related to metallic materials.

*The icon found throughout this book will link you to Internet sites related to topics being covered. The dynamic nature of the Web brings frequent changes, so some of these sites, while available at the time of writing, may not be up now.

The auto and light-truck industry is a major economic player in developed countries. Because these vehicles affect society in so many ways, ranging from ease and comfort of travel to environmental pollution, they continually undergo research and development for improvements. Have you read anything about FreedomCAR?

FreedomCAR isn't an automobile, it's a new approach to powering the cars of the future. The acronym CAR stands for Cooperative Automotive Research. This research program aims to develop fuel technologies such as fuel cells and hydrogen power from domestic renewable sources and improve upon the Partnership for a New Generation of Vehicles (PNGV) program. The transition of vehicles from gasoline to hydrogen is viewed as critical both to reducing carbon dioxide emissions and to reducing U.S. reliance on foreign oil. FreedomCAR will focus on technologies to enable mass production of affordable hydrogen-powered fuel cell vehicles and the hydrogen-supply infrastructure to support them. FreedomCAR also will continue support for petroleum-dependent technologies that have the potential to dramatically reduce oil consumption and environmental impacts. The DaimlerChrysler Natrium minivan uses a sodium borohydride fuel system as seen in Figure 9-1. This version of the Chrysler Town & Country

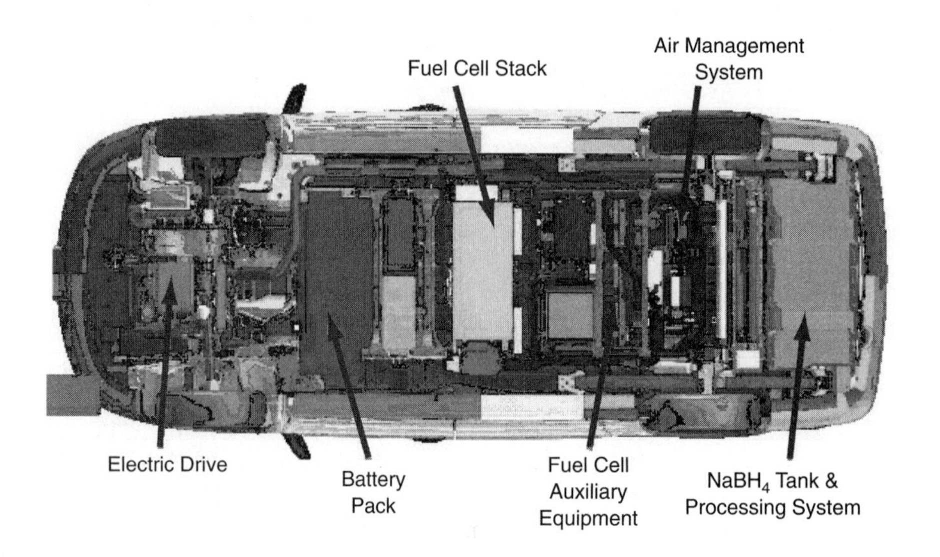

Figure 9-1 Chrysler Town & Country Natrium minivan uses a sodium borohydride fuel system. (DaimlerChrysler)

Natrium is the first fuel-cell concept vehicle to run on clean, nonflammable, and recyclable sodium borohydride fuel, a fuel made from borax, which is a mineral available in abundant supply in the Western United States. In the Natrium minivan, this technology delivers the environmental benefits of a fuel-cell vehicle without the loss of cargo or passenger space, while providing a range of 300 miles, longer than any other fuel-cell vehicle. Hydrogen is extracted from sodium borohydride to power the fuel cell. Sodium borohydride is a compound chemically related to borax, the naturally-occurring substance commonly used in laundry soap.

The unique fuel cell technology could have great benefits for the military: In particular, it is nonflammable, which would greatly improve safety in battle zones, and the main ingredient can be transported as a dry powder, which dramatically reduces the enormous logistical demands of fueling our military in advanced battle settings. Other advantages of the sodium-borohydride fuel-cell powered Chrysler Town & Country Natrium include:

- Zero dependency on oil for propulsion
- Cargo and passenger space is not compromised for on-board storage of hydrogen
- Byproduct can be rehydrogenated and used again as fuel
- Near-silent operation
- Capable of producing 110- and 240-volt electricity
- Greater driving range than other fuel-cell vehicles
- Potential for zero emissions of smog-forming and greenhouse gases

The Hy-wire, a General Motors concept vehicle, gives us a glimpse of the type of vehicles that may emerge from FreedomCAR. GM Hy-wire combines a hydrogen fuel cell with by-wire technology.

PNGV created a unique partnership for a new generation of vehicles. This industry/government consortium, composed of the U.S. auto industry and the federal government, aims to improve fuel efficiency to 80 mpg (miles per gallon). Researchers for PNGV sought to simultaneously reduce vehicle weight by as much as 40%, increase engine efficiency by 40–55%, implement regenerative braking, and increase energy storage by 90%. These goals came from the United States Council for Automotive Research (USCAR). The group promoted a number of precompetitive, cooperative research and development programs at DaimlerChrysler, Ford, and General Motors. Many of the research programs organized under the USCAR umbrella aimed at environmental improvements, fuel economy, and safety for occupants. "Super Car," another name for USCAR's concept vehicles, focused on fuel efficiency and other aims.

Working as a subgroup to USCAR, the United States Automotive Materials Partnership (USAMP) focused on body materials, which encompassed the car's frame and outer skin. The cars in which you ride represent more than a century of evolving automotive technology. Nearly all high-production cars use frames with unibody, stamped-steel construction and skins made mostly of sheet steel. The steel industry aims to keep its dominant role in autobody materials as well as other parts for cars; however, steel faces several challenges from plastics, plastic composites, and aluminum. DaimlerChrysler's Dodge ESX3 PNGV concept car (Figure 9-2a) uses "mybrid," or mild hybrid electric power trains. This technology combines an all-aluminum 1.5-liter diesel engine, electric motor, and lithium-ion battery to achieve a fuel economy of 72 miles per gallon gasoline equivalent.

DaimlerChrysler's injection-molded thermoplastic body technology achieves significant improvements in weight and cost; ESX3 weighs only 2,250 pounds. Its body system is estimated to weigh 46% less and cost 15% less to manufacture than a comparable metal body, while meeting all required safety standards.

If DaimlerChrysler built the ESX3 today, it would cost about $7,500 more than the comparably sized Dodge Intrepid family sedan. However, that's far more than buyers would be willing to pay for high fuel economy according to the DaimlerChrysler engineers who built the car. Their focus is how to make this evolved breed of car affordable.

Consumers desire a mass-produced moon-shot technology but at down-to-earth prices. The challenge: triple the fuel efficiency, cut emissions ever further, yet keep the styling and

(a)

(b)

Figure 9-2 (a) The Dodge ESX3 PNGV concept car aims to apply new materials, processes, and power train technology to satisfy customers who desire mass-produced moon-shot technology but at down-to-earth prices. (DaimlerChrysler) (b) Synergy 2010 concept car is a PNGV approach to a family sedan designed to be capable of 80 mpg with comfort and convenience and comparable cost to today's sedans. (Ford Motor Co.)

performance that our customers love—and make the whole package affordable. The ESX3 should provide some answers to these challenges and converge on reality. Figure 9-2b shows Ford's Synergy 2010 PNGV concept vehicle using an aluminum unibody aimed at one-third weight reduction and 40% less aerodynamic drag. The ultralight and synergy aluminum unibodies are but two of the numerous approaches being studied by USAMP members for reducing the weight in cars.

How does a designer decide on body and frame materials? On the one hand, aluminum can be processed in many ways like steel and it is about 45% as dense as steel. But sheet aluminum costs about $1.50 per pound compared with $0.30 for sheet steel. Also of great significance is the fact that aluminum is about one-third as stiff. These factors hint at the challenges faced by materials competing with steel. Materials substitutions are complex issues that embrace concerns far beyond simple mass reduction. The aerospace technology presently used for many of the candidate automotive materials has proven to be feasible at low production rates for airplanes and spacecraft. But it must be modified for the much higher production rates required for cars and light trucks. This also means the techniques developed for aerospace manufacturing must be modified to permit complex molding or machining, joining of dissimilar materials, and improvements in assembly methods—all at competitive costs in line with consumer demand. You will learn more about some of these approaches in Unit 8 on composite materials.

As you read through this module for insights into metal structures and thermal processing and its references to properties and applications, keep this example of metals for transportation in mind. The concepts explored will prepare you to deal effectively with issues related to metals for transportation and many other applications. Enjoy the ride!

FreedomCAR—*http://www.cartech.doe.gov/freedomcar/freedomcar-remarks.html*
United States Council for Automotive Research (USCAR)—*http://www.uscar.org*
USCAR—*http://www.uscar.org*
AISI Autosteel—*http://www.autosteel.org/cars/default.htm*
Aluminum Association Automotive—*http://www.autoaluminum.org*

9.1 NATURE OF METALS

Metals have been useful to humanity through the ages because they are "strong" when subjected to the external forces encountered under service conditions, yet they become "soft" enough to yield to a machine cutting tool or to a compressive shaping force. Above a certain temperature, they melt and become liquids capable of being shaped by casting. Only in recent times have we realized that the properties of all types of solid materials, including metals, arise from their atomic architecture, that is, from the manner in which their atoms arrange themselves into a crystalline order, from the number and types of imperfections found in this structure, and from the bonding forces that keep the collection or structure of atoms bound or joined together. Most metals occur in nature as compounds. These compounds must be reduced to free the metal (see Module 5 on oxidation and reduction for a further explanation).

The "softness" quality of metals can be explained by the atomic structure and metallic bonds of the metal atoms that form a crystalline structure. We recall that the electrons in the metallic bond are free to move about their positive ions in an electron cloud or gas, which acts as a glue to bond the ions together. This free movement, within limits, also allows for the movement of the atoms under the influence of an external load. This slight movement, visible only under the most powerful microscopes, is called ***elastic deformation*** or ***elastic strain***. Once the external force (load), such as a bending force, is removed, the internal electrical forces that cause the atoms to move will decrease, allowing the atoms to return to their normal position; they leave no sign of ever having been moved. If you bend a piece of spring steel, such as a machinist's rule or vegetable knife, it will return to its original shape, thus demonstrating elastic deformation. If you were not careful and applied too much external force by excessively bending the rule or knife, the atoms might move too far from their original positions to be able to move back again when you released the external force. Consequently, the rule or knife would be permanently bent and no longer fit for use. This permanent deformation is known as ***plastic flow***, ***plastic slip***, ***plastic deformation***, or ***permanent set***. When automakers stamp out a metal car body from low-carbon steel in a huge die press, they use this softness quality of metals. The term *cold working* (defined later) is applied to this stamping operation and many other metalworking processes that produce plastic deformation in a metal. Cold-working operations include rolling, heading, spinning, peening, bending, pressing, extruding, drawing, and others. The steel framing of the skyscraper seen in Figure 9-3 consists of structural steel produced by hot working; steel and other metals, shaped by cold working, serve in the construction and furnishing of the typical office building. Compare the use of the structural members seen here in the Steel Triangle to those shown in the World Trade Center at the beginning of Unit 4.

The microstructure of metals can be modified in a number of ways. By now we know that this last statement can be interpreted to mean that through advances in metals technology, we can affect the atomic structure of metals in a precise, controlled manner to design metal alloys with the desired properties. Throughout this module, much of the text and many of the illustrations refer to the primary metal alloy, steel. However, much, if not all, of the text applies to most metal alloys used in industry. In this module we classify, for learning purposes, the basic methods of changing a metal's properties into the following categories: (1) work or strain hardening,

Figure 9-3 The Steel Triangle, the United States Steel Corporate Center in Pittsburgh. High-strength, low-alloy steel (HSLA), Cor-Ten, forms a tight adherent, dark russet, ferrous-oxide surface coating and requires no other protective coating. The partially completed building exposes steels such as USS "T-1,0," which is nearly three times the strength of carbon steel and bears great loads as interior columns, and USS Ex-Ten, an HSLA steel that serves as floor beams and some core columns. Much carbon steel finds use in areas not requiring maintenance and high strength. (United States Steel Corporation)

(2) thermal processing under equilibrium conditions (solid solution hardening), (3) thermal processing under near-equilibrium conditions (annealing and grain refinement), and (4) thermal processing under nonequilibrium conditions. Of the four categories, only the first does not involve primarily thermal processing. Prior to discussing thermal processing techniques, we need to understand the concepts involved in the internal structure of solids, as presented in Module 2, and have a working knowledge of phase and phase diagrams, covered in the present module.

Work or ***strain hardening*** is a way to change a metal alloy's structure in order to alter its properties by performing work (***cold working***) on the metal itself. Work is a form of energy. If we can find a way to deliver energy to the metal atoms, we can give them the energy necessary to increase their movement and thus their ability to diffuse through the metal structure. Figure 9-4 shows four industrial cold-working methods; the photomicrographs demonstrate grain structure before and after work hardening.

In our opening remarks about the usefulness of metals, we mentioned that metals have the ability to yield to a mechanical force that can form them into desired shapes. An external, mechanical shaping force is a force that causes a metal to exceed its elastic deformation limit and deform plastically without fracturing. This force and the deformation produced are the means for transferring sufficient energy to the atoms to allow them to flow or move plastically. This movement of atoms, one row at a time along planes of close-packed atoms, shifts the positions of the atoms in relation to each other. From our study of crystal structures, we know that the spacing between atoms not only is critical but varies with the particular crystal structure. Fcc structures have the closest-packed atoms and the greatest number of closest-packed planes of atoms, which take less energy to allow for the slip of the atoms. As more and more slip takes place, more dislocations are produced. The greater the slip, the greater the dislocations and the

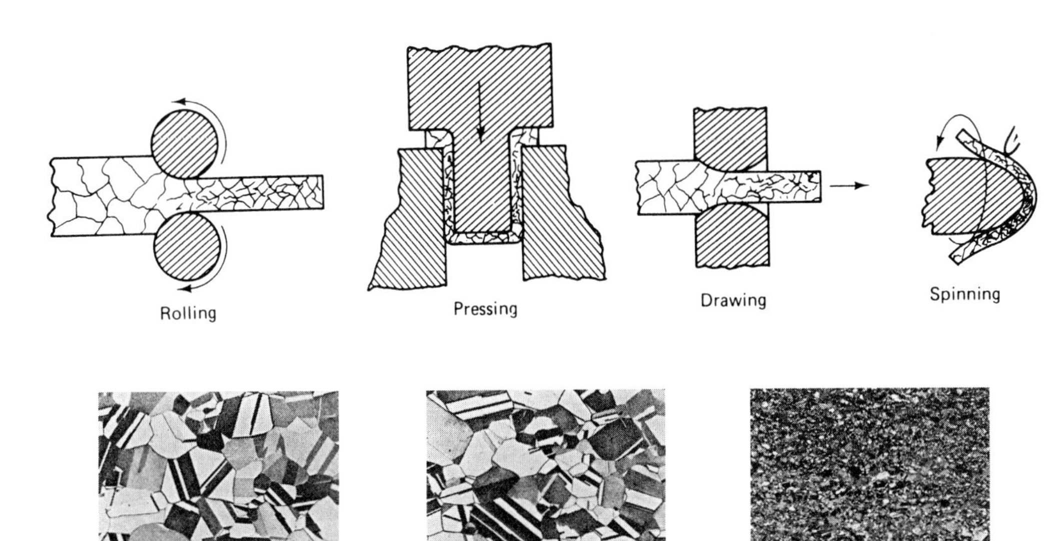

Figure 9-4 Some work-hardening (cold-working) processes.

more distortion of the lattice structure. The end result is that the deformed metal is stronger than the original undeformed metal and offers greater resistance to further deformation.

Figure 9-4 also shows that the processing of materials, particularly the operations of rolling and drawing, can affect the orientation of metal grains (crystals). This is another example of the fact that the structure determines (in the main) the properties of a material. In this example, due to the preferred orientation of the grains, the metals will have different properties in the direction of the orientation than in other directions. This is known as ***anisotropy*** (see Section 9.3.4 for further discussion). Metal grains that are random in their orientation produce ***isotropy***, the same properties in all directions.

In summary, through cold working we have (1) reduced the metal's ductility, (2) reduced the susceptibility of the metal atoms to slip, (3) reduced dislocation movements in the structure, (4) created distortion of the lattice structure, and (5) ended up with a stronger metal that requires a greater force or greater amount of energy to deform it further. At the same time, changes in other properties, such as electrical conductivity, have also occurred. This condition of requiring a noticeable increase in energy for further deformation or an increase in the metal's yield strength is known as *work* or *strain hardening*.

9.2 THERMAL PROCESSING OF STEEL: EQUILIBRIUM CONDITIONS

By controlling the thermal energy, that is, by controlling the supply or removal of heat, we can make metal atoms move. The art and science of controlling thermal energy for the purpose of altering the properties of metals and metal alloys is known as ***thermal processing*** or ***heat treating***.

As with most subjects in materials science, thermal processing concerns a metal's atomic and crystal structure. So far in our study of materials technology, we have accomplished many learning objectives dealing with the "inner workings and hidden mechanisms" in the world of metal atoms. *Crystal structures*, *bonding*, *diffusion*, *crystal imperfections*, *dislocations*, *allotropy*, *solid solutions*, *phases*, *multiphase solid mixtures*, *solute*, *solvent*, and *saturated solid solutions* are but some of the many terms that now have meaning for us. When we studied phases and solid solutions, we used cooling curves for pure metals and metal alloys. These curves are graphic pictures showing the relationships between the variables of time and temperature. Many of these single cooling curves are plotted to give a graphical picture of the possible phases a metal alloy could be in at any particular temperature and at any particular composition. Such phase diagrams are also known as equilibrium diagrams. ***Equilibrium*** means that the time variable is not controlled but is allowed to run its course. In other words, equilibrium implies that phase changes shown on equilibrium diagrams are produced under conditions of slow cooling with no restraints on time. This allows the atoms that are diffusing through a solid material the time to seek and find the equilibrium position of lowest energy level. Phase diagrams show us the following:

1. The phases that are present for a particular alloy composition and temperature
2. The extent of solubility between two metals
3. The maximum solubilities of each metal in the other
4. The alloy composition with the lowest melting point (eutectic)
5. The melting points of the pure metals making up the alloy

But phase diagrams do not show what happens to a solid system of materials in equilibrium when a change in temperature, pressure, composition, or any combination of these variables occurs. We know that the system will seek another state of equilibrium with its lowered level of energy (free energy). We also know that this process takes time. In many cases, the process of arriving at this new state of equilibrium is extremely slow, particularly at low or room temperatures. During this time the system is said to be in a nonequilibrium condition or metastable state. The word *metastable* describes a material changing readily to either a more stable or less stable condition. This metastable state may persist indefinitely. Some steel and nonferrous alloys are in metastable states brought on by heat treatments whose goal is to increase their strength or other desired property. The heat treatments change the microstructures of the alloys not only in a metastable state during the process but also in the final or complete solid material.

To learn how thermal processing can produce phase changes, we use iron and its alloy steel as examples of ferrous metals. We will not discuss other industrial materials that are heat treated, such as glass, but leave this important area of study for independent research. Being an alloy of iron and carbon, steel is also allotropic (existing in several forms). Referring to the iron–carbon phase diagrams (Figures 9-5b and c), which are graphical records of the various phase transformations of steel, and using the eutectoid composition, we see that above 727°C, steel is in a solid-state phase with a crystal structure called *austenite*, or *gamma* (γ) *iron*, a single-phase fcc solid solution. Figures 9-5a and d are simpler versions of the complete diagram shown in Figures 9-5b and c. Austenite is characterized by its ability not only to be deformed but also to absorb carbon. The fcc unit cell of iron atoms contains interstices that are large enough for the small carbon atoms to occupy, producing an interstitial solid solution. Many hot-working operations in industry take place with steel heated to a temperature that produces this austenite phase. Furthermore, the austenite region is the starting point for many of the thermal processes about to be discussed.

The iron–carbon phase diagram is the basis for the heat treatment of steels. It is also known as the "heat treater's road map," which forms the basis for understanding the heat treatment (thermal processing) of steels. Two diagrams are actually shown in this figure, the stable iron–graphite diagram (darker lines) and the metastable $Fe–Fe_3$ (diagram). The Fe–C diagram shows which phases are to be expected at equilibrium (or metastable equilibrium) for different combinations of carbon concentration and temperature. The distinction between the two diagrams is negligible above 2100°C. This equilibrium or stable condition takes a very long time to develop, especially in the temperature and low-carbon ranges that are of interest. Accordingly, the iron–cementite phase diagram is a plot of the unstable or metastable conditions over which most heat treating and the formation of most metallurgical phases take place. Both phase diagrams approximate each other, with only minor differences in temperatures and carbon concentrations, resulting in boundary lines running adjacent to one another. Note that at the highest

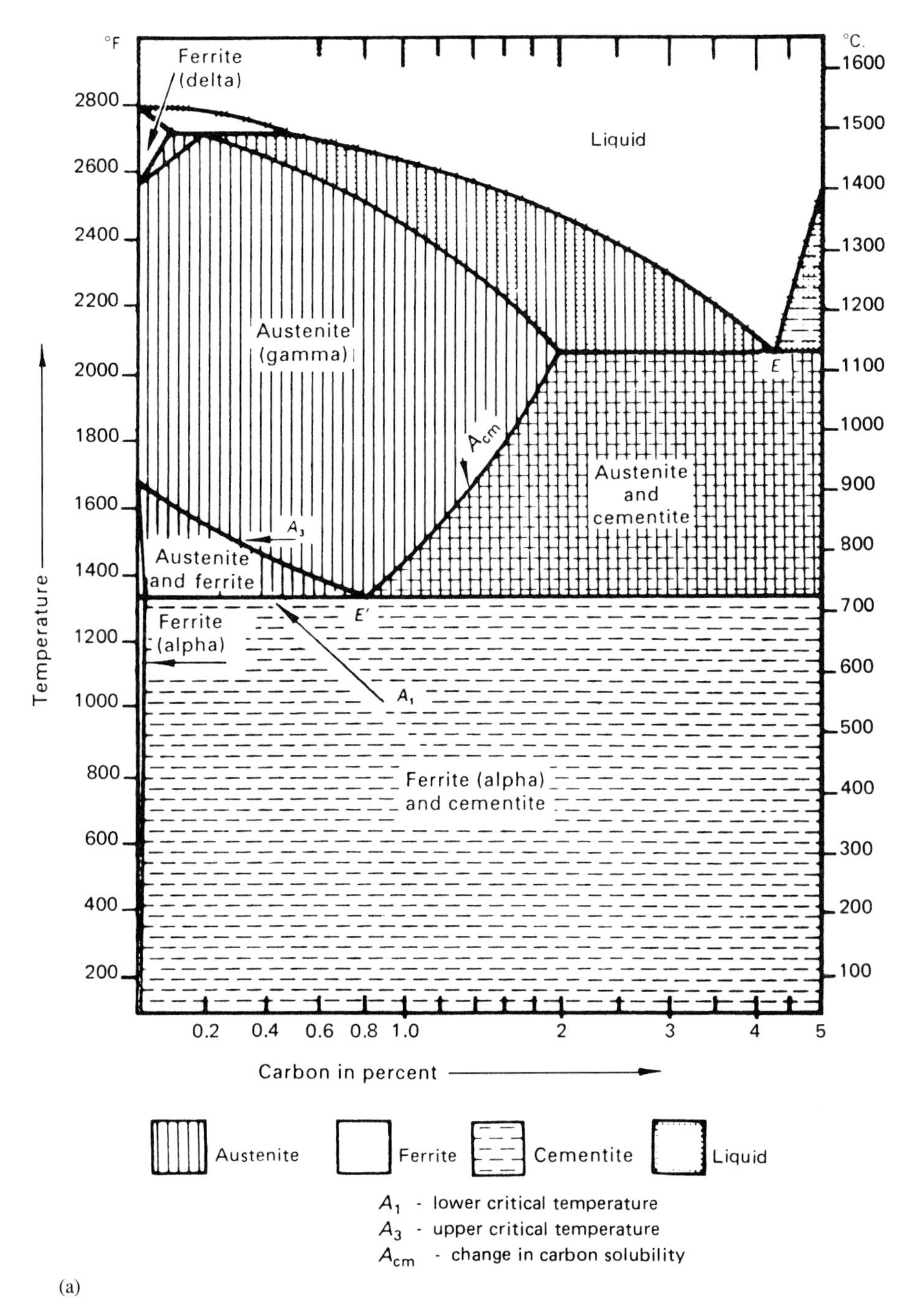

Figure 9-5 (a) Simplified iron–carbon equilibrium diagram. (United States Steel Corporation)

temperatures the liquid phase field can be found, but in the heat treatment of steels the liquid phase is always avoided. Cementite (Fe_3C) is a metastable or unstable compound and under some circumstances it can decompose to form ferrite (alpha ferrite) and graphite. Graphite is a pure form of carbon, that is, it is 100% carbon, which is expressed as 100 wt % C. This latter notation is referred to as weight percentage. Weight percent as used in this text is the weight (mass) of a particular element relative to the total alloy weight. For example, most cast irons (alloys) contain between 3.0 and 4.5 wt % C in addition to other elements. Note in the figure that the arrow for cast irons stops at about 3% carbon.

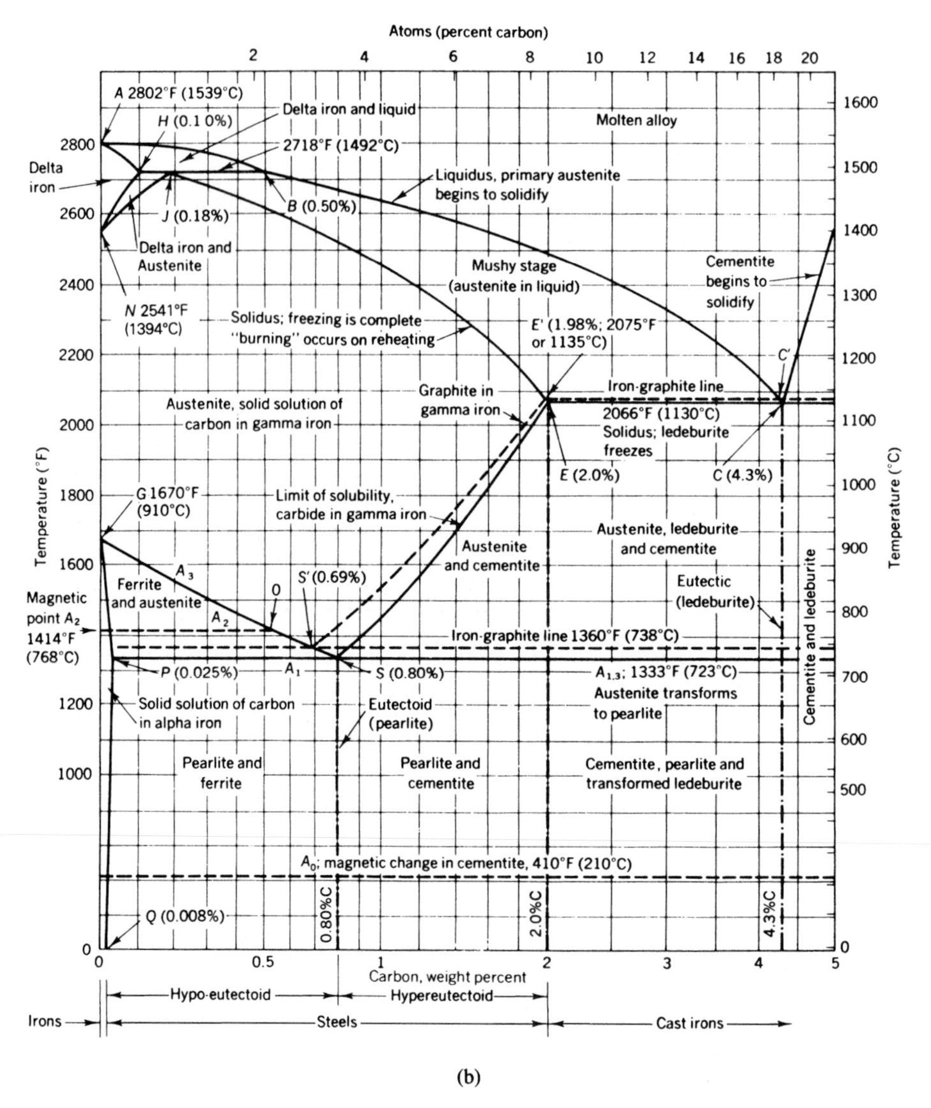

(b)

Figure 9-5 (b) Iron–carbon equilibrium diagram. Dashed lines show true equilibrium of iron and graphite. Solid lines show a metastable phase diagram of iron and iron carbide (Fe_3C), also called cementite. The metastable diagram is used in the same manner as a true equilibrium diagram. The distinction between the two is negligible above 2100°C. (ASM International)

At temperatures below 910°C, pure iron changes to a stable phase called alpha ferrite, alpha (α) iron, or ferrite. This ferrite phase can accommodate up to a maximum of 0.02% carbon (by weight), which produces a solid solution with a bcc structure. The bcc structure contains interstices in its unit cell that are too small to accommodate the carbon atom. Another component in Figure 9-5c is cementite, an interstitial compound with the chemical formula Fe_3C. Cementite is orthorhombic, with twelve iron atoms and four carbon atoms in the interstices among the larger iron atoms in the unit cell. The ratio of iron atoms to carbon atoms is 3 to 1 and it does not change. (Refer to Sections 3.2.10 and 10.1.2 and Table 3-3: Note: cementite is also discussed briefly in Section 9.4). Also called iron carbide, this brittle substance contains 6.69% carbon and has an HB of 700. When steel with the eutectoid composition forms at 727°C, it produces a lamellar two-phase mixture of ferrite and cementite called pearlite (see Figure 9-6). Using tie-lines across the two-phase regions of Figure 9-6, a hy-

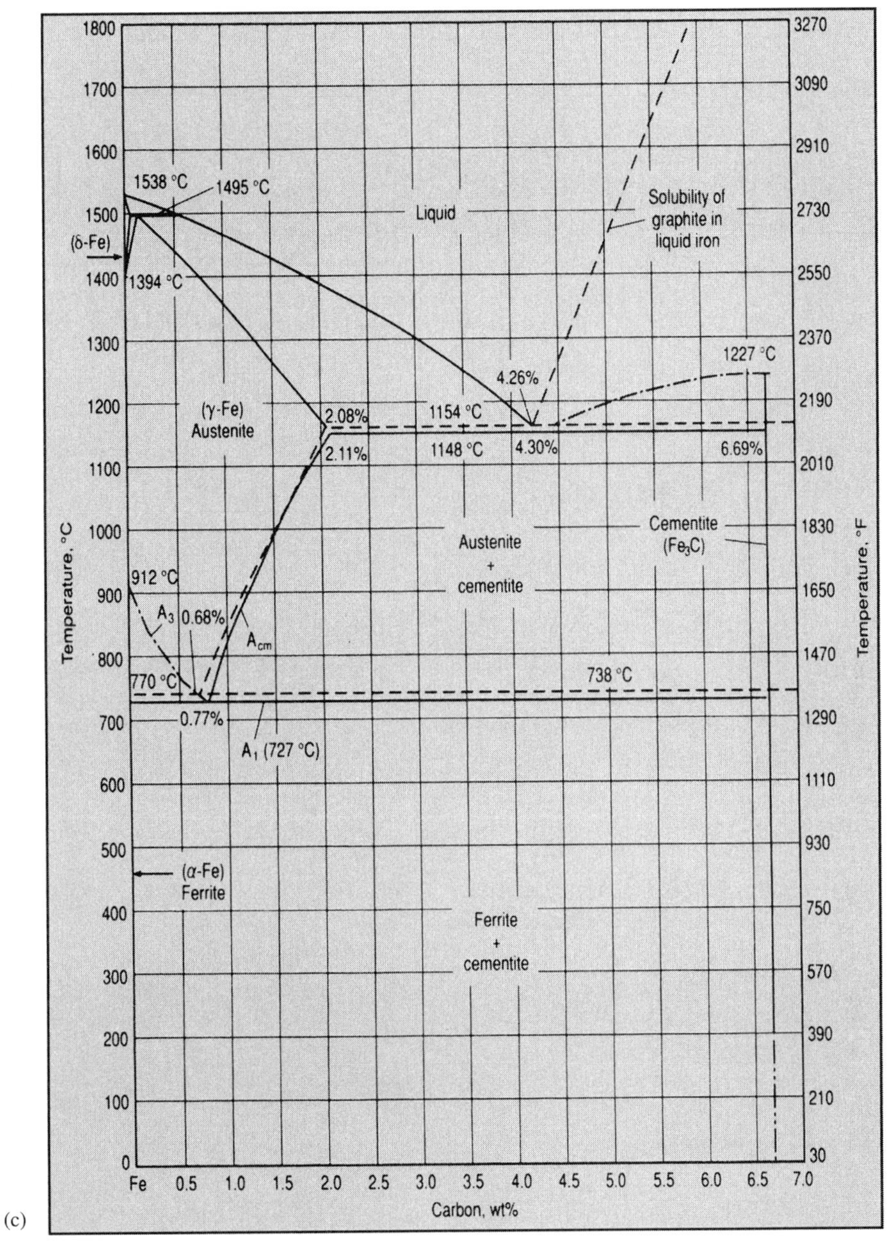

Figure 9-5 (c) This Fe–C equilibrium phase diagram graphs C up to 6.67 wt %. Solid lines indicates the Fe–Fe_3C diagram, and dashed lines, the iron–graphite diagram. (*Heat Treating Progress*, April/May, 2002, p. 45)

poeutectoid steel can be described as a mixture of ferrite and pearlite (Figure 9-7), eutectoid steel as having a pearlitic structure, and hypereutectoid steel as a mixture of pearlite and cementite. Such structures can be discerned under a microscope. At this point it should be evident that there are countless alloys of carbon and iron (carbon steels) with varying amounts of carbon, producing a corresponding variety of steels with different properties to serve the demands of industry. Later in this book other elements will be discussed that also play a part in producing numerous specialty steels (alloy steels) that find ever-increasing use.

It is most important to understand that steel is possible only because iron is allotropic and carbon atoms are small enough to fit between the iron atoms in an fcc austenite structure (interstitial solid solution). The many different properties of carbon steel are produced as a

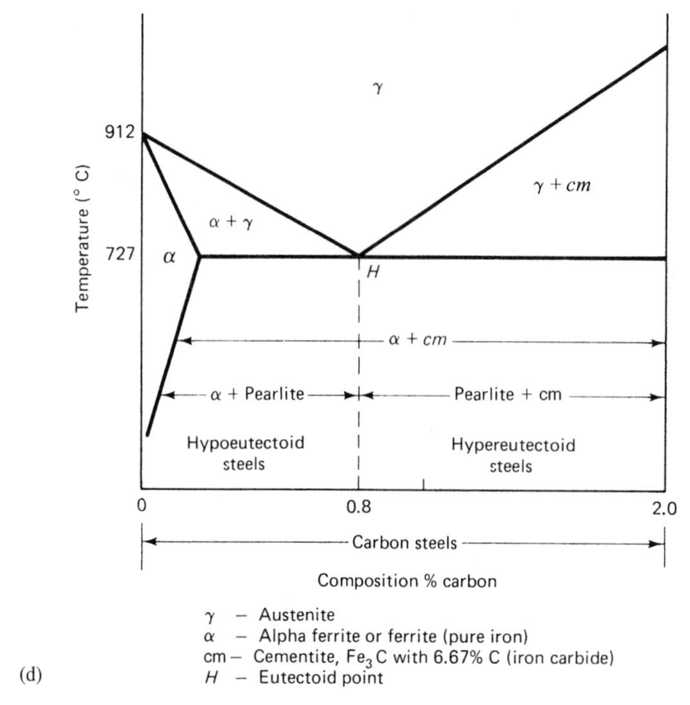

Figure 9-5 (d) Simplified steel equilibrium diagram.

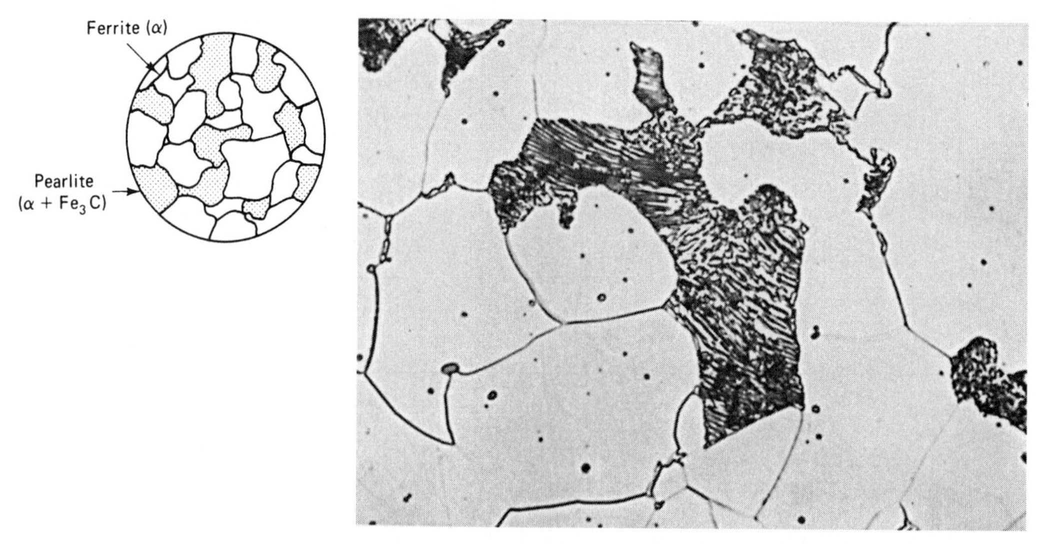

Figure 9-6 Pearlite. Lamellar two-phase mixture of ferrite and cementite. The white areas in this hypoeutectoid steel are grains of proeutectoid alpha ferrite. The dark areas are pearlite colonies consisting of alternate lamellae of alpha ferrite and Fe_3C (cementite). The resolution of the two phases in the pearlite region depends on magnification and orientation and varies from area to area. (Buehler Ltd.)

result of changes in the amount of carbon and the difference in the abilities of ferrite and austenite to dissolve carbon. Austenite can dissolve almost 100 times more carbon than can ferrite. But austenite does not exist at room temperature under equilibrium conditions, whereas ferrite does. It is worthwhile to note on the iron–iron carbide phase diagrams (Figure 9-5a–d) the locations of wrought iron (almost pure iron), steels, and cast irons (greater than 2% carbon). Most important of all is the realization, first, that everything we have discussed about the allotropic forms of iron, the diffusion of carbon atoms, phase trans-

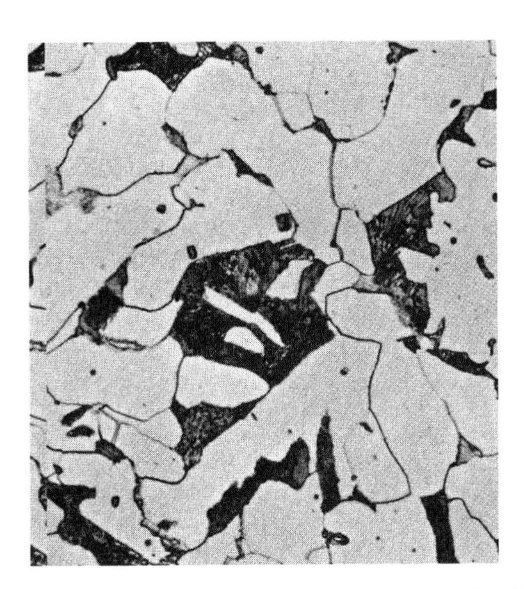

Figure 9-7 Same alloy as Figure 9-6 (1018 Steel [0.18% C, 0.79% Mn, 0.22% P, 0.04% S] at 500X). However, this hypoeutectoid steel specimen was air cooled rather than furnace cooled, yielding a considerably finer structure. Dark areas of alpha ferrite/cementite (Fe_3C) lamella are shorter and closer together. The large white areas are proeutectoid alpha ferrite grains. (Buehler Ltd.)

formations, austenite, ferrite, cementite, and pearlite takes place in the solid state. Not once did we get even close to the melting point of iron, 1538°C. Second, all phase transformations took place under equilibrium conditions.

9.3 THERMAL PROCESSING OF STEEL: NEAR-EQUILIBRIUM CONDITIONS

Several thermal processing techniques used with ferrous metals approach equilibrium conditions, as represented by the iron–iron carbide equilibrium, or phase diagram. By definition, such processes bring about a change in a material's properties. Therefore, they must affect the material's atomic structure. The first thermal process to be discussed is generally known as ***annealing***. Annealing processes bring about changes mostly by producing phase transformations that result in a rearranged, stable atomic structure with less-distorted grains. Therefore, the initial step in most annealing techniques is to heat the steel above its critical temperature to form austenite.

Annealing is broken down into a number of related operations whose overall purpose is to reduce hardness, refine grain structure, restore ductility, remove internal stresses left over from some industrial forming process, or to improve machinability or some other property. These operations—full anneal, normalizing, stress relief, process anneal, and spheroidizing—are shown on a simplified equilibrium diagram for steel (see Figure 9-8). A full anneal consists of heating steel to about 30°C above the line marked A_1 (the eutectoid temperature line), depending on the carbon composition; holding it at that temperature to obtain a homogeneous structure of austenite; and slowly cooling it in a furnace. The result is, for a hypoeutectoid steel, a transformation of austenite to a coarse lamellar pearlite that is soft, stress-free, and a fine ferrite.

The word *annealing* used alone when referring to ferrous alloys implies full annealing. When referring to nonferrous alloys, the word is used to describe a heat treatment designed to soften a cold-worked structure by recrystallization and/or subsequent grain growth. Annealing of age-hardened alloys implies softening of the alloy by a nearly complete precipitation of the second phase. If the annealing is intended for the sole purpose of reducing stresses, it should be designated as *stress relieving*. Forged parts are annealed to refine the microstructure produced by

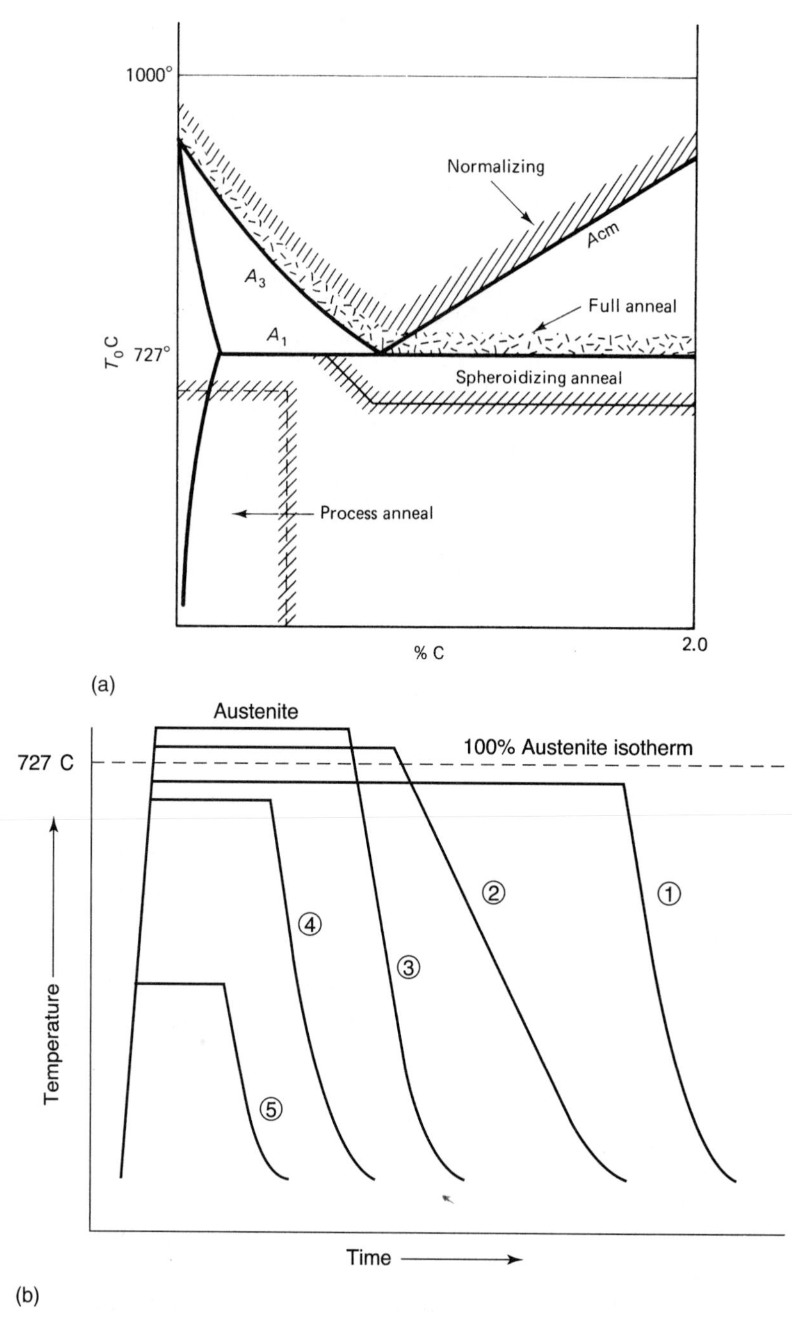

Figure 9-8 (a) Simplified steel equilibrium diagram showing various types of anneals. (b) Schematic of heat-treat annealing processes for steel: (1) spheroidize, (2) full anneal, (3) normalize, (4) process anneal, and (5) stress relief.

forging, which is very hard, so that machining operations such as trimming can be performed. Annealing of gray iron, especially small castings that require a large amount of machining, is noted with an example of ASTM Class 35 gray iron, which ends up as a Class 20 iron due to the reduction in strength as a consequence of the treatment. An example of the annealing treatment taken from the ASM *Handbook* of a heat-resistant casting made of Alloy Castings Institute (ACI) alloy HA or a ASTM A217 iron–chromium heat-treatable alloy used in the oil industry, is cited. These alloys are heated to an annealing temperature of 1625°F, followed by slow cooling in the furnace at about 50°F per hour to below 1300°F. For improved strength these castings are normalized by heating to 1825°F and air cooled to below 1250°F. ***Galvannealing*** refers to galvanized parts that are given a subsequent heat treatment to convert the zinc coating to an oxide to promote paint ad-

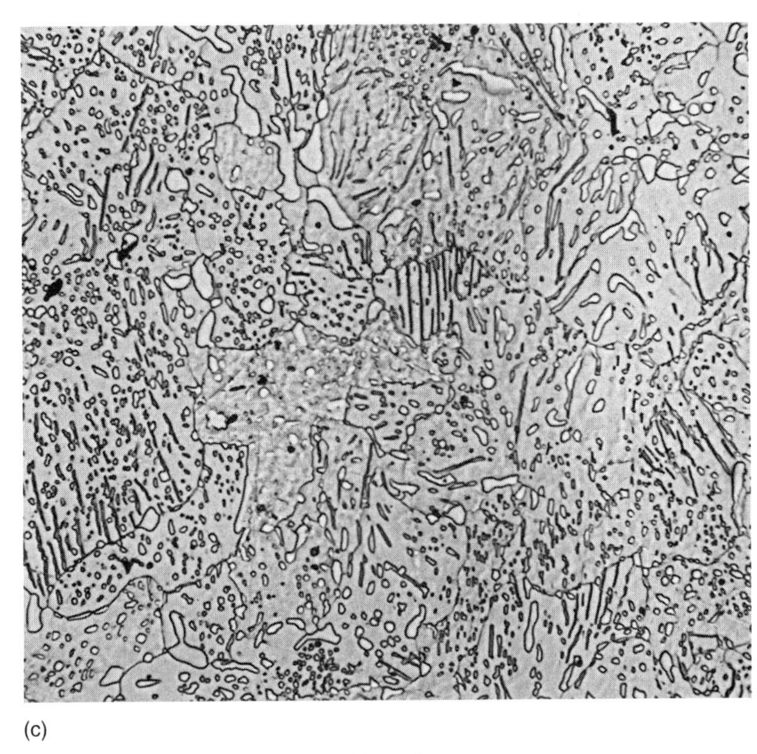

(c)

Figure 9-8 (c) Spheroidized 1045 steel, 4003. (Buehler Ltd.)

hesion. Steel parts are dipped into molten zinc followed by a high-temperature treatment in an annealing furnace. The result is a zinc–iron alloy coating that contains 9-12% iron. This coating provides better paintability and weldability than a hot-dipped zinc coating of equal thickness.

Because it is time consuming and expensive (due primarily to the use of furnace cooling), full annealing in many cases is replaced by normalizing. ***Normalizing*** of steels through using higher temperatures does not require furnace cooling. Instead, all cooling is in still air, and a fine pearlite structure is obtained. Steel is normalized to obtain greater hardness than that from a full anneal. Normalizing does not approach equilibrium cooling conditions as closely as a full anneal. The tensile strength of an annealed alloy is 95,000 psi, versus 107,000 psi for the same alloy if normalized. Nodular cast-iron castings are usually heat treated to improve their properties. A typical heat treatment to obtain a pearlitic structure is to heat to 1650°F, hold at temperature, cool to 1450°F in the furnace, and then air quench. The quench may be followed by tempering to obtain greater hardness and strength. This additional treatment consists of heating to between 1600°F and 1650°F, oil quenching, and tempering. The yield strength for normalized and tempered nodular cast-iron castings is 76,000 psi, compared to 124,000 psi for normalized only, and the HB increases from 255 to 321. Normalizing of steels through using higher temperatures (about 60°F above the A_3 line in Figure 9-8) does not require furnace cooling. Instead, all cooling is in still air, and a finer and more abundant pearlite structure is obtained than from a full anneal. However, its main purpose is to homogenize the microstructure, particularly if further hardening is required. Heating to high temperatures is critical. If too high a temperature is used or if the steel is held for too long at a high temperature, the most likely result is undesirable grain growth. In general, slow heating is more desirable, particularly when working with highly stressed materials (i.e., cold-worked), to avoid distortion. Parts with varying cross sections (thick and thin sections) call for slow heating.

A ***stress-relief anneal*** requires temperatures around 600°C, which are below the critical temperature at which austenite begins to form upon heating. The primary objective of this technique is to relieve residual stresses in all metals and steels as a result of a welding, cold-working, or casting operation. These stresses are eliminated or reduced, even though there is no change in the metal's microstructure.

When a work piece is nonuniformly deformed throughout the part by whatever means, it develops residual stresses. These stresses remain in the part after all external forces have been removed unless the part is made of perfectly elastic material, which is unlikely. A metal part, for example, undergoing bending in order to form a desired shape, is bent permanently. Nonuniform distortion, such as in the bending of a metal part, produces yielding in the outer fibers of the part due to tension, whereas the fibers close to the center of the part acquire compressive stresses insufficient in magnitude to cause any deformation. With the passage of time, the part may experience a lessening or relaxation of the tensile stresses, resulting in dimensional changes that are not desired. Parts subjected to nonuniform temperature change within the body will also undergo dimensional changes, and parts subjected to temperature change, as in the cooling of forged or cast parts, will acquire residual tensile stresses. Residual tensile stresses may be created from the generation of excessive localized heating during grinding. The buildup of residual tensile stress in the surface of parts is undesirable since it lowers the fatigue life and fracture strength of the part. Further, any additional tensile stress imposed on the surface only adds to the residual tensile stresses, which, in turn, limits the ability of the part to sustain external tensile forces. This phenomenon produces stress cracking of manufactured parts over time.

On the other hand, parts that have compressive residual stresses, particularly on their surface, prove to be an asset. Such stresses are desirable because they oppose the applied, repetitive, and undesirable tensile stresses that cause fatigue failures. Hence, the rationale for techniques like shot peening (Module 5), whose primary purpose is to enhance the fatigue life of parts such as gears by inducing a high residual compressive stress at the surface of the flanks and roots of gears. The impact of each shot stretches the surface enough to yield it in tension. Because the surface cannot fully restore itself due to the mechanical yielding that has taken place, it is left in a permanent compressed state. The carburizing process (refer to Module 13) also induces desirable residual compressive stresses throughout the case-hardened layer.

X-ray diffraction (refer to Section 8.4.2) can measure residual stresses at surfaces and subsurface locations. This technique measures strain by measuring changes in atomic distances. Strains can then be converted to stresses using Hooke's Law and the appropriate elastic constant or modulus (see Sections 4.2.3 and 4.2.7). Related subjects are glass tempering (Module 23) and thermal stresses (Module 19). It is extremely important to be always conscious of the importance of understanding how all manufactured processes may affect residual stresses and fatigue performance.

Process anneal is allied with cold working, in which the metal, after heating, is cooled slowly in a furnace down to room temperature. The cooling phase distinguishes this process from normalizing. The end product has higher ductility and lower strength than if normalized. Process annealing is used to "soften" metals, particularly steel sheet and wire products, for further cold working. An electrical transmission line made of copper would become brittle after several drawing operations and, without a process anneal, would fracture. The annealing technique, as well as stress relief, does not involve a phase transformation (no austenizing). The structure of the metal involved, however, is affected. A process called *recrystallization* is involved, which will be explained later in Section 9.3.1.

Spheroidizing anneal is used to improve the machinability of high-carbon steel. Hypereutectoid steels provide good wear resistance. A spheroidizing anneal helps to toughen them by providing more ductility. The hard and brittle cementite network present in hypereutectoid steel also makes machining to close tolerances difficult. Heating the metal for a longer duration near the critical temperature, followed by slow cooling, is one technique used to produce a spheroidal or globular form of cementite in the ferrite matrix. This entire structure is called *spheroidite* or *spheroidized pearlite*; it has a lower hardness, higher ductility, and higher toughness than does the original metal (Figure 9-8).

Most of the preceding processes may be used to improve machinability, but the carbon content of the steel usually is the determining factor. Hypoeutectoid steels with less than 0.3% carbon are normalized, those with up to 0.6% carbon are annealed, and spheroidizing is reserved for steels with 0.6–1.0% carbon.

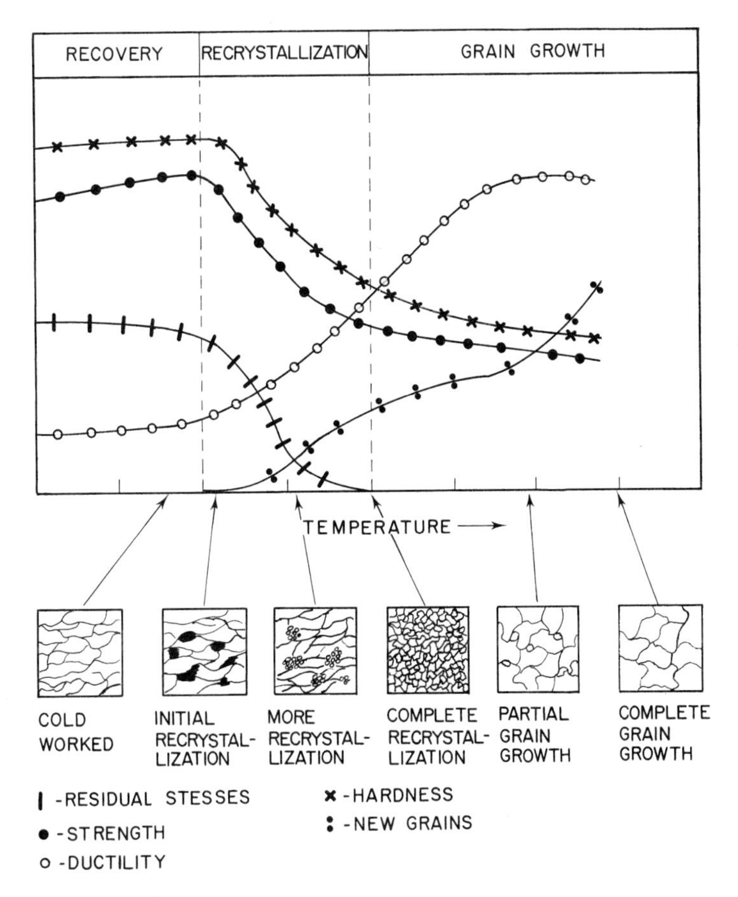

Figure 9-9 Cold-work annealing treatment.

9.3.1 Grain Refinement

Grains that have become plastically deformed as a result of cold working can be given enough energy through thermal processing to permit an orderly rearrangement of the atoms to take place with less-deformed grains. The thermal process involved is annealing, and a certain stage of annealing in which new crystals form, known as *recrystallization*, is worthy of separate comment.

Recrystallization is the formation of a new, strain-free grain structure from that existing in a cold-worked metal upon heating the metal up through a critical temperature for a suitable period of time. When referring to cold-worked metals (mostly all nonferrous metals), annealing can be described as consisting of three stages: recovery, recrystallization, and grain growth.

The recovery stage (see Figure 9-9), the lowest-temperature region, provides the deformed atoms with sufficient energy to rearrange themselves, thus reducing distortion and its attendant stress. The properties recover toward their original values with no visible change in their microstructure. The recovery is due to the cancellation or annihilation of the point defects and dislocations in the deforming metal. A stress-relief anneal, as described previously, requires temperatures in the lowest temperature region of the recovery stage. As temperatures increase, a point is reached where the nucleation of new crystals appears in the microstructure. These new crystals have the same structure and composition as those of the original undeformed crystals before the metal was cold-worked. The grain boundaries and slip planes with the maximum distortions caused by the cold working contain the majority of these new crystals, and eventually they grow and spread throughout the metal structure. A short time after the appearance of the new crystals, the properties of the metal change rather drastically. Hardness and strength drop off, while ductility increases. As the temperatures increase, the region of grain growth is reached, about 0.3 to 0.6 of the melting point of the metal in kelvin, which points out that control of grain size can be exercised to some extent by a wise choice of annealing temperatures. Recrystallization temperatures vary with metals and with the degree of cold working. The number and size of crystals that develop in the recrystallization region result from the amount of cold working done to the metal. If a metal is not cold-worked prior to heating, recrystallization will not occur and the grains will grow in size. To transfer sufficient energy to the deformed crystals, a minimum plastic deformation of about 7% is required before any change in grain size takes place.

From our knowledge of cold working, we know that a highly cold-worked metal contains very little capacity for further plastic deformation. Consequently, the metal is heated to a temperature that permits the atoms to realign and diffuse to a more stable position. The temperatures over which a marked softening occurs (a drop in tensile strength) are in the recrystallization zone. This range or zone of temperatures falls between one-third and one-half of a metal's melting temperature expressed in absolute units (kelvin). See Table A-7A. The average recrystallization temperature is not the same as the critical or transformation temperature. It depends on the amount of cold working, as well as the duration of the heating. The more a metal is cold-worked, the more energy the atoms possess. Therefore, they need less energy transfer from an external source than does a less-cold-worked metal. The recrystallization temperature is used to distinguish between cold working and hot working of metals. The cold working of copper at 95°C is at a higher relative temperature than the hot working of zinc at 20°C, because the recrystallization temperatures of pure copper and zinc are about 120°C and 10°C, respectively.

A point to remember is that these changes do not come about as a result of any phase change. When a metal is not allotropic, several processes are needed to change its properties. First, it must be cold-worked to bring about a minimum deformation of the crystal structure. Second, it must be heated sufficiently to produce recrystallization. Third, after slow cooling the emergence of new, stress-free, small, uniformly dimensioned grains will result in the properties desired. Finally, it must be realized that if, while in service, a metal is subjected to a temperature that is higher than its recrystallization temperature, it will recrystallize, lose its strength and hardness, and possibly cause a failure of some magnitude.

9.3.2 Grain Size

Heat treating can alter the size of grains, which produces corresponding changes in properties. A grain, or individual crystal, varies in size from one metal to another. Zinc crystals in galvanized steel sheet and crystals in some brasses can be seen with the naked eye, but these are exceptions to the general rule. Most crystals are visible only with the use of a microscope.

Even within one metal, the grain size can vary from one region to another. Grain size has a direct influence on the properties of the metal. For example, as grain size decreases, the yield strength of the metal increases. Another fact is that fine-grained metals are stronger and tougher than coarse-grained metals under low or room temperatures. At high temperatures, the reverse is true; coarse-grained metals are stronger and tougher than fine-grained metals. Grain-size ratings, as contained in the American Society for Testing and Materials (ASTM) specifications, are made by comparison with a standard chart of sizes numbered from 1 to 10. When a match is arrived at, the grain size is then designated by the number corresponding to the index number (n) of the matching chart. These matching standard-grain-size charts represent a microstructure projected at a magnification of 100X. The ASTM comparison method (ASTM E112) is an estimation at best. Steel is ordered on the basis of either fine grain or coarse grain. Grain size must be specified for metals that are to be cold-worked. If the grain size is too coarse, the worked surface will be uneven, and if too fine, the metal will be too hard to work satisfactorily. Several different methods, in addition to the ASTM comparison method, are used in industry to estimate grain size.

Temperature alone causes grain growth and not grain size reduction or refinement. If you wish to reduce the size of grains to refine them, you must destroy the original grains and produce new ones with the desired size. One way of doing this is by heating an allotropic metal sufficiently to produce a phase change with its initially small grains. In steel, the fcc grains of austenite are large, but on transforming to bcc crystals, the grains are small. Another way to refine grains is to deform the metal plastically by cold working. Refinement in grain size is not

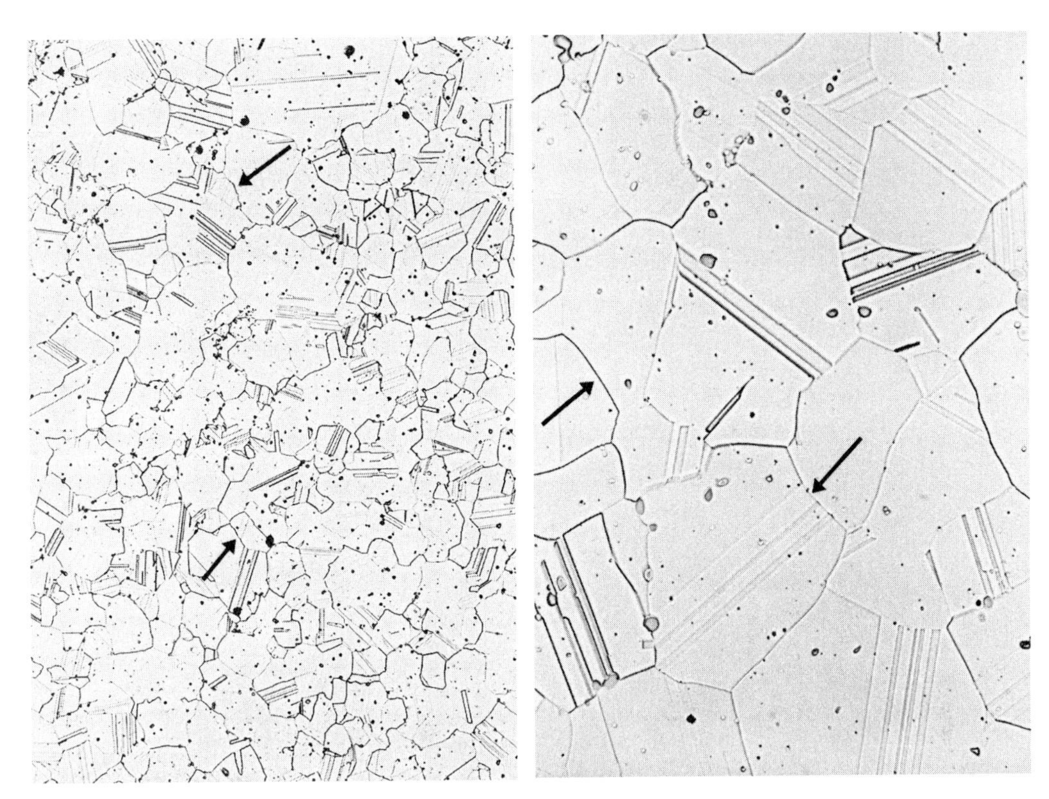

Figure 9-10 Different micrographs showing grain boundaries. (Buehler Ltd.)

the primary purpose of cold working; however, due to the smaller size of the grains, the cold-worked metal is found to possess properties different from those before it was cold-worked.

9.3.3 Grain Boundaries

We know that the atoms that make up the grain boundaries are the last to solidify. As a result, they are not arranged in the same orderly way as are the same atoms in the interior of the crystal. We also know that any time we observe a disordered arrangement of atoms, we look for distortion, increased stress, and resistance to further plastic deformation or slip—all of which result in stronger material in the grain boundaries (Figure 9-10). Arrows point to some of the many grain boundaries shown in Figures 9-10 and 9-11.

The more small grains present in a metal, the more grain-boundary material per unit of volume of metal. The conclusion drawn is that fine-grained metals are stronger than coarse-grained metals at room temperatures. At high temperatures, the heat supplied to the fine-grained metal atoms that make up the grain boundaries increases their energy. This reduces their strained condition, which allows them to spread out and become less densely packed, which in turn results in a decrease in strength. Based on this reasoning and coupled with knowledge gained from experience, coarse-grained metals are required for high-temperature applications where resistance to creep is critical. Can you recall the definition of creep? A good example where knowledge of it is needed is in the selection of the correct metal for the manufacture of turbine blades that revolve at very high speeds and temperatures, with extremely small clearances between the moving tips of the blades and the stationary parts of the turbine.

Cold working results in the formation of more grain boundaries due to the coming together of the large number of dislocations with their disorderly arrangement of atoms. With more grain boundaries come smaller grains; reduced ductility; and increased hardness, tensile strength, and electrical resistance. These conditions describe the terms *work* or *strain hardening* produced by cold working. A good example of cold working is the drawing of copper wire.

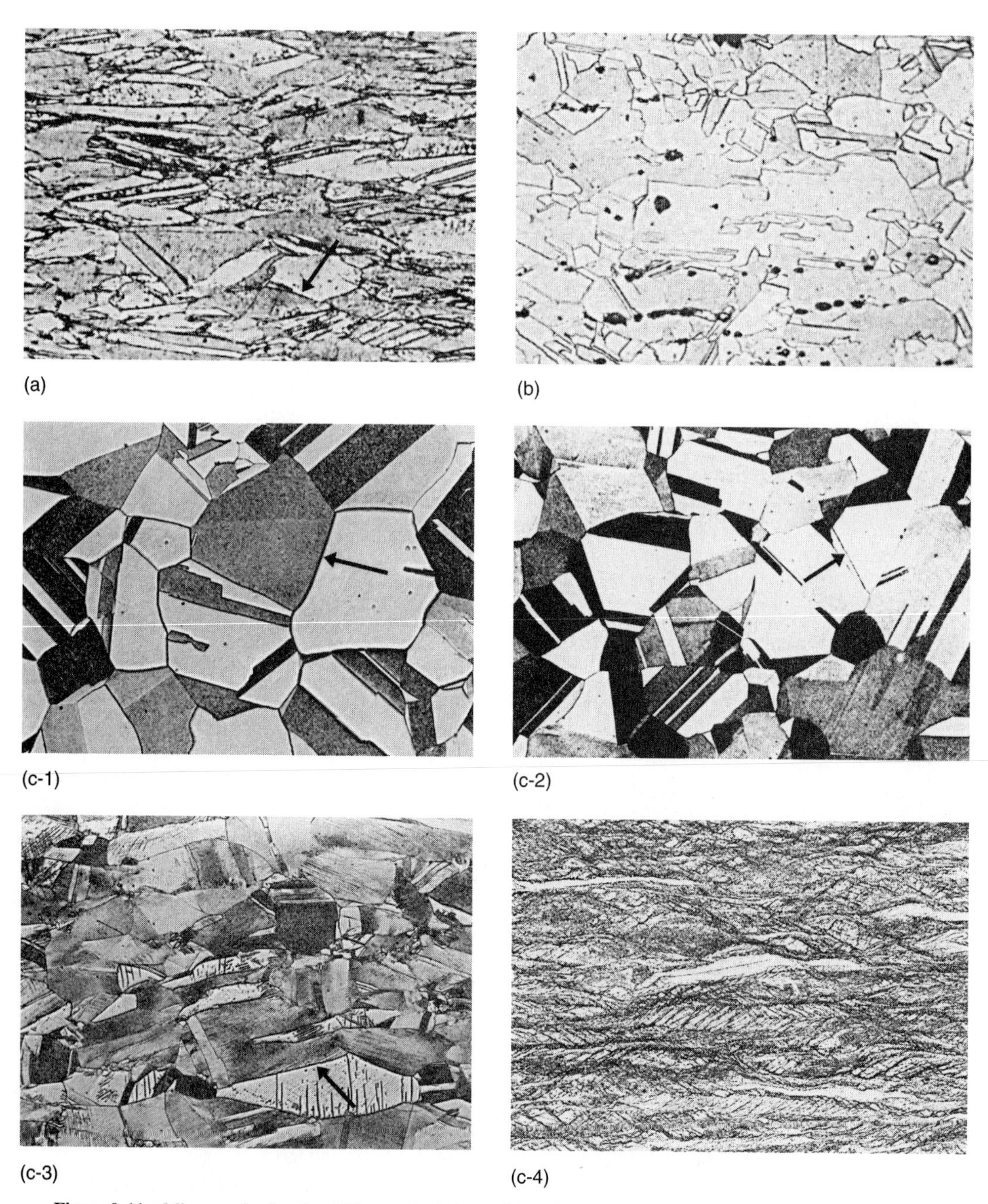

Figure 9-11 Micrographs showing (a) hot-worked, then cold-worked, copper specimen (1003); (b) hot-worked copper specimen; (c) 70/30 brass 3003. Reduction: (1) 0%, (2) 5%, (3) 25%, (4) 75%. (Buehler Ltd.)

Drawing has two meanings in materials engineering. In heat treatment, it means tempering (to be defined). In cold working, the process of drawing refers to the forming of a recess in sheet metal with dies (see Figure 4-10) and the pulling of metal through dies, as in the manufacture of wire. A copper wire transmission line is cold-worked to increase its strength so that it can support itself without excessive sagging over a moderate span.

9.3.4 Grain Shape

Cold and hot working produce a distorted grain structure, that is, the shape of the grain is deformed or becomes elongated in the direction of the metal flow. Figure 9-11a and b compares hot- and cold-worked grains of copper. Such deformed metals no longer possess uniform properties in all directions (isotropic). Instead, the metal generally shows higher-strength properties in certain directions, with a corresponding decrease in those same properties in other directions (anisotropic).

To summarize the preceding discussion, the set of photomicrographs in Figure 9-11c, 1–4 illustrates the concepts of cold working, recrystallization, and grain growth. The specimens are from annealed, commercially available 70% Cu–30% Zn nonleaded brass, initially 0.128 in. thick. Reductions were accomplished by small-reduction rolling passes while maintaining the strip at room temperature at all times. All photomicrographs of the strip samples are shown at 300X for the best possible comparison of microstructures.

9.4 THERMAL PROCESSING OF STEEL: NONEQUILIBRIUM CONDITIONS (MARTENSITE STRENGTHENING)

By varying the amount of carbon dissolved in the single-phase solid solution of iron and carbon (austenite) and cooling the austenite under equilibrium conditions, the austenite transforms into a multiphase mixture known as carbon steel. Depending on the amount of carbon, steel alloys of ferrite and pearlite or cementite and pearlite are formed, which have different hardness and strength properties. But we now reach a limit. Low-carbon steel (0.25% carbon) has a strength of about 44,000 psi; eutectoid steel (0.8% carbon) has about 112,000 psi. To obtain steel with greater strengths, we must depart from equilibrium conditions, primarily in the cooling of the austenite. This departure, in the form of a rapid cooling, is known as ***quenching***. The degree of cooling depends mostly on the quenching medium. For drastic quenching, water or brine is used. Less-severe cooling rates are achieved with such media as oil, molten salt baths, still air, or molten sand. Most liquid media require agitation of the liquid to help reduce the gaseous layer formed adjacent to the metal as a result of the vaporization of some of the liquid in contact with the hot metal, whose temperature is above the boiling point of the liquid quenching medium.

A number of quenching problems are directly related to nonuniform wetting of the surface being treated. The quenching of a heated part by immersion into a fluid whose boiling point is considerably below that of the part being quenched has been researched and divided into three stages or sequences of cooling behavior. First is the development of a vapor blanket or film that literally prevents any cooling from taking place. The second stage is initiated when the vapor collapses. In this stage only localized cooling occurs after the part is randomly wetted by the quenching medium. The final stage is where the surface is completely wetted and convective heating transfer takes place. In the vacuum carburizing of gears an alternative quenching technology is used called *high-pressure gas quenching*. This quenching process has successfully replaced traditional oil quenching. An example is sun gears designed for automotive engines. Made of AISI 4820 steel, the gears are pressure quenched at 12 bar using nitrogen after being vacuum carburized at 1700°F. (Refer to carburizing and surface hardening in Module 13 for further details on carburizing.)

Rapid cooling of steel produces a metastable phase called ***martensite*** (Figure 9-12), which is defined as an interstitial, supersaturated solid solution of carbon in iron having a bct lattice structure. The transformation of martensite results from a three-dimensional mechanical shearing action that distorts the structure, producing hardness and brittleness. Martensite transformations occur in other metals and alloys, such as titanium, lithium, iron, nickel, and copper–aluminum.

To summarize what we have said about equilibrium and nonequilibrium conditions of cooling, our initial step was to heat the basic components of steel (into the austenite-forming region) to produce austenite, a single-phase solid solution of carbon in gamma (γ) iron. Once we formed austenite, we could cool it under equilibrium conditions to produce a two-phase mixture of ferrite and cementite called ***pearlite***. The carbon in the cementite distorts this structure, producing qualities of strength and hardness. The ferrite contributes ductility. If we depart from equilibrium conditions and cool the austenite more quickly, we can form pearlite with greater hardness. If we cool it even more rapidly, we produce a transformation product called martensite, which is extremely brittle, strong, and hard. It far surpasses the hardness and strength of pearlite. Steel with 100% martensite has limited use due to its sensitivity to fracture by impact.

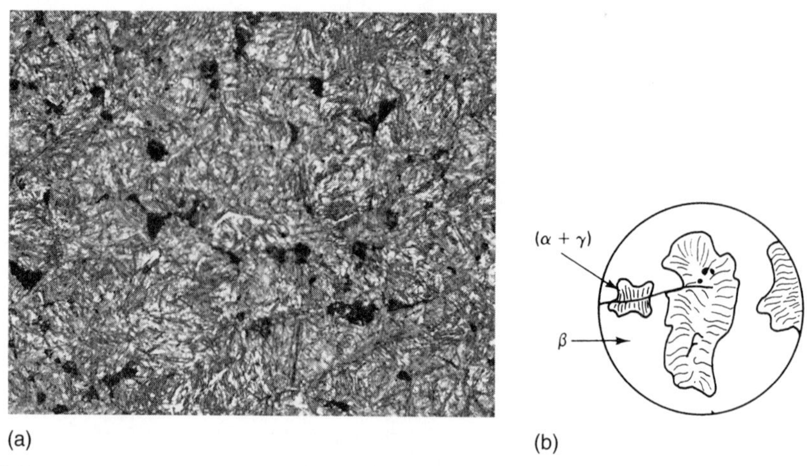

Figure 9-12 Micrograph of 1095 steel martensite. (a) 1095 steel. (b) Microstructure of martensite (1003). (Buehler Ltd.)

Therefore, we call on another heat-treating process known as ***tempering*** to heat the martensite to selected levels and to transform some or all of the martensite to other microstructures with differing properties.

Tempering is the process of heating the as-quenched steel (in the form of as-quenched martensite) to give it some ductility. This as-quenched martensite is a supersaturated solution of carbon in either a bcc or bct ferrite. The bcc structure is producing the strength of the steel and depends almost entirely on the carbon content of the steel. After tempering, the steel is in the quenched and tempered (Q&T) condition of having the as-quenched martensite transformed into stable phases of ferrite and iron carbide (Fe_3C). Tempering is sometimes known by another name, *drawing*, due to the fact that, through controlled heating, hardness can be drawn from the material. Actually, tempering, like any process that supplies energy to atoms through heating, depends on the diffusion of these atoms. The rate of diffusion depends not only on the temperature but also on the time allowed for diffusion to take place. In allowing the carbon atoms to proceed once more on their paths through the atomic structure, varying microstructures and grain sizes are formed with just the correct hardness and strength desired for some industrial application. One application might require a good amount of toughness with very little brittleness; another might demand greater hardness. As in many situations, we cannot have the best of two worlds; to gain greater strength and toughness, we have to trade a little hardness.

In discussing nonequilibrium conditions in the thermal processing of steel, we make use of an ***isothermal-transformation (IT)*** or ***transformation-temperature-time (TTT)*** diagram for a particular steel composition. This diagram is derived by plotting hundreds of isothermal cooling curves for samples of a particular steel composition. The isothermal cooling curve sketched in Figure 9-13 shows the percent of austenite remaining and the percent of the austenite-transformed phase plotted against time. The isothermal temperatures are all below the critical A_e (or eutectoid) temperature at which the austenite becomes unstable and begins to transform (see Figure 9-14). The data thus collected from these isothermal curves are then used to construct the isothermal-transformation diagram for that particular steel. The diagram shows the beginning and end of the austenite transformation as well as the transformed product at various isothermal temperatures.

This procedure is illustrated in Figure 9-14 using just one set of data. A characteristic *S* curve is formed that represents the transformation zone or region at various temperatures. Figure 9-15 is a copy of such a diagram for eutectoid steel, indicating the Brinell and Rockwell hardness readings alongside photomicrographs of the various products formed from austenite at the various temperatures. Figure 9-16, another sketch of a TTT diagram, shows the S-shaped austenite-to-pearlite transformation zone, the austenite-to-martensite transformation zone, and a subdivision of the pearlite into coarse and fine pearlite and bainite, a transformed product of austenite that has characteristics of both pearlite and martensite. In addition, two cooling curves

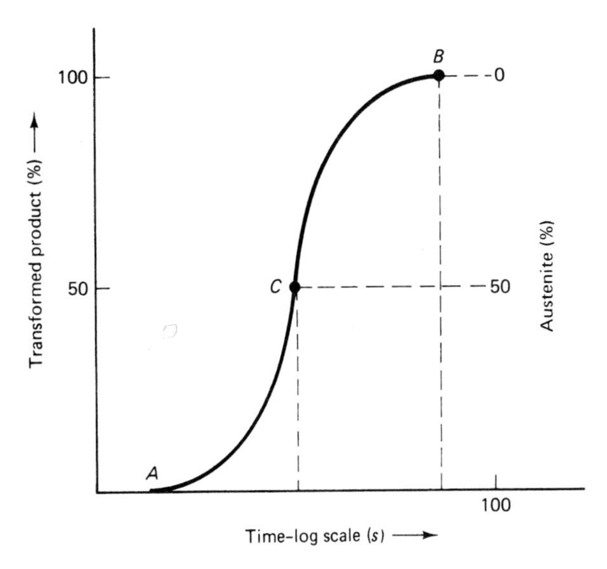

Figure 9-13 Isothermal-transformation curve at temperature T°C.

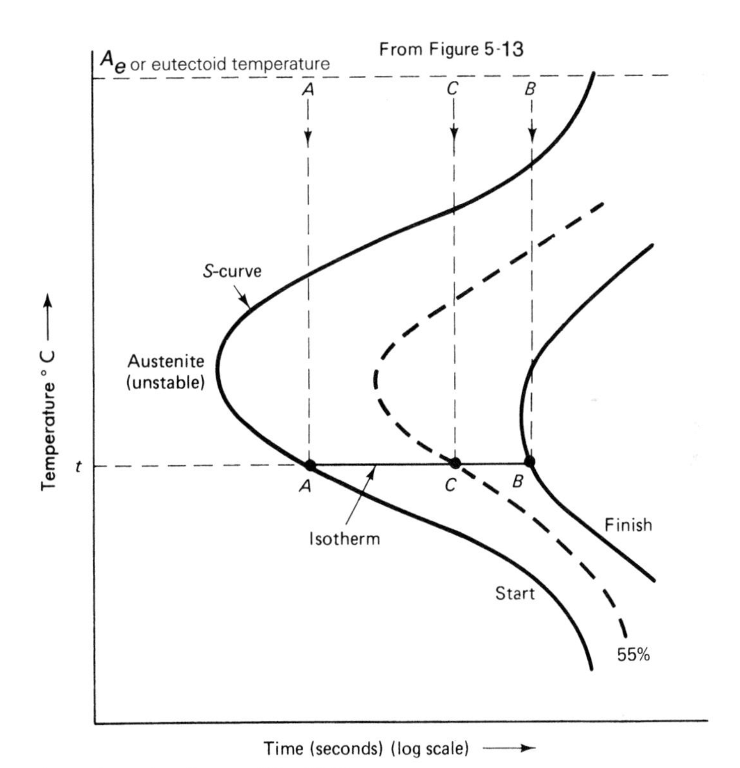

Figure 9-14 Isothermal-transformation diagram (IT or TTT).

are plotted. The first curve, labeled 1 (Figure 9-16), represents the transformation of austenite to coarse pearlite. Such a curve has a relatively small slope, which indicates a slow cooling rate. The second curve has a much steeper slope and consequently a much faster cooling rate. The latter curve is shown passing to the left of the leftmost portion of the pearlite transformation zone, known as the *knee* or *nose*. As a consequence, all the austenite transforms into martensite.

It should be evident at this point that the goal of thermal processing under nonequilibrium conditions is to produce steels with greater strengths than those produced by solution hardening under equilibrium conditions (very slow cooling). The mechanism used to accomplish this goal is a phase transformation. Such a mechanism is possible because steel is allotropic. In the next few pages we will describe briefly some major thermal processing techniques used by industry to strengthen steel under nonequilibrium conditions.

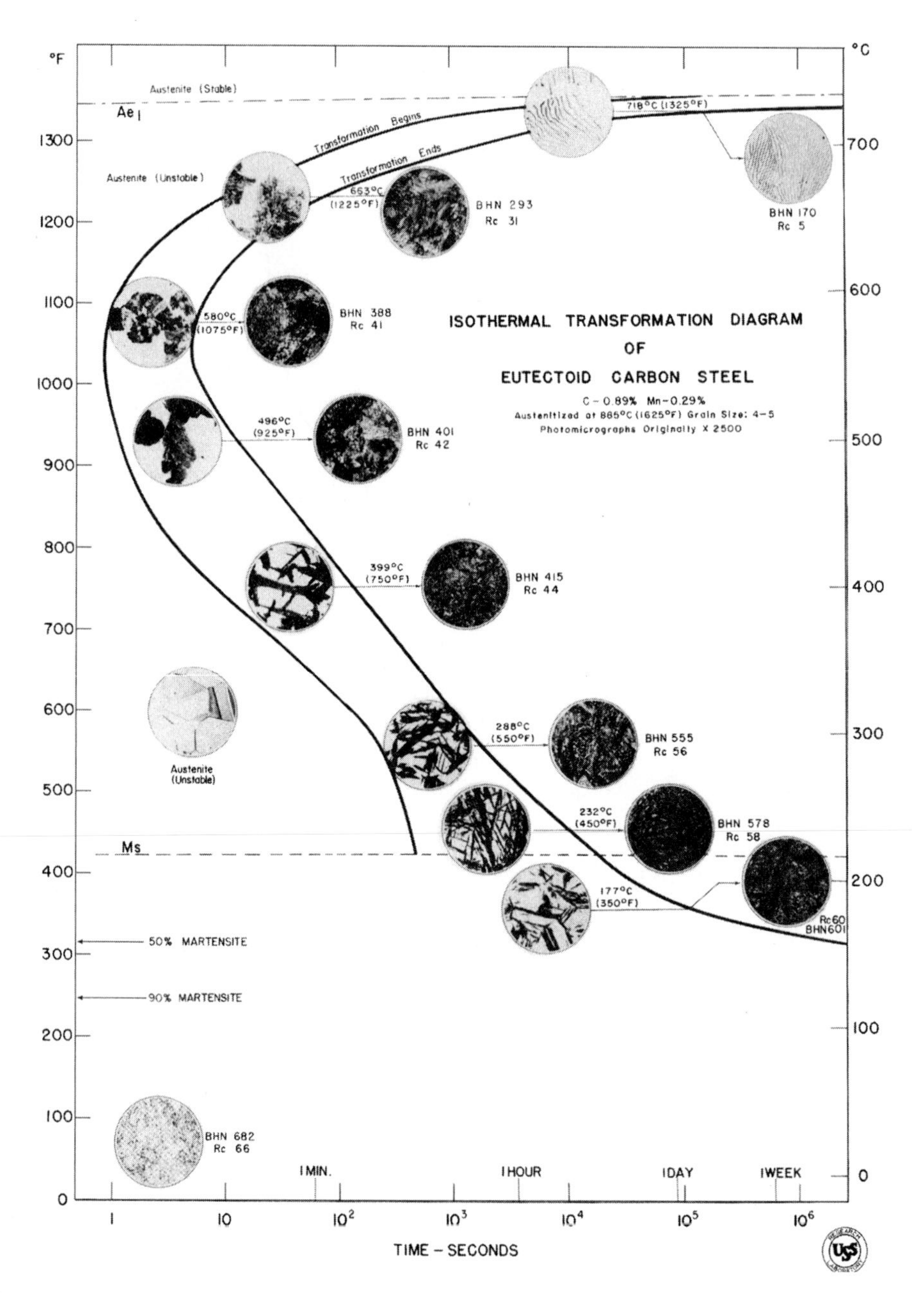

Figure 9-15 Isothermal-transformation diagram of a eutectoid steel. (United States Steel Corporation)

9.4.1 Conventional (Customary) Heat, Quench, and Temper Process

The previous statements on the formation of austenite, quenching to form martensite, and the final heating to temper the martensite describe the conventional process of making steel parts with desired properties. Study Figure 9-17 for the customary quench/temper cooling rates. The log scale along the bottom indicates time. One serious drawback to this process is the possibility of distorting and cracking the metal as a result of the severe quenching required to form the martensite without transforming any of the austenite to pearlite.

As a metal object is quenched, the outer area is cooled more quickly than the center. Thinner parts are cooled faster than are parts with greater cross-sectional areas. What this means is that transformations of the austenite are proceeding at different rates in a single metal object. As we cool a metal object, it also contracts and its microstructure occupies less

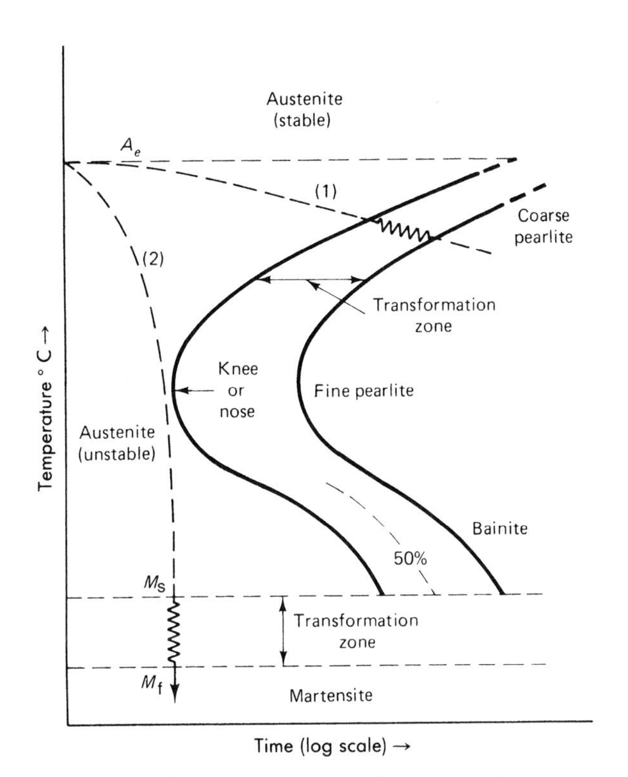

Figure 9-16 IT diagram showing transformation zones, two cooling curves, and products formed.

volume. Put these statements together and the conclusion is that extreme care is necessary to prevent undue distortion and/or fracture. Extreme variations in the size of metal objects complicate the work of the heat treater and should be avoided in the design of metal parts. This means that there is a limit to the overall size of parts that can be subjected to such thermal processing.

9.4.2 Martempering

To overcome these restrictions, two other thermal processes are used. The first, ***martempering*** or *marquenching*, permits the transformation of austenite to martensite to take place at the same time throughout the structure of the metal part. This is shown graphically in Figure 9-17. By using an interrupted quench, cooling is stopped at a point above the martensite transformation region to allow sufficient time for the center to cool to the same temperature as the surface. Then cooling is continued through the martensite region, followed by the usual tempering.

9.4.3 Austempering

A second method of interrupted quenching is called ***austempering***. Figure 9-17 shows this process graphically. The quench is interrupted at a higher temperature (200° to 375°C) than for marquenching, to allow the metal at the center of the part to reach the same temperature as the surface. By maintaining that temperature, both the center and the surface are allowed to transform to bainite and are then cooled to room temperature. The advantages of austempering are (1) even less distortion and cracking than marquenching, due primarily to the higher transformation temperatures, and (2) no need for final tempering. However, austempering has the disadvantage of requiring more time, even though it requires no tempering treatment. Also, parts with large sections of thickness cannot be processed. Sections with a maximum thickness of 1/2 in. can be cooled sufficiently fast to permit the transformation of austenite to bainite without the formation of pearlite.

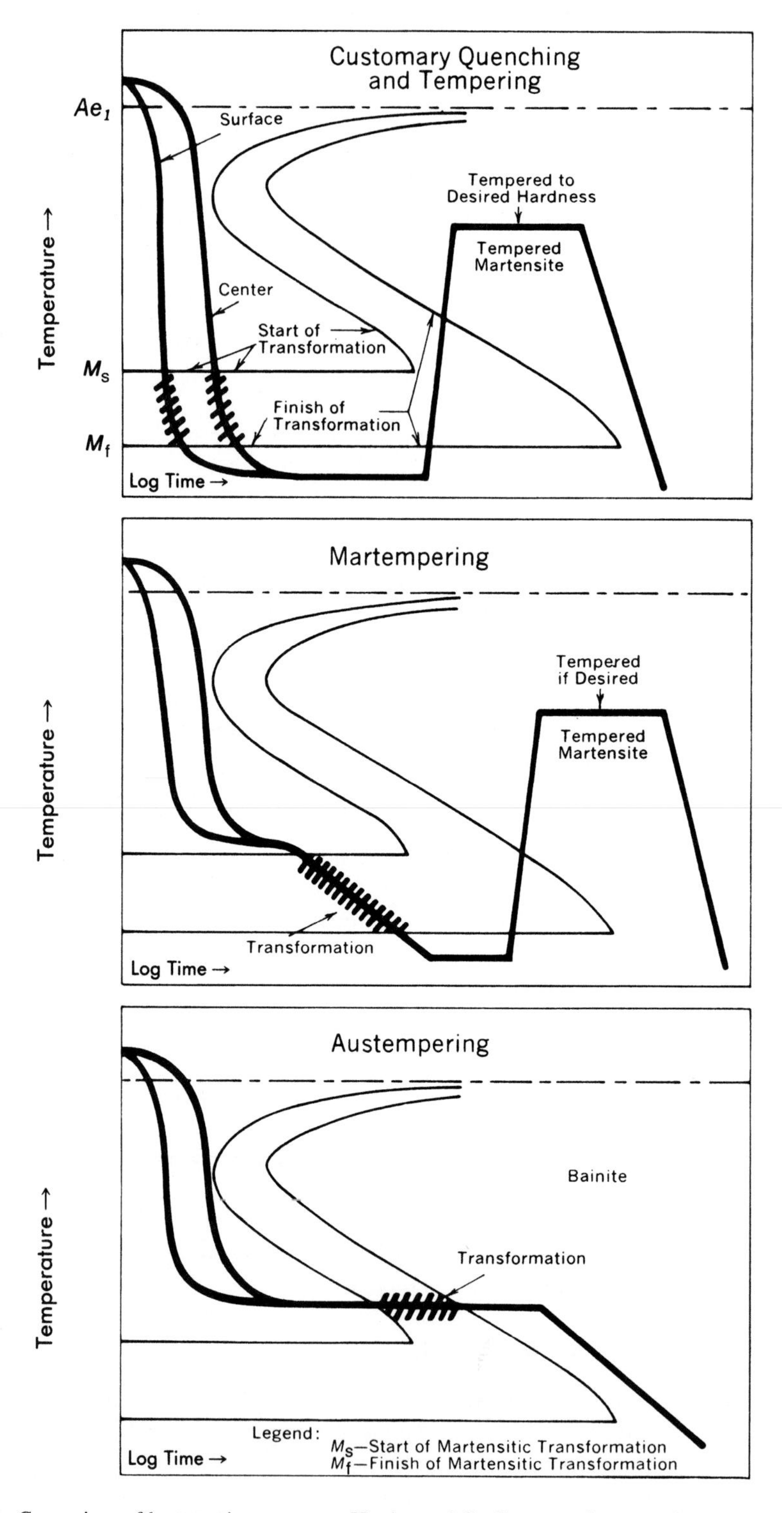

Figure 9-17 Comparison of heat-treating processes. Heating and Cooling operations superimposed on a typical isothermal-transformation diagram. The temperature range at which the transformation to the hard products bainite or martensite occurs has been indicated. (United States Steel Corporation)

9.5 PRECIPITATION STRENGTHENING/HARDENING (AGE HARDENING)

Nonferrous metals do not undergo the significant phase transformations possible with steel. Consequently, most thermal processing of nonferrous metals is used to relieve stresses in a single-phase microstructure and/or to produce recrystallization. These techniques do not result in significant strengthening of a metal's structure. Instead, the most effective thermal processing technique for increasing the strength of such metals is *precipitation hardening*, or *age hardening*. Precipitation hardening involves two steps (see Figure 9-18). The first is solution treatment, the heating (annealing) of an alloy that exists at room temperatures as a two-phase

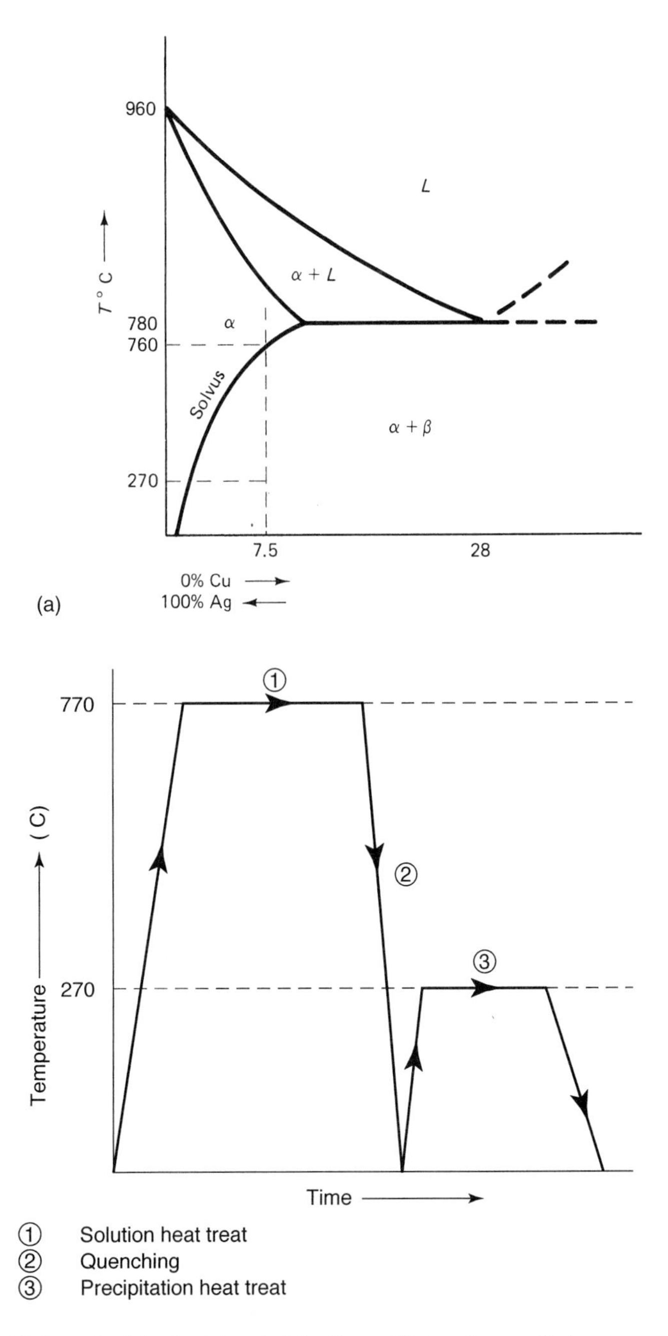

Figure 9-18 (a) Schematic of the Ag-rich end of Ag–Cu equilibrium diagram. B phase is pure copper (Cu) plus a little silver (Ag). (b) Temperature/time plot showing heat-treat cycles for precipitation strengthening.

solid solution to produce a single-phase solid solution at the elevated temperature. The second step is the subsequent rapid cooling (quenching) of this single-phase solid solution to form an unstable supersaturated solid solution. Some alloys require a further heating process to expedite the strengthening of the metal.

To explain how this strengthening comes about, it is first necessary to review earlier material in this module, specifically that concerning diffusion (refer to cold working in Section 9.1) and the effects of solid solubility curves (solvus lines) using Figures 9-18 and 9-19. Such a review will help you understand our explanation of precipitation hardening. As stated previously, the partial solubility of one solid metal in another is determined by the slope of the solid solubility curve (solvus line). The solvus must slope so that it indicates there is greater solubility of one metal in another at a higher temperature than at a lower temperature.

Figure 9-18a, similar to Figure 9-19, is a schematic of the equilibrium diagram for silver (Ag)–copper (Cu) alloys. At each end (the silver-rich and the copper-rich), the solvus indicates the type of partial solubility desired for precipitation hardening. Figure 9-19 shows the silver-rich end of the diagram (the left end of Figure 9-18); an alloy consisting of about 92.5% Ag and about 7.5% Cu (sterling silver) is shown as a vertical dashed line. At room temperature, this alloy exists as a metal consisting of two distinct phases (α and β). Phase α represents a silver-rich solid solution with some dissolved Cu, and β is a solid solution of almost pure Cu with very little dissolved Ag. As the alloy is heated, it crosses the solvus at about 760°C. The β phase dissolves at this temperature and diffuses uniformly to form a solid solution of α. If, then, this same alloy were slowly cooled, the β phase would precipitate out of the α solid solution because the solubility of β (copper) in the α (silver) decreases from about 8.5% at 780°C to less than 1% at room temperature, as shown by the solvus line.

Now if after the solution is heated to ① around 770°C (see Figure 9-18b), this alloy is cooled quite rapidly (quenched ②), there would be insufficient time to permit the Cu atoms to diffuse, which would produce equilibrium-type precipitation of β atoms. The result is a retained α phase that is unstable, with its excess Cu atoms trying to precipitate out to form the β-phase crystal structure. The unstable, supersaturated solid solution formed is quite ductile. With the metal in such a crystal structure, it can be worked, straightened if distorted, or machined much more easily than when in the stable, original form at room temperature. Once the metal is worked, the strengthening process can continue.

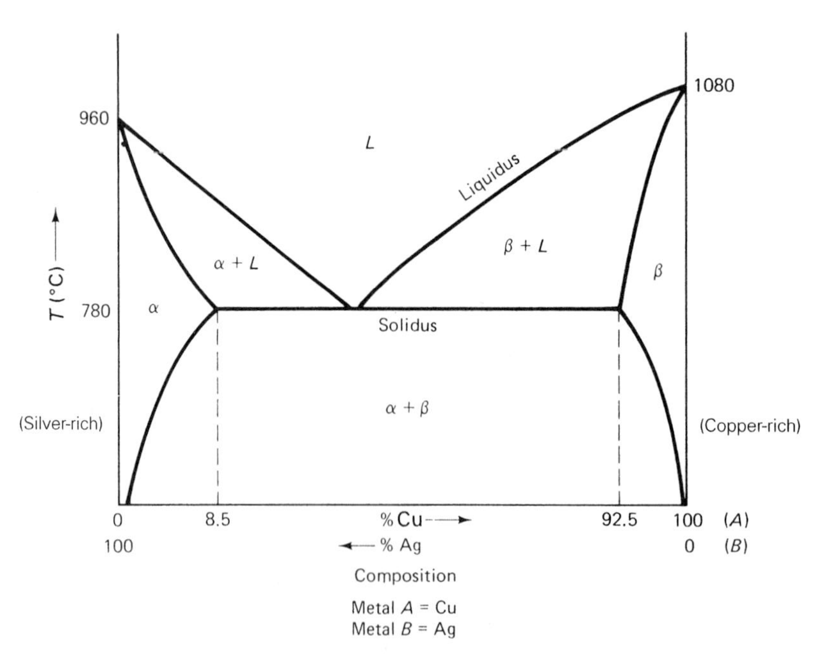

Figure 9-19 Ag–Cu alloys equilibrium-phase diagram.

Other alloys require heating to a temperature below the solvus line to cause this precipitation. Sterling silver can be heated to around 300°C for about 30 minutes and the hardness will double in value. The increase in strength and hardness is explained by precipitation of the trapped β metal out of the α-metal structure at the grain boundaries and at the sites of impurities in the lattice structure of a metal, which produces resistance to the movement of dislocations. If the precipitation occurs at room temperatures within a few days' duration, it is known as ***natural aging***. If additional heating is required, as it is in sterling silver, to cause precipitation and full strengthening, the process is called ***artificial aging***. This additional heating/cooling cycle is also referred to as the *precipitation heat-treating cycle* (Figure 9-18b). At this stage, the supersaturated α solid solution is heated to some intermediate temperature (see the 270°C isotherm in Figure 9-19) to increase the rate of diffusion of the β precipitate phase in the form of finely dispersed particles of metal α or, in this case, copper atoms. It is possible to stop the precipitation entirely by lowering the temperature sufficiently through refrigeration. Once the precipitation has produced a saturated normal dual phase at or near room temperature, the metal reaches its greatest hardness and strength. As fine precipitates form, they have a tendency to grow with time, and with natural-aged alloys this growth of the precipitate will result in a decrease in the strength. Such a phenomenon is known as ***overaging***. Artificially aged alloys are less subject to overaging because the growing process can be stopped effectively by a simple quench.

It must be pointed out that the strength property was used more in this discussion than was any other property. We know that other properties are also affected by distortions of a lattice structure brought about by the precipitation of atoms of another element. The main difference between precipitation strengthening and the thermal processing under nonequilibrium conditions that produces martensite is the formation of a new crystal structure in the transformation of martensite.

To round out our treatment of precipitation hardening, the popular aluminum–copper alloy known as Duralumin (2017) and a part of its equilibrium diagram are used to illustrate the steps taken to bring out the strength of the alloy in industry. Figure 9-20 is a schematic of the aluminum-rich end of the Al–Cu equilibrium diagram. Alloy 2017 contains about 96% Al and 4% Cu. Naturally or artificially aged, this alloy finds many applications, one of which is for aircraft rivets. The vertical dashed line (which intersects the solvus line) in Figure 9-20 represents this particular alloy. Upon solution treatment to a temperature of about 550°C, the single-phase solid solution designated *k* is formed in which all the Cu atoms are diffused into the crystalline structure of the Al atoms. Next, it is quenched to form the unstable, supersaturated *k* phase, and the alloy becomes very ductile and can be maintained if it is refrigerated to retard the eventual precipitation of Cu atoms from the *k* phase. Once sufficient energy is received through heating to room temperature or beyond, the aging process, and hence an increase in strength, take place. Rivets made

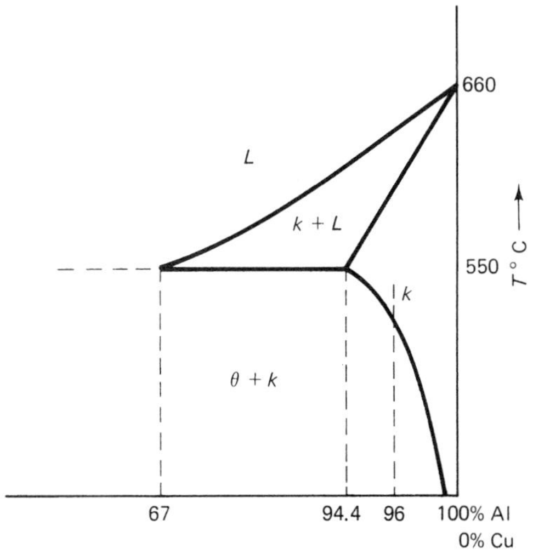

Figure 9-20 Schematic of the Al-rich end of Al–Cu equilibrium diagram.

from this alloy taken from refrigerated storage can then be driven easily. Once driven, the aging process begins as room temperature is reached, which brings on the attendant growth in strength and hardness. Figure 9-21 describes the heat treatment and typical mechanical properties of AerMet® 100 alloy steel produced by Carpenter Technology Corporation. This alloy is more fully described in Module 10 and in Figure 10-9, a photograph of a McDonnell Douglas F/A-18 carrier-based Navy aircraft fighter, which uses this alloy in its landing gear. All procedures and vocabulary in this discussion of the heat treating of this alloy steel should pose no problem for those who have studied the thermal processing of steel in this module. As a result, careful reading of this material can serve you well as a test of your learning progress in the area of heat treating.

9.6 HARDENABILITY

Not to be confused with hardness, ***hardenability*** is the measure of the depth to which hardness can be attained in a metal part using one of the standard thermal processing techniques designed primarily to increase hardness. We know that carbon is vital to steel making—steels are even classified by their carbon content—and hardness, as an indicator of strength, depends on carbon content. The higher the carbon content, the higher the maximum hardness that can be attained at greater depths within a part, assuming that other factors remain the same. See Figure 9-22b. Alloys also play a significant role in producing steels with higher degrees of hardenability.

We know that one way to change a metal alloy's microstructure is to purposely add impurities. The impurities are elements that will alloy or join with the main elements or ingredients to form solid solutions—substitutional and/or interstitial. These alloying elements, such as nickel, boron, copper, or silicon, produce effects similar to the effects of impurities found in the original metal when mined and/or those added, out of necessity, during the processing of the metal. The presence of atoms of the alloying element (1) distorts the lattice structure; (2) increases the metal's resistance to plastic slip; (3) hinders the diffusion of atoms past these areas of distortion; (4) effectively increases the brittleness, hardness, and strength of the metal alloy; and (5) increases significantly the metal's hardenability. Adding alloys or increasing carbon content shifts the location of the *S*-curve (transformation zone) of austenite to pearlite or bainite to the right on the TTT diagram. The result is that a steel need not be quenched as rapidly to achieve the desired hardness. One disadvantage is that alloys are costly, but the benefits usually outweigh such costs. One benefit in using a less drastic quench is the reduction in the danger of the cracking and/or warping of steel parts, particularly the parts that are, of necessity, complex in shape or size or that require great dimensional stability.

It is essential to point out here the necessity of following proper design standards, which help to reduce such things as residual stresses, warping, and cracking. Such techniques as maintaining, if possible, uniform cross sections or thicknesses and using fillets and rounds make the job of the heat treater less difficult.

The Jominy end-quench test performed on steels using standardized specimens and procedures determines the necessary nonequilibrium cooling rates to attain a certain hardenability in a particular steel part (see Figure 9-22a). The 25-mm-diameter specimen with a length of 100 mm represents a metal alloy austenized at a prescribed temperature and for a prescribed time. After removal from a furnace, it is clamped into a fixture and quenched at a specified flow rate and temperature. After cooling to room temperature, shallow flats are machined along its length and Rockwell hardness readings are taken at prescribed intervals over the first 50 mm. Each hardness reading is plotted to produce a hardenability curve similar to Figure 9-22b.

Each steel alloy has its own unique hardenability curve. A steel alloy that has a high hardenability is one that hardens (forms martensite) not only at the surface but generally throughout the interior. Hence, ***hardenability*** is defined as the ability of an alloy to be hardened by the formation of martensite as a result of a given heat treatment. Note that the hardness readings decrease as the measurements are taken farther and farther from the quenched end—the quenched end will be the hardest. In Figure 9-22c, carbon steel (curve 1) and alloy steel (curve 2) have the same maximum hardness at the quenched end. Also the carbon content is the same

HEAT TREATMENT

Decarburization

Like other carbon bearing high strength alloys, AerMet 100 alloy is subject to decarburization during hardening. Heat treatment should take place in a neutral atmosphere furnace, salt bath or vacuum. Decarburization should be determined by comparing the surface and internal hardness of a small test cube for proper response. Metallographic determination of decarburization is not recommended for this alloy.

Normalizing

AerMet 100 alloy can be normalized by heating to 1650°F (899°C) holding for one hour and air cooling to room temperature. Optimum softening for machining is obtained by following the 1650°F (899°C) normalize with a 16 hour 1250°F (677°C) overage anneal.

Annealing/Overaging

AerMet 100 alloy is softened by using a 1250°F (677°C) overage anneal for 16 hours. The optimum annealed hardness of 40 HRC maximum is obtained following this anneal.

Solution Treatment

The solution treatment temperature range is 1625°F +/−25°F (885°C +/−14°C) for 1 hour. The solution treatment temperature must be monitored by a thermocouple attached to the load.

Quenching

Water quenching is not recommended. Proper quenching practice is essential for AerMet 100 alloy. The alloy should be cooled from the solution treatment temperature to 150°F (66°C) in 1 to 2 hours to develop optimum properties. Individual sections larger than 2" diameter or 1" thick (plate) must be quenched with oil in order to obtain 150°F (66°C) in 1 to 2 hours. Individual sections up to 2" diameter or 1" thick (plate) will air cool to 150°F (66°C) in 1 to 2 hours. The cooling rate of the furnace load must be monitored by a thermocouple attached to the hottest spot in the load to insure that the 2 hour cool to 150° (66°C) is obtained.

Refrigeration

Following cooling to room temperature, to obtain the full toughness capability AerMet 100 alloy should be cooled to −100°F (−73°C) and held for 1 hour. The parts can then be air warmed.

Straightening Treatment

AerMet 100 alloy exhibits minimal size change during heat treatment; however, for some parts, mechanical straightening to compensate for distortion during heat treatment is appropriate.

Prior to straightening, a low temperature stress relief at 350/400°F (177/204°C) for 5 hours following the refrigeration operation will provide an optimal combination of ductility and yield strength for the mechanical straightening operation.

Aging

The standard aging treatment for AerMet 100 alloy is 900°F +/−10°F (482°C +/−6°C) for 5 hours. Parts made from AerMet 100 alloy should never be aged at a temperature below 875°F (468°C).

Effect of Aging Temperature on Hardness—AerMet 100 Alloy

Specimens solution treated using 1625°F (885°C) 1 hour, air cooled, refrigerated −100°F (−73°C) 1 hour.

Aging Temperature	HRC
As hardened	51.0/53.0
875°F (468°C) 5 hrs.	54.5/55.5
900°F (482°C) 5 hrs.	53.0/54.0
925°F (496°C) 5 hrs.	51.0/52.5

TYPICAL MECHANICAL PROPERTIES

Typical Mechanical Properties—AerMet 100 Alloy

Heat treatment—1625°F (885°C) 1 hour, air cooled, −100°F (−73°C) 1 hour, aged 900°F (482°C) 5 hours.

Yield Strength		Ultimate Tensile Strength		% Elongation	% Reduction of Area	Charpy V-Notch Impact Energy		Fracture Toughness K_{IC}*	
ksi	MPa	ksi	MPa			ft-lbs	J	ksi in	MPa M
Longitudinal Orientation									
250	1724	285	1965	14	65	30	41	115	126
Transverse Orientation									
250	1724	285	1965	13	55	25	34	100	100

*Per ASTM E399

Shear strength

ksi 175

MPa 1207

Figure 9-21 Heat treatment and typical mechanical properties of AerMet® 100 alloy steel excerpted from alloy data sheet for AerMet® 100 alloy. (Courtesy of Carpenter Technology Corporation—Carpenter Steel Division)

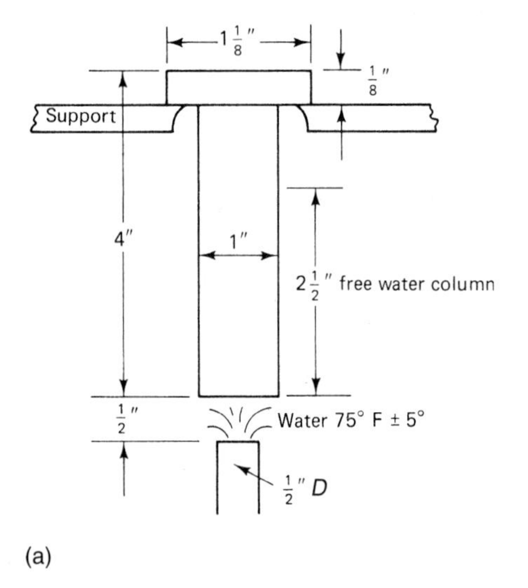

(a)

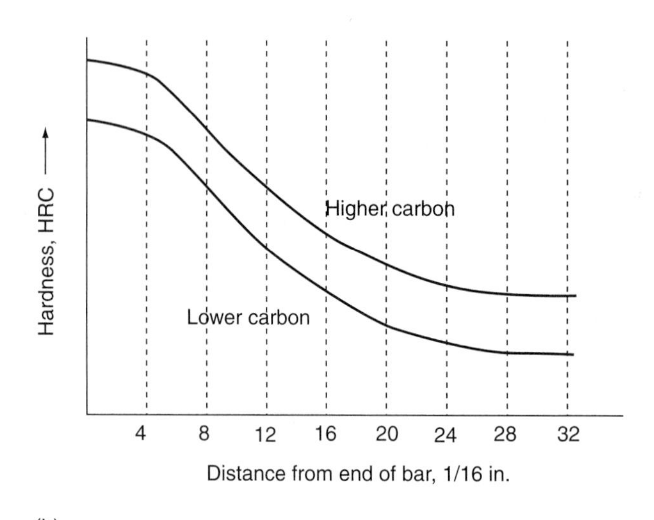

(b)

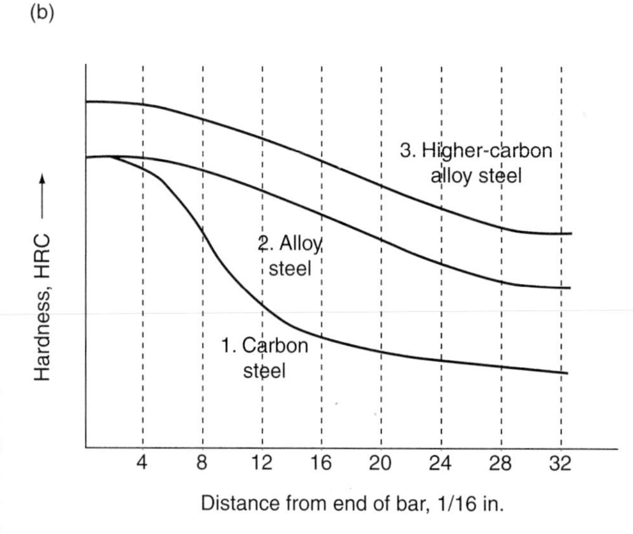

(c)

Figure 9-22 (a) Jominy end-quench hardenability test. Free water pressure must elevate the water column to 2 1/2 in. (b) Hardenability plot of hardness (Rockwell C-scale) versus distance from quenched end of specimen for higher- and lower-carbon steels. (c) Jominy hardenability curves for carbon and alloy steels. (William K. Dalton, *The Technology of Metallurgy*, Merrill, 1994, p. 163, Fig. 7–10 and 11)

for each curve. Curve 2 has the greater hardenability as its hardness is maintained for a considerable distance. The two alloy steels in Figure 9-22c have flatter curves due to the presence of alloying elements, which result in greater hardenability than the carbon steel. Curve 1 has the greater hardness because it has more carbon content than curve 2. Note the shape of such curves, which is the indicator of hardenability. Alloying elements delays the formation of pearlite, which, in turn, permits the formation of martensite at slower rates of cooling.

The purchase of steel cannot be based on its chemical composition because the composition when expressed as a percentage allows for too much variation. This variation in a particular grade of steel will result in a variation in the cooling rate, which will produce a variation in the results of heat treating. This leads to the possibility that this steel will not attain the desired hardness. To ensure full hardness (an indication of strength), a measure was needed that would base the specification for a steel on the response of the steel to heat treating (to produce the austenite transformation to martensite) rather than on composition. Hence, the end-quench hardenability test, or Jominy test, was developed and standardized by ASTM, AISI, and SAE. The SAE identification system for carbon and alloy steels in the United States uses a suffix letter H added to the four digits to indicate that a particular grade of steel will meet certain hardenability requirements, that is, that its hardenability curve will fall within a specified range or band between the minimum values for the particular alloy.

SELF-ASSESSMENT

9-1. Adding elements to replace base metal atoms in a solid solution is called
a. Interstitial **b.** Elutriation **c.** Addition **d.** Substitutional

9-2. The ability of a metal to exist in more than one space lattice is called
a. Isotopic **b.** Amorphous **c.** Allotropic **d.** Equilibrium

9-3. Heat treatment that draws out some (but not all) hardness to improve toughness is
a. Tempering **b.** Spheroidizing **c.** Case hardening **d.** Annealing

9-4. Heat treatment that relieves stress and produces softer and more ductile metal is
a. Tempering **b.** Spheroidizing **c.** Case hardening **d.** Annealing

9-5. Heat treatment that produces a very hard shell but a tough inner core is
a. Tempering **b.** Spheroidizing **c.** Case hardening **d.** Annealing

9-6. The measure of depth to which hardness can be obtained in a piece of steel is called
a. Hardenability **b.** Hardness **c.** Toughness **d.** Ductility

9-7. Steel that is work-hardened has been
a. Hot-rolled **b.** Forged **c.** Cold-rolled **d.** Plated

9-8. The type of heat-treatment process in which the metal is rapidly cooled is
a. Annealing **b.** Equilibrium **c.** Nonequilibrium **d.** Allotropism

9-9. The type of grain structure that provides good creep strength and toughness at high temperatures is
a. Fine **b.** Large **c.** Polymorphic **d.** Serrated

9-10. Regions within a metal that are high energy and form as a result of metal cooling into a polycrystalline solid are
a. Boundaries **b.** Lattice **c.** Void **d.** Interstices

9-11. A homogeneous phase with all solute atoms dissolved into the solvent is
a. Multiphase **b.** Uniform **c.** Single phase **d.** Substitutional

9-12. What is Fe_3C?
a. Cementite **b.** Pearlite **c.** Ferrite **d.** Austenite

9-13. A permanent change in a metal as a result of force is brought about by
a. Elastic flow **b.** Elutriation **c.** Tempering **d.** Plastic flow

9-14. Metals or alloying elements that are insoluble in each other are called
a. Amorphous **b.** Immiscible **c.** Morphologic **d.** Diffused

9-15. A diagram used to determine quenching points for nonequilibrium thermal processing is
a. TTT **b.** Alpha **c.** Gamma **d.** Jominy

9-16. The metal designation system used to indicate the alloy and percentage of alloy in steel is
a. AA **b.** ASTM D2000 **c.** AMS **d.** AISI–SAE

9-17. An element with a radius of less than 0.1 nm is
a. Mn **b.** Mg **c.** C **d.** Fe

9-18. A complex multiphase mixture of different phases, some of which are solid solutions is
a. MMC **b.** Steel **c.** Wrought iron

9-19. A two-phase mixture of ferrite and cementite is
a. Austenite **b.** Ledeburite **c.** Pearlite

9-20. Annealing that is intended for the sole purpose of relieving stresses is called
a. Normalizing **b.** Full anneal **c.** Stress relief **d.** Tempering

9-21. Spheroidizing is used for steels with a carbon content of
a. Less than 0.1% **b.** 0.3% to 0.6% **c.** 0.6% to 1.0%

9-22. A eutectoid steel transformed from austenite to fine pearlite at about 750°F is
a. HRC 31 **b.** HRC 44 **c.** HRC 58

REFERENCES & RELATED READING

AMERICAN SOCIETY FOR TESTING AND MATERIALS. *2003 Annual Book of ASTM Standards*. Philadelphia, PA: ASTM, 2003.

ASM INTERNATIONAL. *ASM Handbook*, Vol. 3, *Alloy Phase Diagrams*; Vol. 4, *Heat Treating*. Materials Park, OH: ASM International, 1992, 1991.

ASM INTERNATIONAL. Heat Treating Library CD-ROM Software. Materials Park, OH: ASM International, 1999.

ASM INTERNATIONAL. SPECIAL, Software for Heat Treating. Materials Park, OH: ASM International, 1992.

DALTON., WILLIAM K. *The Technology of Metallurgy*. Columbus, OH: Merrill, 1994.

DEEVI, SEETHARAMA. "Intermetallics for the Third Millennium," *Advanced Materials & Processes*, September 1999, pp. 44–47.

FOLEY, CHRISTOPHER M. "Why They Fell," ASEE website. cfoley@asee.org, January 12, 2003.

GLANZ, JAMES. "Engineers Closer to Determining How the World Trade Center Towers Collapsed," *The New York Times*, November 11, 2001.

History Channel. *Rise & Fall: Rise and Fall of an American Icon*. December 3, 2002.

Neely, John E., AND THOMAS J. BERTONE. *Practical Metallurgy and Materials of Industry*. Upper Saddle River, NJ: Prentice-Hall, 2000.

EXPERIMENTS & DEMONSTRATIONS IN METALS FROM NATIONAL EDUCATORS' WORKSHOP. EMSET 2 CD-ROM. Over 400 Experiments and demonstrations available in easy to adapt Acrobat Reader format; Department Licensed. ISBN 0–13–030534–0 at http://www.prenhall.com or call 1 800–922–0579.

Periodicals

Advanced Materials & Processes

Journal of Materials Education

Module 10

Ferrous Metals

After studying this module, you should be able to do the following:

10–1. Differentiate among ferrous metal alloys, structures, processing, and properties.
10–2. Cite examples of properties and applications of major classes of ferrous metals.
10–3. Decipher designations for iron and steel alloys.
10–4. Use tables and references on ferrous metals for materials selection.
10–5. Use the website addresses denoted by the icon* to obtain current information and view applications of ferrous metals.

*The icon found throughtout this book will link you to Internet sites related to topics being covered. The dynamic nature of the Web brings frequent changes, so some of these sites, while available at the time of writing, may not be up now.

10.1 FERROUS METALS

Iron and its many alloys, including cast irons and a nearly limitless variety of steels, comprise the ferrous metals group. ***Wrought iron*** is a commercial iron consisting of slag (iron silicate) fibers entrained in a ferrite matrix. It contains approximately 2% slag but very little carbon, and is easily shaped by hot-forming operations such as forging. ***Ingot iron*** contains about 0.1% impurities including 0.01% carbon. It is used in applications where high ductility or corrosion resistance is needed. ***Electrolytic iron,*** about 99.99% pure, is used mostly for research. ***Pig iron*** is iron that is tapped from the base of blast furnace and contains over 4% carbon plus other elements (impurities). It is converted into gray cast iron by heating, along with scrap, in a vertical furnace called a *cupola*. For steelmaking, it is refined into steel in the basic oxygen furnace (BOF) or electric furnace. Even in the face of the wide acceptance of aluminum and polymeric materials, the iron-based alloys dominate all other materials in terms of the weight consumed annually for manufactured products. Ten times more iron (mainly in the form of steel) is used than all other metals combined. Figures 10-1 and 10-2 show the processes and equipment involved in transforming iron ore, coal, and limestone into iron and then into steel in various forms. Coke is made from coal; many other products also come from this source.

10.2 CAST IRON

As shown in the iron–iron carbon equilibrium diagram (Figure 9-5b), cast iron has between 2% and 4% carbon, compared with less than 2% for steel. Other elements in cast iron are silicon and manganese, plus special alloying elements for special cast irons. Many cast-iron products are used as they are cast, but others require changes in properties, which are achieved through heat treatment of the cast parts. Modern casting processes include continuous casting, investment casting, sand casting, ingot casting, lost foam casting, and semisolid casting. Figures 10-1 and 10-2 illustrate some of these casting processes along with their products.

10.2.1 Gray Cast Iron

Gray cast iron is a supersaturated solution of carbon existing in a ***pearlite*** (two-phase structure) matrix. This carbon is mostly in the form of ***graphite*** flakes (a soft form of carbon known as ***elemented carbon***). It is the familiar metal formerly used as the engine block of automobile and other internal combustion engines. Figure 10-3 shows photomicrographs of two "as-cast" gray cast-iron specimens, one of low strength and one of medium strength. The amount of carbon, 3.2%, exceeds the solubility limits of iron, and the carbon precipitates out of solution with ferrite (carbon precipitates out in graphite form). Inoculation with small amounts of iron–silicon alloys helps produce finer graphite flakes, thus improving strength. The graphite promotes machinability and lubricity of this metal. The damping ability of this alloy provides excellent absorption of vibrations and noise, which leads to its selection for piano soundboard frames, engine blocks, and machine tool bases. These combined properties have made gray cast iron a popular gearing material; it is also the least expensive of all metallic materials.

Machinability is a term used to attempt to describe the ease of machining steel to the size, shape, and surface finish required commercially. Machinability describes a material's surface finish, tool life obtained in machining it, and the force/power needed to machine it. ***Machinability ratings*** are based on a tool life of 60 minutes ($T = 60$ min). The standard is free-machining wrought brass bar alloy UNS Alloy C36000, to which is assigned a machinability rating of 100. This brass alloy should be machined at a cutting speed of 100 ft/min. Higher speeds will reduce the tool life and lower speeds will prolong it. Machinability ratings decrease with increases in hardness of the material. Ratings should be considered as a guide and used with caution. Gray cast iron is also available alloyed with nickel, chromium, and molybdenum to improve resistance to wear, corrosion, and heat while improving strength. Flame and induction hardening allow for increased surface hardness with a slightly tougher core.

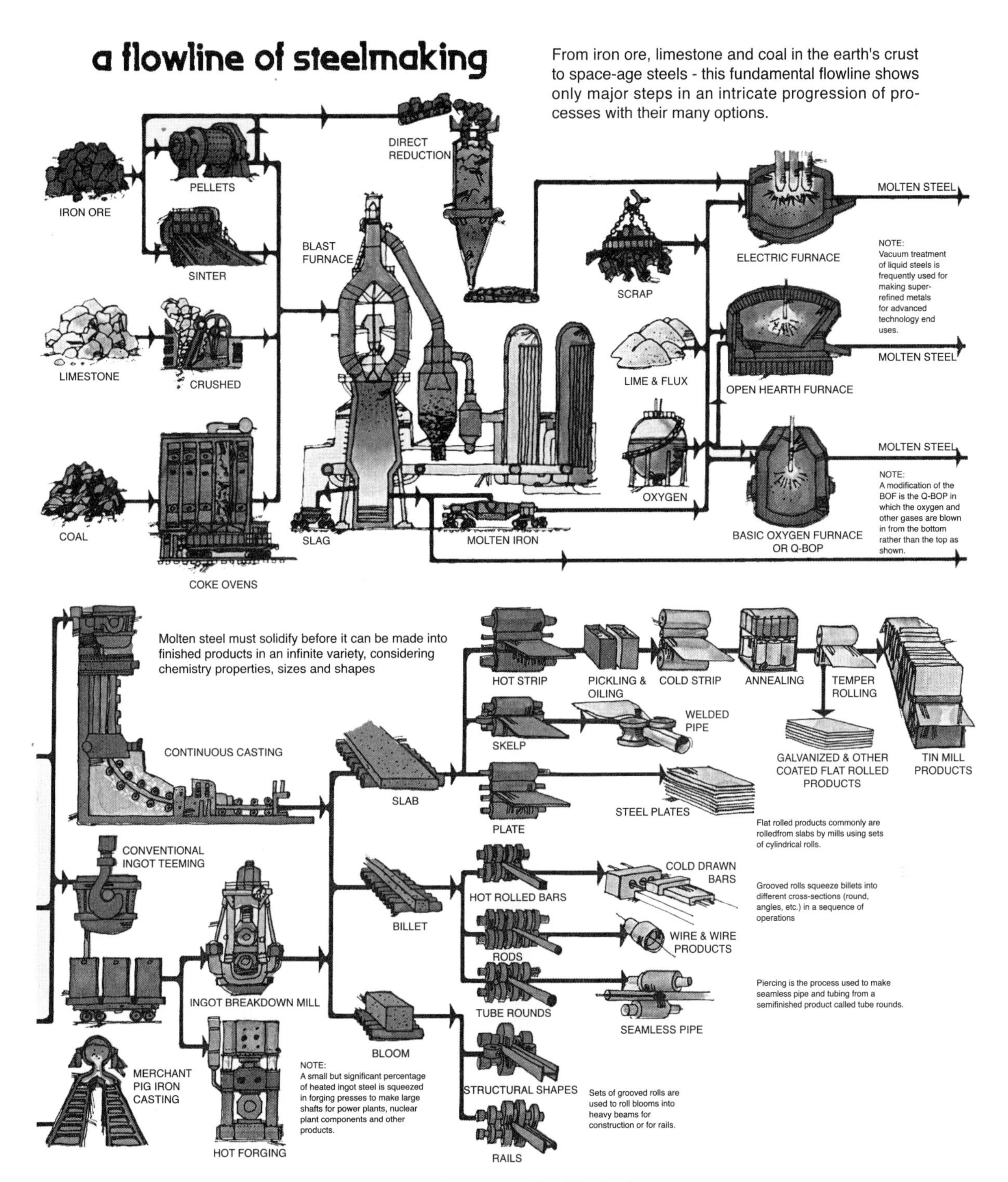

Figure 10-1 A flowline of steelmaking. (American Iron and Steel Institute)

The ASTM system of designation for gray iron places it into classes 20 to 60 based on the minimum tensile strength for each class. For example, class 30 would have a minimum tensile strength of 30,000 psi (207 MPa), while a class 60 gray iron would be 60,000 psi (414 MPa). This classification is often preceded by "48" (ASTM A48 class 40), which designates the specification used to determine the mechanical properties of representative samples. Brinell hardness numbers (HB) range from 160 to 200 for ASTM 48 class 20 to 212 to 248 for ASTM 48 class 60.

 Stainless Steel Dendrites—*http://www.lucent.com/microscapes/13.html*

Figure 10-2 (a) The tapping a heat of steel from a state-of-the-art electric furnace. (MacSteel)

10.2.2 White Cast Iron

Through slow cooling in sand molds, chilling of specific portions of a casting, and alloying, graphitic carbon is stopped from precipitating out of solution with the ferrite, which produces a white cast iron. Most of the carbon in white cast iron exists as cementite instead of graphite, as is the case with gray cast iron. This is due to the low silicon content (less than 1%). The name ***white iron*** comes from the white color produced in the fracture surface of the alloys. Figure 10-4 shows a photomicrograph of white cast iron (compare the photomicrograph of gray cast iron in Figure 10-3, which reveals graphite flakes). The carbon composition of 3.5% for unalloyed white iron has 0.5% silicon. The structure is an interstitial compound of carbon and iron known as ***cementite,*** plus a layered two-phase solution of ferrite and cementite known as pearlite. The castings are very brittle, with Brinell hardness values from over 444 to 712. White cast iron has very good compressive strength, above 200,000 psi (1380 MPa), with tensile strength around 20,000 psi, and good wear resistance; it finds applications as rolls for steelmaking, stone and ore crushing mills, and brickmaking equipment. Heat treatment can reduce brittleness, and as with gray iron, other properties are possible through alloying, but both of these processes add cost to the castings.

10.2.3 Nodular or Ductile Cast Iron

The addition of small amounts of magnesium (Mg), sodium (Na), cerium (Ce), calcium (Ca), lithium (Li), or other elements to molten iron with 3.5% C and 2.5% Si will cause tiny balls,

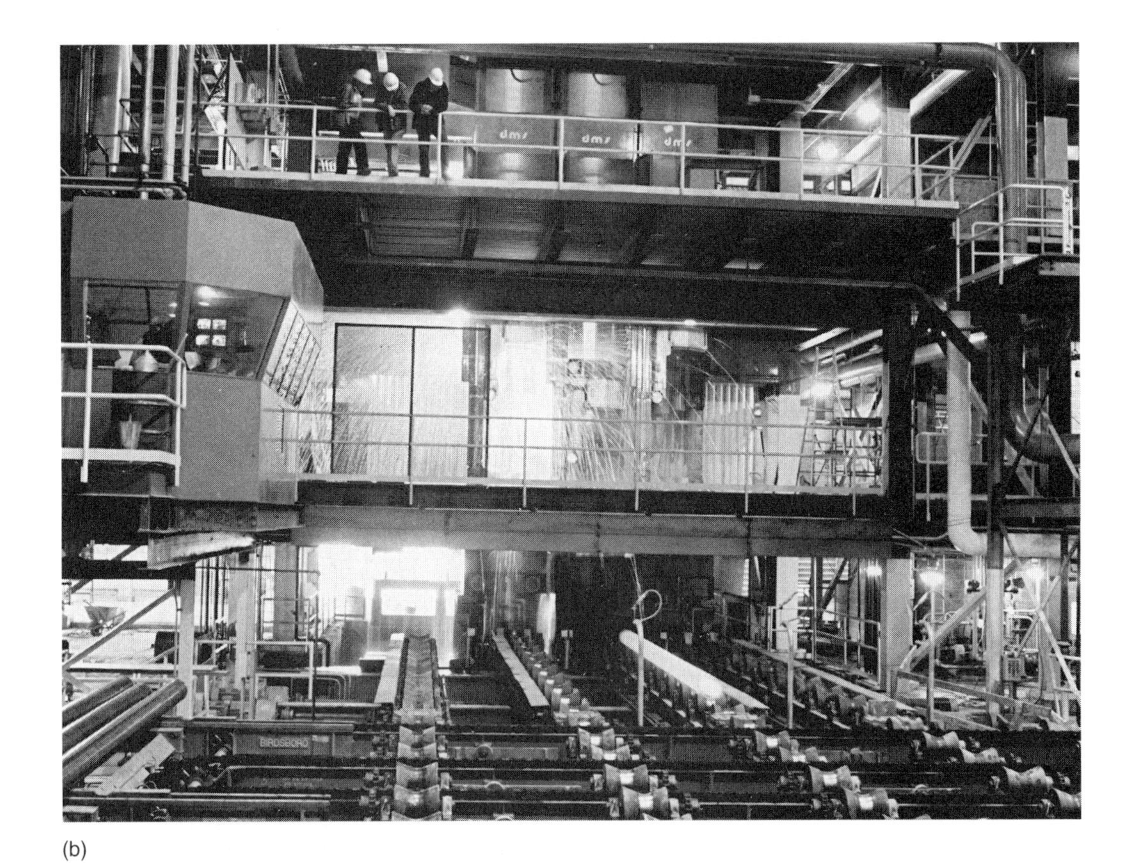
(b)

(c)

Figure 10-2 continued (b) This photo shows a mill line that includes a three-strand continuous caster. (c) MacGold steel bars are precision hot-rolled by a reducing mill installed within the continuous casting and mill production line. (MacSteel, Fort Smith, Arkansas)

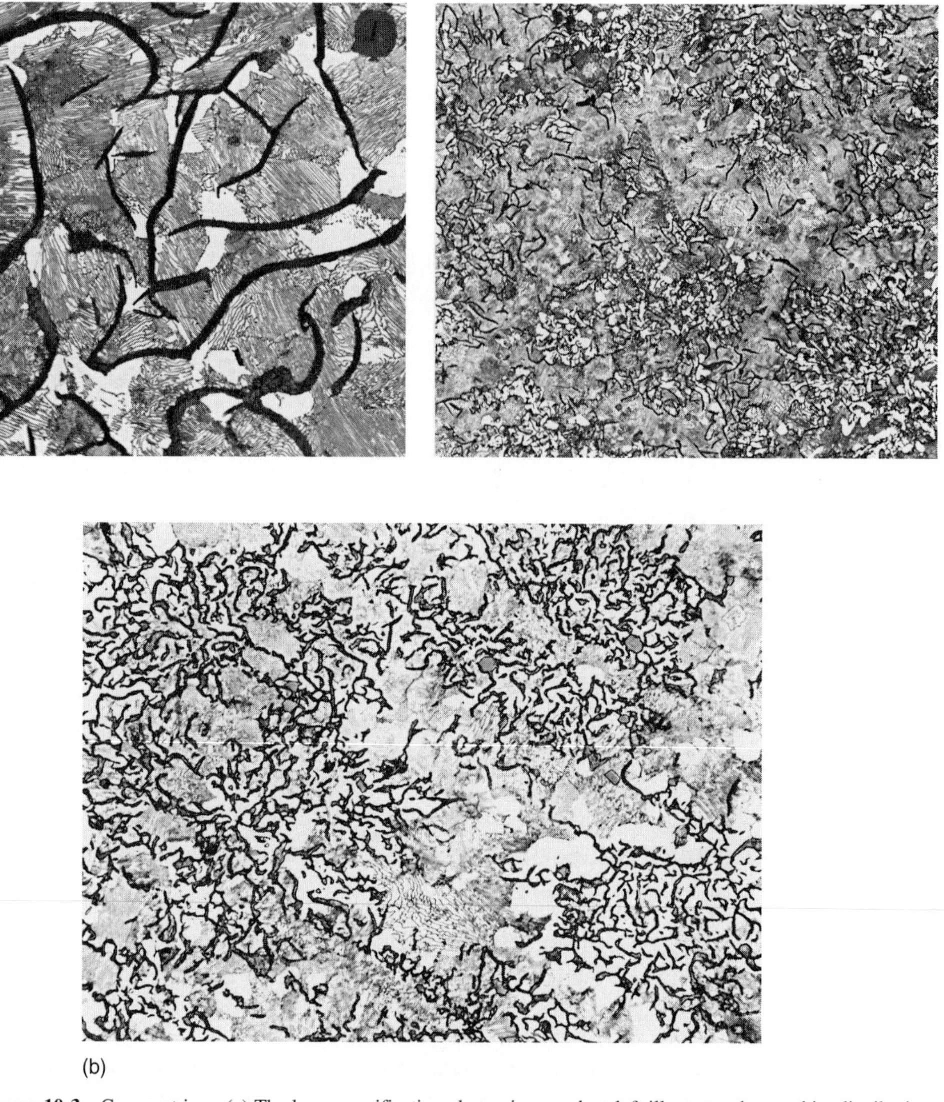

Figure 10-3 Gray cast iron. (a) The low-magnification photomicrograph at left illustrates the graphite distribution, type, and size. At 1000 × (at right), the pearlite colonies and small ferrite grains adjacent to the graphite flakes are clearly distinguishable. The large, round, gray particles are manganese–sulfide inclusions. The surface of the specimen contains products of transformation of a faster cooling rate. (b) The distribution of fine graphite flakes in this sample results in an increase in strength of the iron. (Buehler Ltd.)

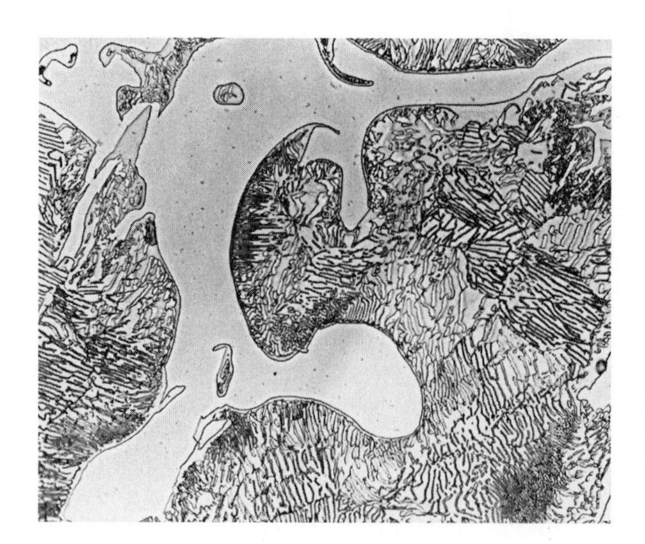

Figure 10-4 White cast iron. This specimen shows a hypereutectoid structure of pearlite and massive cementite. The dark areas are pearlite colonies surrounded by a network of cementite. At higher magnification, the alternate lamellae of alpha ferrite and Fe_3C are clearly resolved. (Buehler Ltd.)

TABLE 10-1 PROPERTIES OF THREE CAST-IRON GRADES

Grade	Minimum tensile strength [psi (MPa)]	Minimum yield strength [psi (MPa)]	Minimum elongation [% in 2 in. (50.8 mm)]
35018	53,000 (365)	35,000 (241)	18
32510	50,000 (345)	32,500 (224)	10
Cupola	40,000 (276)	30,000 (206)	5

or ***spherulites,*** of graphite to precipitate out. As the name implies, the spherulitic structure improves elongation or ductility while yielding superior tensile strength (150,000 psi, or 1034 MPa) and machinability (similar to gray iron).

ASTM specifications for nodular iron indicate minimum tensile strength, minimum yield strength, and minimum percentage of elongation in 2 in. For example, ASTM 120–90–02 would have a minimum tensile strength of 120,000 psi (828 MPa), minimum yield strength of 90,000 psi (621 MPa), and 2% minimum elongation in 2 in. (50.8 mm). Nodular iron finds application in internal combustion engines as crankshafts, rocker arms, and pistons. It is also used for cast gears, pumps, and ship propellers and for equipment that handles caustic substances. It is more expensive than gray iron in terms of weight, but its specific strength makes ductile iron more economical. However, it lacks the damping ability and thermal conductivity of gray iron.

A newer type of cast iron with a *spherical structure* has been developed in Japan that has excellent resistance to corrosion and friction. To the basic composition (iron, chromium, and nickel) of stainless steel (to be discussed later in this module) is added carbon and palladium, which precipitates spherical vanadium carbonate. These spherical carbonates make this new cast iron harder and therefore more useable under harsher conditions.

10.2.4 Malleable Iron

The annealing of white iron castings causes nodules (large flakes) of soft graphitic carbon to form through the breakdown of the hard and brittle cementite (Fe_3C). By varying the heat-treatment cycle, and is possible to produce two basic types of malleable iron. ***Pearlitic malleable*** iron is strong and hard, whereas ***ferritic malleable*** iron is softer, more ductile, and easier to machine. Malleable iron has 2.2% carbon and 1% silicon. In pearlitic malleable iron, 0.3–0.9% of the carbon is combined as cementite and allows for selective hardening of portions of a casting. According to ASTM specifications A47–52 and A197–47, three grades of malleable iron are available: 35018, 32510, and cupola malleable iron. The 35018 and 32510 grades are ferritic, with the latter lower in silicon and consequently more ductile. Cupola malleable iron has a higher carbon and lower silicon content than the other grades, which yields lower strength and ductility. The basic properties of the three grades are shown in Table 10-1.

Applications of the ferritic grades include machined parts (120% machinability rating), automotive power trains, and hand tools such as pipe wrenches that take hard beatings. Applications for the stronger and harder pearlitic malleable iron include parts that require high surface hardness (up to HRC 60 or HB 163–269), such as bearing surfaces on automobiles, trucks, and heavy machinery.

Table A-11 provides a comparison of properties of cast, malleable, and wrought irons. As shown in the photomicrograph in Figure 10-5, wrought iron is an iron of high purity (less than 0.001 parts carbon) with the slag (iron silicate) rolled or wrought into it. The ferrite matrix encloses iron silicate fibers shaped in the direction of rolling, which makes it an easy material to form. It is not a common metal today, but before the development of cast iron and steelmaking, a cruder form of wrought iron served as weapons, tools, and architectural shapes. The Eiffel Tower in Paris was constructed of wrought iron in 1872.

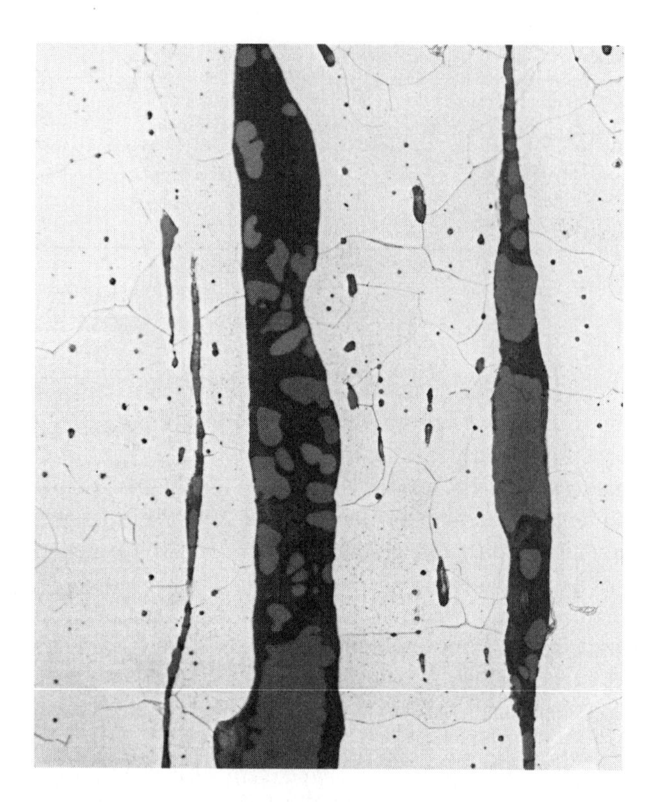

Figure 10-5 Wrought iron. The matrix of this specimen consists of ferrite grains similar to that of ingot iron. The elongated stringers are inclusions of slag composed largely of FeO and SiO_2. At higher magnification, small, dark particles within the ferrite grains are visible. These are finely dispersed impurities, apparent only after etching. (Buehler Ltd.)

10.3 STEEL

The most widely used engineering material, steel is available in an almost limitless variety. Steel contributes to the "greening of manufacturing," for it is the most recycled of all materials. Note in Figure 10-1 that scrap steel is a basic ingredient in the steelmaking process.

Several groups of steel can be used to make other steels, such as cast steel and wrought steel. Wrought (hot- or cold-worked) steel, the largest group, is the steel most known to consumers. The steel is cast into ingots as it comes from such steelmaking processes as the open-hearth furnace or basic oxygen furnace. These ingots are processed further while in the hot "plastic" state to produce a variety of ***wrought*** or ***hot-rolled steel*** (***HRS***) products, such as bars, angles, sheet, or plate. Further working of HRS sheet or bar stock at below the recrystallization temperature of the steel is known as ***cold working*** or ***cold finishing. Cold-rolled steel*** (CRS) is a harder steel because its grains have been work hardened. Other classifications of steel are the carbon steels and alloyed steels.

There are a variety of ways that steels can be classified. A very common system developed by both the Society of Automotive Engineers (SAE) and the American Iron and Steel Institute (AISI) uses four or five digits and certain prefix and suffix letters for many steels and steel alloys. As shown in Figure 10-6, the numbers reveal the major alloying element, its approximate percentage, and the approximate amount of carbon in *hundredths of 1%*, commonly called ***points of carbon.*** Table 10-2 shows the major groupings of carbon and alloy steels under the SAE–AISI classification. The alloy and carbon contents given in this four- or five-digit system are approximations. Complete specifications are available from SAE, AISI, handbooks such as *Machinery's Handbook*, and A-11. For example, the chromium–steel alloy 5120 has the following composition: C = 0.17 to 0.22%, Mn = 0.70 to 0.90%, Cr = 0.80 to 1.10%, P = 0.040%, and Si = 0.20 to 0.35%. Prefixes can indicate the process used in making the steel, such as E for electric-arc furnace; and suffixes clarify further, for instance, H for hardenability guaranteed. Other societies, such as ASTM and ASME (American Society of Mechanical Engineers), have specifications for specialty

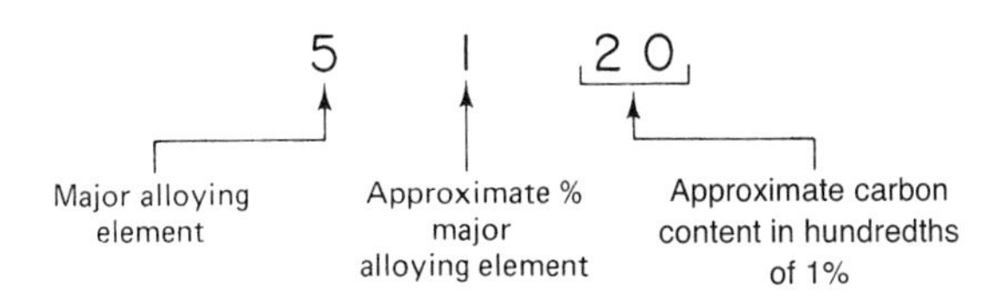

Examples:

Shown above – chromium steel alloy with about 1% chromium and 0.20% (0.002 parts or 20 points) carbon

1015 – plain carbon steel with 0.15% (0.0015 parts or 15 points) carbon

E52100 – chromium steel alloy with about 20% chromium and 1 point carbon produced in an electric arc furnace

(a)

(b)

Figure 10-6 (a) SAE–AISI steel designation. (b) Bicycles take many forms such as seen here: kiddy bike with training wheels and bikes for 1, 2 or 5 people. Bikes use a wide range of steel alloys. What specific alloys do you think would be selected for such parts under different loads as tubular frames and handle bars, spokes, seat springs, chain and sprocket, or ball bearings in the peddle crank?

steels, such as tool steels for dies, cutting tools, and punches or structural bolts and plates. Steel is now classified by ASTM as an ***engineered*** material.

 American Iron & Steel Institute—*http://www.AISI.org*

10.3.1 Carbon Steel

This group of steels, also known as ***plain carbon*** and ***mild steel,*** dominates all other steels produced and is essentially iron and carbon with other elements that occur naturally in iron ore or that result from processing. These elements are held to certain maximum levels: manganese (Mn), 1.65%; silicon (Si), 0.60%; and copper (Cu), 0.60%. Manganese (Mn) AN 25 is present in all types of steel. In fact, the steel industry accounts for about 90% of the world demand for manganese, with carbon steel being the principal market, receiving for up to 70% of total manganese consumption. Carbon steel may be cast or wrought. Typically, the cast steels have more uniform properties since wrought steel develops *directional properties* as a result of its being rolled into shape (see Table A-11).

TABLE 10-2 SAE-AISI SYSTEMS OF STEEL CLASSIFICATION

Digit designation[a]	Types of steel
10xx	Plain carbon
11xx	Sulfurized (free-cutting)
12xx	Phosphorized
13xx	High manganese
2xxx	Nickel alloys
30xx	Nickel (0.70%), chromium (0.70%)
31xx	Nickel (1.25%), chromium (0.60%)
32xx	Nickel (1.75%), chromium (1.00%)
33xx	Nickel (3.5%), chromium (1.50%)
34xx	Nickel (3.00%), chromium (0.80%)
30xxx	Corrosion and heat resistant
4xxx	Molybdenum
41xx	Chromium–molybdenum
43xx	Nickel–chromium–molybdenum
46xx	Nickel (0.75 to 2.00%), molybdenum (0.15 to 0.30%)
48xx	Nickel (3.25 to 3.75%), molybdenum (0.20 to 0.30%)
5xxx	Chromium alloys
6xxx	Chromium–vanadium alloys
81xx	Nickel (0.30%), chromium (0.30%), molybdenum (0.12%)
86xx	Nickel (0.30%), chromium (0.50%), molybdenum (0.20%)
87xx	Nickel (0.55%), chromium (0.50%), molybdenum (0.25%)
88xx	Nickel (0.55%), chromium (0.50%), molybdenum (0.35%)
93xx	Nickel (3.25%), chromium (1.20%), molybdenum (0.11%)
98xx	Nickel (1.10%), chromium (0.80%), molybdenum (0.25%)
9xxx	Silicon-manganese alloys

[a] x's indicate that numerals vary with the percentage of carbon in the alloy.

10.3.1.1 Ultrahigh-Carbon Steel (UHCS). Ultrahigh-carbon steels are hypereutectoid steels ranging in composition from 1.0 to 2.1% carbon. Despite their high carbon content, they have remarkable mechanical properties that are based on desirable microstructures achieved through control of thermomechanical processing and chemistry. They are superplastic at intermediate and high temperatures. At room temperatures, they have ultrahigh strengths with good ductility. When heat treated to achieve martensitic structures, they also can be made exceptionally hard with high compression toughness. These steels may well be a rediscovery of the ancient Damascus sword steels produced by blacksmith at the time of Alexander the Great (325 B.C.). Damascus steels forged from ultrahigh-carbon steel and with beautiful surface markings to this day are not fully explainable. Modern solid-state forging and rolling procedures can almost, but not quite, duplicate this surface marking. ***Laminated steel*** is a modern example of the bonding of low- and high-carbon direct cast strips similar to the approach used for ancient swords, but with much greater control. ***Cladded steels*** are made of stainless steel combined with low-carbon steels. Table A–12 provides a listing of some of these steels with their principal properties.

High- and ultrahigh-strength steel has been used to construct the ultralight steel autobody (ULSAB), which defines high-strength steel as having a yield strength of 210 through 550 MPa and ultrahigh-strength steel as having a yield strength above 550 MPa in material thicknesses ranging from 0.65 to 2.00 mm. The high-strength steels used in the USLAC design were bake-hardenable steels, isotropic steels, high-strength low-alloy steels, high-strength interstitial free steels, and dual-phase steels. Ultrahigh-strength steel was used for most of the lower cross members. Together, the high- and ultrahigh-strength steels made up of 90% of the body structure. Up until this time mild steel was the principal choice for fabricators.

In their quest to create a lightweight, steel autobody, engineers developed a ***steel sandwich*** material consisting of a thermoplastic core sandwiched between two thin steel skins. Its main property is bending stiffness. The core made of polypropylene acts as a spacer between the two outer sheets, separating the outer surfaces from the neutral axis when a bending load is applied. This material can be up to 50% lighter than a comparable sheet of homogeneous material without compromising performance. By definition, this new material could be classified as a true composite. The processing of this material is the same as for sheet, but a consortium of 35 sheet steel producers from 18 countries came together in 1994 to design and validate a ULSAB that would reduce the weight of steel autobody structures while maintaining performance and affordability. In 1998 they published the results, which culminated in a weight savings of up to 36% and substantially improved performance steel with the exception that it cannot be welded. In the 1998 vehicle the steel sandwich material was used for the spare tire tub and the dash panel. The yield strength for the steel used in the spare tire tub is 240 MPa, with a width of 1050 mm, a thickness of 0.14 mm, and a core thickness of 0.65 mm. Note that the steel used for the spare tire tub, with a yield strength of 240 MPa, is considered to be a high-strength steel. ***Adhesive bonding*** is used to bond steel sandwich material. The bonding material provides structure as well as a seal and is a two-component, nonconductive, high-modulus, high-viscous, chemically curing polyurethane adhesive/sealant that cures virtually independent of temperature and moisture. The parts to be joined must be securely held in place while the adhesive bonding cures. Processing temperature is 10° to 35°C and working time is about 10 minutes at 23°C and 50% relative humidity. A very recent U.S. Steel laminated product adequate for automobile sheet applications takes advantage of the distribution of bending stresses just discussed. Called **steel-plastic-steel (SPS),** this sandwich material consists of light-gauge steel skins (the top and bottom layers or laminates) and a plastic core (center) . The polymer core (polypropylene) has sound-dampening characteristics in addition to being low cost and low density. The SPS laminates with the same bending stiffness as solid sheet steel weigh considerably less, corresponding to the weight advantage of aluminum over steel. With such an array of new steel materials, designers and engineers had to become familiar with a steel's capability to spring back when stamped. These materials and advanced processes such as hydroforming (See Module 11) and tailored blanking made it possible to consolidate functions in fewer parts, resulting in mass savings and improved performance.

10.3.2 High-Strength Low-Alloy (HSLA) Steel

HSLA steel is a product of recent technology aimed at producing strong, lightweight steel at a price competitive with that of carbon steels. Although the price per pound of HSLA steel is greater than carbon steel, thickness is reduced due to a higher strength of 414 kPa (60,000 psi) versus 276 kPa (40,000 psi) for carbon steel. Consequently, the overall cost may be cheaper for the HSLA, and significant weight savings are realized. The transportation industry, especially the automotive section, has employed HSLA steel in numerous structural applications. Not as malleable as carbon steel, in the past, sheet HSLA steel could not be used in autobodies, but a modification resulted in a dual-phase steel acceptable for the small bending radii required on autobodies. Some of these HSLA steels, called ***microalloyed steels,*** contain small or microalloying amounts of elements such as nitrogen, titanium, aluminum, or niobium that form carbides or nitrides in the austenite, preventing the growth of the austenite grains. These steels gain their strength without the need for heat treating through the ability to control their grain growth. Weighing less than regular alloy steels but with increased yield strengths (290–480 MPa), these economical steels are ideal for bridges, off-highway vehicles, ships, and machinery. Possessing good weldability, they are produced for sheet, plate, pipe, and forging applications. Stiff commercial competition in this family of steels has led to designating them by trade names. To specify these steels, it is best to use the ASTM designation followed by the strength grade desired, such as ASTM A 242, Grade 60.

10.3.3 Alloy Steel

The classification of alloy steel is applied when one or more of the following maximum limits are exceeded: Mg, 1.65%; Si, 0.60%; Cu, 0.60%; or when specified amounts of the following elements are added: aluminum (Al), boron (B), chromium (Cr, up to 3.99%), cobalt (Co), niobium (Nb), molybdenum (Mo), nickel (Ni), titanium (Ti), tungsten (W), vanadium

(V), zirconium (Zr), or others. Alloy steels are grouped into low-, medium-, or high-alloy steels, with high-alloy steels encompassing the stainless steel group. Refer to Table 10-2 for the SAE–AISI classification systems used for certain alloy steels. Elements added to steel can dissolve in iron to strengthen ferrites or α-iron (bcc) and combining with carbon in the austenite or α-iron phase (fcc) to produce carbides, which improve hardness. High-temperature (heat resistance) alloys must maintain corrosion resistance, mechanical strength, creep resistance, stress–rupture strength, and toughness at temperatures greater than 425°C. These alloys generally contain high levels of chromium and cobalt, which are found outside the United States; consequently, newer alloys are being developed that completely eliminate or reduce the amounts of these two elements.

Chromium is most important for oxidation resistance; cobalt, aluminum, silicon, and the rare earth elements (the 14 elements in the sixth row of the periodic table called the ***lanthanides***, such as cerium, Ce, AN 58) also contribute to producing a stable oxide surface layer. Nickel provides strength, stability, and toughness; and tungsten and molybdenum increase high-temperature strength. Chromium is also effective in increasing strength and hardness, copper forms in austenite to reduce rusting; manganese is an austenite former that, much like carbon, increases hardness and strength; vanadium forms with ferrite to improve hardness, toughness, and strength; molybdenum combines in carbide to improve high-temperature tensile strength and high hardness; silicon dissolves in ferrite to improve electromagnetic properties, plus toughness and ductility; nickel is an austenite former that both improves high-temperature toughness and ductility and provides rust resistance; aluminum is effective as a ferrite former in reducing grain size, thus giving improved mechanical properties.

Many of the alloys described here could easily be labeled *superalloys*. ***Superalloys*** are noted for their outstanding combination of properties, particularly their resistance to creep under prolonged high-temperature conditions and oxidizing atmospheres. Classified according to the predominant metal in the alloy, such as cobalt, nickel, or iron, initially they provided the steels necessary for aircraft turbine components. They find uses in nuclear reactors. These steels are now used in metal matrix composites, serving as matrix materials and with a wide array of reinforcements. Module 25 has further information on these alloys. Table 10-3 provides a comparison of selected alloys.

TABLE 10-3 ALLOY PROPERTIES

	AerMet™ Alloy (875°F Aged)*	Custom 465™ Stainless (900°F Aged)*	17Cr-4Ni Stainless (900°F Aged)*	Ti-6Al-4V Alloy (Annealed)	Ti-6 Al-4V Alloy (900°F Aged)*
U.T.S. MPa	2083	1795	1365	895	1140
(ksi)	(302)	(260)	(198)	(130)	(165)
Y.S. MPa	1780	1655	1260	830	1035
(ksi)	(258)	(240)	(183)	(120)	(150)
Elong. %	14	13	15	10	10
R.A. %	64	58	52	25	20
Fracture Toughness					
MPa ($M^{1/2}$)	109	83	61	77	39
[ksi (in.$^{1/2}$)	(99)	(75)	(55)	(70)	(35)
Density kg/m^3	7889	7833	7806	4429	4429
(lb/in^3)	(0.285)	(0.283)	(0.282)	(0.160)	(0.160)
Modulus of Elasticity					
Gpa	192.5	198.6	196.5	110.3	113.8
(10^3 ksi)	(27.9×10^3)	(28.8×10^3)	(28.5×10^3)	(16.0×10^3)	(16.5×10^3)
Strength-to-Density Ratio					
km	26.9	23.3	17.8	20.7	26.2
(10^3 in.)	(1060)	(919)	(702)	(813)	(1031)
Hardness HRC	54	51	44	30	40

*Except for AerMet alloy aged for 5 hr at 468°C (875°F), all aging done at 482°C (900°F). Custom 465 stainless aged 4 hr 17Cr-4Ni stainless aged 1 hr and Ti-6Al-4V aged 6–8 hr. Ref: Carpenter internal documents and *Aerospace Structural Metals Handbook*. Reading, PA: Carpenter Technology Corp.

A recent example of the advances in the technology of alloy steels is AerMet 100, a nickel–cobalt alloy steel strengthened by carbon, chromium, and molybdenum from Carpenter Technology Corp. The patented alloy, with the designation AMS (Aerospace Materials Specification) 6532, has an nominal composition of 13.4 Co, 11.1 Ni, 3.1 Cr, 1.2 Mo, 0.23 C, with the balance Fe. This alloy steel has the highest fracture toughness of any commercially available steel. Through heat treating it can obtain 1930–2700 MPa (280–300 ksi) tensile strength and exceed a fracture toughness of 110 MPa • $\sqrt{m}$ at 1930 (100 ksi • in. at 280 ksi). Resistance to stress-corrosion cracking and fatigue are two other attributes. The superb combination of high strength and hardness coupled with high fracture toughness and ductility make it a superior alloy steel for applications outside the aerospace industry, for which it was developed. Figure 10-9, later in this chapter, shows one aircraft application of this steel's unique combination of properties that make it lighter and tougher and reduce the size without sacrificing strength.

10.3.4 Corrosion-Resistant Steels (CRES)/Stainless Steel

Although this group of steels contains at least 10.5% chromium, it is more correctly called ***corrosion-resistant steel (CRES)*** rather than alloy steel. These stainless steels find many applications, ranging from tank trailers for transporting many liquids and gases to kitchen appliances (Figure 10-7). The 10.5% chromium does not ensure that the steel will not rust, for a sufficiently high content of carbon or other alloys may negate the passivity of the chromium. As with other steels, stainless may be wrought or cast. Wrought stainless is grouped by its structure as ferritic, martensitic, austenitic, or precipitation hardening (PH). Cast stainless may be classified as heat resistant or corrosion resistant.

A new alloy named ***liquidmetal*** made of nickel, zirconium, titanium, copper, and beryllium, claiming to be twice as strong as titanium, is appearing in the construction of golf clubs. A putter made of liquidmetal costs up to $400; a set of irons costs up to $1998. This metal alloy absorbs less energy than other alloys when a club strikes the ball, allowing more energy transfer from the club to the ball. Its density lies between titanium and steel. Another advanced material, ***alpha maraging metal,*** a stainless steel with high nickel and chrome content, is being used in fairway woods in the popular trimetal clubs produced by Orlimer golf

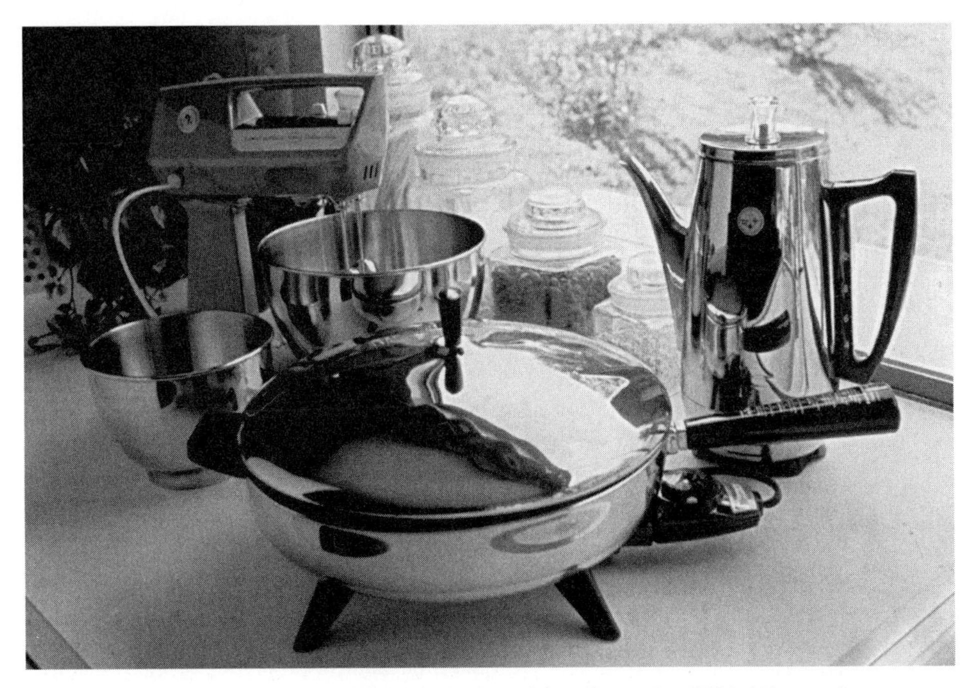

Figure 10-7 Stainless Steel. Cookware of nickel–chrome stainless. (International Nickel Company)

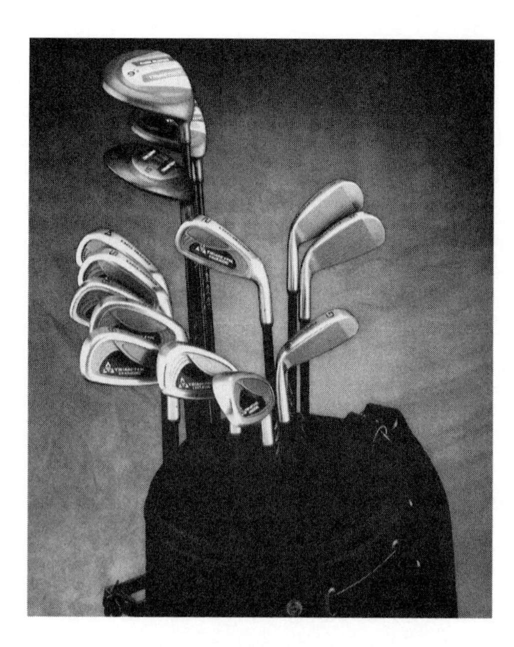

Figure 10-8 New Orlimer irons and woods with face inserts (club head faces) of Carpenter Steel Custom 465™ stainless steel. The club head face is key to flight of a golf ball as its surface is subjected to the all-important impact with the ball. (Carpenter Technology Corp., Reading, PA)

Figure 10-9 A superior alloy steel is the AMS 6532, seen used for the landing gear of the carrier-based McDonnell Douglas F/A-18 because of the alloy's high strength and stress–corrosion cracking resistance.

(see Figure 10-8). ***Maraging steels*** contain large amounts of nickel and iron to produce a soft martensite. Precipitation hardened at low temperatures around 1000°F, they have very high tensile and yield strengths with good toughness. Other clubs are being made with stainless steel combined with strong, shallow faces and copper tungsten weights on the bottom to give them a low center of gravity, which allows the golfer more control without sacrificing distance. The golfing industry also has two next-generation alloys from Carpenter Technology Corp, both of which were originally developed for the aerospace industry. Carpenter AerMet™ alloy was a candidate for the landing gear of U.S. Navy carrier-based F-18E/F jet fighters (see Figure 10-9). It offers a unique combination of high strength and hardness, fracture toughness, exceptional ductility, and resistance to both fatigue and stress–corrosion cracking.

Table 10-3 lists additional mechanical properties of the four different alloy steels contending to be the lead material in the design of golf club heads. Among the many interesting properties is the elongation percentage of the various steels. The higher this percentage, the greater the ductility of the steel; and the more ductility, the greater the steel's toughness in withstanding the impact of, in this instance, a golf ball. Toughness is discussed in several modules of this text. Carpenter Custom 465™ stainless steel alloy is a martensitic, age-hardened stainless steel. It provides an ultimate tensile strength of nearly 1800 MPa, retaining high fracture toughness and yield strength. Note in Table 10-3 that titanium, with its low density, no longer occupies the highest position in specific strength. This means that those three alloy steels that possess higher specific strength will absorb less energy at the point of impact between a club and a ball than titanium. Testing has determined that the impact between the club head and the golf ball can produce a stress of up to 20 MPa. With a high modulus of elasticity and high specific strength, it is possible to design a thinner face that resists damage, allows for a greater sweet spot with no increase in weight, and absorbs less energy at the point of impact (transfers more energy to the ball). The result is that a golfer can drive a ball farther and straighter without swinging harder.

INCO—*www.inco.com*
Carpenter Technology—*www.cartech.com/*

The ASTM and SAE, along with other groups, have developed the Unified Numbering System (UNS), a five-digit designation with an S prefix to replace the AISI designations for stainless steels. Stainless steels in the S30000 (AISI 300) series are nickel–chromium steels; the S40000 (AISI 400) series have chromium as the major alloy; series S20000 (AISI 200) are austenitic alloys, with manganese and nitrogen replacing some of the nickel—far less expensive alternatives when high formability and good machinability are not required.

Austenitic stainless steel is a single-phase solid solution that has good corrosion resistance. Sandvik steel provides a cold-rolled strip steel principally as austenitic stainless steel, which offers high corrosion resistance combined with excellent spring properties. Using the latest rolling technique, this company supplies super-thin strip steel to extremely close thickness tolerances and excellent flatness. The thinnest strip is available in thicknesses down to approximately 0.015 mm. Figure 10-10 shows one wing of a dragonfly made of such steel.

Figure 10-10 The design of the dragonfly wing is unique, with wings that move independently of one another. It can hover like a helicopter, fly slowly backwards, shoot forward, dive suddenly, or dart to one side. As the wing flutters, it also turns to give optimum effect, and it does all that about 30 times a second. The wing pictured, Strip Division of Sandvik, is made of precision strip steel. (Courtesy of Sandvik Steel)

Martensitic stainless steels in the S40000 (AISI 400) series have high carbon content, up to 1.2%, with 12–18% chromium. The higher carbon content allows formation of more δ iron, which quenches to a hard martensitic (up to 100%) steel, but it reduces some of the corrosion resistance. If an austenitic stainless steel is heated sufficiently so that carbon precipitates out of solid solution as chromium carbide (which leaves less than 12% chromium in some segments of the alloy), it promotes intergranular corrosion.

The ***ferritic stainless steels*** have low carbon (0.12% or less) content and high chromium content (14–27%) in a solid solution and do not harden by heat treatment. They are in the S40000 (AISI 400) series and, unlike the martensitic and austenitic steels, are magnetic. The ferritic grades have good formability, machinability, and corrosion resistance exceeding that of the martensitics. Specific properties of each type of stainless steel are found in standard references on steels. Table A–11 lists selected stainless steels and their properties.

10.3.5 Other Steel Alloys

In addition to the high-alloy and stainless steels previously discussed, there are many other specialty alloys. The ASM publishes several volumes of *Metals Handbook*; Volume I, *Properties and Selection of Metals*, gives in-depth coverage of most metals. Included in the steel alloys is a range of tool steels, high-yield-strength (HY) steels, magnetic and electrical steels, ultrahigh-strength steel (see Table A–12 for properties), maraging steels, low-expansion alloys, and ferrous powdered metals.

Nanosteel, a composite material, is made up of steel-alloy clumps with just a few molecules per particle. Each particle is about 50 nm in diameter. This alloy is solidified with an amorphous structure. Crushed into a powder and applied as a coating on a conventional metal alloy such as steel or aluminum, nanosteel bonds with other base metals to form a dense coating with much stronger atomic bonds. Atomic bonds are the ultimate determinate of a material's strength. The best steel alloys with their fine microstructures of ferrite and carbide list yield and tensile strengths that are only about 10% of their theoretical best strength. With stronger atomic bonds in this very fine nanostructure, nanosteel displays a theoretical strength level of between 40 to 45%. In addition, nanosteel exhibits better corrosion resistance because corrosives find it difficult to establish a connection due to the fineness (fine grain structure). Nanosteel coatings under test are about 30% harder than conventional steels yet are fairly flexible, permitting their application before or after the base metal is coated. Further development of this advanced material will allow engineers to approach their goal of using cheap base metals, coating them with essentially a weightless coating to protect the surface from corrosion, and at the same time increase the composite material's strength, hardness, ductility and strength-to-weight ratio. See Module 13 for additional information on surface coatings and surface modification.

10.4 FERROUS AMORPHOUS METALS

Metal alloys possess crystalline atomic structures in which individual atoms are arranged in ordered, repeating patterns. ***Amorphous metals*** differ in that they consist of atoms arranged in near-random configurations lacking any long-range order. These noncrystalline structures are common in nature, being found in nonmetallic solid materials such as silicates, in which, in the viscous liquid state, the individual molecules have limited mobility and crystallization proceeds slowly. This allows the use of modest cooling rates to suppress crystallization completely. Liquid metal alloys, in contrast, have low viscosity and high diffusivity because they consist of loosely bonded atoms rather than bulky clusters or molecules. This indicates that, due to this free movement of atoms, extraordinary cooling rates would be necessary to bypass crystallization.

The discovery of amorphous metals is credited to P. Duwez in 1960, who used a pressurized gas gun to propel small droplets of molten alloy onto a polished copper plate. On contact with the plate, the molten metal droplets solidified into flake or "splat" form. Thus, the splat quenching process was developed for the rapid solidification of amorphous metals. In the early 1970s, the formula for all amorphous metal products was established. It consisted of various percentages of three main constituents: one or more transition elements; a nonmetallic element such as boron, phosphorus, or carbon; and a metalloid such as silicon, aluminum, or germanium. Amorphous metal alloy, called **Liquidmetal,** is a recent alloy said to be twice as strong as steel. It is composed of Ti, Cu, Zr, and Be. As the atoms of these elements are of different sizes, they do not readily form crystals. However, they can join to produce an alloy up to an inch thick that will not shrink when it solidifies and can be cast with a precision down to 1 micrometer. High-elastic-modulus and corrosion-resistant cast alloy VIT–001 with a zirconium base has a yield strength of 1900 MPa compared to 860 MPa for cast 17–4 stainless steel.

The development of amorphous metals is a classic example of the intimate linkage that exists among the three main principles in materials science, namely, the structure, the processing, and the properties of a material. Further developments in processing techniques led to the production of continuous amorphous-metal filaments by ***chill block melt spinning,*** in which a stream of molten metal is directed at a substrate. Two later variations of this process are known as ***free-jet melt spinning*** and ***planar flow casting.*** Using the free-jet process in 1973, Allied Chemical manufactured the first commercial amorphous metal in the form of continuous ribbon 1.7 mm wide and 50 mm thick. Planar flow casting has allowed ribbon width to increase such that widths up to 210 mm are commercially available. (See Module 13 for additional information on sprayed coatings.)

Initially developments with amorphous metals dealt with precious metals such as gold and palladium. Subsequent events proved that amorphous metals could be produced with common, inexpensive metallic elements and, in addition, such metals could display ferromagnetic properties. Figure 10-11 shows typical applications of METGLAS® brand amorphous metals. Liquid Metal is another brand. ***Ferromagnetism*** is one of three forms of magnetism in materials, each form being distinguished by different degrees of susceptibility and permeability, of which ferromagnetism has large values of both. Ferromagnetic materials such as iron, nickel, and cobalt are strongly attracted to magnets. Such magnetic metals are characterized by an electronic structure in which the electronic orbits of an internal shell of bound electrons are not completely full. In iron, the unfilled 3d orbits with five electrons have a positive spin, and

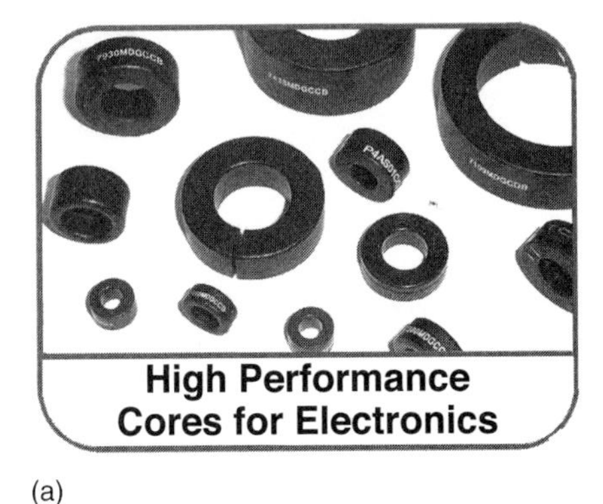

(a)

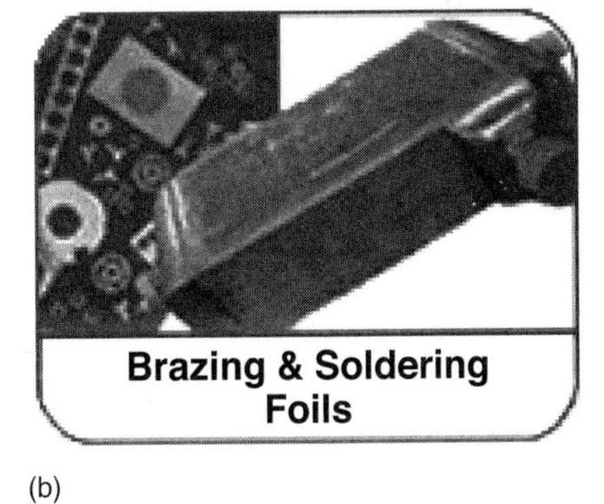

(b)

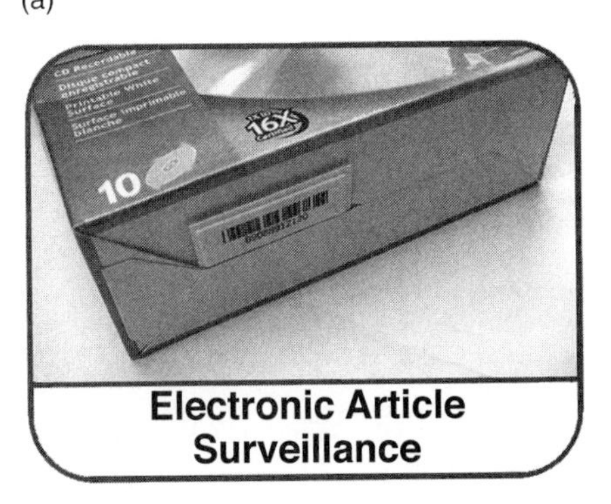

(c)

Figure 10-11 Metglas offers a wide range of amorphous metals for various applications. (METGLAS) Visit their website for more details. (Honeywell—http://www.metglas.com)

one electron has a reverse negative spin. Such a condition is the source of ferromagnetism, the tendency for the spin of these electrons in any given atom to affect the spin of the corresponding electrons in neighboring atoms. The distance between neighboring atoms is critical. (For further discussion of magnetic properties, see Module 6.)

 Honeywell—*http://metglas.com*

Amorphous metals magnetize more easily than crystalline metals because they do not contain the dislocations and grain boundaries that impede the movement of domain walls. ***Magnetic domains*** are regions of ferromagnetic materials in which the spins of all their atomic magnets are aligned. Detectors are under development that will read the status of these domains in a material undergoing load and thus aid in determining when and if the material will experience fatigue failure (see Module 4). Such intelligence would be available long before our present attempts to predetermine when fatigue failure occurs (i.e., when a crack is observed and is accessible for analysis). Even a demagnetized steel bar has magnetic domains that shift positions with respect to each other as the bar is crushed or bent. Such shifting produces detectable changes in the magnetic field. As defects in the bar's steel structure appear, the "flutter" in the magnetic field becomes more pronounced. As the bar begins to crack, spikes appear in the magnetic field. The development of new detectors may result in instruments that will more accurately determine when a part is about to fail. Other metal fatigue is difficult to detect. Once cracks have been discovered, the part undergoing fatigue has been potentially dangerous for an unknown period of time. (Further treatment of domains can be found in Section 6.3.2, "Permeability Classes of Magnetic Materials, Magnetic Domains.")

In the United States, the Electric Power Research Institute (EPRI) is recognized as the major impetus behind the development of amorphous-metal transformers. Several technologies were developed to allow amorphous metal to be used in transformer manufacturing in a manner similar to silicon steel. To simulate the thickness of silicon steel sheet, a process called ***prespooling*** produced multiple-layered packages of amorphous metal ribbon that subsequently formed a strip of multiple-layer thickness. The replacement of grain-oriented silicon steel in the transformer cores with amorphous metal reduced core losses by 75%. In the United States alone, the EPA estimated the equivalent saving in energy at over 70 million barrels of oil while reducing the volume of harmful CO_2, NO_x, and SO_2 emissions. Figure 10-11 shows a variety of amorphous metals for electrical applications.

 Amorphous metals—*http://coatings.liquidmetal.com,http://www.metglas.com/*

SELF-ASSESSMENT

10-1. The carbon content in a 1020 steel is

a. 20% **b.** 0.20% **c.** 2% **d.** 0.20

10-2. Steel with alloying elements to resist corrosion is called

a. CRS **b.** HRS **c.** CRES **d.** TTT

10-3. Metal with no iron is

a. an Alloy **b.** Ferrous **c.** Nonferrous **d.** Magnetic

10-4. As the steelmaking flowline shows, ___________ is a major source of raw material.

a. Scrap **b.** Lead **c.** Aluminum **d.** Silica

10-5. The carbon content of cast iron is

a. 0.6% to 1.0% **b.** 2% to 4% **c.** 4% to 6%

10-6. Graphite in a metal promotes ___________.

a. Lubricity **b.** Malleability **c.** Toughness **d.** Conductivity

10-7. Silicon added to gray cast iron helps produce ___________.

a. Strength b. Malleability c. Ductility d. Magnetism

10-8. The least expensive of all metallic materials is

a. Carbon steel b. Gray cast iron c. Wrought iron d. CRES

10-9. Structural steel shapes are produced from ___________.

a. Billets b. Slabs c. Ingots d. Blooms

10-10. Another name for nodular cast iron is

a. Malleable b. Gray c. Ductile d. Brittle

10-11. What ferrous material was used to construct large buildings prior to the development of steel in the late 1890s?

a. Cast iron b. Wrought iron c. Ductile iron d. Stainless

10-12. Slag is found in ___________.

a. Cast iron b. Wrought iron c. Low-carbon steel d. CRES

10-13. Another name for plain carbon steel is

a. Low-carbon steel b. Mild steel c. Wrought steel d. HSLA

10-14. To compete with lighter metal and plastics; the steel industry offers ___________.

a. Low-carbon steel b. Mild steel c. Wrought steel d. HSLA

10-15. Copper forms in austenite to reduce ___________.

a. Rust b. Spheres c. Graphite d. Strength

10-16. In addition to corrosion resistance, chromium also gives ___________ to steel.

a. Magnetism b. Formability c. Castability d. Strength

10-17. A steel alloy with nickel–cobalt and rated with the highest fracture toughness is

a. AISI-SAE 2430 b. AerMet 100 c. HSLA d. AISI 400

10-18. The metal designation system used to indicate the alloy and percentage of alloy in steel is

a. AA b. ASTM D2000 c. AMS d. AISI-SAE

10-19. SAE 4000 steel contains the major alloying element ___________.

a. Manganese b. Molybdenum c. Nickel d. Lead

10-20. Aluminum, as an alloying element, improves the steel's ___________.

a. Magnetic properties
b. High-temperature properties
c. Mechanical properties
d. Processing properties

10-21. A steel with alloying elements to resist corrosion is

a. CRS b. HRS c. CRES d. TTT

10-22. Two major alloying elements needed for high-temperature steel that are not mined in the United States are

a. Lead and tin
b. Nickel and silicon
c. Chromium and cobalt
d. Iron and aluminum

10-23. Martensitic stainless steels have at least ___________ of Cr.

a. 0.8% b. 4% c. 12% d. 21%

10-24. From Table A–11, what is the hardness capability of type 302 stainless?

a. 85 HRB to 40 HRC b. 75 HRB c. 325 HB d. 180 HB

10-25. From Table A–12, what ultrahigh-strength steel would provide these properties: tensile strength, 2140 MPa; yield strength, 1650 MPa; hardness, 54.5 HRC?

a. UNS G41300
b. UNS K44220
c. UNS G61500
d. UNS T20811 Mod

10-26. Write the chemical symbol for cementite and attach another name to this ingredient of steel.

10-27. List one application, the key structural difference, and the best quality for each of the following cast irons:

a. ductile cast iron,

b. gray cast iron,

c. white iron,

d. malleable iron,

e. wrought iron.

10-28. What is the major function of the following in alloy steel:

a. copper,

b. vanadium,

c. chromium,

d. aluminum,

e. lead,

f. nickel, and

g. tungsten?

10-29. Describe the major advantage of each of the following stainless steels:

a. martensitic,

b. austenitic, and

c. ferritic.

10-30. What is the group name of the 14 elements in Period 6 of the periodic table that contribute to the stable oxide layer on alloy steels?

10-31. From Tables A-11 and A-12, name the two highest hardness steel alloys.

REFERENCES & RELATED READINGS

ASM INTERNATIONAL. *Alloy Finder*. CD-ROM software. Materials Park, OH: ASM International, 1996.

ASM INTERNATIONAL. *Worldwide Guide to Equivalent Irons and Steel*, 4th ed. Materials Park, OH: ASM International, 2000.

DANIELSON, PAUL. "Micro-Images of Metals," *Advanced Materials & Processes*, July 1999, p. 35.

DAVIS, J. R., ed. *Metals Handbook Desk Edition*, 2nd ed. Materials Park, OH: ASM International, 1998.

FELLERS, WILLIAM O. *Materials Science, Testing and Properties for Technicians*. Englewood Cliffs, NJ: Prentice-Hall, 1990.

GILLESPIE, RONALD J., Donald R. Eaton, David A. Humphreys, and Edward A. Robinson. *Atoms, Molecules, and Reactions. An Introduction to Chemistry*. Englewood Cliffs, NJ: Prentice-Hall, 1999.

GROOVER, MIKELL P. *Fundamentals of Modern Manufacturing. Materials, Processes, and Systems*. Englewood Cliffs, NJ: Prentice-Hall, 1992.

HIGH-STRENGTH STEEL (HSS) BULLETIN, 18th ed., Spring 1998. ULSAB: A Lightweight, Affordable, Structurally Superior Steel Auto Body, Auto Steel Partnership (A/SP), Southfield, MI 48075–1137.

INCO. *A QUICK REFRESHER ON STAINLESS STEEL*. New York: INCO, 1999.

KHATAK, H. S. ed. *Corrosion of Austenitic Stainless Steels*. Materials Park, OH: ASM International, 2002.

KUHN, HOWARD A. *METALWORKING TECHNOLOGY UPDATE*. National Center for Excellence in Metalworking Technology, 3rd Quarter, 1992.

LAMB, STEPHEN, ed. *Practical Handbook of Stainless Steels and Nickel Alloys*. Materials Park, OH: ASM International, 1999.

LEFFLER THOMAS J. Manager of Advertising and Projects, Findlay Industries, Crooks Road, Troy, MI 48084. Letter Dated June 7, 1995.

LIGHT TRUCK STRUCTURE (LTS) STUDY. Final Report, Automotive Applications Committee, AISI, AAC-2686 0498 4M HM, Southfield, MI 48075–1137.

MANGONON, PAT L. *The Principles of Materials Selection for Engineering Design*. Englewood Cliffs, NJ: Prentice-Hall, 1999.

MIELNIK, EDWARD M. *Metalworking Science and Engineering*. New York: McGraw-Hill, 1991.

OSTWALD, PHILLIP F. *Engineering Cost Estimating*, 3rd ed. Englewood Cliffs, NJ: Prentice-Hall, 1992.

PARTNERSHIP FOR EXCELLENCE—AUTO/STEEL PARTNERSHIP ACCOMPLISHMENTS, REVISED JULY 1999, Southfield, MI.

PRISM. AMERICAN SOCIETY FOR ENGINEERING EDUCATION, NOVEMBER 1999.

SAMUELS, L. E. *Metals Engineering: A Technical Guide*. Materials Park, OH: ASM International, 1988.

ULTRALIGHT STEEL AUTO BODY (ULSAB). Final Report, 2nd ed. American Iron & Steel Institute (AISI), May 1998.

Periodicals

Advanced Materials & Processes

Journal of Materials Education

EXPERIMENTS & DEMONSTRATIONS IN METALS FROM NATIONAL EDUCATORS' WORKSHOPS

EMSET 2 CD-ROM *Experiment in Materials Science, Engineering and Technology*, 2nd ed. ISBN 0–13–030534–0. Over 300 experiments and demonstrations available in easy-to-adapt Acrobat Reader Format; Department licensed. Call 1 800–922–0579 or go to www.prenhall.com

Module 11 Nonferrous Metals

After studying this module, you should be able to do the following:

11–1. Differentiate among nonferrous metal alloys, structures, processing, and properties.
11–2. Cite examples of properties and applications of major classes of nonferrous metals.
11–3. Decipher designations for nonferrous metals and alloys.
11–4. Use tables and references on ferrous metals for materials selection.
11–5. Describe some of the advances in and barriers to the use of advanced metals and metals processing, such as metal–matrix composites (MMCs), superplastic forming (SPF), and light aluminum alloys.
11–6. Recall how batteries work and describe some advances in battery technology.
11–7. Use the website addresses denoted by the icon* to seek out current developments and emerging technology related to nonferrous metals.

*The icon found throughtout this book will link you to Internet sites related to topics being covered. The dynamic nature of the Web brings frequent changes, so some of these sites, while available at the time of writing, may not be up now.

TABLE 11-1 PROCESSABILITY OF COMMON METALS

	Steel	Iron	Aluminum	Copper	Nickel	Magnesium	Zinc	Titanium	Tin
Castability									
Centrifugal	■			■	■				
Continuous	■			■					
Investment	■		■	■	■				
Permanent mold	■	■	■	■	■	■	■	■	■
Die casting			■	■		■	■		■
Formability									
Cold	■		■	■	■	■	■	■	
Hot	■		■	■	■	■		■	
Machinability	■	■	■	■	■	■	■	■	
Powder metal Compacting	■	■	■	■					
Weldability									
Gas	■		■		◢			■	
Inert arc	■		■		◢	■		■	
Electrical Resistance	■		■	■	◢	■		■	

■ – Common on most alloys

◢ – Used on some alloys

Source: Adapted from *Machine Design*, March 15, 1976.

Although the ferrous alloys are used much more than other metals, nonferrous metals comprise three-fourths of the known elements. However, the problem of extracting certain metals from the earth or ocean and the practical commercial value limits the number of metals used in significant quantity. The most commonly used nonferrous metals include aluminum, copper, zinc, nickel, chromium, tin, magnesium, beryllium, tungsten, lead, molybdenum, titanium, tantalum, and the noble (or precious) metals such as gold, platinum, silver, and rhodium. With the demand for superior properties made by advances in technology, each metal receives careful consideration for its potential value in unique circumstances. Platinum is used in catalytic converters in automobile exhaust systems to aid in the removal of pollutants from our environment. Ruthenium, which has little practical application by itself, is alloyed with platinum for use in thin-film circuitry (ceramic substrates with printed circuits) for solid-state electronics. In addition to the properties of a material, the processability of the material bears heavily on the selection process. Table 11-1 is a simplified listing of metals and how they are commonly processed.

11.1 ALUMINUM

The most abundant of metals, aluminum is locked up with other elements in the form of bauxite ore. In 1852, this metal was only rarely available in usable form and was prized by Emperor Napoleon as dinnerware; at that time, it sold for $545 a pound. In 1886, Charles Hall in Ohio and Paul Heroult in Paris, France, simultaneously developed an electrolytic process for the economical extraction of bauxite. Even though the price had dropped to 24 cents per pound by 1964, the cost of electrical energy kept the price of virgin aluminum high. About 4% of our metal-processing energy is consumed in the production of aluminum. However, the recycling of aluminum has drastically reduced the cost of the metal.

Figure 11-1 shows the steps in the production of aluminum from bauxite. Figure 11-2 shows products that stem from the properties of aluminum; Table 11-1 showed that it is processable by the major methods. The Aluminum Association designates wrought aluminum and cast aluminum by two numbering systems. The meanings of the digits in each system are shown in Figure 11-3. The last digit in cast aluminum indicates cast (0) or ingot (1 or 2). Temper designations following the four digits in the wrought system indicate thermal treatment (T through T10) and solution heat treatment (W), as-fabricated (F), annealed (0), or strain hardened (H).

Aluminum has many favorable properties, including excellent thermal conductivity, good strength, corrosion resistance, and light weight (one-third that of steel). Aluminum of high purity (1000 series), or at least 99% purity, finds applications as thin foils for electronics; as plate and bar stock for chemical apparatus, fuel tanks, and cooking utensils; and as paint pigments and in gasoline production. Aluminum alloyed with copper (2000 series) has extended tensile strength (greater than mild steel), improve machinability, and good specific strength but has reduced corrosion resistance and weldability. Aluminum–manganese alloys (3000 series) provide good strength, workability, formability, and weldability, and the high corrosion resistance makes this a popular alloy for applications such as highway signs, furniture, cooking utensils, architectural shapes, and truck panels. The aluminum and silicon alloys (4000 series) have high wear resistance and relatively low coefficients of thermal expansion; they permit good castings and forgings and find applications as welding and brazing rods, anodized architectural shapes, marine equipment, and automotive pistons. Aluminum alloyed with magnesium (5000 series) yields moderate to high strength, good corrosion resistance in marine environments, and good weldability and formability for uses as extrusions, ship and boat parts, and automotive structural parts. The 6000 series alloys of aluminum, magnesium, and silicon have excellent corrosion resistance, plus good weldability, machinability, and formability; they find uses as bridge rails, automotive sheets, piping, and extrusions. A 6000 series aluminum alloy is replacing 5000 series aluminum alloys, which develop stretcher strains (elongated markings that appear on the surfaces of some sheet materials when deformed just past their yield point). Up until recently, 5000 series alloys were widely used as automotive body panels. Nippon Steel Corp. and Sky Aluminum Co. of Tokyo, Japan, have developed a 6000 series alloy with slightly lower density than a 5000 series alloy that possesses the four key properties required of automotive body panels: strength, formability, bending processability, and corrosion resistance. Nissan Motor Co. will be the first automaker to use the new alloy in the hood of its new Skyline GT-R model. The aluminum–zinc series 7000 alloy is heat treatable with tensile strength up to 606 MPa (88,000 psi) and has good corrosion resistance in rural areas but is poor in marine environments. Uses for the 7000 alloys are chiefly as aircraft structures and other equipment requiring high specific strength. The 8000 series includes alloys of combinations not already mentioned, for example, titanium or zirconium. Table A-11 compares various aluminum alloys. Figure 11-2e is a schematic of applications of aluminum alloys in the Boeing 777 as well as advanced titanium. Figure 11-4a shows an example of a newer breed of hybrid bicycles using aluminum for the frame and other components. The "comfort bike" is a cross between a mountain bike and a beach cruiser with adjustable shock absorbers in the front fork and a spring-loaded seat post for comfort. Figure 11-4b shows the oversized "power" tube frame made of Alpha™ aluminum. Compared to the steel bikes in Figure 10–6b, this bike is much lighter and has many more features, but a much higher price range—from $400 to $750 or more.

Corrosion resistance in aluminum is a result of the natural occurrence of an aluminum oxide film that protects the metal from further corrosion. By making anode of aluminum, the metal is oxidized (i.e., it loses electrons to form metallic ions at the anode of an electrolytic cell). The surface of the aluminum is thereby electrochemically converted from metallic aluminum to aluminum oxide (Al_2O_3), which is an extremely hard and porous ceramic. (Refer to Module 5 for further information on oxidation and corrosion.) For best corrosion protection, the porosity is reduced by a sealing operation by immersing the anodized metal in hot water. This converts the oxide to a hydrated form, which results in a reduced pore size. Dyes can be added to the sealing

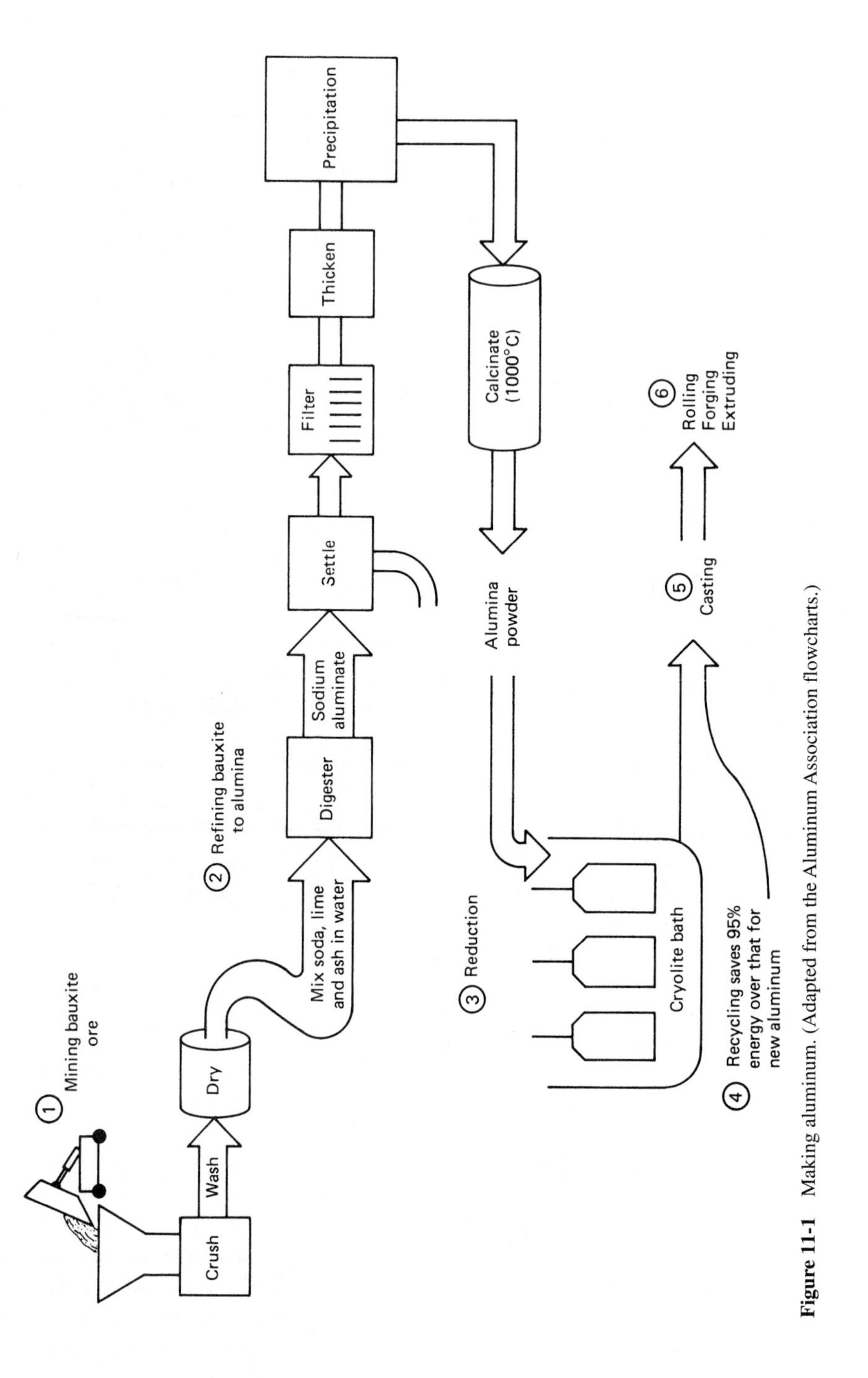

Figure 11-1 Making aluminum. (Adapted from the Aluminum Association flowcharts.)

water that will permeate the coating and thus become an integral part of it. Once formed, the coating resists chipping or peeling. Automobile trim is processed in this way.

Anodizing (the word *anodize* is a contraction of *anode* plus *oxidize*) is mostly applied to aluminum and magnesium alloys, but it may also be used with zinc, titanium, and other less

Figure 11-2 (a) Roof structures for arenas and gymnasiums are usually 6063 or 6061 extruded tube, covered with 5xxx alloy sheet. Alloy 6262 is also used in brake housings and brake pistons. (b) The inner hull stiffener structure of high-speed, single-hull yachts is constructed using 5083 machined plate for hulls, hull stiffeners, decking, and superstructure. (c) Automotive structures are likely to employ increasing amounts of 5754–0 formed sheet for parts such as internal door stiffeners or the entire body-in-white. (d) Aircraft internal structures include extrusions and plate of 2xxx and 7xxx alloys such as 2024, 2124, and 2618. If used as external sheet skin, they may be cladded with higher-purity cladding to provide greater corrosion protection to the Al–Cu alloys that will otherwise darken with age. (Aluminum Association)

common metals. Although primarily decorative, anodized coatings fill a major role in the protection of metal surfaces from corrosion. Most aluminum alloys can be ***hard anodized,*** forming thicknesses that range from 2.5 to 20 mm. Thus, anodizing is one of many surface modifications performed on metals that is helping reduce the cost of corrosion in the United States, amounting to $3 billion a year according to Battelle and NIST estimates. Concerns about the toxicity of materials and their effect on the environment have led to an improved process for anodizing aluminum. The older process used chromic acid and sodium dichromate, which contained hexavalent chromium, a carcinogen. The newer process uses sulfuric acid and nickel acetate, which produces thinner coats (2 to 5 mm thick) that offer better corrosion and fatigue resistance than the older method. Aluminum oxide formed in the anodizing process is quite brittle and rough, creating a poor wear surface. Adding zirconium by ion implantation after anodizing greatly improves its wear resistance (see Section 21.6).

Alloying, cold working, and heat treatment can reduce the corrosion resistance of aluminum. For example, intergranular corrosion can occur in a precipitation-hardened aluminum

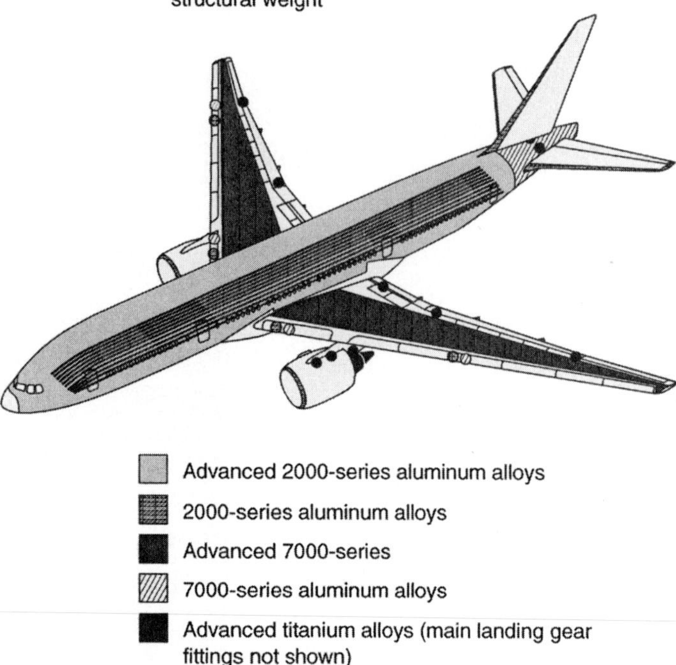

Figure 11-2 continued (e) Advanced nonferrous alloys for the Boeing 777. Various aluminum and titanium alloys (shown with symbols) used on the 777 improve durability while saving on structural weight. (Courtesy of the Boeing Company)

due to the lack of homogeneity (i.e., some areas have a high copper content and become anodic to copper-depleted areas). However, the newer ***spray formed aluminum alloy*** billets being produced by Osprey Ltd., UK—a high-strength aluminum–zinc alloy, a low-density aluminum–lithium alloy, and a wear-resistant aluminum–silicon alloy—provide many microstructural and property advantages from this rapid solidification technology, and at a lower cost because only one step is needed from molten metal to cylindrical billet.

11.1.1 Superplastic Forming (SPF) and Diffusion Bonding (DB)

When deformed in tension at elevated temperatures, some advanced aluminum alloys are also capable of achieving large, uniform elongation (≥500%) at relatively low stresses prior to localized thinning and fracture. The property of superplasticity joined with diffusion bonding produces a space-age composite. ***Superplasticity*** decribes the property of a metal alloy that is not considered very ductile at or near room temperatures but at certain higher temperatures possesses extremely high ductility. Titanium, with normal plastic deformation around 20%, at certain high temperatures can be deformed as much as 2000%. Diffusion bonding or hot pressing is a widely used technique for metal–fiber composite fabrication. Examples of such composites are aluminum and titanium alloys reinforced with boron or borsic fibers. The matrix is usually in sheet form, with the reinforcing filaments or wires mechanically spaced and oriented between the sheets to form alternating layers of filaments and matrix. Surface coatings applied by carburizing, nitriding, and bimetallic castings are also examples of diffusion bonding, but they do not use metal filament reinforcement.

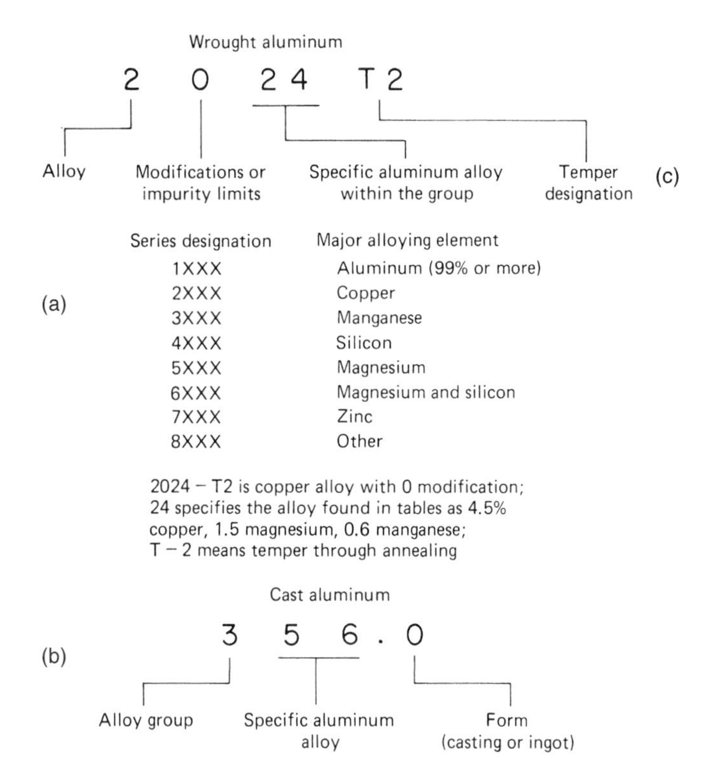

Figure 11-3 Aluminum Association number system: (a) Wrought alloys. (b) Casting alloys.

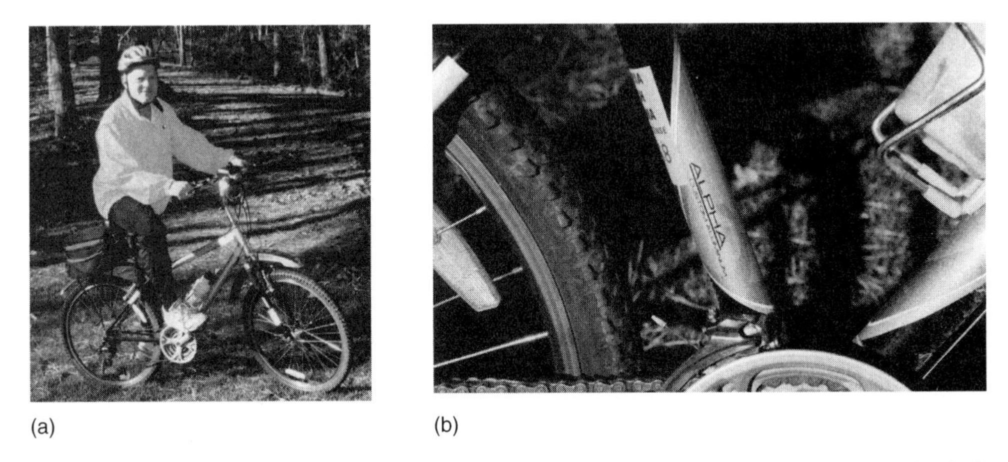

Figure 11-4 (a) Hybrid "comfort bikes" provide an upright riding position with mid-size tires for on-road and off. (b) The oversized "power" tube frame made of Alpha™ Aluminum, described as the highest quality, lightest weight, and lowest density bike-grade aluminum available. The Trek Company processes the tubes to make them thicker and stronger in key locations or thinner and lighter where possible.

The first step in diffusion bonding is to wind the fiber over a metal-foil-covered drum. The resulting mats are cut and made into tapes by diffusion bonding under high temperature and pressure. The tapes, in turn, are cut into plies, stacked in a die, and consolidated by further diffusion bonding. Complex structural shapes can be formed in a single operation by the large forming strains produced by SPF. Superplastic forming/diffusion bonding (SPF/DB) is used, for example, to fabricate major structural components of aircraft out of several alloys. Figure 11-5 depicts the SPF process. The older fabrication method (Figure 11-5a) using machining and chemical milling

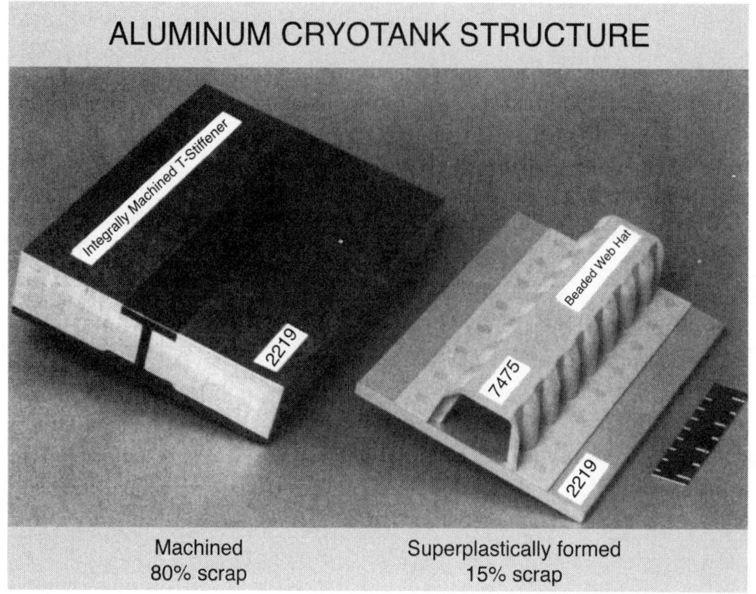

(a)

Roll spot weld pattern
Gas pressure tube
Envelope (sectioned) before superplastic forming
Welded
Gas pressure
Envelope (sectioned) partially formed
Face sheets
Gas pressure
Limiting fixtures
Envelope (sectioned) with face sheets forming complete tank

(b)

Figure 11-5 (a) An aluminum cryotank structure for the space shuttle produced by old machining methods (left) and newer SPF/DB methods (right). Note alloy designations—2219 and 7475. (b) Steps for superplastic forming with diffusion bonding. (NASA)

resulted in 80% of the starting plate machined away as scrap. The newer built-up structure approach, using an SPF beaded-web stiffener that was joined to the skin material by resistance spot-welding techniques, produced only a 15% loss in scrap material while reducing the material cost for fabrication (Figure 11-5b). SPF increases the efficiency of hat stiffeners by as much as 60% by using more complex configurations, such as the beaded web, which increases stiffness. These reductions in cost and increases in efficiencies will permit the more costly aluminum–lithium alloys to be employed, which will lead to further reductions in weight and cost.

A new material for automobiles is even lighter and stronger than aluminum. ***Aluminum foam,*** developed by Karmann Corp., is processed to leave the foam with a smooth aluminum interior and exterior. This new material is reported to produce automobile bodies that are stronger, stiffer, and lighter than existing prototypes. A prototype chassis built by Volkswagen demonstrates the viability of the process.

11.1.2 Tube Hydroforming and Sheet Hydroforming

Hydroforming technology has been used by industry for many years wherever complex shapes of steel or aluminum must be formed with a high degree of accuracy. It has been only recently, however, that designers and engineers have become convinced that a tubular component can replace a stamped assembly with attendant benefits in cost, weight, structural integrity, dimensional capability, and design flexibility. Today, tube hydroforming is recognized as a cutting-edge technology in the automotive industry, where it is being used for spaceframes, engine cradles, roof headers, frame rails, and other subassemblies. In essence, ***tube hydroformed*** parts are replacing stamped and welded components, resulting in lighter, stronger, and fewer required parts.

The tube hydroformed part-making process (Figure 11-6a) incorporates four steps: (1) making the tube, (2) bending the tube, (3) preforming the prebent tube, and (4) hydroforming the

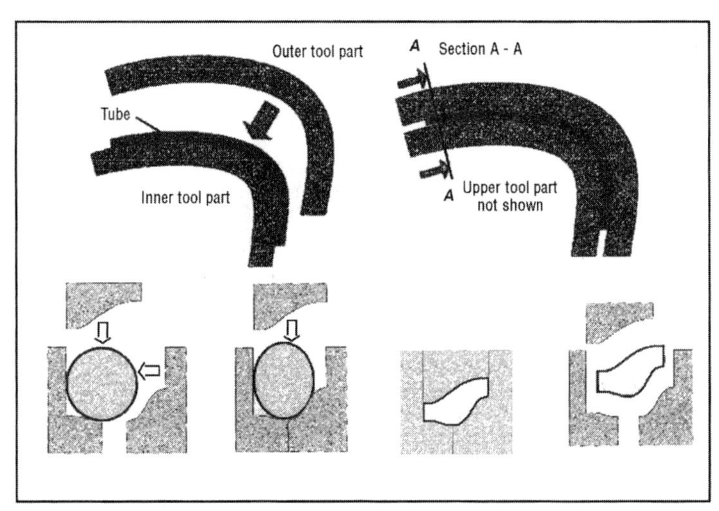

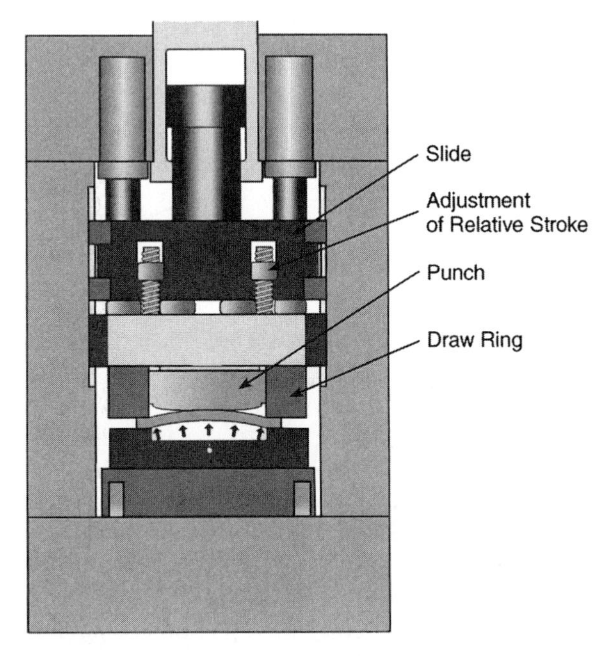

(b) Sheet hydroforming

Figure 11-6 Tube and sheet hydroforming. (a) Rapid prototype tube hydroforming is becoming a fast-growth manufacturing process for automotive exhaust and frame components as well as aerospace components. Benefits include reduced weight, fewer components and assembly operations, and improved strength and quality. (b) Sheet hydroforming involves the forming of large-surface body shells. For example, automobile roofs, hoods, or doors can be formed using large presses that can exert pressures up to 100,000 kN and employ over a dozen programmable axes to provide greater flexibility. (*Advanced Materials & Processes*)

preformed tube into the final component shape. A Computer Numerical Control (CNC) bender is used in the initial steps to bend the tube blank to approximately the final shape. Later steps use a die built to the dimensions of the final component. Fluid pressure (water or oil) is used to force the tube to flow completely into the die. In some cases, punches may be mounted in the die to pierce holes in the forming tube, and slides may be mounted in the die to provide indented surfaces on the tube sidewalls.

Tube hydroforming is the most widely used process, but sheet hydroforming is growing in importance, and aluminum is also emerging as a feasible material especially for hydroforming automobile parts. The ULSAB's roof was ***sheet hydroformed*** (Figure 11-6b) for weight reduction. The sheet chosen had a thinner gauge (0.70 mm) with greater strength (210 MPa) than its replacement due to the work-hardening effect achieved from using fluid pressure to stretch the blanks (plastic deformation) in the opposite direction toward the punch. ***Hydrostretch forming*** is a related process capable of using conventional double-acting presses. Quality is improved because there is no metal-to-metal contact on the outer portion of the part surface.

11.1.3 Tailor-Welded Blanks

Tailored blanks are generally steel blanks of various gauges and grades that are welded into a single blank prior to forming in a press (see Figure 11-7). This technique allows manufacturing

(a) Rear quarter panel

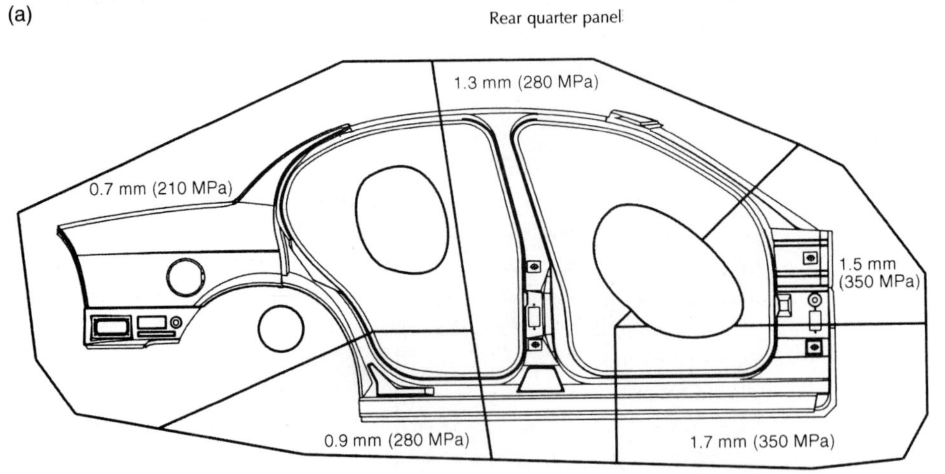

(b) Tailored blank: panel body side outer

Figure 11-7 Tailor welded blanks. A tailor welded blank is composed of several blank, flat sheets of steel each having different strengths, thicknesses, and coatings. A laser blank welding process welds these individual sheets together before they are stamped into a single automobile body panel. (*Advanced Materials & Processes*)

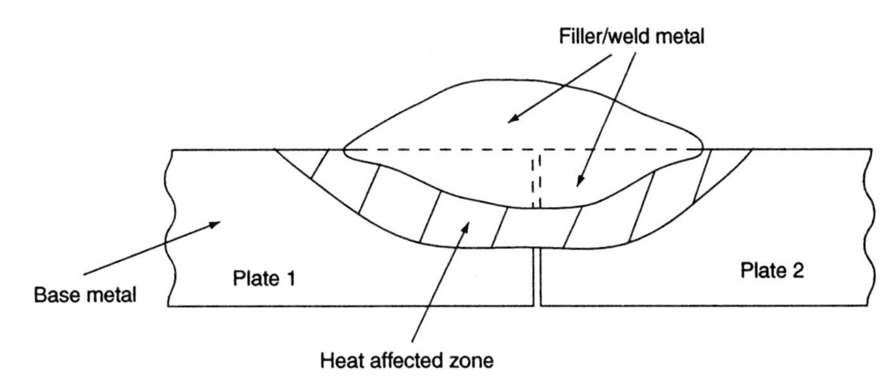

Figure 11-8 Fusion-welded metal joint showing the heat-affected zone (HAZ). High temperatures cause microstructural changes in the HAZ and thus its properties. It is in this region where welding failures occur. Even though the base metal adjacent to the HAZ has not changed in its structure, it is usually left in a state of residual stress. Laser welding produces a much narrower HAZ.

engineers to "tailor" the best properties of different steels so that they are located precisely where they are needed. The result is reduced mass, reduced tooling and assembly costs, dimensional variation reduction, improved crash reduction, reduction in the number of spot welds, part integration, and improved structural integrity, quality, and manufacturing efficiency. The database for aluminum tailor-welded blanks is not extensive, but laser, electron beam, arc, and friction stir-welding techniques can be used to fabricate aluminum blanks with advantages such as less weight, reduced waste generation, and lower stamping and tooling costs. The blanks are laser welded for a smaller heat-affected zone (HAZ). (Figure 11-8 is a schematic representation of a fusion-welded metal joint showing the heat-affected zone.) A small portion of the metal between the HAZ and the filler metal is made up of fused base (parent) metal. The HAZ experiences changes in its microstructure and subsequent changes in its properties due to the high temperatures and subsequent cooling. For steels, the material may have formed austenite, which would produce brittleness. Residual stresses also may form in this region, which would weaken the joint. (Refer to Module 25 for other references to the HAZ.) Nearly half of ULSAB's mass consists of parts using tailored blanks, which promote smooth load flow, reduce structural discontinuities, and allow for the combination of thicker and higher-strength materials within the same part.

 Aluminum—*http://me.edu/2.01/taxonomy/characteristics/aluminum.html*

11.1.4 Aluminum in MMCs

As explained in Module 25, metal–matrix composites (MMCs) can involve fiber reinforcing with boron, carbon, and silicon. Aluminum alloys with ceramic-particle reinforcing are prime candidates for castable MMCs. Ingots of aluminum MMCs can be used for many casting processes (Figure 11-9).

11.2 COPPER

Both as an alloying element in alloys such as steels and aluminum and as a base for copper alloys, copper is a valuable metal. Included in the copper-based alloys are brass (copper and zinc), bronze (copper and tin), cupronickel (copper and nickel), leaded copper, leaded brass, aluminum bronzes, nickel silver (copper, nickel, and zinc), and high copper. In Figure 11-10 a sculpture depicts the wide variety of copper forms and indicates Cu's affordability and many useful properties. Table 11-1 shows that copper and its alloys are processed easily by most common methods, except certain types of welding. There are several systems of designation for copper and its alloys, but these are beyond the scope of this book.

Besides its value as an alloy, copper's greatest value is as an electrical conductor. High purity of over 99.9% is sought to attain the best conductivity and formability. But due to its

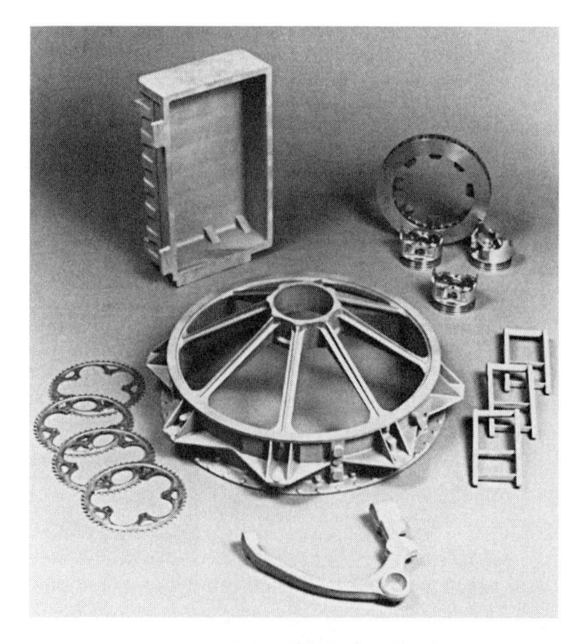

Figure 11-9 Casting alloys. Aluminum alloys with ceramic particle reinforcement can yield near-net shapes. Representative aluminum-alloy parts cast from MMC foundry ingot include, clockwise from top left, an investment-cast pressure vessel; a green-sand cast automobile brake rotor; permanent-mold cast pistons; shell-investment cast airbus components; a dry-sand cast automobile control arm; and high-pressure, die-cast bicycle sprockets. At center is an investment-cast aircraft-camera gimbal. (*Advanced Materials & Processes*, ASM International)

Figure 11-10 This cleaver copper sculpture in the Arizona Science Museum demonstrates the many forms of copper. Note the various copper forms and see if you can recognize what application they were taken from.

scarcity and weight, the trend is, when possible, to move away from copper for this use. Copper and copper alloys develop a protective film that will not corrode in water or nonoxidizing acids. It will corrode, however, if liquids, solids, or gases break down the protective film or if other conditions cause electrolysis to develop. In brass, the presence of oxidizing chemicals can result in dezincification, in which the zinc dissolves, leaving a porous metal sponge. Spray forming of copper is now an established technology along with the production of preforms and near-net shapes from ferrous and nonferrous alloys. This technology will be discussed in other sections on powder metallurgy, rapid prototyping, spray forming of composites, and various laser-forming methods throughout the text.

Copper is an excellent example of new applications for traditional engineering materials to satisfy modern needs as a consequence of the never-ending changes in the technology of materials brought about by industrial research and development efforts. Recent breakthroughs in brazing, external corrosion protection and laser welding are making new copper radiators for automobiles and commercial vehicles competitive with their aluminum counterparts. These new processing techniques, combined with copper's superiority in thermal conductivity, corrosion resistance, and mechanical strength, result in radiators that are significantly lighter, more compact, and more durable than those presently in wide use throughout the world. Although copper has been known throughout history as a coinage metal along with silver and gold (see Figure 11-11), it is the only element that is now primarily used in making the world's coinage. The properties that place it in demand are its durability and corrosion resistance (to air oxidation). It is used to coat (plate) other materials. For example, since 1982, U.S. pennies have been copper-coated zinc; current dimes and quarters are copper–nickel alloy laminates or bimetallics; and U.S. nickels have consisted of 75% copper plating over 24% nickel since 1948. The eight new euro coins (Figure 11-11), the official coinage of the 12 European Monetary Union countries, are made up of various elements plated in copper. For example, the 1-, 2-, and 5-cent euro pieces are made of copper-plated steel. The material used for the 10-, 20-, and 50-cent pieces is "Nordic Gold," a Cu–Al–Zn–Sn alloy that is said to be " difficult to counterfeit." The 1- and 2-euro coins are bimetallic with a center disk and outer ring of different materials. While the 1-euro coin features a three-layer magnimat center surrounded by a brass ring, the 2-euro coin is designed "the other way around" with a brass center and magnimat ring. Magnimat is a Cu–Ni alloy developed in 1969 and was adopted for use because of its "high reliability with vending machines." The 2 euro bimetallic coin consists of two sections, an outer ring and an inner disk. The outer ring is 75% copper–25% nickel; the inner disk is pure nickel (99.2% min) clad on both sides with nickel brass (75% copper–20% zinc–5% nickel), with the nickel layer making up 12% of the clad's total thickness. This coin is 25.75 mm in diameter and weighs 8.5 g. All eight coins may be distinguished by form, color, weight, and thickness. Table A-11 lists several copper alloys with corresponding designations, compositions, and properties. Figure 11-12 shows a popular use of copper-base bronze for casting statues, a practice centuries old.

11.3 NICKEL

Nickel ranks twenty-fourth in abundance among the elements in the earth's crust. The world mine production of nickel totaled 1.1 million tons in 1998, the highest output ever recorded. The principal producing countries were Russia (22%), Canada (19%), New Caledonia (13%), and Australia (12%). Production of refined primary nickel is dominated by three major producers: Together, Norilsk (Russia), INCO (Canada), and the UK account for approximately one-third of world output. Nearly 80% of nickel is used in alloys. About 57% is used in making stainless steel corrosion resistant, some 10% is used in making heat-resistant and low-alloy steels, and 13% is used in nickel-based alloys to improve their heat-resistant and corrosion-resistant properties. While alloyed in steel, copper, and other materials, nickel is also electroplated onto objects for decorative and corrosion-resistance purposes. Nickel-based ***superalloys*** are widely used for the hottest parts in aircraft gas turbine engines, where they

Figure 11-11 On January 1, 2002, eight new euro bimetallic coins went into circulation in 12 European Monetary Union countries. Britain, Sweden, and Denmark did not adopt this currency when it became the official currency of the Europe Union (EU) in January 1999, when stock exchanges and large businesses began using euro notes to keep track of their finances. The coins on the left in the graphic will all have a common design or face on the obverse side (front or main side). This common face represents Europe. On the reverse side (the national face), each country will have its own design. For the coins in the figure the national face is that for Finland. All the coins are round except the 20-cent euro, which has a rib design called the "Spanish flower shape." The euro coins could be called "green" coins since the producer of the coin blanks has reduced the nickel content "as far as possible to avoid the potential risk of provoking allergic reactions." (*Advanced Materials & Processes*)

Figure 11-12 Bronze sculptures often memorialize heroes. Note the detail and realism achieved on the statue of Arthur Ashe in Richmond, Virginia. Ashe, an African-American tennis champion, overcame prejudice and disadvantage to reach greatness. Emphasizing education and good sports ethics, he traveled the world with class and dignity, serving as a role model for all Americans.

Figure 11-13 Incoloy alloy 864 was developed specifically for automotive exhaust system flexible couplings and bellows. These components are attacked by a wide array of corrosion mechanisms. In addition, higher exhaust temperatures and warranties lengthened to 100,000 miles or 10 years of service life are becoming prevalent, requiring a longer life for these parts. (*Advanced Materials & Processes*)

constitute nearly 50% of the weight of these advanced aircraft engines. These alloys can withstand temperatures of up to 1110°C. Directional solidification, perfected in the 1970s, uses nickel-based superalloys (e.g., IN-6203, developed in 1981) that withstand temperatures up to 1020°C.

Common alloys of nickel include Monel (nickel–copper), Inconel (nickel–chromium), Hastelloy (nickel–molybdenum–iron), Duranickel (nickel–aluminum), and Illium (nickel–chromium–molybdenum–copper).

Table A-11 shows the composition and properties of certain nickel alloys. Nickel is magnetic up to 360°C and is alloyed with iron, aluminum, cobalt, and copper in the powerful permanent Alnico magnet. Figure 11-13 shows a flexible automotive exhaust coupling that is

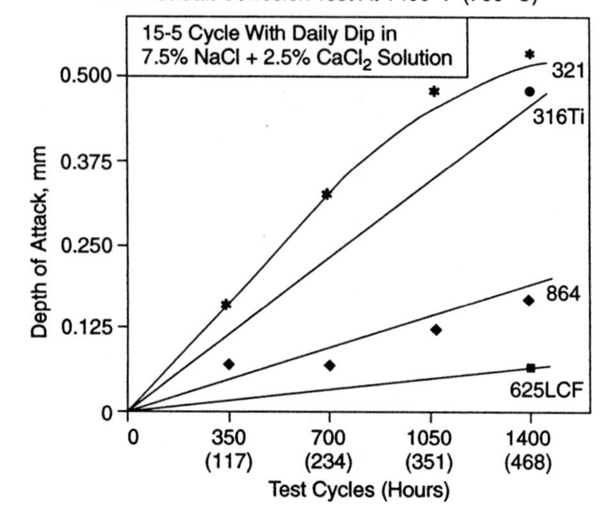

Figure 11-14 A comparison of hot salt corrosion resistance of stainless steel (SS) 321 and 316 Ti and nickel alloys 864 and 625 LCF. Depth-of-attack measurements for nickel alloys and stainless steels in a hot salt corrosion test at 1400°F (760°C).

made of an alloy composed of (in wt %) 34 Ni, 2.1 Cr, 4.2 Mo, 1.8 Si, 0.6 Ti, bal Fe, designated Incoloy 864 by INCO Alloys International Inc. Two of the main effects of higher nickel content in austenitic alloys are an increase in metallurgical stability (maintaining its strength during processing) and an increase in resistance to chloride stress–corrosion cracking. Figure 11-14 is a graph comparing the 760°C hot salt corrosion resistance of stainless steels (SS) and nickel alloys. The presence of de-icing salts and alternate wet and dry conditions support aqueous corrosion mechanisms such as stress–corrosion cracking, pitting, and general corrosion of automotive exhaust system components such as wire braid and exhaust gas recirculation tubes.

www.INCO.com

11.4 MAGNESIUM

A chief source of magnesium is seawater. Magnesium is therefore abundant in supply and 100% recyclable. About 6 million tons of magnesium are available for processing in each cubic mile of seawater, but the electrolysis process required to extract the metal from mineral deposits or seawater increases its cost. The lightest of all structural metals, with a specific gravity of 1.75, magnesium weighs 1.5 times less than an equal volume of aluminum and 4 times less than zinc. It has good strength, stiffness, and dimensional stability, and in pure form and alloyed with other elements, it provides a high strength-to-weight ratio. Magnesium alloys have relatively high thermal and electrical conductivities, good energy absorption characteristics (i.e., its ability to damp vibration compares with cast iron), and nonmagnetic properties. It is often used as an alloying element in engine parts and to produce wheels for racing cars. Rapid technical advances in magnesium alloys and processing have increased its use by automakers and suppliers. A typical vehicle today contains 8 pounds of magnesium. Because it has excellent flow characteristics, magnesium makes possible the manufacture of thin-walled die castings for a variety of automobile parts. Thin-walled parts reduce the weight and cost of castings because less material is needed.

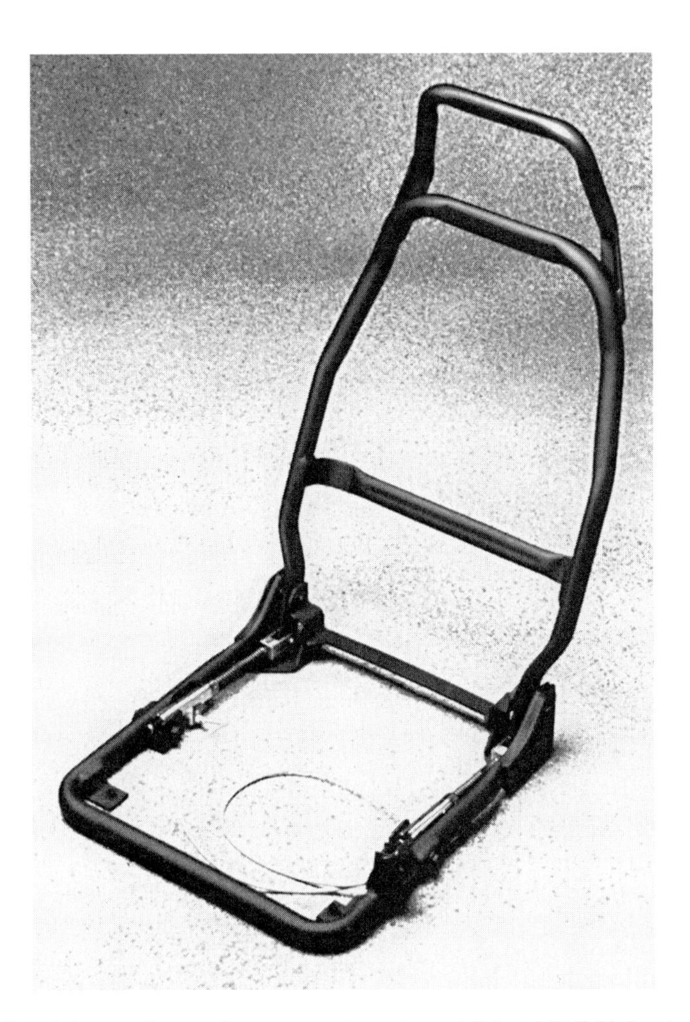

Figure 11-15 Extruded magnesium seat frame represents an advanced, lightweight (3.6 kg) seat system. It is made from magnesium tubular extrusions and stampings, and it reduces seat frame weight by 40%, while retaining its strength integrity. (Findlay Industries)

Findlay Industries developed a lightweight magnesium seat frame made from tubular extrusions and stampings (see Figure 11-15) for use in the automotive and/or aircraft industries. The intent was to design a frame that reduced the overall weight of existing tubular steel frames by 40 to 50%, continue to meet federal safety standards, and satisfied customer requirements. HSLA steel, aluminum, carbon fiber, and Asdel (compression-molded plastic) were also considered, but carbon fiber and compression-molded plastic were eliminated because of the high investment cost for prototype samples. The HSLA steel provided a 25% weight reduction, and a 20% weight reduction if the extruded tube shape was altered. Aluminum provided substantial weight savings but also presented certain manufacturing problems. For example, aluminum tubing of the size specified gave a 35% weight reduction, but the tube was too brittle to bend to the required shape. Heat treating the entire frame, along with increasing the thickness of the tubing 100%, was too impractical, especially when the overall weight reduction would be only 15 to 20%. Welding problems were also encountered. Magnesium AZ61 and AZ31b alloys offered the best weight savings, being 4.51 times lighter than steel. AZ31b won out because of its greater ductility. The new seat frame weighs only 3.6 kg (7.9 lb) in bucket-seat form, which is about half the weight of a steel frame. Other advantages of using magnesium for this application are that it is noncorrosive and weldable, it has greater design flexibility to accommodate customer preferences and styling concepts, it has excellent design and performance/packaging characteristics (i.e., improved comfort, better vehicle packaging, less foam required), and it is applicable to front and rear seats (bucket- or bench-type). Finally, the design meets all federal standards (FMVSS 201, 202, 207, and 208).

11.5 TITANIUM

Like aluminum and magnesium, titanium (Ti)—"the world's most glamorous metal"—is a very abundant metal in its mineral form. As a raw material, it ranks fourth in availability among the structural metal elements. Unfortunately, its fabrication processes are costly. Discovered in 1795, it was not until 1930 that extracting the metal from titanium oxide (TiO_2), its ore, was commercially feasible. This process produces titanium sponge, which is then further refined in an electric furnace to form ingots, which, in turn, are further remelted to form a final, pure titanium. Boeing's 777 uses the largest amount of titanium ever in a Boeing aircraft (Figure 11-2e). With a density between aluminum and stainless steel, titanium is extensively used in the aircraft industry. In fact, commercial and military aircraft now use 80% of American-made titanium. A major effort is under way to reduce the weight of such aircraft by substituting lighter-weight advanced materials such as continuous-fiber ceramic composites (CFCCs) for titanium alloys, thus permitting larger payloads and greater hauling distances.

Compared to steel or aluminum, titanium alloys have a high strength-to-weight ratio, particularly at elevated temperatures. One of their main drawbacks is the difficulty in cutting, milling, and forming them. In the SPF process discussed earlier in this module, one or more titanium sheets are laid on top of each other and held together by a jig while being heated in a furnace (Figure 11-5b). Argon gas is blown through tiny holes drilled in the sheets at desired locations. When the titanium reaches a soft and supple superplastic state, the gas assists the titanium to expand into cells, while the other sheets remain flat. The internal structure takes on whatever shape is intended. Through this process, the outer titanium sheets flow together under pressure to form a complex expanded sandwich unit that has the strength of a single piece.

Titanium alloys offer superior specific strength (up to 26.5×10^3 in.) in high temperatures (over 590°C) and low temperatures (–253°C), which makes them a popular structural metal in ultra-high-speed aircraft and accounts for their use in the space shuttle. The superplastic nature of titanium allows it to deform over 2000% without nicking or cracking when it is heated to around 925°C in the superplastic forming process, which has enhanced the potential of titanium. Titanium is nonmagnetic and has a lower linear coefficient of expansion and lower thermal conductivity than steel alloys or aluminum. Table A-11 lists some titanium alloys, compositions, and properties. Like zirconium and beryllium, titanium is allotropic and exists in a cph structure (α) at below 885°C and in a bcc structure (β) above that temperature (See Table 3-6). The composition of any titanium binary alloy consists of three phase fields in the solid state: α, ($\alpha + \beta$), and β. Various alloying elements alter the effect of the structure and stabilize the alpha or beta phases. For example, aluminum stabilizes the alpha structure, raising the temperature of the alpha–beta transformation and enlarging the α phase field. Other elements are β stabilizers and enlarge the β field. These three distinct phases in titanium alloys produce different properties. Alpha alloys have superior creep resistance and are heat treatable. Beta-phase alloys have excellent forgeability and fracture toughness. Alpha–beta alloys have good superplastic elongation and good strength characteristics. Alloys that are based on the intermetallic compound gamma titanium aluminide (γTiAl) have good stiffness, high-temperature strength, and low density, so their use in advanced gas engines, airframes, and automotive components is quite easy to understand. These alloys are usually fabricated by casting or by powder metallurgy, which makes their processing costs high. Because this alloy can withstand temperatures over 300°C hotter than conventional titanium alloys, the skin of the future X-33 space plane or the engines of the future F-22 jet fighter will be a logical choice for its applica-

Figure 11-16 Black Bird (SR-71) reconnaissance aircraft represents a pinnacle of aeronautical R&D with titanium. The SR 71 flew Mach 3.2+ (over 3 times the speed of sound) and higher than 80,000 feet. The plane could also fly 3200 nautical miles without refueling. All of these capabilities made possible thousands of hazardous reconnaissance missions over North Vietnam, North Korea, and the Middle East without being shot down.

tion. Figure 11-16 shows the SR-71 aircraft with titanium monocoque and some super-high-temperature plastics construction that flew at Mach 3.2+.

Another application of titanium is as a ***biomaterials***, a term for materials that are compatible with human and animal systems and so can be implanted or manipulated in people and animals. Titanium can be used as implants for joint replacements and dental reconstruction (Figure 11-17) due to its ability to osseointegrate (form bonds) with biomolecules of bone.

Titanium can be cast or molded. ***Rammed graphite castings*** are used to cast reactive metals such as titanium because the silica in the sand reacts with these metals. Because of its crystal structure, graphite has poor yield strength and a low coefficient of friction in the presence of air or moisture. It is a misconception to believe that a titanium casting is only good for large, simple shapes whereas investment casting with titanium is only good for small, intricate parts. Probably the limiting factor is wall thickness. Casting wall thicknesses down to 0.25 in. are common. ***Investment casting*** of the versatile titanium alloys results in very complex shapes with excellent surface finishes and unique properties often not possible with other processes (Figure 11-18). Investment castings with wall thicknesses down to 0.40 in. or less are produced depending on how far the metal has to flow before solidifying. Some investment castings weigh well over 100 lb, and some rammed graphite castings of complex shapes are over 1000 lb. The decision between forging versus casting for titanium parts depends on how simple the part is. If the part is somewhat complex, then casting is the economic choice. As far as strength is concerned, a cast part can be equal to a forged part provided it has undergone the HIP process. HIP (hot isostatic pressing), discussed later in this module, subjects the casting to high heat and pressure over time in an inert atmosphere. For example, HIP titanium castings are heated into the temperature range of 1550–1750°F at a pressure of 14,500 psi for a maximum of 2 hours. HIP collapses and bonds the internal bonds that are inherent in the casting metals and produces tensile and yield strengths equal to those of forged products. Because of the casting, the mechanical properties are distributed equally in all directions due to the orientation of the grain structure. And HIP castings will have generally higher fatigue properties than forged parts.

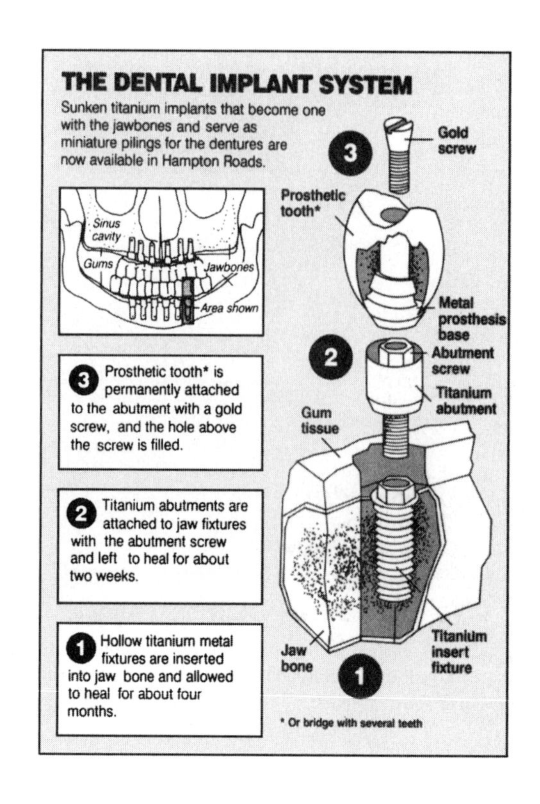

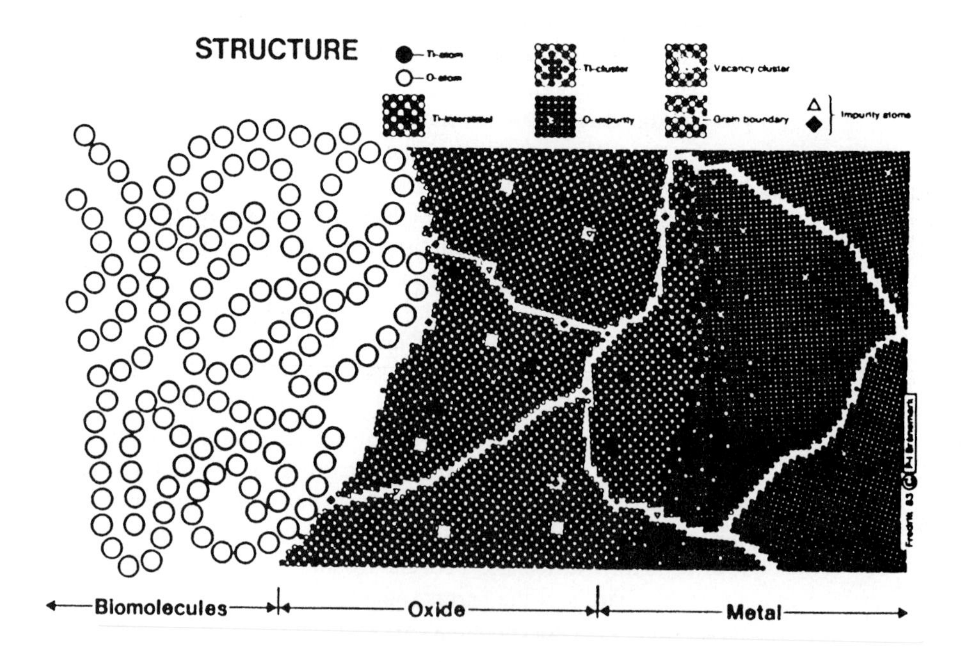

osseointegration

Figure 11-17 Titanium is a biomaterial compatible with the human system for dental and joint implants. The large-scale sketch shows the interface of bone tissue with the titanium and TiO_2 boundary layer. (*Ledger-Star*)

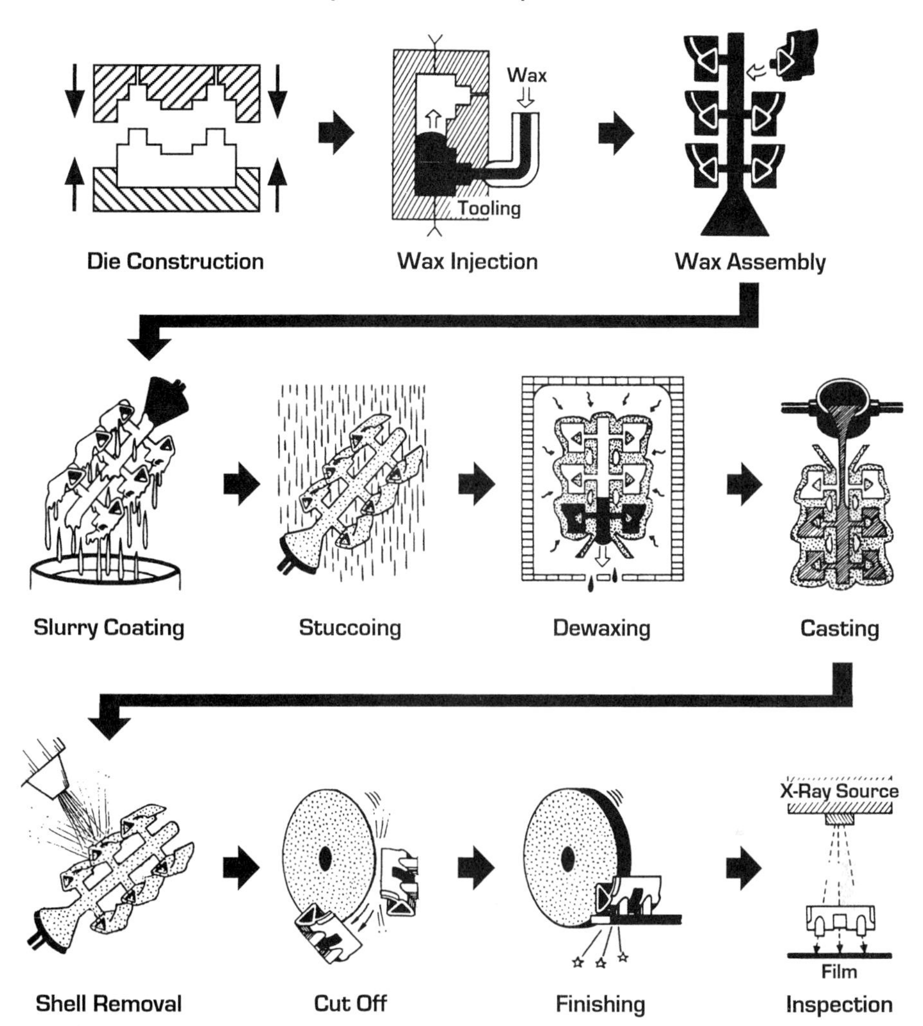

Figure 11-18 Investment casting. From teeth to jet engines come products made from a process that evolved from the ancient art of "lost wax casting." Investment casting serves many fields ranging from medicine for dental and bone implants to intricate aircraft gas turbine engine components. Constant innovations in the process came with improved nonferrous metal alloys such as titanium and vitallium (cobalt, chromium, molybdenum, and nickel), refinements in ceramic mold materials, control of metal grain structure, and the ability to produce larger castings. (Howmet Corporation)

11.6 WHITE METALS

The general term ***white metal*** includes zinc, tin, lead, antimony, cadmium, and bismuth, which all have relatively low melting temperatures (Table 11-2). ***Bismuth*** (Bi, A.N. 83), like lead, is completely insoluble in copper and not toxic to humans. ***Selenium*** (Se, A.N. 34) is also nontoxic to humans and in combination with Bi reduces the amount of Bi needed to achieve machinability and pressure tightness. ***Zinc*** is a readily available and inexpensive metal that can easily be applied to steel to serve as the sacrificial protection coating, as discussed previously. Zinc is also die cast to make housings and decorative trim; used as additives in rubber, plastics, and paint; and used in wrought forms that can achieve a high surface

TABLE 11-2 MELTING TEMPERATURE OF WHITE METALS

Metal	°C	Metal	°C
Antimony (Sb)	631	Lead (Pb)	327
Bismuth (Bi)	271	Tin (Sn)	232
Cadmium (Cd)	321	Zinc (Zn)	420

hardness of over 70 HRC. ***Tin*** has long been used as a coating for "tin cans" made of tin plate and is also a valuable alloying element when used with lead as soft solder, in tin babbitt, and as antimonial tin solder. Solder balls as small as 2 mil (0.002 in.) hold semiconductors onto metal-pinned ceramics used in computers. ***Pewter*** (alloyed with lead) was a popular alloy for cooking and eating utensils, plus ornamental casting, in colonial America. The modern tin-based pewter is lead-free with 91% tin, 7% antimony, and 2% copper and is often used to replicate colonial dishes. ***Cadmium*** is harder than tin and serves as a corrosion-resistant coating electroplated on steels, especially fasteners. A coating of 0.0008 mm provides protection equal to a 0.025-mm zinc coat. It also serves as an alloy in copper to improve hardness.

Lead is not only an important alloy in solder but when evenly dispersed in leaded steels at 0.15–0.35%, it provides built-in lubrication to ease machining and produce tightly curled chips. Traditional leaded brass casting alloys such as UNS Alloy C83600 contain 85% Cu and 5% each of the elements Sn, Pb, and Zn. Lead was originally added to brass casting alloys because it forms discrete globules when the metal solidifies, which then act as a lubricant when brass is machined. Additionally, lead fills the intradendritic spaces that form upon solidification, resulting in pressure tightness that prevents the weeping of water to the outer surface of plumbing castings. The need to replace lead in brass components for potable water systems as mandated by the 1996 amendments to the Safe Drinking Water Act of 1974 resulted in several new brass casting alloys called EnviroBrass. For example, UNS C89510 is a red brass and C89550 is a white brass. Machinability ratings for this new class of brasses range from 97 to 75. Lead is also being removed from foundry sand, increasing sand reclamation, reducing waste disposal, eliminating air purification and worker inhalation safeguards, and lowering health insurance costs. Small portions of lead built up in the human body can cause lead poisoning (plumbism). Consequently, it has become desirable to eliminate lead from paint for children's furniture and toys, in gasoline, and in soldered tin cans and to reduce exposure by industrial workers. Tin is now replacing lead in steel alloys as an environmentally friendly alternative. Dubbed **"green steel,"** this new alloy is being used by Ford Motor Company. Lead has also been important in electrical batteries as terminals and, alloyed with antimony, for grid plates (Figure 11-19). Its resistance to many corrosive chemicals, X rays, and gamma rays and its sound-damping capacities coupled with its high density find unique applications in the medical, chemical, and nuclear industries. It can be compounded with ceramic glazes.

11.6.1 Battery Technology

The desire for greater mobility and longer life in devices as various as heart pacemakers, telephones, portable radios, CD players, space vehicles, and notebook computers with multiple functions (e.g., fax/modem, CD-ROM, and sound) continues to drive improvements in battery technology. State legislatures gave battery technology a boost by mandating a certain percentage of electric vehicles to be made available in their respective states to offset air pollution from gasoline vehicles. A variety of metals in chemical systems serve as both primary batteries and rechargeable batteries, including lithium/alkaline-manganese, manganese dioxide, zinc air, silver oxide, lithium ion (see Figure 11-20), nickel–metal hydride, and nickel–cadmium. Nickel–cadmium batteries have undergone tremendous refinement since their invention at the end of the nineteenth century. These rechargeable batteries are found in handheld electronic calculators, aircraft engine starters, emergency power equipment, and portable medical equip-

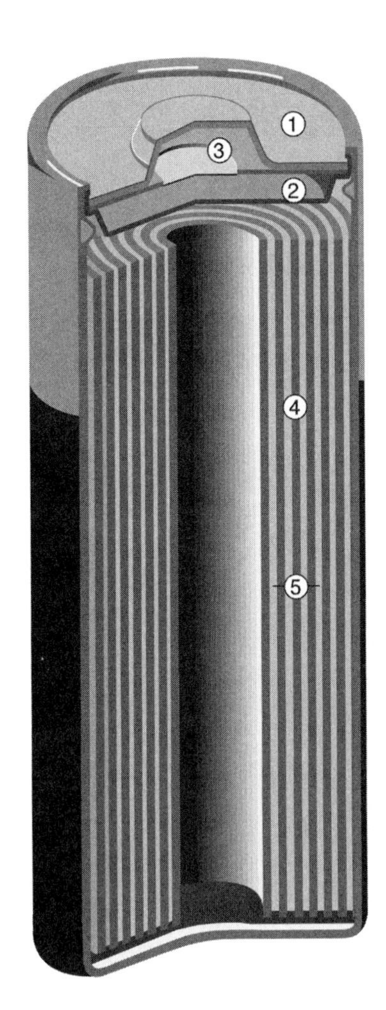

Lithium Cell

1. Positive Cap
Formed nickel-plated steel disk which serves as positive terminal

2. Seal
Molded plastic part which contains cell materials and provides a safety vent

3. PTC Device
Polymeric device which shuts off current in a cell that has been accidentally short-circuited

4. Electrode Jelly Roll
Assembly of positive and negative electrodes which are separated by microporous polymer separator

5. Electrolyte
Solution of complex salts which are dissolved in organic solvents

Figure 11-19 Construction of a standard (AA, C, D, etc.) lithium cell. (Duracell International Inc., *The Story of Packaged Power*)

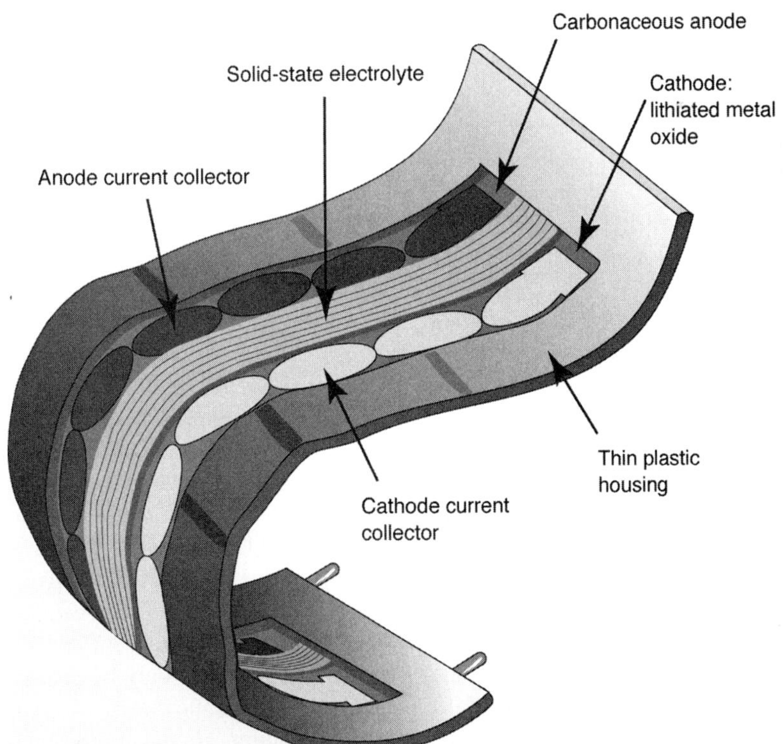

Figure 11-20 Lithium-ion batteries have replaced the nickel-based batteries for many portable notebook computers. (a) These new batteries combine lithium technology with solid polymer electrolyte for a thin, flexible, lightweight, and high-energy-density battery that can be shaped to fit most computer and portable electronics appliances.

Inside the Power Cell

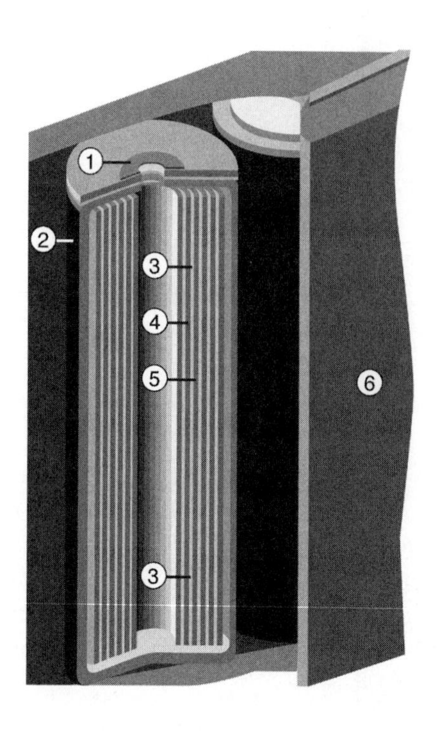

Nickel-Metal Hydride Battery

1. Positive Cap
Connected to the nickel electrode; serves as the positive terminal

2. Can
Connected to the metal-hydride electrode; serves as the negative terminal

3. Separator
Prevents direct internal short circuits

4. Cathode
Composed of nickel oxide; serves as the positive electrode of the cell

5. Anode
Composed of metallic alloy; serves as negative electrode of the cell

6. Battery Case
Plastic container which holds interconnected cells with contacts and internal electronics

(a)

Figure 11-21 (a) Nickel–metal hydride battery.

ment. Figure 11-21a shows the construction of nickel–metal hydride batteries and Figure 11-21b is a schematic of how a typical battery works. One type of nickel–metal hydride battery is being developed, with the involvement of the U.S. Battery Consortium (USBAC), for possible use in the next generation of electric vehicles. Lithium manganese oxide and lithium cobalt are replacing lead and cadmium in the development of lightweight rechargeable batteries.

The ***proton polymer battery*** developed by Japan's NEC Corp. works more like a capacitor, which stores an electrical charge, than a conventional battery. When the battery is fully charged, H atoms with one electron and one proton are bound within the cathode. When a switch is opened, the electrons are released from the atoms and flow through the circuit toward the anode, giving off electricity. The protons flow through an electrolyte to the anode, where they recombine with the incoming electron to form H atoms that bind with the anode material. Recharging the battery reverses the process. Tiny protons can speed through the electrolyte faster than the complex molecules that carry charges in conventional batteries, and the battery generates greater power in a shorter period of time. There is less degradation of the electrodes, resulting in much greater battery life. Bluetooth short-range radio technology would be a likely user of such batteries.

A consortium of scientists is developing a battery to replace the 36-pound battery that U.S. Special Forces carry as part of their 120 pounds of combat equipment. A working prototype is to be developed within a 2½ years. The new design calls for a fuel cell that will provide 20 watts of power for several days of such a size as to fit inside a soda can-sized package and weigh about 4 pounds. A fuel cell creates an electrical current by a chemical reaction between O and a fuel source, in this case diesel fuel. The only by-products would be water and CO_2. The cell is composed of several layers of ceramic material with a middle layer capable of putting O in contact with the fuel. The resulting chemical reaction releases the electrical energy needed to power the connected device. There would be numerous nonmilitary applications for such a battery, especially wherever emergency power is needed.

The wide use of batteries in a huge array of consumer products has created environmental issues concerning their disposal. Some municipalities provide collection stations for

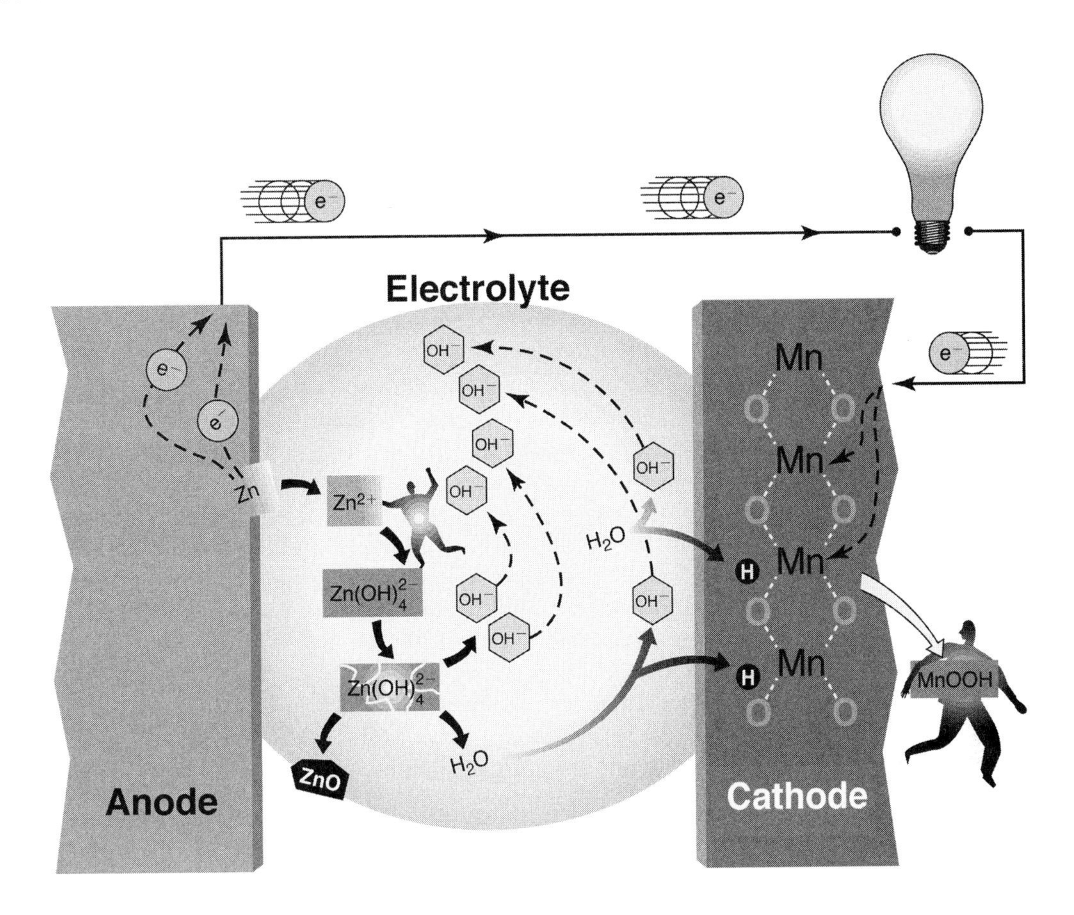

(b)

Batteries may seem simple, but delivery of packaged power is a very complicated electrochemical process. Electric current in the form of electrons begins to flow in the external circuit when the device, in this case a light bulb, is turned on. At that time, the anode material, zinc (Zn), gives up two electrons (e^-) per atom in a process called ***oxidation,*** leaving unstable zinc ions (Zn^{2+}) behind. After the electrons do their work powering the light bulb, they re-enter the cell at the cathode, where they combine with the active material, manganese dioxide (MnO_2), in a process called ***reduction.*** The combined processes of oxidation and reduction could not occur in a power cell without an internal way to carry electrons back to the anode, balancing the external flow of current. This process is accomplished by the movement of negatively charged hydroxide ions (OH^-) present in the water solution called the electrolyte. Every electron entering the cathode reacts with the manganese dioxide to form $MnOO^-$ ion. Then, $MnOO^-$ reacts with water from the electrolyte. In that reaction, the water splits, releasing hydroxide ions into the electrolyte and hydrogen ions (H^+) that combine with $MnOO^-$ ion to form MnOOH. The internal circuit is completed when the hydroxide ions produced in this reaction at the cathode flow to the anode in the form of ionic current. There, they combine with unstable zinc ions, which were formed at the anode when the electrons were originally given up to the external circuit. This produces zinc oxide (ZnO) and water (H_2O).

Figure 11-21 continued (b) How a typical power cell (battery) works. (Duracell International Inc., *The Story of Packaged Power*)

such toxic and hazardous materials. Other means of proper disposal involve suppliers and retailers. Guidelines can be found at the websites of battery manufacturers such as Duracell.

 *http://duracell.com*

11.7 OTHER METALS

Beryllium (Be) is a high-cost metal that serves as an alloying element in copper for age-hardening to enhance oxidation resistance and refine grains. Beryllium is ideal for space telescopes because it is the only metal with a stiffness greater than glass and a density lower than aluminum. Tungsten (W) (3370°C), tantalum (Ta) (2850°C), columbium (Cb) or niobium (Nb) (2415°C), and molybdenum (Mo) (2620°C) are known as ***refractory metals***, those with melting points above 1980°C. They are commonly alloyed with many of the metals discussed earlier to impart strength and hardness at high temperatures—Type 6–6–2 molybdenum–tungsten high-speed steel, for example. Because of the high temperatures and fabrication difficulties, refractories are often formed as powdered metal parts although they can be cast, as well. Their high strength requires chipless-type machining methods. As we have mentioned, numerous nonferrous alloys, known as superalloys, have been developed to meet the requirements of high-technology innovations, such as jet and rocket engines. These superalloys use iron and nickel, nickel, or cobalt as the base metal. Some examples of such alloys are Inconel (nickel base), Incoloy (iron–nickel base), and S-816 (cobalt base). Superalloys contain large amounts of alloying elements to produce a combination of properties that conserve high strength and resist creep and corrosion at high temperatures (exceeding 1100°C). All contain small amounts of carbon, which, in combination with other elements, produces a network of stable carbide particles that interfere with dislocation and grain boundary movements. For example, Stellite, a superalloy that contains 60% Co and 30% Cr, has a tensile strength of 177,000 psi and is used for its abrasive wear resistance at high temperatures. The superalloy industry accounts for almost one quarter of total cobalt consumption (89,500 tons in 2000), which is increasing with the increasing demand for aerospace applications.

Nitinol contains a nearly equal mixture of nickel (55 wt %) and titanium (Ni–Ti). It is one of the memory metal (martensite) alloys, as are Cu–Al–Ni and Cu–Zn–Al. Other elements can be added to adjust or 'tune' the material's properties. The elements form extremely brittle intermetallic compounds that precipitate out into the Ni–Ti matrix structure. Hardness values of R_c 60–62 along with its corrosion resistance make it a good material for making cutting instruments. Nitinol has two phases, austenite, which exists at high temperatures, and martensite, its low-temperature phase. When in the austenite phase, the material is hard to bend, but in the martensite phase the material exhibits flexibility. Even when allowed to cool from its high-temperature phase, the austenite phase can still remain. Nitinol is difficult to machine—extremely tough, a 3000-watt laser is used to cut it. Consumer uses for shape-memory alloys include brassiere underwires, eyeglass frames, high-security locks, gun barrels, rupture disks, and cellphone antennas. Nitinol is being used in bearings that can operate in saltwater without the use of lubricants. A memory metal shirt that rolls up its own sleeves when the temperature gets too warm is woven from fibers of Nitinol interspersed with nylon. The sleeve fabric is programmed to shorten as soon as the room temperature becomes a few degrees warmer. Five nylon fibers to every Nitinol fiber allow these materials to be washable and nonallergenic. Even when the fabric is pleated, creased, or rolled up into a ball, a hair dryer blast can return it to its original shape.

The Oscar statuette, considered the movie industry's most prestigious award, is made of **britannium**, a metal alloy. It is plated with copper, nickel, and silver and topped with 24-karat gold. Each one stands 13½ inches tall and weighs 8½ pounds. The cost of 50 to 60 Oscar statuettes each year is about $300 apiece.

SELF-ASSESSMENT

11-1. Advanced alloys saved __________ pounds of structural weight in the Boeing 777.

a. 1000 **b.** 2000 **c.** 3200 **d.** 5400

11-2. Nonferrous metals comprise __________ of the known elements.

a. 1/2 **b.** 2/3 **c.** 3/4 **d.** 4/5

11-3. The temper designation (W) that follows the four digits specifying wrought alloy groups for the aluminum association stands for ___________.

a. Solution heat treatment **b.** As fabricated **c.** Strain hardened

11-4. Recycling aluminum saves ___________ energy over that used to produce new aluminum.

a. 60% **b.** 20% **c.** 80% **d.** 95%

11-5. Titanium, at certain temperatures, can be deformed as much as ___________ .

a. 100% **b.** 200% **c.** 500% **d.** 2000%

11-6. What metal is added to the surface of anodized aluminum to improve its wear resistance?

a. Beryllium **b.** Silicon **c.** Zirconium

11-7. A pound of aluminum is ___________ as expensive as a pound of steel.

a. Two times **b.** Three times **c.** Four times

11-8. ___________ is a metal and an example of a biomaterial that serves as implants for joint replacements in humans.

a. Titanium **b.** Steel **c.** Copper **d.** Lead

11-9. The lightest of all structural metals, with a specific gravity of 1.75, is ___________.

a. Lithium **b.** Manganese **c.** Molybdenum **d.** Magnesium

11-10. ___________ is added to copper to improve its oxidation resistance.

a. Aluminum **b.** Brass **c.** Niobium **d.** Beryllium

11-11. Brass is a metal alloy made from two elements: ___________ and ___________.

a. Tin, lead **b.** Nickel, copper **c.** Zinc, copper

11-12. Monel is a metal alloy consisting of two elements: ___________ and ___________.

a. Tin, lead **b.** Nickel, copper **c.** Zinc, copper

11-13. Inconel is a metal alloy consisting of two elements: ___________ and ___________.

a. Nickel, copper **b.** Nickel, chromium **c.** Nickel, aluminum

11-14. Which metal listed below is capable of being superplastically deformed 2000%?

a. Nickel **b.** Aluminum **c.** Titanium **d.** Magnesium

11-15. Which metal listed below is a white metal?

a. Zinc **b.** Nickel **c.** Aluminum **d.** Magnesium

11-16. Superalloys are made from a particular base metal, such as one of the following:

a. Cobalt **b.** Iron **c.** Aluminum **d.** Beryllium

11-17. Explain why aluminum, magnesium, and titanium are expensive when they are so readily available as raw materials. Give a reason for recycling aluminum.

11-18. Explain the corrosion-resistance mechanism of copper and aluminum. Which metal offers the widest range of processability?

11-19. Name one barrier and two advantages for advanced metals technologies such as MMCs, near-net-shape forming with SPF and SSF, and light aluminum alloys.

11-20. **(a)** What common characteristic do white metals possess? **(b)** Provide a typical application for each of these metals. **(c)** Why is lead eliminated from paint, plumbing, gasoline, ceramics cookware, and similar uses?

11-21. Explain the reasons for proper disposal of batteries and some means of disposal.

Module 12

Metal Powders, Powder Metallurgy, & More Metals Technology

After studying this module you should be able to do the following:

12–1. Define metal powders and powder metallurgy and describe methods for producing metal powders and other particulates.

12–2. Recall P/M processing techniques to achieve desired densities, shapes, and impregnation of parts.

12–3. Explain the advantages and disadvantages of P/M parts in terms of production lot size, properties, and cost compared to other materials and processes.

12–4. Recall methods of producing amorphous metals and powders and state the advantages of these materials.

12–5. Use the website addresses denoted by the icon* to gain insights into the technology, design considerations, and emerging technology related to powders and powder metallurgy and to keep abreast of advanced metals technology.

12–6. Recall emerging trends and benefits offered by advanced metals technology.

12–7. Explain green design in terms of how metals can contribute or distract from its goals.

12–8. Discuss the concepts included in Cradle to Cradle infrastructure for manufacture/construction systems and the materials cycle.

12–9. Describe some of the issues involved with metals contained in computers in terms of environmental impacts.

*The symbol found throughout the book will link you to Internet sites related to topics being covered. The dynamic nature of the Web brings frequent changes, so some of these sites, while available at the time of writing, may not be up now.

12.1 METAL POWDERS

Powders of selected materials are produced for ***powder metallurgy (P/M)*** applications by a variety of techniques. Although P/M technology existed at least 5000 years ago, only recently has the technology improved significantly. Up until World War I metallurgists were unable to fuse metals such as platinum and tungsten, so they turned to powdered metal, compacted and heated to allow diffusion bonding. This last step is known as ***sintering***, and P/M parts are often referred to as *sintered metals*. As seen in Figure 12-1, powder metallurgy involves most metals and a host of processes and has a wide range of applications.

12.1.1 P/M Production

Figure 12-2 shows the flow of the raw powder through compacting (***briquetting***) to finished P/M parts. Some parts undergo secondary operations to improve tolerances or finishes, to apply lubricants or coatings, or for heat treatment. A major advantage of the P/M part is the reduction or elimination of scrap and machining; too many secondary operations can reduce this advantage, however. Once the part is compacted and sintered, it may undergo such secondary operations as ***coining***, to squeeze the part to closer tolerances; ***infiltration***, in which a lower-melting-point metal is added to close pores and increase density; ***impregnation*** (Figure 12-3), which adds a lubricant for self-lubricating bearings; or ***machining***, to achieve a profile not practical through the pressing operation.

12.1.2 P/M Selection

In addition to the advantages already named, P/M allows the manipulation of a part's density through control of particle size and degree of compaction. Particles range from 0.2 to 2000 mm in size. Smaller particles can pack tightly for increased density; or if open pores are required, as in filters or oil-impregnated, self-lubricating bearings, larger particles can be compacted to-

(a)

(b)

Figure 12-1 (a) P/M parts showing a variety of sizes, shapes, and materials (note quarter for comparison). Carbide (cermet) inserts for metal cutting require P/M. Photomicrographs of the sintered iron sheet in the foreground are seen in Figure 12-4. (b) Two examples of porous P/M parts. A brass gear with carbon on top that will be heated and infiltrated into the porous brass to provide lubrication; a porous brass filter on the right that threads onto the oil nozzle to filter out impurities.

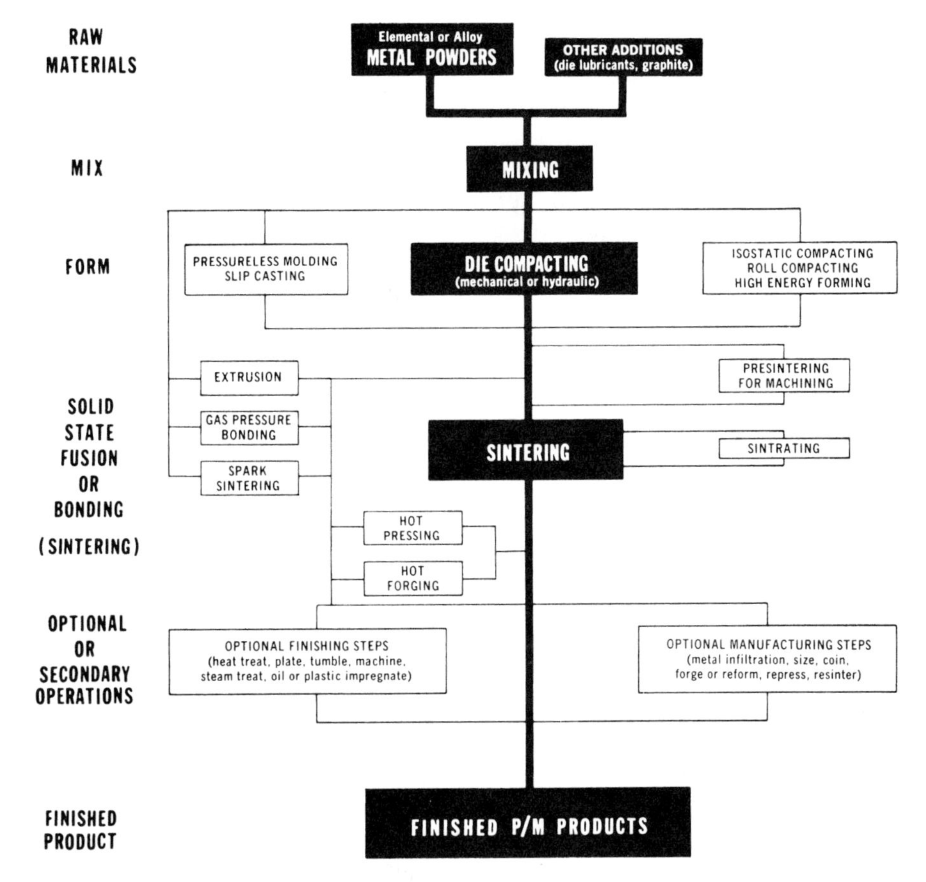

Figure 12-2 Schematic flow diagram of the P/M process. (Hoeganaes)

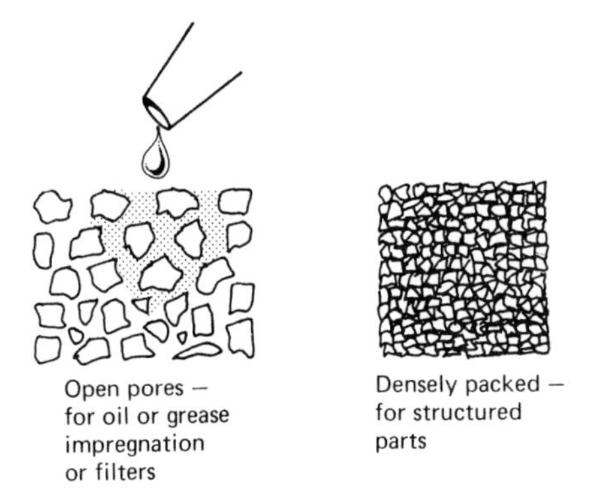

Figure 12-3 Impregnation for lubrication.

gether (Figure 12-4). Powders are getting finer. Cu and bronze powders now average about 10–15 μm; nickel powders approach 1 μm. P/M structural parts of steel powder have tensile strengths from 310 MPa (45,000 psi) to over 1200 MPa (175,000 psi), which makes them competitive or superior in terms of strength with cast iron, steel, and nonferrous alloys in certain applications. But the presence of porosity, or void space, in P/M makes it difficult to achieve full or theoretical density. For example, the density of pure iron is 7.85 g/cm^3. A ferrous P/M part with 90% density would contain 7.87 g/cm^3 $\times$ 0.90 = 7.08 g/cm^3. High-velocity compaction can provide up to 7.5 g/cm^3 density. The density of a P/M part affects its hardenability as well as its thermal conductivity. Denser P/M parts have higher thermal conductivity, which, in turn, ensures rapid heating and cooling, hence good hardenability. Although conventional P/M processing can produce parts weighing up to 35 lb, most are smaller and weigh less than 5 lb.

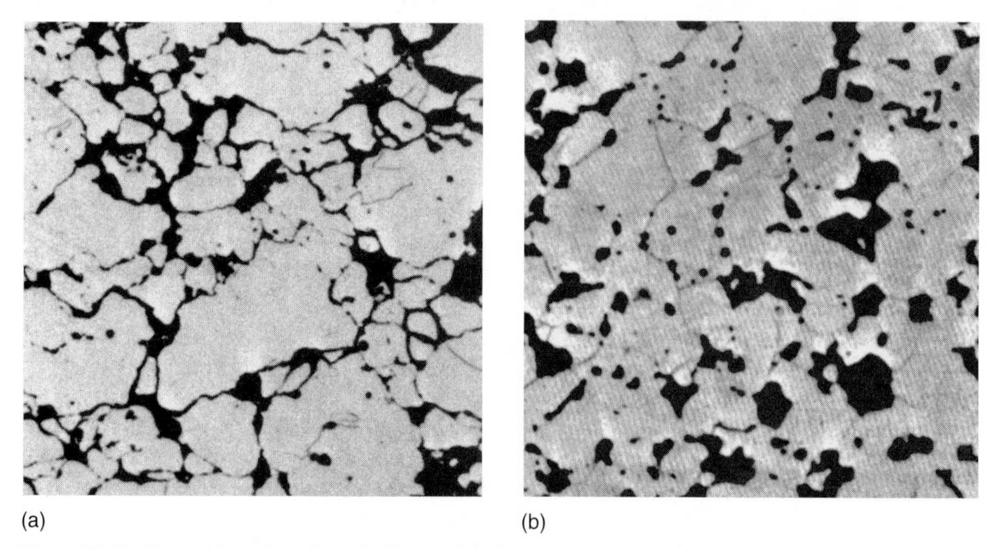

Figure 12-4 Sintered iron sheet shown in Figure 12-1. Iron–carbon sintering involves four steps: (1) establishing particle bonds at their area of contact and the growth of these bonds, (2) grain growth and migration of grain boundaries formed at the bonds, (3) spheroidization of pores and (4) possible growth of larger pores. (a) As pressed. (b) After sintering at 2050°F. (Hoeganaes)

12.1.3 Metal Powders

The most common P/M materials are iron, steel, copper and copper alloys, stainless and other alloy steels, nickel alloys, aluminum alloys, and refractories such as tungsten carbides, but most other metals find certain unique P/M applications. Tungsten light bulb filaments, for example, are made with a P/M process (Figure 12-5). Approximately 70% of P/M parts are made for the automotive industry. Most of these are iron but the use of stainless steel is growing for components of automobile exhaust systems. Powder metal parts are now found in automobile transmissions and engines. The Ford 194R70W and E40Dd transmissions each contain 11 to 127 kg of P/M parts; DaimlerChrysler's 2.7-l engine contains 11 kg. A powder metallurgy composite of sintered and pressed molybdenum–copper powder with outstanding thermal conductivity is used in an electronic package that maximizes voltage and power handling for power hybrid circuits. The base is sealed to the alumina ceramic package with a Kovar seal ring. The high resistance to vibration and mechanical shock failure makes these packages ideal for aerospace applications.

12.1.3.1 Amorphous Powders. Amorphous metals, introduced in earlier modules, continue to gain wider acceptance. As a result of ultrarapid cooling , amorphous metals do not have time to form dendrites and crystallize, as seen in Figures 12-6 and 12-7. Iron- and aluminum-based amorphous metals now in production thus provide plastic behavior without the dislocations common in crystalline materials. In a simulated production process, the Metallurgy Division of NIST used an eddy-current-based density sensor (Figure 12-8) for *in-situ* (within the process) monitoring of hot isostatic pressing (HIP) of amorphous aluminum powder (Figure 12-9). As seen in Figure 12-9, the NIST experiment consolidated amorphous powders by HIP to achieve a relatively dense part (about 80%). Boeing sees possibilities for HIPed amorphous aluminum parts in airframes. Amorphous iron powders offer similar potential for parts production.

Not all metal powder is compacted and sintered into machine parts. Iron powder is added to human and livestock food as a dietary supplement. Iron powders are also used in nondestructive testing (NDT) and magnetic-particle testing, for welding electrodes, and in oxyacetylene cutting and scarfing (an exothermic reaction that increases the heat of a flame to melt the oxides formed on high-temperature alloys or refractories). Platinum powder is a catalyst used in making gasoline, copper powder goes into marine paints, carbon powder is used in automotive ignition wires and in making xerographic paper copies, and aluminum powder aids in paper manufacture.

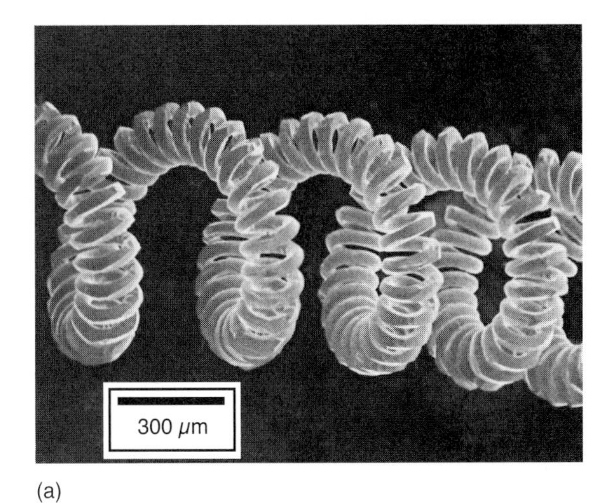

(a)

(b)

Figure 12-5 Scanning electron photomicrograph of a nonsagging AKS tungsten (W) filament after several hundred hours of operation at 2500°C in a common 60-watt light bulb. (a) The "coiled coil" geometry of the filament is possible because of the availability of ductile, Coolidge-process P/M, tungsten wire. (b) A closeup reveals that some of the tungsten evaporated, thus exposing atomic facets of the individual W grains and characteristic interlocking grain boundaries (chevron features). This grain boundary structure prohibits creep deformation (sagging) through grain boundary sliding, which results in longer-lasting filaments. (Jerry P. Wittenauer, Lockheed Missiles and Space Co., and *Advanced Materials & Processes*)

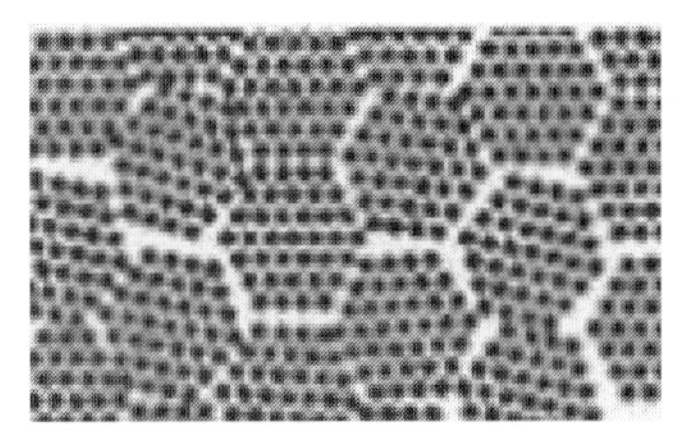

Figure 12-6 Orderly structure of crystalline metals. (www.liquidmetals.com)

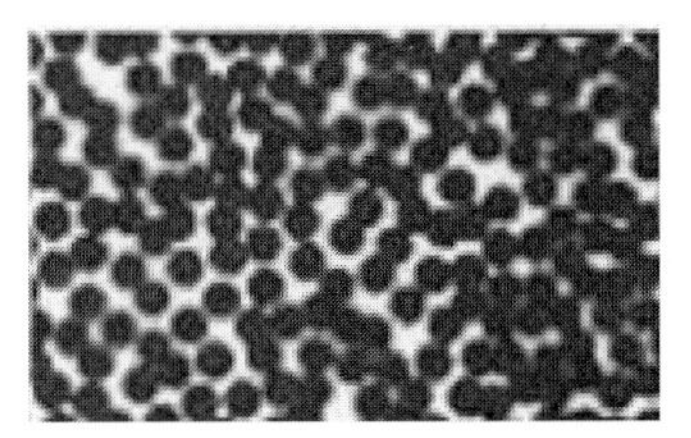

Figure 12-7 Unordered amorphous metal structure. (www.liquidmetals.com)

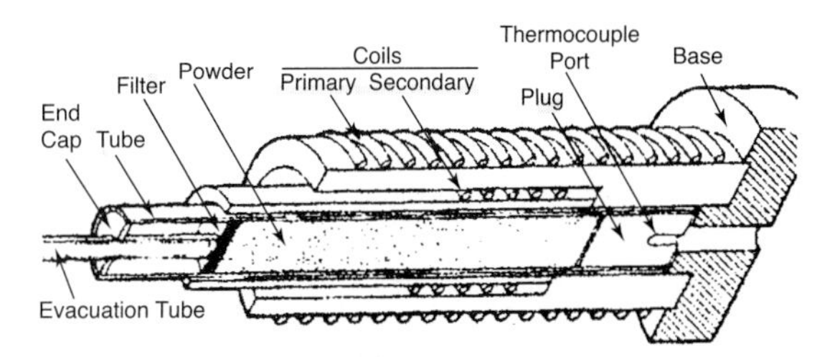

Figure 12-8 Schematic diagram of NIST's eddy-current-based sensor for *in-situ* monitoring of hot isostatic pressing of amorphous metal powders. (Rodney Jiggetts and Thomas Pierce, NIST and Norfolk State University)

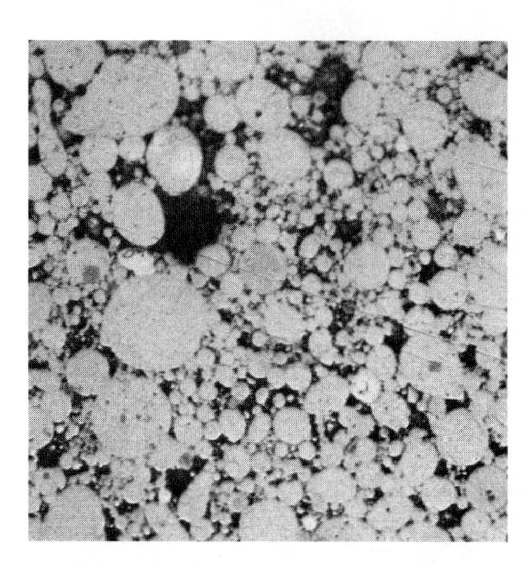

Figure 12-9 Microphotograph of pressed amorphous aluminum alloy showing plastic yielding of powder particle contacts. This compact attained a relative density of about 80%. (Rodney Jiggetts and Thomas Pierce, NIST and Norfolk State University)

12.1.4 P/M Cost

P/M allows production of complex shapes from a wide variety of materials (metals and nonmetals) and with a variety of densities at low cost, assuming large production runs. The parts shown in Figure 12-1 reveal the complexity and small sizes possible with P/M. Some parts, such as carbide cutting inserts, porous metals, and sintered sheet, require P/M for their production. However, due to the cost of presses, tools, and dies, P/M usually requires large lot sizes to justify those expenses. Parts involving smaller lot sizes would be more economically produced by machining.

The Metal Powder Industries Federation website listed here provides a variety of useful insights into metal powders, powder metallurgy, and particulate materials. The Liquidmetal™ website includes animated and still graphics on this emerging technology.

Powder Metallurgy—*http://www.mpif.org*
Liquidmetal™—*www.liquidmetal.com*
http://www.mpif.org/chart.html

12.2 ADVANCED METALS TECHNOLOGY

Throughout this book and in many technical publications, advanced materials receive attention because of their favorable properties and processing possibilities. Figure 12-10, a cross-sectional view of a concept gasoline engine that uses lightweight components, brings together several of the advanced materials components discussed in this book: advanced metal alloys and metal–matrix composites (MMCs), engineering plastics (especially thermoplastics that are easier to recycle), advanced ceramics, and advanced composites. Requests for advanced materials technology come from many disciplines, including fuel efficiency, safer transportation, computers, energy, and on and on. For example, the advanced metal AlBe-Met, an aluminum–beryllium metal–matrix composite (MMC) that is lightweight, high-strength, and oxidation resistant, has been adopted in the computer industry. It combines the low density and high stiffness found in beryllium with the ductility and low cost of aluminum. It reduces weight, removes heat, and increases the lifetime of electrical systems. Used now in rotary actuators for disc drives, it improves operations by allowing a lighter and stiffer arm, which permits the heads to move faster.

A number of examples of advanced materials substitutions for engine parts have already been mentioned in this book to show how materials selection and engineered materials can make more efficient engines. Since the beginning of automobile manufacture, various cast iron

Gasoline Engine
Lightweight Components

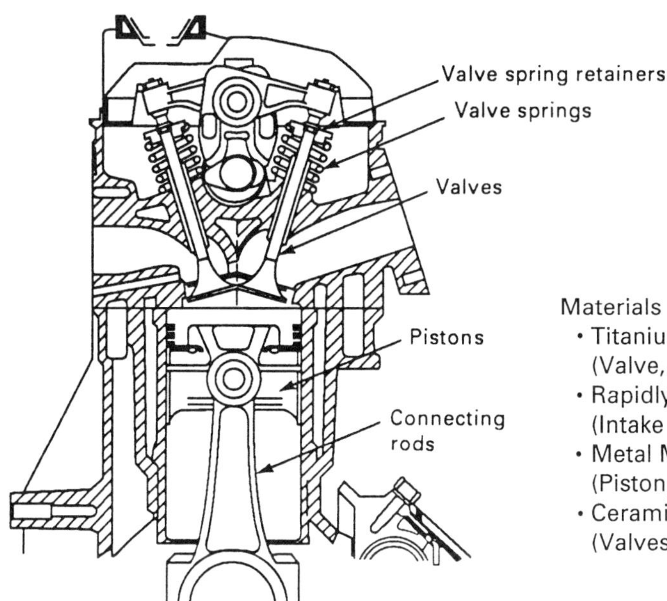

Materials and Applications
- Titanium Alloys
  (Valve, retainers, connecting rods spring)
- Rapidly Solidified Aluminum Alloys
  (Intake valves, connecting rods, retainers)
- Metal Matrix Composites
  (Piston, connecting rods, retainers)
- Ceramic, Aluminides
  (Valves)

Figure 12-10 Gasoline engine: Lightweight components and advanced materials applications. (Ford Motor Company)

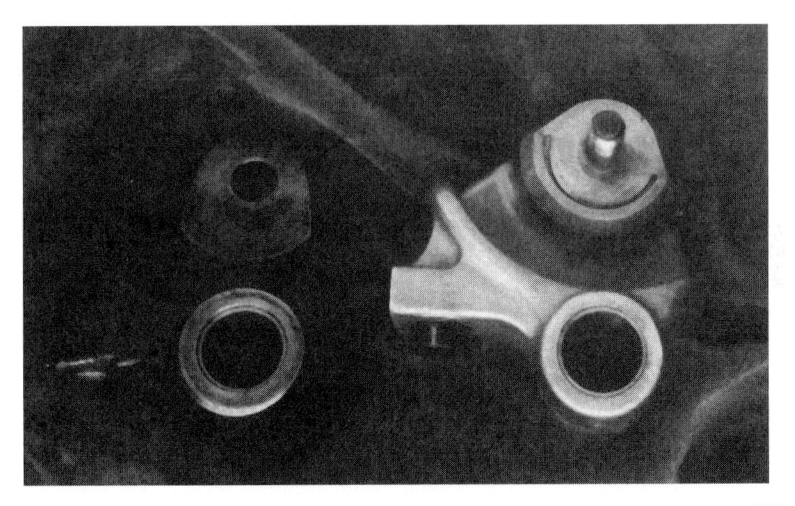

Figure 12-11 The part at left is the original, designed in cast nodular iron. The part at right shows SSF aluminum design. Note the forged-in-place wear pad and pulley mounting stud. The only machining area on the SSF part is the large hole where the pivot bolt mounts. (*Advanced Materials & Processes*)

alloys have been widely used for a wide variety of engines. Figure 12-11 compares two versions of a timing-belt tensioner-pulley pivot bracket—an earlier one made of nodular cast iron and the redesign made of ***semisolid forged (SSF)*** aluminum. SSF uses the principles of thixotropic metal processing (refer to Applications & Alternatives at the end of Module 3) to make alloys for forging. The SSF aluminum bracket was redesigned using CAD with finite element analysis. Weight was reduced from 0.31 kg (0.69 lb) for nodular iron to 0.16 kg (0.36 lb) for the SSF aluminum. The SSF process provides near-net shape since it enables the wear pad and the pulley mounting bolt to be integrally formed into the pivot brackets. This eliminated the machining operation required for the wear pad and pulley mounting bolt areas on the nodular iron version.

Semisolid metal forming (SSF) is a mature, high-volume production method of satisfying the need for lightweight, high-quality, and low-cost aluminum parts. Products made by this method have uniform mechanical properties, near-net shape, and are a cost advantage over those produced with permanent mold processes. A permanent mold is used repeatedly and designed so that the casting can be removed. The molds are made of metal such as cast iron, steel, bronze, and graphite to withstand high temperatures and readily conduct heat away from the casting. Typically, automobile cylinder heads, pistons, and connecting rods are made by permanent mold (hard mold) casting. Die casting and pressure casting are examples of permanent mold casting. The other major classification of casting processes is expendable mold casting, in which the molds are made of refractory materials capable of withstanding the high temperatures of molten metals. When the casting has solidified, the molds are broken up in order to remove the casting. Investment casting (or lost wax process) is one example of such casting. (Refer to Section 3.4, "Single-Crystal Castings" and Module 3 for an example of turbine blade castings.) SSF casting makes use of two alloys, Al–Si–X and Al–Mg–X, which have adequate fluidity in the semisolid state to produce high-ductility, thin-wall castings. A newly developed SSM process called Vexocast also makes use of the rheological properties of a crystallizing melt. Unlike the conventional SSM process, a reheating unit is not needed and there are no limitations in regard to component size or choice of alloy.

12.2.1 Systems

High cost is the main barrier to the use of advanced materials. Usually, they are several times to tens of times more per pound than traditional engineering materials. Any forecast for their use in the twenty-first century will be based on the expectation that these costs will decline as materials producers and product fabricators gain more experience with the new materials, thereby increasing the demand for them. Because advanced technology brings with it lower ***total-systems cost*** and improved systems performance, such forecasts can be made with confidence. However, the changeover to advanced materials technology will not be a sudden one, but will evolve and coexist with traditional engineering materials technology. Although, for example, the auto materials market is driving the development of advanced materials to fill the need for reduced weight and, in turn, reduced emissions and fuel consumption, the gasoline engine made with advanced materials components in Figure 12-10 is still expected to have an engine block of cast iron or aluminum for some time. Hybrid fuel vehicles are only just now appearing on the road, and fuel cells such as the GM Hy-wire shown at the beginning of this unit will continue to be under development for some time.

As a materials user, designer, manufacturer, or tester, or in whatever role you interact with materials, expect to continue to use traditional metals for many years. Appendix A-11 offers only a tiny sampling of the wide variety of traditional metals available. Many handbooks and volumes of references, some listed throughout this book, will enable you to make wise selections. Failure to consult reliable references can have tragic consequences. Twelve people lost their lives when the roof of a Swiss swimming-pool building collapsed. Failure analysis revealed that 207 stainless steel hooks had corroded and could no longer support the concrete roof, which weighed 166 tons. The hooks corroded over 13 years from the pool's aggressive chloride atmosphere, which caused intergranular stress–corrosion cracking. Did you know that stainless can corrode? The section in this book on stainless steels and properties of materials dealt with metals corrosion and degradation; more information will be offered in Module 13. As this swimming pool disaster and many others over the years illustrate, careful materials selection and regular maintenance must be a part of dealing with the materials systems.

12.3 LIFE CYCLE OF MATERIALS AND GREEN DESIGN

Engineering design is a vital activity concerned with, among other areas, a unified, life-cycle approach to manufacturing. Engineering or ***product design*** should involve ***green design,*** taking into account whether a material can be recycled, disposed of as waste, or reused in some other form. In this assessment of a product, the effect of ecology (that branch of biology that studies the relations between living organisms and their environment), the health of human beings, and the world's dwindling resources are concerns not only for the local environment but the global environment as well.

In 1992, the U.S. Council for Automotive Research (USCAR), an umbrella organization of DaimlerChrysler, Ford, and General Motors, was formed to further strengthen the technology base of the domestic auto industry through cooperative, precompetitive research. We have already mentioned the UltraLight Steel Auto Body (ULSAB) Consortium, formed in 1994 with the goal of reducing the mass of steel in autobody structures while maintaining their performance and affordability. The weight-reduction program devised by this consortium of 35 prominent sheet steel producers from 18 countries around the world produced a plethora of new alloyed high- (210 to 550 MPa) and ultrahigh-strength steels (greater than 550 MPa). Over 90% of the ULSAB body structure was composed of these steels, thus saving mass. Also, advanced forming and joining techniques enabled engineers to consolidate functions into fewer parts. Tailor-welded blanks removed masses of metal that did not contribute to performance, and hydroforming both tubular and sheet materials helped reduce part count to 108 as compared to more than 200 for an existing typical body structure. Other features were fewer spot welds (about one-third less) and significantly more laser welding. Steel is the most recycled material in the world and consequently the whole ULSAB structure is recyclable. With less mass, the consumption of gasoline will be less as will its contribution of pollutants to the atmosphere. The aluminum industry also in conjunction with the automotive industry has expanded the opportunities for green design. Because of its corrosion resistance, aluminum is nonbiodegradable; whereas steel rusts and biodegrades. In fact, aluminum and all of its alloys are the most recycled nonferrous metals. The production of aluminum is itself a large consumer of energy, and the process of extracting aluminum from its ore (bauxite) also contributes pollution to the environment. Hence, the low energy required to reprocess aluminum makes its recycling attractive. In addition to recycling, the aluminum industry joined the U.S. Department of Energy's "Industries of the Future" partnership for supporting national objectives for clean and energy-efficient production. The industry engaged in research leading to the development of an advanced aluminum cell to replace the energy-intensive Hall–Heroult process for primary production of aluminum. This century-old process produced significant emissions of greenhouse gases, particularly carbon dioxide and perfluorocarbons.

The Partnership for a New Generation of Vehicles (PNGV) resulted in greater use of high-strength, lightweight automotive aluminum. Alloys 5xxx–Al–Mg are finding wide usage, for example, 5754 for autobody panel and frame applications. Aluminum alloys allow the designer to create special shapes that place the metal where it can carry the required load most efficiently, thus eliminating the need for several separate parts joined together into one, specially shaped part. The many alloys of aluminum (see Figure 12-12) also make it a multipurpose material.

The discussion of recycling/disposal/reuse in Section 1.2.6 included, among other things, the fact that recycling has been "accomplished by requiring deposits on beverage containers to provide financial incentives to people to return cans and bottles." Aluminum has been recycled since it first appeared on the hood of Henry Ford's Model T. In 1996, an average of 252 pounds of aluminum per passenger car was recycled. Figure 12-13 shows the automotive aluminum recycling loop. Aluminum has proven sustained recyclability; that is, it can be recycled many times with no loss of performance or quality. Some automotive parts—aluminum engines, manifolds, motor housings, and other parts—are often made from 100% recycled aluminum. The phrase "recycled content" implies that it is from postconsumer (i.e., its origin is the marketplace) or postmanufacturing scrap. Secondary aluminum producers make aluminum alloys using a high percentage of scrap, primary producers make aluminum from basic raw materials.

Aluminum Association—*http://www.aluminum.org*
Thixotropic Injection Molding Processes with Image Gallery—*http://www.thixomat.com*
Model T Ford T100 Project—*www.ford.com/en/ourCompany/centennial*
http://www.boucherusa.com/1914modelT/

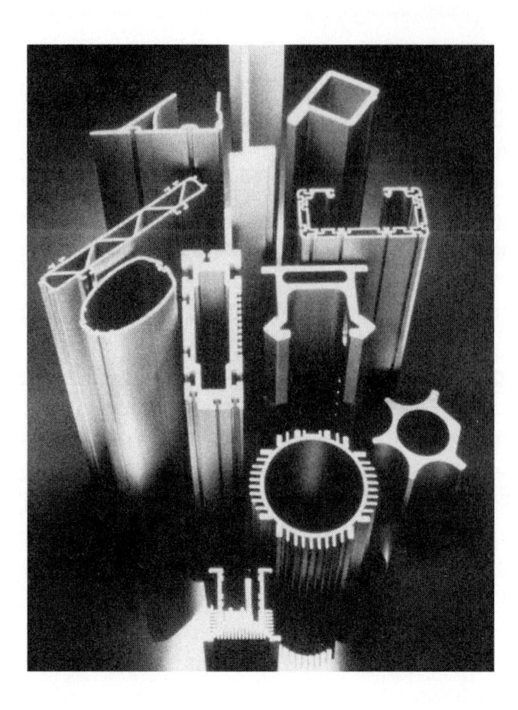

Figure 12-12 6xxx alloys utilizing Mg and Si are heat treatable. They possess great formability and corrosion resistance with high strength. Alloy 6061, a versatile alloy, is used to produce extruded autobody components such as brake cylinders and bumper reinforcements. 6022 is used in outer and inner body panels. These alloys provide a flexibility that can be described as "putting the metal where you want it."

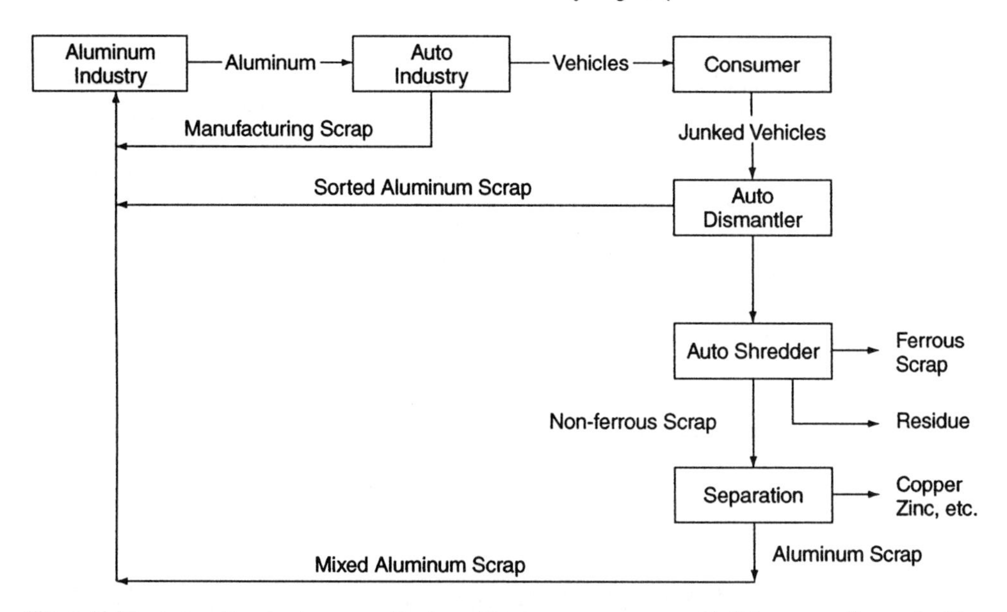

Figure 12-13 Automotive aluminum recycling loop. Aluminum recovery generally follows one of two paths. The first is at the dismantler, where large aluminum parts that are easily separated from an auto hulk are sold to a scrap metal dealer. Such parts as radiators, wheels, and bumpers are examples. The other path is through postshredder recovery. After the hulk passes through the shredder, the nonferrous portion is sent through heavy media separation, where the remaining aluminum is recovered. (Courtesy of the Aluminum Association)

12.4 CRADLE-TO-CRADLE DESIGN

The cradle-to-grave concept in design developed from a need to reduce the environmental impacts of manufacturing. A ***cradle-to-cradle*** concept advanced by McDonough and Braungart goes beyond the 3 R's of eco-efficiency—Reduce, Reuse, and Recycle—by adding Regulation as the fourth R. McDonough and Braungart promote regulation as a way to keep toxic materials out of the materials cycle. Their vision aims at creating laws that regulate design, materials and manufacturing/construction systems to ensure (1) designing buildings to produce more energy than they consume, (2) using manufacturing systems that generate drinking water rather than toxic waste water, (3) producing products that become food for plants and animals at the end of the life cycle, and (4) creating transportation systems that improve the quality of life. Their concepts would build on the materials cycle and life-cycle analysis to create an interconnected system of processes they label *technosphere*, an interconnectivity of industrial systems like the biosphere. For example, one cycle in the biosphere would involve plants. Plants begin from seeds, which grow to maturity, age, then die; after which they decay and are consumed by microorganisms to create nutrients in the soil for the next plant cycle. In the technosphere, a similar cycle should occur, much like that of the biosphere, only *technical nutrients* would come from recycling the materials from obsolete products into new materials for the next generation of products. Unit 5 will address these concepts further and show how plant-based plastics can be technical nutrients.

Some major corporations have implemented aspects of the cradle-to-cradle approach in their industrial operations. For example, Ford Motor built a plant that purifies storm water using the natural cleansing of a local marsh. The design of its new River Rouge plant involved producing more energy than the plant consumes, use of green roofs, and other features you can explore on the website below.

MBDC Cradle to Cradle website—*www.mbdc.com/c2c_home.htm*
Ford Motor's River Rouge Plant—
www.mcdonoughpartners.com/projects/p_ford_rouge.html

12.4.1 Metallics in Computers

Computers and other electronic equipment that become obsolete rapidly offer a good case for applying green design and using life-cycle analysis. Computer monitors and computer system units contain a wide range of metals and alloys, as shown in Figure 12-14. In addition to the recognizable steels, copper, and aluminum, we find other metals in computers and other electronic gear. Heavy (toxic) metals in components include lead, tin, cadmium, silver, antimony, zinc and copper in circuit boards; lead and barium in monitors; and mercury in switches and flat screens. According to the National Safety Council's website, the approximately 300 million obsolete computers in 2004 generated the following amounts of materials:

Lead—1 billion pounds
Cadmium—1.9 million pounds
Chromium—1.2 million pounds

(a)

(b)

Figure 12-14 Computers use a wide variety of materials, including heavy (toxic) metals that have environmental impacts if not disposed of or recycled properly. (a) Inside of a computer monitor showing the back of the CRT and circuit board and (b) a computer system unit revealing a wide range of metals, alloys, and some plastics and elastomers.

Mercury—400,000 pounds
Various plastics—4 billion pounds

Some of these materials will be recycled, but some will find their way into the biosphere as toxic waste.

The Silicon Valley Toxics Coalition (SVTC), among other concerned groups and individuals, wants to ensure that this ***E-waste*** is (1) properly disposed of and (2) reduced or eliminated. In order to achieve a sustainable world, SVTC wants the Soesterberg Principles applied to electronic product life cycle: "Each new generation of technical improvements in electronic products should include parallel and proportional improvements in environmental, health and safety as well as social justice attributes."* The Soesterberg Principles is an example of the type of regulation, the fourth R in the cradle-to-cradle approach, which can help achieve green manufacturing.

The following websites and Internet searches should provide current information on the impact of E-waste and efforts to reduce it.

National Safety Council—*http://www.nsc.org*
Silicon Valley Toxics Coalition—*www.svtc.org*
Toxic Dude—*http://www.toxicdude.com/getTheFacts.asp*
Intel's Education—*www.intel.com/education/recycling_computers*

SELF-ASSESSMENT

12-1. Another name for sintered metals is ___________.
a. Fused **b.** Compacted **c.** P/M

12-2. In P/M the powder is compacted and heated to allow for ___________.
a. Diffusion bonding **b.** Reduced grain growth **c.** Lower density

12-3. Name another use for iron powders (other than P/M).
a. Making steel **b.** Destructive testing **c.** Incendiaries **d.** Welding electrodes

12-4. The primary purpose of briquetting is to ___________.
a. Facilitate handling **b.** Diffusion bond **c.** Sinter

12-5. Which of the following is the main barrier to the adoption of advanced materials?
a. Lack of database **b.** High cost **c.** Lack of experience with material

12-6. Generally speaking, lightweight materials are selected for what particular area of application?
a. Building construction **b.** Machine tools **c.** Transportation

12-7. Name one source of information about materials that serves as a valuable aid in their selection.
a. Handbooks **b.** CD-ROM **c.** Internet **d.** Textbooks

12-8. The following operation is *not* an example of a secondary operation in the P/M process.
a. Mixing **b.** Coining **c.** Impregnation

12-9. A P/M part designed as a self-lubricating bearing requires ___________.
a. Extremely close-packed particles **b.** Loose-packed particles **c.** High-diffusion bonded particles

12-10. Sintering implies ___________.
a. High pressure **b.** High temperature **c.** Both high pressure and high temperature

12-11. Describe an emerging advanced metals technology and the benefits it offers.

12-12. Cite an example of how metals can be a part of green design and manufacturing.

*Adopted by the Trans-Atlantic Network for Clean Production, May 16, 1999.

12-13. Creating laws that regulate design, materials, and manufacturing/construction systems to ensure reduced environmental impacts by establishing a technosphere is described by the term

a. Green design **b.** Cradle to grave **c.** Cradle to cradle **d.** Green chemistry

12-14. The fourth R for green design concept of Reduce, Reuse, and Recycle is

a. Redundancy **b.** Recapture **c.** Reform **d.** Regulate

12-15. Materials used in products that will easily recycle into new products and not harm the environment if put into landfills are

a. Technical nutrients **b.** Biocompatible **c.** PBCs **d.** Amorphous

12-16. A major concern with the disposal of obsolete computers is

a. Heavy metals **b.** CNT **c.** Biosphere **d.** SSF

REFERENCES & RELATED READINGS

ANNUAL BOOK OF ASTM STANDARDS. Philadelphia, PA: ASTM, 2003.

ASM INTERNATIONAL. *Alloy Finder*. CD-ROM software. Materials Park, OH: ASM International, 1996.

ASM INTERNATIONAL. *Powder Metal Technologies and Applications*. Materials Park, OH: ASM International, 1998.

ASM INTERNATIONAL. *Processing and Fabrication of Advanced Materials*, Vol. XI Materials Park, OH: ASM International, 2002.

CAPUS, JOSEPH M. "Advances in Powder Metallurgy Processing," *Advanced Materials & Processes*, September 1999, pp. 33–36.

DAVIS, J. R., ed. *Metals Handbook Desk Edition*, 2nd ed. Materials Park, OH: ASM International, 1998.

LIQUIDMETAL WEBSITE—www.liquidmetal.com, January 12, 2003.

MCDONOUGH, WILLIAM, AND MICHAEL BRAUNGART. *Cradle to Cradle: Remaking the Way We Make Things*. North Point Press, 2002.

MCDONOUGH PARTNERS WEBSITE—http://www.mcdonoughpartners.com/projects/p_ford_rouge.html, January 11, 2003.

MBDC CRADLE TO CRADLE WEBSITE—www.mbdc.com/c2c_home.htm, January 11, 2003.

METAL POWDER INDUSTRIES. *Powder Metallurgy*. Powder Metal Industries Federation, website process diagram—http://www.mpif.org/chart.html, November 19, 1999.

NATIONAL SAFETY COUNCIL WEBSITE— http://www.nsc.org/ehc/epr2/smith/sld020.htm, January 10, 2003.

RIDDER, STEPHEN D., AND RICHARD J. FIELDS. "Instrumented Consolidation of Amorphous Powders," *Advanced Manufacturing Methods*, reprint, January 2003.

WITTENAUER, JERRY P., AND JEFFERY WADSWORTH. "Refractory Metals Forum: Tungsten and Its Alloys," *Advanced Materials & Processes*, September 1992, pp. 28–37.

Periodicals

Advanced Materials & Processes

Journal of Materials Education

EXPERIMENTS & DEMONSTRATIONS IN METALS FROM NATIONAL EDUCATORS' WORKSHOPS

EMSET 2 CD-ROM *Experiment in Materials Science, Engineering and Technology* 2nd ed. ISBN 0-13-030534-0. Over 300 Experiments and demonstrations available in easy-to-adapt Acrobat Reader format; department licensed. Call 1 800-922-0579 or visit www.prehall.com

UNIT 5

Surface Science & Engineering

PAUSE & PONDER

"Surface science" may sound like a strange topic and far from everyday conversation or knowledge. However, we encounter issues of surfaces in our daily lives. Let's explore a few examples.

Buying a new kitchen knife involves knowing something about surfaces. Your choice may be between a good, sharp knife made of carbon steel or a good stainless steel knife. A carbon steel knife will be sharper than a stainless steel but will rust easily; the stainless steel knife will not rust even if put away without proper cleaning and drying. Designers know about the strength and durability of stainless steel and its rust resistance and ability to withstand extreme temperatures. In both commercial and home kitchens, stainless steel finds many uses for large surfaces such as countertops and appliance cabinets. However, polished stainless steel shows fingerprints and will scratch and dent. Enter *brushed stainless steel surfaces*. The lower luster and brushed and satin finishes hide blemishes and cleaning is a breeze by wiping in the direction of the brushed grain.

A silver flower vase costs much more than pewter, plastic, or glass, and silver requires more care to keep it from tarnishing. Traditional polished brass provides shining beauty for lamps, door pulls and handles, bed frames and vases until the protective polymer coatings wear off or are polished off, then it tarnishes to an unsightly state. Advances in surface science with PVD (physical vapor deposition, discussed later) now provide polished brass that remains shiny indefinitely due to an ultrathin transparent protective layer of chrome, such as the faucet in the figure.

Applying paint to wood, metal, or other materials involves working with surfaces. Prior to painting you must do proper surface preparation, which may involve a thorough scraping and cleaning to remove all forms of dirt and evidence of corrosion. Remember the saying "95% of the job of painting must be devoted to preparing the surface."

In an effort to avoid the need to paint wooden sheathing on homes, aluminum siding became popular. But in spite of being anodized even aluminum corroded. Then polymers came to the rescue. Vinyl siding is now in vogue, but even this material requires pressure washing after a number of years to brighten it up. How many new types of home siding or finishes have you seen advertised that promise to reduce or eliminate surface maintenance?

Marine environments provide many challenges to surface protection. Even the composite fiberglass commonly found on boats requires a good working knowledge of surfaces for upkeep whether on a freshwater lake or in contact with saltwater. Barnacles create problems by attaching themselves to boat hulls with remarkable natural adhesion—the envy of adhesive designers. The many fasteners used with boats also require careful materials selection and often protective coatings to avoid rapid deterioration.

These examples describe surfaces we normally see, but what about those critical hidden surfaces like fillings in your teeth, joints within aircraft wings, or those that hold components soldered inside your computer?

To ensure the high reliability of the circuit boards in computers, DVD players, and other electronic gear, the soldering of electrical wires and components must be successful. This means clean surfaces and using fluxes to prevent air from getting to the molten filler metal and to remove oxide films from the faying (contacting) surfaces to be joined. Surface engineering plays a key part in designing the processes for highly automated surface mounting of components on circuit boards.

The considerations for surface science and engineering, as well as interface technology cited here should stimulate you to explore the topics with an appreciation for their value in quality products that we make and use.

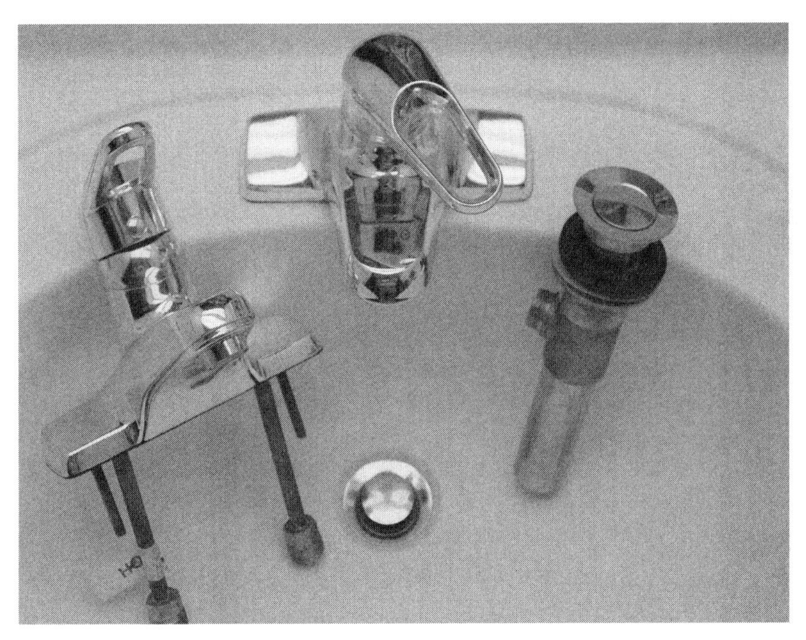

This brass faucet set can carry a 10-year warranty against tarnishing as a result of applying a PVD coating of ultrathin chrome over the brass base metal.

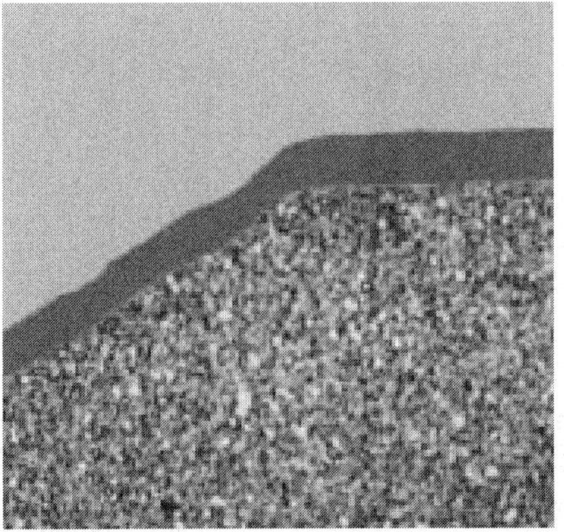

10 µm

Cross section of a PVD coated component

Photomicrograph showing the edge of a PVD-treated surface. Note the scale of 10 µm and the dark outer protective layer (2.5 ± 0.7 µm thickness tolerances) over the base metal shown with its less dense grain structure. PVD coating follows the parts contour, making no dimensional changes; therefore, it does not require subsequent machining. (Balzer Tool Coatings, *http://www.balzers.com/*)

Module 13

Surface Science & Engineering

After studying this module, you should be able to do the following:

13–1. Describe some of the functions performed by surfaces and the key issues in surface science.

13–2. Distinguish between the mechanical reactions and chemical reactions that surfaces can undergo.

13–3. Explain why engineers and designers should observe nature and remain curious about how nature operates.

13–4. Distinguish between the terms *surface treatment* and *surface modification* and provide examples of the problems involved in designing coatings.

13–5. Explain the necessity for good surface cleanliness prior to applying any surface treatment.

13–6. Describe principles and reasons for the use of thermal spraying and cold spraying as well as limitations in the choices for substrate materials.

13–7. Explain the effects of hardening on the ultimate toughness of a part, including the rationale for various forms of surface hardening and the role of the element carbon in hardening steel.

13–8. Use the website addresses denoted by the symbol* to explore sites for new developments and evolving concepts related to surface science and engineering.

*The symbol found throughout the book will link you to Internet sites related to the topics being covered. The dynamic nature of the Web brings frequent changes, so some of these sites, while available at the time of writing, may not be up now.

13.1 SURFACE SCIENCE AND ENGINEERING

An axiom of long standing states that "God made solid materials but the Devil made surfaces." Surfaces shape and define our physical world. The structural elements that make up the solid beneath the surface mainly provide support to the surface. The many functions performed by surfaces include providing an aesthetic appearance, but principally involve interactions with the surrounding environment. Mechanical interactions such as friction and wear, chemical interactions such as corrosion and adherence, electrical, optical and thermal interactions—all require consideration from a design standpoint. The importance of surfaces has led to the development of a new discipline called ***surface science and engineering***, which is gaining greater attention from other engineering disciplines because of the economic benefits that accrue from this new expertise. The International Federation for Heat Treating and Surface Engineering (IFHTSE) is a federation of scientific and technical societies and associations most involved in surface engineering and heat treatment. As in other areas of materials science, nanotechnology will radically transform surface science and engineering, for materials with nanoscale structures have demonstrated significantly improved properties for magnetic and electronic applications as well as structural coatings and superplastic formability.

Module 8 described the instruments and techniques for analyzing structures, mostly the surfaces of materials (even though often a cross section exposing an interior). In Section 8.5 you learned about material characterization, and we repeat the figure here (Figure 13-1) for convenience in this discussion of the characterization of surfaces and their features. Figure 13-2 is a cross section of titanium plate at 400X revealing the interior and plate grains and a contaminated surface layer of light elements (oxygen and nitrogen) from the ambient air. Obtaining images of surfaces for characterization requires careful metallographic techniques, as explained in Module 8.

 IFHTSE website—*www.ifhtse.org/home.htm*

13.1.1 Surface Science

Surface science helps solve problems involving wear, corrosion, hardness, carburizing, nitriding, electroplating, electroless plating, phosphating, vapor deposition, and hardfacing. This new discipline is now incorporated into the broader field of materials science and engineering.

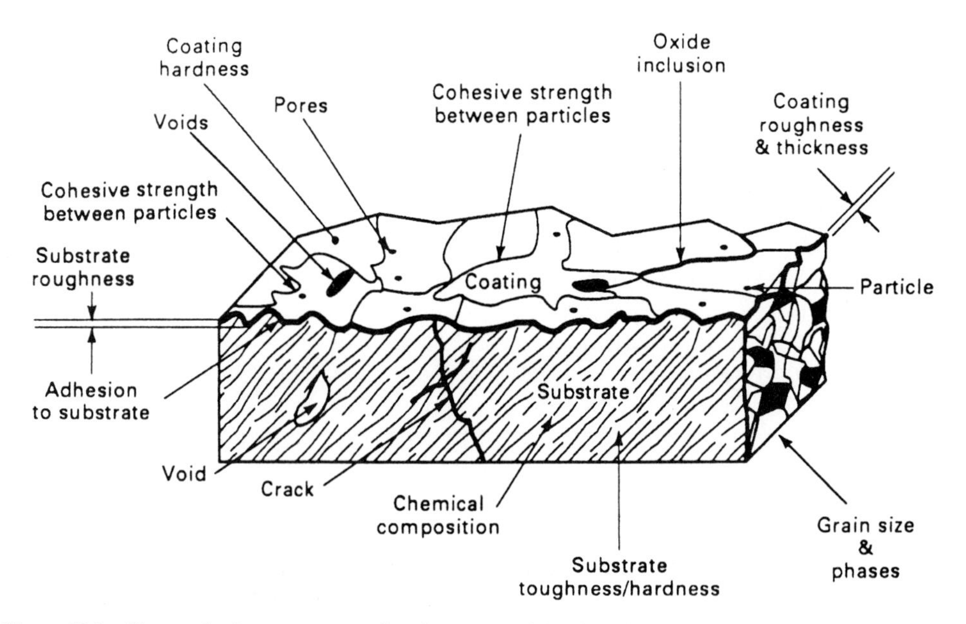

Figure 13-1 Characterization components of a substrate material and coating.

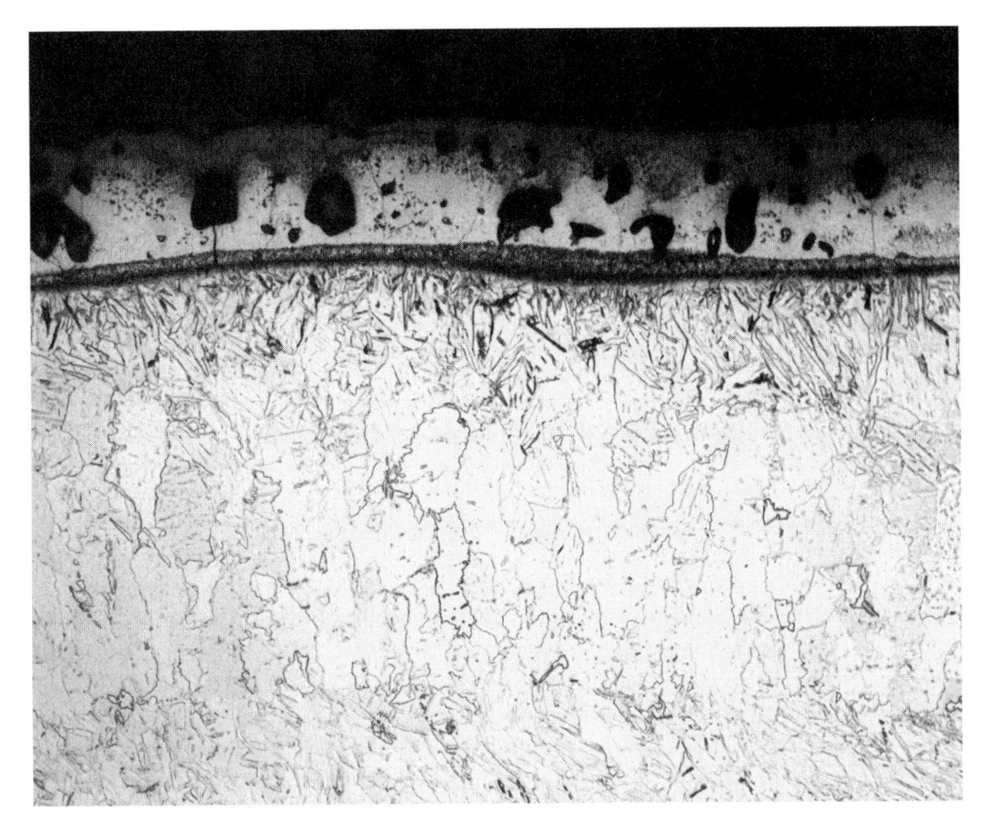

Figure 13-2 Longitudinal view of titanium plate at 400X with surface consisting of contamination film of the light elements oxygen and nitrogen. (*Advanced Materials & Processes*)

One example of an application developed by this field of technology and science is the thin films of polymers and adhesive used instead of paint on USAF aircraft.

13.1.2 Surface Engineering

Engineers have always been encouraged to look to nature to observe how certain tasks are accomplished or to stimulate the mind when involved in problem solving. *Biologically inspired materials,* as defined earlier, result from MSE mimicking living organisms to create advanced materials. Take, for example, the structure of seashells. Observations of birds dropping clams and oysters from considerable heights onto some hard surface to break the shells and of sea otters smashing abalone shells with the aid of a rock attest to the hardness of these shells. But the shells are basically made from a relatively soft chalk. What makes them extremely tough is the manner in which thin alternating layers of calcium carbonate and organic polymers are intertwined—much like a layered composite material. (Two later modules in this text discuss layered composites.) These microscopic layers could be compared to a brick wall, in which each brick is surrounded by mortar. Cracks in an individual brick are stopped from traveling into the adjacent brick by the mortar. Scientists at Sandia National Laboratory, inspired by the properties of abalone shells, have attempted to duplicate this natural process using detergent molecules similar to those of ordinary household detergents. They developed an arrangement of ***micelles,*** a structural unit of molecules, that upon being polymerized, bonded at the surface, resulting in a hard surface that can be used in such applications as automobile finishes or optical lenses.

With emphasis on materials that can withstand high temperatures (greater than about 320°C), whether in space or on earth in turbines or heat engines, new developments in materials are reported almost daily. The search has led in many cases to the field of surface engineering. One of the first discoveries was the use of zirconia to form a surface layer on a metal

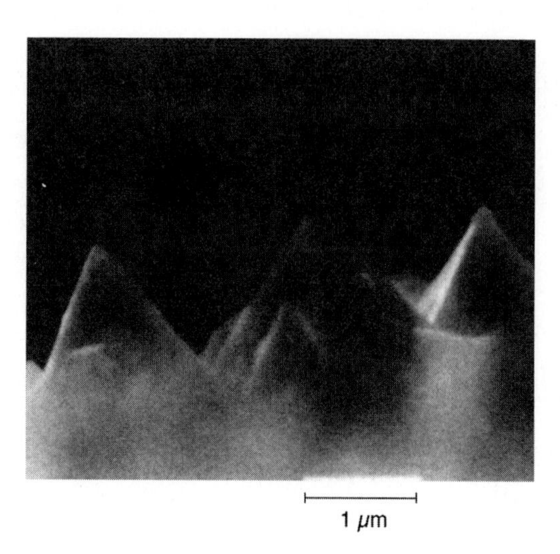

Figure 13-3 SEM photomicrograph of crystal-oriented Pb-8Sn-2Cu. "Side view" of deposit shows the pyramidal surface morphology that enhances wettability by motor oil, which results in improved bearing performance. (*Advanced Materials & Processes*)

substrate. This ***thermal barrier coating (TBC)*** is very effective in resisting high temperatures in a heat engine's moving parts. It fails, however, to protect materials in sliding contact from wear under high temperatures. To add lubrication properties to this TBC, a new, self-lubricating material consisting of 80% silicon carbide, 10% silver, and 10% barium/fluoride eutectic was developed to provide lubricity at high temperatures to sliding-contact-bearing surfaces of internal combustion engines made with conventional iron and aluminum. Many of these engines are now operating at such high temperatures that oil-based lubricants cannot be used. Improved motor oil ***wettability*** (meaning oil adheres to more surface area) of bearing surfaces was achieved with a Pb-8Sn-Cu coating (Figure 13-3) using techniques described in the next section.

13.2 SURFACES

Surfaces, the first layers of a material in contact with the outside environment, play an important role in engineering and scientific applications. Parameters such as adhesion, wear, corrosion, and hardness are directly affected by surface properties. For example, poor surface cleanliness or improper surface chemistry can cause defects or imperfections during subsequent coating operations. Surface atoms, having no neighboring atom on one side, have higher energy and are less firmly bonded than internal atoms of the same solid. Furthermore, the surface finish (rough or polished) or a surface subjected to the effects of corrosion can create additional surface energy that could result in the development of cracks and possible failure of a part. Drops of liquid matter, however, take on a spherical shape, which minimizes the amount of surface area and, hence, ***surface energy.*** For example, small objects like a beetle can float on the surface of water unless it is too heavy. The surface molecules of the water resist being pushed apart. In this case, they are resisting the force of gravity. ***Surface tension,*** the attractive force between surface molecules in a liquid, is a product of surface energy. Like rain, any liquid when sprayed will form drops. Each drop assumes the shape of a small sphere because surface tension causes the drops to have the smallest possible surface area. The smaller the area, the smaller the surface energy. Reducing surface area is the main principle, for example, behind sintering.

13.3 SURFACE TREATMENTS

One of the ways surface treatments can be categorized is by the form of the material being deposited on the surface. Four categories of surface treatments are therefore (1) atomistic deposition, (2) particulate deposition, (3) bulk coating/cladding, and (4) surface modification. Physical vapor and plasma deposition are examples of atomistic deposition; thermal spray and

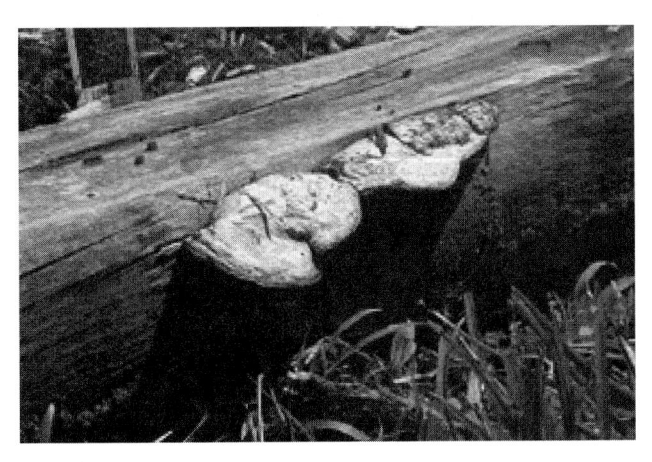

Figure 13-4 Fungus growing on pressure-treated timber.

Figure 13-5 The biological attack overcame the pressure treatment (moisture provided the environment for microorganisms).

cold spray represent particulate deposition; painting and weld overlay would represent the third category, bulk coating; and anodizing, shot peening, and ion implantation represent the ***surface modification*** category. This last category is discussed in Section 13.5. It is important to recognize for all categories of surface treatment that the properties and performance of surfaces, as with other materials, are largely determined by their atomic compositions and microstructure.

Surface treatments or coatings solve problems of corrosion, wear, and compatibility. Module 5 discussed biological attacks on materials and more coverage of these concerns will be presented in Unit 6 on polymerics. Figure 13-4 shows a surface attack by fungus on pressure-treated landscape timbers; Figure 13-5 shows how the timbers have rotted all the way through. The "salt treatment" of these timbers should have provided decades of protection from soil contact. Why did this treatment fail to provide the specified protection? The rest of this module will explore the answers to questions like this.

13.3.1 Thermal Spray Technology

Thermal spray technology can rapidly deposit essentially anything that melts onto a large variety of substrates, including a wide range of metal alloys, many ceramics/glasses and even some polymers. It has the ability to produce consistent, repeatable, and high-quality coatings in the most economical manner. Further, it is flexible enough to allow the user to react quickly to diverse and constantly changing requirements.

The ***thermal spraying*** processes of electric arc combustion and plasma spray apply any material to the surface of any other. Using a low-temperature, high-speed directed spray (applying the coating only where needed), the process called ***kinetic energy metallization (KEM)*** sprays particles of aluminum, titanium, or copper mixed with helium and/or nitrogen at high speeds onto a metal substrate. Upon hitting the substrate, the particles deform, increasing their surface area and bonding metallurgically to the substrate. Nanoparticles such as these in high temperature processes usually grow during consolidation, negating the benefits of the original small cross sections, but the KEM process permits them to consolidate while maintaining their beneficial mechanical properties.

13.3.1.1 Plasma surface engineering. Carburizing and nitriding, to be discussed later in this module, are examples of ***thermochemical*** surface treatments. NO_x emissions from conventional, gaseous nitrocarburizing are 1000 times greater than those of ***plasma technology,*** which is edging out the gaseous methods due for environmental, economic, and energy reasons and the materials being sprayed are no longer limited to traditional metals or alloys. Composites, polymers, both ceramic and polymeric, biomaterials, or just about any material that can be melted is suitable as a coating for just as many different and unique substrate materials. Both thermal barrier coatings (TBCs) and self-lubricating coatings can be applied by the previously mentioned ***plasma-arc spraying*** method. The process produces a true ***laminar composite*** (MMC) material consisting of an aluminum or iron substrate laminated by a TBC and a self-lubricating wear coating superimposed on the surface of the TBC. The ***nickel–silicon–carbide metal–matrix composite coating*** for cylinder bores on Jaguar AJ-V8 engines is 0.08 mm thick yet resists wear and corrosion and reduces the overall weight of the 4.0-liter (L), 290-HP engine by 7 kg (this model has a "best-in-class" weight of 200 kg). The aluminum cylinder blocks used in the Jaguar since 1984 had experienced wear problems at the top ends and center of the bores as a result of high rubbing speeds produced by static-load reversals of the pistons. Cast-iron liners solved these problems but they were too heavy. With the new composite coatings, the Jaguar's fuel efficiency reached 8.9 L per 100 km or 26.4 mpg (miles per gallon) highway and 17.2 mpg in city driving. The ***electroplated nickel–silicon–carbide composite coating (NiCom)***™, developed by U.S. Chrome Corp., Stratford, Connecticut, can be as thin as 0.005 to 0.015 mm and as thick as 1.5 mm. The SiC particles are codeposited with the Ni and are uniformly disbursed in the nickel matrix. The coating hardness averages about 55 to 57 HRC. The company reports that nickel plating is more efficient than chromium and test results indicate that NiCom™ wears better than chromium in all engine bores. Resistance to ***sliding wear*** is its outstanding characteristic, but its frictional resistance properties also rank high.

Adding a small amount of a laminar lubricant such as boron nitride (BN) or graphite to any of these composites can improve their frictional properties. The ***oleophyllic*** (oily) nature of the resulting coatings adds to their ability to reduce friction in internal combustion engine cylinders and hydraulic equipment. In addition, ***oil retention*** (the ability to maintain hydrodynamic lubrication) is better than that provided by chromium or electroless nickel. Prolonged exposure to 540°C does not result in significant loss of these physical properties.

Ceramic barrier coating systems can also be applied with plasma technology. They protect sensitive display devices such as liquid crystal (LCD) and organic light-emitting diodes (OLED) against moisture with a smooth surface combined with a redundant ceramic coating and protective overcoat. In this process a liquid monomer coats the substrate, filling in all the valleys and submerging the peaks and creating a microscopically flat surface. Upon hardening, the monomer transposes into a solid ***polymer*** film. A layer of a transparent ceramic creates the first barrier and is followed by a second flat surface. The procedure is repeated until water and oxygen impermeability is attained. The total coating thickness is less than 3 micrometers (3 μm) and possesses enough flexibility to be rolled up and still retain its properties.

Other thermal spray technologies and coatings are receiving attention as solutions to surface engineering problems involving wear, corrosion, and thermal degradation. Among them is the versatile ***atmospheric plasma spray (APS),*** which is capable of producing coatings from almost any meltable or heat-softenable materials for a wide variety of substrate materials. Oscillating electrical fields called ***atmospheric plasmas*** are used to modify the surfaces of fabrics so that they retain strength over longer periods of time, resulting in the reduction of the number of treatments required during the production process by current chemical processes and the attendant costs. Eventually, this technology will lead to the production of biomaterials in which the plasma generates thin films that will modify the surface of fabrics to make them bioactive, for example, sterile fabrics for bandages, surgical garments, or implants. AK Steel, maker of antimicrobial steel, has coated steel with a compound that kills microorganisms such as mold and fungi. The applications are numerous—doorknobs, handrails, faucets, countertops, and appliances. ***Vacuum plasma spray (VPS)*** and ***low-pressure plasma spray (LPPS)*** are special processes that operate at reduced pressure to produce high-performance coatings having very low porosity and oxide content. One company alone, using eight different techniques, offers more than 300 different plasma-deposited coatings.

13.3.1.2 Cold spray/gas dynamic spraying. An emerging new coatings technology, ***cold spray*** or ***cold gas dynamic spraying,*** reduces or eliminates the effects of high temperatures (oxidation, vaporization, melting, crystallization, residual stresses, debonding, and gas release). Using high-deposition rates, high spray rates, and low-cost feedstock, it improves cost efficiencies. Cold spray can rapidly deposit onto various substrates many metals and some composites at or near room temperature in an ambient air environment. It is a solid-state process in which the coating is generated by mechanical deformation of powder particles. The material to be coated is injected into a supersonic jet, which impinges on the surfaces to be coated. The resulting deposits typically have low porosity and little or no additional oxide, and compressive residual stress. There is minimal heating of the substrate, and the particles do not melt during their brief interaction with the warm gas stream (air, helium, or nitrogen) used in cold spray, nor does the plastic deformation of the particles result in any melting at impact. Normally, thermal spray processes use powders heated to the melting temperature prior to application, and the bonding mechanism is similar to the explosive welding process, in which two metals are impacted with such force that the oxide layers are removed and a metal-to-metal bond occurs. Cold spray technology, however, can now produce high-quality deposits of a wide range of metals and some composites at or near room temperature without melting or even significant heating.

13.3.1.3 Fluidized bed technology. Low-temperature surface engineering treatments that improve wear resistance of metal parts also use ***fluidized bed technology,*** which is based upon diffusion at low temperatures of metallic elements like chromium, nitrogen, and carbon into the base metal. The metal parts to be treated are immersed in a specially designed fluidized bed and heated, typically, to 560°C, at which point the reactive gases containing the basic elements needed for the formation of the required layer are introduced into the bed. The bed itself consists of inert particles—aluminum oxide, for example—coated with the element to be diffused into the surface layer.

13.4 SURFACE HARDENING AND SURFACE MODIFICATION

Throughout the discussions of hardening and strengthening metal alloys in Module 9, the objective was to obtain the same hardness and strength throughout the part that was being heat treated. It would be impractical, if not impossible, in many instances, to harden the entire part, that is, both the interior and the exterior, to that hardness demanded by wear-resistant surfaces. Even if this degree of hardness could be attained throughout, the part would have little *toughness* (the ability to withstand impact loads). There is little application in industry for such a part.

Through hardening is the hardening of a part throughout its cross section, as opposed to ***surface hardening,*** commonly referred to as ***case hardening,*** which hardens only the surface of the part without changing the surface composition from the original steel. Most steel parts require machining, and this operation is best performed when the part is not hard. Once machining is completed and the part is within the required tolerances, it may be hardened on

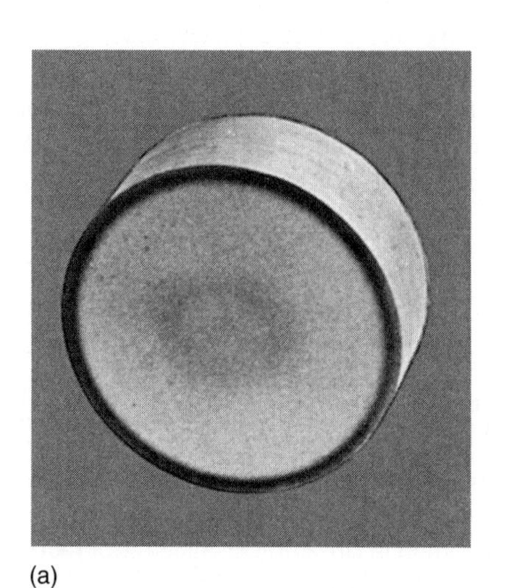

(a) (b)

Figure 13-6 (a) Polished section of 8620 steel bar carburized to a depth of 0.060 in., measured to a 0.40% carbon content. (Republic Steel Corporation) (b) To combat wear, gear teeth get case hardened so the gear body retains its toughness to handle torsion stresses.

Figure 13-7 The ibgNDT Systems Corporation's preventive multifrequency tester (eddyliner® P) is shown with some typical automotive parts that can be automatically tested using eddy current testers. (ibgNDT Systems Corporation)

the surface to provide a strong, hard, wear-resistant surface. Figure 13-6a shows a case-hardened bar; Figure 13-6b shows a gear with case-hardened teeth.

Figure 13-7 shows an eddy current tester that can be used to determine the structure, composition, or properties of parts, such as the automotive parts shown. Included in the test results are a part's hardness, case depth, temper, residual stresses, and structural variations. ***Eddy currents*** are the result of the interaction between a ferrous material and an electromagnetic field. This field is produced by an alternating current flowing in a conductive coil. If a metallic sample is placed near or within this field, then the field of the coil will induce eddy currents in the sample, which, in turn, interact with the original field of the coil. By observing or measuring the effect of the sample on the coil, much information may be gained concerning the structure and composition of the sample. Once optimum conditions are established, such as frequency and magnitude of the current in the coil, changes in the electrical conductivity or magnetic permeability of the sample can be detected. Eddy currents do not penetrate deeply into a sample. Therefore, it is flaws at or near the surface that are readily detected, but through analysis of the hysteresis loop, many characteristics of ferrous materials can be determined (see Module 6 for further explanation of hysteresis loops and magnetic permeability of materials).

13.4.1 Case Hardening

Hardening of the surface only is, as we have said, known as case hardening. The surface layers that are actually hardened make up the *case*, and the inner surfaces that are not hardened are known as the *core* (Figure 13-6a). The first three case-hardening methods mentioned in this section are chemical modifying methods or diffusion-coating methods. They provide the elements that are lacking in the surface layers to permit hardening reactions to take place. We have learned that carbon is an essential element in steel if it is to be hardened (about 0.3% carbon in steel is the minimum for hardening steel). Many hot-working operations result in a removal of the carbon from the surface, however, a process known as ***decarburization*** (Figure 13-8a).

13.4.2 Carburizing

To harden the steel, the carbon content must be restored. As the carbon content in steel increases, the heat treater's ability to create greater hardness also increases. The purposes of carburizing are to restore lost carbon and/or to increase the amount of carbon in the outer layers of the metal part. Basically, during carburizing the metal part is surrounded by a high-carbon-content solid material or gas; with the necessary energy supplied by high furnace temperatures, the carbon atoms diffuse from this high-carbon-content medium into the crystal structure of the metal part to be case hardened.

Carburizing involves not only the absorption of carbon into the surface of the metal but the diffusion of carbon away from the surface and into the interior of the part. Further, this process also induces desirable residual compressive stresses through the case-hardened layer as a consequence of the delayed transformation and volume expansion of the carbon-enriched surface of the metal. After carburization, the steel part is ready for nonequilibrium thermal processing, which transforms the high-carbon-content austenite into martensite. This processing is followed by various quenches and tempers, depending on the degree and depth of hardness as well as the degree of grain refinement desired in both the case and the core (Figure 13-8b).

13.4.3 Nitriding

In our study of the various phase changes produced by heat treating steel, we learned that some iron reacts with carbon to produce a very hard, brittle compound, iron carbide (FeC), that precipitates out during a phase change, giving the steel hardness and strength. Aluminum alloyed with copper forms another compound ($CuAl_2$) with similar results. In addition to Ti, Al, and Fe, there are several other elements, such as chromium (Cr) and molybdenum (Mo), that react with nitrogen (N) in steel to produce compounds called ***nitrides,*** which are also hard and brittle like carbides. The purpose of nitriding is to provide the necessary nitrogen to the steel so that these hard nitrides may form and produce the necessary hardness. The nitrogen-rich medium used may be ammonia gas. As with carburizing, heat is used to provide the energy for the nitrogen atoms to diffuse into the steel case. The heat also helps produce the necessary chemical changes, and therefore additional heat treating is not required.

13.4.4 Cyaniding and Carbonitriding

The methods discussed provide cases that are rich in carbon or nitrogen. ***Cyaniding,*** an old process, might be considered to provide both carbon and nitrogen. Cyanide (CN) decomposes

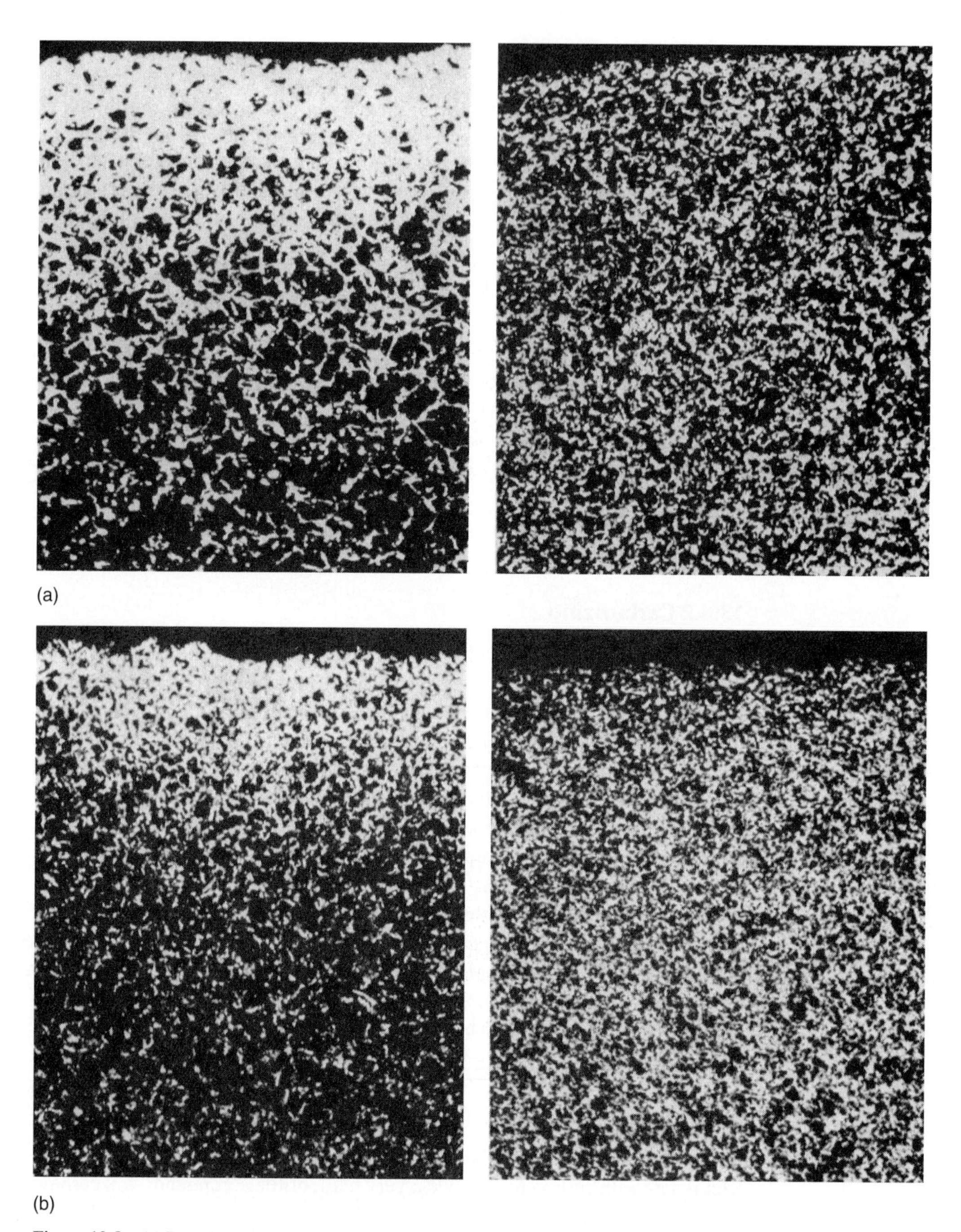

Figure 13-8 (a) Decarburizing. *Left*: Photomicrograph showing "decarb" to a depth of 0.020 in. in the surface of a hot-rolled bar of 1050—magnified 100 times. *Right*: Same bar after carbon correction annealing. Note the restoration of carbon on the outer surface and also the refinement of the grain structure. (b) Carbon correction. *Left*: Photomicrograph of AISI 5046 before carbon correction—magnified 100 times. *Right*: Same bar after carbon correction annealing with surface carbon restored. (Republic Steel Corporation)

in a liquid bath of molten salts containing sodium cyanide, and the nitrogen reacts with iron to form hard iron nitride. The carbon diffuses and assists in further hardening operations, as discussed previously. ***Carbonitriding*** is very similar to gas carburizing in that it uses a medium consisting of a mixture of gases rich in carbon and also nitrogen. The presence of nitrogen in the austenite also permits slower transformation to martensite and therefore lower temperatures and less drastic quenching can be used. These conditions result in less distortion and less danger of cracking.

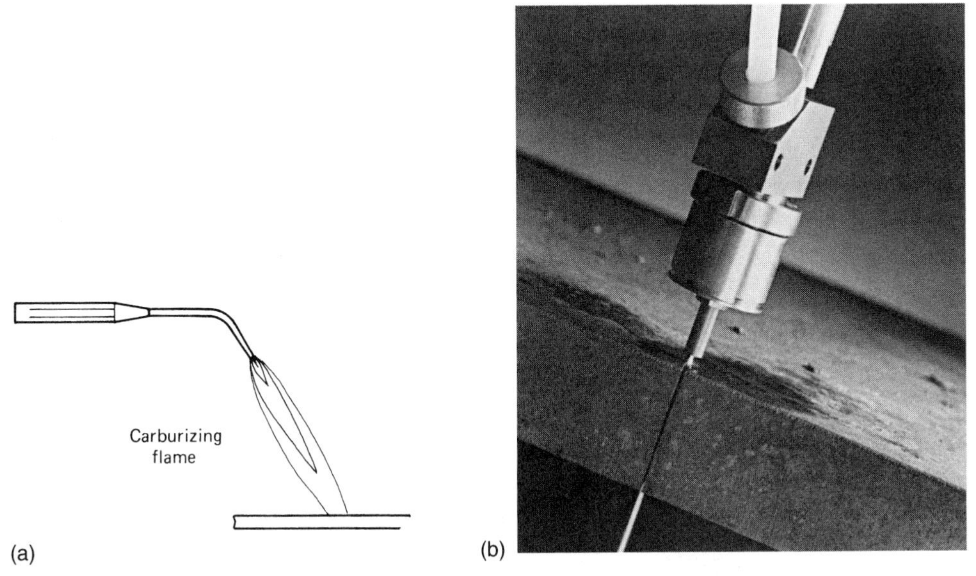

Figure 13-9 (a) Flame hardening. (b) Water-jet cutting allows thick metals to be cut without the thermal effects produced by flame cutting. This Waternife is cutting 3-in. steel plate. (Flow Systems)

13.4.5 Flame Hardening and Induction Hardening

Flame hardening and induction hardening do not change the chemical makeup of the steel case. Through the application of heat, the outer layers of the metal part are first transformed into austenite, and then the austenite is transformed into martensite by quenching. The steel to be case hardened must contain all the hardening ingredients. Any decarburized surface would have to be removed by machining or grinding prior to induction or flame hardening. In ***flame hardening,*** oxyacetylene torches are used as sources of heat, as shown in Figure 13-9a. In ***induction hardening,*** heat generated by the resistance of the metal to the passage of high-frequency-induced electrical currents is used.

Surface-hardened steel products are truly composite materials; they contain two or more integrated components. Reinforced concrete is a composite material whose integrated components are steel rods strong in tension and a concrete matrix strong in compression. They work together in an integrated system to provide strength properties not obtainable by either one acting alone. In surface-hardened steel parts, the integrated components are the hard case and the tough core, which together provide a superior performance not obtainable by either alone. Even steels that are through-hardened, such as fine-pearlite or tempered-martensite steels, are strong and relatively hard as a result of the distribution of fine particles of a second, harder phase throughout a soft ferrite matrix (see dispersion hardening, Section 25.2.4.2). The relatively soft matrix material, for example, the solvent component of an aluminum–copper alloy, though deformable, cannot deform independently of its harder phase. The point is that there is a striking similarity among all human-made materials—steel, aluminum alloys, prestressed or reinforced concrete, fiber-reinforced plastics, or surface-hardened metal products. They all contain integrated components working together to provide the right mix of properties needed for today's highly technological society. (Further details on composite materials have been provided in Module 8.)

13.5 OTHER METHODS OF SURFACE MODIFICATION

Many manufactured metal parts require additional processing of their surfaces to impart certain properties. Zinc-plated steel, jewelry that is plated with precious metals, and gold-plated electrical contacts are but a few examples. Surface treatments, surface modification, and surface engineering are called upon to improve

1. Wear and erosion resistance
2. Friction and fatigue resistance

3. Stiffness
4. Surface roughness
5. Lubrication (lubricity)
6. Texture, color, or decorative features (appearance)
7. Corrosion protection
8. Solderability
9. Electrical conductivity
10. Frayed/damaged surfaces (rebuilding surfaces)
11. Paintability
12. Weldability

Figure 13-10 shows the result of improper lubrication when friction "cold welds" moving parts, resulting in cavities, ***galling*** (welding of high spots causing scuffing), and buildups.

Shot peening subjects a metal surface to a high-velocity stream of small, cast-steel pellets (shot) that effectively cold works the surface, inducing compressive stresses into the surface layers, which improves fatigue strength. ***Electroplating (electrochemical plating)*** suspends the workpiece in a bath containing a water-based electrolyte solution. The metal to be plated on the workpiece is also suspended in the bath. The workpiece becomes the cathode and the plating metal the anode in an electrical circuit. When the circuit is energized, the metal ions from the anode combine with the ions in the solution and deposit themselves on the surface of the cathodic workpiece. A typical thickness of the deposited layer of metal ranges from one or two atomic layers up to a maximum of 0.05 mm. ***Chromium plating,*** also known as ***chromizing, chromidizing,*** or ***chrome plating,*** uses a combination of metals, starting with copper followed by a coating of nickel. Chromium coatings are very thin and porous and normally offer little protection against corrosion. Copper is used as an undercoat particularly on zinc-based die castings. (See Section 11.6 for using cadmium coatings on steel.) Because the by-products of this treatment (specifically chromate) are, in many cases, harmful to the environment, research is actively being undertaken to find secondary uses for these materials. Chromium coatings are used on automobile trim, plumbing fixtures, and appliances. A somewhat similar surface treatment called ***electroless plating,*** uses chemical reactions instead of electricity. Nickel is the predominate plating metal, and nickel chloride is chemically reacted with sodium hypophosphite as the reducing agent to produce nickel for plating. The hardness

Adhesive Wear

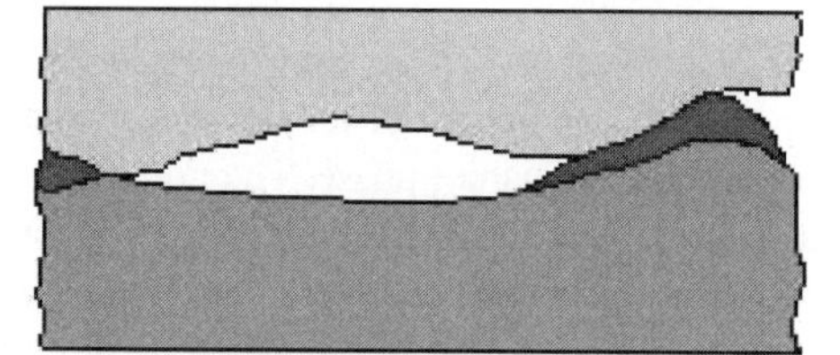

Cause
Cold welding with material transfer

Appearance
Cavities, galling and material build up

Figure 13-10 Adhesive wear. (Balzer)

of nickel plating is about 500 HV (Vickers Hardness), providing excellent wear and corrosion resistance to the workpiece.

The diffusion process, which impregnates the surface of a substrate with foreign elements (atoms), is used in several processes and treatments: carburizing and nitriding (described previously) are examples. The diffusing of impurity atoms (foreign elements) into the surfaces of silicon chips changes the electrical properties of transistors and diodes (see Module 6). ***Ion implantation,*** an alternative to diffusion, embeds foreign elements into a substrate surface using a high-energy beam of ionized particles. ***Hardfacing*** is the welding of alloys to a metallic substrate. ***Thermal spraying*** uses molten and semimolten coating materials. ***Metallizing*** (or metal spraying) refers to the spraying of metal coatings. ***Porcelain enameling*** is a process that applies porcelain (a ceramic material) to a metal substrate. Anodizing, hard anodizing, physical vapor deposition (PVD, Figures 13-11 and 13-12), chemical vapor deposition (CVD), and many other treatments are described in greater detail elsewhere in this book.

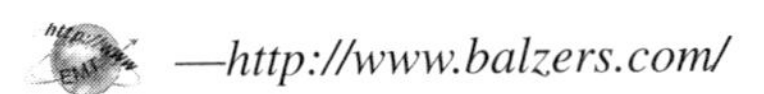
—http://www.balzers.com/

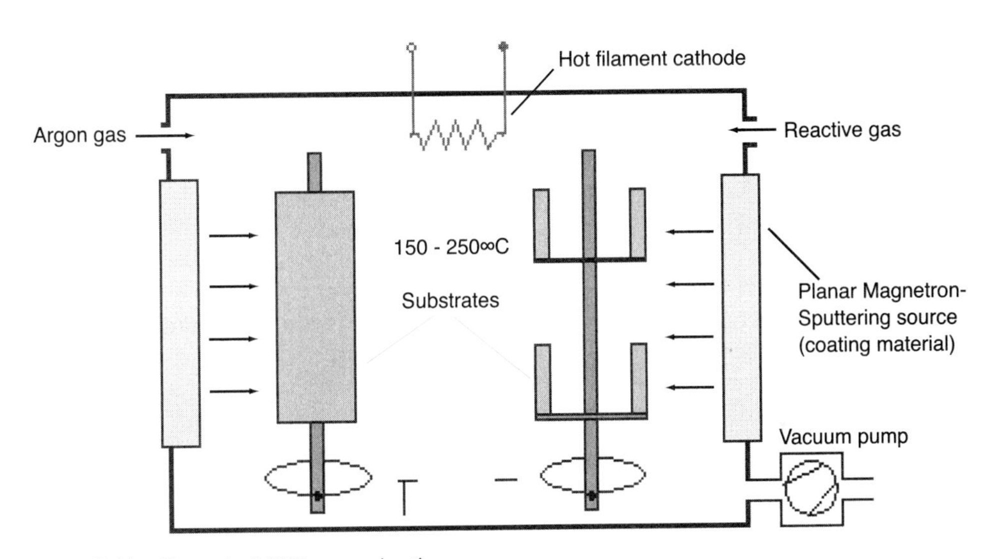

Figure 13-11 "Sputtering" PVD-process/coating.

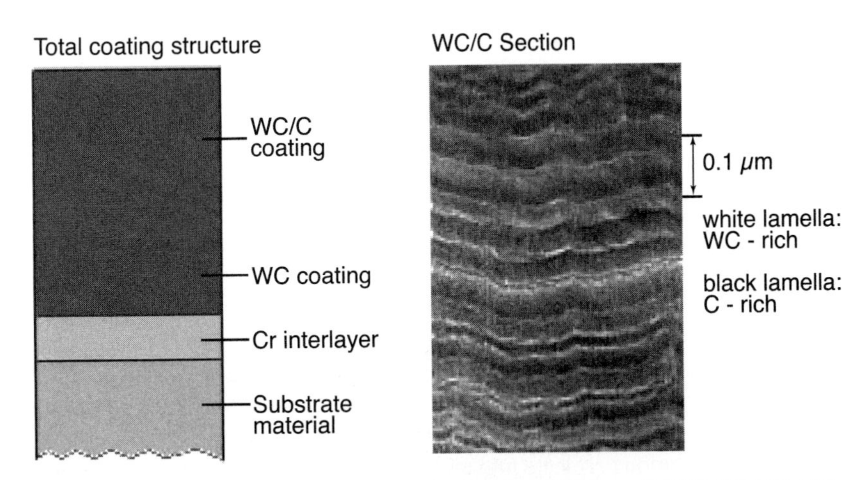

Figure 13-12 PVD Balinit C coating consisting of WC/C with a microhardness of 1´000/1´500 (HV 0.05), a coefficient of friction against dry steel of 0.1–0.2, and a maximum working temperature of 300°C. (Balzer)

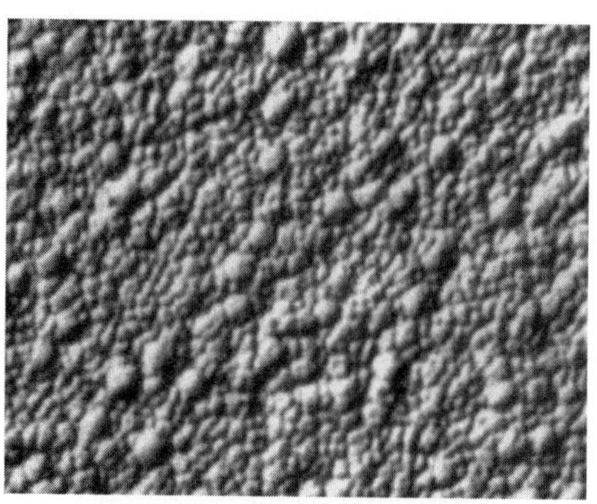

Figure 13-13 Armoloy nodularity provides a ball-bearing-like surface for less surface contact and improved lubrication. *(Advanced Materials & Processes)*

Throughout this book we describe means of modifying or engineering surfaces for tribological reasons in the automotive industry, which concern interfering surfaces that may cause friction, wear, or erosion and corrosion. In the ceramics module you will read about additional surface engineering techniques, such as diamond film and titanium coatings, for cutting tools and other high-friction/abrasion parts. Similarly, discussions in the polymer and composite modules deal with surface modifications to improve characteristics required for other particular applications.

The quest of the automotive industry to improve engine efficiencies continues to lead to modifications in materials and materials surfaces. One example involved improving the tribological aspects of sliding in the plain bearings of automotive crankshafts. The goal was to produce an overlay surface that would best retain lubrication, and the modification involved a lead-alloy electroplated overlay. An optimum surface was produced by controlling the crystallography of the electroplated alloy deposit by designing a surface morphology controlled by the orientation of the deposits into a highly oriented crystal texture in primarily the [200] and [400] planes (refer to Section 3.2.3). As the scanning electron microscope (SEM) photomicrograph in Figure 13-3 revealed, the surface of the overlay is made up of relatively uniform, pyramid-shaped crystals approximately 1.5 mm across with bases approximately 2 mm across.

 *www.aksteel.com*

APPLICATIONS & ALTERNATIVES

In the Pause & Ponder section for Unit 5, we emphasized the importance of proper surface preparation. Numerous examples exist of failures resulting from improper surface preparation on adhesive and welded joints, auto and house painted finishes, and so on. An interesting case study can be made of a 1996 crash of a Sikorsky CH-53E Super Stallion heavy-lifting helicopter in which four Sikorsky employees died. (See the website listed at the end of this section.)

Armoloy coatings use relatively low-tech application processes but produce advanced surface finishes for bearings, valves, pumps, machine tools, molding, and marine applications—to name a few. Low temperatures are used for several stages of an electrocoating process that is based on chrome plating technology, using, however, a proprietary chemical solution instead of more common chrome-plating solutions. Metallic elements of the Armoloy solution permeate into the pores of the base material, integrating the coating with the metal itself. The result is a hard, slippery, corrosion-resistant, nonporous, and crack-free surface.

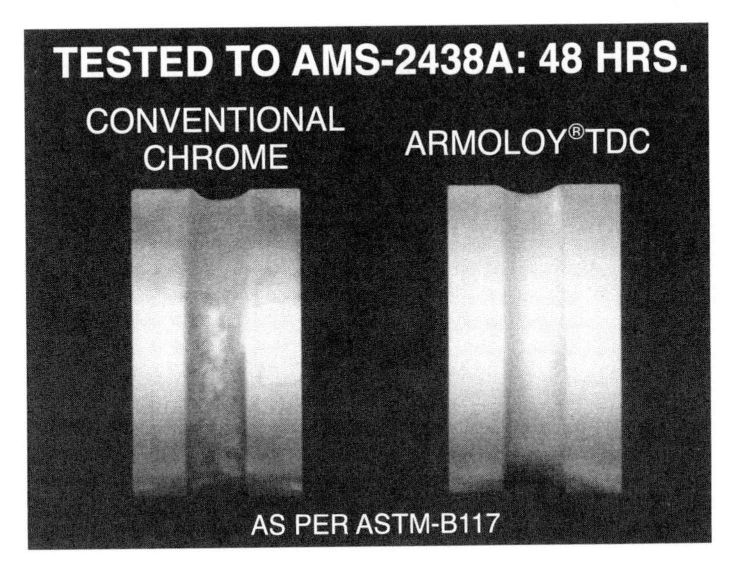

Figure 13-14 Comparison of coatings after ASTM B-117 test. (*Advanced Materials & Processes*)

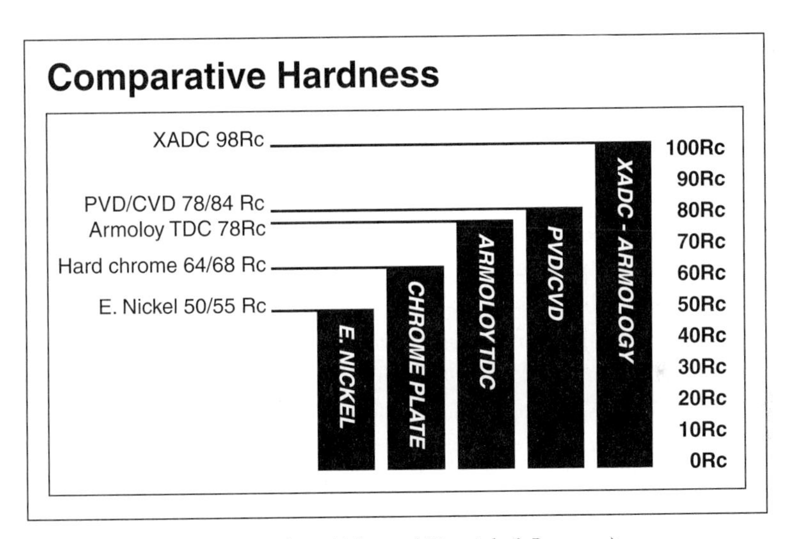

Figure 13-15 Comparison of selected coatings. (*Advanced Materials & Processes*)

Armoloy coatings produce a dense ***micronodular finish,*** as seen in Figure 13-14, similar to the skin of an orange. The nodularity acts like ball bearings, yielding a relatively low contact area to reduce friction while imparting self-lubricating characteristics by retaining available lubrication between the nodules. Figure 13-14 compares a conventional chrome surface to an Armoloy surface. After testing to AMS 2438A for 48 hours, there was no change in Armoloy; the conventional chrome showed corrosion and erosion. Figure 13-15 compares the hardness of several surface coatings. More information about Armoloy can be found at the following website:

 Armoloy—*http://www.azom.com* and *www.armoloy.com*

To complete this case study, use an Internet search engine to find more details on Armoloy and the Sikorsky CH-53E Super Stallion 1996 crash at the following link:

 Sikorsky CH-53E crash—*www.ebearing.com/news2001/news199.htm*

Another example of surface science research is a recent development from MIT's new Institute for Soldier Nanotechnologies that deals with Teflon. This polymeric material, polytetraflouroethylene (PTPF), is truly an amazing material, with properties that make it water repellent, slippery, and biologically inert. This array of characteristics makes it desirable for a wide variety of applications. One limitation has been the inability of manufacturers to attach PTPF satisfactorily to the surfaces of other materials. Starting as a powder in the process of coating cooking utensils, Teflon in the form of particles is melted at high temperatures onto the surface of such devices. However, this process fails to produce a strong bond between the Teflon and the underlying surface. Evidence of such poor bonding is apparent when spots appear on a frying pan where the Teflon has been lost. Another disadvantage is the high temperatures of today's processing that limit applications to non-delicate objects. Further, the excessive thickness of the Teflon coating excludes its use for many intricate devices. Present day waterproofing of soldier's flak vests with a heavy and thick water vapor barrier results in soldiers not wearing their vests. This situation has resulted in research leading to the use of nanotechnology to process molecules of Teflon using a precursor gas under controlled conditions to render a film only 50 um thick, which is deposited on the Kevlar coating of flak vests. It is anticipated that with such success in waterproofing material with Teflon, the attachments of different molecules to the ends of Teflon chains of polymers using advances in nanotechnology will lead to textiles that will protect their wearers from biological weapons.

SELF-ASSESSMENT

13-1. Name an example of mechanical reaction involving a surface.

13-2. What issue does the IFHTSE deal with as a technical society?

a. Ferrous metals **b.** Heat treatment **c.** Internet engineering **d.** Software engineering

13-3. Nanotechnology is involved with surface science in producing surfaces for what applications?

13-4. Name four problems that are being solved by surface science.

13-5. What natural material achieves extreme toughness due to thin, alternating layers of calcium carbonate and organic polymers intertwined much like a layered composite?

a. Abalone shells **b.** Case-hardened steel **c.** Nanoscale film **d.** Surface contaminants

13-6. What approximate high temperature are materials in heat engines required to withstand?

a. 400°F **b.** 1200°C **c.** 200°F **d.** 540°C

13-7. The attractive force between surface molecules in a liquid is

a. Surface tension **b.** Surface compression **c.** Surface contamination **d.** Surface bubbles

13-8. What shape represents the smallest surface area?

a. Triangle **b.** Sphere **c.** Cube **d.** Rhomboid

13-9. Surface atoms are not as firmly bonded as atoms in the interior of a material. This results in surfaces having

a. High energy **b.** Low energy **c.** Neighboring atoms **d.** Dislocations

13-10. Which word would *not* substitute for the word *treatment* in the expression "surface treatment"?

a. Engineering **b.** Modifications **c.** Crystals

13-11. List three problems that surface treatments can solve.

13-12. The performance of surfaces is largely determined by what two characteristics?

a. Atomic compositions and microstructure

b. Microstructure and grains

c. Microstructure and contaminants

d. Contaminants and grains

13-13. Besides metals and alloys, what other material is suitable for spraying?

13-14. What element is replacing chromium due to better wear properties in engine bores?

a. Nickel **b.** Zinc **c.** Lead **d.** Nanoscale

13-15. What is the rationale (the reason for) case hardening?

13-16. A good many steel products require heat treating as well as machining. Which process should come first and why?

13-17. Shot peening enhances

a. Tensile strength **b.** Surface wear **c.** Adhesion **d.** Fatigue strength

---

REFERENCES & RELATED READINGS

ANNUAL BOOK OF ASTM STANDARDS. Philadelphia, PA: ASTM, 2003.

ASM INTERNATIONAL. *Alloy Finder*. CD-ROM software. Materials Park, OH: ASM International, 1996.

ASM INTERNATIONAL. *ASM Handbook*, Vol. 20: *Materials Selection*. Materials Park, OH: ASM International, current edition.

ASM INTERNATIONAL. *ASM Handbook*, Vol. 18: *Friction, Lubrication and Wear Technology*. Materials Park, OH: ASM International, 1992.

DAVIS, J. R., ed. *Surface Engineering for Corrosion and Wear Resistance*. Materials Park, OH: ASM International, 2001.

GLEASON, KAREN, Professor of Chemical Engineering, "Demo: Waterproofing Anything." *Technology Review,* Nov. 2003, MIT's Magazine of Innovation, Cambridge, MA 02142.

PARK, J., ed. *Chemical Vapor Deposition*. Materials Park, OH: ASM International, 2001.

POPOOLA, O., ed., et al. *Surface Engineering: Coatings and Heat Treatments*. Materials Park, OH: ASM International, 2003.

Journals and Websites

Heat Treating Progress

Materials Today

MRS Bulletin

www.technologyreview.com

UNIT 6

Polymerics

PAUSE & PONDER

Pause to recall concerns about the environmental impacts of plastics. A new process involving fermentation is a novel synthesis technique for producing plastic and synthetic fibers from corn and other plants. What does this mean? We now have an ***annually renewable resource*** for plastics and polymer fibers. Will this mean that we will no longer need to wait millions of years for the formation of oil and gas to produce certain synthetic polymers? How does the novel technique work and what are the benefits?

Cargill Dow considers its NatureWorks™, an agriculture-based polymer, as a bridge in the performance gap between petroleum-based synthetic polymers and natural polymers. ***Polylactic acid resin (PLA)*** is intended to compete with synthetic fibers for carpeting and clothing and with synthetic plastics for food wraps, bottles, and packaging foam. The biodegradability of PLA should make it more appealing than the nonbiodegrading polystyrene used in food packaging such as the clamshells for carryout burgers and hotdogs, which often litter the landscape and waterways. Cargill Dow expects to raise production to 500,000 metric tons of PLA by over the next few years for world distribution.

As revealed in the diagram, NatureWorks is promoted as environmentally friendly because it uses crops such as corn that may be returned to the earth or recycled. The crops take CO_2 from the atmosphere as a part of photosynthesis, whereas petroleum-based polymers release CO_2 into the atmosphere. Later, as seen in stage 6 of the cycle, if PLA is composted, as is done with other biodegradables including vegetable waste, it releases carbon back into the soil.

The modified NatureWorks diagram* shows six stages in the cycle of production:

1. Agriculture Resource Annually Renewable—Crops that renew annually such as corn are milled, separating starch from the raw corn. From the starch comes unrefined dextrose. Derived from annually reusable resources, it uses 20–25% less fossil resources than comparable petroleum-based plastics.

2. Fermentation—Here the unrefined dextrose is converted into lactic acid in a process similar to the fermentation of wine and beer. Lactic acid is a common food additive as well as a substance found in the muscle tissue of humans.

3. Intermediate Production—A lactide, a cyclic intermediate dimer, is formed through a special condensation process.

*Adapted from NatureWorks™ Cycle at the website listed at the end of this introduction.

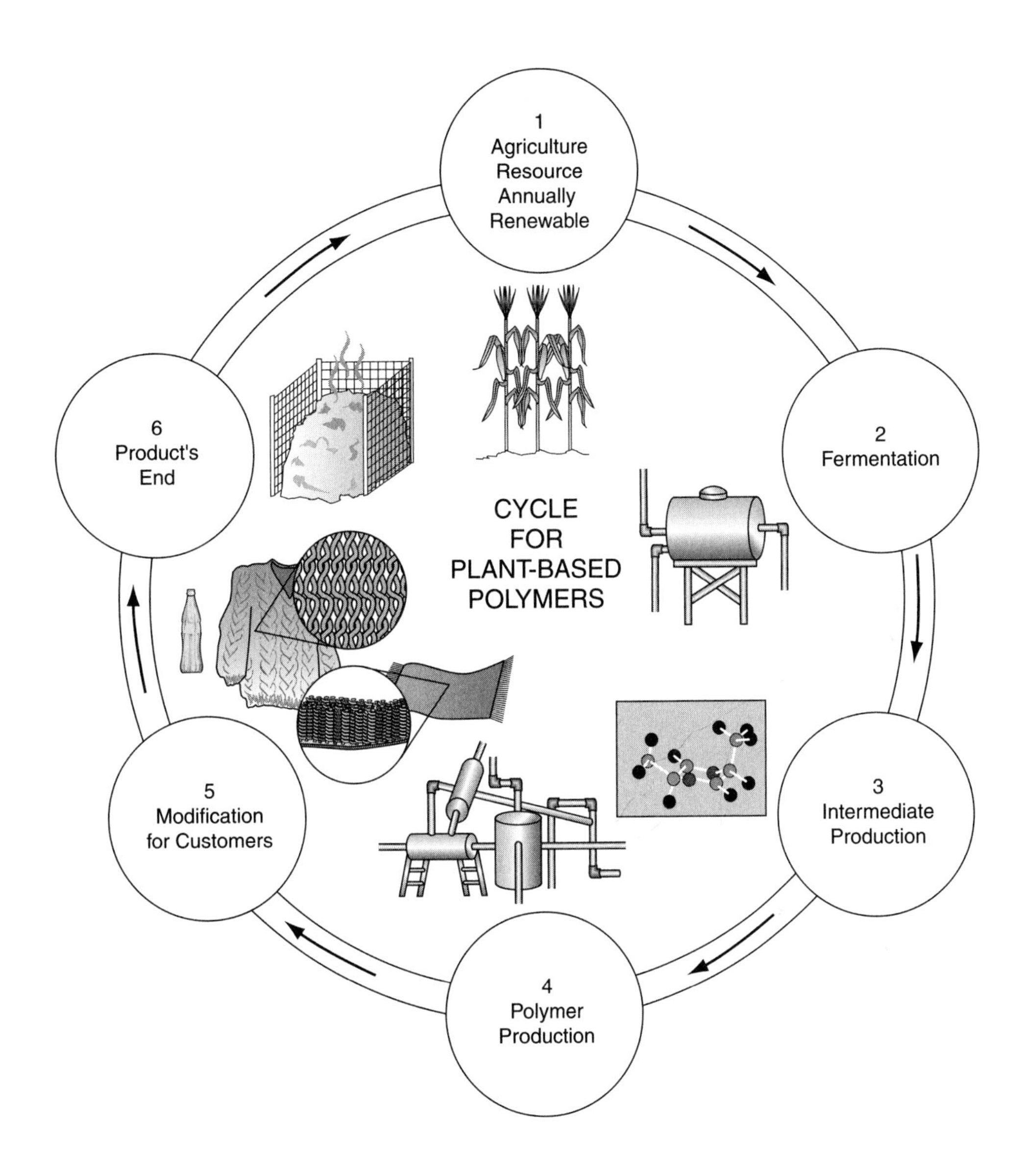

4. Polymer Production—Through vacuum distillation, the monomer lactide is purified. Through a solvent-free melt process, the ring-opening polymerization of the lactide is accomplished.

5. Modification for Customers— By modifying PLA, Cargill Dow can produce a wide range of products that vary in crystallinity and molecular weight: soft drink bottles, carpet, clothing, food wrap.

6. Product End (earth-to-earth)—At the end of the life cycle, PLA can be broken down into its simplest parts so that no signs of the life cycle remain. If the infrastructure were established, and it fits all disposal systems or is compostable (biodegrades), it can be recycled into polymers or to a monomer again. The CO_2 in the atmosphere is potentially reduced because

NatureWorks recycles the earth's carbon. PLA may also be returned to the earth, adding carbon, or it may be recycled to make new PLA. See Table 14-1, which depicts various forms of energy and corresponding air emissions and water effluents used in life cycle analysis.

This cycle can be compared to polymer synthesis in Figures 14-5 and 14-6 and the materials cycles in Figure 14-12. Some producers of synthetic polymers strive to employ ***green chemistry*** whereby processes and products have less adverse impacts on the environment. This approach involves employing *atom economy,* which aims to put all ingredients into the finished material with no by-products.

Use the website address below to learn more about NatureWorks.

NatureWorks™—*http://www.cargilldow.com/s_natureworks.asp*

Module 14

Nature of Polymers

After studying this module, you should be able to do the following:

14–1. With the aid of diagrams, explain the synthesis of polymers in terms of raw materials, bonding, and by-products, and explain how specific types develop their properties.

14–2. Analyze applications of polymers to determine the effects on structure as a result of branching, degree of crystallinity, molecular weight, and networking on the properties required for the finished product.

14–3. Calculate molecular weight for a given compound and explain its significance in polymers.

14–4. Explain the techniques to improve or change polymer properties, such as optical, mechanical, chemical, electrical, color, thermal, and density properties.

14–5. Develop a rationale for emphasizing green design and green manufacturing with regard to practices involving polymers. Describe how life cycle analysis should promote green manufacturing.

14–6. Recall the types of natural and manufactured fibers, and describe their properties and typical uses.

14–7. Use the website addresses denoted by the symbol* to explore sites for new developments and evolving concepts related to polymeric materials.

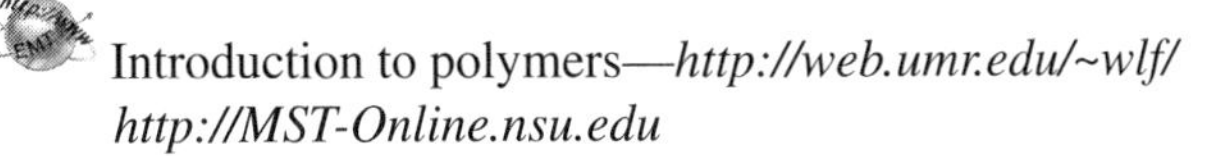

Introduction to polymers—*http://web.umr.edu/~wlf/*
http://MST-Online.nsu.edu

*The symbol found throughout the book will link you to Internet sites related to the topics being covered. The dynamic nature of the Web brings frequent changes, so some of these sites, while available at the time of writing, may not be up now.

14.1 POLYMERICS

Section 2.2.3 provided an introduction to polymeric materials (often classified simply as polymers). The emphasis in this module is on the main polymeric materials used in engineering, manufacturing, and construction. Excluded are natural polymers such as skin, hide (leather), and bones although some coverage of natural fibers such as cotton, silk, and jute along with synthetic fibers is provided. Plastics, woods, elastomers, adhesives and many protective coatings are all polymers. Plastics, adhesives, and synthetic rubber are made by humans; wood and natural rubber are naturally occurring materials that are used in industry after various alterations are made to their natural state. Polymers can include many types of industrial materials, as seen in Figure 14-1. ***Biologically inspired materials*** promise to add to the array of synthetic polymers as they emerge from on-going research.

Figure 14-2 shows many examples of natural and synthetic polymers. Natural polymers seen here include the pine door molding around the doorframe and Connor's hair, skin, and cotton T-shirt. The photo also shows such synthetic polymers as white acrylic paint on the pine molding, the vinyl plastic doorframe with a neoprene rubber gasket to hold the glass, which is a ceramic. Connor's night vision goggles are made of polycarbonate lenses and an acetal frame; the ABS plastics toy robot uses silicon rubber suction cups to climb the glass. The large number of plastic toys demonstrates the low cost and design flexibility offered by plastics, which make it attractive to designers of toys and many other products. But pause a minute—Can you think of any drawbacks to using synthetic polymers such as plastic?

14.2 NATURE OF POLYMERS

Polymers consist of smaller units (***mers***) joined together either naturally or synthetically to produce polymers or macromolecules. This section deals with the structures common to most polymers, and most of the examples cited are plastics, due to their abundant use in industry. The discussion will concentrate on the synthetic methods of producing polymers known as *polymerization* and will not deal with the biological process by which the natural or biological polymers are grown.

The initial step in polymer production is performed by petrochemists or petrochemical engineers, engineering technicians and engineering technologists, or others with education and

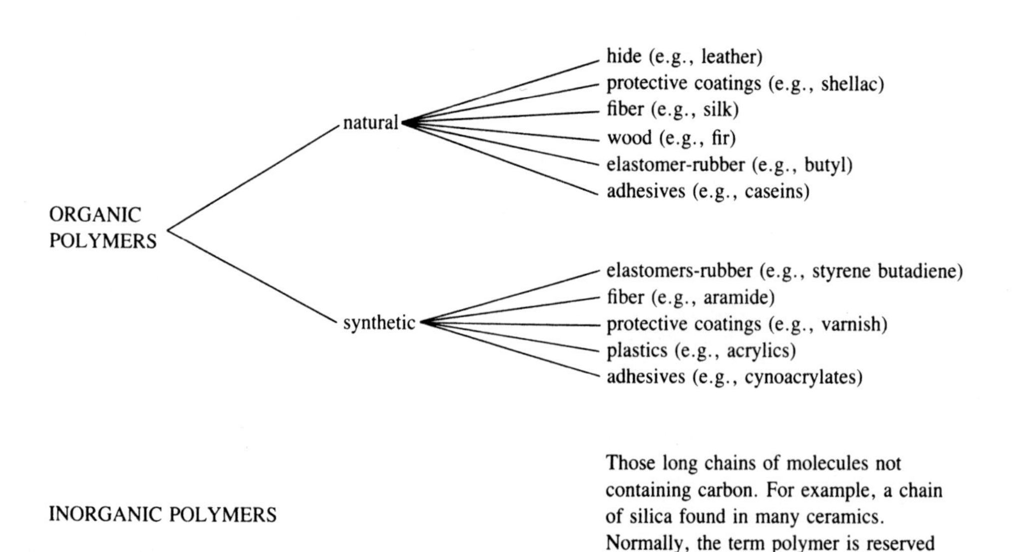

Figure 14-1 Industrial polymers.

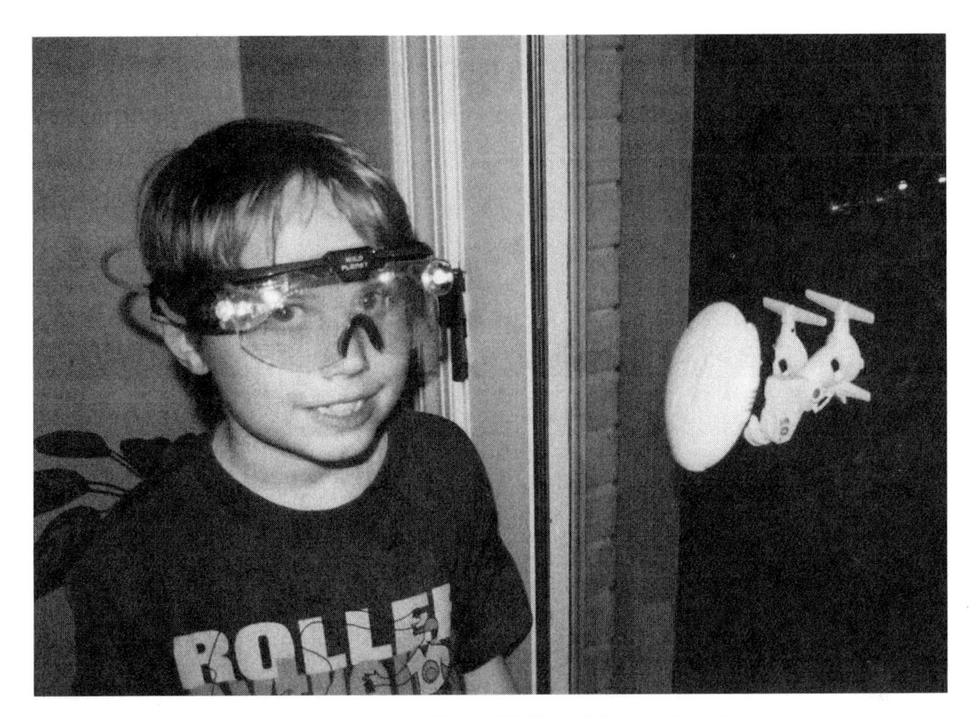

Figure 14-2 Natural polymers—Connor's hair, skin, and T-shirt and the pine door trim.
Synthetic polymers—White acrylic paint, neoprene door gasket, plastic night goggles (polycarbonate lenses and acetal frame), ABS plastic climbing robot with silicon rubber suction cups.

training in chemistry. It involves breaking down raw materials such as crude oil, coal, limestone, salt, natural gas, and air. For cellulose plastics, the raw material is cotton or wood; however, they are not broken down, as are most other polymer ingredients. Whether it be the cracking of natural gas, the distillation of coal or crude oil, or the esterification of cotton or wood, processing the basic organic (carbon-base) raw materials yields the element carbon together with hydrogen. Other basic elements are also obtained in the initial stages, including oxygen, nitrogen, silicon, fluorine, chlorine, and sulfur. Ethylene and naphtha, basic ingredients of polymers, are but two of the many products obtained in the cracking and distilling processes. Another is polyester resin. ***Esters,*** organic compounds that are colorless, liquid, and insoluble in water and that have a very pleasant aroma, find use in perfumes and in the manufacture of flavoring agents for the soft-drink industry. Polyester resin is produced when an acid and an alcohol chemically react (***esterification***).

From 100 gallons of crude oil, at a market value of around $25, it is possible through distillation and further processing to produce gasoline, motor oil, propylene, and a large variety of polymers—styrene-butadiene rubber, polyethylene plastics, polypropylene fibers for fabrics, and others—products that have a combined market value of over $3500. While oil is the major raw material for plastics, plastics take only a very small percentage (about 1.3%) of the world's petroleum production. Yet shortages in oil cause problems in the plastics and polymers industry, so alternative raw materials continue to receive study. Plastics made from corn show promise both because they (1) do not use petroleum and (2) may be synthesized so as to ensure biodegradability.

14.2.1 Polymerization

Scientists observed nature's methods of joining elements into chains and duplicated that natural process to produce macromolecules or polymers. Module 2 covered the principles of bonding, which provide a basic understanding of polymerization. ***Polymerization,*** as we have

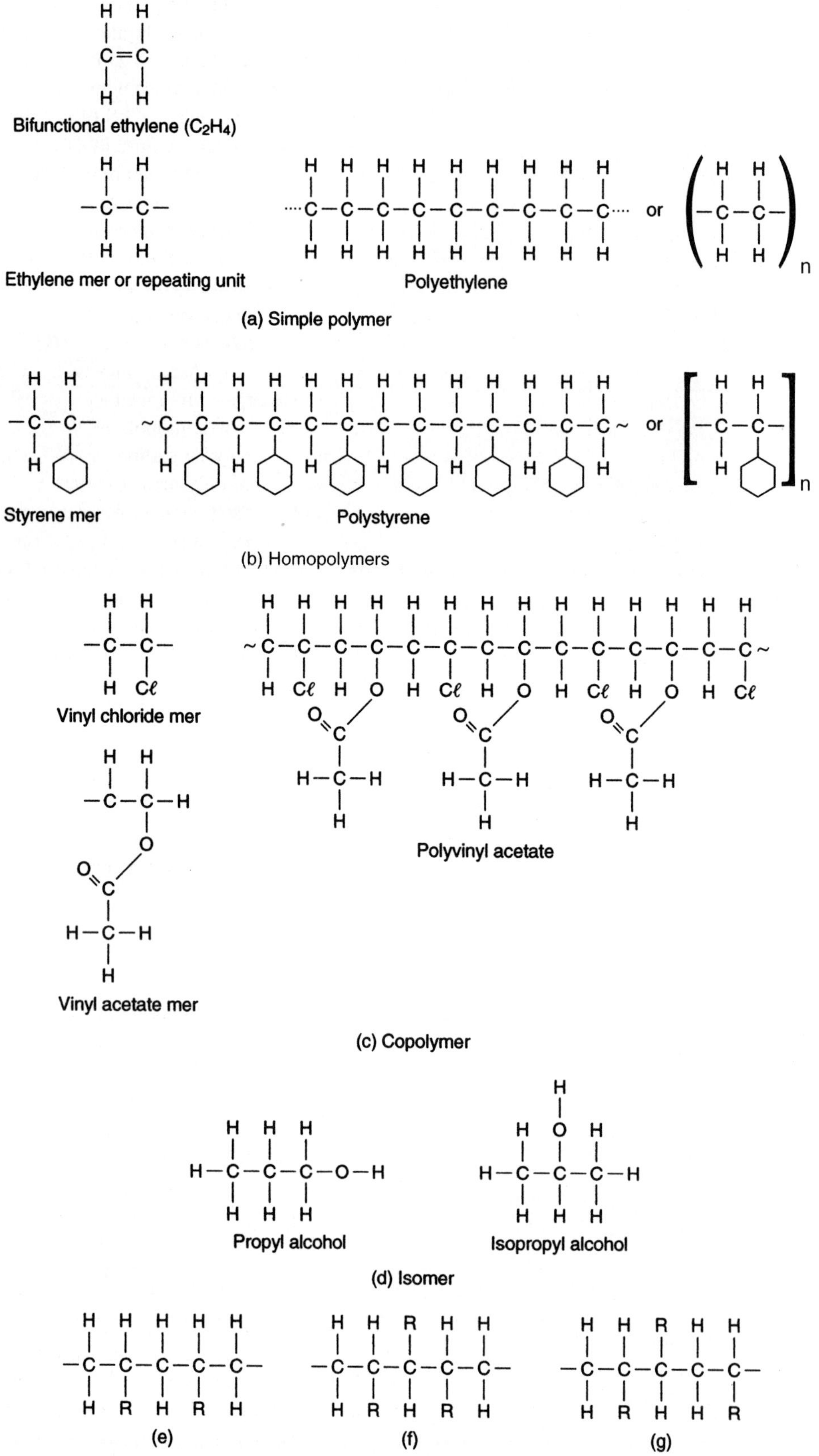

Bifunctional ethylene (C_2H_4)

Ethylene mer or repeating unit

Polyethylene

(a) Simple polymer

Styrene mer

Polystyrene

(b) Homopolymers

Vinyl chloride mer

Vinyl acetate mer

Polyvinyl acetate

(c) Copolymer

Propyl alcohol

Isopropyl alcohol

(d) Isomer

(e)

(f)

(g)

Figure 14-3 Types of polymers: (a) simple polymer, (b) homopolymers, (c) copolymers, (d) isomers, (e) isotactic stereoisomer, (f) syndiotactic stereoisomer, (g) atactic stereoisomer. NOTE: Dotted lines (---) and tildes (~) are like "etc." An n after [ ] or ( ) means to repeat *n* times, where *n* is a very large number, e.g., *n* could be equal to 10,000 or over 10,000 mers in PE (see Fig. 14-8a).

said, is the linking together of smaller units (monomers) into long chains. The repeating units (mers) of some polymer chains are identical, as in polyethylene (Figure 14-3a), polystyrene, and poly(vinyl chloride); these are labeled ***homo*polymers** (Figure 14-3b). *Co*polymers contain two different types of monomers, such as poly(vinyl chloride) mixed with vinyl acetate to produce poly(vinyl acetate) (Figure 14-3c); and *ter*polymers, such as ABS (acrylonitrile–butadiene–styrene), which contain three types of monomers. ***Isomers*** (Figure 14-3d) are variations in the molecular structure of the same composition. Isomers are found not only in hydrocarbons (HCs) but also in polymeric molecules. One class of isomers, known as ***stereoisomers,*** is divided into three categories. The first category is ***isotactic*** stereoisomers. In a single carbon (C) chain of atoms some hydrogen (H) atoms are replaced by another atom or group of atoms, denoted by the symbol **R,** all situated on the same side of the carbon chain. In a ***syndiotactic*** stereoisometric arrangement, the **R** group atoms are located on opposite sides of the carbon chain in an alternating fashion. In the third category, ***atactic*** stereoisomers, the **R** groups are in a random position. (Refer to Figure 6-3e, f, and g.) It is important to note that copolymers and terpolymers consist of units from each contributing mer and are not an alloy of mers.

If polymerization permitted only the production of homopolymers, the properties of polymers would be severely limited. Because of copolymerization, terpolymerization, or other multicomponent polymerizations, a large variety of polymers is available. One example is styrene, a brittle polymer that has limited toughness and poor chemical resistance. Through copolymerization of acrylonitrile with styrene, however, it is possible to obtain a chemically resistant, more rigid, and stronger plastic. Copolymerization of butadiene with styrene yields an elastomer; and terpolymerization of acrylonitrile, butadiene, and styrene produces ABS plastics, which are tough and elastomeric and have good chemical resistance (see Table A-9).

Most polymers are produced by ***unsaturated hydrocarbons,*** which means that they have one or more multiple covalent bonds, such as ethylene:

```
H   H
|   |
C = C
|   |
H   H
```

or adipic acid, which is used in nylon synthesis and has two hydroxyl monomers:

```
     O              O
     ||             ||
HO — C — (CH2)4 — C — OH
```

Saturated hydrocarbons have all single bonds. To achieve polymerization, monomers must be capable of reacting with at least two neighboring monomers or be ***bifunctional.***

Catalysts begin the polymerization process. A catalyst is a substance that increases the rate of a chemical reaction but is not used up in the reaction. In addition, these chemicals have a variety of purposes in chemical processing, ranging from serving as molecular sieves to agents for extracting heating oil, gasoline, and ethane from crude oil to serving as molecular matchmakers holding chemicals together so that a reaction can occur, such as forming long-chain polyethylene polymers from molecules of ethane. ***Zeolites*** are catalysts composed of inorganic grains of alumina (Al_2O_3) and silica (SiO_2) through which molecules of crude oil are strained to gain oil by-products. Activated carbon, platinum, and nickel are also used to catalyze synthetics. ***Aerogels*** are open-cell (often silica) foams that have an ultrafine cell or pore size of less than 50 nm, high surface areas (500 to 1000 m^2/g), and density approaching that of air. They are the lightest, most transparent human-made solids. Known as the world's best insulators, they do not conduct heat. With a density of 15 mg/cm^3 and a melting point of 1552°F, the properties of these materials approach those of glass. Aerogels of organic and inorganic systems of fibrous chains provide catalyst support. To improve fuel cell efficiency, materials scientists tailor silica aerogels by adding minuscule particles of elements such as carbon and platinum to gain a degree of electrical conductivity and catalytic reactions. The development of scanning tunneling microscopes (STMs) has allowed nanoscale examination

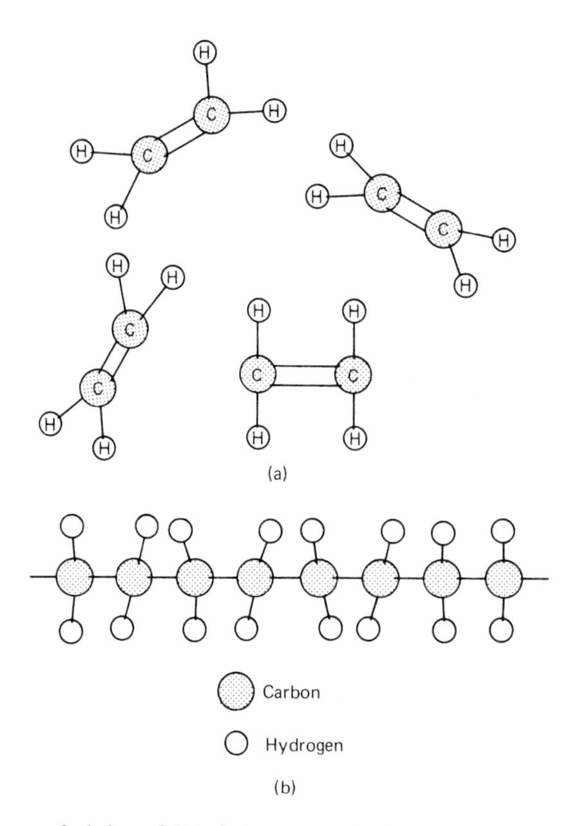

Figure 14-4 (a) Monomer of ethylene (C_2H_4). Carbon atoms with four valence electrons have four arms (shared electrons) for covalent bonding. Two shared electrons hold another carbon atom in a double bond. Two other shared electrons, each covalently bonded, hold one shared electron of hydrogen, which has only one valence electron for single bonds. Double bonds are not stable and form ***reactive sites.*** (b) Polyethylene, a chain of ethylene monomers. During polymerization, heat and pressure break the hold of one shared electron on each of the double-bonded (reactive sites) carbon atoms, thereby leaving each carbon atom of the monomers free to grasp another carbon atom (covalent bond) from other ethylene monomers, which form into a chain of thousands of ethylene monomers (polyethylene). The single bonds satisfy the carbon bond arrangement and produce the most stable saturated polymers because they have no reactive sites.

of catalysts. With this tool new designer catalysts have been developed that can catalyze products more efficiently without creating polluting by-products.

The main polymerization processes are ***addition (chain reaction) polymerization*** and ***condensation (step reaction) polymerization.*** The addition process is the simpler of the two. By use of heat and pressure in an ***autoclave*** or reactor, double bonds of unsaturated monomers (Figure 14-4) break loose and then link up into a chain. These addition reactions (in addition polymerization for unsaturated HCs) are atoms or groups of atoms that attach themselves to the carbon atoms at the sites of multiple bonds. No products other than the polymer are formed. Saturated hydrocarbons undergo substitutional reactions in which hydrogen atoms are replaced by other atoms or groups of atoms. The products of addition polymerization, also referred to as *chain reaction polymerization,* include polypropylene (PP), polyethylene (PE), poly(vinyl acetate) (PVA), poly(vinyl chloride) (PVC), acrylonitrile–butadiene–styrene (ABS), and polytetrafluoroethylene (PTFE).* *Chain polymers* generally fit into the ***thermoplastic*** (soften when heated) group. Most plastics that cannot be resoftened when heated (***thermosetting polymers***) come from condensation polymerization, also known as ***step reaction polymerization.*** This group gets its name from the by-product (***condensate***) of the polymerization, which is often water but may be a gas. Phenolic (PF), polyester (PET), silicon (SI), and urethanes (PUR) are typical thermosets from the step reaction synthesis; nylon (PA) and polycarbonates (PC)* are thermoplastic resins synthesized through step polymerization.

*ASTM abbreviation.

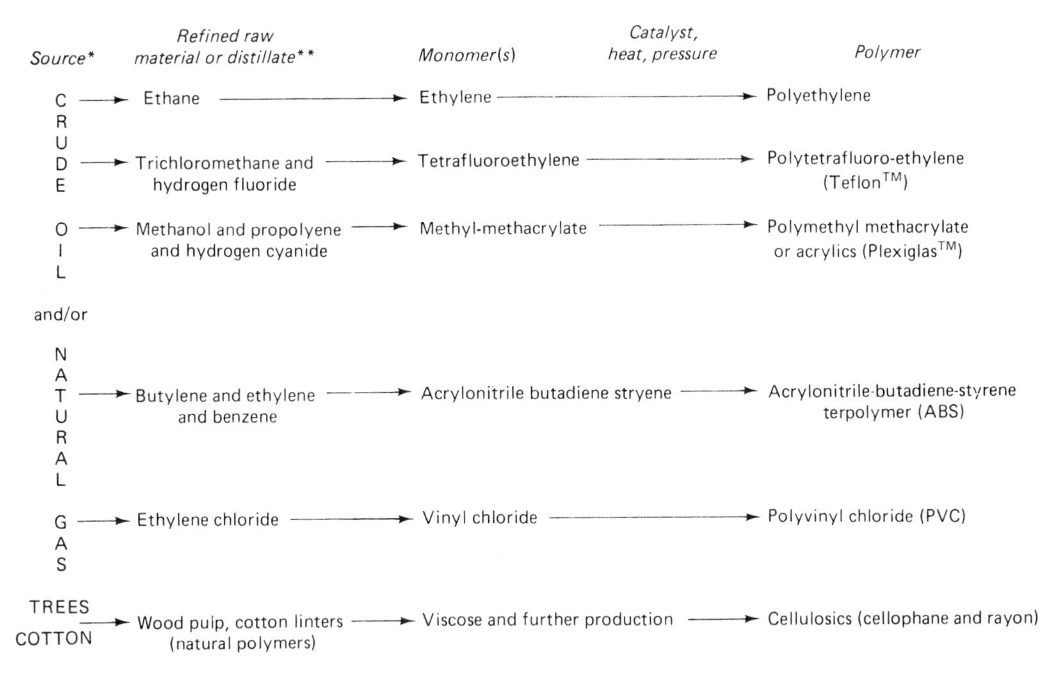

*LRG (liquefied refinery gas) is obtained from refining crude oil and can be further refined to yield the butadiene, ethylene, methane, and propylene. Coal and coke also serve as raw materials for production of gases used to produce polymers.

**In addition to petroleum raw materials which provide the carbon and hydrogen base, other elements such as oxygen, nitrogen, sulfur, chlorine, hydrogen, and fluorine are mixed to obtain the monomer.

(a)

(b)

Figure 14-5 (a) Typical polymers produced through addition (chain) polymerization. (b) A portion of a modern polymer processing plant showing some of the complex apparatus and piping necessary to produce resins.

Figure 14-5a diagrams the addition process for the production of some typical polymers, and Figure 14-5b shows a modern polymer processing plant and the complex equipment used to produce resins. Figure 14-6 shows the flow for the production of nylon through condensation polymerization.

The term ***resin*** applies to both solid and semisolid organic polymers. Resins are often considered to be the uncompounded ingredients or monomers that are mixed but not yet polymerized. For example, thermosetting resins or pellets are molded into thermosetting plastic or

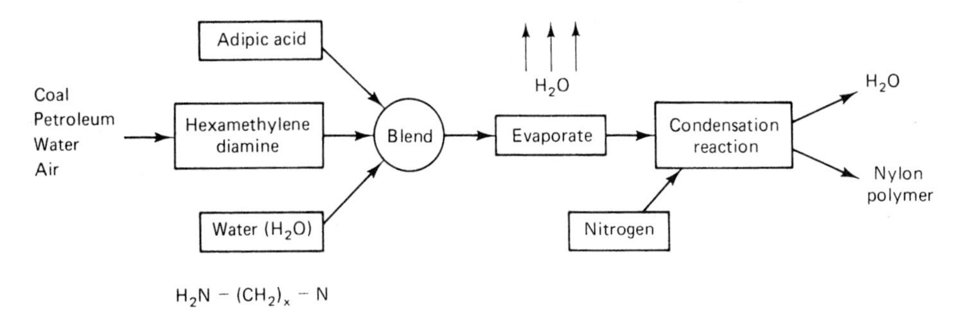

Figure 14-6 Graphical illustration of condensation (step) polymerization of a polyamide (nylon).

elastomeric parts. Sometimes the term *resin* is used synonymously with plastics (e.g., acetal resins of acetal plastics or thermoplastic resin instead of thermoplastic plastics).

14.2.1.1 Polymer chain lengths. Considerable variation in polymer chains results during polymerization. The ***chain lengths*** play an important role in processing polymers and in their final properties. Although we cannot see the chain lengths of the polymers, their molecular weight provides a good indication of length. (Refer to Module 2 for an introduction to molecular structure and bonding.) Molecular weight is the weight of a single unit or molecule in a compound. To determine the ***molecular weight of the compound,*** it is necessary to multiply the weight of a molecule by the number of molecules or monomers. We said polyethylene typically has about 2000 monomers, so we multiply 2000 times the molecular weight of one monomer to obtain 56,000:

Illustrative Problem

Molecule or monomer	Molecular weight		Number of monomers		Molecular weight of compound
C_2H_4	28	×	2000	=	56,000

Involved chemical analysis is required to determine polymer chain lengths, and such analysis reveals that chains are not uniform in length. Consequently, the value given for a particular resin would be either its ***average molecular weight*** or ***degree of polymerization (DP),*** which indicates the number of repeating mers in the material. The degree of polymerization is the ratio of the molecular weight of the molecule to its mer's weight. Long-chained polymers have a greater molecular weight, which causes more entanglement of the chains and thereby increases ***viscosity*** (resistance to flow). The degree of viscosity affects the processing of the polymer, so manufacturers must take it into account.

 Addition polymerization and condensation polymerization—
http://mrcemis.ms.nwu.edu/polymer

14.2.2 Crystalline Structures

In addition, polymerized neighboring chains are held together by weak secondary (intermolecular) bonds known as van der Waals forces, as described in Module 2. As the chains grow during polymerization, they intermingle in a random pattern that will lead to an ***amorphous*** (disorderly) ***structure.*** Bear in mind that the growth of chains is three-dimensional and not simply flat, as normally diagrammed. Visualize a bowl of cooked spaghetti with its uneven strands entangled; that gives a good idea of the arrangement of an amorphous polymer structure.

Actually, polymers are semicrystalline to varying degrees, with the amorphous-structured polymers having only slight regularity and other polymers having a high degree of crystallinity (Figure 14-7). Metals achieve crystallinity due to the uniform nature of the unit cells of their space lattices. The twisting, coiling, and branching of polymer chains cause disorder, and amorphous regions develop in polymers as they solidify. Chain configuration and rate of cooling affect the

Cool Stuff

Materials come in a wide range of structures, which causes them to interact with light for a wide range of colors. Through the ages, objects made from available materials gained value for both their utility and as objects of art. Objects from early times shown here now have value as art and are collected for their beauty and value. In earlier times the very wealthy could afford the colorful objects. Today's objects, composed of an infinite combination of materials, provide colorful and cool "stuff." Due to improvements in materials technology, today and into the future, the average citizen may possess colorful, cool stuff at very reasonable prices. The images shown here represent a very small sampling of the materials available, both natural and man-made, for art and utility. Both macro and micro scale images reveal colors and textures that tell a lot about the properties, structures, and processes.

◄ **Figure C-1 Color Polymer Stuff** Products made of various plastics and elastomers demonstrate the wide ranges of colors, forms, and properties possible with synthetic polymers. Note the white and blue and white and black toothbrushes in the middle with flexible heads, contoured shape, and soft rubber dots and strips for gripping. What types of plastics and elastomers are shown here?

Figure C-2 Dragonfly Lamp A combination of stained glass bonded with copper foil on a bronze base, this 1910 lamp from the Tiffany Studio shows how mixing materials can provide beauty and utility. [Chrysler Museum of Art http://www.chrysler.org]
▼

▲
Figure C-3 Mazda RX7 Timex Racing Car Racing car owners welcome the introduction of newly developed materials and manufacturing processes to help them maintain their competitive edge in racing. Sports Car Club of America's Group 66 car is representative of this trend with its use of lightweight steel engine parts, carbon/Kevlar brake pads, polyurethane bushings, HD flame-resistant foam, and Thermo Tec header wrap. Cryogenic processing of various engine and brake parts reduces wear and costs of operation. [Wally and Liz Hunt]

Figure C-4 Photoelastic plastic and polarized light show stress distributions of a dovetail opening on a turbine disk of a jet engine turbine blade, thus allowing for evaluation of creep properties. [Measurement Group, Inc.] ►

Metals

Figure C-5 Metallography in Color. Electrolytically polished and anodized: (a) Aluminum-2% uranium alloy, as cast, (b) Same alloy after cold working 75%, (c) Same alloy after recrystallized by heat treatment, (d) Oxidized single crystal-tantalum, (e) Oxidized single crystal copper + 1% Ni. [Metals and Ceramics Division ORNL]

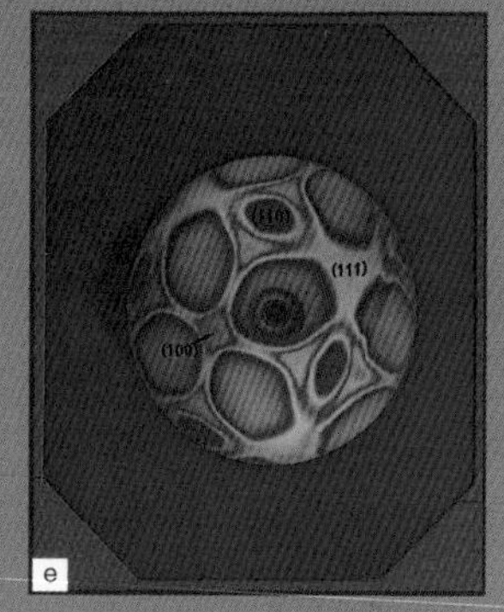

Figure C-6 Photomicrograph shows the longitudinal view of a stain-etched titanium plate. The surface layer is contaminated with light elements (oxygen and nitrogen). It is shown at 400X magnification, bright field, etched with Kroll's etch. [*Advanced Materials & Processes*]

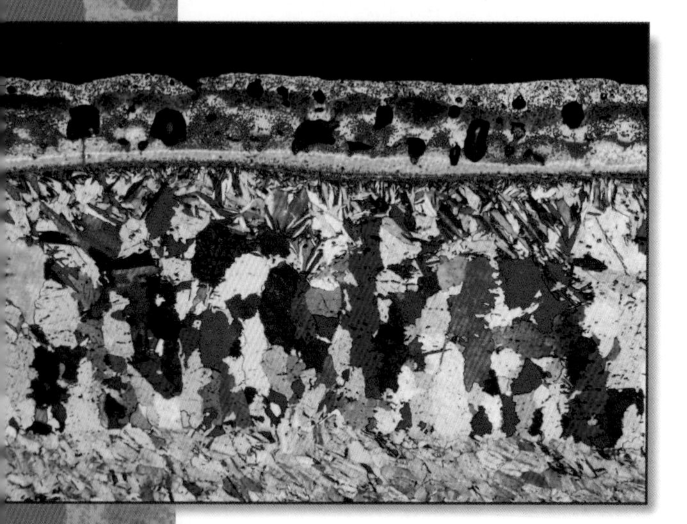

Figure C-7 Zinc alloy die castings, both natural and plated, reflect the intricate shapes possible when this low-melting-point metal and casting process is used. Finishes include nickel, brass, tin, copper, and silver.

Figure C-8 Semi-solid forming produced this as-cast ingot of A356 alloy with anodized microstructure in polarized light at 250X. Refer to Applications & Alternatives in Module 3. [A. Kazakov and *Advanced Materials & Processes*] ▼

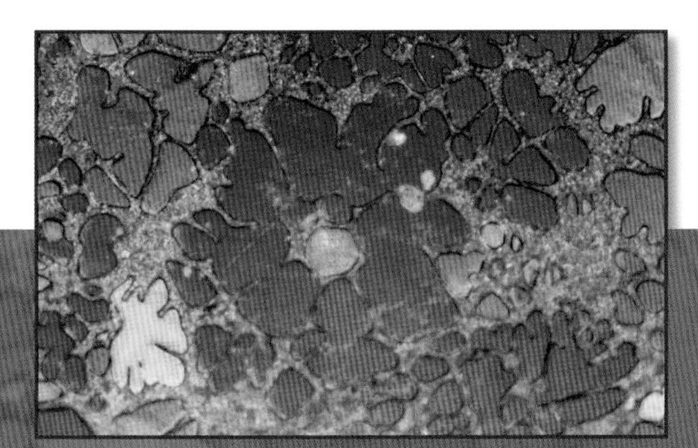

Figure C-9 HAZ (Heat Affected Zone) of a weld. Transverse view of a weld joining Ta-10 W and C103 (NbHfTi) anodized sheet alloys. Blue shows C103 alloys, yellow shows Ta-10W, and red shows weld metal (15X). [P. Danielson and *Advanced Materials & Processes*] ►

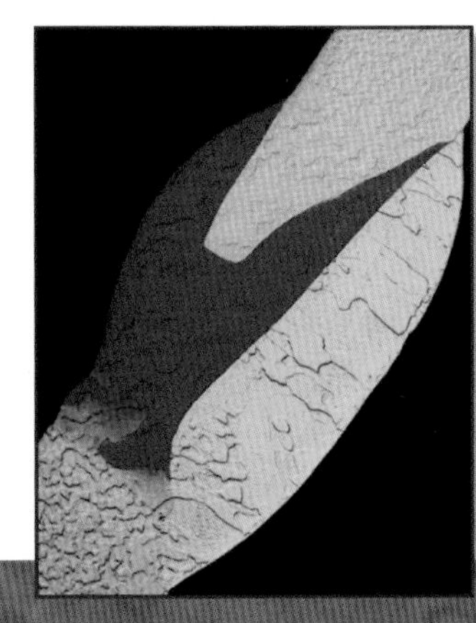

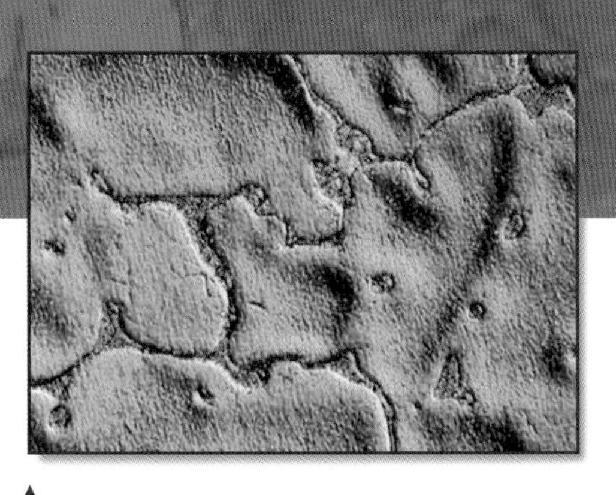

▲

Figure C-10 As-cast A1-3.12 wt% Cu mounted in epoxy, manually polished, and etched with Kellar's reagent and photographed in light microscopy under Normarski differential interference contrast. [P. Crawford and *Advanced Materials & Processes*]

▲

Figure C-11 Bronze has served as the metal of choice for castings of statues through the ages. Contemporary artists apply new processes to add color. Do you know how?

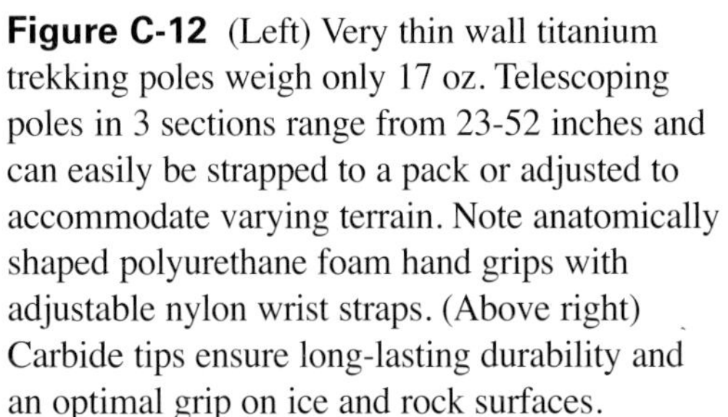

Figure C-12 (Left) Very thin wall titanium trekking poles weigh only 17 oz. Telescoping poles in 3 sections range from 23-52 inches and can easily be strapped to a pack or adjusted to accommodate varying terrain. Note anatomically shaped polyurethane foam hand grips with adjustable nylon wrist straps. (Above right) Carbide tips ensure long-lasting durability and an optimal grip on ice and rock surfaces.

Polymers

◄ **Figure C-13** Many colors, forms. and textures of plastics: 1. Polystyrene foams (gas-filled beads, preformed expanded beads), 2. Polystyrene float, 3. Foamed PS sheet, 4. Laminate insulation foam, 5. Polycarbonate pellets, 6. Acetal sheet, 8. Rigid PVC powder, 9. Rigid PVC pellets, 10. Rigid PVC (window frame, jointed tubing sitting on top of wood grain laminate), 11. Flexible PVC pellets, 12. Elasticized PVC (tubing, stain-edging, coated wire, and electric plug).

Figure C-14 Polycarbonate (Lexan™) provides crystal-clear to translucent to opaque colors for tough, easily molded products [GE Plastics] ►

▲

Figure C-15 Hardwoods and softwoods: 1. Birch, 2. Hickory, 3. Eastern Red Cedar, 4. Southern Yellow Pine, 5. White Oak Group, 6. Western Red Cedar, 7. Douglas Fir, 8. White Pine, 9. Black Walnut, 10. Red Gum, 11. Redwood, 12. Hard Maple, 13. Red Oak Group, 14. White Ash [Heritage Workshop]

▲

Figure C-16 Wood sculptures as examples of traditional African art were carved according to tribal customs and reflected the life cycle of the tribe. [Lois E. Woods Museum, Norfolk State University http://www.nsu.edu/resources/woods]

Ceramics

Figure C-17 Acid-etching process for creating a raised pattern on the surface of art glass was perfected in 1906 by Frederick Carder at Steuben Glass Works. The process is similar to the photolithographic process of creating patterns of electronic circuits on silicon chips for computers and other electronic devices. [Margaret Rasmussen, Paul Vickers Gardner Glass Center, Alfred University] ▶

◀ **Figure C-18** *Intruders*—an English glass: blown, acid-etched, cameo-carved by George and Thomas Woodall circa 1893. [Chrysler Museum of Art http://www.chrysler.org]

Figure C-19 Chapel of the Holy Cross rises out of the rock cliffs in Verde Valley outside of Sedona, Arizona. Designed by Anshen and Stevens Architects and constructed in 1957 of reinforced concrete, glass, and steel. With a 90-foot cross of reinforced concrete, the chapel blends into a rugged canyon setting and marks a sharp departure from traditional churches. ▶

Figure C-20 Airport art—one of numerous pieces of contemporary art found in the Atlanta airport. The African art from Zimbabwe uses native granite and depicts the life cycle of villages. ▶

▲
Figure C-21 The advanced ceramics of silicon nitride and silicon carbide parts such as turbo rotors, valves, and inserts provide superior heat resistance in automobile engines. [American Ceramic Society]

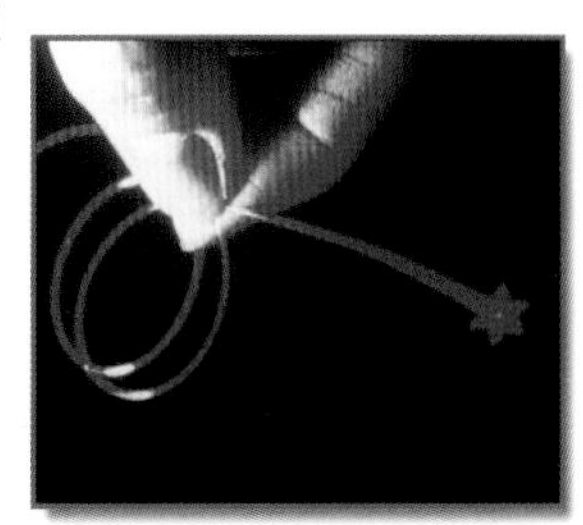

Figure C-22 Glass optical fibers have provided a technological breakthrough in the area of telecommunications. Information (in the form of electricity) was once carried through hundreds of copper wires. Now information (in the form of light) can be carried through just one high-quality transparent silica (glass) fiber. [American Ceramic Society] ▶

▲
Figure C-23 Abrasives have evolved into many forms and are made in a wide variety of ceramics, adhesives, and cloth and paper backings to provide the finishes necessary for the high quality and high performance demanded in today's products.

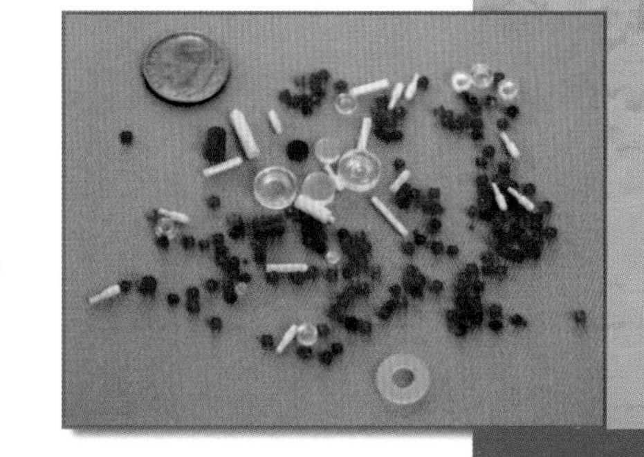

Figure C-24 Materials technologies develop techniques to grow synthetic jewels (rubies and sapphires) and machine these materials to tolerances of .000030" diameter, .000007" roundness, and .000040" concentricity to provide the necessary lenses, fiber optic guides, substrates, and rods for communication devices. ▶

Composites

Figure C-25 Polymer matrix composites and reinforcers: 1. S glass fibers, 2. Carbon fibers, 3. Kevlar fibers, 4. High-temperature performance polymer powder, 5. LARC TP1 resin, 6. Graphite epoxy prepreg, 7. Epoxy fiber (orientation 0/90/90/0 4ply), 8. Nomex honeycomb, 9. Fiberglass pipe, 10. Aluminum/polysulfone pultrusion, 11. Pultrusion hat section, 12. Fiberglass ▸

◂ **Figure C-26** Fiberglass reinforced polymer composite (PPS) thermoplastic, liquid crystalline spheroids in row-nucleation along the fiber interface. [Luther M. Gammon, The Boeing Company]

Figure C-27 Woven graphite composite bike frame is ultralight and stiff. Note: Titanium insert provides rigidity for connecting fork, handlebars, peddle crank and rear wheel. ▾

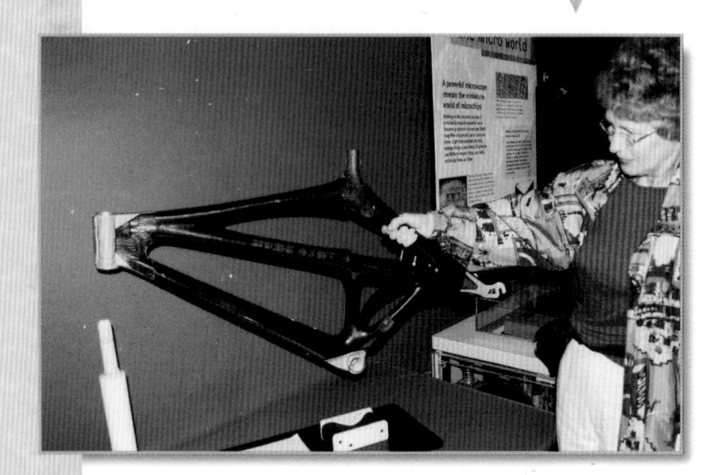

▴ **Figure C-28** Tennis rackets use exotic blends of fibers in various polymer matrices to make unique composite frames. Fibers include tungsten, copper, carbon, graphite, glass, and boron. Resins for the composite matrix may include liquid crystal epoxy.

Figure C-29 Defects seen in glass fiber-filled polyethermide composite. With 500X magnification and ultra-thin section transmitted differential interference contrast, voids are revealed within the matrix filled with short chopped glass fibers. [Luther M. Gammon, The Boeing Company] ▸

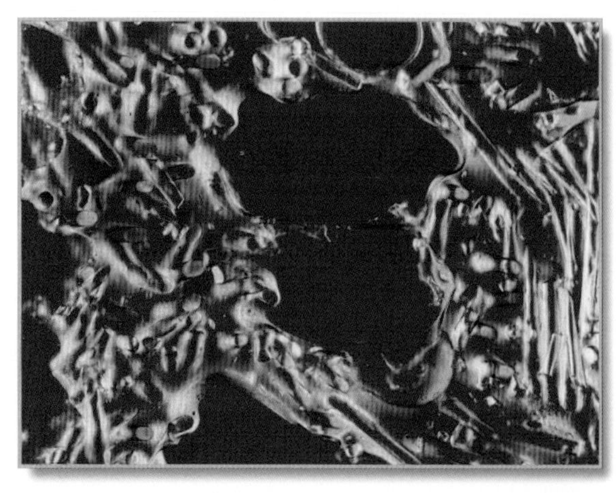

Materials Systems

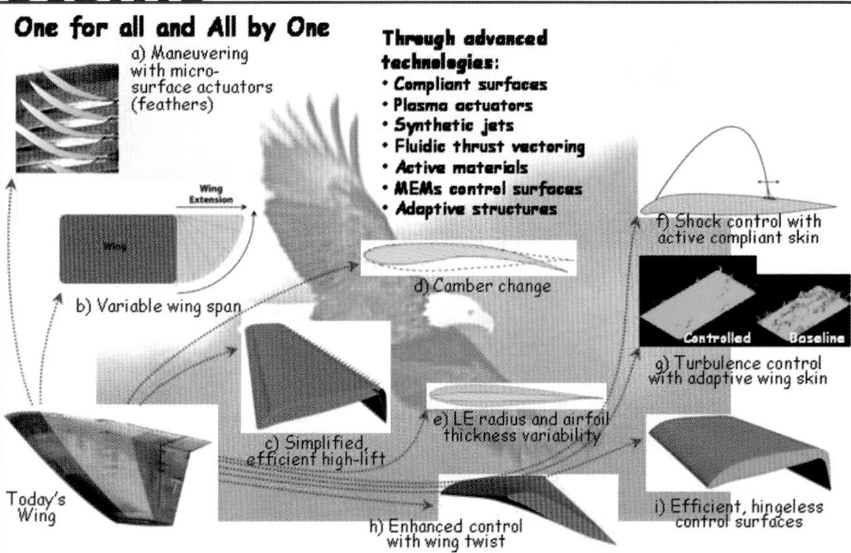

Figure C-30 Figure C-30 Biologically Inspired Systems NASA and its academic partners seek to mimic nature (e.g., eagles) in design of materials systems that will adapt to environments. For example, morphing wings on aircraft and spacecraft will change their shape or "warp" in flight to improve efficiency. This diagram depicts ideas modifying today's complex, multiple component wings with "One for all and All by One" wings that meet all airfoil requirements. The "Wing for all Regimes" will: (a) maneuver by manipulating micro-surface actuators that mimic birds' feathers, (b) change wing length, and (c) have a simplified efficient high-lift, (d) change camber, (e) change thickness, (f) and (g) modify skin or surface, (h) twist or warp and (i) bend without hinges. The materials in these systems should also self-heal much as skin heals from a cut or abrasion. [Adapted from "Materials World—What Can Biology Teach Us?" Lecture by Ilhan Aksay, NASA Langley, June 2, 2003] ▲

◄ **Figure C-31** Sporting equipment. Precipitation-hardened stainless steels and ultrahigh strength steels of different levels of hardness, tensile strength, strength-to-weight ratios, modulus of elasticity, and corrosion resistance. Hardness ranges from HRC41 for Custom 450 stainless, to HRC56 for AeroMet 310 alloy. The unique alloy Aeroloy is seen here used for the face insert on these metal "woods." [Carpenter Technology Corp.]

Figure C-32 In 2003 Ford Motor Company celebrated its 100-year anniversary. The Model T is an early example of materials systems. First built in 1914, it used steel for the frame and other stamped metal components, cast iron for the block and other castings, aluminum for the hood and transmission cover, and brass for the trim. The body frame was made of ash, the wheel spokes of hickory, and the top bows of laminated oak. The upholstery was backed with natural fibers. Paint was baked enamel or "japaned" lead base. [Wally and Liz Hunt] ►

Materials Systems

Figure C-33 In contrast to the Model T, modern cars make use of many types of plastics. Following are examples of plastics use in the 21st century: Speaker Grilles (POM), Instrument Panel, ((LFRT(long-fiber reinforced thermoplastic)), Safety Systems (POM, Polyester, LCP, LFRT), Sun Roof Systems (POM, Polyester, PPS, LFRT), Seating Systems (POM, Polyester, LFRT), Fuel Systems (POM, PPS), Door Lock Systems (POM, Polyester), Window Winders (POM, Polyester, LFRT), Door Module (POM, LFRT, Polyester), Airbag (Polyester-HI), Battery Tray (LFRT, Polyester-HI), Auto Electrical Systems (Polyester, PPS, LCP, POM), Front end (LFRT, Polyester-HI), Engine & Mechanical Applications (PPS, Polyester, LCP, POM, LFRT), Power Distribution Box (LFRT, Polyester-HI), Wiper Systems (POM, Polyester, LFRT), Mirrors (POM, Polyester, LFRT. [*Advanced Materials & Processes;* Alfred E. McKenney]

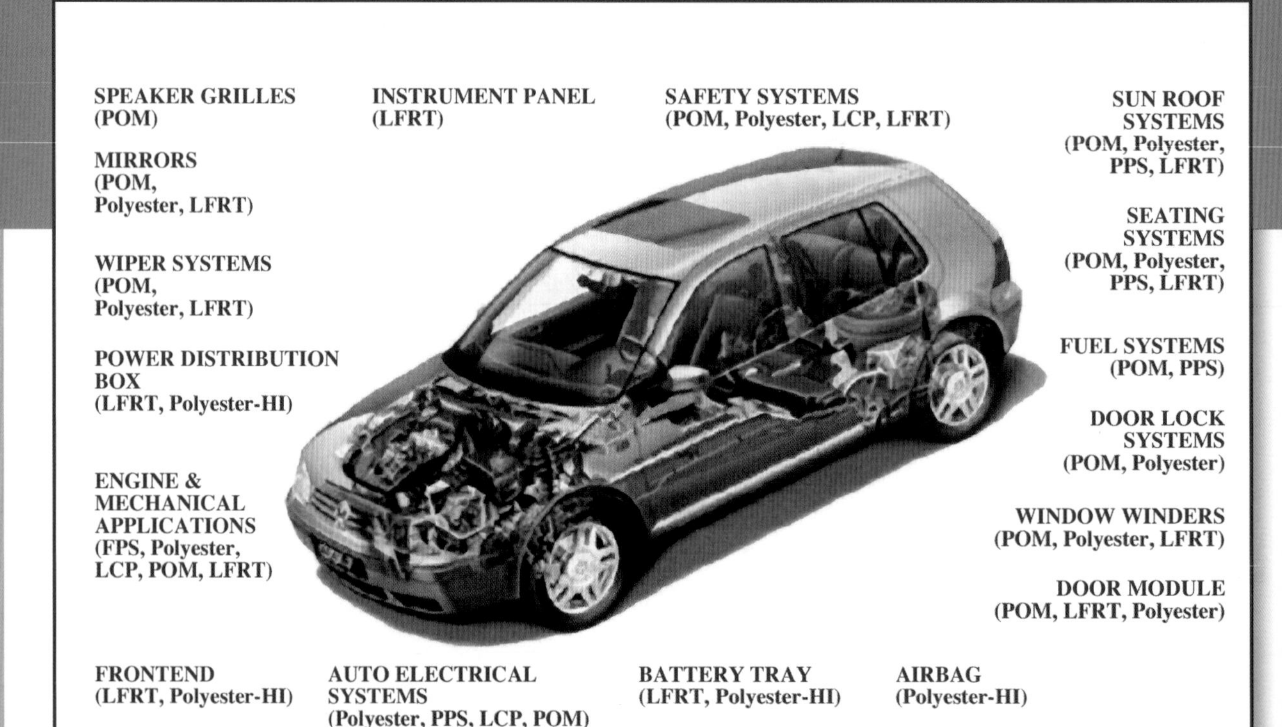

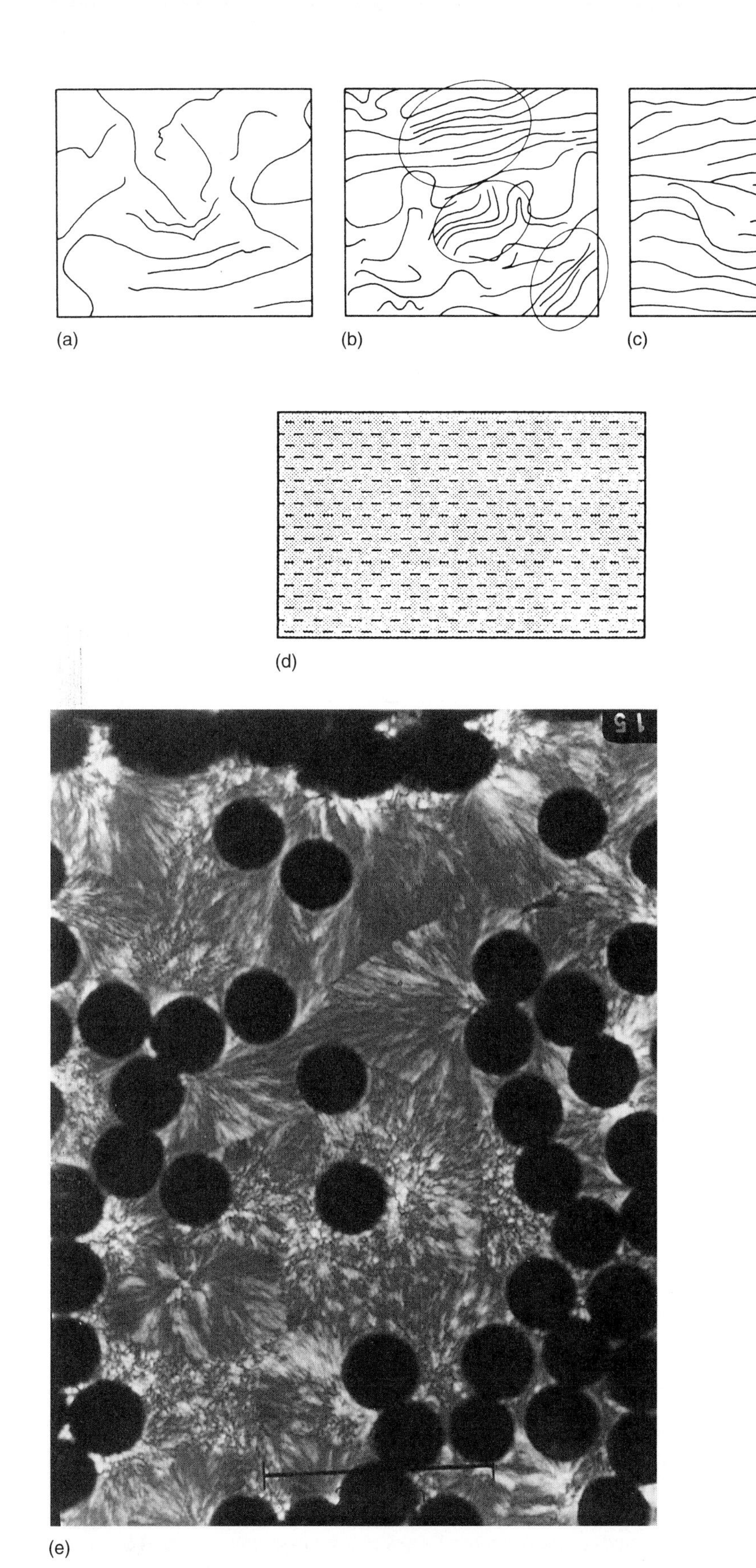

Figure 14-7 Degrees of crystalline structure of polymers: (a) mostly amorphous, (b) semicrystalline (crystallites encircled), (c) mostly crystalline, (d) liquid crystal polymer, (e) transmitted, cross-polarized light circles ultra-thin section of polymer matrix composite—carbon fiber (cut at 90° and seen as black circles) in a matrix of PEEK thermoplastic with liquid-crystalline spheroids at 2000X/100X obj. (Luther Gammon, Boeing Materials Technology)

orderliness. Slow cooling allows chains to move into alignment. While metals are normally fully crystalline, the degree of crystallinity in polymers can range from mostly amorphous to mostly crystalline. Crystallinity in polymers can alter strength and toughness. Mostly amorphous polymers will be less dense than mostly crystalline polymers because the chains pack closely together.

Molecular motion of short polymer chain segments and polymer branches cause amorphous polymers to be stiff, hard, and brittle at room temperatures. Increases in temperature cause greater molecular motion through thermal mixing of the atoms and molecules. This causes an increase in the volume of the material. The thermal mixing from higher temperatures also increases the spacing between molecular segments and permits increased flow of the materials. A discussion of glass transition later in this module deals further with the concept of molecular motion and flow in polymers.

The degree of crystallinity is determined by structural regularity, compactness, and amount of flexibility, which allows packing of chains. Reduction of random chain lengths provides regularity. Stronger secondary forces allow greater compacting. ***Liquid crystal polymers*** (LCPs) develop highly oriented (rodlike) molecules for increased directional strength (Figure 14-7d) and LCP spheroids (Figure 14-7c). Other chemical factors affect crystallinity, including configuration and tacticity (see Figure 14-3e–g). The references at the end of this module provide more information.

Polyethylene (PE) is a good example of a polymer capable of a high degree of crystallinity because (1) the linear structure of the chains is conducive to packing and (2) its molecular pattern is flexible, which provides easy packing even though it has weaker secondary bonding. Secondary bonds (van der Waals forces) normally promote crystallinity. Polyethylene has the potential for a wide range of properties because of technology's ability to control its molecular weight and crystallinity. Low-density polyethylene (LDPE) has a molecular weight below 10,000, while ultra-high-density polyethylene (UHMWPE) has molecular weights much above 1.0×10^6 (see Figure 14-8a) and as shown in Figure 14-3.

Crystallinity in polymers is also achieved through processes such as extrusion and drawing. ***Oriented polymer fibers*** are obtained by drawing an amorphous polymer through a die, which improves strength or ***orientation*** in the direction of drawing (see Figure 14-8b). This concept can be illustrated by stretching a polyethylene sandwich bag. The method of manufacturing orients the polymers in one direction. By pulling on the bag in perpendicular directions you can notice the greater resistance from the oriented direction; also, as the bag is stretched, it becomes stronger due to further orientation.

The cooling rate and processing during cooling affect the crystal patterns of polymers. ***Spherulites*** form as a result of supercooling, with nuclei generating a spherical pattern that grows until several spherulites melt at boundaries. When large spherulites are allowed to grow, weakness results. Heat treatment of polymers can change their crystal structure; for example, the annealing processes reheat the polymer for a specified time to permit crystal thickening. The electron micrographs in Figure 14-9 show polyethylene in the amorphous state, then as a single linear crystal, and then as an annealed crystal.

Polymers and liquid crystals—*http://plc.cwru.edu/*

14.2.3 Branched and Network Structures

As seen in Figure 14-10c, some polymers do not simply grow linear chains (Figure 14-10a and b); rather, the chains develop branches much like those of a tree. As a result, the branched polymers are not as symmetrical and consequently will not achieve the degree of crystallinity obtained in the more regular linear polymers. Numerous techniques are employed by polymer chemists and materials technologists to vary both the length of linear chains and the degree and uniformity of branching.

Networks of polymers develop in branched polymers and primary bonds form between chains. In contrast to the weaker secondary forces in linear polymers, ***cross-linking,*** the development of covalent bonds, occurs between chains, as seen in Figure 14-10d and e. The cross-linked network polymers do not soften when heated as do the ***thermoplastic*** polymers

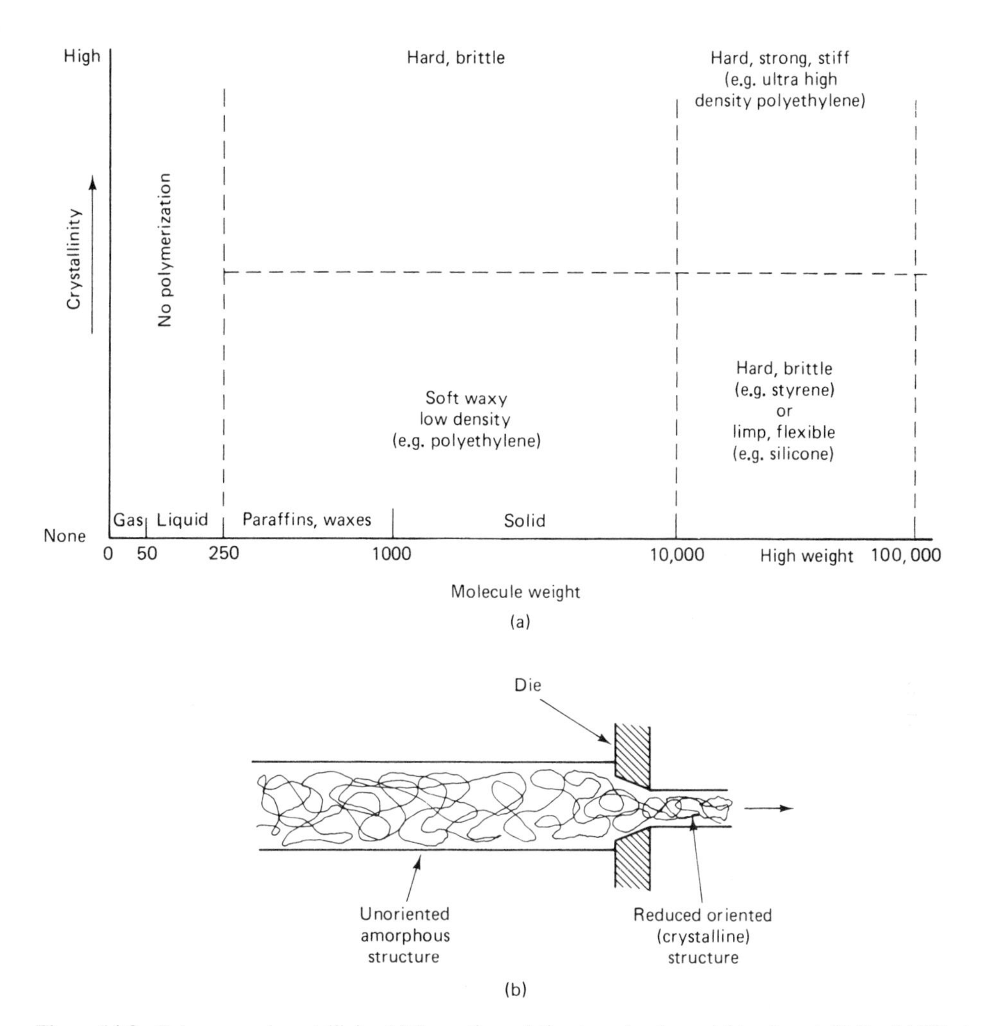

Figure 14-8 Polymers and crystallinity. (a) Properties: relation to molecular weight and crystallinity. (b) Effect of pulling an amorphous polymer through a die.

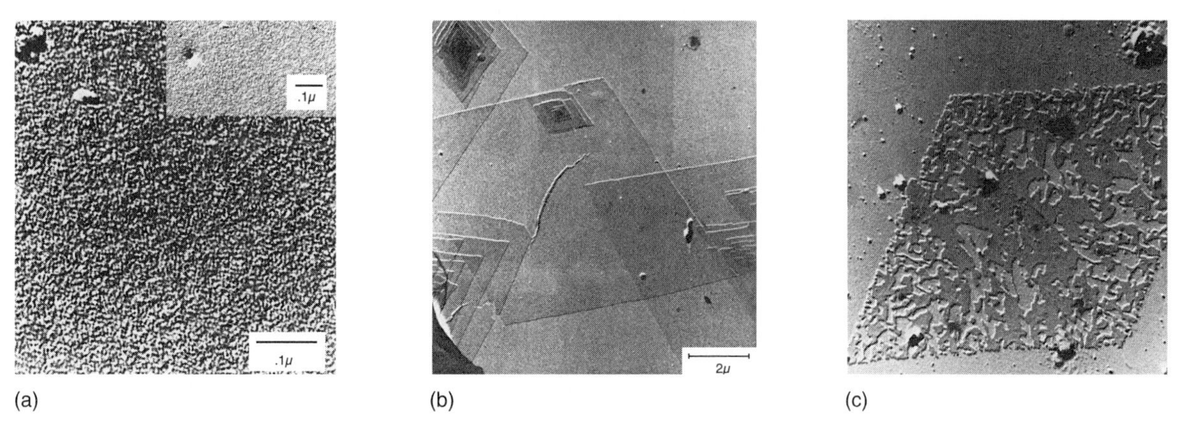

Figure 14-9 Crystallinity in polymers. (a) Electron micrograph of amorphous polyethylene. (b) Single crystal of polyethylene. (c) Single crystal of linear polyethylene annealed at 120°C for 30 minutes. (Dr. Phillip H. Geil, University of Illinois)

that are held together by van der Waals forces (Figure 14-10f). Heating of thermoplastics causes these weaker bonds to lose their hold, and thus movement of the mers and softening of the polymer. Thermosetting polymers are like hard-boiled eggs. Once cooked, the egg cannot be softened to flow into another shape. But a thermoplastic material such as a candle can continually be reheated and molded into new shapes. Normally, polyethylene is a thermo-

(a) Single linear polymer

(b) Multiple linear polymers

(c) Single branched polymer

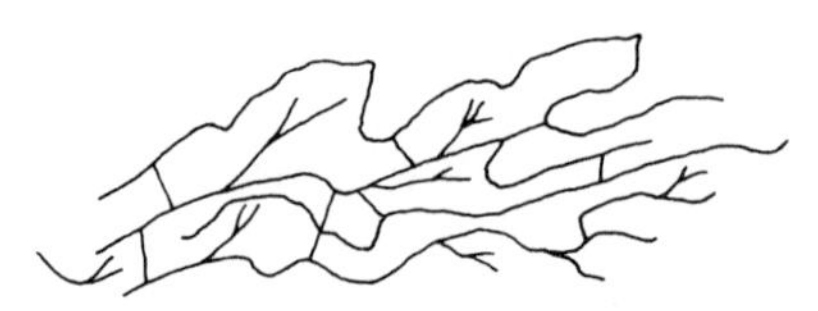

(d) Network – crosslinked branched polymers (e.g., phenolic, epoxy, and silicones)

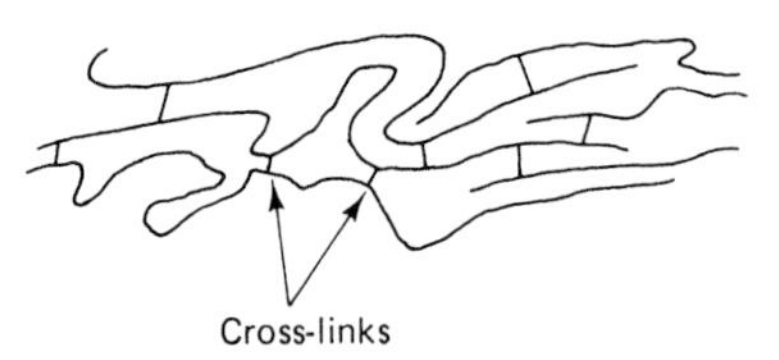

(e) Crosslinked linear polymers (e.g., polyesters and diene elastomers)

Secondary bonds

(f) Thermoplastic polymer

Figure 14-10 Various polymer structures.

plastic material, but through electron irradiation cross-links are developed. Epoxy can exist as a thermoplastic polymer, or, through the connections of end groups, it can become a thermosetting polymer. Sulfur is used in rubber to form numerous cross-links in the ***vulcanization*** process. The more cross-links present, the harder the rubber is. A reference to cross-linking is the ***netting index,*** which designates the number of cross-links per 100 linear bonds. A higher number of double bonds in a polymer provides more sites for reaction than in those with fewer double bonds. Natural rubber has hundreds more double bonds than butyl rubber, which has more tightly placed, single-bonded molecules. A hard rubber has a netting index of 10 to 20, while a hard thermosetting plastic such as phenolic has an index of about 50.

Introduction to polymers—*http://www.umr.edu/,wlf*
Polymers and liquid crystals—*http://abalone.phys.cwru.edu/tutorial/enhanced/main.htm*
Introduction to plastics primer—*http.//www.polymers.com/dotcom/polycon/primer.html*

Table A-9 compares various plastic classification systems. When using the table, it is important to realize that these are general data, and any given plastic may have considerable grade variations due to fillers, additives, and structure. The table also provides the basic molecular structure, chemical formula, American Society for Testing and Materials (ASTM) abbreviations, some selected trade names, and typical uses of the plastics listed. Note that carbon (C) is the backbone for all the plastics except silicone, which has an inorganic silicon (Si) and oxygen (O) backbone. In the polyolefins group, ionomers come from ethylene gas combined with inorganic compounds consisting of sodium (Na), zinc (Zn), magnesium (Mg), or potassium (K) salts that bond both covalently and ionically. Ionomers are resilient, oil resistant, and tough, while having very high transparency. Although the large number of plastics generated from materials technology have a wide array of properties, they create problems for green manufacturing in terms of recycling. The auto industry has sought to reduce the types of plastics it uses in a move to save costs and improve the recyclability of automotive plastics.

Numerous websites provide current information on plastics. For example, the American Plastics Council offers an extensive site, including a glossary of terms used with plastics. Plastic suppliers' sites are a good place to compare types and properties to supplement the information provided in Appendix Table A–9 and the properties tables in this module.

14.3 FIBERS

ASM International's *On-line Engineered Materials Handbook Desk Edition* defines fiber as:

> A general term used to refer to filamentary materials. Often, fiber is used synonymously with filament. It is a general term for a filament with a finite length that is at least 100 times its diameter, which is typically 0.10 to 0.13 mm (0.004 to 0.005 in.). In most cases it is prepared by drawing from a molten bath, spinning, or depositing on a substrate. Fibers can be continuous or specific short lengths (discontinuous), normally no less than 3.2 mm (1/8 in.).

14.3.1 Natural Fibers

Natural fibers come from plants and animals and include jute, hemp, wool, and cotton. These fibers have been used since early times and still are important commercial materials. Fibers are woven into fabrics and used for line, string, and rope (e.g., nylon-monofilament fishing line and tennis string or polypropylene rope). In Module 24, we will further define fibers and related terms and discuss their importance in composites.

14.3.2 Synthetic and Manufactured Fibers

Manufactured fibers are polymer strands classified as either regenerated cellulosics or noncellulosics (i.e., synthetics). Because cellulose does not melt or dissolve, it must be processed (regenerated) into syrup, and from there it is formed into fibers such as rayon and acetate. ***Synthetic fibers*** created by chemical synthesis methods (condensation polymerization and addition polymerization) include polyamide, polyester, and polyurethane fibers. The blending of natural fibers, such as cotton and wool, will provide properties sought in woven fabrics for clothing and upholstery. Engineering of fibers allows desired properties to develop and is accomplished by varying molecular weight, varying cross section, and orienting molecules through drawing and stretching. The ability of fibers to ***wet*** (absorb or ***wick*** moisture) is important in coloring them or in using them in clothing. Olefins wick well and make good sports clothing; acrylic fibers do not absorb moisture well. Fibers treated with poly(ethylene glycol) (PEG), a polytherm coating, are ***enthalpic*** (absorb, store, and release large amounts of heat) and ***hygroscopic*** (release moisture as vapor, not droplets), so as woven fabric they make excellent sporting and cold-weather clothing. ***Microfibers*** represent another polymer advance, the transforming of synthetic fibers such as polyester so that they behave like the natural fibers cotton and wool.

Fiber-reinforced plastics (FRP), along with many laminated and filled plastics, comprise a large group of plastics that fit diverse engineering and general-purpose applications. Used both in woven fabrics and as reinforcing fibers in FRP, the aramid fiber is a plastic of superior qualities. Under tensile loads, this polymer equals steel of similar size but has a superior strength-to-weight ratio. The superior strength develops as a result of the stringing out of the carbon atoms, which line up rather than coiling like a spring as in most polymers. The built-in stiffness of the C—C bond greatly resists stretching under tensile loads. The superstrong, lightweight aramid fibers are woven into bulletproof vests and serve as reinforcers for tires, firefighting and racing helmets, and clothing. DuPont's aramid fiber ***Kevlar*** is a liquid crystal polymer (LCP) fiber. (See Module 24 for more on Kevlar.) Poly-*p*-benzamide can be produced only as fibers; it cannot be molded into solid bulk shapes. Even superior to aramid is the polyolefin fiber ***Spectra,*** which consists of ultrahigh-molecular-weight polyethylene

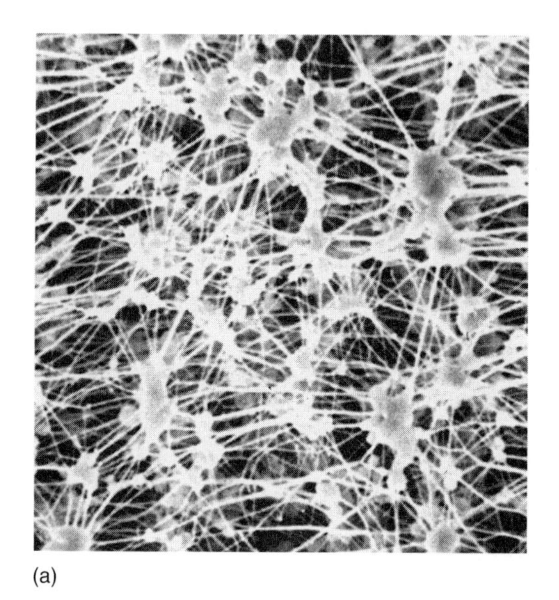

(a)

Figure 14-11 (a) Expanded PTFE Gore-Tex® fiber. The photomicrograph reveals porous microstructure. (b) PTFE fibers for synthetic knee ligament. (W. L. Gore & Associates, Inc.) (c) Cool threads with extruded microfibers. Bundles of scalloped, oval-shaped polyester microfibers create microchannels that allow perspiration to be "wicked" away from the skin by capillary action and evaporation of the moisture. The fabric has a very low moisture content—about 1/2 of 1%. By contrast, normally nylon fabrics have about 4% moisture and cotton around 7%. Fabric designers work to provide a weave that wicks moisture but does not allow rain and cold to enter. For example, bundles of fibers, as seen here, use larger filaments against the skin to create air space. Smaller bundles on the surface of the fabric have greater surface area. (Adapted from "Working Knowledge—Cool Shirt," *Scientific American*, October, 2003, pp. 92–93)

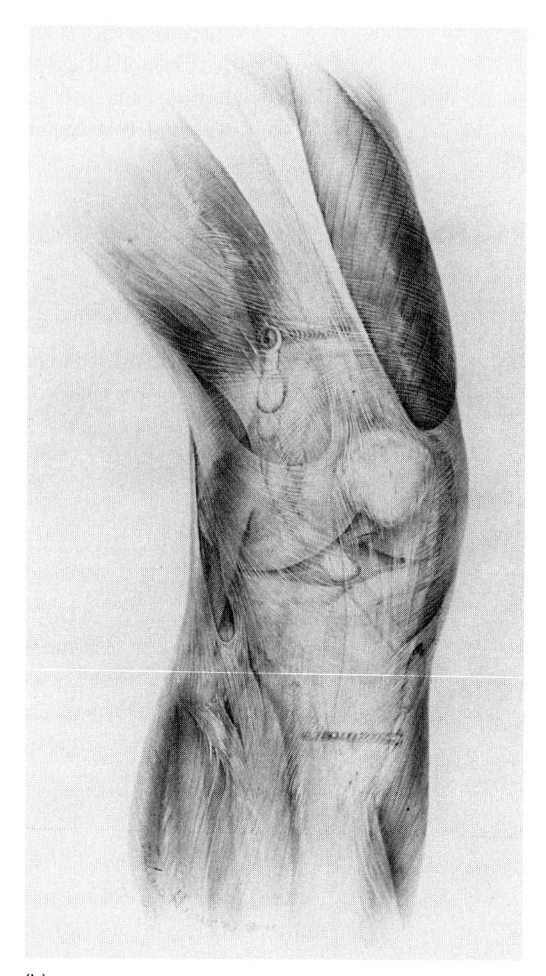

(b)

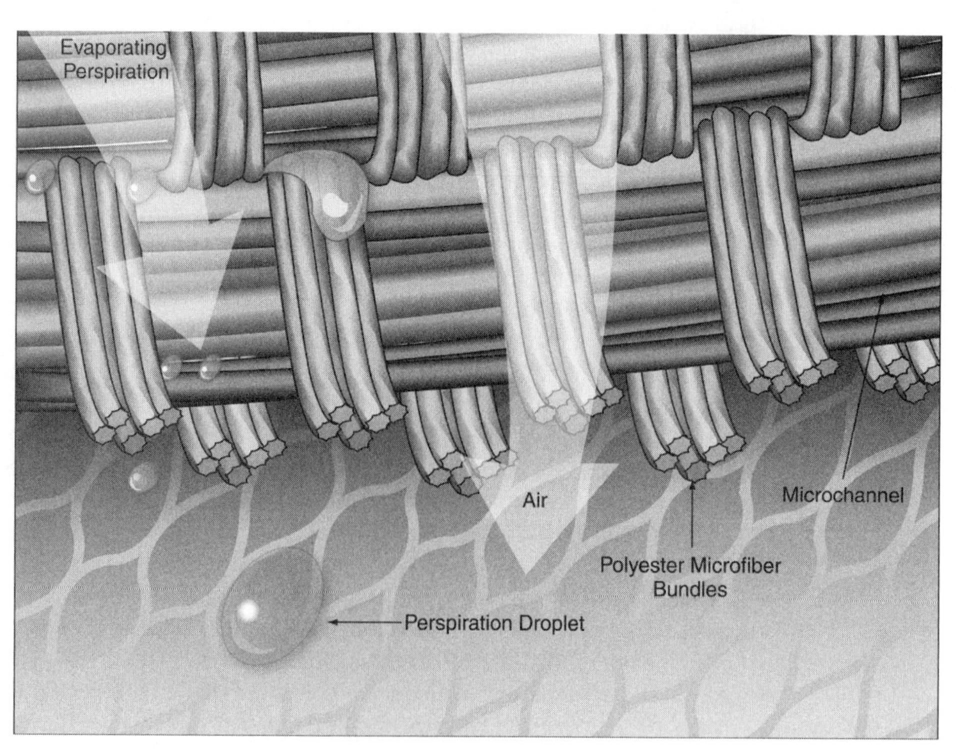

(c)

polymers processed to yield highly oriented fibers. This polyolefin fiber is woven into sails and space suit/boot/glove structures because of its superior abrasion resistance, high flexural endurance, chemical resistance, low moisture absorption, and superior specific strength.

Typical of advanced materials, the engineered Gore-Tex® fiber possesses a string of properties that makes it suitable for a wide range of applications. This expanded polytetrafluoroethylene (PTFE) (Figure 14-11a) is chemically inert (not affected by any common chemical), has a low friction coefficient, can function in a wide temperature range, does not age, and is weather durable. Such properties have resulted in its selection for such wide-ranging applications as clothing, joint sealants, nonshedding filters to provide microcontamination control for computer disk drives and manufacturing clean rooms, and membranes that offer microfiltration of bacteria and other particulates from air or liquids. Because Gore-Tex materials have superior strength and are biocompatible, hydrophobic, porous, and air permeable, they have been selected for clothing used in sports, outdoor activity, and walking in outer space. They are used in surgery for threadlike sutures to sew together tissue, as heart patches, for tubes to replace diseased arteries, and as synthetic knee ligaments to provide stability when replacing the natural anterior cruciate ligaments (Figure 14-11b). Another advanced fiber, polybenzimidazole (PBI), is woven into fabrics that will not burn in air. Sulfar resists melting until 285°C and is an excellent filter that, through reverse osmosis, can separate toxins and heavy metals from water. (***Osmosis*** is the passage of a liquid from a dilute to a more concentrated solution through a semipermeable membrane. This type of membrane allows passage of the liquid but not the dissolved solids in the concentrated solution. ***Reverse osmosis*** is the forced reversal of this natural phenomenon by the application of enough pressure to the concentrated solution to overcome the pressure (osmotic) of the less concentrated solution.)

14.3.3 Nanothreads

An early commercial application of nanotechnology has cropped up in fashion: application of nanopolymers to clothing. Such trade name fabrics as Nano-Dry, Nano-Tex, SoleFresh, and Nano-Touch are results of new molecular-level processes for enhancing fibers used to make threads for clothing. Techniques include applying Teflon on cotton as Stain Defenders with "whiskery molecules" that repel oil and liquids which cause stains and "polymer nets" that wick away body perspiration. ***Nanoengineered textiles*** are available by brand names such as Lee, Gap, Eddie Bauer, and Haggar. SoleFresh socks and underwear use *Phase Change Materials (PCM),* constructed of microspheres that melt when the body heats up to draw away heat. Then as the body cools they freeze again, thereby releasing heat. SoleFresh socks guard against fungus and bacteria as a result of silver nanoparticles which naturally counter these biological organisms. Nanoscale capsules with perfume, moisturizers, insect repellents, and deodorants can also be imbedded in fabrics.

On a more sophisticated level, the military works on body armor fabrics that will harden with the sounds of an explosion and become 50 times stiffer to serve as a splint in case of a fracture. Nanometer scale magnetic particles fill hollow fibers that can be activated by a small device. The new fabrics employ carbon nanotubes to create continuous spinning of nanocomposite fibers with 17 times the toughness of Kevlar. These fibers make superior body armor with battery-like capabilities for storing electricity.

 MIT Technology Review—*www.technologyreview.com*

14.4 GREEN DESIGN AND GREEN MANUFACTURING

Although polymers offer a wide range of benefits to society, they also pose potential burdens for the environment. These burdens present challenges at all stages of the materials cycle, beginning with the synthesis/processes through disposal/recycling. At the end of this unit, we provide a perspective on recycling plastics. To promote ***green design*** (design aimed at reducing burdens on the environment), the recycling stage can become the beginning of the materials cycle for certain polymers. When we introduced the materials cycle we also tied it to life-cycle analysis, also referred to as ***life-cycle assessment (LCA)***, which involves a life-cycle inventory (LCI). LCA seeks to quantify environmental burdens of a product over its entire life cycle. The United States Automotive Materials Partnership (USAMP) states the objectives and intended results for LCI as follows:

Objectives of LCI—Life-Cycle Inventory*

1. Assess the cradle-to-grave environmental performances of vehicles
2. Conduct an LCI for an entire vehicle, in this case, the generic family car (often referred to as PNGV's Super Car)
3. Quantify energy and material flows over a product's life cycle

Use of LCI Results

1. Serve as a benchmark for future vehicles (e.g., PNGVs)
2. Communicate with public and private sector
3. Not intended for materials comparisons
4. Stress that cost and performance must also be considered

The collaborative efforts of the auto industry in such efforts as PNGV, and use of LCA, appropriate legislation, and growing consumer interest will steadily improve green manufacturing. During its life cycle, the generic family sedan (1) consumes 961 GJ of energy, (2) consumes 46,000 pounds of hydrocarbons, and (3) generates 130,000 pounds of carbon dioxide. LCI allows automakers and suppliers to monitor environmental impacts and set goals for reducing these burdens.

To illustrate how LCI works for a product with many environmental impacts, Figure 14-12 depicts a materials cycle developed by the Automotive Composites Alliance, another group working with USAMP and PNGV, for an automobile part. This diagram helps to identify the points where LCI can be applied. Emissions from synthetic polymers occur during the stages of materials acquisition and processing and component manufacture and assembly. Those polymers and hydrocarbon fuels used for powering vehicles, along with the lubricating, hydraulic, and cooling fluids, have potential to pollute water and the atmosphere, contribute to global warming, and affect our ozone layer. More dangers occur at the recovery, reuse, and disposal stage (discussed later). During normal lifetime use and maintenance, synthetic polymers provide many advantages that tend to offset the harmful effects. For example, lightweight, noncorroding plastics and plastic-based composites that provide good design flexibility can yield more fuel-efficient vehicles. Table 14-1 shows the data categories used for LCI as applied to vehicles, including types of energy, water consumption, air emissions, water effluents, solid wastes, and raw materials.

Figure 14-12b, shows the material flow for the Generic Vehicle Product System in more detail, excluding the energy flow shown in Figure 14-12a. This figure follows the description outlined for the materials cycle in Figure 1-4. Beginning on the left, primary materials (steel, aluminum and plastics) along with other materials (glass, rubber, paint, etc.) enter the manufacturing process. Parts are assembled into subcomponents, and the subcomponents join with glass parts, tires, paints, motor oil, etc. for component and vehicle assembly within the vehicle system boundary. After assembly, operation of the vehicle carries the product through its useful life. After the car is taken out of service, the last stage of the materials cycle begins and couples with the first stage of the cycle, that is, disposal/recycling–extracting raw materials, in this case, the raw materials of secondary steel, secondary aluminum, and other recycled materials rather than primary materials. The circled labels A, B, C and D help to connect the

*Adapted from Susan G. Yester, "Life Cycle Inventory," *NEW*: *Update 99*, November 1, 1999, Auburn Hills, MI.

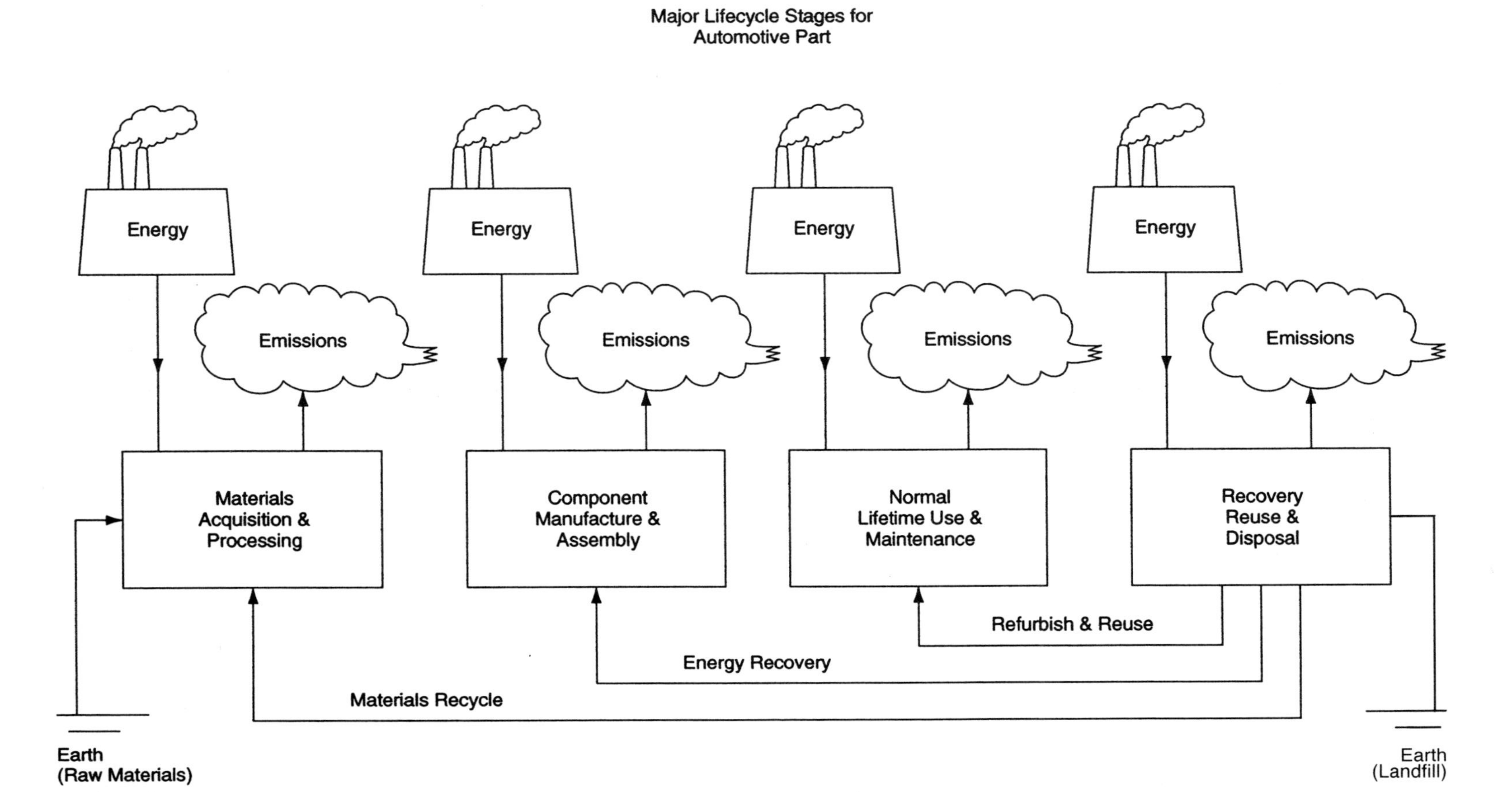

Figure 14-12 (a) Life-cycle inventory for an automotive part. (Adapted from Automotive Composite Alliance)

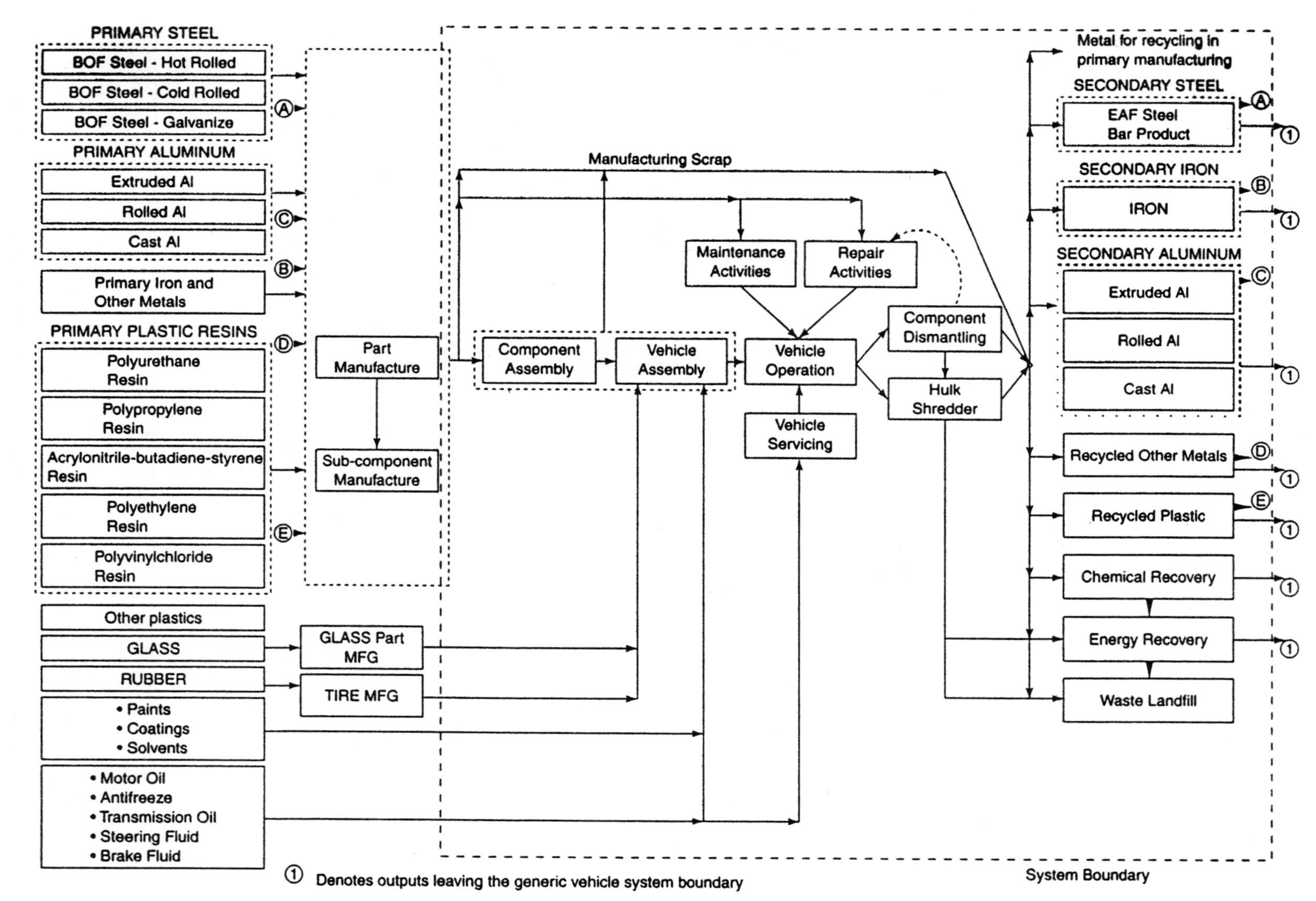

Figure 14-12 continued (b) Material flow for the family sedan. (Generic Vehicle Product System)

TABLE 14-1 DATA CATEGORIES FOR LCI*

Energy	Air Emissions	Water Effluents
Fossil and Nonfossil	CO_2, CO, SO_x	Dissolved Solids
Process	NO_x, CH_4, HCl	Suspended Solids
Transportation	HF, NMHC, Pb	Heavy Metals
Feedstock	Dust and Particles	Oils and Greases
		PO_4 and NH_3
Solid Wastes	**Water Consumption**	
Sanitary and Municipal	Ground	
Total Solid Waste	Surface	
Raw Materials		

*Adapted from Susan G. Yester, "Life Cycle Inventory," NEW: Update 99, November 1, 1999, Auburn Hills, MI.

primary materials group with the secondary materials used in primary manufacturing. Note how scrap moves in the cycle and where materials disposal occurs.

Cost will always influence materials selection. Even though a lighter, more fuel-efficient car will save the owner money over the life of the vehicle, purchase price drives the owner's decision. The following list of general costs for raw materials shows why steel and iron are still the primary body materials used in the average car, even though plastics and other polymer-based materials are increasingly available.

Raw Materials by Dollar per Pound*

Steel	\$0.40	Glass fiber composites	\$1.20
Iron	\$0.50	Magnesium	\$1.50
Aluminum	\$0.90	Carbon fiber composites	\$6.00
Plastics	\$1.00	Titanium	\$10.00

*Adapted from John Fillion, "Trends for Automobile Materials," talk presented to NEW: Update 98, November 4, 1998, Brookhaven National Laboratory, Long Island, NY.

Green architecture connects actions to consequences. It looks downstream to sources of materials and upstream to how the building will impact the environment. Buildings can have life spans up to hundreds of years. Green architecture, defined in terms of a building,

> . . . serves the needs of the people who inhabit it. It supports and nurtures their health, satisfaction, productivity, and spirit. It requires the careful application of the acknowledged strategies of sustainable architecture—non-toxic construction, the use of durable, natural, resource-efficient materials, reliance on the sun for daylighting, thermal and electric power, and recycling of wastes into nutrients. An elegant architectural integration of these strategies produces a building which honors the aspirations of those who use it and engages the natural world. . . . aim for buildings which improve the quality of the air and water, produce surplus power and food, and convert waste into nutrients and useful products. (*http://www.coldhamarch.com/green/what.html,* January 9, 2000.)

Also termed ***green*** or ***sustainable design,*** green architecture involves the issues surrounding change, as explained in Module 1. Construction techniques that apply new design and materials depart from the norm. New building technologies must be integrated with clients' needs using the principles of concurrent engineering while having a positive impact on people and helping sustain a renewable, wholesome environment.

Sometimes building technology that aims to minimize energy consumption and use synthetic polymers results in "sick buildings" that do not provide a wholesome habitat. Carpets, furniture, paints, and other finishes may pollute internal environments. Volatile organic compounds (VOCs) released from such synthetic materials as polyvinyl chloride plastics, dyes, and adhesives combine with particulates such as crystalline silicates in paint and spackle, pollen, and tobacco smoke to make people sick. In the next module you will learn about living environments and the potential of plastics in new building technology.

Green Design Network—*http://Greendesign.net/*
Green architecture—*http://www.coldhamarch.com/green/*
Greening the Government, EPA's Environmentally Preferable Purchasing Program—*http://www.epa.gov/opptintr/epp*
Green design materials—*http://data.oikos.com/products/*
Environment Show—*http://www.enn.com/envshow/*

SELF-ASSESSMENT

14-1. A chemical that serves as a "molecular matchmaker" necessary to begin polymerization reaction is called a

a. Monomer **b.** Isotope **c.** Catalyst **d.** Filler

14-2. Polymers with one or more multiple covalent bonds, such as ethylene, are

a. Unsaturated hydrocarbons **b.** Saturated hydrocarbons **c.** Aerogels **d.** Zeolites

14-3. Adding sulfur to rubber to strengthen it is called

a. Condensation **b.** Vulcanizing **c.** Wetting **d.** Addition

14-4. Multiple covalent bonds that provide active sites for cross-linking are

a. Isotopes **b.** Single bonds **c.** Double bonds **d.** Ablation

14-5. Highly oriented (rodlike) molecules that increase directional strength in polymers are

a. Liquid-crystal polymers **b.** Linear polymers **c.** Amorphous polymers **d.** Branched polymers

14-6. What does the netting index indicate about a polymer?

a. Degree of hardness **b.** Ability to weave fiber **c.** Degree of branching **d.** Level of adhesion

14-7. Allowing large crystals to grow in polymers will usually do what to them?

a. Strengthen it **b.** Weaken

14-8. What effect will drawing polymers through dies yield?

a. Increases in amorphous structure **b.** Increases in crystallinity **c.** Development of cross-link **d.** Development of branches

14-9. A method of quantifying environmental burdens of a product over its entire life cycle is

a. LCA **b.** Green design **c.** Algorithm **d.** Recycling

14-10. Plastics compete best with steel on which factor?

a. Recycling infrastructure **b.** Raw materials availability **c.** Cost per pound **d.** Design flexibility

14-11. Name some natural fibers. Name some manufactured (synthetic) fibers and give typical applications.

14-12. Calculate the molecular weight for poly(vinyl chloride), polystyrene, and nylon-6.

14-13. Calculate the average molecular weight of a polyethylene compound with 10,000 monomers. Would this be classified as a low-, medium-, or high-molecular-weight polymer?

14-14. Justify the extra effort and cost for green manufacturing practices in terms of LCA and burdens on the environment.

14-15. Explain the benefits of LCI for estimating the impacts of a product on the environment; include some categories of data for evaluating.

REFERENCES & RELATED READINGS

American Fiber Manufacturers Association. *Manufactured Fibers.* Video. New York: AFMA, 1989.

ASM International. *Engineering Materials Handbook*, Vol. 2: *Engineering Plastics.* Materials Park, OH: ASM International, 1988.

Brady, George S., and Henry R. Clauser. *Materials Handbook.* New York: McGraw-Hill, 1977.

Marshall, John A. "Liquids That Take Only Milliseconds to Turn into Solids." National Educators' Workshop: *Update 93—Standard Experiments in Engineering, Materials Science, and Technology.* NASA Conference Publication 3259, April 1994, pp. 315–322.

McDonough, William, and Michael Braungart. *Cradle to Cradle: Remaking the Way We Make Things.* Madison, WI: North Point Press, 2002.

Mittal, K. L. *Polyimides: Synthesis, Characteristics, and Applications*, Vols. 1 and 2. New York: Plenum Press, 1984.

Mulhall, Douglas. "Comment—Ban what we don't yet understand?" *Smalltimes*, September/October 2002, pp. 33–35.

NatureWorks™ website, http://www.cargilldow.com/s_natureworks.asp, January 5, 2003.

Pater, Ruth H. "Interpenetrating Polymer Network Approach to Tough and Microcrack Resistant High Temperature Polymers," *Technical Support Package for Tech Brief*, Part 2, LAR-14338, LaRC-RP41.

Peiffer, Robert W. "Solid Imaging: New Opportunities for Photopolymers." *Radtech Europe '91 Conference*, Edinburgh, September 29–October 2, 1991.

Peterson, I. "Geometry for Segregating Polymers," *Science News*, Vol. 134, no. 19, p. 151. Ritter, Stephen K. "Green Chemistry Progress Report", *Chemical & Engineering News.* November 25, 2002. pp. 19–23.

Rogers, Craig A. *Intelligent Materials Systems: The Dawn of a New Materials Age.* Center for Intelligent Materials and Structures, Virginia Polytechnic Institute and State University, 1992.

Periodicals

Advanced Materials & Processes

Automotive Engineering

SME Composites in Manufacturing

Module 15

Plastics

After studying this module, you should be able to do the following:

15–1. Analyze the properties of various plastics to determine how specific plastics would compete with other plastics, wood, glass and other ceramics, and metals for specific design applications and cost factors.

15–2. Use tables and handbooks to select plastics, elastomers, and wood to meet specified environments and stresses.

15–3. Calculate the specific strength of materials to make comparisons.

15–4. Recall the characteristics and applications for general-purpose, engineering, and cellular plastics, and determine selection criteria for plastics.

15–5. Explain with the use of diagrams the basic techniques for processing plastics and how types of plastic affect processing selection.

15–6. Use the website addresses denoted by the symbol* to explore sites for new developments and evolving concepts related to plastics.

*The symbol found throughout the book will link you to Internet sites related to the topics being covered. The dynamic nature of the Web brings frequent changes, so some of these sites, while available at the time of writing, may not be up now.

Figure 15-1 Containers made of steel, aluminum, glass, and a variety of plastics. *Left*—PET, HDPE cleanser bottle; *right*—PP (polypropylene) syrup bottle; *middle*—composite "barrier" plastic catsup bottles; *right front*—two more barrier containers for yogurt and fruit juice

As a result of the many advances in materials science and technology, there seems to be no limit to the variety of materials for use in today's products. How do we choose the right "stuff" or material for a given application? In Figure 15-1 we see containers made of steel, aluminum, glass, and a variety of plastics. On the left is a Coke bottle made of polyethylene terephthalate (PET), next to it is a cleanser bottle made of high-density polyethylene (HDPE), and the syrup bottle on the right is polypropylene (PP). Have you seen these same types of products in glass containers? In the middle of the picture is a "barrier" plastic catsup bottle, and to the right is a glass catsup bottle. Is the one made of glass better than the one made of plastic? Shown at the right front are two more barrier containers—the one in front contains yogurt and the box contains fruit juice. How do barrier containers keep milk and juice products from spoiling in the absence of refrigeration? This module will explore those issues and relate them back to the materials cycle—product life-cycle analysis (Figure 1–4).

How much thought do you give to the materials used in products you purchase? Do you consider plastics as cheap materials and not very durable? What advantages can you name for selecting plastics over metals or ceramics or the natural polymer wood? What recycling issues come into play with polymers? Can plastic be part of green chemistry and green manufacturing? By now you should realize that achieving the goal of materials and processes that have limited environmental impact and low cost requires the use of complex problem solving. Although materials technology has made possible a large number of plastics with a wide array of properties, they create problems for green manufacturing in terms of recycling. The auto industry, for example, seeks to reduce the types of plastics in order to save costs and improve the recyclability of automotive plastics. At the websites listed in this module, note how corporations and technical societies tout their claims to being environmentally friendly.

This module will provide you with much information about plastics, thus enabling you to develop competencies related to their processing, structure, properties, and applications. The information and activities will allow you to be a wiser materials consumer and to deal with polymers in your career.

15.1 CLASSIFICATION OF PLASTICS

The Society of Plastics Industry (SPI) defines ***plastics*** as any one of a large and varied group of materials consisting wholly or in part of combinations of carbon with oxygen, hydrogen, nitrogen, and other organic or inorganic elements that while solid in the finished state, at some stage in manufacture are made liquid and thus capable of being formed into various shapes, usually through the application, either singly or together, of heat and pressure. This definition is very broad and rather awkward, but it points out that plastics come from a wide variety of raw materials and processes, have many varied properties, and take such diverse forms that they almost defy definition. Even natural and synthetic rubbers fit this description.

There are several methods used in classifying plastics. The older and perhaps obsolete system of dividing them into thermosetting plastics and thermoplastic plastics groups is still commonly used; however, many plastics, and someday possibly all plastics, can fit into both groupings. Grouping plastics as thermosetting or thermoplastic is useful as a ***processing classification*** because it indicates what types of processes a certain plastic can undergo. To illustrate, a thermoplastic sheet can be used as a finished product or heated to be reshaped. Once a thermoset part has been molded, however, it can no longer be reheated for further shaping. A more descriptive system is classifying plastics as either ***linear*** or ***cross-linked network,*** which is useful to the designer in determining general properties. For example, linear plastics can continually be remolded and usually have low heat resistance, whereas the cross-linked network plastics have greater heat and chemical resistance. Another grouping system involves the nature of the material: rigid, flexible, or elastic. Closely connected with this classification are the uses of the material: general purpose or engineering. ***General-purpose plastics*** would include the bulk of plastics that we encounter daily, such as cellulosic, acrylic, and vinyl. ***Engineering plastics*** are those that substitute for traditional engineering materials such as steel and wood. Some engineering plastics are considered ***structural plastics*** capable of bearing supporting loads. There is much overlap in these classification systems.

Cellular (foam) plastics are designated by the SPI as a plastics group. Cellular plastics are those that have been foamed or have had gas and/or air entrapped in the polymer resins to reduce the density of the finished product (Figure 15-2a). Foaming agents include gases (nitrogen, carbon dioxide, or air) and liquids (chlorinated aliphatic hydrocarbons, alcohols, or ethers). Hollow spheres of glass or resin are also used to create voids. Whipping action introduces air, carbon dioxide is generated from chemical reactions in the resin, and volatile liquids vaporize into gas through ***endothermic*** reactions (heat is absorbed when resin bonds break). Water is produced through ***exothermic*** chemical reactions (heat is given off) and volatilizes to produce cells. Cellular polymers may be open-cell or closed-cell polymers. Closed cells are discrete, separate cells; open cells are interconnected by openings between cells. ***Aerogels,*** which include both polymer and ceramic open-celled structures, have nanoscale matrices composed of interconnecting colloidal-like fibrous chains with diameters of about 10 nm. They are used as refrigerator insulators and for solar windows and have the potential for many similar uses due to their density (a few times that of air) and low thermal conductivity (1% of full-density glass). Practically all thermosetting and thermoplastic resins can be foamed. The Lawrence Livermore National Laboratory reported developing an entirely new class of aerogels, ***organic aerogels,*** which are stiffer and stronger than silica aerogels and better insulators with extremely high thermal resistance (6 times more resistance than fiberglass insulation). They can also be converted to electrical conductors. Figure 15-2c and d shows transmission electron microscope (TEM) images of aerogels.

Silica aerogels website—*http://eande.lbl.gov/ECS/aerogels/satoc.htm*

The advantages of cellular plastics include weight reduction, increased bulk, improved thermal insulation values, greater shock absorption, increased strength-to-weight ratios, and

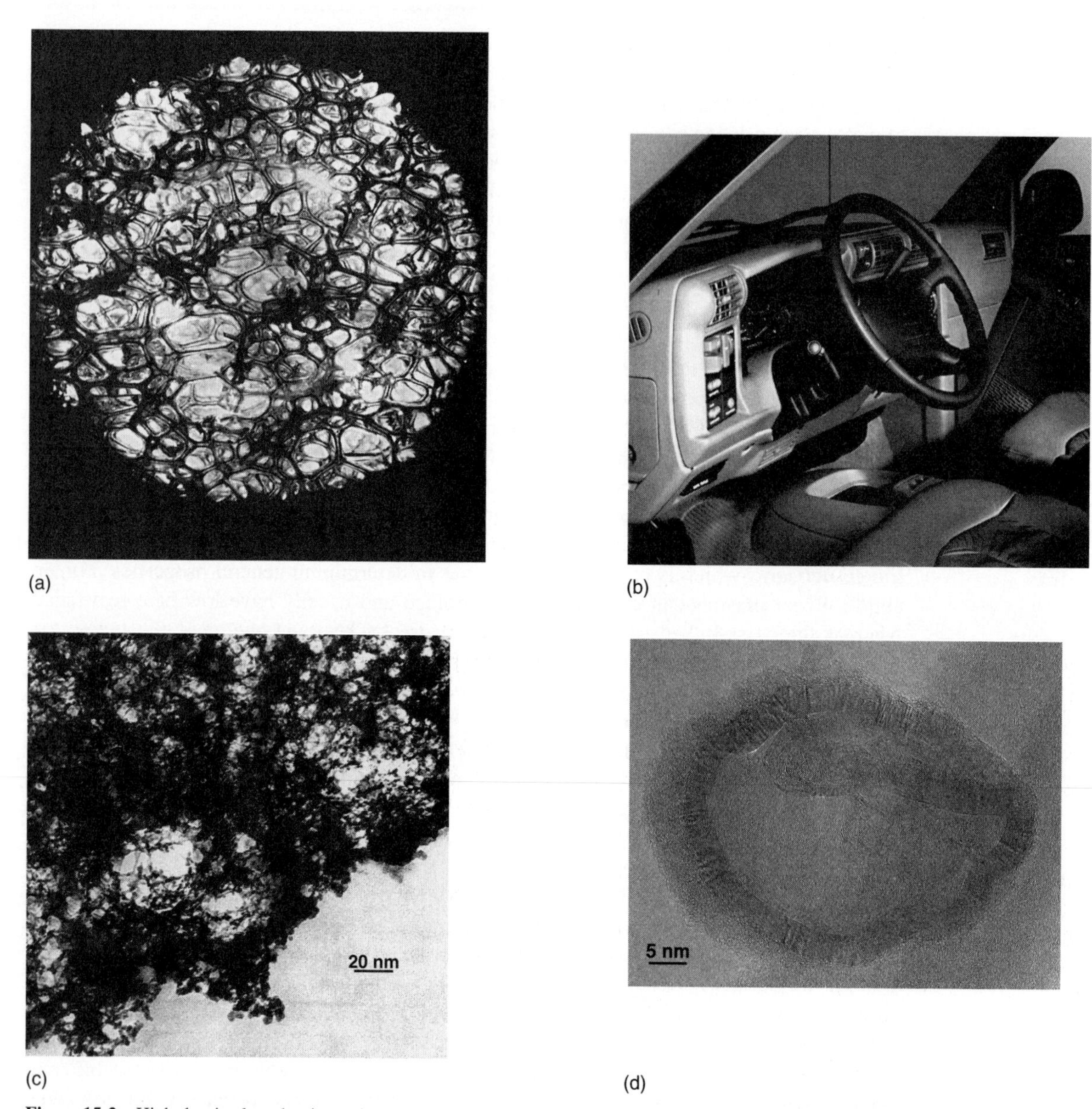

Figure 15-2 High-density, low-density, and energy-absorbing cellular foam polymer. (a) Photomicrograph enlargement of foamed polymer reveals cells created by gas. (b) Cellular plastics and elastomers are found in many vehicles, as seen here for the seats and steering wheel. The heat-resistant, high-impact, dimensionally stable thermoplastic polycarbonate plastics are also seen in the instrument panels for GM pickup trucks and sports utility vehicles. The ML6143H grade of Lexan™ provides ease of processing. NVOCs (nonvolatile organic compounds) and water-based painting are also used (GE Plastics) (c) TEM electron micrograph of a base-catalyzed silica aerogel. (d) TEM showing ring in aerogel impregnated with carbon by the use of chemical vapor infiltration/degradation of acetylene. (E.O. Lawrence Berkley National Lab Silica Aerogel website)

ability to duplicate wood in terms of texture, density, and feel. Cost savings also come with foaming. The addition of glass microspheres or bubbles to ***sheet molding compounds (SMC)****
can reduce weight while improving impact strength because the hollow spheres dissipate energy through shock-absorbing action (dampening). The tiny bubbles, which are from 10 to 200 μm in diameter, also improve mechanical and chemical properties while facilitating processing. ***Cellular plastics*** are available in many densities (1.6 to 96 kg/m^3) as rigid, semirigid, or flexible compounds, and with or without color. The auto industry has accepted many cellular plastics because of their numerous favorable properties, especially their weight-saving property

*ASTM abbreviation.

(Figure 15-2b). The polyurethane (PUR)* foams produced through reaction injection molding (RIM)* and composited with reinforcing fibers (FR)* serve as hoods, seat frames, or doors.

Cellular plastics are everywhere. They include polystyrene foams (Styrofoam™) for building construction, packaging, and appliance insulation; cellular polyethylene for electrical wire insulation; polyamide foams with high heat resistance for aircraft; epoxy foams for flotation devices; cellular silicone for electrical and electronic encapsulation, chemical-resistant fillers, and structural insulation; flexible and rigid poly(vinyl chloride) (PVC)* for cushioning, upholstery, carpet backing, and toys; and foamed acrylics for decorative materials. Syntactic glass-filled foams of polyester are used in sports equipment like composite tennis rackets, bowling balls, floating golf balls, helmets, skateboards, and skis.

Table A-9 compares the various plastic classification systems. When using the table it is important to realize that the data are general and that any given plastic may have considerable grade variations due to fillers, additives, and structure. The table also provides the basic molecular structure, chemical formula, American Society for Testing and Materials (ASTM) abbreviations, some selected trade names, and typical uses of the plastics listed. Note that carbon (C) is the backbone for all those plastics shown except silicone, which has an inorganic silicon (Si) and oxygen (O) backbone. In the polyolefins group, ethylene gas combined with inorganic compounds like sodium (Na), zinc (Zn), magnesium (Mg), or potassium (K) salts bond both covalently and ionically to produce ionomers. Ionomers are resilient, oil resistant, and tough, while having very high transparency.

There are numerous websites that provide current information on plastics. For example, the American Plastics Council has an extensive site, which includes a glossary of terms. Consulting plastic suppliers' sites is a good way to compare types and properties and to supplement the information provided in Appendix Table A–9 and the properties tables in this module. Exploring the suggested website addresses for vendors of plastics will reveal a wide range of classifications, properties, and applications for various plastics. For example, the GE Plastics page** on engineering resins lists the following: ***amorphous materials,*** random entanglements of polymer chains known for very good mechanical properties (strength, stiffness) and dimensional performance, including the GE brands of acrylonitrile–butadiene–styrene (ABS), polycarbonate/acrylonitrile–butadiene–styrene (PC/ABS), acrylic–styrene–acrylonitrile (ASA), polycarbonate (PC), modified polyphenylene ether (PPE), poly(2,6-dimethyl-1, 4-phenylene ether (PPE), expandable polystyrene, and polyetherimide; ***semicrystalline thermoplastics,*** noted for very good electrical properties as well as the ability to withstand both high heat and severe chemical environments, including polybutylene terephthalate (PBT), polyethylene terephthalate (PET), polybutylene terephthalate (PBT), polyamide 6 and 6.6, polyphenylene suphide (PPS); ***amorphous/semicrystalline blends,*** designed to boost specific performance capabilities depending on the blends of polymers and their proportions, including GE's brands of polyamide (PA) reinforced with modified polyphenylene ether (PPE), polycarbonate/polybutylene terephthalate (PC/PBT), and polycarbonate/polyethylene terephthalate (PC/PET). GE's Xenoy™ polyester resin offers good electrical properties, chemical resistance, high-temperature performance, flame retardancy, and fast molding. Typical applications are automotive bumpers and body panels, business equipment housings, cellular phones, and lawn mower decks (see Figure 15-3).

American Plastics Council glossary—*www.plastics.org*
Comparison of polymers including tables and other graphics—
http://www.dupont.com/enggpolymers/europe/techlit/TopTip/toptip/html/e1tip.html
Teflon—*http://www.dupont.com/teflon/*
Gore-Tex—*http://www.gore.com/corp/divisions.html*

*ASTM abbreviation.

**  GE Plastics—*http://www.geplastics.com/resins/devprod/polymer.html*

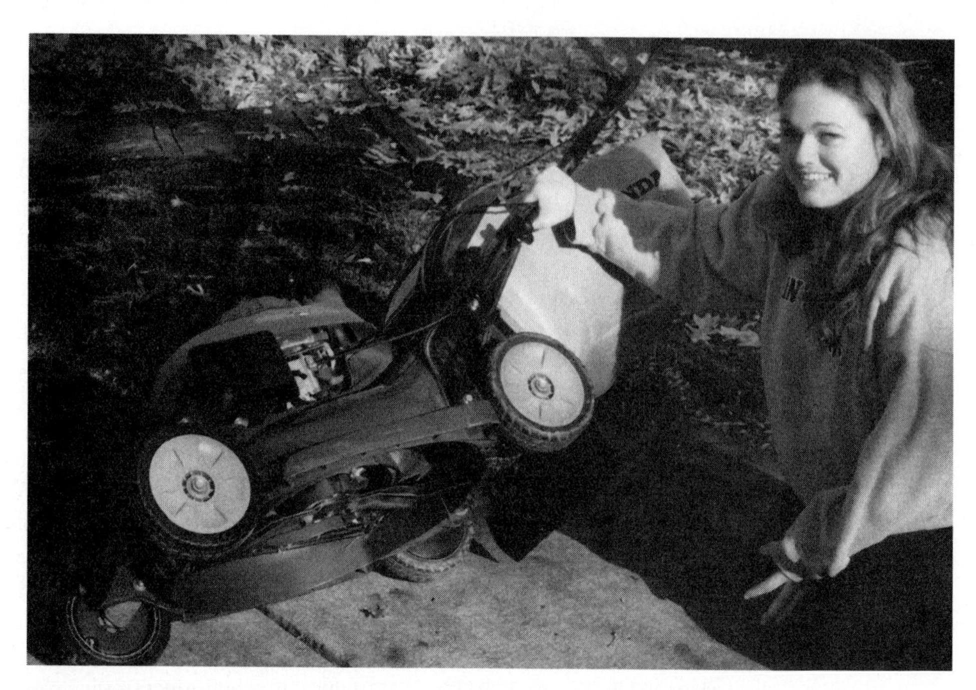

Figure 15-3 Seen on the deck of this lawnmower, GE's Xenoy™ polyester resin helps muffle engine noise and offers good chemical resistance, high-temperature performance, and flame retardancy and allows fast molding. Typical applications are automotive bumpers and body panels, business equipment housings, cellular phones, and, as shown, lawn mower decks.

15.2 PROPERTIES OF PLASTICS

Earlier we saw that the internal structure of polymers dictated their properties. These nearly infinite variations in structure provide a range of properties that continually expands as technological demands prompt new discoveries through the manipulation of polymeric structures. These properties can be categorized as mechanical, chemical, thermal, electrical, and general. The plastic fuel tank seen in Figure 15-4 shows some engineering thermoplastics that require specific properties: resistance to gasoline and various strengths to resist impacts and carry the weight of the fuel while also being lighter than earlier steel fuel tanks. Plastics also provide design flexibility for a complex product.

Additives and fillers so significantly affect the properties of plastics that one must turn to a handbook or manufacturers' specifications for a given plastic to ascertain its properties.

15.2.1 Additives (Modifiers)

Many plastics are true composites because of the many additives that are compounded with the resin to enhance their properties. With the exception of the catalysts (also known as *accelerators* or *initiators*) that are added in minute quantities to start polymerization, additives serve to increase processability, reduce oxidation (corrosion), add color, reduce molecular weight, increase flexibility, reinforce, retard flammability, or increase electrical conductivity. Most polymers will absorb foreign elements that can cause degradation. ***Stabilizers*** are added to prevent this. In addition, ultraviolet radiation and oxygen (ozone) can so alter the chemical bonds of plastics that ***free radicals*** develop within chains. A ***free radical*** is an atom, ion, or molecule with an unpaired electron. Covalents bonds are created only by paired electrons with opposite spins. (The ***Pauli exclusion principle*** states that no two electrons in an atom can have the same four quantum numbers.) The free radicals are segmentations within the polymer that can easily combine with other elements to cause a breakdown of the plastics structure and so affect properties severely. They may form cross-links that will produce a more brittle plastic or disrupt the chain structure, which will limit tensile strength. ***Carbon***

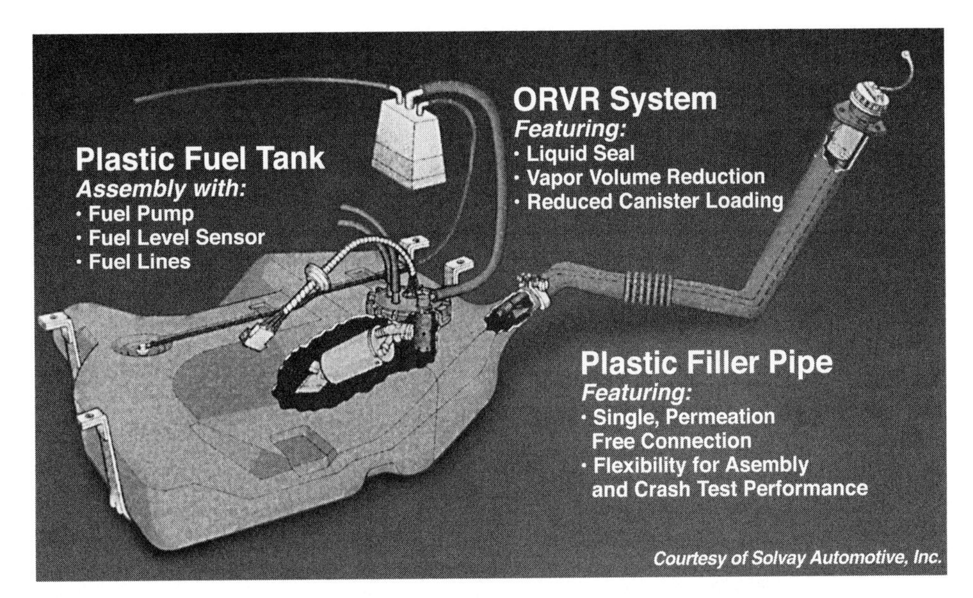

Figure 15-4 Engineering thermoplastics. The complex, integrated, plastic molded fuel system: the tank houses components for a fuel system consisting of six layers of coextruded high-density polyethylene (HDPE) plus two layers of ethylene vinyl alcohol (EVOH). The EVOH acts to provide on-board fueling vapor recovery (ORVR) that emits less than 0.1 gram of hydrocarbons from the gasoline per day. Advantages offered by these systems include reduced hydrocarbon emissions, design flexibility, light weight, and lower cost. (Solvay Automotive, Inc.)

black, an amorphous form of carbon produced through the burning of ***carbonaceous*** materials such as gas or oil in the absence of air, will block out ultraviolet light while strengthening the plastic. Metals such as barium, cadmium, and lead are also used as stabilizers in compounds. Amine and phenol chemicals can serve as sunscreens because they interact with photons.

Colorants consist of both organic ***dyes*** and inorganic (metal-based) pigments that add color to plastics, and some serve dual roles, such as stabilization. Generally, the inorganic pigments will withstand high temperatures without charring or fading. ***Pigments*** disperse rather than dissolve in plastics and reduce the transparency of the material. Pigments also hide flaws such as air bubbles, making it difficult to judge quality. Nearly an infinite range of color, transparency, translucency, and opacity is possible; the Food and Drug Administration (FDA) allows only certain types of colorants for plastics that are in contact with food and drugs.

Plasticizers are additives that increase flexibility, whereas cross-linking agents such as organic peroxides cause hardening and produce free radicals. Plasticizers reduce the attraction (secondary valence bonds) between polymer chains and are normally solvents, such as alcohol. For water and sewage piping poly(vinyl chloride) (PVC) can be very hard and rigid. The addition of a plasticizer such as carbon alphates alcohol produces a resin of low volatility that is used in the slush or dip molding of very flexible products such as tool handle coatings, rain boots, or doll parts.

Flame retardants, which have become more important in the United States as our search for greater product safety broadens, consist of additives such as boron, nitrogen, chlorine, antimony, and phosphorus. They reduce the flammability of plastics by preventing oxygen reactions and by improving ***charring,*** seen when wood burns and leaves a protective residue or ash that slows burning, a process known as ***ablation.*** Flame retardants can cause problems, however, by reducing flexibility, tear resistance, tensile strength, and heat deflection. Fluorocarbons, polyvinyls, and polyimide plastics offer low- or nonburning properties without the use of flame retardants.

Fillers improve plastics by increasing bulk, tensile strength, hardness, abrasion resistance, and rigidity; they also improve electrical and thermal properties, appearance, and chemical resistance while either increasing or decreasing specific gravity. Common fillers are woodflour, quartz, glass spheres, talc, wollastonite ($CaSiO_3$), calcium carbonate ($CaCO_3$), carbon black, clay, and alumina trihydrate (ATH). For example, glass spheres, clay, or calcium carbonate are added to sheet-molded compounds (SMC) or bulk-molded compounds (BMC) to decrease cost while increasing the rigidity of the polyester resins.

Other additives include ***antistatic agents*** and ***coupling agents*** to aid bonding between plastics and inorganic materials in composites, ***foaming agents*** to produce cellular plastics, ***heat stabilizers*** for processing and end product durability, ***lubricants*** to decrease friction and resin melt during processing, ***mold release agents, preservatives*** to retain physical and chemical properties and in some cases prevent growth of bacteria and algae, and ***viscosity depressants*** to reduce viscosity during processing.

15.2.2 Mechanical Properties

As discussed previously (see Figure 14-8a), variations in properties are achieved through variations in molecular weight and crystallinity. In terms of mechanical properties, an increase in crystallinity or density, plus an increase in molecular weight, produces corresponding increases in tensile strength, hardness, creep resistance, and flexural strength, but decreases the impact resistance and percentage of elongation. Additives, fillers, and reinforcers also change mechanical properties to a great extent.

Table 15-1 lists a few selected plastics that reveal the range of tensile strength for engineering plastics; other engineering materials are included for comparison. Because many different varieties and grades are available within a certain plastics group, there is a wide range of tensile strength for plastics of the same name. For example, some flexible cast epoxies have a tensile strength lower than 10 MPa, while a cycloaliphatic epoxy casting resin exceeds 100 MPa. The values in Table 15-1 are for specimens tested at ***standard temperature and pressure (STP).*** Many plastics lose strength rapidly at relatively low temperatures (20°C); others, such as polyimides, can maintain full strength near 500°C over short periods and also withstand cryogenic temperatures around −300°C.

Figure 15-4 illustrates an automative application of engineering thermoplastics. Note the integration of various plastics to form a materials system for the fuel tank. Figure 15-5 shows applications of engineering thermoplastics in building construction. Figure 15-6 is another example of an application of an engineering thermoplastic, polyethersulfone (PES), in this case used for fiber-optic (FO) connectors. Radel®, shown in the figure, a PES blend from Amoco, was selected for FO connectors because it allows precise injection moldings; facilitates concurrent engineering due to its ease of design in that it does not require metal inserts or glass reinforcers; makes tough, chemically resistant, and dimensionally stable components in temperature ranges from 240° to 185°C (240° to 1185°F); and permits use of organic colorant systems to meet standard industry color codes for FO connectors.

Table 15-1 shows that other engineering materials have far greater tensile strengths than the engineering plastics, but the low density of plastics often yields a competitive or better strength-to-weight ratio. Using ***specific strength*** as the comparison factor, we have

$$\text{specific strength}\left(\text{meters}\right) = \frac{\text{tensile strength}}{\text{density}\left(\text{weight}\right)}$$

Polyester pultrusion, fiber-reinforced plastics (FRP) have specific strengths around 35,000 m; stainless steel has a specific strength around 25,000 m. Other FRPs, discussed in Module 8, have even higher specific strength ratios (e.g., up to 90,000 m for epoxy reinforced with carbon fibers). Table 15-6 can be used with the tensile strengths given in Table 15-1 to calculate other specific strengths for comparison.

Illustrative Problem

Use Table 15-1 and Table 15-6 to determine the middle value in the range of tensile strengths and densities of plain-carbon steel and polyimide. Substitute the values into the proceeding formula and determine which has the higher specific strength. Use Table A–7 for conversion of units.

TABLE 15-1 RANGES OF TENSILES STRENGTH AT ROOM TEMPERATURE FOR SELECTED PLASTICS COMPARED TO OTHER MATERIALS[a]

Plastics[b]	Tensile strength [MPa (ksi)]: 0, 10 (1.5), 20 (2.9), 30 (4.4), 40 (5.8), 50 (7.3), 60 (8.7), 70 (10.2), 80 (11.6), 90 (13.1), 100 (14.5), 150 (21.8), 175 (25.4), 200 (29.0), 225 (32.6), 250 (36.3), 500 (72.5), 750 (108.8), 1000 (145.0)
ABS, e	XXXXXX
Acetals, e	XXXX
Acrylics (PMMA)	XXXXX
Alkyds	XXXXXXXXX
Amino	XXXXX
Cellulose acetate (CA)	XXXXXX
Cellulose butyrate (CB)	XXXXXXX
Diethylene glycol bisallyl carbonate	XXXX
Epoxy (EP)	XXXXXXXXXXXXXXXXXXX
Fluorocarbons (TFE), e	XXXXX
Nylon, e	XXXXXXXX
Phenolic (PF)	XXXXXX
Phenylene oxide (PPO)	XXX
Poly (amide/imide), e	XXXXX
Polycarbonate (PC), e	XXX
Polyester, TS	XXXXXXXXXXX
Polyester, TP	X
Polyethylene (LDPE)	XX
Polyethylene (HDPE), e	XX
Polyethylene (UHMWPE), e	XXXXX
Polyimide, e	XXXXXXXX
Polypropylene (PP), e	XXXX
Polystyrene (PS)	XXXXXXXXXX
Polysulfone (PES), e	XXX
Polyurethane (PUR), e	XXXXXXXXXXXXX
Silicones (SI)	XXXXXXXXX
Vinyl (PVC)	XXXXX
Vinyl (PVDC)	XXXXXX[c] XXXXXXXXXXXXX[d]
Comparison Materials	
Alumina	XXXXX
Aluminum alloys	XXXXXXXXXXXXXXXXXXXXXX
Brass	XXXX
Cast iron	XXXXXXXXXXXXXXXX→
Fiber-reinforced plastic	XXXXXXXXXXXXXXXXXXXXXXXXXXXX→
Glass	XXXXXXXXXXXXXXXXXXXX E & S fibers →
Stainless steel	XXXX →
Steel—plain carbon	XXXXX →
Wood	X

[a]Ranges reflect the varieties of the named plastic. Does include certain filled varieties but not fiber-reinforced plastics (FRP). Ranges for other materials are general.

[b]e, Engineering plastic; TS, thermoset; TP, thermoplastic.

[c]Unoriented.

[d]Oriented.

Solution

$$\text{specific tensile strength} = \frac{\text{tensile strength}}{\text{density (weight)}}$$

From Table 15-1,

Plain-carbon steel	680 MPa
Polyimide	76 MPa

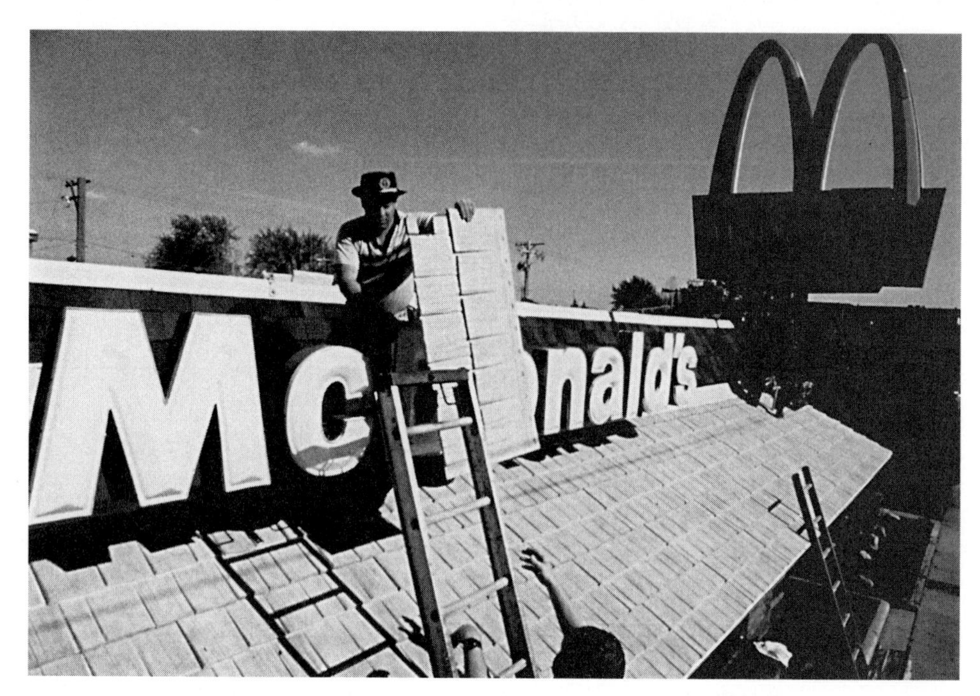

Figure 15-5 Polyphenylene oxide (PPO), Noryl®, offers many advantages over traditional roofing materials, including light weight, easy installation, very good impact resistance, weatherability, walkability, and fire performance. Because of its easy moldability, it can be molded to resemble cedar shake, slate, and barrel tile for the construction industry. (GE Plastics)

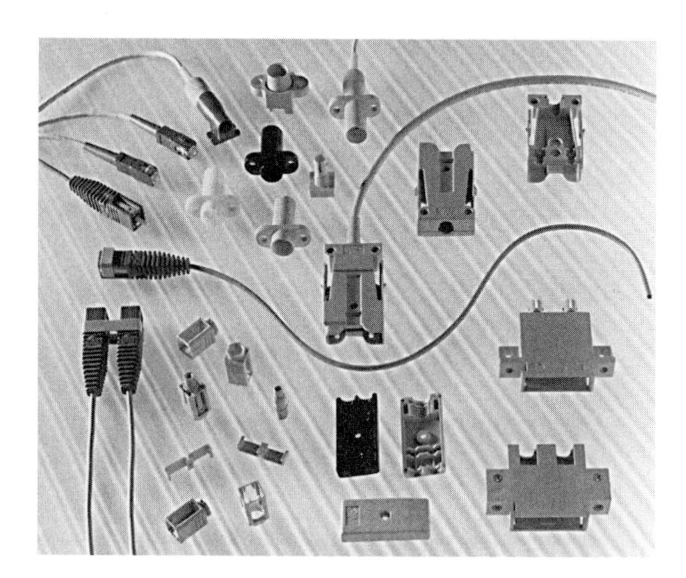

Figure 15-6 Polyethersulfone engineering thermoplastic allows for a wide range of injection-molded fiber-optic connectors and components. Intricate parts can be molded to extremely close tolerances to meet the critical alignment requirements of FO cables. (*Advanced Materials and Processes*)

From Table 15-6,

Plain-carbon steel	8200 kgf/m^3
Polyimide	1260 kgf/m^3

Substituting in the formula, we find that

$$\text{specific tensile strength} = \frac{680 \text{ MPa}}{8200 \text{ kgf/m}^3}$$

Reduce units to meters (m) by using the following unit conversions from Table A–7: 1 kgf = 9.807N, 1 Pa = 1 N/m^2, and M = 10^6. The specific tensile strength for plain-carbon steel is

$$\text{specific tensile strength} = \frac{680\ \text{MPa}}{8200\ \text{kgf/m}^3}\left(\frac{1\ \text{N/m}^2}{1\ \text{Pa}}\right)\left(\frac{10^6}{\text{M}}\right)\left(\frac{1\ \text{kgf}}{9.807\ \text{N}}\right) = 8456\ \text{m}$$

The specific tensile strength for polyimide using the same conversion units is

$$\text{specific tensile strength} = \frac{76\ \text{MPa}}{1260\ \text{kgf/m}^3}\left(\frac{1\ \text{N/m}^2}{1\ \text{Pa}}\right)\left(\frac{10^6}{\text{M}}\right)\left(\frac{1\ \text{kgf}}{9.807\ \text{N}}\right) = 6150\ \text{m}$$

Compared to steel and most other engineering materials, plastics are much softer. Table 4–1 showed an approximate comparison of the hardness of selected materials using various hardness scales. Hardness comparisons of most common plastics are made with the Shore Durometer A and D method and Rockwell M (HRM) tests. The problem with hardness tests of plastics is that they do not closely correlate to wear or abrasion resistance as do most other materials: Polystyrene has a Rockwell M value of 72 but scratches easily; diethylene glycol bisallyl carbonate, marketed as CR39® (allyl diglycol carbonate) by PPG Industries, has a Rockwell M value of 95 to 100 but has abrasion resistance approximating that of glass. This makes it a competitor with glass because it has optical properties approaching those of glass. Methyl methacrylate, under trade names such as Plexiglas and Lucite, has an average hardness of 93 to 98 HRM but abrasion resistance 30 to 40 times less than CR39 using a modified Taber test. ***Abrasion resistance*** and ***wear resistance*** are measured by Taber and other tests that determine the ***percentage of haze*** (loss of clarity) due to marking or ***percentage of material lost*** through rubbing with abrasives such as aluminum oxide. Table 15-2 shows the hardnesses and the results of Taber tests on several plastics.

To enhance plastic substrate properties and to achieve superior surface finishes, films may be applied over injection-molded parts as an in-mold decoration process. Sollx™ film yields a high-gloss, class A, paint-free exterior, as seen in Figure 15-7. It performs better than paint because of its improved weatherability, "scratch and mar" resistance, and chemical resistance. Sollx is coextruded with a pigmented color layer to form a clear, protective polymer film over the color layer. The plastic film is derived from polyester carbonates based on resorcinol arylates. The in-mold decoration process is well known and is used extensively to manufacture prefinished plastic trim parts such as bezels, appliqués, and mirror housings. This process is currently being extended to the production of large parts, such as exterior automotive body panels. Ongoing R&D with Sollx film is aimed at developing a weatherable polymer system and eliminating very expensive paint shops and painting processes for automotive exteriors and interiors.

TABLE 15-2 TABER TEST RESULTS

Plastic	Hardness	Abrasion resistance (mg loss per 1000 cycles)
Nylon 6/6	114 HRR	5
Acetals	95 HRM	16
ABS	100 HRR	84
Polysulfone	120 HRR	20
Polyimides	98 HRM	0.08

Figure 15-7 The plastic film Sollx™ was chosen for the fender of the Segway Human Transporter. This precommercial polymeric film from GE Structured Products is applied using an in-mold decoration process over an injection-molded Xenoy® resin substrate to yield a high-gloss, class A, paint-free exterior trim component in a champagne color. Sollx film offers better performance than paint in terms of improved weatherability, "scratch and mar" resistance, and chemical resistance. (GE Plastics)

TABLE 15-3 COMPARATIVE IMPACT (IZOD) STRENGTH OF SELECTED PLASTICS

	Notched impact strength (ft · lb/in.)										
Plastic[a]	0	2	4	6	8	10	12	14	16	18	20
ABS, e		XXXXXX									
Acetals, e		X									
Acrylics (PMMA)	XX										
Alkyds		XXX									
Aminos—melamines	X										
Cellulose acetate (CA)		XXXXXXX									
Epoxy (EP)	X										
Fluorocarbon (TFE), e		XXX									
Nylon 6/6, e	XX										
Phenolic (PF)	X										
Phenylene oxide (PPO), e			X								
Poly (amide/amide), e	XXX										
Polycarbonate (PC)							XXXXXX				
Polyester (PBT)	X										
Polyester (PET)	X										
Polyethylene (LPDE)										XXX→	
Polyethylene (HDPE)	XXXXXXXXXXXXXXXXXXXXXXXXXXXXXXXX										
Polyimide (GRP), e										XXX	
Polypropylene (PP), e	XXX										
Polystyrene (PS)	X										
Polysulfone (PES), e	XX										
Polyurethane (PUR), e				XXXXXXXX							
Silicones (SI)	X										
Vinyl (PVC)	XXXXXXXXXXXXXXXXXXXXXXXXXXXXXXX										

[a]e, Engineering plastic.

Thermoplastics generally have better ***toughness*** or ***impact resistance*** than thermosets. As seen in Table 15-3, polycarbonate is the toughest of the transparent rigid plastics, at 12 to 14 pounds per inch, and far exceeds polystyrene (10.4 ft·lb/in.) and acrylics (0.8 to 1.6 ft·lb/in.). Polyurethane, with an impact strength of 5 to 8 ft·lb/in., has gained acceptance as automobile bumpers and hoods. Both polyethylene and poly(vinyl chloride) (PVC) have very broad ranges of impact strength, as you may know from the thin-walled, tough, polyethylene milk jugs and PVC piping that have begun to replace much of the copper and cast-iron pipe used for water and sewage. The values given in Table 15-3 are notched specimens. Notching of most plastics severely limits their toughness, so designs should avoid sharp corners or conditions in which parts subjected to impact might develop scratches or cuts. Instead of the notched Izod test, which may not give the most reliable results, some designers prefer data obtained from impacts made with a falling weight on sheet plastic (ASTM D3029), film (ASTM D1709), and pipes and fittings (ASTM D2444). Figure 15-8 reveals that certain plastics compete well with metals when toughness is required.

15.2.3 Viscoelasticity

Viscoelasticity is a property unique to polymers (plastics, elastomers, adhesives, and wood) and is not found in metals or ceramics. Because of the viscoelastic property of plastics, engineering technologists must become involved in ***rheology*** (the study of flow and deformation of matter) to determine why plastics will react to mechanical stresses in a different manner than metals. The property ***viscoelasticity*** has two aspects: viscosity and elasticity. ***Viscosity*** refers to the nature of a liquid's resistance to flow. Motor oil used in an automobile engine is rated by its viscosity. Heavy oil, such as 40 SAE (Society of Automotive Engineers), is thick as syrup and is used in hot weather, while lighter oil (10 SAE) that flows much like water is used in temperatures below zero. Like oil, plastics become less viscous or flow more easily with increased temperatures. This makes them easier to process but also means that they lose strength with heat.

Elasticity (discussed in Module 4) refers to the ability of a material to return to its original size and shape once a load is removed. With plastics, the viscosity and elasticity both come

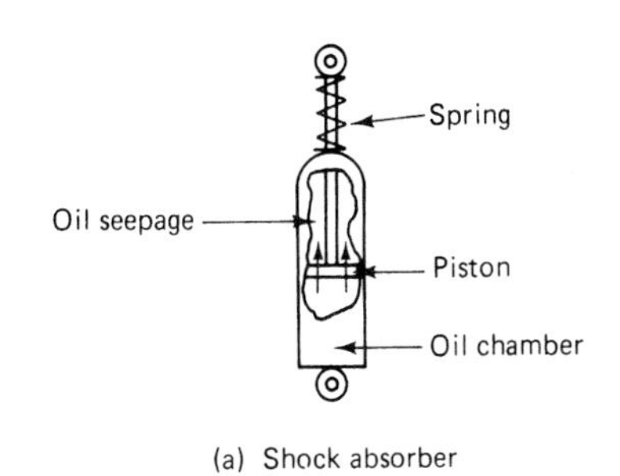

(a) Shock absorber

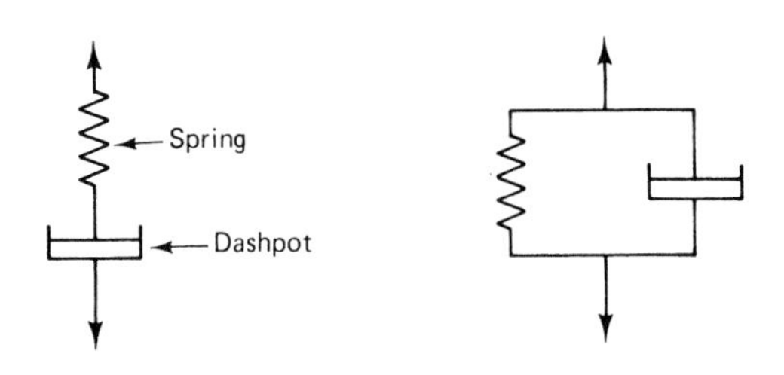

Figure 15-8 Viscoelasticity. (b) Maxwell series model (c) Voigt parallel model

into play, so a sudden impact on the plastic will not result in immediate strain or immediate and full recovery when the stress is removed. Rather, there is a viscous flow internally as the force is absorbed through shearing of molecular bonds, much the way that a shock absorber works in an automobile. The shock absorber is a combination of ***dashpot*** (force gradually released through a slow bleeding of fluid) and spring. Even after all load is removed, the deformation is not fully recovered as with a metal that experiences elastic loading. The effect of heat on viscoelastic properties is discussed under thermal properties in the next section. Figure 15-9 illustrates the mechanical models (Maxwell and Voigt) that simulate viscoelasticity.

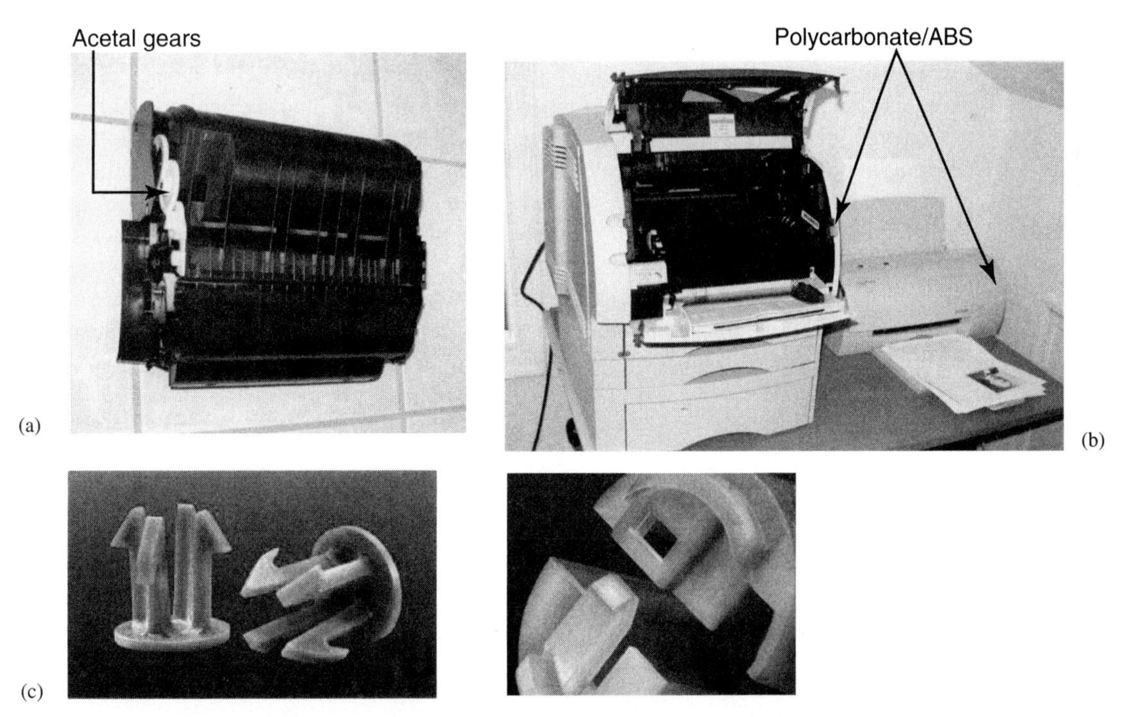

Figure 15-9 Engineering plastics provide toughness for computer applications and appliances. (a) Tough gears used on laser printer toner cartridge made of du Pont's Derlin (acetal) also provide excellent tribological properties (resistance to wear and friction), light weight, corrosion resistance, and quiet operation. (b) Examination of internal and external parts of a computer and peripherals reveals very few nonplastic structural parts; the figure shows cases made of polycarbonate/ABS copolymers. (c) Snap-fit plastic parts made of various engineering plastics eliminate fasteners such as nuts and bolts and enhance robotic assembly. (Adapted from http://www. honeywell-plastics.com/ed/snapfit/beam2.html)

Viscoelastic properties are determined by measuring the stress relaxation (Section 15.2.4) of a material subjected to a constant strain at different temperatures as specified by ASTM D2991. The reduction in stress is measured until it reaches equilibrium. Materials that offer the best strength retention and have good viscoelastic properties are needed for many applications where, for example, a force is required to be maintained for contact purposes or for clip strength. Figure 15-9 exemplifies such an application for the electronics industry.

15.2.4 Thermal Properties

Many plastics lose their strength at relatively low temperatures. Continuous-service temperature comparisons of plastics reveal that most common plastics can endure temperatures of no more than 150°C when under low or no stress. Slight increases in stress would reduce the continuous-service temperature even more. Newer plastics, however, such as the polyimide thermosetting resins, can resist intermittent heat up to 500°C with low stress and up to 250°C for thousands of hours. Graphite-reinforced polyimides can withstand flexural stress to nearly 69,000 Pa at 250°C. Plastic composites have even replaced aluminum pistons in some race cars because the aluminum, with high thermal conductivity, acts as a heat sink and begins to soften at around 150°C. Plastic composites of polyimide and epoxy with glass and graphite reinforcement withstand temperatures approaching 290°C and tremendous mechanical stress, and they offer weight savings over metals.

Glass-transition temperatures or ***glass point*** (T_g) is the point at which polymers act like glass or become viscous liquids (see Figure 15-10). The glass transition is a reversible change that occurs when a resin polymer is heated to a certain temperature (T_g), resulting in a sudden change or transition from a rigid polymer to a flexible, rubbery material or a viscous liquid. Such a material suffers a major loss in strength when subjected to a temperature near its T_g value, leading to catastrophic failure even when the service (use) temperature exceeds the T_g value for only a short time.

In cooling, the thermal mixing of the atoms and molecular chains slows and the volume of the plastic decreases because they pack closer together. In packing, many polymers do not form crystals as do metals; instead, they solidify into an amorphous, glassy structure. Of course, highly crystalline polymers like high-density polyethylene (HDPE) do not follow this pattern. Exact melting points are determined optically in crystalline and semicrystalline plastics when ***birefringence*** (double refraction) is lost due to the change of structure from the crystalline to the glassy state. The degree of crystallinity of the plastics controls T_g. Cooling to the glass point results in noncrystalline polymers becoming brittle. This occurs at or above room temperature for certain plastics, for example, at about 100°C for polystyrene and acrylic. In crystalline plastics with low T_g, such as polypropylene at −10°C, flexibility and impact strength are maintained. In reheating, an amorphous thermoplastic passes from the brittle glassy structure through its T_g point into a tough, rubbery form, progressing from there to a softer and more pliable stage and into a viscous liquid upon reaching the melting point (T_m).

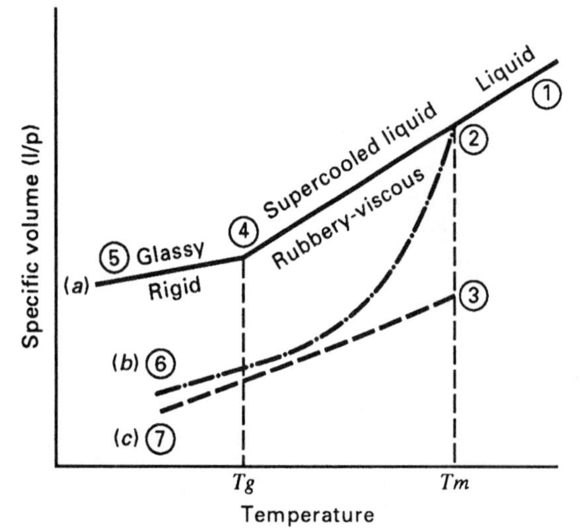

Figure 15-10 Specific volume versus temperature for polymers. (a) Linear amorphous polymer [path (a) 1, 2, 4, 5]. (b) Partially crystalline polymer [path (b) 1, 2, 6]. (c) Crystalline polymer [path (c) 1, 2, 3, 7]. (d) Path through points 1, 2, 3, 7 is a theoretical path approaching a crystalline solid such as a metal. (e) T_g (glass temperature) and T_m (melting temperature) for a crystalline polymer.

T_g provides a way to determine which plastics are suitable for applications that depend on ***pliability*** in a plastic material. Structural applications require those plastics with lower temperatures than the glass transition temperature. A rule-of-thumb for structural applications is that plastics materials should not be used at temperatures above three-fourths of T_g. From our earlier discussion of the mechanical property of viscoelasticity one can see how T_g will affect this property. A linear amorphous plastic will be quite brittle and not exhibit viscous flow below its T_g; it will be rigid and elastic. As the degree of crystallinity increases in plastics, they become more viscous and less capable of recovering from deformation under load, because they possess less elasticity.

Time becomes an important factor in the loading of plastics. A stressed linear amorphous polymer at temperatures above T_g will, over sufficient time, begin to deform without additional loading or heat as the molecular chains become untangled through viscous flow; this is ***stress relaxation.*** A network thermoset at relatively low temperatures under constant stress over a long time will begin to deform and continue deforming with progressively less stress; this is ***creep.*** The softening temperature of polymers can also be increased through an increase in average molecular weight or polymer chain lengths. Thermoplastics are far more prone to creep than thermosets.

Thermal conductivity is low in most plastics, which makes them valuable as thermal insulators. In the cellular or foamed state, air cavities, which do not carry heat or cold, improve the insulating properties even further. However, the rate of *thermal expansion* is quite high for

TABLE 15-4 COMPARISONS OF THERMAL PROPERTIES FOR SELECTED PLASTICS AND OTHER MATERIALS[a]

	10^{-6} m/m/K					
Coefficient of thermal expansion		60	120	180	240	300
Plastics						
ABS,e	X					
Acetals (POM), e			X			
Acrylics (PMMA)		XXXXXXXXX				
Cellulose (CA)			XXXXXXXXXXXX			
Epoxies (EP)		XXXXXX				
Fluorocarbons (PTFE), e			X			
Nylon			X			
Phenolic (PF)		XXX				
Polycarbonate (PC)		XXXXXXXXX				
Polyester (PBT)			XX			
Polyethylene (LPDE)				XXXXXX		
Polyethylene (HDPE), e				XXXXXXXXXXXXXXXXXXXXX		
Polyimide		X				
Polypropylene (PP)		XXX				
Polystyrene (SAN)		X				
Polysulfone, e		X				
Polyurethane (PUR)	X					
Silicone (SI), e		XXXXXXX				
Vinyls (PVC)	XX					
Alumina	X . . .		. . .			
Aluminum	XXX . . .			. . . .		
Cast iron	XX . . .		. .			
Copper	XX . . .				. . . .	
Glass	XX					
Rubber	X					
Silver		X				
Steel	X					
Stainless steel	X					
Wood	I . . .		. . . .			
Zinc		X				
	W/(m · K)					
Thermal conductivity	0 . . . . . . 10 . . . . . . 50 . . . . . . 100 . . . . . . 150 . . . . . . 200 . . . . . . 300. 400					

[a]e, Engineering plastic.

most plastics, generally 10 times as much as metals. Table 15-4 provides a comparison of the rate of thermal conductivity and coefficient of thermal expansion for selected plastics and compares them with other materials.

Plastic films made of select plastics for packaging can withstand heat from conventional cooking ovens; most are not affected by microwaves but some melt from the heat of the food as it is cooked. A polyvinylidene chloride (PVDC) product such as ***Saran Wrap***™ has oxygen protection like glass and will not melt with ***microwave cooking***; crystallized polyester (CPET) will withstand conventional oven temperatures, to 425°F, and microwaves. Polyethylene (PE) products such as Handiwrap and Glad Wrap do not provide the oxygen protection of PVDC or CPET. They can be used in microwave cooking if the fat and sugar content of the food is low enough so that they do not melt the plastic. Polystyrene (PS) foam, often used to package food, is safe for heating liquids such as water, tea, and coffee but will melt if used to heat food such as hamburgers or french fries. As discussed earlier, R&D into new films such as the thermoplastic Sollx film may pave the way for coatings less expensive and more durable than paints.

The range of *flammability* for plastics is quite wide. For plastics that sustain combustion, a burning rate test (ASTM D635), which measures the propagation of flame (in./min) on a specimen of a given size permits a measure of comparability. For instance, phenolics, polyimides, and fluorocarbons without fire-retardant additives are considered nonburning, whereas the cellulosics are highly flammable. Certain plastics, such as polyurethane and vinyls, give off highly toxic fumes when burned. The vinyl siding used on housing ignites above 370°C, whereas wood siding ignites at about 270°C. Generally, thermoplastics are more flammable than thermosets. Thermoplastics can be ground into chips and melted for recycling. Although thermosets have the potential for recycling through the use of chemical solvents, the cost may not be justified; however, thermosets can be ground and used as composite filler. The high thermal energy possible from thermosets, nearly 3 MJ/kg (megajoules per kilogram), has value as an energy source in which scraps are burned to generate heat for steam-turbine generators in the production of electrical energy. As with wood, plastics develop a char layer that serves as an ablative or protective shield, which insulates the unexposed area. Nylon and phenolic resins were used on earlier NASA spacecraft as ablative heat shields for reentry into the earth's atmosphere. Although the polymers did burn when exposed to temperatures over 6500°C for several seconds, the thermal insulating properties of the charred ablative shield protected the spacecraft.

15.2.5 Chemical Properties

Plastics do not corrode like metals do, but they are subject to deterioration and chemical attack. Whereas the corrosion of metals is determined by weight loss, a thermoplastic's deterioration is measured by weight gain, because the attacking chemical combines with the plastic. The result is usually discoloration, swelling, or crazing (fine cracks) with corresponding loss of tensile, impact, and flexural strengths. Polymer-matrix composites (PMCs) have undergone considerable development to arrive at the point where they can be used with confidence in the aerospace industry. Wide use of PMCs on the Boeing 777 (Figure 15-11) attests to the corrosion resistance and many other favorable properties of these polymer-based materials.

As with mechanical properties, ***chemical resistance*** will decrease in plastics as temperatures increase. Table 15-5 provides data for plastics tested at room temperature on their degree of resistance to weak acid, organic solvents, and water absorption. Some common strong acids are hydrochloric and nitric, and weaker acids are lactic, boric, and citric. Common organic solvents include gasoline, alcohol, and acetone. The ***weatherability*** comparison involves several chemical stresses, including heat, moisture, ultraviolet light, and chemicals in the air such as ozone (allotropic form of oxygen) and hydrochloric acid (see Figure 15-12). Notice that the fluorocarbons, due to their inertness, have great resistance to weak acids, organic solvents, water absorption, and weathering. Acetal homopolymer resins have great resistance to acetone at room temperature, but at elevated temperatures (65°C) they are unacceptable for service and do not perform well when in environments with strong acids. The copolymerization of acetal resins greatly improves their resistance to inorganic liquids at elevated temperatures, so they are utilized as plumbing valves and pumps and other plumbing

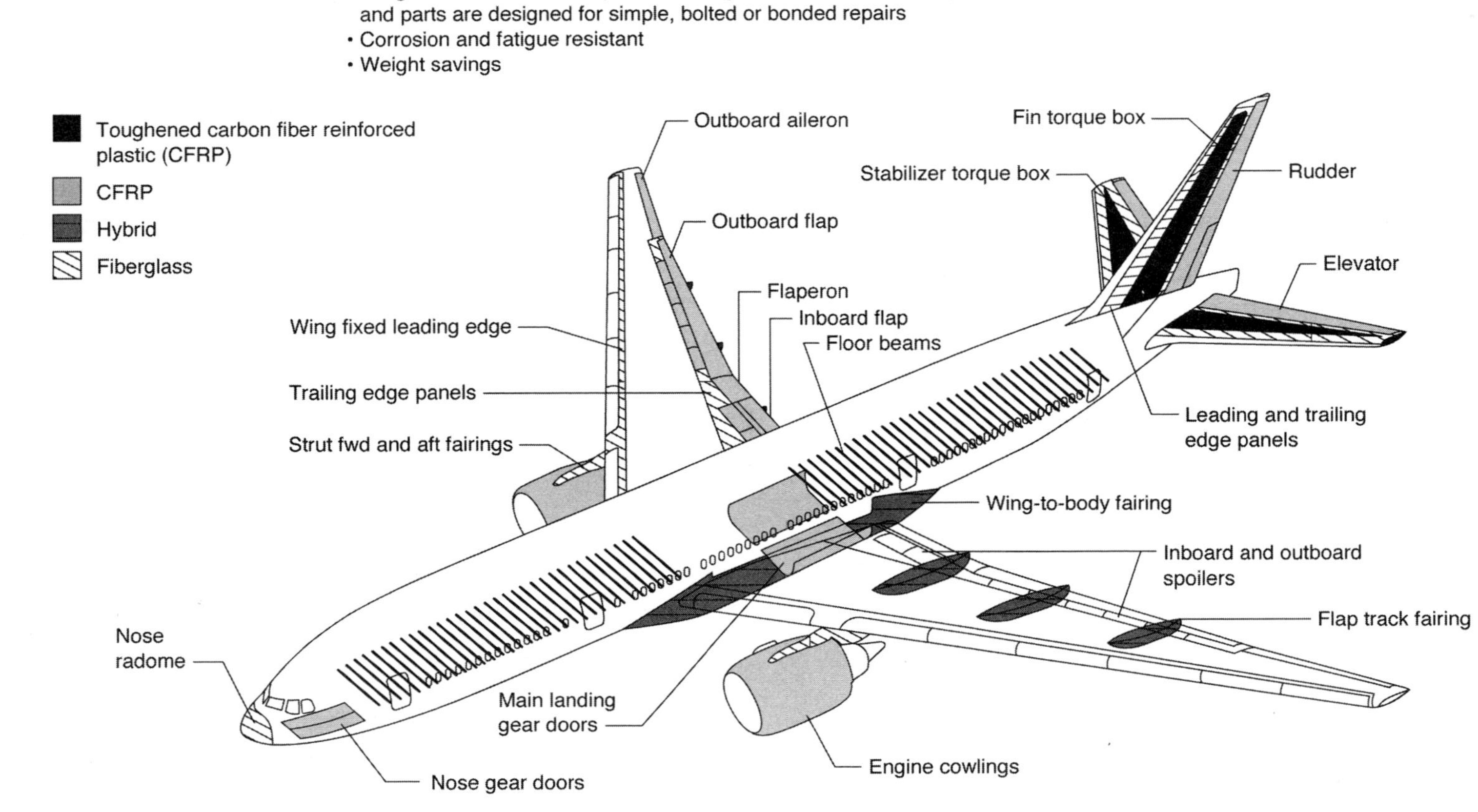

Figure 15-11 Boeing's 777 lightweight composite structure makes extensive use of polymer matrix composites (PMCs). After thorough testing of PMCs in military and research aircraft, polymers proved reliable for commercial aircraft. The "paperless" project utilized concurrent engineering (CE), in which design engineers and manufacturing engineers worked in design–build teams. In addition to advanced metals, the 777 structure used graphite, toughened graphite, fiberglass, and hybrid PMC composites, which were selected for corrosion and fatigue resistance among other properties. (Courtesy of the Boeing Company)

TABLE 15-5 COMPARATIVE CHEMICAL PROPERTIES OF SELECTED PLASTICS AND OTHER MATERIALS[a]

Plastics[b]	Weak acid resistance	Organic solvent resistance	Water absorption resistance	Weatherability
	Little Fair Great	Little Fair Great	Little Fair Great	Little Fair Great
ABS, e	XXXXXXX	XXXXXX	XXXX	XXXXX
Acetals (POM), e	XXXX	XXXXXXXXX	XXXX	XXXXXXXX
Acrylics (PMMA)	XXXXXXXXX	X	XXXXXX	XXXXXXXXX
Alkyds	X	X	XXXXXXXX	XXXXXXXX
Aminos—melamides	XXXX	XXXXXXXXX	XXXXXXXXX	
Cellulose acetate (CA)	XXXXXX	XXXXXX	X	XXXXXX
Cellulose butyrate (CAB)	XXXX	XXXXX	XX	XX
Epoxy (EP)	XXXXXXXXX	XXXXX	XXXXXXXX	XXXXXXXXX
Fluorocarbons (PTFE), e	XXXXXXXXX	XXXXXXXXX	XXXXXXXXX	XXXXXXXXX
Nylon 6/6, e	XXXX	XXXX	X	XXXX
Phenolic (PF)	XXX	XXX	XXXX	XXXX
Phenylene oxide (PPO), e	XXXXXXXXX	XXXXX	XXXXXXXXX	XXXXXXXXX
Poly(amide/amide), e	XXXXXXX	XXXXXXX	XXXX	XXXXX
Polycarbonate (PC)	X	X	XXXXXXXXX	XXXXX
Polyester, TP	XXXXXXXX	XXXX	XXXXXXXXX	XXXXXXXXX
Polyester, TS	XXXX	XXXX	XXXXXX	XXXXXXXXX
Polyethylene (LDPE)	XXXXXXXXX	XXXXX	XXXXXXXX	XXXXXXXXX
Polyethylene (HDPE), e	XXXXXXXXX	XXXXX	XXXXXXXXX	XXXXXXXXX
Polyethylene (UHMWPE), e	XXXXXXXXX	XXXXX	XXXXXXXXX	XXXXXXXXX
Polyimide, e	XXXXX	XXXXX	XXXXXXXXX	XXXXXXXXX
Polypropylene (PP), e	XXXXXXXXX	XXXXXXXX	XXXXXXXXX	XXXXX
Polystyrene (PS)	XXXXXXX		XXXXXXXX	
Polysulfone, e	XXXXXXXX	XXXXXXXXX	XXXXXXX	XXXXX
Polyurethane (PUR), e	XXXX	XXXXXXXXX	XXXXXXX	XXXXXX
Silicones (SI)	XXXX	X	XXXXXX	XXXXXXXX
Vinyl (PVC)	XXXXXXXXX	XXXXXX	XXXXXXXXX	XXXXXXXXX
Vinyl (PVDC)	XXXXXXXXX	XXXXXX	XXXXXXXX	XXXXXXXX
Comparison materials				
Alumina	XXXXXXXX	XXXXXXXXX	XXXXXXXXX	XXXXXXXXX
Aluminum	XXXXXXXX	XXXXXX	XXXXXXXX	XXXXXXXX
Brass	XXX	XXXXXXXXX	XXXXXXXXX	XXXXXXXXX
Cast iron	X	X	XXXXXXXX	XXXXXXX
Fiber-reinforced plastics vary with plastic matrix from little to great.				
Glass	XXXXXXXXX	XXXXXXXXX	XXXXXXXX	XXXXXXX
Stainless steel	XXXXXXXXX	XXXXX	XXXXXXX	XXXXXXXXX
Steel—plain carbon	X	X	X	
Wood	XXXXXXXXX	XXXXXXXXX	X	XXXX

[a]Exposures at room temperatures except weatherability.

[b]e, Engineering plastic; TS, thermoset; TP, thermoplastic.

products. High-density polyethylene (HDPE) has great resistance to acids, water absorption, and weathering, which, coupled with its low cost, makes it a good candidate for numerous packing applications like the blow-molded bottles used for home and industry; it also serves as electrical wire insulation, oil storage tanks, and bodies for recreational vehicles. As seen in Table 15-5, acrylics (Lucite and Plexiglas) have great resistance to ***weathering,*** which is another property that makes them better for exterior windows in addition to their greater impact resistance over that of soda-lime glass; they also transmit light (92%) as well as fine optical glass. Certain plastics with little ***water-absorption resistance,*** such as nylon 6/6, are dried before processing so that the absorbed moisture will not cause corrosion of machinery. Moisture absorption causes swelling in plastics, which creates problems in holding dimensional accuracy. When high accuracy must be maintained, plastic of low moisture absorption

(a)

(b)

Figure 15-12 (a) The chem suit used when entering chemical tanks relies on optical-grade CR39® monomer, which resists solvents, acids, scratching, and impact. (PPG Industries.) (b) Plastics under the hood of a car such as battery cases, cowling, ducts, and reservoirs are subjected to heat and chemicals. Hotter-running engines require newer formulations of thermoplastics, such as nylon electrical connectors.

is required. Polyester used as the resin in fiberglass has great water-absorption resistance and serves well in marine applications such as boat hulls and surfboards.

Aseptic packages use polymer (e.g., paper and polyethylene) and metal (usually aluminum) layers or laminates to protect the package contents from contamination. Milk and juices do not require refrigeration when stored in aseptic packages, as shown in Figure 15-13. ***Barrier containers*** such as the catsup bottle seen on the left in Figure 15-13a employ polymers to prolong shelf life of the contents. The five layers of polymers, diagrammed in Figure 15-13b, consist of a copolymer of polyethylene/vinyl alcohol to form an oxygen barrier, polypropylene for a moisture barrier, polycarbonate to give strength as the hot contents fill the bottle, plus two adhesive layers to bond the other three layers. Toothpaste tubes use a similar construction and include an aluminum layer to enable the tube to be rolled, instead of the earlier tubes that used lead, which had to be eliminated due to its toxicity.

15.2.6 Density

The density (kilograms per cubic meter) of plastic is generally lower than that of other engineering and general-purpose materials. Table 15-6 shows that only magnesium and some wood have as low mass per unit volume as plastics. Even glass-reinforced plastics (GRPs) are lighter than aluminum and steel, but their strength approaches or, in some instances, equals that of these popular engineering metals. CR39® (allyl diglycol carbonate) (Figure 15-12), developed by Pittsburgh Plate Glass for competition with optical glass, has a density about one-half that of glass. This plastic is a practical substitute for glass in larger, stylish eyeglasses and for industrial safety glasses because it combines light weight with impact resistance and is superior to acrylics in wear resistance.

15.2.7 Specific Gravity

Specific gravity is the ratio of the mass of a measured volume of any material to the mass of an equal volume of water at a standard temperature. This ratio is often used for weight comparisons. Specific gravity ratios for plastics range from below 0.06 for foams to over 2.0 for fluorocarbons, compared to 0.5 for softwoods, 0.7 for hardwoods, 2.2 to nearly 4 for glasses and ceramics, nearly 3 for aluminum, and around 8 for steels.

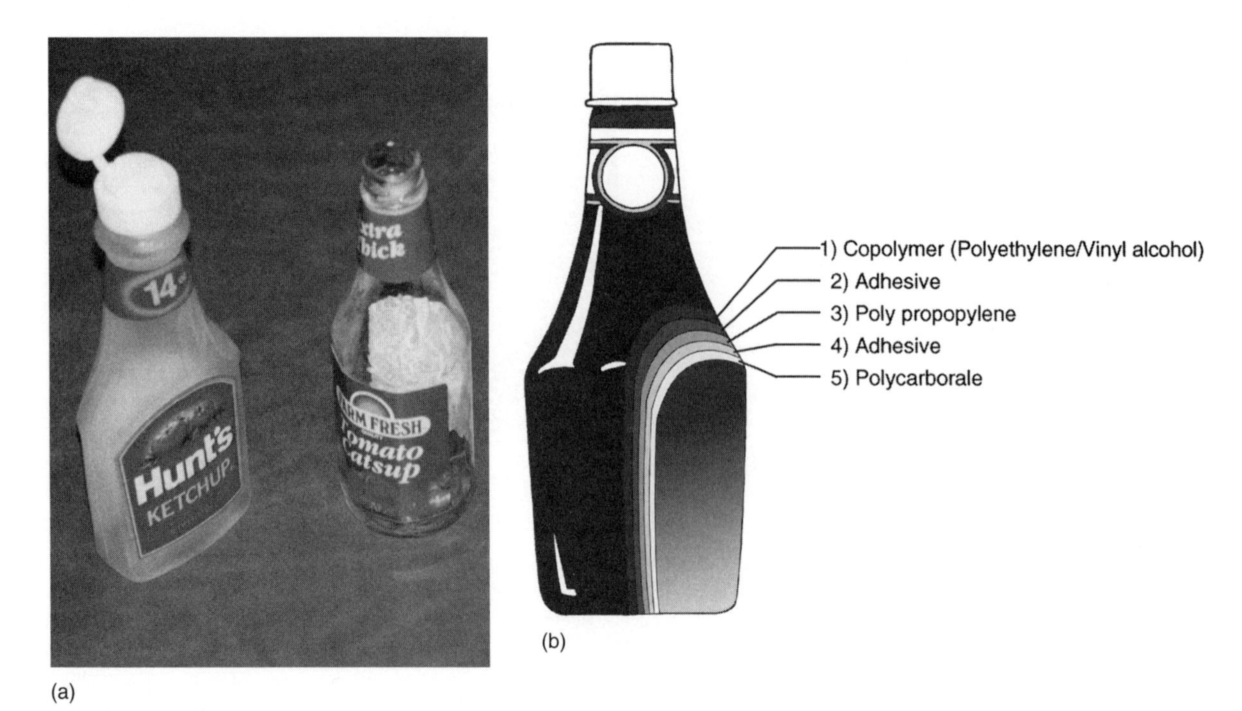

Figure 15-13 Plastic food containers. (a) The tough and lightweight properties of plastics make them very competitive with glass and steel. The barrier plastic ketchup bottle on the left, made of layered composite plastics (see detail in part b), weighs 80% less than the glass catsup bottle on the right. The plastic container is less likely to break than the glass bottle and saves transportation costs. Why would one company choose glass and the other company plastic? (b) Five layers of barrier plastic container.

15.2.8 Optical Properties

A number of plastics have optical properties comparable to glass and offer impact strength superior to most glasses. In addition, they process more easily and, when broken, do not produce sharp splinters or chards. Refer to Section 6.4 for the definition of optical properties used here. Opacity develops in plastics with an increase in crystallinity, whereas an amorphous structure produces transparency. Low-density polyethylene film, such as sandwich bags, is clear; HDPE for detergent bottles is opaque. Acrylics are available in an unlimited range of colors and are as transparent as the finest optical glass (light transmission equals 92%); they can also be made opaque or can be produced with a full range of translucencies. Acrylics have a refractive index of 1.49 compared to 1.52 for soda-lime glass, 1.47 for borosilicate glass, and 1.46 for 96% silica glass. (***Refractive index*** indicates the ratio of the speed of light through the materials compared to light traveling through a vacuum. Higher values indicate greater bending of the light.) The limits of transparency of some other plastics follow:

Acrylics	92%	Polyethylene	80%
Cellulose	88%	Polypropylene	90%
Ionomer	92%	Polystyrene	92%
Polycarbonates	90%		
Amino	29%[a]	ABS	33%[a]

[a]Translucent.

Translucent plastics transmit light but objects cannot be clearly seen through them.

Certain plastics, such as acrylics, polyesters, cellulosics, and polystyrene, show colorful stress concentrations when viewed with a polarized-light filter. This photoelastic effect is used

TABLE 15-6 MASS WEIGHT DENSITY COMPARISONS OF SELECTED PLASTICS AND OTHER MATERIALS

	Kilograms per cubic meter (pounds per cubic inch)																		
Plastics[a]	500 (.018)	1000 (.036)	1500 (.054)	2000 (.072)	2500 (.090)	3000 (.108)	3500 (.126)	4000 (.145)	4500 (.163)	5000 (.180)	5500 (.199)	6000 (.217)	6500 (.235)	6500 (.235)	7000 (.253)	7500 (.270)	8000 (2.89)	8500 (.307)	9000 (.325)
ABS, e		I																	
Acetals (POM), e			I																
Acrylics (PMMA)			I																
Amino (MF)			I																
Cellulose (CA)			I																
Epoxy (EP)																			
Fluorocarbons (PTFE), e								I											
Nylon, e			I																
Phenolic (PF)								I											
Polycarbonate (PC), e			I																
Polyesters			■																
Polyethylene (LDPE)			I																
Polyethylene (HMWPE), e			I																
Poly(ethylene terephthalate) (PET)			I																
Polyimide			I																
Polyphenylene oxide (PPO)			I																
Polypropylene (PP), e			I																
Polystyrene (SAN)			I																
Polysulfone, e				I															
Polyurethane (PUR), e				I															
Silicones (SI), e				I															
Vinyls (PVC)				I															
Plastic foams		■■																	
Glass-reinforced plastics			■■■■																
Comparison materials																			
Alumina										I									
Aluminum					■■														
Brass																			I
Cast iron															I				
Glass					■														
Magnesium				I															
Steel																		■	
Wood	■■■																		

[a]e, Engineering plastic.

as a design tool. ***Photoelastic plastic*** is shaped to a specific design and stressed under polarized light to determine what areas will receive the greatest stress, thus allowing for material to be added or redesign to handle expected loads (Figure 15-14). As with glass, many plastics can effectively bend light and serve as ***light pipes*** for optical fibers used in medical applications, signs, telecommunications, and plastic art. Polycarbonates (PC) possess excellent optical properties as well as hardness and toughness. PC provides the substrate for CD-ROMs, which can store up to 660 MB (megabytes) of information as compared to less than 2 megabytes on floppy disks. The emerging DVDs hold 4.7 GB (gigabytes).

15.2.9 Electrical Properties

Advances in materials have fostered many technological developments. The electrical and electronic developments over the past 75 years owe much to the continuing breakthroughs in plastic materials. The superior insulating properties coupled with good heat resistance of silicones and fluorocarbons led to large reductions in the weight of electrical motors. Epoxies

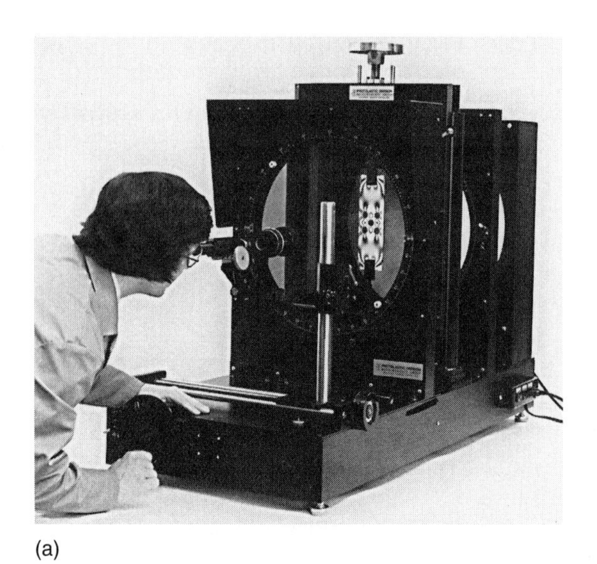

(a)

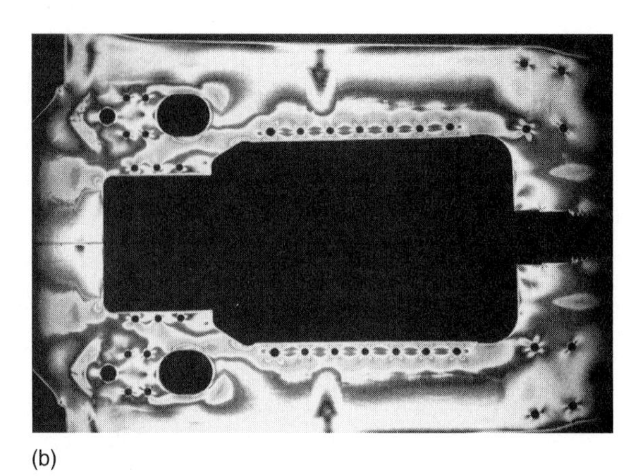

(b)

Figure 15-14 (a) Photoelastic plastics allow stress analysis. (b) Acrylic photoelastic model of a jet turbine engine component under load reveals stress concentrations (closely spaced lines). (Measurement Group, Inc.)

serve to encapsulate electronic components subjected to temperatures as high as 150°C, corrosive chemical environments, and strong vibrations and shocks. Phenolics have the oldest history as electrical insulators and find wide use as housings for automotive ignition parts (coils and distributors), switches, receptacles, terminal blocks, and bulb bases.

The ***dielectric strength*** (maximum voltage that a dielectric can withstand without rupture) of most plastics makes them the logical choice as insulators. They range from 79×10^6 volts per meter (V/m) for fluorocarbons to 12×10^6 V/m for certain phenolics. These values compare to 19×10^6 V/m for porcelain and 12×10^6 V/m for alumina. The ***dielectric constant*** (ability to store electrical energy) of certain plastics puts them into the condenser category. Condensers in electrical circuits maintain voltage with less fluctuation. At 60 hertz, the dielectric constants (K) compare as follows: poly(vinyl fluoride), 8.5; cellulose acetate, 4.0 to 5.0; polyethylene (LDPE), 2.25 to 2.35; and polystyrene, 2.45 to 4.75; as compared to alumina, 8 to 10; mica, 5.4 to 8.7; and glass, 3.8 to 4.6. Sections 6.1 and 6.2 describe the electrical properties discussed in this section.

Arc resistance is the ability of a material to withstand the arcing effect of an electric current. Insulators, terminals, and switches are subjected to arcing effects, which can burn through or damage the material's surface. The arc test (ASTM D495) measures the total elapsed time in seconds that an electrical current must arc to cause failure. The plastic may ***carbonize*** and become a conductor; burst into flames; produce thin, wiry lines or ***tracks*** (surface imperfections in the form of wiry lines) between electrodes; or become an incandescent (glowing hot) conductor. Arc resistance for unfilled polyimide is 125 s; melamines, 110 to 150 s; FEP fluorocarbon, over 300 s; macerated fabric and cord-filled phenolic show tracking but do not conduct; thermosetting cast polyester, 100 to 125 s; and polyethylene (LDPE), 135 to 160 s.

15.3 PLASTIC SELECTION

In the selection of a material, cost is of paramount concern. Assuming that properties are equivalent and other features such as appearance are similar when comparing materials, then cost is usually the overriding factor in the selection of one material over the other. Often, a trade-off is made when costs are higher for one material than another. For example, epoxy reinforced with graphite fibers offers tensile strength nearly 3 times that of stainless steel; however, the cost for stainless steel may be 10 or 20 times less than that of the epoxy composites. If weight is important, as in aircraft, the fact that the stainless steel is over 5 times as dense as the graphite composite can lead to the selection of the reinforced plastic. These comparisons are for exotic materials. More common materials such as plain-carbon steel, which is used more frequently, find strong competition from low-cost plastics such as polyethylene, polypropylene, and vinyls.

High demand for a particular material may result in price drops if suppliers increase their production capabilities and competition results in improved processing methods. In the mid-1970s, graphite fiber composites cost about $24 to $29 a pound. To purchase a fishing rod or tennis racket of this new composite, you would have to pay $150 to $250. Now it is possible to buy even better-designed rods and rackets of graphite composites for as low as $35 to $60. Processing technology driven by demand and competition yields benefits to the consumer. In turn, that demand should lead to even better composite processing technology, which will drive prices down while improving composite quality. Looking at materials from the more mature plastics processing technologies reveals how low prices may go. Plastics in high demand, such as polyethylene and polyvinyls, sell for prices of around $1 to $3 a pound.

Green manufacturing practices affect cost. Demands by the public and legislation for green design and manufacturing, in the short term, increase costs as new technologies must be applied to lessen the burdens of synthesis and processing on the environment. As improved greening technologies evolve and more raw materials come from recycled plastics, the industry should reap a decline in costs just as recycled aluminum and steel have saved on energy and overall costs. Much of our materials innovation results from these market pressures and the resulting legislation. Corporate average fuel economy (CAFE) is a federal legislative concept mandating that automakers achieve a set average fuel efficiency for the entire line of cars they produce. As the miles per gallon (mpg) values continually rise to meet CAFE, plastics come into greater use due to their good strength-to-weight ratios.

Once a material is available, the next problem is informing designers of its availability. Some companies specialize in developing databases for a variety of materials by many producers. Producing companies also provide databases to customers. The PC-based *Plastics Materials DATA Digest* gives information on over 13,000 TP and TS resins, including properties, ASTM test descriptions, and manufacturers and suppliers. To encourage auto and truck designers to use its plastics, General Electric Plastics created the EDD-PC engineering-design database. It is coupled with an Alpha I multimaterial, multiprocessing machine for the development of large, integrated thermoplastic parts such as bumper systems and automotive hoods. Customers can perform materials evaluation 24 hours a day using the latest test and data-acquisition hardware and software. Figure 15-15 shows an example of a large, integrated component that reflects the results of CAFE and industrial design/materials applications of new technology.

Materials selection takes many paths as manufacturers seek profitable products. ***Disposal*** or ***single-use products***, although convenient and often low cost, usually do not follow the principles of reduce, reuse, and recycle. Still, business opportunities drive manufacturers to market this kind of product. How do the tenets of green manufacturing jibe with single-use products?

15.4 PROCESSING PLASTICS

The plastics industry has been very successful in developing a large variety of resins and numerous processing systems, and plastics have become so competitive with other materials that they are now found in most markets. Designers continue to seek lightweight, strong materials that have improved forming capabilities, and plastic–matrix composites, reinforced with fibers and particles and as laminates, are often favored as the engineering material of choice. Reduced costs are often realized through the near-net and net-shape processes inherent in composite processing. Unit 8 provides thorough coverage of plastic–matrix composites.

Intelligent processing of materials (IPM) involves placing nondestructive evaluation (NDE) sensors in the processing equipment to allow computer monitoring and ***adaptive control*** (mixtures, temperatures, pressures, etc., adjustment in "real time" during the process) of the complex variables to be matched against the process model or expert system. IPM lends itself to the trend toward ***total quality management (TQM)*** and ***statistical process control (SPC),*** techniques

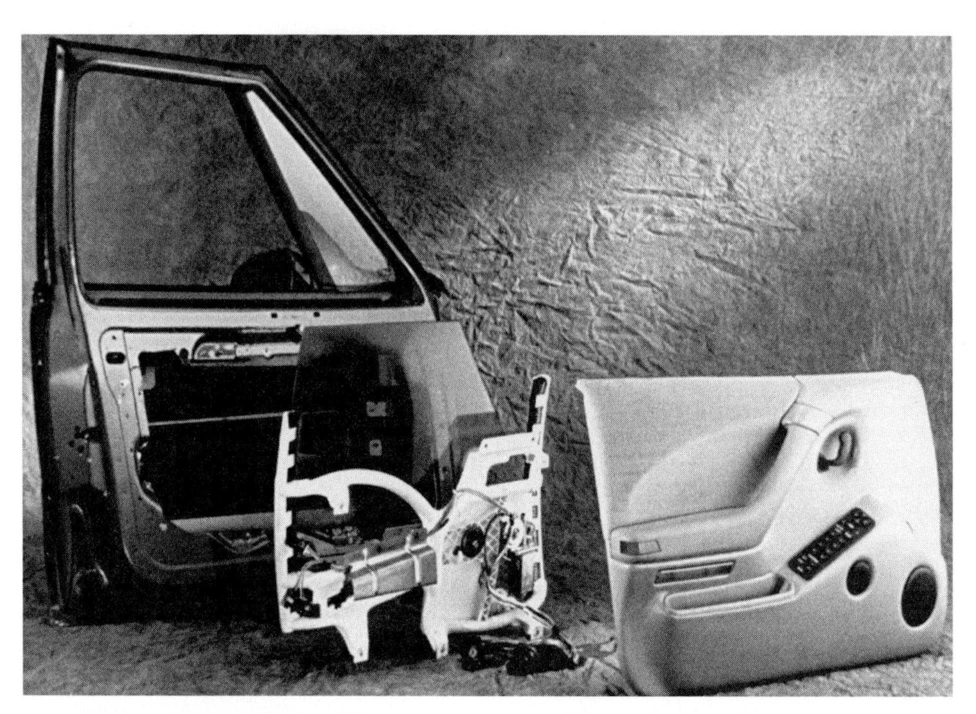

Figure 15-15 The Super Plug is a patented, "next-generation" door hardware module developed as a result of a special, five-year program of GE Plastics and Delphi Interior & Lighting Systems. The module, which can consolidate up to 61 separate door parts into one system, provides original equipment manufacturers (OEMs) with a reduction in system costs and weight and vehicle assembly time. GE Plastics custom-compounded a new XENOY® PC/polyester, 30% glass-fiber blend material for proprietary use in the Super Plug. (GE Plastics)

associated with improved quality in manufactured products and reduced waste and overall costs. World competition is driving the move toward these improvements in ceramics and metals as well.

Manufacturing engineers working with designers, materials engineers, and technicians engage in ***concurrent*** or ***simultaneous engineering,*** which brings together team members from materials, manufacturing, and design during the design stages. The designers then provide ongoing engineering throughout the production stages. The demand for shorter cycles from design to production, often coupled with low-volume products, favors plastics. Materials scientists and engineers are continually developing new resins to meet performance demands.

Thermosetting (TS) plastics are typically processed by casting, spraying, compression molding, transfer molding, and reaction injection molding (RIM) (Figure 15-16). Because thermosets cannot be resoftened, they are not suitable for recycling except as regrinds to be used for fillers. They are burned for energy production. Thermoplastics (TP) plastics are much easier to recycle and are therefore favored as the plastics of choice in more and more applications. ***Liquid crystal polymers*** (LCPs) are a newer class of engineering thermoplastics that include aromatic polyesters and phenylene polyamide. LCPs lend themselves to easier melt processing, which develops highly oriented (rodlike) molecules (Figure 14-7d). The resulting molded LCP parts possess anisotropic properties and chemical and thermal stability at service temperatures up to 190°C. In a manner similar to wood, which is composed of cellulose fibers that generate multidirectional strength, the rod-shaped LCP molecules provide self-reinforcement. Amoco's Xydar, an LCP copolyester resin, represents the newer engineering thermoplastic, which can be molded into complex shapes in injection-molding processes. Xydar has a heat-deflecting temperature of 350°C and a tensile strength of 115 MPa, making it suitable for automotive engine parts. The LCPs can replace ceramics for chemical-processing equipment and find application as engineering plastics in such items as tennis rackets, electric motor insulators, and printed circuit boards, where because the LCP can withstand temperatures about 40°C hotter than the molten solder, it is used to surface mount components.

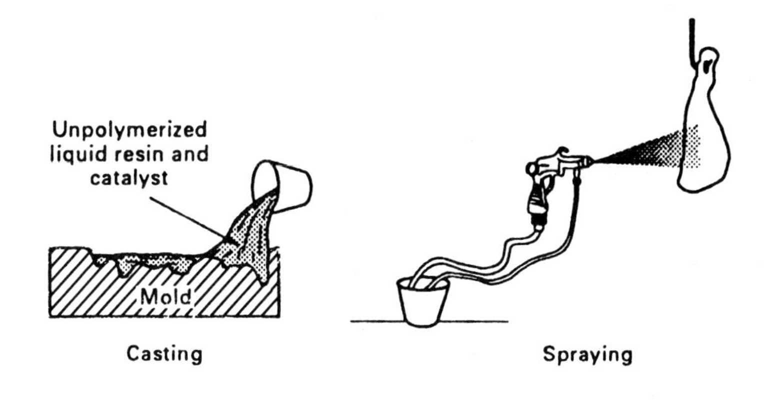

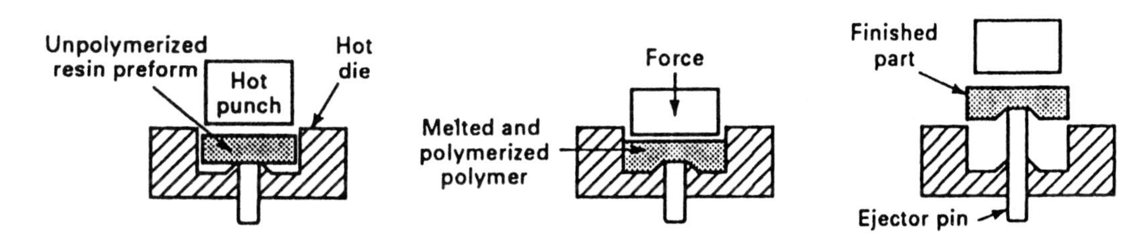

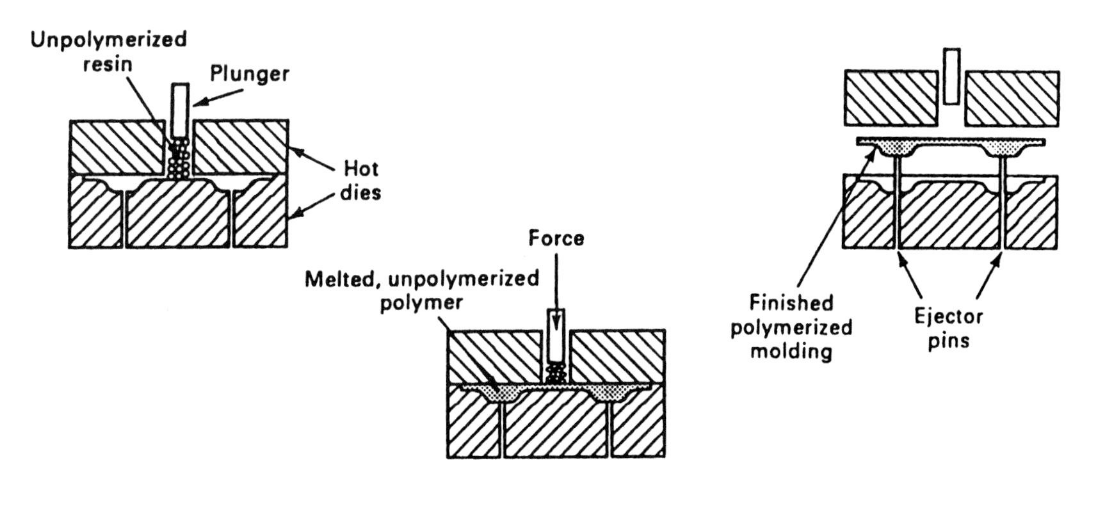

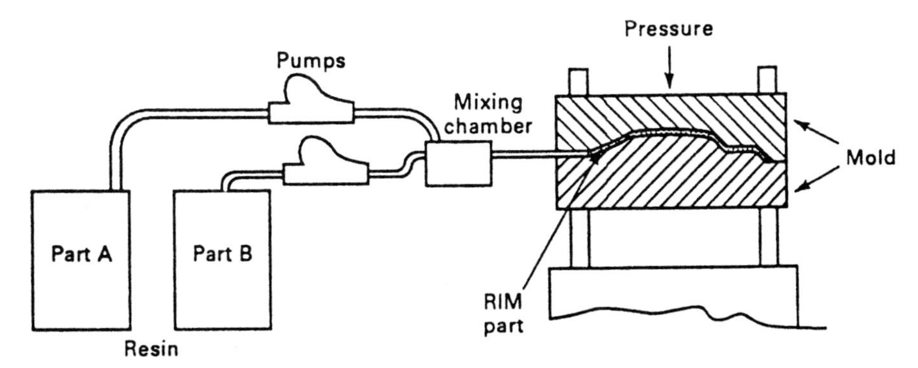

Figure 15-16 Typical thermosetting processes.

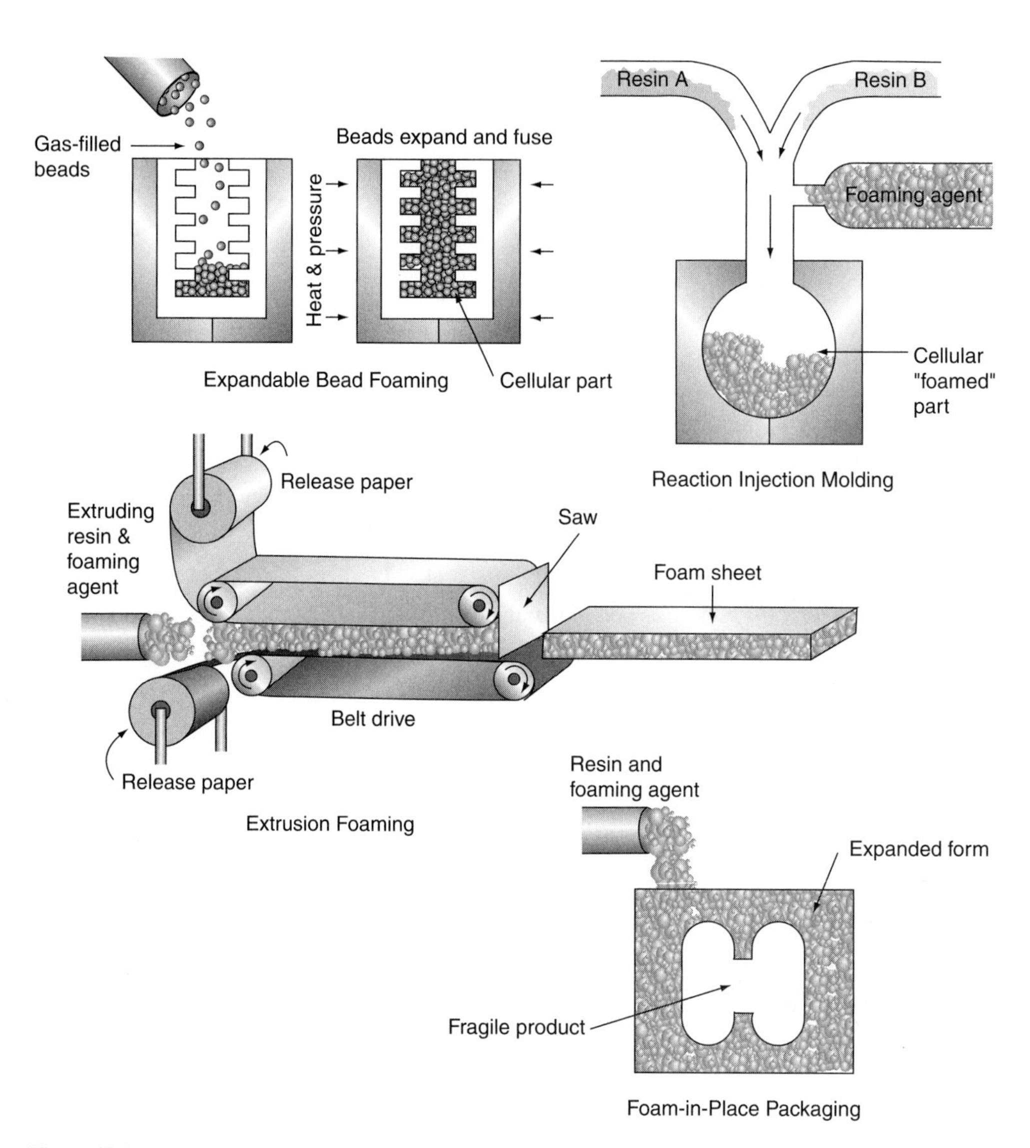

Figure 15-16 continued Typical thermosetting processes.

Thermoplastics are typically processed by casting, extrusion, injection molding, blow molding, thermoforming, drawing, rolling, calendaring, rotational molding, resin transfer molding (RTM), foaming, and spinning (Figure 15-17). Unit 8 discusses other processes for FRP. Because of their resoftening abilities, scrap thermoplastics can often be put back into the process. After service life, recycling works well. ***Cyclic thermoplastic polymers,*** improved in the late 1980s by General Electric (GE), provided a breakthrough for the automobile industry, which was searching for a low-viscosity engineering thermoplastic for high-production structural composites that could easily wet fibers. The resulting FRP provided strength comparable to steel with about a third of the weight to provide significant gains in automobile fuel efficiency. Demonstration of the feasibility of this material, for the Advanced Technology Program, fell to a joint GE-Ford Motor-NIST venture as part of the Big-Three automakers' consortium, the Automobile Composite Consortium (ACC), which sought to improve U.S. automakers' world competitiveness. This advance in cyclic polymer technology made it possible for the first time to use engineering thermoplastics in high-speed, low-cost, resin-transfer-molding (RTM) composite manufacturing processes.

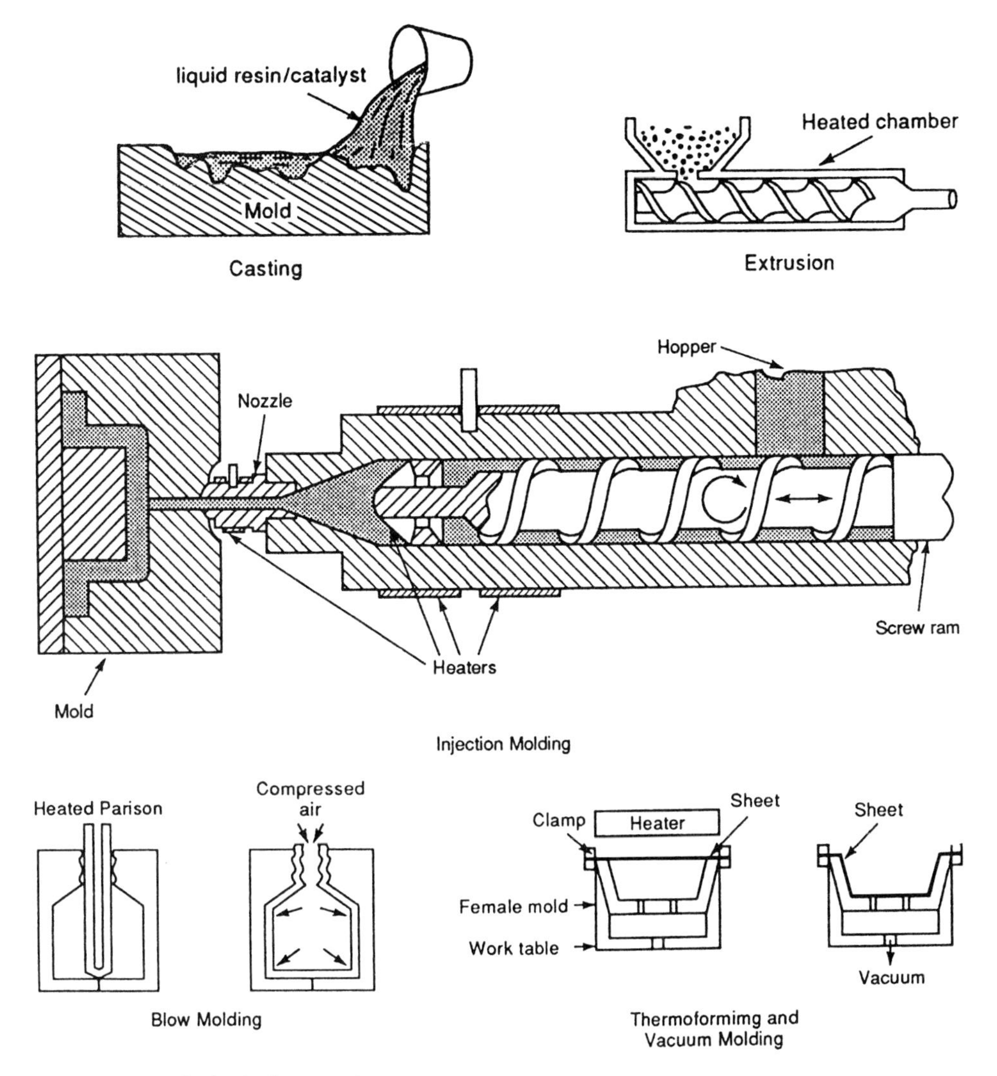

Figure 15-17 Typical thermoplastic plastic processing

To form a structural thermoplastic matrix of bisphenol-A polycarbonate (BPAPC), which is used to impregnate fibers, two stages of polymerization must occur. First, an ***oligomer*** is produced, a low-molecular-weight (2 to 20) polymer, which has a consistency similar to water rather than the viscosity of cold molasses as is normal in polycarbonates. The polymer ring (***cyclic polymer*)**, shown in Figure 15-18a, has a low melt viscosity below temperatures of 300°C and flows like water around the reinforcing polymer preform for complete impregnation. Once that is accomplished, a special "initiator" triggers a polymerization reaction that rapidly transforms the rings into long, linear (Figure 15-18b), high-molecular-weight molecules (M_W 50,000 to 300,000) that give thermoplastics their characteristic strength.

In addition to excellent damage resistance and favorable process characteristics, the cyclic ester material offers excellent resistance to common automobile solvents such as gasoline and brake fluids. This BPA polycarbonate is also environmentally friendly because it does not give off volatile or nonvolatile by-products in the condensation polymerization, and it

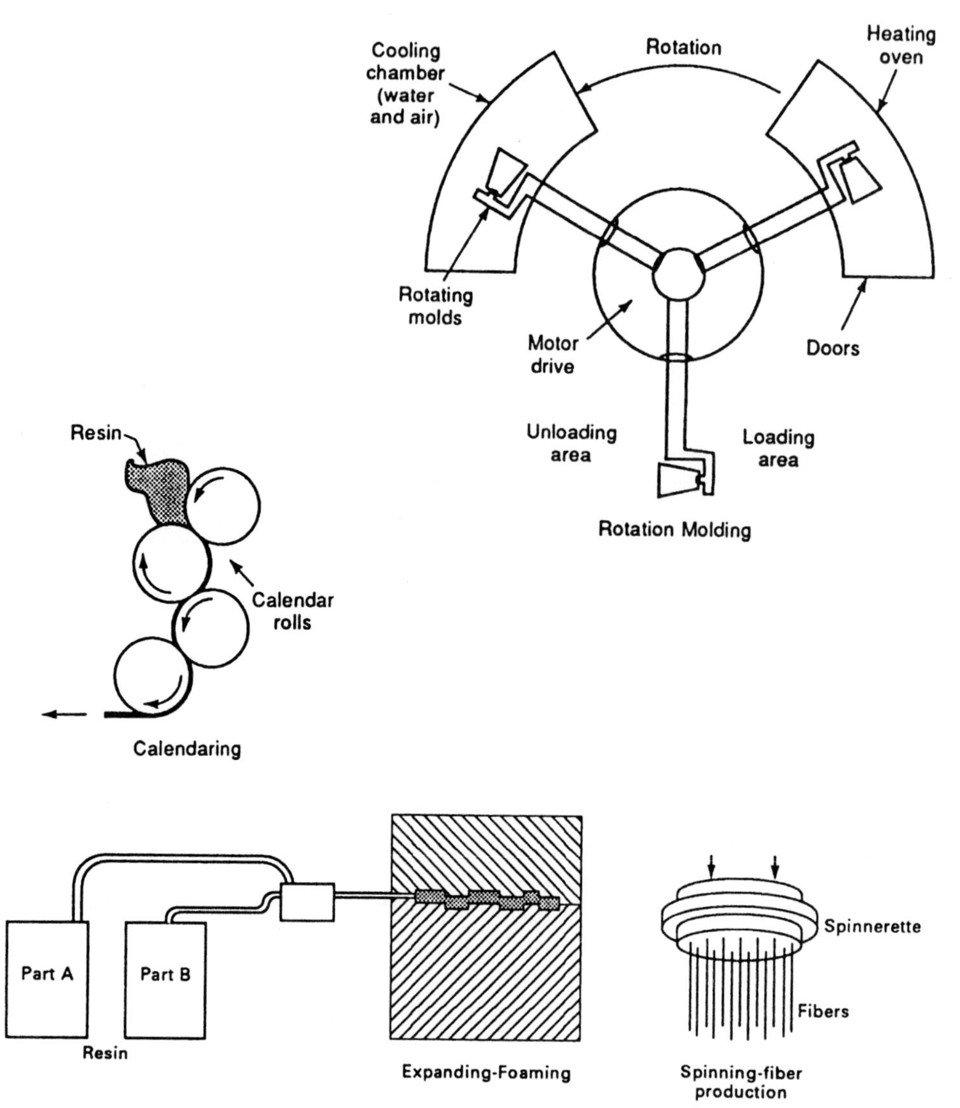

Figure 15-17 continued

offers recycling opportunities because it is thermoplastic. An example of resin-transfer molding (RTM) in the auto industry is Chrysler's choice of the process for its Viper sports car (see Figure 15-19a and b) because it offered a 40% weight savings over sheet steel, with reduced tooling time and tooling costs. Compared to sheet-molding-compound (SMC) composites, RTM gives lighter, stronger panels at lower molding pressure with less labor. With only minor hand finishing, RTM provides a class A finish. Some problems in holding tolerances on the large, complex hood slowed the production rate.

Applying new engineering materials technology that couples new analysis technology such as ***magnetic resonance imaging (MRI)*** with the ability to engineer materials can result in remarkable design and production innovations. For example, Toyota engineers produced a super olefin polymer out of a desire to change the conventional bumper fascia material, ***elastomer modified polypropylene (EMPP).*** Figure 15-19c is a comparison

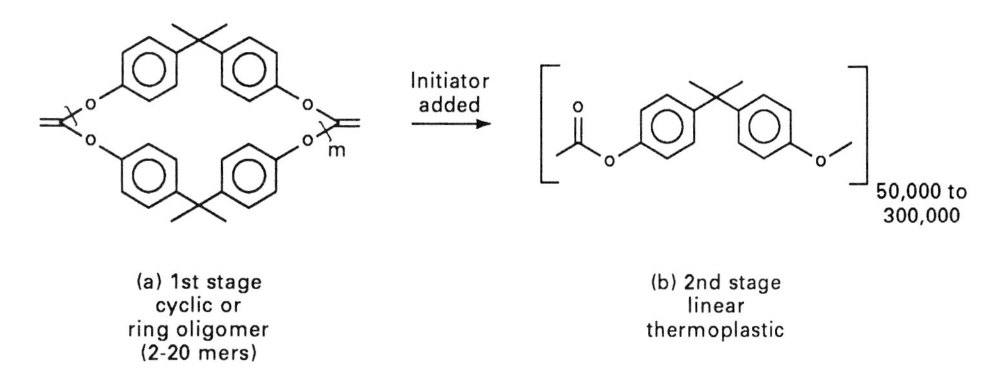

Figure 15-18 Steps in making bisphenol-A polycarbonates (BPAPC).

(a)

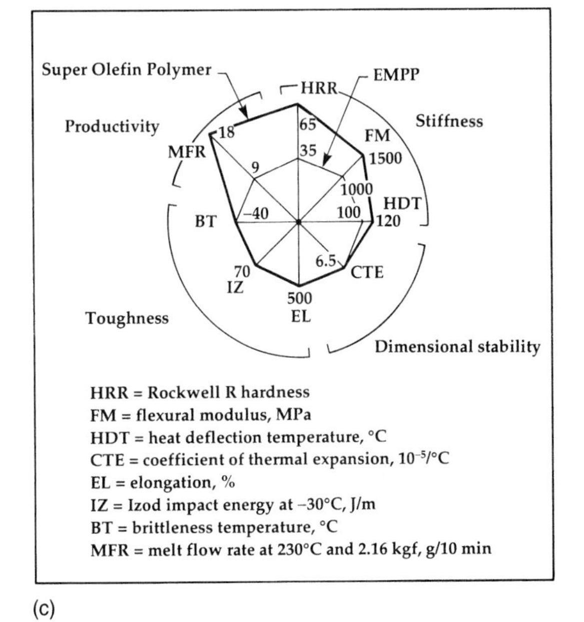

(b) (c)

Figure 15-19 (a) The Dodge Viper was an experiment in new technologies, including concurrent engineering, low-volume and low-investment production, unique design, and numerous materials and processes innovations. (b) An exploded view of the Viper showing RTM plastic panels as the majority of outer and inner body panels; only the lower front body enclosure used SMC. (Chrysler) (c) Major advantages of Super Olefin Polymer over conventional EMPP are dramatized in this property comparison. Note the toughness values of brittleness, temperature, impact strength, and elongation were maintained. Productivity values increased with improved flow rate, as did stiffness values (flexural modulus and heat deflection temperature). There was a slight improvement in dimensional stability, which is a combination of the coefficient of thermal expansion (COT or CTE) and stiffness. (*Advanced Materials and Processes*)

of the properties of the two modified polymers. The inner octagon is the EMPP and the outer polygon depicts the improved properties of the ***Super Olefin Polymer.*** EMPP is a blend of polypropylene (PP) and ***ethylene propylene rubber (EPR).*** Toyota's newly engineered polymer met the goal of a super olefin polymer with a 50% minimum improvement in flexural modulus, melt flow rate, and hardness over the conventional EMPP, while retaining a low coefficient of thermal expansion (CTE) and high impact resistance at −30°C (−20°F) and below. The reengineered polymer composite used an EPR elastomeric matrix to provide the following basic performance characteristics required of a bumper fascia:

- Low-temperature impact resistance and a low CTE
- Increased hardness, toughness, and stiffness through the introduction of hard segments into the EPR phase forming a microscale interpenetrating network (IPN) structure
- Reduced molecular weight and maximum crystallization of the PP component to increase melt flow rate and hardness
- Enhanced formation of the IPN-like structure by controlling its crystallization rate as well as the crystalline component of the PP
- Recyclability of the polymer component

The following comparison reveals the dramatic improvements:

	Super Olefin Polymer	EMPP
Wall thickness, mm	3.8	4.8
Fascia weight, kg/car	6.8	8.7
Molding cycle time, s	85	115
Relative production cost, $	89	100
Molding temperature, °C	210	230

Because plastics offer so much to the designer, new processes coupled with improved resins will continue to emerge. One recent development in plastics processing, ***solid imaging,*** uses lasers to polymerize plastic for net-shape processing (Figure 15-20). ***Photopolymer*** plastics are based on acrylic monomers and oligomers and are generally inferior to typical engineering plastics. Photoactive polymers have been used for decades in applications such as printing plates and inks, ultraviolet (UV) and electron-beam (EB) curable coatings and photoresists, and for the production of holograms. As one of several ***rapid prototyping*** techniques, UV laser beams follow paths generated by computer-aided design (CAD) data to create solid parts layer by layer as the lasers polymerize the resins. As each layer is cured, a very accurate, solid-plastic model emerges on an elevator from the vat of resin. This and other techniques of rapid prototyping have been developed to allow designers to shorten the time required to go from drawing to a solid model and avoid the need for numerous molding, sculpting, cutting, joining, and similar processes traditionally involved in model making. However, although photopolymers are suitable for model making, fully functional parts await new polymer development. Rapid prototyping is now being integrated with solidification modeling and computer tomography to produce patterns for investment-cast rapid tooling for the die-casting, permanent-mold, and investment-casting industries. Existing parts can be scanned with computer tomography to produce a CAD file. The CAD data can be used with solidification modeling software to determine the best method for casting a tool. Still another promising technique for processing plastics is the free electron laser, discussed in Module 6.

Conduct Internet search on "Direct Manufacturing" and "Rapid Prototyping"

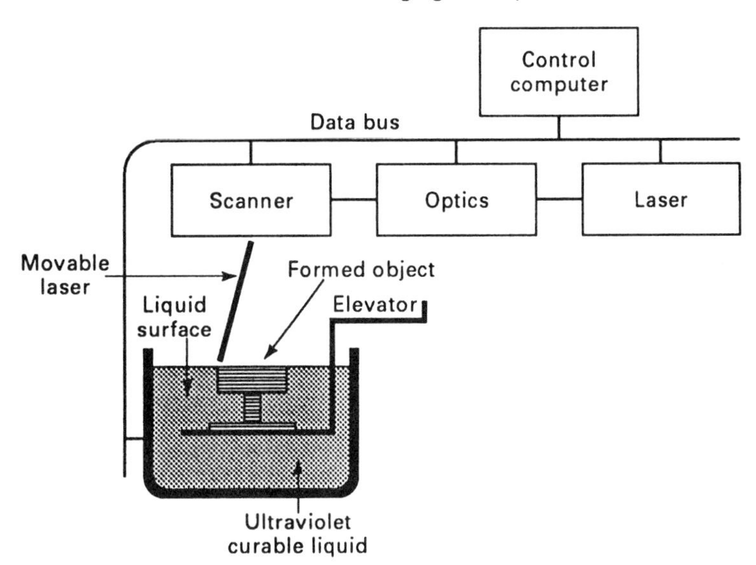

Figure 15-20 Photopolymers permit a solid imaging technique, known as rapid prototyping, that uses laser beams to process resin into solid models and prototypes.

Dendrimers, another new class of *oligomers* (Figure 15-21), are spherical or rod-shaped nanostructures with hollow interiors that allow easier engineering of desired properties. They offer the potential for superior protective coatings, nonlinear fiber optics, adhesives, conductive polymers, lubricants, and reinforcers for composite matrices. Interpenetrating polyimide-network (IPN) plastics represent research aimed at having the best of both worlds: in this instance, thermosets (TS) and thermoplastics (TP). NASA research has developed a tough, high-temperature polyimide that resists microcracking. The IPN ***neat resin*** (no additives or reinforcers) makes a valuable composite matrix for graphite reinforcement suitable for aircraft engines and aerospace structural applications. By combining cross-linked (TS) polyimide with linear (TP) polyimide (Figure 15-22), the ease of processing TS is retained while the toughness of TP is achieved. ***Block copolymers*** (identical mers clustered in blocks along the backbone chain polymer) are also valuable as composite matrices because of their ability to combine the stiffness of a plastic with the toughness of an elastomer. One example of this synergy is with plastic-resin polystyrene, which combines the stiffness of a block copolymer with the toughness of the elastomer polyisoprene, as illustrated in TP rubber (refer to Figure 16-8).

Figure 15-21 Computer model of carbon-based dendrimers, which resemble the structure of atoms, but at the molecular level. The symmetry provides special physical and chemical properties. (George R. Newkome, University of South Florida)

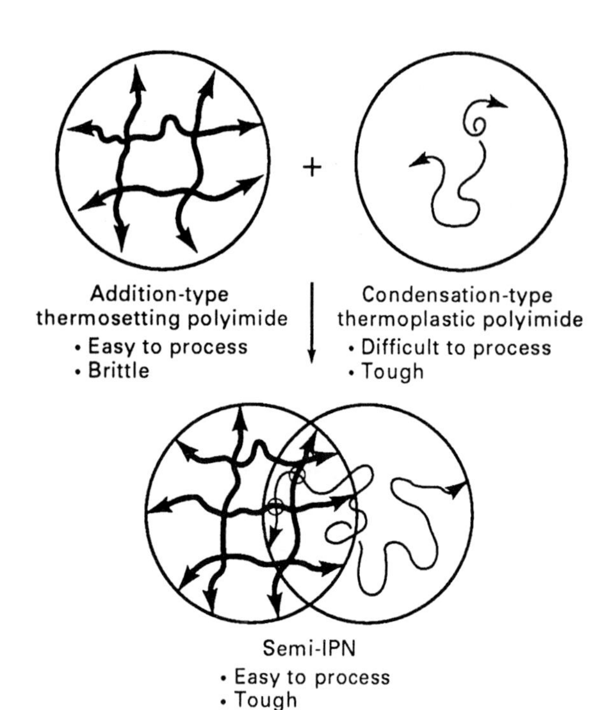

Figure 15-22 Formation of an IPN plastic matrix.

15.5 SMART POLYMERIC MATERIALS

Because polymer chemists can synthesize a wide range of organic polymers, many opportunities for smart polymers wait to be exploited. Current examples of smart polymers include multigrade and synthetic lubrication oils. A 10W40 SAE motor oil will become more viscous with increased heat and less viscous as the temperature drops because of the polymers that expand and contract with heat variations. ***Electrorheological (ER)*** and ***magnetorheological (MR) fluids*** also change viscosity, plasticity, or elasticity. Variations in the application of either electrical current or magnetic fields cause ERs and MRs to go from liquids to gels in milliseconds. These ***smart fluids*** consist of polymers (oils and additives) and ceramic particles suspended in liquids. The liquid carrier must be a good insulator and the particles either good conductors or ferromagnetic. When the field is applied, the particles link together to form a gel. The particles reassume their random pattern with removal of the external field (Figure 15-23). Smart fluids have potential in robotics, hydraulic valves, power transmissions, and other automated equipment that can react to computerized controls.

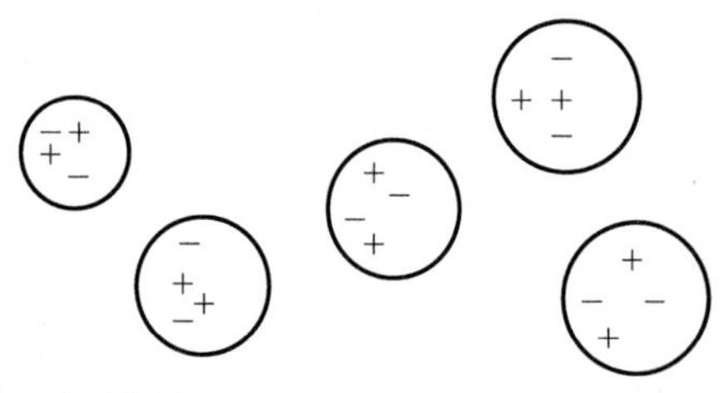

No electrical field present, charges are random, no particle attraction.

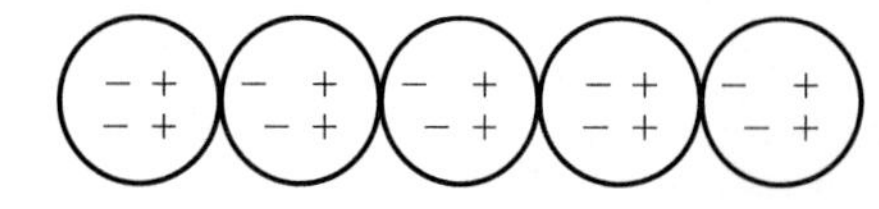

Probe energizing the fluid, particles linked together.

Figure 15-23 Smart polymers—electrorheological (ER) fluids. (From John A. Marshall, "Liquids That Take Only Milliseconds to Turn into Solids," National Educators' Workshop: *Update 93—Standard Experiments in Engineering, Materials Science, and Technology*. NASA Conference Publication 3259, April 1994, pp. 315-322)

Shape-memory polymers, much like the shape-memory alloy Nitinol, offer control after polymerization. Polynorbornene, a shape-memory polymer, is molded to a desired shape to which it will return after being subjected to loads. Other smart polymers, including liquid crystals, change colors, opacity, and other characteristics with variations in external forces. These polymers find applications in such diverse fields as medicine (color thermometers), sports (smart tennis rackets), and building construction (LCD and smart windows) (see Figure 15-25). Polyvinylidene chloride (PVDC) film is another smart polymer. It is piezoelectric (see Module 27) and senses motion. PVDC sensors can be used in automatic lights and auto alarm systems. Liquid crystal polymers (LCPs) in thermometers change colors with changes in temperature (Figure 15-24a) and are used in novelty drinking cups, cereal bowls, and straws (Figure 15-24b, c, d).

APPLICATIONS & ALTERNATIVES

Plastics and Building Construction Technology. Shelter is one of our basic needs. For most people, their homes will account for the largest outlay of money they will make over their lifetime, whether they buy or rent. Affordable shelter, including homes, is a major goal of all societies. So it is wise to give careful consideration to your home, especially if you plan to purchase a house. The materials selected are very important in terms of their initial cost, and they will also affect the ease and long-term cost of living in and maintaining the house.

House construction technology has evolved over the history of civilization. Currently, the average house in the United States is a custom home, framed on its site (stick construction) with wood and having plastic or masonry siding, glass windows, and wood or metal doors. House designers and materials manufacturers have come to believe that it is possible to make tremendous improvements in traditional building techniques to achieve more cost-efficient and enjoyable homes. The "living environments" smart house shown in Figure 15-25 reflects the desire of teams of specialists in building technology, computer technology, materials engineering, industrial design, and architecture to utilize advanced technologies to improve home construction. Prototype homes are designed to demonstrate to homeowners energy-efficient, high-tech comfort and convenience. This "living environments" concept house was built to convince the home construction industry that plastics can have a beneficial impact on houses in a manner similar to that for automobiles. Built by General Electric (GE) and other business partners, the concept house serves as a research tool and a showplace for design, materials engineering, manufacturing and construction processes, and home systems (e.g., heating, air conditioning, security, entertainment, and appliances).

Approximately 30% of the "living environments" concept house is plastic, including the roof, windows, siding, plumbing system, foundation, and electrical and mechanical systems. Plastics are used in harmony with the traditional materials (wood, metal, and ceramics), and many are engineering thermoplastics. Because plastics generally do not biodegrade and their processing may lead to pollution, the GE project also focused on demonstrating that plastics can be environmentally friendly. Recycled engineered plastics, such as those used in automotive parts, can be used for house construction and should keep many plastics out of municipal landfills.

Factory-manufactured houses rather than traditional site-built houses can also be more affordable, of higher quality, and less expensive to run and maintain when built with new materials and techniques. They should also have features that make for more enjoyable living. Among the unique features in the house shown in the cutaway (Figure 15-25b) are the modular wall and flooring systems. Walls include foam-polymer sections that radiate heat into the room and are part of the system for heating, air conditioning, humidity control, and air filtration, with the capacity to recover wasted heat. Hollow composite-plastic-sandwich floors, held in place by Velcro, provide easy access, encase plumbing and electrical components, and provide storage bins. The roofing shingles are of lightweight fiber-reinforced plastic that do not need underlayment. Plastics have low emissivity and therefore conserve energy that might

(a)

(b)

(c)

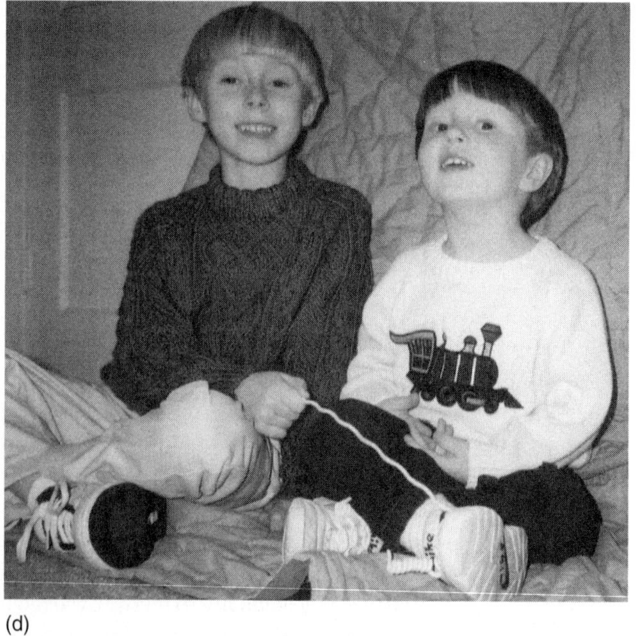

(d)

Figure 15-24 (a) LCPs serve as thermometers. (b) Child's cereal bowl made of a smart LCP changes from green to yellow with hot meal and (c) drinking straw turns from white to red upon contact with cold liquids. (d) Smart shoe strings—Luke straightens Sean's polymer shoe laces; note the one not under tension stays curled. The laces have a memory and stay coiled, thus not requiring tying. What a relief for the child's parents!

(a)

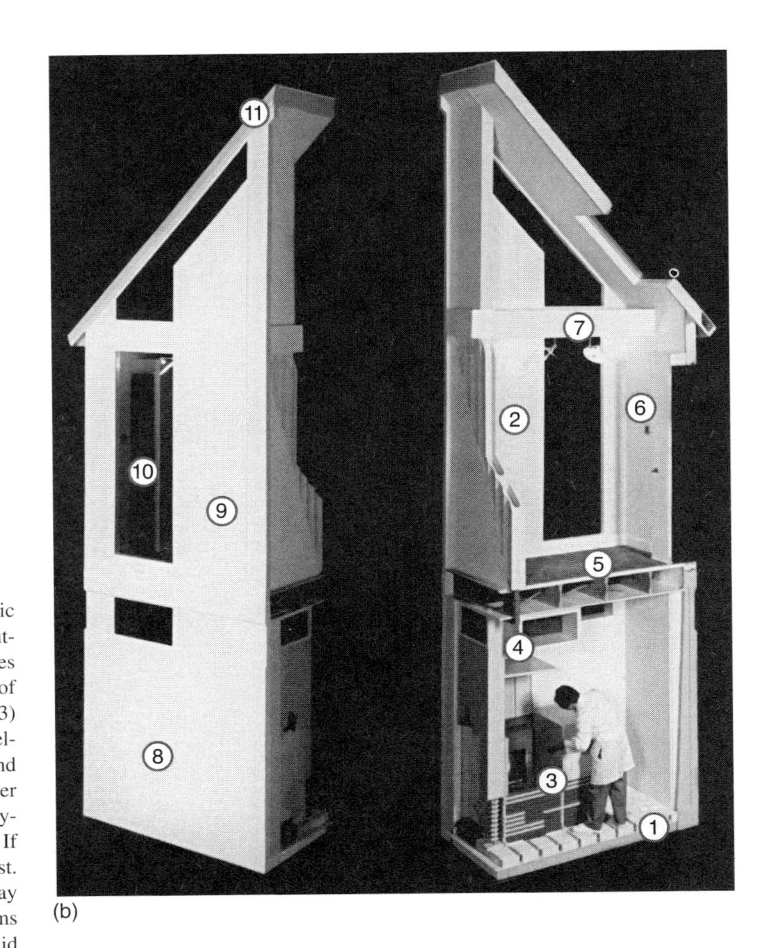

(b)

Figure 15-25 (a) Smart house. This prototype house uses smart plastics, intelligent structures, and smart electronic systems to provide the homeowner with energy-efficient, high-tech comfort and convenience. (GE Plastics) (b) Cutaway model of "living environments" showing unique features. On the right is interior Technopolymer Structures (TPS) of high-strength composites (glass fiber mat and Azdel resin). TPS is used for (1) the distribution floor made of Azdel skin and low-density foam and (2) radiant wall panels. Azdel is a glass-fiber-reinforced thermoplastic resin. (3) Total Environmental Control (TEC) are modules made of Noryl (modified polyphenylene oxide resin that has excellent mechanical and electrical properties) and Alloy (composite of amorphous polymers or semicrystalline alloys and blends, reinforced with glass fiber) for heating, cooling, hot water, air purification, and energy recovery; the number of modules depends on house size and climate. (4) Heating and air-conditioning ducts made of TEC Altum (polyetherimide resin, structural foam, sheets, and film). Altum is exceptionally strong and flame and heat resistant. If burned, it produces a low amount of smoke; it comes in clear or opaque. (5) Floor panels of Azdel and Alloy floor joist. (6) Doors of Noryl low-density foam. (7) Plastic extrusions form lighting tracks that fit into wall panels. Left cutaway of exterior: (8) precast concrete foundation wall panels with integral low-density plastic foam wall inside; plastic forms for casting concrete can be left in place. (9) Vinyl siding. (10) Windows of Lexan (polycarbonate resin) with liquid crystals (LCD) are energy-efficient and photochromic (change from transparent to translucent). (11) Roof of low-density plastic foam can look like cedar shakes or ceramic tile and can also make use of photovoltaic solar panels to generate electricity. Silicone rubber sealing creates an air and moisture barrier. Modular components such as roof systems could be fabricated in a factory to assemble solar panels, skylights, gutters, and roofing, which would then be transported to the building site and placed over roof trusses. (GE Plastics)

otherwise escape through roofs and walls. Other components, such as doors, drawers, and cabinets, snap together, which allows for easy assembly and disassembly. Acrylic windows can be changed from clear to opaque because of their embedded liquid crystals. This affords privacy and blocks the sun to avoid the bleaching of fabrics and carpets. Instead of the traditional baseboards, there are hollow raceways that carry all wiring: telephone, coaxial cable, and low-voltage electrical.

Other green architectural aspects of "living environments" include special features in the kitchen: the polished surface of the breakfast counter with the look of marble but engineered from recycled computer housings and plastic automotive parts, and a built-in bottle and food package recycling unit with a bar code reader to segregate engineering plastics from general-purpose or commodity plastics. Engineering plastics can be ground and reused; commodity plastics can be incinerated to reclaim their BTU (British thermal units) value.

Tour GE's Living Environments house with movies, floor plans, and novel applications for shelter—*http://www.crd.ge.com/esl/cgsp/projects/le/endtour.html*

Plastic Container Coding System. Symbols on packaging containers aid in sorting plastics. These symbols, usually found on the bottom of packages and other products, are there as part of the Society of Plastic Industry, Inc. (SPI) voluntary ***plastic container coding system.*** Figure 15-26 shows the seven symbols used in this coding system and the legend of the code. Note that "1—PETE" represents polyethylene terephthalate. The American Society for Testing and Materials (ASTM) abbreviation is PET, and the ASTM acronym for poly(vinyl chloride) or vinyl is PVC, whereas the SPI code is V. These departures from ASTM standards are made to avoid confusion with registered trademarks. But the coding system required for packaging by most U.S. states may soon be replaced due to the confusion it has caused. Much package labeling has given the erroneous impression that the plastic will be recycled, when in fact it will not be. Only about 3% of plastic garbage actually makes it back into new products. Recycling centers can use only select materials, such as HDPE from blow-molded milk bottles or PETE from soda bottles. HDPE used for injection medicine bottles has a different composition that cannot be mixed with the milk bottle resin.

Other organizations have also developed coding systems to aid plastics recycling, including the Society of Automotive Engineers (SAE), the American Society for Testing and Materi-

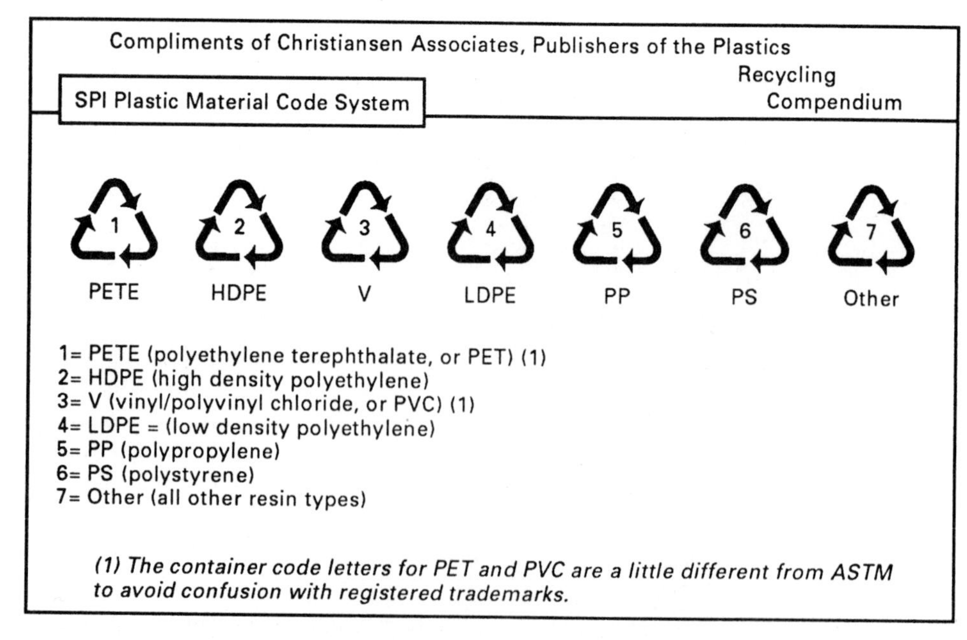

Figure 15-26 SPI's voluntary plastic container coding system. (Dow Chemical)

als (ASTM), and the International Standards Organization (ISO). When laws do not dictate the use of a particular code, manufacturers use the coding system most appropriate for their product.

Although some communities report as much as 75% participation in recycling paper, aluminum, steel and certain plastic bottles (PETE, HDPE), there is much skepticism about plastic recycling. Even though most plastic containers have symbols, only PETE and HDPE have a reliable recycling infrastructure. However, the auto industry recycling infrastructure has made good use of plastics from junked cars. Because of the savings realized from using recycled rather than virgin plastics, more new cars carry parts with recycled plastics. For example, PET from soda bottles recycles into parts such as luggage rack siding, headliners, and reinforcement panels. Ford estimates that nearly 95% of scrapped cars make it to recycling centers, where more than 75% of the contents are recycled. Research reveals that many plastics can be scrapped, shredded, pelletized, and remolded into new parts up to a dozen times.

From the Pause & Ponder section at the beginning of this unit, how do you think a plant-based polymer like NatureWorks PLA will affect the plastic industry, recycling, and progress in green manufacturing?

SPI container coding system—*http://www.plastic.org/outreach/recycling/2124.htm*

The following websites will provide additional insights into plastics:

DuPont homepage—*www.dupont.com*
Society of Plastics Engineering—*http://4spe.org/*
Introduction to plastics—*http://www.polymers.com/dotcom/polycon/primer.html*
GE Plastics Education through cartoons on history, properties and terminology, design and processing—*http://www.ge.com/plastics/resins/education/cartoon.html*
Nylon—*http://me.mit.edu/2.01/Taxonomy/Characteristics/Polyamides.htm l*
Dow Chemical—*http://www.dow.com*

SELF-ASSESSMENT

15-1. What main advantage do cellular plastics offer?
a. Reduced density **b.** Improved crystallinity **c.** Greater density **d.** Improved strength

15-2. The objective of adding microspheres to plastics is which of the following?
a. Increased weight **b.** Improved impact strength **c.** Improved processability **d.** Reduced crystallinity

15-3. Which group of polymer fibers provides superior strength and resistance to biological attack?
a. Manufactured **b.** Cotton **c.** Jute **d.** Hemp

15-4. In transportation vehicles, structural plastics are often not used because they lack good specific strength.
T (true) **F** (false)

15-5. A property unique to polymers that gives them the ability to return to their original size and shape once a load is removed, much like a shock absorber handles bumps, is
a. Viscoelasticity **b.** Voigt **c.** Specific strength **d.** Branching

15-6. A problem with many plastics is that they often creep at room temperature over time with fairly small loads.
T (true) **F** (false)

15-7. Paper is a better packaging material because, in the modern landfill, it will biodegrade.
T (true) **F** (false)

15-8. Thermosetting plastics are not good for recycling because they cannot be softened for reprocessing.
T (true) **F** (false)

Pg 468

15-9. Common processes for thermoplastics are casting, extrusion, blow molding, and thermoforming. **T** (true) **F** (false)

15-10. The first stage of a cyclic thermoplastic that makes processing easier is
a. Oligomer **b.** Linear **c.** Hevea **d.** RIM

15-11. A property found in wood that results from grain patterns is
a. Anisotropy **b.** Isotropy **c.** Homogeneity **d.** Interface

Pg 450

15-12. A ratio of strength to weight is
a. Isotropism **b.** Tensile strength **c.** Specific strength **d.** Density

Pg. 474

15-13. Which advantage do smart polymeric materials offer designers?
a. Elimination of selection problems **b.** Reduction in raw materials costs
c. Adaptation to environmental changes **d.** Reduction in processing

15-14. Which application would benefit most from the properties of electrorheological fluids?
a. Multigrade engine oil **b.** Robotic power transmission **c.** Superconductors
d. Telephone switching

15-15. What does "organic" refer to in the SPI definition of plastics? What are the other major ingredients in plastics?

15-16. What advantages do most plastics offer compared to linear plastics?

15-17. Describe the following plastic classifications:
a. general purpose **b.** engineering **c.** cellular

15-18. List additives for plastics that moderate the following service conditions:
a. aircraft storage compartments subject to fire **b.** coating for a wrench handle
c. vegetable storage drawers in refrigerated bins **d.** flexible gloves
e. stiff, lightweight automobile interior door panels

15-19. Determine whether thermosetting or thermoplastic plastics are *generally* best for the following type of service conditions:
a. high impact **b.** creep **c.** chemical exposure

15-20. Use Tables 15-2 and 15-6 to calculate the specific strength of aluminum, and compare it to the specific strength of an acetal plastic.

15-21. From the tables in this module and current prices listed in the reference issues of periodicals such as *Machine Design and Modern Plastics*, calculate the specific strength and cost per unit volume for the following materials:
a. acrylic **b.** polycarbonate **c.** plain-carbon steel **d.** wood **e.** epoxy
f. aluminum.

REFERENCES & RELATED READINGS

American Society for Testing and Materials. *Annual Book of ASTM Standards*, Volume 08.01, *Plastics* (D-20). Philadelphia, PA: ASTM, 1989.

Brown, R. P., and J. H. Greenwood. *Practical Guide to the Useful Life of Plastics*. Detroit, MI: Plastic Design Library of William Andrew Publishing, 2002.

Coalition on Resource Recovery and the Environment. "Incineration of Municipal Solid Waste: Scientific and Technical Evaluation of State of the Art." Washington, DC: The United States Conference of Mayors, February 1, 1990.

Constance, J. "Dendrimers: Materials by Design," *Compressed Air Magazine*, June 1992, pp. 12–17.

Crane, F. A. A., and J. A. Charles. *Selection and Use of Engineering Materials*, 2nd ed. Woburn, MA: Butterworth, 1989.

Data Business Publishing. *Plastics Materials DATA Digest*. Englewood, CO: DATA, 1996.

Fine, Leonard W., and Herbert Beall. *Chemistry for Engineers and Scientists*. Orlando, FL: Saunders College Publishing, 1990, pp. 698–702.

Fischer, David. "Turning Trash into Cash," *U.S. News & World Report*, July 17, 1995, p. 43.

GOBSTEIN, SAUL. "Introduction to Plastics and How to Choose Them." ASM International continuing education short course materials. Materials Park, OH: ASM International, 1993.

HOCKING, MARTIN B. "Paper Versus Polystyrene: A Complex Choice," *Science*, June 7, 1991, pp. 504, 506.

INSTITUTE OF SCRAP RECYCLING INDUSTRIES. *http://www.recycle.net/recycle/assn/rs000103.html,* January 4, 2000.

JACOBS, JAMES. "Recycling Plastics," *The Technology Teacher*, September/October 1990, pp. 15-22.

LABANA, S. S. "Recycling of Automobiles—An Overview." Paper presented to NEW: Update 93, NASA–LaRC, November 4, 1993.

Lawrence Livermore National Laboratory Silica Aerogel website. *http://eande.lbl.gov/ECS/aerogels/satoc.htm,* January 2, 2003.

SALEM, A. J., et al. "Fabrication of Thermoplastic Matrix Structural Composites by Resin Transfer Molding of Cyclic Bisphenol—A Polycarbonate Oligomer," *SAMPE Journal*, January/February 1991, pp. 17–22.

SEYMOUR, RAYMOND B. *Polymers for Engineering Applications*. Materials Park, OH: ASM International, 1987.

SHEER, ROBERT J. "Incorporating Intelligent Materials into Science Education." National Educators' Workshop: *Update 94—Standard Experiments in Engineering, Materials Science, and Technology*. Gaithersburg, MD, Nov. 7–9, 1994.

SOCIETY OF PLASTICS INDUSTRY. *Plastics Today—SPI's Fax-on-Demand Service*. Phone 800–774–4614 (U.S. only), January 10, 2003.

WIRKA, GENE. *Wrapped in Plastic*. Environmental Action Federation, 1990.

Wypch, George. *Handbook of Material Weathering*, Detroit, MI: Plastic Design Library of William Andrew Publishing. 2003.

YESTER, SUSAN G. "Life Cycle Inventory." NEW:Update 99, November 1, 1999, Auburn Hills, MI.

ZAJONC, AUGUST. Green building website. *http://www.coldhamarch.com/green,* January 17, 2000.

Periodicals

Modern Plastics

Plastics Technology

Plastics World

Module 16

Elastomers: Natural & Synthetic Rubbers

After studying this module, you should be able to do the following:

16–1. Diagram thermoplastic and cross-linked rubber and describe the effects of rubber bonds (lengthening and straightening).

16–2. Recall types, properties, and comparative cost of elastomers.

16–3. Describe rubber additives and hardness.

16–4. Use the website addresses denoted by the symbol to link to sites related to elastomers.*

16–5. Cite examples of new rubber technology including nanotechnology.

*The icon found throughout this book will link you to Internet sites related to topics being covered. The dynamic nature of the Web brings frequent changes, so some of these sites, while available at the time of writing, may not be up now.

Elastomers refer to the group of polymers that exhibit rubbery or elastic behavior, including *natural rubber* and *synthetic rubber*. The American Society for Testing and Materials (ASTM) defines ***elastomers*** in ASTM D1566–66T as "macromolecular material that returns rapidly to approximately the initial dimension and shape after substantial deformation by a weak stress and release of the stress." Elastomers will stretch from 200% to over 900% of their original length at room temperature and return to their original state when the tensile stress is released. Natural rubber, known as ***hevea rubber,*** comes from the rubber tree *Hevea brasiliensis*. Columbus brought this bouncy material back to Europe with other curiosities from the New World, but nearly 300 years passed before it was given serious attention. The name ***rubber*** resulted from the material's ability to erase black pencil marks from paper by rubbing. Synthetic rubber, as with plastic, is a hydrocarbon with petroleum as its major source. Development of the processing of synthetic rubber into a method suitable for high-volume production did not fully materialize until the beginning of World War II, when the United States was cut off from the natural rubber supply in India and was forced to use large quantities of synthetic rubber.

Figure 16-1 provides an idea the widespread applications of rubber. We often don't give much thought to elastomers, unless our radiator hose springs a leak out on the freeway or if we experience a tire blowout at high speeds. This module will provide you with basic concepts for natural and synthetic elastomers.

16.1 NATURAL RUBBER AND VULCANIZATION

Natural rubber (NR), in the form of liquid resin secreted from the inner bark of the *Hevea brasiliensis* tree, is known as *latex*. Latex is not tree sap; it consists of isoprene molecules (Figure 16-2a). Polyisoprene, or rubber, is formed through a natural polymerization in the tree (Figure

Figure 16-1 Elastomers in cars. How many types of rubber can you name in the typical car? This diagram depicts applications of rubber. The rubber reduces vibration and isolates against noise, as seen in (1) engine mounts, (2) strut mounts, (3) bushings, (4) couplings, (5) air sleeves, (6) insulation against impact, (7) insulating foams, (8) gaskets, (9) seat padding, (10) trim, (11) cushions for headrests, (12) door cushions, and (13) roof trim. It's easy to notice the rubber tires, which involve the blending of various natural and synthetic rubber with reinforcement wire and fibers. Rubber is also used to make cooling and brakes hoses, floor mats, and fan belts. Each of the applications has special requirements that result in many rubber compounds. Other important polymers include sealants, lubricants, oils, natural leather, and numerous adhesives. Of course, many plastics are also used in the modern automobile. (Bridgestone Firestone)

16-2b). The liquid latex dries into a thermoplastic material. The structure is an amorphous mass of coiled and kinked chains having constant, thermally induced motion. Rather than stretch when heated, rubber shrinks because the thermal agitation causes the chains to entangle and draw up. The structure provides the ***resilience*** for rubber to spring back into shape when compressed or stretched (Figure 16-3a and b). When a tensile load is applied, the structure changes as the chains straighten (bond straightening) and stretch (bond lengthening) and crystallinity is achieved in varying degrees depending on the amount of stress. With increased crystallinity comes a greater strength, increased rigidity, and increased hardness. The structural changes in rubber make it good for tires because the portion of the tire under high stress is crystalline and provides support for the vehicle it is carrying, while at the same time the portions of the tire not under high stress are still resilient and absorb shock from bumps in the road. Figure 16-4a and b shows examples of other products that take advantage of the tough, flexible nature of rubber.

The problems with natural rubber are that it is too soft and has too many reactive sites (double bonds), which cause rapid oxidation and ***dry rot.*** It is also somewhat plastic and will not recover from high stress. In 1839, Charles Goodyear discovered, through the addition of sulfur to natural rubber compounds, that it was possible to increase hardness and reduce susceptibility to oxidation and reaction with other chemicals. The sulfur ***vulcanizes*** the rubber, or changes it into a thermosetting polymer, by linking together the molecular chains at their double bonds (Figure 16-5a). Two sulfur atoms per pair of isoprene mers are needed, and with 5% of the mer pairs cross-linked,

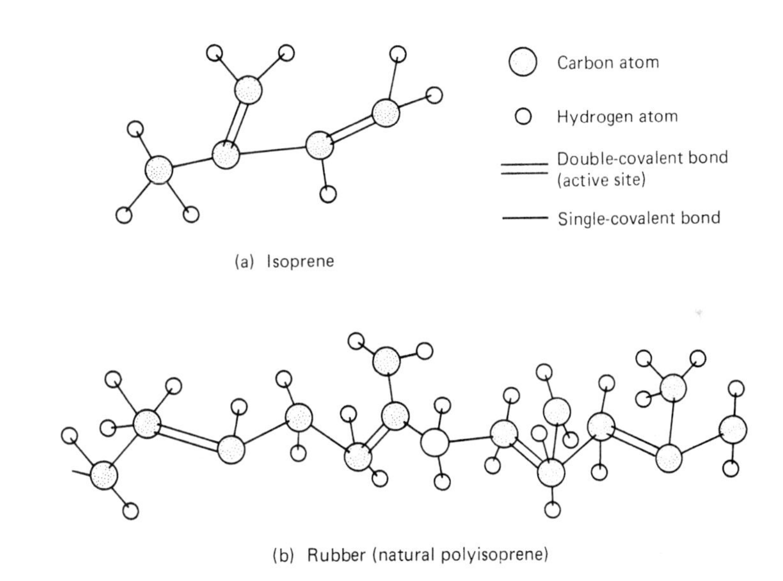

Figure 16-2 (a) and (b) Natural rubber.

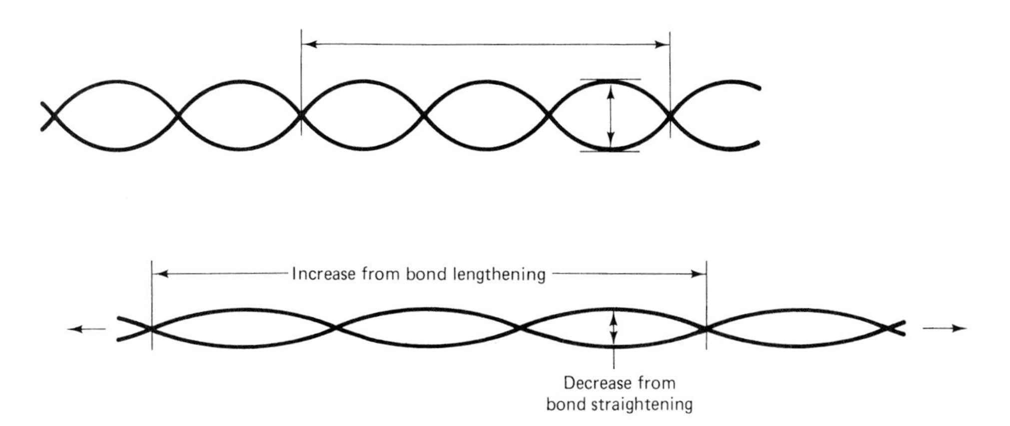

Figure 16-3 Effect of tension on bonding: (a) no load, (b) tensile load.

(a)

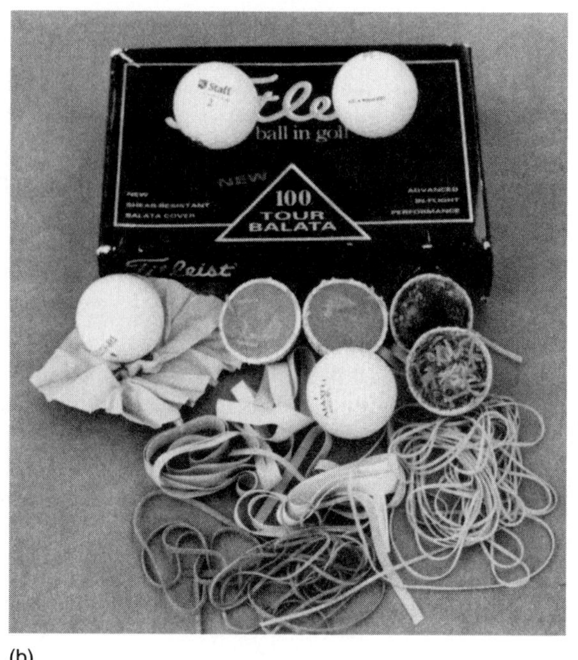

(b)

Figure 16-4 (a) Tennis, running, and other sport shoes vary in density and stiffness as a result of blending rubber compounds. (b) Golf balls exemplify the variety of properties achieved with synthetic and natural rubbers. Note the variety in core rubber, which ranges from solid core to thin band winding and liquid center. Ball covers of synthetic balata rubber or rubber blends provide a soft feel as the ball is struck, yet are tough and scuff resistant. Considerable effort goes into the dimple design on the covers in order to achieve optimum aerodynamics.

a flexible, resilient rubber results. Further cross-linking increases hardness. Vulcanization of only sulfur to rubber requires several hours and temperatures around 145°C. However, accelerators and activators added to the compound will result in vulcanization within minutes.

In addition to natural rubber are the sulfur cross-links polybutadiene (BR), polyisoprene (IR), and acrylate (ACM) elastomers. Along with sulfur, other cross-linking agents vulcanize certain synthetic elastomers. Butyl rubber (IIR) and ethylene–propylene copolymers (EPM) rely on sulfur and phenolic agents; zinc oxides form cross-links in polysulfide (PTR) and polychlorapene (CSM). Vulcanized natural rubber has excellent flexural strength or deformability, tensile strength, and abrasion resistance, but it is attacked by petroleum oil, greases, and gasoline. Its value in automobile tires comes from low heat buildup, which is discussed later. Natural rubber has superior overall engineering capabilities compared to synthetics. The choice of a synthetic rubber over natural rubber or of one synthetic over another boils down to the specific properties required, price, and processability.

Because natural rubber comes from a renewable source and because it has excellent properties, the study of natural rubber science and technology continues. Deproteinized natural rubber (DPNR) is an example of the developments in improving this natural elastomer. By removal of ingredients with an affinity for water, such as protein, polyols, and inorganic salts, fatigue resistance and other mechanical properties are enhanced. Research has been renewed in the United States into the use of guayule shrubs to produce rubber; they were grown in California during World War II. Only 5% to 10% yield was possible then; a 20% yield is required to justify the use of this raw material.

16.2 SYNTHETIC RUBBER

Even with the improved properties obtained through vulcanization, natural rubber has poor resistance to aging. It is attacked by ultraviolet light, oxygen, and heat because it still has many reactive sites. Due to these shortcomings and because the rubber tree grows only in

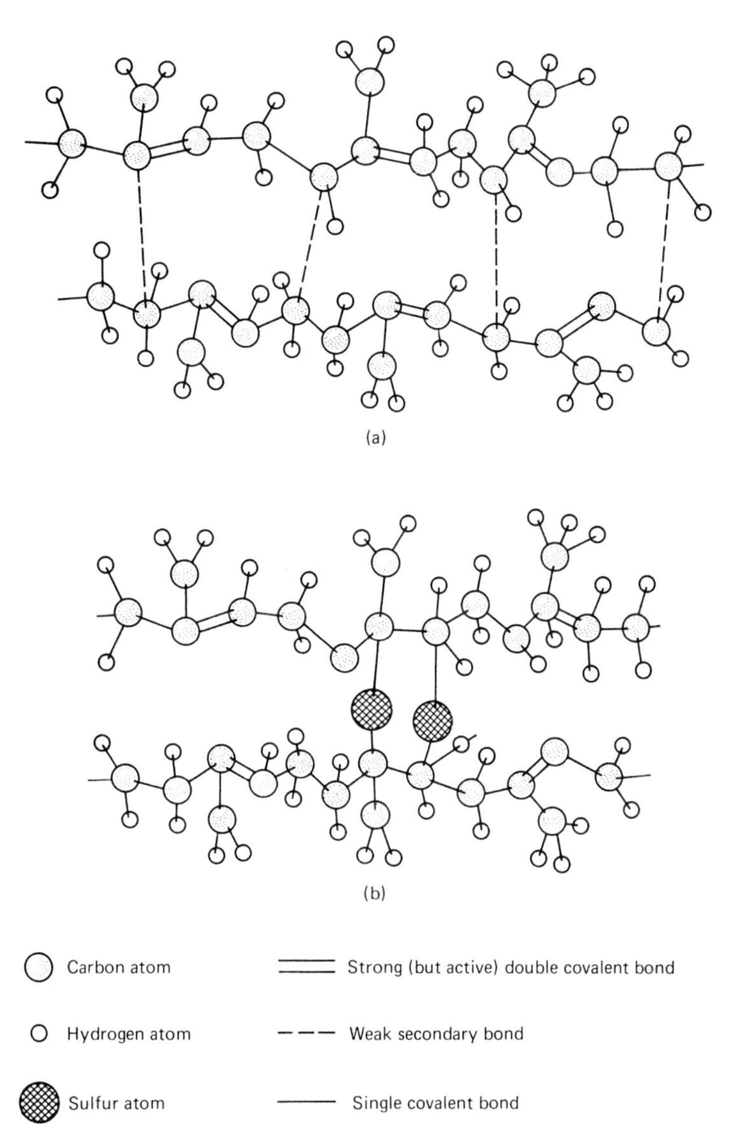

Figure 16-5 Vulcanized rubber. (a) Natural rubber—thermoplastic. (b) Vulcanized (cross-linked) rubber—thermoset.

special environments that were vulnerable to political sanctions, the search for a substitute produced several synthetic rubbers in the early to middle part of the twentieth century. The number of synthetic rubbers has grown to include many special-purpose elastomers. Improvements in synthetic elastomers have also brought improvement in the additives and fillers for natural rubber (NR). Almost identical to NR is synthetic polyisoprene (IR), except that it has greater stretching ability. Synthetic rubbers have the same raw materials used in plastics. Starting with crude oil and natural gas, many formulations are possible. Styrene, the monomer for styrene–butadiene rubber (SBR), is derived from coal as vinyl benzene, $CH = CH_2$, a product of benzene and ethylene. Butadiene is derived from petroleum. The hydrocarbon is obtained in the fractionation cracking of petroleum used for olefins, polymers, or gasoline. The 1-butene is separated and catalytically dehydrogenated in the vapor phase to produce butadiene.

Figure 16-6 shows the steps in modern automobile tire production. During the final processing step, heat and pressure are used to form the final tire shape, then the rubber that was in a thermoplastic condition during forming is vulcanized. Note, at the beginning stage, the mixing of natural and synthetic rubbers with carbon black, sulfur, and other additives.

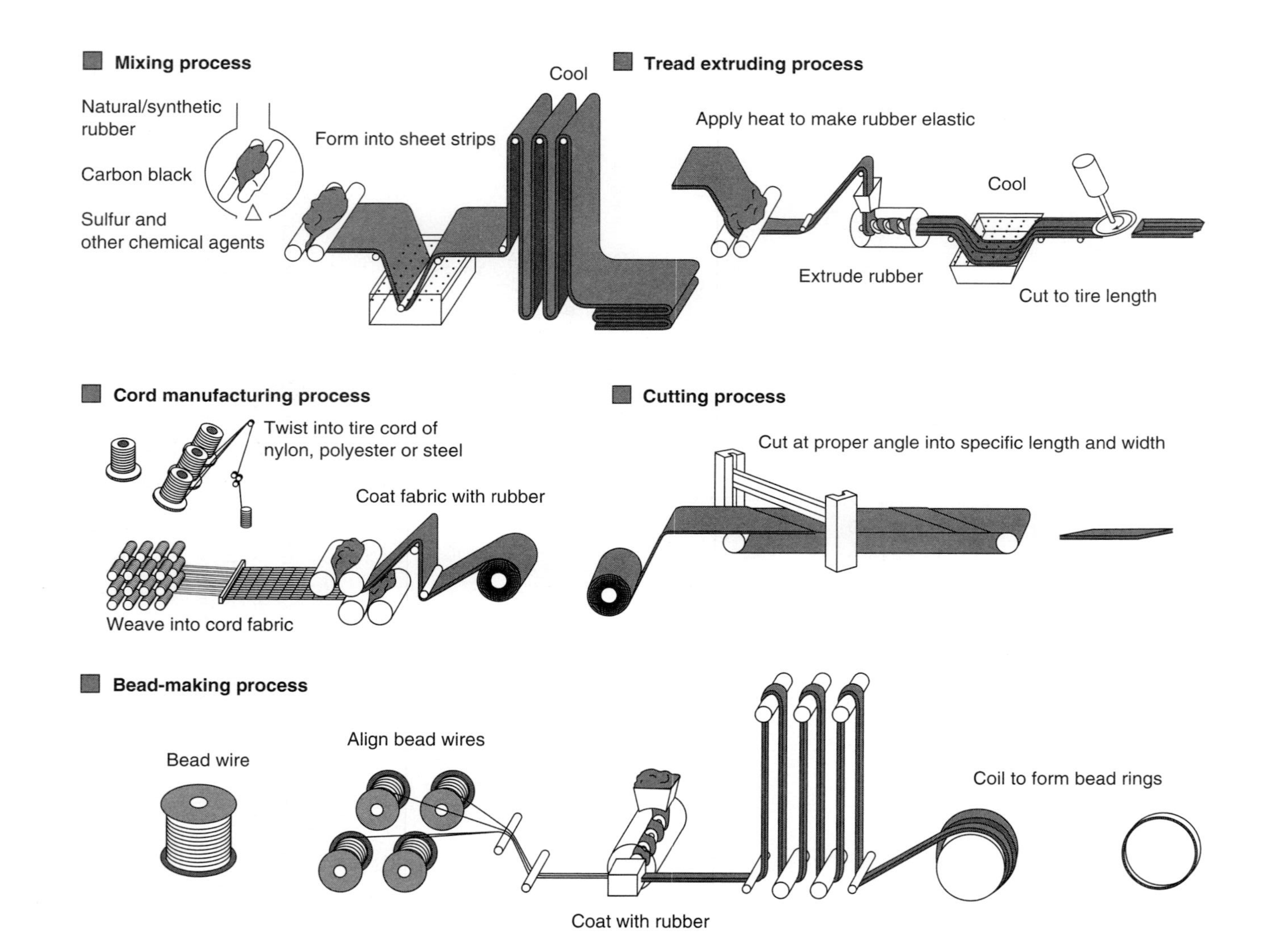

Figure 16-6 (a) Steps in tire production. (Bridgestone Firestone)

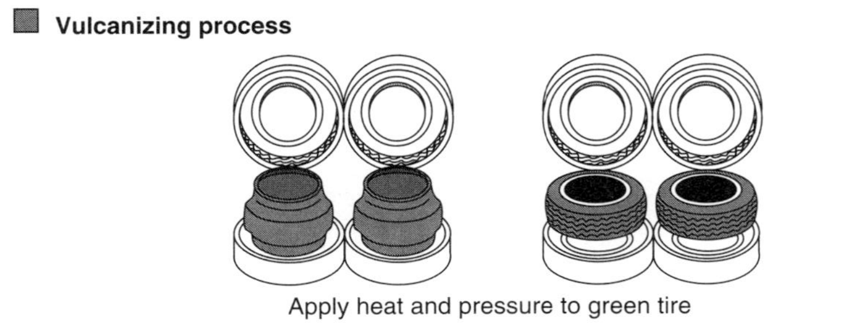

Figure 16-6 continued (b) Steps in tire production.

16.2.1 Additives (Modifiers)

Additives (modifiers) to synthetic rubber are introduced at the initial processing stage and also during compounding. Soap and water are added to styrene and butadiene in the making of raw styrene–butadiene rubber (SBR) to produce an emulsion that keeps droplets of the monomers from separating out. ***Accelerators*** and ***activators*** speed sulfur vulcanizing, while ***retarders*** prevent vulcanizing before it is required. For those saturated synthetic elastomers that have no double bonds, ***peroxides*** promote sulfur vulcanizing. ***Peptizers*** soften raw rubber, ***pigmenters*** add color, ***abrasives*** provide abrasing action for products such as erasers, ***hardeners*** increase rigidity, and ***antioxidants*** and ***antiozonants*** prevent aging or dry rot from the sunlight and ozone.

Fillers help to increase the bulk of the compound while holding down cost. Inert fillers include talc, chalk, and clay. Reinforced rubbers use many of the same fillers as plastics, including metals, glass, and polymer fibers. ***Carbon black*** also serves as a strengthening filler and imparts hardness.

16.2.2 Thermoplastic Rubbers or Elastoplastics

Thermoplastic rubbers (TR) or elastoplastics do not cross-link as do most elastomers, even though their properties are similar to many rubbers. A copolymer of styrene and butadiene was introduced in 1965 as a thermoplastic rubber. The elastoplastics have the ability to soften when heated for processing and, upon cooling, become solid yet maintain their elastic behavior. Vulcanization is eliminated. The thermoplastic behavior develops as a result of the structure, in which the chains of one monomer form links in the center of block molecules of other monomers. Figure 16-7 shows that the domains in the middle have a hard linking effect, while the other end of the segments has elastomeric behavior. Upon heating through Tg, the hard centers soften to plasticize the rubber for processing. When cooled, the hard blocks re-form to again link the copolymers. This linking is especially effective at low temperatures.

Two-block, copolymer, thermoplastic elastomers manufactured under the trade name Kraton include styrene–butadiene styrene (SBS) and styrene–isoprene styrene (SIS). The styrene (S) occurs in two thermoplastic blocks, as shown in Figure 16-8a, on either end of a rubber block of butadiene (B) or isoprene (I). Rather than having randomly distributed monomers, as do most copolymers, block copolymers form uniformly distributed sections as blocks. Figure 16-8b shows that a network of block chains has formed. It consists of a highly uniform distribution of ***polystyrene domains*** and the separate butadiene rubber chains. The ***rubber network***, which is physically joined, links the domains and cannot move because the domains are immobile, thus providing a physical rather than a chemical cross-linking. The physical cross-linking can be broken when heated, which makes the compound a thermoplastic. Kraton is therefore capable of continuous molding and remolding. The polystyrene domain consists of actual particles that become hard and glasslike at room temperature; they take on cylindrical, spheroidal, and planar configurations.

The major elastoplastics are polyester copolymers, olefinics, polyurethanes, and styrene copolymers. The wide service-temperature ranges, low-temperature flexibility, and ease of

Figure 16-7 Polystyrene domains of thermoplastic styrene butadiene rubber.

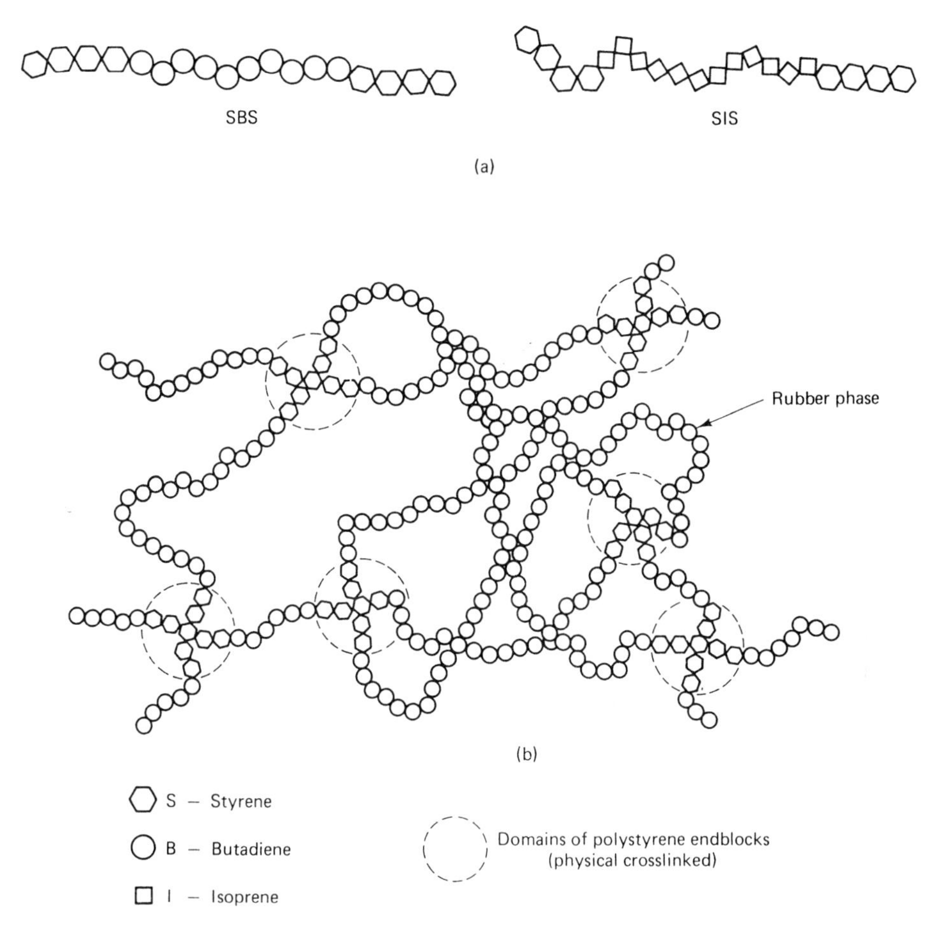

Figure 16-8 Thermoplastic rubber "network" structure. (a) Block copolymers. (b) SBS.

painting place thermoplastic rubbers in contention as a substitute for thermosetting rubbers. Their ease of processability has given them wide acceptance in the auto industry as flexible bumpers and other exterior panels, steering wheels, hose covers, housings for seat belts and horns, wire covering, and oil seals. Other applications include housewares, toys, sporting equipment, adhesives, rainwear, footwear, and skate wheels. As with plastics, scrap thermoplastic elastomers can be ground and remolded. Because of the complex nature of the modern automotive tire, its "after-life" has been a real problem. Many methods have been employed to recycle the millions of discarded tires, including grinding them into materials for road paving and forming them into tiles for patios and tennis courts and fence posts and for burning them to produce electricity.

As we have seen, ***shape-memory polymers*** are a class of "smart materials" that return to their original fabricated shape under the influence of temperature. ***Bioelastics*** are synthetic fibers that behave much like muscles. Based on the natural protein elastin, these biomaterials offer medical applications for surgical repairs as well as industrial uses such as for robotic hands and automatic valve controls to replace metal springs and hydraulic and pneumatic devices. Bioelastics are nontoxic, biocompatible with human tissue, and biodegradable. Because bioelastics can convert chemical energy into mechanical energy, they offer potential as sensors or transducers.

16.3 CLASSIFICATION AND PROPERTIES OF ELASTOMERS

Designations of elastomers by the American Society for Testing and Materials (ASTM D2000 standard) and the Society of Automotive Engineers (SAE) group them by type and class (Figure 16-9). ***Type*** reveals the maximum service temperature, which ranges from A (70°C) to

Type	Test temp (°C)	Class	Maximum swell (%)
A	70	A	No requirement
B	100	B	140
C	125	C	120
D	150	D	100
E	175	E	80
F	200	F	60
G	225	G	40
H	250	H	30
J	275	J	20
		K	10

(a) ASTM D 2000/SAE J200 type and class

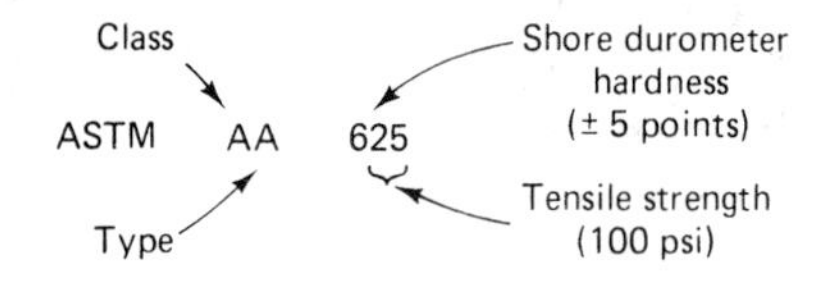

(b) ASTM designation example

Figure 16-9 Elastomer designation. (a) ASTM D2000/SAE J200. (b) ASTM designation example.

J (275°C). ***Class*** indicates the maximum percentage of swell when immersed in oil, and ranges from A, with no requirement, to K, at 10% swell. Table A-10 shows the class and type for selected elastomers. ASTM D1418 establishes abbreviations for elastomers, also shown in Table A-10. To illustrate, PTR, AK designates a polysulfide elastomer with a maximum service temperature of 70°C that can swell as much as 10%. Additional systems of designation indicate hardness and minimum tensile strength (Figure 16-9). ASTM AA625 designates (AA) natural rubber with Shore durometer hardness (6), meaning 60 plus or minus 5 points on the scale and 25 (2500) as the minimum tensile strength.

Table A-10 compares properties for common elastomers. NR is the standard of comparison for the synthetics, as can be seen by its overall superiority, with excellent resistance to abrasion, tear, impact, electrical current, and water absorption, plus a tensile strength of 3100 Pa and a wide hardness range of 30 to 100 Shore A. By means of designed formulations, the synthetic rubbers can surpass NR for special properties. NBR, ACM, FPM, and PTR offer excellent resistance to oil, kerosene, and gasoline; NR is not recommended for service when exposure to these chemicals is expected. For high-temperature service, SI and UE are excellent. Neoprene resists deterioration from weathering, oxygen, ozone, oil, gasoline, and greases and has good tear and abrasion resistance. It finds applications as soles and heels for work shoes and liners for chemical tanks and pipelines, serves as an adhesive, and is applied like paint as a protective coating. Fluoroelastomers such as FPM are space-age elastomers that meet the demands of the aircraft and aerospace industry because of excellent resistance to heat and fluids, but they carry a high price tag. NR has only fair ***gas permeability,*** which means that it will allow some gas to pass through it. Butyl rubber (IIR) resists gas permeating and serves well as inner tubes, gas hoses, tubing, and diaphragms.

Rubber plays an important role in health care. Silicone elastomers, the most stable group of all elastomers, have polymer chains composed of silicon and oxygen backbones. Silicone rubber is used for various body replacement and enhancement components. Silicone is often used for plastic, reconstructive surgery of body tissue; recent problems with breast implants have caused concern about silicone as an implant. Silicone is a derivative of the second most abundant material on earth—silicon. Silicon composes about 28% of the earth's crust and is usually found in combination with other elements. One combination is silica, a naturally occurring silicon–oxygen compound known as silicon dioxide. Sand is probably the best known silica. Silicones are a family of extraordinary synthetic materials made from polymers. Of high molecular weight, their molecular structure consists of spines of oxygen alternating with silicon. By attaching various organic, carbon-containing groups of elements to the spines, liquids, pastes, and rubberlike materials can be produced. The liquids and pastes can be cured into solids with application of heat or exposure to air to make adhesives and sealants. In other forms they are used as elastomers, coatings, oils, lubricants, and resins. If the attachment of these compounds to the spines of the SiO_2 polymer is short, the result is a volatile liquid that has the appearance of water and evaporates in air. If the attachment is long and ropelike, the physical properties change and the materials are more viscous, ranging from syrupy to a "solid" that flows if left unattended. You can stretch it, pound it, roll it into a ball and bounce it, and hit it with a hammer so that it shatters. This material has the trade name Silly Putty.

Uses for silicone include baby bottle nipples, heat-resistant gaskets, cable insulation, spark-plug boots, encapsulations, bumper gels, brake fluid, adhesives for exterior tile on space shuttles, sealants, and weatherproof caulks. Theoretically, millions of silicone materials can be made just by altering the structure of the silicone chain. Only a few thousand of these have been produced and studied so far.

Natural-rubber latex has gained considerable attention recently because of concern over the transmission of AIDS and other sexually transmitted diseases. Latex surgical and examination gloves and condoms must meet the highest standards of manufacturing and testing to ensure the highest reliability. ASTM standards D3577, D2578, and D3492 address these procedures.

Many synthetic elastomers, when heated, lose energy, a phenomenon called ***hysteresis.*** If you stretch a rubber band and hold it against your cheek, you can feel the release of heat. The constant flexing of automobile tires generates enough heat in elastomers such as SBR that, if a tire were made wholly of this compound, it would quickly deteriorate. NR, on the other hand, converts the stressing into elasticity and quickly rebounds with less heating. The high hysteresis of SBR has an advantage in preventing slippage on wet surfaces and providing better abrasion resistance than NR, so it is used for the tread of tires while the NR goes into the sidewalls.

Elastomers possess extremely diverse properties, as do plastics. Thousands of additives and fillers are available and as many as 30 may be mixed into a single elastomeric compound. The range of hardness includes soft foam rubbers used for pillows to hard-rubber (ebonite) battery cases. Figure 16-10 compares the hardness of selected elastomers and plastics. ***Foam rubbers*** have been formulated to respond to body temperature and pressure.

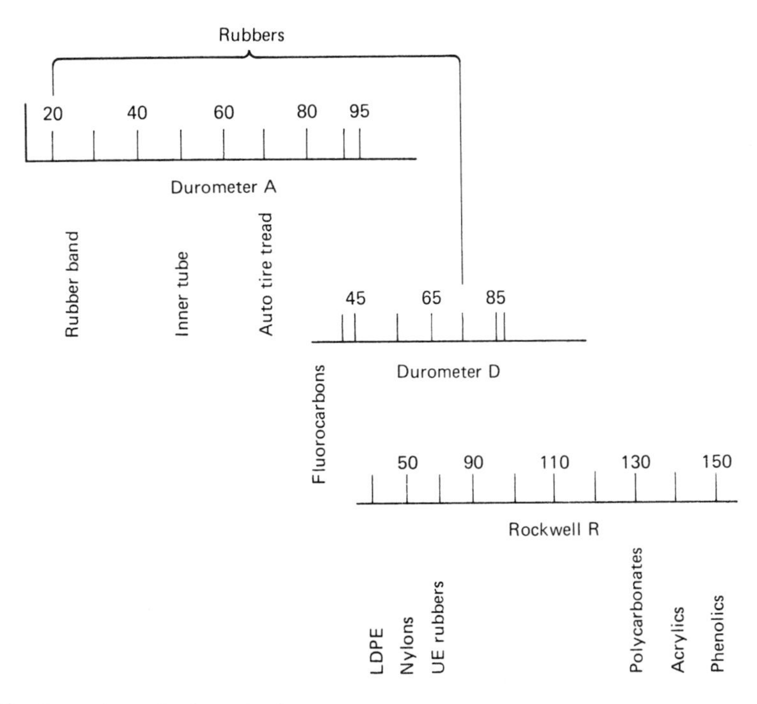

Figure 16-10 Comparisons of polymer hardness.

Developed for astronauts, these foams now serve as orthopedic support for mattresses and chairs because of their ability to provide even support, which reduces fatigue and soreness caused by prolonged body contact. Upon impact, the foam firms up, and when heated, it softens. Joggers and other athletes use foam inserts in their shoes to reduce shock from the impact of running and jumping.

EPDM is a class of 13 types of materials that are a miracle of rubber technology because of their low cost and ability to be painted. Through compounding of carbon black, clay, processing oils, and cross-linking agents, these elastomer compounds can be molded or extruded as thermoplastics or thermosets. Temperature insensitivity, impact resistance, high deformation recovery, dimensional stability, and distortion resistance make them good choices for flexible exterior parts on automobiles, including bumpers and fascia.

Heat resistance is an important property in certain elastomer applications. Fluoroelastomers (FPM) can withstand temperatures over 340°C and continuous service temperatures up to nearly 290°C in exposure to oils, steam, and certain fluids. They have been improved to handle low temperatures down to around −35°C, at which point they become brittle. Among the silicon elastomers, room-temperature vulcanizing (RTV) compounds have stable properties through a range of temperatures, from nearly −53° to 250°C. Ethylene acrylics used for belts, seals, rollers, hoses, and gaskets can withstand hydraulic fluids, engine coolants, and hot oils for over a year at temperatures over 200°C. Chlorinated polyethylene elastomer, which is a random compounding of chlorine with high-density polyethylene, has good weather, hydrocarbon fuel, and oil resistance down to −15°C and up to nearly 150°C in continuous service. Grades of thermoplastic elastomers (TP), such as polyurethane, provide good weathering as low as −15°C and as high as 120°C. Appendix Table A-10 provides general ranges of maximum service temperatures for the groups of elastomers listed.

16.4 SELECTION OF ELASTOMERS

Whether natural or synthetic, rubber offers distinguishing properties. It is flexible, airtight, tough and resilient, and waterproof, and it resists corrosion caused by most common chemicals. Its ability to stretch in any direction and return to its original shape makes it valuable in a variety of applications, from tires, to ball bladders, to hoses and gaskets. Reinforcing fibers can make it a nonstretch composite that retains the other valuable attributes of elastomers. Some high-performance elastomers, such as polyurethane (PUR), can be synthesized as thermosets or thermoplastics (TPUs); TS PUR will elongate 650% and has a tensile strength up to 100,000 kPa; TPU can elongate 600% and has a tensile strength of about 40,000 kPa.

The ***cost of elastomers*** varies widely and has a tremendous influence on selection of one over another. The fluoroelastomer mentioned previously that is capable of temperatures over 340°C sells for over $4500 per kilogram on finished parts, whereas the price for SBR would range from $0.60 to $0.80 per kilogram. Natural rubber sells for about $0.90 per kilogram, SI for $3.75 to $11 per kilogram, and heat-resistant ethylene acrylics sell for $3 to $4 per kilogram. All fluoroelastomers do not cost as much as $4500 per kilogram, but they are generally the most expensive elastomers, selling at around $28 per kilogram, and as copolymers with silicones, most fluorosilicones reach prices of $65 to over $90 per kilogram. Of the thermoplastics, the following indicate general prices per kilogram: copolyesters, $3.20 to $3.65; olefins, $1.37 to $1.75; and styrenes, down to $0.90 to $1.37.

Environmental concerns regarding rubber center on disposal of reinforced rubber, especially vehicle tires. Many approaches have been applied to the disposal issue, as can be seen in the categories listed with links to the *Recycler's World* website *http://www.recycle.net/recycle/Rubber/index.html* (January 15, 2000), as follows: Scrap Tire Disposal and Recycling, Used Tire, Retread Tires and Casings, Tire Parts Recycling, Non Tire Rubber Recycling, Granulated Rubber (Tire Crumb), Tubes and Flaps Recycling, Tire Pyrolysis & Gasification. Recycling approaches range from (1) burning the tires as fuel for heat and power production to

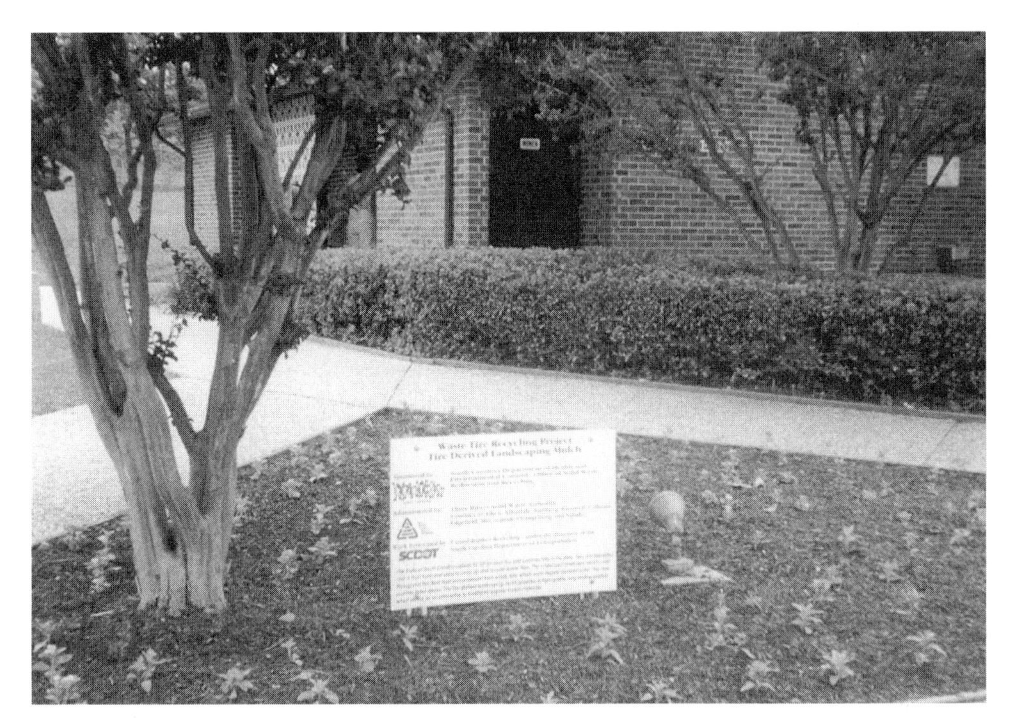

Figure 16-11 A unique recycling idea involves grinding rubber from discarded tires to produce long-lasting mulch to keep out weeds and retain moisture for plants while keeping tires from polluting our environment.

(2) using them as filler and encasing whole tires in very large concrete blocks used as bulkheads to prevent soil erosion to (3) shredding them for filler for roads and gardening mulch, as seen in Figure 16-11.

16.5 NANOTECHNOLOGY

Some applications are beginning to emerge from the slow, but steady, progress of nanotechnology R&D. Rubber provides an excellent material when applications demand a high degree of flexibility and compressive force resistance in product substrates such as tennis balls, soccer balls vehicle tires, and survival rafts. However, we all know that sooner or later these products lose their air pressure. One of the reasons given for rollovers of SUVs with Firestone tires related to low tire pressure. A runflat tire and wheel system is shown in Figure 16-12a. Figure 16-12b shows a relatively new tire that suddenly blew the tread at highway speed. While there was no other damage from this blowout in a sedan, numerous tread separations of a Firestone tire in SUVs did result in death and injury, which resulted in a worldwide recalls, lawsuits, and a major study of quality control in tire manufacture. Anyone who plays tennis knows the balls begin to go "flat" and become "dead" due to a loss of pressure. A rubber nanocomposite has been developed by InMat to create an impermeable core for tennis balls and to have potential for other balls and for tires. Tennis balls known as Double Core reportably hold pressure twice as long as regular balls. As seen in Figure 16-13, the layered composite structure consists of a layer of latex containing suspended butyl spheres about a micrometer in size; then nanoscale ceramic (vermiculite) platelets are introduced into the inner core layer. This forms a layer of about 20 μm that is impenetrable to air.

Making rubber—*http://www.negromex.com.mx/english/Learn.htm*
Firestone rubber products—*http://www.firesyn.com/*
Enterplast rubber products—*http://www.enterplast.com/rubber.html*

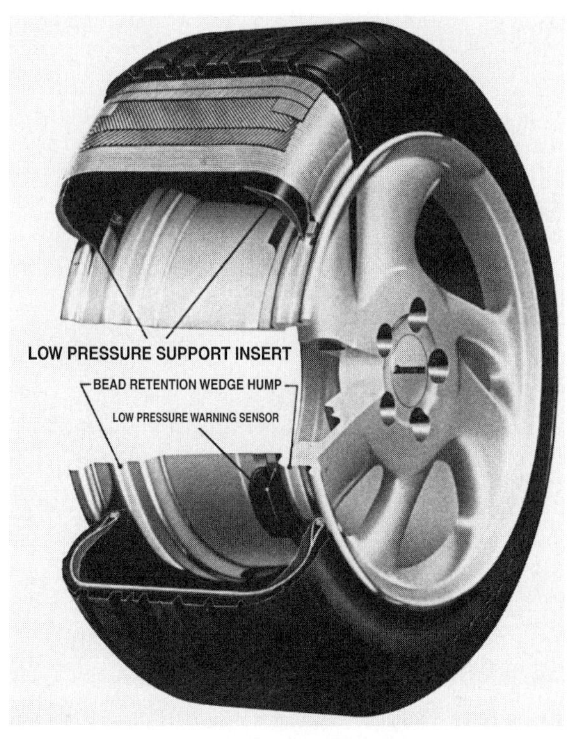

(a)

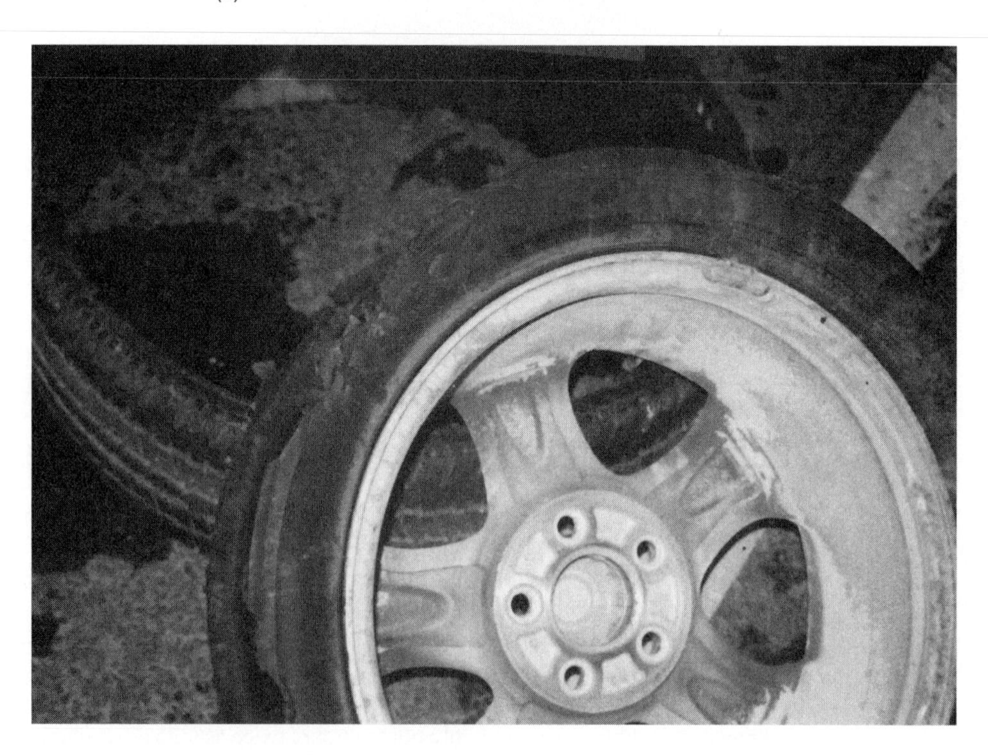

(b)

Figure 16-12 (a) Runflat tire and wheel system. (Bridgestone Firestone) (b) With only a few seconds of warning noise, a relatively new tire threw its tread at highway speed.

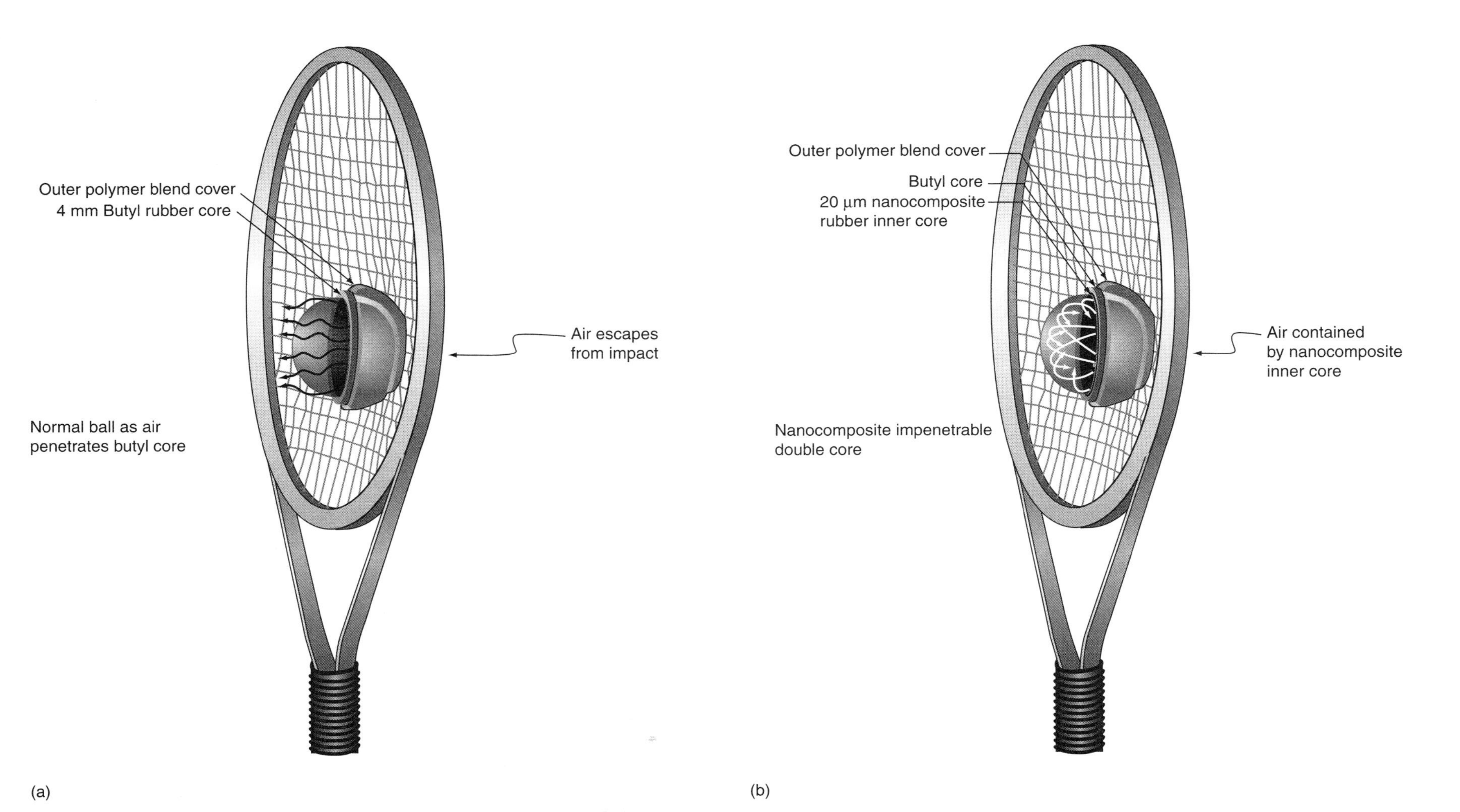

Figure 16-13 (a) Normal ball as air penetrates butyl core.(b)An ultrathin nanocomposite layer of latex, butyl, and vermiculite offers an impenetrable air barrier. (Adapted from AMS International. "Double-Core tennis balls hold bounce twice as long," *Advanced Materials & Processes,* September, 2002, p. 31)

16.6 ARTIFICIAL MUSCLE

Another example of biologically inspired materials is ***electroactive polymers (EAPs),*** polymers that move in response to an electric current. For decades designers have sought materials that would replicate muscles, to replace heavy and complex motors, actuators, gears, and linkages. Muscles of all types (ranging from those of fleas to frogs to kangaroos) are mechanisms that work equally efficiently regardless of size. Muscles change size in response to nerves to make minor, subtle movements such as a twitch of the nose, or large, complex movements, as when a pole-vaulter propels herself 16 feet over the crossbar. EAPS in the form of elastomers are beginning to evolve that may replace small motors in robots or on aircraft, or in other applications where small, light mechanisms will provide appropriate controlled movement.

Figure 16-14 depicts a simple electroactive polymer mechanism, constructed as rubbery capacitors with two parallel carbon plates with silicone (dielectric) sandwiched between them. With the application of voltage, positive and negative charges build up on opposite plates, thereby causing attraction. The polymer insulator is squeezed and elongates. Figure 16-15 shows a complex robotic hand opening and closing. EAP could simplify this type of device. Dielectric elastomers, as seen in Figure 16-14, can elongate up to 400 percent.

Visit the websites below for more details.

Jet Propulsion Laboratory, NASA—*http://ndeaa.jpl.nasa.gov/nasa-nde/lommas/cap*
Muscle demonstrations—*http://www.erg.sri.com/projects/miscle/receipe*

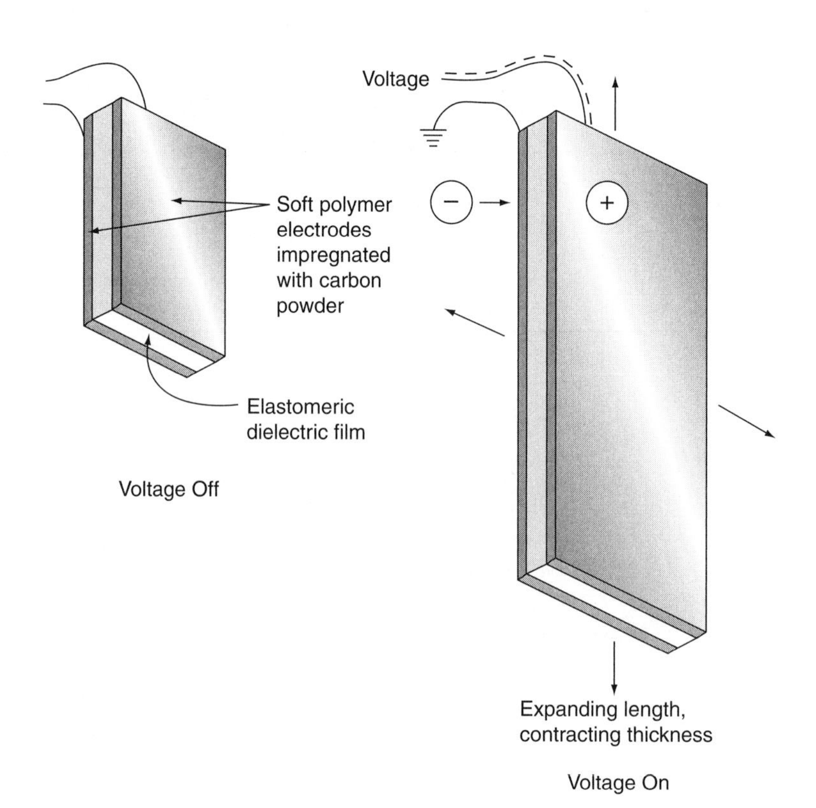

Figure 16-14 EAP mechanisms are being developed as one form of artificial muscle. By laminating thin (30–60 μm) films of silicone or acrylic between soft polymer sheets impregnated with carbon powders, current applied to the oppositely-charged plates causes them to squeeze the film and elongate it up to 400%. (Adapted from "Artificial Muscles," *Scientific American* October, 2003, pp. 52–59)

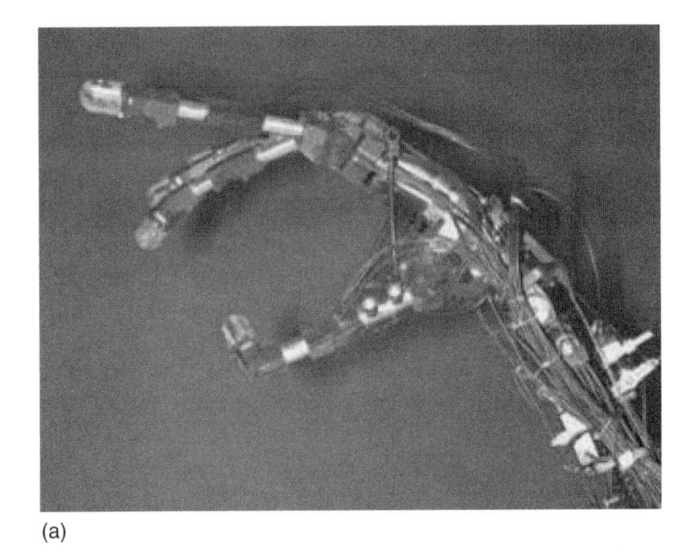

(a)

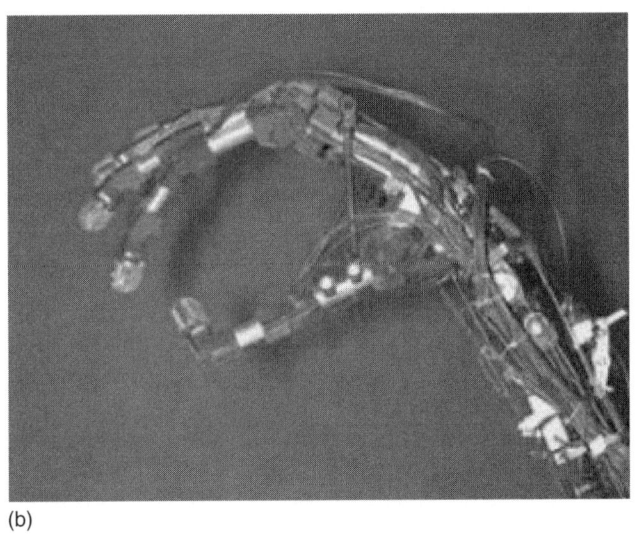

(b)

Figure 16-15 Robotic hands now require complex mechanisms to attempt to replicate human motions. EAPs offer the potential to create the artificial muscles for simpler mechanisms. (NASA, Jet Propulsion Laboratory)

SELF-ASSESSMENT

16-1. Diagram the polymer chain structure of rubber at rest and with tensile stress applied.

16-2. The objective of vulcanizing rubber is

a. To improve processability **b.** To increase hardness and chemical resistance
c. To re-duce materials cost **d.** To increase amorphousness

16-3. Vulcanization of rubber uses sulfur to transform it into a thermosetting elastomer.

a. T (true) **b.** F (false)

16-4. Before vulcanization, natural rubber has many double covalent bonds known as

a. Reactive sites **b.** Benzene rings **c.** Isoprenes **d.** Copolymers

16-5. The major source of raw materials for synthetic rubber is

a. Hevea **b.** Oil **c.** Trees **d.** Corn

16-6. The purpose of an antioxidant additive is to

a. Prevent aging **b.** Speed processing **c.** Accelerate curing **d.** Strengthen

16-7. The purpose of carbon black as an additive is to

a. Prevent aging **b.** Speed processing **c.** Accelerate curing **d.** Strengthen

16-8. Use Table A-9 to choose a suitable elastomer for the following containers:

a. hydraulic hose **b.** hot water bottle **c.** diver's wet suit **d.** bicycle-tire inner tube

16-9. What additives to synthetic polymers accomplish the following:

a. retard aging from ultraviolet rays **b.** soften raw rubber **c.** increase rigidity
d. strengthen

16-10. Select elastomers and specify their ASTM abbreviation and their favorable properties for the following applications:

a. fire hose **b.** lawn mower wheels **c.** stopper on chemical test tubes **d.** golf-ball winding

e. racket ball **f.** tennis-shoe soles **g.** rubber band

h. seal on automobile brake master cylinder **i.** carpet backing

j. work-shoe soles **k.** pad for bicycle caliper brakes **l.** wrestling mat

16-11. Thermoplastic rubbers process easier than most thermosetting rubbers.

a. T (true) **b.** F (false)

16-12. An application for shape-memory polymers is

a. Robotic hands **b.** ATMs **c.** Tire reinforcement **d.** Netting

16-13. The energy loss through heating in elastomers, which creates problems in applications such as car tires, is called

a. Ablation **b.** Oligomer **c.** Cohesion **d.** Hysteresis

16-14. The elastomer used most often to help prevent the transmission of AIDS is

a. Neoprene **b.** Fluoroelastomers **c.** Latex **d.** IIR

16-15. A most stable elastomer used for human implants is

a. Latex **b.** NR **c.** Fluoroelastomers **d.** Silicone

16-16. Elastomers offer a wide range of properties from very soft to quite hard, they resist most chemicals, they provide good traction and abrasion resistance, and they are quite flexible.

a. T (true) **b.** F (false)

REFERENCES & RELATED READINGS

AMERICAN SOCIETY FOR TESTING AND MATERIALS. *Annual Book of ASTM Standards*, Volume 09.01, *Rubber, Natural and Synthetic*. Philadelphia, PA: ASTM, 1991.

AMS International. "Double-Core tennis balls hold bounce twice as long," *Advanced Materials & Processes.* September 2002, p. 31.

BHOWMICK, A. K., and H. L. STEPHENS. *Handbook of Elastomers*. Materials Park, OH: ASM International, 1988.

BRADY, GEORGE, and H. R. CLAUSER. *Materials Handbook*, 12th ed. New York: McGraw-Hill, 1986.

HARPER, CHARLES A., ed. *Handbook of Plastics and Elastomers*. New York: McGraw-Hill, 1975.

INTERNATIONAL INSTITUTE OF SYNTHETIC RUBBER PRODUCTS. *Synthetic Rubber: The Story of an Industry*. New York: IISRP, 1973.

PUGH, BRADLEY. "The Role of Standards in the Development of the Condom," *ASTM Standardization News*, August 1992, pp. 23–29.

REISCH, MARC S. "Following the Bouncing Ball," *C&EN*, December 2, 2002, p. 21.

SHELL. *Kraton Thermoplastic Rubber*. Houston, TX: Shell Chemical Company, Polymer Division, 1972.

Module 17

Adhesives, Caulks, Sealants, & Other Important Polymerics

After studying this module, you should be able to do the following:

17–1. Use appropriate terms to describe adhesive and interface technology.

17–2. Use Marra's adhesive joint model to explain the principles of bonding, name the conditions and events required for an adhesively bonded joint, and describe the main methods by which adhesives work.

17–3. Use adhesive selection guides and suggested websites to determine appropriate adhesives for specified applications.

17–4. Explain why interfacial technology is gaining wider study and application.

17–5. Recognize the value and uses for protective coatings, sealants, smart polymers, and other important polymers.

17–6. Use the website addresses denoted by the symbol* to explore sites for new developments and evolving concepts related to.

*The symbol found throughout the book will link you to Internet sites related to the topics being covered. The dynamic nature of the Web brings frequent changes, so some of these sites, while available at the time of writing, may not be up now.

17.1 ADHESIVES

Until the mid-1940s, the selection of adhesives for joining metallic materials together normally included either mechanical fasteners, such as bolts, rivets, and pins, or a thermal bonding method, such as welding, brazing, or soldering. Today, the knowledgeable designer learns that bonding with adhesives must receive equal consideration with the traditional methods of joining parts together. The technician and crafter involved in maintenance and repairs can choose adhesives as an alternative technique for fastening broken or unjoined components. Generally, welding, riveting, and adhesive joining are classified as permanent joining methods; mechanical fasteners such as bolts, screws, nuts, and pins allow disassembly.

The ideal material, described in Module 1, does not exist, and neither does the ***ideal adhesive.*** If it did exist, it would (a) stick to any material, (b) need no surface preparation, and (c) maintain its strength and other properties forever under all types of conditions. An adhesive bond has a great advantage over a mechanical joint, such as a nailed, screwed, bolted, or riveted joint. The strength of mechanical joints depends on the combined strengths of the materials joined together and the fasteners involved. This interaction takes place where these fasteners come in contact with the material to be joined, producing a concentration of stress that may cause one of the materials to exceed its ultimate strength and the connection to fail. Not so with an adhesive joint, which permits an even distribution of stress over the entire bond area.

Adhesives have the benefit of being able to join dissimilar materials. Adhesives are also lightweight and provide joint sealing, sound and vibration dampening, thermal and electrical insulation, and uniform strength and require only low-cost, low-skill techniques. These advantages mean that adhesives find many uses in building construction, including replacing nails with mastic adhesives to bond plywood or particleboard to flooring joists and drywall panels to wall studs (see Figure 17-1). ***Elastomeric adhesives*** are good for joining wooden furring strips to concrete. Various other adhesives are used to fasten vinyl and ceramic tiles.

Adhesives are widely used by the aerospace industry to join thin sheets of aluminum, plastics, or composites to frameworks because they can help maintain smooth, aerodynamic surfaces and light weight and provide uniform strength and good fatigue resistance. ***Epoxy*** adhesives stick together structural members formerly joined by rivets, which caused problems due to their nonuniform stress distribution, which created stress concentrations; rivets are also heavy and prone to corrosion. Among the major reasons mechanical or thermal joining is preferred to adhesive bonding in some cases is heat sensitivity to normal service temperatures ranging from 160°C to over 500°C. Most common adhesives can only withstand a lower range, from −90° to 290°C. The special adhesive bonding system developed for the space shuttle's ceramic heat shield tiles however, can withstand temperatures up to 1260°C. Epoxy adhesives in the form of microcapsules serve as locking systems for mechanical fasteners on the shuttle and mix and cure quickly with the shearing force of a nut on an epoxy-coated bolt. Adhesive-

Figure 17-1 Adhesives in construction. Low-cost, lightweight panel resulting from a strong honeycomb paper core bonded with adhesives between plywood sheeting. (U.S. Forest Products Laboratory)

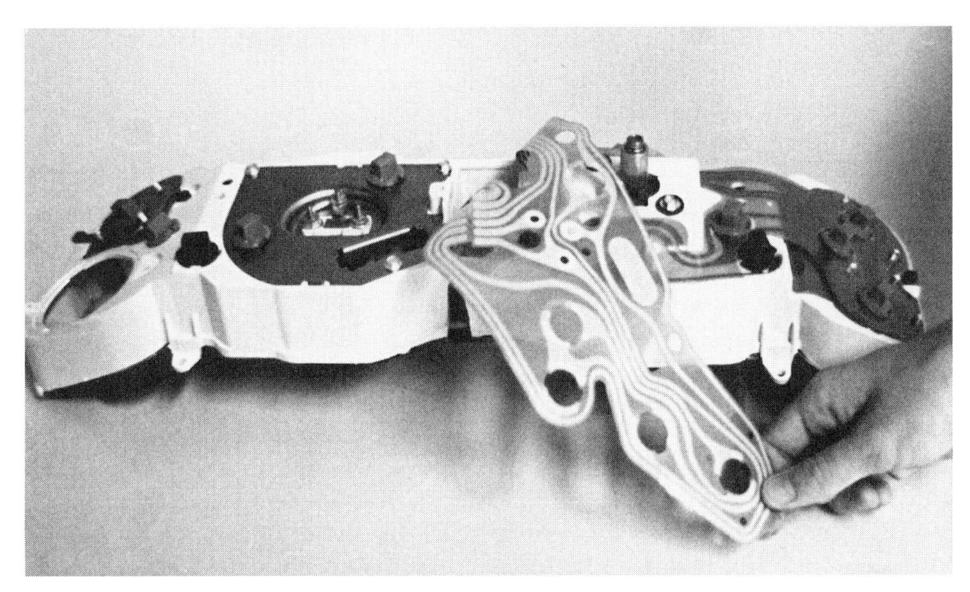

Figure 17-2 Cyanoacrylics set rapidly for electronic-component assembly. (General Motors)

Figure 17-3 This array of adhesives and substrate cleaners shows just a few of the hundreds of adhesive types and forms available. Included here are two-part epoxies, contact cement, water-based vinyl adhesives, acrylic and PVC solvents, elastomeric cement, cyanoacrylates, and loss-of-solvent and anaerobic adhesives.

joining technology provides many unique joining methods that help the automotive industry achieve more fuel-efficient vehicles through reduction of weight (see Figure 17-2 and Figure 17-3). Certain ***structural acrylic*** and ***plastisol*** adhesives offer auto and other assembly lines the advantage of compatibility with oily metal surfaces; they also bond many dissimilar metals and plastics in addition to wood, glass, hardboard, and asbestos board. These acrylics resist moderate temperatures, have high impact and peel strength, and cure rapidly with minimal shrinkage. ***Urethane adhesives*** bond together steel and glass sunroofs. Elastomeric and plastic bumper systems and plastic body side molding use adhesives that dissolve the contaminating oils and films found on production lines, thus providing a dependable joint. ***Cyanoacrylates,*** or "***superglues,***" which are instant adhesives, are widely used on electronic, electrical, appliance, and instrument assembly lines. Advances in microprocessors and computers resulted from improvements in semiconductors and ceramic substrates, but many of those advances were possible because of continually improved ***interface technology,*** which focused on the joining of materials. Multilayered ceramics (MLCs) are examples of improved integration, one product of interface technology. Very-large-scale integration (VLSI) of semiconductors also presents interface challenges, as solder balls only 0.002 in. thick join microthin metal circuitry deposited on ceramics. Printed circuit boards use ***surface-mount technology (SMT)*** for electronic components (see Figure 17-2). Figure 17-3 shows a just a few of the many types of adhesives and substrate cleaners and the various forms available.

17.1.1 Adhesive Systems and the Bond Joint

An adhesively bonded joint results from the adhesive's ability to flow, wet, and set. ASTM defines an ***adhesive*** as a substance capable of holding materials together by surface attachment. The materials held together are *adherends* or *substrates.* ***Substrate*** is the broader term, defined as a material upon the surface of which an adhesive-containing substance is spread for any purpose such as bonding or coating. As liquids of varying viscosity, ranging from below that of water to a semisolid (viscoelastic) or syrup consistency, including hot-melted solids, adhesives must wet the surfaces to which they bond. ***Wetting*** involves flowing into (1) large openings such as pores of woods or cells of foam plastics or (2) the microscopic hills and valleys that exist on the smoothest of surfaces, such as polished metals or glass. (Refer to Figure 18-1 for a microscopic view of wood to see the difference between this substrate and smoother surfaces.) A key factor in adhesive selection is the ability of the adhesive to wet all or most of the available bonding sites of the substrates (surfaces making the bond joint) (Figure 17-4).

The free spreading of the adhesive on the substrate surface may be impaired by such conditions as air entrapment, moisture or oxide buildups, and contaminants like oil and dirt. Figure 17-4 illustrates a nonwetting condition—water beading up on a waxed surface, for example—and a wetting condition—for example, water absorbed in tissue paper. To achieve wetting of an adherend, the adhesive must not have ***cohesive forces*** (attraction of molecules inside the adhesive) that provide a very strong ***surface tension*** (forces contracting the liquid into a droplet). The surface tension must be less than the adhesive forces between the adherend and the adhesive. A high-surface-energy liquid such as epoxy adhesive will not wet a low-energy polyethylene solid, but the polyethylene liquid with lower surface energy will wet and form good adhesion to an epoxy solid.

Cyanoacrylate adhesives consist of simple molecules that upon contact with moisture polymerize instantly. Hydroxyl ions (OH) create single bonds in the adhesive molecules, which lead to the formation of macromolecules. As seen in Figure 17-5, substrates comprised of nonpolar (+ + + +) materials are nonadhering, and substrates and adherends of polar (+ − + − + −) materials provide good adhesion. Waxy materials such as polypropylene and polyethylene are nonpolar and do not adhere to cyanoacrylates nor most other adhesives. Because most materials such as glass and many metals are polar, they adhere well with cyanoacrylates, which, since the opposite charges attract, form polar plastic upon contact with moisture.

17.1.1.1 Contact angle.

The wetting or contact angle is a measure of how a liquid will spread on a solid. Many liquids spread easily on solids and have zero or very small contact angles. If the contact angle is greater than 90°, it is considered nonwetting. Water on clean glass would have a contact angle of 0°; on polyethylene the water would form an angle of over 100°. The analysis of contact angle has many applications in interface technology, ranging from the spread of adhesives, lubricants, printing inks, and protective coatings to achieving maximum coverage with insecticides and herbicides. Once the wetting has occurred, the adhesive must ***cure*** into a solid (occasionally a viscoelastic) or ***set*** so that it develops adhesion,

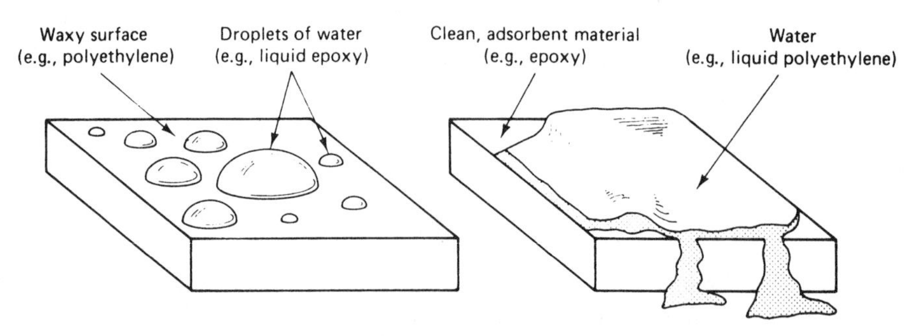

Figure 17-4 Wetting of substrate. Poor wetting: liquid beads up on the surface. Good wetting: liquid flows into micro hills and valleys on the surface and porous materials absorb the liquid.

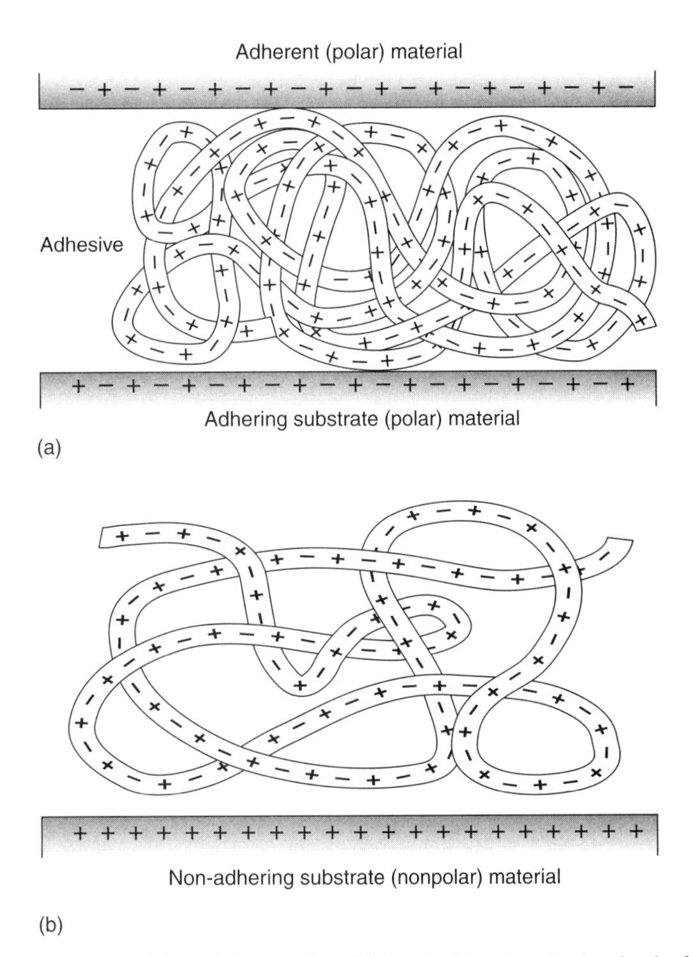

Figure 17-5 Cyanoacrylics bond through the attraction of their polar (+ − + − + −) molecules for the polar molecules of the substrate and adherend, such as glass and steel, as seen in (a), whereas, nonpolar materials (b) such as polyethylene won't bond due to nonpolar (+ + + +) molecules.

thus bonding materials (***adherends***) together into an adhesive joint. Synthetic resins and natural adhesives ***cure*** through the various polymerization processes covered in Module 14. Addition and condensation polymerization develop in the thermosetting adhesives that form permanent bonds; thermoplastic adhesives, on the other hand, can be softened to reposition parts or for disassembly.

There are four ways to look at the forces involved in the adhesion of an adhesive to its substrates. Both ***chemical*** and ***physical bonds*** are involved in adhesive bonding, but the exact nature of all forces is not fully known. Covalent bonds and physical attraction through van der Waals forces seem to operate in adhesives as they do in other polymers, but scientists differ on how the adhesion develops. On porous materials such as wood, ***mechanical interlock*** (penetration of adhesives into openings) provides added surface area for the adhesive to attach, but some scientists believe that mechanical interlock has only a minor role in adhesion. Testing of adhesive bonding systems must deal with ***interfacial adhesion*** (attraction between adhesives and substrates) and ***practical adhesion*** (breaking strength of a joint).

17.1.2 Adhesive-Bond Chain Links

The complexities of adhesive bonding are outlined in the model developed by Alan A. Marra (Figure 17-6), which illustrates the nine links in the adhesive-bond chain. The chain-link model reveals the interrelations of factors in the bonded product that include (1) adhesive composition, (2) adhesive application, (3) the bonded materials' properties, (4) their preparation,

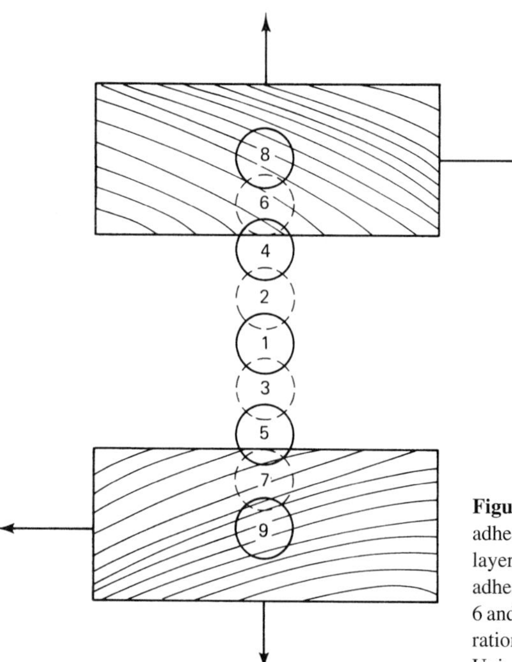

Figure 17-6 Marra's adhesive-joint model. Link 1—the adhesive film; links 2 and 3—intra-adhesive boundary layer, strongly influenced by the adherend; links 4 and 5—adhesive-adherend interface, site of adhesion forces; links 6 and 7—adherend subsurface, partially fractured in preparation; links 8 and 9—adherend proper. (Dr. Alan A. Marra, University of Massachusetts)

and (5) the stress and environment that the product will encounter in actual service. The physical and chemical theories and principles associated with the chain links become quite involved and require adhesive specialists to deal with them. The user of adhesives must recognize that proper adhesive bonding depends on the choice of the correct adhesive for the intended environment and on application according to specifications. The best adhesives will fail if not used properly.

Beginning with link 1 of the model, the adhesive film must have proper ***cohesion*** to stick together when in service. The intra-adhesive boundary layers at links 2 and 3 are strongly influenced by the chemical and physical properties of the adherend; for example, extractives in wood may have an undesirable chemical interaction with the adhesive and change the composition and/or action of the adhesive as it is curing. Links 4 and 5 are critical links in the system because this is where the adherend and adhesive engage. Problems such as surface contamination can prevent the anchoring of the adhesives at the necessary bonding sites. Here, the compatibility of adhesive to adherend is required so that proper wetting takes place. As we have mentioned, several theories attempt to explain the actual forces operating in these two links at links 6 and 7; the adherend/substrate is critical in a material such as wood or ceramic, where a damaged surface must be repaired through penetration of the adhesive into the pores. Mechanical interlocking may help with the repair. Finally, links 8 and 9 of the adherend material must possess sufficient cohesive strength and compatibility with the bonded joint. For example, a porous material such as a cellular plastic or elastomer may absorb and transport harmful liquids to the bonded joint where the chemical could degrade the adhesive.

Marra's model leads to the four ***main factors in adhesive selection:***

1. Materials to be bonded and type of joint
2. Conditions under which the bonding must be done
3. Durability and strength levels required during service of the product
4. Cost of the bonding process, including joint preparation, material cleaning, cost of adhesives, and so on

These four factors will vary considerably and require close attention to specifications by those making adhesive selections. Technical sales representatives, handbooks, and other sources of technical information guide the user, but as with any industrial production system, the manufacturer of the product must evaluate the adhesively bonded product throughout its development and maintain quality control over the adhesives and the application of those adhesives to products.

17.1.3 Types of Adhesives and Their Properties

Adhesives fit into various classifications, depending on factors such as the substrate, physical form (solid, liquid, viscoelastic), and bonding characteristics. A basic classification is whether the adhesive is structural or nonstructural. ***Structural adhesives,*** by ASTM definition, are those used to transfer required loads between adherents exposed to service environments typical for the structure involved. Some authorities classify structural adhesives as those whose failure could result in a threat to public safety, loss of life, or substantial property damage. These adhesives were formulated to replace nuts, bolts, rivets, and welds. A U.S. government study called the Primary Aircraft Bond Structure Test (PABST) demonstrated that adhesive bonding is feasible and economical for replacing mechanical fasteners and welds for the main structural components of aircraft. Structural adhesives are used to repair metal automobile bodies for after-market use. They are used as an option to welding to bond replacement body panels (door, deck lid, hood, and roof skins). Such adhesives are covered by a lifetime guarantee of bond integrity and noncorrosion. Adhesives must also possess ***serviceability*** (the ability to withstand design conditions such as heat, cold, stress, moisture, or chemical attack) and ***durability*** (the ability to maintain serviceability of the bond joint over its expected life).

Structural adhesives for wood include ***phenol–resorcinol–formaldehyde*** polymers, which are thermosetting, all-weather exterior adhesives used on laminated structures. ***Casein*** adhesives are structural water-resistant adhesives (moisture content of wood less than 16%) commonly used on softwood laminated beams and doors. ***Nonstructural adhesives*** include ***urea–formaldehyde,*** which is used on particleboard and furniture; ***poly(vinyl acetate)*** (e.g., Elmer's) for furniture and interior woodwork; and ***starch*** and ***dextrin*** for paper products. ***Protein*** adhesives include animal glues made from bone, hide, and fish (they are used for woodworking and gummed labels); casein from skimmed milk; and ***blood*** and ***soybean flours*** formulated with phenolic and formaldehyde resins, which cross-link to serve as water-resistant adhesives on interior plywood.

Thermoplastic adhesives are generally serviceable at temperatures below the thermosets. Among the thermoplastics, cyanoacrylates (Superglue) have a service range of −5° to 80°C; polyamides, 15° to 100°C; poly(vinyl acetate), up to 45°C; and butadiene styrene, 5° to 70°C. These compare with thermosets such as epoxy, −10° to 140°C; phenolic epoxy, −25° to 260°C; and phenolic neoprene, −20° to 95°C. Some synthetic adhesives come as thermosets and thermoplastics, such as polyimides (TS up to 370°C and TP −160° to 350°C), silicones (TS −25° to 250°C and TP −75°C), and polyurethanes (TS and TP −60° to 175°C). Elastomeric adhesives are formulated with natural or synthetic rubber or blended with other resins to provide impact resistance and to withstand wide temperature variations. These adhesives include neoprene, polysulfides, butyls, and nitriles. A phenolic–rubber blend adhesive bonds brake linings that withstand severe shock and temperature and saltwater exposure.

Structural bonding tape provides a bonding system that combines the advantages of pressure-sensitive tapes and structural adhesives. These double-coated tapes are tacky at room temperature and possess a highly elastomeric nature. Advantages of structural bonding tapes include ease of use, immediate adhesion, and uniform bond thickness, which makes them good replacements for liquid adhesives and other joining techniques.

Except for pressure-sensitive tapes that remain tacky, adhesives change from the liquid state to the solid state upon curing. Adhesive tape, duct tape, and transparent tape do not develop the strength found in solid adhesives; however, double-coated, pressure-sensitive tapes can serve as hold-down tapes on the machine tables of milling machines and surface grinders. The curing of adhesives involves the cooling of hot melts, loss of solvents, anaerobic environments, and two-part mixtures. ***Hot-melt adhesives*** usually begin as a solid that becomes liquid in heat guns or furnaces, or with other heat sources that develop temperatures around 150°C; most cool and set within seconds after removal of the heat source. ***Loss-of-solvent adhesives*** employ volatile liquids, including water and organic solvents, that dissolve the adhesive base material to form the fluid; upon application to the joint, the adhesives harden as the solvents evaporate or penetrate porous adherends. With solvent adhesives, especially, as well as other adhesives, one must determine whether the adhesive will attack the substrate so severely that it may produce weaknesses in the material being joined. ***Anaerobic adhesives*** retain their fluid state when exposed to oxygen, but when squeezed into thin joints that block the oxygen, they set up into hard, strong adhesives. ***Two-part mix adhesives*** are stored with the resin and catalyst (or hardener) separated; when mixed, the catalyst causes cross-linking of the resin.

Loss-of-solvent adhesives often have long setting times, and except for water-solvent types, the solvents present health and fire safety problems; they also rely on our short supply of petroleum. As a result, the trend is toward water-based adhesives. Anaerobics come in one part, cure quickly, and develop high shear and impact strength, and some have gap-filling abilities. Other adhesives such as cyanoacrylates cure as a result of exposure to the moist alkali environment found on most surfaces. Contact cements are usually rubber-based structural adhesives that form an instant bond when substrates coated with the cured adhesives are brought into contact, thus eliminating the need for clamping. Most other adhesives require the application of pressure long enough for the adhesive joint to cure.

17.1.4 Adhesive Selection

The wide range of adhesives and sealants available is partially the result of the difficulty of finding suitable matches of adhesive to adherend. Many general guides exist, but they should be used with caution. For important production jobs, it is best to consult suppliers to obtain the best choice of adhesive. For general-purpose jobs, some broad guidelines help.

As discussed earlier, plastics present the most difficulty because of wetting problems. Table 17-1 lists some common adhesives and their typical applications and provides an overview of selected adhesives, adherents, and applications. The below contains the locations of some useful adhesive selectors. The list included here provides a basic guide for adhesive selection in selected materials groups. Read the technical data for each adhesive to learn

TABLE 17-1 COMMON ADHESIVES

Adhesive	Typical Adherents and Applications
Acrylics (two-part)	Plastics (ABS, phenolic-melamine, polyester, polystyrene, polysulfone, PVC) to metal, rubber to metal, and plastics to plastics
Casein	Interior wood components
Cyanoacrylates (anaerobic)	Thread locking, gear and bushing mounting, aluminum, copper, steel, ceramics, glass, rubber, polycarbonate, phenolic, acrylics
Epoxy	Most adherents except some plastics such as acetals, polypropylenes, polyurethanes, and silicones; many modifications including nylon epoxies and epoxy phenolics
Melamine formaldehyde	Paper, textiles, hardwood, plywood for interior use
Natural rubber	Cork, foam rubber, leather, paper-building materials, pressure-sensitive tapes
Neoprene rubber	Many plastics (ABS, fluorocarbons, polystyrene), aluminum, ceramics, copper gasketing, and laminating
Nitrile rubber and phenolic	Many plastics (ABS, cellulose, fluorocarbons, PVC, polyester), aluminum, ceramics, glass, magnesium, microwave isolators
Phenol–formaldehyde	Cork, cardboard, softwood, plywood for exterior use
Poly(vinyl acetate)	Textiles, polystyrene, wood, other porous and semiporous materials, interior furniture and construction
Resorcinol–formaldehyde	Asbestos, cork, paper, rubber, wood furniture, laminated exterior wooden beams, wooden boats
Silicone	Aluminum, ceramics, glass, magnesium, phenolics, polyester, acrylics, rubber, steel, textiles, integrated circuits, sealants, gasketing
Urethane elastomer	Many plastics (ABS, polyester, acrylics, PVC), aluminum, ceramics, copper, glass, magnesium, steel-bond solid propellants, cryogenic applications, insulation

more specific details on compatibility. For important jobs, conduct a small sample test before doing the complete bonding. If adhesive tapes will not stick to an adherent, you must abrade the surface to remove coatings or apply another coating that is compatible with the adhesive.

General Materials Group	General Adhesive Groups
Metals	Epoxy, cyanoacrylates, urethane
Wood	Vinyl emulsion, resin, epoxy
Ceramic	Cyanoacrylates, epoxy, urethane
Plastic	Numerous options—read packaging information

Adhesives have been developed to meet the most varied of circumstances. In medicine, methylmethacrylate resins join artificial hip sockets to natural hip bones and repair broken, damaged bones; cyanoacrylates bond eye tissue and other skin tissue. Thermoplastic polyester resins bond copper foil to flexible backing on printed circuit boards and, with epoxy and phenolic adhesives, withstand etchants and the heat of ***weld bonding,*** which combines spot welding and adhesives. The weld immobilizes the adherends to allow the adhesives to cure. The increased use of resin-based composites promises even wider use for adhesives. The duPont website page on adhesives, http://www.dupont.com/industrial-polymers/adhesives/index.html, January 15, 2000, lists an array of polymers to meet various applications, with data sheets on properties, uses, and procedures. Among the types of polymers listed are

- Hot-melt and wax-blend adhesives
- Extrusion and laminating adhesives and sealants
- Water-based adhesives
- Binders and film formers

The websites listed below provide similar data on adhesives, and some also provide activities.

Adhesive and sealants digital reference—*http://www.adhesivesandsealants.com*
DuPont industrial polymers—*http://www.dupont.com/industrial-polymers/index.html*
Devcon adhesives—*http:/www.devcon.com/*
Elmers products—*http://elmers.com/*
Loctite Corporation—*http://www.loctite.com/*

17.1.5 Biomimetics and Biologically Inspired Adhesives

Knowledge of adhesives and adhesive forces can profit from biomimetics (copying nature), as in the case of research by Autumn and Full* into the ability of gecko lizards to walk across ceilings. These creatures possess toes with millions of tiny hairs (*setae*), and each *seta* consists of hundreds of projections known as *spatulae* (Figure 17-7). This gives the lizards enormous surface area, which enhances bonding. The toes are ***hydrophilic*** (having strong affinity for water) like the toes of insects and frogs, but they can also adhere to ***hydrophobic*** (having little or no affinity for water) surfaces, indicating that van der Waals forces, not capillary action, give geckos abilities that seem to defy gravity. ***Capillary attraction*** or ***adhesion*** involves a combined force of adhesion and cohesion that causes liquids to flow between tightly spaced, solid surfaces, even against gravity. Understanding how the gecko's toe structure couples with van der Waals bonding can lead to improvements in adhesives and applications where adhesion aids in design, such as climbing robots. Further research along these lines should lead to the development of effective dry adhesives.

*  *http://www.lclark.edu/~autumn/gecko.html,* January 3, 2003.

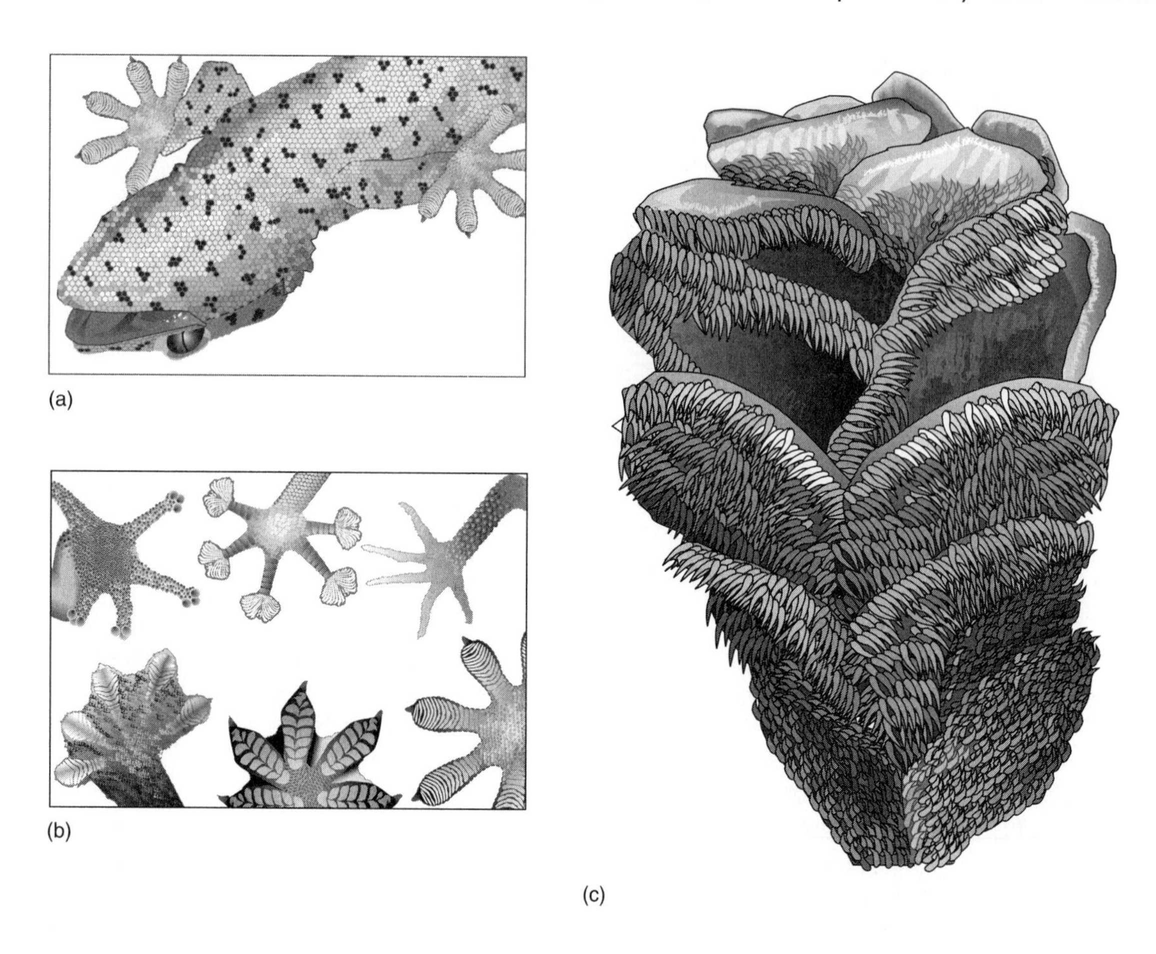

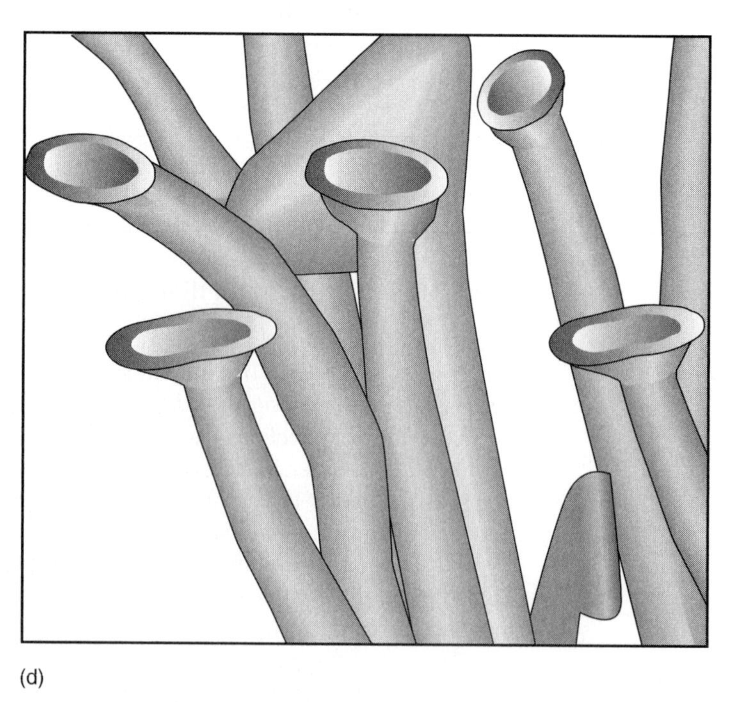

Figure 17-7 (a) Geckos can easily cling to glass and walk on ceilings, seeming to defy gravity. (b) Feet of various geckos—note the variety of toes. (c) Gecko toe consisting of millions of *seta* (hairlike) strands which have hundreds of (d) *spatulae,* which form van der Waals bonds with hydrophobic and hydrophilic surfaces. (Adapted from Kellar Autumn, *http://www.lclark.edu/~autumn/gecko.html)*

17.2 CAULKS AND SEALANTS

Caulks and sealants, like adhesives and protective coatings, fill gaps and seal against weather and liquids and have many commercial applications. ***Polymer caulks*** are classified by their expected service life, known as *range*. Oil-based caulks are low-range and low-priced. Acrylics, butyls, and latexes are medium-range caulks and serve many construction needs. High-range caulks are the most costly; they include silicone (for considerable joint movement, ultraviolet-light exposure, waterproofing, and wide temperature variations), various copolymers, and urethane. Some high-range caulks such as urethane are also available as foam to fill large gaps. Industrial suppliers and home improvement centers have hundreds of caulks and sealants; some even allow patching and joining under water. An advanced material, Gore-Tex joint sealant, is 100% PTFE in expanded, continuous cord form that allows it to be used as an "asbestos-free" sealant. It can serve as gaskets for engine blocks (Figure 17-8) that will never leak or need replacing and can be used on irregular or damaged surfaces because it fills in and conforms to surface imperfections to provide an ideal seal against gas or liquid. ***Sealants***, applied as liquids or pastes, cure to form seals against gases or liquids. Often, the sealants also act as adhesives, especially in building construction. One-part silicone sealants combine with glass and other materials to make effective sealed systems (Figure 17-9). Caulking/sealant use has increased in building construction due to a demand for better-sealed glazing for complex insulated glass systems, as seen in Figure 17-10 and 17-11, as exemplified in Biosphere2, discussed in the Applications and Alternatives section at the end of this module.

17.3 OTHER IMPORTANT POLYMERICS

Polymer science and technology continue to generate innovative materials. In the medical field, research and development efforts encompass replicating human tissue, such as skin, and producing a polymer that will stimulate cartilage regeneration. Polymers that will not burn yet are easy to process, lightweight polymer batteries, water-soluble polymers, and controlled-release polymers are some materials of continuing interest. The following groups of polymers are other significant engineering polymers.

17.3.1 Protective Coatings and Preservatives

Because polymers include synthetics, an endless variety of them can be synthesized. Protective coatings, a category of polymers not discussed in this module, include natural polymers such as shellac, oil-based paint, and many synthetic polymers, including the same general types used for plastics, elastomers, and adhesives (e.g., epoxys, silicones, latex, urethanes, and acrylics). These coatings develop their protective skins in a number of ways, including polymerization,

Figure 17-8 Gore-Tex® gaskets of PTFE provide excellent seals for rough surfaces where oil, fluids, and temperatures may damage other sealing materials. (W. L. Gore & Associates, Inc.)

Figure 17-9 One part insulating glass, silicone sealant offers exceptional hydrolytic stability, fast cure, high strength, high modulus, and primerless adhesion to glass and many substrates. (G.E. Plastics)

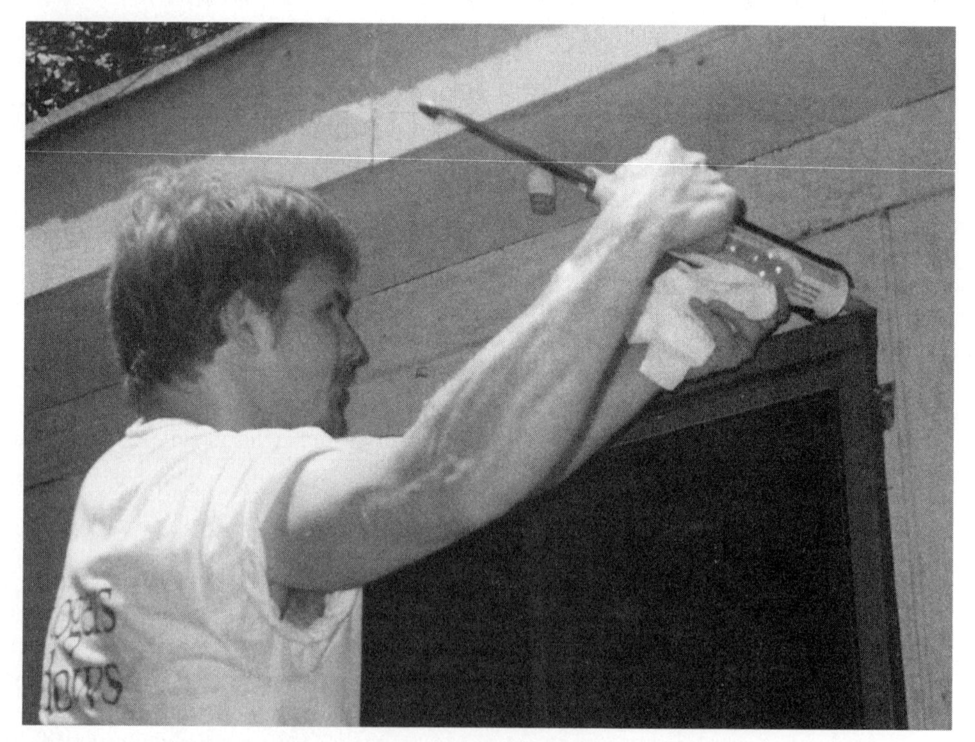

Figure 17-10 A technician applies 100% silicone caulk to seal window unit against fir framing.

cooling of hot melts, and evaporation of solvents or water. Evaporation leaves behind a film of the polymer that was dissolved in the solvent or suspended in water.

Environmental concerns about ozone layers, water quality, and overall air quality have required the petrochemical industry to put considerable effort into eliminating the traditional materials-related uses of chlorofluorocarbons (CFCs) and ***volatile organic compounds (VOCs).*** VOCs create air-quality problems at the ground level (troposphere) by forming ozone and photochemical oxidants, the primary ingredients in smog. CFCs create problems in the upper atmosphere (stratosphere) by reacting with the delicately balanced ozone layer that protects the

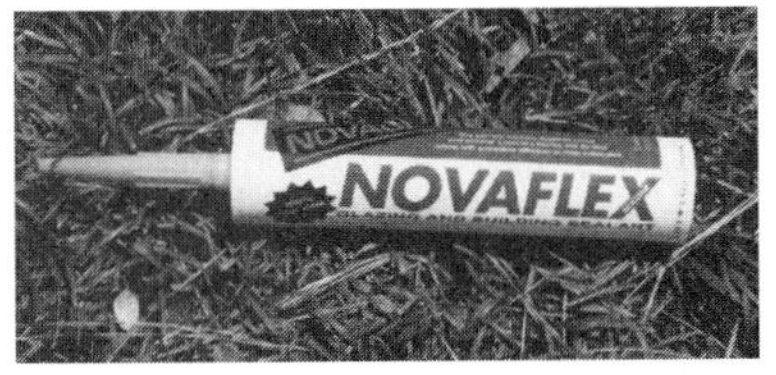

(a)

(b)

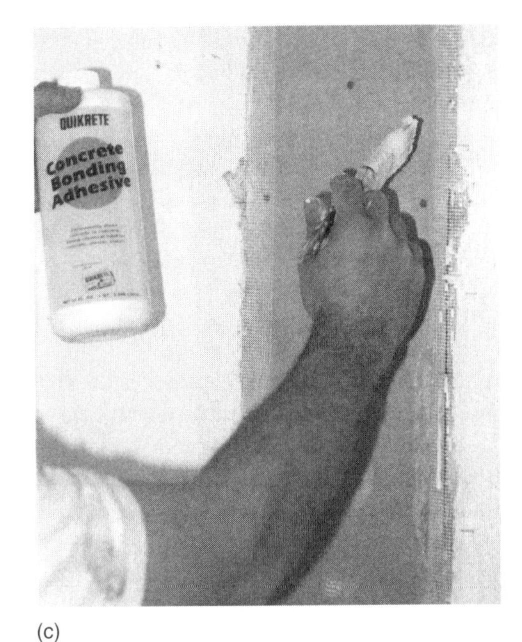

(c)

(d)

Figure 17-11 (a) Caulks and sealants come in cartridges that allow their use with manual and power caulking guns. (b) Note that the tapered polyethylene tip allows for cutting to various size openings to change the bead size. PE is nonpolar and does not bond with the sealant. (c) Bonding agent penetrates pores of old concrete to form chemical bonds with new coatings. (d) Acrylic latex undercoat forms a breathable film to "kill" stains and form bonds for final coating.

earth from the sun's ultraviolet rays. CFCs deplete the "good" ozone. Much of the current development in protective coatings focuses on eliminating VOC solvents because they create health hazards to workers and harm the atmosphere. By replacing VOCs with water as the vehicle to carry the polymer dispersions, they evaporate with few harmful effects. Much of the paint used today is water based. As a part of the ***green movement,*** paints based on natural substances (plants and minerals) have been developed without pesticides, preservatives, and antimildew agents. Some paint manufacturers separate the solvents from the solids in excess paint, leaving pigment and resin that is recycled as primer or utility paint. Acrylics have been developed using a polymerization process known as ***group-transfer polymerization (GTP),*** which permits unlimited polymer-chain growth and has the potential for use as high-performance auto finishes that will cure at lower temperatures (180°F versus 250°F to 300°F). The process also uses lower concentrations of polluting solvents.

Wood offers natural beauty and warmth, but it is susceptible to weathering, moisture, insect attack, and stress on weaker sections that have suffered during the life of the tree. Many protective coatings and polymer enhancements, such as diffusion of ***poly(ethylene glycol) (PEG),*** have been developed to preserve archeologically significant wooden pieces. Cross-linking chemicals can permanently bond cell walls of wood for stability and resistance to fungi, but their use results in decreased strength. The enhancement of the mechanical properties of wood is achieved with vacuum impregnation of liquid monomers such as epoxy and methyl methacrylate into openings that result from sap leaving the tree during curing. The resins are polymerized after impregnation to make harder, stronger modified wood for boat hulls, billiard cues, knife handles, and parquet flooring. Large structural timber and lumber are both pressure and nonpressure treated (brushing, soaking, dipping, and heating) using oil-type preservatives: creosote, pentachlorophenol (penta), copper naphthenate, and waterborne inorganic arsenic compounds.

 Frequently Asked Questions about finishes—*http://www.fpl.fs.fed.us/faqs.htm#finishing*

17.3.2 Oils, Lubricants, and Fluids

Polymers play an important role in many types of lubricating, cutting, and hydraulic applications as solids and as additives or suspensions in liquids. For example engine-oil viscosity in multigrade oils, such as SAE 10W-40, changes when the ambient and engine temperatures change because polymers control the oil thickness. New long-lasting engine oils are made from synthetic polymers. ***Electrorheological (ER) fluids*** (introduced in Section 15.5 as smart polymers) congeal from a liquid into a solid mass in a few milliseconds when an electrical current is passed through the liquid. When the current is removed, the solids return to liquids. ER fluid has promise as a hydraulic fluid for robots, automotive clutches, and artificial human limbs. Experimentation with combinations of selected oils with particles (microscopic solids of polymethacrylic acid and water in long, tangled chains) will continue in the quest for the ideal ER fluid. What possibilities might this type of solid lubricant hold for problems encountered with high-temperature ceramic engines or automotive brakes?

17.3.3 Conductive Polymers

A major disadvantage of lead–acid and other metallic batteries is their weight. The development of lighter, conductive polymers, such as polyacetylene electrodes, could lead the way to the long-sought electric automobile. Remember that polymers do not normally serve as conductors because their electrons are locked in covalent bonding, unlike the free electrons in metals. Doping polyacetylene with iodine, however, results in oxidation and the free movement of electrons. Reduction in the same polymer, achieved by reaction with the metal sodium, turns polyacetylene into an anion while the sodium becomes a cation. Disulfide polymers, long chains of sulfur–sulfur bonds, make battery cathodes that can act with lithium anodes to produce a depolymerization–polymerization process that yields electrochemical energy. The polymeric form of sulfur nitride becomes conductive at low temperatures without doping, and poly(*p*-phenylene), or PPP, which is made up of single and double bonds, can achieve electron mobility with doping. Polypyrrole also has conductor potential because of bonding similar to that of PPP. Both can be switched back and forth between conductors and insulators. Each of these polymers offers potential as conductors, but there are major problems involved in converting them to reliable engineering materials.

17.3.4 Geosynthetics

ASTM defines a ***geosynthetic*** as "a planar [lying in one plane] product manufactured from polymeric material used on soil, rock, earth, or other geotechnical engineering related material as an integral part of a man-made project, structure, or system." Geotextiles, geogrids, geofoams, geofilters, and geocomposite drains serve many purposes in roads, dams, landfills, and similar applications where earth erosion, pollution, or containment of liquids is a factor. Civil, building, and ***geotechnical engineering*** (related to the earth's crust) often call on impermeable membranes to contain water in dams, to keep oil and other liquid chemicals in reservoirs, or to line sanitary landfills to prevent leached liquids from leaving the landfill and polluting groundwater. Porous woven polymer fabrics hold soil on river or lake banks until rocks and vegetation stabilize the banks. Polymer fabrics go into road construction to develop laminar composites that prolong the life of macadam (asphalt-and-rock composite) roads. A major concern of the geotechnical community is how long a geosynthetic will last. A landfill has many potential pollutants. Once the geomembrane liner is in place, will it prevent pollution of the groundwater indefinitely? ***Geomembranes*** of PVC, PE, and ethylene copolymers undergo extensive testing based on ASTM Committee D35 standards. Geotechnical engineers also use expanded polystyrene (EPS) blocks, also known as ***geofoam,*** to rebuild highway embankments. Increased awareness of the fragile nature of the earth has prompted use of innovative geosynthetics, and continued improvements in these synthetic polymers are inevitable.

APPLICATIONS & ALTERNATIVES

The widespread use of polymers such as PVC plastic piping, caulks and sealants, and numerous elastomers have made possible structures like Biosphpere2 (Figure 17-12a–c). Located in the desert outside Oracle, Arizona, Biosphere2 serves as a research facility designed to replicate various earth environments such as rain forests, deserts, plains, and oceans. It contains 175,000 square feet of glass surface made up of approximately 6500 panes of glass in

Figure 17-12 Biosphere2 serves as a research center to study the earth's environments. (a) The main housing consists of approximately 6500 panes of glass sealed in steel struts by silicone sealant. Small-scale environments including (b) rain forests, plains, deserts and (c) oceans are manipulated to study the impacts of pollutants. (Note that recycled HDPE milk jugs provided the raw materials for the simulated wood walkways, steps, and handrails instead of short-lived pressure-treated wood materials). (d) The 1/2-in.-thick glass unit consists of three layers: 1/8-in. insulating layer sandwiched between two 3/16-in. pieces of white glass. This glass transmits 85% of visible light, 62% of solar energy, and less than 1% of ultraviolet light. (e) Maintaining balanced environments requires complex systems of pumps, PVC water pipe, a large rubber bellows for air exchange, air-conditioning units, and complex computer controls.

steel and glass frames. The glass transmits 85% of visible light, 62% of solar energy, and less than 1% of ultraviolet light. A special silicone sealant was used to seal the glass glazing to the metal struts (Figure 17-12d). Maintaining balanced environments requires complex systems of pumps, PVC water pipe, air-conditioning units, and computer controls (Figure 17-12e). To learn more, go to the Biosphere2 website listed below.

www.bio2.edu

SELF-ASSESSMENT

17-1. Which advantage(s) would adhesive bonding have over other joining methods?

- **a.** Ease of assembly and disassembly
- **b.** Light weight, load spreading, joint sealing
- **c.** Withstands very high temperatures normally above those withstood by the base material
- **d.** Easy selection for joining any type of material

17-2. A key factor in adhesive selection is the ability of the adhesive to __________ the substrate material.

a. Wet **b.** Cover **c.** Cure **d.** Entrap

17-3. Attraction of molecules within an adhesive or substrate is called

a. Adhesion **b.** Contact **c.** Cohesion **d.** Interface

17-4. Penetration of adhesives into the openings of porous materials is called

a. Interface **b.** Mechanical interlock **c.** Cross-linking **d.** Surface tension

17-5. Which group of materials presents the greatest challenge for adhesive selection?

a. Plastics **b.** Wood **c.** Metal **d.** Ceramics

17-6. How does Marra's adhesive joint model aid in dealing with bonding?

- **a.** Illustrates the relationships between the adherents and adhesives
- **b.** Shows how to design the substrate
- **c.** Diagrams methods to select plastic adhesives
- **d.** Provides data on curing times

17-7. Loss-of-solvent curing of adhesives usually occurs more quickly than cooling of hot melts.

T (True) **F** (False)

17-8. **a.** In the adhesive system, what three events must occur to obtain a bonded joint?

b. What types of forces are involved in adhesion?

17-9. What is the difference between adhesion and cohesion?

17-10. In a study of the gecko's ability to walk on ceilings, what was the force that allowed adhesion to hydrophobic and hydrophilic surfaces?

a. Capillary **b.** Cohesive **c.** Ionic **d.** Van der Waals

17-11. Findings from the gecko research on attraction forces and surface area hold promise for

a. Dry adhesives **b.** Stronger elastomers **c.** Plastics **d.** Smart polymers

17-12. Select an adhesive for the following joining applications:

- **a.** aluminum electronic component to fiberglass printed circuit board
- **b.** PUR-foam bumper fascia to steel bumper
- **c.** ABS control mount to porcelain-enamel dishwasher panel
- **d.** plywood veneers
- **e.** aluminum identification label to PC plastic electric drill housing
- **f.** aluminum towel rack to ceramic tile wall
- **g.** glass walls to chrome-plated steel aquarium
- **h.** PS-foam pad to wooden seat
- **i.** exterior laminated beam

j. copper foil to flexible printed circuit board

k. brake liners to steel brake shoe

17-13. Materials used to work like adhesives, provide protective coatings, and keep out liquids and gases are

a. Adhesives **b.** Caulks **c.** Sealants **d.** Stains

17-14. Considerable emphasis has been placed on eliminating _____ in protective coatings to improve air quality and other aspects of our natural environment.

a. Silicone **b.** Water-based coatings **c.** ER **d.** VOCs

17-15. The term for synthetic polymers used on soil, rock, and earth in construction is

a. Aramides **b.** Geosynthetics **c.** Styrofoams **d.** Earth screens

17-16. Visit the Biosphere2 website listed in applications & alternatives and list examples of applications of plastics, elastomers, adhesives, caulks, and sealants.

REFERENCES & RELATED READINGS

Adhesives

http://www.adhesivesage.com/mag/aa1.html

Adhesives Age On The Web. http://www.adhesivesage.com/, January 3, 2003.

Adhesive and Sealants Digital reference. http://www.adhesivesandsealants.com, December 16, 2002.

ASM International. *Engineering Materials Handbook, Volume. 3: Adhesives and Sealants.* Materials Park, ott: ASM International, 1990.

Autumn, Kellar. Gecko van der Waals forces. http://www.pnas.org/cgi/content/abstract/192252799v1

Consumers Union. "The Right Glues," Consumer Reports, July 1995, pp. 470–473.

Valero, Greg. "Gecko Technology Climbs to New Heights" *Adhesives Age on the Web,* December, 19, 2002.

Periodicals

Adhesive Age

Journal of Adhesion Science and Technology

Module

18

Wood & Related Products

After studying this module, you should be able to do the following:

18–1. Analyze wood structure and explain how its physical and mechanical properties develop.

18–2. Cite typical uses for hardwood, softwood, composite wood, and wood by-products.

18–3. Give reasons why wood is being rediscovered as a valuable engineering material and state some concerns for managing forests.

18–4. Explain why wood, a renewable resource, contributes to "greening" the materials cycle.

18–5. List some examples of newer wood-based products and the benefits those products bring to building and construction.

18–6. Explain how wood fits in with trends in materials technology with regard to improved building construction, design, materials selection, and maintenance of a sustainable environment.

18–7. Use the website addresses denoted by the symbol to seek out current developments in wood technology.*

* The icon found throughout this book will link you to Internet sites related to topics being covered. The dyanamic nature of the Web brings frequent changes, so some of these sites, while available at the time of writing, may not be up now.

18.1 WOOD

Wood—one of nature's unique, natural composites—and rock were humanity's first materials. As a ready weapon or fuel for the fire, the stick required no major processing. Since those early beginnings, wood has been the target of much technology and is still being researched not only for improving the yield of forests but also for ways of better using the tree and the many by-products of wood. In Figure 18-1, a model of the Wright brothers' "03 Flyer" hanging in the Smithsonian Air & Space Museum shows that the main structural component (struts) consisted of Sitka spruce and ribs made of ash. Wood provided the Wright brothers the best lightweight structural material available in the early twentieth century. They covered the wooden wings with muslin cloth, which is made from another natural polymer, cotton. Although wood won't be found in modern aircraft, it finds many contemporary uses because it still has much value as a structural material that suits current needs.

 Library of Congress digital files—*http://lcweb2.loc.gov/pp/wriquery.html*

As with other living organisms, trees are subject to environmental conditions. Acid rain, lack of biodiversity in tree farms, overharvesting, fire, and similar problems present constant challenges to maintaining the fragile balance necessary for the world's forests. Many underdeveloped nations have harvested their trees too heavily to sell the timber for needed income and/or for cooking and heating because of the scarcity of other fuels. Where forests are lost, regions turn to deserts, causing disaster to all living things. As with most other materials in our family of materials, all aspects of wood must be closely monitored in the materials cycle.

Wood must be considered a valuable engineering material and should be exploited to its fullest extent because it is environmentally friendly. Fossil fuels and minerals required an eon of time to develop, so civilization faces or will face shortages of oil, gas, iron, chromium, and other raw materials used to produce many of our engineering materials. Forests are key to a sustainable environment. Responsible forest management must be a part of wood technology because trees, while renewable resources and valuable construction materials are also crucial to the quality of the air we breathe, stabilize soil, provide shelter and nutrition for many species, and add to the natural beauty of our landscape. Photosynthesis is a series of chemical reactions by which green trees and plants, using energy from the sun and the presence of chlorophyll, convert CO_2

Figure 18-1 The Wright brothers' 1903 Flyer with most structural components made of Sitka spruce. The brothers constructed the biwinged aeroplane out of wood, steel wire, and fabric in their bicycle shop in Dayton, Ohio. The Flyer serves as another example of wood's value as a fairly lightweight structural material.

and water vapor (H_2O) into carbohydrates such as sugar, glucose, and cellulose—the principal structural material of trees and plants. Through photosynthesis, trees and other green plants consume CO_2, which helps combat the ever-increasing CO_2 content of the atmosphere. CO_2 in the atmosphere absorbs some of the infrared radiation emitted by the earth and prevents it from returning to space, thus keeping the earth's surface warmer than it would be in the absence of CO_2. This phenomenon is referred to as *global warming*. CO_2 is produced primarily by the burning of fossil fuels (coal, petroleum, natural gas, and gasoline) and has been increased by the destruction of vast areas of tropical rain forests, which can no longer do their part in consuming it.

It is possible to grow wood in a variety of ways that yield a full range of construction and consumer products. If grown in plantations using new planting and harvesting techniques, then processed with new technology to maximize all of the plant, and finally recycled efficiently, wood will serve humanity well. In our modern era, wood has been replaced by other materials, such as steel, plastic, and concrete; however, our experience with these alternative materials has not always been completely positive.

Many plants possess woody or cellulosic substances, but the tree is our major source of wood. ***Extractives*** from trees that contribute to the properties of wood—such as decay resistance, color, odor, and density—are also useful in a number of industries. These extractives include tannin (tannic acid), used in processing leather, and polyphenolics, used to make phenolic plastics, plus coloring agents, resins, waxes, gums, starch, and oils. In addition to the fruit and nuts that are harvested, trees serve as the raw material for paper, cellulosic plastics, explosives, rayon fibers, films, lacquers and drilling muds, ethyl alcohol, food flavoring, concrete additives, and rubber additives. Wood was our first fuel, but with the technological development of coal, oil, and gases, wood became only a romantic fuel for most of the technologically advanced people of our world. However, tree products once again are being given serious consideration as a fuel. The ability to renew wood supplies through forestry increases humanity's interest in this material as nonrenewable resources such as oil, coal, and minerals rapidly become depleted.

18.1.1 Structure

Wood develops through ***photosynthesis***—the chlorophyll in trees converts carbon, hydrogen, and oxygen to sugar, starches, and cellulose. The dry wood used as an engineering material is composed of the following approximate percentages by weight: cellulose (50%), lignin (16–33%), hemicelluloses (15–30%), extractives (5–30%), and ash-forming mineral (0.1–3%). ***Cellulose*** ($C_6H_{10}O_5$) (Figure 18-2a), a high-molecular-weight linear polymer, forms fibers that make up the cell walls of vessels and ducts. ***Lignin*** (Figure 18-2b), an amorphous polymer, forms a matrix around the cellulose, much like the plastic matrix in fiberglass. Through removal of lignin, wood can be broken down into the fibers that are used in making paper and other synthetics. The cellulose forms cellular networks of ducts, vessels, fiber rays, and pits, which transport and store the extractives and minerals throughout the living tree. The network differs depending on whether the tree is a hardwood or a softwood. Softwoods are not always softer than hardwoods; Douglas fir, a softwood, is about twice as hard as basswood, a hardwood. ***Softwood*** species (Figure 18-3) include firs, pines, and spruces; they bear cones and have needles or scalelike leaves. ***Hardwoods*** are broadleaved and are not normally evergreen; they include maples, oaks, birches, and mahogany. ***Hemicelluloses*** is another polymer closely akin to cellulose; like cellulose it breaks down into sugars when chemically treated. The ash-forming minerals include calcium, potassium, phosphate, and silica. Extractives, which have a significant commercial value, are removed by heating the wood in water, alcohol, or other chemicals.

Most common engineering materials, such as steel and concrete, are ***isotropic,*** which means that generally their strength is the same in all directions because the materials are homogeneous. Wood is ***anisotropic,*** meaning that it has greater strength in some directions than others. The anisotropy develops as a result of the way a tree grows, with various factors influencing the wood structure: cellular structure, branches, and bending by prevailing wind. A full discussion of all these factors is beyond the limits of this book, but the resulting tensile strength and stiffness of wood are greater in the radial direction than in the tangential direction (Figures 18-4 and 18-5). Rays that radiate from the pith outward on the tree tend to tie together the layers of

(a)

(b)

Figure 18-2 (a) Cellulose formula. (b) Lignin formula.

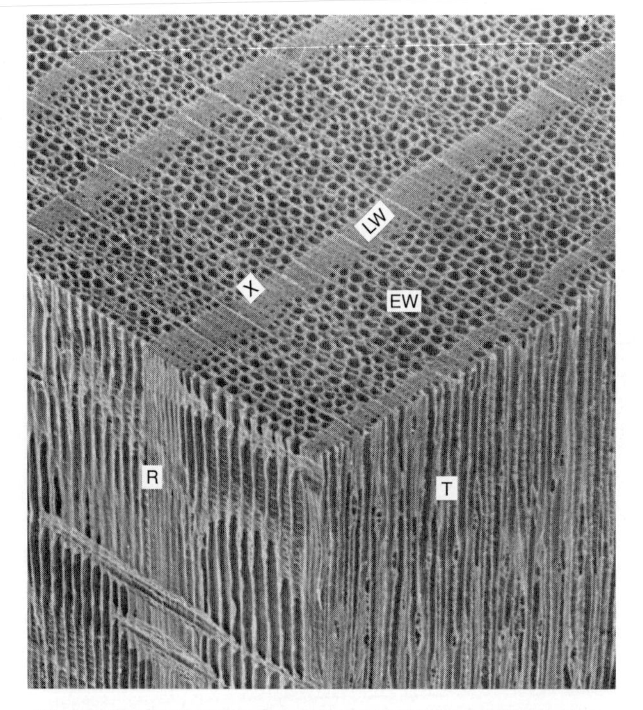

Figure 18-3 Softwood block showing three complete and two partial growth rings in the cross-sectional view (X). Individual cells can be detected easily in the earlywood (EW), whereas the smaller latewood cells are difficult to distinguish in the latewood (LW). Note the abrupt change from earlywood to latewood. The two longitudinal surfaces (R—radial; T—tangential) are illustrated. Rays, which consist of food-storing cells, are evident on all three surfaces. 70X. (Dr. Wilfred A. Côté, SUNY College of Environmental Science and Forestry)

cells (tracheids) growing longitudinally. This ***directional strength*** results partly from the complex structure of cell walls, which have long polymer chains of cellulose that form layers running in varying patterns to reinforce each other. Primary covalent bonding holds together these ***micro-***

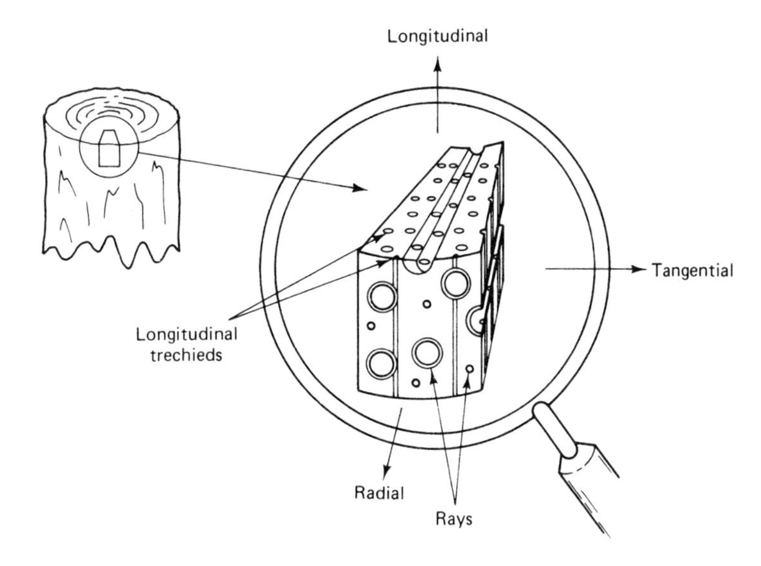

Figure 18-4 Section cut from tree shows patterns from Figure 18–3.

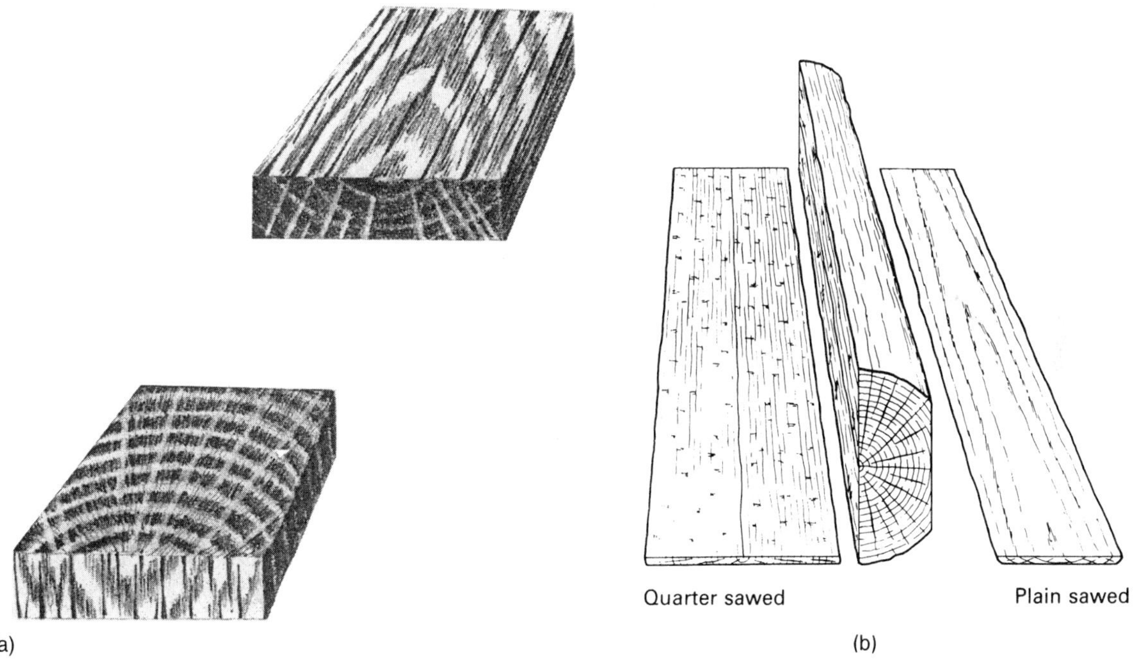

Figure 18-5 (a) Grain direction. (U.S. Forest Products Laboratory, Madison, Wisconsin) (b) Sawed boards.

fibrils, which are almost parallel to the cell axis. Weaker, secondary bonds operate on the perpendicular axis.

Lumber consists entirely of dead cells. The cross section of a white oak tree trunk in Figure 18-6 shows the various layers. The ***cambium layer*** produces all the growth of cells (wood and bark). ***Sapwood*** is a transitional wood of both living and dead cells. The sapwood's main role is to conduct sap and store food. ***Heartwood*** evolves from sapwood as the cells die and become inactive. The heartwood cells no longer transport sap but rather become storage cells. The sapwood of all species is highly susceptible to rotting; the heartwood of some species is protected from decay because it has taken on infiltrating materials that serve as

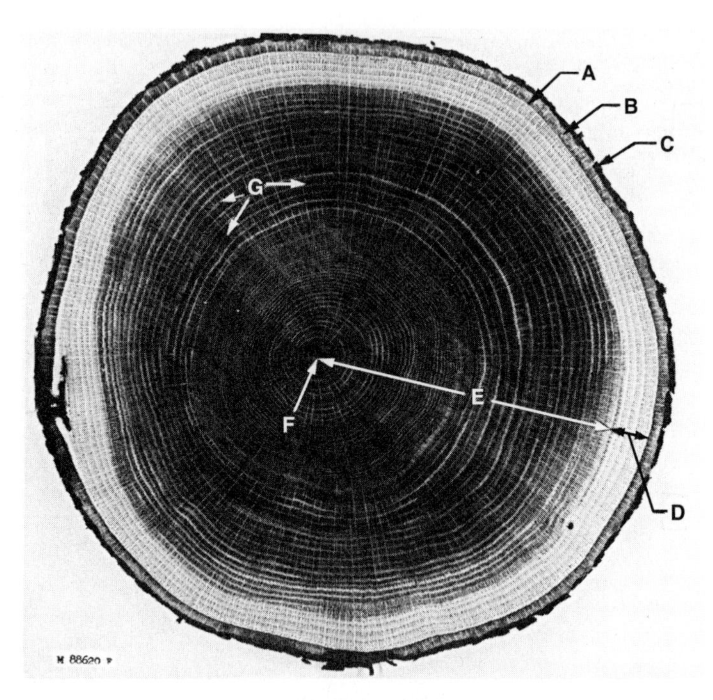

Figure 18-6 Cross section of white oak tree trunk. (A) The cambium layer is inside the inner bark and forms wood and bark cells. (B) Inner bark is moist and soft and contains living tissue; it carries prepared food from leaves to all the growing parts of the tree. (C) Outer bark containing corky layers is composed of dry, dead tissue; it gives protection against external injury. (D) Sapwood, which contains both living and dead tissue, is the light-colored wood beneath the bark; it carries sap from the roots to the leaves. (E) Heartwood (inactive) is formed by a gradual change in sapwood. (F) Pith is the soft tissue about which the new growth takes place in newly formed twigs. (G) Wood rays connect the various layers from pith to bark for storage and transfer of food. (U.S. Forest Products Laboratory, Madison, WI)

preservatives. Additional information on wood structure, plus other information on wood, is available in the excellent source *Wood Handbook*: *Wood as an Engineering Material* (see the References section). The handbook defines ***lumber*** as the product of the saw and planing mill that is not manufactured further; ***timber*** is a classification of lumber that is nominally 5 or more inches in its least dimension. Figure 18-7 shows some applications of logs and lumber in building construction.

18.2 WOOD PRODUCTS

18.2.1 Commercial Lumber

Lumber is graded to provide the buyer with an indication of its strength and appearance. Unlike most engineering materials, wood is not homogeneous. Grading specifies conformity of the lumber to its end use. The grading thus is an averaging of the quality of boards in a given grade. Visual grading spots defects such as knots, stains, bark pockets, decay, shakes (cracks), pitch pockets (resin buildup), and checks (splits). ***Softwood lumber*** is a vital material for construction and manufacturing and is graded as construction and remanufactured lumber based on the American Lumber Standard. ***Construction grade*** includes stress-graded, nonstress-graded, and appearance lumber. The ***stress-graded*** category gives an indication of engineering properties and covers pieces of 2 to 4 in. nominal size, larger timbers (5 or more in. nominal thickness), posts, beams, decking, stringers, and boards (less than 2 in. nominal thickness). ***Nonstress-graded*** lumber includes a combination of allowable properties and visual defects and is used for boards, lath, batten, and planks. ***Appearance lumber*** that is nonstress graded is cut to patterns such as trim, siding, flooring, and finished boards. ***Remanufactured*** grade is

(a)

(b)

(c)

Figure 18-7 Wood structures built to weather time and the ocean. (a) Reconstructed slave homes at Carter's Grove (a portion of Colonial Williamsburg) represent both log and lumber home construction that has been for centuries. Much of current building construction technology relies on the same basic "stick construction." (b) The U.S. Coast Guard lifesaving station, built at Virginia Beach on the ocean's beach in 1903, reflects early twentieth-century wood construction. (c) A modern wood structure also built on the Atlantic Ocean has an imposing roof support made of laminated beams and wood trusses using southern pine. This structure departs from "stick construction" techniques by taking advantage of design flexibility with laminated wood beams.

lumber that will receive further processing or secondary manufacturing, such as cabinet stock, molding, and pencil stock. ***Hardwood lumber*** is graded as factory lumber (amount of usable lumber in a piece) to be used for dimensional parts and finished market products (maple flooring, stair treads, trim, and molding). The furniture industry is the prime user of most hardwood grading lumber. There is also a system for designating the degree of dressing or finish ***surfacing:*** one side surfaced (S1S), two sides surfaced (S2S), one edge surfaced (S1E), two edges surfaced (S2E), and various combinations, such as one side and one edge surfaced (S1S1E) or all sides and edges surfaced (S4S).

18.2.2 Composite Wood

To gain maximum use of trees and to achieve properties not possible from solid wood, composite woods have been developed for use in construction and manufacture. Among composite woods are laminated timber, impreg-wood, wood and recycled plastics, plywoods and veneer, particleboard, hardboard, sandwiched materials, and insulation board.

18.2.2.1 Laminated timber. Laminated timber is a product of adhesive-joining technology. Through adhesive bonding of pieces of lumber so that the grain of all pieces is parallel to the length of the timber, it is possible to produce straight to curved structural-wood members of large size and outstanding strength. ***Glulam*** bonds sawn lumber laminations with waterproof structural adhesives, allowing nearly unlimited depth, width, and length in a variety of shapes. One glulam product, Parallam, achieves higher stress ratings and uniform strength than those of conventional materials through a patented process in which thin strands of veneer are glued using waterproof adhesives, then cured with microwaves. Laminated beams offer the architect flowing lines and the warmth and beauty of wood. Custom-ordered laminated beams have the capability to span more than 100 m of unobstructed space for sports arenas, convention centers, entertainment halls, and churches (Figure 18-7c). Seasoning lumber for laminated beams prior to gluing provides improved strength, dimensional stability, and elimination of unsightly surface cracks and shrinkage. The ablative nature of the large beams is an advantage in protection from the hazards of fire, since the thin layer of char protects the interior of the beam and the wood does not become plastic at high temperatures, as do metal and plastic members.

18.2.2.2 Impreg-wood. ***Impreg-wood*** is a very stable lumber achieved by the bonding of phenolic resins to the cell wall microstructure. This is accomplished through the saturation of thin veneers with phenolic resin that is polymerized before the veneers are stacked into thick laminates. Applications of impreg-wood include sculptured models for the huge metal dies used to stamp automobile sheet metal parts. ***Compreg-wood*** also employs veneers impregnated with phenolic resin, but the polymerization is accomplished when the veneers are stacked and compressed into the desired final shape. The dimensionally stable, high-density, hard-compreg laminate has a glossy finish throughout the material. Knife handles, bowls, jigs for manufacturing, and parts for textile looms use the compreg laminates.

18.2.2.3 Wood and recycled plastics. The progress in municipal recycling has resulted in a dependable flow of recycled plastics. Plastic lumber made of recycled plastics offers many benefits—it resists biological attack (rotting and insects) and has no need of protective finishes. For example, the Smartdeck® is a synthetic lumber system that includes rails, trim, fascia, and posts. Wood is eliminated, and the system allows simple deck installation due to such features as molded channels in the posts and rails. These allow wires and cables to be run for electricity and sound equipment without drilling. Combinations of recycled plastics and wood offer the benefit of wood's beauty, are usually lower in cost, and have the durability of plastics. Durawood EX®, a wood–polymer composite (approximately two-thirds oak fiber combined with recycled plastic resins), combines the look of wood with highly detailed design, durability, and low maintenance. Like other decking material, Durawood will not rot, splinter, or show other effects of moisture typical of natural wood, and is backed by a 10-year limited warranty. After installation, it turns a light "driftwood" gray. Eaglebrook also makes a 100% recycled plastic lumber product called Durawood PE®, which comes in several durable colors such as redwood, cedar, gray, and oak. Durawood PE®, which carries a 20-year limited warranty, absorbs no moisture and therefore provides high durability and low maintenance. Fibrex™ exemplifies the result of years of R&D aimed at combining reclaimed wood, a renewable resource, with polymer resins to achieve superior building materials. Figure 18-8 shows Fibrex constituents, beginning with pine (e.g., Ponderosa) reclaimed from wood milling, then further processed into wood fiber or *wood flour*. The fine wood flour is blended with virgin (not recycled) PVC, then extruded into pellets that will be extruded again to make various framing pieces for windows and doors. The second extruding is a coextrusion to provide a nonporous vinyl for interior and exterior surfaces of windows and doors, a PVC laminate (Perma-Shield®) for exteriors, or a thin layer of pine for interior surfaces. The thin pine veneer interior laminate allows staining and/or finishing. A PVC external laminate can serve as the interior layer as well, with the advantage of through-coloring on windows, doors, and frames that should not require further finishing (Figures18-9 and 10).

Composite woods made from recycled milk jugs (HDPE) and recycled or reclaimed woods take many shapes and come in a variety of colors. They serve as replacement for wood

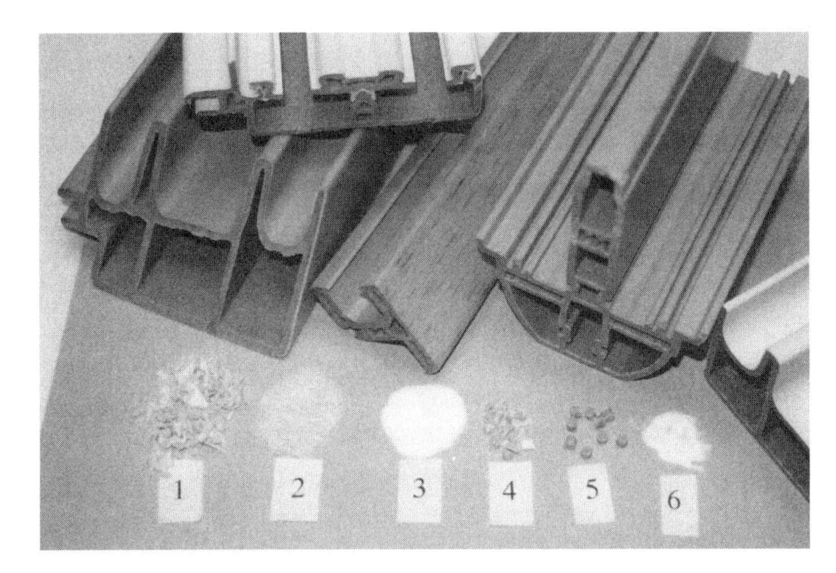

Figure 18-8 Fibrex™, a high-end composite wood/plastic material consists of (1) pine (such as Ponderosa reclaimed from the milling of wood for solid wood products). Next, the reclaimed pine is processed to produce (2) fine wood fibers (flour) that is mixed with (3) virgin (not recycled) powder for making a (4) blend of wood fiber in a PVC matrix. This blend is extruded into (5) pellets. The pellets go into another extruder that coextrudes either (6) nonporous PVC (Perma-Shield®) or pine veneer (seen on the middle sample) to make the complex, cross-sectional, structural components seen in the background and in Figures 18–9 and 18–10.

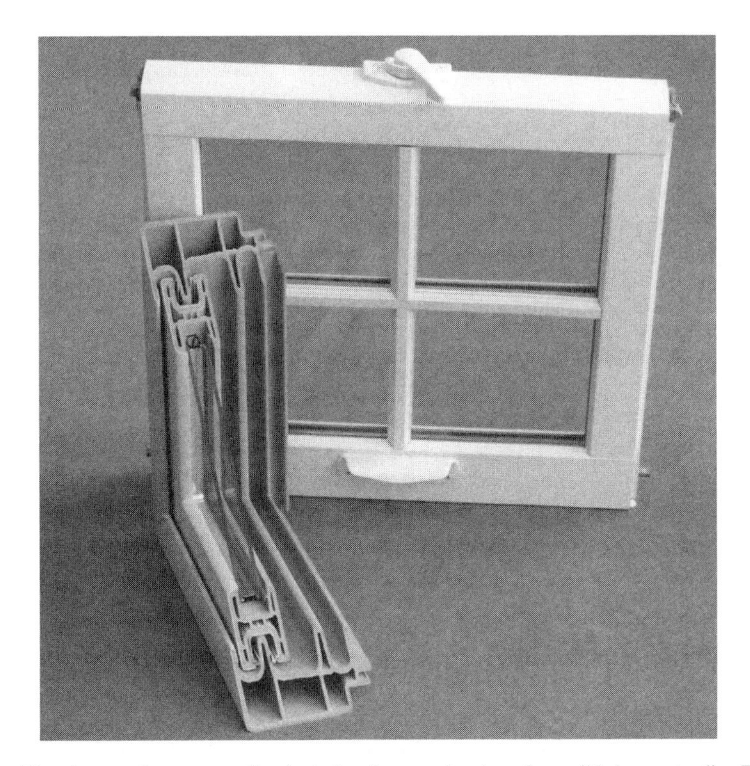

Figure 18-9 Note the complex cross-sectional window frame and sash made possible by coextruding Fibrex™ with nonporous PVC. Also note the double-glass panes that seal inert gas to provide superior insulation in the Renewal by Anderson® replacement window system.

decking, stairs, and rails (Figure 18-11). These composite woods aid in "greening" the materials cycle by keeping plastic out of municipal landfills. They resist a wide spectrum of weather extremes and do not need to be refinished as do pressure-treated boards (which also harm the environment during processing and in use). Available in various brands such as TecWood®,

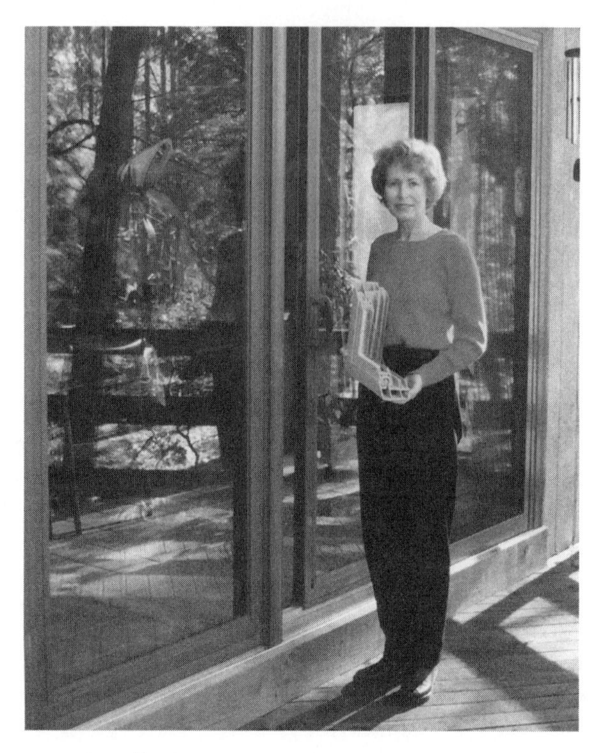

Figure 18-10 Renewal by Andersen® patio door frame made of Fibrex™ composite wood components and sealed glass.

Figure 18-11 Various extruded composite wood forms for decking made from reclaimed wood and recycled milk jugs (HDPE).

Fiberon®, and ChoiceDek®, these products do not require constant refinishing and they neither splinter nor warp. All of these benefits do come at a cost however, for these alternatives are many times more expensive than pressure-treated wood.

 Recycled plastics for wood—*http://oikos.com/products/wood-plastics/eaglebrook/index.html*

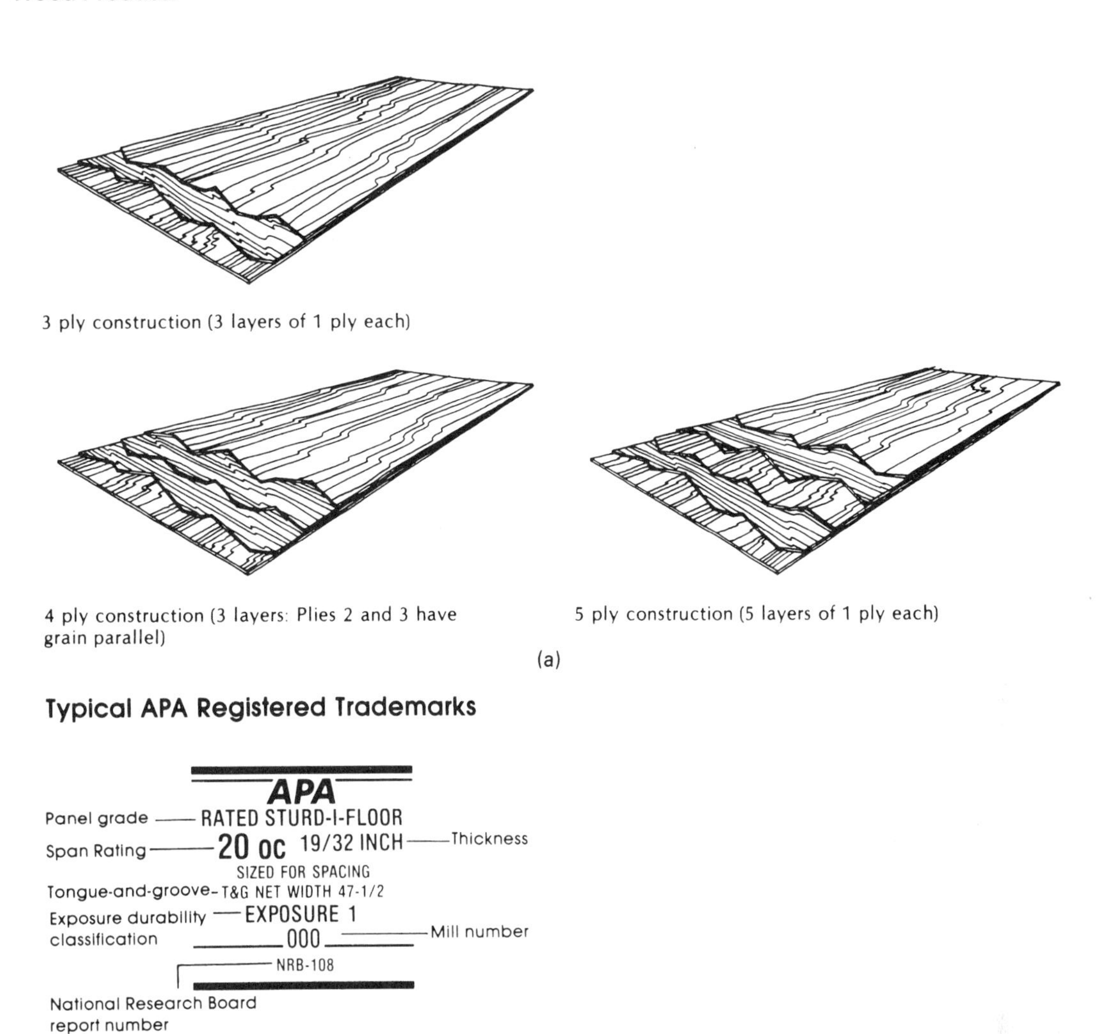

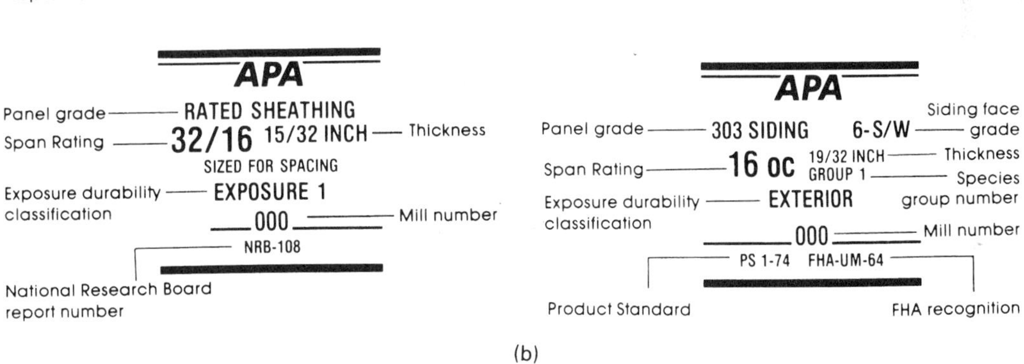

Figure 18-12 (a) Plywood. (b) Marking used by American Plywood Association.

18.2.2.4 Plywood. Plywood is another form of laminate common in building construction. It is produced by stacking layers of veneer with the grain direction in each layer at right angles to the next, beginning with the grain running the length of the panel (Figure 18-12a). This cross-laminating takes advantage of the fact that wood is much stronger in the direction of the grain than across the grain. Thus, plywood yields equalized strength in the plane directions while also being shear, puncture, and split resistant. Numerous grades of plywood are available; Figure 18-12b shows the marking applied to three grades of plywood using the American Plywood Association

Grade	Description
N	Smooth surface "natural finish" veneer. Select, all heartwood or all sapwood. Free of open defects. Allows not more than 6 repairs, wood only, per 4 × 8 panel, made parallel to grain and well matched for grain and color.
A	Smooth, paintable. Not more than 18 neatly made repairs, boat, sled, or router type, and parallel to grain, permitted. May be used for natural finish in less demanding applications.
B	Solid surface. Shims, circular repair plugs and tight knots to 1 inch across grain permitted. Some minor splits permitted.
C Plugged	Improved C veneer with splits limited to 1/8-inch width and knotholes and borer holes limited to 1/4 × 1/2 inch. Admits some broken grain. Synthetic repairs permitted.
C	Tight knots to 1-1/2 inch. Knotholes to 1 inch across grain and some to 1-1/2 inch if total width of knots and knotholes is within specified limits. Synthetic or wood repairs. Discoloration and sanding defects that do not impair strength permitted. Limited splits allowed. Stitching permitted.
D	Knots and knotholes to 2-1/2 inch width across grain and 1/2 inch larger within specified limits. Limited splits are permitted. Stitching permitted. Limited to interior (Exposure 1 or 2) panels.

Figure 18-13 Veneer grades. (American Plywood Association)

standards. Letters assigned to veneer grades specify the quality of exposed plys (Figure 18-13). Plywood is divided into ***interior*** and ***exterior*** plywood. Interior plywood has its laminates glued with moisture-resistant glue (phenol formaldehyde, urea, and melamine). Interior plywood veneers may be of lower quality than the exterior type. Exterior plywood is glued only with phenol formaldehyde to provide a waterproof bond. The waterproof glue can withstand aging and boiling water. Marine plywood is the superior-quality exterior plywood. Both appearance- and engineering-grade plywood can be either exterior or interior grade.

Thin appearance-grade plywood, in the form of paneling, is very popular for walls in houses, mobile homes, recreational vehicles, and boats. There is a nearly unlimited variety of hardwood-faced paneling. An inexpensive grade of Philippine mahogany known as ***lauan*** receives many types of coating and textures for decorative paneling. Thicker appearance-grade plywood is used for furniture and cabinets. Engineering-grade plywood is used to make forms for casting concrete, underlayment for flooring and carpets, structural panels, roofing, decking, walls, and cabinets. Plywood siding has gained wide acceptance in contemporary-style buildings. Siding offers warmth, low maintenance, and good thermal and acoustical insulation, among other properties, and is available in various species with textures including smooth, deep grooved, brushed, rough sawn, and overlays. ***Veneer*** in thin sheet form is used in a single thickness for baskets, for boxes, and as ornamental inlays.

18.2.2.5 Wood-based fiber and particle panel materials. Wood-based fiber and particle panel materials include a variety of panels and boards used in building construction, packaging, furniture, and other manufactured products. ***Particleboard*** panels, also known as *reconstituted panels*, are produced through the use of thermosetting resins, such as urea–formaldehyde and phenol formaldehyde, which serve as a matrix to bind together wood residues or shavings in the form of small wood flakes, wood flour, and additives (Figure 18-14). Water resistance is improved through the addition of wax. The resins account for 5% of the dry panel's weight and polymerize as the particles are pressed in flat presses or extruded through thin die presses. Many particle sizes and shapes are manufactured to yield panels with specific properties. Particleboard serves as backing for plastic and wood laminates or as a decorative panel, which is normally painted. ***Hardboard*** is produced by processing wood chips with steam and/or pressure; the lignin bonds the fibers together. Medium-density, high-density, or special-density

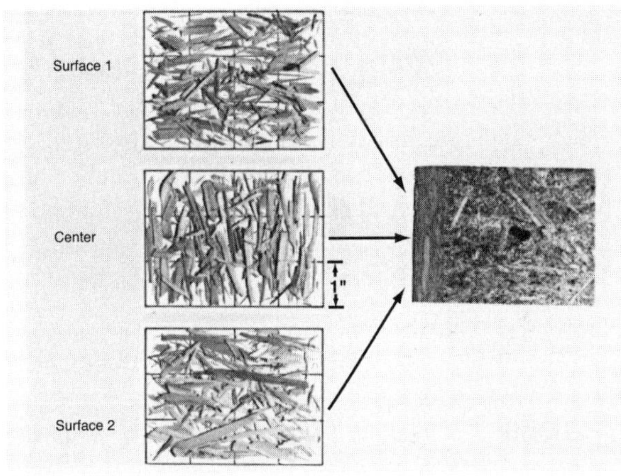

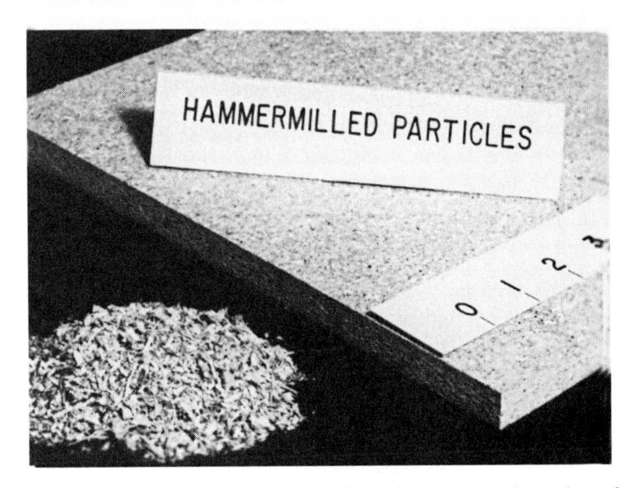

Figure 18-14 Particleboard (also called *reconstituted panel*) products consist of a variety of compositions and organizations. Flakeboard uses relatively square chips; oriented strandboard uses long, narrow chips that can be aligned for maximum directional strength. (U.S. Forest Products Laboratory)

hardboard offers high strength, wear resistance, moisture resistance, and resistance to cracking and splitting and has good working qualities. ***Tempered hardboard*** has oils added to increase water resistance, hardness, and strength. Hardboard finds use as painted house siding, diestock, and electrical panels; it is also used with plastic and other laminates and as components in furniture, such as drawer bottoms, mirror backs, and television sides and backs.

18.2.2.6 Sandwiched materials. Sandwiched materials include treated paper honeycomb faced with laminating veneers. Doors and panels employ these materials. Fiberboards made of wood fibers bonded with rosin, asphalt, alum, paraffin, oils, fire-resistant chemicals, and plastic resins are used as insulation panels on walls and roofing to which some other material is added, such as siding or tar and gravel.

18.3 WOOD BY-PRODUCTS

The most common by-product of lumber and plywood processing is *chips* used in the production of pulp for various paper products. Pulp is obtained through chemically defibrating the chips by dissolving the lignin to obtain fibers. Other by-products include naval stores (turpentine and rosin), bark for mulching plants, charcoal briquettes, and various resins from the tree.

18.4 PHYSICAL PROPERTIES

18.4.1 Appearance

When selected for furniture, house trim, doors, floors, wood turning, or other applications in which wood is exposed, appearance is paramount because it lends warmth and beauty. Color, figure (patterns produced by grain, texture, and machining), luster, and the manner in which finishes affect wood all become important factors in selection.

18.4.2 Sawing Effects

The Forest Products Laboratory of the U.S. Department of Agriculture defines grain and texture as follows:

Grain: direction, size, arrangement, appearance, or quality of fibers in wood or lumber
Texture: structure of wood finer than annual rings

Twenty grain specifications cited in the *Wood Handbook* provide precise definition for various grain patterns. The angle at which a log is cut is one factor that determines grain pattern. Figure 18-15 shows the difference in quarter-sawed and plain-sawed boards. Also see the photographs of wood specimens in the color section of this book.

Lumber is used in many stages, including logs with just the bark removed and rough-sawn timbers, but finished lumber is the most common form encountered. Standard dimensions include 2 × 4, 1 × 12, and 4 × 4, which are nominal measurements in inches. Nominal measurements indicate the approximate size of the lumber when it is in the rough state. A finished 2 × 4 is actually 1 1/2 × 3 1/2 in. Following rough sawing, lumber is ***seasoned*** to remove the moisture, reduce warping and cracking, reduce weight, and increase strength and many other properties. There are a variety of seasoning methods, including air drying and kiln drying. Air drying is a slow process in which rough timber is stacked in a specified manner and allowed to dry naturally. Kiln drying employs ovens, in which the rough lumber is stacked and hot air is forced over the wood. Research into solar kilns shows promise for low cost and quick seasoning. Beyond seasoning, many methods are employed to improve the physical, chemical, and mechanical properties of wood. ***Chemical modification*** involves a chemical reaction (covalent bonding) between the cellulose, lignin, and hemicellulose and a chemical that acts as a preservative. Woods are also impregnated with polymers, oils, salts, and other solutions. A wide variety of coatings, from natural clear finishes to stains and paints, provide protective and decorative value.

18.4.3 Moisture Content

Moisture content is a measure of the water in wood. Normally expressed as a percentage of the amount of water weight to the weight of oven-dry wood, moisture content is an extremely important factor in the mechanical and other physical properties of wood. Trees have moisture

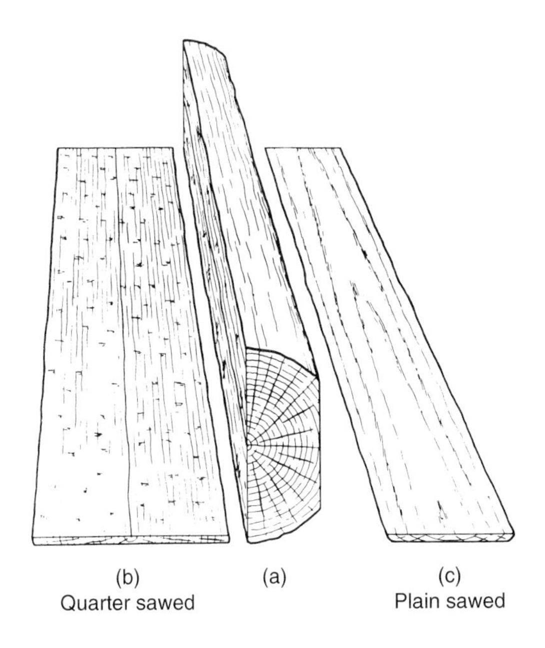

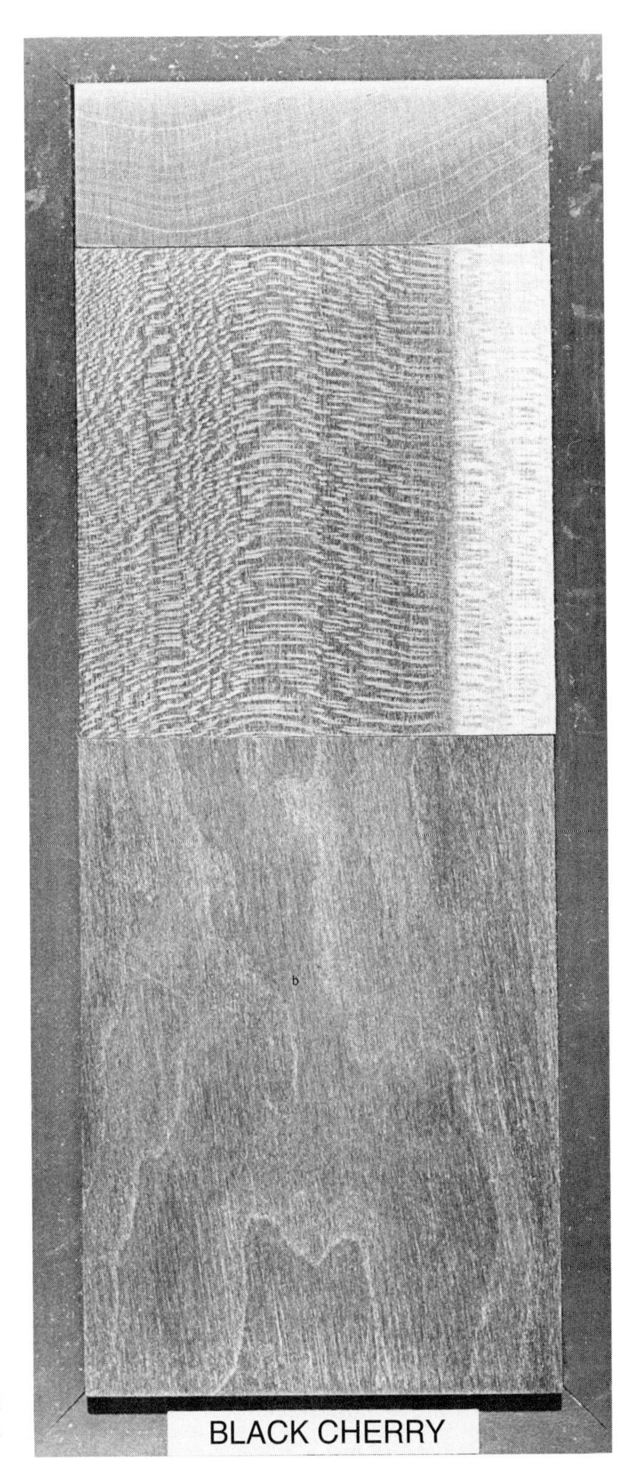

Figure 18-15 (a) The top segment in the photo illustrates end grain. (b) The middle segment is quarter sawed. (c) The bottom segment is plain sawed. These match the 3-D sketch above.

content ranging from 30 to 200%. Drying of freshly cut wood removes moisture, but the hygroscopicity of wood substance allows it to pick up moisture from the surrounding atmosphere. For this reason, seasonal and environmental changes in relative humidity cause a constant shift in moisture content. You may recall that during humid and rainy seasons, doors and drawers become hard to open, while during dry days the same doors or drawers become easy to use. Treatment of wood must include steps to minimize moisture content changes, weathering, chemicals, and biological attack. Expanding and shrinking or improperly finishing wood cause ***warping, splitting, checking*** (lengthwise separation across the annual rings), and other problems. Damp wood also provides opportunity for bacteria and fungi to ***decay*** or ***rot*** wood. Seasoned lumber has been dried in varying degrees. For shipping and to avoid molding, moisture is reduced to 25% or below. Thoroughly dried lumber has from 12 to 15% moisture. The key to seasoning lumber is to attain a moisture content corresponding to the average atmospheric conditions in which the lumber will be used.

18.4.4 Weatherability and Decay Resistance

Similar to the reaction of plastics, woods change color and structure due to heat, light, moisture, wind, and chemicals in the environment. Ultraviolet (UV) light causes chemical degradation and graying of color, while metal fasteners (screws, nails, etc.) and hardware (hinges, etc.) also cause color changes. The UV light breaks down cellulose in the fibers. However, the degradation of the surface fibers, which is rapid when wood is first exposed, slows to a negligible rate of about 1 mm per 100 years. Other effects of weathering include warping, checking, abrasion, and surface roughening. While the even gray color is attractive, the appearance of dark gray and blotchy colors indicates attack by biological organisms and the probability of rotting. Table 18-1 shows the varying resistances of woods to this sort of decay. The cut of the

TABLE 18-1 PHYSICAL AND MECHANICAL PROPERTIES OF IMPORTANT COMMERCIAL WOODS GROWN IN THE UNITED STATES (AVERAGE VALUES OF SMALL, CLEAR, STRAIGHT-GRAINED SPECIMENS

Specific gravity Common species	Hardness[a] (12% moisture)	Modulus of elasticity[b] (N)	Decay (MPa)	Resistance[c]
Hardwoods				
Ash, white	0.60	5,900	12,000	S
Basswood	0.37	1,800	10,000	S
Beech, American	0.64	5,800	11,900	S
Birch, yellow	0.62	5,600	13,900	S
Cherry, black	0.50	4,200	10,300	V
Elm, American	0.50	3,700	9,200	S
Maple, sugar	0.63	6,400	12,600	S
Oak, northern red	0.63	5,700	12,500	S
Oak, white	0.68	6,000	12,300	V
Walnut, black	0.55	4,500	11,600	V
Yellow, poplar	0.42	2,400	10,900	S
Softwoods				
Bald cypress	0.46	2,300	9,900	V
Cedar				
Eastern red cedar	0.47	4,000	6,100	V
Northern white cedar	0.31	1,400	5,500	V
Western red cedar	0.32	1,600	7,700	V
Douglas fir				
Coast	0.48	2,200	13,400	M
Interior west	0.50	2,300	10,300	M
Fir				
Eastern species	0.38	2,200	10,300	S
Western species	0.36	1,800	8,500	S
Pine				
Eastern white	0.35	1,700	8,500	M
Sugar	0.36	1,700	8,200	S
Western white	0.38	1,900	10,100	S
Redwood				
New-growth	0.35	1,900	7,600	V
Old-growth	0.40	2,100	9,200	V

Source: Wood Handbook: Wood as an Engineering Material, by Forest Service, U.S. Department of Agriculture.

[a]Newtons required to embed an 11.26-mm ball one-half its diameter in a direction perpendicular to the grain.

[b]Measured from a simple supported, center-loaded beam, on a span/depth ratio of 14:1.

[c]V, very resistant or resistant; M, moderately resistant; S, slight or nonresistant.

boards, such as vertical grain rather than flat grain, can aid weatherability, but special treatment with preservatives is normally required. Composite wood overcomes some disadvantages of natural wood, as seen in the graphs in Figure 18-17. Generally, heartwood provides more resistance to decay than does sapwood. The naturally occurring wood extractives determine the resistance of a particular species of heartwood to attacking fungi. Formerly, the southern and eastern pines were slower growing and older and thus had more heartwood, which made them more resistant to decay. New forestry methods produce faster growth but more sapwood, which reduces these species' decay resistance. Bald cypress has outstanding natural durability.

18.4.5 Density

The variation in a wood's moisture content depends on the environment and the preparation of the wood and will greatly affect the density, weight, and specific gravity. The range of density for most woods falls between 320 and 720 kg/m^3. Calculations of ***specific gravity*** are determined on either kiln-dried wood or wood with a specific percentage of moisture. Twelve percent is frequently used; at a 12% moisture level, most species of trees yield specific gravities in the range of 0.32 to 0.67, as shown in Table 18-1.

18.4.6 Working Qualities

In hand working, normally the wood with the lower specific gravity works best. Machinability of wood depends on characteristics such as ***interlocked grain, hard deposits*** of minerals such as calcium carbonates and silica, ***tension wood,*** and ***compression wood*** in addition to density. High density and hard deposits will dull tools. Reaction wood is tension wood and compression wood. Reaction wood and interlocked grain cause binding of boards as they are fed through saws and planers. Other considerations in the workability of wood include nail splitting, screw splitting, and the surface results of sanding.

18.4.7 Ablation

An advantage offered by the use of wood timbers and boards in building is their ability to char when burned. This ablation property allows woods to retain much of their strength because ***charring*** slows the burning. Conversely, metals act as good thermal conductors and uniformly transmit the heat of a fire throughout the metal. This can cause metal supports to lose much of their strength and bend under their loads. Figure 18-16 shows the ablative value of wood; the steel beams failed (softened) under the heat, but the wood structural members held some of the load.

18.5 MECHANICAL PROPERTIES

The mechanical properties of woods can vary widely because of the lack of homogeneity in wood structure. Knots, cross grain, checks, and growth rings provide varying properties within boards and between boards from the same tree. The properties shown in Table 18-1 reflect values obtained from small, clear, straight-grained specimens. They are average values for comparing common woods and for comparing wood with other materials on the basis of modulus of elasticity.

Figure 18-17* compares wood, wood/plastic-based products, and competing materials used for windows and doors. As described earlier, Fibrex is one of a growing class of building construction materials that combine the benefits of wood fibers and plastic resins. Note that Fibrex has a much lower rate of thermal expansion than PVC (the most common material for residential and small business windows). The benefit of Fibrex for double-insulated, sealed windows includes improved durability since its lower expansion and contraction rate will cause less distortion and cracking. Fibrex's greater stiffness over vinyl allows thinner components to be used for larger window openings. The composite wood fiber in a thermoplastic resin matrix also provides excellent insulating and decay-resistance properties.*

*Based on ASTM D-1413 testing methods, Fibrex provides comparable decay resistance to vinyl.

Figure 18-16 Ablative nature of wood: Steel beams became plastic and draped over wooden beams in a factory fire. (U.S. Forest Products Laboratory)

The wooden bridges shown and described in Figure 18-18 are demonstrations of how materials may go out of favor but then may return. Laminated timber, like laminated beams, produces a very durable bridge material.

Advances in the development of polymer materials like those discussed in this unit are taking place every day. If you understand the basic concepts covered in Unit 6, you will possess the ability to keep pace with these developments. To stay current with the innovations, make it a regular habit to visit the websites listed in this module.

Wood products—*http://www.weyerhaeuser.com/woodprod*
Sustainable Forestry Initiative™ (SFI)—*http://www.afandpa.org/forestry/forestry.html*

18.6 LABORATORY ACTIVITIES AND CLASSROOM DEMONSTRATIONS WITH POLYMERS

The *Experiment in Materials Science, Engineering, and Technology (EMSET2)* CD-ROM provides numerous experiments and demonstrations for polymers as a result of a cooperative effort by scores of educators and materials specialists from industry. See Prentice-Hall's website for ordering information on the latest edition of *Experiments in Materials Science* (CD-ROM), at *www.prenhall.com.*

SELF-ASSESSMENT

18-1. In light of the diminishing supplies of many raw materials, what makes wood such a promising engineering material for future generations?

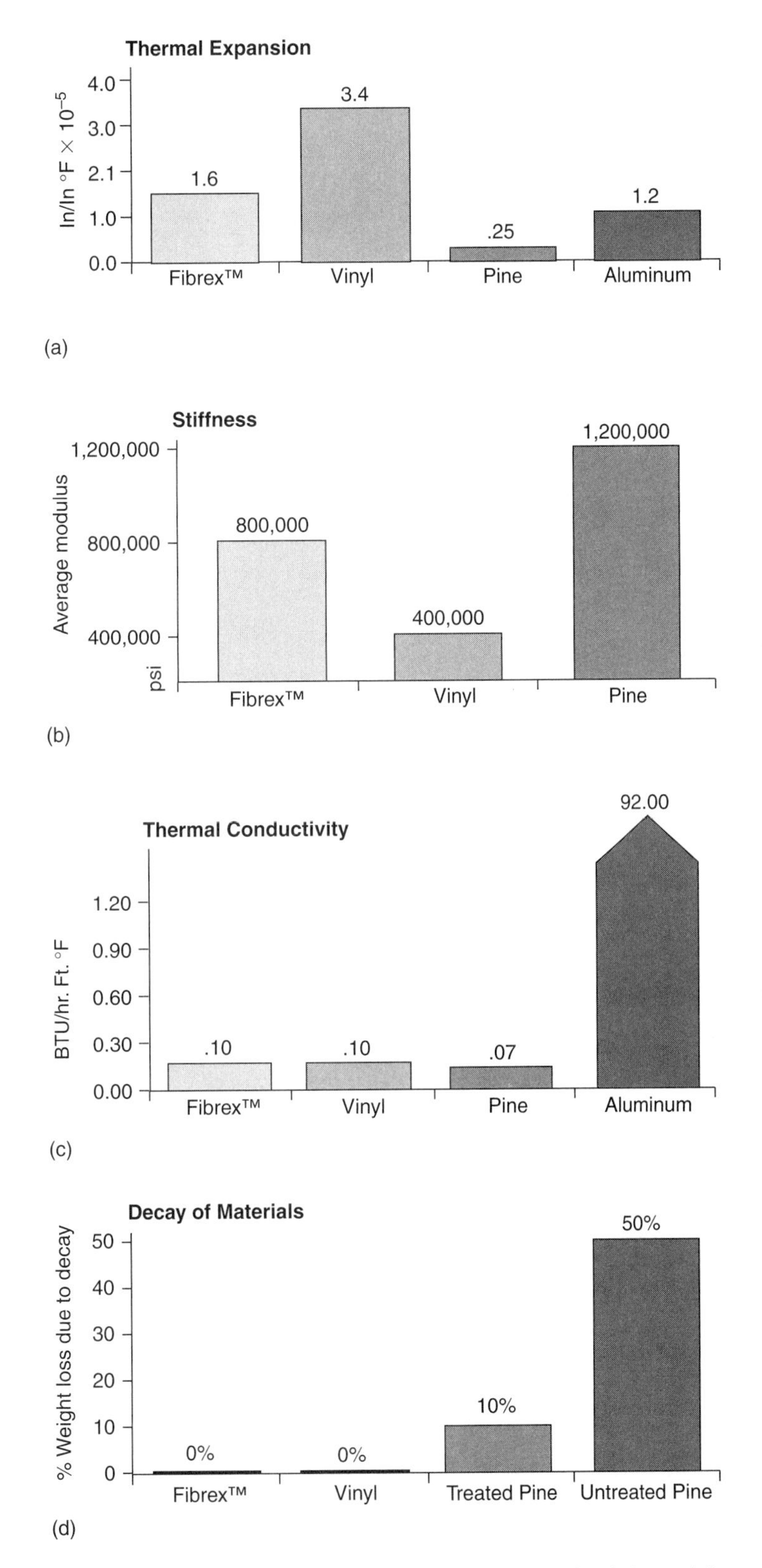

Figure 18-17 Graphs provide a comparison of properties for materials used in window and door construction. (a) Thermal expansion for durability and reliability. (b) Stiffness allows for more glass. (c) Thermal conductivity should be low for best insulation. (d) Decay resistance provides long product life. (Adapted from Andersen® Renewal by Andersen® product literature)

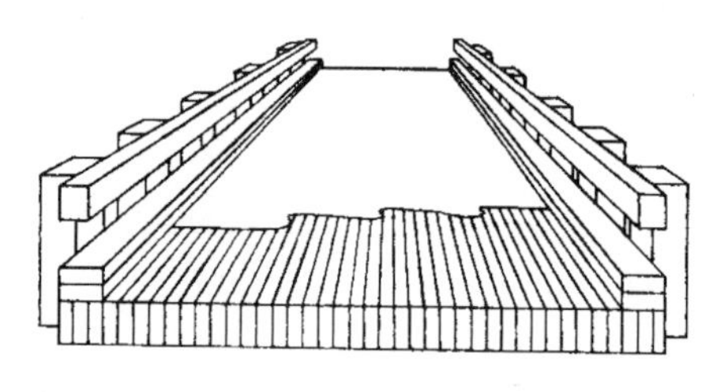

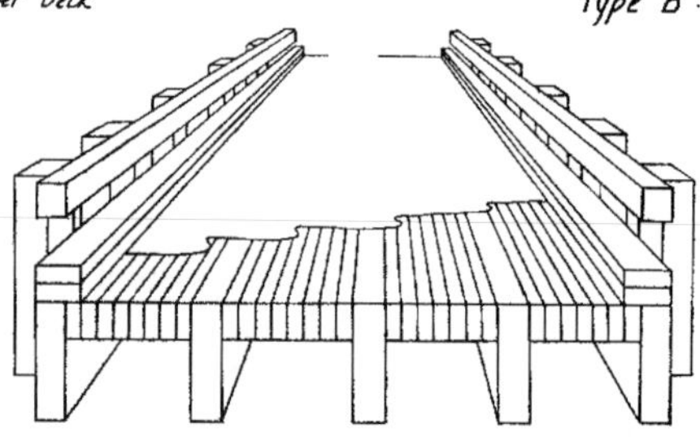

Figure 18-18 The rebirth of wooden bridges. For decades, concrete and steel replaced the old timber bridges that had served humanity since early civilization. Recently, we have learned that acid rain, road salt treatments, and other environmental conditions have rendered nearly half of the 600,000 short-span bridges in the United States unsafe and/or unusable. New technology applied to wood has resulted in timber bridges that should provide much longer service life than steel and concrete bridges. Timber for bridge construction treated with wood preservatives and laminated into stress timber decks, T-sections, and box sections offers the following advantages: pleasing aesthetics, good shock resistance, high strength-to-weight ratios, sound and thermal insulation, and weather and chemical resistance (see photo). Using wood, a renewable resource, localities should be able to harvest regional materials and labor for economical alternatives to those hundreds of thousands of defective concrete and steel bridges. (Barry Dickson, Construction Facilities Center, West Virginia University)

18-2. Name typical applications for the following grades of lumber:

a. nonstressed graded **b.** construction appearance **c.** remanufactured **d.** stress graded

18-3. Name a wood to fit into the following categories:

a. high hardness, high decay resistance, high specific gravity
b. very low hardness, low specific gravity, good modulus of elasticity

18-4. Which polymer serves as a matrix in natural wood to bond the cellulose fibers?

a. Sulfur **b.** Hemicellulose **c.** Lignin **d.** Isoprene

18-5. As a natural composite, wood's grain structure gives it __________ properties.

a. Uniform **b.** Anisotropic **c.** Isotropic **d.** Poor

18-6. The main difference between hardwood and softwood is __________.

a. Polymer structure **b.** Hardness **c.** Economics **d.** Growing environment

18-7. The reason for grading lumber is its lack of __________.

a. Value **b.** Processing **c.** Homogeneity **d.** Visual defects

18-8. A means of overcoming the problem in Question 18-7 to ensure uniform strength is

a. Seasoning **b.** Curing **c.** Heating **d.** Laminated timber

18-9. The key difference between interior and exterior plywood is

a. Number of plys **b.** Direction of grain **c.** Glue **d.** Appearance

18-10. A material developed to maximize the yield of wood from forest products is

a. Plywood **b.** Paper **c.** Glulam **d.** Particleboard

18-11. A major reason wood is prized as an engineering material is its

a. Natural beauty **b.** Low cost **c.** Plentiful supply **d.** Uniform properties

18-12. Wood should be considered as a valuable engineering material because the raw materials are __________, whereas those for most other materials are not.

a. Recycled **b.** Cheap **c.** Renewable **d.** Limited

18-13. To prevent wood from warping, splitting, and checking, it is important to control

a. Growth **b.** Weather **c.** Moisture **d.** Lignin

18-14. A problem with wood not found in most other engineering materials is

a. Rotting **b.** Weight **c.** Processing **d.** Supply

18-15. Because of the way that wood grows, most wood of the same species has identical properties.

T (true) **F** (false)

18-16. Wood timber may offer better protection in a fire than steel beams.

T (true) **F** (false)

18-17. Wood's ability to char when burned is called

a. Ablation **b.** Isotropy **c.** Homogeneity **d.** Checking

18-18. Most wood has very good workability because of the uniformity of its structure.

T (true) **F** (false)

REFERENCES & RELATED READINGS

AMERICAN SOCIETY FOR TESTING AND MATERIALS. *Annual Handbook of ASTM Standards*, Volume 04.09, *Wood* (D-7); Volume 06.03, *Paints, Related Coatings and Aromatics*. Philadelphia, PA: ASTM, 1992, 1991.

ASSEMBLY MAGAZINE. "Plans for the 1903 Flyer." *http://www.assemblymag.com/execute-vSection-Articles-vSub-Detail-vSideBar-Active-vrID-BA3845A396184D31B9B2E08721ECDC24-vDate-October+2002.htm*, January 3, 2003.

CASSENS, DANIEL L., and WILLIAM C. FEIST. *Exterior Wood in the South: Selections, Applications and Finishes,* FPL-GTR-69. Madison, WI: Forest Products Laboratory, 1991.

FOREST PRODUCTS LABORATORY. *Wood Handbook: Work as an Engineering Material.* Washington, DC: U.S. Department of Agriculture, 1974.

INTERNATIONAL COUNCIL FOR MATERIALS EDUCATION. Volume I, *Wood: Its Structure and Properties;* Volume II, *Wood as a Structural Material;* Volume III, *Adhesive Bonding of Wood and Other Structural Materials;* Volume IV, *Wood: Design of Structural Elements.* State College, PA: MEC, 1983, 1988.

RITTER, MICHAEL A. *Timber Bridges: Design, Construction, Inspection and Maintenance.* EM 7700–8. Washington, DC: U.S. Forest Service, 1990.

ROWELL, R. M., and P. KONKOL. *Treatments That Enhance Physical Properties of Wood.* FPL-GTR-55. Madison, WI: Forest Products Laboratory, 1987.

UNIT 7

Ceramic Materials

PAUSE & PONDER

Early people learned to convert natural materials into useful products mostly by trial and error. Adobe bricks were made from wet mud, shaped, and dried in the sun then used to assemble buildings. In the earliest building construction technology, primitive people used sticks, stones, and mud for their shelters. Eventually they progressed from the hunter–gatherer stage to living in a community with domesticated animals and plants, which supplied their needs for food and clothing. This advancement in living style allowed them greater time to develop better and more efficient materials. With this great change came the discovery that fire could be used to dry the adobe bricks much more efficiently than the sun. Firing clay and bricks produced stronger objects that made cooking and storing of food much easier. Molds made of these natural materials also could withstand higher temperatures and, with the discovery of metals, these molds could be used to shape the metals. This led to the making of bronze tools and, later, tools and utensils made from iron. Pottery and porcelain made by heating aluminosilicate clays to high temperatures at which the particles partially melt and fuse together are examples of *traditional silicate* ceramics (see figure). Traditional ceramics are derived and processed from clay or nonclay minerals from the earth's crust or outer shell, which consists

(a) Modern building construction technology ranging from modest homes to skyscrapers to dams and bridges relies on traditional ceramics, as seen in the constructed seawall and surrounding structures in Virginia Beach. Note the decorative brick walkway, concrete sculptures, and reflective glass glazing in the building. (b) The seawall is made of reinforced concrete with huge, integrated concrete pipes that connect to pumps to carry storm-surged seawater back out to sea, thereby reducing flood damage.

of rocks and minerals. Each of the crystalline substances in a rock is called a *mineral.* Minerals that have a commercial value are called *ores.* Bauxite ore is a mineral from which the metal aluminum and alumina are produced. Limestone is a common rock made of *calcium carbonate* ($CaCO_3$). The word *carbonate* indicates compounds made up of a metal combined with C and O_2, the word *silicate* refers to compounds formed of metals joined with Si and O_2, and the word *oxide* refers to compounds formed of metals combined with oxygen. $CaCO_3$ or limestone is found in many different forms, including chal , coral, marble, and eggshells.

Traditional products made from these minerals include refractories, white wares, cement, porcelain, and structural clay ceramics. Included in the definition of traditional ceramics are the products produced by the use of silicate glass. Concrete is included in this category because two of its ingredients are ceramic materials—sand/gravel and cement. Cement in various forms is a ceramic we see regularly. As concrete it provides the roads, storm water transport systems, bridges, sidewalks, buildings, and numerous other products and systems of building construction technology (see figure).

Due to the immense size of cement and concrete projects such as dams and highway systems, these projects require extremely large volumes of raw materials. Because of the large volumes of ***industrial by-products and waste (IBPW)*** produced annually, there has been considerable interest in developing new construction materials incorporating IBPW. Transforming IBPW such as granulated blast furnace slag, fly ash, and glass cullet or ceramic shards into useful building materials reduces environmental impacts while offering economic benefits. Twenty-five years of research has yielded a high-volume mineral additive with

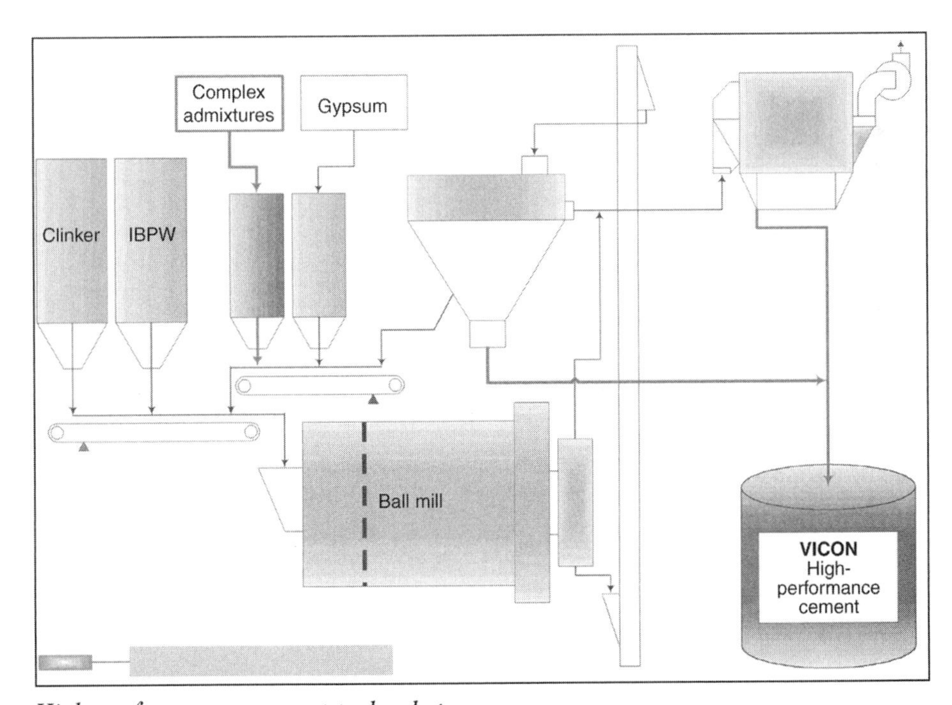

High-performance cement technology.

The process for making Vicon high-performance cement for use in eco-cement. Note the use of IBPW. (K. Sobolev and Metin Arikan, "High Volume Mineral Additive for Eco-Cement", *American Ceramic Society Bulletin,* Volume 81, No. 1, pp. 39–43)

higher compression strength than standard cement while making use of IBPW as the additives to make ***eco-cement,*** a cement good for the ecology while being economical.*

Eco-cement is a high-tech, ***high-performance (HP) cement.*** As seen in the figure, HP cement involves the addition of a new, reactive, silica-based complex admixture (supersilica) during the grinding of normal portland cement. In the case of HP cement, the clinker is ground in a ball mill with mineral additives, gypsum, and supersilica. The resulting cement can then be made into a wide range of concrete, including HP concrete. By using high-volume and inexpensive mineral additives (sand, limestone, or various industrial by-products), HP cement can be made to order, ranging from low-cost cements with up to 70% mineral additives to superstrong cements with high durability. Eco-cement serves as a model for "win–win" products for a sustainable environment—economical and ecologically friendly.

This unit will help you to gain perspective on the structure, properties, selection, and basic processes in ceramics, both traditional and advanced. In your study of ceramics, ponder their nature and their ability to compete with metals, polymers, and composites. Be alert to news articles, advertisements, and Internet sites that describe emerging applications of advanced ceramics. Also, imagine where there may be new opportunities for the use of ceramics. This knowledge will serve you well in preparing for the dramatic changes in this era of new materials.

American Ceramic Society—*www.ceramics.org*
NIST's *Ceramic Webbook—www.ceramics.nsit.gov/webbook/webbook.htm*
NASA TechBriefs—*www.nasatech.com*

*Sobolev, K., and Metin Arikan. "High Volume Mineral Additive for Eco-Cement," *American Ceramic Society Bulletin,* Volume 81, No. 1, pp. 39–43.

Module 19

Nature & Design of Ceramics

After studying this module, you should be able to do the following:

19–1. Name some goals of industry and society that caused a demand for greater use of ceramics.

19–2. Describe attributes of ceramics that make them competitive with plastics and metals.

19–3. Contrast traditional and advanced ceramics and sketch common ceramic microstructures.

19–4. Relate bonding, structure, and processing of ceramics to their properties (electrical, thermal, optical, and mechanical).

19–5. Compare ceramics with ceramics and ceramics with metals in terms of structure and properties.

19–6. Use diagrams and explanations to show stress concentrations plus methods of toughening ceramics, including composite matrix ceramics.

19–7. Define work of fracture and compare work of fracture versus tensile strength in glass, cements, and biomaterials.

19–8. Differentiate among empirical, deterministic, and probabilistic design.

19–9. Make sketches to show methods of improving ceramic structure for greater toughness.

19–10. Express fracture toughness in MPa and psi and density in SI and U.S. customary units.

19–11. Calculate the thermal shock resistance of given materials.

19–12. Use the website addresses denoted by the symbol* to explore sites for new developments and evolving concepts related to traditional and advanced ceramic materials.

*The symbol listed throughout the book will link you to Internet sites related to topics being covered. The dynamic nature of the Web brings frequent changes, so some of these sites, while available at the time of writing, may not be up now.

Traditional ceramics are usually heavy, fragile, and not suitable for use as structural or load-bearing materials. This modern age demands lighter materials that have great strength at high temperatures and maintain their strength at these high temperatures while subjected to impact forces. Figure 19-1 shows samples of what you may find in a typical home and what most people consider to be ceramics. The great sculptures of marble or granite (Figure 19-2) done through the ages also fit many people's idea of what ceramics are.

Figure 19-1 A mix of typical applications of ceramic materials that illustrate the vast time over which ceramics have benefitted the human race—from the Stone Age to the Space Age. Shown is one example of the use of soda-lime glass; two examples of Pyrex™, borosilicate glass by Corning that are oven and microwave proof with thermal-shock resistance; two examples of clay pottery with a glazing into which has been incorporated colors and other chemicals to produce a colored, glassy, impermeable coating; and two pieces of Corelle tableware by Corning, which can be placed in a preheated oven or microwave directly from the freezer.

Figure 19-2 Statuary sculpted from marble, a ceramic material that has withstood the ravages of time and weather, even modern-day air pollution. Marble is a hard, crystalline form of limestone ($CaCO_3$). Sulfuric acid (H_2SO_4) in acid rain causes marble buildings and statuary to disintegrate, forming a crumbly compound of calcium sulfate ($CaSO_4$).

Today, we can use our expanding scientific knowledge and technologies to help us develop new materials with the properties necessary to meet new applications demanded by society, the greatest demand for which comes from the land, sea, air, and space industries. These industries require construction materials that are lightweight so that the energy required for the movement of vehicles is minimized. The definition of traditional ceramics does not fully describe ***new, technical, engineering,*** or ***advanced ceramics.***

Are advanced ceramics inorganic, nonmetallic, nonmolecular solids? Yes, some are. Table 19-1 compares some of the properties of ceramics and metals. At one time, the word ***organic*** when used with terms such as ***compounds*** or ***chemistry*** described materials that contained carbon. Today, some of these (e.g., materials that contain CO_2 and $CaCO_3$) are no longer classified as organic materials. All other compounds that lack carbon are now defined as ***inorganic compounds. Ceramic compounds*** can be defined as inorganic compounds made by heating clay or other mineral matter to a high temperature at which they partially melt and bond together. Adobe bricks are ceramics that are made of mud that has been dried in the sun. They include both crystalline and amorphous materials. New ceramics are synthetically developed from rather simple chemical compounds, but advances in processing have provided greater control over their structures and resulted in vast improvements, particularly in their electrical, magnetic, and optical properties. One characteristic these new ceramics possess that traditional ceramics do not is their high purity, which is the basis for their improved properties. Some examples of advanced ceramics include ***oxide ceramics,*** such as alumina, beryllia, and zirconia, and ***nonoxide ceramics,*** which are represented by carbides and nitrides such as boron carbide, silicon carbide, and silicon nitride.

Ceramic is a unique material. Although developed in the Stone Age, it has found numerous uses as a space-age material in the third millennium. Much of the development in ceramics resulted from aerospace ***research and development (R&D);*** gradually the cost of advanced ceramics has declined, making them feasible for more earthly applications, such as in the automobile, sports, and machine tool industries. Most of these are actually synthetics produced from fine, relatively poor powders using new technology, including microwaves, electron beams, and polymer chemistry. Pictured in Figure 19-3 are a few of the newer ceramics that you will learn about in this module. The silicon nitride (Si_3N_4) turbocharger rotor makes possible more efficient engines that have smaller displacement and operate at higher temperatures. Si_3N_4 valves are 50% lighter than steel valves and therefore reduce friction in the valve train; they also run faster and hotter. Si_3N_4 roller bearings replace the complex steel needle bearings and cage and save about 40% in cost. They also operate at higher temperatures, which means reduced cooling needs. Partially stabilized zirconia (PSZ) has also found

TABLE 19-1 COMPARISON OF METALS WITH CERAMIC MATERIALS

Metals	Ceramics
Crystal structure	Crystal structure
Large number of free electrons	Captive electrons
Metallic bond	Ionic/covalent bonds
Good electrical conductivity	Poor conductivity
Opaque	Transparent (in thin sections)
Uniform atoms	Different-size atoms
High tensile strength	Poor tensile strength[a]
Low shear strength	High shear strength
Good ductility	Poor ductility (brittle)
Plastic flow	No plastic flow
Impact strength	Poor impact strength
Relatively high weight	Lower weight
Moderate hardness	Extreme hardness
Nonporous	Initial high porosity
High density	Initial low density

[a]Small flaws; pores act as stress concentrators that are not reduced by ductility/plastic deformation.

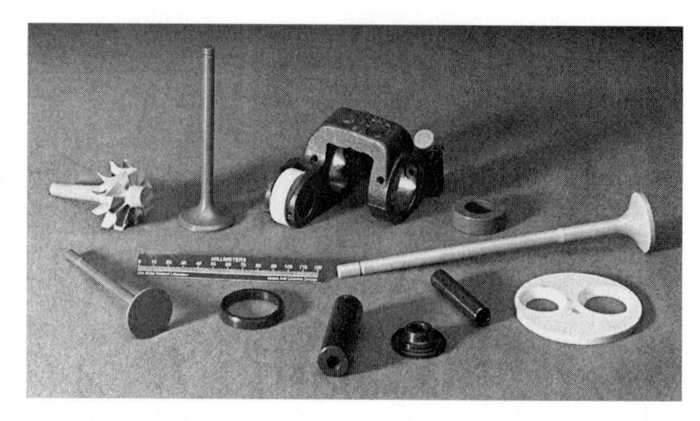

Figure 19-3 Examples of newer advanced ceramics for engines. *Top, left to right*: silicon nitride (Si_3N_4) turbocharger, Si_3N_4 valve, cast steel diesel engine rocker arm with partially stabilized zirconia (PSZ) cam follower and wear button, Si_3N_4 cam follower, and valve. *Bottom row, left to right*: valve; silicon carbide (SiC) water pump seal, piston pin, valve spring retainer, and valve guide of Si_3N_4; and PSZ diesel head plate with integrated valve seats. (ORNL)

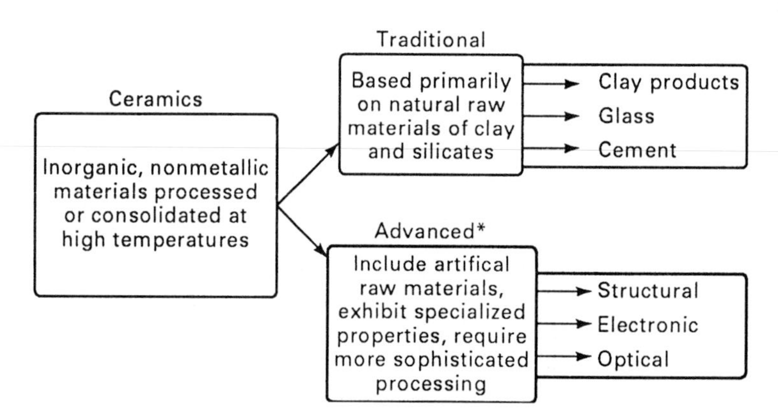

Figure 19-4 Broad classifications of ceramics. Traditional ceramics are contrasted with the newer advanced ceramics, which are also referred to as fine ceramics, engineered ceramics, new ceramics, and value-added ceramics. (ORNL)

applications in heat engines, as shown by the diesel engine rocker arm, head plate, piston pins, and valve guides in Figure 19-3.

ORNL Advanced Ceramic Technology Center—
http://www.ornl.gov/orcmt/specific/advcer.html
Materials Science and Technology Gateway—*http://MST-online.nsu.edu*

Figure 19-4 lists some differences between traditional and advanced ceramics, including some applications for each category of ceramic. Advanced ceramics have become the focus of attention in sci-tech applications because they offer the designer properties such as light weight, good strength at elevated temperatures, and wear resistance. Many of these ceramics are composites that are reinforced with whiskers and fibers to improve their fracture toughness, so they do not break catastrophically as we generally expect dishes and coffee cups to do when we drop them. The wear resistance of ceramics makes them valuable as coatings for cutting tools, surgical instruments, punches, and dies. For a modest cost, it is possible to apply a TiN coating only a few micrometers thick to tools, which can extend their

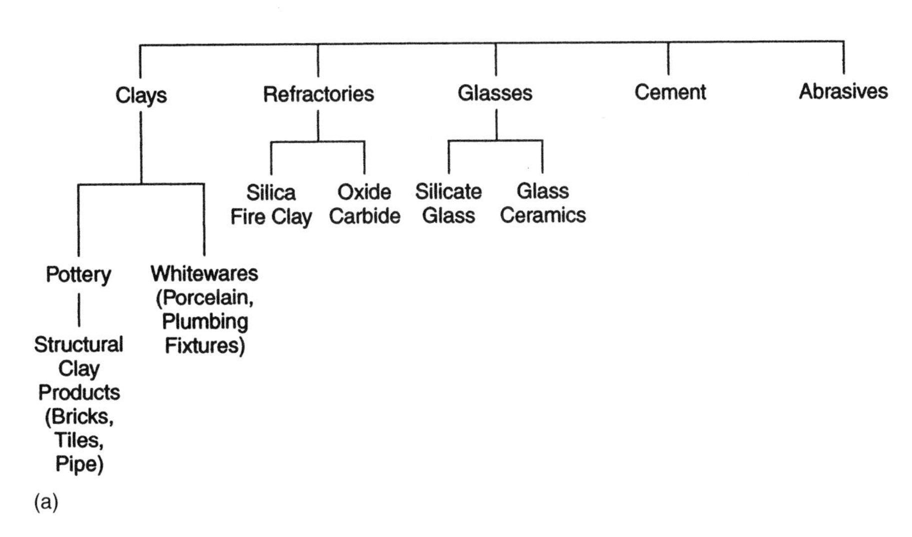

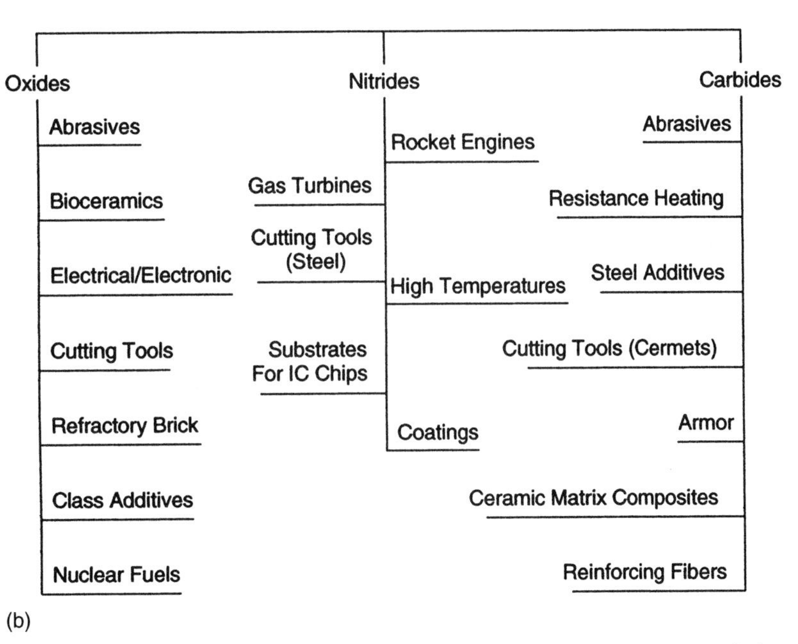

Figure 19-5 (a) Taxonomy of traditional ceramic materials based on applications. (b) Taxonomy of advanced (new) ceramics materials based on applications.

life 7 to 15 times. You may have seen golden TiN coatings on twist drills in the hardware store.

Figure 19-5 is a taxonomy of ceramic materials based on their applications. Note that ceramics can also be classified in other ways based on such differences as the way they are processed. New or advanced ceramics do not possess the plasticity and formability of clay when mixed with water, which necessitates new techniques in processing, discussed later in this module.

In the ongoing search for superior materials that can be produced with a minimum of energy (low temperature, low pressure, and minimum number of steps), researchers have replicated the processing techniques in nature, an approach termed ***biomimickry*** or ***biomimetics.***

Seashells, for example, are intricate, lamellar ceramic composites made with low energy and at a nanometric scale, yet yield complex patterns that hold promise for ceramic electronic devices and similar advanced applications.

With this discussion of advanced ceramics, could you expand on your list of ceramic materials found in the home? To sum up, ceramics encompass a wide variety of materials for which humans have found applications over the centuries.

19.1 NATURE OF CERAMICS

Ceramics are loosely defined as crystalline solids composed of metallic and nonmetallic materials. Some ceramics are intermetallic compounds (see Section 3.3.2.1). Ceramics are inorganic materials; that is, they do not contain carbon (with a few exceptions) and are derived from mineral sources. The crystal structures of ceramics are made of metallic ions in which bonding is either partially or completely ionic. The variables that govern bonding are (1) the magnitude of the electrical charge on the ions and (2) the relative size of the ions. The resulting crystal must be neutral; that is, all the cation positive charges must be balanced by an equal number of anion negative charges. As we mentioned earlier in this book, the ***oxidation state*** or ***oxidation number*** of an element designates the number of electrons an atom loses or gains or otherwise uses in joining with other atoms in compounds. The total of the oxidation states of all atoms in a molecule or formula unit is zero. Na has an oxidation state of +1; Cl has an oxidation state of −1. When they join to form the compound expressed by the formula unit NaCl, their oxidation states must equal zero.

Originally, ceramics were clay-based materials. The word *ceramic* comes from the Greek ***keramos,*** meaning "burnt earth." Glass, with an amorphous structure (a rigid liquid), can be classified either as a traditional ceramic, which includes the glasses made from silicates, or as a separate category due to its amorphous structure. When glass is transformed into a largely crystalline structure by heat treatment to form glass ceramics, it can be considered an advanced ceramic. ***Silicon carbide*** does not occur as a mineral and is produced synthetically. Its crystalline structure is diamondlike, with alternating Si and C atoms covalently bonded together to produce a material that is almost as hard as diamond. In addition, its crystals fracture in a way that leaves them with sharp edges. These properties make SiC an important industrial abrasive.

Granite is a mixture of quartz and the silicate mica and feldspar. The main source of quartz is sandstone, which is mixed with other minerals and clay to improve the final product. Crystalline quartz is a three-dimensional array of silicon dioxide (SiO_2), also known as ***silica*** (see Figure 23-1 for a sketch of the crystalline pattern formed by the Si and O atoms in silica). Crystals of pure quartz are colorless, but various impurities produce colored quartz; for example, traces of Fe (an impurity) in the crystal structure produce amethyst with its purple color.

Ceramics are inexpensive compared with competing materials. Consisting primarily of forms of silicon, aluminum, and oxygen—the most abundant elements in the earth's crust—ceramics can be produced at less cost than that of competing metal-alloy components, many of whose strategic minerals, including cobalt, that are needed to produce sophisticated metal alloys, such as the superalloys presently used in high-temperature applications, are not found in the United States. What sets ceramics apart is the set of properties common to nearly all its forms: (1) extreme hardness, (2) heat resistance, (3) corrosion resistance, (4) low electrical and thermal conductivity, and (5) low ductility, or brittleness. Table 19-1 compares these and other properties of ceramics to those of metals. Table 19-2 provides an additional comparison of the properties of ceramics and those of competing materials.

Characteristics—*http://me.mit.edu/2.01/Taxonomy/Characteristics/Ceramics.html*
Yahoo's materials science site and search engine—
http://dir.yahoo.com/science/engineeringmaterial_science/

TABLE 19-2 TYPICAL PROPERITIES OF SELECTED ENGINEERING AND TECHNICAL CERAMICS

	Ceramics				Comparison materials		
Properties	Alumina	Beryllia	Silicon carbide	Zirconia	Mild steel	Aluminum	Nylon
Melting point (approximately °C)	2050	2550	2800	2660	1370	660.2	215
Coefficient of thermal expansion (m/m/K) $\times 10^{-6}$	8.1	10.4	4.3	6.6	14.9	24	90
Specific gravity	3.8	—	3.2	—	7.9	2.7	1.15
Density (kg/m^3)	3875	2989	3210	9965	7833	2923	1163
Dielectric strength (V/m) $\times 10^6$	11.8	—	—	9.8	—	—	18.5
Modulus of elasticity (MPa) $\times 10^4$	34.5	39.9	65.5	24.1	17.2	6.9	0.33
Hardness (Mohs)	9	9	9	8	5	3	2
Maximum service temperature (K)	2222	2672	2589	2672	—	—	422

The several examples of ceramics that follow illustrate the properties listed in Tables 19-1 and 19-2. The Knoop hardness of cubic boron nitride (CBN) is 5000 kg/mm^2; that of diamond, the hardest known material, is about 7500 kg/mm^2. Partially stabilized zirconia (PSZ), to be discussed in more detail later, can withstand a pressure of 78,000 psi for about 2000 hours without any appreciable effect. Its ***coefficient of friction*** with steel is only 0.17 at room temperature, and its coefficient of thermal expansion has very nearly the same value, between 0° and 1000°C. It is because of properties like these that ceramic composites are used in electronic package casings. Casings need to have superior heat transfer characteristics with low weight, high thermal conductivity, and matching coefficients of thermal expansion. For example, silicon ***integrated circuit (IC) chips*** must be joined out of necessity to metal. Si has a coefficient of thermal expansion of 2.6 × 10 in./in./C and specific heat (C_p)of 702 J/kg-K. Si_3N_4 comes closest to Si, with a range of thermal expansion between 2.3 and 2.9 × 10 in./in./C. Aluminum is used to fabricate the metallized circuits on the surface of the chips to provide circuits between devices and to bring circuits to contact pads where outside contact is established. Aluminum and its alloys have an expansion coefficient of about 25, which is much higher than that of Si. The IC chip itself must be packaged so that it may be bonded to a printed circuit board. All the materials mentioned must be compatible—they all need to be equal in thermal conductivity to help dissipate the excess heat generated by the passage of electricity in such a small area and such extremely small circuits. Second, these materials should expand and contract together as much as possible when subjected to changes in temperature. The choice must be to use a material with matching coefficients and thermal conductivities. Some nonferrous alloys have been designed to fulfill this need, such as ***KovarB,*** a registered trademark of Carpenter Technology Corp. for its nominal 29Ni–17Co–54Fe glass-sealing alloy UNS K94610. **Kovar** is a ternary alloy of Fe–N–Co with a coefficient of thermal expansion of 5 × 10^{-6}in./in./C and a specific heat of 460 J/kg-K. It is important to note that materials that have excellent thermal conductivities are in demand, particularly for applications like the one described here. Designing a metal composite such as silicon carbide aluminum composite (a CMX–ceramic matrix composite) with a matching coefficient of thermal expansion and lighter weight than Kovar would be a worthwhile project, for a 10°C rise in temperature reduces the life of an integrated circuit chip by a factor of 3.

Applications of advanced ceramics have received major media attention in recent years, particularly as parts in a future ceramic heat engine. However, corrosion resistance, chemical inertness, thermal shock resistance, and other properties that materials scientists and engineers can design into ceramic materials make both traditional and advanced ceramics highly attractive in a large number of applications. This combination of properties makes ceramics good candidates for wear-resistance applications. Their electrical properties place ceramics in great demand as solid electrolytes in experimental batteries and fuel cells. Other uses include automotive sensors, packaging for integrated circuits, electronic/optical devices, fiber optics, microchips, and magnetic heads. In the marriage of the computer and communications technologies, ceramics also play a major role. Two additional outstanding characteristics of ceramics that could be added to

Figure 19-6 Ceramics are increasingly being used to repair the human body. This porous ceramic material is used as a bone void fiber and for optical implants. It is produced by converting natural coral into hydroxyapatite—the main component of human bone. (Produced by Interpore Cross International, Irvine, CA)

Table 19-1 are wear resistance and chemical inertness. Their wear resistance, mentioned briefly, can make ceramics replacements for human bone, particularly where bone moves in close contact with other bone components. Figure 19-6 is one example of a ceramic used in biomedicine. One not so obvious reason for using ceramics is that if excessive wear were to occur, the residue would be carried off by body fluids and cause serious pollution problems within the human system. Thus, the chemical inertness of ceramics is finding many applications in the medical field where contact with body fluids is more of a problem than with other materials. Finally, thermal and mechanical stability of ceramics play a big role in the machine tool industry. This stability allows ceramics to retain their smooth, accurate cutting surfaces longer than metals do. Ceramic-coated cutting tools and inserts, some with as many as 12 extremely thin coatings, each designed to serve a special function, can run productively at faster cutting speeds and at faster feed rates than can any metal-alloy tool in the machining of hard steels, superalloys, and ceramics. What is remarkable about these multicoated carbide inserts is that with slightly over a dozen thin ceramic layers, the total thickness of the coatings is only 10 to 12 μm.

High-tech companies for materials science and engineering ceramics—
http://doll.eng.sunysb.edu/other4.html#ceramics

19.1.1 Porosity and Density

When referring to a solid material such as a part made from copper or stainless steel, the word ***density*** takes into consideration a microstructure that contains no ***porosity,*** by which term we do *not* mean the voids or vacancies in the atomic structure. By *density* in speaking of a solid material we mean the material's theoretical density. ***Mass density,*** which is the mass of a material divided by its volume (Section 3.2.6), refers to this theoretical density. Such a density may be expressed in terms of weight instead of mass, as will be discussed in this section. Atomic weight is a major factor in determining the density of a material. Low-atomic-weight elements have low densities. A second factor is how the atoms/ions are stacked in the microstructure. Close-packed metals are more dense than open-structured materials. Our study of crystalline materials in Module 3 revealed that a bcc unit cell with a maximum coordination number of 8 is only 68% of full density, whereas fcc and cph unit cells with a coordination number of 12 can have a density of 74% (of a maximum of 100%). Ceramics are composed primarily of different phases, each with its own density. As an example, a ceramic that has a porosity of 20% is described as being 80% of theoretical density.

Green density is the bulk density of a compact prior to its densification. In preparing particles for the production of powdered metals or ceramics, the packing of spherical-shaped, monosized (0.1 mm) particles, even when vibrated to settle the particles, does not achieve a density level much higher than 60%. Despite tapping or vibrating, these particles do not rearrange themselves completely to eliminate pores. The optimum density of spherical particles can be reached by varying the size distribution of particles to permit smaller particles (50 nm or less) to locate in the interstices of the larger ones. One disadvantage with this procedure is that the larger particles tend to grow excessively during high-temperature sintering, which takes place later in the process. After particle sizing is complete, the pores remaining must be reduced by additional steps, such as mixing and kneading of the initial green mixture prior to the making of a green compact.

Ceramics are, by nature, generally porous materials with varying degrees of porosity. The term ***bulk density*** is used to refer to a ceramic's density, as opposed to that of solid material, and it includes the material's porosity and the fact that most ceramics contain both a crystalline and a noncrystalline phase. ***Open porosity*** refers to the network of pores that is open to the surface and into which a liquid such as water can penetrate if the part were submerged in it. ***Closed porosity*** refers to those pores that have become sealed within the grain structure. New technologies are under development to reduce, if not eliminate, pores in ceramic materials. Pores affect the strength of ceramics in two ways. First, they produce ***stress concentrations.*** Once the stress reaches a critical level (see Section 19.1.5), a ***crack*** will form and propagate. Because ceramics possess no plastic-deformation attributes to absorb any energy transferred to these materials once a crack is initiated (started), it propagates (grows) until fracture occurs. Second, pores (i.e., their size, shape, and amount) reduce the strength of ceramics because they reduce the cross-sectional areas over which a load can be applied and, consequently, lower the stress that these materials can support.

The amount, size, distribution, and the continuity of voids in a ceramic are the principal factors that determine its properties. In general, porosity is detrimental. Grain boundary sliding, creep rate, and chemical reactions that attack the ceramic or reduce its melting point are adverse conditions produced by impurities and porosity. However, modern processes have been developed to minimize porosity to a great extent. One exception to this goal of minimum porosity is ceramic insulators. Porosity increases the insulating capacity of ceramics, and at the same time, increased porosity makes ceramics lighter in weight. These two characteristics, when combined, make an ideal ceramic for use in such applications where the component or object is part of a moving system propelled by the consumption of energy, such as an automobile or spacecraft. The bulk density of complex geometric parts as well as their various porosities can be determined by using ASTM procedure C373, which is based on Archimedes' principle,* which states that the weight of an object in a fluid equals its dry weight minus the buoyant force (or the weight of the fluid displaced). Taking only three weight measurements—the object's dry weight, its saturated weight in a fluid, and its net saturated suspended weight—several physical properties relating to the density and porosity of an object can be calculated.

*Refer to the experiment by Gail Jordan at the end of this module.

For example, the following five relationships expressed in terms of these three weights can be calculated:

$$1.\quad d_B = \frac{W_D}{V_B} = \frac{W_D \cdot d_L}{W_S - W_{SS}}$$

$$2.\quad V_B = \frac{W_S - W_{SS}}{d_L}$$

$$3.\quad V_A = \frac{W_D - W_{SS}}{d_L}$$

$$4.\quad d_A = \frac{W_D}{V_A} = \frac{W_D \cdot d_L}{W_D - W_{SS}}$$

$$5.\quad \%P_A = \frac{W_S - W_D}{W_S - W_{SS}} \times 100\%$$

where

W_D = dry weight
W_S = saturated weight
W_{SS} = saturated suspended weight
d_L = density of saturating liquid
d_B = bulk density
d_A = apparent density
V_A and V_B = apparent volume (bulk volume)
P_A = apparent porosity

Notes:

1. If water is the saturating liquid, its density may be assumed to be 1 g/cm^3; the ratio *W/d* results in units of volume (cm^3).
2. The apparent density is its dry weight divided by the difference between its bulk volume and its open-pore volume.

Illustrative problem

A specimen of sintered silicon carbide has a dry weight of 3.1 g, a saturated weight of 3.5 g, and a saturated suspended weight of 2.3 g after being suspended and soaked in water. Calculate the bulk volume, bulk density, apparent density, and apparent porosity.

Solution

$$V_B = \frac{3.5 - 2.3}{1} = 1.2 \text{ cm}^3$$

$$d_B = \frac{3.5}{1.2} = 2.6\text{g/cm}^3$$

$$d_A = \frac{3.1 \cdot 1}{3.1 - 2.3} = 3.9\text{g/cm}^3$$

$$\%P_A = \frac{3.5 - 3.1}{3.5 - 2.3} \times 100\% = 33.3\%$$

Note: The saturated weight is the weight of the object after it is removed from the water.

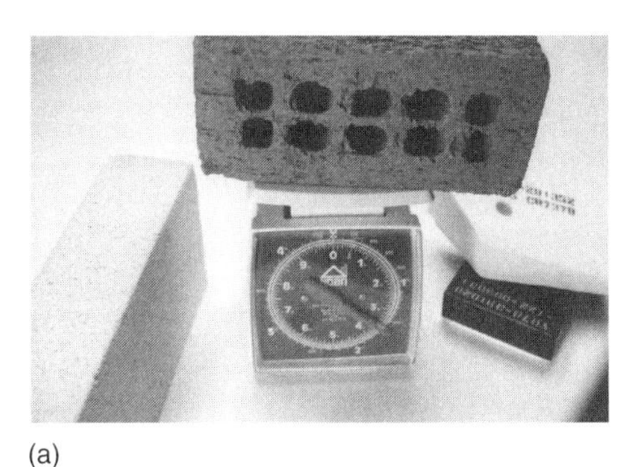

(a)

(b)

(c)

Figure 19-7 Examples of ceramics of varying densities: (a) The common architectural brick with holes weighs about 3.5 lb; (b) an insulating brick used on glass furnaces that weighs less (about 2 lb) than the brick in (a) due to the high porosity achieved during processing, which also improves insulation value; (c) the weight of the Space Shuttle tile (silica fibers with glass coating) processed to achieve minimum density and maximum insulation value—note that while about the same size of the two bricks in (a) and (b), it only weighs a few ounces.

American Ceramic Society—*http://www.acers.org*

Figure 19-7 provides examples of ceramics of varying densities. In Figure 19-7a, the common architectural brick with holes weighs about 3.5 lb. In Figure 19-7b, an insulating brick used on glass furnaces, while solid, weighs less (about 2 lb) than the brick in (a) due to the high porosity achieved during processing, which also improves insulation value. The Space Shuttle tile (silica fibers with glass coating) in Figure 19-7c was processed to achieve minimum density and maximum insulation value; note that while it is about the same size of the two bricks, it weighs only a few ounces.

19.1.2 Structure, Bonding, and Properties

The structure of most ceramics varies from relatively simple to highly complex. Being compounds, ceramics are made of different types of atoms of varying sizes. ***Clay*** is a complex mixture of ***silicates,*** which are complex ionic compounds that are exceedingly numerous. Approximately 90% of the earth's crust consists of silicates, the starting material for a wide variety of products including tile, bricks, cement, glass, and china. The basic unit of the silicate structure is the SiO_4 tetrahedron (see Figure 19-8a). Si, one of seven semimetals in the periodic table, is a hard, gray, semiconductive solid that melts at 1410°C. In nature it is generally found combined with O in ***silica*** (SiO_2) and in various silicate minerals. Oxygen (O) and silicon (Si) together account for about 75% of the earth's crust. More than 1000 different silicates occur naturally, depending on the different ways the SiO_4 tetrahedra can be linked. Sand consists of fine particles of quartz. Other common forms of silica are ***mica*** and soapstone (a form of talc). Particles of mica

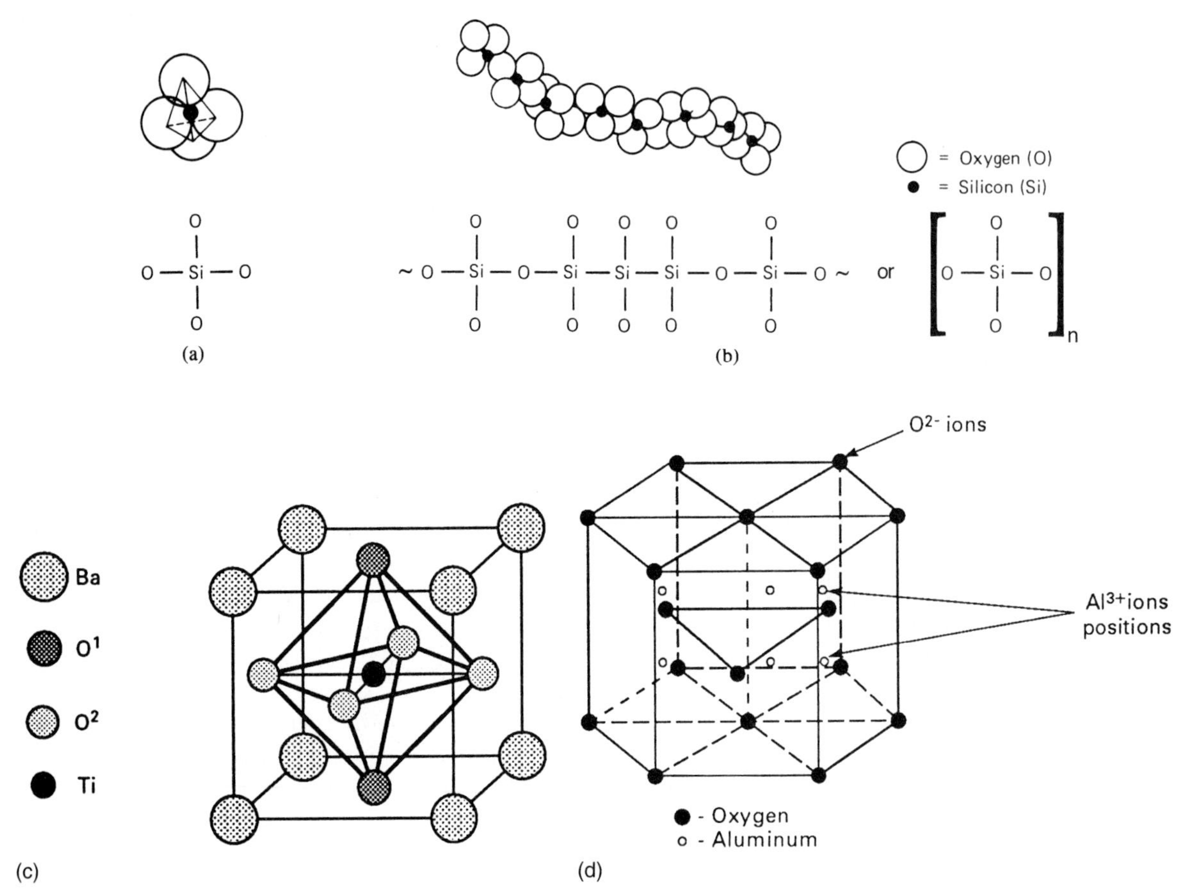

Figure 19-8 Silica structure. (a) Silicate tetrahedral. (b) Single chain of silicate. (c) Cubic cell of ideal perovskite structure of $BaTiO_3$ above 130°C with the origin on a barium ion. The ideal cubic perovskite structure, abbreviated ABX_3, has a large cation (A); a smaller cation (B); and an ion (X). Various arrangements or views of this structure might show the B cation in the center of the cube or the A cation in the center. The figure depicts this structure using $BaTiO_3$. The A ions are barium (Ba); the B ions are two oxygen ions, 01 and 02; and the X ion is titanium (Ti). Superconductivity occurs in oxides with this same structure, and the new high-temperature, superconducting copper oxides closely resemble it also. Such a ceramic possesses large values of dielectric constant as well as ferroelectric properties. (*Journal of Materials Education,* Volume 13, No. 3). (d) Crystal structures of Al_2O_3 showing O^{2-} ions occupying the hexagonal close-packed (hcp) unit cell sites and Al^{3+} ions occupying about two-thirds of the interstitial sites to maintain electrical neutrality.

produce the small flecks in metallic paint used on some automobiles. Clays and mica have sheets of linked SiO_4 tetrahedra that slide over each other, giving their minerals a characteristic soft feel. This layered structure can be penetrated by water molecules, which act as a lubricant and produce a material that is pliable and slippery. When clay is heated to over 1100°C in a kiln, the water is removed and the layered sheets bond or lock into a rigid structure called a *ceramic*.

Silicates include sand, clay, feldspar (K-Al-Si_3O_8), quartz (SiO_2 silicon dioxide), and the semiprecious stone, garnet. Pure silica (SiO_2), the most numerous silicate compound, has three common polymorphs: cristobolite, tridymite (high temperature), and quartz (low temperature). The silica structure is the basic structure for glasses and many ceramics. Its internal arrangement of pyramid (tetrahedral or four-sided) units, as shown in Figure 19-8a, has four large oxygen (O) atoms surrounding each smaller silicon (Si) atom. The silicon atoms occupy the openings (interstitials) between the oxygen atoms and share four valence electrons with the oxygen atoms through covalent bonding. This bond is one of the strongest single bonds—it is considerably stronger than the C—C bond. These SiO_4 tetrahedra are joined to one another through the O_2 atoms, each of which is bonded to two Si atoms.

The tetrahedra can be arranged in a linear manner to form fibers, rings, chains, and planar sheets or complex three-dimensional arrays. An example of the simple anion SiO_4 is the mineral ***zircon*** ($ZrSiO_4$), which is an inexpensive gemstone. More common are larger anions in which two or more O atoms bridge between Si atoms to produce various structures. Two-dimensional layer anions ($Si_3O_{10}^{5-}$) are found in clay minerals, micas, and talc. Micas can be easily cleaved into thin, transparent sheets and were once used as lantern panels. The bonds between the layers made up of cations, principally Al^{3+}, are relatively weak and easily broken. Asbestos, a magnesium silicate, is one of many forms of fibrous silicates that have double chains of tetrahedra. Kaolinite [$Al_2(Si_2O_5)(OH)_4$] is a type of clay valuable for pottery making because its layer structure can be easily penetrated by water. When Al ions are partially substituted for Si ions, the aluminosilicates called ***feldspars*** are formed, which are the most abundant of all minerals. ***Zeolites*** are aluminosilicates with three-dimensional network structures containing tunnels. They are being used as catalysts.

The silicate structures (Figure 19-8b) can link by sharing the atoms at two corners of the SiO_4 tetrahedrons, forming chain or ring structures. By sharing three corner atoms, they produce layered silicates (talc, kaolinite clay, mica); or by sharing four corner atoms, they produce framework silicates (quartz, tridymite). The alternating Si and O covalent bonding is most common in ceramics and compares to the carbon-to-carbon (C—C) bonding in organic materials. It is their unique atomic structures that lead to the properties possessed by most ceramics. Figures 19-8c and d indicate some similarities and differences between ceramic and metal unit cells. (See Section 3.2.1 for a discussion of metal unit cells.)

As mentioned in Module 6, silica, in the form of inorganic aerogels, has the potential to improve materials structures through its many microstructure and nanocomposite applications. In the 1980s, an aerogel became the world's lowest-density solid material, with a density of 0.003 g/cm^3, only three times that of air. Unique surface chemistry, pore structure, and optical and thermal properties have propelled aerogel technology. The following website offers good insight into this unique silica material.

 Silica aerogels—*Http://eande.lbl.gov/ECS/aerogels/satoc.htm*

Ionic and covalent bonds are the primary bonding mechanisms for ceramics. The crystal structure of bulk ceramic compounds is determined by the amount and type of the bonds. The percentage of ionic bonds can be estimated by using electronegativity determinations (refer to Figure 2-8). Being compounds, ceramics have different types of atoms. Hence, their resistance to shear and high-energy slip is extremely high. Because their atoms are held (bonded) so strongly compared to those of metals, there are fewer ways for the atoms to move or slip in relation to each other. Thus, the ductility of ceramic compounds is very low and these materials act in a brittle fashion. Fracture stresses can initiate a crack buildup before there is any plastic deformation, and once started, a crack will grow spontaneously. The combination of high shear stresses and reduced ductility produces high compressive strength but low tensile strength. The maximum bending stress for ceramic materials in tension at failure is called the ***bend strength*** and is often referred to as the ***modulus of rupture (MOR).*** Ceramics are noted for their heat resistance. At room temperature, metals and ceramics are often competitive, but at temperatures above 1500°F, metals weaken while ceramics retain much of their strength. The maximum service temperature for alumina is 3450°F, and for silicon carbide, it is 3000°F. Heat-resistant nickel alloys are considered unserviceable above 1500°F. (Table 19-1 compares some characteristics of metals with those of ceramic materials.)

19.1.3 Ceramic Phase Diagrams

The statement "the properties of materials depend on their structure" is emphasized more than once in this text. The questions that need to be asked are where the different microstructures originate, and how the optimum microstructure that will render the properties we are seeking in a material can be obtained. Microstructures of a material depend on their compositions as well as the variables of temperature and pressure. As the composition becomes more complex, the greater the number of phases that present themselves. The desired microstructure comes from ***processing.*** In processing a material, we melt, refine, alloy, shape, heat treat, and finally fabricate

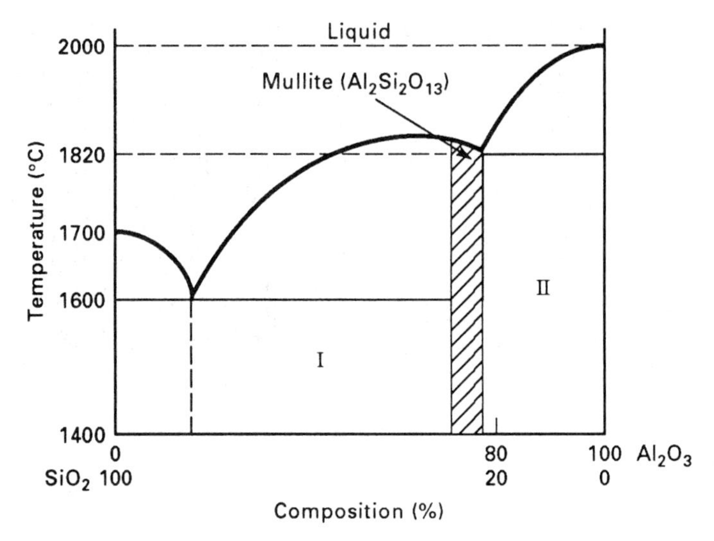

Figure 19-9 Rough sketch of SiO_2–Al_2O_2 binary system used to produce many refractories. Mullite ($3Al_2O_3$–$2SiO_2$), an intermetallic compound with a small range of compositions, is a ceramic with many high-temperature applications.

the structure within a component, which involves forming, machining, or welding. These processing steps have an effect on the microstructure, and before we can start on processing, we need to know the details of the melting, solidification, and the various phases that the material goes through as it is heated and cooled. Without this information we are unable to control processing in order to obtain a material with the desired characteristics. The phase, equilibrium, or constitutional diagram is the map that shows us the different compositions at different temperatures under equilibrium conditions. Module 20 provides more explanations of ceramic processing.

Phase diagrams having two components were discussed in relation to metallic systems in Module 4. Ceramics materials have at least two components, but more than two is the norm. Figure 19-9 is a phase diagram for a binary ceramic system that shows the development of mullite ($3Al_2O_3$–$2SiO_2$), an intermetallic compound much in demand for high-temperature applications. Ceramicists dealing with three-component or ternary phase diagrams need to construct three-dimensional diagrams, which are difficult not only to construct but also to interpret. However, the concepts developed in the study of simple binary, two-component systems can be directly applied to the more complex ternary systems found in ceramic systems.

19.1.4 Wear Resistance

Wear can be defined as the removal of surface material. Wear is a facet of tribology, introduced in Module 5. Several mechanisms are involved in wear. One is adhesion, or the bonding of two surfaces with subsequent removal of material. Inertial welding is an example of how this mechanism is used in the welding of metals. The roughness of the surface results in removal of part of the protrusions and depressions. When there is a large difference between the hardness of two materials in contact, gouging can occur. Gouging is how an abrasive grinding stone removes material from a workpiece. ***Impact,*** also known as ***erosion-corrosion,*** occurs when abrasive particles in a fluid impinge on a surface. Other variables at play are abrasion, oxidation, and contact stress. Wear is responsible for many equipment breakdowns and the attendant downtime. With wear comes friction and with friction comes a large expenditure of financial resources to overcome it. Aside from the loss of material from wear, the dollar value of the energy lost each year due to friction is in the billions of dollars. Ceramics are likely candidates to help reduce this exorbitant expenditure of energy and funds. A quick perusal of their properties should be a convincing argument for greater use of ceramic materials. Sapphire and ruby spheres have ideal wear

(a)

(b)

Figure 19-10 (a) Ruby (red) color and sapphire (white) jewel bearings and related precision products such as orifices, nozzles, pins, valve seats, bushings, balls, ferrules, and split sleeves are manufactured from raw silica materials with tolerances such as roundness within 7×10^{-6} in., or concentrically within 4×10^{-5} in. Ruby and sapphire are both forms of alumina (Al_2O_3). The presence of chromium ions (Cr^{3+}) in alumina gives the materials its red color, and small amounts of Fe and Ti give a white color to sapphire. Corundum is the name given to alumina that occurs in nature. (Microlap Technologies, Inc.) (b) Ceramic industrial wear parts are used as mechanical seals, erosion-resistant liners, metal processing dies, slurry pump rotors, and more. Materials include alumina, silicon nitride, zirconia, silicon carbide, boron carbide, and boron nitride. (American Ceramic Society)

resistance. Figure 19-10 shows an assortment of these miniature pieces of synthetic alumina from which are manufactured precision parts for hundreds of uses. In addition to hardness, these materials feature low coefficients of friction, shock resistance, a high index of refraction, low specific gravity, and nonmagnetic and high dielectric strength. For example, ferrules with inside diameters (ID) that range from 125 to 750 μm can be manufactured with great accuracy.

Materials can be ranked according to their wear resistance by a variety of wear tests. Many are single-purpose tests with a particular application in mind. Some concern themselves with only one type of wear factor. A recently developed test method uses flat rectangular specimens ($100 \times 25 \times 6$ mm) mounted on a cylinder that rotates inside a container filled with an abrasive medium tailored to simulate different types of wear. This test, called the *rotating-cylinder wear test,* is supposed to be a better indicator of a material's performance under service conditions. Specimens are weighed before and after the test, and the percent weight difference is an indicator of the material's wear resistance. The test conditions must be included when reporting the results. Figure 19-11 illustrates the test results of a ceramic metal composite (Alanx CG 896) produced by Alanx using a new process called Dimox (discussed later in this module). This composite, designed specifically to provide maximum wear resistance, consists of densely packed silicon-carbide particles within a matrix of aluminum oxide and a metal alloy. The particles and the alumina resist wear, and the metal alloy provides the bonding between the two ceramic phases. The accompanying graph compares this wear-resistant composite to the service lives of other, competitive materials by use of a wear-life factor: The larger the number, the longer the wear life. A slurry erosion test was used to measure the wear resistance of a test specimen. The material loss was compared to a standard aluminum test specimen and the wear-life factor was determined.

19.1.5 Stress Concentrations and Fracture Toughness

When selecting a material that might be subjected to ***cyclic loading,*** or where the load cannot be distributed uniformly over an area, the subjects of stress concentration and fracture tough-

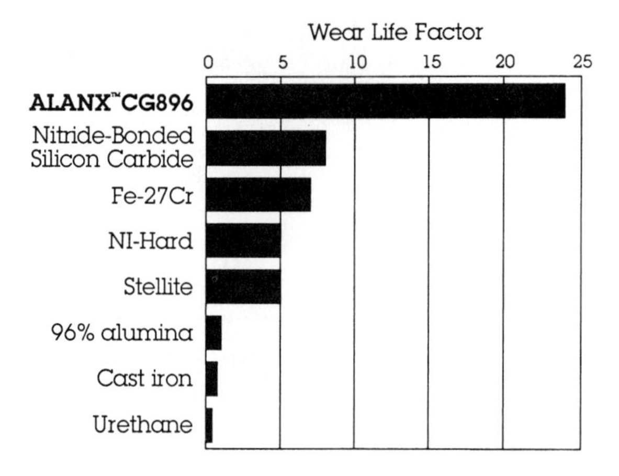

TYPICAL PROPERTIES OF ALANX™ CERAMIC/METAL COMPOSITES

	ALANX™ CG896	ALANX™ CG273	ALANX™ FGS	Developmental Metal Matrix Composite (MMC)
Wear Life Factor	24	24	12	9.2
pH Range	5-8	1-14	5-8	5-8
Density (g/ml)	3.32	3.42	3.32	2.95
Hardness (R_A)	80	80	80	57
Flexural Strength (MPa)	137	100	216	216
Fracture Toughness K_{Ic} (MPa · m$^{1/2}$)	6.0	4.4	5.5	11.3
Shear Modulus (GPa)	131	144	144	79
Young's Modulus (GPa)	313	357	366	204
Poisson Ratio	0.19	0.23	0.27	0.29
Thermal Expansion (10^{-6}/K)	5.43	5.20	5.89	10.24
Thermal Conductivity (W/m · K)	160	160	160	160

Figure 19-11 Typical properties of ALANX ceramic/metal composite.

ness come into play. For example, when a hole is drilled in a flat plate, such as that sketched in Figure 19-12, and is loaded with a force *P*, the stress in the plate is assumed to be uniformly distributed (Figure 19-12b) across the cross section of the plate everywhere but at the cross section containing the hole. Figure 19-12c shows the distribution of stress as it finds its way around the hole, thus producing a concentrated stress at the edges of the hole. Such a buildup of stress may cause the material to crack. It is important to point out that stress concentrations depend on the geometry of the flaw (hole, notch, or crack) and the geometry of the specimen or component (shape and thickness). *They have nothing to do with the properties of the material.* In materials that may contain cracks due to flaws, voids, inclusions, or oxides in their microstructure, the stress concentration is greatest at the tip of the crack. Flaws that cause the fracture of ceramic materials usually are smaller than 50 μm. This size of flaw is not readily detectable by most nondestructive testing (NDT) methods.

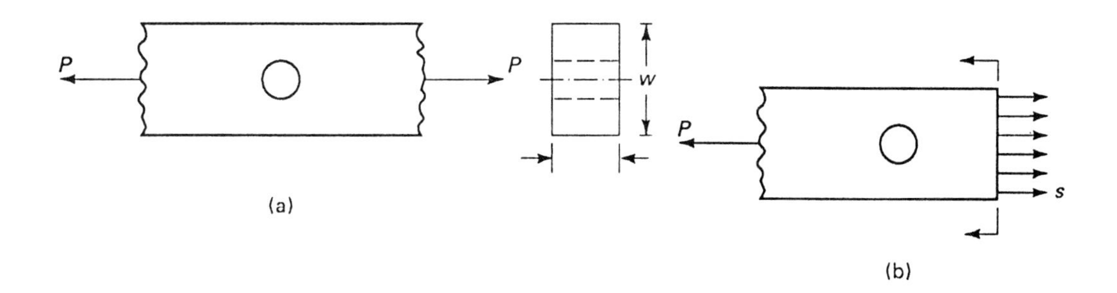

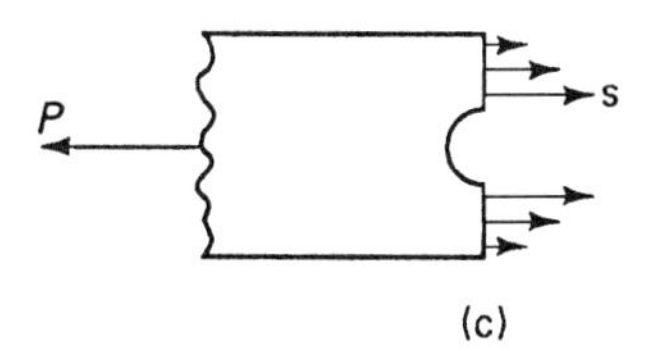

Figure 19-12 A stress concentration resulting from a through hole in a flat plate. (a) Sketch of the plate loaded with a tensile force *P*, showing a side view of the plate. (b) Cross section of the plate at some distance from the hole, with uniform distribution of tensile stress indicated by small vectors. (c) Cross section through the center of the hole, with the largest vectors portraying the maximum stress concentrations at either edge (side) of the hole.

The ***fracture toughness*** of ceramics, often perceived as the limiting factor for ceramics applications, is being improved to make ceramics competitive as engineering materials. A stress–strain diagram (Figure 19-13a) illustrates the effects of ceramic brittleness compared to those of relatively ductile metals. Figure 19-13b graphs the contrasting fracture strengths of metals and ceramics; the graph also shows how this property increases using the toughening techniques discussed in this module—to the point where advanced ceramics approach metals in their fracture toughness.

Using fracture mechanics techniques, the critical stress–intensity factor (K_{1c}), or fracture toughness of the material, is determined independently from the nature of the material. The "1" in the symbol K_{1c} refers to the mode of loading the specimen. In this instance, the load is parallel to the 1 or x material axis, producing a normal stress. When used with the symbol for fracture toughness, the numerals 2 and 3 refer to the shear loads acting on the y- or z-axes, respectively. Figure 19-14 is a sketch of a fiber-reinforced part being stressed parallel to the 2 axis. Several mechanical tests, including indentation methods, are used to measure the fracture toughness of ceramics. Which one is chosen depends in part on the type of information needed. One popular test is the single-edge, notched-beam (SENB) test (Figure 19-15a). It is similar to a four-point bend test except that an artificial crack is placed in the specimen before testing. As Figure 19-15b illustrates, the chevron-beam test specimen is similar in loading to the SENB test but differs in that it has a chevron notch rather than a straight-through notch machined or cut into the surface. The reason for this type of notch is to force the crack to propagate slowly through the specimen. It does so because it is always passing into a zone of greater load-bearing area. Both tests give a direct measure of the fracture toughness of a material. The SENB test uses the following equation:

$$K_{1c} = \frac{3P(S_1 - S_2)a^{1/2}Y}{2BW}$$

where

P = applied load
a = crack length
S_1, S_2 = dimensions between the outer and inner roller spans
B = specimen thickness
W = specimen height
Y = dimensionless calibration factor

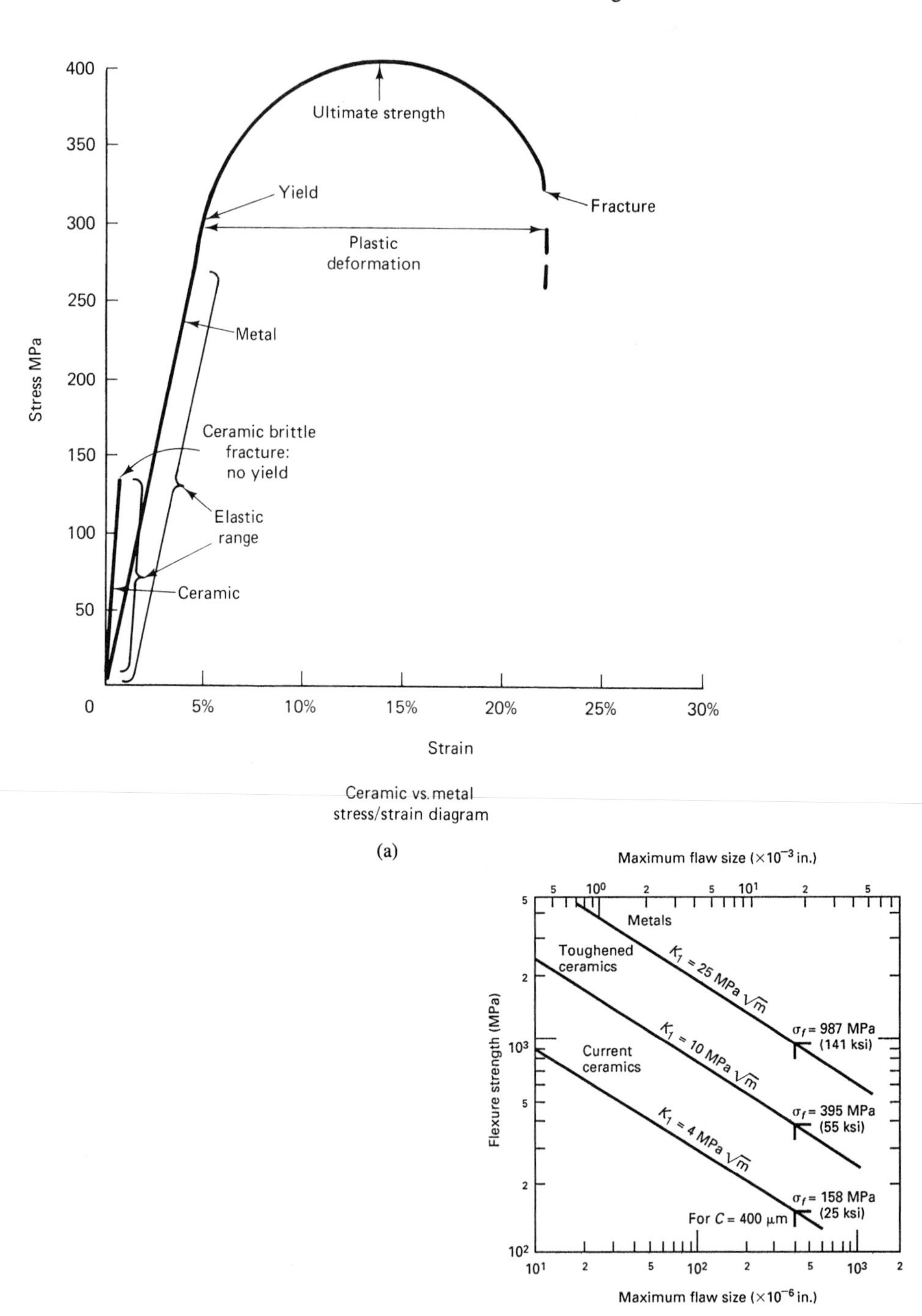

Figure 19-13 (a) Brittle fracture in ceramics results from their inability to yield like metals do. This diagram shows a large percentage of plastic deformation in metals that occurs because localized stresses are relieved by localized plastic deformation. No yielding point shows on the ceramic curve, so localized stress eventually causes catastrophic failure when the elastic range is exceeded. The percentage of strain before failure on ceramics can be measured in hundredths of a percent compared to tenths of a percent for metals. (b) A plot of flexural strength versus maximum flaw size for metals, toughened ceramics, and traditional ceramics, which translates into (represents) toughness values ranging into the 20s. (ORNL)

The units for expressing fracture toughness are either psi $\times \sqrt{\text{inches}}$ or MPa $\times \sqrt{\text{meters}}$. It is of interest to note that once this critical stress–intensity factor is exceeded, the crack will propagate and the material will fracture, producing a catastrophic failure. The failure can be caused by an increase in the applied stress (load) or in the crack length. Once the fracture

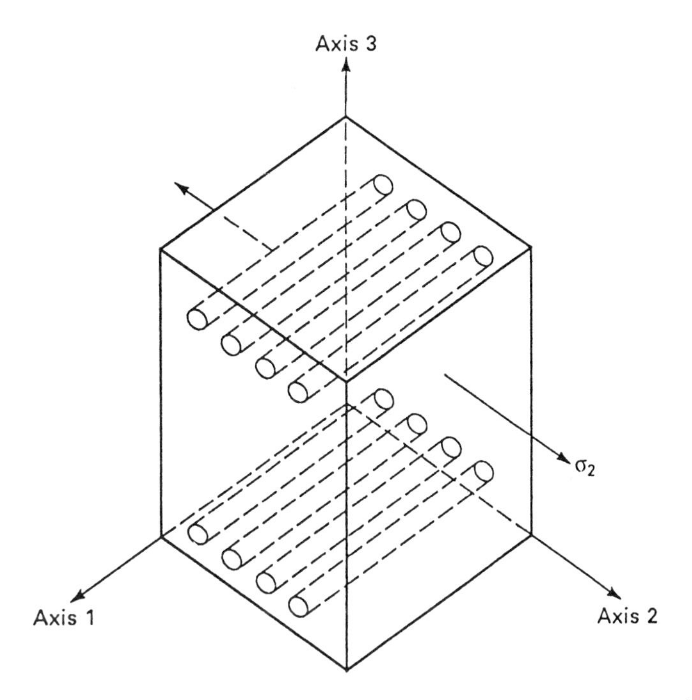

Figure 19-14 A composite material with fiber reinforcement. The axes (1, 2, and 3) are shown with a normal tensile stress acting in the direction of axis 2.

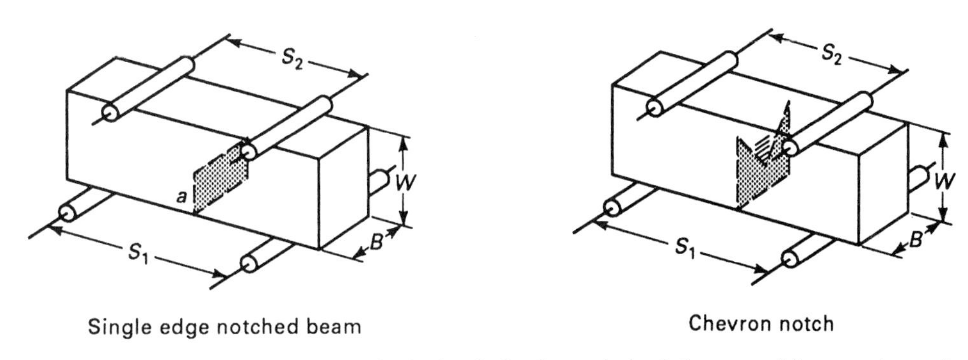

Figure 19-15 Fracture toughness is determined using single-edge notched and chevron notch beam specimens. An artificial cut is placed in both specimens. (a) A straight-through notch in the SENB is cut with a diamond-impregnated blade and (b) a chevron is machined into the chevron notch (CN) specimen.

toughness is known, the fracture stress or the stress at which the crack will propagate can be calculated. Also, the stress at failure may be less than the yield strength. Also to be noted is that flaws in almost all silicate and oxide-based ceramics and glasses will grow at stress levels lower than the calculated stress that produces a catastrophic or fast fracture. This environmentally assisted crack growth is produced by a reaction between the strained bonds at the tip of the crack and water or other molecules. See Figure 7-19 for a sketch of a crack zone toughened as a result of the transformation of ZrO_2 and a micrograph of the area. Transformation toughening is covered in greater detail in Section 19.3. Toughened ceramics, including ceramic armor for military land vehicles such as tanks and personnel carriers, consist of tiles toughened by the hot pressing of B_4C, SiC, AlN, and TiB_2 and sintered SiC and Si_3N_4. More can be seen at the following Ceradyne website.

Flaws in ceramic materials that produce failure are usually smaller than 50 μm, but most NDT procedures do not detect flaws smaller than 50 μm. In summary, the fracture toughness, a property of a material, is an indicator of the ability of a stress concentrator (crack or flaw) to produce a catastrophic failure. To predict the ability of a material to resist fracture (fracture strength), knowledge of both fracture toughness and the presence of ***stress concentrators*** in a material is vital.

MSEL of NIST—
http://www.nist.gov/item/About_NIST_Materials_Science_and_Engineering_Laboratory.html
Advanced ceramics—*http://www.ceradyne.com/*

19.1.6 Thermal Shock Resistance

Thermal shock is a mechanism that can produce mechanical failure in materials. It involves the buildup of thermal stress in a material as a result of exposure to a temperature difference between the surface and the interior. If the temperature difference occurs rapidly, the possibility of failure increases. Testing of ceramic turbine vanes for thermal shock subjects them to repeated recycling from 340° to 1230°C in less than 2 s. At 1230°C, the vanes glow as a result of the heating. Future automobile gas-turbine engines will require materials that can withstand a rapid temperature change, from room temperature to 1200°C or more in a few seconds, and can be exposed to stresses in excess of 207 MPa. Some engine designers are contemplating future engines capable of operating for 3500 hr with internal operating temperatures as high as 1745°C.

Some of the material properties involved in producing thermal shock are the elastic modulus (E), linear coefficient of thermal expansion (α), thermal conductivity (k), tensile strength (σ), and fracture toughness (K_{1c}). Also, if a material is allotropic and undergoes a phase transformation as a consequence of the temperature change, this factor must be considered. In the selection of materials for good resistance to thermal shock, each of these variables must be analyzed to determine how it affects the material's ability to limit and/or reduce the thermal shock produced. For example, a material with high thermal conductivity, high strength, a low coefficient of thermal expansion, and a low elastic modulus would be a good candidate for resisting thermal shock. Lithium aluminum silicate (LAS) has an extremely low coefficient of expansion along its hexagonal structure ($2.0 \times 10^{-6}K^{-1}$) and, despite its low tensile strength, low thermal conductivity, and fracture toughness, is a good candidate for thermal-shock-resistance applications. A new ceramic, titanium silicon carbide (Ti_3SiC_2) combines the heat resistance of a ceramic with the machinability and ductility of a metal. Heated to 1400°C (2550°F) then plunged into cold water, it withstands a thermal shock that would shatter any other ceramic. And, although even the best engineering ceramics such as silicon nitride and silicon carbide are brittle at room temperature, the new ceramic can be deformed as much as 50% at room temperature before failure. It also machines like ductile iron or graphite.

Due to their ability to deform both elastically and plastically, most metals have few problems with thermal shock. The presence of free electrons, which conduct heat as well as electricity in metals, results in good thermal conductivities. Organic materials have low thermal conductivities due to their covalent bonding, which ties up their electrons. Ceramic materials exhibit a diverse range of thermal conductivities due to their complex microstructures. Those ceramics with the least complex structures, such as silicon carbide (SiC), composed of atoms of similar size and shape, achieve the highest thermal conductivities. Although not a ceramic material, diamond has the simple crystal structure of pure carbon, with a thermal conductivity at room temperature (RT) of 900 W/m · K. Copper's conductivity is half that of diamond's. Graphite, another polymorph of carbon, has a layered microstructure, which makes it isotropic. In addition, graphite undergoes different fabrication techniques that result in different orientations of the crystal structure, which only compounds the problem of assessing its ability to conduct heat or electricity. Finally, porosity, dispersions, noncrystallinity, grain size, and grain boundaries all affect the thermal shock resistance of materials. Pores and microcracks are probably the most important of these.

Several parameters have been developed that take into consideration the factors mentioned to assist in the selection of likely candidate materials for thermal shock resistance. One

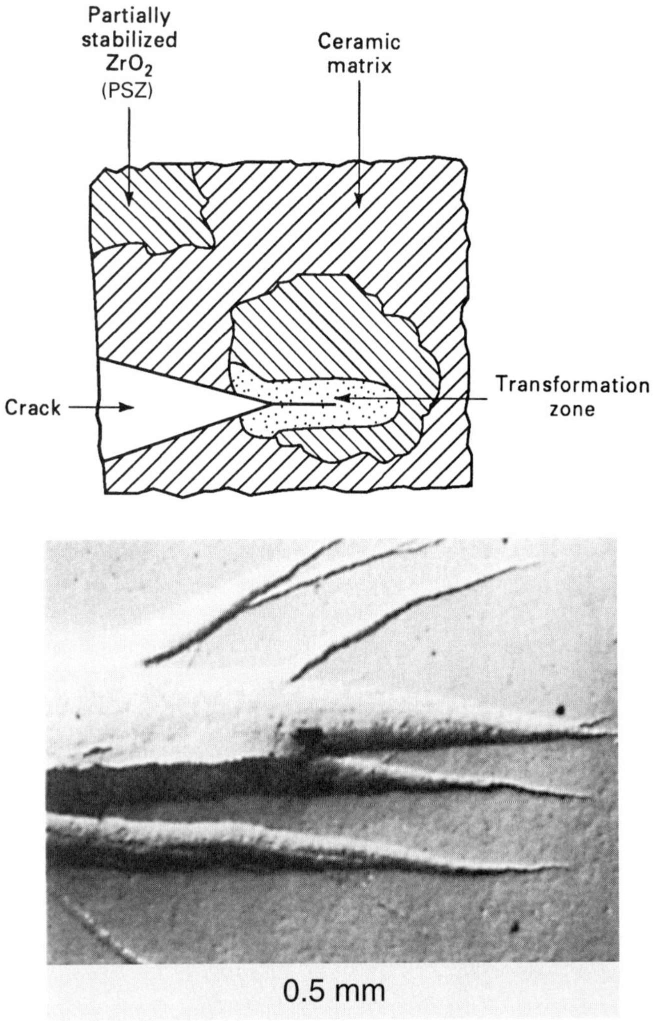

Partially stabilized ZrO_2 (PSZ)

Ceramic matrix

Crack

Transformation zone

0.5 mm

Figure 19-16 A diagram showing the formation of a transformation zone surrounding the tip of a crack due to the polymorphic transformation of ZrO2, including a micrograph of the same area. (ORNL)

such parameter is the ***thermal shock index (TSI),*** also known as *thermal fatigue resistance,* expressed in equation form as follows:

$$\text{TSI} = \frac{\sigma \times k}{\alpha \times E}$$

where
σ = tensile strength
k = thermal conductivity
α = linear coefficient of thermal expansion
E = modulus of elasticity (Young's modulus)

TSI is a ratio of the products of four variables. High values for the variables in the numerator and low values for the variables in the denominator render large or increased values for TSI. When comparing different materials based on the value of TSI, it is essential to use the same units of measure for each of the variables. For example, if you use MPa to express tensile strength for one material, you must also use MPa (not psi) for the material you are comparing it to. Table 19-3 lists the TSI and some representative property values for selected materials. Table 19-3 reveals that soda-lime glass has a low TSI of 2.1 W/cm, while fused silica glass has a TSI of 94 W/cm. Figure 19-17 provides a dramatic illustration of 96% silica glass's ability to handle thermal shock.

TABLE 19-3 THERMAL SHOCK INDEX (TSI) FOR SELECTED MATERIALS

Material	σ(MPa)	k(W/cm · °C)	α(°C^{-1} × 10^{-6})	E(GPa)	TSI[a](W/cm)
Fused SiO_2	68	6×10^{-2}	0.6	72	94
Al_2O_3	204	3×10^{-1}	5.4	344	33
Graphite	8.7	1.4	3.8	7.7	416
Soda-lime silica glass	69	2×10^{-2}	9.2	68	2.1

[a]TSI units in the U.S. customary system are Btu · in./hr · ft^2 · F°; in the SI system, they are W/cm.

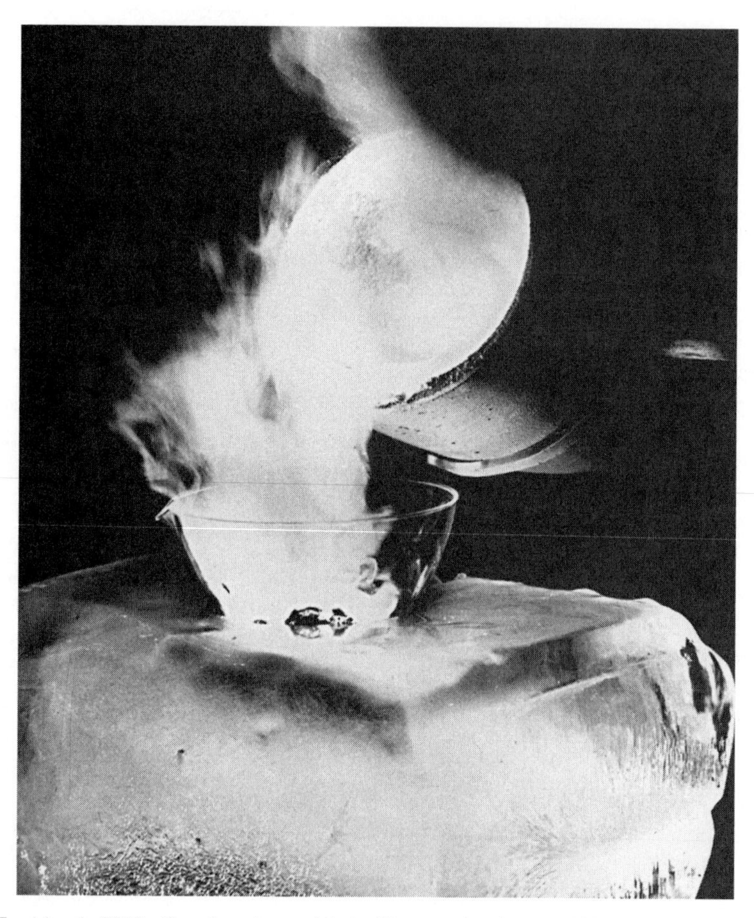

Figure 19-17 A bowl of 96% silica glass sits on a block of ice as molten bronze (2000°F) is poured. Failure to break attests to the ability of this *hard* glass to withstand thermal shock. (Corning Glass Works)

Illustrative Problem

Determine the TSI for aramid fiber and E-glass fiber, and indicate which has the best thermal shock resistance based on these calculations. Selected properties for each of these materials are provided in the table that follows:

Aramid fiber	E-glass fiber
$E = 18 \times 10^6$ psi	$E = 10.5 \times 10^6$ psi
$\sigma = 400 \times 10^3$ psi	$\sigma = 500 \times 10^3$ psi
$k = 3.5$ Btu · in./hr · ft^2 · °F	$k = 7.9$ Btu · in./hr · ft^2 · °F
$\alpha = 0.8 \times 10^{-6}$ °F^{-1} (long.)	$\alpha = 1.6 \times 10^{-6}$ °F^{-1}

Solution

$$\text{TSI}_{\text{aramid}} = \frac{4 \times 10^5 \times 3.5}{0.8 \times 10^{-6} \times 18 \times 10^6} = 9.7 \times 10^4 \text{ Btu} \cdot \text{in./hr} \cdot \text{ft}^2$$

$$\text{TSIE}_{E-\text{glass}} = \frac{5 \times 10^5 \times 7}{1.6 \times 10^{-6} \times 10.5 \times 10^6} = 2.1 \times 10^5 \text{ Btu} \cdot \text{in./hr} \cdot \text{ft}^2$$

E-glass fiber was the greater and therefore has the best shock resistance.

 Virtual Library: Technical Ceramics —*http://www.ikts.fhq.de/v1/ceramics.html*

19.1.7 Emissivity

Emissivity is an important property, particularly in high-temperature ceramics. When a surface is exposed to radiation, it may absorb part or all of the incident radiation (energy). The fraction of the energy absorbed is the emissivity (e) of the surface. A blackbody, the perfect absorber, has a value of $e = 1$; for the perfect reflector, $e = 0$.

Thermal barrier coatings (TBCs) on the metal components of gas turbines have low thermal conductivities and low emissivity. Having a coefficient of thermal expansion similar to that of the base metal allows the coating to adhere to the metal as the metal undergoes changes in temperature. The newer, energy-efficient window glass, which is more costly to the customer, has low emissivity, which means that a much greater percentage of the sun's radiation (heat) will be reflected than from older types of window glass.

Architectural flat glass for applications such as glazing in windows and doors has good optical qualities, strength, and weathering resistance. Additional requirements imposed on flat glass mandate sophisticated industrial treatments that do not involve changing the composition of the glass. Some of these additional requirements include heat reduction, light and sound protection, and increased resistance to breakage. One example of such treatment is an atoms-thick coating. Known as *low-emissive coating*, which provides an emissivity of approximately 0.14 (83% better than a normal glass surface). (See Section 6.4.2 for definition of emissivity). This low-emissive coating inhibits the absorptivity of the glass surface, resulting in more of the long-wave heat energy being reflected. All such glass must meet rigid ASTM test standards E773-83, E774-81, and E6,P-1. To determine whether a window glass is low-emissive requires a special detector available from window contractors and utility companies. A film designed to protect astronauts from the sun's radiation in outer space has found use in climate-controlling windows. Suspended between two panes of krypton gas, the film called *Heat Mirror* provides R-values (a measure of insulation effectiveness) up to 13. Average R-values in North America fall between 1.0 and 4.0. The krypton gas supports the windows in reducing noise, heat loss, energy consumption, and fuel expenses. The window blocks solar heat by 50% and keeps 99% of damaging ultraviolet light from penetrating and fading fabrics and furniture.

Another technique to reduce solar heat gain in glazing applications is shown in Figure 19-18. These new windows are another example of ***green manufacturing.*** The benefits that accrue to all of us are staggering. With more efficient windows the most obvious benefits are

less burning of fossil fuels for heating and cooling
less greenhouse gases produced
less dependence on foreign oil
less money spent on utility bills
less discomfort from drafts near windows
fewer medical bills due to better living conditions
less outside noise
less fading of fabrics

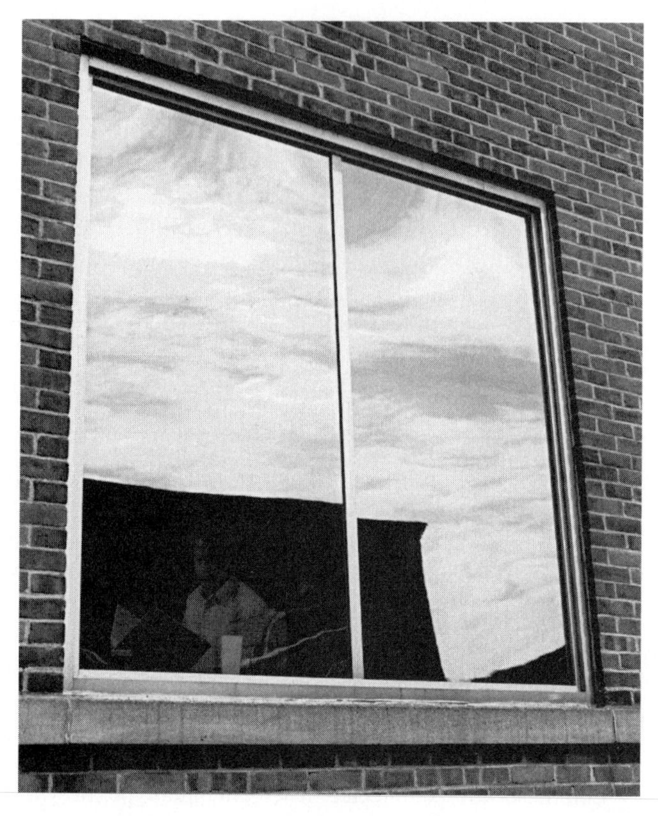

Figure 19-18 Reducing heat gain through its reflective nature, Lexan® polycarbonate sheet material (either MR® or XL-1®) is laminated with a proprietary, metallized reflective film for commercial glazing applications. The plastic also provides natural light and the high impact strength of polycarbonate sheet. It is a competitor to glass glazing. As a materials selection issue, what are the pros and cons for each type of glazing, beyond their emissivity properties? (G.E. Plastics)

19.2 DESIGNING CERAMICS

Because of their properties—their inherent brittleness; inability to yield, change shape, or plastically deform; sensitivity to defects; tendency to break at any point of high stress concentration; microstructures that possess flaws which can cause failure without being detected beforehand—ceramics require a design model that calls for greater degrees of accuracy in design and manufacture. Anyone who has broken a piece of glass knows about the sensitivity of brittle materials to defects. A slight scratch on the surface of the glass and a small bending force to produce a tensile stress are all that is needed to break a relatively thick piece of glass. This flaw sensitivity means that a ceramic cannot be considered as having only one strength but rather a distribution of strengths, depending on the flaws in the material that were added during processing, final machining, or testing, or while in service. The three general design models used by engineers are the empirical, the deterministic, and the probabilistic.

Empirical design can be defined as the trial-and-error method of selecting a material for a particular application. ***Deterministic design*** is based on mathematical analysis of average properties of ductile materials, together with a factor of safety to accommodate any errors. ***Probabilistic design*** is required for designing ceramics, particularly when the parts are complex in form and when there is a complex set of applied stresses to contend with. Using the tools of fracture-mechanics theory, highly refined stress-analysis techniques, and statistical representations of material properties, this probabilistic design model allows the engineer, aided by numerous computer programs, to determine the stresses statistically, and parts are then designed not to exceed a specified probability of failure. The density, size, and severity of flaws are also determined statistically. Because ceramic parts have slight variations in their structure, which can have a severe effect on their properties, each part must be designed individually.

The mathematical means for representing the statistical nature of brittle failure is the Weibull equation, which relates the cumulative probability of failure to the failure stresses of all the elements of the sample. The graphical plot of the data along with a best-fit curve is known as a ***Weibull plot*** of strength for a material. The slope of the best-fit curve, known as the ***Weibull modulus,*** indicates the degree of variance in the strength data. A high value for the modulus (m) is desired because it indicates low variance or a narrow range of failure stresses. Many critical parts are ***proof tested*** at a predetermined proof-stress level. Those parts with large flaws will fail the test. Several assumptions are made at this point. One is that parts that pass the test will perform satisfactorily in service. It is also assumed that the testing did not initiate cracks or cause cracks to propagate that will cause failure later in service.

 Diamond film—*http://www.nortondiamondfilm.com/*

19.3 IMPROVING TOUGHNESS

Toughness is an indication of how much resistance a material offers to a crack starting to grow through it. Most materials either contain cracks already or have structural flaws that, under stress, can readily turn into cracks. The work to be done in propagating a crack, also known as the ***work of fracture,*** is measured by the energy absorbed in creating a unit area of new crack ahead of an existing crack tip. This work is a measure of toughness (G_O) and is expressed in joule/m^2. Figure 19-19 shows a graph of specific strength versus work of fracture for some typical ceramics. Work of fracture is the preferred measurement of toughness for fiber-reinforced ceramics.

Another procedure for assessing toughness is to measure the area under the stress–strain curve up to the point of fracture. This procedure is more suitable for a brittle material due to the difficulty in identifying the area under the curve that represents the energy attributable to plastic deformation (see Figure 7-20). The units for expressing toughness by this procedure are joule/m^2. ***Impact*** is defined as the sudden application of a load confined to a localized area of a material. Exemplified by the striking of a material with a hammer, this relatively quick application of a

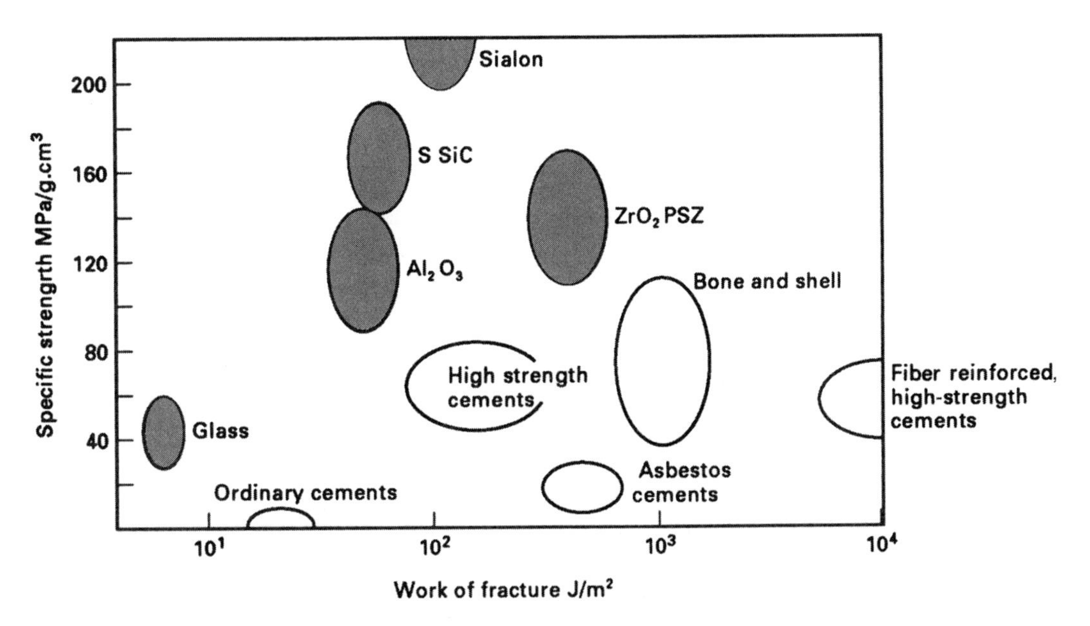

Figure 19-19 Plot of the specific strength versus the work of fracture for some typical ceramic materials—glass, cements, and biominerals. (JME)

force (as opposed to a slow or static loading of a material) can cause considerable damage to a material that cannot adequately redistribute the stresses caused by the impact. Figures 4-13a and 4-13b showed a picture of an impact machine and some fractured specimens. ***Ductile materials*** usually survive impact due to their microstructure, which allows slip to take place. Metals are thus considered tough, or having good impact resistance. Due to their inherent nature as compounds of metals and nonmetals, ceramics do not possess the ability to redistribute stresses and deform plastically. Consequently, they have poor toughness, impact resistance, and fracture toughness.

To improve the impact resistance of ceramics, measures are taken to increase their fracture toughness. One such procedure is ***fiber and particulate reinforcement.*** Common examples are the use of glass fibers to reinforce plastics, carbon fibers to reinforce carbon–matrix composites, zirconium–diboride platelets to reinforce a zirconium–carbide matrix, and Al_2O_3 fibers in an aluminum matrix. Figure 19-20a is a micrograph of precoated SiC fibers reinforcing a SiC matrix by using the forced CVI process. Figure 19-20b is an accompanying stress–strain diagram, showing the effect on toughness of the composite by modifying the interfacial bond between fibers and the matrix material. In ceramic–matrix composites, the bond between the fibers and the matrix must be weakened to allow the fiber to absorb more of the energy that is trying to initiate or propagate a crack. This is the opposite strategy to fiber bonding with a polymeric matrix, which calls for an increase in the interfacial bonding. For example, silicon fibers are coated with carbon before they are used in a matrix of silicon carbide (Figure 19-20a). This coating increases the toughness of the composite by reducing the interfacial bonding (friction) between the fibers and the matrix material (Figure 19-20b). Doing so allows the fibers to debond from the matrix material and pull out before breaking. All of these displacements require energy expenditures and thus reduce the energy that can propagate as a crack until there is a failure of the composite. Figure 19-21a is a sketch of fiber bridging (fibers span across the crack) and pullout (fibers pull out of the matrix material) as the fibers resist the propagation of a crack; Figure 19-21b is a micrograph of this same phenomenon. Figure 19-22 is a stress–strain plot showing the point of fiber pullout and matrix cracking as well as the progressive fracturing of the fibers. Each of these individual events takes energy away from that required to propagate a crack and leads to total failure of the ceramic. In addition to precoating the fibers, the toughness of the materials is increased partly because a crack that tries to propagate in the matrix encounters fibers that block its movement. The crack is diverted to run along the fiber–matrix interface. The stress applied in this direction is much lower, so the crack might stop propagating. Figure 19-23 is a micrograph of the crack pattern surrounding a diamond-hardness indentation in a polished surface of a Lanxide™ zirco-

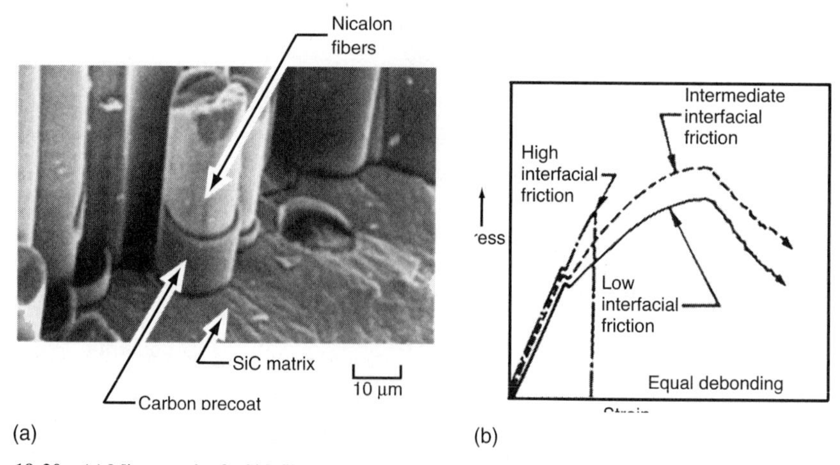

Figure 19-20 (a) Micrograph of a SiC fiber precoated with carbon reinforcing a SiC matrix in a *continuous fiber ceramic composite* (CFCC). (b) Stress–strain curve showing the dramatic results of such a coating on the toughness of the composite. Note the differing results of reducing the friction or interfacial bonding between the fiber and its matrix material. (ORNL)

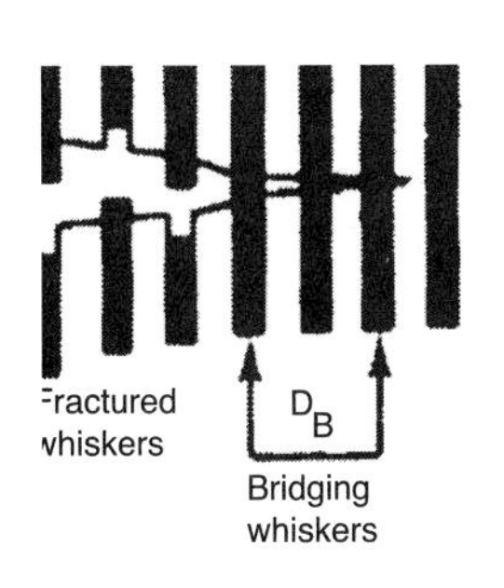

(a)

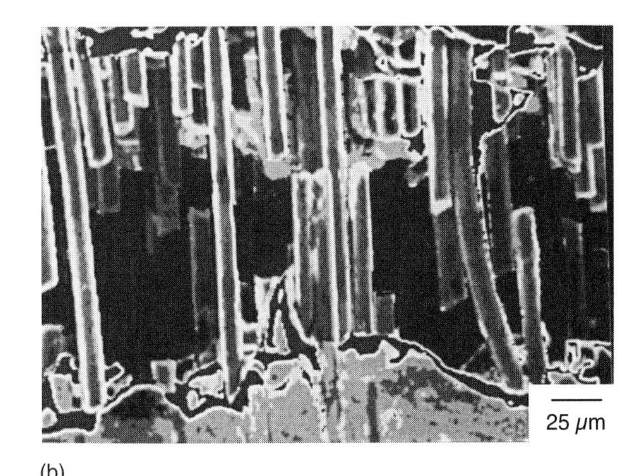

(b)

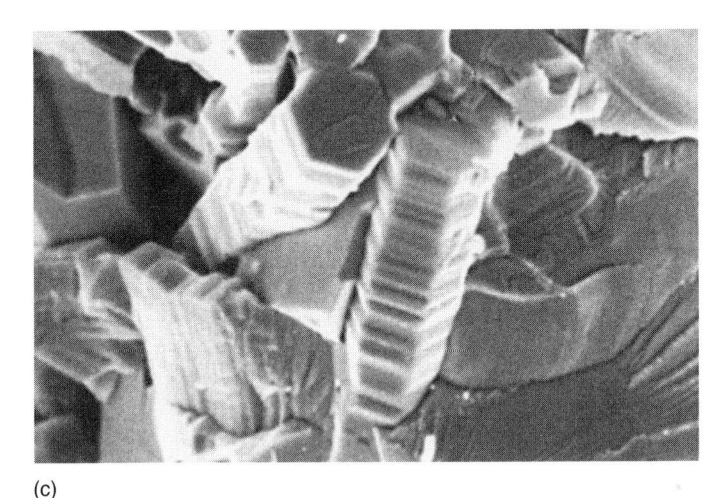

(c)

Figure 19-21 (a) A crack zone showing the effects of whiskers (or fibers) as they bridge the crack and pull out of the surrounding matrix material before fracturing. (b) Photomicrograph of a fractured, fiber-reinforced Lanxide ceramic composite showing the fibers spanning the fracture to resist crack propagation. By designing an interfacial bond with the correct strength between the fibers and their matrix, the fibers can expend energy, overcoming friction and pulling out of the matrix, thus bridging the fracture instead of fracturing and assisting in crack propagation. (c) Ceramic composite fracture surface. (ORNL and Lanxide)

nium–diboride platelet, zirconium–carbide/zirconium composite. The matrix material of the composite, written or abbreviated $ZrB_{2p}/ZrC/Zr$, contains up to 20% metallic zirconium in addition to ZrC. These composites have fracture toughness as high as 23 $Mpa \cdot m^{1/2}$ and flexural strength exceeding 1 GPa, properties that are the result of a combination of crack deflection and pullout of the platelets, as shown in the micrograph. Additionally, the excess zirconium provides some ductile yielding (plastic deformation) in the crack zone. These reinforcing platelets have successfully localized the damage, as shown by the numerous small cracks surrounding the indentation.

Another method for toughening materials is the use of ***second-phase reinforcement.*** Examples are the addition of small quantities (about 10%) of cobalt (Co) to tungsten carbide (WC), a ceramic (a hard refractory material), while both are in the particle stage. After additional steps in processing (pressing and sintering), a cermet is produced that has the toughness needed by cutting tools. A third method for toughening materials is called ***transformation toughening.*** We discussed the formation of martensite in the study of steels, in Section 9.6. The martensite reaction was once believed to be confined only to steels, but now a number of other alloy systems,

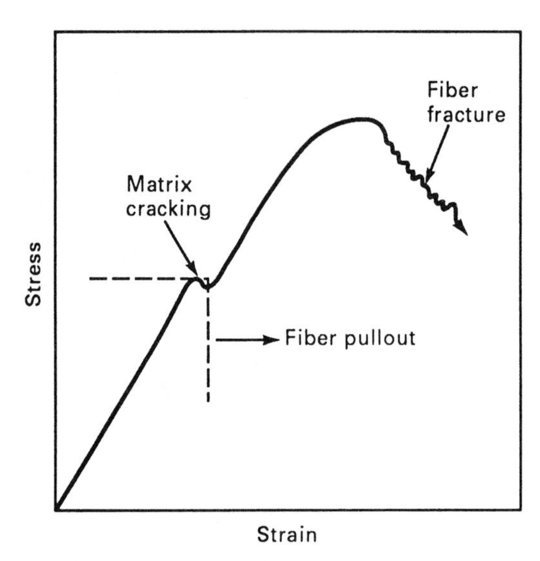

Figure 19-22 A fiber-reinforced ceramic composite has increased toughness due to the added strain energy of the fibers and the work done in debonding the fibers from the matrix.

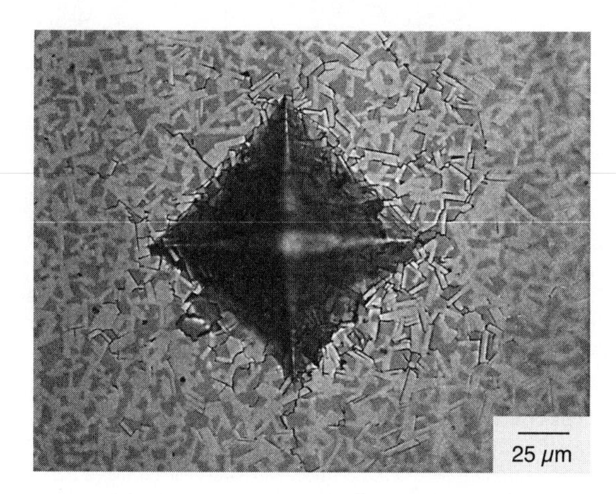

Figure 19-23 A micrograph of a platelet-reinforced Lanxide® ceramic composite showing the crack pattern produced by a diamond stylus indentation that effectively stops the propagation of a critical crack, which could have led to failure. (Lanxide)

such as iron–nickel and copper–aluminum as well as titanium, produce similar phase transformations. In the field of ceramics, ***zirconia*** or ***zirconium oxide (ZrO_2)*** has been found to possess some unique properties (see Figure 19-24). Zirconia possesses the lowest thermal conductivity of any ceramic material. Pure ZrO_2 has a stable, tetragonal crystal structure of temperatures above 1000°C but changes to a monoclinic crystal structure below 1000°C. This displacing transformation produces an abrupt change in the dimensions of the crystal structure, accompanied by stress buildup and possible cracking. A similar transformation occurs when silica allotropically changes from α-quartz to β-quartz, and from tridymite to cristobalite. When zirconia is cooled through 1000°C, the accompanying volume change (3.25% expansion) causes the material to fall apart. Figure 19-25 is a rough sketch of the effects of temperature on the volume of ZrO_2. However, if a small amount of yttria is added to the ZrO_2, the tetragonal phase can be stabilized down to room temperature. In addition to yttria, manganese and calcium can be used. The stabilizer size is critical. The yttria must be uniformly distributed throughout the microstructure of the ZrO_2. This is accomplished by mixing the two materials in powder form, with particle sizes varying from 0.05 to 1 μm. Note that this does not require a phase transformation.

As stated, specific microstructures are crucial for optimum properties. High-temperature sintering (firing) is essential for the formation of tetragonal zirconia precipitates

Figure 19-24 Three differently designed hand scissors show the cutting surfaces made of zerconia, a ceramic that never needs sharpening. Note the differences in the way this extremely hard material (white color) is supported to augment its flexural strength.

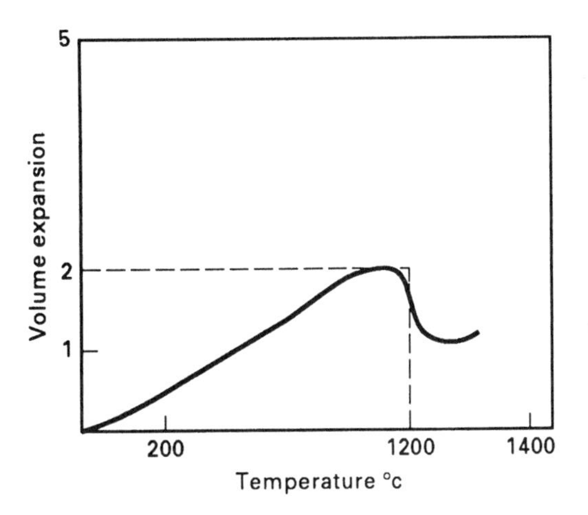

Figure 19-25 Volume expansion curve for ZrO_2 showing the effects of the displacing polymorphic transformation.

within a matrix of cubic-stabilized zirconia. The morphology (shape) of these tetragonal precipitates can be controlled by various stabilizers. The ability of these precipitates to be transformed to monoclinic zirconia (with a 4% volumetric dilation on application of a stress or passage of a crack) imparts increased fracture toughness to these materials. The zirconia is now referred to as *stabilized* zirconia. By not adding enough yttria to stabilize the zirconia and with adjustments in the particle sizes and processing controls, mixtures of the stabilized cubic phase and the unstable monoclinic phase that possess high fracture toughness are obtained. This metastable mixture, known as ***partially stabilized zirconia (PSZ),*** will undergo transformation if it is disturbed sufficiently. If a part is made of this ceramic material and the part is subjected to impact, vibration, or another tensile-stress-producing condition, the stress will cause the material to absorb energy sufficient to bring about transformation of the monoclinic phase. The transformation takes place ahead of the tip of any crack growing in the material, resulting in a diminishing of the stress and in turn causing the crack to propagate and hence an increase in the toughness of the ceramic material. Figure 19-16 was a sketch of the PSZ transformation zone surrounding a crack. It has been shown that Mg–PSZ, one of many PSZ systems, has the highest fracture toughness. As with other materials, it can be heat treated in a variety of ways to modify its microstructure for particular applications.

The last toughening method to be mentioned is ***surface compression.*** If the surface of a material is placed under compression, any applied tensile stress must exceed the compressive stress before a stress concentration can begin to build up and lead to a crack propagation. Surface compression or ***prestressing*** can be accomplished by ion exchange, quenching, or layering. ***Ion exchange*** can be brought about by exchanging ions of greater size with those of the parent material (see Figure 19-26). The larger ions substitute for some of the smaller ions when the material is heated, and after the material is cooled, the larger ions near the surface cause a compressive stress. This phenomenon comes about because, at the higher temperature, the unit cells of a crystalline material or the atoms in an amorphous material have moved outward, making room for the larger ions that still find room with the smaller ions squeezed, producing the compressive stress.

When a glass material is heated sufficiently to a temperature that produces some viscous flow and then quickly cooled, the surface will cool more quickly than the interior, forming a solid, hard case around the still somewhat viscous interior. This process is known as ***glass tempering.*** Note the difference between glass tempering and tempering of martensitic steel, discussed in Module 9. Figure 19-27 shows an example of tempered glass that has undergone an

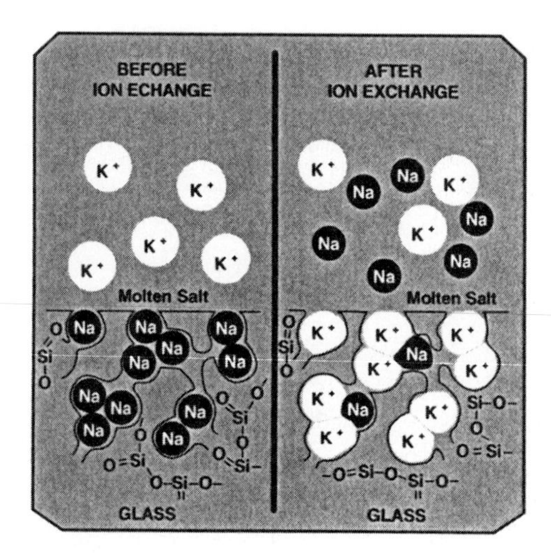

Figure 19-26 Chemically strengthened glass. Ion exchange causes relatively large potassium ions (K^+) to replace smaller sodium ions (Na^+), thus crowding the surface and producing compression (prestressing).

(a)

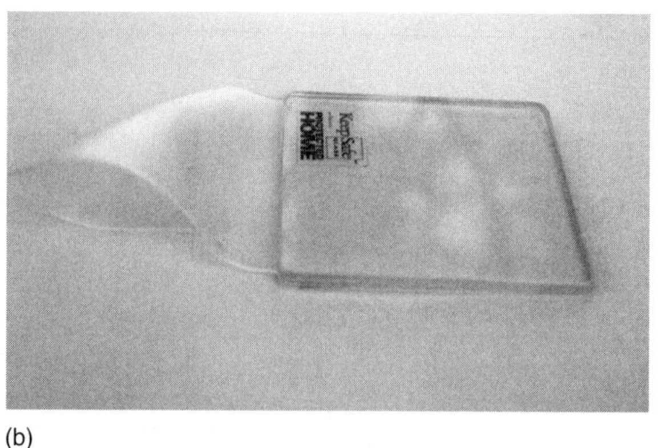

(b)

Figure 19-27 (a) Tempered glass is found in some automobile windows, large glass doors, and in eyeglass lenses. Note the fractured pieces of the tempered glass are small, rounded, blunt shaped, and relatively smooth and are incapable of causing injury. A piece of laminated safety glass, with an enclosed polymer adhesive joins two pieces of annealed glass held in position by the adhesive. The broken pieces of this glass, held in position by polyvinyl butyral (PVB), shown in a darker color at the pointed end, are somewhat long, narrow, and dagger-shaped shards that emanate from the origin of failure but adhere to the tough PVB interlayer. (b) Another example of laminated safety glass showing an extended piece of the PVB resin sheet material. Note that the polymeric membrane between the two layers of glass is relatively thick. Annealed glass, broken comparatively easily by impact or wind loads, produces shards similar to the laminated glass, but an automobile windshield's transparency is retained.

impact contrasted with safety glass that incorporates a laminated polymer sheet (PVB) to restrain the glass shards. Automobile windshields use a laminated, annealed glass sandwich composite with a PVB sheet between two sheets of glass. This composite material keeps the glass from flying about on impact yet maintains the transparency of the windshield so that the driver can maneuver, if possible, the automobile to a safe stop. By producing a surface compression with a different coefficient of thermal expansion (layering), heating to a temperature to produce some plasticity, and then cooling, the surface layers of the glass experience a compressive ***prestress.*** Figure 19-28e is a graphic step-by-step portrayal of this process.

This technique is used to not only toughen tempered glass but also other composite materials, including ***prestressed concrete.*** The technique involves fiber-reinforcing composites and reinforced concrete and places the matrix materials under compressive stress to prevent a tensile

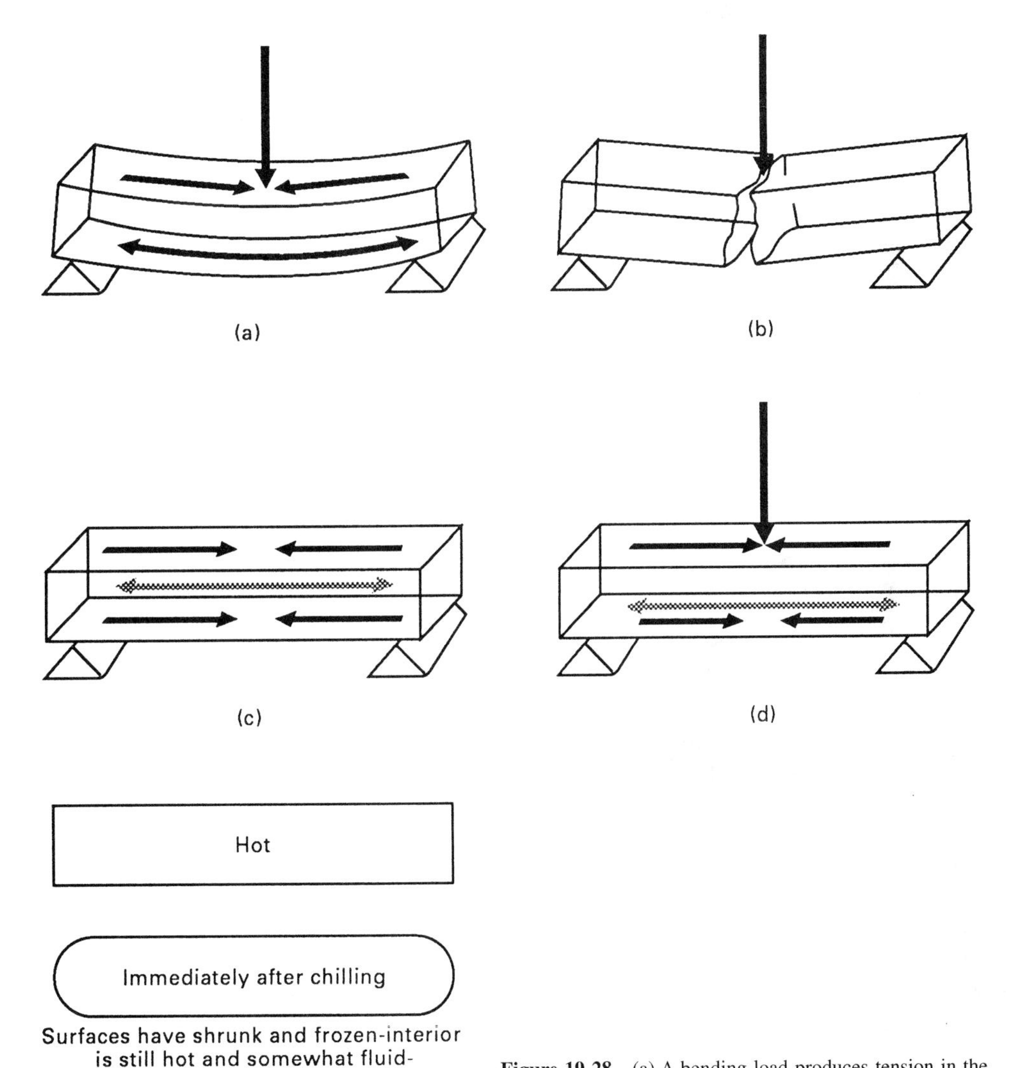

Figure 19-28 (a) A bending load produces tension in the lower surface of the test bar and compression in the upper surface. (b) The bar breaks when the tension on the lower surface exceeds the ultimate strength of the glass. (c) A prestressed bar shows compression in all surfaces. Reactive tension is buried within the bar. (d) A bending load applied to the prestressed bar must first overcome built-in compression before the surface can be put in tension. (e) Prestressing by physical tempering consists of heating glass until it begins to soften. An abrupt chill then shrinks and freezes the surfaces. When the interior of the glass cools and shrinks, the surfaces are compressed. (Corning Glass Works)

stress from acting to cause crack propagation in the matrix. Prestressing is accomplished in a multistep process. First, the fibers are tensioned mechanically. Second, they are encapsulated by the matrix material, which is allowed to solidify. Third, the tensioning force is released, causing the fibers to contract elastically toward their original length, placing the matrix under a compressive stress. Once an external tensile force (or load) is applied to the composite, the compressive stress in the matrix must be exceeded before the matrix feels the presence of the tensile stress.

SELF-ASSESSMENT

19-1. A broad classification of ceramic materials that use artificial raw materials, exhibit specialized properties, and require sophisticated processing is

a. Traditional ceramics **b.** Advanced ceramics **c.** Cements **d.** Glass

19-2. What appeal does biomimetics represent for the selection of ceramic materials? What other attributes make ceramics competitive with metals and polymers?

19-3. Ceramics based primarily on the natural raw materials of clay and silicates are

a. Traditional ceramics **b.** Advanced ceramics **c.** Cements **d.** Glass

19-4. A major objective in designing new ceramics is to overcome

a. Hardness **b.** Brittleness **c.** Reactivity **d.** Conductivity

19-5. Generally, ceramics compete well with metals when there is a need for lower weight, lower density, high shear strength, and good tensile strength.

a. T (true) **b.** F (false)

19-6. Metals are usually superior to ceramics in areas such as impact strength, ductility, and plastic flow.

a. T (true) **b.** F (false)

19-7. The term describing the density of ceramics that includes its porosity and both crystalline and noncrystalline phases is

a. Bulk density **b.** Open density **c.** Closed porosity **d.** Green density

19-8. A polymer with a backbone of SiO and whose ASTM designation as Si is

a. Silicon **b.** Silica **c.** Silicone **d.** Silicates

19-9. The element Si is

a. Silicon **b.** Silica **c.** Silicone **d.** Silicates

19-10. The ceramic SiO is called

a. Silicon **b.** Silica **c.** Silicone **d.** Silicates

19-11. Because of their bonding and their different atoms and unit cell structure, ceramics offer high resistance to shear and are good insulators.

a. T (true) **b.** F (false)

19-12. Using Figure 19-11, place in order from good to poor the following materials in terms of wear resistance:

a. Cast iron **b.** Silicon carbide **c.** Urethane **d.** 96% alumina

19-13. Transformation zones such as those created by ZrO_2 serve what purpose in ceramics?

a. Thermal shock resistance

b. Fracture toughness

c. Improved conductivity

d. Tensile strength

19-14. Bonding in ceramic materials is

a. Ionic **b.** Covalent **c.** Secondary

19-15. Oxidation numbers designate the number of ____________ taking part in the bonding of elements during a chemical reaction.

a. Protons **b.** Atoms **c.** Electrons **d.** Ions

19-16. Ceramic armor, initially developed for the Vietnam War and also used in the Gulf War, was a composite made up of boron carbide (B_4C) and fiberglass plus aramid fiber matting. The ceramic has a density of 2.4 g/cm^3. Is this a low- or a high-density ceramic? Explain why you think this ceramic was chosen for this application.

19-17. Match the correct ceramic to its favorable characteristics. Some will have the same characteristics.

a. Boron nitride (1) Low cost
b. Graphite (2) Machinable
c. Al_2O_3 (3) Best lubricity
d. Glass ceramic (4) Maximum hardness

19-18. Express a fracture toughness of 186 MPa in terms of psi.

19-19. The measure of energy absorbed in creating a unit area of a new crack ahead of an existing crack is

a. Impact **b.** SENB **c.** Flaw **d.** Work of fracture

19-20. Using the data from Table 19-3, determine the SI units of TSI for graphite.

REFERENCES & RELATED READINGS

American Society for Testing and Materials. *Annual Book of ASTM Standards,* Part 17. *Refractories. Glass and Other Ceramic Materials; Carbon and Graphite Products.* Philadelphia, PA: ASTM, 2003.

Anderson, Richard M. "Testing Advanced Ceramics," *Advanced Materials & Processes.* Volume 135, No. 3, 1989.

ASM International. *Engineering Materials Handbook,* Desk Edition. Materials Park, OH: ASM, 2000.

Besmann, T. M., B.W. Sheldon, R.A. Lowden, and D.P. Stinton. "Vapor-Phase Fabrication and Properties of Continuous-Filament Ceramic Composites," *Science,* Volume 253, September 6, 1991.

Bonds, James W. et al. "Wire-Arc Spraying of Metal onto Insulating Foam." Rockwell International Corp. for Marshall Space Flight Center, *NASA Tech Briefs*, May 1995, p. 84.

Bunnell, L., Roy. "Tempered Glass and Thermal Shock of Ceramic Materials," NEW: Update 91, Oak Ridge, TN: Oak Ridge National Laboratory, November 12–14, 1991.

Corning Glass Works. *All about Glass.* Corning, NY: CGW, 1968.

Corning Glass Works. *Properties of Glasses and Glass-Ceramics.* Corning, NY: CGW, 1973.

Craig, Douglas F. "Structural Ceramics," NEW: Update 91, Oak Ridge, TN: Oak Ridge National Laboratory, November 12–14, 1991.

Cselle, Tibor. "New Directions in Drilling," Applications Research & Development, Guhring, Inc., Brookfield, WI, *Manufacturing Engineering*, August 1995, pp. 77–80.

Double, David D. "Chemically Bonded Ceramics; Taking the Heat Out of Making Ceramics," *Journal of Materials Education*, Volume 12, No. 5/6, 1990, p. 353.

Evans, James W., and Lutgard C. DeJonghe. *The Production of Inorganic Materials.* New York: Macmillan, 1991.

George, Laurie A. "Dental Applications of Ceramics," in *Bioceramics*: *Materials and Applications Ceramic Transactions*, Volume 48. Westerville, OH: American Ceramic Society, 1995.

Harper, Scott. "New-Age Concrete Acts Like a Big Filter," *The Virginian-Pilot*, September 7, 1995, p. B3.

Jordan, Gail W. "Adapting Archimedes' Method for Defining Densities and Porosities of Small Ceramic Samples," NEW: Update 90, Gaithersburg, MD: NIST, November 12–14, 1990.

Lehmann, Richard L. "Primer on Engineering Ceramics," *Advanced Materials & Processes*, Volume 141, No. 6, 1992.

MacChesney, J.B. "The Materials Development of Optical Fiber: A Case History," *Journal of Materials Education*, Volume 11, No. 4, 1989, p. 321.

McColm, I. J. *Ceramic Hardness.* New York: Plenum Press, 1990.

MPR. "Net and Near-Net Shaped Ceramic Composites Formed by Lanxide Technology," *Materials and Processing Report,* Volume 3, No. 3, 1988.

Ometite, Ogbeni O., Mark A. Jannaya, and Richard A. Strehlow. "Gelcasting: A New Ceramic Forming Process," *American Ceramics Society*, Volume 64, No. 12, 1985.

Pfaender, Heinz G. *Schott Guide to Glass.* New York: Van Nostrand Reinhold, 1983.

Reed, James S. *Introduction to the Principles of Ceramic Processing*. New York: Wiley, 1988.

Richerson, David W. *Modern Ceramic Engineering: Properties, Processing and Use in Design*, 2nd ed. New York: Marcel Dekker, 1992.

Richerson, David W. *The Magic of Ceramics*. Westerville, OH: The American Ceramic Society, 2000.

Rogers, Craig A. "Intelligent Materials Systems: The Dawn of a New Materials Age." Center for Intelligent Materials and Structures, Virginia Polytechnic Institute and State University, 1992.

Roy, Della M., ed. "Instructional Module in Cement Science," *Journal of Materials Education*. Materials Education Council, 1985.

Swain, Berta. "Microwave Processing of Ceramics," *Advanced Materials & Processes*, Volume 134, No. 3, 1988, p. 76.

Urquhart, A. W., G. H. Schiroky, and B.W. Sorenson, "Ceramic Composites for Gas Turbine Engines via a New Process," ASME 89-GT-316, *Gas Turbine and Aeroengine Congress and Exposition*, Toronto, Canada, June 4–8, 1989.

Periodicals

Advanced Materials & Processes

Journal of Adhesion Science and Technology

Ceramic Bulletin

Ceramic Monthly

Manufacturing Engineering

MRS Bulletin

Module

20

Ceramic Processing

After studying this module, you should be able to do the following:

20–1. Name the major processing and raw materials used to produce ceramics.

20–2. Describe some advanced processes and their benefits in terms of high and low technology.

20–3. List and describe the favorable major properties of ceramics (e.g., wear and heat resistance and low thermal expansion) and the units in which to express them.

20–4. Name the common tests to measure the properties of ceramics.

20–5. Compare typical properties such as wear, density, hardness, flexural strength, fracture toughness, and thermal expansion and thermal shock resistance of CMCs with other ceramics, metals, and polymers.

20–6. Use the website addresses denoted by the symbol to explore websites for new developments and evolving concepts related to traditional and advanced ceramic processing*.

*The symbol found throughout the book will link you to Internet sites related to topics being covered. The dynamic nature of the Web brings frequent changes, so some of these sites, while available at the time of writing, may not be up now.

A major problem in ceramic processing is achieving the same properties every time a process is repeated. As materials become more complex in composition and structure, it becomes more difficult to make a uniform product. ***Intelligent processing*** of materials is an attempt to solve the problem of making complex materials more uniform in their structure and properties. The National Institute of Standards and Technology (NIST), in cooperation with industry, has an ongoing program to automate ***quality control*** to allow quality to be built in during processing rather than attempting to obtain it by inspection after the fact. Sensors analyze a material as it is processed to warn the computers controlling the process to correct the errors that are beginning to occur, a method known as ***adaptive control.*** By improving existing processes and developing more efficient ones that reduce the size and number of flaws and even out the distribution and size of particles, attaining tougher, stronger, more reliable ceramics is a goal rapidly being realized.

In the initial preparation of ceramic materials for shaping or molding into some useful product, the raw materials must first be reduced to a powder form. The term ***comminution*** describes the operations of crushing and grinding that accomplish this task. ***Crushing*** reduces the raw material to particle size, whereas ***grinding*** takes the particle-size ceramic and further reduces it to a powder form. Grinding is accomplished by confining the ceramic powder with extremely hard balls and pebbles (known as ***media***—see Figure 21-11) or rods in a revolving, cylindrical drum called a *ball mill.* This grinding action causes the reduction to a fine powder by a combination of impact and attrition. Often, water is added to the powder, which then changes into a slurry. Yttria-doped tetragonal zirconia polycrystal (Y-TZP) ceramics, which have high strength and high wear resistance, are widely used as milling media for making powders for more advanced ceramics.

New and emerging processes offer the wide range of ceramic products discussed in this unit. Figure 20-1 shows a prototype ceramic automotive engine. A fully ceramic engine eliminates the need for cooling water, which not only saves weight but allows a hotter, more efficient engine.

20.1 THERMAL PROCESSING—SINTERING/DENSIFICATION/FIRING

Sintering is a critical factor in the production of advanced ceramics. A green, or unsintered, ceramic material is softer than chalk, consisting of millions of powder particles. When sintered at high temperatures, usually between 1000° to 2000°C, the particles fuse to become a solid material with properties superior to many metals.

Figure 20-1 Prototype ceramic engine (Kyocera, American Ceramic Society)

Densification, also referred to as ***firing,*** is a process by which a particulate compact is transformed into a ceramic part that has adequate properties to satisfy the needs of a specific application. In sintering, the powder compact is heated to a temperature *below* the melting temperature of its components. (The melting temperatures of most ceramics make it impractical to actually melt the materials and thus bond the grains of powder together.) The bonding between the grains that takes place is due to the action of many complex mechanisms, including diffusion. In sum, sintering removes by a variety of mechanisms, the initial pores, and causes the growth of grains and strong bonding forces to provide strength to a powder compact. The main effect is the reduction in surface area (i.e., the surface energy of the item). In general, this results in a part with the same shape as the original or ***"green"*** shape but reduced in size. The density of the part as well as the particle size also increases, although particle size must be controlled because final strength is determined partly by the size of the final grains. Relatively large particles have a greater number of interstices, which are filled with air, a good insulating material. This may be satisfactory if the ceramic is being designed for heat resistance. Usually, a ceramic requires strength, and strength comes with a homogeneous microstructure that has a fine grain structure. However, the initial pores between the original particles are not completely eliminated in sintering. Further, the final density reaches only some 95% of the maximum *theoretical density* (see Section 19.1).

Sintering furnaces usually heat ceramic compacts externally by radiant and convection thermal transfer. Figure 20-2 shows graphs of the effects of sintering alumina and zirconia using microwave sintering versus standard furnace heating.

Microwave sintering produces uniformly heated ceramic parts of large volume and irregular shape rapidly at temperatures in excess of 1600°C in a vacuum or under atmospheric pressure. Microwave sintering causes the molecules throughout the workpiece to vibrate, which produces uniform heating without gradients in temperature. The key to this method is

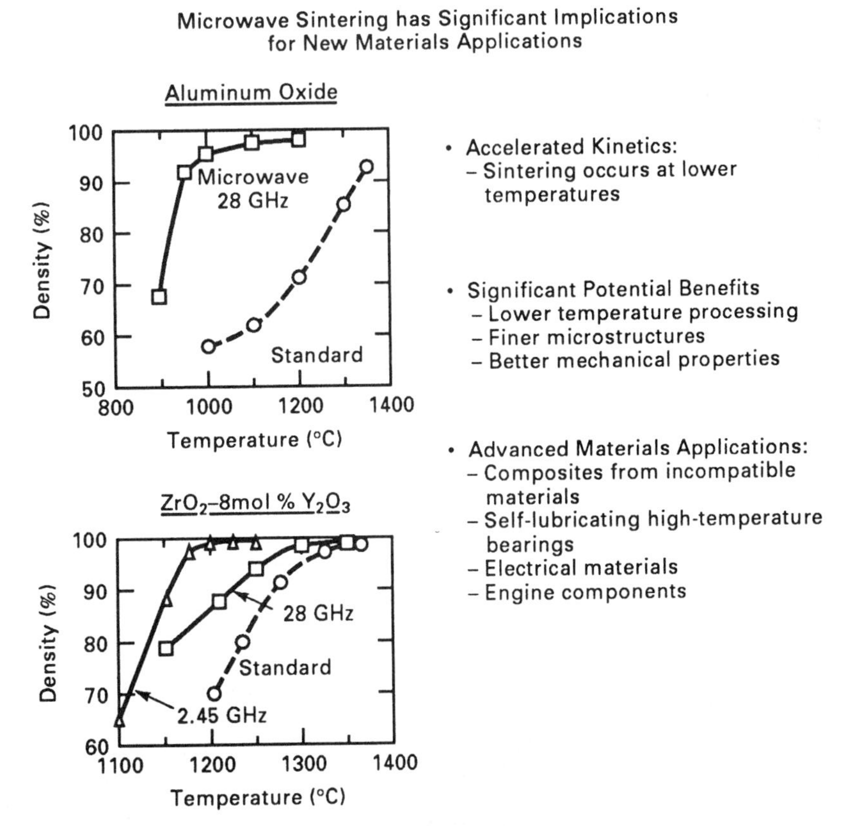

Figure 20-2 Plots of density percentage versus temperature for Al_2O_3 and ZrO_2, showing the vivid effects of sintering these materials with microwaves versus standard heating procedures. Note the significant increases in attained density as a result of microwave heating. (ORNL)

the specialized, insulated sintering chambers, which trap the heat in the greenware and allow the temperature to rise rapidly. In contrast, in radiant or convection heating, energy is absorbed only at the surface of the workpiece and must be transferred into the bulk of the material by conduction.

Ceramics that have rather high dielectric-loss factors (a measure of electric energy lost as heat energy by a capacitor in an ac circuit) can benefit from the lower temperatures used in the microwave process. Figure 20-2 also lists other significant benefits and a few applications for the new ceramic materials produced by the microwave process, which achieves finer microstructures and better mechanical properties. In addition to the conventional and microwave sintering processes, there are other sintering processes, such as liquid-phase and solid-state sintering, used to densify compacted powders.

Many other processes employed by industry improve on the pressureless sintering process just described. Some involve the application of pressure and heat simultaneously, for example, (1) hot pressing, (2) overpressure sintering, and (3) hot isostatic pressing (HIP). In ***hot pressing,*** also known as ***pressure sintering,*** powders are compacted using a die within an evacuated chamber. Temperatures are about one-half of the absolute melting temperature (T_m+273°)* of the powdered material. Due to the uniaxial techniques used to apply the pressure in the die, hot-pressed materials may have a preferred orientation in their grain structure, which produces anisotrophy. ***Hot isostatic pressing*** (HIP) combines compounding (compressing and densifying loose powders into a desired shape) and sintering using temperatures as high as 2500°C and pressures as high as 200 MPa in a furnace within a pressure vessel. This process relies on plastic deformation to achieve densification. Using argon or helium gas to exert pressure from all sides, HIP results in near-shape forming of ceramic parts, particularly with cemented-carbide particulates. As with the other densification processes mentioned, HIP can be used in combination with other methods to achieve highly densified parts. Enhanced densification rates have brought on reduced sintering times and temperatures, which means that this innovative technology produces a higher quality product and does so at a lower cost.

In another novel process, graphite–carbon particles with controlled electrical properties are used as both the pressure-transmitting medium and as an electrical resistor for simultaneous heating and consolidation of the powder preforms. Cyclic times are very short compared to other methods, such as HIP (minutes versus hours), and both high temperatures (about 1000°C) and relatively low pressures are involved. This process has the ability to densify complex shapes without "***canning***," which refers to the need to encapsulate (for HIP) preforms with glass to seal them in a gas-impermeable envelope. The rapid densification possible with this new process has the potential to inhibit grain growth, resulting in fine-grained microstructures with improved mechanical properties.

CERAM Research Ltd.—*http://www.ceram.co.uk*

20.2 TRADITIONAL PROCESSING

The first ceramic materials produced by processing were bricks and lime plaster. Limestone was roasted to produce lime, which was mixed with water. To this mixture was added clay, straw, or fibers to provide building materials that were used widely and are still in use today. When added to the clay materials, water provides the clay with plasticity or workability, in addition to providing the bonding mechanism that gives the material its strength. Traditional ceramics are still produced with basically the same materials (Figure 20-3). ***Slip casting*** (Figure 20-4), an age-old process for making vases and other hollowware, is a production process still used for many ceramics. Many other traditional ceramics are produced by blending, pressing, and firing (Figure 20-5a–b). Often, traditional ceramics develop a vitreous or glassy phase upon cooling after firing, which accomplishes densification (Figure 20-6). Glass phases must

*T(K)=T(°C)+273.15. K is kelvin. T_m=T°C

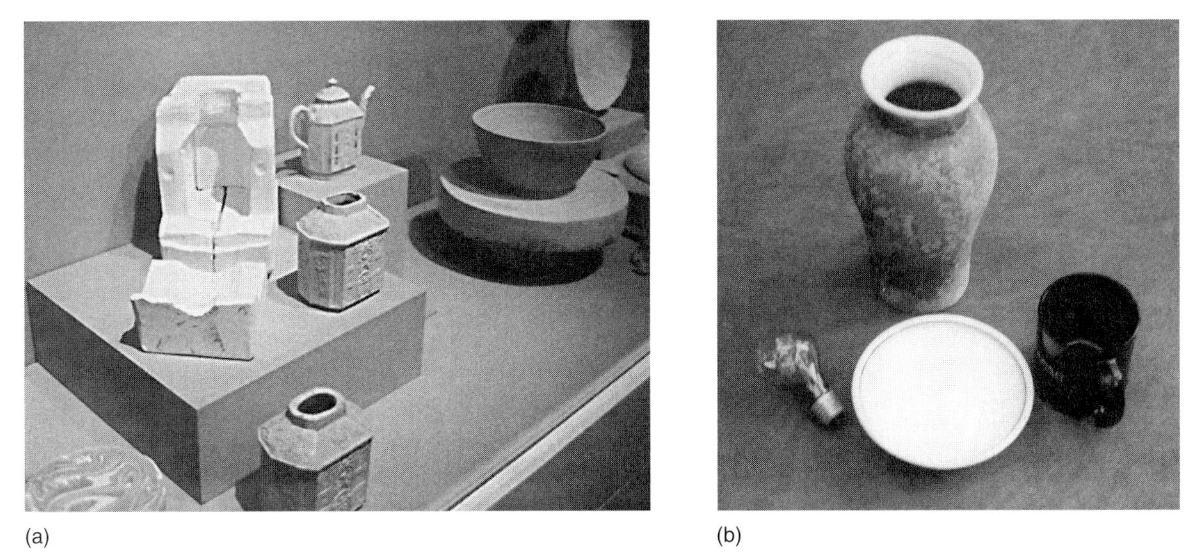

Figure 20-3 (a) Examples of traditional ceramics from Colonial Williamsburg Museum exhibit. *Left to right*: Glazed clay pitcher, split mold made of plaster of paris for slip casting, samples of slip-cast/glazed pots, potter's wheel with sample bowls and platters. The ancient ceramic processes seen here used for traditional ceramics are still used today with some aspects automated. (b) Examples of modern ceramic artifacts—pottery, glass cup, flatware, and an incandescent light bulb produced by an automatic glass-blowing machine that can produce over 1000 bulbs per minute. Fluorescent lights use a phosphor coating on the inside of the sealed glass tubes, a ceramic that reemits white, soft, cool light.

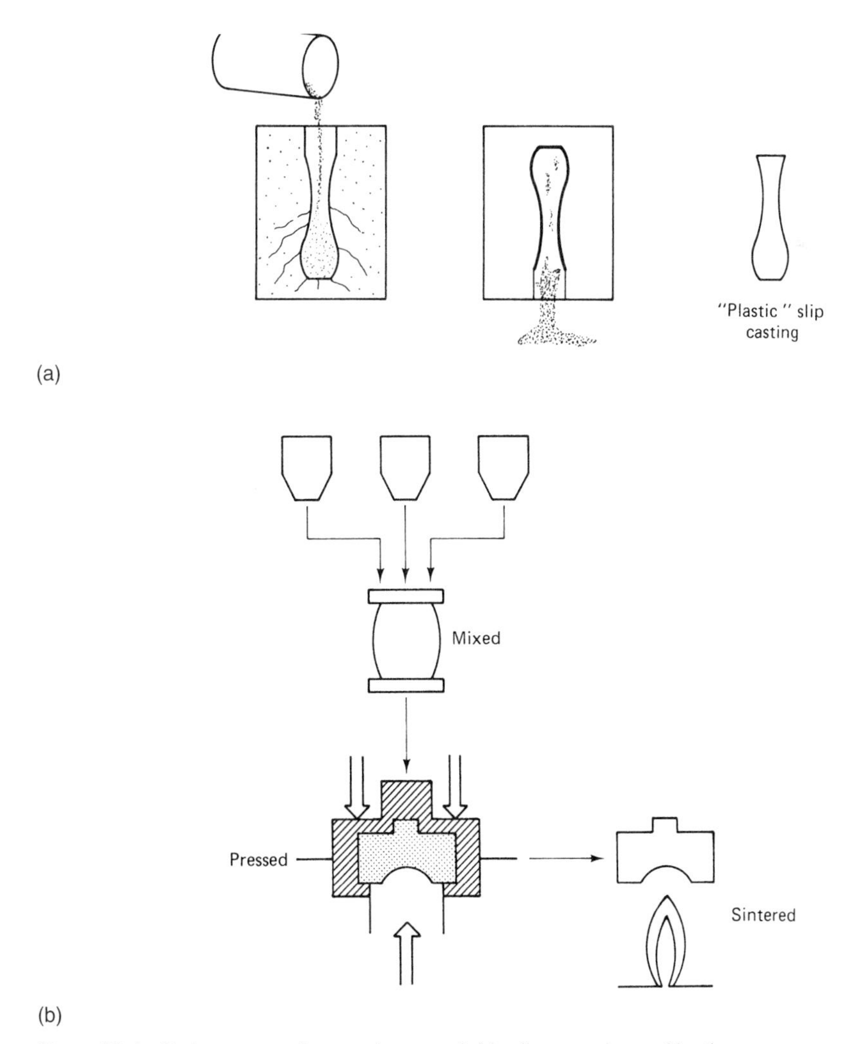

Figure 20-4 Basic processes for ceramics: ceramic blending, pressing, and heating.

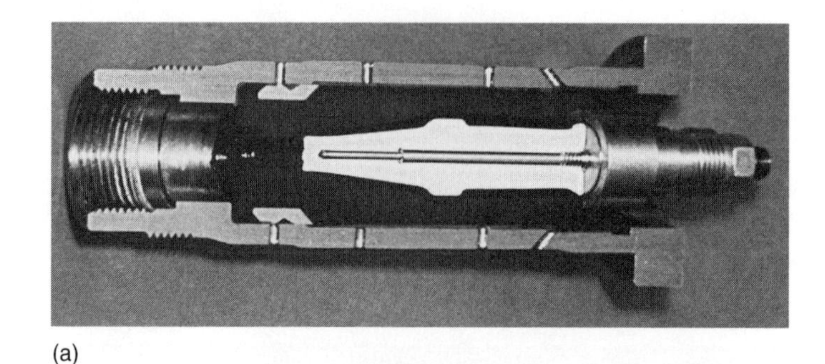

(a)

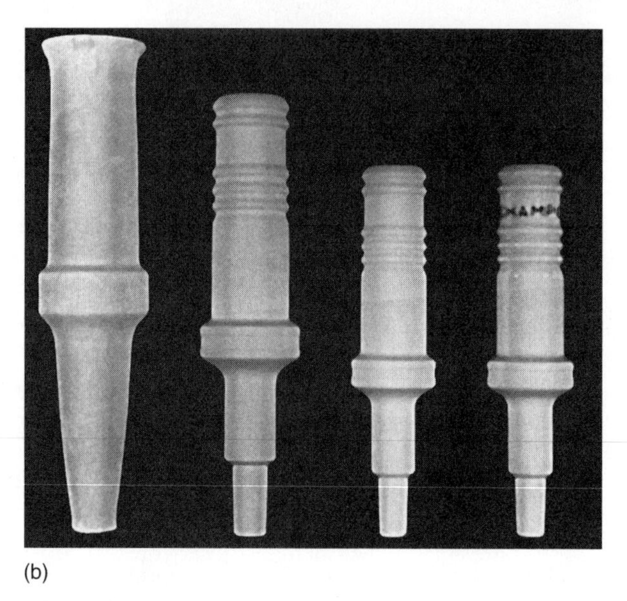

(b)

Figure 20-5 (a) Isostatic pressing mold in cross section showing insulator blank. (American Ceramic Society) (b) Stages of insulator body development (left to right): pressed "green" blank, turned "green" insulator, bisque-fired insulator, glass-glazed and decorated refired to finished insulator. (American Ceramic Society)

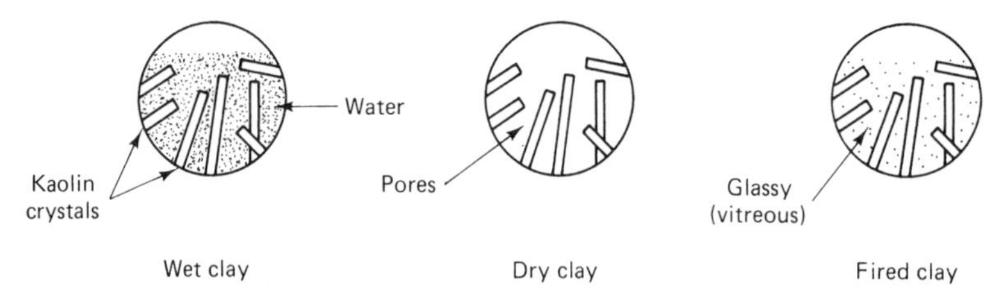

Figure 20-6 Stages of sintering (firing) clay.

be avoided, however, in advanced ceramics produced from ceramic particulates. Ceramic materials processed to produce refractories, ***structural clay*** products (brick, pipe), or ***whiteware*** (dinnerware, tile, pottery) can be considered traditional ceramics. In early times, whiteware, such as pots or vases, was formed by pressing one's hands against material supported on a rotating wheel. This very old process, called ***jiggering,*** is still practiced by people engaged in making hollowware with the aid of a potter's wheel. In industry, the same process is used in the fabrication of hollowware, but it is highly automated.

20.3 ADVANCED CERAMIC PROCESSING

Modern ceramics with the unique and more sophisticated properties needed in this highly technological age are referred to using a variety of terms. ***High-tech, high-performance, fine, new, value-added, engineered, engineering,*** and ***advanced*** are some of the terms seen in the literature. In this book we use the word *advanced* to describe such ceramics. The raw materials for advanced ceramics are not new or revolutionary. What is new are the many processes developed in recent years to produce them. For example, Figure 20-7 shows a hybrid sol–gel process for fiber optics.

The multistage fabrication process for ceramics includes powder production, powder conditioning, shaping (forming), drying, and densification (also called *firing* or *sintering*). Note the similarity of ceramic processing to the processing of powder metals (Module 12). Tungsten carbide (WC), a synthetic compound with extremely brittle characteristics, was developed in the late 1890s. In the 1920s, WC powders (particulate) were combined with cobalt (Co), a metal, acting as a binder and sintered, producing the first modern ***cemented carbide*** or ***cermet*** (ceramic plus a metal). A true composite material, it had the trade name of Carboloy (General Electric Co.). Cermets can be classified as ***large-particle composites.*** Relatively large-volume fractions of the particulate often exceed 90%. (See rule of mixtures in Module 25.) Carboloy was used primarily as a cutting tool for cutting cast iron and nonferrous metals. Later, other cemented carbides using titanium carbide (TiC), tantalum carbide (TaC), and chromium carbide (Cr_3C2) were developed. Their powders (particulate) were added to either Co or Ni as metallic binders to produce cemented carbides. Carbon black provided the carbon for these carbides, which composed 80–95% of the total weight of the composite materials. Another class of cermets are oxide based and are made from Al_2O_3 or MgO using Cr as a metallic binder. Some applications for these materials, which possess great hardness and wear resistance, are bits, dies, mining tools, valve liners, spray nozzles, gauge blocks, and indenters for hardness testers. The carbides are increasingly being used in ***ceramic matrix composites (CMCs),*** discussed in Module 25. These advanced materials use particulates, fibers, or whiskers embedded in a matrix of another ceramic. The source minerals for these advanced ceramics are metallic elements combined chemically to form oxides, carbides, nitrides, or borides. It is common practice to refer to oxides by a special name ending in *-ia* or *-a.* For example, Al_2O_3 is alumin***a,*** MgO is magnes***ia,*** Cr_2O_3 is chrom***ia,*** ThO_2 is thor***ia,*** and ZrO_2 is zircon***ia.***

Ceramic processing can be simply described as consisting of two steps: (1) a cold step, in which the ceramic part is formed or shaped into a "green" part or preform and (2) a hot step, in which the green compact is first subjected to heat to dry up any liquid phase formed during the processing and then subjected to higher heating, known as firing, sintering, or densification. The firing consolidates the particles and bonds the ceramic particles together permanently to give the ceramic its strength and other final properties. Some processes do not require drying, nor do some require the final step in most processing—that of machining to final specifications.

Some, but not all, engineering-ceramics fabrication processes begin with minerals ground into fine powders before they are transformed into useful products. The processing of the powder is itself a major undertaking, involving steps such as milling and sizing to produce a powder with the desired particle size, shape, and particle-size distribution. Either physical flaws (i.e., porosity or inclusions) or chemical flaws (i.e., unwanted or unknown chemical elements that combine to produce second phases such as liquids) can occur at any of these stages and may be difficult, if not impossible, to correct or eliminate. It must be stressed that the microstructure of the final ceramic product is dependent on the range of particle sizes and the inclusion of impurities in the starting materials. The use of clean rooms is evidence of the importance of cleanliness in the processing of advanced ceramics. We also know that the microstructure of any material determines the material's properties. Thus, much importance has been placed on developing low-technology techniques that make the conversion of mineral ores into industrial ceramic products more efficient with the expenditure of less energy and with ever-lower temperatures. Some advanced ceramics require that the powders used in the process be produced by chemical means, which allows greater control over both impurities and the final grain structure.

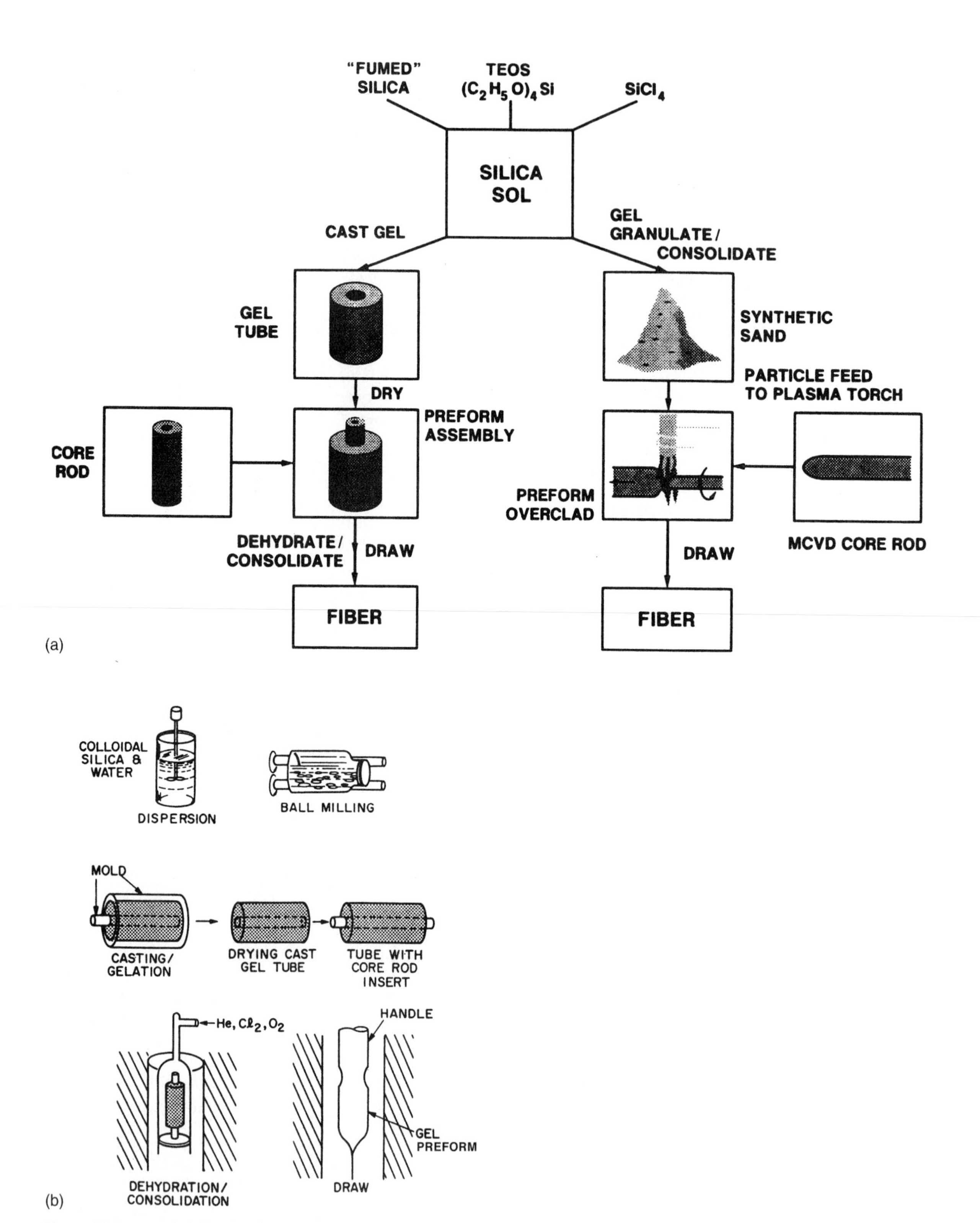

Figure 20-7 (a) A hybrid sol–gel strategy in which gel is cast as a tube to overclad a core rod. As an alternative, gel is granulated then fusion-sprayed on a preform to accomplish overcladding. (b) Sequence of steps to prepare fiber from core rod and cast gel body. (JME)

20.3.1 Chemical Processes

Sol-gel processing of ceramics and glass is a chemical process for producing powders for use in advanced ceramics. This process and chemical-vapor-deposition techniques (discussed later) provide the means for designing and controlling the composition and structure of a ceramic at the molecular level. Their benefits include increased purity, lower processing temperatures, finer grain size, and more homogeneous microstructures. To gain further understanding of these techniques requires a brief description of some terms. A ***sol*** is a stable dispersion in a liquid of particles less than 0.1 μm in diameter. In more general terms, it refers to a mixture of solid colloidal particles in a liquid. A ***colloidal*** substance consists of very small particles that remain in suspension in a liquid for some time, without settling. When a sol loses liquid, it becomes a gel. ***Gels*** are noncrystalline solids formed by chemical reaction rather than by melting. To describe them in other terms, gels are a jellylike substance formed by the coagulation of a sol (causing a liquid to become a soft, semisolid mass). Through a chemical reaction, the sol–gel process turns the solution of organometallic compounds (sol) into a gel by the formation of polymerlike bonds. The final step is raising the temperature to convert the dehydrated gel into a ceramic material. Such temperatures are much lower (less than 900°C) than those used in the conventional processes of pressing, casting, or plastic forming, which involve high-temperature sintering. Applications for sol–gel processing include preparation of glasses for coatings, optical fibers, fabrication of glass ceramics, forming of oxide matrices for fiber composites, coatings on fibers to lessen the interfacial bonding between the fibers and the matrix to increase the toughness of reinforced fiber composites, and monolithic ceramics. Sol–gel and reaction-bonded ceramic processes, ***green manufacturing,*** are low-temperature, relatively high-energy processes that produce many advanced ceramic materials without the many difficulties (i.e., chemical interactions and residual stresses) associated with the use of high-temperature processes. Figure 20-8 contrasts them with chemically bonded ceramics (CBCs) such as cement and concrete, discussed later in this module.

As mentioned, ***gelcasting*** is a new ceramic-forming process in which a slurry (a fluid suspension) of ceramic powders in a solution of organic monomers is cast in a mold. The monomer mixture is polymerized in the molds to form gelled parts. Both monolithic and composite ceramic parts and complex-shaped and near-net-shaped parts that require minimal final finishing are produced. Figure 20-9 show a gelcasting process flowchart, and Figure 20–10 show an application as well as a list of some materials that are gelcast.

The formation of the microstructures of ceramic materials is the result of chemical changes, interactions, or reactions. Chemically bonded ceramics (CBCs) can be distinguished by the manner in which consolidation of the components is brought about. In one process, consolidation results from sintering (densification) or fusion. In another, consolidation results from cementation by chemical reaction and bond formation. Figure 20-11 is a graphical representation of these two consolidation processes. Traditional sintering, depicted in Figure 20-11a, is a solid-state diffusion process that results in shrinkage; Figure 20-11b illustrates that the volume between reactants and products can be conserved, resulting in net- or near-net-shape forming. Generally, these chemical processes take place when the compacted powders are heated to their sintering temperature. On other occasions the compacted powders react with a gas or liquid.

Reaction-bonded silicon carbide (SiC), or reaction-sintered SiC, is one such process. A powder mixture of Si and C is formed into the desired shape and exposed at high temperature to Si vapor or molten Si. The reaction with Si produces SiC, which bonds to the original SiC compact, and the excess Si infiltrates the compact's pores at a controlled rate. The result is a nonporous composite (SiC/C) with varying degrees of strength and elasticity. By using carbon fibers to form laminae of fibers woven within the molten Si, a range of fiber composites can be developed having the characteristics desired.

A similar process is the ***reaction-bonded Si_3N_4 (RBSN)*** process, which starts with a compacted shape of Si in a furnace under an atmosphere of N, N/He, or N/H gas. Heated initially to about 1250°C, the N permeates the compact, forming Si_3N_4. The temperature is then raised to just below the melting temperature of Si to increase the reaction rate. After about one week in the furnace, the process is completed, producing a ceramic that has good dimensional

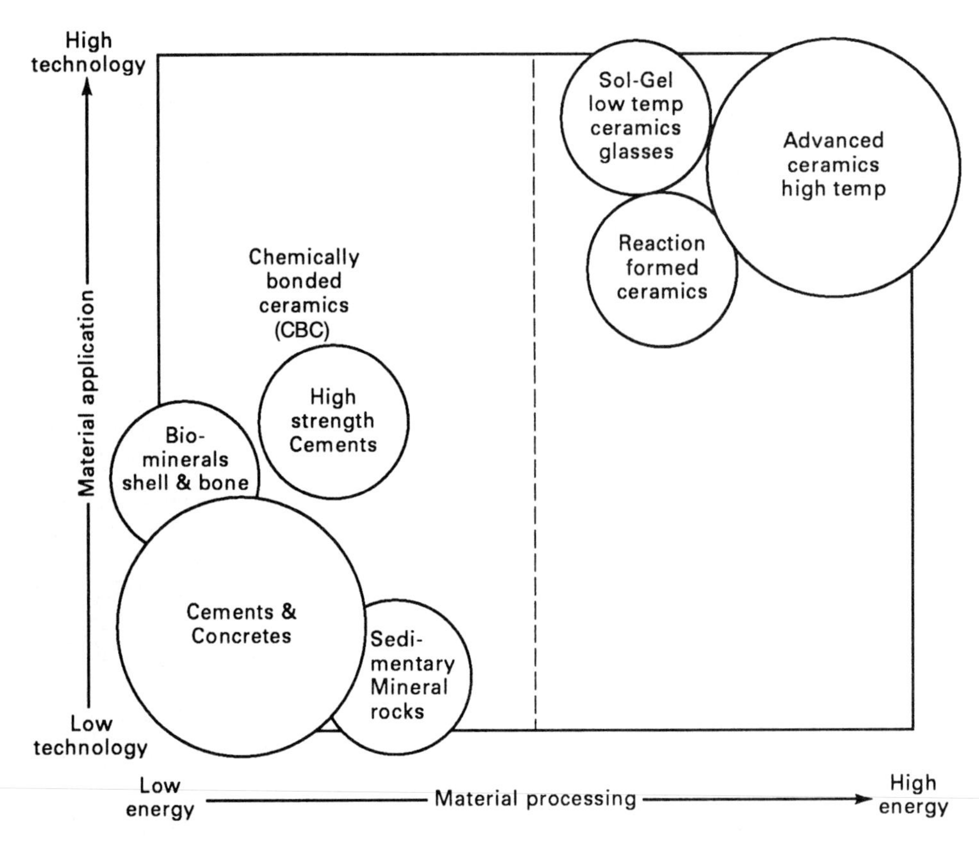

Figure 20-8 The compromise between using the world's materials for an ever-increasing number of technological materials applications and the need to reduce the energy required to process them. The greater the need for complex technology in processing, the more energy is consumed. An underlying theme in this graphic is the never-ending search to duplicate the methods that nature uses to accomplish many of our technological goals. (JME)

stability, thermal-shock resistance, and thermal conductivity. Its one disadvantage is its low oxidation resistance due to its interconnected porosity (open pores). ***Oxidation*** describes the chemical reaction of oxygen when it combines with another element. If sufficient oxygen is present, it will react with the surface of the silicon nitride, producing silicon dioxide (SiO_2), which forms a protective layer at the surface that prevents further reaction. This is known as ***passive oxidation***. If not enough oxygen is present, the gas silicon monoxide (SiO) is formed. Known as ***active oxidation,*** this type of oxidation can be continuous, resulting in complete destruction of the ceramic. Any oxygen-poor environment, such as outer space, would be conducive to such attack. Other factors, such as temperature and porosity, influence the degree of reaction. By using oxidation initially, the surface may be coated with a layer of SiO_2, which will protect the interior from further oxidation attack. This is known as *flash oxidation,* which is useful for protecting $MoSi_2$ heating elements.

20.3.2 Melt Processing

Most glasses and many ceramics are melted before being processed. Abrasives, glass shapes, and bricks of Al_2O_3 are made by casting from a melt. Glass fibers for fiberglass composites are drawn or spun from a molten state. Molten droplets made from ceramic particles are sprayed onto surfaces through a variety of spray techniques that provide corrosion, wear, and thermal protection to a wide variety of metal, plastic, and ceramic products. Figure 20-12 shows photographs of various parts made of silicon nitride. ***Coatings***—barrier, inhibitive, sacrificial, organic, inorganic, and metallic—are used to control corrosion as a part of surface engineering.

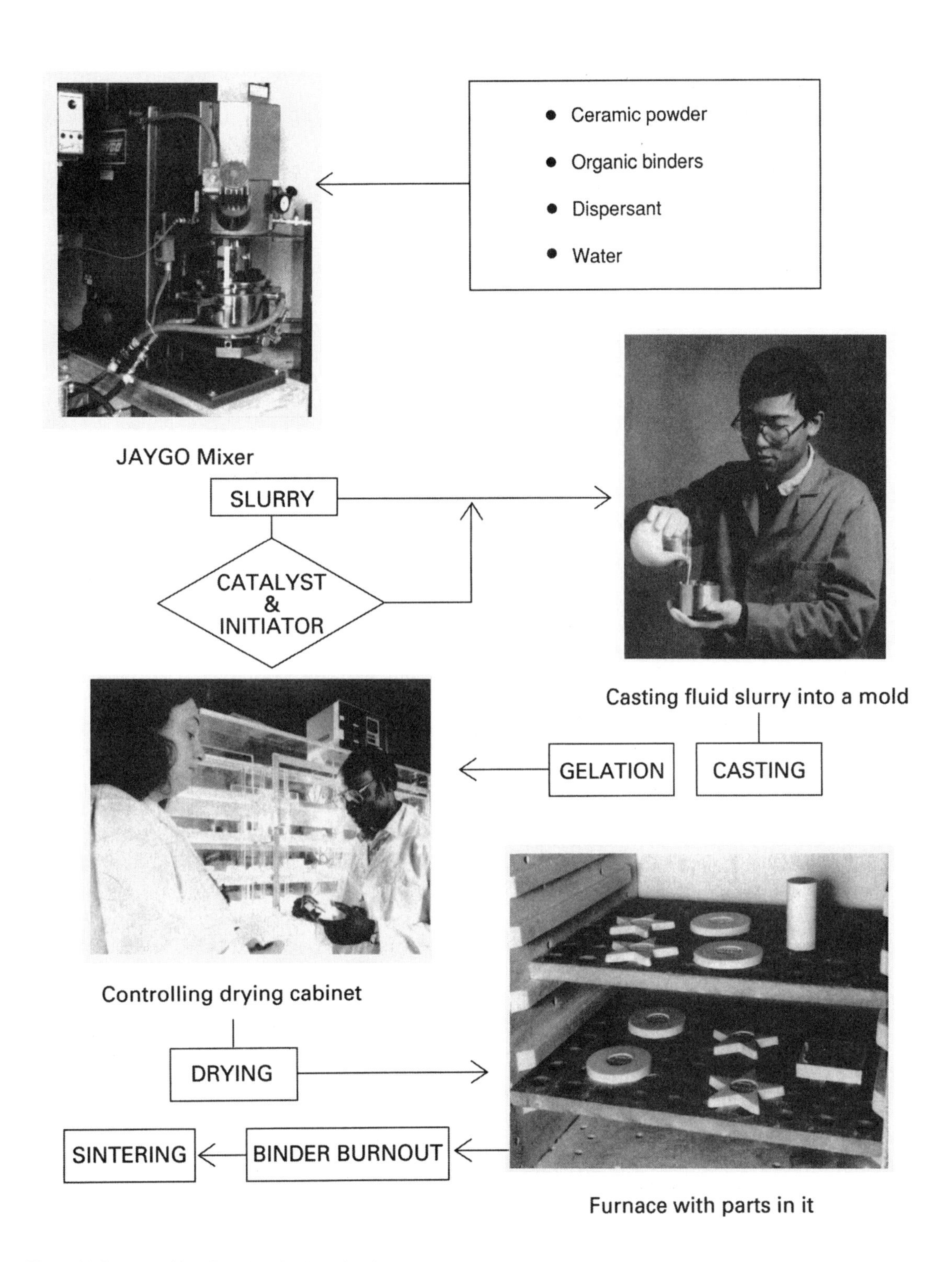

Figure 20-9 A graphical flowchart for the gelcasting process.

20.3.3 Producing Single-Crystal Materials

Single-crystal materials solidified from a melt are developed to obtain materials with special properties not possessed by polycrystalline materials. Such materials are formed by several different techniques, of which two are mentioned here. Figure 3-62 showed a typical application of this technique as it relates to metals. In the semiconductor industry, great advances

Gelcasting Process Applications

Ceramic gears (Alumina)

List of Materials gelcast

Monoliths

- Alumina
- Silicon
- Silicon carbide
- Zirconia
- Fused silica
- Sialon

Composites

- Nicalon fiber-reinforced reaction-bonded silicon nitride
- Alumina-zirconia

Ceramic rotor fired (Alumina)

Figure 20-10 Some applications for the gelcasting process along with a partial listing of materials that could be selected for their manufacture.

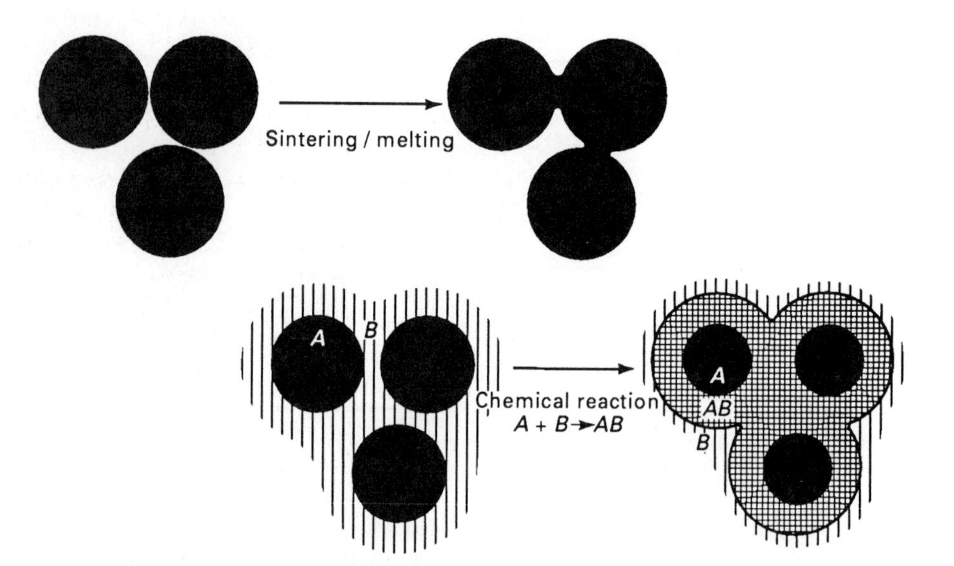

Figure 20-11 How the densification of particles is accomplished under (a) conventional sintering and (b) reaction bonding sintering processes. Conventional sintering depends, in the main, on high temperatures (over 2000°C) to cause solid-state diffusion, whereas reaction bonding sintering requires temperatures around 1400°C. (JME)

(a)

(b)

Figure 20-12 (a) Silicon nitride (Si_3O_4) rolling bearings have many advantages over high-performance steel bearings: lower wear, lower heat generation, reduced noise and vibration (RNV), and excellent performance under marginal lubrication. Applications of ceramic hybrid bearings include machine tool spindles, pump mainshaft bearings, dental drill shaft bearings, in-line skate and mountain bike hub bearings, and the space shuttle main engine oxygen fuel pump. (Produced by Norton, East Grandy, CT) (b) Turbocharger rotors and valves used in automobile engines. The development of silicon nitride gas turbine components in 1988 in Japan permitted the use of turbocharger rotors in passenger car engines because they weighed less than metal rotors, permitting the driver to accelerate more quickly without feeling a lag in power. Ceramic valves have provided consistent excellent performance. Their relative cost of manufacture keeps them from widespread use. (American Ceramic Society)

have been made in growing silicon single crystals, which are now produced with few, if any, structure dislocations. An *edged-defined, film-fed growth (EFG) method* uses alumina with a molybdenum die that is immersed in molten alumina. The molten alumina is wicked to the top of the die, where it comes in contact with a seed crystal that is slowly pulled away to start the growth process. The growing crystal assumes the shape of the cross section of the die. Plates, tubes, and various-size filaments are produced in this way. For example, some continuous filaments are produced at a rate of 62 m/hr. A second technique is the *heat exchanger* or *gradient furnace,* which produces larger single-crystal components. A seed crystal is kept cool by a helium flow in a crucible filled with molten alumina. The temperature of both the melt and the seed crystal is controlled such that the contents of the crucible grow into a single crystal.

20.3.4 Vapor Processing

Unit 5, Surface Science and Engineering, provided an introduction to surface modification, including vapor deposition. ***Vapor processing*** involves heating a solid substance to a temperature that transforms the solid into a vapor, which is then deposited onto a surface. When the vapor contacts a cold substrate or when a gas reacts on contact with a hot substrate, the solid produced on that surface solidifies so quickly that the atoms fail to form a crystalline structure. Such noncrystalline, fine-grained, nonporous coatings are difficult, if not impossible, to produce by other means. ***Physical vapor deposition (PVD), chemical vapor deposition (CVD), chemical vapor infiltration (CVI), metallo-organic chemical vapor deposition, molecular beam epitaxy (MBE), liquid-phase epitaxy (LPE), sputtering,*** and ***ion plating*** are some of the methods that use this technology. As a group these methods represent a wide range of ***surface modification*** techniques used to achieve surface properties in a controlled manner for the ***engineered materials*** needed by today's advanced technologies. In addition to providing high-temperature materials for turbine blades or metallic coatings for plastic parts, these techniques are now involved in producing high-quality optical glass for fiber-optic communications as well as bulk forms in single shapes with structures developed on an atomic scale. These new materials that have designed structures are known as ***artificially structured materials.*** Until recently, most developments in this field were somewhat restricted to the semiconductor industry, but as each day passes, new applications are found for such materials.

PVD vaporizes a target material by using a high-energy source such as ions or electrons. The vapor is deposited atom by atom on a substrate, building up the required layers of atoms. ***Laser ablation and deposition*** uses a pulsed laser process to produce high-temperature superconducting thin films. This new process is also based on PVD principles. Using multiple target pellets, each containing the required chemical composition, a thin film is reproduced with an exact composition. The thin-film coatings so made, typically of zirconium, chromium, and carbon, are often used to solve wear problems. Before a particular coating is used, a complete tribological profile of the conditions in which the part is to operate must be determined. Recently, titanium carbonitride (TiCN) has been replacing titanium nitride (TiN) and chromium nitride (CrN) coatings on ***cutting*** and ***forming tools***. Such coatings are applied in thicknesses of 1 to 4 μm by a low-temperature, reactive ion plating PVD process. Figure 20-13a shows a schematic of the process, and Figure 20-13b is a photomicrograph of the edge view of the coating and substrate. Such coatings (Figure 20-13c) combat abrasive wear and adhesive welding when metals like stainless steel, cast iron, brass, titanium, and aluminum alloys are machined. The TiCN coating, for example, is stable to temperatures of 400°C with a hardness of 3000 HV (25 g load).

Another PVD-based process is the electron-beam vacuum evaporation process, developed primarily to coat turbine blades. This method generally produces a higher-quality coating than that produced by its competitor—the plasma spray technique—and at a cost advantage. Two other versions of the electron-beam evaporation method—the arc method and the plasma-assisted evaporation method—are also finding many applications.

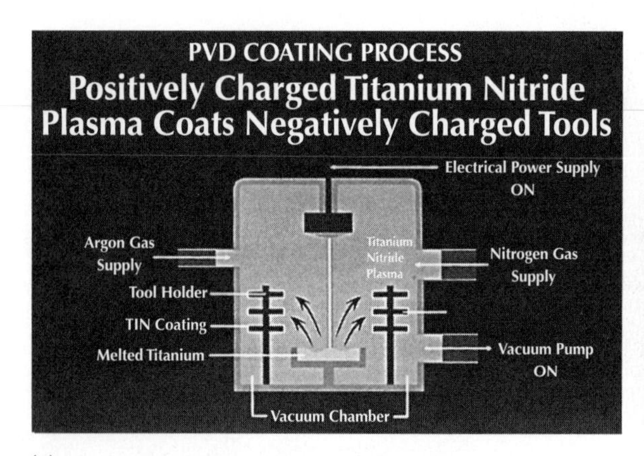

(a)

(b)

(c)

Figure 20-13 (a) Physical vapor deposition (PVD) process used for thin-film ceramic coatings. (b) Photomicrograph showing edge view, with a uniformity of 1×10^{-4} in., of a thin coating of titanium nitride (TiN) on high-speed steel substrate. (c) TiN and other thin-film coatings, sometimes as many as 13 layers, are used on cutting tools and tool inserts.

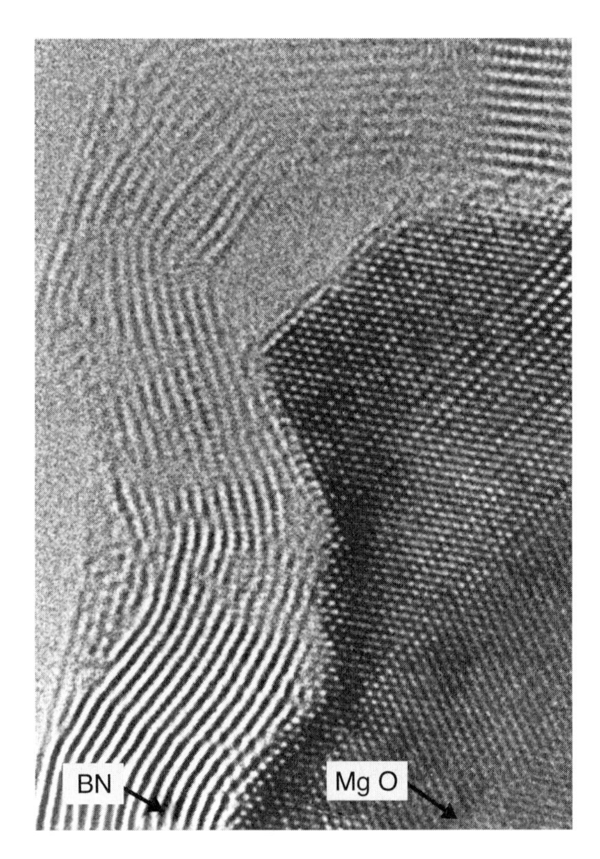

Figure 20-14 A photomicrograph of a layer of boron nitride (BN) deposited on a substrate of magnesium oxide (MgO)—magnification 7,200,000×. (ORNL)

Chemical vapor deposition (CVD) is a process in which a mixture of gases is passed across a heated surface. The temperature causes the gases to react or decompose to form a solid upon contact with the surface. In an example taken from the literature, a TiN coating that protects a surface from wear is produced by CVD with a mixture of gases, $TiCl_4$–N_2–H_2, heated to a deposition temperature between 900° and 1000°C. CVD can be used to manufacture parts such as a boron–nitride (BN) crucible with wall thicknesses of several millimeters. Figure 20-14 is a micrograph of a boron–nitride layer deposited on a substrate of magnesium oxide (MgO). In the composites field, CVD is used to coat fibers that will be embedded in a metal or ceramic matrix to increase the toughness of the fiber composite (discussed in Section 19.3). For example, silicon fibers are coated with carbon before they are used in a matrix of silicon carbide. This coating increases the toughness of the composite by reducing the interfacial bonding strength between the fibers and the matrix material. Doing so allows the fibers to debond from the matrix materials and pull out before breaking. All of these displacements require energy, decreasing the probability of catastrophic failure from a single mechanical overload. Figure 19-21 showed a toughness model sketch (a) and a micrograph (b) of fiber bridging and pullout as fibers resist the propagation of a crack. A ***near frictionless carbon (NFC)*** coating many times slicker than PTFE (see Gore-Tex) developed by Argonne National Laboratory has a coefficient of friction less than 0.001 when measured in a dry nitrogen atmosphere. This is 20 times lower than the previous record holder, molybdenum disulfide. Thin NFC films can be deposited on virtually any substrate by vacuum deposition without risking damage to heat-sensitive materials. The ultrahard coatings have a host of ideal properties that make them suitable for a variety of applications requiring exceptional wear resistance and durability, fast processing with large amounts being produced in only a few hours, adherence to many surfaces including polymers, finished products requiring no secondary processing, lower cost, and improved environmental compliance.

Synthetic diamonds were first made in 1950 by General Electric using extremely high pressures and temperatures. The same technology is in use today. A more recent development is the making of ***synthetic diamond films*** and coatings using low-pressure processes. One of

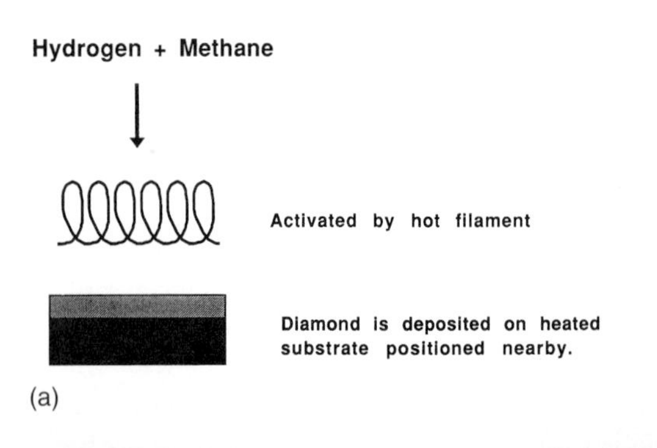

(a)

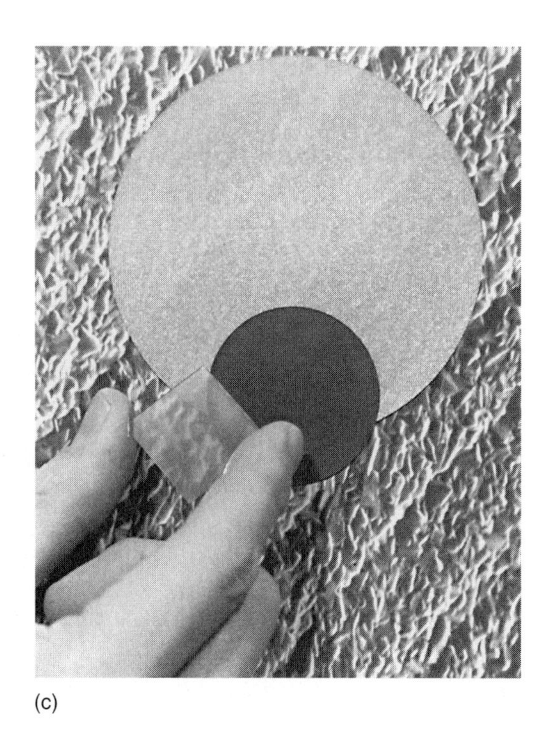

(c)

(b)

Figure 20-15 (a) Sketch of CVD process for making diamond film. A mixture of hydrogen and a hydrocarbon gas, usually methane (CH_4), is heated to about 2000°C using an electric filament or other heat source. At this temperature, the gases disassociate, freeing the hydrogen and carbon atoms. The carbon atoms are deposited on a substrate heated to around 1000°C. Most other CVD processes are variations of this technique. (b) The chamber used in the CVD process. (c) A diamond film micrograph magnified more than 100× serves as background for several grades of thick-film CVD diamond, including optical (white and clear) and cutting tool quality (black). (ORNL and Norton Diamond Film)

many vapor deposition techniques used is CVD, which deposits the film from a gaseous mixture (CH_4–H_2) onto a substrate, such as molybdenum, silicon, ceramics, and natural diamond. The largest obstacle to successful film growth with this method is the need to heat any substrate to temperatures above 600°C. To do this, hydrocarbon gases, of which methane is most common, are used as the precursor gas. There are various methods for forming the vapor, such as microwave activation, thermal activation, and plasma-jet processes. Figure 20-15 shows a simple sketch of the process, a photograph of a typical reactor, and several grades of diamond film. Raman spectroscopy is used to detect diamond; it consists of looking for a single, well-defined peak that is distinctive for the diamond bond.

Natural diamond has a set of unique properties: It has the highest hardness reading on the Mohs scale; the elastic modulus is about five times as large as most steels, at 160×10^6 psi; its tensile strength is higher than that of steel but lower than that of ceramic fibers, which can go as high as 10×10^6 psi; natural diamond's 0.1 coefficient of friction ranks about the lowest for a solid material; the coefficient of thermal expansion for synthetic diamond is only about $0.8 \times 10^{-6} - C^{-1}$, a consequence of its strong covalent bonds; and its electrical resistivity is about

10^{14} · cm. Diamond, one of the two forms of carbon, possesses another great characteristic in addition to those mentioned—its high thermal conductivity, which is 5 to 10 times greater than metals.

To obtain a better picture of its hardness, consider the following statement: It takes 500,000 psi of pressure to indent metals and 0.75 million psi will indent the softest ceramic, whereas it requires more than 10 million psi to indent diamond. Diamond ***cutting tools*** are produced via P/M (powdered metals processes) by hot pressing to full density a mixture of diamond grit and metal powders in a graphite mold. The metals serve as a binder to bond and hold the diamond in the tool during the cutting process. If synthetic diamond could be produced in sufficient quantities at an affordable price, it would replace silicon as a semiconductor material.

Diamond film—*http://www.nortondiamondfilm.com/*

A comparison of the various forms of diamond is interesting. Natural diamond is a single crystal made up of 99% carbon 12, with 1% carbon 13 isotope. General Electric's synthetic diamond contains more atoms per cubic centimeter than does natural diamond and can be made to any desired C_{12}/C_{13} ratio. CVD polycrystalline natural diamond has no binder, whereas polycrystalline sintered diamond contains up to 10% binder. ***Polycrystalline diamond (PCD)*** and ***polycrystalline cubic boron–nitride (PCBN),*** referred to as ***superabrasives,*** are finding more applications in the turning of hard materials, once the sole province of the grinding field. Using tipped turning inserts from the materials mentioned, material removal rates were 4 to 10 times greater than rates for grinding with cleaner surfaces. As with all materials, diamond-film properties can vary significantly, not only with deposition techniques but with substrate temperature, precursor-gas composition, and other factors that can or cannot be controlled in the process. Current research is focused on producing diamond films at lower cost and lower deposition temperatures, with larger film areas (present limits are 100 mm square and about 1 mm thick) and better adhesion between the film and the substrate.* Applications for diamond films are as cutting-tool inserts, wear surfaces for dies, heat sinks for electronics, and infrared (IR) and microwave windows (diamond is transparent to both radiations).

Chemical vapor infiltration (CVI) is somewhat related to CVD, the major difference being that the deposition occurs inside a porous preform rather than only on a surface. CVI originated in efforts to densify graphite bodies by infiltration with carbon. Infiltration is the densification of a porous preform by filling the pores with liquid, vapor, polymer, or a sol. The CVD-type coating grows with continued deposition to form the composite matrix. A wide variety of matrix materials are used, including borides, carbides, nitrides, and oxides. A great advantage of this process is that it allows the formation of high-melting-point materials such as TiB_2 (T_m = 3200°C) at relatively low temperatures, which reduces the risk of damage to the fiber reinforcement. Also, CVI does not use high pressures, which translates into less residual stress in the final composite materials.

CVI is now used to produce carbon–carbon composites as well as ceramic–ceramic composites (see Module 25 for additional details). When fabricating fiber-reinforced ceramic composites, such as SiC high-modulus ceramic fibers with diameters of only 10 μm or less in a matrix of dense alumina, the fragile fibers cannot withstand the high stresses and handling of more conventional processes. With its low stress and gaseous precursors, CVI can infiltrate around the bundles of continuous fibers and deposit the matrix material around them, filling in all pores and thus encapsulating the fibers with the matrix material and knitting them together in a strong interfacial bond. The coating of the fibers is a slow molecular process that builds up molecule by molecule. Figure 20-16 shows (a) a sketch of the CVI furnace, (b) a micrograph of SiC fibers reinforcing a SiC matrix composite material, and (c) a dramatic demonstration of the increased toughness of reinforced CVI ceramic materials. (Liquid-phase epitaxy and molecular-beam epitaxy methods are discussed in Section 27-5.)

*EMSET CD-ROM cited in References & Related Readings provides experiments.

FIBER-REINFORCED CERAMIC COMPOSITES HAVE BEEN FABRICATED USING A FORCED CHEMICAL VAPOR INFILTRATION PROCESS

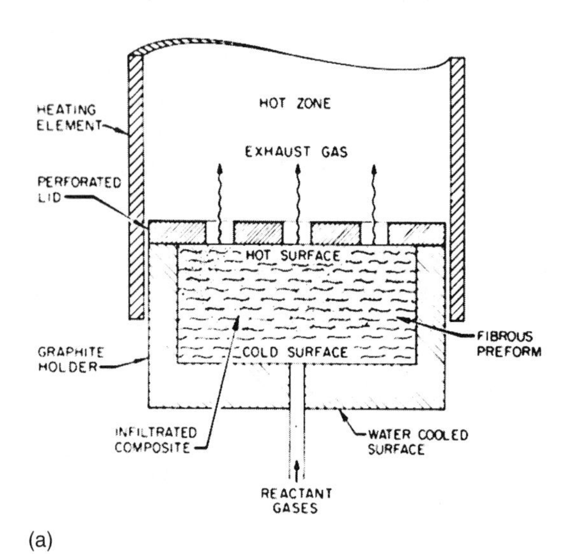

(a)

(b)

(c)

Figure 20-16 (a) Forced chemical vapor infiltration (CVI) process. (b) Photomicrograph of SiC ceramic produced by CVI. (c) Dramatic demonstration of the toughness obtained in ceramics as a result of the reinforcement achieved in CVI. (ORNL)

Reactive melt infiltration is a new process that produces silicon–carbide-based ceramics and composites faster and more economically than do processes such as CVI or hot isostatic pressing (HIP). In comparison to CVI, this process takes minutes instead of days and costs about half as much. In the reactive melt infiltration process, a microporous carbon preform is infiltrated with molten silicon or a molten silicon alloy. The liquid and solid react to form a solid ceramic or composite. If the end product is to be a composite, the preform contains fibers. Products can be formed into complex shapes with full density, controlled microstructures, and tailored thermal and mechanical properties. A composite material consisting of a silicon–carbide matrix reinforced by silicon–carbide fibers has been produced. This ability to produce a composite material of complex shape and nearly final dimensions prepares the way for the production of silicon–carbide cutting tools.

Ion plating and sputtering involve plasmas. A ***plasma*** is a cloud of gas at high temperature made up of ions and electrons. Extremely high temperatures in the plasma can be reached because the cloud of gas is electrically conductive. This allows electrical heating (i.e., radio-frequency induction or direct contact with a metal electrode that supplies electrical current). The various factors involved in maintaining a plasma are somewhat complex, but in this brief treatment of the subject, they may be said to involve the temperatures of the ions in the plasma, energy input (heat), and the gas pressure. At high pressures, the plasma temperature can reach several thousands of degrees Celsius. Under these conditions, a powdered ceramic can be fed into the plasma, producing tiny droplets that can splatter onto a substrate as in plasma-arc spraying. At lower pressures, the gas atoms in the plasma remain cold and permit high electron temperatures to produce chemical reactions in the cold gas, the products of which can be deposited on a substrate.

Sputtering is a basic coating process that uses a plasma, a target positioned above the substrate that is to be coated, and a sputtering gas. The target acts as the negative electrode, and the substrate acts as the positive electrode. The positive ions (usually argon ions) in the plasma are accelerated toward the target at high speed, resulting in the knocking off of desired atoms from the target, which are subsequently deposited on the substrate. This is known as the *sputtering action.* There are numerous variations to this basic process that depend, in part, on the type of substrates involved, how the coating atoms are introduced, the manner in which the plasma is maintained, or the differences in the applied voltages between the target and the substrate. One final comment is a reminder that the temperature of the substrate is extremely important to the final coating achieved. A recent innovative process for depositing diamondlike surface films on a substrate is the *microwave-assisted chemical vapor deposition (MACVD)* process. It can handle wafers up to 4 in. in diameter with a microprocessor-based programmable process control system.

20.3.5 Directed Metal Oxygen Process

The reaction of a molten metal with a gas to form near-net-shape metal and ceramic–matrix composite parts is exemplified by a Lanxide Corporation process. In the directed metal oxidation process, DIMOX™, ceramic matrices are grown around preplaced fibers (reinforcements) to produce a ceramic–matrix composite (CMC) to net or near-net shape without limitations on size or shape. Figure 20–17 is a schematic of the DIMOX™ process showing an example of

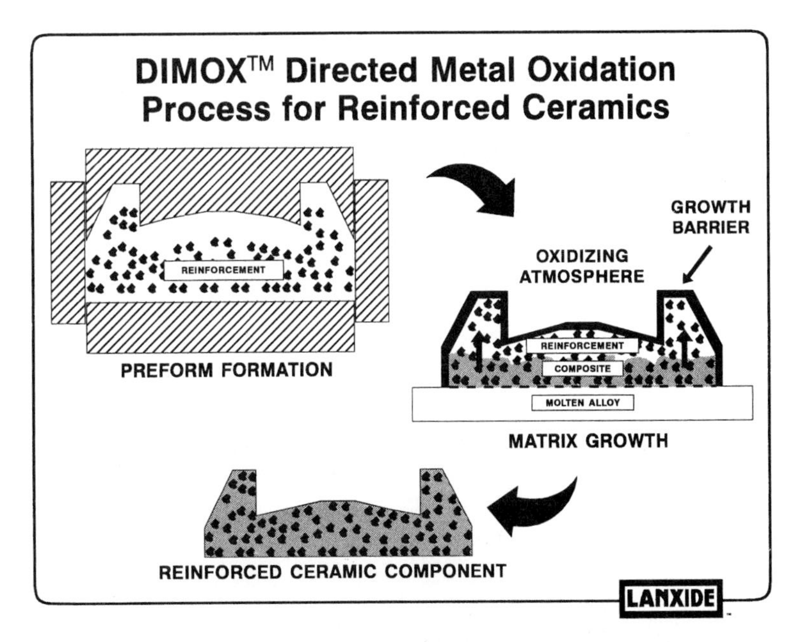

Figure 20-17 DIMOX™ process, demonstrating preform infiltration technology with unidirectional growth. A filler preform coated with a growth barrier is placed on top of a solid metal ingot in a refractory container. Temperatures (850°–1300°C) initiate growth, which proceeds through the filler until either the molten metal is consumed or the growth reaches the barrier. No shrinkage occurs.

unidirectional matrix growth. The preform is shaped like a filler preform of the same size and shape as the desired part. Preforms are made of particulate, which can be produced by any of the ceramic green-body-forming methods, such as pressing, isostatic pressing, slip casting, extrusion, or injection molding. The preform is covered with a special barrier material that limits matrix growth, ensuring that the part shape matches that of the preform. The parent alloy and preform are then heated to the growth temperature, at which the metal begins to oxidize rapidly in the presence of a gas oxidant. Oxidation occurs outward from the metal surface and into the preform such that the reaction product becomes the matrix surrounding the reinforcement in the preform.

Oxidation of a metal generally forms a solid ceramic layer that limits further reaction. However, in one method, additives such as Mg or Si promote wetting of the ceramic by the molten aluminum alloy and reduce the stability of grain boundaries in the ceramic, which allows sustained reaction of the metal with the gas oxidant (oxygen). The metal is continuously made available to react with the gas oxidant, providing a growth rate that is independent of the thickness. In other words, the molten metal is drawn through its own oxidation product. The growth of the matrix into the preform occurs with little, if any, change in dimensions. This characteristic sets this process apart from traditional ceramic processing, with its inherent densification shrinkage.

Figure 20-18 is a photomicrograph of the microstructure of a CMC made by the DIMOX™ process, composed of a matrix of Al_2O_3 reinforced by woven SiC fibers with some unreacted metal still visible. In low-temperature applications, the residual metal provides added toughness. For high-temperature use, that amount of metal can be removed, leaving a small amount of porosity, or treated to prevent melting or further oxidation during service. Figure 20-19 is a photograph of a Lanxide CMC specimen bar, consisting of an alumina matrix reinforced with CVD-coated SiC fibers, undergoing a flexure test. Table 20-1 lists some mechanical properties of SiC_p/Al_2O_3 CMC for use at high temperatures. The subscript "p" in the notation for the composite material refers to platelets or particles. Other abbreviations, "W" for whiskers and "f" for fibers, are used here.

Shape forming can be done by metal-shape replication and/or infiltration into a preform of the desired filler (reinforcement). Metal-shape replication used in making screw threads is shown in Figure 20-20. The shaped parent-metal cast or machined part is inserted into the central portion of the figure surrounded by the silicon–carbide bedding material. The growth will extend outward from the metal-shape preform toward the barrier bedding. In the DIMOX™ process, the silicon–carbide bedding materials are incorporated into the ceramic matrix by the directed oxi-

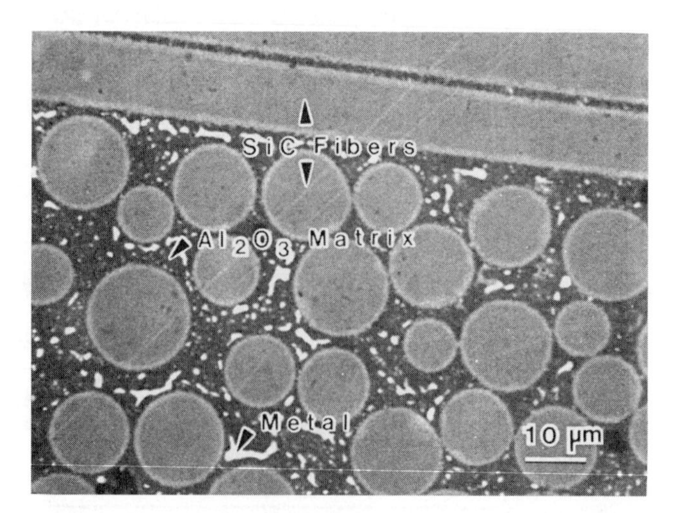

Figure 20-18 A photomicrograph of a typical microstructure of CMC and MMCs made by Lanxide's DIMOX™ process. Al_2O_3 reinforced by woven SiC fibers is shown with some unreacted metal also visible. Impurities in the uncoated fibers reacted with the molten metal to form a reaction layer on the fibers. In practice, fibers are given a protective coating prior to matrix formation, as found in continuous fiber ceramic composites (CFCC). (Lanxide)

Figure 20-19 A photograph of an alumina composite containing CVD-coated Nicalon® fibers produced by the Lanxide DIMOX™ process undergoing a four-point bend test. Despite a deflection greater than 1 mm, sufficient to fail an unreinforced ceramic catastrophically, the bar continues to support the load. (Lanxide)

TABLE 20-1 MECHANICAL PROPERTIES OF SIC_P/AL_2O_3 COMPOSITES PRODUCED VIA DIRECTED METAL OXIDATION (DIMOX™) PROCESS

Property	Degrees Celsius	90-X-008
Flex strength MPa (ksi)	25	179(26)
	1000	141(21)
	1400	155(22)
	1550	130(19)
Toughness (MPa · m-)	25	5.5
	1000	4.0
	1400	3.0
	1550	3.8
Young's modulus (GPa)	25	391
Shear modulus (GPa)	25	153
Poisson's ratio		0.28
CTE (ppm/·C)	25–1400	6.0
Thermal conductivity (W/m-K)	25	116
	500	54
	1000	31
Bulk density (g/cc)		3.30
Carbide loading (v/o)		73
Residual metal (v/o)		6

dation of molten aluminum alloy. The matrix reaction product occupies the spaces between the filler without displacing it, so that densification is achieved without shrinkage. Of course, for growth to occur, the filler used—be it fiber, whisker, particle, or platelet—must be compatible with the gas atmosphere and the parent metal. Coating of the filler materials is used to improve compatibility. Lanxide's PRIMEX™ process, a pressureless molten-metal-infiltration technique, is discussed in the composite module, Module 8. Both of these Lanxide processes employ bulk processing techniques that have been traditional in foundries and ceramic processing for many years. This is in contrast with the expensive equipment and materials encountered in the production of many other advanced ceramics.

20.3.6 Finishing (Machining)

In some instances, the shaped compact (preform) has not acquired the final shape or met the tolerances specified. Machining of the surfaces of ceramics is done to meet dimensional tolerances, improve surface finish, or remove surface flaws. It is to be assumed that all machining will introduce flaws of some size on the surface of ceramic parts. The machining of fine-grained ceramics such as Si_3N_4 significantly reduces their strength—more than that of ceramics of large

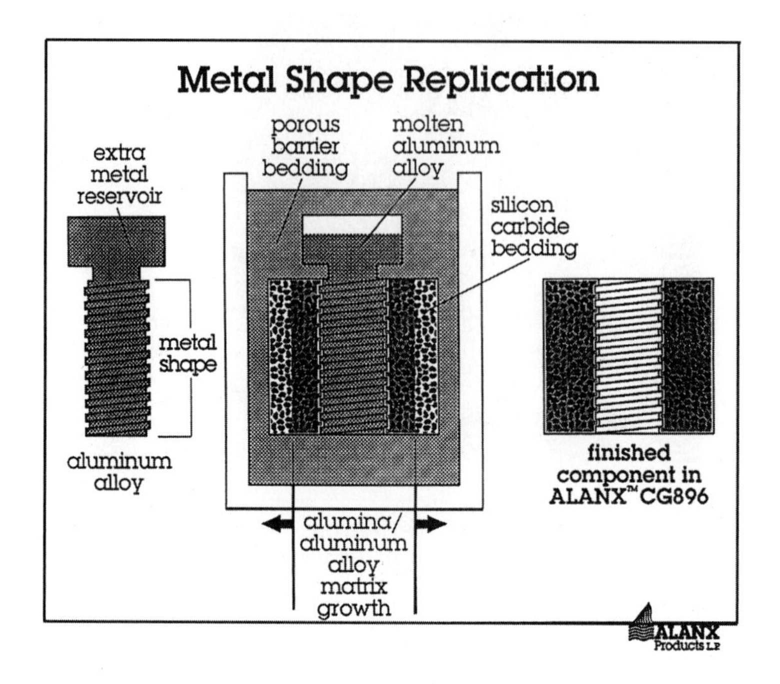

Figure 20-20 The Lanxide DIMOX™ process in the making of screw threads with a ceramic composite material through metal shape replication. (ALANX)

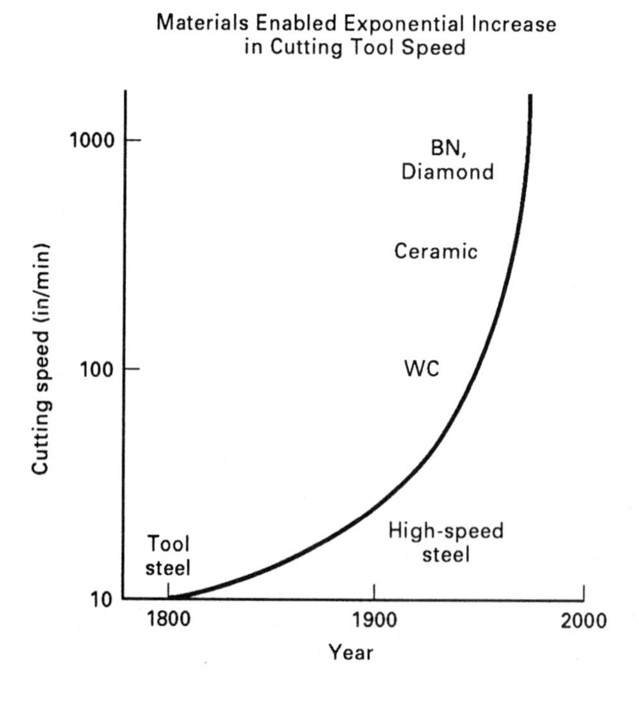

Figure 20-21 Plot of cutting speed versus time (years), showing the exponential rise in speed as a result of the advances in materials development.

grain size. After forming, the compact can be "green" machined (i.e., machined before densification). In other instances, final machining is required after densification.

Regardless of the type of machining (i.e., turning, milling, or wheel or profile grinding) or when the machining is done, tool materials must be made from hard, wear-resistant materials to avoid damaging or lessening the toughness of the relatively fragile ceramic workpiece. Internal flaws of less than 100 μm or surface flaws of less than 50 μm will produce fractures due to stresses in heat-engine ceramic parts. Although more costly initially, ***diamond inserts*** in machining tools usually result in significant cost savings in terms of reduced risk of damage to the

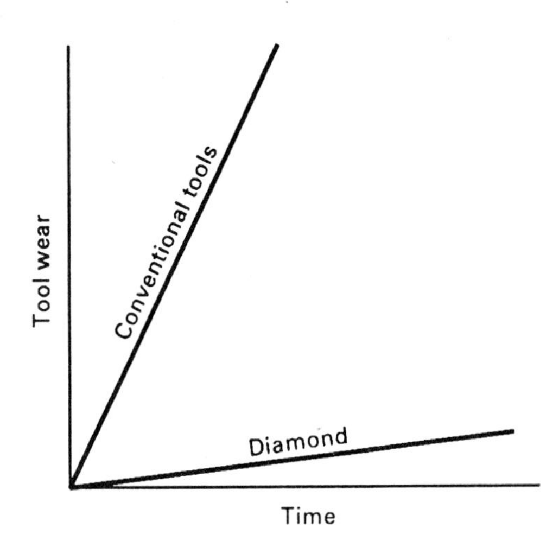

Figure 20-22 Rough plot of tool wear versus time, emphasizing the drastic reduction in tool wear as a result of the introduction of diamond-coated tools and inserts.

workpiece, reduced time expended in tool changes, greater cutting speeds, and increased tool life. Figure 20-21 illustrates the phenomenal rise in cutting-tool speed over the past 200 years as a result of the research and development in metals and ceramics. Figure 20-22 is included to emphasize the advantage of synthetic diamond as a cutting tool, particularly for ceramics.

SELF-ASSESSMENT

20-1. Ceramic that has been molded but is still in a plastic condition is called
a. Frit **b.** Bisque **c.** Glaze **d.** Greenware

20-2. Ceramic that has been fired but is porous and will absorb liquids is called
a. Frit **b.** Bisque **c.** Glaze **d.** Greenware

20-3. The glassy finish such as porcelain fired onto porous ceramic is
a. Frit **b.** Bisque **c.** Glaze **d.** Greenware

20-4. In ceramic processing, the term *green* is used to describe a particular part. What condition does this term describe?

20-5. Normal sintering depends on convection heating as well as heating by ___________.
a. Contact **b.** Radiance **c.** Conduction **d.** Microwave

20-6. Spraying a protective metal on foam uses the technique of
a. Plasma arc **b.** HVOF **c.** Wire arc

20-7. The ceramic with the lowest conductivity is
a. Thoria **b.** Al_2O_3 **c.** ZrO_2

20-8. The process in which high temperatures cause some ceramics to melt into a glassy matrix is called
a. Sintering **b.** Atomization **c.** Vitrification **d.** Annealing

20-9. The process in which clay is heated to change a soft material to a hard thermoset is called
a. Sintering **b.** Atomization **c.** Vitrification **d.** Annealing

20-10. A ceramic metal–matrix composite (densely packed SiC particles in Al_2O_3) produced by the DIMOX™ process that gives maximum wear resistance is
a. CVI **b.** RHM **c.** ALANX **d.** CFCC

REFERENCES & RELATED READINGS

AMERICAN CERAMIC SOCIETY. *CERAMICSOURCE* 2003.

BEHRENDT, DONALD R., and MRITYUNJAY SINGH. *Melt-Infiltration Process for SiC Ceramics and Composites.* Lewis Research Center (LEW-15767), *NASA Tech Briefs,* August 1994, p. 59.

BESMANN, T. M., B. W. SHELDON, R. A. LOWDEN, AND D. P. STINTON. "Vapor-Phase Fabrication and Properties of Continuous-Filament Ceramic Composites," *Science*, Volume 253, September 6, 1991.

BONDS, JAMES W. et al. "Wire-Arc Spraying of Metal onto Insulating Foam." Rockwell International Corp. for Marshall Space Flight Center, *NASA Tech Briefs,* May 1995, p. 84.

MPR. "Net and Near-Net Shaped Ceramic Composites Formed by Lanxide Technology," *Materials and Processing Report,* Volume 3, No. 3, 1988.

NIST. Webbbook site. www.ceramics.NIST.gov/webbook/webbook.htm, February 2, 2003.

OMETITE, OGBENI O., MARK A. JANNAYA, AND RICHARD A. STREHLOW. "Gelcasting: A New Ceramic Forming Process," *American Ceramics Society,* Volume 64, No. 12, 1985. PSH website. www.psh.ca, January 23, 2003.

Saint-Gobain Industrial Ceramics website. *www.saint-gobain.com,* January 18, 2003.

SWAIN, BERTA. "Microwave Processing of Ceramics," *Advanced Materials & Processes,* Volume 134, No. 3, 1988, p. 76.

URQUHART, A. W., G. H. SCHIROKY, AND B. W. Sorenson, "Ceramic Composites for Gas Turbine Engines via a New Process," ASME 89-GT-316, Gas Turbine and Aeroengine Congress and Exposition, Toronto, Canada, June 4–8, 1989.

Module 21

Refractives, Cement, Concrete, Clay, Abrasives, & Protective Coatings

After studying this module, you should be able to do the following:

21–1. Define refractories in terms of structures, properties, and applications.

21–2. Explain the difference between cement and concrete and between setting and curing.

21–3. Discuss the advantages offered by cement and concrete for developing nations.

21–4. Explain the advantages accrued from improved cements and why the "low-tech" aspects of cement and concrete make it environmentally friendly.

21–5. Describe abrasives and ceramic protective coatings in terms of their structures, properties, and uses.

21–6. Use the website addresses denoted by the symbol to explore sites for new developments and evolving concepts related to refractories, cement, concrete, clay, protective coatings, and abrasives.*

*The symbol found throughout the book will link you to Internet sites related to topics being covered. The dynamic nature of the Web brings frequent changes, so some of these sites, while available at the time of writing, may not be up now.

21.1 REFRACTORIES

Refractory metals are among the few materials with significant mechanical strength above 1371°C (2500°F), but they do not have sufficient strength-to-weight ratios for many sought-after applications. A ***refractory material*** is one with a very high melting point and other properties that make it suitable for uses such as furnace linings and kiln construction. A broad range of ceramic materials qualify as refractories. In the traditional ceramic group are sintered clay products, consisting of crystals in a glassy matrix. These materials find use in ovens, kilns, furnaces, and melting pots; in welding and cutting; and as engine parts. Some common refractory ceramics are alumina, alumina–silica, chromite, bauxite, zirconia, kaolin, silicon carbide, magnesite, and graphite. Operating temperatures of some furnaces are extremely high. For example, the basic oxygen furnace (BOF) for making steel subjects its linings for about 1 hour to a temperature of about 1700°C; sintering furnaces may reach 2500°C. An arc furnace heats blocks of alumina at temperatures well above 2000°C to be cast for use as liners in glass-melting furnaces. These linings are highly resistant to high-temperature corrosion. See, for example, the refractory brick in Figure 19-7. In addition to heat resistance, chemical properties bear heavily on the selection of refractories that come into contact with molten metals and glasses. Ceramics with weak ionic bonds have relatively low resistance to chemical attack, whereas those possessing strong covalent/ionic bonding are stable as linings for furnaces used for melting metal alloys and glasses at temperatures above 1200°C. The acidic or basic nature of a ceramic will also determine how it reacts with a given molten material. The reaction between materials can change the properties of the metal or glass being processed.

Coors Technical Ceramics—*http://www.coorstek.com/coorstek/home.asp*
Glass PPG Industries—*http://www.ppg.com*
*ceramic*source online —*www.ceramicsource.org*

The fact that ceramic cutting tools can have more than a dozen thin coatings added to their surfaces, each with its own function, suggests that this technique may be useful for refractory ceramics as well. In the case of furnace linings, liners may be installed that have special functions to perform in addition to basic high-temperature resistance. Such functions could be structural or include specialized insulating properties with low thermal conductivity in addition to high-temperature melting resistance. Finally, the possibility of using additives in these ceramics to increase their specific properties must be considered. In addition, fibers can be added to refractories to reduce thermal conductivities as well as to provide more open space within the material to decrease the material's ability to store up heat as a heat sink.

High-temperature ceramic materials must, therefore, possess a varied assortment of properties. For example, the electrodes must have (1) resistance against corrosion and erosion from high-velocity gases, (2) good thermal-shock resistance, (3) high bulk density for good chemical resistance and erosion resistance, (4) good resistance against electrical-discharge arcing, and (5) good thermionic emission. Notice that strength properties are not mentioned, nor are those properties principally related to insulation against high temperatures.

Many of our present engineered materials are designed for specific purposes. As seen in Figure 21-1, NASA's space shuttle *Orbiter,* a reusable cargo-carrying space vehicle that began its space missions in the early 1980s to service space vehicles in orbit, was protected with a thermal protection system that would protect it not only in space but also during flight to and from space on a repeat basis. Though primarily a high-temperature protection system, the tiles were also required to be, among other things, lightweight, resistant to moisture penetration, and able to withstand the stresses and vibrations of flight. Figure 21-2 shows tiles that were used in two of the four different protection systems on the shuttle that would undergo different conditions of temperature and stress. (See Figure 19-7 for a more complex comparison of tile shape and density.) Some of these tiles have densities as low as 0.14 g/cm^3.

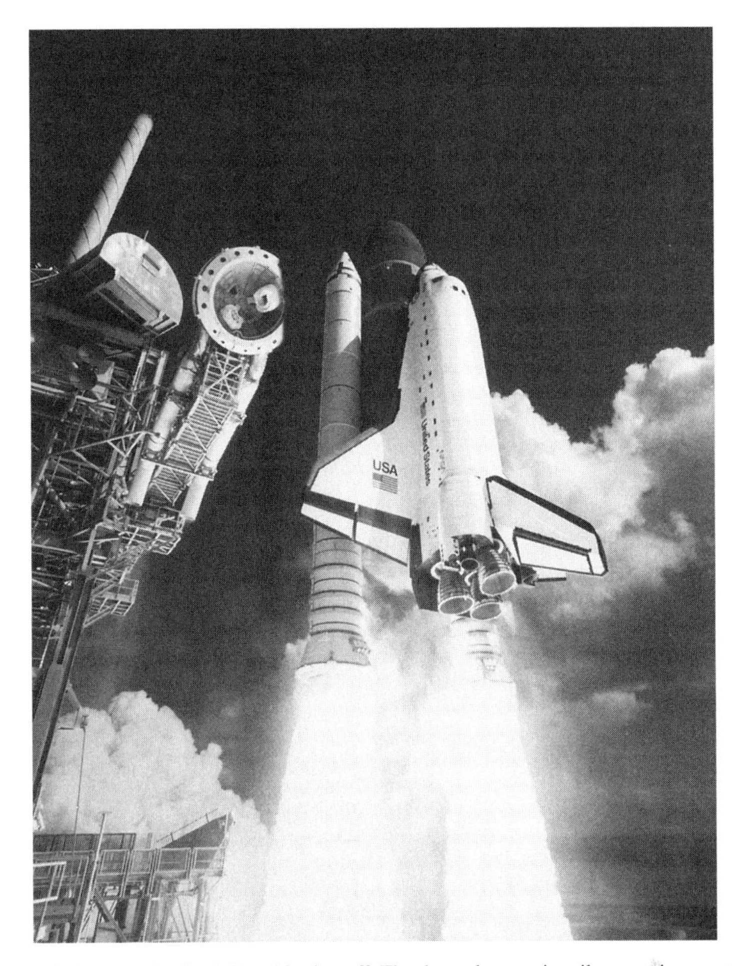

Figure 21-1 NASA's space shuttle *Orbiter* blasting off. The thermal protection tiles covering most of the *Orbiter* are white; those on the leading edges and bottom have black glass coatings. See Applications & Alternatives.

21.2 CEMENT AND CONCRETE

21.2.1 Chemically Bonded Ceramics

Ceramic cements are most familiar to us in the form of ***concrete*** and ***mortar.*** Concrete and mortar are similar because both use cement to bond together the ***aggregate*** (rocks and sand, which combine only through adhesion with cement), which adds bulk to the material. Concrete is more correctly classified as a composite, especially when rocks and reinforcing metal rods and wires are added to it. ***Wheat concrete*** is a new "green" product that uses starch—a granular, solid, complex carbohydrate found in rice, corn, wheat, and many other vegetable foods. ***Masonry mortar,*** applied by brick masons between cement blocks, brick, and tile, consists of masonry cement plus a fine aggregate sand. Masonry cement is composed of portland cement with hydrated (water-incorporating) lime, silica slag, and other additives to improve plasticity and slow setting time.

Traditional cements, both those developed many years ago and those produced in recent years, are actually a group of chemical materials in which consolidation occurs by means of chemical reactions at low temperatures rather than by firing or sintering at high temperatures. When mixed with water in suitable proportions, portland cement, for example, sets and hardens. Consequently, it is used in concrete to bind the sand and coarse aggregate into a solid mass. Figure 21-3 illustrates the use of cast concrete at an oceanfront park. Setting and hardening are caused by the chemical reactions between the cement and water. ***Setting*** (stiffening)

(a)

(b)

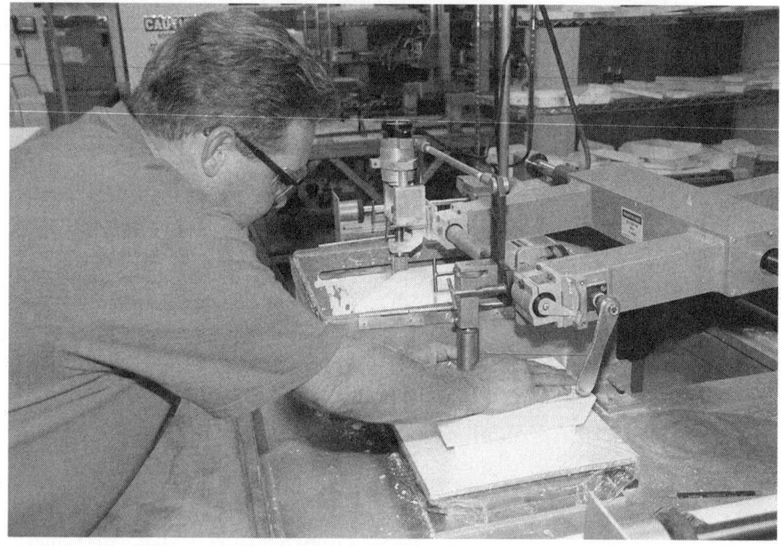

Figure 21-2 (a) NASA's space shuttles are dependent on a thermal protection system to protect them under various adverse conditions, including extremes in temperatures ranging from −130°C to 1260°C. Examples of two different tiles used to cover the exterior of space vehicles are shown. Some exterior surfaces are covered by blankets of quartz fiber batting sandwiched between woven quartz fabric and stitched together using quartz and glass threads. The blankets are further protected with a ceramic coating. This glassy coating protects against moisture penetration and erosion. For those surfaces exposed to higher temperatures than the above-mentioned temperature range, low-density tiles were developed made of low-density silica fiber material bonded together by a sintering system (fused silica). Silica melts at about 1700°C and has good heat insulation qualities due to its large percentage of trapped porosity. Some 24,000 tiles cover about 70% of the exterior of the *Orbiter*. (Lockheed Martin Missiles & Space and American Ceramic Society) (b) As seen in (a), each tile has its own shape and ID number. Tiles are machined to size and shape prior to application of glass glazing. (Lockheed Martin Missiles & Space and American Ceramic Society)

occurs within hours; with most normal varieties of cement, ***hardening*** (the subsequent development of strength) is largely complete in a month. If the water–cement mixture (paste) is kept moist, the strengthening process can continue for many years. If high strength is required in less than 1 month, alumina cements, which can achieve high strength in 48 hours, are used. Figure 21-4 is a schematic of the setting and hardening process.

Figure 21-3 Two examples of cast concrete being used to maximize the main characteristic of concrete—weight. The tables are sitting on a cast-concrete platform. Both objects have withstood the ravages of hurricanes without any damage.

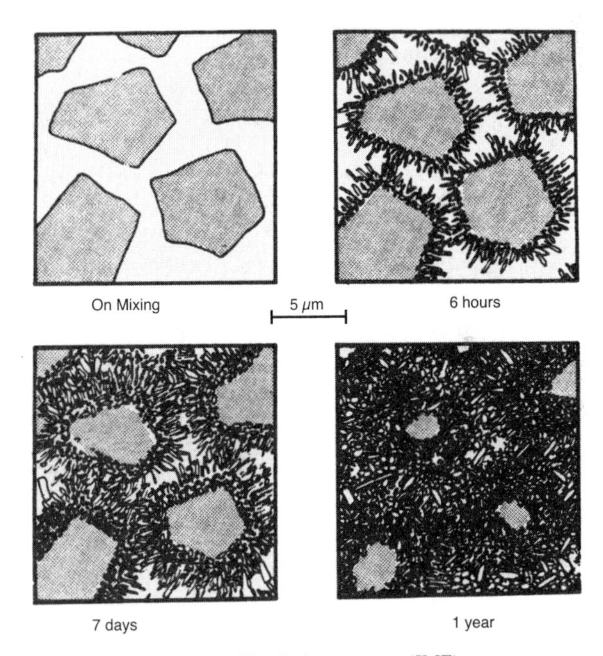

Figure 21-4 Schematic picture of the setting and hardening process. (JME)

Setting occurs when the gel coatings formed around individual grains join, and hardening results from densification of the contact areas in the gel. As seen in Figure 21-4, crystal calcite spindles (long, slender stalks or stems) grow from grains into a network strong enough to support extremely large loads. The drying times used by builders are, at best, conservative due to temperature and humidity variations during setting. However, specified drying times must be complied with in determining when the concrete is ready to support the next step in construction. To speed construction, a more definitive measure is used. A piezoelectric crystal device (see Section 27.4 for the definition of *piezoelectric*), long used to test concrete bridges for internal cracks, is being adopted in other types of construction to obtain greater accuracy in calculating drying times. The operation of this device is based on the fact that the harder the concrete becomes, the faster sound travels through it. By mounting a beep-emitting piezoelectric crystal on one end of a concrete mass and a receiver on the other end to listen for the

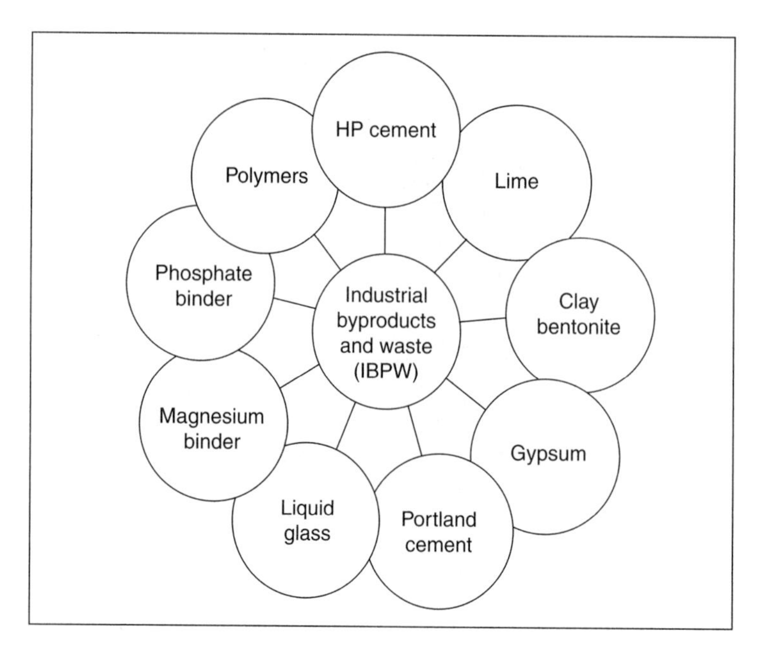

Figure 21-5 Binder materials used with IBPW (industrial by-products and waste) for stabilization and immobilization. (ACerS Ceramics Bulletin)

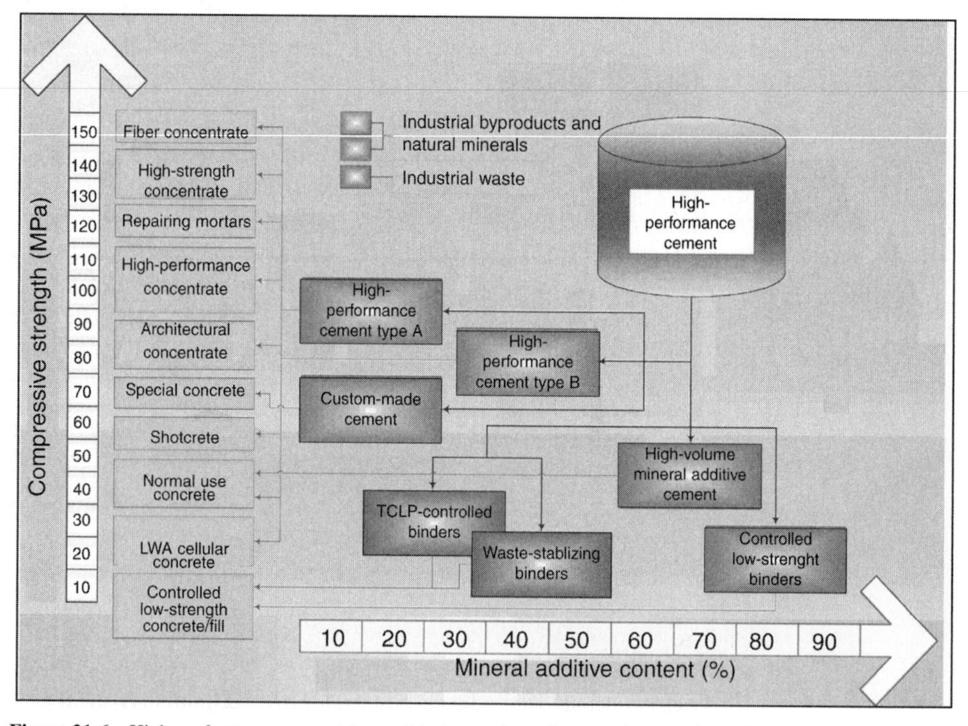

Figure 21-6 High-performance cement is possible in a variety of compositons with varying compressive strength. (ACerS Ceramics Bulletin)

beep, the speed at which the sound travels through the drying concrete can be calculated; and an even more accurate prediction of when the cement will reach a given strength is now reducing the time required in many construction projects. In addition, the high-performance cement (HP) discussed in the Pause & Ponder for Unit 7, which uses industrial by-products and waste and a variety of binders for stabilization and immobilization, as seen in Figure 21-5, offers a variety of products with various compressive strengths (Figure 21-6, arrow on left) resulting from the type of additives and reinforcers used.

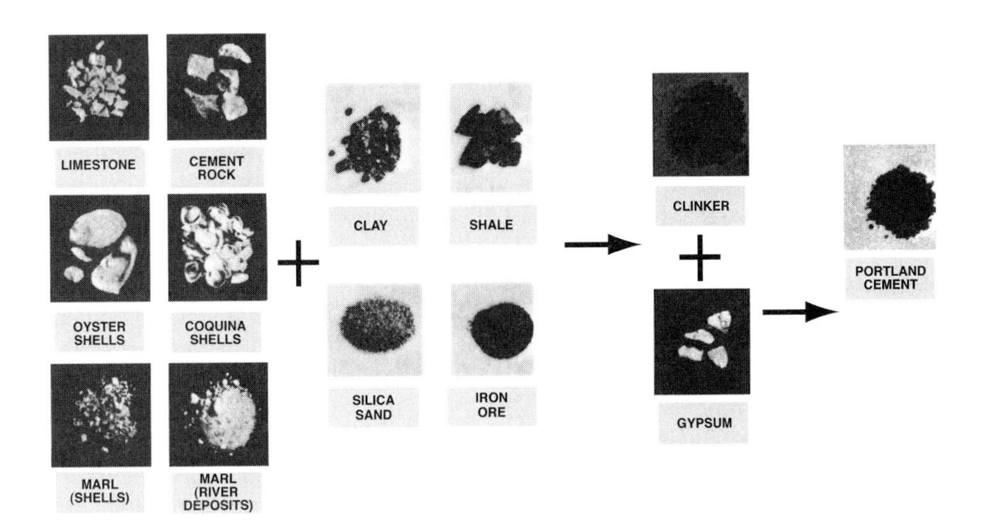

Figure 21-7 Raw materials to make portland cement. (From *Design and Control of Concrete Mixtures* (EB001.11T and Portland Cement Association, Skokie, IL)

Cements can be classified according to their bonding mechanism: hydraulic bonding, reaction bonding, and precipitation bonding. ***Hydraulic cements*** cure or set by interaction with water. ***Portland cement,*** the most common hydraulic cement, is composed primarily of anhydrous (without water) calcium silicate. An example is tricalcium silicate ($3CaO \cdot SiO_2$). Other ingredients are alumina, gypsum, and iron oxides. ASTM C150 recognizes eight types of portland cement, each with different chemical compositions and properties. Figure 21-7 depicts the raw materials used to prepare this cement. Portland cement sets by a complex reaction with water to form a hydrated composition. This reaction is ***exothermic*** (gives off heat), which means the heat can dry out the cement, thus stopping the reaction prematurely. Keeping the cement damp until setting is almost complete precludes this from happening. A compromise is needed here because the lower the ratio of water to cement when mixing, the higher the final strength of the mix that will be attained. When hardened, portland cement can contain a porosity of over 30%, which results in flexural strengths in the range of 4 to 10 MPa.

The strength of cement is affected by the size of the pores arising from the entrapment of air and the incomplete packing of cement particles. Another hydraulic cement is gypsum cement, such as plaster of paris. These cements set using the same hydraulic reaction as portland cement but recrystallize to a higher crystalline structure that has a low degree of adhesion. This property makes them good candidates for components such as wallboard and plaster molds made for slip casting of metals and ceramics. To give wallboard more strength, both sides are covered with paper, another example of a stressed-skin composite material.

Traditional hydraulic cements, as well as reaction and precipitation cements and the concretes made by combining them with various aggregates, are among the most widely used materials. The term ***cement*** can include any materials that act as an adhesive to bond components together. ***Cementitious*** materials include pozzolans (containing natural hydrating or water-incorporating material), such as shale, lime, gypsum, asphalt, tar, synthetic plastics, and cement. ***Asphalt*** is the by-product of the ***catalytic cracking*** that separates gasoline and heating fuels from crude oil. The combination of asphalt and gravel has been used for centuries to build roads, but the amazing fact is that very little was known about the engineering properties of this combination of materials. Only one test, a viscosity test for high temperatures, was ever used to determine how well asphalt would ward off rutting by vehicles in the summer, and only one grade (a certain proportion of the two ingredients) was used regardless of temperature and/or traffic conditions. This ***empirical*** approach (based on experiment and observation without reference to any scientific principles) was the one taken in the art of road building for nearly 94% of U.S. road pavements. Do you know why there were so many potholes and ruts in our street pavements?

The Strategic Highway Research Program was federally funded in 1987, and engineers from universities and industry began the six-year job of finding a better way to build pavement that resulted in Superpave. Actually, *Superpave* is not only the name for the new, superior performing asphalt product but also includes a new set of testing equipment and protocols for mixing materials in the proper proportions to hold up to the weather and traffic at a specific location. New tests now include both the old reliable viscosity test for high temperatures and a new viscosity test for low temperatures. In addition, a shearing test to measure deformation under a shearing type load and a load test to determine the material's ability to resist cracking are now standard practices. These tests now provide the data to determine the engineering properties of this new road-building material. The ingredients that combine with asphalt were also investigated. Data have been collected about the correct proportion of aggregate (mix of rock as to its size, shape, coarseness, and fineness) such that each road is custom designed with the ideal grade of asphalt and the correct aggregation. One final factor determined by research is the optimum procedures for laying down and rolling out Superpave. Researchers are collecting new pavement data to determine how many years a new pavement will last before rutting or cracking from temperature and fatigue. These data are from several test tracks where vehicles are run continuously to simulate many years of traffic on a particular pavement under certain weather conditions.

The "greening" of concrete has been achieved with a "new-age," pervious (porous) concrete that can trap contaminants as it filters runoff water from parking lots, driveways, and other large concrete pads. ***Pervious concrete,*** as seen in Figure 21-8, was developed by Tarmac America, Inc. It is a pervious material that permits water, oil, sediments, and other solids and liquids to flow through the pores and slowly seep into the ground. With normal, impervious concrete and asphalt pads, water and contaminants carried by the water must be directed to catch basins and storm drains, where they are piped through piping systems to

Figure 21-8 Pervious or porous concrete allows runoff water to seep through, disperses water naturally, and traps some pollutants. (John Corbitt, The Virginian-Pilot)

streams, rivers, lakes, and oceans. Any waterborne contaminants flow into the bodies of water and contribute to pollution. Pads made of pervious concrete provide for the slow, natural filtration of water, which works its way back into groundwater reservoirs deep in the earth. To keep the pervious concrete clear, it is flushed by water. ***Eco-cement*** describe in the Pause & Ponder for Unit 7 described how industrial by-products and waste serve as raw materials in high-tech concrete.

New legislation to control storm water runoff has prompted new concrete technology. When introduced, pervious concrete cost more than twice impervious concrete. As with any new technology, this "green" concrete must prove its ability to withstand stresses from changing weather conditions, traffic loads, and aging. Its higher cost may be offset by savings in storm systems and improved water quality.

21.2.2 Improved Cements

Two recent advances in hydraulic cements have improved their tensile strength, primarily by decreasing their porosity. In one technique, known as DSP (densified with small particles), SiO_2 particles of very fine size (ultrafine) are added. Because they are smaller than the cement particles, they fill spaces between the cement particles, decreasing the overall porosity and pore size. In addition, less water is required to produce plasticity. These factors reduce the permeability of the mix, which translates into longer service life for the many structures built with reinforced concrete that contain metal bars. However, these mixes suffer corrosive attack from water seeping through the concrete.

A second technique is referred to as MDF (macrodefect-free) cement, in which additives such as small amounts of polymers and other ingredients are added to achieve better particle dispersion and packing. The absence of pores, whether due to initial particle size, particle distribution, or the elimination of entrapped air by kneading, results in the reduction of stress concentrators and a dramatic increase in tensile strength (over 200 MPa in some instances) and fracture toughness, with a corresponding decrease in permeability, all with little increase in overall cost.

High-temperature cements produced by chemical reactions not involving water are yet another improvement. One common reaction cement is monoaluminum phosphate, produced by the reaction between aluminum oxide and phosphoric acid. An example of these cements is the dental cement based on zinc oxide. Precipitation cement is acid resistant and has good resistance to abrasion. Sodium silicate, which is formed by gels that are precipitated from colloidal solutions, is a very common, inexpensive cement in this category of cements.

More recent chemically bonded ceramics (CBCs) find uses as tooling mediums, for military armor, and in containers for storing hazardous and radioactive materials. Metal aggregates and fiber reinforcements are added to improve tensile strength and fracture toughness. These low-temperature CBCs are improving to the point where they have nearly 20 times the strength of concrete. As for cost, when raw materials and energy are considered, CBCs are nearly 6 times less expensive than polymer materials, 20 times less expensive than steel, and 25 times less expensive than aluminum.

Improved cements, such as the CBCs, will no doubt be produced using techniques that call for low temperatures and low energy consumption and thus require fewer high-technology materials and processes. The significance of this new, "low" technology is that the resources needed to produce this new class of engineering materials are universally available. Third-world nations will benefit greatly, as will more developed countries that now rely on petroleum-based raw materials. As mentioned in the Module 1 discussion of the materials cycle, there is much to be said for utilizing ***low technology*** as the world struggles to reduce consumption of limited natural resources and energy. Reducing environmentally threatening waste (by-products of materials processing and energy consumption) and the disposal of used products will be some of the results of the new technology. While we study high-technology materials and processes, we must continue to reexamine the viability of older materials such as wood (a renewable resource) and cement (from readily available resources) where it is possible to apply new processes.

21.3 CLAY

Clay is classified as either residual or sedimentary. ***Residual clay*** results from the wearing down of rock from the mechanical and chemical action of wind, water, earth movements, and chemicals in the soil. Such clays are found where they were formed. ***Sedimentary clays*** also form through mechanical and chemical erosion of rock, but they are moved by wind or water to places other than where they have formed. Common clays include kaolin ($Al_2O_3 \cdot 2SiO_2 \cdot 2H_2O$), ball clay, fire clays, stoneware clay, and slip and flint clays. In various forms, clay finds many applications: as ingredients in most traditional ceramics, structural products of brick and tile, raw materials for sculptured art, china dishes, and electronic components. Clay is also an important ingredient in paper making. Silicon and aluminum, comprising 28% and 8%, respectively, of the earth's upper crust, are prominent in clay. Most other elements are present in only small percentages, such as carbon, with 0.009%.

21.4 RADIOACTIVITY AND NUCLEAR FUEL

Radioactivity is the spontaneous emission of radiation from certain unstable elements. Over 1100 unstable radioisotopes are known, of which about 65 occur naturally and the remainder are laboratory-made under controlled nuclear reactions. The spontaneous emission is the disintegration or decay of the unstable nucleus. From our study of atoms and their nuclei in Module 2, we learned that the nucleus of atoms may contain the same number of protons but different numbers of neutrons. In these atoms, known as isotopes, nucleons (protons and neutrons), like electrons, spin and have angular momentum. ***Stable*** isotopes bind the nucleons in the nucleus by a combination of forces, some of which are not well understood. Certain isotopes, however, are not stable and they decay, with the emission of some type of radiation. The disintegration of unstable isotopes occurring in nature is called ***natural radioactivity,*** whereas that of unstable isotopes produced in a laboratory by nuclear reactions is called ***artificial radioactivity.***

When unstable nuclei undergo radiation decay, they change into other nuclei and emit alpha (α), beta (β), or gamma (γ) particles, each of which has a different degree of penetrating power. For example, α rays are weak and could be blocked by a single piece of paper, whereas γ rays can penetrate several centimeters of lead. Gamma rays, like X rays, are high-energy protons. The only difference between them is that X rays are produced by an electron–atom reaction and a gamma ray is produced in a nuclear process. The energy released by nuclear reactions will appear as kinetic energy of the outgoing particles (radiation). Einstein's equation ($E = mc^2$)states that mass and energy are two different aspects of the same thing. During radiation, the mass of the nucleus is diminished by a tiny amount, and this loss of mass is translated into radiation energy. This energy then can be used or converted to other forms. A hydrogen bomb explosion converts less than 1% of the mass to energy. Radiation energy can be transformed into heat in a nuclear reactor, which is used to boil water (to make steam) flowing through the core of a heat exchanger. Steam drives a turbine that generates electricity and is then cooled in a condenser that converts the steam back to its liquid form.

All forms of radiation can ionize atoms by knocking out electrons. Radiation passing through materials can do considerable damage. Metals and other structural materials become brittle and lose their strength if the radiation is intense. An isotope of strontium, at. no. 38, a product of radioactive fallout, concentrates in human bone causing bone cancer and destroying bone marrow (see Figure 21-9). Gamma radiation can adversely affect polymeric materials by rupturing bonds that hold molecular chains together, a degradation called ***scission.*** This same radiation is used to cross-link PE molecules to increase the materials' resistance to softening at high temperature.

The applications of nuclear radiation are numerous. Smoke detectors in homes and offices use a weak radiative source to ionize air molecules. When these molecules combine with smoke particles, the difference in the speed of movement is detected electronically and an alarm is triggered. Radioactive isotopes are used to detect leaks in pipes and to study corrosion and

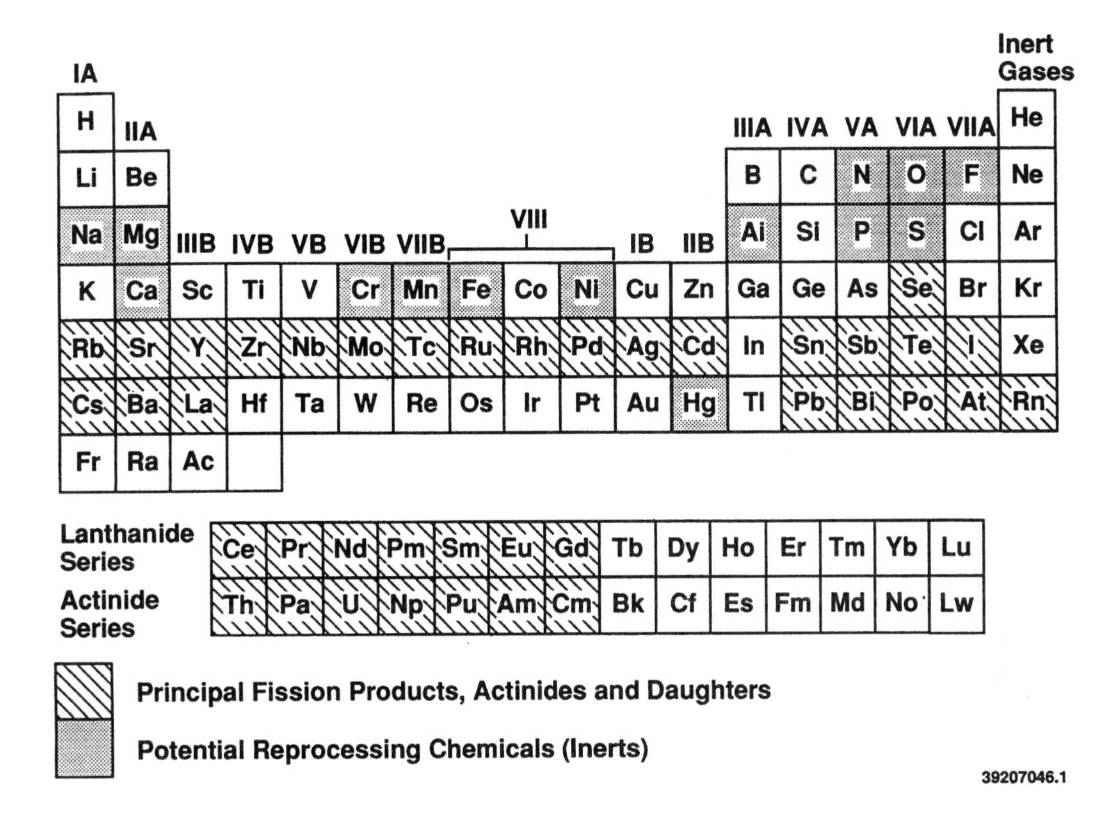

Figure 21-9 Elements typical in nuclear waste. (U.S. Dept. of Energy—Pacific Northwest National Laboratory)

wear. In the medical field, there are numerous applications of radioactivity and radiation to human beings. The treatment of disease, mainly cancer, through radiation therapy utilizes radiation because in large doses it will kill cancer cells. ***Microspheres,*** a special glass composition of S, Al, O, plus rare earth elements (lanthanum series, known as the lanthanides) are also being used in medicine to combat cancer (see Figure 21–9 for a listing of the lanthanide series of elements in the periodic table). By radiating this material in a nuclear reactor, the lanthanides become radioactive and give off beta, not gamma, radiation. By controlling the amount of radiation and placement of the radiative elements, the penetrating distance can be adjusted to attack only cancer cells, not healthy cells, in an organ or tumor. Radioactive tracers are also used in medicine for diagnostic purposes, such as locating the spread of disease tissues in the human body. Co, I, and the metastable state of Tc (technetium) are some of the radionuclides used in medicine. (See Module 2 for a review of the notation used in designating isotopes.) Gamma radiation is used to kill bacteria and viruses and to preserve food (irradiation).

A ***radioactive isotope*** consists of a vast number of radioactive nuclei that do not decay all at once. They decay randomly over a period of time. The rate of decay of any isotope is often specified by giving its ***half-life,*** which is defined as the time it takes for half the original amount of the isotope in a given sample to decay. The longer the half-life of an isotope, the more slowly it decays. Spent fuel from nuclear reactors is extremely radioactive, so it must be kept isolated from people, animals, and the food chain. A radioactive fuel such as plutonium (Pu, at. no. 94), for example, has a half-life of about 24,000 years. Starting with a pound of plutonium, after the passage of about 24,000 years, one-half pound of this element will decay to a stable condition. After another 24,000 years, an additional one-quarter of the original pound will become stable, leaving one-quarter of the original pound still radioactive.

The ***radioactive decay*** of the fission products also produces heat, so cooling must be provided for the first few years that the spent fuel is stored. The usual way to provide cooling is by storing the spent fuel in water. Because of the current lack of a permanent place to dispose of spent fuel, the material is now stored on-site at power reactors. It is estimated that by the year 2020, nearly 100,000 tons of uranium-dioxide fuel will be awaiting a more permanent way of isolating it from the biosphere. The electrical power to operate the average household for three years generated by burning 22.5 tons of coal would produce about 82.5 tons of carbon dioxide and about 2.3 tons of ash. The ash represents its own disposal problem, and the carbon dioxide would be released irretrievably into the atmosphere, where it accumulates. Carbon dioxide is one of the greenhouse gases, which impede the earth from radiating energy into space. Many scientists believe that these gases could cause global warming, changing the climate worldwide.

Uranium dioxide, a ceramic material, is used as a fuel in the current generation of nuclear reactors for the production of electric power. This oxide was chosen for its chemical stability and high melting temperature (2750°C). Figure 21-10 shows a cross-sectional view of a typical reactor fuel rod. The uranium oxide is made into pellet form by pressing and sintering; these pellets are then inserted into tubes made of a corrosion-resistant alloy of zirconia called *zircaloy,* which are then welded shut. A nuclear reactor contains thousands of these tubes, held into frameworks that allow water to flow among them. The water acts as both coolant and moderator. It slows the neutrons produced by splitting the uranium atoms so that they can then split other uranium atoms in a controlled fission process. The fuel inside each tube produces an amazingly large amount of energy, each foot of length producing about 2 kW, enough electric power to run an average household.

Because the ceramic fuel is not a very good conductor of heat, the release of energy produces high temperatures, 600° to 1400°C, at the center of the oxide pellets. The water outside the fuel is at a fairly low temperature, about 300°C, so the temperature gradient is very steep. As the nuclear-fission reactions proceed, elements of lower atomic weight are produced and accumulate inside the solid oxide fuel. Eventually, these elements, called *fission products*, accumulate to the point that the fuel must be removed and new fuel inserted. The fuel produces a lot of energy before this is necessary; the spent fuel that produced all the electrical power for a household for a period of three years weighs about 4 oz. In a typical power reactor, a third of the fuel will be exchanged for fresh fuel every year during a planned outage. Refueling is sometimes done on a two-year schedule, so fuel usually produces power for three to six years.

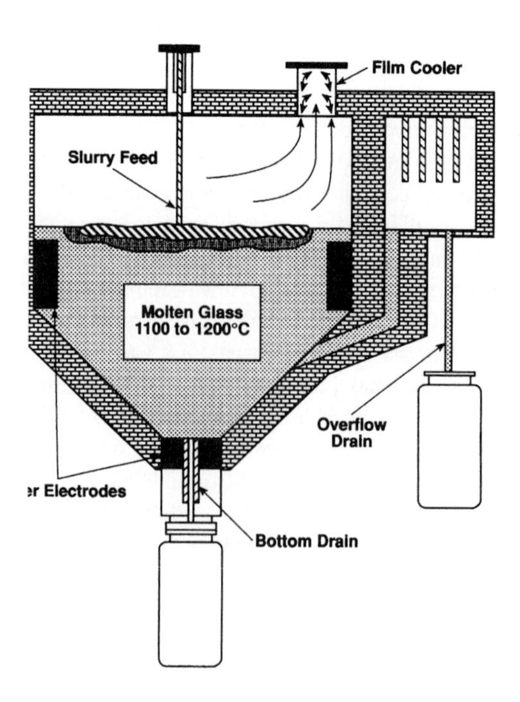

Figure 21-10 Nuclear-waste glassmelter. (U.S. Dept. of Energy—Pacific Northwest National Laboratory)

21.4.1 Disposing of Nuclear and Chemical Waste[1]

Nuclear materials are an excellent example of the importance of understanding the total materials cycle. These materials represent some of the highest levels of technological development, but they also present huge problems in terms of their safe processing, use, maintenance, and disposal. As with other chemicals, biological organisms, and materials, materials science and engineering is often called on to immobilize the end products.

When in a molten state, glasses tend to be highly corrosive and dissolve other oxides readily. Glasses are nonselective in this dissolution because their noncrystalline nature permits them to incorporate other atoms without much regard for size, valence, or crystal form. One way to take advantage of this characteristic is to use particularly durable glasses as a matrix material for disposal of high-level nuclear wastes. The various elements present in a typical nuclear waste are shown in Figure 21-9: These elements can be dissolved by nuclear-waste glass, although solubility limits are occasionally reached for noble metals and elements such as ruthenium. Development of this concept started in the mid-1960s, and operating nuclear-waste glassmelters exist in several countries at present. The glass type used for nuclear-waste disposal is roughly similar in composition to Pyrex, a borosilicate glass developed by Corning Glass Company. This glass is very durable but can be processed at moderate temperatures.

Processing of glass at temperatures of about 1100°C inside a shielded, remote-control-operated enclosure called a ***hot cell*** is a substantial engineering challenge. The most common way of heating nuclear-waste glass is to pass electricity directly through it, heating the glass by its own resistance. Glass behaves like a semiconductor, meaning that the electrical resistance drops as temperature increases, opposite to the behavior of a metal. Figure 21-10 shows a typical nuclear-waste glassmelter. Nuclear waste is combined with glassmaking ingredients and fed to the top of the melter as an aqueous slurry. The water evaporates and the ingredients combine to form a homogeneous black glass, in which the radioactive elements are dissolved at the molecular level. After a period to allow a homogeneous glass to form, the product is poured into a stainless-steel canister approximately 2 feet in diameter, where it cools and hardens to form a "log" about 10 feet long and weighing 4500 lb. After a lid is welded on and the canister is cleaned of external contamination, the waste glass is ready for transport to a permanent repository.

21.5 ABRASIVES

Abrasives are very hard particles used for grinding, sanding, and polishing. ***Abrasive materials*** include particles of flint, garnet, diamond, aluminum oxide, silicon carbide, emery, pumice (pulverized volcanic lava), rottenstone (shale rock), and corundum. Aluminum oxide and silicon carbide find the greatest industrial application, although garnet is often used as sandpaper—an abrasive paper or belt used for sanding wood. See Figure 21-11 for examples of abrasives, including media used in ball mills. Adhesive papers include paper and cloth sheets, belts, disks, and drums onto which an abrasive is cemented. Wet or dry abrasive paper employs waterproof adhesives. Grinding wheels consist of abrasive particles (grains) held together by a tough bonding material such as rubber or organic resins. The particles constantly fracture on the wheel to expose new, sharp cutting edges, or they break off as the wheel becomes smaller in size. Abrasive particles are also used in loose form for sand blasting, hand rubbing, or abrasive drilling. Selection of the ***grit*** (particle size) is based on a screening method that determines the number of particles per inch, such as 60 grit or 600 grit. A 120 grit would have 120 abrasive grains per linear inch or 14,400 (120^2) per square inch. Sialon (silicon–aluminum–oxygen–nitrogen) is a hot-pressed cermet that works well on nickel-based superalloys.

[1] Written by Roy Bunnell, Kennwick, WA

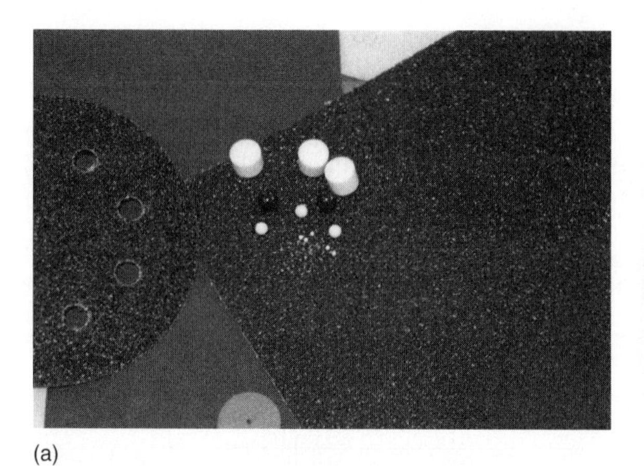
(a)

(b)

Figure 21-11 Ceramic materials are used to make abrasive products that process other ceramics as well as to make abrasive materials that help in the shaping, grinding, and finishing of many other types of materials. (a) A grouping of white-and-black objects represents milling or grinding media used in ball mills to reduce ceramic materials to a fine powder or slurry. Note their spherical or cylindrical shapes. (b) A sampling of abrasives such as adhesive-backed sanding paper, sanding sleeves, grinding wheels, disks, disks with holes for more efficient removal of dust, sanding drum sets, diamond burr sets with assorted shanks and head shapes, a sanding flutter (flap) head, and a file.

Superabrasives, synthetic diamond and cubic boron nitride (CBN), are used for grinding wheels or sintered, polycrystalline cutting tools for milling and turning. Diamond abrasives are four times as hard and have three and one-half times the wear resistance of common abrasives such as aluminum oxide. CBN is two and one-half times as hard and has two and one-half times more wear resistance than aluminum oxide. In addition, superabrasives are excellent thermal conductors. The combined superior properties of hardness, wear resistance, and thermal conductivity provide sharper, longer-lasting tools that can operate at higher temperatures and speeds. When grinding and machining high-speed steel, cemented carbide, high-silicon aluminum alloys, cast iron, and superalloys, superabrasives offer increased productivity through high material-removal rates, achieving better quality (tolerances and surface finishes), less residual stress, reduced scrap, and lower overall manufacturing costs.

21.6 PROTECTIVE COATINGS

Coatings solve problems of corrosion, wear, and compatibility. The thermal spray technology of electric arc combustion and plasma spray can apply any material to the surface of another. Ceramic particles melted into a ***glassy vitreous state*** are used to coat and decorate pottery and metals. ***Glazes*** and ***enamels*** impart color, hardness, and corrosion resistance when they are suspended in a slurry and fired to a vitreous state; as the ***substrate*** (base material) cools, the coating bonds tightly to the substrate surface. ***Flame spraying*** uses a high-temperature heat source to melt the ceramic oxide powder and then sprays or blasts the coating onto the surface of a metal. The oxide material cools and bonds to increase surface hardness and offer protection against heat and oxidation.

To make mirrored sunglasses, a layer of silver or other metallic compound, which is thick enough to retain its reflective properties but thin enough for the wearer to see through, is applied to the glasses. Some light is reflected from this "mirror," but some passes on through to the eye. In addition to this coating there are up to eight other coatings that add extra properties to the glasses depending on the desires of the consumer. These coatings, listed in the reverse order in which they are applied, are (1) silver or metallic, (2) scratch resistant,

(3) UV light absorbing, (4) haze resistant, (5) polarizing filter, (6) haze resistant, (7) UV light absorbing, (8) scratch resistant, and (9) antireflective. The polarizing filter absorbs over 98% of the glare created by light bouncing off horizontal surfaces such as an ocean or a road surface. The multiple coatings correct, clarify, or refine the light as it passes through them. Vertically aligned light is allowed to pass through to the eyes. Both UV-A and UV-B light radiations are eliminated.

APPLICATIONS & ALTERNATIVES

On February 1, 2003, the Orbiter *Columbia* broke up during reentry and scattered into parts across the western states of the USA, resulting in the deaths of all on board. This tragedy was investigated by the Columbia Accident Investigation Board (CAIB), headed by retired Admiral Harold Gehmen, and top scientists, engineers, and other experts. The extensive CAIB report provides an excellent case study of many of the concepts covered in this book, such as the failure analysis of the shuttle's thermal protection systems (TPS). The TPS apparently failed as a result of damage to the tiles during blast-off, when frozen chunks of the foam polymer coating on the external fuel tank struck the tiles, compromising their insulation value. The delicate silica tiles with their glass protective coating are discussed on p. 602 with Figures 21-1 and 21-2. The report also points to issues of safety. Included with these issues were the absence of effective lines of communication.

This disaster also points to the need for people to have a working knowledge of materials science and technology in order to function well in our society. In this case, the needs for this knowledge involved NASA managers, member of the CAIB, the news media, and the general public who have to decipher accounts of complex technology involving aerospace travel.

Read the CAIB Report at *http://caib.us/news/report* to learn about corrective action to improve the TPS.

SELF-ASSESSMENT

21-1. A ceramic, such as masonry mortar, used to bond bricks together is called

a. Cement **b.** RHM **c.** Cermet **d.** Abrasives

21-2. Ceramic particles floating in a liquid such as water is called

a. Slurry **b.** Cement **c.** Amorphous **d.** Composition

21-3. Explain the difference between the setting and the hardening of cement.

21-4. Discuss the advantages of improved low-tech cements, such as CBCs, now and in the near future for developing nations and a greener materials cycle.

21-5. Match the name of the ceramic to its chemical formula or composition.

a. Portland cement — (1) Al_2O_3

b. Soda-lime silica glass — (2) SiO_2

c. Alumina — (3) 99% SiO_2

d. Quartz — (4) $Al_2(Si_2O_5)(OH)_4$

e. Fused silica glass — (5) 74% SiO_2, 15% Na_2O, 10% CaO, 10% Al_2O_3

f. Kaolinite — (6) C_3S, C_2S, C_3A, C_4AF

21-6. What is the element, atomic numbers, and weight for T_c?

REFERENCES & RELATED READINGS

BANERJEE, SUBRATA. *Monolithic Refractories*. Westerville, OH: American Ceramic Society. 1998.

BENNETT, J. P., and J. D. SMITH. *Fundamentals of Refractory Technology*. Westerville, OH: American Ceramic Society. 2003

ROY, DELLA M., ed. "Instructional Module in Cement Science," *Journal of Materials Education.* Materials Education Council, 1985, pp. 47–92.

Saint-Gobain website, *www.refractories.saint-gobain.com*, January 24, 2003.

SOBOLEV, KONSTANTIN AND MERTIN ARIKAN. "High Volume Mineral Additives for Eco-Cement," *American Ceramic Society Bulletin,* January, 2003, pp. 39–43.

Module

22

Electronic and Magnetic Ceramics & Smart Ceramics

After studying this module, you should be able to do the following:

22–1. Describe types and applications of electronic, magnetic, and smart ceramics.

22–2. Discuss how nanotechnology involving ceramics will contribute to advanced technology.

22–3. Use the website addresses denoted by the symbol* to explore sites for new developments and evolving concepts related to smart ceramics.

*The symbol found throughout the book will link you to Internet sites related to the topics being covered. The dynamic nature of the Web brings frequent changes, so some of these sites, while available at the time of writing, may not be up now.

22.1 ELECTRONIC AND MAGNETIC CERAMICS

Typical electronic ceramics include some representative compounds and mixtures, such as ferrites (ZrFeO), silicon (SiO_2), zirconia (ZrO_3), steatite (SiO_2MgO), porcelains (Al_6SiO_{13}), and alumina (Al_2O_3). Nearly 1.2 billion pounds of ***hard ferrites*** were produced in 1997 for use in such applications as speakers, household magnets, memory cores, windshield wiper motors, and automobile power seats. ***Soft ferrites,*** which can be magnetized or demagnetized by changing the direction of the applied magnetic field, are used in television, radio, telephone, electric ignition systems, transformer cores, and tape and disk recording heads. Some 20 million pounds were produced in 1997. Module 9 discussed some of these and other electronic materials.

22.1.1 Ceramic Magnets

Ceramic magnets that contain a sufficient number of ***dipoles*** (atoms with electron spin in the same direction, see Section 6.3) are ferromagnetic. They consist of iron oxides containing ions of elements such as zinc (Zn^{2+}), iron (Fe^{2+}), magnesium (Mg^{2+}), and nickel (Ni^{2+}) that combine with oxygen. Ceramic ferromagnetics can be either soft or hard ferrites. The high electrical resistivity of ceramic magnets gives them an advantage over metal magnets in high-frequency devices, and their permanent magnetic behavior finds application in microwave devices. Ferrites are also commonly found in the computer industry in memory cores because computers process data magnetically. Sheets of rubber and ferrite are employed as well as "stealth" coatings on some steel suspension bridges that cross navigational waterways. The coatings partially absorb radar and minimize clutter on ships' radar screens as they navigate under the bridges. Figure 22-1 shows an assortment of ce-

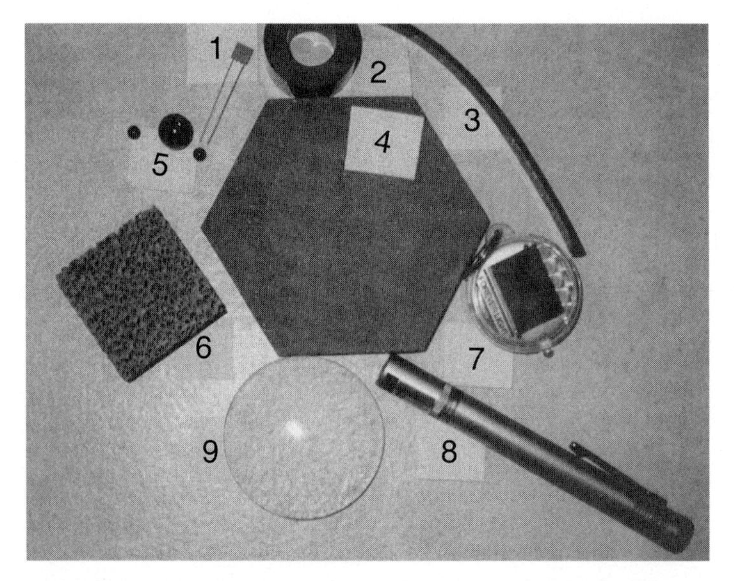

Figure 22-1 American Ceramics Society specimen kit. (1) Ceramic capacitors become polarized when under the flow of electrons. The electric charge distribution shifts such that one side of the ceramic becomes more positive and the other side more negative. This polarization is called the *dielectric behavior* of ceramics. Capacitors are required in all electrical circuits. Those that have a high degree of polarization (defined by the dielectric constant) are used to store electric charge. (AVX Corp.) (2) Cam follower roller. (Kyocera Industrial Ceramics Corp.) (3) Strip magnet-iron on elastomeric strip. Ceramic magnets are generally less dense than metallic magnets and are less costly. For high-frequency applications (e.g., microwave devices), ceramic magnets are made of rare earth ferrites, a ceramic material. (4) Armor tile. Armor to protect personnel and aircraft was developed for the Vietnam War. A ceramic material was chosen because of its hardness. Boron carbide (B_4C) was bonded to a Kevlar backing and covered with a fabric spall shield to make a composite material that defended against armor-piercing projectiles. Boron carbides' hardness on the Knoop scale is 4.5×10^6 psi. (Coors Ceramics) (5) Silicon-nitride ball bearings. (Cerbec, Inc.) (6) Ceramic foam filter. (Selee Corp.) (7) Photovoltaic cell on light that never requires new batteries. (8) Ruby crystal used in laser light pointer. (9) Chemically toughened lens blank, which will be ground to form an eyeglass lens. (American Ceramics Society)

ramic parts, some of which depend on their ability to polarize their electric charge; others are chosen because of their hardness.

22.1 Thick-Film and Thin-Film Ceramic Devices

Thick- and thin-film ceramic devices are of major significance to the computer industry, particularly as ceramic substrates (mostly alumina) onto which semiconductor integrated circuits are mounted. These substrates are generally single-layer or multilayer ceramic (MLC) devices that employ thick- or thin-film metal layers to serve as circuitry. A substrate is usually a base material, but in these devices it includes the composite of ceramic with the metal layer.

For ***thick-film*** devices, the layers begin as continuous slip-cast sheets of greenware, which are blanked to size, then silk-screened. The ceramic substrates consist of 92% alumina and frit made up of alumina, calcia, magnesia, and silica. (Frit is a form of glass that has been melted, cooled, then broken into small particles.) The Al_2O_3 substrate receives a layer of metal paste, such as gold, platinum, palladium, or silver. The paste is printed by silk screen in a circuitry pattern (designed to carry the electron signal) onto either the greenware or bisque. The bisque substrates are stacked in layers (5 to 35) and sintered at about 850 °C, which changes the paste structure into resistors and conductors; greenware and paste are fired together (cofiring). Computer-controlled lasers may trim resistors to exact resistance values before the chip is soldered onto the composite substrate. ***Thin-film*** circuits composed of microthin layers (about 150 nm) of material, such as gold or nickel–chromium alloys, are produced through vacuum deposition onto ceramic substrates such as alumina, glass, or beryllia. The references listed at the end of this module contain information on specific techniques, such as sputtering, evaporation, and electron-beam deposition.

22.2 SMART CERAMICS MATERIALS

Electrorheological (ER) and ***magnetorheological (MR)*** fluids, covered in other parts of this book, use inorganic particles and (LCD) liquid crystal displays that change their structure with the application of an electrical current. Optical fibers are used to construct ***artificial nerves*** and ***sensors,*** part of fiber-optic technology. The glass optical fibers can be embedded in a host of materials or bonded to surfaces and are connected to instruments in order to receive signals. In a smart materials system, the movement of embedded fibers breaks a light path to reveal movement or rupture due to various external forces, such as earthquakes shaking a building or bridge.

Optical fibers embedded in composites with "smart skins" employ various light-sensing systems to indicate material damage due to variation in pressure, such as may be experienced by the surface of a model aircraft in a wind tunnel. A similar approach can be used to test designs of automobiles. ***Piezoelectric materials,*** explained in Module 9, include lead zirconate titanate (PZT), which is the most popular piezoceramic. PZT acts as a control device in intelligent materials systems for active acoustic attenuation (reducing sound), active structural damping (reducing vibration), and active damage control (see Figure 22-2).

Another of the smart ceramics, ***magnetostrictive actuator materials,*** respond to magnetic fields in a manner different than PZT, which changes its microstructure upon the introduction of electrical currents. This group of intelligent materials includes Terfenol-D, a compound with the rare earth element terbium, at. no. 65. Terfenol-D is controlled with magnetic fields that bring on an alignment of magnetic domains, thereby causing controlled expansion of the magnetostrictive material. It can be used in vibration damping systems; high-torque, low-speed rotary motors; and hydraulic actuators.

Smart skis—*http://www.acx.com./cool_smartski.html*

Figure 22-2 "Smart Skis." All structures are subject to vibrations, which move or deform them. Snow skis are structures similar to a cantilever beam. Unwanted vibrations cause the edges of skis to lift off the snow, resulting in reduced control and responsiveness. Piezoelectric ceramics convert mechanical vibration energy into electrical energy that is dissipated through a tuned circuit in a shunt, in the form of light and heat, reducing the vibrations. Active Control eXperts, Inc. (ACX) has designed and developed a line of piezo control modules for damping vibrations for military applications and for the commercial market, such as mountain bikes and Formula One race cars. (American Ceramic Society)

22.3 ELECTROCERAMICS

Highly conducting electroceramics are comprised of inorganic single-crystal and polycrystalline compounds as well as glasses that can be used in a variety of electrical, optical, and magnetic applications. Oxide ceramics are the materials of primary interest, but carbides, fluorides, and sulfides are also finding many commercial applications.

Such ceramics are generally classified as electronic conductors, ionic conductors, mixed (electronic/ionic) conductors, and insulators. The electronic conductors include superconductors, metallic and ionic conductors, and semiconductors. This grouping represents materials with the highest conductivities. The metallic conductors exhibit high conductivities that decrease with increasing temperature; the conductivities of oxide semiconductors increase with temperature. Ionic conductors generally exhibit conductivities in the range 10^{-8} to 100 S/m (siemens per meter) that increase exponentially with temperature. A ***siemens*** (refer to Table A-5) is defined as the quotient of amperes divided by volts (A/V). Insulators such as high-purity alumina are at the lower extreme of the conductivity spectrum, with conductivities of 10^{-13} S/m. Figure 27-3a summarizes the electrical conductivity characteristics of these classes of ceramic materials.

22.4 NANOTECHNOLOGY

The drive for ever-smaller, faster, and cheaper processors for computers, smart materials, morphing aircraft, and computers will involve ceramics at the nanoscale. Figure 22–3 depicts the format of carbon nanotubes forming zinc-oxide nanowires. Figure 22–4 shows another

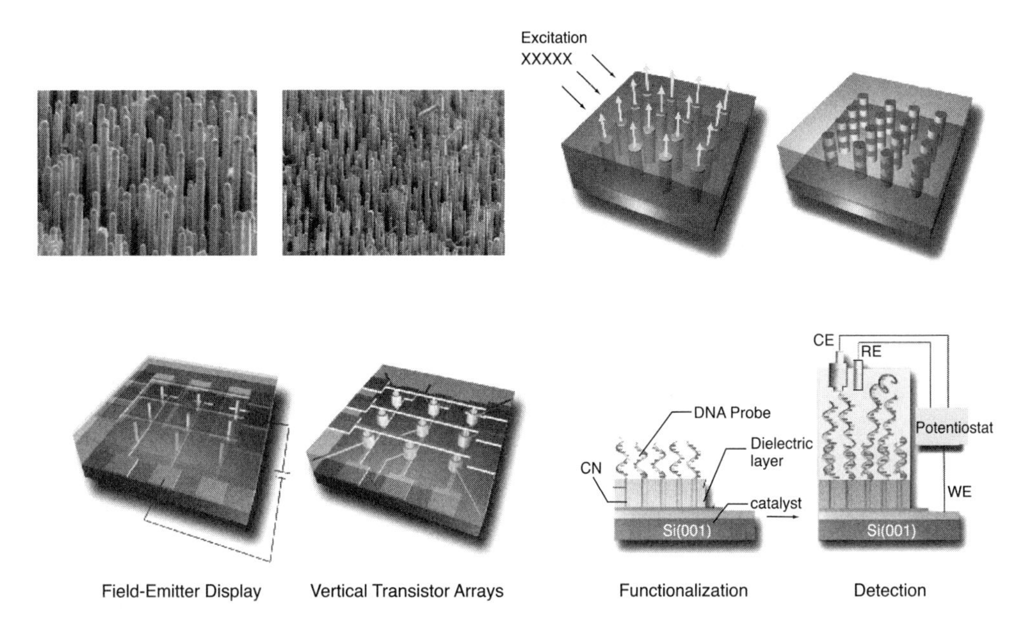

Figure 22-3 Zinc-oxide nanowires used for field-emitter display. Note the alignment of CNT nanowires. (NASA Ames)

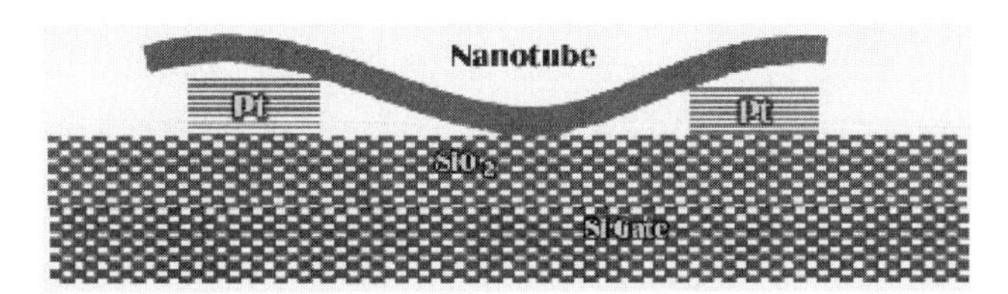

Figure 22-4 Carbon nanotubes used for quantum wire interconnects on a SiO_2 substrate forming a NOT gate for computer logic circuitry. The nanowire connects p-type transistors. (NASA Ames)

nanotube being used as a connecting wire in a semiconductor device; note the SiO_2 substrate with the CNT serving as quantum wires. CNT applications under R&D include diodes and transistors for computing, capacitors, data storage, field emitters for instrumentation, and flat-panel displays. Coupling biology with materials science, researchers attached a CNT to a specific site on a DNA strand. Proteins on the DNA strand and nanowires ***self-assemble*** (bind to each other without human intervention).

SELF-ASSESSMENT

22-1. List at least 10 applications for ceramic materials within a home.

22-2. A category of ceramics, such as piezoelectric materials, that act as sensors and for acoustic attenuation is

- **a.** Glass ceramics
- **b.** Hard ceramics
- **c.** Traditional ceramics
- **d.** Smart ceramics

22-3. Describe a smart ceramic, including examples of materials and applications.

22-4. Citing examples of applications, explain how use of ceramic nanotechnology can assist in advancing technology.

REFERENCES & RELATED READINGS

MEYYAPPAN, MEYYA. "Nanotechnology: Opportunities and Challenges", *National Educator's Workshop,* NASA/CP-2003, June, 2003.

ROGERS, CRAIG A. "Intelligent Materials Systems: The Dawn of a New Materials Age." Center for Intelligent Materials and Structures, Virginia Polytechnic Institute and State University, 1992.

PC Magazine. "Tiny Triumph," January 20, 2004, p. 25.

Module 23

Glass, Radioactive Materials, & Nuclear Fuels

After studying this module, you should be able to do the following:

23–1. List four types of glass and state their major properties and typical applications.

23–2. Explain, with the aid of sketches, methods of strengthening glass.

23–3. Explain the process of making polycrystalline glass and applications of this material and draw the heating curve for glass ceramics.

23–4. Discuss some benefits of ceramics with regard to the goals of green manufacturing.

23–5. Use the website addresses denoted by the symbol to explore sites for new developments and evolving concepts related to traditional and advanced ceramic materials.*

*The symbol found throughout the book will link you to Internet sites related to topics being covered. The dynamic nature of the Web brings frequent changes, so some of these sites, while available at the time of writing, may not be up now.

23.1 GLASS

Glass is both an ancient and a modern material. We see it in art museums as beautiful vases (Figure 23-1), in our kitchens as drinking glasses and cooking surfaces, on skyscrapers as reflective glazing, and in communication systems as optical fibers. This module will explore the many types and applications of glass. The American Society for Testing and Materials (ASTM) classifies ***glasses*** as inorganic products of fusion that have cooled to a rigid condition without crystallizing. They differ from glass-ceramics in that they lack polycrystalline structure. Some authorities prefer not to consider glass as a ceramic because of its structure, and thus give it a separate classification.

23.1.1 Nature of Glass

The base raw material of glass is the very pure, white silica sand found in abundant supply in the central United States and other parts of the world. Although there are approximately 750 different glasses and glass-ceramics, most can fit into six groups: soda-lime, lead-alkali, borosilicate, aluminosilicate, 96% silica, and fused silica.

The ***vitreous state*** of glass is mechanically rigid, like crystalline materials, yet it has an atomic space lattice that is amorphous, much like a liquid. Even though raw materials such as quartz sand (SiO_2) have a naturally crystalline structure (Figure 23-2a and Figure 19-8), when they melt the lattice breaks up (Figure 23-2b). Slow cooling will allow some crystals to form, but normal cooling produces an amorphous vitreous or glassy structure (Figure 23-2c), such as the SiO_4^{4-} tetrahedron (silicon oxide ion with a net charge of –4).

The discussion of phase diagrams in Section 10.3 pointed out that amorphous materials do not have clearly defined freezing or melting points as do crystalline materials; rather, they harden or liquefy over a range of temperatures. Molten glass becomes more viscous (Figure 23-3) and thickens to a working or softening point after cooling. Figure 23-3 is a plot of viscosity versus temperature for several different materials. (The centimeter, gram, and second (cgs) metric system of units uses the ***poise (P)*** as the unit for viscosity. The smaller multiple of the poise, the centipoise (cP), which is 1/100 of a poise, is commonly used instead of the SI system of units. $10P = 10^3 cP = 1\ Pa \cdot s$.) ***Viscosity*** is the tendency of a material to resist flow. As the temperature increases, this internal resistance to flow decreases in most materials, as indicated in the graphic. SAE multigrade oils such as 20–50 are designed to maintain viscosity as temperature increases. 20–50 means that the oil is 20 weight (wt) when cool but is like 50 wt pure oil when it is hot (engine running temperature). Figure 23-4 plots the heating of crystalline raw materials and glass. Curve *a–b* shows an increase in volume until the crys-

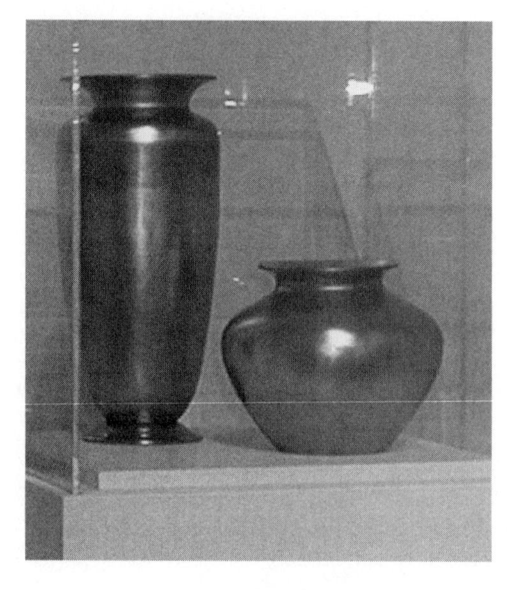

Figure 23-1 Vase of Gold Aurene glass, created around 1905 by Frederick Carder. Silver oxide added to glass batch rises to the surface in an oxidizing atmosphere to form a mirrorlike surface; it produces an iridescent glow when sprayed with stannous oxide. The glass was studied for electrical conductivity in efforts to produce technical glasses. See more at *PVGResearch@alfred.edu.* (Photo from Paul Vickers Gardner Class Center by Margaret Rasmussen)

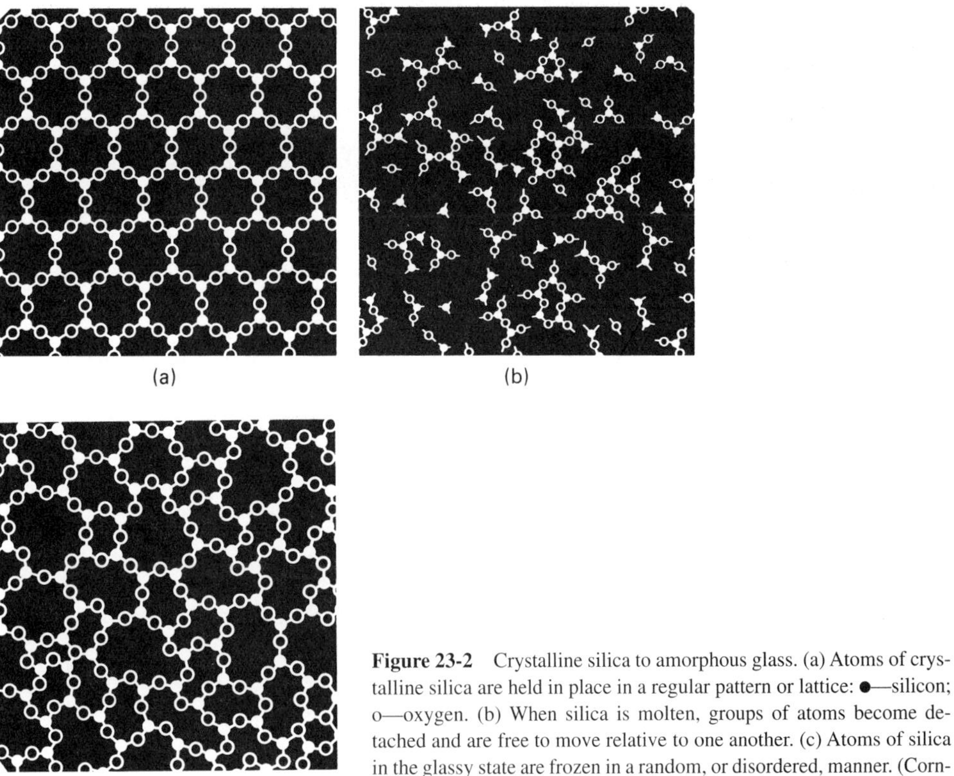

Figure 23-2 Crystalline silica to amorphous glass. (a) Atoms of crystalline silica are held in place in a regular pattern or lattice: ●—silicon; o—oxygen. (b) When silica is molten, groups of atoms become detached and are free to move relative to one another. (c) Atoms of silica in the glassy state are frozen in a random, or disordered, manner. (Corning Glass Works)

talline material becomes molten (liquidus temperature); the increased temperature yields a corresponding increase in volume as the atoms expand. Glass is formed as the raw materials cool through curve *d–c–e,* but the volume remains higher at point *e* than when the raw materials were in a closely packed crystalline state. This curve represents a normally cooled glass.

Slower cooling permits some crystallization in the process known as ***devitrification,*** the conversion of glass or other noncrystalline solids into polycrystalline solids. This chemical reaction is opposite to that of ***vitrification*** (producing amorphous structures). Used commercially, devitrification produces high-quality, homogeneous, fine-grained, nonporous ***glass-ceramics*** (see Section 23.2). Devitrification can occur only below the liquidus, so to control the glass structure, it is important to know the location of the liquidus line (point), as with metals. High resistance to devitrification is desirable and prominent in commercial glasses.

A variety of raw materials yield glass compositions of assorted oxides. Oxides such as SiO_2 (silicon dioxide), B_2O_3 (boron trioxide), GeO_2 (germanium dioxide), and V_2O_5 (vanadium pentoxide) are glass formers. ***Glass formers*** or ***network formers*** are those oxides that promote the ionic linking or polymerization of oxide molecules in the glass compound. Network modifiers or fluxes such as lead, zinc, and alkali ions lower the liquidus temperature, improve workability, and change thermal and optical properties. Other stabilizers, such as CaO (calcium oxide), improve chemical properties.

One technique for producing glass objects is hand ***glass blowing*** using a blow tube and molds, which dates back to around 200 B.C. Glass blowers were in evidence among the earliest settlers in Virginia. Hand blowing of glass is now used primarily by hobbyists to produce glass articles for their aesthetic value. Most modern shaping of glass by blowing is performed by highly automated equipment that uses a mold to shape jars, bottles, drinking glasses, incandescent light bulbs, and other pieces. Other shaping processes are continuous processes for making sheet and plate glasses for windows and tubing, and fiber-making processes for making fibers for such items as fiberglass composite materials, fiber optics, and insulation.

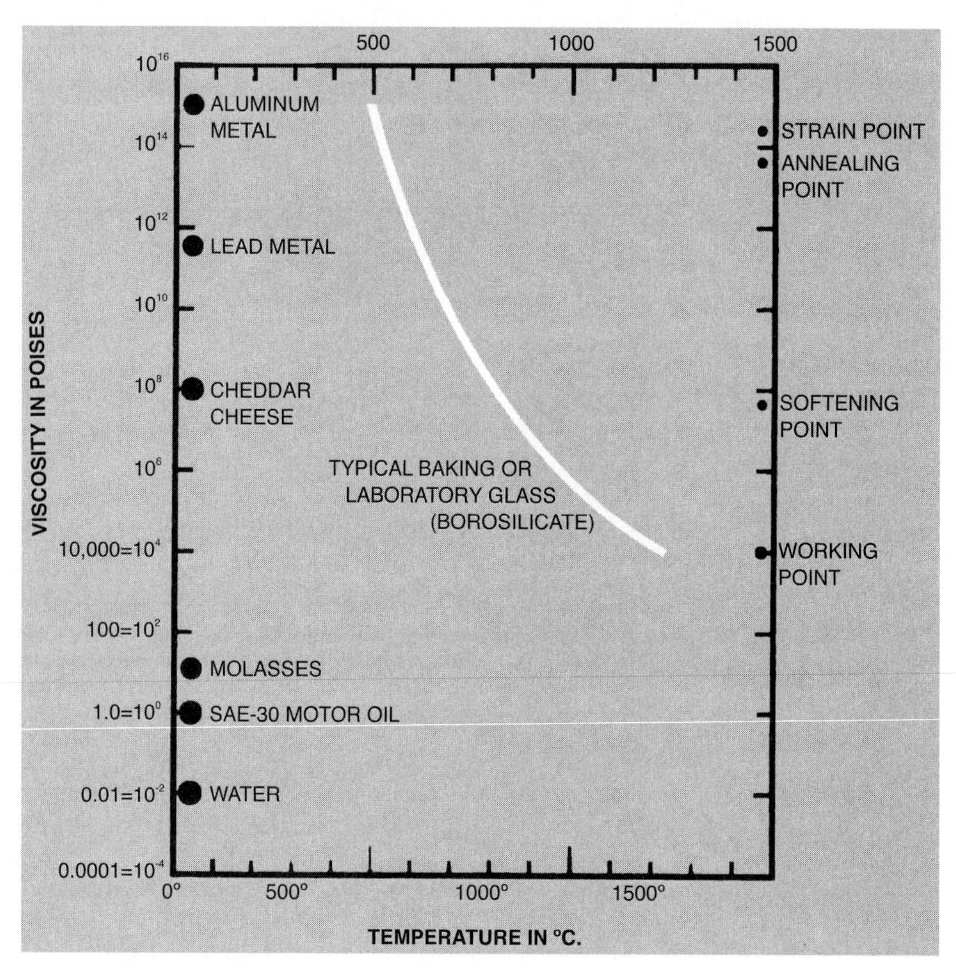

Figure 23-3 Glass viscosity. As glass is heated it becomes gradually less viscous. (Poise—absolute viscosity in Pascal seconds [Pa • s]. The measure of force required to overcome resistance to flow.) (Corning Glass Works)

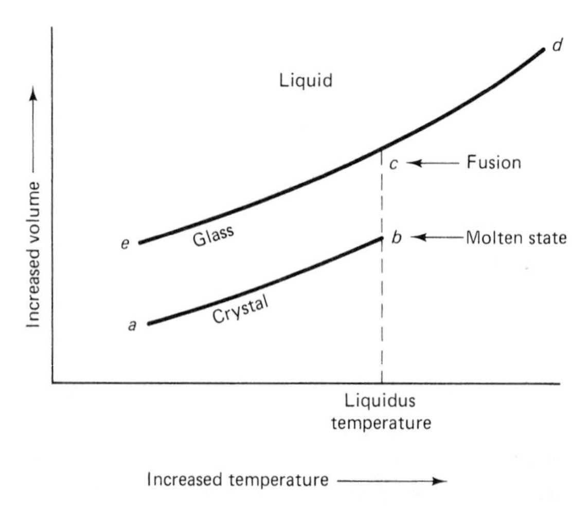

Figure 23-4 Glass formation.

Glass is a nearly perfect elastic material. At any stress below breaking stress, glass will return to its original shape when the stress is removed. Some plastic flow can be achieved, but the amount of stress must nearly equal its ultimate strength. Except when internal flaws are present, glass will fail in tension on the surface opposite the compressive stress. Whereas crystalline materials fracture along planes of slippage that may not be normal with the tensile stress, the amorphous structure of glass causes fracture to be normal to the stress. Heat treatment of glass can either reduce internal stress or create high internal stress. High stresses could lead to breakage from relatively minor forces. ***Annealing*** of glass through slow cooling provides a homogeneous structure by reducing internal stresses to give isotropic (equal in all directions) properties. ***Tempering*** involves rapid cooling of the outer surface of glass while it is still in the plastic state. (See the discussion of the tempering of steel in Section 9.4 to note the differences in meanings when these techniques are applied to steels.) The tempering results in compressive stress on the surface and tensile stresses in the core (see Figure 19-28), which must be overcome before the tensile stresses can act on the surfaces. This condition occurs because the slower-cooling core tries to contract but is restrained by the rigid outer glass. The nonequilibrium condition causes tempered glass to fracture into small pieces rather than large splinters. Tempered glass is now required on doors in buildings and homes because of this safety feature. This is similar to prestressing concrete, which is also weak in tension.

Table 23-1 provides a comparison of the electrical properties (see Module 6) of various insulating materials. The last material in the table, *steatite,* is an alumina-based ceramic used in spark plug insulators. Most of the glasses and ceramics have higher dielectric constants than those of plastics and rubber. Except for soda-lime glass, they offer greater volume resistivity than that of organic polymers. Table A-13 provides a comparison of the mechanical, physical, and chemical properties of selected glasses. With petroleum getting scarcer, it is conceivable that glass and other ceramics will replace plastics in many applications.

TABLE 23-1 COMPARISON OF ELECTRICAL PROPERTIES OF INSULATING MATERIALS AT ROOM TEMPERATURE

	Intrinsic dielectric strength[a]					
Thickness material	mm	in. $\times\Omega 10^{-3}$	kV/cm	kV/in.	Dielectric constant	Volume resistivity ($\Omega\cdot$cm)
Cellulose acetate	0.025–0.12	0.98–4.7	2,300[b]	5,840	5.5	10^{12}
Glass						
Borosilicate code 7740	0.10	3.9	4,800[c]	12,200	4.8	10^{16}
Soda-lime	0.10	3.9	4,500[c]	11,400	7.0	10^{12}
Soda-lead	0.10	3.9	3,100[c]	7,880	8.2	10^{14}
Mica, muscovite clear	0.020	0.79	3,000	7,620	7.3	10^{17}
ruby	0.10	3.9	8,200[b]	20,850		
Phenolic resin	0.012	0.47	2,600	6,600	7.5	10^{11}
	0.04	1.6	3,300[b]	8,380		
Porcelain, electrical			380[b]	965	4.4–6.8	10^{14}
Silica, fused			5,000[c]	12,700	3.5	10^{18}
Rubber, hard	0.10	3.9	2,150[b]	5,460	2.8	10^{13}
	0.30	11.8				
Porcelain, steatite—low loss			500[b]	1,270	6.0–6.5	10^{15}

Source: C.J. Phillips, *Glass, the Miracle Maker,* Pitman Publishing Co., New York, 1987.

[a]Intrinsic dielectric strength can be realized only under special test conditions and is very much higher than the working dielectric strength attainable in ordinary service. These data are listed for purposes of comparison.

[b]Values of S. Whitehead, *World Power,* p. 72, Sept. 1936.

[c]Values of P. H. Moon and A. S. Norcross, *Trans. AIEE* 49, 755 (1930).

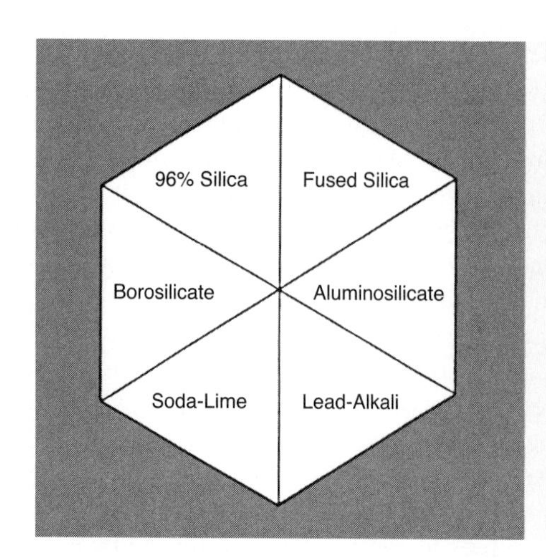

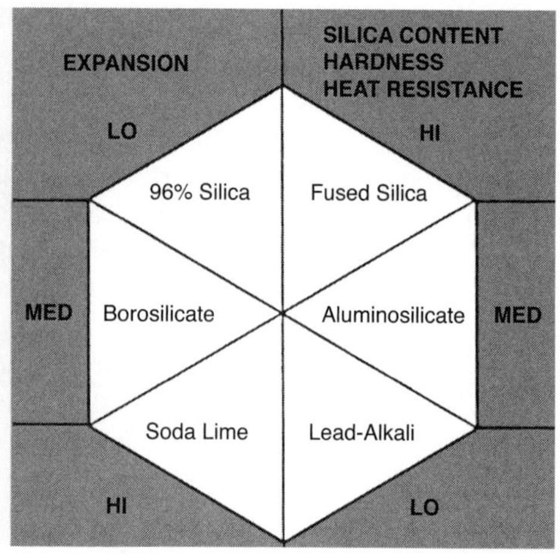

Figure 23-5 Categories of glass. Hard glasses are lower in thermal expansion and higher in both heat resistance and silica content than soft glasses. (Corning Glass Works)

23.1.2 Types and Properties of Glass

The six categories of glass are diagrammed in Figure 23-5, where they are divided into soft glasses and hard glasses. The classification of hard or soft does not denote mechanical hardness but, rather, the ability to resist heat. ***Soft glasses,*** those that soften or fuse at relatively low temperatures, include soda-lime and lead-alkali. Soft glasses have lower heat resistance and a higher coefficient of thermal expansion than hard glasses. ***Hard glasses*** are borosilicate and aluminosilicate; the hardest high-silica glasses are 96% silica and fused silica.

Soda-lime silica glass is the oldest glass, dating back 4000 years, and is most familiar to you in the form of windowpanes and bottles. It accounts for more than 90% of the glass produced. The typical composition is 74% silica (SiO_2), 15% soda (Na_2O), 10% lime (CaO), and 1% alumina (Al_2O_3). The least expensive glass, soda-lime glass has a composition of oxides of silicon (from silica sand), calcium, and sodium. Although easy to form and cut into many designs, soda-lime glass has only fair chemical resistance and cannot endure high temperatures or rapid thermal change (thermal shock). Because it shatters so easily into dangerous splinters, it is not recommended for doors or other glazings where persons might damage it. Thinner soda-lime glass, such as that used for small windowpanes, is ***single-strength glass.*** Increased thickness is referred to by strength, for example, ***double-strength glass.***

Lead-alkali silica glass is slightly more expensive than soda-lime; both are considered soft glass and neither will endure high temperatures or thermal shock. ***Borosilicate glass*** is considered a hard (thermally) glass; it is the oldest type of heat-resistant glass. Pyrex is its familiar Corning trade name. Although three times as costly, ***aluminosilicate glass*** can sustain higher service temperatures than borosilicates and is similar in its ability to handle thermal shock. The ***96% silica glass,*** a valuable industrial glass, is capable of setting on a block of ice and having molten metal poured over it without breaking, attesting to its superior thermal hardness (see Figure 19-17). ***Fused silica glass*** is composed of at least 99% silicon dioxide (SiO_2). This pure glass is the most costly and most resistant to heat. It is capable of service temperatures of from 900° to 1200°C and has the highest corrosive resistance. It is superior for the transmission of ultraviolet (UV) rays.

Transparency colors are achieved in soda-lime, lead, borosilicate, and 96% silica glasses by the addition of elements that change the structure to absorb certain bands of the light spectra. For example, manganese ions (Mg^+) can produce purple glass, and iron yields green and brown bottle glass. ***Colorants*** such as iron are very undesirable in clear glasses and are kept at low levels by selecting clean sand deposits as a silica source.

 More lighting, color, and materials—*www.sci-ed-ga.org*

23.1.2.1 Specialty glasses. Using the six basic groups just discussed, it is possible to develop glasses with unique properties to meet special needs. Some of the specialty glasses are tempered, optical, colored, sintered, glazing, fibrous, laminated, cellular, photosensitive, and light-sensitive glasses. Advances in optical glass technology have resulted from several societal and economic demands, including the need for greater transmittance (light transmission), safety (impact resistance), styling, and convenience, and competition from the plastics industry. Consumer groups and the federal government are requiring more energy efficiency in heating and cooling buildings, and specialty glass and glazing manufacturers are responding with a host of new materials and products.

Aluminosilicate is a common ceramic packaging material for IC chips because it hermetically seals the chip. Al_2O_3 is dispersed in a liquid binder, formed into sheets, and dried. The sheets are then laminated by pressing and sintering to form a monolithic body. Another process sometimes chosen to seal chips uses a refractory glass ($PbO–ZnO–B_2O_3$) with a melting point around 400°C. Burn-in testing may include elevated temperature testing in an oven at 125°C for 24 hours.

The hard, stable nature of glass permits grinding and polishing of optical lenses for telescopes to 1/6000 of 1%. Glass is also used in corrective lenses; and photochromic lenses provide eyeglasses that darken in bright light but lighten up in subdued light for style and convenience. This ability to change the transmittance of light results from the use of silver ions. When photons of light strike the silver particles, they change from silver ions to metallic silver and absorb more light. When the light source is removed, they revert back to silver ions (Ag^+) (see ***transmittance*** in Section 6.4.5) and the lenses lighten in color. ***Photosensitive (PS)*** glass also changes its chemical characteristics when exposed to light. The distinction between PC and PS glasses is that in PS glass the change is permanent; that is, it acts like a photographic film. In addition to silver (Ag), PS glass contains halogens (Br and Fl). As the glass cools during manufacture, minute crystalline droplets (5–30 nm) of silver halide precipitate to form a dispersion so fine-grained that it does not affect the transparency or color of the glass. The surface interacts with the UV and violet wavelengths from the sun to form a tiny speck of pure Ag, which absorbs a portion of the light. As light disappears, the silver recombines with the halogen to form a transparent silver halide and the glass fades to colorless. Dissolving atoms in the glass, exposing it to UV light, and heat treating produce many PC glasses with remarkable properties.

Lead lanthanum zirconate titanate (PLZT) discovered by Haertling at Sandia National Laboratory in 1969 is a transparent polymeric crystalline form of an ***electrooptic ceramic.*** Used in a lens that allows light to pass through under normal lighting conditions and coupled with a photodiode sensor (***photodiode sensors*** detect light and send out an electrical signal), an electrical signal is transmitted to the lens, which changes the electric field in the lens, making it opaque in less than 150 μs. Gallium arsenide, a compound that can convert light directly into electricity (***photoconduction***), is used in LEDs and transistors.

23.1.2.2 Chemically toughened glasses. This type of ***tempered glass,*** also known as chemically tempered glass, is produced by using a molten salt bath in which eyeglass lenses are soaked for hours to achieve appropriate impact strength. During the long-term soaking, larger potassium ions from the salt bath replace sodium ions, as shown in Figures 19-26 and 19-28. This crowds the surface layer and causes compressive stresses. The impact strength of the tempered glass is then a product of its untempered strength and the amount of surface compression developed in chemical toughening. A chemically tempered glass is stronger than a heat-tempered glass.

23.1.2.3 Glazing. Glass is an important architectural material; it adds beauty and it can improve energy efficiency if properly used in design. In Module 21, the word *glaze* or *glazing* described a hard, glossy finish or coating on an object for purposes of adding color and/or making the coated object impervious to moisture. Here these words refer to window glass. For example, ***glazing point*** or ***glazing compound*** pertains to setting glass in a window frame. A variety of types of glass serve as glazing, or glass windows. Most older buildings utilize single soda-lime sheet glass. Because the cost of energy to supply heating and cooling has increased, improved glass technology has resulted in ***insulating glasses*** that are constructed of two or three panes, which are often fused together on the edges. Spaces between the sheets of glass are sometimes filled with gases to provide low thermal conductivity and improved sound dampening. Additives to glazing glass and reflective metallic/plastic films are also techniques for reducing the transmission of ultraviolet (UV) rays (see Section 6.4 and Figure 19-18).

23.1.2.4 Glass fibers. Glass fiber can be woven and matted into sheets or used in loose form. The composition of glass fibers differs according to function. E-glass, used as reinforcing fiber, consists of calcium alumino-borosilicate, which produces high tensile strength (3.45 GPa). Along with excellent electrical properties, dimensional stability, and low cost, it is the most widely used fiber and is found particularly in printed circuit boards and as reinforcing fiber for plastic materials. S-glass which contains magnesium aluminosilicate, has greater tensile strength than E-glass (4.59 GPa) and is more expensive. Its tensile strength, combined with high thermal stability, makes S-glass suitable for use in the aircraft/aerospace industry. Glass fibers are produced by various techniques, including drawing and blowing. (See Glossary 26.6.)

Fibers can be produced as thin, continuous filaments or discontinuous fiber segments. A 5/8-in.-diameter glass marble can produce 0.97 mile of filament. These thin fibers are used as reinforcers in standard, woven, and other forms of plastic resins, such as fiberglass. (Fiberglass is discussed extensively in Module 24.) Segmented or discontinuous glass fiber is also used as a very effective thermal insulation in buildings, in refrigeration and heating units, and in land and air vehicles. Its light weight and good insulation properties, coupled with low cost, make glass fiber very popular.

Optical fibers, used to transmit light, are gaining wide acceptance in the communications field because they can be produced from plentiful silica sand and replace heavier, bulkier, and more expensive copper and aluminum conductors. A single optical fiber the size of a hair has the potential to transmit several thousand voice signals, compared to the fewer than 50 voice signals that can be carried on a copper wire of the same size (more than 600 MB/s). For this reason, telephone, television, computer, and other communications systems are moving to smaller, coaxial fiber-optic cables to replace copper-stranded cables. Optical fibers of glass and plastic have been used for many years in the medical profession and by engineers as inspection tubes. Flexible fibers on probes or endoscopes are inserted into the human body or into a motor and attached to a television system or magnifying lens to allow viewing of these otherwise inaccessible places. Photons of electromagnetic radiation (light) replace electrons in signal transmission through these transparent optical fibers. The speed, the distance, and the amount of information (density) in the signals have been increased beyond all imagination. Figure 23-6 shows some optical fibers and their connectors.

Several forms of optical fibers exist, including glass cores clad with silica and cores coated with plastic (Figures 23-7 and 23-8). The high-purity glass fibers can transmit more than 95% of the light beamed into them for over 1 km, have a tensile strength of over 4137 MPa, and possses good flexural strength.

Optical properties and electronic structure —
http://www.lrsm.upenn.edu/frenchrh/opes.htm

The cladding of fiber-optic cables with a transparent covering whose index of refraction is lower than that of the fiber prevents leakage of light. The transatlantic cable TAT-8 uses two pairs of glass fibers simultaneously. Because of the lower quality of the older fibers, it is necessary to boost the signal in the cable every 50 km using repeater stations. With the use of ***laser light*** and thinner cable (10 μm in diameter), little energy is lost and boosting stations are seldom needed. Figure 23-8b is a sketch of an optical fiber showing the cladding and the coating of the cable, both of which consist of sophisticated glass-fiber materials designed to enhance and protect the signal transmission through the core. Present-day technology limits the laser transmission of data over

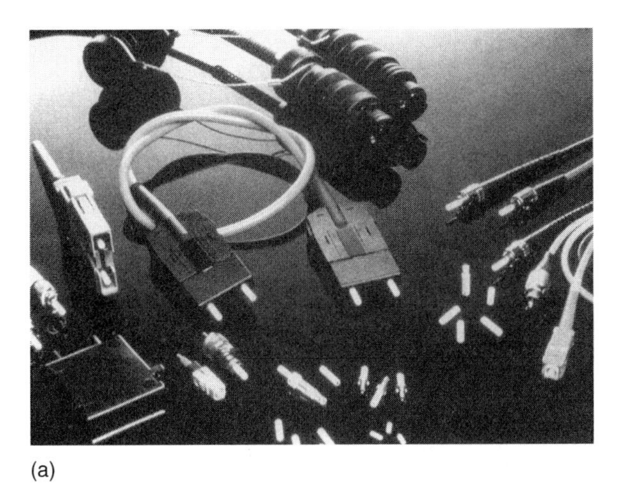

(a)

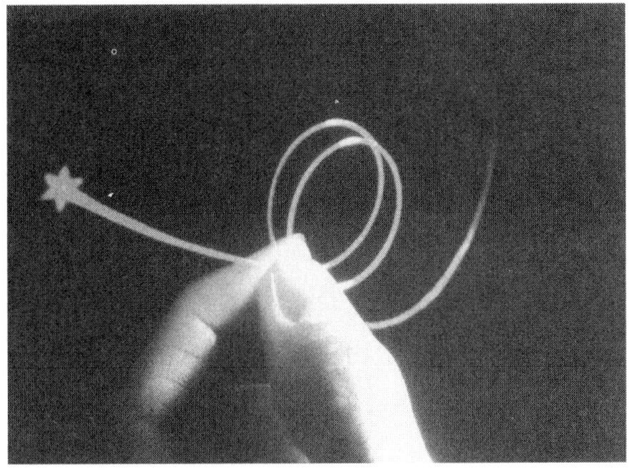

(b)

Figure 23-6 (a) Zirconia ferrules and examples of fiber-optic connectors in which they are used. (Coors Ceramics Co.) (b) Glass optical fibers have provided a technological breakthrough in the area of telecommunications. Information that was once carried electrically through hundreds of copper wires can now be carried through just one high-quality transparent silica (glass) fiber. Using this technology has increased the speed and volume of information that can be carried by orders of magnitude over what is possible using copper cable. The reliability of the transmitted information is also greatly improved with fiber-optic cables. (American Ceramic Society)

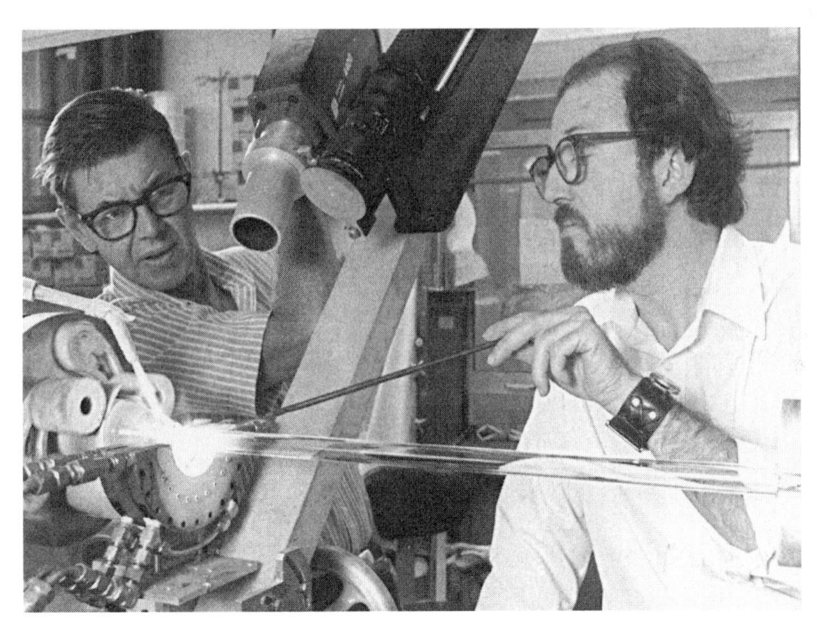

Figure 23-7 Optical fibers. The fibermaking process uses highly purified glass layers that form as chemical vapors react within the silica tube; the tube and its contents are heated until they collapse into a solid glass rod, from which nearly 10 miles of fiber are drawn in the modified chemical vapor deposition process. Single-strand, optical waveguide has the potential, through fiber-optic technology, to transmit 10,000 simultaneous telephone conversations when used in pairs. (Corning Glass Works)

fiber-optic cable to some 50 mi and numerous lasers are required to keep the signals stable. This represents a data capacity of about 40 gigabits per second (40 Gbps or 40×10^9 bps) and a cable capacity of 10 terabits per second (10×10^{12} bps). Silkroad of San Diego, California, has developed a single-laser transmission that is now capable of transmitting 200 Gbps over 200 mi. This new technique, called *refraction synchronization communications*, is bidirectional, or capable of being used interactively. Fiber-optic sensors are based on optical signal transmission through sensitive optical fibers. Environmental changes such as the weight of a passing vehicle may cause

(a)

(b)

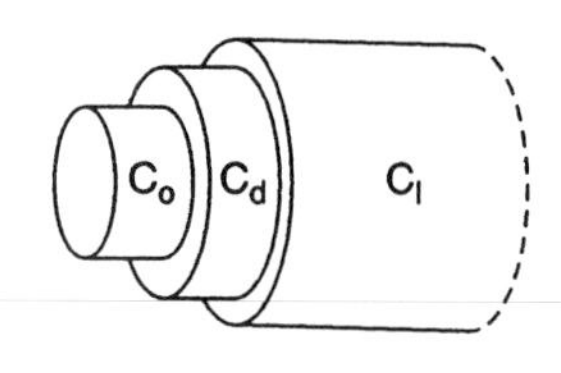

Figure 23-8 (a) Spool of optical fiber with laser light exiting. (American Ceramic Society) (b) The core consists of extremely high purity (flow-free) silica glass with diameters that range from 5 to 100 μm. The containment of the light rays within the core is the job of the cladding while the outer coating protects both the core and the cladding. Each of these components is designed to accomplish certain objectives, and each is the product of sophisticated and advanced processing techniques. This rendering is not to scale.

an induced strain, which, in turn, affects the intensity of light transmitted through the fiber. Another design may be based on micro-bending effect in which the fibers are bent, causing a leakage of light and thus reducing the total internal reflection of the light.

23.2 GLASS-CERAMICS

The glass-ceramics group includes ß-cordierite ($2MgO \cdot 2Al_2O_3 \cdot 5SO_2$) (Figure 23-9) and ß-spodumene ($Li_2O \cdot Al_2O_3 \cdot 4SiO$), which are specific phases of glass that possess very low expansion coefficients and superior resistance to oxidation when subjected to high heat. As discussed earlier (Section 23.1.2), glass will crystallize under a range of temperatures, including slower cooling processes than are normally used in glass manufacture. Crystals form on the surfaces of glass with impurities and will cause failure of the glass part. However, controlled heat treatment can produce a fully crystallized material known as ***glass-ceramic.***

Glass-ceramics are polycrystalline glasses that have four to five times the strength of glass and a mechanical hardness about equal to that of tool steel. This polycrystalline glass is capable of developing structures that resist extreme thermal shock (rapid change from cold to hot, or vice versa). Figure 23-9 shows an application of a glass-ceramic that was selected and designed for an array of sophisticated properties. Cookware is another familiar form of glass-

Figure 23-9 Automobile internal engines contribute to photochemical smog by exhausting unburned and partially burned hydrocarbons (HC), compounds containing carbon and hydrogen. Most (65%) HC vapor in the atmosphere comes from exhaust gas. Catalytic converters reduce HC and carbon monoxide (CO) emissions in auto exhaust by coating a substrate with a catalyst such as platinum (Pt) and allowing oxidation of the exhaust products to take place, producing harmless water, nitrogen (N), and carbon dioxide (CO_2). Cordierite, a glass-ceramic and one of over 1000 different silicates of a magnesium aluminum silicate (MAS), is the main ceramic material used as a substrate for this application. One reason for its selection is its high temperature capability. (Corning Inc., Corning, NY)

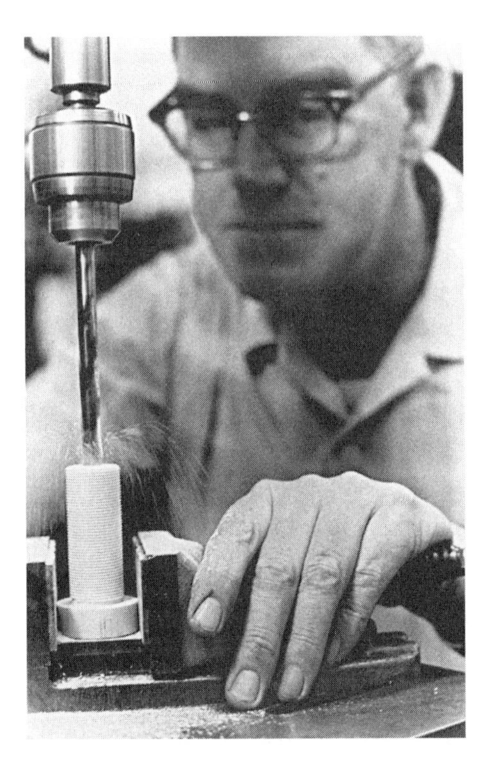

Figure 23-10 Machinable glass ceramic. (Corning Glass Works)

ceramic. One polycrystalline glass-ceramic consisting of mica crystals in an opal-glass matrix (known by the trade name ***Macor***) can be machined with standard metalworking tools (Figure 23-10). Randomly oriented crystals about 20 μm in length and width keep fracturing locally instead of propagating (spreading), as would a crack in an amorphous glass or plastic. The fine crystalline structure of glass-ceramics is achieved by the introduction of a nucleating agent into the ceramic compound (Figure 23-11). Typical nucleating agents include phosphorus oxide

Figure 23-11 A radome is a protective covering or fairing as well as a window for electronic guidance and protection equipment in aircraft, missiles, and spacecraft. It provides protection against high-velocity impacts of various particles in the atmosphere as well as rain. It functions as a window to allow the transmission and receipt of electromagnetic radiation from and to the craft (high transparency to radar). Most radomes are made of Pyroceram 9606™, a glass-ceramic produced by Corning Glass Works.

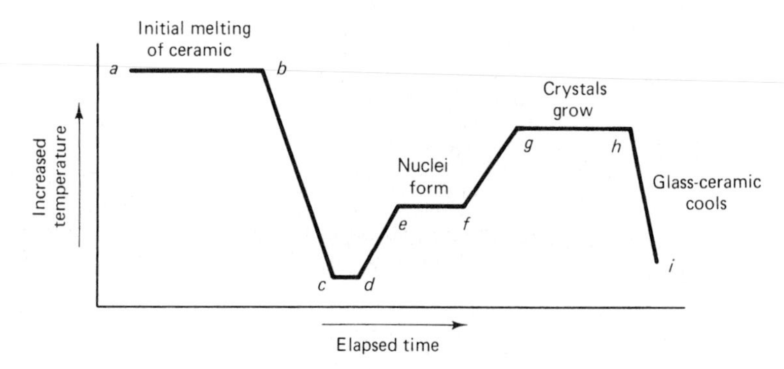

Figure 23-12 Heating curve for crystal formation in glass ceramics.

(P_2O_5), fluorides, titanium oxide (TiO_2), platinum, and zirconium oxide (ZrO_2). These nucleating agents are barely soluble in glass and remain in solution at high temperatures. At lower temperatures the agents precipitate out of solution to become seeds or nuclei around which crystals grow. The process is shown in Figure 23-12. A ceramic composition is allowed to cool down as a normal glass *(b–c),* is raised slightly to nucleation temperature and held there *(d–e–f),* and then the temperature is raised and held for crystal growth *(f–g–h)*. Figure 23-13 shows the formation of a glass crystal from an amorphous opal glass with fluorine-rich droplets dispersed throughout. Through heat treatment, it is transformed to a fully crystalline structure. Such crystallization of glass-ceramics exceeds 95% and forms grains between 0.1 and 1 μm in size, smaller than in typical ceramics. Glass-ceramic parts are made using typical glassmaking processes, including pressing, blowing, drawing, and centrifugal casting.

23.3 GLASS AND GREEN MANUFACTURING

The rules of (1) reduce, (2) reuse, and (3) recycle for green manufacturing apply to glass. Because glass is very hard and usually easily cleaned, it has many secondary uses. How often have you washed a mayonnaise or pickle jar and reused it to store another food in the refrig-

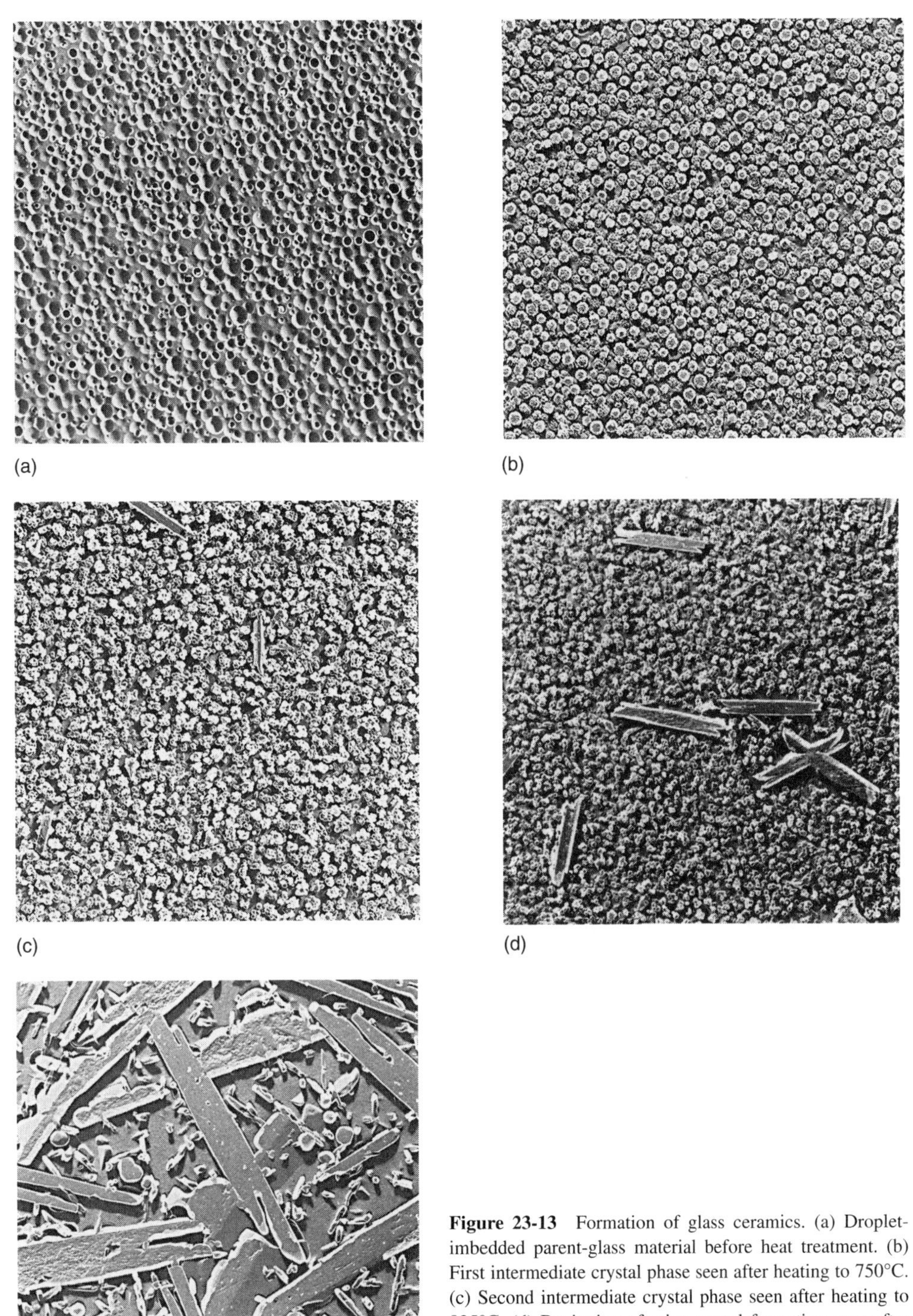

Figure 23-13 Formation of glass ceramics. (a) Droplet-imbedded parent-glass material before heat treatment. (b) First intermediate crystal phase seen after heating to 750°C. (c) Second intermediate crystal phase seen after heating to 825°C. (d) Beginning of mica-crystal formation seen after heating to 850°C. (e) Fully crystallized Macor glass ceramic seen after heating to 950°C. (Five micrographs from Corning Glass Works)

erator? The auto scrap industry finds uses for the glass in junked cars. Like steel, glass will last forever in the materials cycle. The recycling of glass, much like that of steel, involves a well-established infrastructure. Commercial glass such as bottles and jars requires recycled glass for its production. About 25% of these products consist of recycled glass.

Resources conserved by recycling glass include about 28% of the processing energy as a result of not having to extract glass from raw materials and also the raw materials for glassmaking, including sand, soda ash, limestone and feldspar. About a ton of raw materials is saved for every ton of recycled glass used. At the same time, recycling reduces the air pollution that would result from processing the raw materials, a reduction of about 28 pounds of particulate from the air for every ton of glass made, or a savings of about 17%. Mineral waste from extracting the raw materials is also avoided, and recycling keeps glass out of municipal landfills.

Recycling glass—*http://envirosystemsinc.com/glassfac.html*

23.4 MATERIALS SELECTION

A material(s) is needed with good thermal-shock resistance to serve as a liner for a high-temperature vacuum furnace. The liner will be in contact with molten aluminum. Strength requirements for the liner are minimal, but resistance to crack propagation is desired. The requirements are as follows:

1. Based on your knowledge of structures, bonding, and properties of ceramics, make a list of properties that the selected material(s) should have, based on the preceding information.
2. For each property listed in (1), explain why each property was chosen (i.e., justify your choice in terms of how the property adds to the overall characteristics of the material selected).
3. List suitable candidate materials that possess all or some of the properties listed in (2).
4. Select the material(s) and justify your selection. Consult the textbook tables of data or outside sources for appropriate data to include with your justification.

Hint: Multicoatings and additives such as fibers can be considered in making your selection.

23.5 LABORATORY ACTIVITIES AND CLASSROOM DEMONSTRATIONS WITH CERAMICS

The *Experiment in Materials Science, Engineering, and Technology (EMSET2)* CD-ROM, 2nd ed., includes numerous experiments and demonstrations of ceramics provided through the cooperative effort of scores of educators and materials specialists from industry. See Prentice-Hall's website, www.prenhall.com, for ordering information on the latest edition of *Experiments in Materials Science* (CD-ROM).

APPLICATIONS & ALTERNATIVES

Failures analysis in glasses and ceramic materials includes ***fractographic analysis*** or ***fractography***—the analysis of fractures under laboratory conditions of parts being tested. Fractography also serves as a powerful forensics tool for parts that break in service. The Ceramics Division of NIST contributes to fractography standards, for example, the revised ASTM C 1322 Fractography, a standard for interpreting and reporting fractographic analysis, and ASTM C 1161 (Flexural Strength Specimens). Scientists and technicians at NIST apply these standards and many more in conducting tests of new materials and in investigating failure of parts.

By studying fractured glass and ceramics with tools such as stereo binocular microscopes, a fractographer can determine points of stress at failure, as seen in the macroscopic views of shattered tempered glass (Figure 23-15) from the side window of a car involved in a tornado. See the origin of impact (arrow). Note that due to tempering the glass broke into relatively small pieces rather than the large, dangerous shards that one normally sees with untempered glass. Figure 23-14, the microscopic view of the window in figure 23-15, shows the origin of impact and the elements of the photomicrograph.

Figure 23-16 shows the steps for preparing flexural strength specimens according to ASTM C1161, as discussed in Modules 4 and 7 and shown in Figure 19-15. Note the detail

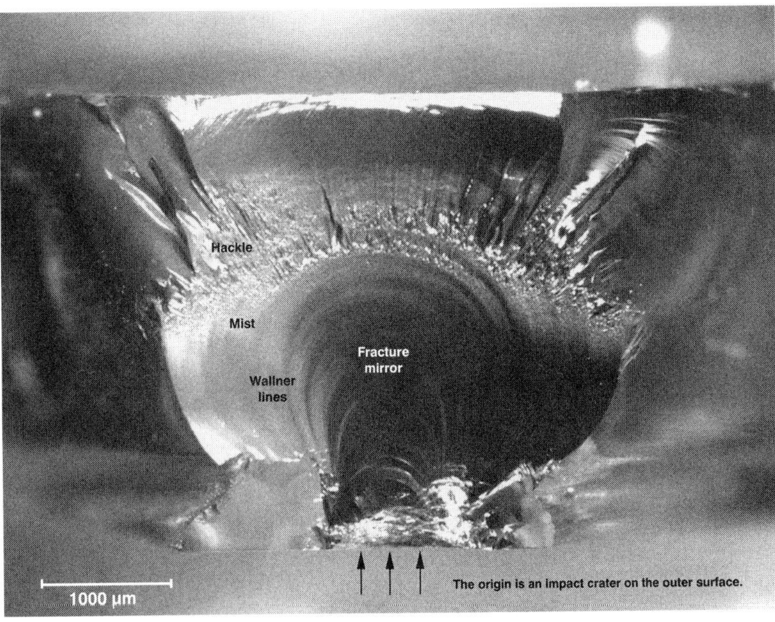

Figure 23-14 Result of tornado-thrown debris on tempered auto glass. Note in increasingly enlarged macroscopic views (a and b), the origin of impact (arrow), and (c) microscopic view. See 1000 µm scale bar. (George Quinn, NIST)

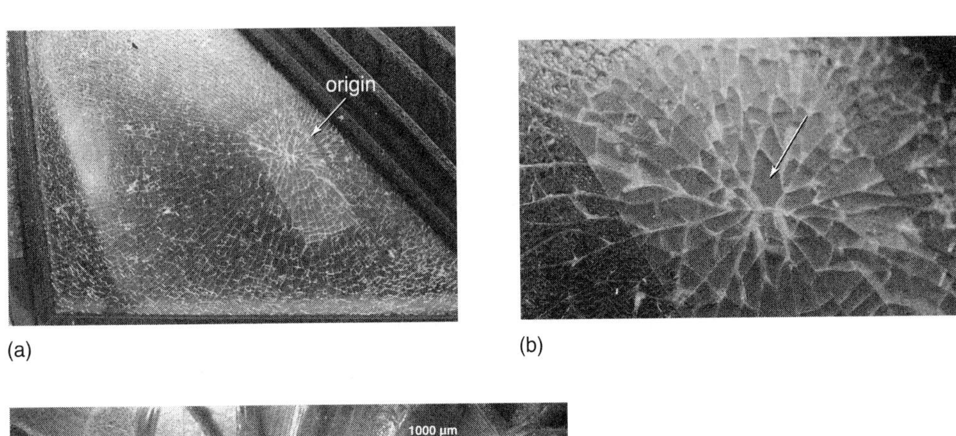

(a) (b)

(c)

Figure 23-15 Further magnification of origin of impact crater (note 1000 µm scale bar) on the outer surface of the window with fractography notations for elements of the fracture. (George Quinn, NIST)

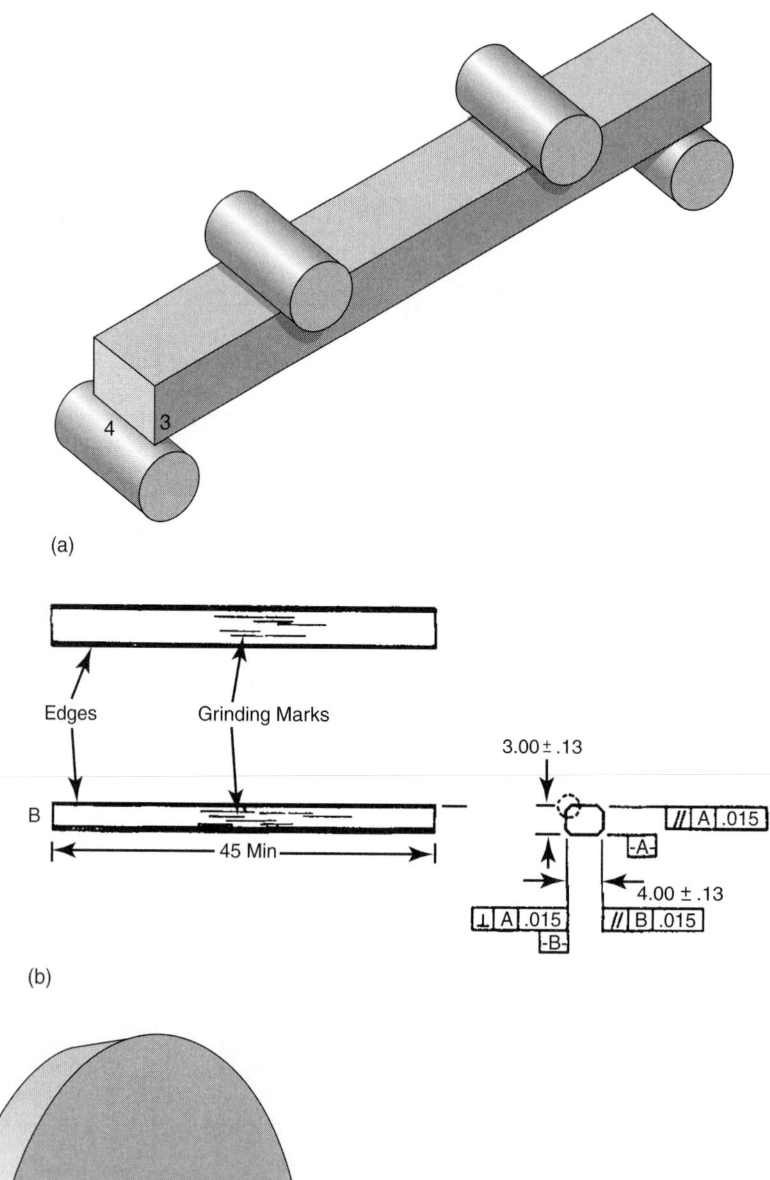

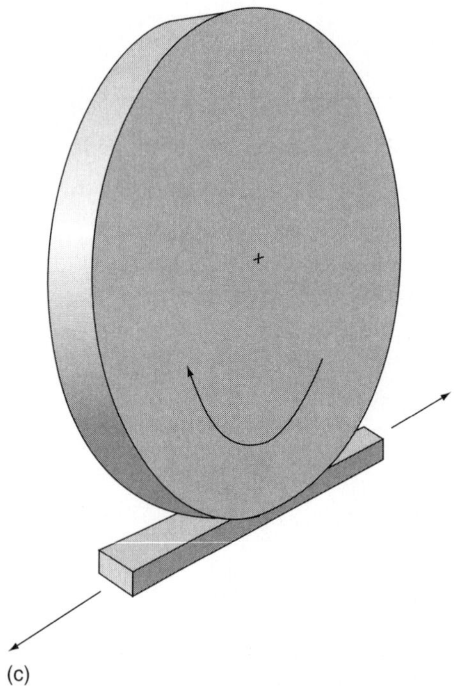

Figure 23-16 ASTM C1161 Flexural Strength Specimens. (a) Schematic of four-point flexural fixture. Note four rollers held by rubber bands in Figure 23-17. (b) Detail drawing of "B" specimen in C 1161. (c) Diamond abrasive grinding wheel for final finish of specimen with 400–600 grit. (George Quinn, NIST)

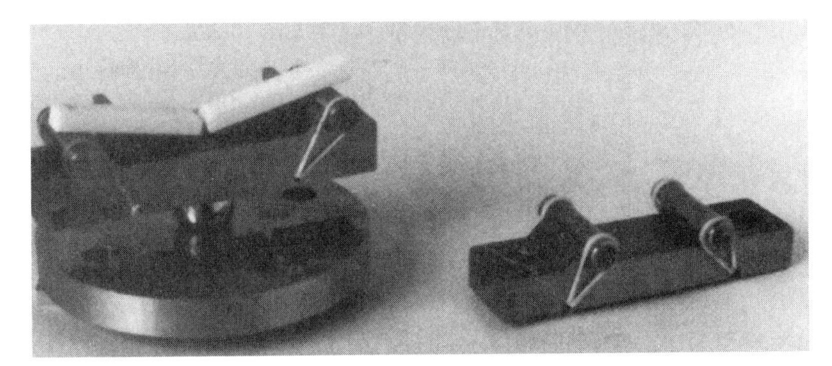

Figure 23-17 Actual four-point flexural strength test fixture with fractured ceramic specimen. (George Quinn, NIST)

drawing with the geometrical and positional tolerance symbols, direction of grind, and the specification for using a diamond abrasive grinding wheel. Figure 23-17 shows the ***four-point flexural test*** fixture and fractured part. Fractography on the specimen after failure will provide clues to improved materials and part design.

The two standards discussed here represent a very small fraction of those used in materials science, engineering, and technology.

SELF-ASSESSMENT

23-1. A very hard glass that withstands extremes of thermal shock is

a. Soda-lime glass
b. Polycrystalline glass
c. Lead glass
d. 96% silica glass

23-2. A ceramic-structured material produced by controlled heat treating of glass that has nucleating agents added to the mixture is

a. Soda-lime glass
b. Polycrystalline glass
c. Lead glass
d. 96% silica glass

23-3. Glass that was rapidly cooled while in the plastic state to place the surface in compression and the core in tension is

a. Annealed glass
b. Tempered glass
c. Soft glass
d. Chemically toughened glass

23-4. Most often the structure of which of the following materials is amorphous?

a. Diamond
b. Cement
c. Ceramic
d. Glass

23-5. In terms of green manufacturing, which is NOT an advantage that glass containers have over plastic containers?

a. Saves weight in transportation
b. 25% of waste glass used to make glass
c. One ton of raw materials saved for every ton of recycled glass used to make glass
d. Glass will last forever in the materials cycle

23-6. **a.** What effect does tempering have on glass?

b. How does light cause photochromic lenses to darken?

c. How does toughened glass obtain improved impact strength?

23-7. Match the ceramic to the application for which it is best suited:

a. Tempered glass	(1) Grinding and polishing
b. Fused silica	(2) Furnace windows
c. Ferrite	(3) Protect metal from oxidation
d. Glaze (porcelain)	(4) Computer memory disks
e. 96% silica glass	(5) Storm door glazing
f. Diamond	(6) Contain hot acids

23-8. Describe how glass fits into the concepts of green manufacturing.

REFERENCES & RELATED READINGS

AMERICAN SOCIETY FOR TESTING AND MATERIALS. *Annual Book of ASTM Standards* Part 17. *Refractories, Glass and Other Ceramic Materials; Carbon and Graphite Products.* Philadelphia, PA: ASTM, 2003.

BUNNELL, L. ROY. "Tempered Glass and Thermal Shock of Ceramic Materials," *NEW:Update 91,* Oak Ridge, TN: Oak Ridge National Laboratory, November 12–14, 1991.

CORNING GLASS WORKS. *All about Glass.* Corning, NY: CGW, 1968.

CORNING GLASS WORKS. *Properties of Glasses and Glass-Ceramics.* Corning, NY: CGW, 1973.

General Atomics Education Foundation website *www.sci-ed-ga.org,* January 20, 2003.

HOLAND, WOLFRAM, and GEORGE BEALL. *Glass-Ceramic Technology.* Westerville, OH: American Ceramics Society, 2002.

MACCHESNEY, J. B. "The Materials Development of Optical Fiber: A Case History," *Journal of Materials Education,* Volume 11, No. 4, 1989, p. 321.

PFAENDER, HEINZ G. *Schott Guide to Glass.* New York: Van Nostrand Reinhold, 1983.

QUINN, GEORGE. "ASTM Fractogapy Standard C 1322 Gets Major Overhaul," *American Ceramic Society Bulletin*, January 2003, pp. 28–29.

QUINN, GEORGE. "Ceramic Fractography: Broken Pieces tell the Story," *NEW: 2001—Standard Experiments in Engineering, Materials Science and Technology*, NASA/CP-2002–211735, June 2002, pp. 28–29.

QUINN, GEORGE. "Overhaul of ASTM C1161 Flexural Strength Specimens," *American Ceramic Society Bulletin*, May 2002, p. 65.

RASMUSSEN, MARGARET. "Glass Art Reflects the Future," *American Ceramic Society Bulletin,* September, 2002, pp. 30–32.

Periodicals

Advanced Materials & Processes

Ceramic Bulletin

Ceramic Monthly

MRS Bulletin

UNIT 8

Composite Materials

PAUSE & PONDER

The initial growth of composite technology came largely from the demands of the industrial sector for such items as pultruded industrial ladders and railings, fiber-reinforced plastic (FRP) sheet and structural components for aerospace and automotive purposes, and graphite and Nomex honeycomb for aerospace use. The sports equipment industry quickly saw opportunities for using the advanced composites designed for the Space Age. Soon athletes had an array of newly designed equipment—tennis racquets, snow skis, helmets, and vaulting poles—that made use of the design flexibility of advanced composites.

Large, stiff, lightweight tennis racquet frames provided larger sweet spots, increasing the coefficient of restitution and translating into more power and accuracy. Conventional, smaller, wood racquets required a fair degree of skill to hit the tennis ball with accuracy. New racquet designs developed around the 1970s featured steel and aluminum, but the breakthrough achieved with large graphite-impregnated epoxy racquets meant the less skillful recreational player could hit a ball off-center but still have a fair degree of accuracy and improved power. As a result, the popularity of tennis increased and that provided the funding to the sporting industry to improve research into even better designs for composite tennis racquets. Now there is a wide array of racquets constructed of various glass, carbon, and metal fiber-reinforced plastics. At the professional level, some argue for new rules to require the return to wood racquets so that the player's skill, not just the equipment, determines the best player. The entire sports and recreational equipment industry, not just racquet manufacturers, turned to advanced composites. One now finds fishing rods, skateboards, water skis, drag racers, speed boats, and softballs, to name a few, using a wide range of fiber-reinforced plastics. The increase in the use of composites dramatically drove down the cost of the materials and products. For example, the first Prince large-head graphite racquet sold for about $250. Now a casual tennis player or someone interested in buying a child's racquet can find many composite racquets for around $50. However, state-of-the-art racquets designed for maximum performance still sell for over $200.

Soon do-it-yourself suppliers to homeowners, along with suppliers to the building construction industry, began using polymer–matrix composites to replace traditional

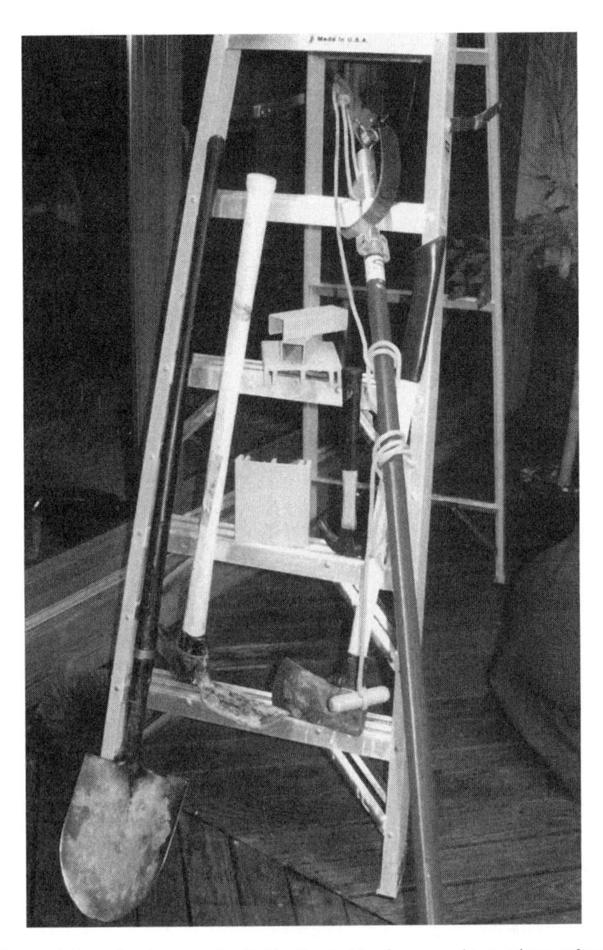

Pultruded parts made with glass-fiber-reinforced polyester plastic for the ladder frame and tree trimmer handle. The shovel, hammer, pick ax, and ax handles are made of cast-glass-reinforced plastics. On the aluminum steps of the ladder are several forms of decking made from scrap wood fibers and recycled PETE plastic. The advantages of plastic–matrix composite decking are the recycling of what would otherwise be waste, superiority to pressure-treated wood in terms of weather resistance, and eliminating harmful chemicals used to pressure-treat wood. Fiber-reinforced plastic tools come in a wide range of colors, reduce the danger of electrical shock, and reduce weight while increasing strength and eliminating the splintering and cracking experienced with wood handles and frames, thus providing improved functionality and longer service life.

wooden tools such as ladders, ax and shovel handles, and even fence posts. The photo shows typical fiber-reinforced plastic composite and polymer tools one will find around the home and construction sites today. As in the sports equipment industry, the increase in use of composites for industrial applications brought about a decline in materials cost and an increase in their availability for the general market. As reported in the *Composites Technology* August 2002 issue, homeowners purchased 10 times more pultruded composite ladders than did the industrial sector.

Fiber-reinforced plastics have also become popular with artists. Artists soon learned what auto designers knew: FRP provides considerable flexibility in design. The second photo shows a FRP mermaid located in the Norfolk Botanical Gardens. It is one of a large number of mermaids created from the same mold but in different colors and decorated by individual artists from around Hampton Roads, Virginia. You may have seen the FRP Painted Ponies around Albuquerque, New Mexico and Kitty Hawk, NC, or other theme FRP sculptures of pigs, fish, and so on used elsewhere as part of community art projects.

Unit 8 will provide extensive study of a wide range of composites. Pause for a minute to think about the many composites you encounter daily.

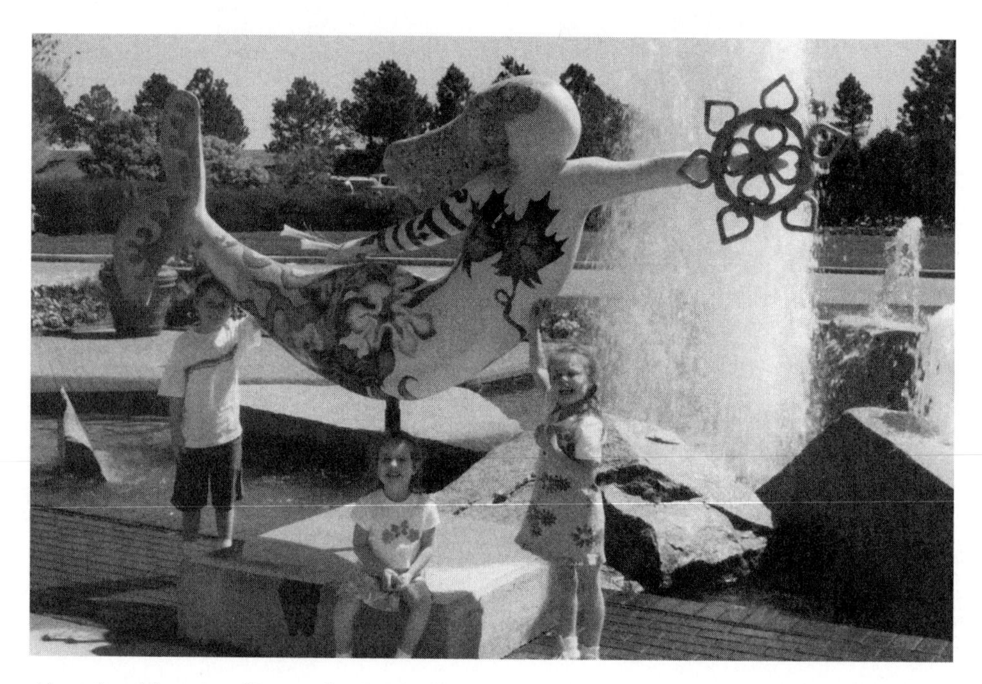

Painted mermaids are found throughout Hampton Roads, Virginia as part of a community art theme. While the molds provide the same basic shape from FRP, artists employ add-ons and colors to achieve unique expressions.

Module 24

Fiber Structures & Properties

After studying this module, you should be able to do the following:

24–1. Sketch the Venn diagram that shows how the sets of the three main family groups of materials intersect with reinforcers to form subsets of polymer–matrix composites (PMCs), ceramic–matrix composites (CMCs), and metal–matrix composites (MMCs) and how the PMC subset expands into a wide array of composite materials.

24–2. Select examples of composites and advanced composites, sketch their macrostructure and label the constituents (matrix, interface zone, etc.), and justify their use over monolithic materials. Recall reasons to use composites and cite the applications of advanced composites.

24–3. Discuss carbon and graphite fibers in terms of their definitions, production, and important properties, including coefficient of thermal expansion (CTE), specific strength, and wear (erosion) rates.

24–4. Calculate the specific strength and specific stiffness of given materials and determine when these properties are major reasons for the selection of materials.

24–5. Use the website addresses denoted by the symbol to explore advances in composites, fibers, matrices, applications and research.*

24.1 INTRODUCTION

A ***composite material*** or ***composite*** is a complex, solid material composed of two or more materials that together, on a macroscopic scale, form a useful product. The composite is designed to exhibit the best properties or qualities of its constituents or some properties possessed by neither. The combining of the two or more existing materials is done by physical means, as opposed to the chemical bonding that takes place in forming alloys of monolithic solid materials. A true composite might be considered to be made up of a matrix material that completely surrounds a reinforcing material, the two phases acting together to produce characteristics not attainable by either constituent acting alone. Note that the insoluble phases or main constituents in a composite do not lose their identity. This is not true of solids that are metallic alloys or copolymers, whose phases are invisible to the naked eye, because they are formed as a result of natural phenomena. In other words, steel is a multiphase metal alloy but is not considered a composite. A metallic alloy of Al and Cu, with the addition of Al_2O_3, is considered a dispersion-strengthened copper composite in which a dilute copper/aluminum alloy powder is exposed to an oxidizing environment, with the aluminum being oxidized within the copper matrix. The aluminum oxide particles are

*The icon found throughout this book will link you to Internet sites related to topics being covered. The dynamic nature of the Web brings frequent changes, so some of these sites, while available at the time of writing, may not be up now.

inert and act as inhibitors of dislocation movement and thus strengthen the alloy at high temperatures. By adding niobium (10%) to this same extruded composite in the form of a uniformly distributed powder, even greater hardness and strength are produced with only a minimal reduction in electrical conductivity. Two-phase alloys are not considered composites by some because their phases are not formed in separate processes but originate in a single manufacturing process.

This discussion reveals that the definition of a composite is still somewhat arbitrary (see Section 2.2.5). A broad definition of composite materials includes the naturally occurring composites, such as wood, as well as the synthetic or human-made composites. A recent example of this broad definition is a composite developed for use in the medical field. A mineral paste composed of monocalcium phosphate monohydrate, α-tricalcium phosphate, and calcium carbonate, dry mixed, to which a sodium phosphate solution is added to form a paste. The paste is surgically implanted into acute bone fractures and hardens within ten minutes after injection. It holds fractured bones in place while the native bone remodeling process replaces the implant with living bone. This process causes the in situ formation of the mineral phase of bone. Attaining a compressive strength greater than that of long bones, the paste's tensile strength of about 2.1 MPa is comparable to long bones. This new material eliminates the need for heavy and uncomfortable casts.

There is an urgent need for a knowledge of new materials on the part of all who are in a position to not only recommend, select, and design materials for the civil or military infrastructure but for those who already live in such structures or are about to purchase such structures. For example, composites based on unsaturated polyester resins do not burn easily and produce low levels of smoke and toxic fumes. However, many traditional materials that burn fiercely are often selected as structural fire-safety components. In answer to the question "Why use composites?" one can reply, in part, as follows:

1. To increase stiffness, strength, or dimensional stability
2. To increase toughness (impact strength)
3. To increase heat-deflection temperature
4. To increase mechanical damping
5. To reduce permeability to gases and liquids
6. To modify electrical properties (e.g., increase electrical resistivity)
7. To reduce costs
8. To decrease water absorption
9. To decrease thermal expansion
10. To increase chemical wear and corrosion resistance
11. To reduce weight
12. To maintain strength/stiffness at high temperatures while under strain conditions in a corrosive environment
13. To increase secondary uses and recyclability and to reduce any negative impact on the environment
14. To improve design flexibility

The objective of this module is to impart a fundamental knowledge of composites and to present the terms used in this specialized field; the nature, structure, and properties of composites; and an arbitrary classification of the major types. Examples of the applications of some composites in a limited number of industries are included with a view toward providing readers, many of whom occupy positions in industry that require a degree of expertise in the selection of materials, with a base of knowledge of composites from which to make intelligent selection decisions.

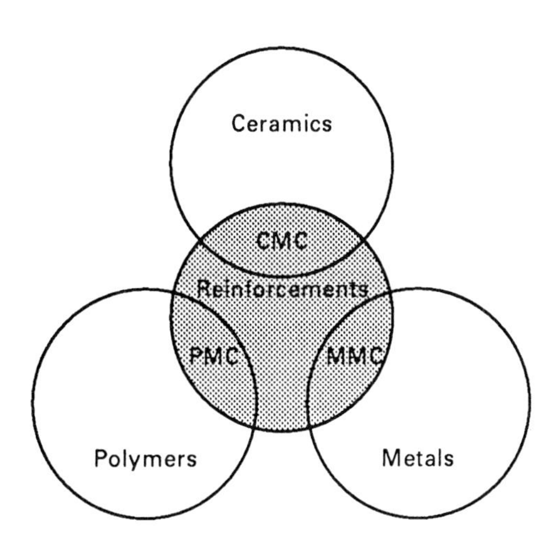

Figure 24-1 Composite materials. The design process for engineering materials involves the selection of components from major members of the family of materials, which, when combined into a composite material, will possess a mix of desired material properties that will satisfy the requirements of a particular application. A ceramic selected as a matrix material combined with a reinforcement in any form and quantity produces a ceramic–matrix composite (CMC). A similar combination produces a metal–matrix composite (MMC) or a polymer–matrix composite (PMC).

24.1.1 Development of Composites

Composite materials have been used from earliest times. Mongol bows of the thirteenth century utilized a materials system consisting of animal tendons, wood, silk, and adhesives. The ancient Israelites used straw to reinforce mud bricks. The early Egyptians fabricated a type of plywood. Medieval swords and armor are examples of laminated metal materials. Nature has provided composite materials in living things—seaweed, bamboo, wood, and human bone.

Developments made through the collective efforts of government, NASA, industry, and universities gave rise to ***advanced composites.*** Figure 24-1, a Venn diagram, shows how ceramics, metals, and polymers are integrated into sets and subsets of advanced composites. The sets of CMC, MMC, and PMC multiply further into subsets based on the form of reinforcement and the general structural arrangement, as illustrated in Figure 24-2 for fiber-reinforced composites, a subset of PMC. ***Nanocomposites,*** a new subset of CMC, MMC, or PMC, are composites with a substrate that is reinforced with nanometer (10^{-9} m)-size particles that enhance the substrates' physical properties. For example, nanocomposite, a subset of PMC, with a thermoplastic olefin (TPO) substrate uses submicron particles of clay. An addition of only 5% of this natural clay provides significant benefits over traditional thermoplastic materials and fillers such as talc, mica, and $CaCO_3$. The stiffness characteristics are equivalent to 25 to 35% additions of talc-filled materials. In addition, the stiffness is achieved without increasing weight or losing temperature properties or opacity. Developed specifically for automobile applications, this new material should improve the recyclability of automobiles in the future.

The major ingredients needed for successful production—engineering design and manufacturing—kept pace with developments in materials and resulted in the continuous introduction of new composite materials for use throughout industry. Figure 24-1 also represents the continual search for materials systems that can withstand even greater loads under hostile environments. Composite materials with ceramic and metal matrices reinforced with similar materials are being developed and tested for use in many high-temperature applications. The Advanced Composites Technology (ACT) program (see Figure 24-3) has the

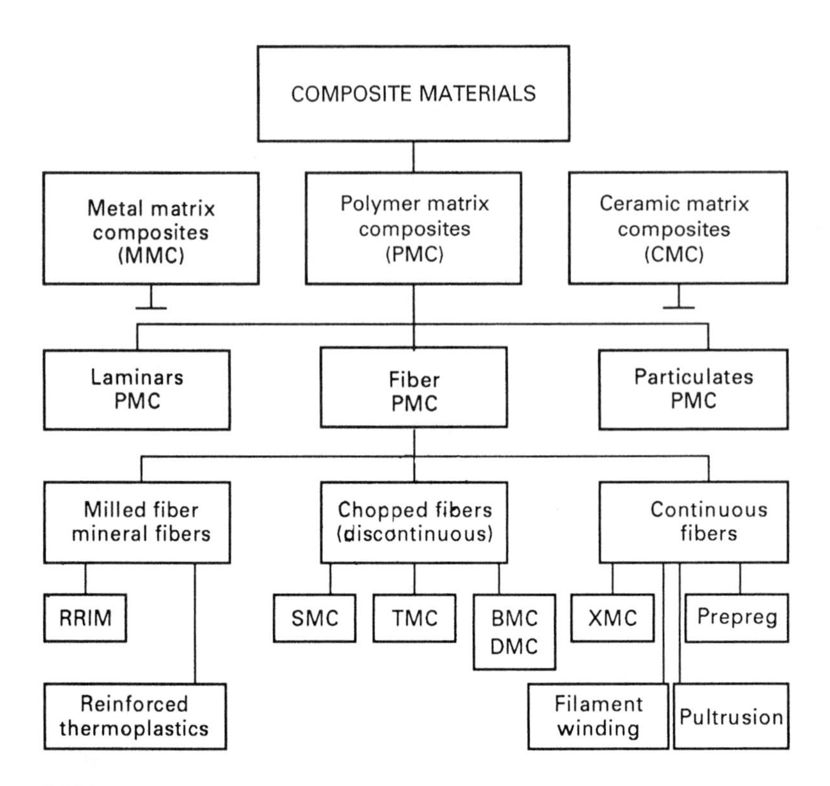

RRIM: reinforced reaction injection molding
SMC: sheet molding compound
TMC: thick molding compound
BMC: bulk molding compound
XMC: x-pattern molding compound (trademark of PPG ind.)

Figure 24-2 Composite materials—their matrices and reinforcements produce an infinite array of materials.

broad objective of developing the technology to design and manufacture a cost-effective and structurally optimized, lightweight, composite airframe. In its first phase, the program developed three major manufacturing methods for fabricating these cost-effective composite structures, namely, ***stitched dry preform, textile preform,*** and ***automated tow placement.*** Production costs of a composite structure, once twice those of a corresponding aluminum airframe structure, continue to creep down as new technology evolves. It is reasonable to speculate from this example that materials designers will eventually use materials that do not yet exist. It also demonstrates that the effort to develop new materials calls for a "synergistic advancement" of technology in which a partnership between materials scientists and materials users will, for the first time, be put into practice.

Down on the ground, designers of a range of diverse products, including clothing, athletic gear, appliances, furniture, office supplies, home entertainment systems, packaging, computer systems, boating, camping gear, and housing, have called on materials technology for new composite systems that make better products. Shoes are one example of a line of products making use of new composite systems. Look at inflatable basketball and tennis shoes and cut open a pair of jogging shoes to find the air chambers and wide variety of polymers in the form of rigid plastics, soft foams, dense foams, nylon mesh, and hard rubber. Designers used CAD technology coupled with materials technology to keep this market humming with ever-improving products. The customer should be the winner.

Another example of advances in materials technology is a composite consisting of 80% advanced polymer and 20% aluminum developed for automotive applications by XCORP, Beverly Hills, California, at a cost said to be comparable to that of stamped steel. Designated "Composite X," it is reportedly resistant to abrasion and wear and offers superior crashwor-

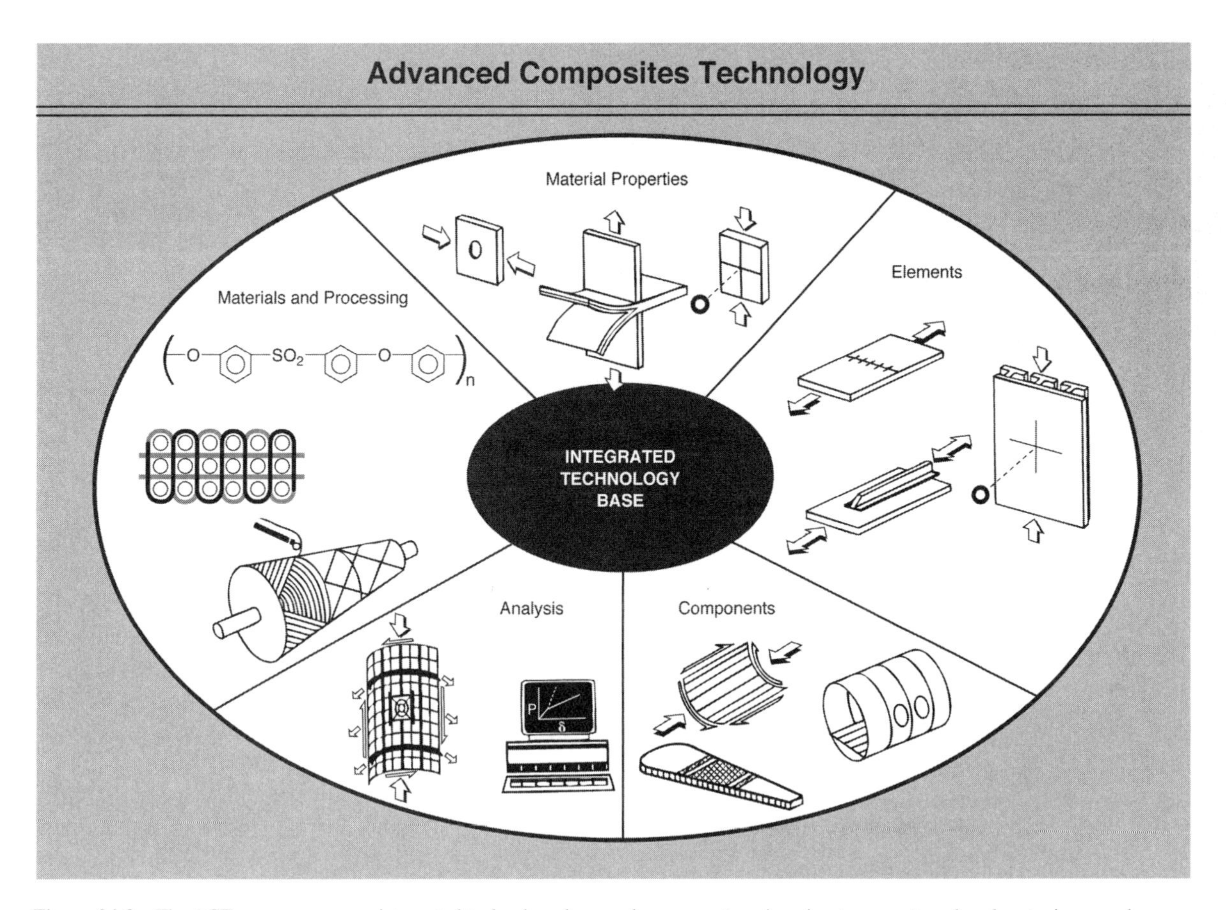

Figure 24-3 The ACT program uses an integrated technology base and concurrent engineering to promote reduced costs, improved manufacturing techniques, and high-quality products. (NASA-LaRC)

thiness. The material is one of a group XCORP calls ***environmental composites,*** because they are easily recycled and are made using low energy and a clean manufacturing system.

The National Composites Center (NCC) in Kettering, Ohio, studies new methods for building (a) infrastructure (construction related to roads, bridges, etc.), (b) automobiles, and (c) aerospace vehicles. The NCC and others who develop new building technology continue to look to composite materials. For example, a new building material—a fiber-reinforced polymer composite (graphite fibers embedded in an epoxy resin)—was used to build a vehicle bridge that can support loads heavier than a comparable bridge made of traditional materials. First tested in 1995 on a 26-foot span, the "Tech" (Materials Technology for the 21st Century) vehicle bridge (33 feet in length and capable of supporting more than 36 tons) opened in Butler County, Ohio, in 1997. It is the first all-composite vehicle bridge in Ohio and the first fully instrumented bridge in the United States—its performance can be remotely monitored by computer. There are 230,000 bridges across the country that are structurally deficient or functionally obsolete. Replacing or reinforcing these bridges with new building materials would prolong the lives of these structures—tripling their life expectancies, improving highway safety, and saving billions of tax dollars.

The many recent developments in advanced composite systems, not only as spin-offs from the nation's space program but also as a consequence of growing concern for energy conservation and the depletion of the world's nonrenewable materials resources by more and more citizens, make the emphasis placed in this module on composite materials appropriate. This is not meant to downgrade the importance of the many other composite materials without which our economic well-being would suffer. In industries' new thrust for "design for manufacturability" and "design for automation," designers seek to reduce the number of components in a product, eliminate mechanical fasteners, and design to facilitate assembly by robots. Composites offer great promise in helping to meet at least some of these goals.

Automotive Composites Alliance—*http://autocomposites.org/*
National Composite Center—*http://www.compositecenter.org*

24.2 CONSTITUENTS OF FIBER-REINFORCED COMPOSITES

A ***fiber-reinforced composite*** is a materials system made primarily of varying amounts of a particular fiber reinforcement embedded in a protective material called a *matrix*; a coupling agent is applied to the fiber to improve the adhesion of the fiber to the matrix material. Many of the terms in the modern materials technology field originated in the textile industry. A few such terms are selected here for brief definition to facilitate an understanding of these highly important engineering materials. See the glossary at the end of Unit 8, Section 26.9, for other definitions.

High-performance fiber-reinforced composites exhibit higher structural performance than that of commercial E-glass and polyester composites. The degree of performance over a fiberglass composite depends on the fiber, its orientation and loading, and the matrix. A significant processing aspect that affects performance is the degree to which the fibers are saturated with wet matrix, leaving few or no voids. The matrix's ability to adhere to the fibers, as well as its inherent toughness, also make a contribution. As for fibers, their loading (density); continuous lengths (high aspect ratios); and orientation, including straightness, degree of knitting, and interlacing all play a role in contributing to the final properties.

In the study of atomic structure, we learned that a material's strength in a certain direction or plane is directly proportional to the type of atomic density and the atomic bonding existing in that particular direction or plane. If the atoms are tightly packed and are bonded with strong primary bonds, the material will possess a high degree of strength. We also learned that properties, particularly in brittle materials, are adversely affected by defects or flaws. The larger the flaw, the less the ability to withstand any degree of load. As an example of the application of this knowledge to the production of carbon fibers (to be discussed), we recall that carbon, an element, is polymorphic. The sketch of the structure of graphite in Figure 2-14 showed two basal planes consisting of atoms covalently bonded that were connected in a lateral direction by weak secondary bonds. In the production of carbon fiber, for example, the original precursor material is in a similar state. As part of the process, the basal planes are oriented along the axis of the fiber, thus placing the atoms with the strongest bonding in the best position to withstand a longitudinal force. Defects in the atomic structure or macrodefects on the surfaces of the fibers are dealt with by reducing the amount of cross-sectional area in the fiber. The chance of a defect remaining when the diameter of a fiber decreases is minimized considerably. This explains the ongoing efforts to produce very fine fibers, down to the nanometer range.

24.2.1 Fiber

A ***fiber*** is the basic individual filament of raw material from which threads or fabrics are made. Fibers constitute one of the oldest engineering materials; jute, hemp, flax, cotton, and animal fibers have been used from the earliest days of history. Today, fibers can be organic (plant, animal, and mineral), synthetic (human-made from a polymer or ceramic material), or metallic. Manufactured polymeric fibers were discussed in Module 14. ASTM further defines a fiber as having a length at least 100 times its diameter, with a minimum length of at least 5 mm. A fiber can be a filament or a staple. ***Filaments*** are long, continuous fibers, whereas ***staple*** fibers are less than 150 mm in length. Most natural fibers are in staple form, whereas synthetic fibers may exist in both forms.

Whiskers are formed from extremely thin single crystals of high perfection with a large length-to-diameter ratio for reinforcing composite materials. They are virtually flaw-free and possess exceptionally high strength. Whiskers are, however, extremely expensive to produce and difficult to handle in forming composite materials. SiC and Si_3N_4 ceramic whiskers inhibit crack propagation (fracture toughness) and strength. These two properties increase with an increase in whisker/fiber content in composite materials such as metal–matrix composites. Whiskers are examples of discontinuous reinforcements, as are chopped fibers of alumina or carbon and particulates of SiC and alumina. ***Continuous fibers*** are essentially infinite in length and extend continuously throughout the matrix in a composite material. ***Discontinuous fibers*** are not less than 3 mm in length and tend to orient themselves in the direction of resin flow as a composite material is processed. As we have mentioned, particulate composites with reinforcements of SiC and alumina are an example of discontinuous fibers. Whisker-reinforced alumina is used to make inserts for cutting tools. Discontinuous reinforcements in metal–matrix composites (MMC) do not cause any problems in the shaping of these composites by metal-forming operations such as rolling. Other terms relating to fibers and fiber composites can be found in Section 26.6.

The strength of fiber-reinforced composites depends on the ability of the fibers to transmit their load-carrying properties to the surrounding matrix. If a fiber ends in a matrix, then the fiber cannot transmit its stress to the matrix and thus provides little strength reinforcement. A formula has been devised to determine the ***critical length of a fiber (l_c).*** The calculation uses the smallest of the fiber diameter, its tensile strength, the shear strength of the matrix, or the fiber–matrix bond strength.

$$l_c = \frac{\sigma_f d}{2\tau_c}$$

σ_f = tensile strength
d = fiber diameter
τ_c = shear yield strength

A fiber that is longer than the critical length (l_c>15) is termed a ***continuous fiber.*** Some examples of continuous-fiber metal–matrix composites (MMCs) are the matrix of alumina and continuous boron fiber reinforcement used in the space shuttle *Orbiter* and the carbon fibers in an alumina matrix used in the Hubble Telescope. Other examples of continuous fibers are SiC, alumina, and refractory metals.

24.2.2 Matrix

The functions of a ***matrix,*** the binder material, whether organic, ceramic, or metallic, are to support and protect the fibers, the principal load-carrying agent, and to provide a means of distributing the load among and between the fibers without itself fracturing. When a fiber breaks, the load from one side of the broken fiber is first transferred to the matrix and subsequently to the other side of the broken fiber and to adjacent fibers. The load-transferring mechanism is the shearing stress in the matrix. Typically, the matrix has a considerably lower density, stiffness (modulus), and strength than the reinforcing fiber material, but the combination of the two main constituents (matrix and fiber) produces high strength and stiffness, while still possessing a relatively low density.

A ***coupling agent,*** also known as a *bonding agent* or *binder,* provides a flexible layer at the interface between fiber and matrix that will improve their adhesion and reduce the number of voids trapped in the material. ***Voids,*** air pockets in the matrix, are harmful because the fiber passing through the void is not supported by the matrix. Under load, the fiber may buckle and transfer the stresses to the weaker matrix, which could crack.

The matrix serves as the structure, keeping the reinforcing fibers at the same distance from the bending axis, and the filler provides the desired properties. There is no significant chemical reaction between the two phases except for the bonding action at their interface. If this were not true, any reaction between the two materials would have a varying negative effect on the inherent properties of the filled composite. The structure or matrix of a filled composite can take a porous or spongelike form. This network of open pores can be impregnated

TABLE 24-1 OPERATING TEMPERATURES FOR COMMON POLYMER RESINS

Resin	Maximum temperature (°C)
Polyester	Room temperature (RT)
High-performance thermoplastic polyesters	150
Epoxies	200
Phenolics	260
Polyimides	300
Polybenimidazole	Above 300

with a variety of materials, from plastics to metals to ceramics, and then the final shape of the filled open-pore composite is formed during the processing. The matrix can also have a predetermined shape and size; for example, the open honeycomb core made of metal impregnated with a ceramic filler for high-temperature applications.

Maximum operating temperatures for resins used as matrices in fiber-reinforced composites are listed in Table 24-1 to give some idea of their limitations. No attempt has been made to introduce the element of time in these data; we simply state that the time a material is subjected to these extremes in temperature is a critical factor that must be taken into consideration in any design. An example taken from the literature illustrates how the heat-resistant property is described for a series of polyimide molding compounds, Tribolo PI-600, available from Tribol Industries: Such compounds perform at 316°C continually and 427°C intermittently; a typical molding retains 56% of its mechanical strength after 100 hr at 316°C; weight loss is 2.1% after 500 hr.

Both thermosetting (TS) plastics and thermoplastic (TP) plastics are used with hybrid composites (defined later). Epoxies, which have a wide range of properties, are the most used TS plastics, but thermoset polyester is a strong contender, particularly for automotive applications. Intermediate modulus epoxies are used in hybrids. They have good high-temperature properties but suffer from moisture absorption when exposed to temperatures near 180°C. Poly(ether ether ketone) (PEEK), a thermoplastic that is tougher and potentially less expensive to manufacture than conventional resins, is used for the doors of F-15E fighters and T-38 trainer aircraft. Polyimides are used for extended operations near 260°C and above. TS polyester, formulated to be hard, brittle, or tough, is used with chopped glass in many applications (see Figure 24-6).

Polyimide molding compounds capable of withstanding continuous-use temperatures ranging from 480° to 600°F and intermittent service at 800°F are now available. A typical molding retains 56% of mechanical strength after 100 hr at 600°F, and weight loss is 2.1% after 500 hr. All of these compounds are compatible with reinforcing fibers.

TP resins are the newest matrix materials; an example of TP use for automotive components is shown later in Figure 24-7. Nylon; polysufone (PSU); poly(phenylene sulfide) (PPS); and poly(phenylene terephthalate) (PET), better known as thermoplastic polyester, are used in hybrids and advanced composites. Bismaleimide (BMI), which is manufactured like conventional composites, is able to withstand higher operational temperatures. Improved resin-transfer molding (RTM) resins have a longer pot life and lower viscosity. User-friendly prepreg resins have better-controlled reactivity and an extended out-time.

Resins can be divided into two main groups: structural pastes and structural film adhesives. Pastes can be stored for a year at room temperature (RT) and cure at RT. Four versions are available: one is aluminum filled for easy handling and high compressive strength, another is a thixotropic nonmetallic for composite surfaces and random structures, another is an unfilled version designed for wet-layup composite repairs, and the fourth version has high shear modulus at elevated temperatures and can be stored six months at RT. (Thixotropic metals are discussed in Module 3. Figure 3-64 contains a micrograph of a semisolid, thixotropic alloy.) Structural film adhesives resist moisture in the bond line and are extremely resistant to severe environmental exposure. They can be cocured with the composite structure. ***Cocuring*** describes the adhesive bonding that is carried out at the composite layup stage during the laminate process schedule. High-temperature resin systems for structural applications are exemplified by use in supersonic aircraft involving long exposures to temperatures in the range of 150 – 250°C.

Glass ceramics, polycrystalline solids produced by the controlled crystallization of glass, have highly desirable properties for service as matrix materials. Dimensional stability, thermal stability, heat resistance, corrosion resistance, and the ability to maintain strength at high temperatures make these materials attractive despite their inherent brittleness. Their coefficient of thermal expansion is zero up to about 450°C. Glass-ceramic matrices act just the opposite to polymer– and metal–matrix materials because the glass-ceramic matrix fails first, having the lowest strain to failure. Failure occurs when the matrix viscosity becomes too low to transfer the load to the reinforcement. Thus, thermal compatibility between the matrix and fibers must be good. If the coefficients of thermal expansion are not similar, the brittle glass-ceramic matrix will not adjust to any differences. Also, the need for high forming temperatures to bring about matrix flow among the reinforcing fibers requires thermochemical compatibility between the fibers and glass-ceramic matrix material to prevent reactions at the fiber–matrix interface that produce unwanted interphases, which degrade the fibers and reduce their strength. Some of the most common glass-ceramic matrices are litha–alumina–silicate (LAH; Li_2O_2–Al_2O_3–SiO_2); silica, SiO_2; and BaO–SiO_2–Al_2O_3–Si_3N_4 with SiC (Nicalon), the most commonly used reinforcing fiber for glass-ceramic matrices.

An ***intermetallic compund*** or ***phase*** (see Section 3.3.2.1 for a discussion of intermediate-phase materials), formed from two or more metals, has a mixture of metallic and ionic bonds. ***Intermetallics*** will certainly be the key to the future of technology development. Their high-temperature properties will find future applications in road vehicles, in aviation, and in industrial equipment. One such intermetallic compound is molybdenum–silicon–boron (Mo_5Si_3B), which is likely to replace ***superalloys*** in the next generation of aircraft and automobile engines. Cementite, Fe_3C, is a compound or alloy that serves as the backbone for strengthening steel. Titanium aluminides—high-strength, low-density intermetallics—are superseding nickel-based superalloys. The density of $Ti_2 \cdot Al \cdot Nb$ (titanium–aluminum–niobium) alloys is less than two-thirds that of nickel-based superalloys such as UNS N–07718. High specific strength at high temperatures, improved ductility, and fracture toughness are some of the improved properties of such intermetallics that are making these materials feasible for gas-turbine applications at temperatures approaching 1000°C.

24.2.2.1 Interface and interphase. By definition, composites contain a combination of mutually insoluble components (constituents). With fiber-reinforced plastics (FRPs), the fibers and the polymer resin are the main constituents. A wide variety of other material additives, such as coupling agents, fillers, pigments, catalysts, and fire retardants, are also used to satisfy specific needs. Consequently, there exists an ***interface(s),*** a surface that forms a common area or boundary similar in many respects to grain boundaries between any two constituents in a monolithic material. An ***interphase*** is the region formed between two interfaces. It is therefore a distinct phase in itself with its own identity, forming a region contiguous to two phases. Figure 24-4 sketches these main constituents of an FRP. In metal–matrix composite materials (MMCs) the reactions between the metal–matrix and the reinforcement takes place in the interface zone outlined in Figure 24-4b.

24.2.3 Fiber/Matrix Consolidation

Fibers can be saturated with resinous material through a process called ***preimpregnation.*** The resinous material is subsequently used as a matrix material. The preimpregnated fibers are called ***prepregs.*** Unidirectional preimpregnated fibers in a roll with a removable backing that prevents the fibers from sticking together are known as ***prepreg tape. Prepreg cloth*** or ***mats*** are also available. Prepregs are ready-to-be-molded resin and fiber laminates (see Figure 24–8a). The resin-impregnated fibers are partially cured (B-staged) and ready to be softened and formed into a permanent shape with the application of heat and/or pressure. Some important advantages of using prepregs are the elimination of the handling of liquid

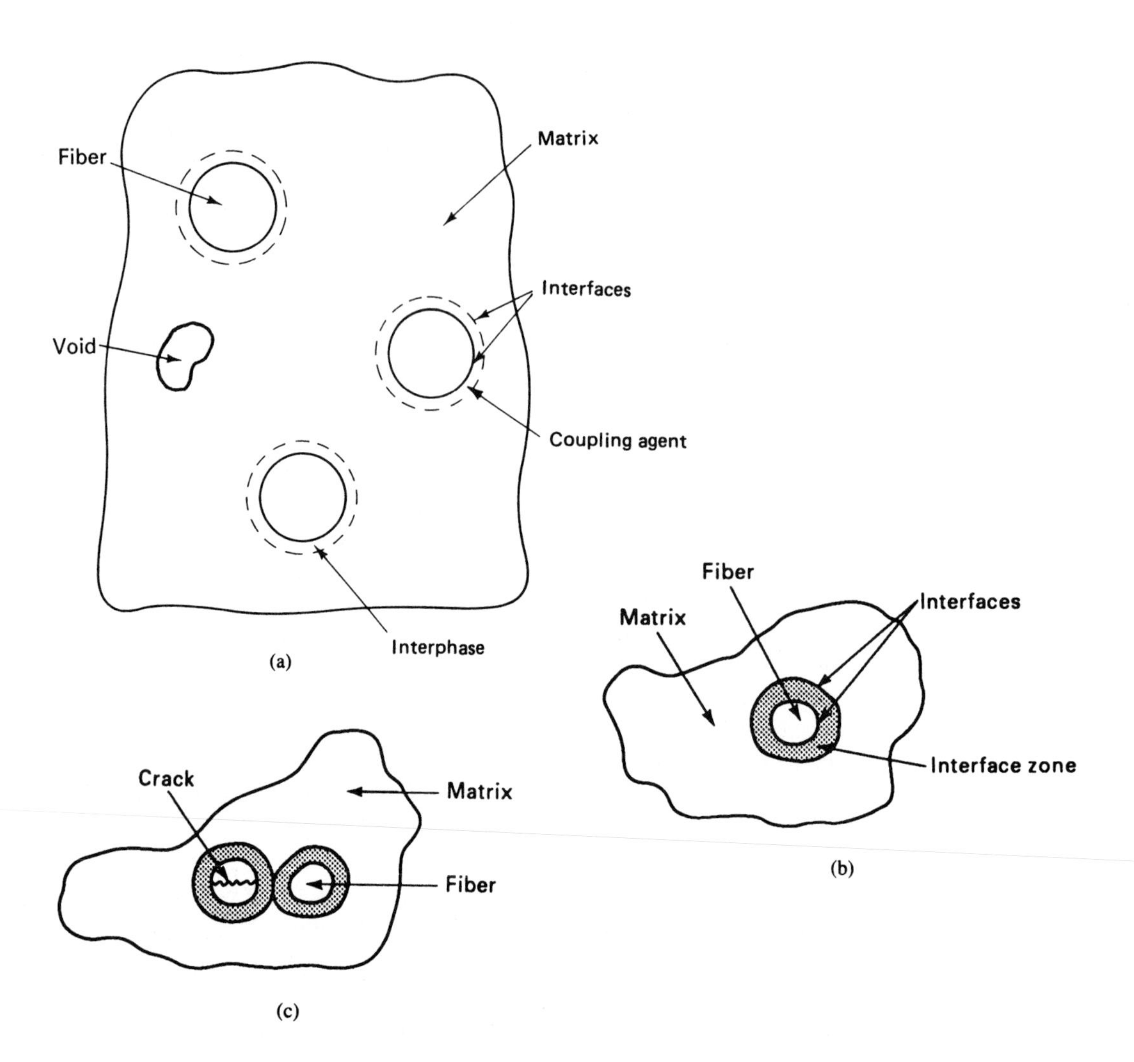

Figure 24-4 (a) Cross section of fiber-reinforcing fibers in a matrix. (b) Coated fibers create two interfaces that act like grain boundaries inhibiting crack propagation. (c) Two fibers prevented from making direct contact with each other stop crack propagation.

resins, the simplification of the manufacturing of reinforced plastic forms and shapes, and the savings in cost with small production lots. A few disadvantages are the elimination of room-temperature cures, the need for refrigerated storage, and the additional cost over dry-blended or undispersed materials. The last disadvantage must be weighed against the use of higher mold pressures and longer molding cycle times for undispersed material to effect the required dispersion of the reinforcement throughout the matrix material to attain the optimum properties of the composite.

Polyimides are attractive to the aerospace industry, as we have mentioned, because of their toughness, thermal and thermo-oxidative stability, resistance to solvents, and excellent mechanical and electrical properties over a wide range of temperatures. One disadvantage is that they have poor melt-flow characteristics during processing, which leads to voids in the prepreg laminates (see Figure 24-5). By blending linear, high-molecular-weight polyamic acid solutions with semicrystalline thermoplastic polyimide powders, a slurry is formed that can make void-free prepregs. The slurry can be applied to film, fabric, metal, polymer, or composite surfaces. As discussed in Module 15, cyclic thermoplastic resins developed recently have also improved wet-out of reinforcing fibers.

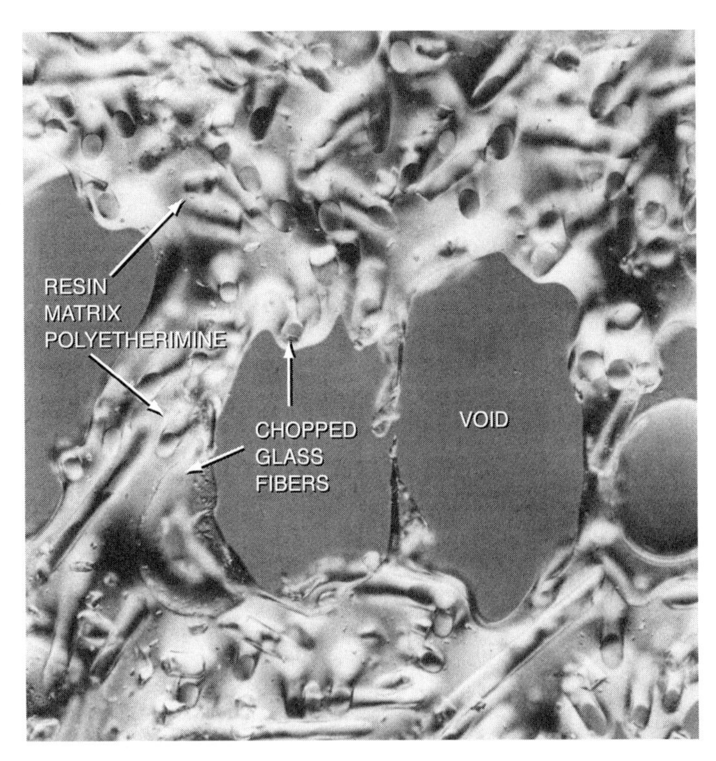

Figure 24-5 Short chopped-glass fibers and voids (produced by air entrapment) in a polyethemide ultrathin section transmit differential interference. (Contrast −500×/20× obj.) (Luther M. Gammon, Boeing Commercial Airplane)

Much development work is involved in using intermetallics as matrix materials for metal–matrix composites. Intermetallics such as Ni_3Al, TiAl, and $NbAl_3$ reinforced with Al_2O_3 or ZrO_2 fibers form composites that are finding use at temperatures from 1000° to 1400°C. Room-temperature ductility of titanium aluminides containing niobium is improved for these composites by the addition of small quantities of magnesium. A special "pressure-casting" technique, in which the molten intermetallic is poured onto preheated fibers, is used to fabricate these composites. Argon gas pressure of 3 MPa ensures complete infiltration of the ceramic fibers. Titanium or yttrium is added to promote fiber wetting.

The ***particulates*** are a large, diverse group of materials that consist of minute particles. Particulate density will affect the composites' final density, as in the case of hollow spheres. The amount used (loading) and their distribution in the matrix greatly influence a composite's performance. Many of their functions overlap, in that they can function as reinforcers, fillers, or both. Carbon black, for example, is used to fill space in the matrix and thus reduce cost; but it also is a reinforcing agent for elastomers. This dual role is possible because of the essentially spherical shape of carbon black particles, which range in size from 20 to 50 μm.* The length-to-diameter ratio (aspect ratio) for spherical particles is 1, which means that the particles can be easily moved or offer no resistance to being moved or realigned. Aspect ratios determine resistance or realignment in the matrix. Particles that have an aspect ratio other than 1 offer resistance to movement. They may be spherical, saucer, needlelike, or nodular or have a platelet or short-fiber shake.

Some of the most important forms of prepregs are the following:

Preforms are custom-shaped, resin-bonded mats for reinforcement of molded parts with complex shapes. In other words, a preform is a bundle of fibers having the desired volume and architecture to fill a mold cavity for a one-shot molding.

A ***lamina*** is a flat arrangement of unidirectional fibers or woven fibers in a matrix.

*See Index for carbon black and buckyballs.

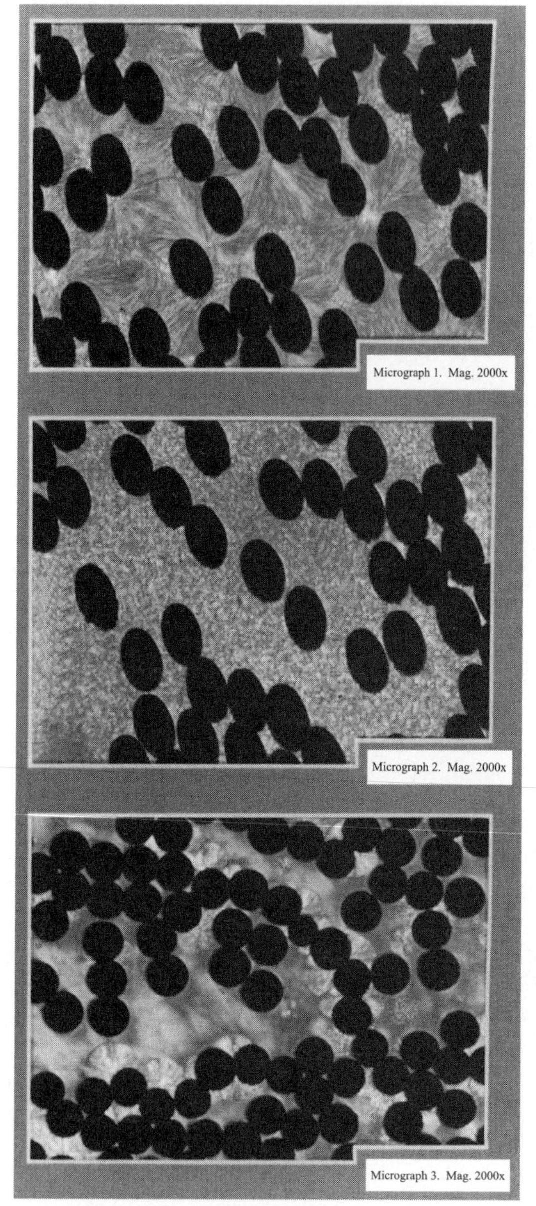

CLASS 2

MICROSTRUCTURES OF THERMOPLASTIC COMPOSITES

L.M. GAMMON, M.Y. ALMASSY, C. H. SHEPPARD, S.L. STANEK. A.M. THORSTENSON

BOEING MATERIALS LABORATORY

Objective

Establish a qualitative relation between mechanical behavior, microstructure, and solidification kinetics in high temperature thermoplastic composites, with an emphasis on Poly Ether Ketone (PEEK)

Background

It has been found that high temperature thermoplastic composites behave like metallic structures, in that mechanical properties change when these composites solidfy at different rates. A precise correlation has not been established between mechanical properties and the amount of polymer crystallinity, which is affected by solidification rates determined by X-ray diffraction, thermal analysis, or density measurements. By extension of the analogy with metals, microstructural phenomenon, such as crystal size, shape, distribution, etc., should affect mechanical properties and may be controlled by the cooling rate.

Procedure

Composite specimens made from PEEK and carbon fiber solidified at various cooling rates (low, moderate and fast [followed by annealing]) were prepared for optical microscopy using transmitted, polarized light. These specimens were 2- to 4-micron thick wafers prepared by mounting and polishing one side, then bonding the polished surface to a Lucite slide. A specially designed hand vise was used to obtain parallel polished surfaces.

Specimen preparation by polishing minimizes internal damage (cold work). Controlling specimen thickness is essential to 1. resolve spherulites that can be stacked behind each other in thick sections, and 2. avoid ghost-like images due to lack of refractive material in thin sections. Thickness may be determined geometrically (refer to sketch).

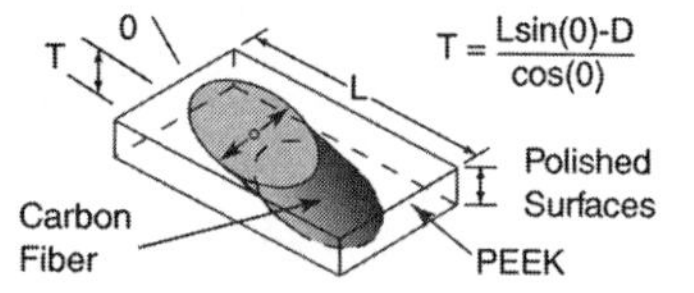

Results

Micrographs 1 and 2 show PEEK/carbon fiber composites with comparable percentages of crystallinity. While both of these matrices are uniformly covered with spherulite, those in Micrograph 1 are larger than those in Micrograph 2 (In metallurgical terms: coarse grain structure versus fine grain structure.) The large spherulites of Micrograph 1 were obtained by cooling slowly, while the smaller spherulites of Micrograph 2 were formed by quenching (rapid cooling) and subsequent annealing.

Matrix crystallinity can be eliminated by rapid cooling which produces a thermodynamically unstable system. A more stable system can be obtained by annealing to nucleate some crystals which then incorporate the surrounding noncrystalline material into what can be viewed as spherulites. Micrograph 3 indicates that the most favorable nucleation sites for crystals tend to be located along the carbon fiber surfaces.

Conclusions

The techniques of metallography described here are proving invaluable for researching matrix crystallization. Mechanical properties of semicrystalline thermoplastic/fiber composites are related to percent of crystallinity and microstructure analogous to metals. These attributes can be controlled during processing by varying the cooling rate from the polymer melt. Material characterization of these materials should include metallography, along with thermal analysis and X-ray diffraction.

Figure 24-6 Class 2 microstructures of thermoplastic composites. (Luther M. Gammon, Boeing Commercial Airplane)

A ***laminate*** is a stack of lamina with various orientations of the directions of the principal materials in the lamina. Laminates can be built up with plates or plies of different materials or of the same material, such as glass fibers. Shear stresses are always present between the layers of a laminate because each layer tends to deform independently of its neighboring layers due to each layer having different properties. These shear stresses, including the transverse normal stresses, are a cause of delamination.

Bulk molding compounds (BMCs) are a premixed material of short fibers (chopped-glass strands) preimpregnated with resin and various additives. A *dough molding compound (DMC)* is an alternative term for a BMC. Some thermoset resins are quite thick and are called *molding doughs.* Parts made by BMCs are limited to about 16 inches in their longest dimension due to problems with the separation of the components of the molding compound during molding.

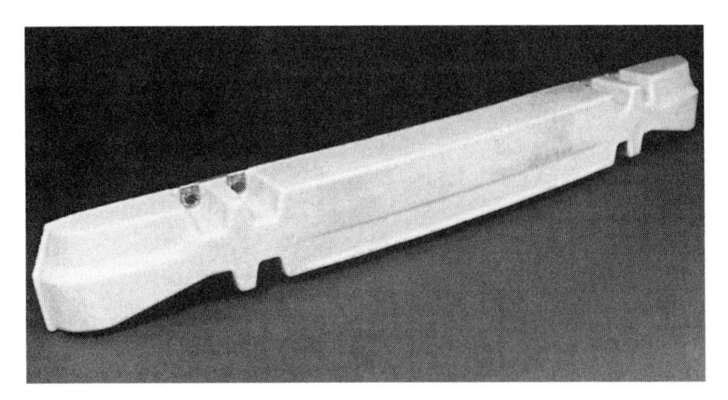

Figure 24-7 TP resins serve as a matrix for glass fibers in the new thermoplastic composite used in Ford's Lincoln Continental bumper beams. AZDELPlus® is a high-strength, lightweight, impact-resistant composite available in a layered structure of glass mat or chopped glass impregnated with thermoplastic resins. Its superior properties to polypropylene-based glass mat help meet higher bumper collision standards (5 mph versus 2.5 mph). Demonstrating up to 90% higher peak loads, these new bumpers weigh 5 pounds less than their steel counterparts. This joint development by PPG Industries and AZDEL, Inc. has led to a 40% improvement in the performance of molded parts. (G.E. Plastics)

Sheet molding compounds (SMCs) are impregnated, continuous sheets of composite material. SMCs cut to proper size and stacked to provide the required thickness before heat curing in matched metal molds are used in automotive bodies and large structural parts (see Figure 24-7). There are three types, depending on the length of the glass fibers used in the reinforcement. Fibers can be as short as 25 mm and arranged in random fashion in the resin. Longer fibers (200 to 300 mm) can be oriented in one direction (directional fibers). Continuous fibers laid in only one direction make up another category. Various combinations of these SMC types are used with fibrous glass reinforcements, reaching 65% by weight of the composite (see Figure 24-8). A recent molding compound, designated XMC, can contain up to 80% glass fiber by weight and uses continuous fibers running in an X pattern (see Figure 24-19). Thick molding compounds (TMCs) are similar to SMCs but are up to 2 inches thicker.

The more common fibers used in composites are described on the following pages. As research continues, more materials (organic, polymeric, ceramic, and metallic) are becoming sources of fibers for different composites designed for ever-increasing applications in an energy-conscious age.

24.3 REINFORCING FIBERS

24.3.1 Glass Fiber

Glass fibers are made by letting molten glass drop through minute orifices and then attenuating (lengthening) them by air jet. The standard glass fiber used in glass-reinforced composite materials is E-glass, a borosilicate type of glass. The glass fibers produced, with diameters from 5 to 25 μm, are formed into strands having a tensile strength of 5 GPa. Chopped glass used as a filler material in polymeric resins for molding consists of glass fibers chopped into very short lengths. ***E-glass*** is the first glass developed for use as continuous fibers. It is composed of 55% silica, 20% calcium oxide, 15% aluminum oxide, and 10% boron oxide. It is the standard grade of glass used in fiberglass and has a tensile strength of about 3.45 GPa and high resistivity. Fiber diameters range between 3 and 20 μm. ***S-glass*** was developed for high-tensile-strength applications in the aerospace industry. It is about one-third stronger than E-glass and is composed of 65% silicon dioxide, 25% aluminum oxide, and 10% magnesium oxide.

Automakers are finding more uses for ***fiberglass*** fiber reinforcements for both thermosets and thermoplastic polymers, which are appearing in both exterior and interior parts as

MAGNAMITE GRAPHITE PREPREG
X-AS4/1904

Magnamite graphite prepreg X-AS4/1904 is a 250°F curable epoxy resin reinforced with unidirectional graphite fibers. The reinforcement is Magnamite continuous AS4 graphite fiber that has been surface-treated to increase the composite-shear and transverse-tensile strength. The 1904 resin matrix was developed to cure under tape-wrap or vacuum-bag conditions at 250°F and to operate at temperatures up to 180°F.

	At 77°F	
Typical Composite Properties	Without Glass Scrim	With Glass Scrim
0° flexural strength, psi × 10^3	225	215
0° flexural modulus, psi × 10^6	18.4	16.1
0° tensile strength, psi × 10^3	240	—
0° tensile modulus, psi × 10^6	20.1	—
Shear strength, psi × 10^3	14.5	15.0
Cured ply thickness, mils	5.6	6.4
Fiber volume, %	58	58
Typical Prepreg Characteristics		
Tape width, in.	12	12
Resin content, %	40 ± 3	40 ± 3
Fiber content, [a] g/m^2	145 ± 4	145 ± 4
Glass scrim[b]	—	104
Yield, ft/lb	19	18

(a) Various fiber contents are available between 95 and 195 g/m^2.

(b) 104, 108, and 120 glass scrim have been used.

(a)

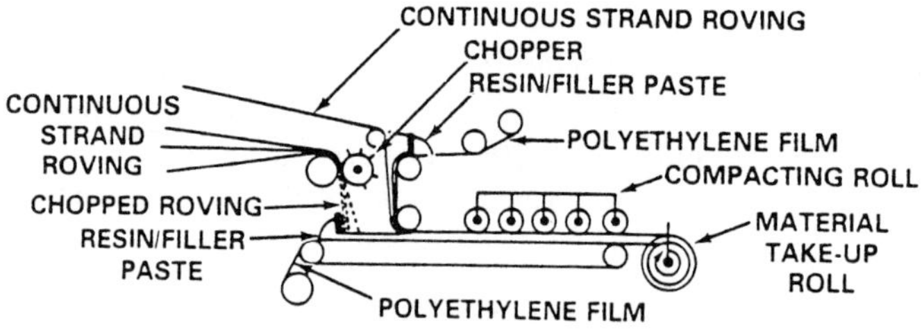

(b)

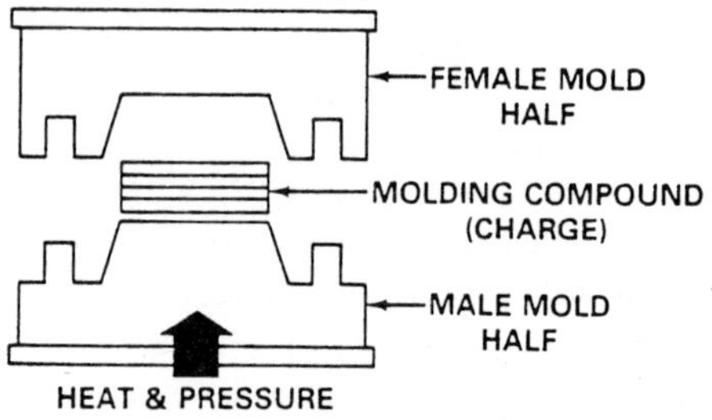

(c)

Figure 24-8 (a) Hercules prepregs. (Hercules, Inc.) (b) Schematic of machines used to make SMC sheet and (c) mold SMC plaques.

well as under the hood. Air-intake manifolds, radiator caps, cables and connectors, bumpers, fenders and hoods, seats, and instrument panels increasingly contain fiberglass. Air-intake manifolds, in particular, represent a dramatic growth area for glass fiber. The demand for this reinforcement originated from the desire of manufacturers to build cars that cost less, are visually appealing, and are easy to maintain.

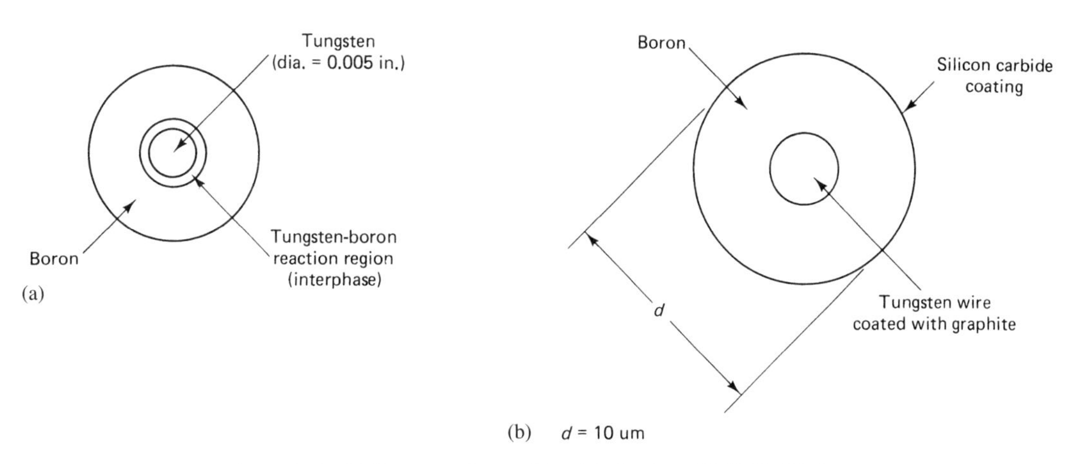

Figure 24-9 (a) Structure of boron fiber. (b) Structure of BORSIC™ fiber.

24.3.2 Boron Fiber

The year 1963 marked the birth of boron technology. Advanced composites with boron fibers were developed at that time by the U.S. Air Force. Boron fibers are composites of the substrates tungsten, silica coated with graphite, or carbon filaments upon which boron is deposited by a vapor-deposition process (CVD). The final boron fiber has a specific gravity of about 2.6, a diameter between 0.01 and 0.15 mm, a tensile strength of about 3.45 GPa, and a modulus of around 413 GPa. Figure 24-9a is a representation of a boron fiber.

Boron is more expensive than graphite and requires expensive equipment to place the fibers in a resin matrix with a high degree of precision. Problems with chemical reactions between boron and other metals continue to be a source of concern to materials technologists, who are exploring ways to overcome this limitation. One approach to this problem was the coating of the BORSIC fiber with silicon carbide (see Figure 24-10), which provides a fiber that is more compatible with metal–matrix materials, particularly at high operating temperatures. Figures 24-9 and 24-10 show a sketch of a BORSIC™ filament and an aluminum matrix in which boron is deposited on a graphite-coated tungsten wire, which makes the fiber itself a composite.

24.3.3 Carbon and Graphite Fibers

Carbon is a nonmetallic element. Black crystalline carbon, known as graphite, has a specific gravity of 2.25. Transparent carbon, diamond, has a specific gravity of 3.5. Graphite and diamond are allotropes of carbon, as are amorphous forms of carbon such as coke and charcoal. Both graphite and diamond have very high melting points of over 3732 K, the reason for which becomes clear upon examining the crystal structure and seeing the nearly infinite network of carbon–carbon covalent bonds that must be broken to melt these materials.

The terms ***carbon*** and ***graphite*** are often used interchangeably. However, a line of demarcation has been established in terms of modulus and carbon content. Carbon fiber usually has a modulus of less than 344 GPa and a carbon content between 80% and 95%. Graphite fibers have a modulus of over 344 GPa and a carbon content of 99% or greater. Another distinguishing feature is the pyrolyzing temperatures. For carbon, this temperature is around 1315°C, and for graphite, it is around 1900° to 2480°C. ***Pyrolysis*** is the thermal decomposition of a polymer. This heat can vary from 1000° to 3000°C, depending on the desired fiber properties. A carbon-containing polymer can decompose to carbon in an inert or reducing atmosphere and thus produce carbon fibers and/or carbon–carbon composite materials. If elements other than carbon are present in the original polymer, these elements will form a residue as a result of pyrolysis of ceramic compounds. An example of pyrolysis is the production of SiC fibers from polycarbosilane polymers having tensile strengths of 2.7 GPa.

One authority classifies carbon as a ceramic. Another says the pyrolized organic material is a polymer, even though it is mostly carbon. They admit that carbon itself is an element,

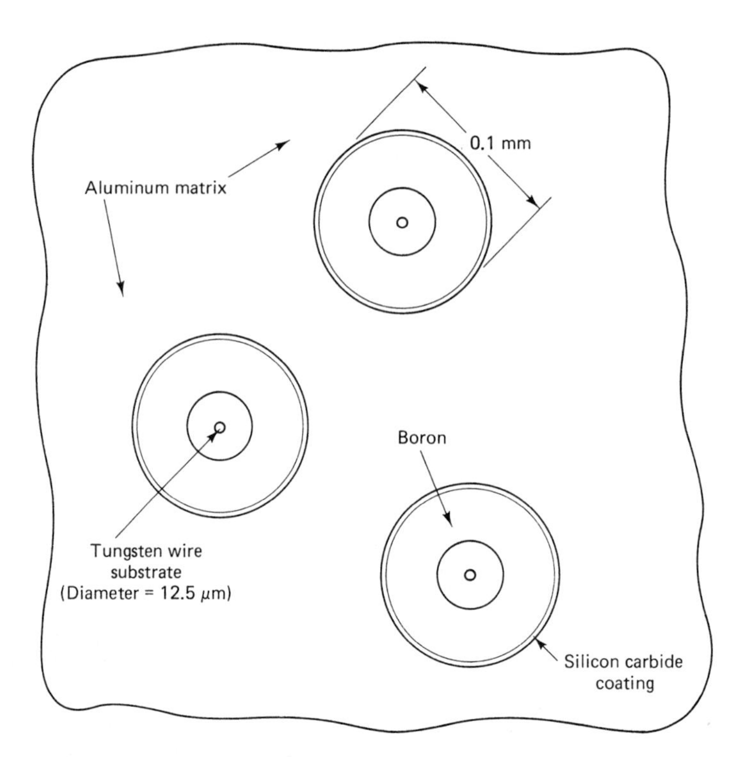

Figure 24-10 BORSIC™ fibers in an aluminum matrix would produce a MMC. The boron fiber itself is a composite, having been deposited on a Ti substrate and coated with SiC.

but if you believe that the word *carbon* really means "graphite," then carbon is a ceramic. Another authority uses the word *carbon* to describe fibers because the fibers are composed of crystalline graphite regions. In general, *carbon* describes both carbon and graphite fibers.

As with composite materials, graphite fibers are not a new development. They were first produced in small quantities in the nineteenth century for use in incandescent lamp filaments. Several methods are used today to make carbon fibers of varying length and diameter and having the versatility of glass fibers. The oldest method of producing these fibers, used in the late 1960s, was ***graphitization*** of organic fibers such as a rayon at temperatures up to 3250 K. An acrylic fiber, polyacrylonitrile, abbreviated PAN, is also used as a source of such fibers, which the Royal Air Force (RAF) produced in England in 1961. Heated (pyrolized) under tension to stabilize the molecular structure at temperatures between 920 and 1140 K, the noncarbon elements (O_2, N_2, H_2) are driven off, leaving a fiber high in carbon content. A more recent process for producing high-modulus carbon fibers uses low-cost pitch; the pitch is converted to a liquid crystal, or mesophase, state before it is spun into fibers. Although carbon fibers have a maximum-use temperature range from 250° to 2000°C, the use temperature of a composite is controlled by the use temperature of the matrix.

Graphite fibers are available to the fabricator in the following forms:

1. Continuous fiber
2. Unidirectional prepreg as tow or tape in widths from 3 to 36 inches to form laminations
3. Chopped fiber
4. Pultruded shapes
5. Woven fabrics as dry cloth or prepreg for laminations

Twenty different carbon fibers with cross-sectional configurations varying from circular to kidney shaped, in diameters of 0.008 to 0.01 mm, with strengths ranging from 1.72 to 3.1 GPa, and modulus values from 193 to 517 GPa are now available from six manufacturers. One carbon yarn comes in plies from 2 to 30, with each ply composed of 720 continuous filaments about 8 μm in diameter. This fiber is 99.5% carbon and maintains dimensional stability to 3420 K. In addition to carbon, there are glass and aluminum fibers, supplied in the form of tows (rovings or strands) consisting of many individual continuous filaments consisting of 3000 to 30,000 filaments. Another grade of fiber consists of 2000 filaments gathered together into a uniform strand that provides a modulus of 379 GPa and a tensile strength of 2.0 GPa. Thornel 300 carbon fiber, produced by Union Carbide, consists of 6000 filaments in a one-ply construction. The surface of the fiber has been treated with sizing to increase the interlaminar shear strength in a polymer–matrix composite in excess of 89.6 MPa. The tensile strength is listed as 2.69 GPa, with an accompanying modulus (tensile) of 229 GPa. Carbon and graphite fibers and yarns are used to produce various fabrics for a variety of industrial uses (see Figure 24-11).

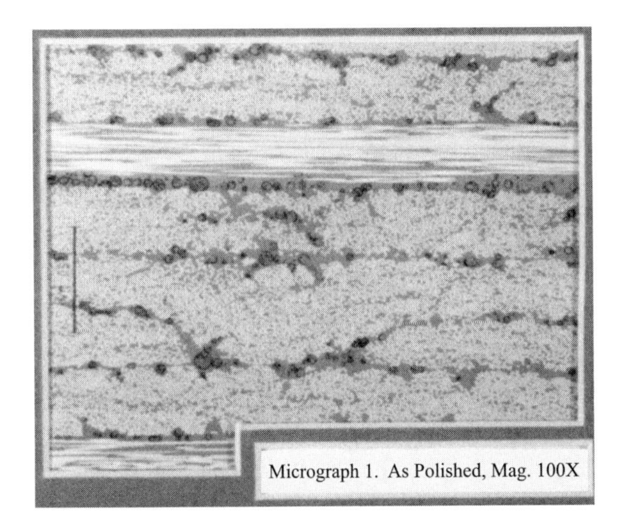

Micrograph 1. As Polished, Mag. 100X

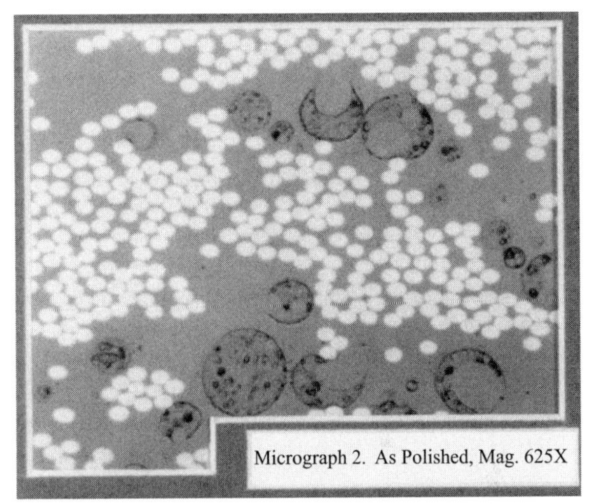

Micrograph 2. As Polished, Mag. 625X

Figure 24-11 Class 8 microstructure of carbon-fiber-reinforced composites. (L.M. Gammon, J.S. Chen, R.A. Grove, A.M. Thorstenson, Boeing Materials Technology)

The use of composite, *Carbon Fiber Reinforced Plastic* (CFRP), for airplane components has resulted in significant weight savings while yielding improved performance and fuel efficiency. Low damage resistance and fracture toughness have commonly limited the application of CFRP throughout the industry.

The Boeing Company, during the past several years, has actively engaged in the development and evaluation of tougher and less prone to damge CFRP composites. A promising composite candidate was researched and identified in 1987. It was the result of a breakthrough in resin technology that this new composite exhibited significant improvements in damage resistance without compromising other vital properties. The main objectives within the laboratory has been to come to a full understanding of those key microstructural features which relate directly to damage resistance, fracture toughness, and process and formulation changes.

Boeing Materials Technology research has led to the development of several specimen preparation methods, and has resulted in an accurate and broader understanding of relationships between microstructure and material performance. It was found that size and distribution of individual resin phases would directly affect damage resistance and the changes in formulation and manufacturing processing could also be evaluated from a microstructural viewpoint. The latter benefit allowed the development of a "fingerprint" of the morphology for maintaining material consistency.

This presentation reveals three successive specimen preparation techniques for CFRP laminates developed with a complex, multiphase resin system. Micrographs of the same field are shown for each of the following techniques: 1. Polishing. 2. Etching; and 3. Epi-fluorescent dying/microscopy. For each integral specimen preparation method, the correlation between key microstructural features and material performance is being determined.

Micrographs 1 and 2: Specimen preparation. Silk wheel primary polished at 1200 rpm, high pressure and alumina/H_2O slurry. Used 5.0 u alumina followed by 0.3 u alumina. Final polished with medium nap wheel, high pressure and 0.05 u alumina/H_2O slurry. **Observation.** In the "as polished" condition, a stacking sequence of fibers (white) can be seen and spheres (Phase 1) are distributed within the resin (Phase 2). Although this method provides a superior polish, this is what many microstructural investigations of composites have been limited to in the past.

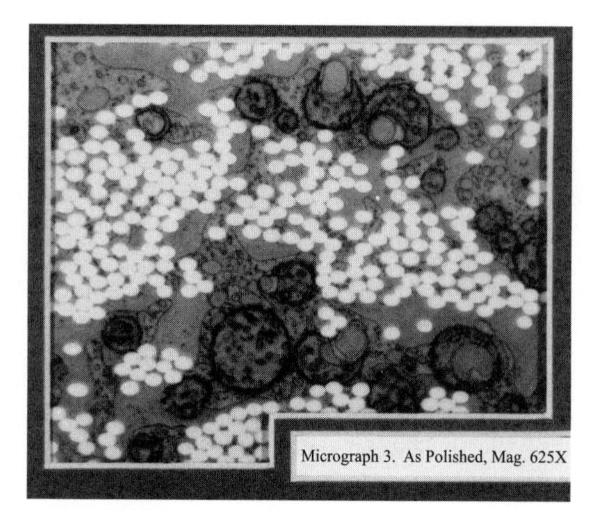

Micrograph 3. As Polished, Mag. 625X

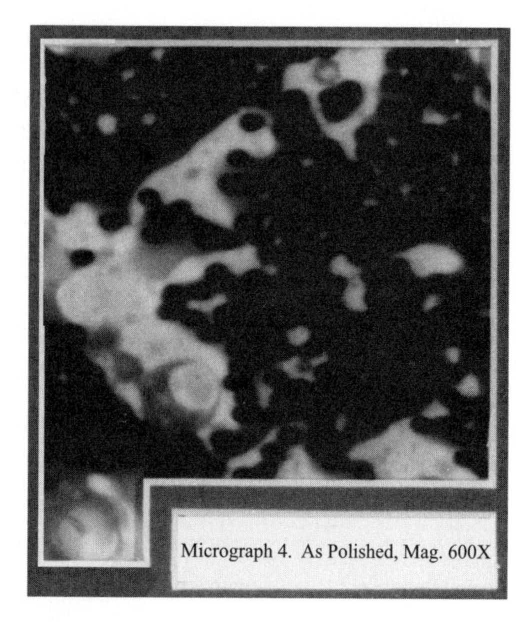

Micrograph 4. As Polished, Mag. 600X

Figure 24-11 continued

Micrograph 3: Specimen Preparation. Polished specimen is etched for 30 seconds using [CrO_3 (12.5 g): Conc. HNO_3 (50 ml): H_2O (50 ml)] solution. **Observation.** The resin (Phase 2) gives indications of two sub-phases: as continous phases (2a) and a discrete phase (2b). Precise detail can be seen within each phase, especially within the spheres (Phase 1). This information allowed Boeing to visually characterize the microstructure and correlate material performance with formulation and processing changes and also enabled them to understand how each of the phases contributed to damage resistance and fracture toughness.

Micrograph 4: Specimen preparation. Specimen is immersed for 30 minutes in a solution of Rhodamine B/methylene chloride dye. Ultrasonically cleaned excess dye using methylene chloride, followed by a bake at 250°F to remove solvent. **Observation.** Using epi fluorescent microscopy with a 190 110 mg excitation range each individual phase becomes distinctly colored. This breakthrough allows investigation of image analysis for quantitative morphology characterization (where more contrast is required) and also more accurately allows definition of the chemical nature of each phase with finer detail.

First introduced in 1959 at a price of over $500 per pound, carbon fibers have steadily dropped in price. By 1984, the price had declined to $13 to $15 a pound for lower-grade graphite fiber and about $19 a pound for aerospace-grade fibers. Generally, the decline in prices of materials can be expected to continue with an increase in use, as long as there is a good availability of raw materials and improvements in production technology. Graphite fibers are linear elastic and, like other carbon materials, are anisotropic. Graphite possesses excellent creep and fatigue resistance, as well as dimensional stability at high temperatures. It is one of the few truly efficient structural reinforcements, with both high specific modulus and strength. Figure 24-12 compares the strength and modulus of two grades of graphite with some other fiber materials.

The thermal properties also must be considered unusual. The most important properties are low ***coefficient of thermal expansion (CTE*** or ***COT),*** high thermal conductivity, and dimensional stability at temperatures of 2000°C and above. The CTE of graphite is negative at 0°C, but increases with temperature (see Table 24-2). However, its generally negative CTE, coupled with the ability to tailor the orientation of the fibers, permits the designer to adjust the thermal coefficient over a wide range of values. Its thermal conductivity is similar to that of many metals in the same temperature range, including steel and nickel. Aluminum is the only metal that has a significant advantage in thermal conductivity over graphite. However, aluminum's high CTE and low-temperature capability make it useful only in low-temperature composite curing applications. Finally, graphite is essentially free from residual stresses and never needs stress-relief annealing, as is common with many metals.

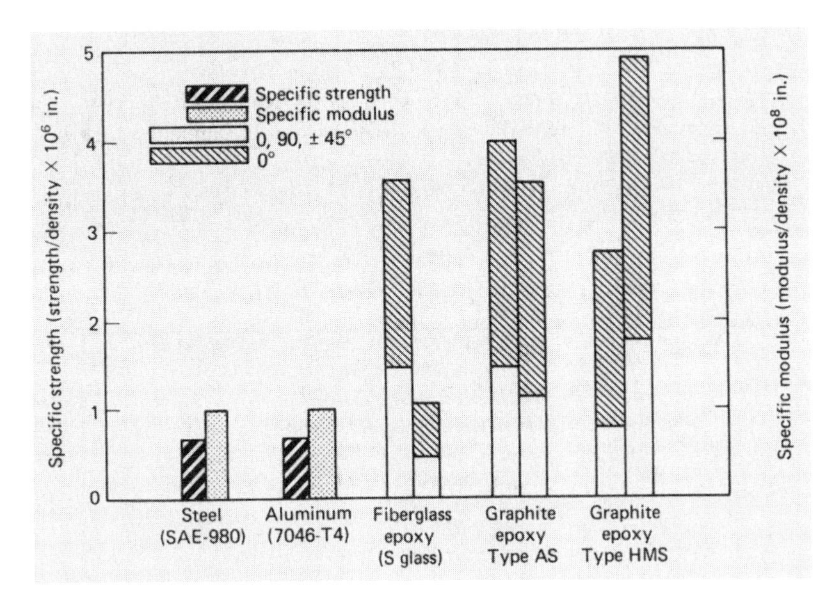

Figure 24-12 Comparison of specific strength and specific modulus (specific stiffness) of commonly used fiber materials. (DuPont)

TABLE 24-2 DIMENSIONAL STABILITY OF SOME FIBER MATERIALS (CTE)

Material	Thermal Expansion (in./in./°F) × 10^{-6} 0°	0°, ±45°, 90°
Graphite AS	−0.20	1.01
Graphite HTS	−0.25	0.80
Graphite HMS	−0.30	0.45
Fiberglass	3.50	6.00
Aluminum	13.0	—
Steel	7.0	—
Nylon (6/6)	45.0	—

Graphite is commercially available in three different types: high modulus, intermediate modulus and tensile strength, and high stiffness and strength. All three have low density but also low impact resistance. The strength of graphite fiber/polycarbonate matrix composites is increased by precoating (sizing) the fibers with a thin layer of polycarbonate. Sizing can promote the wetting of fibers by matrix polymers, thereby enhancing fiber/matrix adhesion. The COT of graphite increases with temperature. Graphite fibers can also be coated with other materials to extend the usefulness of their properties to higher temperatures. Titanium carbide (TiC) is used to coat graphite or carbon/carbon composites with unusual thicknesses (about 150 mm) to provide exceptional resistance to erosive wear and to corrosion by chemical agents. In addition, this coating has demonstrated very high thermal-shock resistance. Table 24–3 lists a few properties of TiC-coated graphite and includes a graph showing the erosion rates of various materials, including TiC-coated graphite.

Bromine-intercalated graphite fibers in an epoxy material are ***electrically conducting composites*** used for many applications, such as grounding planes. The electrical conductivity of graphite fibers (derived from pitch blend) has been increased from about 4000 to about 50,000 $\Omega^{-1}cm^{-1}$ through ***intercalation*** ("insertion between layers"). Common metals have conductivites that range from 10,000 to 550,000 Ω^{-1} cm^{-1}. The process of intercalation introduces donor/acceptor atoms, such as bromine, between the layers of graphite atoms in their

TABLE 24-3 ENGINEERING PROPERTIES OF TiC-COATED GRAPHITE

Property	Value	
Flexural strength [MPa (ksi)]	90	(13)
Elastic modulus [GPa (Msi)]	11	(1.6)
CTE [ppm/K (ppm/°F)]	8.4	(4.7)
Thermal conductivity [W/m · K (Btu/hr · ft · °F)]	121	(70)
Density [g/cm³ (lb/ft³)]	1.81	(113)

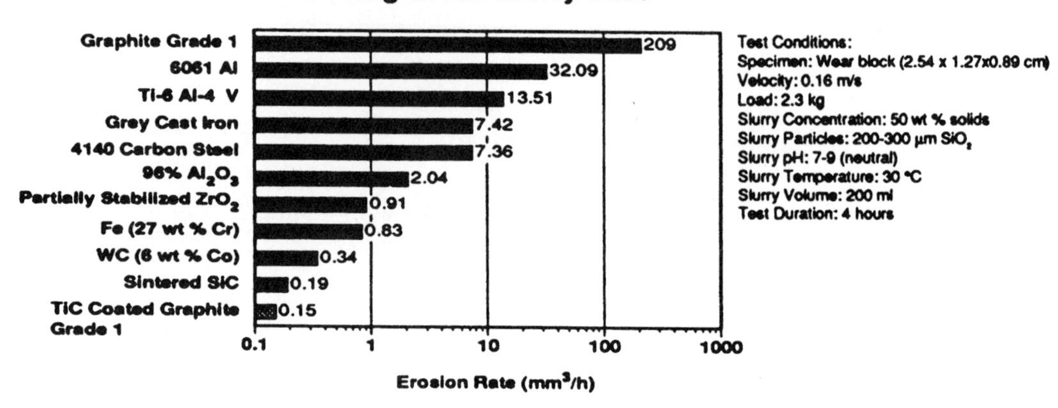

Source: Courtesy of Lanxide.

characteristic layer arrangement. A cloth made of graphite fibers immersed in bromine is stacked between epoxy films making up the composite, which is then pressed and cured. Bundles of a thousand thin carbon fibers are pultruded into a connector and coated. These connectors can conduct electricity as well as copper but at a lower cost and higher reliability (they never need replacement). About 1000 fibers make a rod 0.3 nm in diameter. Carbon connectors are far less susceptible to corrosion and contamination.

One disadvantage of graphite is its poor oxygen resistence. It begins to oxidize in air at about 700 K. Unlike metals, graphite does not form a protective film. Instead, the graphite oxide is volatile. To protect graphite from oxidation, several coatings have been developed, one of which, silicon carbide, protects graphite for a limited time at temperatures approaching 1920 K. Another limitation of carbon fiber is that it is a brittle, strain-sensitive material that cannot be depended on to offer much impact resistance.

24.3.3.1 Nano composites. Potential future reinforcements for composite materials are high-modulus and high-elastic-strain nanotubes. ***Nanotubes*** are carbon molecules constructed as a single sheet of graphite rolled into a tube with fullerine caps on the ends. (See discussion of fullerines in Module 2 and the molecule in Figure 2-14.) These single-walled elongated ***fullerine molecules*** contain millions of carbon atoms that form defect-free hollow strucutures just a few atoms in circumference. Nanotubes are many thousands of times longer than their diameters (see the discussion of aspect ratio in Section 24.4.1) and they possess exceptional mechanical, electrical, and thermal properties. For example, their tensile strength is at least 10 times stronger and their weight is less than half that of conventional carbon fibers and their thermal conductivity is as high as that of diamond. These properties can be tailored through processing to fit a multitude of applications.

Nanosteel, a composite material, is made up of steel alloy clumps of just a few molecules per particle. Each particle is about 50 nm in diameter. Solidified, with an amorphous structure, this alloy is crushed into a powder and applied as a coating to a conventional metal alloy such as steel or aluminum. The nanosteel bonds with the other base metals to form a dense coating of much stronger atomic bonds, the ultimate determinates of a material's strength. The best steel alloys with their fine microstructures of ferrite and carbide have listed yield and tensile strengths that are only about 10% of their theoretical best strength. With the stronger atomic bonds of the very fine nanostructure, nanosteel displays a theoretical strength level of between 40 to 45%. In addition, corrosives find it difficult to establish a connection due to the fineness (fine grain structure of nanosteel). Nanosteel coatings under test are about 30% harder than conventional steels yet are fairly flexible, permitting their application before or after the base metal is coated. Further development of this advanced material will allow engineers to approach their goal of using cheap base metals to produce, by coating them with essentially a weightless coating to protect the surface from corrosion, composite materials of greater strength, hardness, ductility, and strength-to-weight ratio.

24.3.4 Kevlar™ Fiber

Kevlar, an organic fiber introduced in 1972 by DuPont for use in radial tires, is an aramid, or aromatic, polyamide fiber. This organic fiber is melt-spun from a liquid polymer solution. The aromatic ring structure (see Figure 2-13) results in high thermal stability. The rodlike nature of the molecules classifies ***Kevlar*** as a liquid-crystalline polymer characterized by its ability to form ordered domains in which the stiff, rodlike molecules line up in parallel arrays. These domains orient and align themselves in the direction of flow during processing, causing a high degree of alignment parallel to the fiber axis and resulting in anisotropy, with high strength and tensile modulus in the fiber-longitudinal direction. Refer to Figure 14-7 and the accompanying text for additional discussion of polymer crystallinity. These overall properties plus a density of 1.44 g/cm^3 (about one-half that of glass) are the key to Kevlar's use in weight-limited applications.

There are three grades of Kevlar. Kevlar 29 provides high toughness with a tensile strength of about 3.4 GPa for use where resistance to stretch and penetration are important. Kevlar 49 has a high-tensile-strength modulus of 130 GPa and is used with structural composites. Kevlar 149 has an ultrahigh-tensile-strength modulus of 180 GPa. Figure 24-13 is a stress–strain diagram showing the relation of Kevlar 49 to other common reinforcing fibers. Kevlar fibers incorporated into high-strength composite panels are being used by several U.S. air carriers to reinforce cockpit doors and bulkheads. On Airbus commercial aircraft the cockpit doors are being reinforced with composite armor made of S-2 glass

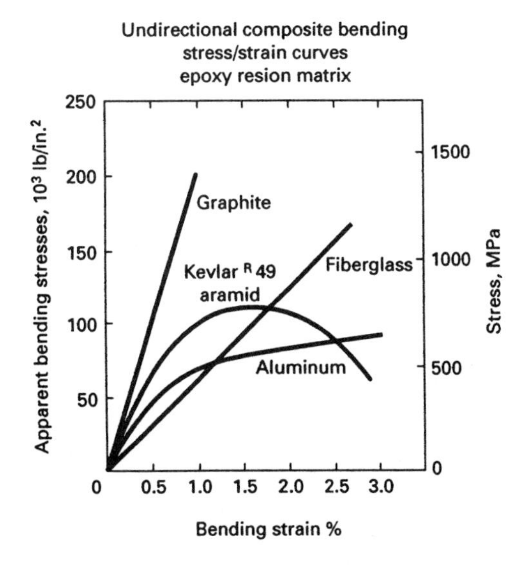

Figure 24-13 Bending stress–strain curves of some fiber materials. (DuPont)

roving and phenolic resin that exceeds the protection and fire safety regulations of the U.S. Federal Aviation Administration (FAA). A flexible textile that conducts electricity developed by DuPont researchers is made of fibers with a core of Kevlar covered with a layer of electrically conductive material such as silver or nickel. Bundles of these fibers are coated with a second polymer for protection during washing. This advance in wearable electronic devices is the first to combine high electrical conductivity with flexibility sufficient for weaving and embroidery. It is commercially available, and a Finnish sportswear company has made a prototype cold weather survival suit with this advanced textile complete with an embedded heart monitor, a body temperature sensor, and a Global Positioning System device.

24.3.5 Ceramic Fibers

The development of both continuous and discontinuous ceramic fibers based on oxide, carbide, and nitride compositions was undertaken due to the need for high-temperature reinforcing fibers in composites for the aerospace industry. Most oxide fibers are compositions of Al_2O_3 and SiO_2, although a few are almost pure oxides of aluminum and silicon (alumina and silica). The average properties of continuous oxide fibers are as follows: density, 3 g/cm^3; diameter, 12 μm; tensile strength, 2 GPa; tensile modulus, 200 GPa; and use temperature of around 1300°C.

Carbide fiber properties differ significantly from those of the oxides. Representative of continuous carbide fibers is SiC. ***Silicon carbide (SiC),*** a ceramic fiber, is made by one of two methods. The first method uses CVD to deposit Si and C onto a pyrolytic graphite-coated carbon core. The resulting fiber, designated SCS-6, is similar in size and microstructure to boron fiber and has a diameter of about 140 μm. The second method uses controlled pyrolysis of a polymeric precursor. The filament produced (trade name Nicalon) is similar to carbon filaments in size (14 μm) and microstructure, with a diameter of about one-tenth that of an SCS-6 fiber. It is more flexible and its fibers are arranged in rows of 250 to 500 filaments per row. A new coating for SiC particulates is SiO_2, produced from rice hulls (SiC_w is the designation for this treatment). SiC is available in three different forms. Monofilaments are 140 μm in diameter; yarns consist of 500 fibers and each fiber has a diameter of 14.5 μm. SiC whiskers have diameters ranging from 0.1 to 1 μm. Allied Signal Inc.'s SiC yarn fabrication process is similar to those used for manufacturing fiber from PAN.

SiC fibers are the reinforcement of choice for use with aluminum, magnesium, titanium and titanium aluminide, and toughened ceramics materials because such fibers offer better resistance to oxidation and galvanic corrosion than graphite fibers.

Approximations of SiC's properties are as follows: density, 2.5 g/cm^3; diameter, 10 to 20 mm; tensile strength, 3.0 GPa; tensile modulus, 100 to 400 GPa; melting point of 2830°C; and use temperature of 1200°C.

Both SiC and Si_3N_4 are available as discontinuous (whisker) forms. The principal advantage of carbide fibers over oxide fibers is the great increase in the modulus of elasticity.

24.3.6 High-Performance Manufactured Polymeric Fibers

Included in the high-performance category are fibers based on polyester, nylon, aramid, and polyolefin. As discussed in Module 15 (Section 15.4.1), fibers such as aramid (Kevlar and Nomex), polybenzimidazole (PBI), Sulfar, and Spectra have increased the range of choices for materials engineers in designing materials with tailor-made properties. However, most do not possess the thermal properties found in ceramic- and metal-based fibers. Kevlar, a trademarked material, has been discussed separately in this module. A polyolefin fiber, ***Spectra*** (ultrahigh-molecular-weight polyethylene) has a low specific gravity (0.9), which allows 40% more fiber per pound than aramid (Kevlar) fibers. It also has higher specific strength and specific stiffness, and has seven to ten times more abrasive resistance than aramid, with low moisture absorption and good chemical resistance.

24.4 FIBER PROPERTIES

24.4.1 Fiber Strength

A material's brittleness greatly affects its strength, for brittle materials have little resistance to the propagation of cracks. Pure metals possess varying degrees of ductility (the opposite of brittleness), which allows for some yielding in the face of longitudinal stress concentrations, which in turn prevents fracture. Brittle fractures, on the other hand, are usually catastrophic and come without advance warning. The alloying of metals and thermal treatments are two ways of improving the resistance of a material to plastic deformation (strength) by providing for the dispersion of harder particles within a matrix of softer material, which tends to limit the motion of dislocation through the matrix. Another way to increase the strength of a material is to add another material that has greater load-carrying capacity. The most prevalent form of this added material is fiber. How well fibers strengthen a material depends on the efficiency with which the relatively soft matrix material interacts with the fibers to transfer the load between them.

The strength of an individual fiber is dependent on the absence of surface defects. The presence of microcracks anywhere in the small fiber causes localized stress and load concentrations and eventual failure of the individual fiber. The properties of a fiber are mainly dependent on the fiber's length, diameter, and orientation. The optimum reinforcing fiber for a composite material has a ratio of length (L) to diameter (D) of about 150. As this ratio, known as the ***fiber-aspect ratio,*** increases, the strength of the composite increases (see Figure 24-14). A single polymer chain has the highest aspect ratio of any "fiber." Incorporating molecular fibers into composite polymers is a technique with great potential for the future. Even with fibers of small diameter, 1 to 25 μm, aspect ratios range from 100 to 15,000.

Illustrative Problem

Given a fiber that is 25 μm in diameter and has an aspect ratio of 15,000, find the length of the fiber expressed in meters.

$$\text{Length}(L) = \text{diameter } (D) \text{ of fiber} \times \text{aspect ratio (AR)}$$

Solution

$$L = 15 \times 10^3 \times 25 \times 10^{-6}$$

$$L = 0.375\text{m}$$

The shape of the reinforcing phase, rodlike for fibers, plays an important role in determining the performance of a composite, as does the shape of the unit cell in a homogeneous solid. Their size and distribution control the texture of the material and determine, in part, the interfacial area between the fibers and the surrounding matrix. The topology of the fibers (i.e., their spatial relation to each other) is important because certain properties of the composite may be affected by how much the individual fibers or filaments touch each other (see Figure 24-15).

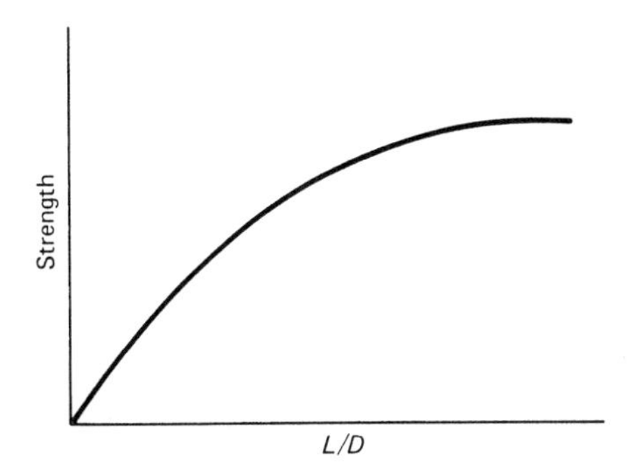

Figure 24-14 Strength versus fiber-aspect ratio (L/D).

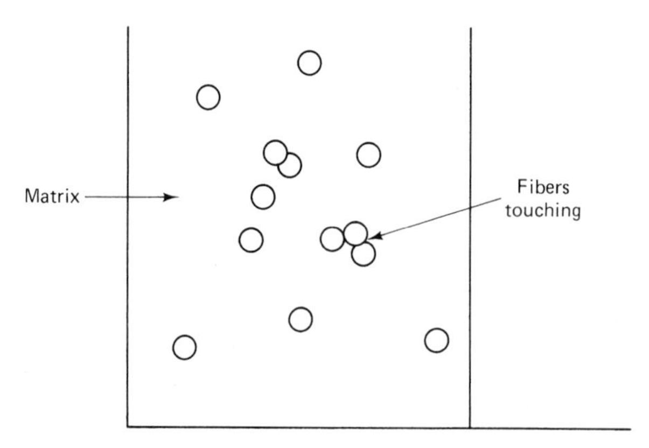

Figure 24-15 Cross section of a fiber-reinforced material showing a poor spatial arrangement of fibers within the matrix material.

24.4.2 Specific Tensile Strength

Fiber composites find use in many weight-sensitive applications in space exploration, general aviation, and the automobile and sporting-goods industries. One indicator used to portray the effectiveness of the strength of a fiber is specific strength. ***Specific tensile strength,*** the ratio of the tensile strength of a fiber material to its weight density, is an indicator of structural efficiency, given the relative load-carrying ability of equal weights of material. Typical units of specific strength are millimeters (mm), inches (in.), or kilometers (km). Table 24-4 and Figure 24-16 list the specific strength for some typical materials from which fibers are formed for use in composite materials.

24.4.3 Specific Stiffness—Specific Modulus

Stiffness is the ability of a material to maintain its shape when acted upon by a load. Another indicator of the special properties of fiber composites—in particular, the effectiveness of a fiber—is ***specific stiffness.*** It is a ratio of the modulus of elasticity (or tensile stiffness) to the weight density of the fiber. Table 24-5, using the same materials as in Table 24-4, and Figures 24-12 and 24-17 show the average specific stiffness of some fiber materials, starting with the highest values. Note that graphite, boron, and carbon have values almost six times that of steel. Graphite is known to have six times the strength of steel with six times less weight. These values of both strength and stiffness of materials in fiber form over the same materials in bulk form are most significant. As an example, the strength of glass in bulk form, such as plate glass, is only a few megapascal, yet in fiber form this figure rises to around 3.5 GPa. Structural steel (plain-carbon steel) in bulk form has a tensile strength around 0.5 GPa, but in fiber form the value is 4 GPa. This difference is due to structural differences between the two forms of the material. In fiber form, the microstructure approaches a nearly perfect structure, with the crystals aligning themselves along the fiber axis in ordered fashion. As a result, there are fewer internal defects or dislocations than would be present in bulk form. As we recall from our previous discussions, the presence of dislocations explains many properties of a material. In this situation, the movement of dislocations permits the material to yield, changing the internal structure accordingly and accounting for ductility and accompanying decreases in strength and stiffness. Figure 4-5 showed the increase in stiffness gained by reinforcing materials (aluminum and titanium) with boron fibers.

Specific stiffness and specific strength can be expressed in terms of the mass density rather than the weight density. Such a practice is more compatible with the SI. Typically, units of mass density are expressed in $kg{\cdot}m^{-3}$. Modulus of elasticity and tensile strength can be expressed either in $N{\cdot}m^{-2}$ or Pa. The specific ratio will have units of m $N{\cdot}kg^{-1}$. For example, a

TABLE 24-4 SPECIFIC STRENGTH

Materials	Weight density, ρ (kN/m^3)	Tensile strength, *S* (GN/m^2)	Specific Strength, *S*/ρ (km)
S-glass	24.4	4.8	197
E-glass	25.0	3.4	137
Boron	25.2	3.4	137
Carbon and graphite	13.8	1.7	123
Beryllium	18.2	1.7	93
Steel	77	4.1	54
Titanim	46	1.9	40
Aluminum	26.2	0.62	24

Note: The specific strength of fibers can be more than 16 times that of structural metals. E-glass has a tensile strength of about 500 ksi (3.44 GPa). Its modulus of elasticity (E) is 10.5 Msi (72.3 GPa) and its density is 2.54 g/cm^3. S-glass has a higher strength-to-weight ratio and is more expensive than E-glass. Its tensile strength is about 650 ksi (4.48 GPa). Its modulus of elasticity is about 12.4 Msi (85.4 GPa). E-glass fibers have diameters that range from 3 to 20 μm.

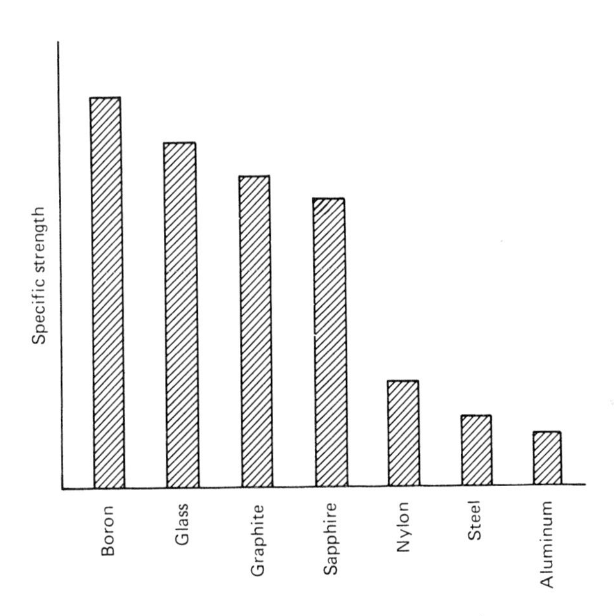

Figure 24-16 Relative specific strengths of typical materials used in composites.

TABLE 24-5 SPECIFIC STIFFNESS/MODULUS

Materials	Weight density, ρ (kN/m^3)	Tensile elasticity, *E* (GN/m^2)	Specific Stiffness, *E*/ρ(mm)
Graphite	13.8	250	18
Beryllium	18.2	300	16
Boron	25.2	400	16
Carbon	13.8	190	14
S-glass	24.4	86	3.5
E-glass	25.0	72	3.0
Steel	77	207	2.7
Titanium	46	115	2.6
Aluminum	26.2	73	2.8

Note: The specific stiffness of fibers can be more than 13 times that of structural metals.

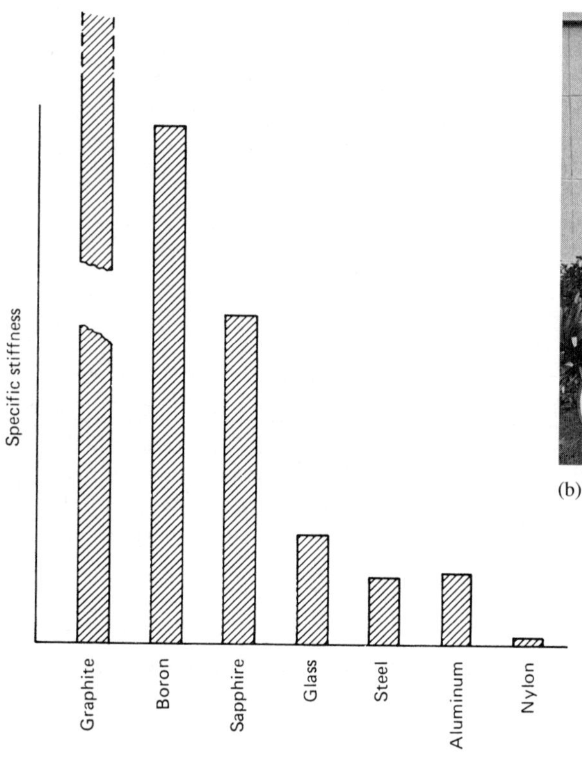

Note: Graphite has a specific stiffness of 5000 versus steel with 25 in units of 10^6N 'm/kg (using mass density in kg/m^3)

(a)

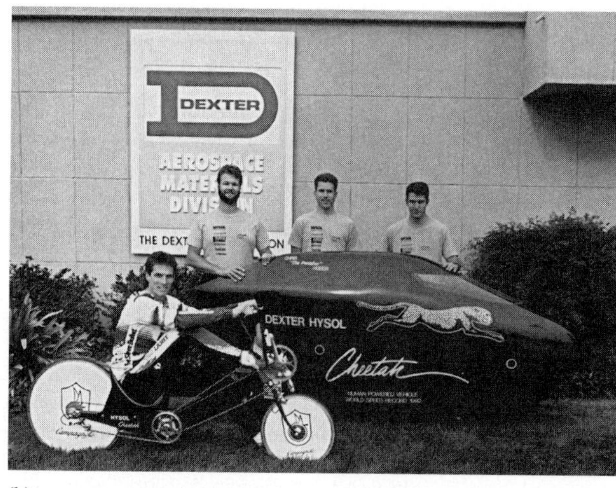

(b)

Figure 24-17 (a) Relative specific stiffness of typical materials used in composites. (b) A light, strong, and rigid aerodynamic composite shell surrounded the rider of the Dexter Hysol Cheetah bicycle as it traveled through a 200-meter speed trap at San Luis Valley, Colorado, going 68.73 mph (110.6 km/h) to capture the new human-powered vehicle speed record. *(Advanced Materials & Processes)* Internet Search—.deciMach Prize for Human Powered Speed.

wood's longitudinal modulus (E) is $1.2 \times 10^{10} \cdot \mathrm{Nm}^{-2}$ and its density measured at 12% moisture content is 0.68×10^3 kg·m^{-3}. The specific modulus of this wood is

$$\frac{E}{\rho} = \frac{\dfrac{1.2 \times 10^{10}\mathrm{N}}{\mathrm{m}^2}}{\dfrac{0.68 \times 10^3\,\mathrm{kg}}{\mathrm{m}^3}} = 17.6 \times 10^6 \mathrm{N} \cdot \mathrm{m} \cdot \mathrm{kg}^{-1}$$

Values of the stiffness–mass density parameter (stiffness modulus) for some other materials are

Graphite 5000×10^6 N·m·kg^{-1}
Boron 190×10^6 N·m·kg^{-1}
Steel 25×10^6 N·m·kg^{-1}
Nylon 3×10^6 N·m·kg^{-1}

Figure 24-17b is a striking example of how the specific tensile strength and stiffness values of composites give designers the opportunity to select and design engineered materials to form materials systems for specific applications. The human-powered, record-setting vehicle in the figure used aerospace-grade adhesive bonding to join the lightweight, rigid shell. Paste adhesives were used to bond the frame, seat mold, fairing mounts, aluminum inserts, steering assemblies, and composite parts of the bicycle system. ***Syntactic core material*** (reinforced plastic foams made with microspheres instead of entrapped gas voids) was used for the bucket seat and as a laminar core stiffener. The windshield and fairing were attached by epoxy patch kits. It should be noted, however, that the strength and stiffness properties of fibers are reduced significantly when they are used with a matrix material to form a composite. Nevertheless, the significant weight savings achieved through the use of fiber composites in engineering structures is, in the majority of cases, the primary consideration. The weight of some Cannondale mountain bikes, for example, is less than 25 pounds due to design innovations such as "an internal magnesium spine with carbon-fiber skin."

24.4.4 Fiber Loading

Fiber loading refers to the amount of reinforcement in a composite material. The strength of the composite is directly proportional to the volume of fiber (***volume fraction***) present. In addition to the fiber loading, the arrangement or orientation of the fibers plays a major role in the strength of the given product. Using the analogy of filling a shoebox with pipe cleaners, the maximum number of fibers that can be placed in a given volume (shoebox) is determined by the arrangement of the fibers. If all fibers are placed parallel to each other, a maximum number can be attained. In fiber loadings, a maximum of about 85% can be achieved. A load range of between 50 and 75% can be reached if half the fibers are arranged at right angles to each other (fabrics), while a random arrangement (chopped strands) permits a range of only 15 to 50% (see Section 24.4.5).

A composite made of a fiber, designated *Securus,* made by Allied Signal, uses fiber-loading technology to create an automobile seat belt fiber that increases occupant safety. Working in conjunction with the air bag in a collision, this "smart response" fiber replaces the mechanical load-limiting device with fibers that stretch or relax as needed to limit the force imposed in a collision. The fiber, made from a category of synthetic fibers known as Pelco fibers, is based on a polyester–caprolactone block copolymer. The fiber begins to stretch at 400 pounds of force and then stiffens again to prevent the occupant from hitting the dashboard. Costing 25% less, these seat belts provide protection for a wider range of occupant sizes.

24.4.5 Fiber Orientation

As the direction of an applied load moves 90° to the fiber orientation, the strength of a directionally oriented fiber composite decreases to about 20 to 30% of the longitudinal direction. Several techniques are therefore employed to orient the fibers in a fiber composite. If continuous fibers are oriented such that their length is in the direction (longitudinal) of the loading, this type of arrangement is known as ***directionally oriented.*** If the applied loads and/or their directions are not known, a random-oriented arrangement with continuous fibers running at various directions may be used. Alternating layers of continuous fibers at various angles provide full strength at these various directions. However, this calls for additional layup time, with its increased costs. Discontinuous fibers less than 30 cm long, which tend to orient themselves in the direction of the resin–matrix flow, provide another partial solution to providing balanced strength in several directions. This technique is a compromise between the continuous and random orientation of continuous fibers. By sandwiching random short fibers between continuous fibers, a multiple-configuration fiber composite results that is superior to a fully randomly reinforced material. The effect of adding some continuous fibers to a random, chopped-fiber-reinforced SMC is highlighted in Figure 24-18a, which compares, among other things, SMC-R50, a 50% (by weight) of 25-mm chopped E-glass fibers, 16% calcium–carbonate filler, and 34% polyester–resin–matrix formulation, to SMC-C20/R30, a hybrid of chopped (30%) and continuous (20%) E-glass fibers in a polyester-resin matrix. Figure 24-18b shows the effect of a change in the fiber orientation on the fatigue failure of XMC, an X-pattern molding compound produced by PPG Industries. All of these arrangements are available in SMC form. Figure 24-19 contains sketches of some of these arrangements of fibers in a matrix.

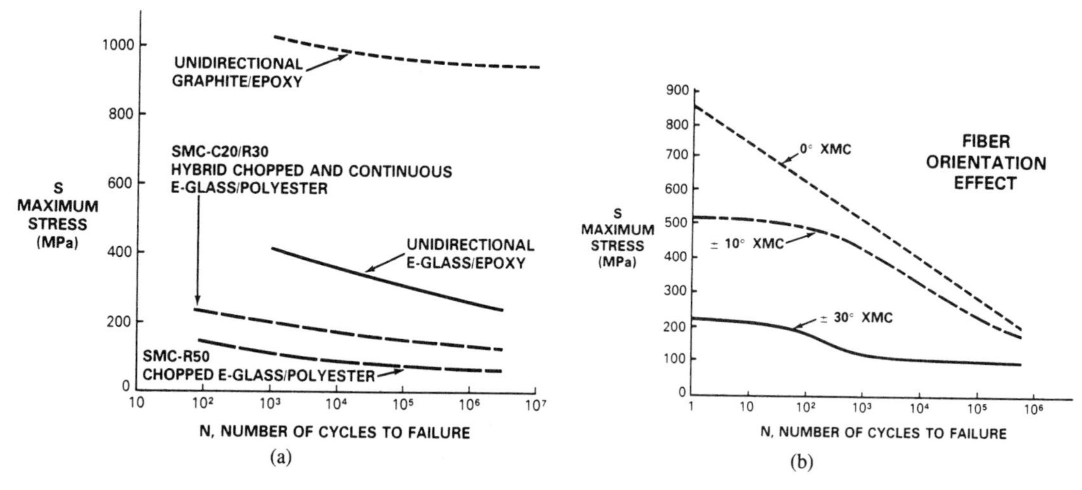

Figure 24-18 SN Diagrams. (a) Range of fiber-composite tensile fatigue performance. (General Motors) (b) Effect of fiber orientation on fatigue of XMC. (General Motors)

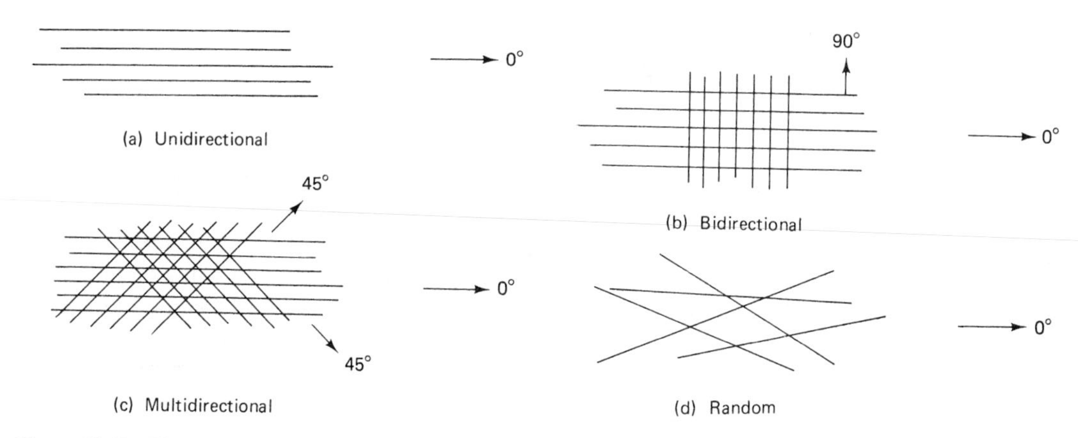

Figure 24-19 Fiber orientations.

24.4.6 Smart Fibers

There is an increased demand from various sources for fabrics that "react to our environment and provide greater degrees of comfort or performance." These fibers that are being developed not for their strength characteristics but for a multitude of other reasons will find future, beneficial applications for consumers, the military, and our environment. ***Antimicrobials,*** which prevent the spread of bacteria and reduce odors caused by them, are being investigated by textile manufacturers particularly for applications in sportswear. Fiber-based antibacterials have the advantage of conserving their germ-killing properties even after numerous launderings. The National Textile Center has funded research into ***"smart" fibers*** that can alter themselves under different conditions. Athletic wear, for example, can be made to react to environmental conditions by warming or cooling as needed. Electrically conductive polymers that act like metals can conduct electricity. Microcircuits within fabrics would permit the fabrics to store information. Future applications for such a fabric are extensive. Other smart fibers using ***photonics*** could respond to the environment by changing colors like a chameleon or, when exposed to a source of radiation such as light or gamma radiation, could respond by identifying the source of such radiation. Other fibers could be developed that would be inert to radar. Also see Section 14.3.2.

Photonic crystals are intricate microscopic structures sprinkled with regularly spaced holes similar in a way to Swiss cheese. The holes create a barrier to light of certain wavelengths. With the right arrangements these holes can force photons along designated paths.

Present-day optical fibers leak light when bent too far. Hence, research with photonic crystals is directed toward developing crystals that would lead photons along pathways with great precision. If successful, a new generation of far faster telecommunications and computing devices would be possible. Photonic crystals might revolutionize the information age by making it possible to make optical-based computer systems.

APPLICATIONS & ALTERNATIVES

Economics, often the overriding factor in product and system design, takes into account all aspects of the materials cycle outlined in Module 1. To review some of these considerations, we turn to two programs involving the use of advanced composites: the Advanced Composites Technology (ACT) program and the Boeing 777 program. ACT, a scitech partnership between NASA and the U.S. aeronautical industry, has the broad objective of developing appropriate technology for the design and manufacture of the primary structure for commercial aircraft. The emphasis is on cost-effective use of lightweight advanced composites, with the goal of reducing the costs 25% below that of aluminum structures.

Gaining enough confidence in advanced composites to use them on commercial aircraft requires considerable research and development. Among the tools of research, microscopy provides a means of examining the microstructure of failed composites in an attempt to determine the reasons for failure. Initially, carbon-fiber-reinforced plastic composite materials had low damage resistance and fracture toughness. These properties severely limited their application, particularly in the aircraft industry. One effort to rectify this problem led to a study of the relationship between microstructure and materials performance. One outcome was the development of several specimen preparation methods that revealed the size and distribution of individual resin phases and subsequently the role played by these resin phases in the improvement of damage resistance. As discovered in the preparation of metallic specimens for observation by a metallograph, good microscopy is 99% sample preparation. Preparing a flat, scratch-free, strain-free cross section without introducing artifacts (human-made changes) is both a science and an art. To grind/polish a material such as a fiber composite material with its relatively hard fibers (carbon, Kevlar, etc.) embedded in a soft, polymeric matrix certainly requires special care. One final result of these new microscopic techniques was changes in formulations and manufacturing processing that could be evaluated from a microstructural viewpoint.

The technique of ***thin-film microscopy**** was one development that provided greater morphological detail than traditional reflected-light microscopy. Thin-section microscopy uses transmitted, polarized light. Since the composite materials are opaque, specimen sections only 1 to 5 μm thick are used; at these thicknesses, the materials become optically transparent. The result is that transmitted-light methods reveal more details of the morphology of polymers and similar materials than are observable using any other microscope techniques available. Ultrathin microscopy also reveals information about fracture processes in structural laminates, which leads to improved composites that resist delamination. It also supplements other failure analysis techniques. Fiber-reinforced laminates with multiple plies, honeycomb-core sandwich structures with laminated face sheets, and polymer matrix composites with finely dispersed particulates are also likely candidates for thin-section techniques.

Most composite microscopy involves looking for porosity and microcracks. ***Microcracking*** is the result of stresses that surpass the ultimate strength of the polymer or

*A short course on "Microscopy of Fiber-Reinforced Polymer Composites" containing numerous experiments and demonstrations as well as color photomicrographs, one of many valuable composite resources, is available on the *CD-ROM Experiments in Material Science, Engineering and Technology, Version 2* from Prentice-Hall at 1-800-526-0485 or *www.prenhall.com*

polymer/fiber interface bond. From a careful study of microcracking via micrographs, scientists can determine the durability of composites in service and investigate the relationship of process parameters and environmental exposure. High-temperature thermoplastic composites behave like metallic structures in that mechanical properties change when these composites solidify at different rates. Controlling the cooling rates and microstructural phenomena such as crystal sizes, shape, distribution, and so on determines, to a great extent, the mechanical properties of the composite.

Several photomicrographs taken with thin-section microscopy, provided by Luther M. Gammon, Boeing Materials Technology, Boeing Commercial Airplane Group, Seattle, Washington, were instrumental in the development of this technique. Figure 24-20a through d, entitled "Stress/Strain Before Your Eyes," consists of five photomicrographs of various defects that illustrate the technique of using ultrathin sections with polarized light microscopy to study a composite material undergoing impact testing. Such studies, among other R&D efforts, led to decisions to make significant use of composites in commercial aircraft.

The Boeing 777 was the first of a new family of wide-body airplanes to make extensive use of carbon-fiber-reinforced plastic (CFRP) advanced composites (Figure 24-20). Although

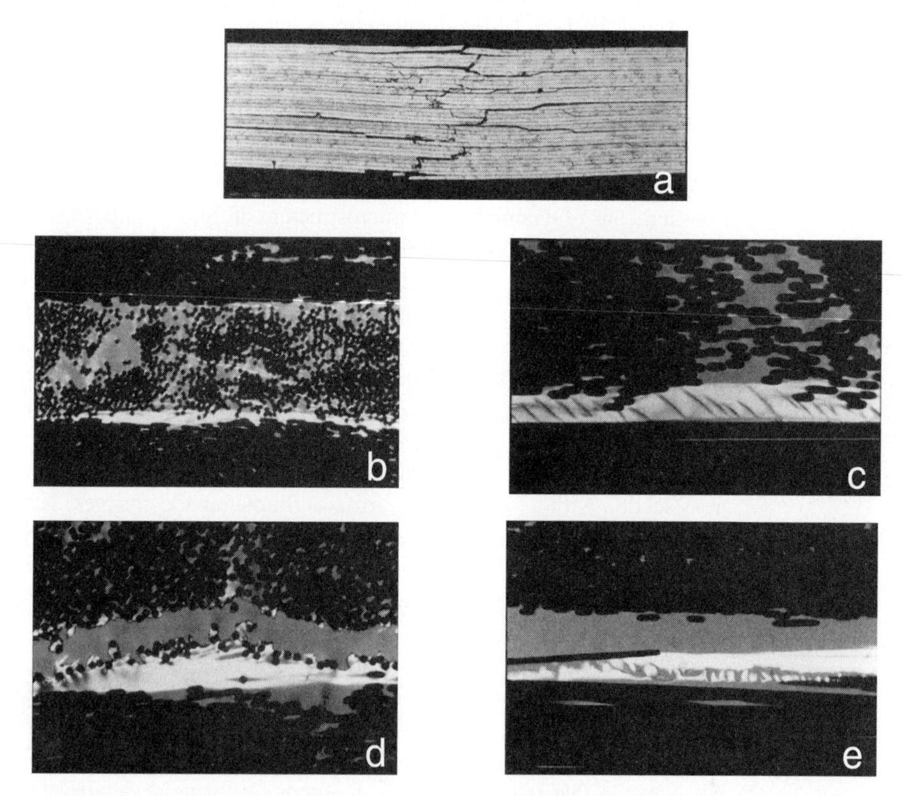

Figure 24-20 Stress–Strain Before Your Eyes. A thermoplastic resin with fiber reinforcement under impact loading. Ultrathin sectioning and polarized-light microscopy reveal in a series of photomicrographs (a through d) the effects of delamination and other fracture processes in this structural laminate composite material. (a) The composite material was impacted at 3000 in.lb/in. with a 0.625-in.-diameter spherical indentor. The specimen dimensions were 6 in. by 4 in. The characteristic fracture path propagates into the laminate like a radially expanding spiral staircase. (Magnification 8.4X.) (b) Shear deformation along the interplies is the main energy absorption mechanism. In contrast, the intraplies absorb less energy due to fiber spacing constraints. In this photo, spreading of the damage zone in the intraply inhibits delamination. (Magnification 250X.) (c) As shear strains in the interply reach a critical value, the resolved tensile strains initiate cracks 45° to the shear plane, becoming hackles as the crack tips link together. Hackle spacing gives qualitative information about resin toughness and strength. (Magnification 500X.) (d) A large interply deformation zone runs parallel to the intraply fracture. Fibers alter the local strain field by restricting polymer alignment. (Magnification 500X.) (e) In contrast to K3B laminate, a toughened bismaleimide laminate exhibits a deformation region that does not span the interply. The hackles are generally smaller and more jagged in appearance due to the more brittle nature of the resin. However, the specific strength of the bismaleimide resin is higher. (Luther M. Gammon, Boeing Materials Technology)

weight savings, and the corresponding fuel savings, from the use of composite materials would provide large economic benefits over the life of the aircraft, the poor economic health of the airline industry demanded that the initial cost of the finished airplane be reduced if composites were to be selected. New design and innovative manufacturing techniques using polymer matrix composites were required to make them competitive because the cost per pound of composites is higher than aluminum, and the current production costs of composites were twice that of aluminum aircraft structures. The industrial partners building the 777 engaged in ***concurrent engineering,*** which involved parallel developments in materials and their processing, determination of the properties for new materials, design and manufacture of individual elements and the structural components they form, and continual analysis of elements and components. This integrated approach to advanced composite technology for primary aircraft structures will carry through the next generation of commercial transport aircraft.

Boeing's decision to use advanced materials for its totally redesigned 777 resulted in the application of various lightweight composites, amounting to ten times more mass of composites than in its 757. These composites are approximately 9% of the new plane's structural weight. Uses of fiberglass, as seen in Figure 24-21, ranged from the fixed leading edge of the wing to the strut fairings. Graphite composites continued to be used, as on the Boeing 767 and 757, on rudders, elevators, and flaps. Hybrid composites served as wing-to-body fairing. *Toughened graphite* (Figure 24-21), a new, toughened epoxy composite, was used on the fin torque box, stabilizer torque box, and floor beams. Toughened graphite addresses a major concern about composites because of its outstanding resistance to impact: nearly seven times tougher than the previous CFRP.

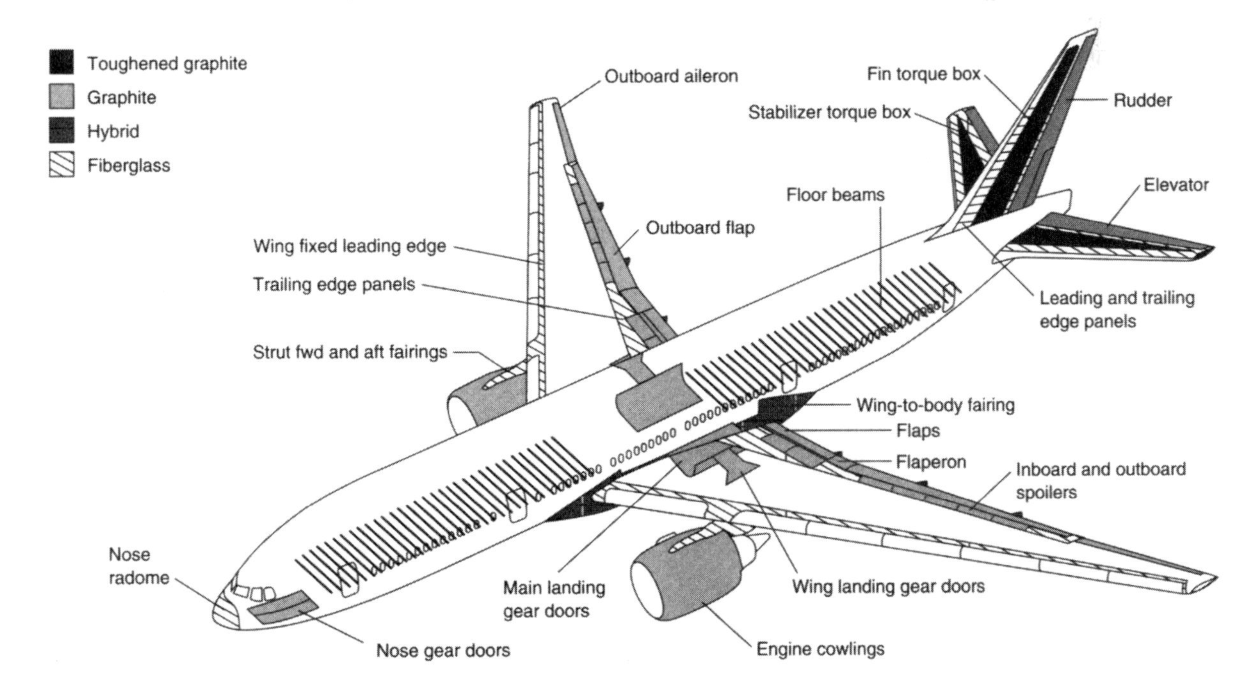

Figure 24-21 The application of composites showing the advantages of reduced-weight, corrosion-free, and durable structures. (Courtesy of The Boeing Company.)

The costs of all aspects of design, tooling, materials, and manufacturing for the 777's wing were greater than those of a competing aluminum wing. Why? One reason is that in design and its related testing, there are additional costs from the extra analysis required because of a lack of experience with the material and the lack of established databases, such as those available for aluminum. Of course, more experience with CFRP should bring reduced costs. The ACT program, mentioned earlier in this module, faces a formidable challenge to reduce the cost of advanced composites to be used in aircraft. The benefits of using advanced composites, as made clear in this module, could be enjoyed by the entire aerospace industry worldwide. In addition, the auto industry, building construction, road construction, and other fields could also benefit.

Boeing's next generation commercial airplane, 7E7 (see Figure 2-30 on p. 67), uses in its design more advanced composites to improve efficiency.*

SELF-ASSESSMENT

24-1. A composite of plastic reinforced with glass is

a. GPa
b. FRS
c. SMC
d. GRP

24-2. Impregnated, continuous sheets of composite materials such as those used for autobody panels are called

a. GPa
b. FRS
c. SMC
d. GRP

24-3. From the Pause & Ponder section, list one advantage of using an engineered polymer composite to replace natural polymers.

24-4. List one type of advanced composite used on the Boeing 777.

24-5. The binding portion of a composite that distributes the load among fibers or particles is the

a. Matrix
b. Reinforcer
c. Coupling agent
d. Interface

24-6. Organic, ceramic, synthetic, or metallic material with a length 100 times the diameter and a minimum length of 5 mm is a

a. Strand
b. Fiber
c. Staple
d. Particulate

24-7. Bonding agent of binder that provides a flexible layer between the fiber and the matrix is

a. Matrix
b. Reinforcer
c. Coupling agent
d. Interface

24-8. The terms *specific strength* and *specific modulus* are often used in describing composites. What does specific modulus refer to, and what does it tell us about a particular material?

*  Search for 7E7's use of composite materials—*www.Boeing.com*

24-9. Using Figure 24–13, which material listed would produce a bending strain of 2% with a bending load that produces a stress of 750 MPa?

24-10. A rod made of polyester under a tensile load that is parallel to the length of the rod complies with the rule of mixtures. If another rod with the same strain under the same load conditions and composed of glass fibers in a polyester matrix is substituted for the rod of polyester, compare the two rods by finding:

a. The density

b. The elastic modulus of the composite rod

c. The difference between the cross-sectional areas of the rods

d. The diameter of the composite rod in millimeters if the area of the polyester rod is 0.7854 in^2 Use the following property values:

E(p), modulus for polyester = 4 GPa

E(f), modulus for glass fibers = 68 GPa

p(p), density of polyester = 1.36 g/cm^3

p(f), density of glass fibers = 2.54 g/cm^3

V(f), volume fraction glass fibers in composite rod = 0.094

V(p), volume fraction for polyester

E(c), modulus of the composite material

F(p), load or force on polyester rod

REFERENCES & RELATED READINGS

AMERICAN SOCIETY FOR TESTING AND MATERIALS. *Annual Book of ASTM Standards*, Part 36. Philadelphia, PA: ASTM, 2003.

AMERICAN SOCIETY FOR TESTING AND MATERIALS. *Composite Materials, Testing and Design*, ASTM STP 617. Philadelphia, PA: ASTM, 2003.

AMERICAN SOCIETY FOR TESTING AND MATERIALS. *Fracture Mechanics of Composites*, ASTM STP 593. Philadelphia, PA: ASTM, 2003.

ASM INTERNATIONAL. *ASM Engineered Materials Reference Book*. Materials Park, OH: ASM International, 1990.

ASM INTERNATIONAL. *ASM Handbook on CD-ROM*, Volume 7, Volume 14, Volume 15, Volume 16, Volume 17, *Manufacturing Processes*. Materials Park, OH: ASM International, 2003.

ASM INTERNATIONAL. *Engineered Materials Handbook*, Volume 1, Composites. Materials Park, OH: ASM International, 1987.

BROSTOW, WITOLD. *Science of Materials*. New York: Wiley, 1979.

CARLSSON, L. A., AND R. B. PIPES. *Experimental Characterization of Advanced Composite Materials*. Englewood Cliffs, NJ: Prentice-Hall, 1987.

CATLIN, CHRIS. "Aluminum/Polymer Composite Developed for Automobiles," *Advanced Materials & Processes*, Volume 147, No. 5, May 1995, p. 8.

HARRIS, C. E. "Industry-University-NASA Partnerships in Aeronautics Research and Technology Development," 4th Technical Conference: New Century Partnerships for Material Systems. Blacksburg, VA, April 23–25, 1995.

LEWIS, III, D. ET AL. "In-Situ Composites," *Advanced Materials & Processes*, Volume 148, No. 1, July 1995, pp. 29–31.

MIRACLE, D. B., AND S. L. DONALDSON. *ASM Handbook, Volume 21, Composites*. Materials Park, OH: ASM International, 2001.

Modern Plastics Encyclopedia. New York: McGraw-Hill, 1993.

QUINN, KEVIN R., AND CARLOS A. CARRENO. "High-Temperature THERMO-PLASTIC Composites," *Advanced Materials & Processes*, August 1991.

RICHARDSON, TERRY. *Composites: A Design Guide*. New York: Industrial Press, 1987.

SCHWARTZ, M. *Composite Materials Handbook*. New York: McGraw-Hill, 1983.

SEYMOUR, R. B. *Engineering Polymer Sourcebook*. New York: McGraw-Hill, 1990.

STRONG, A. BRENT. *Fundamentals of Composites Manufacturing*. Dearborn, MI: Society of Manufacturing Engineers, 1992.

ZHOU, SUWEI, ZAFFIR CHAUDRY, AND CRAIG ROGERS. "Review of Particle Tagging Methods for NDE of Composite Materials and Structures," SPIE North American Conference on Smart Structures and Materials, February 26, 1995. Blacksburg, VA.

Periodicals

Advanced Composites
Automotive Composites
CI on Composites
CDA-Composites Design & Application
Composites in Manufacturing
Composites Technology
High Performance Composites
Journal of Composite Materials
Journal of Composite Technology and Research
Journal of Materials Research
Journal of Reinforced Plastics
*NASA Tech Briefs**

*CD-Rom Experiments in Material Science, Engineering and Technology, Version 2
0-13-030534-0. Order from *www.prehall.com*

Module 25

Composite Types, Structures & Properties

After studying this module, you should be able to do the following:

25–1. Recall types of composites and their structures, properties, advantages, and applications.

25–2. Make sketches of the structures of broad groups of composites.

25–3. Apply the rule of mixtures to determine properties of composites.

25–4. Describe hybrid composites and give examples.

25–5. Define "smart" composite materials and cite examples and their applications.

25–6. Use the website addresses denoted by the icon* to seek out current developments related to composite materials and processes.

*The symbol found throughout the book will link you to Internet sites related to topics being covered. The dynamic nature of the Web brings frequent changes, so some of these sites, although available at the time of writing, may not be up now.

25.1 STRUCTURE AND PROPERTIES OF COMPOSITES

Throughout this book we have stressed that the properties of a material are dependent on the material's structure. In monolithic materials such as steel or polymers, the structure commonly referred to is that of the crystalline or molecular material microstructure. Many times we peer beyond this atomic or molecular level to the subatomic level for further understanding of the behavior of such materials. With composite materials, we can identify with the naked eye (macroscopic) the major ingredients or constituents in the composite material, such as metal particles or glass fibers and different matrix materials, involved in the collective performance of these components. Just as with steel alloys, where we can change the microstructure to obtain different properties by varying the amount of the basic ingredients (elements) or by changing the number, size, and shape of the ingredients through appropriate thermal processing, we can vary the properties of composite materials by varying their composition and their structure. We know also that the final behavior of materials such as steel depends on how well the various phases interact with each other. This is vitally important with composites. As human-made materials, composites can be designed to have isotropic properties, and if so designed, they probably would be homogeneous materials with their components distributed and arranged in a uniform pattern. Or the components can be distributed in a nonuniform manner, and if so, the composites would have anisotropic properties. Third, similar to forged metals, a composite material can have its components distributed and arranged in a particular orientation that would result in directional properties. These last statements help explain some of the interest of materials scientists in composites in recent years. With the new technologies of materials, scientists can design their own material to fit the specific requirements of a particular application that cannot be met by existing technology or conventional homogeneous materials.

Composite technology is continually generating new processes for various structures. Representative of the advanced composites emerging to meet aerospace needs, are the elements involved in the construction of ***unconventional composite tubes structures,*** depicted in Figure 25-1. These structures are comprised of single- and multiple-component monolithic composites. As seen in (a), individual constituents of the monolith composite include, left to right, (1) high-performance insulation, (2) multiple fiber layers, (3) coolant ducts, (4) a heat conductor, (5) an evaporated metallic layer, and (6) a fibrous elastomeric layer that combine to form the FyreRoc monolith used for various components, such as rocket nozzles, missile casing frames, and jet turbines (Figure 25-1b).

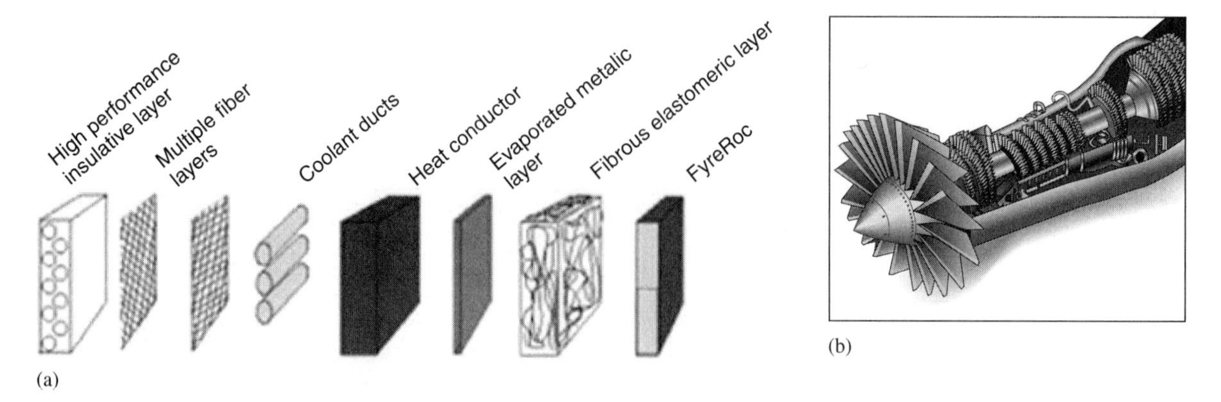

Figure 25-1 Unconventional composite tube structures. (a) Monolithic composite FyreRoc, on right, consists of the six elements to the left and forms (b) tube structures such as this turbine engine. Theses monolithic structures using new RTM processes allow 0 to 90° fiber orientation, processing with advanced resin systems, multilayer/microstructural capability, design flexibility, and automated production. (Cornerstone Research Group, Inc.)

The characteristic properties of individual constituents of a composite interact in various ways to produce the collective properties or behavior of the composite. Some properties obey the ***rule of mixtures (ROM),*** because the composite properties are the weighted sums of the values of the individual constituents. In other words, properties are a function of the amounts and the distribution of the contributing materials. Density, specific heat, thermal and electrical conductivities, and some mechanical properties, such as modulus of elasticity, follow this rule. In some composites the properties of the components are somewhat independent but supplement each other to produce a collective performance by the composite. Other composites have properties that are the net results of the interaction or interdependence of the components with each other. This latter type of composite behavior is by far the most important because the end result of combining materials into a materials system is a performance or set of properties that far exceeds the individual properties possessed by the components acting alone. An example of this type of composite, one of the fiber-reinforced plastics (FRP) composites, is glass fiber embedded in plastic resin. The glass fibers possess extremely high tensile strength, but being very brittle they cannot be used alone. A plastic resin, on the other hand, is relatively weak but very ductile. Once a sufficient number of glass fibers are embedded in a plastic–resin matrix in a unidirectional manner parallel to the direction of the load, the two components act together as a unit to withstand the load by deforming equally and sharing the load proportionately, so that the composite or materials system achieves higher tensile strength than is otherwise possible based on the individual components.

25.1.1 Rule of Mixtures

Using fiberglass filler and a polyester–resin matrix for a composite, we can sketch the stress–strain curves for each material, as shown in Figure 25-2. The modulus of elasticity (E) for glass is about 69 GPa, and for polyester, around 69 MPa. The tensile strength of glass (single-fiber) is about 3.45 GPa. The curves show the great differences in the stiffness (slope of the curves) between the two materials. Glass is extremely brittle and shows little evidence of any elastic strain up to the point of fracture. As indicated by its modulus values and the slope of its stress–strain curve, polyester is a very ductile material. If each material were deformed an equal amount, the stress carried by the glass would be over 1000 times greater than the stress on the polyester. If these two materials were combined in one ideal material, the glass would act as continuous fibers completely surrounded by a matrix of polyester, which forms a bond between the fibers and the matrix. The bond would allow both to deform (elongate) the same amount under a uniaxial load.

Research shows that the modulus of a unidirectional composite material in which (1) the load is applied parallel to the continuous fiber and (2) the bond between the fiber and the matrix material is strong enough to permit both materials to deform equally is the weighted sum of

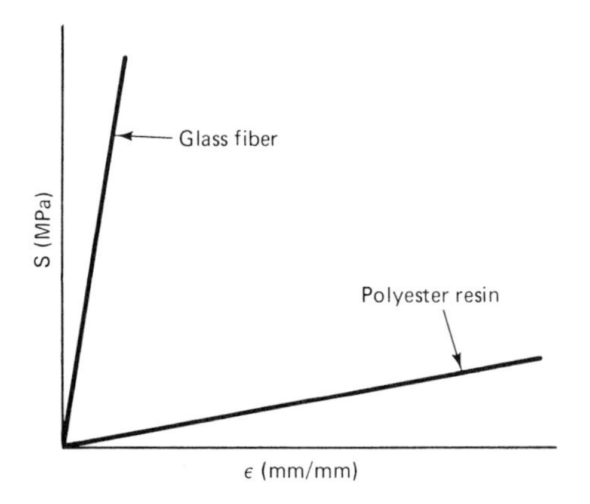

Figure 25-2 Stress–strain curves for glass fiber and polyester resin.

the modulus of the fiber and the modulus of the matrix. This last statement is an example of a ***simple mixture law.*** Mixture laws for physical properties such as specific heat, density, and conductivity for a composite material can also be developed. For such composite materials, most of the flow of thermal energy or electrical current will be through the component that is the best conductor.

The preceding discussion assumed that the chemical bond between the fibers and the matrix was greater than the strength of the matrix materials. This is not always the situation. Much research goes into determining the conditions necessary to improve the bonding so that the stress transfer between the matrix and the fibers can be made without fracture. Studies of metal–matrix composite materials have led to the development of composite materials that can withstand temperatures in excess of 670 K and retain great stiffness. Such materials are represented by tungsten, boron, graphite, or silicon carbide fibers in aluminum, cobalt, or nickel matrices.

As we have said, certain properties of a composite depend on the relative amounts and properties of the components of the composite material. Using the law or rule of mixtures (ROM), properties such as mass, density, modulus of elasticity, thermal conductivity, or electrical conductivity can be estimated provided that the reinforcing fibers are continuous, unidirectional, and have the same length as the matrix. Also, the load on the composite must be a tensile load parallel to the longitudinal axis of the fibers. It is to be emphasized that not all properties of a composite follow the rule (i.e., some composite properties are a simple proportion or are weighted averages of their components). Our discussion of the ROM will extend to expressing the elastic modulus in tension for a fiber-reinforced matrix material conforming to these constraints. Figure 25-3 is a sketch of a composite loaded with a tensile load, F(c). The following abbreviations represent the quantities involved in this example and in our ongoing discussions of the ROM:

V = volume fraction

s = stress

e = unit deformation, strain

E = modulus of elasticity in tension

A = area

(f) = refers to fibers

(m) = refers to matrix

(c) = refers to the composite material

S = tensile strength

F = load or force acting on a material

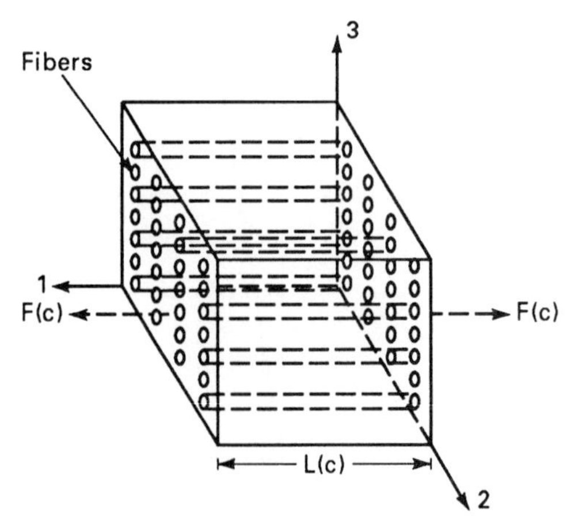

Figure 25-3 Fiber-reinforced-composite material undergoing isostrain loading conditions.

Referring to Figure 25-3, the load (F) is carried by both the fibers and the matrix acting to share the burden:

$$(1) \qquad F(\mathrm{c}) = F(\mathrm{f}) + F(\mathrm{m})$$

Assuming that the bond between the fibers and the matrix is good, we can say that both the fibers and the matrix will stretch or deform the same amount in withstanding the tensile load.

$$(2) \qquad e(\mathrm{c}) = e(\mathrm{f}) = e(\mathrm{m})$$

Using the definition of stress (Section 4.2.1), $s = F/A$, and Hooke's law (Section 4.2.3), $s = E/e$, equation (1) can be written

$$(3) \qquad s(\mathrm{c})A(\mathrm{c}) = s(\mathrm{f})A(\mathrm{f}) + s(\mathrm{m})A(\mathrm{m})$$

Dividing equation (3) by A(c) yields

$$(4) \qquad s(\mathrm{c}) = \frac{s(\mathrm{f})A(\mathrm{f})}{A(\mathrm{c})} + \frac{s(\mathrm{m})V(\mathrm{m})}{A(\mathrm{c})}$$

The ratios A(f)/A(c) and A(m)/A(c) are area fractions. Because the composite and its components are arranged in parallel, areas A(f) and A(m) are proportional to their volumes. Thus, volume fractions can be substituted for area fractions:

$$(5) \qquad s(\mathrm{c}) = s(\mathrm{f})V(\mathrm{f}) + s(\mathrm{m})V(\mathrm{m})$$

Returning to equation (2) with Hooke's law and substituting for s yields

$$(6) \qquad E(\mathrm{c}) = E(\mathrm{f})V(\mathrm{f}) + E(\mathrm{m})V(\mathrm{m})$$

Equation (6) expresses the ROM in terms of the elastic modulus for the ***isostrain*** situation described in Figure 25-3. Note that if one volume fraction is known, the other volume fraction can be found using the relationship $V(\mathrm{f}) + V(\mathrm{m}) = 1$.

Illustrative Problem

A laminar composite of epoxy resin reinforced with graphite fibers is loaded similarly to the model in Figure 25-3. Assume that there are equal volumes of graphite and epoxy sustaining the load. The modulus of elasticity for the graphite fibers is 58×10^6 psi and for epoxy it is 0.6×10^6 psi. Find the modulus for the composite.

Solution

$$\begin{aligned} E(\mathrm{c}) &= E(\mathrm{f})V(\mathrm{f}) + E(\mathrm{m})V(\mathrm{m}) \\ &= 58 \times 10^6 \times 1/2 + 0.6 \times 10^6 \times 1/2 \\ &= 29.3 \times 10^6 \text{ psi} \end{aligned}$$

Figure 25-4 illustrates a model of the ***isostress*** loading of a composite material. The load, F(c), acts transversely as opposed to longitudinally or parallel to the long axis of the fibers. The fibers and the matrix each resist the load equally. Stated in equation form, $F(\mathrm{c}) = F(\mathrm{f}) = F(\mathrm{m})$. This also means that the stress in the fibers and in the matrix is equal. However, the strain or deformation in each material is different. The weaker material will deform more than the stronger component. In equation form,

$$e(\mathrm{c}) = e(\mathrm{f}) + e(\mathrm{m})$$

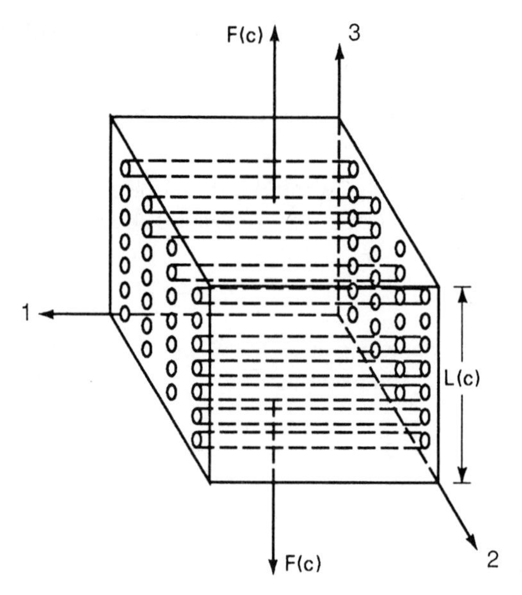

Figure 25-4 Fiber-reinforced-composite material undergoing isostress loading conditions.

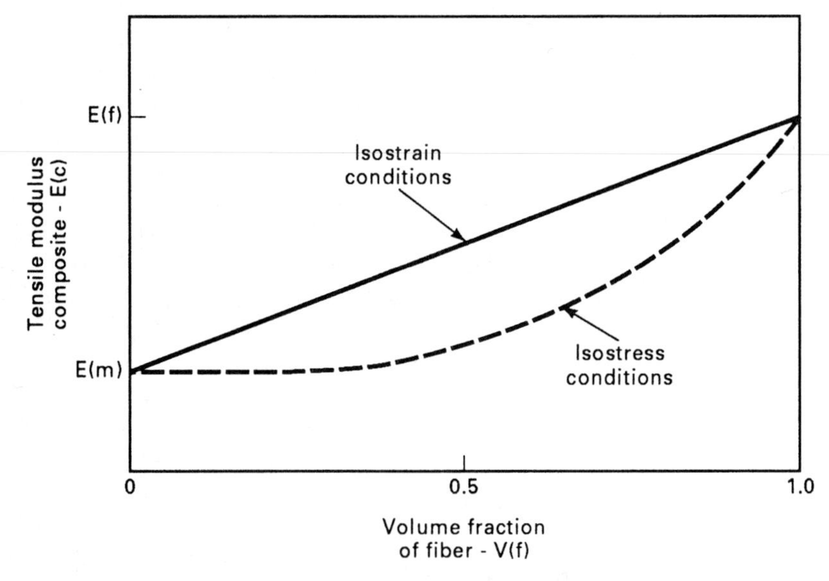

Figure 25-5 Graphical plot of the tensile modulus of a unidirectional composite material loaded under isostrain and isostress conditions as a function of volume fraction of reinforcing fiber.

The equation that expresses this type of loading, known as the ***inverse rule of mixtures,*** reveals that this isostress loading condition does not follow the rule of mixtures. In equation form, the inverse rule of mixtures is expressed arbitrarily in terms of modulus:

$$\frac{1}{E(\mathrm{c})} = \frac{V(\mathrm{f})}{E(\mathrm{f})} + \frac{V(\mathrm{m})}{E(\mathrm{m})}$$

Figure 25-5 contrasts these two types of loading and helps illustrate that, for a given volume fraction of fiber in a composite, the composite loaded under isostrain conditions is more efficient, having the higher modulus of elasticity for the composite. In fact, when looking at the curve in Figure 25-5, you can see that the modulus for the composite is not increased much over the modulus for the weaker component in the composite. On the contrary, the composite modulus in the isostress model using the same materials components and with the same volume fractions, is about 10% less than that obtained in the isostrain model.

When determining how to arrange the components in a composite, it is important to recognize how a composite material is to be loaded. If the direction and approximate size of the loading are known beforehand, this simplifies the problem. If these conditions are not known, the correct components must be chosen and best arranged to support the load regardless of its magnitude (within limits) and direction.

25.1.2 Design and Fatigue

The design of a new or improved product for an industrial and/or technical application entails the detailed analyses of many factors. The design process links the scientific knowledge of materials with modern industrial methods to reach an overall objective of producing a better product at a lower cost. The interrelationship among materials, process, and design is nowhere more pronounced than in the use of composite materials to replace traditional materials such as steel in both structural and nonstructural components. Figure 25-6 is a flowchart showing the interaction of design, materials, and manufacturing processes that must take place to ensure the success of any design project.

Composite materials owe their success in large part to the fact that they provide the engineer with considerable design latitude compared to metals. This fact is in addition to their many other desirable properties. Fatigue in metals, discussed in Unit 4, differs from fatigue in composites (see Figure 25-7 and Figure 25-10). The mechanisms of crack initiation and crack propagation result in the failure of metals under cyclic, repeated loads at stress levels well

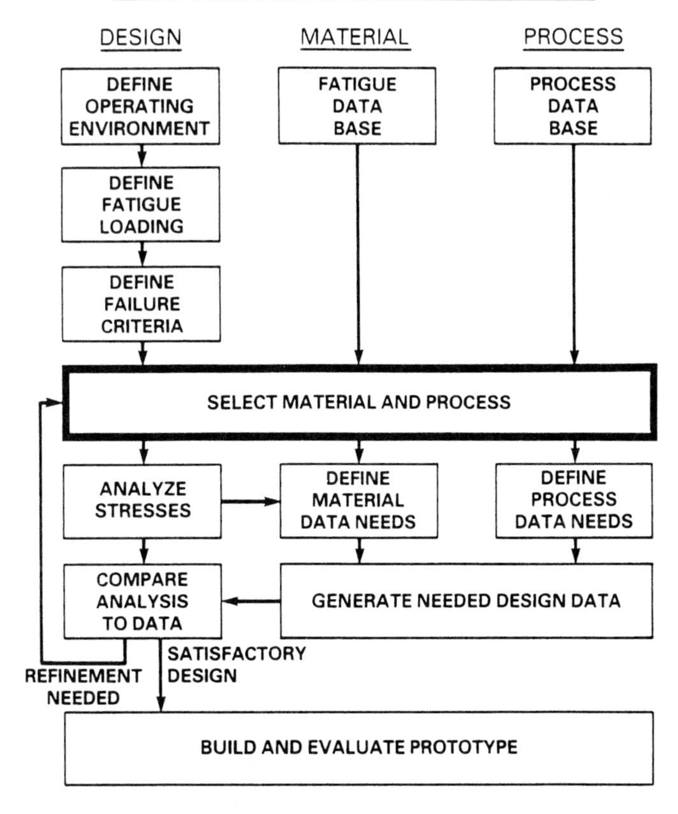

Figure 25-6 Flowchart for fatigue design with fiber-reinforced PMCs showing need for databases. (General Motors)

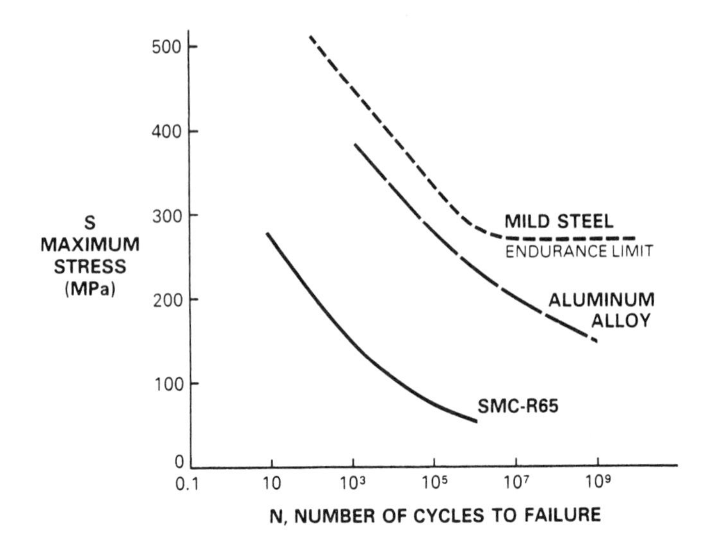

Figure 25-7 Comparison of metals and PMC typical flexural fatigue *s–N* diagram curves. (General Motors)

below their ultimate strengths. Crack initiation is caused by stress concentrations, inclusions, or voids, which are then acted on by the mechanism of crack propagation, reducing the net-load-bearing area to the point where the fatigue stress exceeds the metal's ultimate strength. Final failure in metals usually results in a relatively clean fracture surface. In many instances the failure is of catastrophic proportions. Finally, metals exhibit an ***endurance limit,*** a stress level below which fatigue failure does not occur, regardless of the number of fatigue cycles applied.

Engineers have always been encouraged to look to nature to observe how certain tasks are accomplished or as an aid to stimulate the mind when involved in problem solving. We have referred previously to the example of seashell structure. In composites, crack initiation and propagation produce simultaneous growth of cracks that may (1) extend through the matrix, (2) be stopped at a fiber, or (3) propagate along a fiber–matrix interface. Cracks are initiated by factors such as filler or fiber debonding, voids, or fiber discontinuities. The crack propagation results in cracks joining each other to the extent that the matrix is unable to perform its basic function of transferring the load from one fiber to the next in fiber composites. The fracture surface usually shows evidence of a complex assortment of matrix failure, fiber failure, and fiber pullout. Other ***fatigue–failure characteristics*** of FRP composites are as follows:

1. A gradual decrease in the slope of the stress–strain curve (modulus of elasticity, E) occurs during cyclic loading.
2. Elevated temperatures normally reduce the performance of matrix-dominated composites (stress carried mainly by the matrix material). Fiber fatigue properties are better than those of the polymer matrix.
3. Moisture and chemical exposure can greatly affect the polymer matrix.
4. Notch sensitivity in high-cycle tensile-fatigue conditions is less than for metals.
5. There are considerably fewer matrix-dominated properties than fiber-dominated properties. Hence tensile, compressive, and shear stresses in the interlaminar (through-the-thickness) direction, being matrix dominated, should be kept to a minimum.

25.1.3 Toughness and Impact Strength

As measured by the ASTM D256 test on unnotched Izod specimens, the impact strength at room temperature for "tough" plastic molding materials is at least 8 J/cm (15 ft·lb/in.).* A value of at least 10 J/cm^2 (50 $ft{\cdot}lb/in^2$) using the high-speed tensile impact test (ASTM D1822) also

*Test values and other data presented in this module and elsewhere are typical values for the material. They are offered as an aid to understanding. The properties of parts fabricated from a particular material are too contingent on many factors, such as part configuration, molding techniques, or curing times, to guarantee reproducibility of the data.

qualifies such materials as being tough. Considerable impact testing of such materials using glass- or carbon-fiber reinforcement in varying amounts with different resin materials has been done. Only recently, however, have such tests been carried out at low temperatures (down to 200 K). From our study of other materials, we know that the impact strength of a notched specimen is less than that of an unnotched specimen because notches produce stress concentrations. We also know that it takes energy not only to start a crack but also to propagate one. With notched specimens the crack is already provided, and depending on the sharpness of the notch, less energy is usually needed to propagate it. The tremendous effect of surface imperfections such as cuts, tool marks, and scratches on the toughness of materials is well known. For many reinforced thermoplastics, toughness drops off and the notch sensitivity increases at low temperatures. However, there are exceptions. Polypropylene as well as acetal copolymer become tougher, as evidenced by higher test values for unnotched specimens. Glass-reinforced polycarbonates show little change in toughness, although their notch sensitivity is greater at lower temperatures. With DuPont's "super-tough" nylons (Zytel ST resin), notch sensitivity decreases with increased glass- or carbon-fiber content, while at low temperatures, notched Izod impact-strength readings increase with fiber content. Using unnotched specimens, the toughness of these nylon resins increases with increasing fiber content at both room temperature and low temperatures.

Impact testing measures the material's resistance to fracture under certain prescribed test conditions when a standard specimen is struck at high velocity. In addition to the Izod and Charpy notched and unnotched specimens using a swinging pendulum, other impact tests, such as the falling ball, falling dart, high-velocity tensile stress, and fracture-toughness tests, are also used to measure toughness. ***Fracture toughness,*** in units of stress times the square root of a crack length ($MPa \cdot m^{1/2}$), is an indicator of a material's resistance to the extension of a preexisting crack. The high-speed tensile test uses unnotched specimens and defined impact strength as the area under the stress–strain curve or the energy required to break a material (toughness). Impact strength using this test setup is expressed in units such as kilojoule per square meter (kJ/m^2). Those tests using notched Izod or Charpy specimens express impact toughness in terms of the energy per length of notch, or kJ/m, as one example.

25.1.4 Fire Resistance

Much research effort has been undertaken to provide fire-resistant materials particularly for use in mass transit vehicles. Design requirements for greater fuel economy dictate low-weight reinforced polymeric composites. The complexity of molded structures further limits these complex structures to the use of reinforced plastics. Materials for buses and trains are subject to strict fire safety requirements. The United States and European countries each have their own standards. ASTM E–162 (flame spread), E–662 (smoke density), and National Fire Protection Association (NFPA) 130 are just a few of some of the U.S. regulations. Along with test requirements and standards, the industry has developed sophisticated test equipment to assess the behavior of materials in a fire.

The fire triangle, heat–oxygen–fuel (a material that will support combustion) illustrates the components needed for a fire. Further, all fires give off carbon monoxide even when burning a phenolic material; this oxide of carbon kills more people than any other agent. Resins are attracting the most attention in terms of fire reaction, for organic composites contain oxygen within their chemical structures and are inherently combustible. The factors involved are not only fire resistance but the principal reactions of materials to fire, namely, smoke, heat release, flame spread and toxic gas release. ***Phenolic resins*** are slow to ignite, extinguish easily, and when burned create up to 50% by volume a carbonaceous char that tends to hold in volatiles. They emit little smoke during combustion and heat release rates are low. Accordingly, phenolics are widely used in aircraft interiors and automobile applications, such as engine compartment firewalls. But the downside of phenolics is they tend to be brittle and lack peel strength (as in peeling a banana or pulling a piece of sticky tape off a spool), which is a drawback to their use in sandwich composites. Lastly, they are expensive partly due to difficulty in holding a surface finish due to their high degree of cross-linking, which severely limits the number of reactive sites left for bonding.

In spite of these limitations, companies have developed a successful ***surface coating technique*** using a modified polyester gel-coat that adheres to phenolics.

Resin suppliers have used and incorporated ***additives*** such as ***halogen*** compounds, which, like bromine and chlorine, are highly effective in neutralizing the ***free radicals*** released when composites burn, retarding the spread of fire. Present health and safety concerns have made the use of halogens unacceptable, but the wisdom behind this prohibition is being questioned since they are highly effective fire retardants.

Some alternatives to the halogens, such as phosphorus, melamine, aluminum trihydrate (ATH), antimony trioxide (ATO), magnesium hydroxide, and zinc borate, are being used but with limited results.

As with surface coating techniques, incorporating fire-retardant ***fillers*** into composites is difficult in that too much may impact the processing and/or the mechanical properties of the composite. A more recent alternative to making composites fire resistant is to protect them with a ***fire barrier.*** This outer layer can either be separate or integrated into the laminate as a surface veil or intumescent coating (one that swells or enlarges with heat). Technofire, a product of Technical Fibre Products, has a thin surface layer that swells in a fire to provide a thick insulating *char* that protects the underlying composite. Fire-retardant paints are also now available.

Advanced Materials Instruments and Analysis Inc.—*www.amia.com*
Detroit Testing Laboratory Inc.—*www.Dettest.com*
Materials Research & Engineering Inc.—*www.henge.com/mre*
Polymer Diagnostics Inc.—*www.polymerdiagnostics.com*
University of Dayton Research Institute—*www.udri.udayton.edu/*
Wyle Laboratories Inc.—*www.wylelabs.com*

25.2 TYPES OF COMPOSITES

Products requiring different manufacturing techniques and different reinforcing components, and the specialized nature of the parts themselves, have resulted in many types of composites. We have divided this ever-increasing variety of composites into fiber, laminated sandwich, particulate, flake, and filled (see Figure 24-1).

25.2.1 Fiber-Reinforced Composites

25.2.1.1 Polymer–matrix composites (PMCs) and glass-reinforced plastics (GRPs). GRPs represent the earliest and, because they comprise over three-fourths of the total fiber-reinforced composites produced, the most widely used fiber–resin composites. Using various forms of glass fibers coupled with either a thermosetting or thermoplastic resin, these composites can be produced without the need for high curing temperatures or pressures. The product contains a good balance of properties, has high corrosion resistance, is low in cost, and has a multitude of uses in structural, industrial, and consumer-related products, ranging in size from minute circuit boards to boat hulls. With 20–40% fiber loadings, the composites will, in general, double the strength and stiffness of the plastic resin used alone. Continuous fibers in the form of monofilaments or yarn increase these properties fourfold, with accompanying desirable decreases in thermal expansion and creep rate and with increases in impact strength, heat-deflection temperatures, and dimensional stability. Figure 25-8 shows one application of GRPs.

PMCs are being used to make turbine blades for a German offshore power plant. The wind turbine blade is almost 56 m in length and it will be the largest blade in the world in serial production. The competitiveness of offshore power generation is the driving force for large turbines and blades of 45 m and above. Crucial to the success of these new blades will be carbon-fiber-reinforced plastics. Stiffness is the desired mechanical property and carbon fiber offers about three times the stiffness of glass. For the new 5-MW blade, carbon fiber, in the

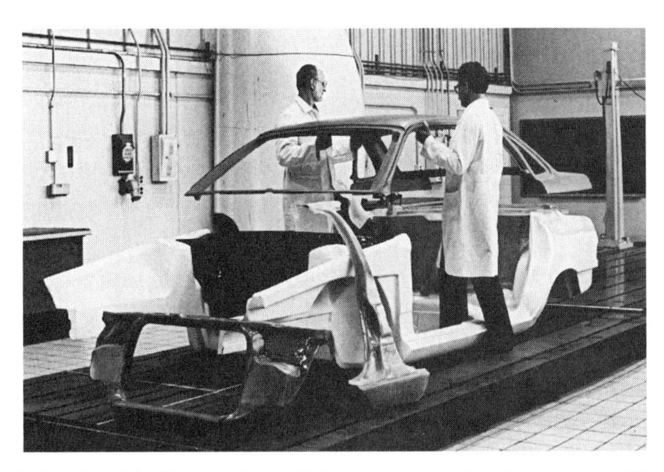

Figure 25-8 A redesign of the Ford Taurus chassis. This prototype chassis reflects designers' keen interest in using reinforced-plastic composite technology. Using GRPs allows 8 or 10 parts to replace over 400 stamped steel parts. A GRP chassis reduces the car weight by about 30% because a lighter chassis means engines, brakes, and other components could be downsized. Additional benefits are realized through more efficient manufacturing techniques. (Courtesy of Ford Motor Company)

form of unidirectional carbon prepregs, is specified for the entire spar structure. The possibility of lightning strikes is dealt with by incorporating a copper mesh layer upon the laminate. Hybrid glass/carbon will be used to fabricate 43.8-m blades for 2.5-MW onshore turbines. A German company that makes 33-m all-carbon blades claims a weight little over half of what a comparable glass blade would weigh. A reasonable conclusion is that carbon blades above a certain size will be no more expensive than glass blades due to the savings in weight. This will be certainly true if carbon prices continue their downward trend. Figure 25-9 shows a similar windmills-for-energy concept developed by the U.S. Department of Energy.

 DOE wind energy program—*http://www.eren.doe.gov/wind/wttr.html*

The disadvantages of these composites arise, in the main, especially from the fact that they are two-phase structures. This leads to a degree of environmental degradation greater than that of either component material alone. Residual stresses and electrochemical effects result from the marriage of two dissimilar materials. Furthermore, the diffusion of fiber materials into the matrix materials, and vice versa, may take place at several stages. Variation in the thermal expansion of these two materials leads to thermally induced stresses that result in warping, plastic deformation, cracking, or combinations thereof.

Naturally, the advantages of such composites outweigh the disadvantages. Some of the major advantages can be summarized briefly as follows:

1. Greater matrix strength and stiffness, which permits little or no cross-plying and greater joint and stress concentration load strength
2. Increased operating temperatures and increased durations of time spent at those temperatures
3. Increased resistance to high temperature and humidity environments
4. Better conformability with existing metal fabrication techniques

A recently developed FRP composite rebar, called ***C-bar,*** is a likely candidate to replace epoxy-coated steel reinforcing bar (rebar) in many applications. C-bar is composed of unidirectional glass fibers wetted by recycled terephthalic acid resin in the center core. Unidirectional mat and 1/2-in. chopped fibers with a polyester surfacing coat are added to the exterior. The base resin, used only on the outside, is a polyurethane-modified vinyl ester. What sets C-bar apart from other

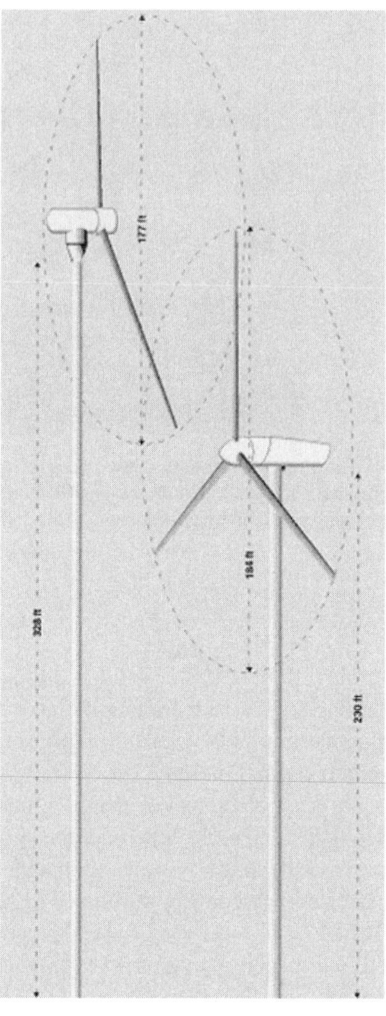

Figure 25-9 Windmills for energy are cropping up all over the world. Composite technology is helping to create huge turbine generators such as the concepts shown here developed with U.S. Department of Energy support. Each windmill generates up to 1 MW of electricity, and hundreds can be placed in fields located on land and at sea. (DOE)

forms of FRP rebar is the combined use of both mat and fibers, plus the mechanical and chemical bonding between the acid resin in the center core and the vinyl ester used on the outside. Its tensile strength is reported to be more than twice that of steel rebar, with four times less weight. C-bar is nonmagnetic and noncorrosive and has a more compatible coefficient of thermal expansion than that of concrete, attributes that can lead to many potential uses. The hybrid manufacturing process combines pultrusion with compression molding (see Section 25.3) and applies the fiber circumferentially to the inner core of the rebar. The result is a consistent, fully cured cross section produced at a rate of 20 ft/min, which is unattainable with any straight pultrusion process.

One of the promising uses of such composites is reinforcing and prestressing concrete intended for waterfront applications. Composite bar (reinforcing or C-bar) is immune to destructive moisture and salt attack, which can corrode conventional steel bar causing it to swell and crack the concrete around it. Even epoxy-coated steel rebar fails to hold up under these conditions. Using a patented process that combines pultrusion and compression molding of GRP, the resulting C-bar core is given an outer skin of a urethane-modified vinyl ester that has high resistance to moisture and chemicals, including the alkaline surroundings of concrete. C-bar has twice the tensile strength of conventional epoxy-coated steel rebar at only a quarter the weight. And the coefficient of thermal expansion is closer to concrete than to steel. The FRP industry is also competing with the manufacturers of railroad cross ties made of treated wood. The traditional market uses 75 million ties per year with 20% of that volume sold in the United States. Seaward International, Inc. has developed ***thermoplastic cross ties*** reinforced with ***fiberglass rebar*** based on the company's successful marine timber technology. This technology uses a continuous, coextruded process, with the company's Duralin Plastic as a matrix and fiberglass rebar as reinforcement. A high-strength outer skin of plastic covers the inner core of matrix and rebar. Another company, TieTek, Inc., is using fiberglass as a chopped strand throughout the matrix, which also includes other materials such as rubber from recycled tires. Both companies feature recycled high-density polyethylene in their designs. Accelerated aging tests of TieTek's composite cross ties show that they lose only 25% of their performance properties in 15 years, whereas the properties of wooden ties drop off by 50%.

The recent switch from steel to FRP composites for ***dowel bars*** was brought about after almost 15 years of service on heavily traveled roads in Ohio. The dowel bar serves as the transfer joint mechanism connecting adjacent slabs of concrete roadway. When vehicles travel over the joint, dowel bars restrain vertical movement between slabs by transferring loads from one slab to another. Carbon steel (mild steel, general-purpose grade) bars are susceptible to corrosion from moisture and deicing salts, limiting their life span to 7 to 15 years depending on their design and location, whereas concrete slabs can last 35 to 40 years (see Module 4 for more details on corrosion and refer to the American Concrete Paving Association (ACPA) for further details). When a steel dowel corrodes, it loses its ability to transfer loads and can "lock" in the joint, which produces excessive loads and causes the concrete to break down. This, in turn, produces a familiar "thump-thump" as wheels pass over the joints. Traditional joints use a steel dowel typically 1.5 in. diameter by 18 in. long, coated with epoxy. After 13 to 15 years of service during the testing period, the epoxy coating on the steel bars delaminated and could be removed as a thin film shell. The composite dowels used in the test were 1.00 to 1.25 in. in diameter and were pultruded by Strongwell (Bristol, Virginia) of 78% E-glass fiber-reinforced vinyl ester. When these composite dowels were removed for testing after the testing period, they "looked like new dowels that just came out of a pultrusion machine." (See Section 26.1.5 for details on pultrusion.)

Carbon–Carbon Composites. Carbon–carbon composites are noted for their light weight and strength, and thus are likely candidates for applications in the aircraft and auto industries. Carbon–carbon has been used on nose cones and leading edges of high-performance space vehicles subjected to very high temperatures (3000°C). Another high-temperature application is brake discs for racing cars. Such composites are stronger at high temperatures than at low temperatures, have high specific strength and good resistance to thermal shock, and are self-lubricating. Composed of a carbon fabric and an organic resin, the fabric is placed in a mold that is infiltrated (one method is CVD, described in Module 20) with a resin that is subsequently ***pyrolized*** (heated to a high temperature), driving off the noncarbon atoms in the polymer and thus forming a carbonized matrix. A coating of silicon carbide is added to reduce oxidation, which limits the high-temperature applications of these composite materials. Recent research in the field of damage detection, where sensors are either attached to or embedded in the structural material, led to the discovery of the electrical properties of carbon composite materials. When subjected to changes in temperature, the composite material undergoes changes in its electrical properties similar to that of semiconductors. A most promising application for this technological advance is using carbon composites to act either primarily or secondarily as a sensor to eliminate the need for the embedding of conventional sensors in aircraft parts.

Carbon–carbon composites possess a unique set of properties that make them desirable materials for high-temperature applications such as in rocket propulsion components, brakes, hypersonic leading edges of flight vehicles, heating elements, and structural components. These composites are stronger and stiffer than steel and less dense than aluminum. In addition, they maintain their mechanical properties above 3000°C and their properties improve with heating as the nonordered carbon is converted to the ordered graphite structure (see graphitization elsewhere in this text). Of limited use up to now due to the high cost of production, a recent development called ***in situ rapid densification*** greatly reduces the processing time and thus the cost. This novel process rapidly densifies the composite uniformly. Such a composite can be produced from any type of fiber preform (e.g., woven, braided, one-dimensional to n-dimensional, and felt). Further, there appears to be no size limit; diameters up to 45 cm and lengths up to 183 cm have been uniformly densified. An additional advantage of the technique is the ability to join parts made of the same material with a mechanically strong, seamless joint. This process does not need to use graphitization, thus saving time and energy. Reduced cost in processing this new material will result in many new applications.

Carbon–carbon composites are also replacing graphite electrodes, which are difficult to use in fuel cells. Using a chemical vapor infiltration technique, a light carbon composite electrode just 2 mm thick with good electrical conductivity and low permeability has been developed.

25.1.1.2 Metal–matrix composites (MMCs) and ceramic–matrix composites (CMCs). Metal–matrix composites (MMCs) and ceramic–matrix composites (CMCs) consist of a base (matrix) that is reinforced with continuous fibers or discontinuous materials in particulate form. ***Continuous reinforcement*** is defined as fibers, usually oriented in one direction, that produce an anisotropic composite in which the fibers predominate, with the matrix serving as a vehicle for transmitting the load to the fibers. ***Discontinuous reinforcement*** refers to whiskers, short or chopped fiber, or particulates that produce a composite exhibiting a blend of the reinforcement and matrix properties.

As with plastic resins, many fibers are used with metal matrices to form composites. The volume fraction of reinforcing materials plus interface–interphase reactions plays a leading role in determining the ultimate success of any metal–matrix composite in meeting the demanding requirements of a particular design. Figure 25-10 shows photomicrographs of metal-reinforcing fiber embedded in a metal matrix. The major rationale for the development of such composite materials is to satisfy the important need for tough, strong materials capable of maintaining their special properties under high-temperature operating conditions. Figure 25-11 is a plot showing how the specific strength of some metal alloys and composites varies with temperature. Originally developed for the aerospace industry, these composite materials are now finding many uses in other industries.

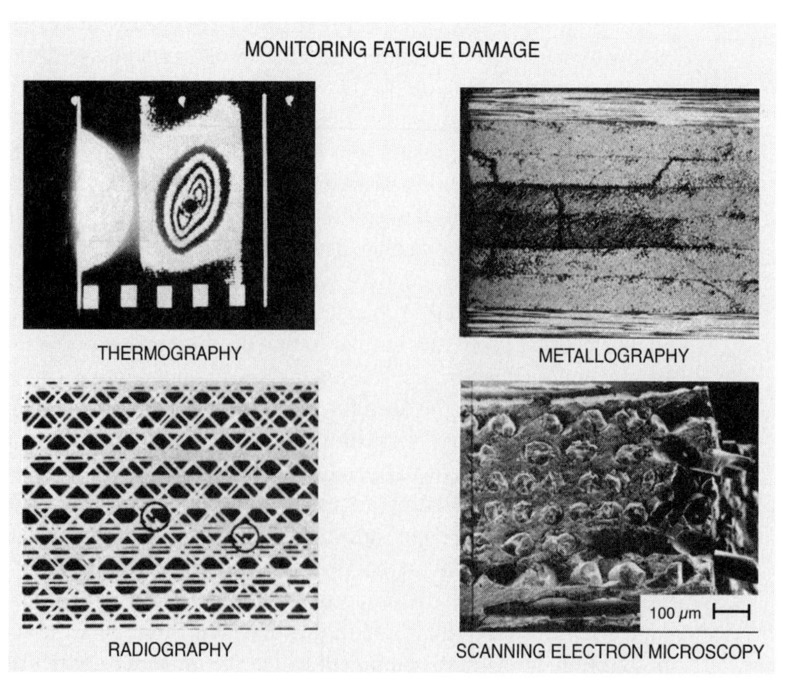

Figure 25-10 Photomicrographs of metal–matrix composites (MMCs). (NASA)

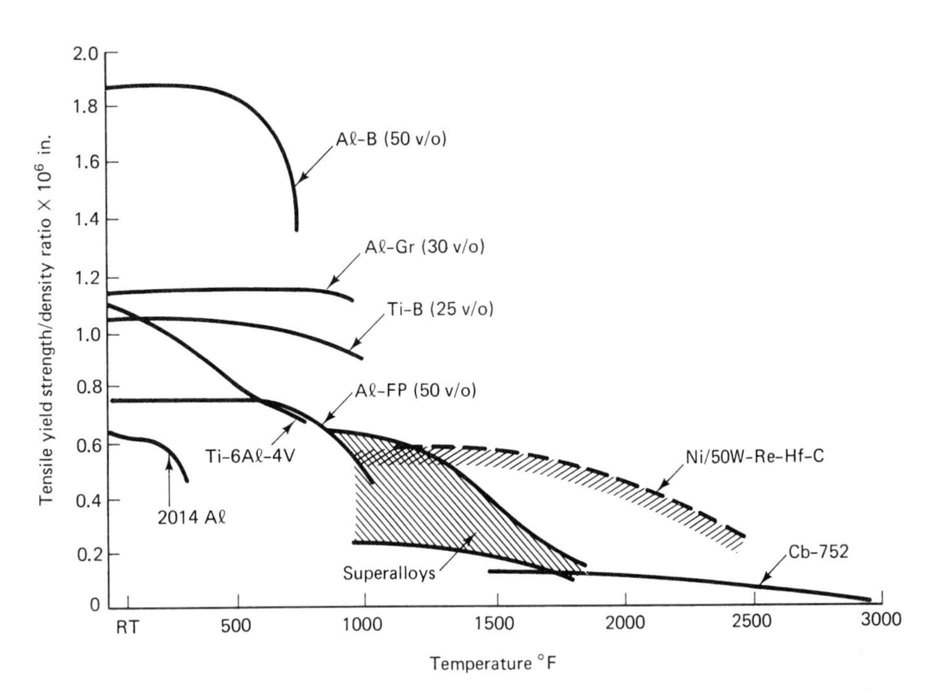

Figure 25-11 Specific strength versus temperature of metal alloys and composites. (NASA)

MMCs are metals, often superalloys, that are reinforced with particulates or filaments of other metals or nonmetallics that provide properties not possible with metals alone. These properties include higher temperature operating limits than the base metal; MMCs can also be tailored to give improved strength, thermal conductivity, stiffness, abrasion and creep resistance, and dimensional stability. A monolithic alloy forms a continuous matrix, and the reinforcement is a high-performance carbon, metallic, or ceramic addition. Being a metallic alloy, the matrix is not attacked by organic fluids such as fuels and solvents. The continuous or discontinuous reinforcements may represent up to 60 vol % of the MMC. The continuous fibers may be carbon, silicon carbide (SiC), boron, alumina (Al_2O_3), and refractory metals. Discontinuous reinforcements may be SiC in the form of whiskers or particulates, alumina in the form of particulates or short or chopped fibers, and titanium diboride (TiB_2) in the form of particulates. Many reinforcing agents are not compatible with their matrix metals—either a negative reaction causes one or both components to suffer degradation or a bonding failure occurs. To correct these interface problems (refer back to Figure 24–4 showing the interface in a fiber-reinforced composite), procedures involving coatings, surface treatment of the reinforcement, or additions to the matrix to make it more compatible are performed.

Processing techniques for MMCs are divided into two stages—primary and secondary processing. Primary processing involves the introduction of the reinforcement into the matrix in the proper amount, location, and orientation and the bonding of the reinforcement to the matrix. Secondary processing involves transforming the composite into a finished hardware component. There are six primary processing methods for fabricating MMCs. Solid-state processing involves diffusion bonding of materials and powder metallurgy techniques, stir casting, squeeze casting, rheocasting (also known as compocasting), pressureless metal infiltration (see the Lanxide Primex process in this module), and spray deposition.

The following are types of matrix composites that incorporate different materials to provide properties for numerous applications.

- ***Aluminum–matrix MMCs.*** Aluminum is the most popular matrix material because of its melting point. Most commercial applications do not require too high a melting point for their matrix materials. A relatively low melting point simplifies composite processing and makes it relatively less complex and less costly. Aluminum also imposes few compatibility problems with reinforcements.

A second family of aluminum metal–matrix composites, ***ceramic-reinforced aluminum MMCs,*** adds a new dimension to tailorability. Structural, thermal, and abrasion-resistant properties are now adjustable to provide customized materials that are castable and extrudable into complex to near-net shapes for a variety of applications.

The designation for aluminum MMCs, approved by the Aluminum Association Inc., uses a four-part system to identify the matrix alloy and the composite's amount and type of reinforcement. For example, $6063/Al_2O_3/15_p$-T6, 6063 designates an alloy–6063 matrix composite reinforced by 15 vol % alumina particles and heat treated to T6 temper.

- ***Boron–aluminum MMCs*** are continuous boron (B) fiber composites. Made by a hot-press diffusion bonding or plasma spray process, boron is expensive and finds use primarily in the aerospace industry. The midfuselage structure of NASA's *Orbiter* is made of tubular truss members that employ B–Al composites.
- ***Silicon carbide–aluminum*** composites can maintain their strength up to 500°F. SiC_p/Al are discontinuous silicon carbide composites materials that use SiC whiskers, nodules, flakes, platelets, or short fibers in an aluminum matrix. (The subscript following the abbreviation for carbon indicates the reinforcement is a particulate of some form. The letter *p* is also used to signify this type of reinforcement.)
- ***Graphite–aluminum MMCs,*** introduced in the 1960s, are an example of the negative effect between the matrix material and its reinforcement. Molten aluminum does not wet the carbon fibers; it attacks the carbon and the carbon itself will oxidize. Combined, these situations make processing difficult. Liquid metal infiltration, vapor deposition, diffusion bonding, pultrusion, and squeeze casting are some of the processes used in fabricating this composite.
- ***Alumina–aluminum MMCs*** using inexpensive ceramic fibers have improved resistance to wear and thermal fatigue. Continuous-fiber alumina–aluminum composites use a preform made of alumina tapes wherein the fibers are arranged in the required direction.
- ***Magnesium–matrix composites*** have similar properties to aluminum composites, which include low coefficient of thermal expansion (CTE), high stiffness, and light weight (Mg is nearly two-thirds as dense as Al). Continuous-fiber C/Mg composites are used in the aerospace industry; Al/Mg composites with alumina short fibers are used in making automobile engine components; and SiC_d or B_4C/Mg composites find use in engine components and low-expansion electronic packaging materials.
- ***Titanium–matrix composites.*** Titanium retains its strength at moderately high temperatures and has excellent corrosion resistance. Titanium reacts negatively with many reinforcements, however. To overcome this problem, titanium composites are hot isostatic pressed and vacuum hot pressed rather successfully. Another method is plasma spraying the titanium matrix onto SiC fibers. These processes result in a titanium composite that is significantly stronger than unreinforced titanium. Figure 25-12 is a photograph of the cross-sectional area of a titanium–matrix composite consisting of continuous SiC fiber reinforcement that was sputter coated with a beta titanium alloy and vacuum hot pressed to form the composite material. This coating is necessitated by the adverse reaction between the titanium matrix and the SiC fibers. One of many film deposition techniques, ***sputtering*** is a coating process in which ions in a plasma are used to knock off atoms of the coating material (the target) in a vacuum chamber, which then diffuse to the adjacent substrate, effectively coating it with the target material. Continuous-fiber reinforcements are used in aerospace for structural applications and as compressor blades for advanced turbine engines.

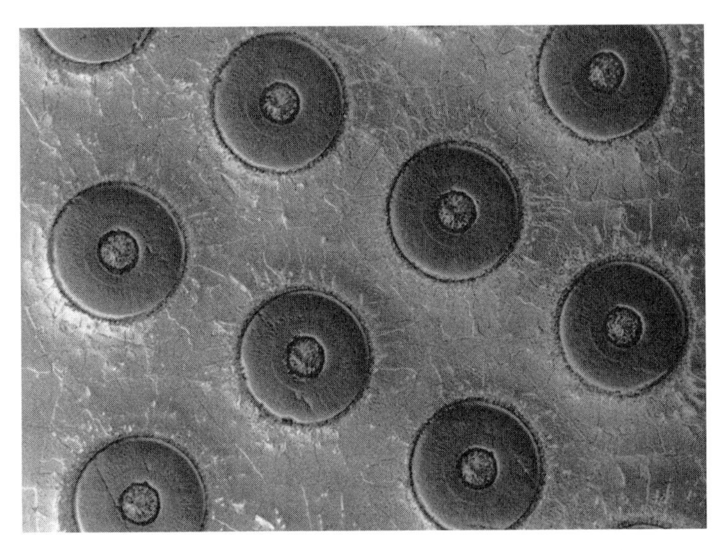

Figure 25-12 A titanium–matrix composite reinforced with silicon carbide continuous fibers coated with a beta titanium alloy by sputtering. Magnification 200X. (Courtesy of James Baughman, Analytical Services and Materials, NASA Langley Research Center, Langley, VA.)

- ***Copper–matrix composites.*** Copper is a better conductor both electrically and thermally than aluminum. As a composite, copper has higher temperature-strength properties than aluminum. W/Cu fiber composites with their high-strength properties up to about 925°C are finding more uses in rocket engine designs. C/Cu MMCs that are less dense, stiffer, and better thermal conductors than copper have been developed. One result of this effort is graphite fibers coated with pure copper that are flexible enough to be made into a fabric.
- ***Superalloy–matrix composites.*** Since the 1950s superalloys have found use in turbine engines. In the 1960s, superalloy composites were developed for similar applications. Using refractory metals reinforcement, in which tungsten, molybdenum, tantalum, and niobium fibers received the most attention, the strongest fiber developed was tungsten with a strength of more than 2070 MPa at 1095°C. Iron, nickel, and cobalt–matrix materials permit the enclosing of tungsten reinforcements without adverse effects. A W/FeCrAlY superalloy composite has been used to fabricate a hollow composite blade.

Due to their improved mechanical properties over traditional alloys, metal–matrix composites are being used as shaped charged liners, military armor, and cylinder liners for piston engines. A novel method of producing MMCs is ***spray atomization*** and ***codeposition.*** There are several advantages of this new processing technique: (1) Low processing temperatures can be used due to the highly efficient heat convection during atomization; (2) spray-deposited materials exhibit some characteristics of rapid solidification, which produces increased solubility of solids and nonequilibrium phases, among others; (3) the process has the potential of being utilized for near-net-shape manufacturing of difficult-to-form materials such as ***refractory metals;*** and (4) the process has the potential of being a cost-effective manufacturing technology because it avoids powder handling.

Most MMCs and CMCs can be synthesized using standard metal-working practices such as extrusion, hot forging, hot molding, superplastic forming, squeeze casting, powder metallurgy, liquid-melt infiltration, and rolling. A recent Mixalloy process uses turbulent eddies and jet impingement to mix two or more liquid metal streams at turbulent velocities in a chamber. The resulting mixture is then solidified fast enough to preserve the required microstructure. This process is unique because it is successful in generating nanoscale particles (e.g., 50-nm refractory boride) for use in a copper matrix. Coarser particles can also be obtained by careful control of the mixing conditions. This new process is replacing older methods of producing a wide variety of dispersion-strengthened alloys (see Sections 25.2.4 and 25.2.4.2 for further information). In addition, brass and bronze can be strengthened by the Mixalloy process using boride particles. A coating of MMC consisting of a nickel matrix in which silicon carbide particles are uniformly disbursed has been developed for engine cylinder bores by U.S. Chrome Corp. Designated NiCom, the coating can be as thin as 0.005 to 0.0015 mm and as thick as 1.5 mm. Average micro hardness ranges from 57 to 55 HRC. The SiC particles can range from as large as 20 μm to submicron size. According to the company, nickel coating is more efficient than chromium. Test results show that NiCom wears better than chromium in all engine bores. Other properties of NiCom are excellent frictional characteristics, oil retention, and environmental friendliness.

Whisker-reinforced MMCs have distinct advantages over conventional alloys. These MMCs, classified by some as dispersion-strengthened alloys, provide the highest-strength discontinuous-reinforced MMCs. Typical fibers are SiC, Al_2O_3, and Si_3N_4, which are combined with some common metal matrices, such as Al, Mg, Ti, and Cu. Powder metal (P/M) or liquid-melt infiltration methods are used to synthesize the fine-mesh metal-alloy powders and the whiskers by thorough mixing, blending, and consolidating to produce a near-net-shape component. SiC whiskers reinforcing an Al matrix have transverse tensile properties that are nearly equal to the longitudinal properties, making them capable of outperforming fiber-reinforced composites in a situation requiring multidirectional reinforcement.

Lanxide Corporation fabricates MMCs by using a pressureless, molten-metal infiltration process (trademarked Primex™) that features excellent wetting of the reinforcement by the matrix alloy. (See Module 19 for information on Lanxide's fabrication of CMCs.) Net and near-net-shape parts, small to large, having complex shapes can be produced economically. Figure 25-13a is a schematic of the process for making an electronics package. To produce an

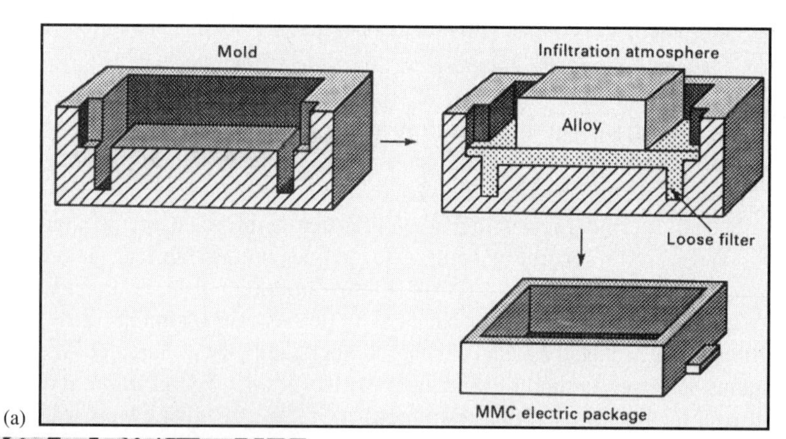

(a)

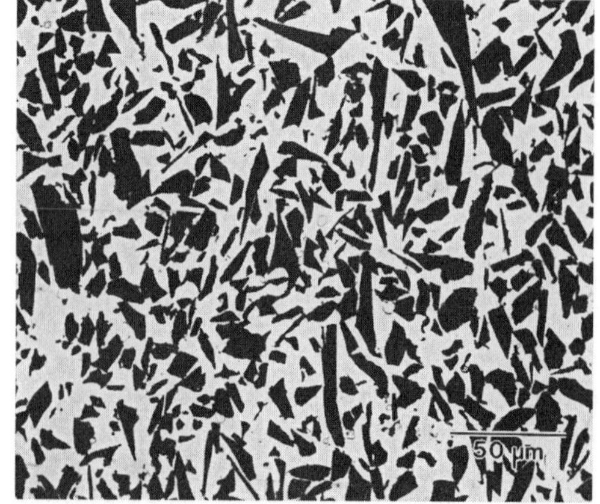

(b)

Figure 25-13 (a) Primese® pressureless metal infiltration process for producing MMC parts such as an electronics package, to net and near-net shape. (Lanxide Corp.) (b) A photomicrograph of Al_2O_{3p}/Al MMC showing a high and uniform particle loading. (Lanxide Corp.)

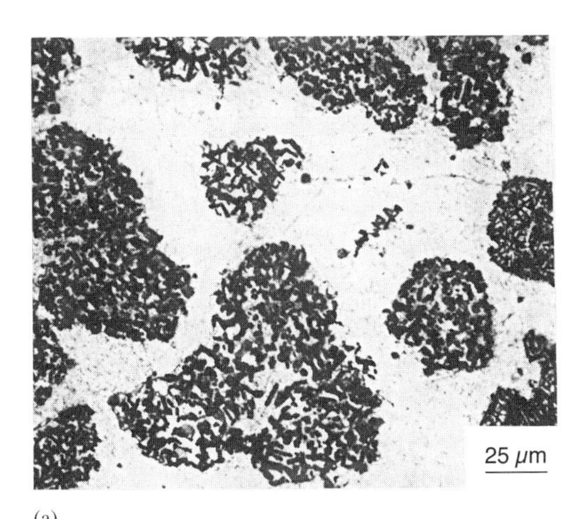

(a)

(b)

Figure 25-14 (a) A photomicrograph of an aluminum–matrix MMC filled with Al_2O_3 agglomerates. The molten metal completely infiltrates both the interagglomerate spaces and the minute openings within the agglomerates. (Lanxide Corp.) (b) The Primex® process for aluminum MMCs, with 60–70% (volume change) particulate reinforcement, can be used in engine components such as connecting rods, piston pins, brake parts, brackets, and stiffeners. (Lanxide Corp.)

aluminum-alloy MMC, for example, the infiltration is carried out in a nitrogenous atmosphere at a temperature somewhat above the melting point of aluminum. No external pressure or vacuum is needed. The MMC produced is completely infiltrated with metal alloy and features high and uniform loadings of particles or fibers. Figure 25-13b is a photomicrograph of an aluminum–matrix MMC filled with fused Al_2O_3 particles, and Figure 25-14 is a similar photomicrograph showing Al_2O_3 agglomerates (clusters of particles). These figures illustrate complete infiltration of the spaces between the particles by the molten aluminum alloy. One application of this SiC/Al MMC is for support structures for electronic components. Typical requirements include a low coefficient of thermal expansion (CTE) to reduce mechanical stresses imposed on the component during attachment and operation, high thermal conductivity for heat dissipation, and low density for minimum weight. Table 25-1 compares the physical properties of SiC_p/Al MMCs to a typical aluminum alloy. In addition, the dimensional stability of this MMC can be more than 3.2 times that of unreinforced steels, aluminum alloys, and titanium alloys. Combined with high thermal conductivity and low CTE, this array of properties provides this MMC with excellent rigidity and dimensional stability for use in space stations, automotive brake calipers, or optical system components. Table 25-2 lists some properties of (Al_2O_3)p/Al MMC in which the particles are more coarse, resulting in a composite that has higher strength and fracture toughness than those of the parent alloy. Despite the lower ductility, these MMCs

TABLE 25-1 PROPERTIES OF SiC-PARTICLE-FILLED Al MMCs FOR ELECTRONIC APPLICATIONS

	Composite examples		Typical Al alloys (for comparison)
SiC loading (vol%)	55	75	—
Coefficient of thermal expansion [$\times 10^{-6}$ ft^2 K ($\times 10^{-6}$ ft^2 °F)]	8.5 (4.7)	6.2 (3.4)	22–24 (12–13)
Thermal conductivity [W/m · K (Btu/hr · ft^2 · °F)]	160 (93)	170 (99)	150–180 (87–104)
Density [g/cm^3 (lb/in.^{3})]	2.95 (0.106)	3.0 (0.108)	2.7 (0.097)
Elastic modulus [GPa (msi)]	200 (29)	270 (39)	70 (10)

Source: Courtesy of Lanxide.

TABLE 25-2 PROPERTIES OF AN AL_2O_3-PARTICLE-FILLED Al MMC

	Composite (51% Al_2O_3)	Base alloy (AA 520)
Tensile strength [MPa (ksi)]	531 (77)	331 (48)
Elastic modulus [GPa (msi)]	161 (23)	65 (9)
Fracture toughness[a] [$MPa \cdot m^{1/2}$ ($ksi \cdot in.^{1/2}$)]	18.5 (16.8)	N.A.[b]
Elongation (%)	0.63	16

[a]Chevron notch method.

[b]N.A. = not applicable.

TABLE 25-3 PHYSICAL PROPERTIES FOR SILION-CARBIDE FIBER-REINFORCED ALUMINA $SiC_F(Al_2)O_3$[a]

	Temperature					
Property	25°C	(73°F)	1000°C	(1832°F)	1200°C	(2192°F)
Fiber content (eight plies)(%)	35	—	35	—	35	—
Density [g/cm^3 ($lb/in.^3$)]	2.9	(90.10)	2.9	(0.10)	2.9	(0.10)
Tensile strength [MPa(ksi)]	283	(41)	—	—	—	—
Modulus [GPa(Msi)]	122	(17.8)	—	—	—	—
Poisson ratio	0.10	—	—	—	—	—
Flexural strength[b] [MPa(ksi)]	450	(65)	400	(58)	350	(51)
Modulus [GPa(Msi)]	200	(29)	—	—	—	—
Poisson ratio	0.29	—	—	—	—	—
Fracture toughness[c] [$MPa\sqrt{m}(ksi\sqrt{in.})$]	28	(26)	23	(21.5)	23	(21.5)
Interlaminar shear strength[d] [MPa(ksi)]	63	(9)	—	—	—	—
Shear modulus[e] [GPa(Msi)]	90	(13.1)	—	—	—	—
Coefficient of thermal expansion [ppm/°C (ppm/°F)]	5.8	(3.2)	—	—	—	—
Thermal conductivity $\left[\frac{W}{m \cdot K}\left(\frac{Btu \cdot in.}{hr \cdot ft^2 \cdot °F}\right)\right]$	8.7	(60)	5.7	(39)	5.5	(38)

Thermal shock
(% retained flexural strength after quenching into RT water)

1000 to 25°C	92.7%	
1000 to 25°C	85.2%	. . . after 5 cycles
1200 to 25°C	83.6%	

Source: Courtesy of Lanxide.

[a]Silicon-carbide fiber (Nicalon); 12-harness satin weave, 0/90 two-dimensional perform layup.

[b]Four-point bend.

[c]Chevron notch beam.

[d]Short-beam method, ASTM D2344-84.

[e]Sonic method.

retain good fracture toughness. Table 25-3 compares properties of a CMC with a MMC. In particular, note the retention of strength properties at high temperatures for this CMC.

A major inhibition to the use of MMC parts is the difficulty in joining them. A newly developed fabrication process produces MMC parts with integral inserts. Parts consisting of a magnesium matrix and graphite fibers with titanium inserts are formed in final shape in a single casting step using a pressure-casting machine. No finish machining is required. Molten magnesium is infiltrated into the graphite fiber preform reinforcement, which is inserted into a mold cavity along with the titanium insert. Pressures, temperatures, and times are calibrated to produce a complete infiltration of the densely packed preform and rapid solidification of the melt. Subsequent tensile testing produced fracture, at around 10,000 N, at a point well away from the insert.

A carbon fiber material called ***graphite fibrils,*** each piece about 1/10,000 the size of a human hair, added to a resin creates a conductive resin for electrostatic painting of automotive side mirrors made of ABS. This GE and United Technologies Automotive Inc. development may lead to a better way of painting plastic parts.

To satisfy the need for materials that are lighter, stronger, more corrosion resistant, and capable of performing at elevated temperatures, the U.S. Department of Energy established a comprehensive program to develop a relatively new family of materials known as ***continuous-fiber ceramic composites (CFCCs).*** Applications in the near term will be for mid-temperature (500° to 1000°C), secondary aerospace structures such as exhaust components, hot gas ducting, and leading edges. In the longer term, there is great potential for high-temperature applications in the automotive and industrial fields, provided the costs of fabrication can be lowered to be competitive. Several methods are used for fabricating these composites. Among them is the use of preceramic polymers for the low-cost fabrication of large, complex-shaped CFCCs. After forming in the polymer condition, the composite is pyrolized to form the ceramic matrix and then densified through repeated reinfiltration/pyrolysis cycles. Regardless of what fabricating process is finally determined to have the most merit, the deciding factor will be the development of a cost-effective, oxidation-stable fiber interface that can withstand high strain fracture. Continuous-fiber ceramic composites (CFCCs) were developed to overcome the brittle nature of ceramics and to capitalize on their favorable properties (see Section 19.3). Low-cost, low-density ***ceramic fibers*** that exhibit high-temperature mechanical properties comparable to superalloys and existing ceramic fibers are a recent development of the Max Planck Institute in Germany. Known as ***Siboramic***, the fibers, composed of borosilicon nitride and borosilicon carbon nitrate, are designed to replace ceramic materials in aerospace and power generation components. Their thermal conductivity is 0.4 w/mK at 1500°C; CTE is 3×10^{-6}/K, and they have oxidation resistance to molten silicon up to 1600°C. ***Discontinuous reinforced aluminum (DRA),*** which can refer to a technology, a Defense Department program, or a specific metal–matrix composite material, is now capable of competing with conventional aluminum and titanium alloys as well as organic matrix composites. DRA is an aluminum metal–matrix composite in which the reinforcement is a particulate, flake, whisker, or short fiber typically of a strong, stiff ceramic material such as Al_2O_3 or SiC. Highly versatile, this isotropic composite has a unique combination of strength, stiffness, and affordability as well as the ability to be processed and finished by conventional metal processing. It has good wear and corrosion resistance and can be anodized. Because its density is similar to conventional aluminum, its specific characteristics (specific compressive strength and stiffness) are superior to conventional aerospace materials. Figure 25-15a is a graph of DRA's compressive strength compared to other aerospace materials. Military applications are numerous, especially in fighter aircraft. Figure 25-15b shows a U.S. Air Force F–16 fighter aircraft with ventral fins made of a discontinuous reinforced-aluminum MMC. These composite fins are projected to last more than 6000 flight hours as opposed to less than 400 hours for the less stiff conventional aluminum alloy sheets bonded to honeycomb panels. The Hubble Telescope antenna, wave guide mast and electronic packages for satellites are further aerospace applications for this new composite material. Automotive applications include Chevrolet Corvette and GMS/T truck driveshafts, Honda Prelude cylinder liners, Plymouth Prowler brake rotors, and GM Electronic Vehicle–1 brake drums. The last two applications take advantage of DRA's low density and high wear resistance.

Table 25-4 compares the physical properties of two aluminium MMCs reinforced by particulates, emphasizing the difference in properties due primarily to the type of reinforcement. Most of these composites tend to be isotropic. MMCs and CMCs are composed of three constituents: the fibers, having high strength and stiffness; the matrix, which holds the fibers together and distributes the applied load; and the interface zone between the fibers and the matrix, which determines the wetting and bonding (coupling between the fiber and the matrix) as well as the transfer of the load between the fibers and the matrix (see Figures 24-4 and 24-10). The formation of a shallow interface zone is generally desirable to ensure a strong bond.

Figure 25-16 depicts how the Chrysler Corporation used the Sullivan process to produce a single-piece, CMC Si_3Ni_4, cam roller follower. As mentioned in Module 7, this CMC component replaces 18 small steel bearings, potentially costs less to mass-produce than steel parts, and offers longer wear life. Nickel-based superalloys are slowly being replaced by

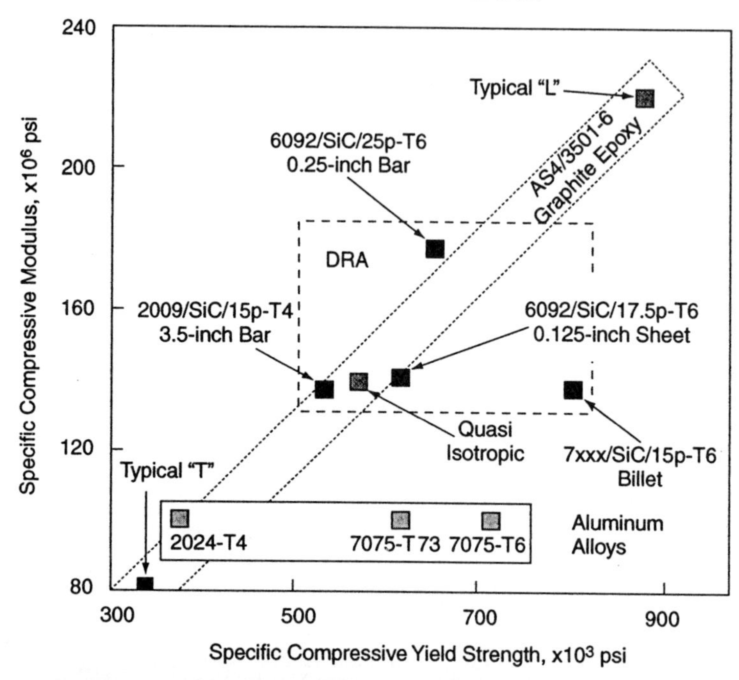

(a)

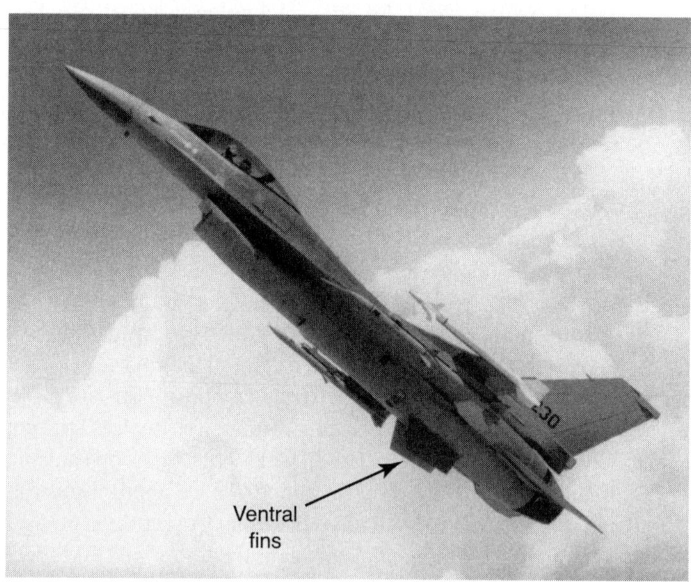

(b)

Figure 25-15 (a) DRA metal–matrix composite is an isotropic composite with a broad range of properties. With a density similar to conventional aluminum, its specific properties are superior to conventional aerospace materials. (Courtesy of DWA Aluminum Composites) (b) The U.S. Air Force F–16 Fulcan fighter aircraft uses ventral fins located under the aft section of the fuselage to provide added stability during tight, high-speed turns. This material replacement program is expected to save the Air Force approximately $21 million in life-cycle costs. (Courtesy U.S. Air Force)

composites in high-temperature applications for aircraft engines. SiC fiber reinforcement added to a matrix, which is itself a composite material, makes for another interesting material. The matrix consists of molybdenum disiliside ($MoSi_2$), which contains a reinforcement of 30 to 50 vol % of Si_3N_4 particles. This composite material has a higher temperature oxidation behavior with a lower density and a higher melting point temperature than super alloys. It retains excellent strength and toughness improvements at temperatures up to

TABLE 25-4 PHYSICAL PROPERTIES[a] FOR SILICON-CARBIDE- AND ALUMINA PARTICULATE-REINFORCED (SiC_P/AL AND AL_2O_3/AL) METAL–MATRIX COMPOSITES

Property	Reinforcement			
	Alumina (Al_2O_{3p}/Al)		Silicon carbide (SiC_p/Al)	
Filler content[b] (%)	50	—	55	—
Density [g/cm^3 ($lb/in.^3$)]	3.2	(0.12)	2.9	(0.10)
Tensile strength [MPa (ksi)]	530	(77)	250	(36)
Modulus [GPa (Msi)]	160	(23)	200	(29)
Elongation (%)	0.6	—	0.3	—
Compressive strength				
0.2% yield [MPa (ksi)]	520	(75)	460	(66)
Ultimate [MPa (ksi)]	900	(130)	490	(70)
Fracture toughness[c] [$MPa\sqrt{m}$ ($ksi\sqrt{in.}$)]	18.5	(130)	12	(11)
Coefficient of thermal expansion [ppm/°C (ppm/°F)]	13.5	(16.8)	8.5	(4.7)
Thermal conductivity $\left[\frac{W}{m \cdot K}\left(\frac{Btu \cdot in.}{hr \cdot ft^2 \cdot °F}\right)\right]$		(7.5)	160	1100

Source: Courtesy of Lanxide.

[a]Room temperature (25°C) properties. Elevated temperature properties available.

[b]Filler content can be tailored (20–80%) to obtain desired properties. Properties as a function of filler content available.

[c]Chevron notch beam.

Figure 25-16 Ceramic–matrix composite (CMC) parts, such as this cam roller follower, are made from a liquid process, developed by Sullivan Mining, instead of the conventional powder process. The process is less damaging to reinforcing fibers and capable of mass-producing complex shapes at about one-tenth the cost of parts made by conventional processing. At least as durable as steel rollers, they may run more quietly. (*Advanced Materials & Processes*)

1400°C. The addition of Si_3N_4 to $MoSi_2$ doubled the room-temperature toughness and reduced the high-temperature creep rate by about five orders of magnitude. The addition of SiC fibers to the matrix materials increased the room-temperature fracture toughness and impact resistance.

25.2.2 Laminar Composites

Laminar composites (Figure 25-17) consist of layers (***lamina***) of at least two different solid materials bonded together so that the fiber orientation runs at different angles (e.g., 0°, 45°, 90°). Lamination allows the use of the best properties of each layer to achieve a more useful

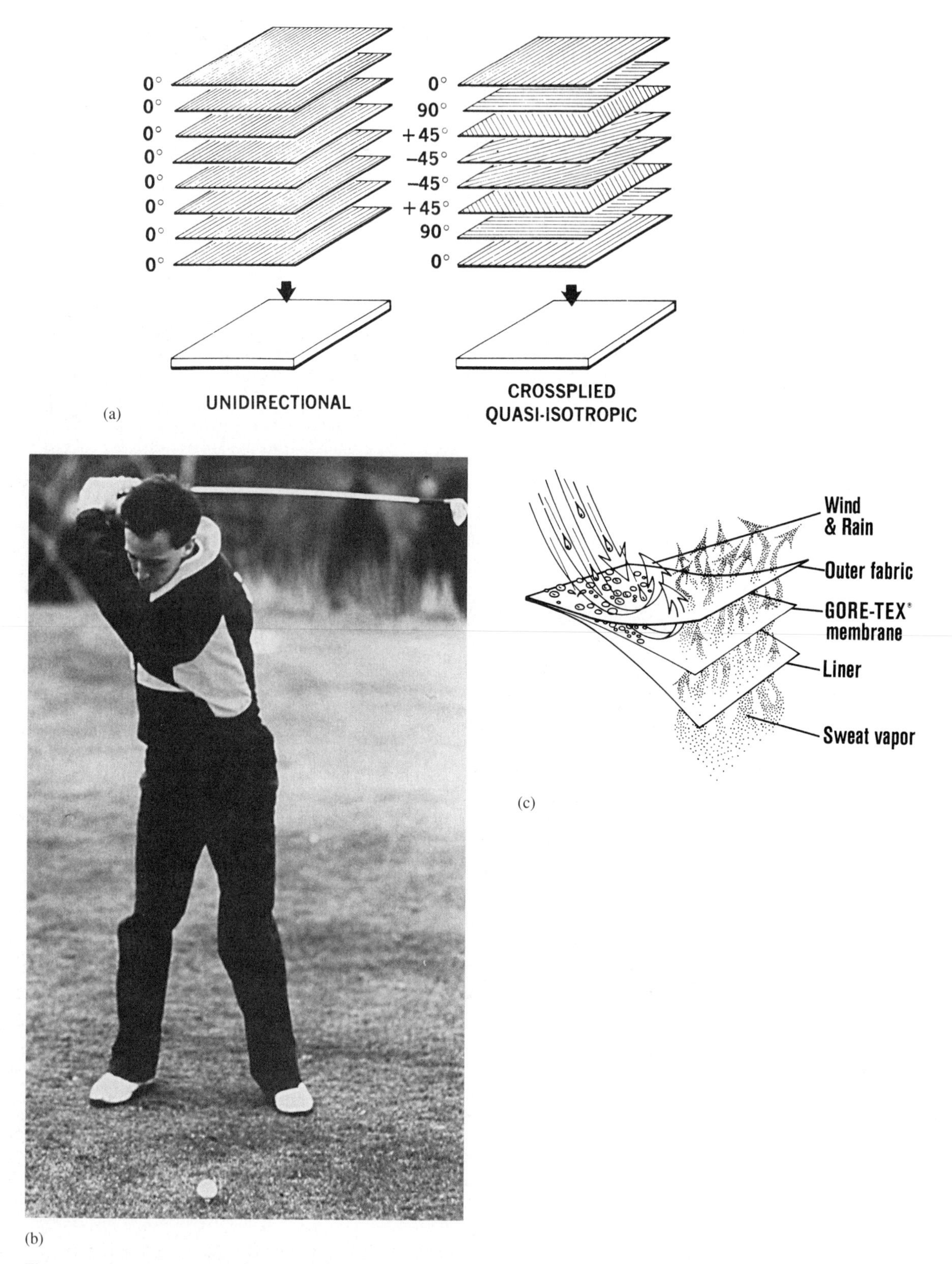

Figure 25-17 (a) Fiber orientation within laminates. Laminates can be tailored to meet specific requirements, with savings in time and material. (Hercules, Inc.) (b) Gore-TEX® laminated membrane used by manufacturers of outdoor clothing and footwear for use in all types of weather. (W. L. Gore & Associates, Inc.) (c) Gore-TEX® lamina between liner and outer fabric allows sweat vapor to escape but keeps large water drops out of foul-weather gear.

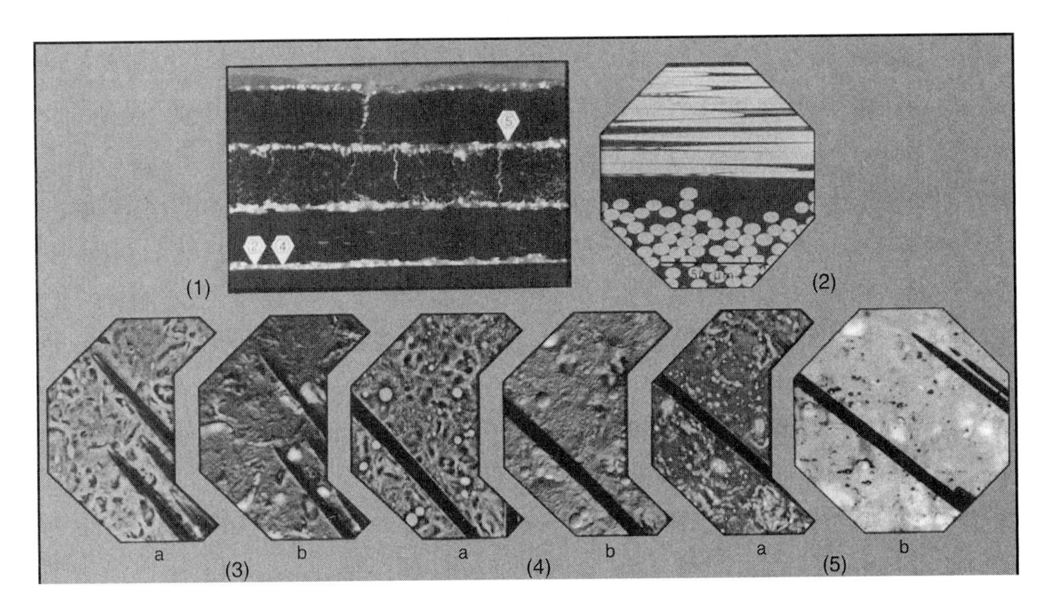

Figure 25-18 Figure 2 is a traditional reflected-light bright-field microscopy technique. The carbon fibers are seen as white rods, and the polymer resin is gray. It is important to note that the soft polymer matrix encapsulating the hard carbon fibers must be polished, with no relief at the interface. An optical wedge created by thinning the polymer just under the surface adjacent to the fiber results in the observed prism. The multiphase resin morphology is difficult to see with reflected-light microscopy. In contrast, thin-section micrographs of baseline interior (3a–3b), aged interior (4a–4b), and aged surface (5a–5b) reveal changes in resin morphology caused by thermal and oxidative processes. Since thin-section microscopy utilizes transmitted light, carbon fibers are seen as black rods and the resin morphology is visible in high resolution. (Luther M. Gammon, Boeing Commercial Airplane)

material. Properties such as wear resistance, low weight, corrosion resistance, strength, stiffness, and many more can be accented by a wise selection of different constituent layers. A fiber-reinforced composite made up of layers of tape, fabric, or mat can be considered a laminar composite. A well-known example of a laminar composite is plywood. Plywood has isotropic properties in the plane of each sheet due to the layers of wood bonded with a thermoset resin such that the longitudinal direction is at right angles in adjacent plies. Only their orientations differ. Figure 25-18 shows thin-section micrographs of degraded laminated thermoplastic and thermoset composites. Analysis of specimens by, for example, isothermal overtesting designed to replicate operating conditions aids in understanding multiphase systems and how they change due to environmental conditions.

Some example of laminar composites are

- ***Clad metals*** such as bimetallic strip or copper-clad stainless steel are composed of two different materials. In the former the two sheets of metal bonded together have different amounts of thermal expansion, and in the latter the thermal conductivity property is the rationale for the design of the composite material. Similar treatments with paper produce isotropic properties. Paper laminated with plastic film or metal foil—or polymers such as nylon fabric as a base and laminating the polymer with layers of metal—produces an endless variety of plastic-based laminates. Roofing paper, Formica, and Kevlar fibers laminated with a resin to produce a material for bulletproof vests are additional examples of these composite materials on the market today.
- ***Safety glass*** is a laminated glass consisting of a layer of poly(vinyl butyral) bonded between two layers of glass. Glass alone is quite brittle and dangerous due to its proclivity to shatter into many sharp-edged pieces. Poly(vinyl butyral) is very tough but weak in scratch resistance. The glass lamination protects the plastic from scratching and gives stiffness, while the plastic contributes its toughness. By protecting each other in different ways, each material allows the other to contribute to the composite material's vastly improved properties.

- A layered composite known as ***Graftskin*** is an artificial human skin replacement tissue that replaces both the outer epidermal and inner epidermal skin layers. It blends dermal cells with collagen, the protein that forms a fibrous framework for skin cells. The collagen is taken from cows. Graftskin has been used successfully in treating chronic skin diseases, burns, and deep wounds, the results of which show quick wound closure, accelerated healing, and reduced scarring.
- Laminates formed by high pressure and heat are referred to as ***rigid laminates.*** Some examples of rigid laminates are countertops, rods, tubes, and printed circuit boards (PCBs) for the electronics industry.

Unsolved manufacturing problems associated with the processing of bulk-laminated ceramic or metal composites limit their potential for use in structural applications. Ceramics have inherent properties of high-temperature strength, hardness, and wear and oxidation resistance, but they also possess low toughness. Unit 7 contained much information on how materials engineers and scientists are combating this problem of brittleness in ceramics by such mechanisms as crack bridging and transformation of residual stresses, to mention but a few. ***Ceramic surface coatings*** have also had great success in providing other materials with some of the properties of ceramics. However, the application of coatings is expensive. A processing time of 500 hours to achieve a laminated structure 1 cm thick is mentioned in the literature. Metallic laminated structures made of ultra-high-carbon steel and iron–silicon alloys using roll bonding would have limitations in producing a bulk ceramic/metal laminated composite due to concern about adhesion between the layers.

Refractory metals are among the few materials with significant mechanical strength above 2500°F, but they do not have sufficient strength-to-weight ratios for many applications. A material with a high strength-to-weight ratio and good mechanical strength at service temperatures above 3000°F is sorely needed for the many high-temperature applications, such as space shields, rocket engine fuel turbines, and gas turbines. Combining a dense refractory metal such as tungsten, molybdenum, vanadium, or tantalum with a lighter-weight matrix material creates a unique material—a lightweight ***refractory composite.*** A fabricating process, ***explosive welding,*** is used to join these incompatible metals. Explosive welding or ***explosive bonding*** is a solid-state welding process that uses controlled explosive detonations to force two or more metals together at high pressures. The contaminant surface films are plastically jetted off the base metals as a result of the collision of the two metals. The time duration is so short that the heat-affected zone (HAZ) is microscopic. During the process, the first few layers of each metal become plasma due to the high-velocity impact (200 to 500 m/s). The remaining thickness of each metal maintains a near-ambient temperature and acts as a large heat sink, cooling the plasma rapidly. The resultant composite system is joined with a high-quality metallurgical bond. Some explosive-bonded materials in industry include niobium/copper, copper/aluminum (for high-voltage busbars), aluminum/stainless steel, titanium/stainless steel (for pacemakers), and zirconium/stainless steel.

Microinfiltrated macrolaminated composites (MIMLCs) offer a new, economically feasible approach to the production of tough composites using cermet/metallic, ceramic/metallic, and ceramic/intermetallic compound bulk-laminated composites. Figure 25-19 illustrates the conceptual structure of an MIMLC. These materials attempt to duplicate a case-hardened metal with a ductile interior. Such a metal has the surface hardness and strength needed for many applications, such as a cutting tool, along with the ductility to resist great impact (toughness). One problem that must be overcome is the proper adhesion between the layers of the bulk composite. Figures 25-20a and b illustrate poor and good solid-state bonding with a W–Ni–Fe/Ni MIMLC. W–Ni–Fe is a hard ceramic alloy that has high strength and modulus but low toughness. It could be compared to the fiber reinforcement in a metal– or ceramic–matrix composite. The nickel (Ni) is a material having low strength, elastic modulus, and hardness but high toughness; it could be compared to the matrix material in a fiber-reinforced composite. Figure 25-20a shows a crack-like discontinuity between the bottom side of the Ni foil and the W–Ni–Fe heavy sheet when the composite was subjected to a sintering temperature of only 1400°C. Figure 25-20b shows proper solid-state bonding and consolidation of the layers when pressure was added to the process.

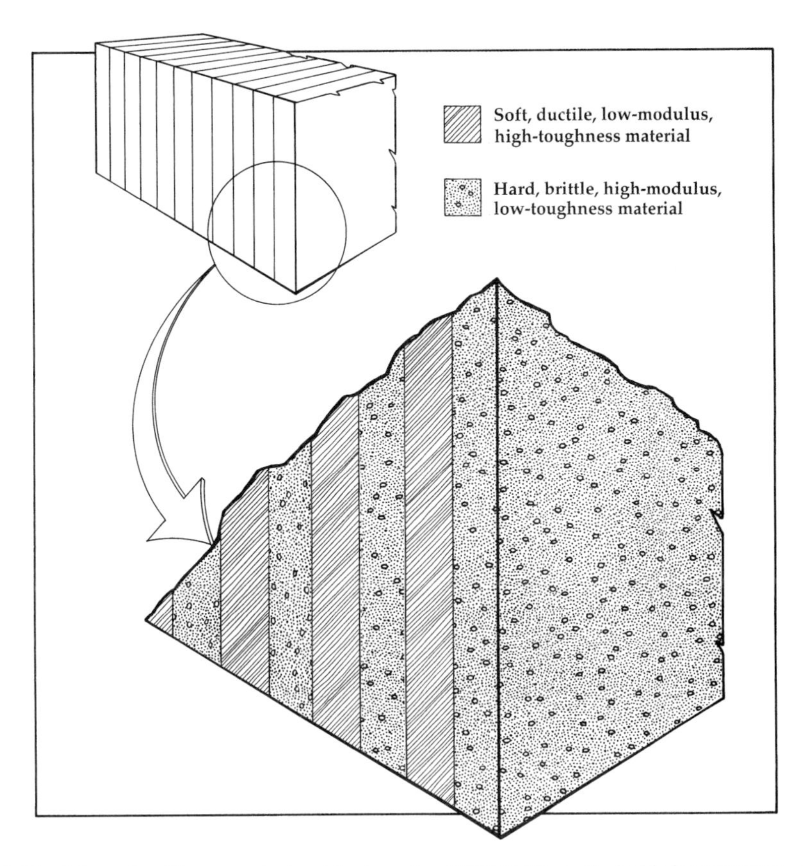

Figure 25-19 Schematic of the conceptual architecture of an MIMLC. The basic double-layer structure, consisting of a soft, ductile material (low modulus and strength and high toughness) and a hard, brittle material (high modulus and strength and low toughness), is repeated as many times as necessary to form a bulk composite material. The brittle material is also infiltrated with the ductile component, tending to increase the brittle material's toughness. (Bose and Lankford, *Advanced Materials & Processes*)

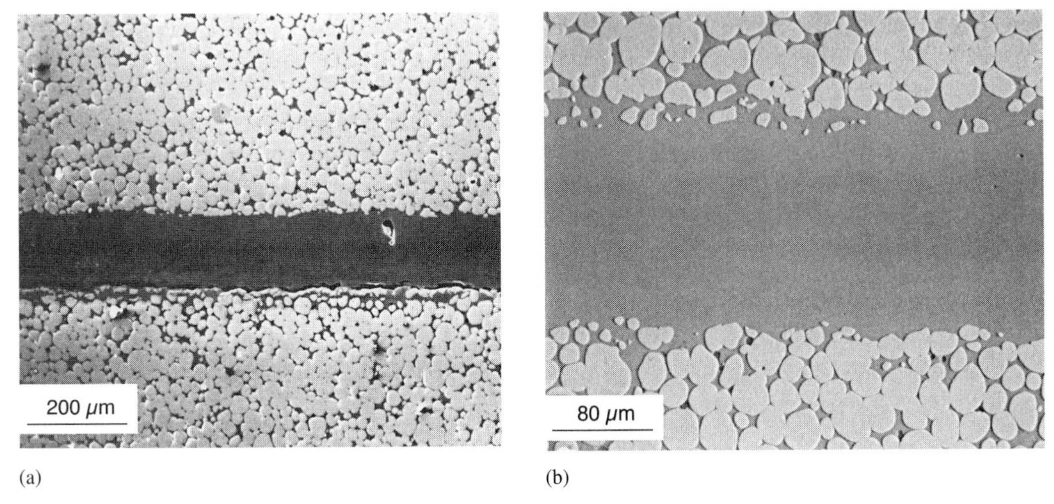

Figure 25-20 (a) W–Ni–Fe/Ni MIMLC composite that was sintered at 1400°C without pressure showing improper bonding between layers. Note a crack along the bottom side of the nickel foil. (Bose and Lankford, *Advanced Materials & Processes*) (b) W–Ni–Fe/Ni MIMLC composite that was sintered under pressure at 1400°C showing good bonding between the nickel foil and the heavy alloy sheet, which creates one continuous matrix that interpenetrates the tungsten grains. (Bose and Lankford, *Advanced Materials & Processes*)

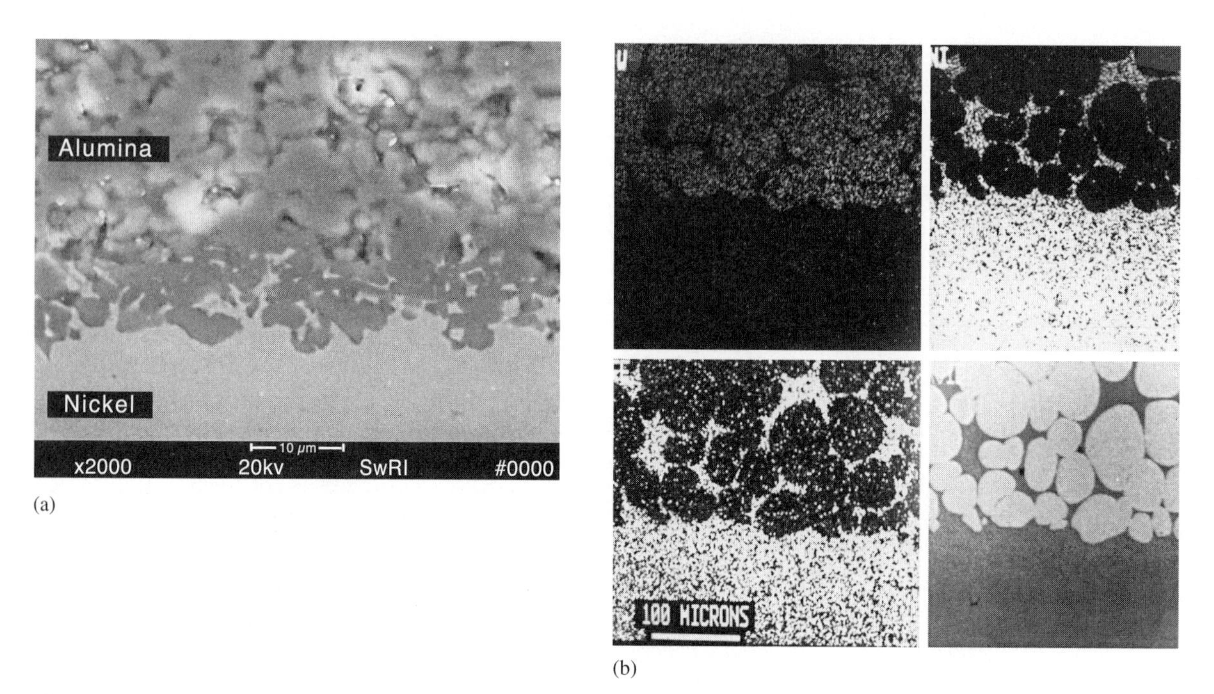

Figure 25-21 (a) Photomicrograph of the microstructure of a hot-pressed alumina/nickel MIMLC where the nickel layer has infiltrated to only a small depth in the alumina sheet, which remains porous. Additional increases in time, temperature, and pressure would be necessary to achieve adequate bonding. (Bose and Lankford, *Advanced Materials & Processes*) (b) An X-ray dot map of a portion of a properly bonded W–Ni–Fe/Ni sheet MIMLC showing the tungsten spheroids embedded in a matrix of nickel and iron. (Bose and Lankford, *Advanced Materials & Processes*)

Many variations of the MIMLC fabrication process are necessary to accommodate the differences in solubility of the materials combinations that can be used. Figure 25-21a illustrates a situation in which two materials, a ceramic and a metal, that have very little or no solubility in one another are combined. Porous alumina (Al_2O_3) and Ni are tape cast (to be defined) to produce an alumina/nickel MIMLC. Using a temperature just below the melting point of Ni and applying pressure, the Ni is extruded into the fine pores of the alumina structure. With further controls of variables such as time, temperature, and pressure, more adequate bonding can be achieved. Additional details of the fabrication of the W–Ni–Fe/Ni MIMLC may assist in understanding this bonding process. To bond them, both materials are tape cast. ***Tape casting*** consists of mixing powders with a suitable binder to form a slurry, which is then cast as a thin tape (sheet) of material. The material is then cut in the green state when it is relatively soft or after it is sintered. The W–Ni alloy is tape cast and sintered to some 60% of theoretical density to produce an interconnected porosity. An 80:20 ratio of Ni to Fe powders is also tape cast and sintered to full density. Laid up in alternating layers and heated to about 40°C above its melting temperature, the molten Ni–Fe alloy infiltrates the porous W sheet and takes into solution a fraction of the W. In Figure 25-21b, the lower white portion (Ni–Fe alloy) shows the infiltration of the porous tungsten sheet. The top, darker portion shows the W spheroids surrounded by the diffusing Ni–Fe matrix, which takes into solution a fraction of the W (darker portion).

There are many process control variables as well as a large number of material combinations of a hard, brittle phase interpenetrated by a soft, ductile material that can be modified to produce a multitude of MIMLCs. With a plethora of different properties, they will find applications as armor plating; high-temperature composites; tough cutting tools; wear-resistant composites; and components with low density, high strength and high modulus, good fracture toughness, and impact resistance for use in aerospace and automotive structures and components.

25.2.3 Sandwich (Stressed-Skin) Composites

Sandwich composites can also be classified as laminar composites. Their outer surfaces, or ***facings,*** are made of some material higher in density than the inner material, or ***core,*** which supports the facings. The primary purpose of sandwich composites is the achievement of high strength with less weight, specifically a high ***strength-to-weight ratio*** or ***specific strength.*** A sandwich composite may be compared to a structural I-beam. The high-density facings correspond to the I-beam flanges that carry most of the applied load, particularly the bending loads. Like the I-beam web, the low-density core allows the facings to be placed at a relatively large distance from the neutral plane to produce a large section modulus. The core carries the shear stresses. Overall, sandwich composites are more efficient than I-beams, because they possess a combination of high section modulus and low weight.

Typical sandwich facing materials are aluminum, wood, vinyl, paper, glass-reinforced plastics, and stainless steel. Core materials, representing all families of materials, are primarily cellular in form and take on the configurations known as honeycomb, waffle, corrugated, tube, and cone. These rigid cores provide the greatest strength and stiffness, with honeycomb possessing isotropic properties. Metals that can be made into very thin sheets capable of being diffusion bonded are finding greater use as cellular-core materials. Solid cores made of plywood, as well as foamed cores of polystyrene or ceramics, are examples of other types of core materials. An ***aluminum honeycomb material*** is finding applications in automobiles to improve head support protection due to its unique shock-absorbing ability in thicknesses down to 10 mm. The nested hexagonal cell material is very light (weight savings of up to 30%), easy to form, simple to install, and is 100% recyclable. It is less sensitive to temperature changes and environmental degradation than foam materials. ***Ceramic honeycomb panels*** that withstand temperatures to 1800°C serve as lightweight structural panels. Using CVD, a ceramic is deposited on a fabric substrate woven in a honeycomb pattern to make a honeycomb structure. This substrate is ultimately eliminated. The fabric can be made of a loosely woven polymer such as polyacrylonitrile (PAN), which is impregnated with an organic binder such as phenolic resin for stiffness. The ceramic materials are not limited to SiC—others are silicon boride, silicon nitride, or boron nitride. Heated to temperatures between 500° and 1000°C in an atmosphere of 2–5% oxygen, the fabric and binder oxidize and pass through the pores of the ceramic as a gas, leaving a microstructure of voids. The voids can be filled with ceramic by additional CVD, as can the hexagonal holes of the honeycomb.

The strength of the honeycomb structure or shape (see Figure 25-22), mathematically the strongest possible and the most economical for a mass of adjacent cells, is dramatically demonstrated in Figure 25-23. ***Nomex*** honeycomb paneling provides superior specific modulus and finds many applications as structural elements in aircraft, spacecraft, racing vehicles, and watercraft. Dupont's cellular honeycomb composite Kevlar N636 offers up to three times the stiffness of an equivalent Nomex honeycomb. A ***honeycomb composite*** sandwich material is used by Boeing Co. to build a one-piece satellite body that is 40% lighter than a traditional aluminum structure, four times stiffer, 40% stronger, and weighs only 14 kg. Fabricating time is halved by reducing the number of parts and eliminating the number of manufacturing and assembly steps to one. The composite face sheets are made of an ultra-high-modulus graphite fiber and polycyanate ester resin composite that is about 20% lighter than conventional composites. This material releases fewer contaminants in space than conventional epoxy resin systems and has good resistance to microcracking. As good a heat conductor as aluminum and an excellent thermal conductor, the honeycomb composite permits passive thermal control, which is impossible with low-conductivity composites.

25.2.4 Particulate (Particle) Composites

Particulate composite materials contain reinforcing particles of one or more materials suspended in a matrix of a different material. A reinforcing particle can be defined as having all its dimensions nearly equal. The particles, either metallic or nonmetallic, by definition do not chemically combine with the metallic or nonmetallic matrix material. As with nearly all materials, structure determines properties, and so it is with these composite materials. The size, shape, and spacing of particles, their volume fraction, and their distribution all contribute to the properties of these materials. How particles influence the properties of a material are explained in the discussion of dispersion-strengthened alloys in Section 25.2.4.2. Particulate composites are many times divided into subclasses by using some characteristic or combination of particular characteristics.

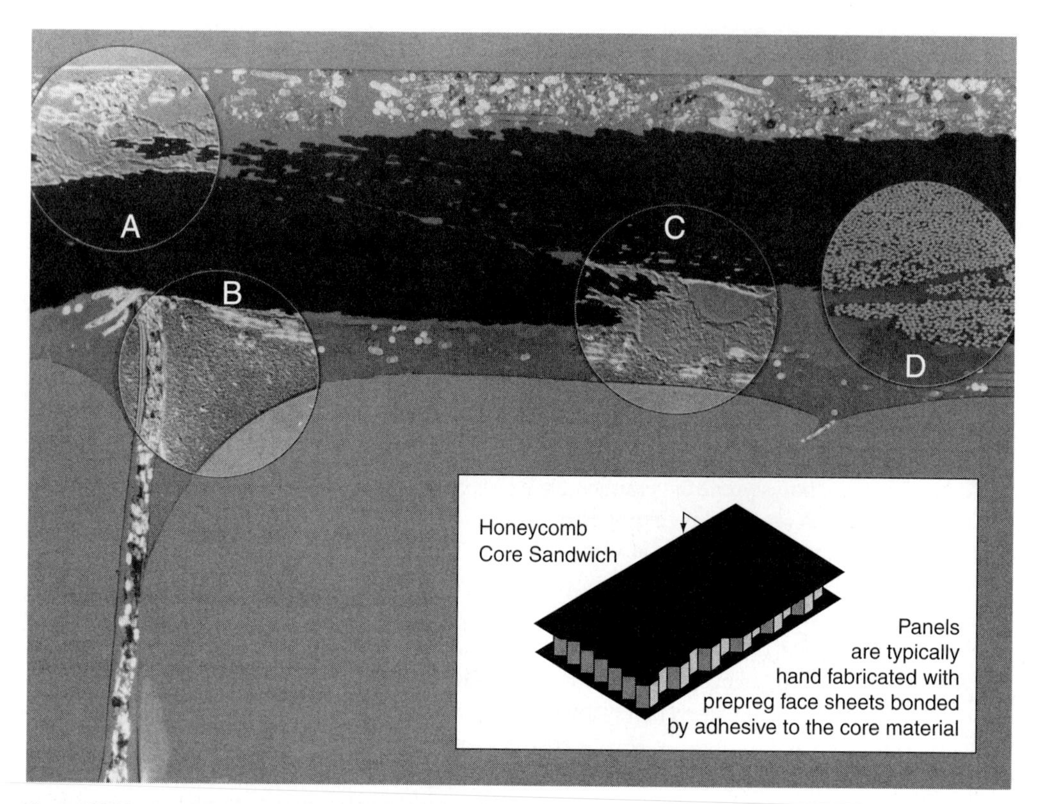

Figure 25-22 The honeycomb sandwich composite is a very efficient yet complex structure widely used in the aircraft industry. By imitating the natural geometric structure of a beehive, the honeycomb core imparts strength and light weight to sandwich panels, while supporting the prepreg skins. The bond between honeycomb core and the carbon/epoxy skin, shown in this cross section of ultrathin sections (1–10 μm), is created by the adhesive film layer. A surfacing film is often co-cured with the composite to improve the appearance of the panel and provide a smooth, uniform, and nonporous surface. With the appropriate technique, transmitted light microscopy allows us to see differences in morphology between the prepreg, adhesive film, and surfacing film. Understanding the intermingling of the resin systems enables the proper selection of materials and/or processing steps. The base figure was produced using transmitted cross-polarized light with a 530-nm compensator plate, 200X, 10-micron thin cross section. The 3K–70-PW carbon fibers are visible as black rods, and the amorphous resins are first-order magenta. The randomly oriented, birefringent polyester scrim fibers are observed in the 3M, AF310 surfacing film, BMS 8–245 adhesive, and in the Nomex BMS 8–124 core cell wall. A filler mineral, talc, appears as birefringent flakes and needles in the surfacing film. Inserts A, B and C: Transmitted light Hoffman contrast, 200X. This technique reveals the subtle amorphous morphology of the ultrathin section. With this contrast mode, a three-dimensional appearance is given, indicated by different contrast due to different refractive indices of the resins. Therefore, one can easily trace resin intermingling. Figure A shows the resin intermingling of the AF310 surfacing film and the BMS 8–256 carbon/epoxy skin. Figure B shows the Nomex core scrim and phenolic dips and the BMS 8–245 adhesive fillet which bonds the core to the skin. Figure C shows the resin intermingling of the BMS 8–256 skins and BMS 8–245 adhesive. Inset D: Reflected light bright field, 200X. This technique shows the fibers in the carbon/epoxy skin as light gray rods, but does little to enhance the morphology shown in transmitted light. A white reflective surface placed on the back of the slide allows one to see a three-dimensional view of the fiber. (Luther M. Gammon, Boeing Commercial Airplane)

This category of composites does not include particulates that are flat in shape, which possess sufficiently different properties to warrant a special classification.

Up to 30% titanium diboride (TiB_2) incorporated into the steel matrix by powder metallurgy has produced a ***particle-reinforced steel composite.*** Hot isostatic pressing consolidates the powders to 100% density. Tensile strengths are in excess of 1000 MPa and the specific modulus is double that of titanium alloys. Several matrix materials are being used to form this composite, such as pure iron, plain carbon steel, and gear steel. Titanium diboride is the particulate reinforcement due to its high modulus, low density, and good chemical compatibility with steel. To date, the composite has been hot forged and extruded and a gear component has been manufactured. This new composite should help meet the demand for higher strength for high-performance structural applications.

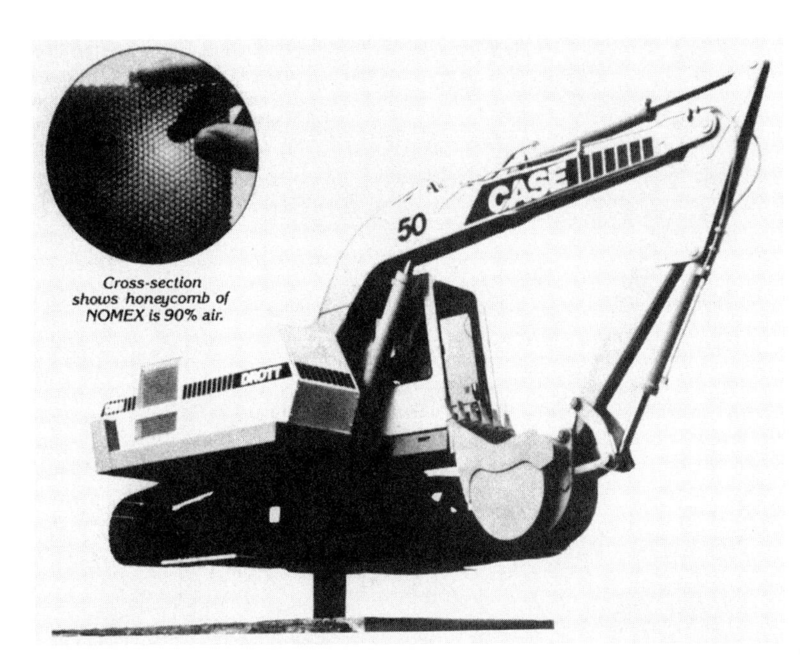

Figure 25-23 A 10-in.-square by 22-in.-high column of honeycomb of Nomex™ aramid weighing 11 pounds supports a load of 27 tons. (DuPont) (Luther M. Gammon, Boeing Commercial Airplane)

Advanced Ceramics Corp., Cleveland, Ohio, has developed a process in which small particles of ceramics, graphite, and metal are encapsulated with ***pure metallic coatings.*** Particle size ranges from less than 5μm to more than 5000 μm, depending on the particle size and shape. Coatings include copper, tin, cobalt, zinc, nickel, or a combination of these metals. Successfully coated particles include graphite, boron nitride, silicon carbide, tungsten carbide, zirconia, titanium diboride, and stainless steel. Figure 25-24 includes photomicrographs of some of these particles. The ability to put thick coatings on a variety of particles will easily lead to applications in thermal spray coatings and as reinforcements in metal–matrix composites.

25.2.4.1 Cermets. Cermets are excellent examples of a particulate composite material. They are produced by sintering a mixture of ceramic and metal powders to form a structure that consists of a dispersion of ceramic particles in a continuous metallic matrix. Carbides of tungsten, chromium, and titanium are widely used in combination with cobalt, nickel, and stainless-steel matrices to provide materials with very high hardness for wire-drawing dies, very high corrosion resistance for valves, or very high temperature resistance for turbine parts. Uranium-oxide and boron-carbide cermets suspended in stainless steel find several uses in nuclear fuel elements and control rods.

25.2.4.2 Dispersion-strengthened alloys/composites. Dispersion-strengthened alloys are similar to cermets and to precipitation-hardened alloys. Differences among these three categories lie in their constituent particles. The particles in dispersion-strengthened alloys are smaller (between 0.01 and 0.1μm) and of a lesser volume fraction (at 3% by volume) than those found in cermets. Dispersion-strengthened alloys are formed by the mechanical dispersion of particles, unlike precipitation-hardened alloys, in which compounds are precipitated from the matrix by heat treatment. Cold solder, for example, is a metal powder suspended in a thermoset to provide a metal that is not only hard and strong but also a good conductor of heat and electricity.

High-temperature structural silicides combine the desired qualities of ceramics with the high-temperature ductility and fracture resistance of metals. New patents have been issued in this field that describe the dispersing of silicon carbide throughout a matrix material of molybdenum disilicide, which produces a molybdenum disilicide–alloy matrix composite with eight times the high-temperature strength of molybdenum disilicide by itself. Such a composite has

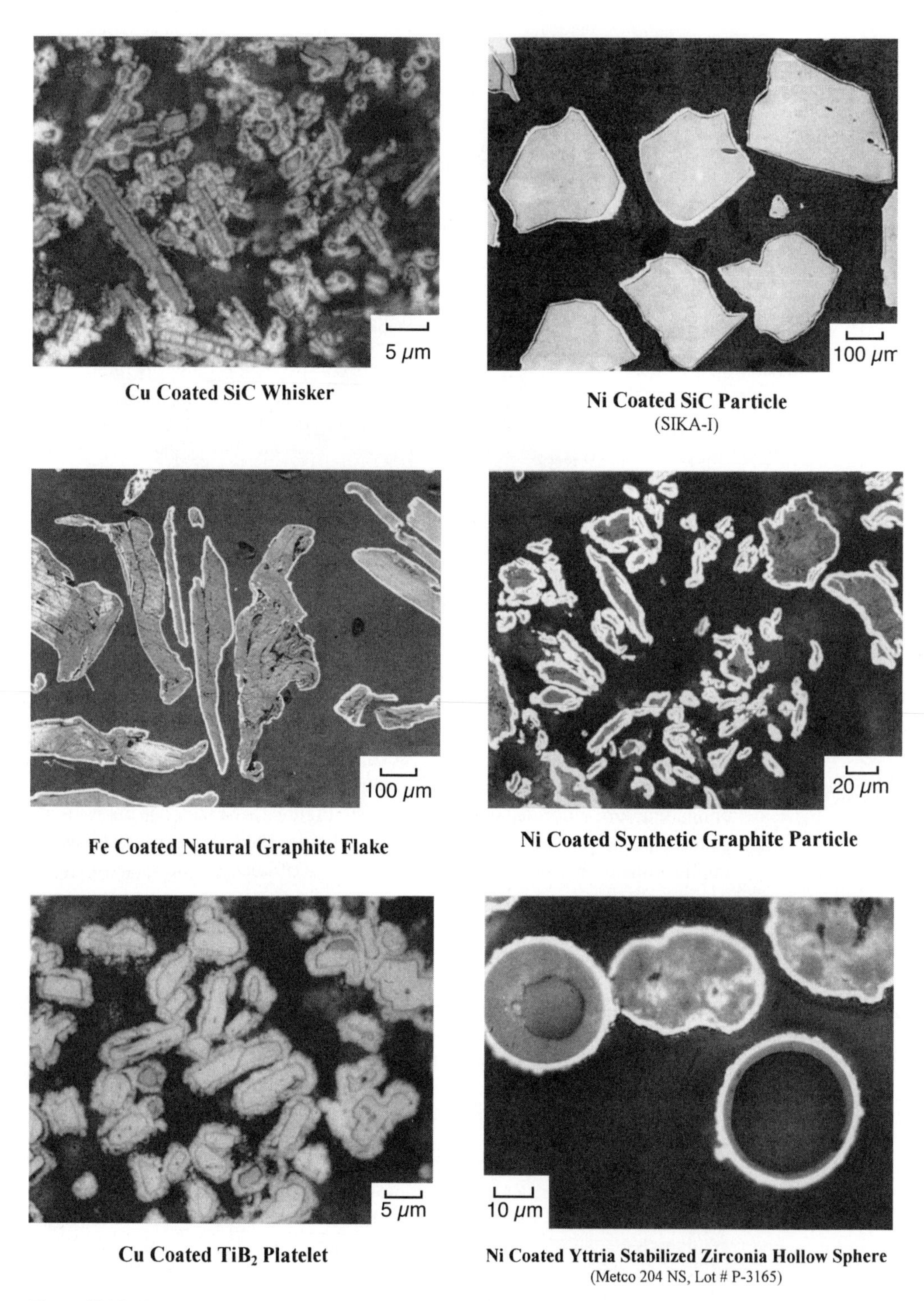

Figure 25-24 Metal-coated particulates used in MMCs can improve mechanical properties by 30–100% or more. Particles can range in size from less than 5 μm to more than 5000 μm. Coatings can be as thick as 50% or more by volume or as thin as 0.5% by volume. Virtually any particle that can be wetted by water, does not react with water, and has a specific gravity greater than 1 is a candidate for coating. (Courtesy of Advanced Ceramics Corp.)

the potential for meeting the demands of advanced high-temperature structural applications in the range of 1200° to 1600°C in oxidizing and aggressive environments. In addition to possessing specialized properties, the composite material is cost-effective in its manufacture compared to other advanced composites, such as silicon carbide ceramics materials.

At this point, it is timely to discuss the differences between ***precipitation hardening (PH)*** and ***dispersion strengthening (DS)***. Most copper alloys achieve their properties through the addition of alloying elements and cold rolling; beryllium copper (BeCu) achieves its properties through alloying elements (Be 1.60–1.79% plus trace amounts of other elements), cold rolling, and a precipitation-hardening heat treatment. The Be precipitates out of a solid solution to form microscopic copper beryllides. Due to their submicron size, very fine distribution, and interaction with the copper matrix, these precipitates are responsible for the unique properties that set apart BeCu from other copper alloys. (See Section 9.5 for further details on precipitation hardening.) PH metal alloys are strengthened by the in situ precipitation of a hard, second-phase particle in a soft, ductile matrix through a multistage heat-treat process. The size and amount of the second-phase particles can be controlled minimally by adjusting the alloy composition and the aging temperature, but certainly not as well as the DS process can. The second-phase particles are *not* stable at high temperatures, which means that as temperatures rise, these particles, including those under service conditions, will coalesce and agglomerate, reducing the number of particles and thus reducing the hardness and strength of the alloy. DS materials (composites) are produced by mechanically combining a hard and strong oxide particle with soft, ductile, metallic matrix particles. Various combinations of these particles can be produced in varying proportions, sizes, and shapes by the powder metallurgy process, yielding a composite material with many desirable properties. The oxide particles chosen for these composite materials are chemically inert. Hence, they do not affect electrical properties. Since they do *not* grow or dissolve in the matrix at high temperatures, they contribute to excellent resistance to softening and to mechanical properties that do not change with changes in service operating temperatures.

The phrase *in situ* means "in its original position." In relation to composites, the term refers to the formation of material components (i.e., the reinforcements, matrices, and coatings of the interfaces) during processing from the elements that make up the material. Because these reinforcements are produced within the material, (i.e., a metal or intermetallic), they are thermodynamically stable in the matrix (they will not degrade or disappear when heated) and they form as single crystals. These reinforcing structures are different from the reinforcements in more conventional composites in the form of fibers or particles deliberately added to the original material. ***In situ composites*** are proving to be more cost-effective and to have higher mechanical strength, particularly at high operating temperatures. The cost savings come from reduction or elimination of processing steps and of the need to add reinforcing components. The many chemical processes that normally occur when a material system is thermally processed, such as phase changes, recrystallization, or nucleation, produce uniform distributions of the reinforcing phases much more efficiently than composites produced by more conventional means—the processing of conventional aluminum alloys versus precipitation-strengthened alloys is an example. Also, it is extremely difficult to handle very fine particles smaller than 1 μm either as reinforcement or matrix contributions. In fact, fine powders, whiskers, or fibers are extremely reactive and present toxicity/health problems. With in situ composites, the reinforcements produced are well within the nanometer range and require no handling. With such a scale of reinforcement comes greater mechanical strength.

Exothermic dispersion (XD) composites, under development since 1983 by Martin Marietta, are good examples of in situ composites. ***XD technology*** deals with composite fabrication in which reinforcing components such as particulates, short fibers, or whiskers develop in situ within metal or intermetallic matrices. These XD materials are produced by exothermic (heat given off) reactions of precursors, and they include metals and intermetallics reinforced by dispersions of intermetallic or ceramic particles or whiskers. One such composite is a titanium aluminide (Ti_3Al) matrix composite made from three starting materials in powder form: Ti, Ti_3Al (titanium aluminide), and TiAl. To these is added titanium diboride (TiB_2), forming TiB_2 dispersions in a matrix of Ti_2Al, which in turn is formed by reactions of titanium, aluminum, and boron. This composite is replacing nickel-based superalloys in several aerospace applications. With a significantly lower density than stainless steel, titanium aluminide composites have greater strength and stiffness at elevated temperatures. For example, the yield strength at 750°C for 17-PH stainless steel is about 345 MPa; for TiAl (Ti–47Al + 7 vol% TiB_2), it is 415 MPa. The modulus at 920 K for steel is 145 GPa, compared to 162 GPa for the in situ composite.

A titanium diboride–alumina composite material with superior high-temperature wear resistance is being produced by a new process called the ***self-propagating high-temperature synthesis (SHS)*** process, developed by the Georgia Institute of Technology. The new chemically based process yields titanium diboride material with smaller (submicron) particles and allows the composite to be produced in final form in molds. The older process using solid-state diffusion left carbides as contaminants in the powder, which accumulated in the grain boundaries. The SHS process mixes powder metal, titanium oxide, and boron oxide; places them in a high-temperature crucible; and ignites them to start a self-sustaining reaction between the components. This produces titanium diboride dispersed within either magnesium oxide or alumina, depending on the choice of metal. Finished products can be hot pressed at 1500°C in less time (2 to 4 hr) than the conventional materials. By varying processing conditions for the titanium diboride–alumina composite, either a dense composite material or a porous form (into which materials such as molten metals can be infiltrated) is produced. Some physical properties for this new material are a melting point of 3000°C, hardness superior to tungsten carbide, thermal conductivity better than cubic boron nitride, fracture toughness greater than silicon nitride, and a good stiffness-to-weight ratio. Applications for this new material are cutting tools, dies, and electrodes.

A "particle-dispersed" steel made by mechanical alloying is said to have a Young's modulus of 265 to 285 GPa, the highest ever achieved for steel. Sumitomo Metal Industries, Japan, reports that it has developed the high levels of stiffness through three stages of processing. The first step was mechanical alloying, in which oxides of iron, carbon, and yttrium are comminuted (pulverized or reduced) to ultrafine particles in a ball mill. The second step was the sealing of the powder in a capsule and hot extruding it. The third step was hot recrystallization, followed by heat treatments. The composition of the steel is Fe, 13–15 Cr, 1–3 Al, 0.5 Y. Its properties are tensile yield strength of, 600 to 900 MPa; ultimate tensile strength of 700 to 900 MPa; and elongation of 10 to 20%. The steel is capable of being fabricated in bar, pipe, or sheet form.

25.2.5 Flake Composites

Flakes of mica or glass in a glass or plastic matrix form a composite material that has a primarily two-dimensional geometry with corresponding strength and stiffness in two dimensions. Flakes tend to pack parallel to and overlap each other, and they provide properties such as decreasing wear, low coefficient of thermal expansion, and increased thermal and electrical conductivity, which are dependent on higher densities in materials. Aluminum in flake form suspended in paint provides good coverage. Silver flakes similarly prepared provide good electrical conductivity.

Metal particles suspended in metal matrices such as lead added to copper and steel alloys lead to increased machinability or reduced bearing wear resistance. These usually brittle metal particles are not dissolved in their metal matrix as in the alloying of metals, but instead, the metal–matrix material is infiltrated around the brittle particles in a liquid-sintering process. When extremely small flakes of aluminum are disbursed into an aluminum metal matrix, a material called ***sintered aluminum powder (SAP)*** is produced. The flakes (0.1 to 0.2 μm thick) are coated with alumina prior to disbursement. (See metal-coated particles in Section 8.6.4.) Nickel alloys can gain high temperature strength by disbursing thoria (ThO_2) particles into a nickel metal matrix, producing ***thoria-disbursed (TD) nickel.***

25.2.6 Filled Composites

A ***filler*** is a material added to another material to alter its physical and mechanical properties significantly or to decrease its cost. Some fillers when added to polymer materials improve their strength by reducing the mobility of the polymer chains, much as an appropriate amount of gravel, as a filler, improves the strength of concrete. Fillers are added to polymer materials for several other reasons (e.g., to improve frictional characteristics, control shrinkage, improve moldability, reduce dielectric properties, lower resistivity, reduce the moisture absorption characteristics, or enhance wear rate). Celluloid and Bakelite are filled polymers, as are electric circuit boards and countertops made from phenolics, a thermosetting resin. See Section 14.2.2 for more details.

Since the accidental discovery early in the twentieth century that natural rubber would accept large amounts of carbon black, thus improving its mechanical properties, great advances in the technology of fillers have continued. Particles of ***carbon black,*** an inexpensive material, are very small (20 to 50 nm) and spherical in shape. When added to rubber, carbon black enhances its tensile strength, toughness, and wear resistance. Its reinforcing effect on such properties is the result of its particularly good adhesive bonding with the rubber molecules. Additional reinforcement effects are attributable to the uniformly distributed particles acting as barriers to plastic deformation. Automobile tires (see Sections 6.4.2 and 6.4.3) may contain up to 30 vol % of carbon black. Just as with multiphase microstructures such as steel, whose structures (micro) can be changed by the relative amounts of the phases present, the phase distribution, and the size of the individual grains making up the phases, the properties of other materials can be changed by the addition of fillers. Glass fibers, for example, are added as fillers to plastics. Not only the type of filler but its size, shape (e.g., spheres), and distribution all play an important role in determining the desired properties of the filled composite.

25.2.7 Hybrid Composites

Hybrid composites, which combine two or more different fibers in a common matrix, greatly expand the range of properties that can be achieved with advanced composites. They can cost less than materials reinforced only with graphite or boron. The common matrix can be either a thermoset or a thermoplastic. Using combinations of continuous as well as chopped fibers also qualifies the material as a hybrid.

There are four basic types of hybrids. One is ***interply,*** which consists of plies from two or more different unidirectional composites stacked in a specific sequence. A second is ***intraply,*** which consists of two or more different fibers mixed in the same ply. A third hybrid is ***interply/intraply*** hybrids, in which the plies of interply/intraply hybrids are stacked in a specific sequence. ***Superhybrids*** are resin–matrix composite plies and metal–matrix composite plies stacked in a specific sequence. The curing procedure must be compatible with the matrix resin, which places something of a restraint on using plies with different matrix materials. In the case of superhybrids, the resin matrix and the prepreg tape must have the same curing cycle as the adhesive used to bond the entire composite. The fabrication of hybrids is the same as that for other composites.

The potential number of fiber/resin combinations for hybrids is vast. With two resins and three types of fibers, an unlimited number of different hybrid combinations can be produced. One design problem is how to make the most effective use of each fiber. Graphite fibers have poor impact resistance but high tensile strength and a high modulus of elasticity. Aramid fibers have good impact but a low modulus. The combination of the two fibers is a natural where each fiber contributes its best properties. A hybrid containing 50% graphite and 5% aramid reportedly shows flex strength on the order of three times that of a straight aramid composite, and both fibers have coefficients of thermal expansion that are similar; in fact, the CTEs are slightly negative. This minimizes internal thermal stresses. Another combination is graphite or aramid and glass. Both fibers increase the stiffness, but with aramid there is some loss of compressive strength. Another combination is graphite and glass in polyester. The more glass that is used, the less the cost; and the more graphite replaces glass, the higher the mechanical properties. For example, 25% carbon/75% glass fibers gives a modulus of 6.39×10^4 MPa; the reverse of this proportion gives a modulus (in tension) of 12.3×10^4 MPa. Using SMCs and continuous graphite fibers with a typical loading of 30% graphite fiber and 45% chopped glass, the composite's properties replicated those of the closest all-glass system.

The B–1 bomber stabilizer was made of a hybrid composite—the largest hybrid ever produced, having an area of some 45 m^2—made of an aluminum/graphite–epoxy/boron–epoxy hybrid. The saving in weight over the traditional aluminum/titanium structure was about 320%; the stabilizer consisted of 108 parts compared to 270 in the metal version. With fewer parts needed, there are fewer holes to drill, countersink, and inspect, and fewer fasteners, all of which reduced the expense of production. Alcoa has developed a hybrid composite, ARALL™, consisting of layers of aramid/epoxy laminated with layers of aluminum. Compared to an all-metal structure, this hybrid has better fatigue life and higher specific strength and stiffness. It is machined similarly to a solid metal, and the outer layers of aluminum provide a moisture barrier plus lightning protection.

The auto industry will be the biggest user of hybrid composites in the form of hybrid driveshafts made of graphite fibers for stiffening and glass fibers to carry the torsional loads—another example of the intelligent selection of materials to meet the requirements of the application. The sports industry uses graphite/aramid and graphite/boron hybrids. Graphite/boron golf club shafts are more rigid than an all-graphite shaft. A project in Australia has taken a new approach to composite bridge design. The bridge, 10 meters long, is based on a hybrid beam concept that combines the high compression capacity of plain concrete with the high-strength/low-weight characteristics of fiber composites. With slender concrete beams, almost all of the load bearing takes place in the top 25% of the beam, which is under compression. The lower 75% of the concrete is under tension and contributes very little to the load-carrying capacity of the beam even though this portion of the beam contributes greatly to the beam's weight. The incorporation of reinforcing bars (r-bars) made of steel, glass fibers, or carbon composites into the lower portion of the beam does little to reduce the weight of the bridge. The new design uses concrete and fiber-reinforced plastic (FRP) composite beams that weigh one-third of a traditional concrete beam. Seven concrete / FRP beams are bonded together forming a section some 2.5 meters wide, which is reinforced with a strong composite laminate in the transverse direction along the bottom of the beams to provide extra stiffness. This section of beams can be trucked to the building site.

This brief description of bridge design has been included here to emphasize the importance of learning about the many new materials that are now available to designers in all areas of endeavor. In this case, the new materials were fiber-reinforced plastic composite materials and the some type of adhesive material to form a strong bond between different materials. These and many more advanced materials are discussed later in this text along with their abilities to withstand the many varied forces that they have been selected and designed to resist. Certain QC/QA issues with composites involving defects, such as hidden voids and delamination of composite layers, remain major concerns because composites are becoming more widely accepted in buildings, roads, and other construction as replacements for traditional structural materials.

In the area of high-temperature materials, where composites are beginning to replace metals, a final example illustrates the ***intelligent design*** of hybrid composites. It must be remembered that the polymer–resin matrix is the determiner of the high-temperature performance of the composite. Carbon- and glass-fiber-reinforced amorphous composites lose their flexural strength as they approach their glass transitional temperatures (T_g). Any exposure to temperatures above T_g, even for a brief moment, could result in a catastrophic failure. Despite this, amorphous thermoplastic–resin composites are stronger than crystalline-resin composites near their T_g values; crystalline–resin composites retain measurable strength above these temperatures. By using a hybrid matrix blend of resins (in which one resin has a higher T_g value), high-temperature properties can be raised enough to meet the application requirements for a particular service temperature. Refer back to Figure 25-18 for micrograph analysis of heat-aged specimens.

25.2.8 Smart Composite Materials

As explained in Module 1, materials engineers and designers seek to mimic biological organisms and to place control and feedback in materials and structures. The infinite variety of structures possible with hybrid composites led to the opportunity to design ***smart composites*** to fulfill many goals. Such materials are currently available, with more varieties to follow.

Hybrid PMCs incorporating ***shape-memory alloys (SMAs),*** such as Nitinol, give the designer more choice in achieving functionality. As a load-bearing armor material, the PMC with embedded Nitinol improves low-velocity impact and perforation resistance compared to monolithic PMC. Nitinol's superplastic nature allows it to deform up to a strain of 7% with fully reversible recovery when the load is removed, thus allowing for greater impacts on the host material. Hybrid composites designed for structural and armor application include various compositions of tough polyethylene, Kevlar, nylon, glass, aluminum, and graphite using the Nitinol SMA to improve strain and impact resistance. The Pause & Ponder section for Unit 1 discussed morphing aircraft and showed a photo of an actuator consisting of ***shape-memory alloy hybrid composite*** with embedded Nitinol ribbons changing the shape. After layup of composite laminations and curing, Nitinol ribbons bond in precise locations allowing the shape of the flexible structure to be manipulated for structural shape control, active noise control, and vibration control of flexible structures.

Figure 25-25 shows one of the evolving group of shape-memory polymers (SMP) contributing to smart composites. A polymer network forms by physical or chemical cross-linking and the cross-links determine the permanent shape. Temperature changes and loading trigger shape changes, as illustrated in Figure 25-26. The tubular part can be compressed and rolled up for storage, as in NASA's *Orbiter* storage bay as a component for constructing orbiting

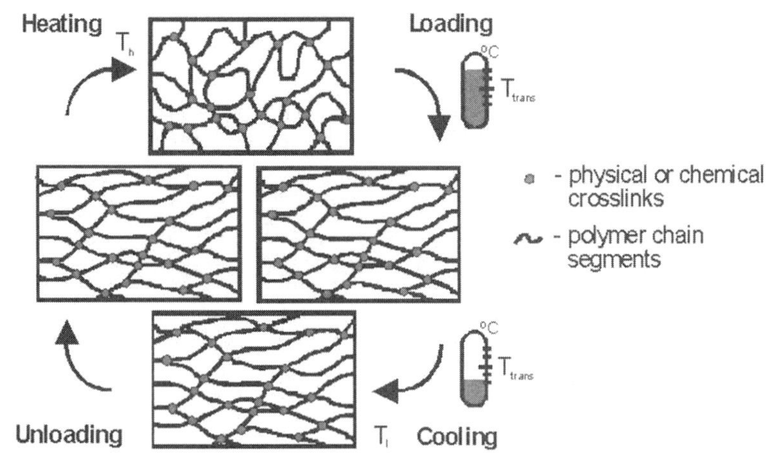

Figure 25-25 Shape-memory polymers (SMP) will change shape based on T_g or other thermal transitions, and pressure. (Cornerstone Research Group, Inc.)

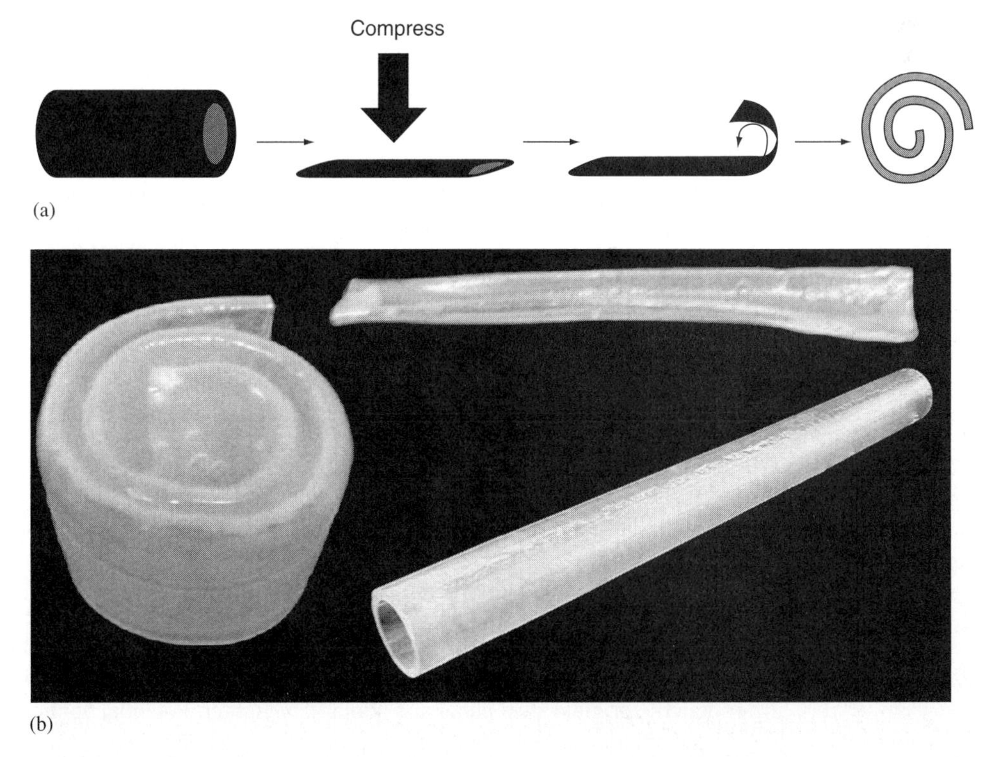

Figure 25-26 (a) Schematic of SMP tube compressed and rolled up for storage; (b) actual part. (Cornerstone Research Group, Inc.)

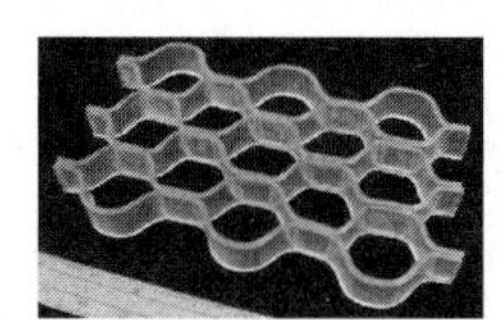
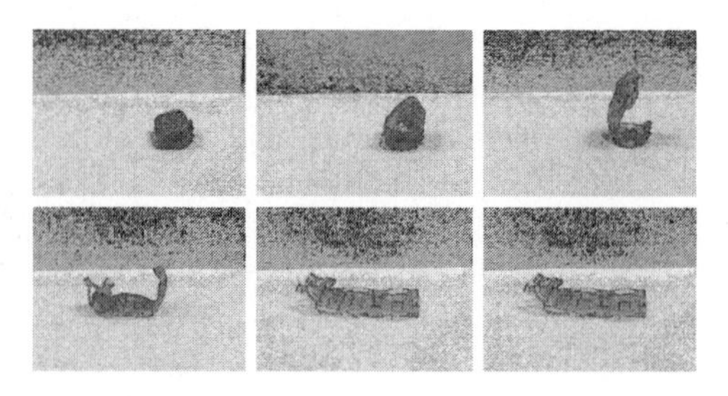

Figure 25-27 SMP for a honeycomb deployable support structure. Photo sequence shows compression-molded SMP rolled, then with temperature and pressure changes deploying back into its original shape. (Cornerstone Research Group, Inc.)

structures. Figure 25-27 shows use of a SMP as a deployable support structure (more complex then the tube in Figure 26–26). The compression-molded SMP has been compressed and rolled up; when removed from storage the honeycomb part will deploy back into its original shape with the application of a change of temperature and pressure.

NASA Technology Center—*http://www.teccenter.org/*
NASA Langley Research Center—*http://www.larc.nasa.gov/*
Cornerstone Research Group, Inc.—*www.crgrp.net*

25.2.9 Foam Composite Materials

To make a ***foam*** or ***cellular plastic*** (both open and closed cell), a foaming agent is added to the resin stream and mixed in the impingement chamber (in-line mixer). Foams are of two types—***rigid foams*** whose walls are stiff, and ***flexible foams*** whose walls will collapse when pressed. Hollow glass spheres (microballoons) are used to produce ***syntactic foams.*** The glass spheres have a diameter of 0.03 mm. Cellular plastics are used in flotation devices, noise abatement, and where thermal insulation is required. A stained and deeply grained composite door (carbon or fiberglass) can now replace a real wood door. Unlike real wood doors, these doors are virtually maintenance free, as they will not crack, warp, swell, or chip. They come with a 25-year warranty. With rigid foam between the indoor and outdoor composite skins, the insulation level is about six times higher than solid wood. Decorative glass in the door is usually triple pane and tempered. Super-secure glass is an option with a lifetime never-break warranty. New stainable, carbon composite doors offer advantages even over fiberglass-skin doors. Carbon expands less than fiberglass with temperature changes so there is less thermal bow. Carbon skins also require less forming pressure during molding so the graining remains more true. Hollow glass microspheres are being used to produce new low-density SMC body panels for today's automobiles. The panels have specific gravities of 1.3 and 1.5 and a thickness of 2.5 mm.

Metal foams are metallic, cellular materials that have a high porosity fraction, typically ranging from 40 to 90 vol %. With their high stiffness and low specific weight, they are used in construction, packaging, noise and vibration damping situations, and filtering. They are considered by many to be a new class of engineering materials. A new powder method for fabricating these foams makes it possible to build metallic foam parts that have complex geometry. Sandwich structures composed of a metallic foam core and metallic face sheets can be produced. Common aluminum alloys, 2XXX and 6XXX, Al-Si casting alloys, and pure aluminum are used for foaming. Tin, zinc, lead, bronze, and steel can be foamed.

Foamed materials are capable of absorbing large amounts of energy at relatively low stress. Automotive applications for these stiff, light structures made of aluminum foam and foam sandwich panels are hood and trunk lids. A foam sandwich panel is about eight times stiffer than the original steel panel and about 25% lighter. Potential applications are bumpers, pillars, and other locations where impact protection is required. Foam structures could replace honeycomb structures based on their reduced cost, isotropy, and fire resistance. Lead and

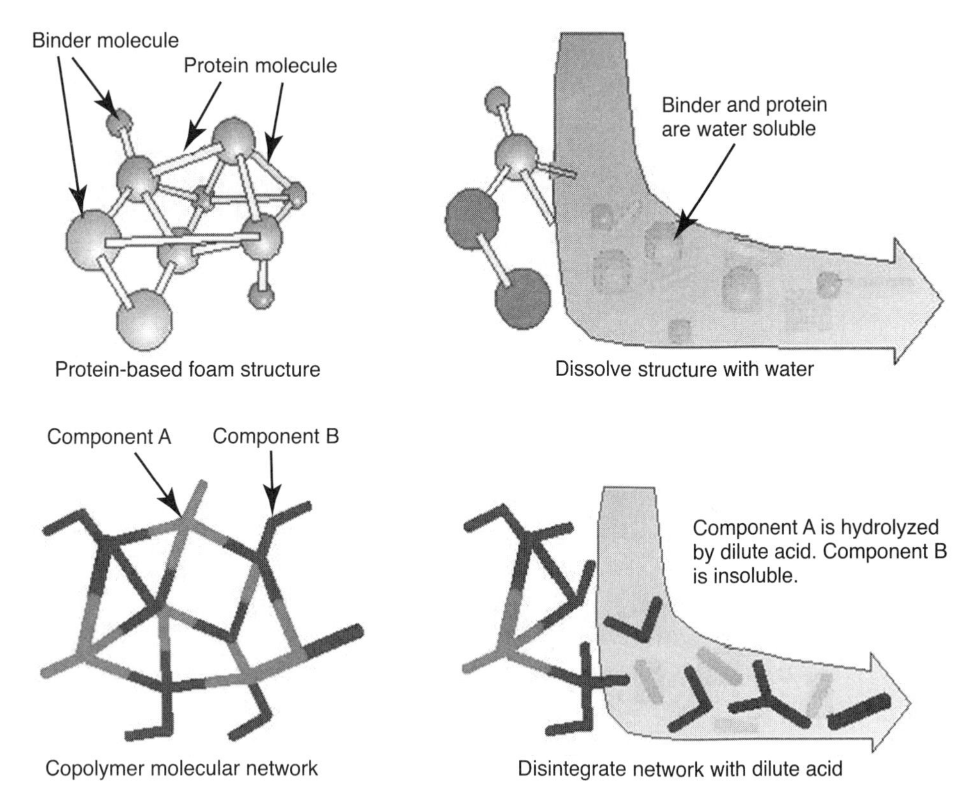

Figure 25-28 Hollow core composite with soluble structural foam. Top: protein-based foam technology. Bottom: polymer-based foam technology. (Cornerstone Research Group, Inc.)

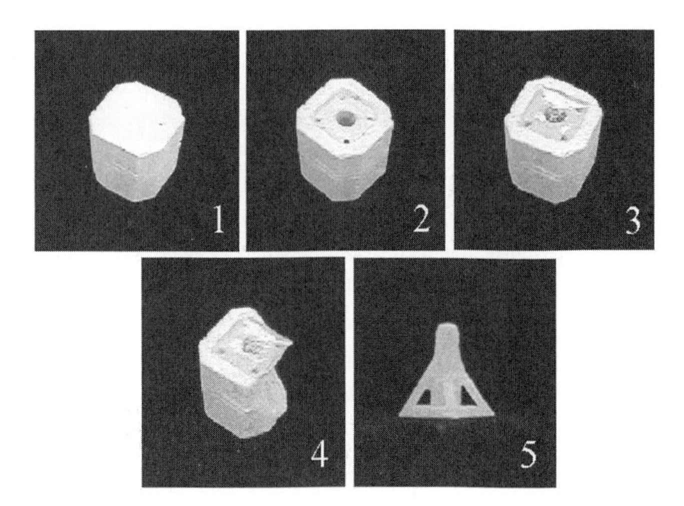

Figure 25-29 Hollow core composite with soluble structural foam stages in dissolving foam to product final shape. (Cornerstone Research Group, Inc.)

nickel foams may even find use in batteries. Open-cell foams would be excellent for several applications such as heat exchangers and filters. Nickel and titanium foams should readily find applications in the aerospace and biomedical fields.

Figure 25-28 depicts protein-based and polymer-based foam technology. The protein-based resin uses water to dissolve the foam into a final shape; the polymer-based foam is dissolved by a dilute acid (component A dissolves leaving the shape of component B), as seen in Figure 25-29. Which would be the greener product?

SELF-ASSESSMENT

25-1. Derive the inverse rule of mixtures in terms of modulus and volume fraction as stated in Section 25.1.1.

25-2. Solve for $E(c)$ using the inverse rule of mixtures equation in Section 25.1.1.

25-3. Calculate the specific strength of S-glass that has a tensile strength of 4.8 GN/m^2 and a weight density of 24.4 kN/m^3.

25-4. Sketch a Venn diagram that shows how sets of the three family of materials groups intersect with reinforcers to form the wide array of advanced materials.

25-5. Sketch a typical fiber-reinforced composite and label the following: matrix, reinforcing fiber, interfaces, interphases.

25-6. Fibers such as rayon that are pyrolized to drive off organic elements, leaving a very stiff, high-strength carbon fiber are

a. E-glass **b.** Kevlar **c.** Borsic **d.** Graphite

25-7. The trade name for aramid fiber, which is used for bulletproof vests and advanced composites for aerospace applications, is

a. E-glass **b.** Kevlar **c.** Borsic **d.** Graphite

25-8. Directional properties in fiber composites are achieved mainly due to

a. Fiber orientation **b.** Laminate plane

c. Interphase **d.** Prepregs

25-9. Air pockets that reduce the strength of the composite structure are called

a. Voids **b.** Laminates **c.** Particles **d.** Honeycomb

25-10. A hollow core material that provides stiffening but very little weight is called

a. Voids **b.** Laminates **c.** Particles **d.** Honeycomb

25-11. One advantage of composites over monolithic cast materials is

a. Low material cost **b.** Anisotropic properties **c.** Isotropic properties

25-12. An advantage of reinforced plastic composites over sheet steel is

a. Lower material cost **b.** Common fabrication methods

c. Isotropic properties **d.** Flame resistance

25-13. The reinforcing effect of carbon black, used as an inexpensive reinforcing filler in tires, is the result of the __________ of its particles.

a. Large size **b.** Irregular shape **c.** Uniform distribution

25-14. Electrically conducting composites use components such as

a. Metal matrix **b.** Polymer fibers **c.** Graphite fibers

25-15. CFCCs for near-term applications will perform in the temperature range of

a. 250° to 500°C **b.** 500° to 1000°C **c.** 1250° to 2000°C

25-16. In precipitation-hardened alloys, what is the state of the second-phase particles at high temperatures?

a. Stable **b.** Unstable **c.** Do not grow or dissolve

REFERENCES & RELATED READINGS

ABRAMOWITZ, HARVEY. "Determining Pull-Out Depth of Composite Reinforcing Bars in Concrete," *2003 National Educators' Workshop,* NASA Langley, October 20, 2003.

ASM INTERNATIONAL. *Engineered Materials Handbook*, Volume 1, *Composites*. Materials Park, OH: ASM International, 1987.

BITTENCE, J. C., ED. *Engineering Plastics and Composites*. Materials Park, OH: ASM International, 1990.

EASTERLING, KEN. *Tomorrow's Materials*. Dorchester, England: Henry Ling Limited, Dorset Press, 1988.

GORDON, J. E. *The New Science of Strong Materials*. New York: Penguin, 1976.

KUMAR, K. S., ET AL. "XD Titanium Aluminide Composites," *Advanced Materials & Processes*, Vol. 147, No. 4, April 1995, pp. 35–38.

LANGLEY RESEARCH CENTER. *Research and Technology*. NASA Technical Memorandum 4243. Hampton, VA: NASA Langley Research Center, 1993.

LEE, ARTHUR K., LUIS E. SANCHEZ-CALDERA, S. TURKER OKTAY, AND NAM P. SUH. "Liquid–Metal Mixing Process Tailors MMC Microstructures," *Advanced Materials & Processes*, Vol. 142, No. 2, 1992.

LINDSAY, KAREN. "FRP Rebar: The Next Generation," *CI on Composites*, April/May 1995, p. 7.

LUCE, S. *Introduction to Composite Technology*. Dearborn, MI: Society of Manufacturing Engineers, 1988.

PAINE, J. S. N. "Enhanced Damage Resistant Composites Incorporating Shape Memory Alloys." 4th Technical Conference, New Century Partnerships for Material Systems. Blacksburg, VA, April 23–25, 1995.

POSTON, IRVIN E. "Recycling of Automotive Polymer Composites," Society of Plastics Engineers Recycling Meeting, Detroit, March 9, 1992.

SUMITOMO METALS INDUSTRIES, LTD. "Particle-Dispersed Steel Has Highest Modulus," *Advanced Materials & Processes*, July 1994, p. 7.

"Synthetic Skin," *Popular Mechanics*. January 1944, p. 14.

Periodicals

Advanced Composites
Automotive Composites
CI on Composites
CDA-Composites Design & Application
Composites in Manufacturing
Composites Technology
High Performance Composites
Journal of Composite Materials
Journal of Composite Technology and Research
Journal of Materials Research
Journal of Reinforced Plastics
*NASA Tech Briefs**

*CD-Rom Experiments in Materials Science, Engineering and Technology, 2nd ed. 0–13–030534–0. Order from *www.prenhall.com*

Module

26

Composite Processing, Fastening, Machining, Repair & Evaluation

After studying this module, you should be able to do the following:

26–1. Use sketches as aids in describing processing techniques for advanced composites.

26–2. Describe the relationship among methods of fabricating composites and the properties that develop from the processes.

26–3. Explain the factors in machining and repairing advanced composites.

26–4. Explain the uses of advanced composites in automobiles and issues concerning recycling composites.

26–5. Use composite databases to determine the applications for select composites. Describe NDE techniques for evaluating composite designs and manufacture.

26–6. Use the website addresses denoted by the http://www EMT symbol to explore new developments and evolving concepts related to composite materials.*

*The http://www EMT symbol found throughout the book will link you to Internet sites related to topics being covered. The dynamic nature of the Web brings frequent changes, so some of these sites, although available at the time of writing, may not be up now.

26.1 COMPOSITE PROCESSING

Although ordinarily not a part of the study of materials, the numerous manufacturing processes involved in the production of composites are intimately connected with the materials, because design affects manufacturing and the particular process affects the properties of the final product. Because of the one-piece forming capability of composites, in which the material is made simultaneously as the end product is manufactured, a brief mention of the major manufacturing processes as a primary factor in evaluating the costs is justified. The accomplishment of near-net-shape (NNS) components with composite manufacture epitomizes this technological development of recent years.

Unlike manufacturing methods for metals, the processes related to the manufacturing of composite materials have limitations based on shape, microstructure, and materials. The basic processes for converting the many polymeric materials into solid shapes include molding, casting, thermoforming, and extrusion. Secondary processes are concerned with finishing, coating, and machining the solid shapes and will be discussed in less detail. Molding processes shape or mold the product by some force other than gravity or atmospheric pressure. Heat and pressure are normally used depending on whether a thermoplastic (TP) or thermosetting (TS) resin is being used as a matrix material. The major processes to be discussed briefly are hand layup, compression molding, injection molding, filament winding, and pultrusion.

Two stages in the processing of most fiber composites are the ***layup*** or combining of the reinforcement and matrix materials and the molding or curing stage. ***Curing*** is the drying or polymerization of the resinous matrix to form a permanent bond between the fibers and the matrix material. It occurs unaided, as with contact molding, or by the application of heat and/or pressure using vacuum bags, autoclaves, pressure bags, or conventional metal stamping machinery (see Figure 26-1).

26.1.1 Contact Molding, Open Molding, and Lamination Molding Processes

Hand layup is the simplest of the processing methods. Using a single, inexpensive, open mold, reinforcement mat or fabric is placed in the mold and saturated with resin by hand. Layers are built up to the desired thickness to form a laminate that is cured at room temperature. This form of curing without the application of heat is called ***contact molding.*** Spray-up molding using a single mold combines short lengths of reinforcing fibers and resin in a spray gun, which deposits them simultaneously on the mold surface. Contact molding follows. Hand layup is a general term to describe a number of procedures using a single mold and a gel coat, followed by the placement of various layers or reinforcements. By using prepreg materials, the careful placement of the fibers in the mold can be ensured. Additional layers can be placed on top of previous layers; the prepregs are generally tacky so that the layers stick to each other. No force or pressure is used to form the composite shape. Hence, it is not considered a molding operation or technique but, rather, contact molding or open molding.

Open-mold processes for shaping FRP laminates use a single-mold surface to produce laminated FRP structures. Open-mold processes include hand layup, spray-up, bag molding, and automated tape-laying machines. These methods differ mainly in the manner in which the laminations are applied to the mold. Hand layup, the oldest and most labor-intensive method, consists of five steps that are also common to the other methods: (1) application of the mold-releasing agent, (2) application of the thin gel coating to the mold, (3) application of the dry fiber reinforcement and impregnation of the fiber with resin through the use of hand rolling, (4) curing, and (5) removal of the fully hardened part from the mold. In step 3, prepregs may be used to more closely control the placement of fibers and gain greater efficiency in adding the laminations. Molds can be made of metal, plaster, GFRP, or other materials depending on economics and other technical factors. Spray-up reduces layup time by spraying liquid resin and chopped fibers onto an open mold to build up successive laminations. The U.S. automotive aftermarket parts and accessories industries use open molding, primarily hand layup and spray-up methods, to produce low-volume products with which vehicle owners can affordably

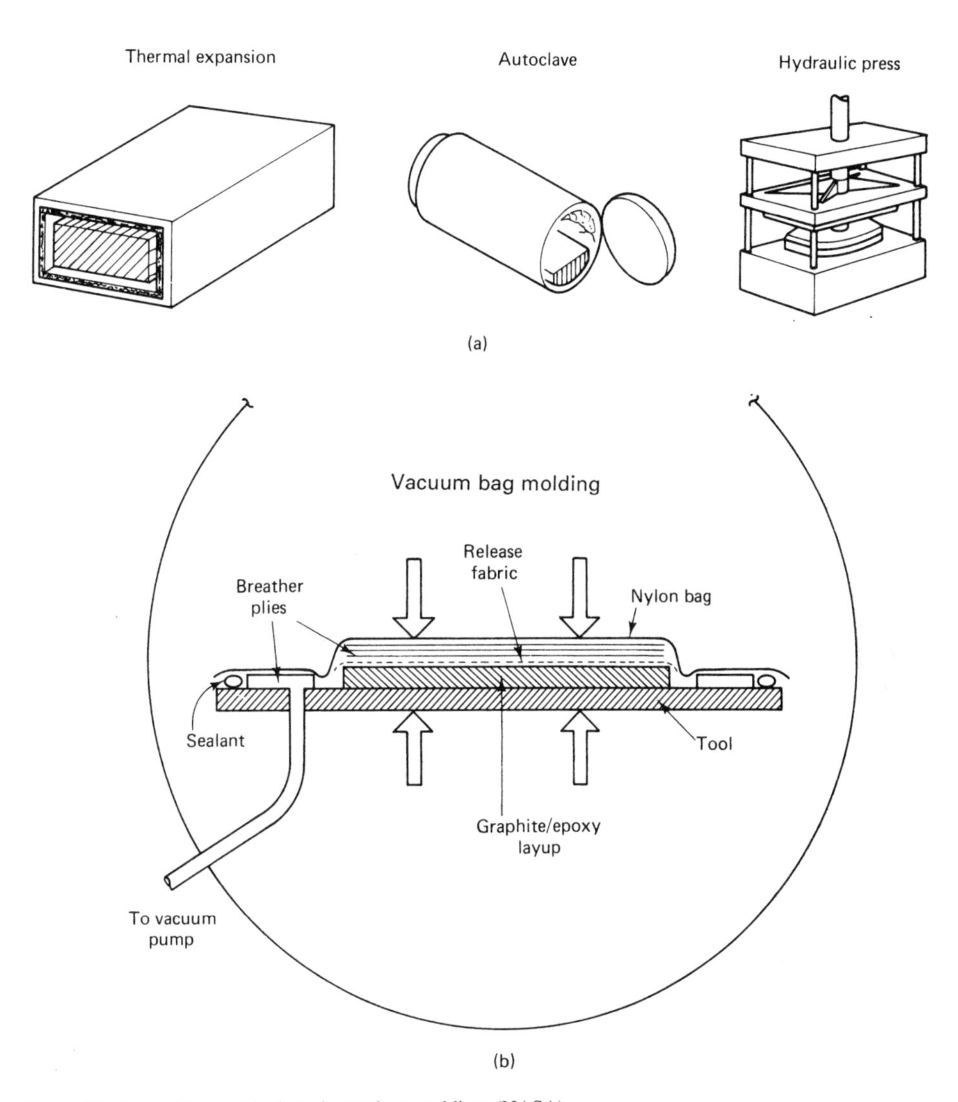

Figure 26-1 Molding methods and autoclave molding. (NASA)

customize their cars, trucks, or vans. Examples of such products are rear deck spoilers (also known as wings), body fairing kits, fender flairs, hoods with aerodynamic air-induction scoops, fenders, bumpers, trunks, doors, running boards, and truck caps and bed covers. Plaster casts, flexible rubber molds, epoxy resins, and composite plates that allow for varying thicknesses without adding weight have also invaded the tradition-bound area of violin making. The plates are layered sandwiches of wood veneer, carbon fiber, and aeronautical foam.

26.1.2 Closed Molding

Compression molding or ***matched metal die molding*** is used to mass-produce high-strength parts that are limited in size by the press equipment. Pressures used in this equipment are about 1.72 MPa and temperatures are about 120°C. The materials are in the form of SMCs or BMCs. The mold consists of two sections that open and close during each molding cycle. Tooling cost is a determining factor due to the complex equipment required. These processes are employed when a good finish is required on all surfaces, when high production rates are involved with smaller tolerances, and where more complex three-dimensional shapes are involved.

Closed molding can be divided into (1) compression molding, also known as matched die molding, (2) injection molding, and (3) transfer molding. Compression molding or matched die molding uses TS resins placed in a cavity of a matched mold in the open position.

The mold is closed, pressure is applied, the material is heated to cause cross-linking, and when complete, the mold is opened. This process is aptly described as waffle making. Parts that have hollow sections are very difficult to mold using this process; however, longer lengths of chopped fibers (approaching 100 mm) can be used. If fibers are used as reinforcement and are to remain in place, a preform can be made and positioned in the mold. The resin is then poured on top, and the pressure from the mold closing distributes the resin.

Preforms are granules pressed together into a particular shape for easy handling. Granules are TS, such as a phenolic that liquefies when heated and then cures. A "green compact," a name given to a ceramic part before it is fired, is an example of a preform. TS materials may be removed from the mold while hot since the curing creates a dimensionally stable form. Carbon-fiber reinforcement can be used in this molding operation. Compared to injection molding, compression molding uses simplified tools and can accommodate high fiber volume fractions and long fiber lengths, thus making parts with better physical and mechanical properties. Compression molding also is suitable for the production of parts with large surface areas. More common is the use of chopped fibers that have been combined with a resin and a filler to form a molding compound. This compound is then used as a charge material. Molding compounds are materials in which the fibers have been combined with the resin and a filler such as $CaCO_3$.

In addition to preforms, SMC, a sheet molding compound, is another starting material for compression molding. Figure 26-2 shows a typical formulation for a SMC. The process begins with the making of a stack of SMC plies that are die-cut to the required size and shape and placed in a preheated mold. This stack is called the *charge*. The top mold is lowered slowly onto the charge, both the top and bottom molds are heated, and the rising temperature reduces the viscosity of the charge. As the mold continues to close, the SMC flows toward the extremities, forcing air out of the mold cavity. Curing times vary from 1 minute to several minutes. The top mold is then opened and the part is removed with the aid of ejector pins. The cycle time is long compared to injection molding since the matrix material must be cooled in the mold below its melting temperature before removal. Compression-molded SMC composites find many uses in the automobile industry. High-strength automobile road wheels formerly made of SAE 1020 steel have been compression molded using SMC 50 polyester composite. Figure 26-3a is a graph of this composite's strength versus use temperatures ranging from 250° to 125°C. Weight reductions of up to 50% over standard steel have been realized. Due to the composite's uniform weight distribution, the need for the balancing of SMC wheels is nearly eliminated, and its corrosion resistance tops that of steel wheels.

Material	Percent	Purpose
Calcium carbonate	48%	Filler
Fiberglass (1-in. lengths)	28%	Strength & physical properties
Polyester resin	13%	Plastic binder
Low-profile additive	9%	Shrinkage compensation & class "A" surface
Additives	2%	Catalyst, mold release, pigment, thickening agent

Figure 26-2 Typical SMC formulation. The presence of fiberglass as a reinforcement indicates that this is a reinforced SMC. High-strength reinforced composites have a larger percentage of fiberglass reinforcement than conventional SMCs. High strength means ability to withstand high impacts (dent resistance). The low-profile additive is present to ensure a smooth, paintable finish for automotive panels.

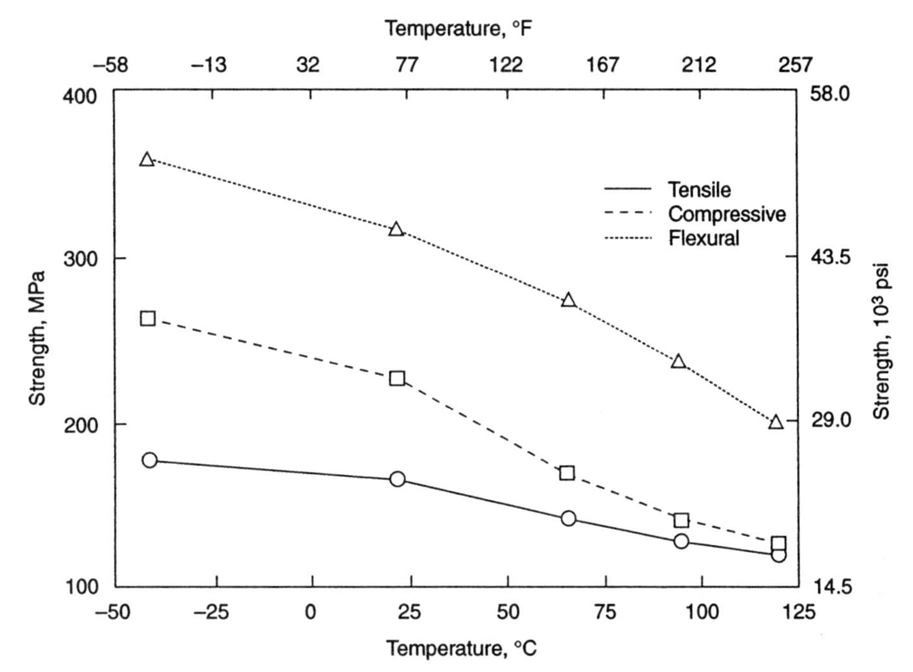

Tensile, compressive and flexural strengths vs temperature.
(a)

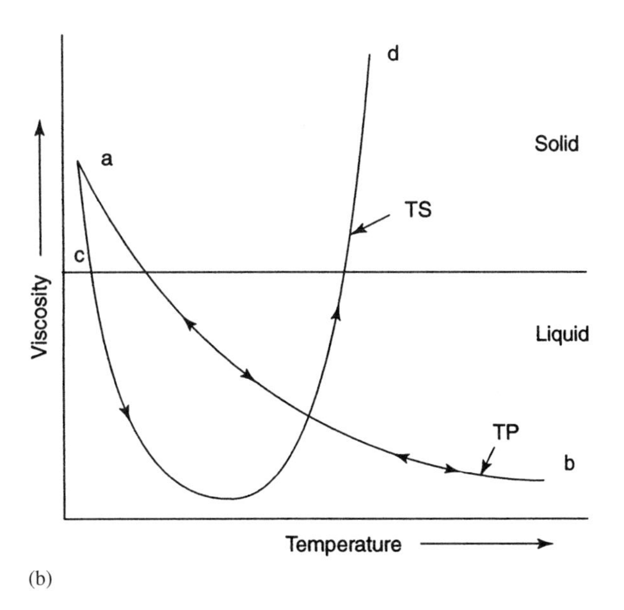

(b)

Figure 26-3 (a) SMCs are thin sheets of fiber precompounded with a thermoset (TS) resin such as polyester or vinyl ester. Low-profile additives are added to compensate for any shrinkage during the cooling period outside the mold, thus assuring a fine finish, as on automobile bodies, that is free from surface defects. SMC R50 polyester composites, which maintain their strength at high temperatures, are finding many automotive applications. (Courtesy of The Budd Co.) (b) A plot of viscosity versus temperature for thermoplastic (TP) and thermosetting (TS) resins. Point *a* on the TP curve represents a resin that is in a solid state at the coldest or room temperature. Point *b* is the same TP resin at its warmest temperature in a definite liquid form. Point *c* represent a TS resin around room temperature at which it could be in either a solid or liquid state. Point *d* represents a TS resin at the warmest temperature in a definite solid state. Note the double arrows on the TP curve, which indicate that these resins can change their form or state depending on temperature; the single arrow direction on the TS curve signifies that as these resins are heated they begin to cure and solidify to a final solid state.

Injection molding (similar to die casting of metals) is used to form molten thermoplastic (TP) and thermosetting (TS) matrix materials in a closed mold, where they are held until the materials solidify. High-volume injection molding is the best example of the manufacturing of TP parts. Modifications to the injection molding equipment allowed the subsequent molding of TS parts, including reinforced TS parts. TS materials are forced into heated molds where they cure (cross-link) into a solid. Conversely, cooler molds are used for TP materials to increase the rate of production. It takes longer to mold and cure a TS material than a TP material, although preheating a TS material before it is placed in the mold reduces the curing time. Both types of materials at or around room temperatures can easily exist as a liquid, as shown in the plot of viscosity versus temperature in Figure 26-3b. This makes the mixing of reinforcements, fillers, colorants, and so forth much easier, with simple stirring or mechanical mixing. Note also that the TP materials have a higher viscosity in the liquid form than TS materials. Filler, reinforcing agents, and other additives reduced to powder or granular form are fed into the injection machine from a hopper. RIM can be used to encapsulate an insert or core. When the core is made of fiberglass, the process is called ***structural reaction injection molding (SRIM).*** Two types of injection machines are the plunger type and the reciprocating screw. Injection molding of metals is discussed in Module 3. See Figure 3-65 for a schematic of an injection molder (screw type).

Reaction injection molding (RIM), the word *reaction* refers to the use of TS polymers in the injection molding process. During the process, two or more TS monomers are forced into an impingement chamber and the mixture is then injected into a mold cavity. Remember that to cure, these resins must be heated, so temperature must be controlled throughout the process. Polyurethane is the most commonly used resin system, and the two-part reaction system begins to cure and form a plastic typically in less than a minute. The two components of polyurethane, polyol and isocyanate, are metered and mixed in-line so that they begin to polymerize or cross-link.

When reinforcement in the form of very short fibers (chopped fibers usually less than 100 mm in length) or particulate such as flakes is added to the mold, a more isotropic solid is produced. To accomplish this, the mold is closed and the resin is injected into it in a process called ***reinforced reaction injection molding (RRIM).***

Mainly due to fiber breakage, there are limits to the strength of the fiber reinforcement that can be achieved using both TP and TS resins. Breakage occurs as the resins with their reinforcing phase are forced through narrow gates and openings in the mold. This restriction limits the degree of reinforcement obtained.

If a filler has an aspect ratio (the ratio of the largest to the smallest dimension) close to 1, then it makes a fairly good reinforcement for a fiber composite part. A ***reinforcement*** might be defined, therefore, as a filler with a ratio greater than 1. A variety of fillers are used with TS resins. Glass filler in the form of microspheres (both hollow and solid) and mineral fillers such as clay, talc mica, and metallic powders used to coat particulates such as aluminum-coated flakes (see metallic-coated particles) are some examples. Some of the fibers used are thermoplastic, glass, aramid, and carbon fibers. Carbon and glass hybrid composites also are produced by these processes.

One of the automotive of parts made by RIM is a glass-reinforced radiator tank for automotive radiators. The plastic tops and bottoms of these radiators must resist temperatures up to 130°C and have chemical and mechanical resistance. Cooling fans for radiators are made of a similar material. FRP composites of both thermosets and thermoplastics continue to find new and growing applications in sporting goods and recreation equipment. Bow hunters are now going to compression-molded composites to gain greater arrow speed. Rollerblade, Inc., the in-line skate market leader, has introduced an innovative children's skate that adjusts over four shoe sizes to accommodate growing feet and thus extend the skate's usable life. The Xtenblade and Xten Plus skates feature upgraded wheels and bearings and hard shell boots with a toe section that adjusts forward as needed. Three injection-molded parts, which provide torsional rigidity, stiffness, and strength, account for 28 % of the glass-filled nylon thermoplastic composites used in the skates. SRAM Corp. (Chicago), a specialist in mountain bike components, uses thermoplastic composites with carbon fibers and/or glass to produce shifters, front and rear derailleurs, braking systems, and related accessories for the $1.7 billion U.S. bicycle and bicycle components market.

26.1.3 Resin Transfer Molding

Resin transfer molding (RTM) differs from RIM in that a reinforcement is molded with the resin; the equipment, however, is basically the same. Both long and short fibers can be used with RTM as well as innovative weaving and knitting techniques that allow for greater latitude in laying out the fibers. This dry reinforcement, cut and shaped, is first placed in a mold cavity. The resin is added to wet out the reinforcement and eliminate air trapped in pores or voids in the preform. The mold is then closed and the resin is cured. This technique allows precise control of the fiber reinforcement in high concentrations and is capable of producing large, integrated sections with complex geometries. A variety of resins and reinforcements are used, particularly low-cost E-glass fibers and mats and woven fabrics. There are numerous variations of RTM, several of which pertain to the application of a pressure differential on the resin flow into the mold. Some of these variations are high-speed resin transfer molding (HSRTM), Seeman process, flexible resin transfer molding (FRTM), continuous resin transfer molding (CRTM), and vacuum-assisted resin transfer molding (VARTM).

Structural resin injection molding (SRIM) uses a preform and mold preparation similar to RTM. Once the mold has been closed, the SRIM resin is rapidly introduced into the mold and reacts quickly to cure fully within a few minutes. This requires the wet out of the preform to occur rapidly and so resins that have low viscosities and are highly reactive are used. The two-part resins must undergo high-pressure impingement mixing to ensure thorough mixing prior to injection.

A new process, ***vacuum-assisted custom resin injection molding (VACRIM)*** contains styrene emissions rather than releasing them into the atmosphere, an example of industry's efforts to reduce the emissions of dangerous solvents (see Section 17.1 for a discussion of VOCs). ***Thixomolding*** is a new one-step process for the production of near-net-shape injection-molded magnesium parts. Current applications include computer cases, digital camera bodies, auto shift cams, and mechanical components of handheld power tools.

General Motors has used the RRIM and SRIM processes to manufacture a pickup truck box that is lightweight, corrosion proof, and virtually indestructible made from high-strength urethane molded with glass fibers. Figure 26-4 shows the composite pickup box assembly with all the major parts broken out. The box's fenders and outer tailgate panel are made of RRIM materials; SRIM materials comprise the one-piece inner panel and the tailgate liner. These two processes decrease the total weight of the truck by about 50 pounds and improve fuel econ-

Figure 26-4 Composite pickup box assembly. The cargo box was fabricated using HD-SRIM, a high-density, structural injection molding process and polyurethane as the matrix resin. Glass fiber mats and glass fiber preforms comprised the reinforcement. The wall thickness is 3 mm. This one-piece inner panel replaced four steel panels, resulting in a reduction in the number of fasteners and easy disassembly and aftermarket repair. (National Composite Center)

	SMC	LD-SMC	RRIM	SRIM
Weight (Specific Gravity)	1.9	1.3	1.3	1.5
Investment (Tooling)	~	~	Lowest	~
Total Program Cost	Lowest	~	~	Highest
Design Flexibility	~	~	Most	Least
Complexity Reduction	Most	Most	Least	~
Heat Resistance (degrees F)	400+	400+	375	350
Dimensional Integrity	Best	Best	~	~
Thermal Expansion	Steel	Steel	Higher	Higher
Corrosion Resistance	High	High	High	High
Dent Resistance	~	~	High	~
Structural Strength	~	~	Lowest	Highest
Stiffness (Modulus)	1.8	1.5	0.3	1.2
Recyclable	Yes	Yes	Yes	Yes
Surface Quality	Class A	Class A	Class A	Poor

(a)

Figure 26-5 (a) Comparison of composite materials. Compression molding uses SMC and low-density SMC as resin materials. RRIM and SRIM are molding processes in which the matrix resin and reinforcements are injected into molds. Note that all materials are high on corrosion resistance and recyclability. Both SMC materials have the same coefficient of expansion as steel, which adds to their advantage when mating composites to a steel structure. This simple comparison reveals that the selection of materials requires an extensive background in materials.

Weight Comparison

Steel	19 lb	Base
SMC (sp. gr. 1.9)	14.5 lb	−4.5 lb
SMC (sp. gr. 1.3)	10 lb	−9.0 lb
Aluminum	10 lb	−9.0 lb

Cost Comparison

	Pc. Cost	Tooling	Amort.	Total Cost	Cost Diff.
Steel	$40.00	$8.9 Mil	$14.85	$54.85	Base
SMC (sp. gr. 1.9)	$45.00	$4.0 Mil	$ 6.65	$51.65	−$ 3.20
SMC (sp. gr. 1.3)	$50.00	$4.0 Mil	$ 6.65	$56.65	+$ 1.80
Aluminum	$60.00	$8.9 Mil	$14.85	$74.85	+$20.00

Note: The above costs are for engineering purposes only and do not represent a cost quotation. Tooling amortization assumes 2-year program at 300,000 units per year.

(b)

Figure 26-5 (b) Cost/weight comparison—small fender. The weight and cost of four candidate materials for the production of small automobile fenders are compared to those of steel. A recent Ford Motor Co. automobile using 100 pounds of SMC components reduced vehicle mass by about 30 pounds versus steel.

omy and payload capability. Stronger than steel, the flexible body panels are dentproof. This structural composite box was used as a test bed for future composite structures. The Silverado 1500 Series truck was the first with the new composite pickup box.

Figure 26-5a compares composite materials and processes for making various automotive components and lists applicable properties. The materials are compared to steel as well as to each other. It must be noted that all do not possess the same properties to the same degree, an observation that points up the fact that with most materials and their manufacturing processes, a compromise based on experience and knowledge is needed. Selecting the best material for any vehicular component involves many steps: Data banks are consulted, models produced, tools developed, and projected costs estimated and analyzed, to name a few. Figure 26-5b is an analysis of producing small fenders for automobiles. Four materials are

Figure 26-6 Ghia Connecta. Ford Motor Company's electrically powered concept vehicle has a carbon and Kevlar fiber composite body, double-skinned glass roof, and a solar panel in the top surface powering the rear roof vent. (Ford Motor Co.)

compared, with steel being the base material. The two SMC materials are low-density materials containing glass microspheres (see Section 25.2.9). If weight were the only factor, then aluminum would win out. If the cost of the components is considered, however, including tooling and materials costs, then aluminum would lose out to SMCs. This example also shows that selection of materials is mostly a compromise, as materials possess neither the same kind of properties nor the same degree of properties. Automakers' concept cars explored a variety of techniques for maximizing use of composites, as seen in Figure 26-6.

26.1.4 Filament Winding and Tape Winding

Filament winding and tape winding, which produce the highest specific strength and glass content by weight of composite parts (fiber loading up to 85% by weight), are generally limited to parts with round, oval, or tapered inner surfaces, although external shapes are unlimited. The continuous glass strand or filament is usually passed through a resin bath prior to winding onto a revolving mandrel. The mechanical as well as other properties of a filament-wound product can be changed by altering the ***wind angle*** (Figure 26-7a), an angle measured between the axis of the mandrel and the lay of the filaments. The tangent of this angle equals the ratio of the circumference of the mandrel and the pitch of the filaments. As the angle increases to 90°, the hoop tensile strength increases and the axial tensile strength decreases. Figure 26-7a shows the angle, the pitch, and the axial and hoop directions. The change in modulus of elasticity in the two principal directions is sketched in Figure 26-7b. Figure 26-8 shows some internal shapes possible with filament winding, along with a few special design and winding problems. Filament winding developed in 1946 under contract with the U.S. Navy's Bureau of Ordnance. The first computer-controlled filament winding machines appeared in the mid–1970s. Filament-wound concrete pipes with diameters up to 20 ft and 65 ft in length competed with concrete as the material of choice for nuclear power stations. Since that time advances in filament-winding machines have included resin content control, vacuum resin injection, 360-degree fiber placement, and up to five axes of motion.

26.1.5 Pultrusion

Brant Goldsworthy, recognized as the father of the reinforced-plastic automobile for his pioneering work on the Corvette in 1953 (Figure 26-9) and other FRP sport cars, is regarded as the father of modern pultrusion, a process he invented in 1950. Pultrusion is well suited to high-volume production of high-fiber-content lineal components with a constant cross section, such as beams, channels, and tubing. Glass rovings, multidirectional glass reinforcements, and a synthetic surfacing veil provide strength and smooth finish to FRP utility poles. Pultruded components with equivalent strength can weigh 50% less than aluminum and 75% less than steel. Most pultruders use liquid thermoset resins but thermoplastic resins are gaining a foothold. Powdered thermoplastic resins are being used in the pultrusion of structural components. Many fiber reinforcements can be combined in the pultrusion process. Fiber content up

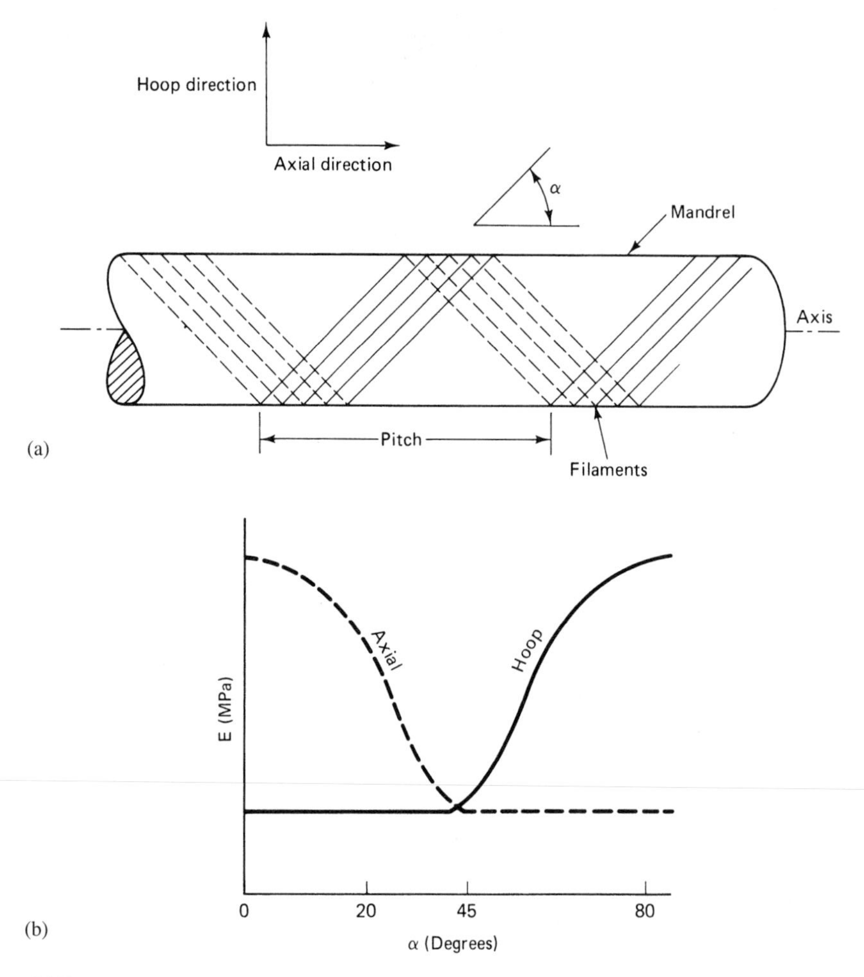

Figure 26-7 (a) Filament winding, showing the wind pitch and the wind angle (α). (b) Wind angle of a filament-wound product.

to 85% by weight is possible, providing high strength and stiffness in the longitudinal direction. Thermoplastics can be readily joined using plastic welding processes, which eliminate adhesives and fasteners. Low-viscosity resin technology by Dow Plastics has solved the problem of the higher viscosity of thermoplastic resins, especially for pultrusion.

 *www.mccleananderson.com*

Pultrusion is a composite fabrication method by which an extremely long, fiber-reinforced, polymer–matrix material can be produced by pulling a fiber or a bundle of continuous filaments through a resin system for impregnation and then through a heated curing die. As the fiber/resin bundle travels through the curing die, polymerization occurs, and a composite structure emerges from the exit end of the die. Figure 26-10a is a profile view of the pultrusion process that produces a graphite-reinforced, epoxy composite T beam for use as a lightweight, high-strength aircraft stiffener. Figure 26-10b is a photograph of a section of the T beam, which can be produced to any desired length. Figure 26-11a is a profile view giving the dimensions of the T beam in inches, and Figure 26-11b shows the fiber orientation used in the T beam. The fiber, supplied in 12-K tows, was sized with a coupling agent for epoxy–resin compatibility. The fiber layup consisted of eight plies oriented in the 0°, 90°, 145°, 245°, 145°, 245°, 90°, 0° directions (in that order). These tailored fiber orientations were obtained by locking two plies in their positions by stitching with a polyester thread; that is, the 0° (longitudinal) and 90° (transverse) fiber orientations were obtained by knitting the two plies together,

THESE INTERNAL SHAPES CAN BE FILAMENT WOUND

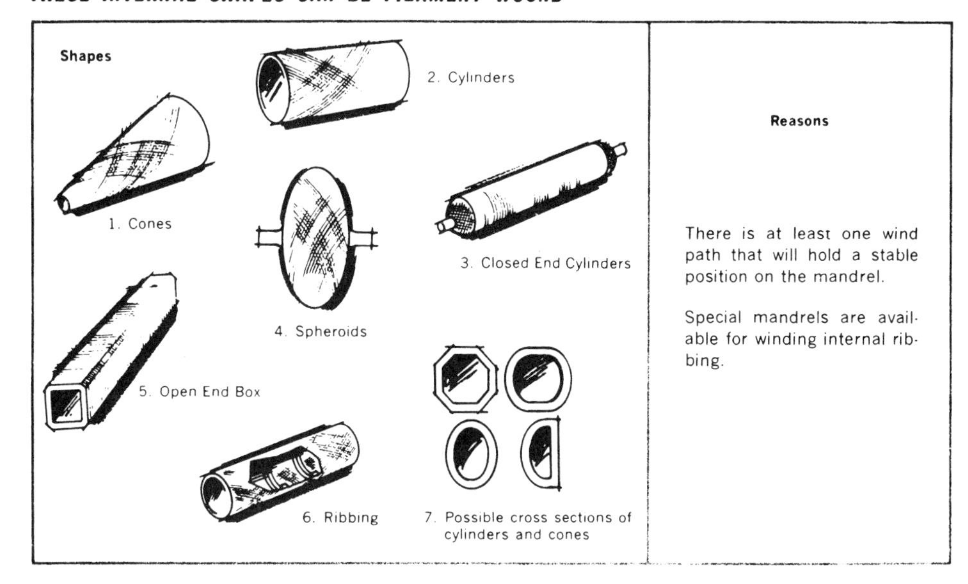

THESE INTERNAL SHAPES CREATE SPECIAL DESIGN AND WINDING PROBLEMS

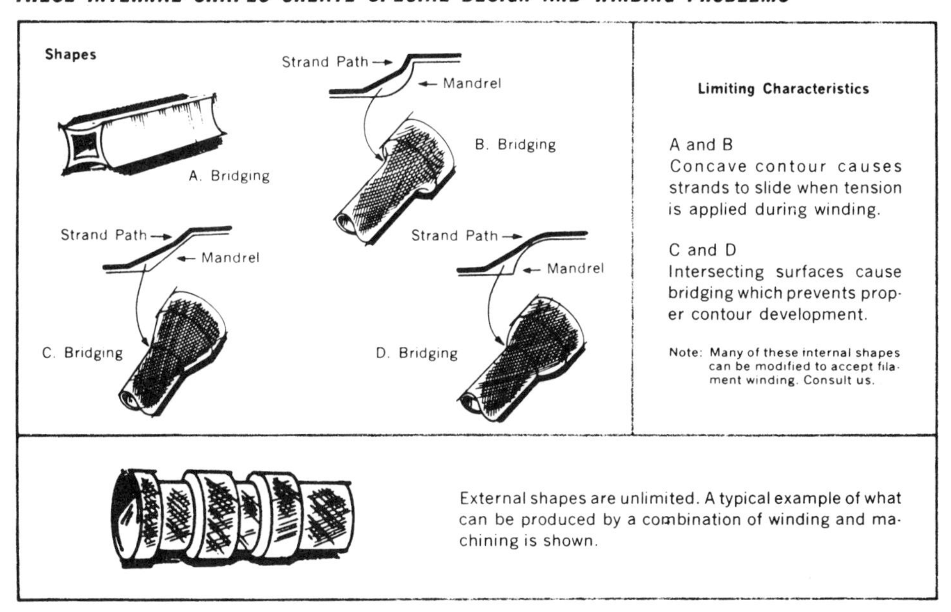

Figure 26-8 Winding problems and some internal shapes of filament-wound products. (Permali, Inc.)

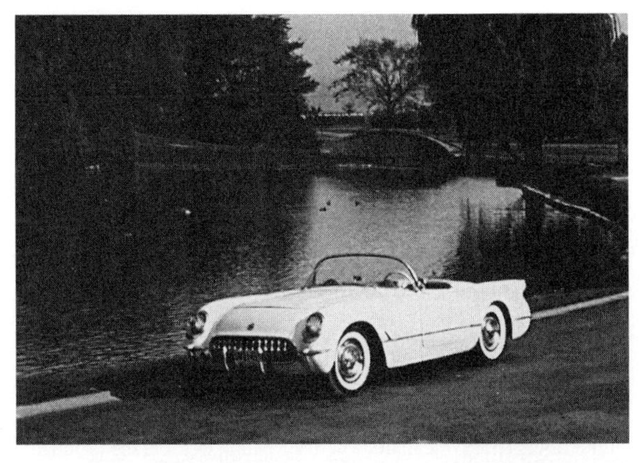

Figure 26-9 Composite body. The 1953 Chevrolet Corvette was the first fiberglass production body. (Morrison Molded Fiber Glass Co.)

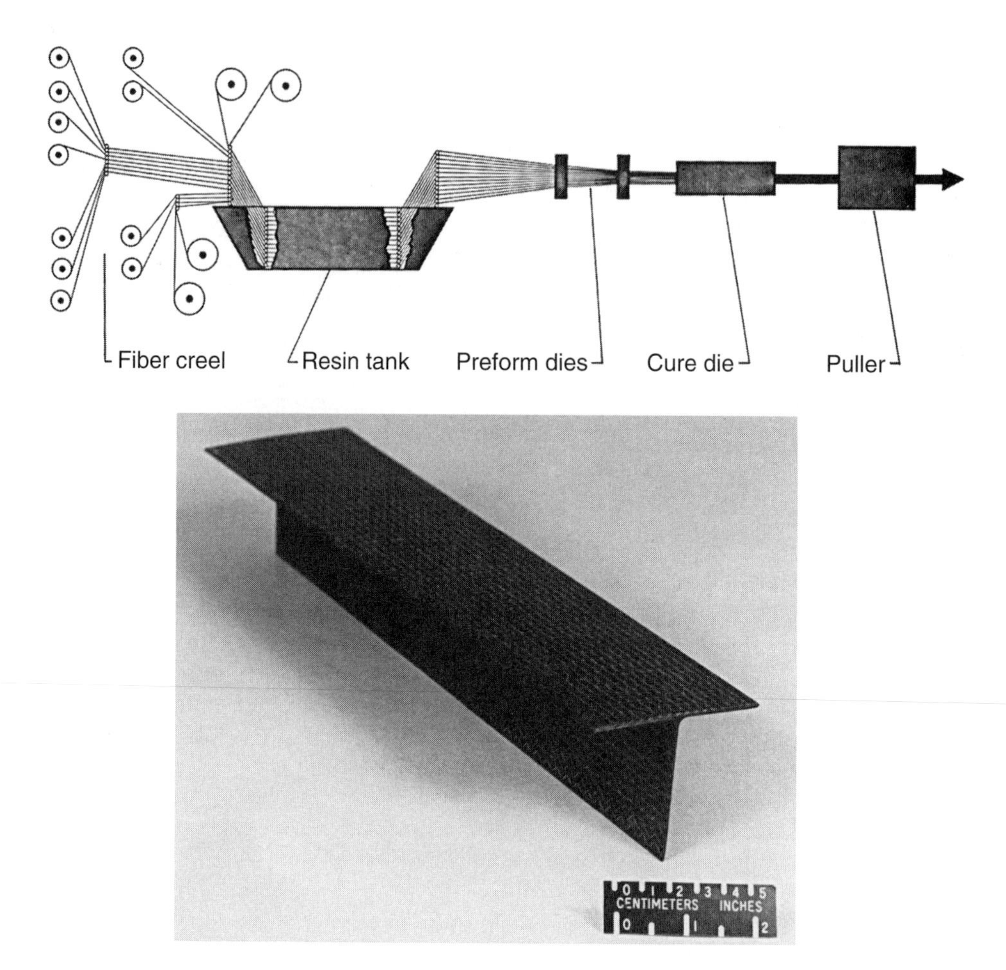

Figure 26-10 A profile view of the pultrusion process. (NASA) (b) Photograph of a portion of a T beam produced by the pultrusion process.

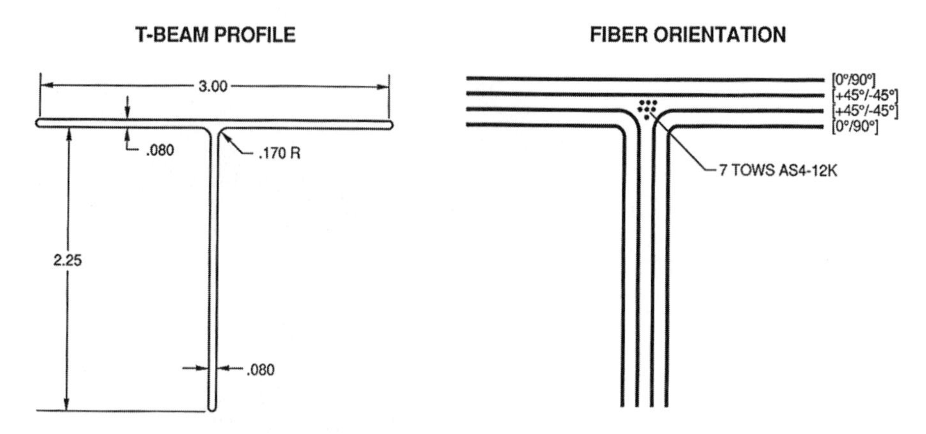

Figure 26-11 T beam profile showing dimensions in inches. A cross section of the T beam showing the placement of the fibers (fiber orientation) in the T beam. (AS4 is the producer's designation for a graphite fiber.)

thus locking them in their respective positions. The fiber volume was 54%. Preforming, preimpregnation, and layup of the plies to form the T beam were performed simultaneously in the 47-inch. resin tank.

This new advance in the pultrusion method illustrates the need to minimize, if not eliminate, the labor-intensive manufacturing processes required for any advanced composites. The knit-locked, tailor-made reinforcement systems that produce tailor-made properties by pultrusion contrasts with some older fabrication techniques that require the manufacture and stowage of preimpregnated materials, laborious layup and fixturing methods, vacuum bagging, autoclave curing, and debagging, which produce composite structures that are limited in size and length to the available autoclaves and drying ovens. Parts being pultruded can be inspected as they are fabricated by ultrasonic techniques to detect the presence of porosity and delaminations. This real-time feedback allows changes to correct or discontinue the process before the production of a large number of defective parts that must be rejected.

26.2 FASTENERS

Current fasteners for composites are not meeting all the requirements of structural designers. Fasteners for composites are needed that will eliminate the problems associated with installing metallic fasteners in nonmetallic structures. Material compatibility is a necessity if problems with corrosion, lightning strikes, and low radar signatures are to be avoided. Fiberlite fasteners made by the Tiodize Company (Figure 26-12) have low creep, water absorption, and thermal expansion and are half the weight of aluminum with equivalent strength. Used to bolt composite structures together, these fasteners come in a variety of composite nuts and bolts with shear strengths of about 30,000 psi. They resist temperatures to 600°F. Machined from composite rod stock, the reinforcement in the rod is a three-dimensional, woven graphite-fiber structure with a polyimide–resin matrix. The cost is equivalent to titanium.

 Fasteners—*http://www.tiodize.com/mainflash.html*

26.3 MACHINING OF COMPOSITE MATERIALS

Machining of composites demands special skill and knowledge: Composites cause severe tool wear through abrasion; dust from the machining of graphite is a health hazard and causes fouling of the machine tools, producing maintenance problems; graphite and carbon fibers are conductive and pose problems with numerical-controlled equipment; coolants used with graphite can cause additional problems by combining with graphite dust to produce a slurry that is difficult to clean up; some coolants also tend to be absorbed and retained by the composite. The Suppliers of

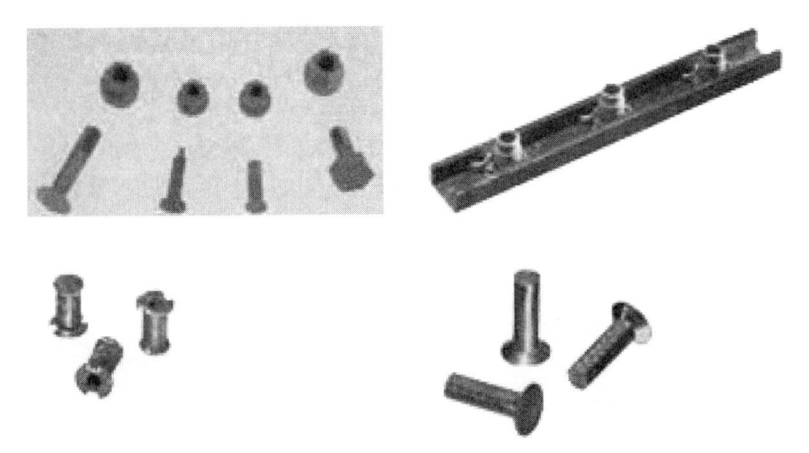

Figure 26-12 Fiberlite fasteners. (Tiodize Co.)

Advanced Composite Materials Association (SACMA, 1600 Wilson Blvd., Arlington, VA 22209) has reported on the workplace hazards of advanced composite materials. Many manufacturers are abiding by the Occupational Safety and Health Administration (OSHA) guidelines for the handling of asbestos and confronting the possible hazards of composite particulates released during fabrication. The best remedy for eliminating airborne particles is adequate ventilation provided by a high-efficiency particulate air (HEPA) filtered vacuum. **Conventional machining** poses several problems, including excessive tool wear, frictional-heat generation, crazing effects when using lubricants, and localized fiber breakage and delamination due to tool impact and vibration. Drilling bolt holes to allow the joining of composites to other materials creates a multitude of special problems, particularly when aligning prefabricated holes with precision tolerances. Drilling holes in laminated composites requires special tools and techniques. The highly abrasive nature of these composites and their low interlaminar strength often results in splintering and delamination of the hole. Without some kind of backside support or frequent sharpening, innovative drill designs fail to produce clean, repeatable holes. There is a new diamond drill, however, that can be used when backing materials or backside support is not cost-effective.

Solid ceramic tools boost production rates 5 to 10 times over carbide tools when composites are machined. Their metal-removal rates are up to 200 times higher than with diamond-coated tools. Tools of fine-grained alumina matrix reinforced with SiC fibers or whiskers are also harder and more resistant than carbide tools. Drills, end mills, and routers made of ceramic materials can be used to cut graphite/epoxy, fiberglass, Kevlar, polyimide, graphite/graphite, and other advanced composites.

Single-layer diamond tools, made of a layer of diamond bonded to a formed metallic preform representing the negative of the shape desired, are manufactured for machining composites. Known as metal-single-layer tools, they are improving precision and lengthening tool life by remaining cool throughout the cut.

Laser machining has some advantages in cutting and drilling holes in epoxy composites. The laser beam can focus on a very small spot, minimizing the heat-affected zone. Its vaporizing of material eliminates fraying. Kevlar and glass/epoxy composites present no laser-machining problems. Because of the high melting temperature and electrical/thermal conductivity of graphite, graphite/epoxy composites do not lend themselves to laser machining.

Waterjet machining, with or without the garnet abrasive added to the stream, is also being used in the machining of composites. Kevlar is cut with a waterjet without an abrasive. Due to safety requirements, only automated waterjet systems are used. Figure 26-13 shows a

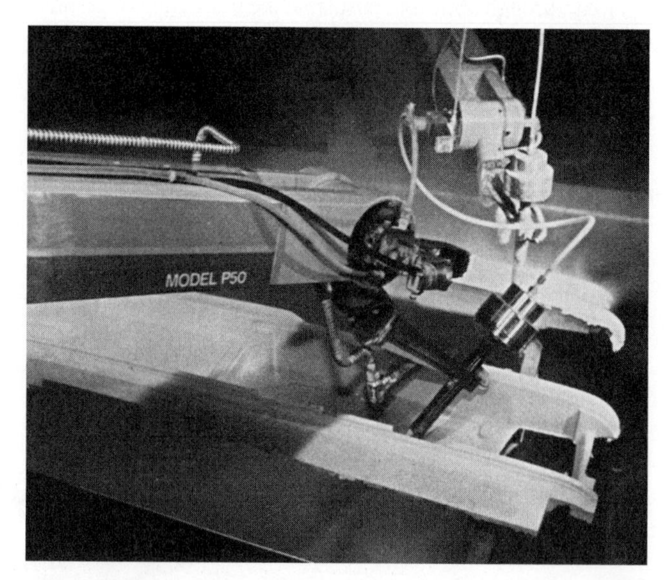

Figure 26-13 The Robotic Waternife® waterjet cutting system is used here to finish-cut and trim automotive Class A surfaced rear-end panels with numerous clearance holes for taillights and a license plate mount. (General Electric)

GE waterjet system used in the automobile industry. Its biggest limitation is the inability to effectively machine material thicknesses greater than 2.54 mm. Otherwise, the abrasive jets cut quickly and leave an excellent edge without tearing, fraying, or delamination. Several companies now specify that composites and other heat-sensitive parts must be cut with abrasive waterjet. Flow International's Paser II waterjet can cut composites at 15 to 30 in./min, compared with mechanical cutting of the same materials at 1 in./min. The same waterjet is capable of cutting 0.8-in. Kevlar at 5 in./min. Both waterjet and laser machining cause very little tool wear.

26.4 REPAIR OF COMPOSITE MATERIALS

Most composite-repair techniques depend on an adhesive with a certain shelf life, requiring that the adhesive be used prior to its expiration date. Additional restrictions are imposed on its disposal to ensure compliance with existing environmental laws and regulations. Many adhesives require environmentally controlled storage, usually at low temperatures, and their outtimes are limited, requiring close monitoring. A fast-repair kit produced by Ferro Corporation for fiberglass parts in the aerospace industry consists of an epoxy–resin system—a two-component thermosetting patch and repair material that, when mixed with an activator, produce a hard plastic requiring no external heat to cure. The activated resin saturates glass cloth, mat, or tape for a structurally sound composite with good dimensional stability, impact resistance, and tensile strength. Patches are abrasion and corrosion resistant and not affected by moderately high or subzero temperatures. The resin adheres to most surfaces, including steel, plastics, wood, and ceramics. Figure 26-14 is a line drawing of a typical honeycomb-core repair. Other types of repair patches for laminar composites are illustrated in Figure 26-15.

Traditionally, the joining of dissimilar materials has been a difficult task. A functional gradient material joint is one in which the materials gradually change in chemical composition and properties across a dissimilar joint, bridging mismatches and reducing thermally induced stress. A recent exercise demonstrated the joining of silicon carbide and a nickel-base superalloy with a functional gradient material made up of elemental powders of nickel, titanium, aluminum, and carbon. Such a process may lead eventually to the efficient and effective joining of hybrid structures consisting of mixed material systems.

Friction stir welding (FSW), another joining technique presently undergoing development, is based upon the high rotational speed of a pin and the resulting frictional heat to forge a bond between two metal alloys. This new technique is more reliable and maintains higher material properties than conventional welding methods. A pin is slowly plunged into the joint between two materials and rotated at high speed. At the end of the weld the single-piece pin tool is retracted. However, in so doing a "keyhole" is left, which is unacceptable, especially for cylindrical objects such as drums, pipes, and storage tanks. Another limitation of the "keyhole" welding method is the need to change to different single-pin tools when welding materials of different thicknesses. An adjustable automatic retractable pin tool that will leave a smooth hole closure at the end of the weld is being developed and the benefits will be many,

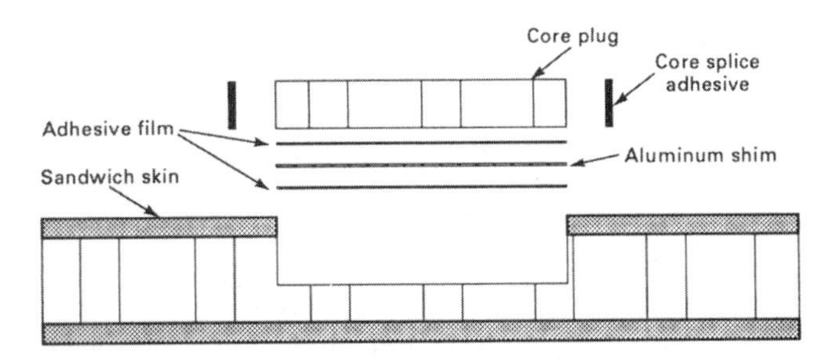

Figure 26-14 Honeycomb-core repair. (USAF)

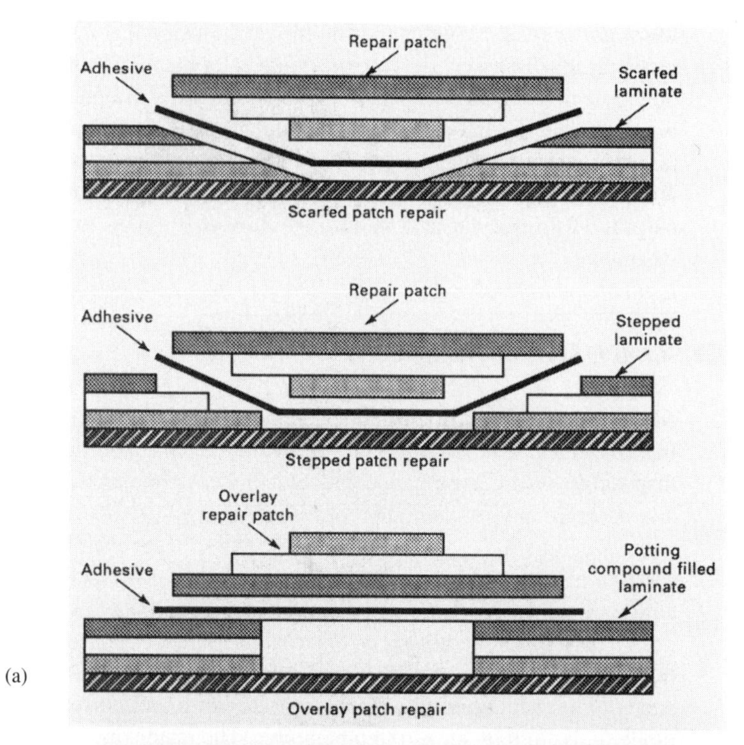

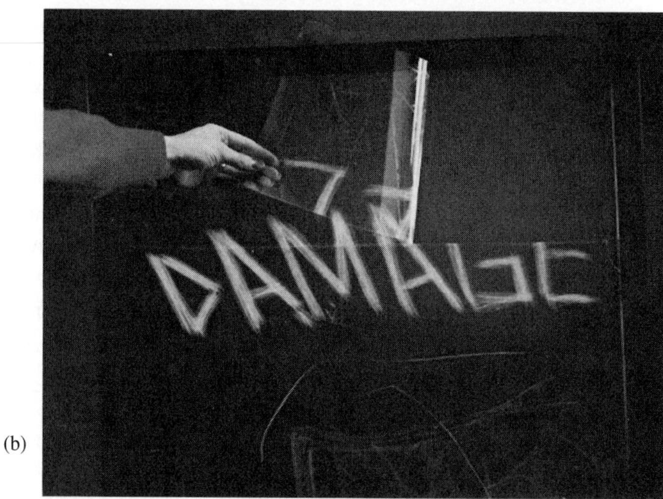

Figure 26-15 (a) Scarfed, stepped, and overlay patch repairs. (USAF) (b) The new material, N-View, is a sheet of Lexan MR5 polycarbonate that receives an abrasive-resistant Margard surface treatment and is then laminated with a thin, abrasive-resistant film of Lexan HPH to avoid damage to plastic glazing. The thin film can be removed if damaged, leaving the MR5 thick sheet and thereby doubling the glazing life. (G.E. Plastics)

including the ability to weld a wide range of materials including composite materials, producing welds with greater fatigue resistance and joint strength, minimal material distortion, lack of contamination, and no welding fumes. The welding of high-performance aluminum alloys previously thought unweldable will also be possible.

26.5 NONDESTRUCTIVE EVALUATION

Composites have extended the lifetime of helicopter blades threefold. Older blades, made with aluminum sheet and extrusions bonded with adhesives, were inspected by X-ray equipment, which took about eight hours. New analysis and characterization techniques (introduced in

Module 8) have drastically reduced the time and improved the reliability of this nondestructive evaluation (NDE) procedure. Using computerized infrared ***thermography,*** inspecting the blades now takes about one hour. The NDE of composites during the manufacturing process will be the key to 100% reliable inspection. Due to the variety and complexity of the material's composition and fabricating processes, composites are a severe challenge to existing evaluation procedures. Also, there is no clear definition of what a defect is. Ultrasonic nondestructive evaluation plays an important role in determining properties and detecting defects in composite materials, specifically in materials that are classified as anisotropic laminates of composites and smart materials. ***Computer tomography (CT)*** using X rays provides high-resolution two- and three-dimensional images of the internal structure of advanced composites for identifying stress-induced damage in small electronic or mechanical components. Unlike radiography, this system makes measurements from many angles, which a computer program then uses to reconstruct a slice of the object's internal structure. One obstacle to the adoption of polymer matrix composites (PMCs) for use in marine structures, including surface ships, submarines, and marine platforms, has been the absence of effective test methods for electronic or mechanical components.

26.6 COMPOSITE SELECTION: DATABASE FOR COMPOSITES

In the selection of composite materials for high-temperature applications, particularly resin–matrix composites, it is imperative to have detailed knowledge of a resin's glass transition temperature (T_g) (see Section 15.4.2.4). An up-to-date materials database is also essential. Replacing traditional engineering materials with composite materials requires a database that reflects varying service conditions, including combinations of conditions such as high temperatures with or without a load and/or an adverse environmental condition such as a corrosive atmosphere. One such database, for automobile materials, was compiled by P. K. Mallick, University of Michigan-Dearborn. In Section 25.1.2 we showed a general flowchart for designing an FRP composite to withstand fatigue. Note the inclusion in the flowchart of databases for both materials and processes. The adoption of new materials by many design personnel is inhibited by the lack of good databases with which to locate sufficient data about materials that would be more efficient in many applications.

The importance of the foregoing two points can be emphasized by using the example of fiber-reinforced thermoplastic composites. The amorphous resin poly(ether sulfone) (PES) and the crystalline resin poly(ether ether ketone) (PEEK), both thermoplastics, are used as matrix materials reinforced with glass or carbon fibers. These reinforcements increase the strength and modulus (stiffness) of the polymer resins. In general, carbon-fiber reinforcements provide a 20% improvement in tensile strength and a 30–100% improvement in flexural modulus over glass-fiber reinforcement at room temperatures, due to the ability of the matrix material to transfer the stresses produced by the load(s) from itself to the fibers. Under high-temperature conditions, the matrix plays another important role—it is the determiner of the high-temperature performance of the composite. How well it is able to maintain its strength and stiffness at high service temperatures dictates the service temperature for the composite. There are innumerable high-temperature applications where the critical properties desired are dimensional stability and stiffness. One is the need to reduce mold shrinkage. PES has a T_g of 230°C, whereas PEEK, being crystalline, has a melting temperature (T_m) of 336°C. Testing these materials at various high temperatures shows that the amorphous polymer PES has higher strength at just below its T_g than does the crystalline polymer PEEK, but the crystalline PEEK resin composite maintains its strength above the T_g of the amorphous PES composite. Figure 26-16 illustrates GE SUPEC® poly(phenylene sulfide [PPS]) resin, which has high ductility and good thermal capabilities and finds applications in **surface-mount technology** as high-performance connectors with good pin-insertion/retention qualities. This example illustrates that when selecting thermoplastic composites to replace metals for high-temperature applications, a knowledge of glass transition temperatures and a database containing the results of performance testing at conditions other than ambient room temperatures are essential. Figure 26-17 illustrates a situation in which a high-temperature

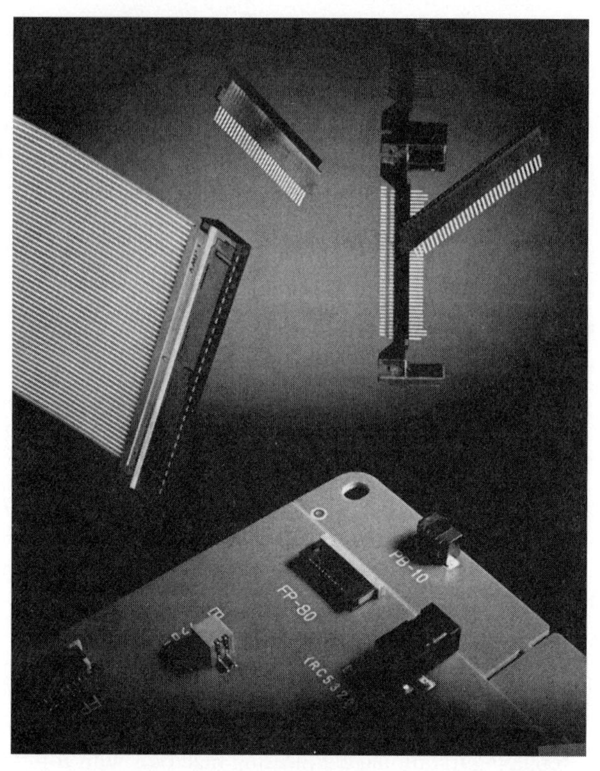

Figure 26-16 High-performance connectors used in the electronics industry's surface-mount technology are made of GE Plastics' SUPEC® resins with high ductility and low-flash grades. (General Electric Plastics)

Figure 26-17 Two impellers used in the chemical industry. The metal impeller on the right has an average service life of only two months in an acid environment at 230°C compared to a carbon-fiber-reinforced PEEK (the high-temperature thermoplastic composite impeller on the left) with an extended service life of 18 months. ("ICI Advanced Materials," *Advanced Materials & Processes*, Volume 140, No. 2, August 1991, ASM International)

thermoplastic composite having an array of properties required to combat not only high temperatures but also a combination of stressful and corrosive environments was substituted for a metal and performed better than the metal.

The submersible craft seen in Figure 26-18 serves as a case study of how the many concepts of materials selection presented in this book are applied. Battelle Columbus's two-person, human-powered *Spirit of Columbus* uses varying-density PVC foam in an epoxy-bonded composite hull. Hysol 9460 structural adhesive offers high tensile lap-shear strength (24 MPa or 35,000 psi) when bonded to the aluminum frame. This aluminum-to-composite system is the same one used in Indy 500 race car bodies and reflects a movement away from the mechanical fastening systems of bolts, rivets, etc. The hull sections consist

Figure 26-18 *Spirit of Columbus*, Battelle Columbus's award-winning entry in the International Submarine Race for human-powered submersible craft allowed its designers to gain experience with emerging composite systems technology. (*Advanced Materials and Processes*)

of PVC foam of different densities sandwiched between layers of fiberglass. High-density, 0.1 g/cm^3 (6.23 lb/ft^3), foam is concentrated on the load-bearing hull bottom, to which air tanks and the drivetrain are permanently bonded. Low-density, 0.06 g/cm^3 (3.75 lb/ft^3) foam is used on the hull's top half, where no other structural support is needed. The propulsion system uses a 455-mm (18-in.)-diameter drive sprocket, connected by a chain to a cable takeup pulley. Pulley-cable ends are attached to the driveshaft and connect to two hinged flappers. The flappers are made of 6061-T6 aluminum alloy with lightening holes, polystyrene foam, glass fiber, two layers of carbon fiber, and a top layer of drag-reducing riblet tape. The drag-reducing tape inhibits the transverse movement of turbulent eddies across the flapper surface. The lightweight hull must be both very rigid and very strong because the innovative, articulated linear-thrust propulsion system causes the drivetrain to move back and forth.

Composite databases—
www.wwcomposites.com/www.plastics.about.com/library/blmatweb.htm
Matweb database—*www.matweb.com*
http://composite.miningco.com/library/data/blglass.htm

26.6.1 Integrated Design for Manufacturing

Textile structural composites are composite materials that are reinforced by textile structures for load-bearing applications. An ***integrated design for manufacturing (IDFM)*** approach has been established based on textile-fiber architecture to facilitate communications between structural designers and composite manufacturers. There are several software packages available based on IDFM, including a cost-analysis module.

26.6.2 Composite Design and Selection

Composites offer an unprecedented flexibility that allows design engineers to "design the material." The development of a standard design methodology and a common standard for design approaches and allowables is vital, in addition to more reliable analytical techniques for designers and manufacturing engineers, such as advanced FEA systems. Detailed standards and methodology applicable to all types of composite materials are needed; their lack is a major drawback to more rapid adoption of composites. The reluctance of design engineers to specify materials with which they are unfamiliar despite an abundance of data showing superior performance of new materials will only be overcome by long-term performance data and industrywide standards. (Further discussion of materials design was contained in Module 8.) One major cooperative effort is underway in the industry. Several U.S. aerospace manufacturers are engaged in developing software to reduce the cost of composite manufacturing. The ***Composite Affordability Initiative (CAI)*** is a collaboration between government and industry to reduce the cost of aerospace structures so that engineers can take full advantage of their unique benefits. Under this initiative, entirely new processes for composite manufacturing are being developed.

26.7 MICROSCOPY OF COMPOSITE MATERIALS

Initially, carbon-fiber-reinforced plastic composite materials had low damage resistance and fracture toughness. These properties severely limited their application, particularly in the aircraft industry. An effort to rectify this problem led to a study of the relationship between microstructure and materials performance. One outcome was the development of several specimen preparation methods that revealed the size and distribution of individual resin phases and, subsequently, the role played by these resin phases in improving damage resistance. The overall result was that changes in formulations and manufacturing processing could be evaluated from a microstructural viewpoint.

26.7.1 Lightning Strike Analysis

Laboratory analysis conducted at Boeing's Materials Technology group involved techniques for evaluating the macro and micro effects of lightning strikes and the resulting high-current, high-energy arcing on aerospace polymeric composite materials. An excerpt from the report follows.*

> Modern aircraft are constructed of a wide range of structural materials, which include a variety of metals and composites. Because of the diversity in these materials, it is important to understand and document all the microstructural changes that can occur when there is an arcing event. For example, when an arc occurs involving annealed steel, an untempered martensite marbled in a matrix of ferrite due to the instantaneous heating and rapid quenching by the surrounding material is seen.
>
> Composite conductivity, however, is not as homogenous as in metals. Although there are entrance and exit zones (as is the case for metals) and carbon fibers are very conductive, they are surrounded by an insulator (the polymer resin). Therefore the current travels differently through the material. Unique to carbon fibers there is an interply arcing as the current jumps from ply to ply.
>
> The macro effects of a lightning strike on composites are usually apparent and can be documented with macrophotography but the micro effects can only be observed with a clean cross section of the area. The zone in and around the strike is very fragile due to the vaporized matrix as shown in Figure [26-19a] and [b], respectively. Care must be taken in cross-sectioning and mounting in order to preserve much of the critical information. Mounting and sectioning lightning strike composite specimens are best accomplished with a two-stage mount. It is advisable on the primary mount to add a dye to the mounting epoxy, such as rhodamine-B laser dye. The best method for creating an artifact-free specimen is to first vacuum impregnate the strike area with the epoxy followed by a pressure cure. This encapsulated area will allow a section through the center of the strike and thus hold the fragile material in place, minimizing artifacts. The specimen can be remounted to provide adequate handling of the microscopic sample.
>
> The microstructural analysis can be observed with the full range of optical micro-tools: bright field, polarized light, dark field, and epi-fluorescence. In polymer composites, however, you may observe hot spots or interply and intraply arcing [Figure 26-19c], unlike metals with their

*Excerpt with permission from reprint "The Effects of Lightning Strikes and High Current on Polymer Composites," L. M. Gammon, and B. S. Hayes, *Boeing Materials Technology*, January 25, 2003.

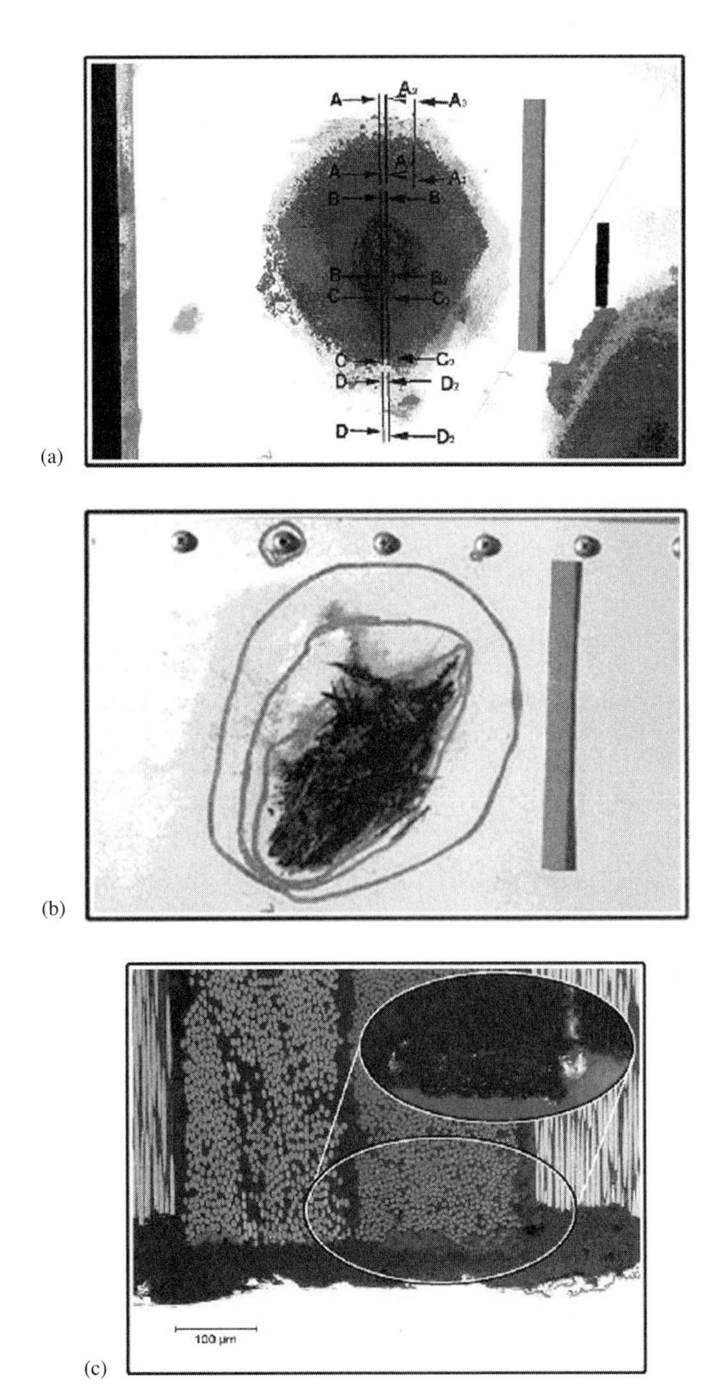

Figure 26-19 (a) A lab-induced lightning strike on a carbon fiber composite. (b) A section map of a lab-induced lightning strike on a honeycomb carbon fiber composite structure that contained an expanded aluminum mat. (c) A carbon fiber composite showing a hot spot due to intraply arcing. It has been backfilled with rhodamine B laser-dyed epoxy. This shows the voided zone adjacent and around the center of the hot spot. Polarized light illumination. (Luther Gammon, The Boeing Company)

multiphase eutectic melted zone. The microstructure of polymer composites that have been exposed to high-energy arcing is quite different and can vary widely depending on the polymer. The strike effects in thermoset polymers usually have a central hot spot zone where the fibers are degraded and surrounded by a zone where the polymer is vaporized. The bulk structure of the surrounding polymer can also change depending on the polymer matrix (single phase or toughened multiphase). The use of a laser-dyed backfilled epoxy is very important so that the vaporized zone is preserved and apparent with the fluorescence, leaving no doubt which are the composite polymer and the mounting polymer [see Figure 26-19c].

26.8 GREEN MANUFACTURING

The auto industry's successful three-tier recycling program (salvage dealers, shredders, and secondary-metal recoverers) has been quite effective in keeping iron, steel, aluminum, and other metals out of the ***municipal solid waste (MSW)*** stream, as we discussed in the Metallics and Polymers units. This tiered system recovers more than 80% of the ferrous metals and aluminum and 66% of the copper from scrap automobiles. Most batteries are salvaged to recover lead plates and the polypropylene case. Figure 6-22 showed the increased use (past and projected) of major plastics for automobile use. PMCs are also on a steady increase. As discussed in Module 9, there is a declining trend in iron use, which is expected to continue.

Setting up the ***recycling infrastructure*** for PMCs is a major challenge. Currently, much of the polymer content ends up in landfills. Not only is this unacceptable from an environmental perspective but from an economic point of view it is unwise to bury polymers for ever-increasing "tipping fees" at landfills when the materials have value as recyclable resources. In the United States, Chrysler, Ford, and GM have established a vehicle recycling partnership to promote and conduct research for improved recycling technology.

26.8.1 Composites Recycling/Green Manufacturing

Composites benefit the environment because they are energy efficient to manufacture and their inherent light weight results in fuel savings over the life of a vehicle. Weight savings in vehicles mean less fuel consumption and thus a reduction in emissions. Energy savings are also realized in the production of raw materials for composites and in the processing and manufacturing of composites when compared to similar steel parts. Composites are recyclable at the end of their own life cycle, and today's composites can contain substantial amounts of recycled materials. The ***Automotive Composites Alliance (ACA),*** which assists automakers in the use of composites, has reported that more than 10 million cars and trucks produced annually have composite components. Reinforced thermosetting polymers, specifically SMCs, are making attractive alternatives to steel and aluminum, because their styling flexibility permits automakers to make frequent styling changes and get incorporate part consolidation and weight savings at low program costs. The ACA has promoted the research and development necessary to establish economically viable processes for recycling these materials. To date, many automotive parts containing recycled SMC filler are used on production vehicles. A recycled part can contain recycled content as high as 25% by weight without adverse effect on physical properties. SMC is an easily recycled plastic, more so than thermoplastics, and there is no need to remove paint and adhesive from the surface prior to processing as recycled filler. SMC is less sensitive to contamination by dissimilar materials and involves no degradation of properties. The resin portion of SMC also can contain recycled PET from pop bottles. In view of these factors, SMC will be the material of choice for auto components when compared to other plastic materials. As this use grows, the recyclability of SMC will be one of its greatest advantages.

Figure 26-20 is a schematic of the recycling of composite thermosetting plastics. The current recycling sequence is shown in this figure via a scrap car delivered to a dismantler, where useful or recyclable parts are removed. The "hulk" is passed through a shredder; ferrous materials are magnetically removed, leaving the remainder, or automotive shredder residue (ASR). This residue is passed through a nonferrous separator where aluminum, copper, zinc, magnesium, and other metals are removed. The balance of the ASR is sent to a landfill. Typical recovery rates for some dismantled materials are: thermoplastics, 1% dismantled of which

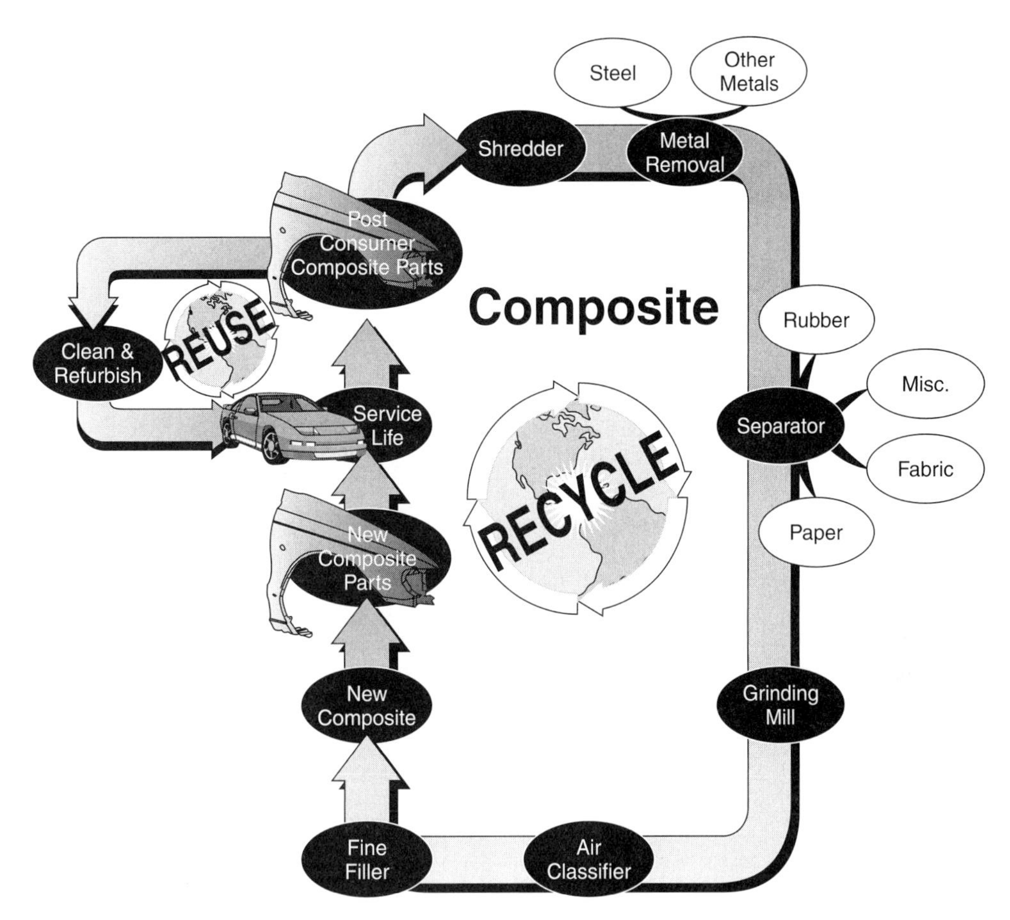

Figure 26-20 Recycling of SMC composite parts in automobiles. Scrapped parts (end-of-life parts or in-plant scrap) made of thermosetting plastics (SMCs) are ground up to provide a material that can replace the filler in new formulations. New composites can contain substantial quantities of recycled materials, including resin, filler, and additives. (Adapted from Automotive Composites Alliance)

93% ends up in landfill; aluminum, 15% dismantled of which 20% ends up in landfill; SMC/BMC, 15% dismantled of which 85% ends up in landfill. The percentage of organic materials (plastics, rubber, fabrics, etc.) in an average vehicle is 15% by weight but 55% by volume. The recycling of SMC and its reuse by manufacturers (***green manufacturing***) will considerably reduce this large influx of materials to landfills.

Figure 26-21 lists the recycling options for five materials used to produce external automobile body panels. Recycled content can be used in several ways. Preconsumer content refers to postindustrial use other than in an automobile. Internal reuse refers to the manufacturing process for vehicles. Primary recycling refers to the reuse of material to make a similar or identical part, and secondary use is reusing material to make a lesser-value part. ***Life-cycle assessment*** is the "cradle to the grave" analysis, or ***life-cycle inventory (energy),*** as described in Module 9. This inventory of total energy consumption begins with raw materials and includes final disposal of the part. Accompanying this analysis is an ***environmental impact assessment,*** which involves natural resources depletion, toxic emissions, pollution, and other forms of negative attack on the earth from initial manufacture to final disposal.

 United States Council for Automotive Research (USCAR)—*http://www.uscar.org*
Automotive Composite Consortium (ACC)—*http://www.uscar.org/consortia/con-acc.htm*

RECYCLING OPTIONS
Material Comparison For External Autobody Panels
(Hood, Fender, Applique, Front-End Panel, Bumper, Etc.)

	Thermosetting Composite SMC/BMC RRIM SRIM	Thermoplastic Composite Azdel GFN/GFPE	Engineering Thermoplastic PET/PBT Polycarbonate TPO	Sheet Steel	Sheet Aluminum
Recycled Content in New Part					
Post-Consumer	No (a)	No (a)	No (a)	No (a)	No (a)
Pre-Consumer	No (a)	No (a)	No (a)	Yes	Yes
Internal Reuse	No	Yes (b)	Yes (b)	Yes	Yes
Recycleable at End of Service Life					
Refurbish and Reuse	Yes (d)	No (e)	No (e)	Some (f)	Some (f)
Primary Recycling	No (a)	No (x)	No (x)	No (a)	No (a)
Secondary/Tertiary Recycling	No (a)	No (a)	No (a)	Yes	Some
Life Cycle Assessment					
Life Cycle Inventory (Energy)	Low	Low	Low	High	Med.
Environmental Impact	Med. (g)	Med. (g)	Med. (g)	Med. (g)	Med. (g)

Notes:
(a) Feasible, but not currently done because of cost penalty
(b) Internal scrap used, but generally into "lower grade" parts
(d) Bolt-on parts commonly dismantled and reused (if not damaged)
(e) Generally not economical to reuse
(f) Bolt-on parts can be reused if free of corrosion and/or damage
(g) Generally no major differences between materials
(x) Generally not feasible or large cost penalty

Figure 26-21 Recycling options for automotive materials. Five materials used to produce external autobody panels are compared showing their potential for various recycling techniques. SMC automobile panels reduce usage of fuel reserves. A Lincoln Continental, for example, uses 1200 pounds of SMC components, reducing vehicle mass by about 30 pounds versus steel and thus reducing fuel consumption. (Adapted from Automotive Composites Alliance)

26.9 COMPOSITES GLOSSARY

Areal weight —the weight of fiber per unit area (width × length) of tape or fabric.

Autoclave —a closed vessel used to consolidate and cure composites using heat and/or pressure.

Bag molding —a technique in which the composite material is placed in a rigid mold and covered with a flexible bag with pressure applied by vacuum, autoclave, press, or by inflating the bag.

Basket weave —a woven reinforcement in which two or more warp threads go over and under two or more fill threads in a repeating fashion or pattern. This weave is less stable than the plain weave (refer to **Weave** in this glossary) but produces a flatter, stronger fabric.

Bidirectional laminate —fibers oriented in more than one direction in the same plane.

Braided fabrics —more expensive than woven materials due to the more complex manufacturing process, but typically stronger by weight. Braiding is a hybrid of filament winding and weaving. The strength comes from intertwining three or more yarns without twisting any two yarns around each other. This mechanical interlocking imparts a natural mechanism to the braid that evenly distributes load throughout the structure. Braided structures exhibit higher impact strength, improved shear properties, superior fatigue properties, and optimal torsional strength. Braids are available in both flat and tubular configurations. Flat braids are used primarily for selective reinforcement such as to strengthen specific areas in pultruded parts. Tubular braids can be pultruded over a mandrel to produce cross sections in a variety of parts such as lamp poles or hockey sticks.

Braiding —the weaving of fibers into a tubular shape as opposed to a flat fabric.

Cavity —the location in a mold where the resin enters and fills to form a solid molded part replicating the features and size of the cavity.

Chopped strands —produced by cutting continuous strands into short lengths for use in mats or molding compounds.

Commingled yarn —a hybrid yarn made with two types of materials intermingled into a single yarn.

Conductive composites —electrically conductive composites that use metal fibers embedded in a nonconductive matrix to provide materials for such uses as antistatic coatings, fuel cell catalysts, solar electrical cells, or lightweight, inexpensive batteries.

Core —the central component in sandwich construction to which the inner and outer skins are attached. Honeycomb, foam, paper, and wood are common core materials.

Cowoven fabric —a reinforced fabric woven with two different types of fibers in individual yarns, such as thermoplastic fibers woven with carbon fibers.

Crazing —a region of ultrafine cracks that may develop on or under a resin surface.

Creel —a device for holding the required number of roving balls (spools) in the desired position for unwinding onto the next step in the process (i.e., weaving, braiding, or filament winding).

Curing agent —initiators, hardeners, or any other materials that promote cross-linking. It differs from a catalyst in that it is consumed by the cross-linking reaction, whereas a catalyst is recoverable once the reaction is complete.

Drape —the ability of a fabric or prepreg to conform to a contoured surface.

Fabric —made by interlacing yarn in a variety of ways, including weaving, knitting, and braiding. Fabrics such as 2-D woven fabrics are currently used for prepregs. There are nearly a dozen types or styles of basic weave such as plain, 2/2 twill, 8HS, and triaxial wave. In 3-D woven fabrics, only z yarns interlace with the filling yarns. 3-D woven fabrics have high interlaminar strengths and find use in impact-resistant structures.

Fiber —filamentary materials synonymous with **filament**, which has an extreme length and very small diameter, usually less than 25 μm. A continuous fiber has an indefinite length, in contrast to a discontinuous fiber, represented by whiskers (chopped fiber in short specified lengths) or particulates.

Filament —slender thread or fiber. Filaments are individual fibers of indefinite length used in tows, yarns, or roving.

Flash —the overflow or squeeze-out of cured resin from a mold.

Harness satin —a weaving pattern that produces a satin appearance. "Eight-harness" means that the warp tow crosses over seven fill tows and under the eighth in a repeating manner.

Impregnate —to saturate the voids and interstices of a reinforcement with a resin.

Knitted fabrics —fabrics produced by interlooping chains of yarns.

Knitting —a process of interlooping chains of tow or yarn.

Laminate —two or more layers (plies) bonded together. Laminates are composite materials made from several layers of materials.

Macrocomposites —single parts that incorporate more than one composite. An example of the incorporation of more than one composite in a single part would be a CMC and an MMC in a composite-fabricated driveshaft.

Mat —a fibrous reinforcing material comprised of chopped filaments, along with a binder to maintain form, available in blankets of various lengths, widths, and weights.

Mold release agents —barriers between the mold or tool and the part being molded primarily to prevent the part being molded from sticking in the mold when curing is completed. They also hold the part in place during cure so that the proper shape and surface finish are attained. The various types of release agents include solvent based, water based, paste, semipermanent, and internal (blended into the molding resin). Silicones, fluorocarbons, and beeswax serve as release agents.

Neat resin —a resin to which nothing (additives, reinforcements, etc.) has been added.

Out-time —the time a prepreg is exposed to ambient temperature (i.e., the total amount of time the prepreg is out of the freezer).

Particulate composites —materials consisting of one or more components (particles) suspended in a matrix of another material. Such particles can be either metallic or nonmetallic (see Section 25.2.4).

Pitch —the residue left from the distillation of coal and petroleum products. It is a base material for the production of some carbon fibers, as well as a matrix precursor for some carbon–carbon composites.

Ply —a single layer of tape or fabric.

Pot life or **working life** —the length of time that a catalyzed thermosetting resin retains a viscosity low enough for it to remain suitable for processing.

Preform —a fibrous reinforcement preshaped on a mandrel to the approximate contour and thickness desired in a finished part.

Prepregs —a precoating of fiber or cloth sheets with resin mixed with a hardener to help keep the fibers in place. Fibers may be in the form of continuous roving, mat, or woven fabric. The resin is allowed to cross-link slightly until it is viscous enough to stay on the fibers. This process is called B-staging, and the result is tacky. Placed on paper to keep from sticking together, the prepreg is rolled up and stored under low temperature until used. Epoxy resins are mainly used in prepreg especially in the aerospace industry, primarily due to the high strength and modulus (stiffness) that can be obtained. Rolls of prepregs have widths from 4 to 40 inches. Roving can be prepregged. Prepregs are subsequently cut into required dimensions and cured into final shape by molding operations.

Prototype —the process that establishes design, material, and fabrication parameters for a new product.

Ramping —the gradual, programmed increase or decrease in the temperature or pressure used to control the cure or cooling of composite parts.

Release agent —a substance usually sprayed or painted on a mold to prevent cured matrix materials from bonding to tooling.

Resins —mostly polymers; in reinforced plastics, the matrix is a resin that binds together the reinforcement. Resins are solids or pseudosolid organic materials, usually of high molecular weight, that will flow when subjected to stress.

Roving —a collection of bundles of continuous filaments, either as twisted yarns or as untwisted strands. Rovings are used in continuous molding operations such as filament winding or pultrusion. They can be preimpregnated with a thin layer of polymeric resin to form prepregs.

Scrim —a low-cost reinforcing fabric made from continuous filament yarn in an open-mesh construction.

Shelf life —the length of time a material can be stored and continue to meet specification requirements by remaining suitable for its intended use.

Sizing —a compound that binds together and stiffens yarn and fibers, providing resistance to abrasion during weaving and handling. The ingredients, such as gelatin, wax, or oil, provide surface lubricity and binding action. Unlike a *finish*, which contains a *coupling agent* to improve the bond of resin to the fibers, it contains no coupling agent. The size is usually removed by heat cleaning and a finish is applied. A finish can also contain a binder. This solution of chemical additives lubricates filaments and reduces static electricity.

Skin —the relatively dense laminate comprising the outer surfaces of the core in a sandwich construction.

Space-grown composites —organic composites grown in space to provide repair materials for a space station. They obviate the need to deliver repair materials from earth at great expense.

Strand —an untwisted bundle of continuous filaments used as a unit, including tows and yarn. Sometimes a single filament or fiber is referred to as a strand. Also, a narrow wood flake on oriented strandboard.

Tape —a material in which the fibers are laid in a single direction within a resin matrix.

Tool —a mold, either one- or two-sided, open or closed, in or upon which composite material is placed to make a part.

Tow —an untwisted bundle of continuous filaments. A tow designated as 100K has 100,000 filaments. This thin, unidirectional prepreg comes in widths up to 12 inches.

Warp —yarn running lengthwise in a woven fabric. *Warp* refers to the 0° yarns; the 90° yarns are *fill*. The method of 2-D weaving involves interlocking two sets of yarns perpendicular to each other. The warp or the 0° direction is the continuous fabric direction and refers to the long, machine, or fiber direction. The fill direction is also called the *weft* or *woof* direction.

Weave —the particular manner in which a fabric is formed by interlacing yarns. Weave consists of interlacing two sets of yarn perpendicular to each other. The 0° set is called *warp* and the 90° set is called *fill*. The warp axis is the continuous fabric direction. In plain weave, warp and fill fibers alternate to make both fabric faces identical. Satin weave has a pattern that produces a satin appearance, with the warp roving crossing over several fill rovings and under the following one. Eight-harness satin has warp roving over seven fill rovings and under the eighth. Other weaves such as harness satin and basket weave allow the yarn or roving to cross over and under multiple warp fibers (e.g., over two, under two). These weaves tend to be more pliable and conform more easily to curved surfaces than do plain weaves.

Weft —the transverse threads or fibers in a woven fabric. They run perpendicularly to the warp. Also called *fill*, *yarn*, or *woof*.

Wet layup —a method of making a reinforced product by applying the resin system as a liquid when the reinforcement is put in place.

Wet-out —the saturation of all interstices between bundles and filaments of reinforcement with resin.

Whiskers —single crystals ranging in size from about 0.5 to 10 μm in diameter and 3 μm to 3 cm in length. They can be metallic or ceramic. Their aspect ratios are between 100 and 15,000. (See Section 8.4.1 for a definition of aspect ratio.) A popular whisker in use is one made of silicon carbide. It varies in diameter from 3 to 10 μm, is 10 to 100 μm in length, and has a Young's modulus of 800 GPa and an ultimate strength of 21,000 MPa. Another popular whisker made of silicon nitride varies in diameter between 0.2 and 0.5 μm, is 50 to 300 μm in length, and has a modulus of elasticity in tension of 300 GPa. The size of small defects in silicon carbide, such as voids, inclusions, and surface roughness, is 0.1 to 0.4 mm. The secret to the strength of whiskers is the smaller the fiber, the smaller the defect. Adding encapsulation to the fiber, which helps protect it from surface damage, results in fibers (whiskers) maintaining their inherent strength.

Woven roving —a heavy, coarse fabric produced by the weaving of continuous roving bundles.

Yarn —an assembly of twisted filaments, fibers, or strands, either natural or manufactured, that forms a continuous length suitable for use in weaving or interlacing into textile materials.

SELF-ASSESSMENT

26-1. A fiber composed of 55% silica, 20% calcium oxide, 15% aluminum oxide, and 10% boron oxide is a(n)

a. SMC **b.** GRP **c.** SCAB **d.** E-glass

26-2. Fibers such as rayon that are pyrolized to drive off organic elements, leaving a very stiff, high-strength carbon fiber are

a. E-glass **b.** Kevlar **c.** Borsic **d.** Graphite

26-3. The method of manufacturing structural shapes, such as channels and angles, in which prepregs are drawn through a die and heated until they set is called

a. Pultrusion **b.** Extrusion **c.** Laminating **d.** RTV

26-4. A means of inspecting materials for flaws without adversely affecting them is

a. NDE **b.** SPF/DB **c.** RRIM **d.** RCC

26-5. Explosive welding is used to

a. Join incompatible metals

b. Remove surface contaminants

c. Weld projectiles

26-6. Regardless of the name given to the processing method for producing fiber-reinforced plastic products, two stages are included in most of them. Name these two stages.

26-7. The key step automakers have taken to promote green manufacturing in the use of thermosetting resin–matrix composite parts is

a. Using ASTM symbols

b. Using only thermoplastics resins

c. Using only CMC matrices

d. Grinding parts to use as fillers

26-8. From the table on recycling options in Figure 26-21 and using a High, Medium, and Low rating, how would the life-cycle assessment of automotive materials rate the following key materials?

Life Cycle Inventory (Energy)

a. Thermosetting composite (e.g., SMC/BMC) ___________

b. Thermoplastic composite (e.g., azdel) ___________

c. Sheet steel ___________

d. Sheet aluminum ___________

Environmental Impact

e. Thermosetting composite (e.g., SMC/BMC) ___________

f. Thermoplastic composite (e.g., azdel) ___________

g. Sheet steel ___________

h. Sheet aluminum ___________

26-9. Environmental composites are so named because they are

a. Easily recycled

b. Easily manufactured

c. Cheap to fabricate

26-10. The chance of a defect remaining in a fiber as the diameter is decreased is

a. Minimized **b.** Unaffected **c.** Increased

26-11. A molding method for producing FRP in which parts are placed in a cylinder that is subjected to high heat and pressure is

a. Filament winding

b. Injection molding

c. Autoclave

d. Pultrusion

REFERENCES & RELATED READINGS

ASM INTERNATIONAL. *Engineered Materials Handbook*, Volume 1, *Composites*. Materials Park, OH: ASM International, 1987.

BITTENCE, J. C., ED. *Engineering Plastics and Composites*. Materials Park, OH: ASM International, 1990.

Manufacturing Engineering. "Cutting Tool Material Makeover," August 1995, p. 26.

MARC ANALYSIS RESEARCH CORP. "Creep-Forming Simulation Helps Form Al Wing Skins," *Advanced Materials & Processes*, July 1994, p. 15.

SCHAEFER, R. J., AND M. LINZER, EDS. *Hot Isostatic Pressing: Theory and Applications*. ASM Conference Book. Materials Park, OH: ASM International, 1991.

STRONG, A. BRENT. *Fundamentals of Composites Manufacturing*. Dearborn, MI: Society of Manufacturing Engineers, 1992.

TRACESKI, E. T. *Specifications and Standards for Plastics and Composites*. Materials Park, OH: ASM International, 1990.

URQUHART, ANDREW W. "Molten Metals Sire MMCs, CMCs," *Advanced Materials & Processes*, Volume 140, No. 1, 1991.

WILSON, MAAYWOOD L., GARY S. JOHNSON, AND ROBERT MISERENTINO. "Pultrusion Process Development for Multidirectional Graphite/Epoxy T-Beam," *SPI Paper*, NASA Langley Research Center, Hampton, VA, 1992.

Periodicals

Manufacturing Engineering

Plastics Engineering

Plastics Technology

Plastics World

UNIT 9

Electronic and Other Important Materials & Materials Systems Technology

PAUSE & PONDER

Fuel cell technology offers the promise of clean, efficient energy. Concerns over the long-term availability of the world's known oil reserves, the fluctuating costs of oil due to armed conflicts, and the potential effects of harmful emissions from gasoline-powered vehicles and coal- and oil-powered electric power plants has spurred the development of fuel cell technology (FCT).

The 1881 Cape Henry Lighthouse in Virginia Beach, Virginia (note the fuel cell building in the lower right corner of the photo) shares the site with the still-standing original Cape Henry light, commissioned in 1791 by the U.S. Congress as the first public works project. The two lighthouses are on the site of the "first landing" on April 26, 1607 by the English colonist's, who later founded Jamestown, Virginia. The light is one of hundreds of lighthouses and other lighted aids to navigation operated by the U.S. Coast Guard, most of which have aging equipment and unreliable sources of electrical power. As well, these installations are inefficient and sources of high energy consumption and pollution. The Coast Guard, set a goal of reducing facility energy costs by 12% and consumption by 20% by fiscal year 2005 over 1995 levels through its research into fuel cells as power sources for these facilities. Traditional power sources, like diesel electric generators or gas turbines, have high operating and maintenance costs and may also produce high levels of pollutants. Underground and submarine cables, such as those that power some lighthouses, can be unreliable and are subject to damage. These factors, along with recent power shortages, have driven the need to explore alternative energy sources. Innovation and new technology have become necessities.

As a test of fuel cells at a remote, unmanned site in a marine environment, a 3-kW direct methanol fuel cell was installed at the Cape Henry lighthouse. The fuel cell had potential for this location because of its promise of reduced maintenance and greater reliability compared to other systems. A fuel cell is an electrochemical power generator, similar to a battery that never loses its charge. As long as fuel and air are supplied, a fuel cell will continue to produce electricity and heat without the pollutants normally associated with burning fuel. Fuel cells only give off heat, water, and carbon dioxide, not the pollutants from traditional fossil fuels. Fuel cell technology is being developed for mobile applications (automobiles, trucks, and buses) as well as for stationary and portable ones. Stationary applications include data communication companies, hospitals, and houses; portable applications could be replacements for small generators and battery substitutes for notebook computers and cell phones.

The chosen fuel for fuel cells is hydrogen. On a weight basis, hydrogen has more energy than any other fuel with the exception of nuclear fuel. However, hydrogen must be produced from a hydrogen compound, which requires energy. Water is by far the cheapest and most abundant source of hydrogen. The purest hydrogen is obtained by electrolysis of water, the passing of an electric current through water to decompose it into hydrogen and oxygen. (See Module 5 for further description of this electrochemical process.) But pure hydrogen is not readily available and, in addition, requires storage in highly pressurized tanks. To power vehicles with fuel cells, the hydrogen tanks would have to be strong, lightweight and-inexpensive. And they would have to be small to fit in the available space. A new materials research effort would be required to design such a tank. Moreover, to replenish the hydrogen in these individual tanks would require a national hydrogen fueling infrastructure— a condition that practically rules out such a system. Therefore, current research is focused on generating hydrogen as it is used in a fuel cell. Since there is no ready source of pure hydrogen, many fuel cell systems convert natural gas into hydrogen-rich gas (reformate) using a fuel reformer or processor as part of the fuel cell package to extract the hydrogen. These reformers are heavy and occupy vital space. Though not as efficient as pure hydrogen fuel cell systems, reformated hydrogen systems have 50% better fuel economy than present internal combustion engines.

Cape Henry Light, built in 1881. Note fuel cell building behind Martha.

Stationary 3-kW methanol fuel cell inside building.

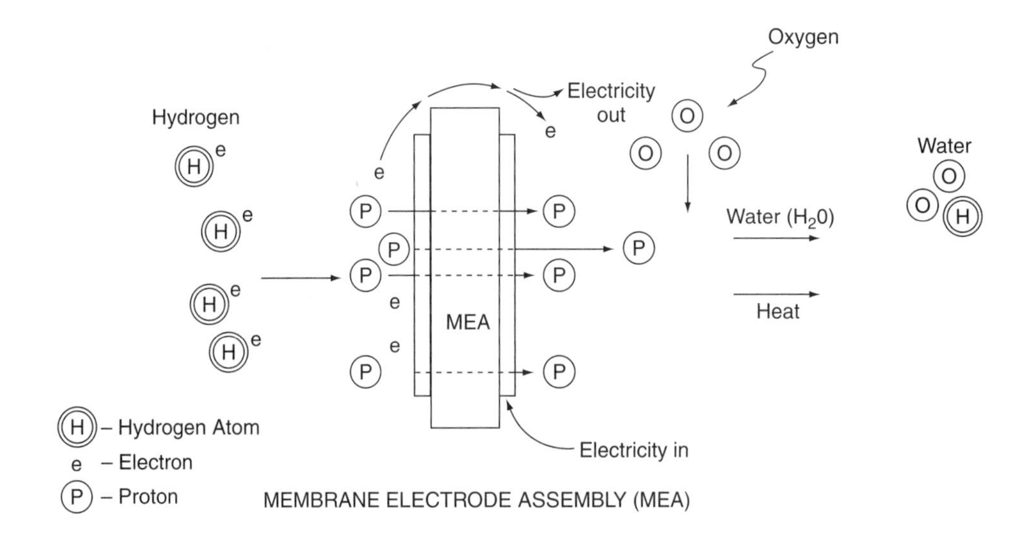

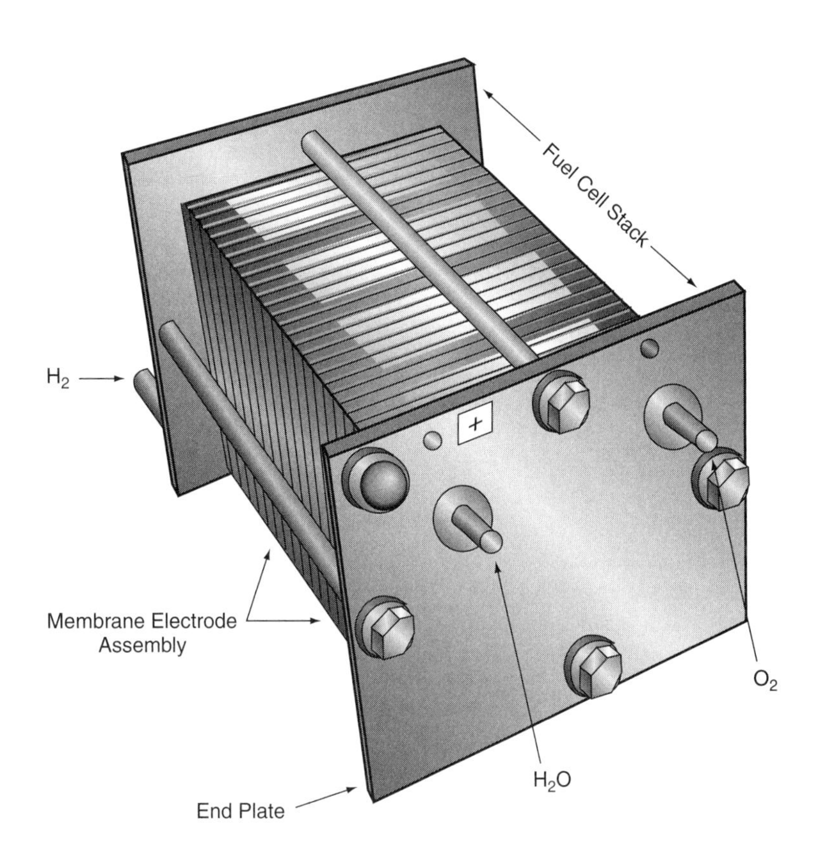

Materials as always are the essential elements in any advance in technology and so they are in the forefront of the development of fuel cell technology. Particularly for fuel cell vehicles (FEVs), materials must meet strict specifications for light weight, corrosion resistance, strength to withstand required operating pressures, and, where needed, the ability to conduct electricity or act as insulators—and provide these characteristics at less cost. In accordance with the automobile industry's increasing use of composites and other polymeric materials, these are the likely choice for mobile fuel cell applications. The Partnership for a New Generation Vehicle Program (PNGVP), now superseded by the ***FreedomCAR*** (CAR standing for Cooperative Automotive Research) and foreign automakers have been encouraging the development of FEVs. Prototypes of vehicles capable of getting more than 70 mpg have been produced and a handful have been sold to the public starting in 2003. Several automobile companies planned to market a "limited" number of fuel cell vehicles by 2004. More than likely, these vehicles will have a proton-exchange membrane fuel cell system that operates at moderate operating temperatures and provides adequate power at affordable costs with near-zero emissions.

The ***fuel cell process*** can be briefly described with the aid of the accompanying figure. Composite materials make up the fuel cell stack, which is central to the system. The stack consists of a series of fuel cells composed of a membrane electrode assembly (MEA) sandwiched between two bipolar composite end plates. When the reformate enters the fuel cell, the catalyst surface of the MEA splits the hydrogen gas molecules into protons and electrons. The protons (a positive hydrogen ion which originate in the nucleus of the hydrogen atom) pass through the membrane to react with oxygen (from the air), forming water and heat. Note that this is a chemical reaction. The electrons, unable to pass through the membrane, go around it, creating a source of direct current (dc), which can be converted to alternating current (ac) if needed. This electricity is generated via a electrochemical process.

The individual fuel cells that make up the fuel cell stack are separated by bipolar plates. Each plate serves as the anode for one fuel cell and the cathode for the next cell. The composite bipolar plates thus perform three basic functions: They contain the cells where the reaction takes place, manage the reactant and coolant flow through the fuel cell, and conduct the electrical current produced in the cells. Incorporated into the plates are a series of flow channels to distribute the hydrogen on one side and the oxygen on the other. This requires good corrosion resistance and chemical compatibility to the fluids present as well as a high level of electrical conductivity. They also must have good thermal conductivity to remove the heat that is generated by the reaction of the hydrogen and oxygen. The plates represent a large percentage of the overall weight as well as the overall cost of the fuel cell stack. They are made from bulk molding compound (BMC), a highly conductive vinyl ester resin loaded with a combination of a proprietary particulate graphite and modifiers. The end plates are made of the same material but without the conducting additives and so are a nonconductive element that provides strength to support the stack assembly.

This brief introduction into fuel cell technology serves as an introduction to the concepts of the emerging technologies discussed in the next two modules. Pause for a minute and consider what you have read and heard about fuel cells , nanotechnology, and other emerging technologies. This unit will introduce you to the materials science and technology behind these new technologies.

National Fuel Cell Center—*http://www.nfcrc.uci.edu/*
How Stuff Works, fuel cells —*www.howstuffworks.com/fuel-cell.htm*
U.S. Coast Guard —*www.uscg.mil*(search for lighthouses)

Module
27
Electrical, Electronic & Optical Materials

After studying this module, you should be able to do the following:

27–1. Explain the connection between semiconductor materials and (1) photoconductive devices and (2) solar cells.

27–2. Explain the physical and economic limits of using silicon in smaller yet more powerful computing devices and in the development of solar cells.

27–3. Give an example of a semiconductive device that functions without electricity.

27–4. Discuss the dependence of piezoelectric devices on the basic characteristics and properties of materials, including bonding, ions, crystal structures, and electric dipoles.

27–5. Discuss briefly how materials scientists have developed semiconductive materials not found in nature and given them tailor-made atomic structures, including some present-day applications.

27–6. Discuss some differences between a liquid crystal and an ordinary liquid.

27–7. Outline the path to the discovery of superconducting materials and state the major goal of modern scientists in this area.

27–8. Define the common thread that runs through advanced technologies, such as the technologies involved in information systems, multimedia, computer chips, and solar cells.

27–9. Recall how properties of materials such as electrical, magnetic, and optical properties play a significant role in the design and development of advanced materials for new instruments, optoelectronic devices, and systems.

27–10. Use the website addresses denoted by the symbol to explore advances in electrical, electronic and optical materials.*

*The icon found throughout this book will link you to Internet sites related to topics being covered. The dynamic nature of the Web brings frequent changes, so some of these sites, while available at the time of writing, may not be up now.

Rapid innovations in materials science and technology gave the information age its momentum and key technologies involving electronic and optical materials continue to propel information technology. To illustrate the impetus that materials science has provided, we will use the example of CD-ROM and DVD technology. Although the DVD and CD-ROM discs may look the same, a standard digital video disc holds seven times more information, 4.7 GB (gigabytes), than a CD-ROM. More advanced DVDs, consisting of two layers, for recording on each side, hold up to 17 GB of binary code.

With the adaptation of optical discs to storage media around 1986, ***CD-ROMs*** (compact disc—read-only memory) revolutionized the storage and retrieval of digital information. Offering a capacity of 500 times more storage of data than the ordinary magnetic diskette, CD-ROM technology and its associated laser technology have already moved the music business away from audiotapes and now offer both the computer and entertainment businesses new variations of ***multimedia technology.*** CD-ROMs provide high-density data storage, durability, and accessibility and allow for the storage and retrieval of text, graphics (both still and motion), and sound. ***Information technology,*** the term often used to label the combination of systems of electronics, optics, computers, telecommunications, graphics, entertainment, and a wide range of other forms of communication owes much of its evolution to materials science and engineering, which spurred its progress.

Once the master disc is produced by lasers, replication of a limitless number of discs is accomplished by injection molding of polycarbonate (see Figure 27-1a). With recordings on magnetic media such as cassette tape, a recording head had to pass over the entire surface to transfer data. The 4.75-in. (121-mm) CD-ROMs contain up to 2.8 billions pits, which spiral for approximately 3 miles as measured from the hub to the outer rim. The cross section (Figure 27-1b) shows the structure of a recordable CD. A regular CD uses aluminum instead of gold and has no dye layer.

A recent application of amorphous metals is the new technology that is reinventing the ***compact disc (CD).*** This new CD is a medium that is capable of being recorded onto—not merely played back. The ***digital video disc (DVD),*** for video recorders and personal computers with capacities over 5 GB, uses a red laser to record digital data by heating spots on the disc's recording layer. The laser changes the structure of the heated portions of the material's surface from a shiny crystalline state to a dull, amorphous state. To erase the data, the laser heats the selected spots on the surface, changing the structure of the heated spots on the surface back to a crystalline structure. A red laser has a wavelength of about 660 nm. A recently perfected blue semiconductor laser has a shorter wavelength, of about 440 nm, which increases the amount of readable data on a DVD-size disc to about 15 GB per side. In essence, this allows for the packing of pits so closely together that one disk will be able to hold four times as much information. This core technology will be the backbone of the industry, producing video products such as high-definition DVD that require large amounts of data storage. Sony markets a high-definition DVD player that utilizes a blue laser licensed from Nichia Chemical Industries that stores as much as 20 GB of data on a double-sided 12-cm disc—more than double that of current players. Blue lasers will be used to produce more effective blue LEDs (see Section 27.3.3.1).*

A study of CD-ROM materials processing and manufacturing techniques provides insight into modern production technology. The processing of a metal master and the subsequent replication of polycarbonate CD-ROMs, shown in Figure 27-1a, involves eight basic steps:

1. Preparing the coated glass substrate
2. Applying a photoresistant coating to the glass substrate
3. Using a laser beam recorder to "cut" pits in the photoresist layer and then developing the glass master
4. Electroforming the glass master to produce a metal "master," or "father"
5. Additional electroforming produces the "mother" and "stamper"
6. Injection molding of polycarbonate plastic discs from the "stamper"

* DVDs—http://entertainment.howstuffworks.com/dvd

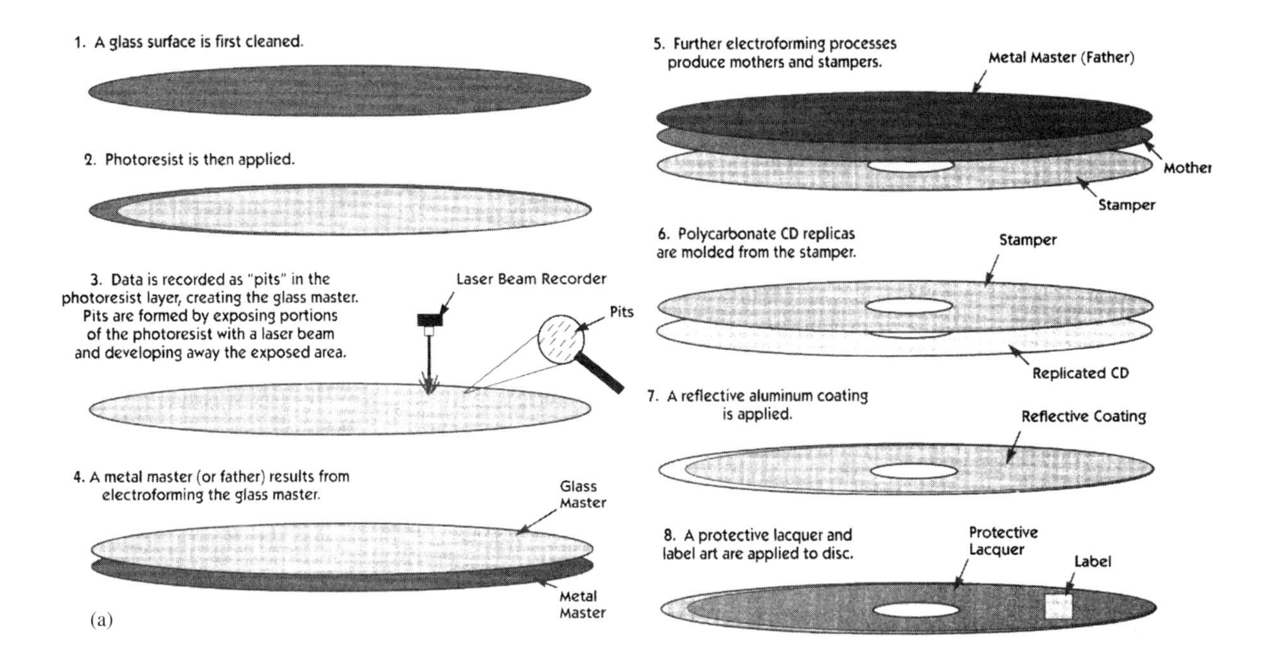

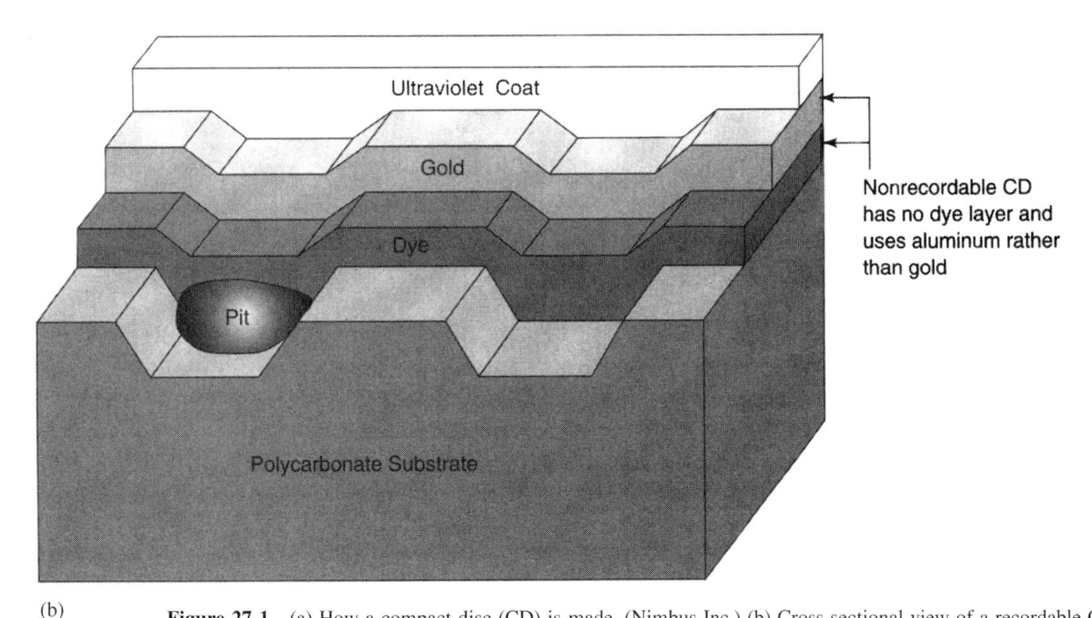

Figure 27-1 (a) How a compact disc (CD) is made. (Nimbus Inc.) (b) Cross-sectional view of a recordable CD.

7. Aluminizing the disc
8. Adding a protective coating to the aluminized surface and applying a label

With multimedia computers becoming commonplace, the demand for CD-ROMs will continue. Most PC computers use these optical storage media in the form of ***OROM (optical read-only memory)***; however, the demand for recordable CD-ROMs, known as ***WORM (write once–read many),*** has already caused an increase in production and a lower price for these more versatile optical storage discs.

Pause for a minute and consider all the rapid developments in information you have witnessed in your lifetime. As with CD-ROMs, many of those developments were a direct result of advances in engineering materials technology. This module will introduce you to some of that technology.

Department of Energy National Laboratories—*http://home.doe.gov/people/labsmap.htm*
Thomas Jefferson National Laboratory—*http://www.jlab.org/*
MIT's *Technology Review—www.techreview.com*
Encyclopedia Britannica—http://www.britannica.com/

27.1 MICROELECTRONICS/SEMICONDUCTORS/DONOR DOPING/ BAND THEORY/ COMPUTER CHIP TECHNOLOGY

Our discussion of energy levels in Module 2 dealt primarily with the action of electrons in a single atom. When atoms come together to form a solid material, the electrons in one atom come under the influence of other atoms. The energy levels that may be occupied by electrons merge or broaden into bands of energy levels. The ***Pauli exclusion principle*** states that only two electrons in the entire solid have the same energy. According to the energy band theory, there are two distinct energy bands in which electrons may exist: (1) the valence band and (2) the conduction band. Separating these two bands is an energy gap in which no electrons can normally exist known as the ***forbidden gap.*** A sketch of these bands is shown in Figure 27-2.

Electrons in the conduction band have escaped the atomic forces that hold them to their nucleus and therefore may move about within the material with the application of minimum energy. Electrons in the valence band, in normal orbit around their nucleus, are tightly held or restrained by the attractive forces originating in the nucleus. Much larger amounts of energy must be applied to extract an electron from or move it within this band. The basic differences among insulators, conductors, and semiconductors can be visualized using the energy band diagram sketched in Figure 27-2. Conductors have no forbidden gap and the valence and conduction energy bands overlap. This permits large numbers of electrons to move (carry electrical current), even at extremely low temperatures. Metals have a partially filled valence band, and insulators have a wide forbidden gap with practically no electrons in the conduction band. Diamond, for example, has a large bandgap. Almost all electrons are in the valence band. Large

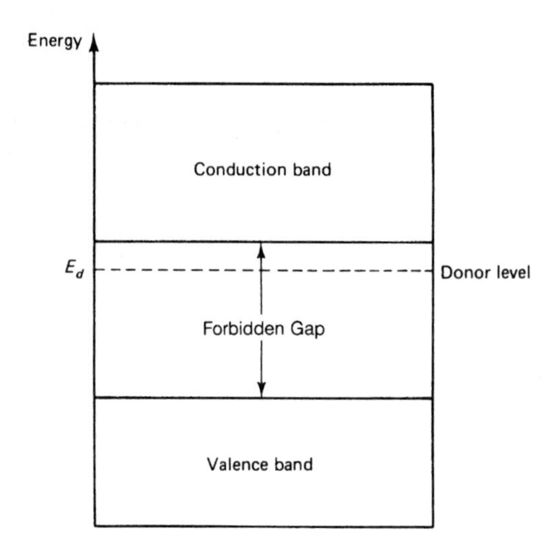

Figure 27-2 Electron energy band diagram showing the forbidden gap.

amounts of energy (about 6 eV) are required to cause an electron to cross from the valence band to the conduction band, but a semiconductor material has a narrow forbidden gap, more narrow than that of an insulator but with electron distribution similar to that of an insulator. However, since the forbidden gap is narrower, only small amounts of energy (0.785 eV for germanium or 1.2 eV for silicon) can raise electrons from the valence band to the conduction band.

Thermal conductivity is carried out by the free motion of electrons and phonons. As a semiconductive material, silicon, in addition to the properties just mentioned, allows the integration of both mechanical and electrical properties in one device. Review Figure 27-3 to see if you can observe any similarities to Figure 27-2. The properties of semiconductor materials make possible the transistor, solar cells, tiny lasers, and the multitude of devices that use integrated miniature circuits. The basis of this new technology is human-made materials that permit the generation and precise control of electron motion within the confines of a tiny crystal of semiconductor material to which has been purposely added (in most instances) a minute amount of impurity atoms, forming a solid solution. ***Transistors*** consist of a crystal of one type of doped semiconductor material sandwiched between two crystals of an opposite type. Minuscule transistors are the basis of all modern computing. Like electrical switches, these devices switch on and off billions of time per second. As they become smaller, they use more power and release more heat due to the leakage of current. This results in short battery life and temperatures too hot for computers to work. One solution is the building of a thin layer of silicon on top of an embedded layer of insulator. This solution lowers the leakage rate considerably. Another solution involves the use of a new material, a high k gate dielectric, that replaces silicon dioxide between the gate and the active area of a transistor. (Review the topics of resistivity and dielectric properties in Module 6 for further descriptions of the terms used here.) Transistors are the basic elements in modern electronic amplifiers and also serve as switching devices for the processing and storage of information. A new ***complementary metal oxide semiconductor (CMOS)*** can provide more than 400 million transistors on a single chip, interconnected with up to seven layers of wiring.

The ***microelectronic circuits*** of semiconductors, when interconnected, are called ***integrated circuits.*** Very-large-scale integrated (VSLI) circuits, when placed on silicon, form a

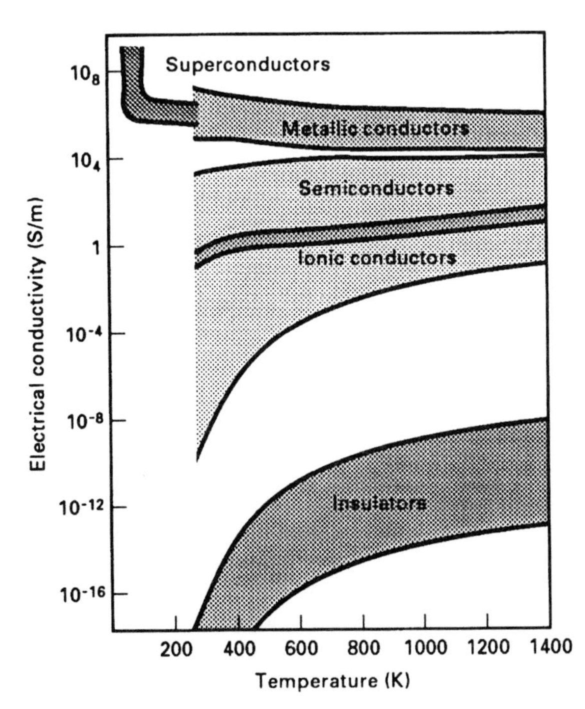

Figure 27-3 Typical conductivity ranges for various classes of electroceramics as a function of temperature. See explanation of units of conductivity in Section 27.2.

Figure 27-4 Silicon wafer with many chips to be cut by diamond saw.

chip, which is rectangular in shape, typically 6 mm on a side (Figure 27-4). In 1999, the maximum number of transistors that could be placed on one chip was about 120 million; the transistors were 0.18 mm long. A knowledge of the electrical properties of materials, makes it somewhat easy to picture conditions that arise when transistor sizes are scaled down, interconnection wiring diameters shrink, and wiring is more closely packed. Resistance rises and capacitance increases, to mention some changes with negative results for performance. Texas Instruments hopes to offset these effects by replacing aluminum wiring with copper, with resistance one-third that of aluminum. If development proceeds as planned, it will make possible complete computer systems on a single chip with operating frequencies exceeding 1 GHz and internal voltages as low as 1 V or lower. The result will be extremely high performance combined with ultralow power consumption and high design flexibility.

Carbon nanotube field-effect transistors (FETs) are replacing silicon-based transistors due to their higher switching rate and the amount of current they can carry per width of conductor. They are designed with a top gate, an electrode situated above the nanotube that controls the flow of electricity through the device. Figure 27-5a is a simple schematic of a FET. Both *p*-type and *n*-type devices can be used in building these FETs, which increases their usefulness (The *n*-type and *p*-type nomenclatures are explained in Section 27.1.2. When the size of silicon-based circuits can no longer be reduced due to the limitations imposed by their fundamental properties of the material, carbon-based FETs will replace them in the ongoing race to shrink electronic circuits.

 Texas Instruments—*http://www.ti.com/sc/docs/news/1998/98079.html*

Pure semiconductor materials are referred to as ideal or ***intrinsic*** semiconducting materials. They possess a minimum of impurity atoms, certainly none added purposely. We recall that it is a very rare occurrence to find a pure substance anywhere in a natural state. Second, we have learned that impurity atoms can effect profound changes in the properties of materials. So it is with semiconducting materials. If selected impurity atoms in accurately measured amounts are added to certain pure semiconducting materials, highly significant changes in behavior will occur. These materials are now known as ***extrinsic*** semiconducting materials. One example is the igniter made of an extrinsic semiconductor material that ignites gas appliances in only 7 seconds by heating up when electric current is passed

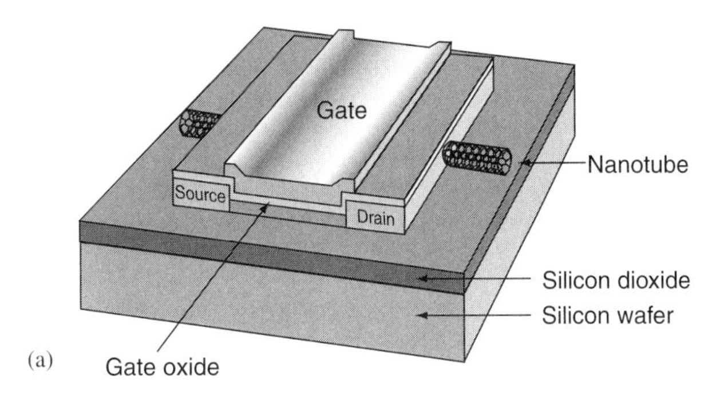

(a)

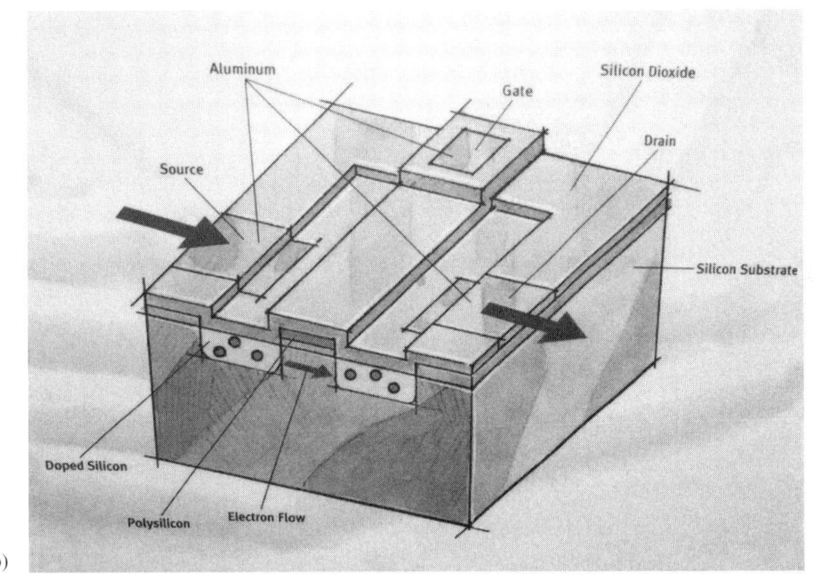

(b)

Figure 27-5 (a) Carbon nanotube field-effect transistors (FETs) with a design that places a gate (electrode) above a carbon nanotube. This design, including very thin insulating layers, outperforms FETs made from a silicon wafer that supports the device as one of its electrodes. (b) Transistors grow smaller as chips become more complex. This cross section of a typical metal-on-silicon (MOS) transistor clearly shows the layers and patterns. As chip makers design more complex chips, the transistors need to be smaller. To do this they make the layers thinner and the patterns smaller.

through it. This invention took nearly 20 years to develop. Intrinsic semiconducting materials use the elements in Group IV of the periodic table, germanium (Ge) and silicon (Si). These elements form crystal structures like that of carbon in diamond (Figure 27-6). Each atom is bonded to its four neighboring atoms with covalent double bonds. There are no free electrons in the conduction band. How can these bound electrons be excited to act as charge carriers?

Pure element semiconductors—http://wwwdmse.mit.edu/wom/elementalsemiconductors.html
Semiconductor applets—http://jas2.eng.buffalo.edu/applets/

27.1.1 Electrically Charged Particles

Electricity is defined in terms of the flow, or movement, of electrically charged particles. This movement to other locations results in increased kinetic energy, which can be converted to some form of work for our benefit. Several forms of charged particles can satisfy these conditions, although some are not as efficient as others. The word *charged* refers to the electrical charge carried by these particles. All particles have this same charge, which is equal to 1.6×10^{-19} coulomb. Its symbol is e^- and it is the standard unit for measuring electrical charge. The magnitude of the

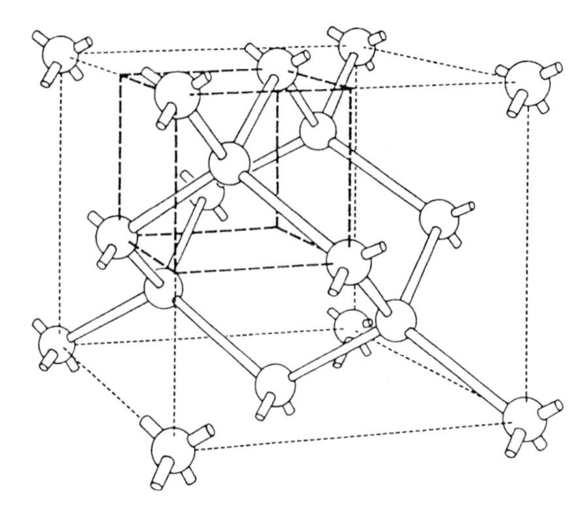

Figure 27-6 Representation of the silicon-crystal lattice arrangement. (U.S. Dept. of Energy)

charge in one proton is the same. However, the proton is a positive charge; its symbol is e^+. The measure of electric current (I) is the amount of charge (q) that passes a given point per unit of time (t):

$$\text{Current } (I) = \frac{\text{charge } (q)}{\text{time } (t)} \text{ or } I = \frac{q}{t}$$

An **ampere** (A) is defined as an electric current of 1 coulomb (C) of charge passing a point in 1 second (s). Knowing that each electron contains an electric charge of 1.6×10^{-19} C, and using a proportion to determine the number of electrons in 1 C of electric current, the results are 6.25×10^{18} electrons in 1 C. Therefore, 1 A of electric current is the passage of 6.25×10^{18} electrons in 1 s.

A ***hole***, the absence of an electron, is considered to have the equivalent of an electron's charge, only with a positive sign. As you recall, holes in the valence band are produced by supplying electrons with energy sufficient to cause them to be excited to the conduction band. Free electrons have greater mobility than holes. For comparison purposes, the mobility of electrons in pure germanium at 300 K is about 3950 cm^2 per volt second, and for holes it is about 1950.

27.1.2 Donor Doping

If a Group V element such as phosphorus (P), with its five valence electrons, is added to germanium (Ge), then only four of its outer electrons are needed to perfect a covalent bond with the neighboring germanium atoms to form a substitutional solid solution. The extra electron not needed for bonding will be attracted to the impurity atom (phosphorus) by a weak coulomb force of attraction of the nucleus. Because this electron is held much less tightly than the four bonding electrons, it and similar electrons contributed by other impurity atoms can be raised into the conduction band by the absorption of much smaller energies than those required for raising electrons from the valence band into the conduction band. This situation is depicted in FiguresFigure 27-7a and b. The extra electron is shown lying just below the conduction band at an extra energy level (E_d) called the ***donor level.*** This type of impurity atom makes the material n-type (possessing free electrons). The process of adding impurities is called ***doping.*** Figure 27-8 is a sketch of one phosphorus atom substituting for a germanium atom in the crystal lattice of an n-type germanium semiconductor material. Other donor impurities that can be used for donor doping from Group V are arsenic (As) and antimony (Sb). Note that these atoms, having five valence electrons, are referred to as ***pentavalent*** atoms. Initially discussed in Section 3.2.8 and again when describing surface-hardening

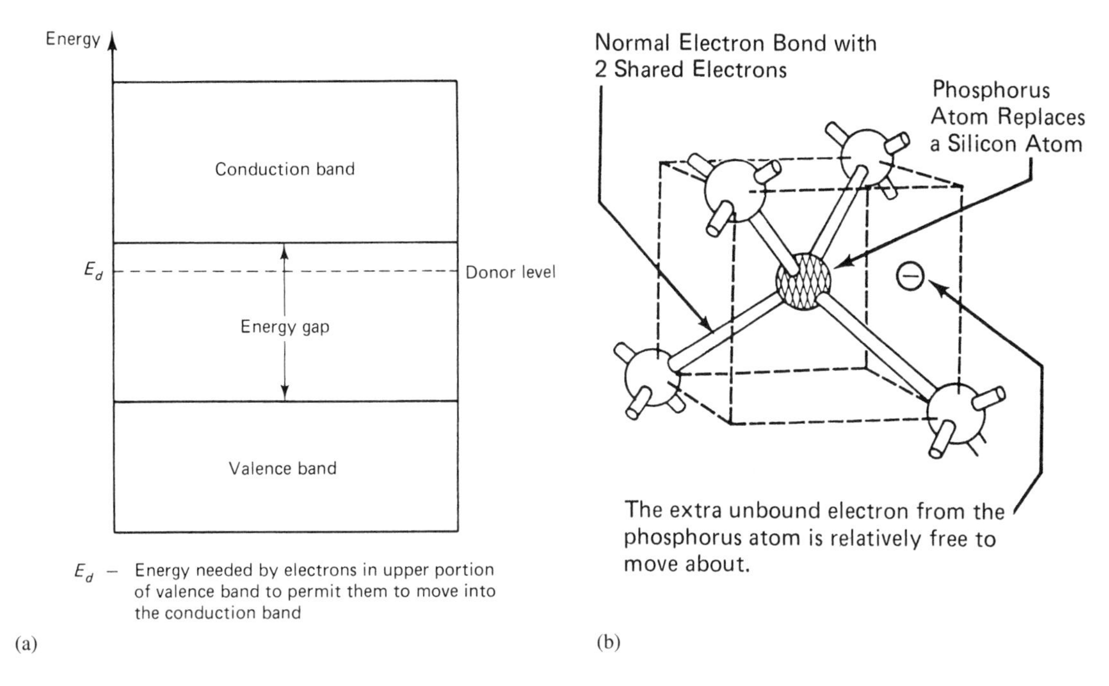

Figure 27-7 (a) Energy-band diagram of an *n*-type semiconductor. (b) Substitution of a phosphorus atom having five valence electrons for a silicon atom leaves an unbonded electron. (U.S. Dept. of Energy)

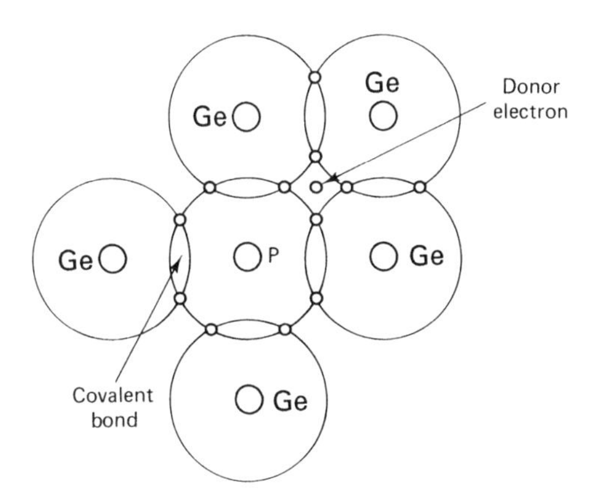

Figure 27-8 Effect of a dopant phosphorus atom substituting for a germanium atom in the crystal structure of an *n*-type semiconductor material.

processes such as carburizing and nitriding in Section 13.1, the ***diffusion*** of atoms and molecules across a boundary between two materials in contact is called ***mass diffusion.*** The two materials must share a common interface or boundary. In metals, the atomic motion is facilitated by vacancies and other imperfections in the crystal structure (see Sections 3.3 and 11.1.1). In semiconductor processing, diffusion is used to dope the silicon substrate with the desired amount of impurity atoms. The dopant enters the microstructure of the silicon substrate, substituting for the silicon atoms until a maximum concentration limit is reached. Temperatures around 1800°F accelerate the diffusion of dopant atoms.

27.1.3 Holes

From our study of the electronic structure of atoms, we have learned that any orbit of most atoms, other than the K shell or first shell (which is completely filled with two electrons), is stable if it contains eight electrons. Silicon (see Figure 27-9a) has only four electrons in its outermost shell (M shell). Being the outermost shell, it contains space for four additional electrons.

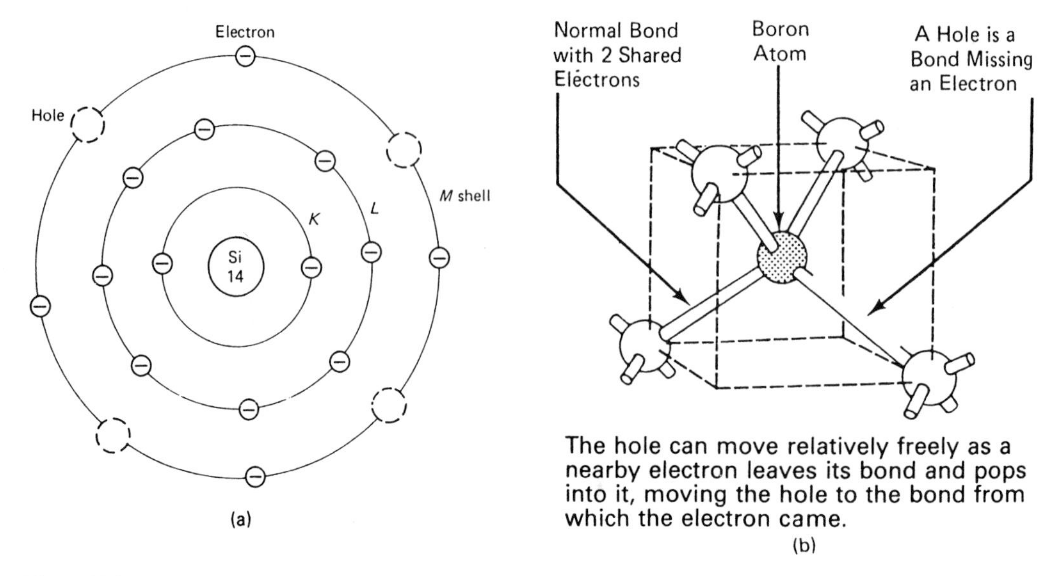

Figure 27-9 (a) Two-dimensional representation of the silicon (Si) atom. (b) Boron, a three-valence-electron atom in a silicon crystal, is normally bonded, except that one of the bonds is missing an electron, creating a hole. (U.S. Dept. of Energy)

These vacancies or absences of an electron in the valence shell are called ***holes.*** Holes are created in the valence band when valence electrons receive sufficient energy to move up into the conduction band. With intrinsic semiconductors, the valence electrons with the highest energies (outer-shell electrons have different energies) will be freed first when energy is added to the material as a result of an increase in temperature. The transfer of these electrons to the upper band creates holes in the lower (valence) band, which can then be occupied by electrons lower down or deeper in the valence band. When an electron receives an input of energy from some outside source, the electron will be loosened from its bonded position in an atom and be free to move about in the crystal structure (in the conduction band). If the energy received is insufficient to loosen the electron, it may just vibrate, which gives off heat. Once the electron breaks loose, it leaves behind a bond that is missing an electron. This bond may also be called a hole (see Figure 27-10b). Electrons and holes freed from their positions in the crystal structure are said to be ***electron–hole pairs.*** A hole, like a free electron, is free to move about in the crystal (Figure 27-10b).

27.1.4 Conduction in Solids

Electricity flow (conduction) occurs in a solid when an applied voltage (electric field) causes charge carriers within the solid to move in a desired direction. In the absence of an electric field, the movements of charge carriers are random and result in zero charge transport. If a metal is placed in an electric field (in an electrical circuit), the free electrons in the conduction band will move toward the positive terminal. As they do so, they receive additional momentum (mass × velocity) and hence more energy. Those electrons moving toward the negative terminal lose momentum and reduce their energy.

Hole transfer involving electrons takes place when an electron from one atom jumps to fill the hole in another atom. This electron jump leaves a hole behind it. Or we can describe this movement, not in terms of what the electron does, but in terms of the hole movement. The hole moves in the direction opposite to the electron. Therefore, the flow of electric current is brought about by the movement of free electrons in the conduction band and/or holes in the valence band. Free electrons are easier than holes to move through a semiconductor material, hence they have greater mobility (***drift velocity*** of the carrier). With the help of Figure 27-10a, the movement of one hole can be clarified. The hole starts with atom *A* as an electric field (E) is applied, and a valence electron breaks free and moves to its left to fill the vacancy (hole)

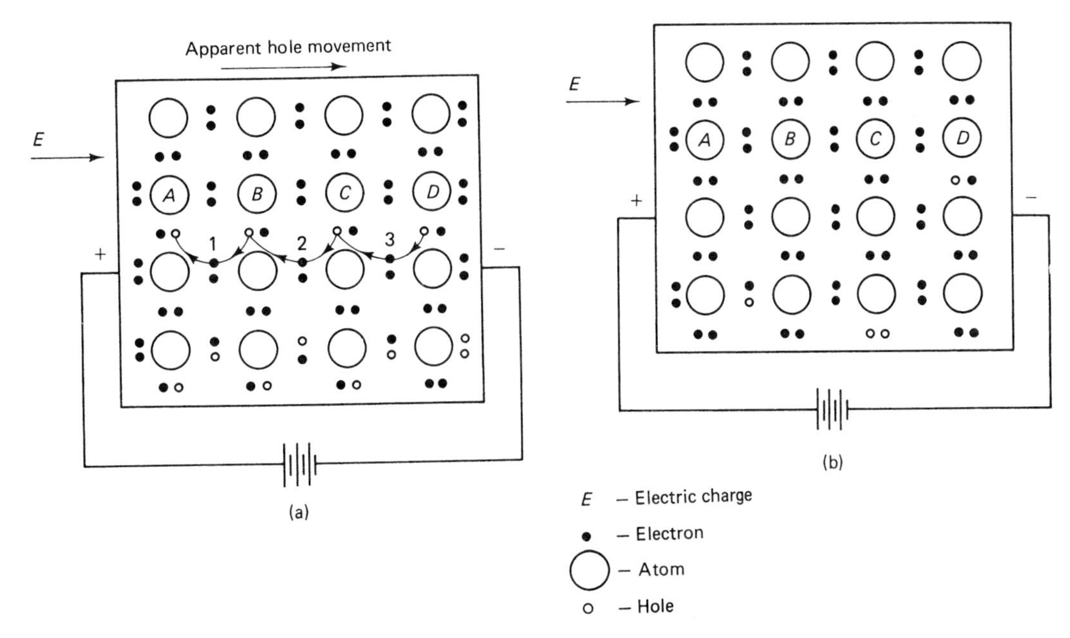

Figure 27-10 Conduction by holes in the valence band.

shown by arrow 1. The hole is now at atom *B*. A similar effect is felt by one of the valence electrons at atom *C*, which also moves to its left, and fills the hole at atom *B*. Arrow 2 shows this movement. Each time an electron moves, it creates a hole. Figure 27-10b shows the final location of the hole at atom *D*. Note that the hole movement, in this instance from left to right, is toward the negative-charged terminal of the battery, opposite to the flow of electrons as they move toward their left to fill the holes. In other words, this hole movement is a measurement of positive charge in the direction of the applied field (*E*). The action of the electrons in their moves, labeled 1, 2, and 3, in filling the just-created holes is known as *recombination*, which will be discussed later.

27.1.5 Acceptor Doping

When a Group III, or trivalent, element such as boron (B), aluminum (Al), gallium (Ga), or indium (In) is introduced as an impurity into an intrinsic semiconductor such as silicon (or other Group IV elements), a mismatch in the electronic-bonding structure occurs. Using aluminum as an example of the dopant, this element lacks one electron to satisfy the covalent bonds of a Group IV matrix element such as silicon. This mismatch is illustrated in Figures Figure 27-11 and 27-9a. One covalent bond near each dopant atom (Al) is incomplete; that is, it is missing an electron or it contains a hole. If an external electronic field is applied to this solid, one of the neighboring electrons from another covalent bond can acquire sufficient energy to move into the hole. This, in turn, causes the hole to move to the position formerly occupied by the electron. In this manner the hole moves through the solid as a positive-charge carrier. The electric current produced is mainly the result of hole movement (positive-charge carrier). This structure is called *p*-type because of the presence of free positive charges (the moving holes).

In terms of the energy band diagram for this *p*-type structure (Figure 27-12), the aluminum atom has provided an energy level that is only slightly higher than the upper limit of the valence band. This puts it in the forbidden gap. Thus, a nearby valence electron could easily be excited into this intermediate level, called an ***acceptor level.*** As with *n*-type extrinsic semiconductor materials, the extremely small concentration of dopant is measured in parts per million (ppm). In summary, the addition of acceptor atoms to a semiconductor material increases the number of holes in the valence band without an increase in the number of electrons. These positive-charge carriers are termed ***majority carriers***; the electrons are called ***minority carriers*** in *p*-type semiconductors, just the reverse of their designation in *n*-type semiconductors.

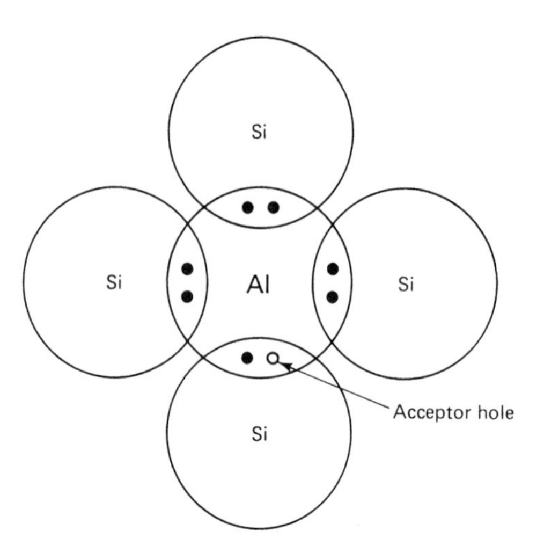

Al = Acceptor atom

Figure 27-11 Structure of silicon with aluminum atom added as an impurity (dopant).

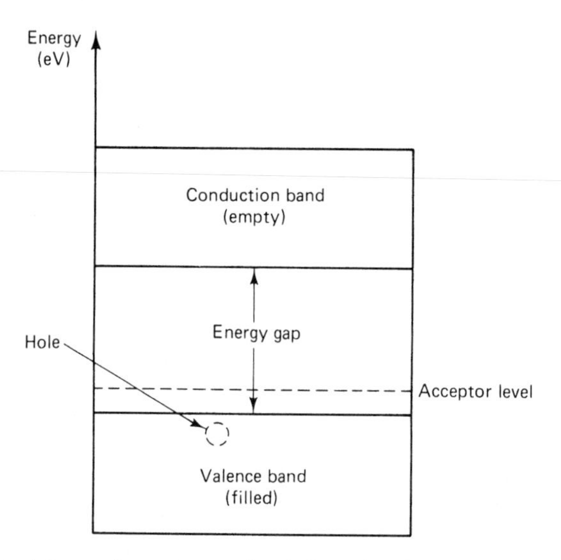

Figure 27-12 Energy-band diagram for a *p*-type semiextrinsic semiconductor material.

Figure 27-13 represents an *n*-type semiconductor material in a circuit with a switch in the open position. No electrons will flow in the external circuit. However, within the doped *n*-type semiconductor material, the free electrons contributed by the dopant atoms and the electrons that break away from their parent atoms in the valence band are diffusing throughout the material in a random fashion. In addition, each time a valence electron breaks free of its atom, an electron deficiency in the valence shell of that atom is created, which makes the atom a positive ion with an electric charge of equal magnitude to that of the free electron. The vacancy in the valence shell (energy band) of the electronic structure of the atom is called a hole and the phrase *electron–hole pair* refers to this energy transfer process, as we have discussed. When a free electron collides with a hole, it is captured by it, which produces a balanced atom in its lowest equilibrium state. This process is known as ***recombination.*** Taking into consideration the remaining atoms that comprise the material, we recognize that overall equilibrium must be maintained, and thus a new electron–hole pair is generated with every recombination (Figure 27-14).

Next we close the switch in the external circuit in Figure 27-13. This causes electron flow to occur from the negative terminal of the source through the semiconductor material

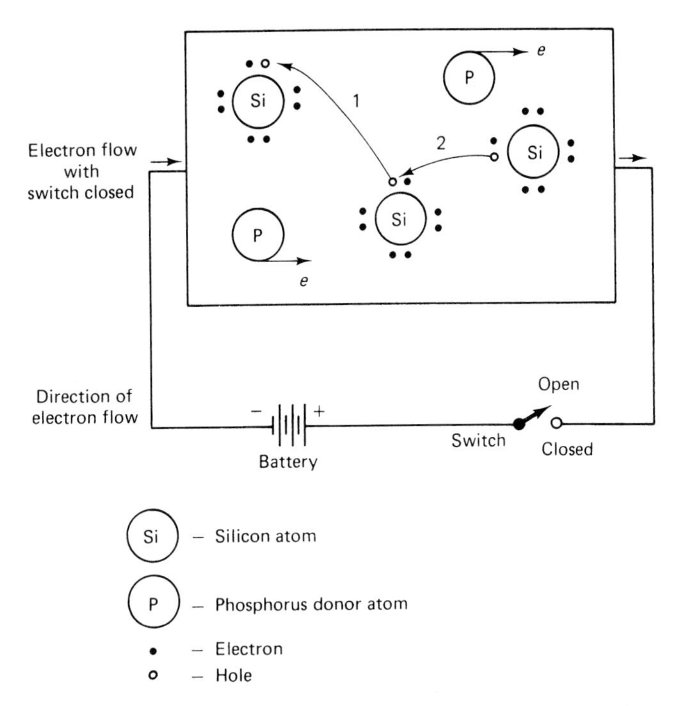

Figure 27-13 Random movement of electrons and holes in an *n*-type semiconductor material.

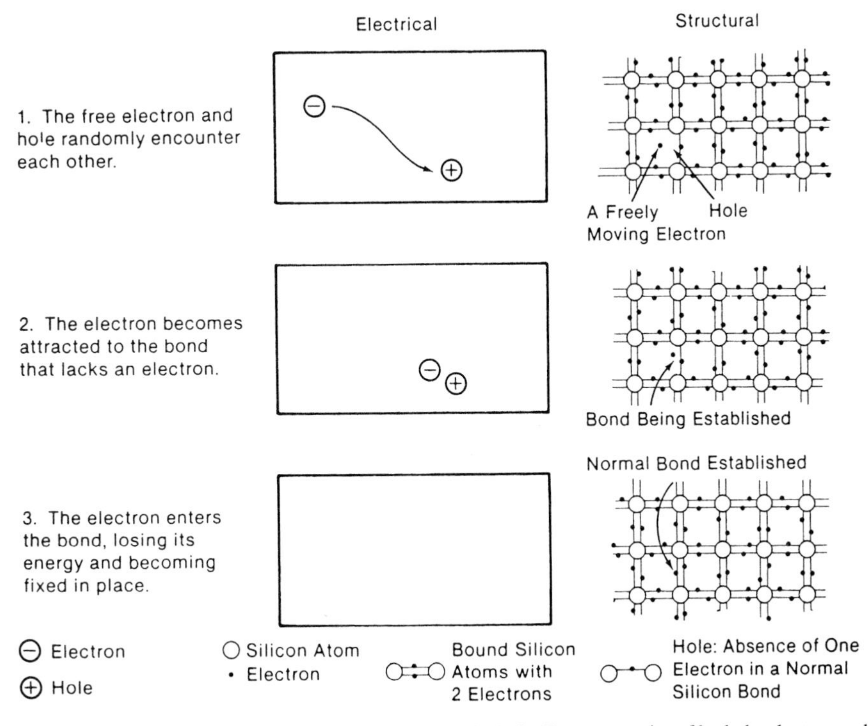

Figure 27-14 Direct recombination of an electron and a hole. Excess energies of both the electron and hole are lost to heat. (U.S. Dept. of Energy)

and back to the positive terminal of the source. Figure 27-13 attempts to show that only electrons are flowing in the external circuit within the material, but the action of the electric field gives direction and movement (drift) to the flow of both free electrons, donated by the phosphorus atoms in this instance, and the holes (minority carriers) in their movement from right to left. Both the free electrons donated by the donor atoms and the electrons supplied through

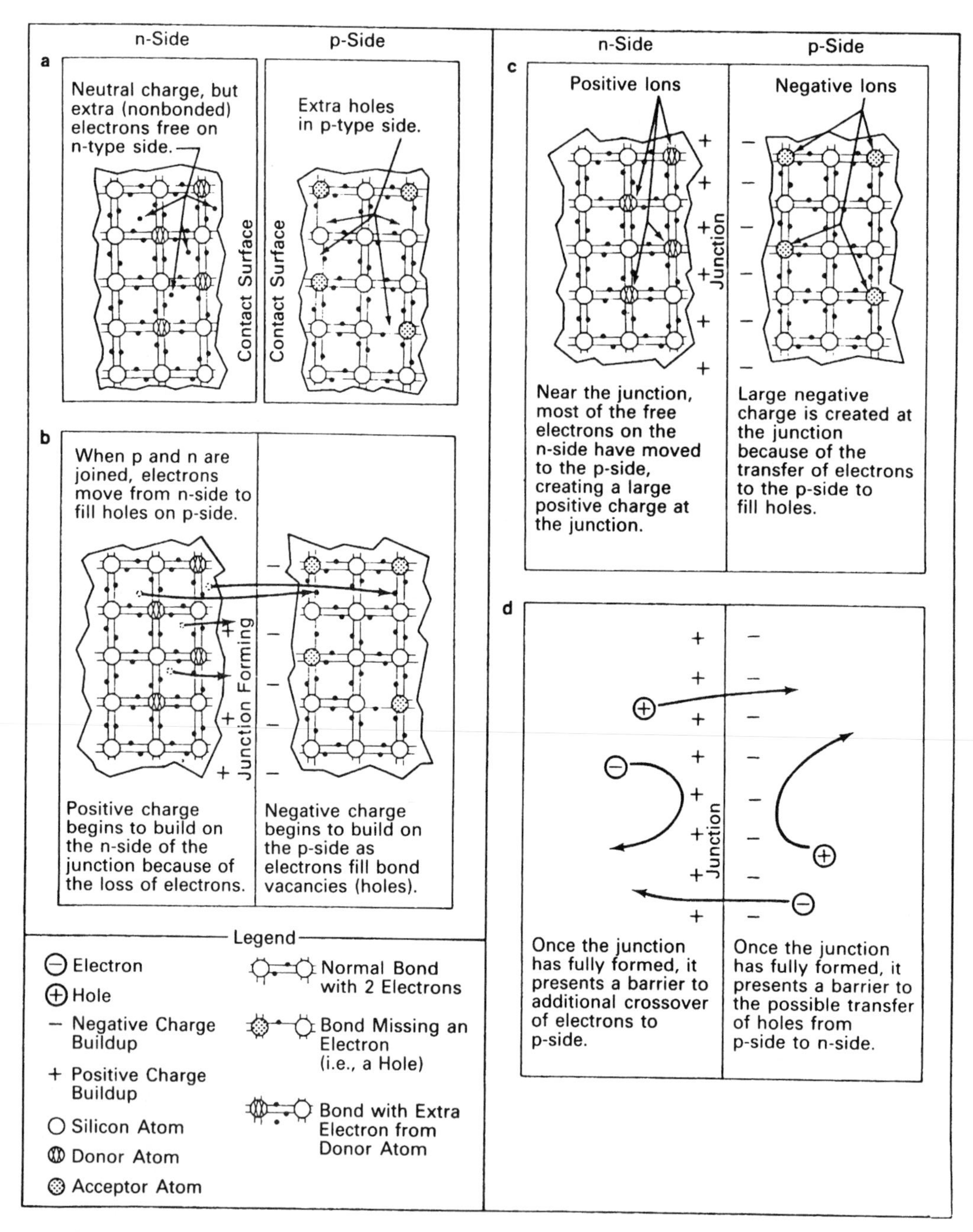

Figure 27-15 Junction formation. Additional movement of charge carriers is stopped and equilibrium is established. (U.S. Dept. of Energy)

the external circuit combine with the holes to accomplish the conduction process through the processes of generation and recombination.

When a piece of *n*-type material is brought into contact with a piece of *p*-type material, a ***pn junction*** is formed (Figure 27-15). A junction of this type produces a ***diode,*** which is the simplest of the semiconductor junction devices. Before observing the diode in a simple circuit, it would be helpful to study the interaction of the electrons at the *pn* junction. The *n*-type material contains an excess of electrons, whereas the *p*-type material has an excess of holes. When the *pn* junction is formed, there is a diffusion of charges between the two materials. The electrons from the *n*-type material will move across the junction to the *p*-type. This will create holes or positively charged particles in the *n*-type. The reason for this diffusion of charges is to counteract the electrical imbalance in the two materials caused by the formation of the junction. The diffusion process will stop once an equilibrium of charge is established between the two materials. This equilibrium is reached when no additional charge carriers have the energy to overcome the electric field that has been built up at the junction. It can be seen in Figure 27-15 that a boundary is formed across the *pn* junction. This boundary is usually referred to as the ***depletion region*** because all mobile charge carriers have been depleted from this area. The electric field set up by the stationary charges in the depletion region creates a difference of potential between the two materials and represents the equilibrium potential difference from *n*-type to *p*-type in direction. This voltage is called the *contact potential* of the junction. For silicon, it is about 0.6 V.

The previous discussion has been focused on the *pn* junction (a diode) as a closed system. We will next observe the junction with an externally applied voltage (Figure 27-16). The *p*-type side of the diode is called the *anode* and the *n*-type side is called the *cathode*. Applying a voltage to the diode is called ***biasing.*** If, as in the second diagram in Figure 27-16, a battery is connected across the diode so that the positive lead is attached to the anode and the negative lead is attached to the cathode, a large current will flow. This is the forward-biased condition of the diode. The reason for the large current is that the junction potential (contact potential and applied potential) has been lowered due to the forward bias. Electrons from the battery that enter the anode of the device will see little resistance and can flow easily across the *narrowed* depletion region to the cathode.

In the third diagram in Figure 27-16, the leads of the battery have been reversed so that the positive side of the battery is across the cathode and the negative side is across the anode. This is the *reverse-biased* condition of the diode, and very little current (only leakage current) will flow. The reason for the extremely small current in this condition is that the battery is acting in the same direction as the electric field across the junction; thus, the junction potential is larger. When the junction potential increases, the width of the depletion region and the amount of stationary charges will increase. This means that the charge carriers must gain an extremely large amount of energy before they can cross from cathode to anode.

The *pn* junction is the basic building block for all semiconductor junction devices. The preceding discussion can be applied to the transistor, silicon-controlled rectifier (scr), field-effect transistor (FET), or any other junction device.

pn junction diode at equilibrium—
http://jas2.eng.buffalo.edu/applets/education/pn/pnformation/pnformation.html

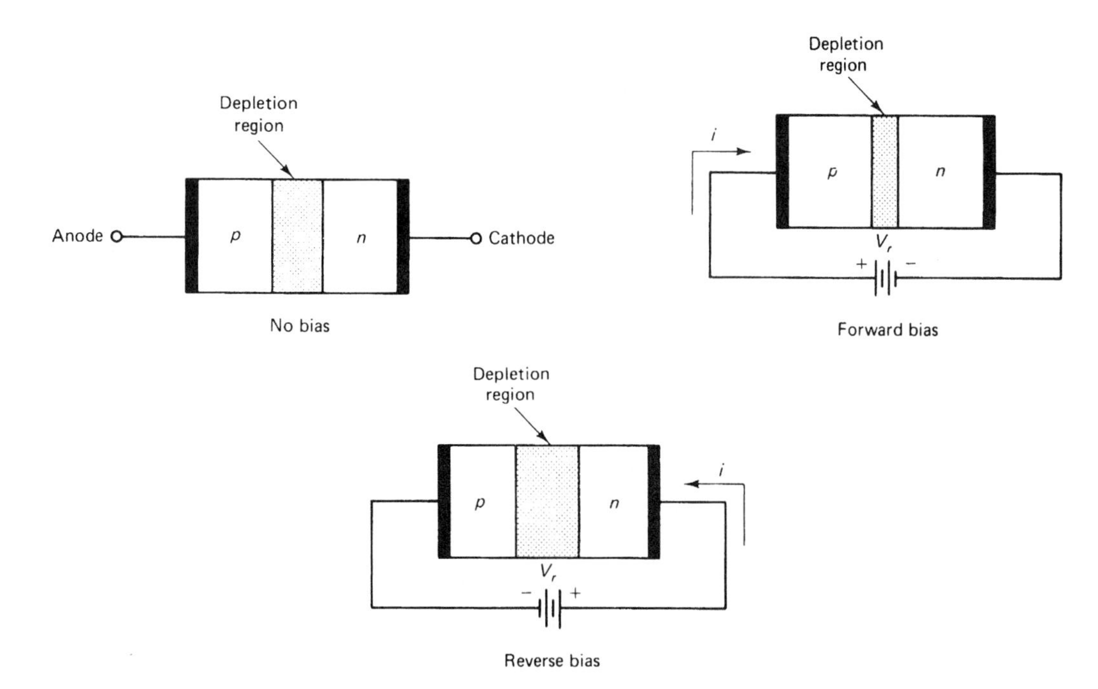

Figure 27-16 A *pn* junction (diode) showing no-bias, forward-bias, and reverse-bias conditions.

27.1.6 Computer Chip Technology

Semiconductors and their properties form the basis for the design, production, and application of electronic microscopic devices. As each year passes, these devices become smaller, such that by the year 2001 the average size was 180 nm. Examples of such microelectronic devices are the *pn* junction diode and the *npn* bipolar junction transistor. These devices and many others are linked to fabricate microelectronic circuits known as ***integrated circuits (ICs)*** or ***chips***.

A simple IC can be an amplifier and a complex IC a complete digital computer (or microprocessor). These devices must be connected by paths that can readily conduct electric currents. In most cases, metal wires are used made by filling tiny trenches in the surface of semiconductor wafers. Multiple layers are created by penetrating a layer to make contacts with the layers above and below. The metal used for these wires has been aluminum, and for interlayer connections, titanium. As these devices get smaller, copper is succeeding aluminum because of its greater conductivity, among other properties. A new method of laying copper wire for ICs is called the ***ion-assisted trench*** method. This technique can produce thin and uniform layers that can conform to the shape of the patterned wafer. As is often the case with other materials combinations, however, copper is not compatible with silicon: It readily diffuses in silicon but causes defects in the silicon structure. The metallization process can deposit many layers of metal, each insulated by a dielectric. The layers of metal are connected by ***vias***. Contacts provide access to the devices on the substrate. The width of this new circuitry is reported to be 1/400 the width of human hair, which varies in diameter from 40 to 120 μm. Expressing this width in micrometers and using the smallest value for the width of a human hair, the new width for circuitry would be about 0.1μm. The wires require connections to other wires using soldered joints. As the size of these devices becomes smaller, the numerous tiny, very fine, tightly packed connections become more difficult to solder. The failure of one joint only a few thousandths of an inch wide can cause failure of the entire device. A computer model has been developed that predicts when solder joints are likely to fail by calculating their precise shapes.

Silicon an excellent natural oxide with a remarkable set of properties is still the favored semiconductor material. Some of its advantages are lower cost; larger bandgap, which permits operations at high temperatures; and integration of mechanical and electrical properties in one device. The fabrication of ICs begins with a thin piece of silicon (see Figure 27-6), an intrinsic semiconductor on which well-defined surface areas are processed in a way that will allow them to contain certain amounts of electrically active dopants. These areas are created by first removing the SiO_2 surface layer, which blocks the diffusion of many types of dopants. Next, the dopant is added, which thermally diffuses into the substrate. When a layer of insulating materials is required, such as in a capacitor (two conducting layers separated by a thin layer of insulating material), the SiO_2 layer is retained. Thus, SiO_2 is the primary choice for most microelectronic materials. Gallium arsenide (GaAs) is preferred for high-speed and high-frequency applications because of its greater electron mobility. Silicon carbide (SiC), with its high-temperature stability (withstands temperatures up to 325°C) and its resistance to chemical attack, is used in aggressive environments. Its one disadvantage involves the difficulty of growing its inexpensive, simple crystals. An alloy made of silicon, carbon, and germanium has been developed for use with high-temperature, high-power, and high-frequency microelectronics and ***microelectromechanical machines (MEMs)***. The germanium (Ge) permits higher speeds than Si, but it is difficult to add much Ge to Si because the Ge atoms are larger than Si atoms and consequently strain the Si lattice structure. By adding of 1–2% carbon, however, greater amounts of Ge can be used, and by switching from Si to SiC, additional Ge can be used (up to 4%) to produce twice the current as pure SiC. Another technology, IBM's ***silicon-on-insulator (SOL),*** for chip manufacture, which involved some 15 years of effort, deposits a layer of dielectric on a silicon wafer. Giving better protection to the metal circuitry allows for closer placement of transistors and more efficient flow of electrical current. Combined with the improvement in chip manufacture made possible by the replacing of aluminum with copper, these advances have boosted chip performance 35%, cut power usage by one-third, and increased the service life of silicon-based chips.

The flow of electrical current, no matter how small, generates resistance, which manifests itself in the form of heat, so the constant challenge for electronic chip manufacturers is to decrease resistance and thus (1) reduce the amount of heat to be dissipated, (2) reduce the amount of dielectric needed to insulate the individual layers of the circuit, (3) increase the ***chip speed*** (electron flow through the circuits), (4) decrease power consumption so that smaller, battery-type energy sources may be used, and (5) allow for the greater placement of more and smaller semiconductor devices such as transistors in circuits of smaller and smaller width. The copper-chip technology that is replacing aluminum wiring to connect transistors increases processor speeds and decreases power consumption. Power PC 740 and 750 processors in workstations now have this technology, and personal computers are next in line to receive it. Present cell phones use 200 to 250 chips to send and receive calls, commonly on one frequency (900, 1800, or 1900 MHZ) and in one encoding format (GSM CDMA, or TDMA). This means the traveling public would need several phones to maintain communications in different parts of the world. IBM has developed a new chip technology using silicon germanium (SiGe) that will allow future cell phones to use multiple frequencies and encoding formats, an integration made possible by SiGe's compatibility with the other cellular components that act on the digital signal and reproduce the sound of the caller's voice. Personal digital assistants, watches, and jewelry are all possible future applications that will benefit from the reduced power consumption and longer operational capability of more densely packed transistors. Philips Electronics the standard development time for new chips in half when it produced its new base-level processor called Nexperia, which uses 32 million transistors. The processor has a variety of options, such as Internet access and telephony, and it contains a customer intellectual property module. Its initial use is for set-top boxes that control TVs'. The processor will allow the viewer to move from a TV show to a website and back again because the disc recorder automatically continues to record the TV program.

IC chips are attached to substrates (***electronic package casings***) so that they may be safely attached to structures or supported for shipment. Normally these two materials are not compatible in that they differ in properties such as thermal conductivity and thermal expansion (expressed by the coefficient of thermal expansion, ***COT***) Table 19–2 compared some metals with a variety of ceramics in terms of COT. Silicon has a COT of 2.6×10^{-6} °C^{-1}, whereas aluminum has a COT of 23.6×10^{-6} °C^{-1}. Materials with such a wide difference in COTs, if attached to one another, will experience a buildup of thermal-induced stresses as temperatures change, a condition that could lead, in the case of ICs, to warping and more likely a loss of adhesion between the layers of conductive metal and dielectric ceramics or polymer materials of which most are composed. One alloy of iron–nickel–cobalt, called ***Kovar®***, has a coefficient of 5×10^{-6}°C^{-1}, which is a better match for silicon chip mounting. (Kovar is the registered trademark of Carpenter Technology Corp. for its nominal 29Ni–17Co–54Fe glass-sealing alloy UNS K94610.) An optoelectronic package incorporating advance materials to maximize voltage and power handling for hybrid power circuit applications is made of a powder metallurgy composite of sintered and pressed molybdenum–copper powder, which has outstanding thermal conductivity. The base is sealed to the alumina ceramic package with a Kovar seal ring. The high resistance to vibration and mechanical shock failure makes these packages ideal for aerospace applications.

27.2 ELECTROCERAMICS

Inorganic single-crystal and polycrystalline compounds as well as glasses are included in the group of highly conducting electroceramics that can be used in a variety of electrical, optical, and magnetic applications. Oxide ceramics are the materials of primary interest, but carbides, fluorides, and sulfides are also finding many commercial applications. Additional information on similar ceramic materials can be found in Module 19. Such ceramics are generally classified as electronic conductors, ionic conductors, mixed (electronic/ionic) conductors, and insulators (Figure 27-3). The electronic conductors include superconductors, metallic and ionic conductors, and semiconductors and represent materials with the highest conductivities. The metallic conductors exhibit high conductivities that decrease with increasing temperature; the conductivities of oxide semiconductors increase with temperature. Ionic conductors generally exhibit conductivities in the range 10^{-8} to 100 S/m, which increase exponentially with temperature.

(The units of conductivity used in this discussion and in Figure 27-3 are expressed using the special SI unit name ***siemens***, with symbol S. A siemens is defined as the quotient of amperes divided by volts: A/V.) Insulators such as high-purity alumina are at the lower extreme of the conductivity spectrum, with conductivities of 10^{-13} S/m. Figure 27-3 summarizes the electrical conductivity characteristics of these classes of ceramic materials.

27.3 OPTOELECTRONIC DEVICES

Optoelectronic devices (1) are operated by light (photoelectric), (2) produce or emit light, or (3) modify light. Photoelectric devices, in turn, can be categorized as photoconductive, photovoltaic, or photoemissive. A review of optical properties in Section 6.4 is appropriate prior to reading the following discussion of these devices. ***Spectroscopes***, or ***spectrometers*** measure accurately the wavelength of light emitted by atoms after they have been excited by the use of diffraction gratings. A ***diffraction grating*** is a large number of equally spaced slits. These gratings can be made by precision machining of very fine parallel lines on a glass plate. The untouched spaces between the lines serve as the slits from which light is reflected and analyzed. These devices can also be used to identify atoms and molecules whose chemical analysis would be difficult. Spectroscopy is initially discussed in Section 8.4.1.

27.3.1 Photoconductive Devices

In photoconductive devices, the conductivity of the semiconductor material will vary, with the energy supplied by the light (visible, infrared, or ultraviolet), which must be sufficient to raise the electrons into the conduction band. A photomultiplier is a device that can produce an visible image when an object is illuminated by a weak light source. A light meter is an example of a device that operates on this basic principle. Gallium arsenide (GaAs), a compound, can convert light directly into electricity, an example of ***photoconductivity*** that is used in LEDs and other solid-state devices such as transistors. Photoconductivity is also discussed in Section 6.4.4.

27.3.2 Photovoltaic Devices

Photovoltaic (PV) cells are semiconductor junction devices that convert electromagnetic energy in the form of light directly into electrical energy. The amount of electrical current (flow of electrons) is directly proportional to the amount of light that is incident on the semiconductor material. A ***solar*** or ***photovoltaic cell*** is a photodiode used to extract electrical energy directly from sunlight. This direct conversion of sunlight to electricity differs from the solar thermal conversion process used in the solar heating of homes and offices, which uses panels to absorb the energy from sunlight to heat water or other media, including air, to heat a building. Photovoltaic cells embedded in the glass of an automobile sunroof are a recent application of these devices. To help keep the interior cool, the cells power two fans that pull in cooler outside air and exhaust hot interior air through vents in the roof when the temperature inside the vehicle exceeds 85°F. If it starts to rain, sensors shut the vents.

Typical solar cells contain two extremely thin layers of silicon connected externally by a wire to the load, where light energy is converted to work (see Figure 27-17a). When light is absorbed in silicon, it creates electron–hole pairs (Figures 27-17b and c). Electron–hole pairs that reach the junction are separated—holes going to the p side and electrons to the n side. This buildup of charge on either side of the junction creates a voltage that drives current through the external circuit. Reflection of sunlight from an untreated silicon cell can be as great as 30%. By chemical treatments and texturing of the surface, this value can be reduced to around 5%. Light with a certain wavelength (energy) is required to interact with specific materials before the optimum electron–hole generation is achieved (Figure 27-18c). This characteristic energy is called the material's ***band energy***. (Silicon requires 1.1 eV; gallium arsenide, 1.4 eV.) Overall, these mismatches of light with a solar cell's material waste some 55% of the energy from sunlight. Research is proceeding to find better ways to process sunlight to make it mono-

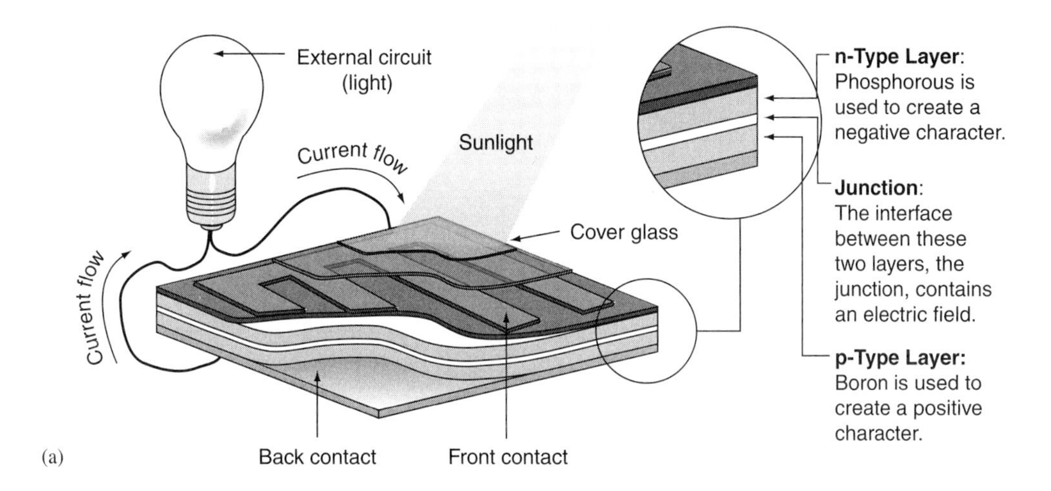

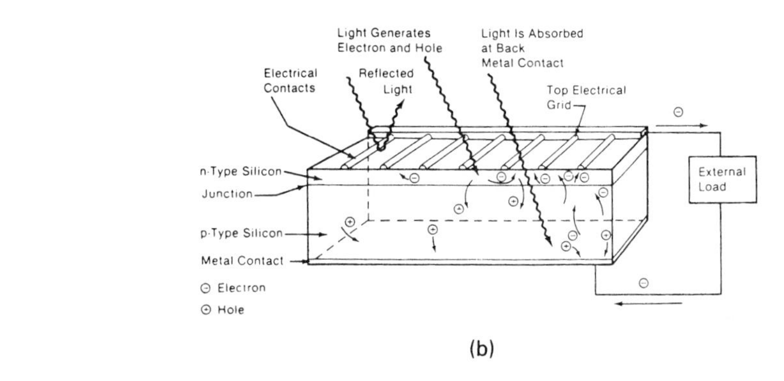

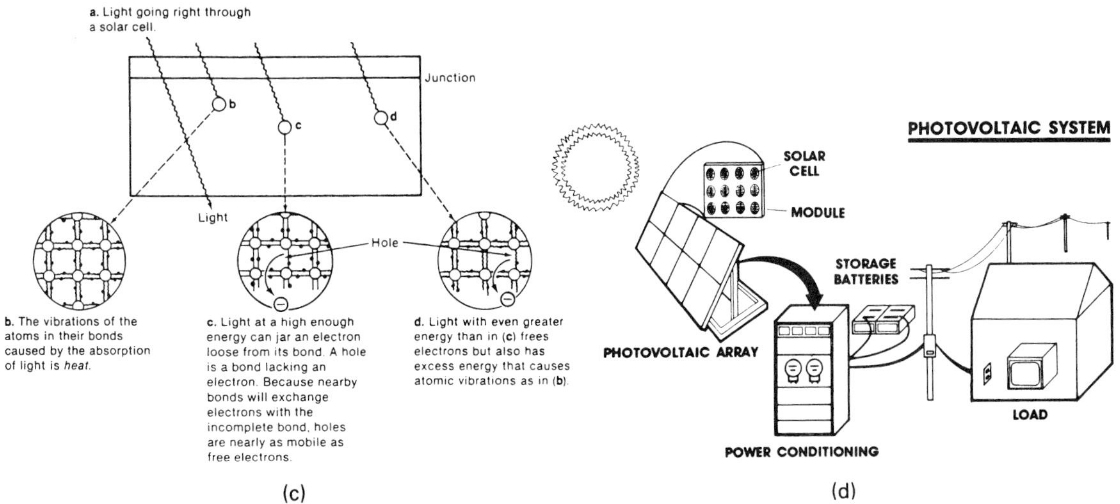

Figure 27-17 (a) A typical *pn* junction, single-crystal, silicon solar cell, outlining its components. (b) Light striking a cell creates electron–hole pairs that are separated by the potential barrier, creating a voltage that drives a current through an external circuit. (c) What happens to light entering a cell? (d) Schematic of a photovoltaic system. (U.S. Dept. of Energy)

(a)

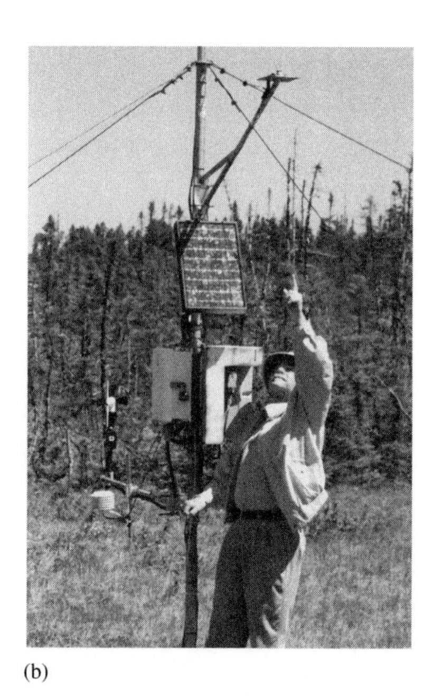

(b)

Figure 27-18 (a) Devices that require electricity in remote areas make solar cells affordable. The U.S. Forest Service fee collection booth for wilderness parks seen here gains reliable energy from the sun in desert areas. Similar applications in developing countries that lack resources for complete electrical power transmission lines also justify photovoltaic systems. (b) Here we see a remote environmental monitoring station designed by Dr. John D. Jacobs in the Salamonic Park in Newfoundland, Canada. Photo cells provide power for the recording instruments, and data from some stations are transmitted to satellites for collection and distribution.

chromatic (one wavelength) and possess an exact energy match with that required by the cell material to make the conversion of incoming sunlight more efficient. (See Figures 27-18a and b).

Theoretically, the silicon solar cell should be able to convert about 25% of the sun's energy into electricity. To achieve greater conversion of sunlight to electricity, two different solar cells can be used, one of aluminum–gallium–arsenide and the other of silicon. These two cells absorb a wider range of wavelengths from the solar spectrum, but various optical losses reduce the overall system efficiency to about 25%. The term efficiency refers to the percentage of energy in sunlight striking the cell that is converted into electricity. Individual solar cells have limited power. To produce electricity for most applications, they must be joined together electrically to form modules that become building blocks for larger arrays (Figure 27-17d). Superthin solar cells with a semiconductor layer thinner than a human hair and 1/100 the thickness of silicon solar cells have been developed in the laboratory. Traditional cells require relatively large amounts of semiconductor material, which drives up their cost. These new cells use but a penny's worth of material. The new semiconductor material is made of copper indium diselenide, which is deposited in thicknesses of only 2 to 3 μm on a material such as glass, and can reach conversion efficiencies (sunlight to electricity) of 18%, which rivals the best silicon cells. As yet, the ability to manufacture these cells outside the laboratory does not exist. The know-how to produce them must be developed and passed on to industry.

Solar cells are very expensive primarily because of their fabrication costs. Many breakthroughs in materials technology are needed before solar cells can be developed to compete on the mass market to produce electricity costing about $0.50 per peak watt or about $0.6 to $0.8 per kilowatthour. Ultrapure (99.999% purity) semiconductor-grade silicon costs around $15 a pound, compared to $0.15 a pound for metallurgical silicon with a purity of 99.5%. The processing cost represents approximately 30% of the cost per peak watt of solar-produced electricity. ***Peak watt*** is a term used to express the amount of power produced by a solar cell in full sunlight at 25°C. A 4-in. cell, for example, can produce 1 peak watt at noon on a sunny day in the U.S. Southwest, that is, under ideal conditions. A more relevant term is ***average wattage,*** which is about one-fifth the peak wattage. Invented at Bell Laboratories in 1954, most common solar cells are made from a single crystal of pure silicon grown artificially in the form of an ingot that is then sliced into wafers 0.012 in. thick. These wafers are then polished, trimmed, and doped in an oven. The finished cells are mounted in arrays or panels containing dozens of individual cells with diameters ranging up to 4 in. One manufacturing innovation produces cells continuously in the form of thin ribbons. This thin-film technology, similar to that used in making electronic integrated circuits, could cut manufacturing costs substantially and permit the fabrication of large-area cells that would make solar cells more cost-effective. (See Figures27–18a and b.)

Today, the Bell Labs process, referred to as the crystalline process, uses scrap silicon from the semiconductor industry. The silicon is melted at 1450°C and poured into 4.5-in-square ceramic molds to form a foot-long silicon crystal ingot. This ingot is sliced into 300- μm-wide cells, about the width of a piece of construction paper. Thirty-six cells are soldered together into a standard 1.5-ft × 3.5-ft module, wires and electrical leads are attached to connect individual cells, and the module is placed between two pieces of glass with a plastic backing and installed into bulky frames for shipment to customers.

A ***thin-film deposition*** process, now called the continuous process, uses 12 silicon gas deposition chambers through which single sheets of 2 × 4-ft glass panels, laser etched with conducting filaments, are coated with a photovoltaic silicon layer that has a thickness of 0.5 μm. Outside layers of glass are added for protection. This process uses less materials and less labor and has wider commercial application than the older process; however, comparative-size crystalline cells are still more efficient in converting sunlight into electricity. In spite of this loss in efficiency, the new thin-film cells cost from 10 to 15% less than cells produced by the crystalline process.

Bulky and less-efficient crystalline and amorphous Si thin films are becoming obsolete with the development of two new thin-film photovoltaic (PV) materials. These materials make it possible to attain efficiencies of 10% or more as opposed to the 4% efficiency of the older processes. Cadmium telluride and copper indium diselenide are both more efficient and less costly than current materials. The latter material is being used in the production of PV roof tiles. The highest efficiency rating for any mass-produced solar cell is held by the HIT Power 21 cell produced by Sanyo, with an efficiency rating of 15.2%. This cell is made of a sandwich with a single crystal layer between thin, amorphous silicon layers. The combination improves the flow of electrons through the cell, cutting internal energy loss and boosting conversion efficiency. The U.S. government has spent some $1.5 billion in research money for the solar industry over the last 25 years and is taking steps to keep some of the benefits at home. ***The Million Solar Roofs Initiative*** has the goal of putting 1 million solar systems on the roofs of commercial and residential buildings by 2010, including both photovoltaic units and thermal systems for space heating and hot water. A drawback to more use of PV devices is the conversion of the generated electricity from direct current to alternating current, but another technological advance, the ***microinverter,*** mounts on the back of PV panels and allows the panels to generate alternating current directly at the roof. As with all innovative technological advances, with greater use comes lower prices, which will permit the technology to be more cost-effective for residential use. New PV panels featuring built-in microinverters are being developed. The small 250-watt microinverters allow the panels to act as their own ac generators, thus eliminating the need for dc wiring and an in-home inverter. Homeowners can now start with a small system and add on since the microinverters are built into each panel.

27.3.3 Photoemissive Devices

In a photoemissive device, the light (or photons) generated by the recombination of electron–hole pairs is emitted from the surface of the device. Electrons striking a television screen, which is coated with a semiconductor material doped with copper, raise the electrons in the coating into the conduction band, where they recombine and emit energy in the form of light (photons). This is known as ***electroluminescence.*** Fluorescent lamps depend on ultraviolet light striking electrons in the inner coating of the lamps, which then emit light in the visible region of the electromagnetic spectrum. We round out our coverage of these devices by the following detailed discussion of light-emitting diodes. It is important to repeat at this point that regardless of the device, the underlying mechanisms involve the mutual interaction of electrons with electromagnetic radiation.

27.3.3.1 Light-emitting diode. Photoconductive materials (see photoconduction in Section 6.4) must be formed into a thin film atop of a substrate. They are deposited by a novel metal–organic chemical vapor deposition (CVD) process. A light-emitting diode (LED) is essentially a sandwich of conductive materials with electrodes attached at the top and bottom. A more technical definition would say that a LED is a device that luminesces visible light (see definition of luminescence in Section 6.4.1). Figure 27-19 shows a cross section of a LED. Passing an electric current through the material forces electrons and positive-charge carriers (holes) to combine in the middle of the sandwich to emit photons of light. When forward bias is placed across the junction, electrons cross from the *n* side of the junction to the *p* side and recombine with the holes, giving off energy in the form of heat and light. The light will be emitted, assuming that the semiconductor materials are translucent and the gold-film cathode effectively reflects the light to the surface. The color of the light depends on the characteristics of the materials that comprise the sandwich. Gallium phosphide (GaP) and gallium arsenide phosphide (GaAsP) are typical semiconductor materials used in LEDs to produce red, yellow, or green light. Gallium nitride is a new material that produces bright blue light.

Present-day applications of LEDs are as tiny "on" lights on computers, stereos, and VCRs. Because they are wafer thin and emit no heat, they can be integrated into building surfaces. The cost of electricity use with LEDs is a fraction of that for incandescent lamps. With LEDs surfaces such as ceilings can be continuously and uniformly illuminated, and "smart" surfaces such as a desktops fitted with LEDs can display e-mail or other messages.

The invention of the blue laser will lead to further uses for LEDs. The bright-blue, room-temperature LED from Nichia Chemical Industries is the first LED that can achieve a 10,000-hour life (deemed necessary for a commercial product). Its present output is 5 milliwatts, which is enough for read-only applications. For read-write applications, 30 milliwatts are required. These newly developed LEDs are currently used mostly for digital displays. Scoreboards and stadium displays are using the new blue and green LEDs. Their next applications will be for emergency lighting, flashlights, and traffic lights. The new goal of researchers is to produce highly efficient LEDs that will replace incandescent lighting.

Organic light-emitting diodes (OLEDs) use organic compounds rather than liquids, for greater ruggedness. They are being used in low-power, small-screen devices, including cell phones and car stereos. They activate a row of pixels at a time; active matrixes that address each pixel individually provide full color displays. Three other types of OLEDs are the transparent OLEDs, the flexible OLEDs, and the stacked OLEDs. Transparent OLEDs have organic molecules between two layers of a transparent conductor made of indium tin oxide. They provide excellent viewing under bright light. Flexible OLEDs have organic molecules mounted on a conformable substrate like plastic or metal foil and are designed to be rolled up like a poster. Stacked OLEDs combine red, green, and blue organics in a single pixel, which produces a full range of color in the space of a single pixel and much higher resolution.

Because consumers quickly adopt products made with optoelectronic devices, designers and manufacturers are eager to create more advanced products. As a result, tech-

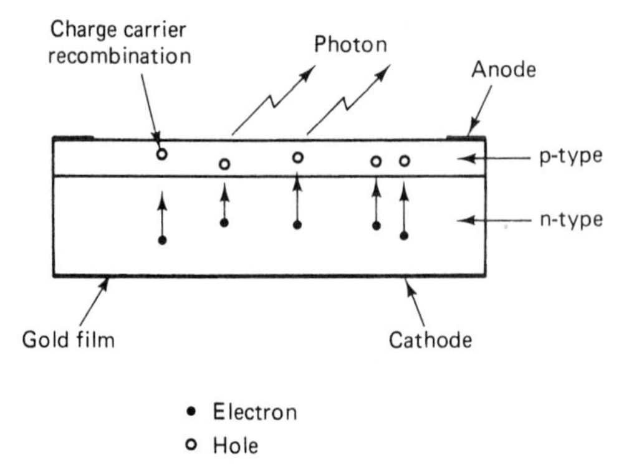

Figure 27-19 Cross section of an LED.

Figure 27-20 Advances in electronic and optical materials have spawned a wide array of toys and fun devices such as this programmable, remote-controlled Spybotic from Lego that kids like Julian can build and program.

nology advances quickly, often providing improvements at reduced prices. The educational toy seen in Figure 27-20 typifies these advances. For decades, high-density-polyethylene Lego bricks offered children an inexpensive toy that had merit of teaching manipulative and problem-solving skills. Now with the addition of an optical detector and a microcomputer that couples with miniature motors, children can learn to assemble robotic toys and create a wide range of configurations that employ gears, belts, wheels and many new bricks.

27.3.4 Silicon Technology

Silicon, the second most abundant element in the earth's crust, exists as silica or silicon dioxide (SiO_2). The processing of quartzite, a mineral (almost 99% silica) and the source of pure silicon, raises the cost of silicon to about $70 per kilogram. Pure polycrystalline silicon is further processed into single-crystal or large-grain (5 mm) polycrystalline silicon (Figure 27-21). The single-crystal silicon technology, the oldest and most well established, is slowly giving way to newer technologies using polycrystalline and amorphous silicon.

The latest fabrication techniques form silicon directly into usable wafers (0.5 mm thick) without intermediate steps, such as sawing, thus reducing processing costs considerably. More advanced methods produce ribbons (long, ultrathin, rectangular sheets). One such method produces ribbons 33 m long, 5 cm wide, and 0.5 mm thick at a linear rate of 55 cm/min, with cell efficiencies of over 10% (Figure 27-21b). ***Polycrystalline silicon*** is cheaper to produce and fabricate than single-crystal silicon, but at present single-crystal cells have maintained their lead in efficiency. In polycrystalline silicon, the grain boundaries impose numerous restraints on the movement of charge carriers (Figure 27-21c). However, new technologies are reducing the effect of the grain boundaries.

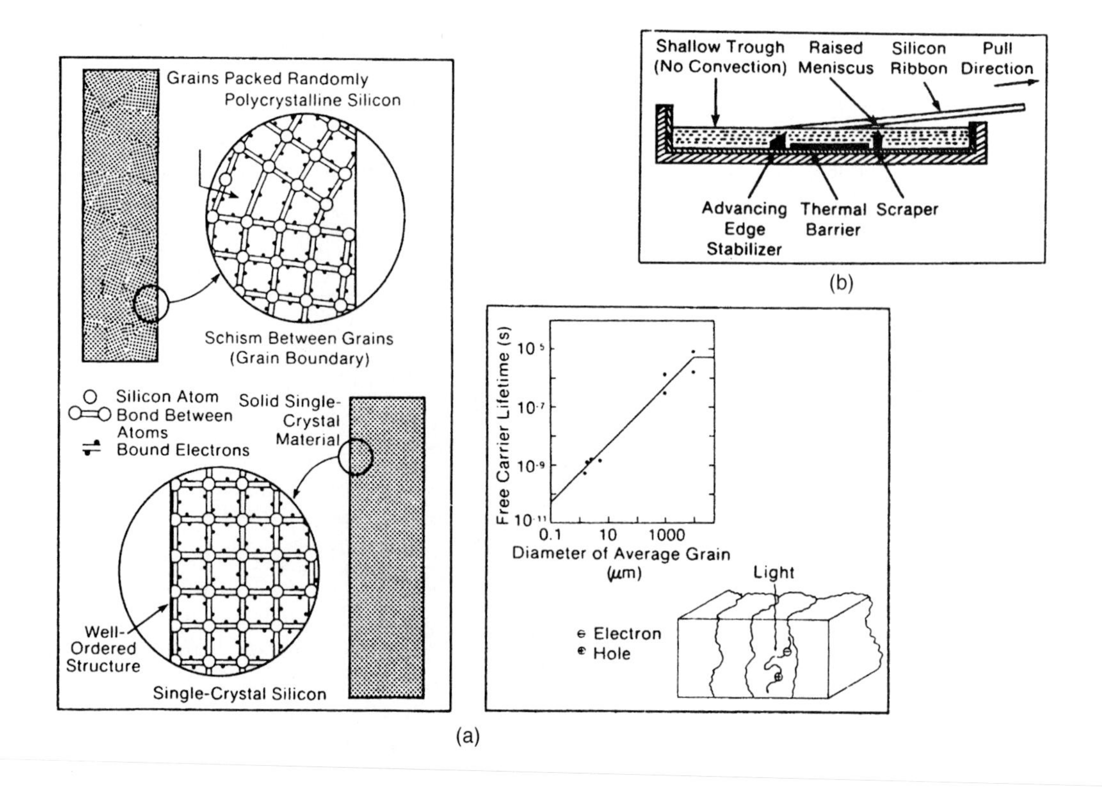

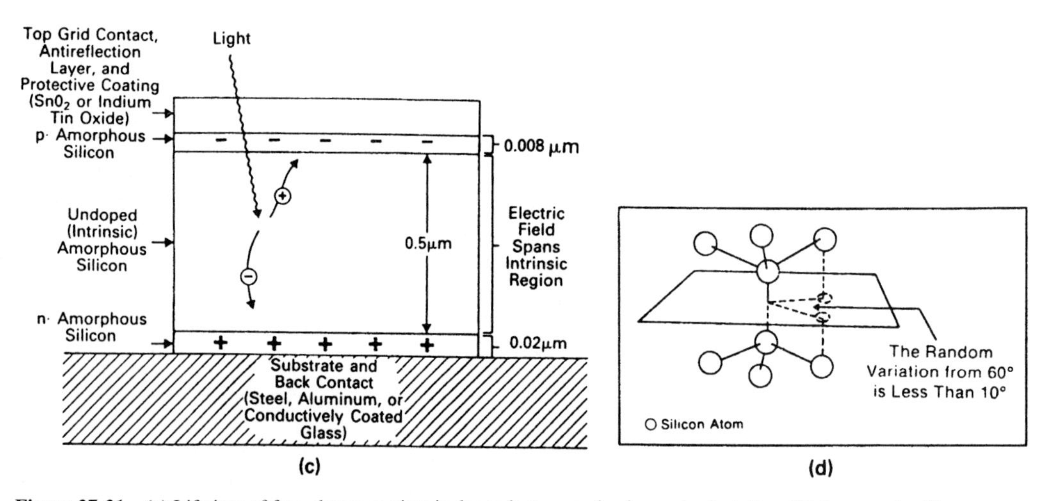

Figure 27-21 (a) Lifetime of free-charge carriers is dependent on grain size and orientation. (b) Low-angle silicon sheet (GLASS) growth is a fast method of drawing a silicon ribbon from a shallow trough of molten silicon. (c) A *p-i-n* device. An electric field sweeps the charge carriers to opposite ends of the cell. (d) Silicon atoms build a three-dimensional tetrahedral structure that, in amorphous silicon, is rotated randomly, producing a disordered atomic structure. (U.S. Dept. of Energy)

A third form of silicon, ***amorphous silicon*** (abbreviated aSi), has been used since 1974 to produce cells with high output voltage (0.8 V), currents greater than 10 mA/cm^2, and efficiencies of 6% using *p-i-n* cells (*p*-type, intrinsic, and *n*-type layers). Figure 27-21c, a schematic of this cell, points out the ultrathinness of the layers of amorphous silicon. To overcome amorphous silicon's disordered atomic structure (Figure 27-21d) with its incomplete bonds, hydrogen is added (doped) to complete the bonding and increase the cell's efficiency. A distinct advantage of amorphous silicon is its ability to absorb light about 40 times more strongly than crystalline silicon. Amorphous silicon cells are therefore used to power handheld calculators because they are more efficient and cost-effective under fluorescent light than either single-crystal or polycrystalline silicon. This characteristic of amorphous silicon allows such cells to be extremely thin, 1/40 of the thickness required by single-crystal and polycrystal silicon to absorb the same light. The low mobility of the charge carriers provides relatively rapid recombination, which means that the charge carriers must be separated by the *p-i-n* junction during the short time of illumination. In addition, thinness translates into less material, and depositions of the material tend to be easier.

27.4 PIEZOELECTRIC MATERIALS

Piezoelectric materials, first discovered by Jacques and Pierre Curie in 1880, were used in crystal radios in the form of slices of quartz crystal to select (receive or block) different frequencies and thus tune the radio. World War I introduced quartz crystals and their piezoelectric properties in the detection of submarines. During World War II, scientists discovered that ceramic materials at very low temperatures could exhibit more power than natural piezoelectric crystals, which led to the development of ferroelectric materials (see Section 6.2.2, "Dielectric Constant" for further data on ferroelectric materials). Now piezoelectrics find widespread use in speakers, microphones, and sensors for wheel balancers, alarms, pagers, motors, transformers, ignitors, ink-jet printers, fish finders, watches, ultrasonic NDTs, and automobile air bags. By controlling the size and electrical input to the crystal, a range of audible sounds can be produced by the frequency of the vibrations. Piezoelectric noise sources are used in tops, telephones, noise alarms, seat belt buzzers, and pagers. Other modern applications for piezoelectric materials are in push-button ignitors for gas barbeques, crystal microphones, vibration dampers in guitars and skis, fish finders, and most digital clocks and timers. The newly established center for Piezoelectrics by Design at the College of William and Mary will engage universities and research laboratories in the search for new piezoelectric materials, which up until recently has been on a trial-and-error basis. The center will house 60 computers to tackle the difficult job of running through the complex equations that represent the similarly complex crystal structures of materials under study.

Piezoelectric materials can be explained most easily through a discussion of dielectrics with an emphasis on polarization. Piezoelectric crystals are physically uniform solids that are bonded together by ionic bonds. We have learned that ionic bonding is a result of the electrostatic forces of attraction between ions of opposite charge. Normally, the number of positive-charged ions equals the number of negative-charged ions. Figure 27-22a represents the symmetric, crystalline structure of a crystal using three representative positive ions located equidistant from each other (at the vertices of an equilateral triangle). Similarly located are three negative-charged ions. The centers of charge both occupy a common center of symmetry. Being of equal magnitude, these centers of charge not only coincide in location but cancel each other, leaving the electric charge of these six ions neutralized.

Figure 27-22b is a similar representation of the same structure of a crystal with the addition of a mechanical load or force *(P)* that produces a stress assumed to be uniformly distributed over the face of the crystal on which it acts. This force, in this instance, acting along the *y* or mechanical axis, tends to compress the crystal, resulting in a compressive, elastic strain or unit deformation. The elastic deformation, in turn, results in the offsetting of the centers of electric charge from each other. In effect, this offsetting of centers of charge creates electric dipoles throughout the material, which combine to produce a measurable electric potential along the *x* or electric axis of the crystal. Conversely, if an electric potential were to be applied to the *x*-axis, a detectable deformation would be observed along the *y*-axis. The prefix *piezo-* comes from the Greek word meaning "to sit on, or press." The word ***piezoelectricity*** refers to electricity generated by exerting pressure on an ionic crystal.

Figure 27-23 is a model of a crystal with an electric field impressed along the x-axis. This particular arrangement of electric charge acting as the input of energy would produce a mechanical deformation (contraction) similar to that of the force *(P)* in Figure 27-22. Note that the ions of negative charge are attracted to the right, toward the positive terminal; similarly, the positive ions and their center of charge move to the left, thus increasing the separation of the

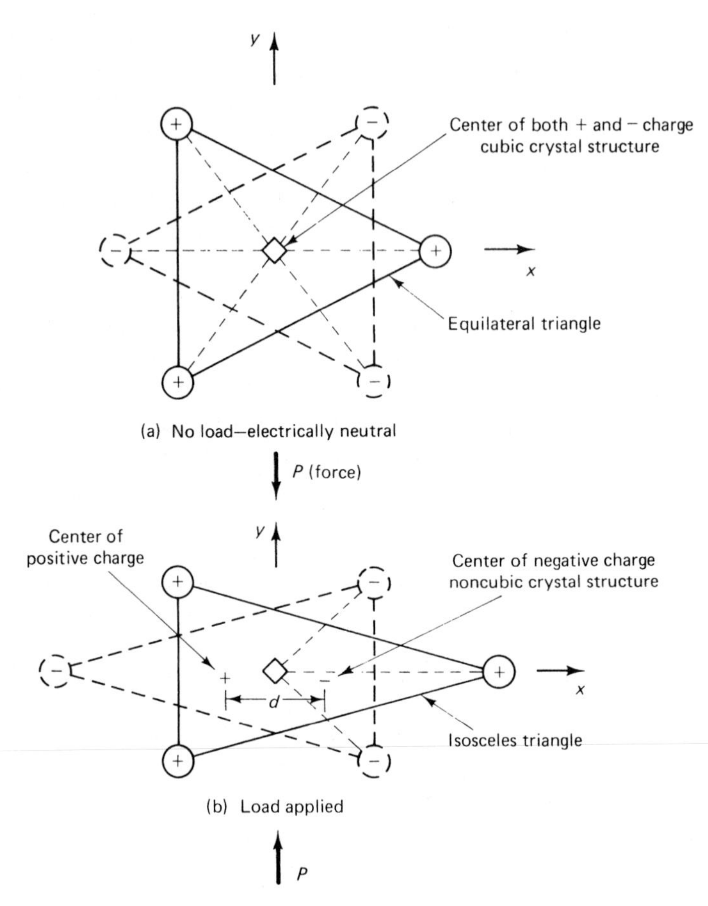

Figure 27-22 Model of piezoelectric crystal structure.

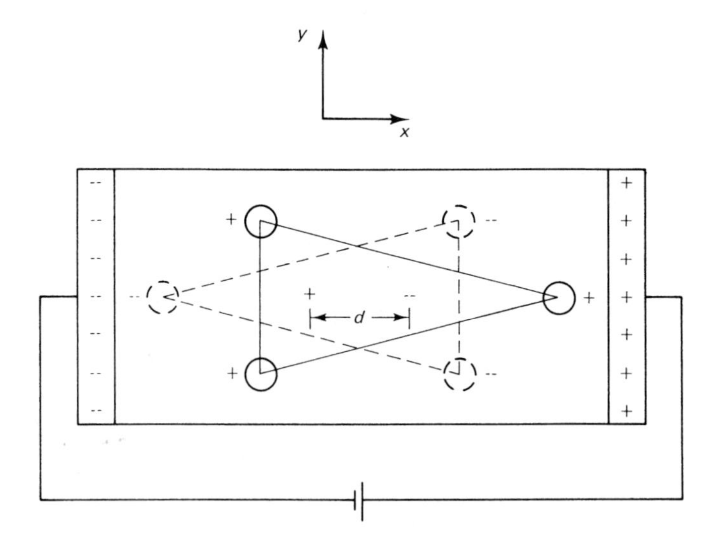

Figure 27-23 An electric field across a piezoelectric crystal.

centers of change (increasing the dipole length). If the polarity of the terminal were reversed, the deformation produced would be distorted in the opposite direction (expansion). An ac voltage, when impressed on the crystal, would cause the crystal to expand and contract (oscillate) at its driven frequency (maximum amplitude at resonance), which could be transmitted as a wave of sound energy into the surrounding medium. In other words, the potential produced by an alternating current would cause the crystal to vibrate. Each crystal has its own natural mechanical frequency and the ac potential can be adjusted to match this frequency to produce resonance (maximum amplitude of vibration). Ultrasound applications in medicine involve high-frequency sound to create an echo image. ***Sonar*** (sound navigation and ranging) uses low-frequency sound to travel long distances in water. The ***ceramic spinning*** technology that is producing ***piezoelectric composites*** in acoustic components for the U.S. Navy make sonar more versatile, because these new fibers can be bent to conform to curved surfaces and thus produce a clearer signal.

In summary, if a mechanical stress is applied to two opposite ends of a crystal (mechanical energy supplied), the remaining ends of the crystal are charged with electricity (output in electrical energy). If an electric voltage is supplied to one set of faces of a crystal, the other two faces of the crystal are deformed (either they contract or they expand); and if an ac potential is supplied, the output is crystal oscillation at its driven frequency, which can be transmitted into a surrounding medium as sound energy at a constant wavelength. A piezoelectronic crystal is called a ***transducer*** because it not only transfers energy from one system to another but converts energy from one form into another.

Quartz crystals (SiO_2), which posses piezoelectric properties, are both naturally occurring and grown commercially. In its uncut state, the crystal is in the form of a hexagonal prism, as sketched in Figure 27-24. The x-axis passing through the corners of the hexagonal cross section is the *electrical axis*; the *y*-axis perpendicular to the faces of the hexagonal cross section is the *mechanical axis*, and the *z*-axis is the *optical axis*. The crystal is cut in a variety of ways, depending on the characteristics desired. The sections cut from the crystal are known as *blanks*. One such blank, sketched in Figure 27-24, is called an *x cut* and has a thickness that is parallel to the *x*-axis and a length parallel to the *y*-axis. A mechanical stress applied to its faces along the *y*-axis produces a voltage along the *x*-axis. The dipoles created in quartz crystals do not rotate; thus, there is a permanent orientation. This characteristic permits a quartz crystal, once cut and polished to specified dimensions, to maintain a resonance frequency with an extremely high degree of accuracy. This feature is of great use in controlling the frequency of radio broadcast signals and in chronometers (time pieces of great accuracy). A quartz crystal vibrates 32,768 times a second when electric current is passed through it. Today's digital watches contain a battery, a tiny computer, and a quartz crystal.

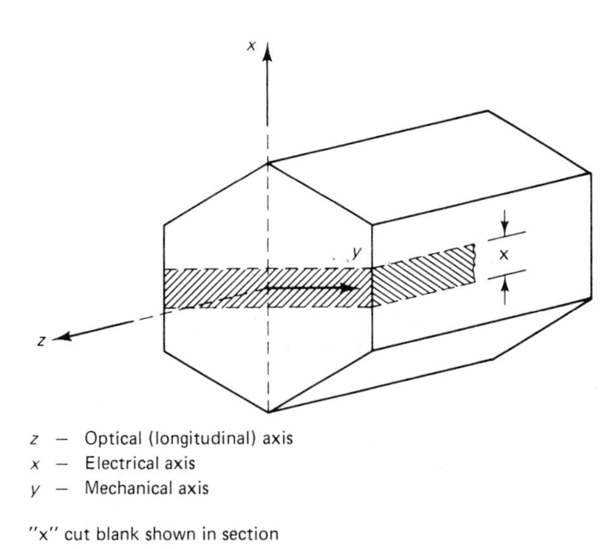

Figure 27-24 Uncut hexagonal quartz crystal.

27.5 LIQUID-PHASE EPITAXY AND MOLECULAR-BEAM EPITAXY

Liquid-phase and molecular-beam epitaxies (MBEs) are methods developed by Bell Laboratories to produce new semiconducting materials not found in nature having tailor-made atomic structures *that* providing an array of new, built-in, electronic, mechanical, and optical properties. The word ***epitaxy*** describes the overgrowth in layers of crystalline material deposited in a definite orientation on a base material of different crystal structure and chemical composition. In this world of human-made materials, this scientific research and development effort by Bell Laboratories demonstrates once more our ability to exert precise control over the creation of a new material by methods that ensure the exact composition and structure of the alternating atomic layers that make up a material.

Starting with a substrate of gallium arsenide, a semiconductor material, atomic beams of controlled intensity are aimed at the base material in an ultrahigh vacuum (Figure 27-25a and b).

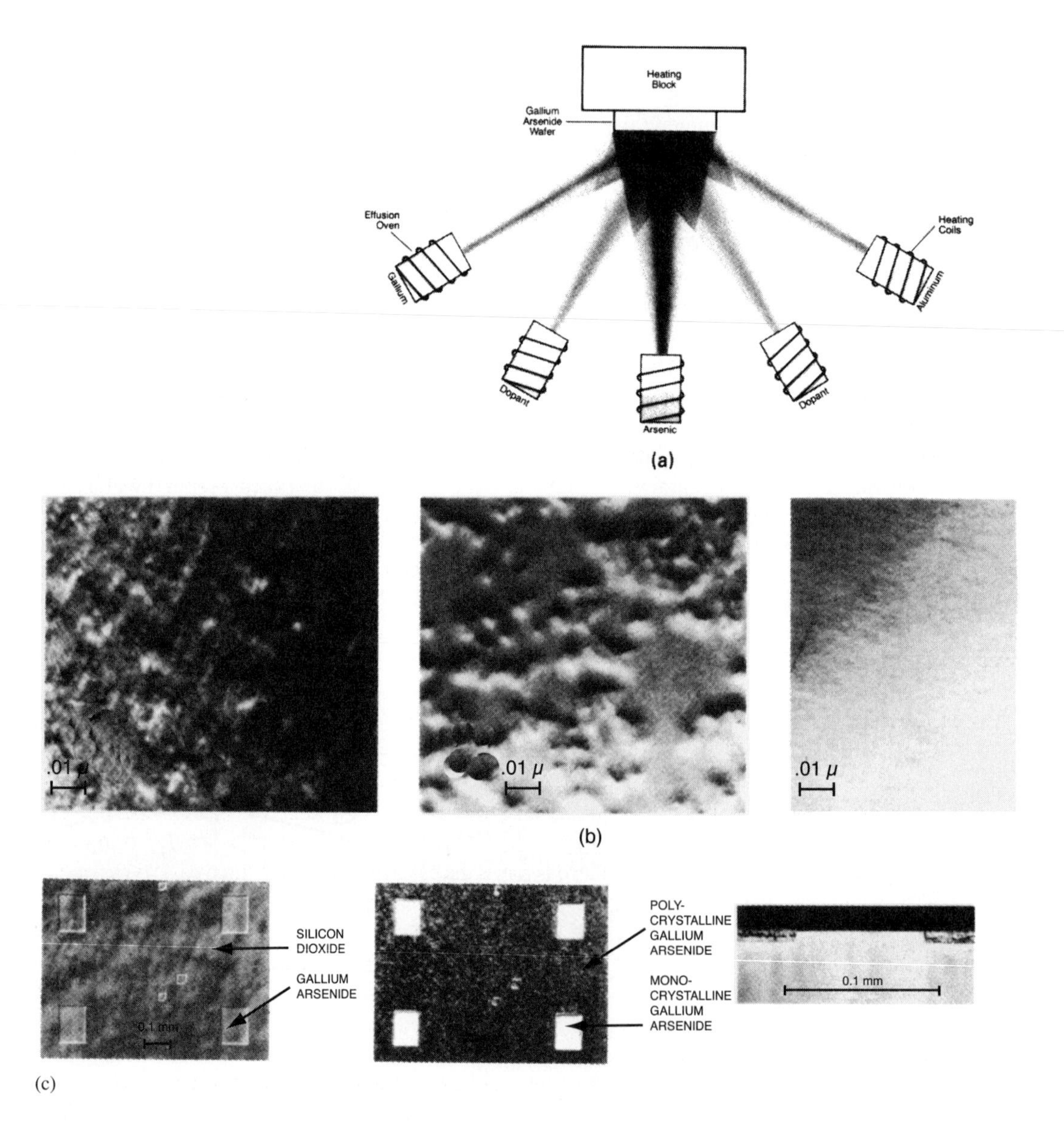

Figure 27-25 (a) Atomic or molecular beams formed by heating elements in effusion ovens aimed at a base metal. (Bell Labs) (b) "Rough" surface of wafer (left) smoothed after MBE growth of one layer (right), permitting the growth of layers of uniform thickness. (Bell Labs) (c) Patterned layer. An island of gallium arsenide doped with elements with desired properties can be grown in a "sea" of semi-insulating gallium arsenide. (Bell Labs)

Shutters are used to turn the beams on and off as they are directed at the base from a heated oven. A layer of gallium atoms followed by layers of arsenic, aluminum, arsenic, then gallium atoms, repeated many times, produces a crystal resembling a thin, highly polished mirror. Liquid-phase epitaxy, an older technique, produces satisfactory wafers, but MBE appears to be more promising. Using MBE, ultrathin, multilayer crystals with different structures have been grown. Quantum-well structures with up to 100 alternating layers represent one class of such crystals, with each layer as thin as several hundredths of a micrometer. Monolayer structures with alternating layers of atoms in a stack, each layer two atom planes thick, are being produced. Multilayer structures consisting of several thousands of individual monolayers have been grown with a total thickness of 1 μm. Research has revealed that because of the thinness of these layers, the electrons and holes normally confined to each layer will interact with their counterparts in adjacent layers and build ***superlattice*** structures with even more desirable properties (Figure 27-25c).

27.6 LIQUID CRYSTALS

Liquid-crystal displays (LCDs), currently used in watches, signs, and similar applications, do not require as much energy as LED (light-emitting diode) displays. A common LC display requires about 125 μA, whereas a similar-size LED display requires about 500 μA. The image familiarly seen on the LCD is a silver display; the LED is the familiar red display. Images that are generated by lasers for the LCD have the potential to create multicolored video displays for information systems. The LCD offers many possibilities for easier information retrieval.

Most known substances can exist, given certain conditions, in a gaseous, liquid, or solid state. Very few compounds can exist in a fourth state between the liquid and solid phases known as the *liquid-crystal phase*. This mesomorphic (intermediate) phase, or anisotropic liquid, was discovered over 100 years ago (in 1888) but remained a physical oddity until recent advances in electrooptics and thin-film technologies made possible several important applications of these materials, particularly in the visual display of information. Liquid crystals (LCs) are organic compounds that flow like a liquid in a small temperature range above the melting point while maintaining the long-range orderliness of a solid. Normally, crystals of a pure compound when heated show a well-defined and characteristic melting point at which the ordered crystalline lattice structure breaks down and the material becomes a liquid. In the liquid phase, the individual molecules show no preferred spatial orientation. The feature of liquid crystals is that during the melting process, the well-ordered three-dimensional crystalline structure transforms into a one- or two-dimensional state of order. This results in a material that has some optical properties of a solid combined with the fluidity of liquids, resulting in a mix of unique properties.

Figure 27-26 is a sketch of the molecules (shown as cigar- or rod-shaped) of ordinary liquid and of a liquid crystal between transparent glass sheets or plates. These cells are known as ***transmittive-type cells.*** When only one glass plate is transparent and the other has a reflective coating, the cells are called a ***reflective-type cells*** (see Figure 27-33a later in this section). Cholesteryl nonanoate and *p*-azoxyanisole are two organic compounds that exhibit these optical properties. The orientation of the molecular aggregates (collection of molecules) of an LC compound in the form of long, cigar-shaped polar rods forms the basis for classifying three basic types of liquid crystals: smectic, nematic, and cholesteric. The ***smectic*** phase consists of flat layers of cigar-shaped molecules with their long axes oriented perpendicular to the plane of the layer. This is the most ordered phase. The molecules within each layer remain oriented within each layer and do not move between layers. Figure 27-27 shows a structural model of this smectic mesophase. The molecules lie in layers with their long axes parallel. The molecules can move relative to each other within the layers, and consequently several types of smectic structures can be formed, depending on the inclination of the long axes of the molecules to the plane of the layers. The molecules in the ***nematic*** phase also have their long axes parallel, but they are not separated into layers. Rather, their structural arrangement is similar to the ordinary packing of toothpicks in a box. Figure 27-28 depicts such a nematic mesophase structure. While maintaining their orientation, the individual molecules can move freely up and down. The ***cholesteric*** mesophase (Figure 27-29) can be described

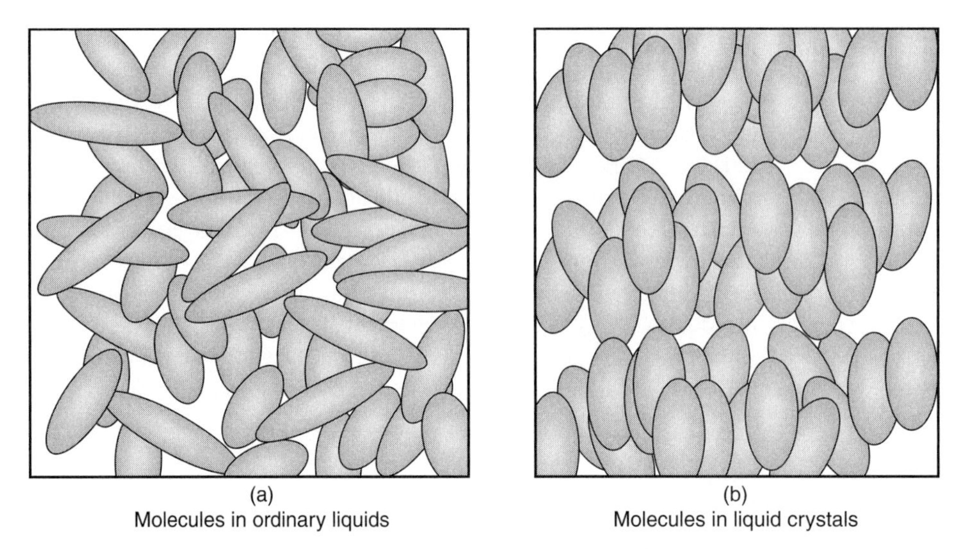

Figure 27-26 Ordinary liquids versus liquid crystals. (a) Ordinary liquids have molecules (shown here as cigar-shaped figures having the same size) that are randomly oriented, constantly changing, irregularly spaced, and free to move about at will. When heated, some compounds (substances) will change to an intermediate state between liquid and a crystalline solid, sometimes described as a cloudy state, before they completely change to a clear liquid state. (b) The liquid-crystal state has molecules (cigar-shaped figures of the same size) that can align themselves preferentially in certain directions. After transforming from a solid state, these substances contain "floating islands" or clusters with a high degree of molecular order. Liquid crystals usually have long molecules that are cylindrical (rodlike) or large and flat (platelike). Rodlike molecules can rotate around their own axis and can slide past each other, but they tend to retain their parallel orientations. Another type of liquid crystal has molecules arranged in layers. Molecules can rotate and move around each other but the layers are retained.

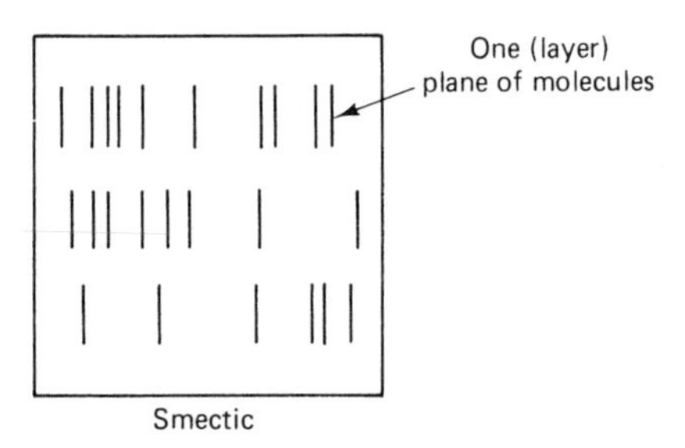

Figure 27-27 Model of the smectic mesophase.

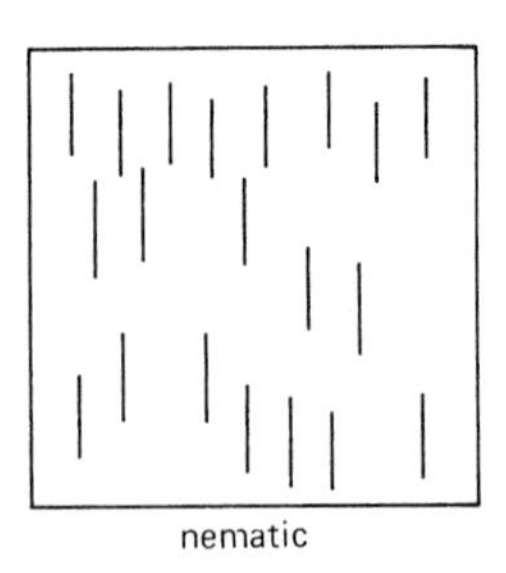

Figure 27-28 Model of the nematic mesophase.

as a special type of the nematic phase in which the thin layers (one molecule thick) of mostly parallel molecules have their longitudinal axes twisted (rotated) in adjacent layers at a defined angle. Each layer is basically a nematic structure. The axes of alignment of contiguous layers differ by a small angle and produce a helix or progressive rotation of many layers.

Because changes in their optical properties, can be produced by aligning their molecules with electric and magnetic fields, the *nematic* liquid crystals are increasingly used in electrooptical devices. Changes in the structure of nematic liquid crystals similar to those brought about by

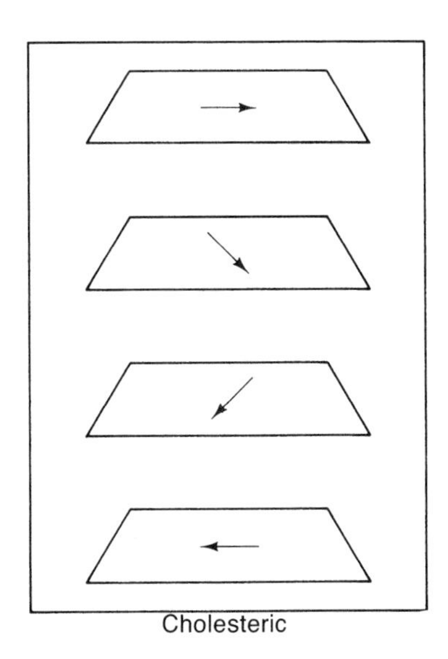

Figure 27-29 Model of the cholesteric mesophase.

electromagnetic fields can also be induced by several mechanical influences such as pressure, impact, or temperature. The cholesteric structure exhibits anisotropic optical properties. When such a material is illuminated with white light at certain temperatures, iridescent colors (shifting or changing) brought about by a fraction of the incident light are observed; the much larger portion of light is transmitted by the LC material. This reversible color phenomenon functions over a temperature range of about 220° to 250°C. These properties of LC materials are finding many applications in heat-transfer studies (thermal mapping); nondestructive testing (NDT); toys and games; holography; medical diagnosing of vascular diseases; and flat-panel, full-color display panels.

Switchable privacy glass, produced by Marvin Corporation, uses a technology similar to that found in laptop computer screens. The windows contain a thin film of liquid crystals made by the 3M Company sandwiched between layers of tempered glass. When a small electric charge of about 1 watt per square foot of window is applied, the crystals align, and the glass appears clear. With power off, the crystals arrange themselves in a random fashion, scattering light. The result is that the glass turns a frosty opaque white. This phenomenon is known as ***dynamic scattering,*** i.e., the cells do not generate light; they transmit or reflect light from external sources. Therefore, the only energy required by the cell is that necessary to produce the dynamic scattering. Refer back to Figure 14-1 and the related text discussion of privacy glass used in the smart house.

27.6.1 Liquid-Crystal (LC) Cell

A LC cell consists of a layer of nematic mesophase LC material between two glass plates that are glued or fused together. The thickness of the LC material is about 10 to 25 μm. The two glass plates have electrodes made of a transparent and conducting material such as tin or indium oxide deposited on their inside faces. Figure 27-30 shows a sketch of the construction of LC cell.

27.6.2 Homeotropic and Homogeneous Orientations

Two preferred orientations of the LC molecules, ***homeotropic*** and *homogeneous,* are used within LC display devices (LCDs). Figure 27-31 shows the homeotropic orientation with the long axes of the molecules perpendicular to the surface of the glass plates and electrodes. This orientation can be achieved by chemical doping of the nematic phase. The long, cigar-shaped rod orientation of the molecules allows the groups of molecules to act like dipoles in the presence of an electric field.

Figure 27-32 shows the molecules oriented parallel to the glass plates, the homogeneous orientation. This orientation can be produced by mechanically rubbing, unidirectionally, the

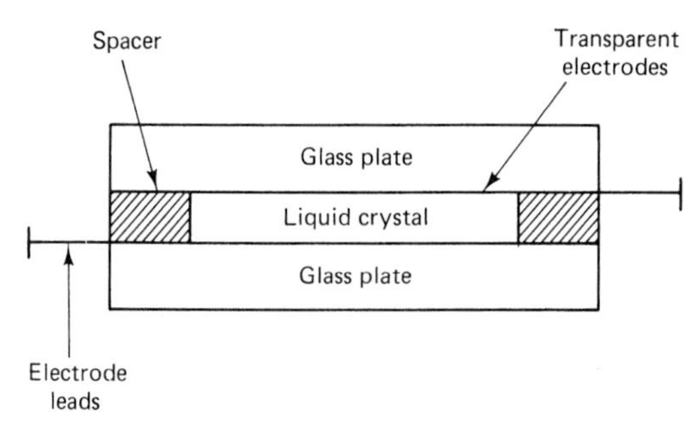

Figure 27-30 Cross section of an LC cell.

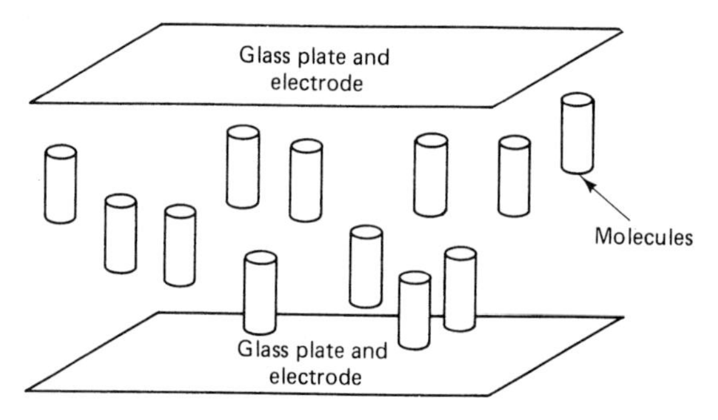

Figure 27-31 Homeotropic molecular orientation.

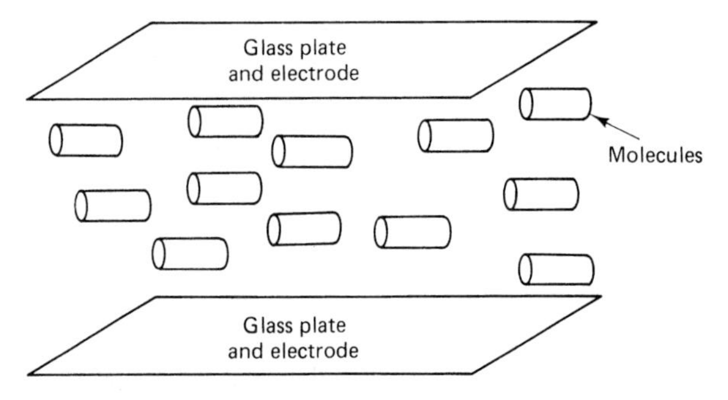

Figure 27-32 Homogeneous molecular orientation.

glass plates with a leather cloth prior to assembly of the LCD. Another means of accomplishing this orientation is by the deposition of a layer of dielectric over the transparent electrodes. Each plate is then rotated 90° in relation to the other. In effect, the LC material now acts as a set of polarizers that causes the light passing through the LC cell to be rotated 90°. Another way of altering the alignment of the layers in the LC material is to apply an electric field across the material. The molecules then align themselves in the direction of the electric field, which prevents the 90° twist of the molecular layers and the transmission of the incident light through the cell. By judicious selection of LC material, polarizers, glass-plate treatment, and electric field, incident light through the cell can be controlled; that is, it may be reflected, rotated, transmitted, or extinguished.

27.6.3 Reflective-Type LC Cell Operation

We have chosen the reflective-type LC cell rather than the transmittive-type to explain the operation of a numerical-display LC device. Figure 27-33a is a sketch of the components of a reflective-type LCD. The main components of the LC cell consist of the two glass plates, electrodes, and the LC material. A vertical polarizer and a horizontal polarizer, plus a reflector, are

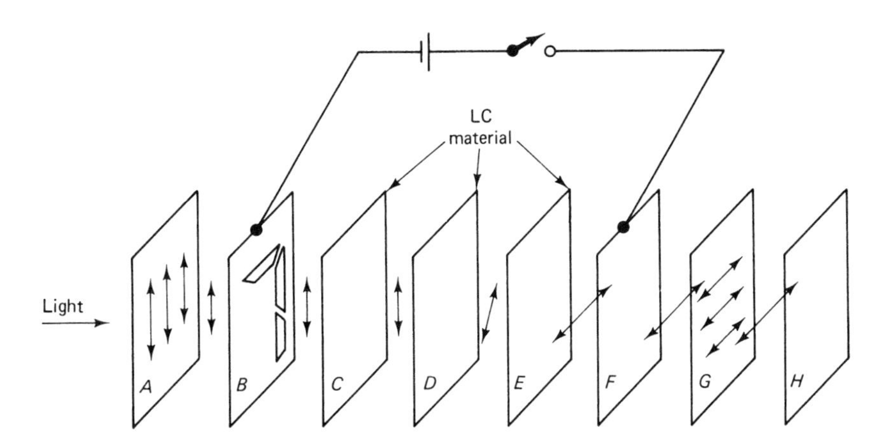

A – Vertical polarizer
**B* – Front glass plate electrode
C – Liquid crystal field effect cell
D – Liquid crystal field effect cell
E – Same as *C* and *D*
**F* – Back glass plate electrode
G – Horizontal polarizer
H – Reflector

* Numerical character segments contained in *B* and *F*.

(a)

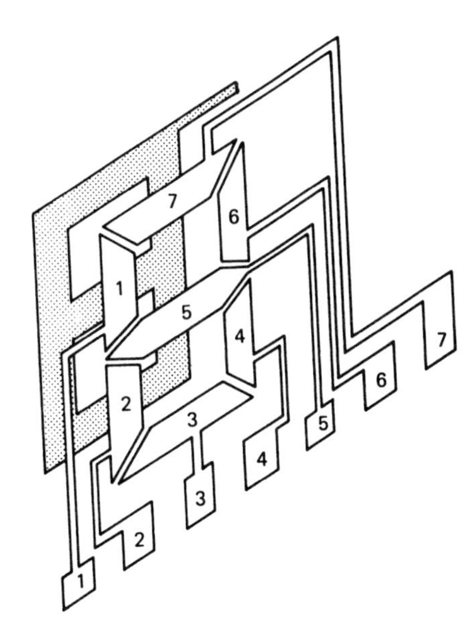

(b)

Figure 27-33 (a) Reflective-type LCD (electric field not applied).(b) Seven-segment numeric display.

added to the cell. With the plates treated and arranged to produce the 90° twist described previously, unpolarized light enters the vertical polarizer, rotates 90° through the LC material, and passes through the horizontal polarizer and on to the reflector. The reflector reflects the light back through the same path. The face of the reflector appears to be a uniform, light color.

To show one numeral in the display, several LC cells are placed together in a particular configuration; each controls one segment of the digital display, as shown in Figure 27-33b. An electric field activates the selected segments, which act together to form the particular numerical character desired. The digital display is impressed on the electrode on the front glass plate in Figure 27-33b; the electrode on the back plate acts as the base of the common electrode. The molecules in the area of the energized segments will align with the electric field and prevent rotation of the polarized light. Consequently, the vertically polarized light passes through the cell everywhere but in the region of the energized pattern elements. The end result is that the energized display elements appear as black images against a light background. By properly choosing the correct combination of segments to be activated, any digit from 0 to 9 can be displayed.

27.7 SUPERCONDUCTORS

As the world seeks cleaner, more efficient, and readily available energy, many developments, ranging from improved solar collectors to cold fusion to high-temperature superconductivity, quickly grab the public's attention. High-temperature superconductors (Figure 27-34a) achieved just that effect back in 1986. A superconductor is a material that allows an electric current to flow without resistance. Most conductive materials, usually metals, restrict the flow of electricity to some degree; this is known as resistance. If metals were compared for their resistance on a scale of 1 to 10, with 1 being the most *resistant* and 10 being the least, iron would be close to 1 and gold and silver near 10. Copper and aluminum are somewhere in the middle. Resistance of an electric current causes energy losses due primarily to heat generated from the resistance. Superconductors have a theoretical resistance of 0 Ω. This means that there is no energy loss when an electric current passes through the superconductor. Because of this feature, superconductors can be of great benefit in electrical transmission, from the very small millivolt signals in computers and other instrumentation to the high-powered, electrical transmission lines that bring electricity to our homes and businesses. This brings up the question, why aren't superconductors utilized more? This question is best answered with a little historical perspective.

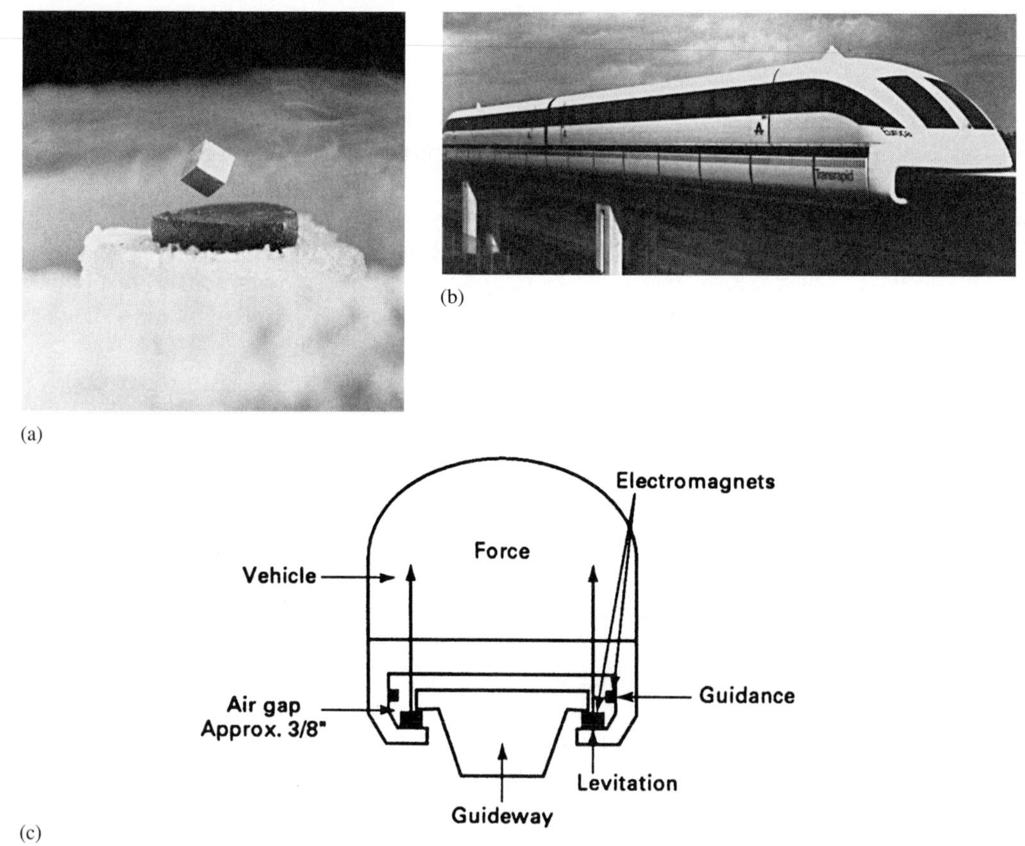

Figure 27-34 (a) High-temperature superconductors from ceramic materials. The Meissner effect, demonstrated here and explained later in this module, shows how superconductivity permits levitation of objects. (NIST) (b) High-speed travel exceeding 300 mph over long distances with Maglev trains is a desirable alternative in some cases to air and road travel. (Transrapid) (c) Electromagnetic Maglev, depicted in this simple diagram, is an alternative suspension system whereby electromagnets, perhaps high-temperature superconductors, will levitate a train over a track. (Federal Railroad Administration)

Superconductivity was discovered in 1911 by a Dutch physicist, Heike Kamerlingh Onnes, while he was observing mercury at liquid-helium temperatures (−452°F, almost absolute zero). Because of the difficulty and expense of working with superconducting materials at such low temperatures, uses for it have been very limited. Research on superconducting materials continued, and by 1973, the achievable superconducting temperature had risen to −418°F with the discovery of superconductivity in a niobium–germanium alloy. Because metals are conductive, the most logical choice for a superconductive material would be metal. The discovery in 1986 of superconducting metal oxides by K. Alex Muller and J. Georg Bednorz, physicists at the IBM research laboratory in Zurich, Switzerland, made quite a stir in the scientific community. Metal oxides are normally nonconductive, but certain oxides, when mixed in the correct ratios, have been found to be superconducting at temperatures that are reasonably achievable.

Since the Muller and Bednorz discovery in 1986, research around the globe has boomed as people seek to find ways to commercialize a product using this so-called high-temperature superconductivity. The number of materials has increased dramatically. One of the amazing things about the field is that it has been the only major discovery in the twentieth century that requires only common laboratory equipment and therefore can be duplicated almost anywhere, including the high school classroom. This means that not only researchers are experimenting with high-temperature superconductivity but people with an interest—engineers, inventors, industrialists, and tinkers—are trying their hands at making a usable product and furthering the research.

This discovery will have a tremendous impact on society. Besides wires for conducting electric currents, we are witnessing attempts to use these revolutionary concepts in medical equipment, trains, batteries, machinery, computers, and so on. Figure 27-34b and c shows a magnetic-levitation (Maglev) train that uses superconducting magnets to suspend the train over the rails, not touching them. The magnets and rails act together as a long-stator, linear induction motor to propel these high-speed supertrains. In the United States the Maglev project seeks to augment air and road travel with the supertrains. Electromagnetic levitation of trains using superconducting magnets is one application of electronic materials that is sure to continue to receive attention.

Superconducting metals act as normal metals and offer resistance above the ***critical temperature (T_c)*** but have zero electrical resistance below T_c. Figure 27-35 depicts electrons flowing through a normal metal conductor (a) and a superconductor (b) from the negative end to the positive end. In the normal conductor, electrons collide with phonons. These quanta, or

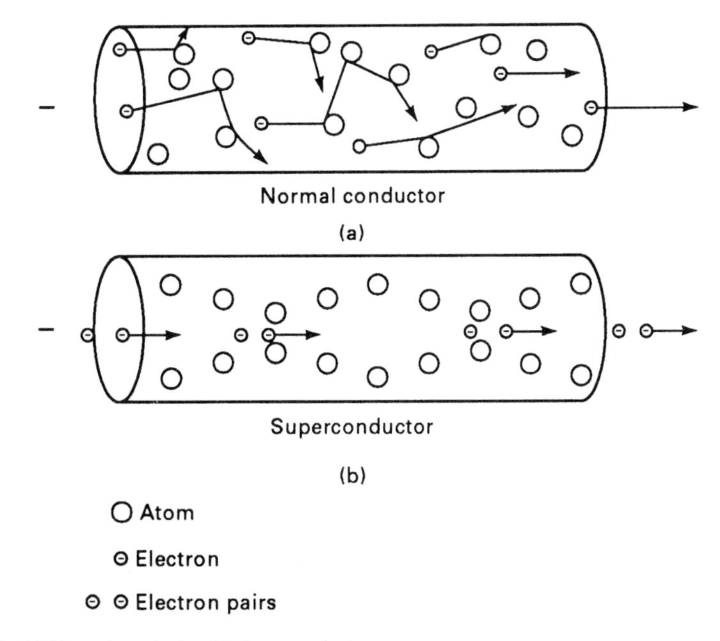

Figure 27-35 (a) Normal conductor. (b) Superconductor.

packets of energy, are produced by the vibration of the atomic space lattice, in which atoms form into crystalline solids. As materials become hotter, the vibration of their atomic lattices increases, causing a corresponding increase in electrical resistivity. Silver has a resistivity of 1.6 μΩ/cm at 20°C, but that resistance increases to 4.6 at 500°C; copper goes from 1.7 μΩ/cm to 4.5 with the same increase in temperature. As long as metals are maintained at moderate temperatures, they are *fairly* good conductors.

A theory known as ***BCS theory*** (after Bardeen, Cooper, and Schrieffer, 1957) holds that, in superconducting metals, pairs of electrons move with ease through the vibrating lattice (Figure 27-35b). As one of the paired electrons passes between two atoms, it causes a drawing together of the positive-charged atoms, which results from their attraction to the negative-charged electron. This is much like a speeding truck sucking cars into its draft. Once the electron passes the two atoms and they are still momentarily close together, they generate a positive attraction for the second electron in the pair and pull it into their region. This seems to set up a pulsing effect that causes masses of valence electrons to move through the superconducting material without resistance. The flow of electrons through the superconductor is perpetual, as long as the critical temperature is maintained.

High-T_c superconductors established some new lines of thought. Ceramics, for example, with their covalent and ionic bonding, are normally insulators, but a group of superconducting ceramic oxides posseses superconducting properties at temperatures high enough to use relatively cheap liquid nitrogen as a coolant to achieve superconductivity. Muller and Bednorz's now-famous 1:2:3 superconductor was comprised of a ratio of 1 part yttrium to 2 parts barium and 3 parts copper. Copper oxides in these ceramics have positive- and negative-charged ends (dipole) in the same manner as a bar magnet. It is believed that valence electrons passing through the lattice cause a shifting in the polarity of the charged ends, thereby developing a strong positive charge as they pass (see Section 6.2). Again, as with a superconducting metal, the second electron in the pair is pulled in by the positive attraction. This phenomenon is repeated many times to provide superconducting current. Figure 27-36 is a diagram of a unit cell for the new $YBa_2Cu_3O_7$ (yttrium–barium–copper oxide) ceramic-oxide superconductor. A ***plasma-spraying technique*** can coat a variety of large and small objects, including large copper panels and spherical vessels, with the ceramic oxide. The ceramic-oxide-superconductor

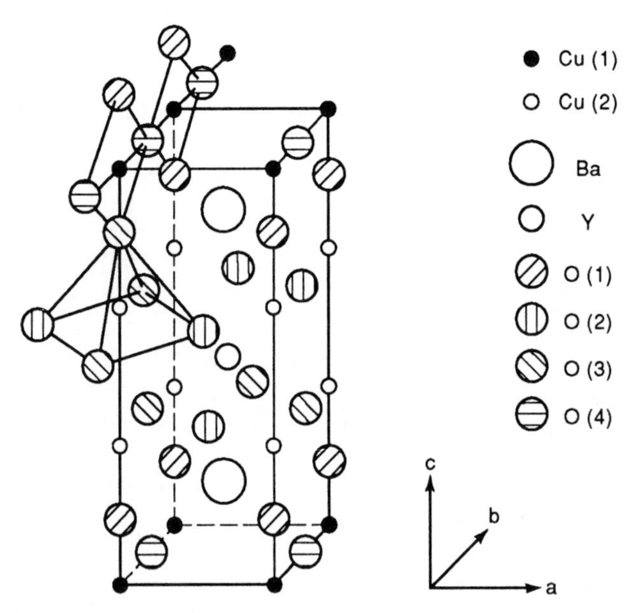

Figure 27-36 A single unit cell from the atomic space lattice of a ceramic-oxide superconductor composed of one atom of the rare earth element yttrium (Y), two atoms of barium (B), 17 atoms of copper (Cu), and 27 oxygen (O) atoms or vacancies. This unit cell is repeated millions of times in the solid material. The superconductor is abbreviated YBCO.

materials present a major challenge, however, in processing because of their brittle nature and the problem of keeping the composition in proper proportions.

Many applications involving superconducting materials call for powerful magnets, for example, magnetic-resonance-imaging (MRI) diagnostic devices (Figure 27-37). MRI uses the hydrogen in tissue to find tumors and other abnormalities. All molecules in the body contain hydrogen, which amounts to 60% of the atoms in one's body. Tumors contain much water and when a superconducting magnet creates a powerful magnetic field that aligns all the hydrogen ions in the same direction, the tumors appear as bright spots in the image created. Harmless radio waves cause some of the hydrogen ions to move off their axes. A scanner then measures the time it takes the hydrogen ions to return to their original positions, which is unique for different types of tissue. A computer uses this movement of ions to reconstruct a detailed picture of the inside of the body.

The superconducting magnets used for these devices consist of doughnut-shaped field-coil windings made of tightly wound niobium–titanium wire. They are cooled to near 0 K and high amperage is applied. As long as the coil stays cold enough, the magnets continue to conduct the electric current, free of resistance, at a highly stable rate. Niobium (Nb, at. no. 41) and its alloys exhibit not only superconducturty at low temperatures but high strength at extremely high temperatures as well as corrosion resistance. In aerospace applications, niobium alloys such as C-103 performed at temperatures exceeding 1300°C, its melting point is about 4260°F (2468°C). Niobium is perhaps the most practical superconductor. It not only enables such technologies as magnetic resonance imaging but is essential in particle accelerators/colliders, which are used in the search for a better understanding of basic matter.

The magnetic characteristics of high-T_c superconductors exclude all magnetic flux from a material. This phenomenon, known as the ***Meissner effect,*** is illustrated in Figures 27-34 and 27-38. The figures compare a superconductor to a normal ferromagnetic material. The drawings show the lines of flux (H) as arrows penetrating the normal conductor. A critical field (H_c) is reached for the superconductor when the temperature is below the critical temperature. Then no lines of flux can penetrate the superconductor. An external field created by a permanent magnet can move the superconductor. Figure 27-34 is the familiar picture showing a permanent magnet levitating over a pallet of ceramic-oxide superconductor. The magnet floats in the air above the superconductor because its lines of flux are repulsed by the superconductor and the magnetic field is excluded from the magnet's interiors. In effect, the superconductor acts as a perfect magnetic mirror to the magnet. The superconductor material and magnet are firmly attracted to one another but never touch, as would a normal magnet attracting iron. The cloud in the picture beneath the pallet is a result of the liquid nitrogen coolant boiling away at room temperature.

High-T_c superconductors can offer many benefits to society, such as cost savings in power generation. Applying superconductivity to electric-power generators, would make it

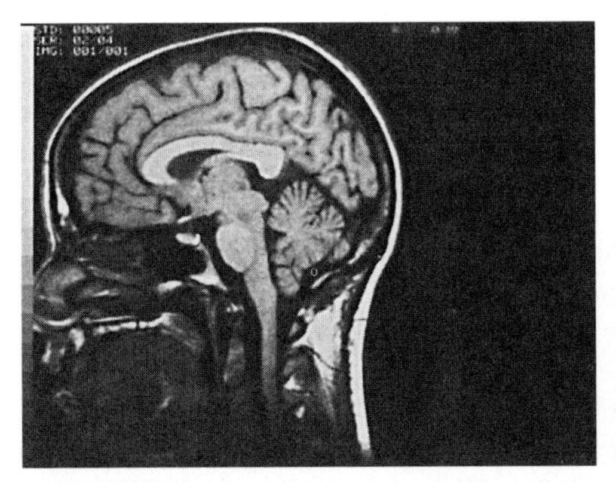

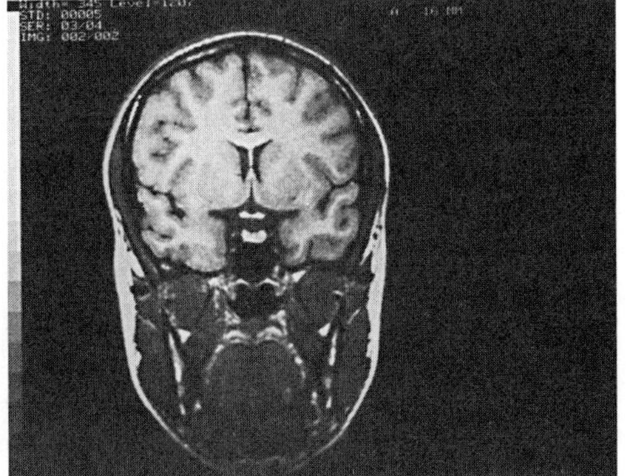

Figure 27-37 Two prints of a set of many slices taken through a human head by a magnetic-resonance-imaging (MRI) scanner to reveal bone and tissue for diagnosis of suspected cancer. (General Electric)

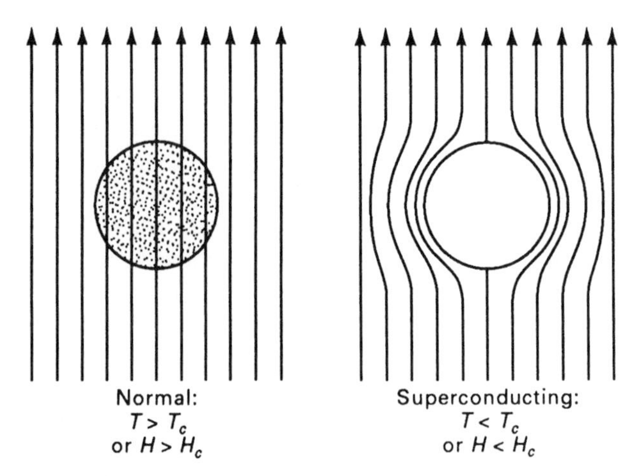

Figure 27-38 Comparison of the effect of the magnetic field (H) on normal conductors and superconductors. Critical field (H_c) is reached when the temperature (T) falls below the critical temperature.

Figure 27-39 The General Electric Company has built a superconducting generator that produces 20,600 kilovolt-amperes of electricity, or enough for a community of 20,000 people. This is about twice the amount of electricity that would be generated by a conventional generator. Field windings of conventional generators are made of a copper–silver alloy. The low-temperature superconductor generator employs modular field windings consisting of hundreds of filaments of a metal composite of a niobium–titanium alloy in a copper matrix. When available, high-T_c ceramic superconductor materials will provide even more cost savings, but no one can say when those breakthroughs will occur. (General Electric)

possible to generate electricity without the losses normally associated with the flow of current in the rotor windings of conventional generators, thereby saving millions of dollars (Figure 27-39). The demand for electricity in the United States is expected to double by the year 2030. Large industrial motors delivering 10,000 hp consume almost 15% of all U.S. electricity. If their coils could be made out of a superconducting wire, the motor could be reduced in size by one-half, with a corresponding 60% reduction in weight and the power required to operate them. The international competition to produce such a wire is fierce. To date, laboratory samples of such wire are small, consisting of tapes about 1/2 in. wide, 2 in. long, and a fraction of an inch thick. These samples are extremely brittle and very difficult to make into wire. However, researchers have used an ion-beam-assisted, deposition spray process to produce a thin film of a proven high-temperature, superconductor metallic-oxide compound, YBCO (yttium, barium, copper, oxygen—see Figure 27-36). In this process, one beam of ions lays down a few hundred thousandths of an inch of zirconia on a nickel-alloy tape; a second beam orients the crystals of YBCO as they form on the nickel substrate. The zirconia prevents the YBCO from reacting with the nickel substrate, and the other ion beam forces the YBCO into a regular crystalline order. In effect, a sandwich is produced that is malleable and retains much of its superconductivity. Much time and effort will be required to transform the lab process into a low-cost volume manufacturing process. A new superconducting tape invented at Los Alamos National Laboratory is made from a nickel alloy coated with fine layers of zirconia using a pulsed laser. Subsequent layers of nickel are applied to create a superconducting film as thin as 6 μm. The film carries 10^6 A/cm^2 of current, which is about 14 times the capacity of present bismuth-based superconducting tape and 200 times better than copper wire. The tape still requires liquid nitrogen cooling but only to −196°C versus −269°C for existing high-capacity tapes. Its faster processing times also make it far more economical.

Superconductors—*http://www.lucent.com/microscapes/16.html*

SELF-ASSESSMENT

27-1. The size of the forbidden gap in conductors is
a. Large **b.** Zero **c.** Small

27-2. An electric current of 1 coulomb of charge passing a point in 1 second is known as a(n)
a. Micron **b.** Ampere **c.** Angstrom

27-3. The p-side of a diode is called the
a. Cathode **b.** Anode **c.** Pentode

27-4. Solar cells are expensive primarily because of the cost of
a. Material **b.** Fabrication **c.** Tooling

27-5. A YBCO is a
a. Small yard tug
b. Superconductor
c. Computer memory code

27-6. High-T_c superconductors have an advantage over normal conductors because they offer no ___________ to the flow of current.
a. Strain **b.** Resistance **c.** Dielectric **d.** Permeability

27-7. Explain why the electrical conductivity of semiconductor materials increases as temperature increases.

27-8. Integrated circuits use many types of solid-state devices. Name three such devices.

27-9. What is the purpose of doping a semiconductor material?

27-10. Compare the energy required to raise a donor electron to the conduction band with the energy required to raise a valence electron.

27-11. Name one method by which an ideal semiconductor having no impurity atoms can be made to conduct electrons.

27-12. Once a donor electron has been raised into the conduction band by the receipt of energy from some external source, conduction occurs as a result of the combined actions of what charged bodies?

27-13. A Group VI element, selenium, is a semiconductor. From what group in the periodic table would you choose a dopant to convert selenium to (a) a *p*-type material; (b) an *n*-type material?

27-14. Identify what are termed the majority carriers in an *n*-type semiconductor.

27-15. Explain what is meant by the term *biasing*.

27-16. The depletion zone in a *pn* junction differs in size when a forward bias is applied from a condition of zero bias. Specify what this difference is in terms of area.

27-17. What device operates in opposite fashion to a solar cell and produces light from electrical energy?

27-18. What creates the voltage in a typical solar cell that drives electrical current through the external circuit?

27-19. Reflection of sunlight from the surface layers of a photovoltaic cell is maximized by treating the surface with some form of antireflective coating. True or false? Explain.

27-20. Down through the ages, humans have dreamed of converting sunlight directly into electricity. How efficient is this process today? Define *efficiency* as it applies to this process.

27-21. Name four possible things that can happen to sunlight entering a solar cell.

27-22. Explain the terms *solar cell, module,* and *array*.

27-23. What is the major cause of the poor efficiency of polycrystalline silicon cells in relation to single-crystal silicon cells?

27-24. Amorphous silicon used in handheld calculators is more efficient under fluorescent lighting than either single- or polycrystalline silicon. Name one characteristic that amorphous silicon has that explains this advantage.

27-25. Define the word *epitaxy* and explain why it is used in the semiconductor industry.

27-26. Both LED and LCD devices produce light. Which one requires more energy and consequently needs a battery source of power to function?

27-27. Optoelectronic devices all operate in accordance with a basic underlying scientific principle. Explain this principle.

27-28. The three types of liquid crystals are based on what characteristic of liquid crystals?

27-29. Name three influences that can change the structure of liquid crystals.

27-30. A homogeneous orientation of LC molecules can be accomplished in several ways. Name two such methods other than the use of an electric field.

27-31. In a reflective-type LC cell, the light source and the scattered light are on the same side of the cell. One glass plate is given a reflective coating. A transmittive-type cell's light source is on one side of the cell and the scattered light appears on the other. For this to happen, how are the glass plates modified, if at all? Draw a sketch of a transmittive-type cell similar to Figure 27-33.

27-32. Referring to Figure 27-33, to display the numeral "7," which segments need to be activated?

27-33. Referring to Figure 27-24, which axis (*x* or *y*)is the electric axis?

27-34. Name several applications of piezoelectric crystals.

27-35. The highest frequency of sounds that can be detected by the human ear is about how many cycles per second? Express in units of hertz.

27-36. A pickup for a record player converts the movements of a stylus into corresponding electrical signals. Other devices, such as a loudspeaker, use a similar element, a crystal. What material property peculiar to crystals do these devices depend on to perform their functions?

27-37. Superconductors have great potential for many technological applications in this energy-dependent world. What is one difficulty that limits their development?

REFERENCES & RELATED READINGS

ALLEN, EMILY. "Exploring Solar Cells," *National Educators' Workshop: Update 2002,* NASA/CP–2003–212403, May 2003, pp. 213–226.

ASM INTERNATIONAL. "Smart Materials Alter Shape of Adaptive Aircraft Wings," *Advanced Materials & Processes,* September 1995, p. 9.

"Don't Let the Sun Shine in." Webasco Sunroofs of Maumee, OH, *Popular Science,* June 1994, p. 12.

DREXLER, K. ERIC. *Nanosystems—Molecular Machinery, Manufacturing, and Computation*. New york: John Wiley and Sons, inc., 1992.

HUMMEL, ROLF E. *Electronic Properties of Materials,* 2nd ed. New york: Springer-Verlag, 1992.

MALLARDI, JOSEPH L. *From Teeth to Jet Engines*. Hampton, VA: Howmet Corporation, 1992.

"Manufacturing Parts Drop by Drop," *Compressed Air Magazine,* March 1995, pp. 38–44.

Moving America: New Directions, New Opportunities, Report to Congress, June 1990.

National Maglev Initiative: Annual Report—November 1992. Washington, DC: Federal Railroad Administration, 1992.

"Nimbus Information System." *CD-ROM Replication Guide*. Nimbus Manufacturing, Inc., Charlottesville, VA, 1995.

PAI, D. M. and N. R. SUNDARESON. "The Use of Piezoelectric Materials in Smart Structures," *National Educators' Workshop: Update 2002,* NASA/CP–2003–212403, May, 2003, pp. 213–226.

UENOHARA, MICHIYKI. "Electro-ceramics and Applications in C&C Era," *Journal of Materials Education,* Volume. 10, No. 3, 1988, pp. 259–280.

WEBER, WILLIAM J. "Highly Conducting Electroceramics," Pacific Northwest Laboratory, October 1992.

Periodicals

Invention and Technology

Reinforced Plastics

Technology Review

Heat Treating Progress

Module

28

Advanced Materials, Processes & Manufacturing

After studying this module, you should be able to do the following:

28–1. Describe the goals of materials science in developing smart materials and intelligent structural systems. Recall some current and potential applications of various smart materials.

28–2. Describe the goals of materials science in research and development with biologically inspired materials. Explain the properties and some applications of biomaterials.

28–3. Name some of the techniques, explain the needs, and list some developments in nanotechnology and microtechnology.

28–4. Gain an appreciation for the role of materials science and technology and their value to you in the future.

28–5. Use the website addresses denoted by the symbol* to explore new developments and evolving concepts related to electronic, photonic, smart materials, materials systems, and other advances.

*The symbol found throughout the book will link you to Internet sites related to topics being covered. The dynamic nature of the Web brings frequent changes, so some of these sites, while available at the time of writing, may not be up now.

28.1 SMART MATERIALS AND INTELLIGENT STRUCTURES TECHNOLOGY

Hardly a day goes by that we don't read or hear about some "smart" technology: smart cars, smart highways, smart credit cards, smart bombs, smart explosives, smart pills, smart structures, smart batteries, smart skins, smart wool and smart fluids. For each major group of our family of materials, we have provided you with examples of how smart materials within each group work and explained some of their uses. This technology shows great promise; it represents the potential for materials systems that involve electronic and optical materials in conjunction with metals, ceramics, polymers, and composites. As Figures 28-1 and 28-2 show, smart materials may be used as shape-memory alloys in rather simple products like eyeglass frame temples or for complex intelligent structures such as bridges and aircraft wings.

An example of the use of a shape-memory alloy is the CryoFit® permanent fittings and Cryolive® dematable end fittings in many USAF aircraft hydraulic systems. These fittings, made by Advanced Metal Components, Inc., are made from nitinol barstock, which is first machined at room temperature to an inside diameter (ID) that is smaller than the outside diameter (OD) of the tube to be joined. The fitting is then cooled in liquid nitrogen below its martensite transformation temperature. Next, it is expanded in the cold condition to an ID larger than the OD of the tube to be joined. The finished fitting is placed over the tube end, where it warms and recovers onto the substrate, forming a sealed joint.

The fastest-growing areas of use for binary NiTi alloys exploit this shaping memory or superelastic property. The alloys will restore up to 8% strain repeatedly in this pseudoelastic manner. NiTi wires are used, for example, to make very flexible eyeglass frames (Figure 28-1), which are essentially unbreakable. Highly flexible antenna wire for cellular phones is also produced from these alloys, whose service life is inversely proportional to the amount of deformation. If deformed close to 8%, a NiTi alloy's life span may be only a few hundred cycles. On the other hand, if the magnitude of the deformations is less than 1%, the alloy's life may be several million cycles.

The U.S. Army is developing a smart skin for tanks that uses sensors much like those seen in the bridge in Figure 28-2a. The armor skin is intended to detect incoming projectiles, such as shells made of depleted uranium, and trigger a repulsive explosion in microseconds to counter the shell's impact. An aerospace intelligent structure is the adaptive wing shown in Figure 28-2b. It employs piezoelectric materials (discussed in Module 27) to see if they can deform or morph wings in response to changes in air speed and air pressure and thus provide for more efficient cruising. Shape-memory alloys and fiber-optic systems are also being evaluated by the aerospace industry.

Research on smart materials is also being conducted at NASA Langley Research Center. It has focused on the feasibility of a smart material actuator system driven by a microwave. If such a system proves to be viable, it could dramatically reduce the cost of distributed shape-control systems and make possible many of the technologically aggressive science instruments

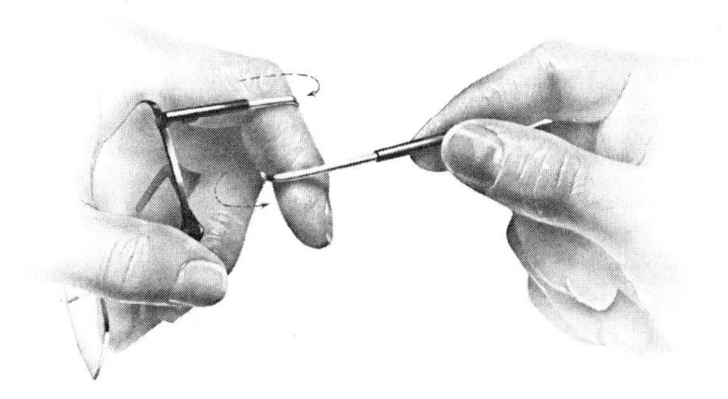

Figure 28-1 Shape-memory alloys, used for the Nitinol eyeglass frame temples, "remember" their original shape and immediately return to that shape when the bending force is removed.

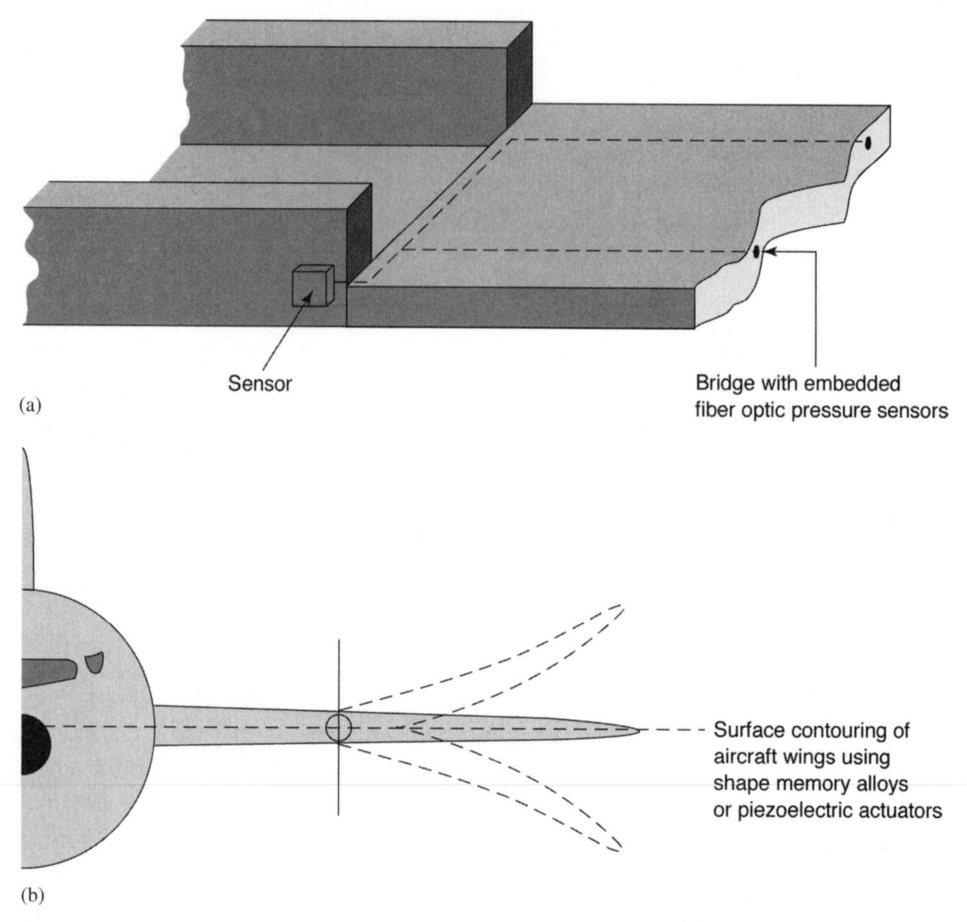

Figure 28-2 (a) Using optical fibers, sensors, and recording devices embedded in concrete or attached to steel structural members, it is possible to monitor stresses on bridges, walls, and other structures. (b) Current research may one day yield aircraft equipped with intelligent structures that can deform adaptive wings in reaction to variations in air pressure and air speed and thus improve high-speed cruising. *http://avst.larc.nasa.cog/morphing*

NASA plans for the future, such as the Next Generation Space Telescope. The Rectenna in Figure 28-3 has piezoelectric materials that receive signals from microwaves to move "smart strings" and "smart membranes," control devices for earth-orbiting devices or microrobots on Mars. Many exotic technologies first developed for NASA space applications soon find their way into broad commercial applications. What down-to-earth applications can you think of for the smart material actuator system driven by a microwave?

Another development, the "smart battery," an intelligent materials system, integrates battery technology with semiconductor technology. The battery pack designed for laptop computers includes a microprocessor inside each lithium-ion battery. The system keeps track of battery use and notifies the user of the time remaining on the battery. Some major computer manufacturers are teaming up with major battery manufacturers to develop standard sizes and configurations for these battery packs, which would ease the burden on consumers. Just as with the standard AAA and D cells, it will become possible to purchase standard replacement batteries for laptop computers in many stores.

In Module 1, we defined ***intelligent materials systems***, ***smart materials,*** and ***smart structures*** as materials and materials systems designed to mimic biological organisms and to offer an ultimate system that can place control and feedback into a material structure. These *biologically inspired materials* take their cue from biological elements, such as muscles, nerves, and bodily control systems, that can adapt to environmental changes. Where will biologically inspired and smart materials and intelligent structures go from here? Much of the focus of smart materials research is based on mimicking nature. We will examine that technology next.

RECTENNA

An antenna that captures and converts RF or microwave power to DC power.

Advantage:

Tailored to generate either high voltage/low current or low voltage/high current DC power.

Piezoelectric: high V - low I
Magnetostrictive: low V - high I

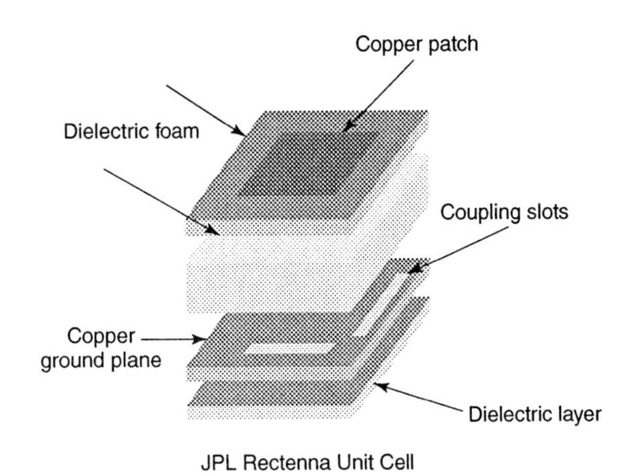

Figure 28-3 Adapted from Kyo D. Song, *Development of Microwave-Driven Smart Materials Actuators*. NASA project overview, March 1, 2000.

Smart materials and structures at Sandia Lab—*http://endo.sandia.gov/9234/smas_expand.html*
Shape-memory alloys—*http://www.uni.uiuc.edu/,richlin/sma.html*

28.2 BIOTECHNOLOGY

Biotechnology is an emerging field that promotes the coming together of the traditional inorganic sciences of physics, engineering, and chemistry with biology. Its immediate goal is to counter bioterrorism and biowarfare. Some of its weapons or capabilities to counter such threats rely strongly on sensors, electronics, mathematical algorithms, and computational ability. Each of these capabilities have been developed as a result of new and complex types of research that have lead to greater scientific discoveries and the creation of new materials and devices—many of which are *bioinspired.* Biotechnology has linked up with nanotechnology to make these new materials and devices. This combined research primarily deals with matter on the nano scale. In biology, the nano-scale objects that can be manipulated by man are cells and molecules, whereas in nanotechnology these nano items can be inorganic particles made from metals or ceramics or polymeric molecules. The influx of biotech discoveries in materials science and technology has brought untold benefits in the form of new and advanced materials.

28.2.1 Biologically Inspired Materials

From the beginning of civilization, technology has reflected human attempts to imitate nature. In the early attempts at human flight, efforts to build flying machines that imitated the structure and motion of birds met with failure. On the other hand, imitating the structure used by bees has produced advanced honeycomb composites suitable for modern aircraft. As mentioned earlier, NASA, among others, is inspired to replicate biology and as a result is putting considerable research into the effort.

Transforming raw materials into finished products usually requires large amounts of energy and many different processes. Because of concerns over the limitations of natural resources, considerable research and development is aimed at reducing the input of both raw materials and energy in processing materials. Synthesis and processing determine the structure and properties of a material. How can we conserve resources and synthesize and process materials in a way that is friendly to the environment and still obtain the desired structure and properties of materials?

Thin-film coatings offer many possibilities if they can be manipulated (1) to achieve unique structures that possess desired properties and (2) to be applied to a selective variety of surfaces. Organic thin films are found everywhere. Antirust primer on an automobile, Mylar film that holds the iron oxide on audiotape, and soap film formed around dirt are but a few of the films that make our lives simpler in the twenty-first century. The ability to orient the crystal structure of coatings offers possibilities for developing special optical, electronic, magnetic, and mechanical properties. Tough ceramic coatings provide increased durability for many products. The Pacific Northwest Laboratory used biomimetic synthesis to develop an organic interface that controlled crystal growth and formed desired ceramic coatings on polymer, glass, and metal surfaces of complex shapes at lower cost than the current vapor-deposition techniques. As depicted in Figure 28-5, the coating was deposited at temperatures below 100°C in a water solution that produced no hazardous waste.

Materials technology has turned to nature in its search for low-energy (low-pressure and low-temperature) and low-environmental-impact synthesis. ***Biomimetics*** is a technology that seeks to duplicate some of the processes that living organisms employ to construct bones, tissues, shells, webs, and other materials found in nature. The cross section of a seashell in Figure 28-4 reveals the intricate structure that mollusks achieve as they construct their hard, durable homes. Some doubt the value of pursuing the biomimetics route for engineering materials, but others feel that it is useful to study nature's way of producing biomaterials. For example, biomaterials for dental and bone transplants have been produced using biomimetics that replicate sea coral. Biomimetic research has enhanced the development of the smart materials already discussed.

Engineers are taught to be curious about the world around them, to observe how Mother Nature does things. Some examples of the questions raised by this curiosity are the following. How do honey bees construct their nests? Why doesn't a blade of grass break when stepped on by an animal or human? How does a tree withstand the force of wind? What makes the adhesive produced by barnacles so strong that the considerable expense of sand blasting is required to remove them from underwater hulls of ships and pilings? How does a bat manage to capture insects as it flies in the dark of night at high speeds? Some 20 years ago NASA researchers were studying how to use networks of strings to fabricate connecting platforms in outer space. Part of its studies involved an experiment with a space-borne spider. As the spi-

Figure 28-4 Biomimetics uses techniques that imitate nature to synthesize materials. Mollusks are examples of complex structures at the nanoscale that nature produces at low pressure and low temperature and without harmful effects on the environment. (DOE–Pacific Northwest Laboratory)

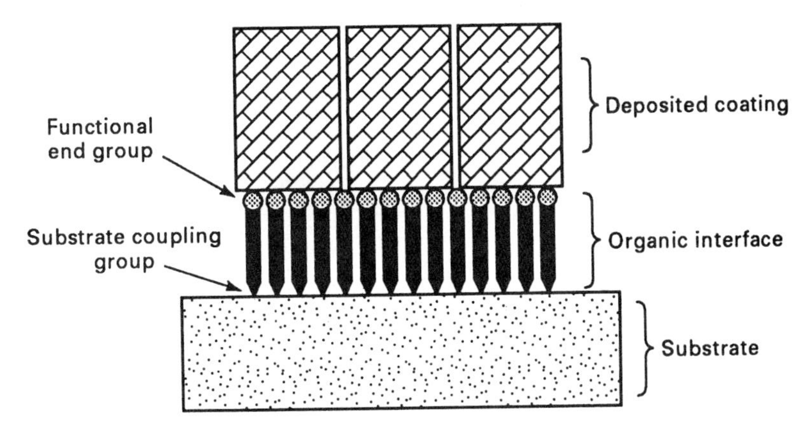

Figure 28-5 Biomimetic synthesis allows thin ceramic films (deposited coating) such as magnetic iron oxide to be deposited on substrates such as polyethylene and polystyrene plastics in a very controlled manner so that the functional end groups covalently bond (discussed in Module 2) to the substrate. (DOE-Pacific Northwest Laboratory)

der spun a web, it used threads of nonuniform size. NASA researchers recently used this information to help design an improved tennis raquet with greater power, feel, and control.

Two characteristics of tennis rackets, as well as other sports equipment such as baseball bats, must first be explained to clarify the significance of NASA's discovery. (This discussion builds on the information about testing sports equipment in Module 4.) The point that reaches the maximum speed when a racquet is swung is located at the toe of the racket, or the furthest location from the player's hand. Maximum power is generated when a ball hits that point, which is the desired point of impact when hitting a tennis ball. The second characteristic is the point of optimum vibration. Known also as the ***center of percussion,*** or ***sweet spot,*** this location produces the least amount of vibration on impact of a ball. Impact vibration, kick, or jar is transmitted through the racket to the player's hand and, in some cases, can cause chronic arm problems. Many tennis racket designs have enlarged the sweet spots to reduce this vibration. Sweet spots are usually found at the geometric center of the string area, but if the center of percussion could be moved outward toward the location of the position of maximum power, the power delivered to the ball would be enhanced because most of the impact would now be on or close to the sweet spot. Fewer vibrations would be felt by the player, with attendant reduction in stress on the player's arm and hand. The player would also have better control because of the increase in the racket's stability. (The racket is less likely to twist in the player's hand when impact occurs at the center of percussion.)

Solving the problem of moving the sweet spot was greatly helped by the knowledge received in the spider experiment. Using a computer and a genetic algorithm and the laws of physics dealing with vibrating strings, an optimum design that involved distributing the required mass along each string (or tapering the individual strings) was obtained. Any string material could be used, and the tapering could be achieved by spinning the string thicker or bonding different string materials along a central braid in the string. A ***genetic algorithm*** is one implementing a mathematical evolutionary process that resembles biological evolution in some respects to arrive at an optimized design by following a sequence of random design changes and preserving those changes that improve performance.

28.2.2 Biomaterials

A ***biomaterial*** is any substance other than food or drugs contained in a ***therapeutic*** (serving to cure or heal) system that is in contact with biological tissues or fluids. The most common biomaterial is dental fillings, which have been in use for more than 2000 years. A new biomaterial is the synthetic polymers being developed as implantable (in humans) dispensing machines to deliver drugs at the proper levels and rates and to direct the release of drugs to a particular cell type or tissue. This is an important development because one conservative estimate is that up to 100,000 deaths each year in the United States can be attributed to adverse drug events

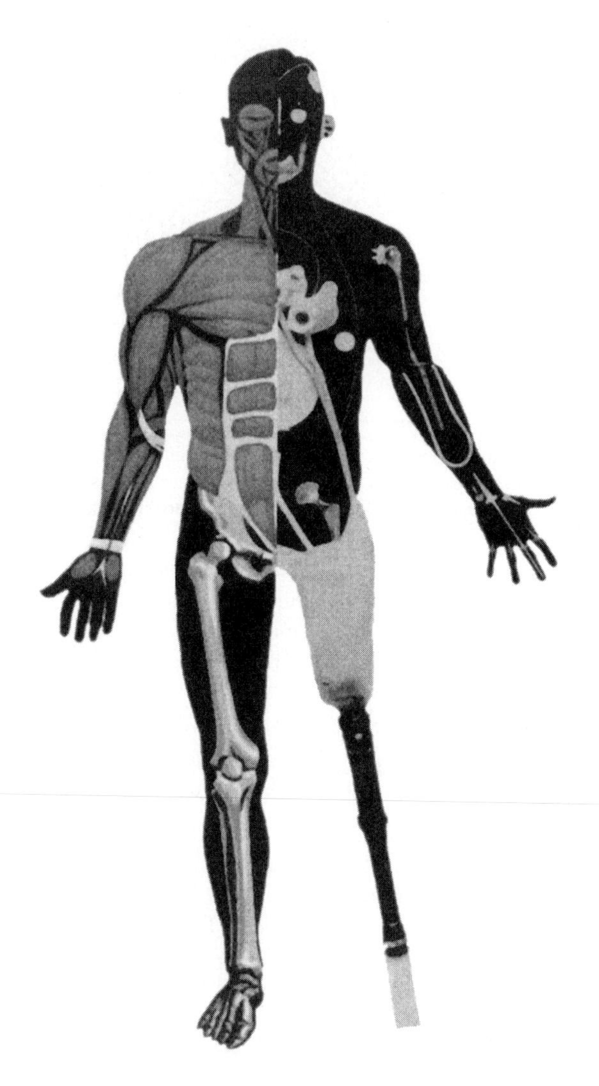

Figure 28-6 Tissue engineering. Living cells grown on synthetic polymers produce body parts; drug delivery via polymeric microspheres, dendrimers, magnetic dendrimers, polyanhydrides, or microchips. Biomimetics and biocomplexity themes are pervading polymer science. (NIST) *http://NIST–Online.nsu.edu*

such as overdose. As a consequence, the manufacture of drug-delivery systems has become a multibillion-dollar industry. Synthetic polymers have also been used as engineered tissue in successful attempts to grow replacement body parts (Figure 28-6). The greatest challenge to the research teams involved—chemical engineers, materials scientists, and surgeons—is to produce a material that is not harmful to the human body and that will degrade once inside a human body after fulfilling its task. Roger Langer, professor of chemical and biomedical engineering at MIT, has been instrumental in molding materials that turn biological discoveries into medical cures. He has discovered a family of polymers called ***polyanhydrides*** that among its other properties is degradable. These polymers are now being used to treat brain cancers. Another avenue of development is the use of ***smart systems*** that can deliver drugs in response to specific signals. For example, an implantable biosensor that detects glucose levels and tells a drug-delivery implant to release insulin when it is needed may be the next "crazy thing" that proves successful. Better still, an implantable microchip that can release specific drugs or vaccines into the body on demand by telemetry is not far off.

The field of ***tissue engineering*** integrates knowledge from cell biology, materials science, and process science to create new tissue (Figurre 28-7). The first engineered tissue was skin. Synthetic polymers have been used as artificial skin to replace burned or otherwise damaged skin.

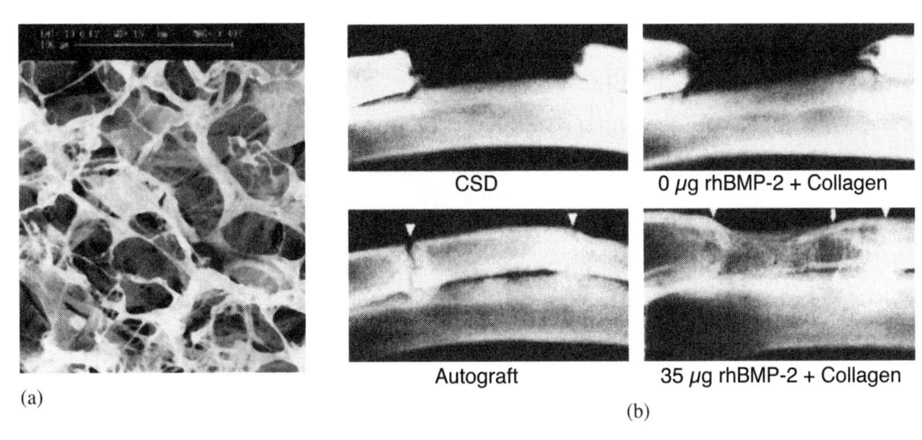

Figure 28-7 Materials for tissue engineering. (a) New functional tissue will be fabricated using living cells and a matrix or scaffolding, which can be natural, polymeric, or a composite material. $500 billion dollars per year in patient care is for tissue loss and organ failure. The American Academy of Orthopaedic Surgeons and NIH recommended the development of new bone graft materials. (b) Injectable composite bond grafts comprise biocompatibility and controlled release of bone morphogenetic protein. (NIST)

The "plastic" skin, sometimes in conjunction with skin grafts, performs like real skin to protect the wound, allowing new skin to grow naturally. Grown on a two-dimensional surface, synthetic polymers are used to construct three-dimensional structures that provide a framework or template for skin cells to grow and take the shape of the template. ***Cartilage cells*** (cells that eventually grow into bone) and degradable polymeric foam are used to make a human nose (see Figure 28-8). The cells are placed throughout the foam and allowed to grow, producing an extracellular matrix that provides strength while the polymer dissolves. The result is pure cartilage in the shape of a human nose. A very recent innovation in this technology is called ***injectible tissue engineering.*** More than 700,000 patients undergo joint replacement surgery every year in the United States. This type of surgery requires the implanting of an artificial joint, a highly invasive procedure. Biomedical engineering researchers at Johns Hopkins University have developed a way to inject joints with specially designed mixtures of polymers, cells, and growth simulators that solidify and form healthy tissue. This research is at the forefront of efforts to grow new tissues on polymer scaffolds. Small amounts of cartilage and skin could be grown successfully. The difficulty was in keeping cells alive on larger scaffolds, and these new tissues would have to be implanted by a surgeon. However, the new tissue developed at Johns Hopkins could be injected by syringe rather than by implant. As the research led to working with ***stem cells,*** the growing of tissue for cartilage and bone could be done simultaneously. Stem cell mixture infused each polymer layer where researchers had also incorporated special chemical signals that triggered the cells to develop into either bone or cartilage. Hybrid materials like this would, for example, simplify knee surgeries, which require a surgeon to replace the top of the shin bone and the cartilage above it.

Scientists are looking for a material that is strong enough to withstand the strain of a human joint and be so resilient that it can stretch dramatically and still return to its original size. Such a material could be a breakthrough in helping to repair human tendons and ligaments. More than 50 companies have received licenses based on recent inventions using such materials as polyanhydride polymers. Like human tendons and ligaments, the tough, stretchy ***collagen threads*** produced by marine mussels possess extraordinary properties. However, these threads are 5 times stronger (tensile strength) and 16 times as resilient as human tendons. Inspection of the molecular structure reveals that the threads are made up of two types of collagen, a fibrous protein commonly found in bone, cartilage, and connective tissue. Further, nature has seamlessly woven this tendonlike material to achieve a gradient between stiffness and elasticity. Considerable investigation will be required before any developmental work for human application can be undertaken.

Materials science is interested not only in mimicking nature but in contributing bioengineering materials and components that work with human and animal systems. Biomaterials must be compatible with human and animal systems so that they can be implanted or

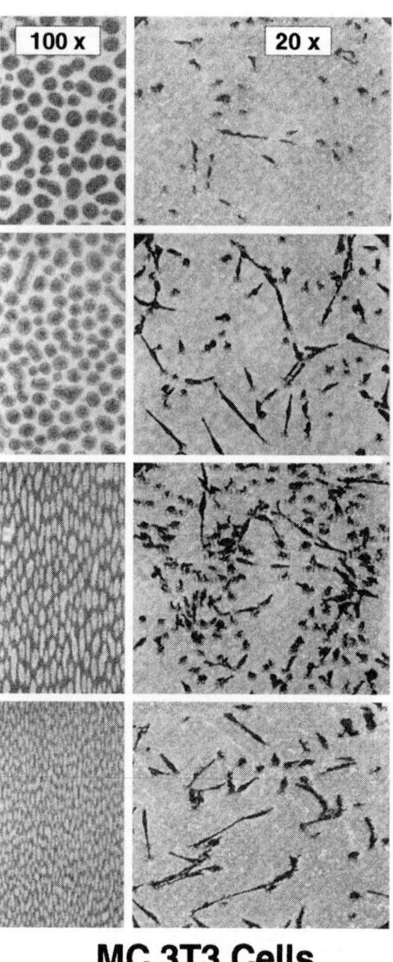

Figure 28-8 Metrology for tissue engineering. Test patterns and cell function indicators shown are to develop a set of measurement methodologies for assessing cell–biomaterial interactions to facilitate the development of tissue engineering and aid research in understanding cell growth and differentiation. (NIST)

manipulated in these systems. The principal requirement for each biomedical alloy is that it be corrosion resistant when inserted into the body and possess optimum mechanical properties. Bone has a modulus of elasticity of $E = 17$ GPa (2.5 Msi; 2.5 mega pounds per square inch = 2.5 million psi). A corrosion rate of less than 0.01 mpy (mils per year) is therefore required for a material to be considered a biomaterial. (A mil is one-thousandth of an inch, or 0.0254 mm.) All types of corrosion have been observed in biomaterials in the body. ***Surface passivity*** is the most important criterion for corrosion resistance. BioLAST is a coating that passivates the surfaces of medical devices that come into contact with body fluids, thus preventing the surface impurities from causing thrombosis. Other qualities of this coating are its high degree of tissue compatibility and its possible resistance to bacterial infection. (*Thrombosis* is a coagulation of blood in the heart or a blood vessel, which forms a clot. The clot is caused by converting fibrinogen to fibrin. By coating the surface with albumen fibrinogen, adhesion will not occur and no thrombi can form at the device surface or farther downstream.) Stainless-steel and cobalt–chromium alloys depend on the presence of chromium and its ability to render the alloys passive. Over the long term, stainless steels are not sufficiently corrosion resistant, but additions of other elements enhance its resistance to nonuniform types of corrosion such as pitting. (Refer to passive oxidation in Section 20.3 and prevention of corrosion in Section 5.1.4 for further information on the passivation mechanism.) In addition, surface finish also may af-

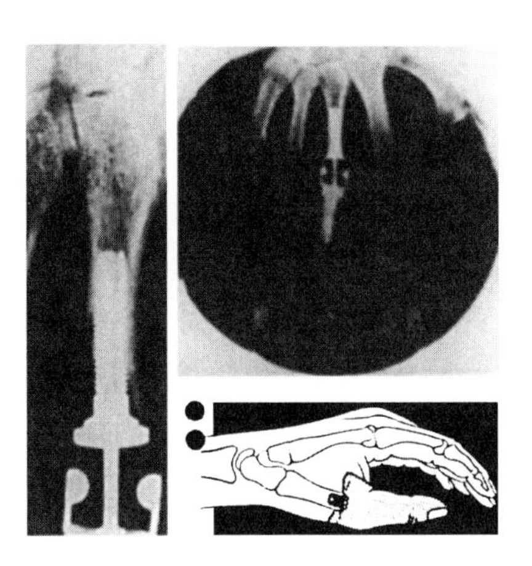

Figure 28-9 Because of its biocompatibility and ability to osseointegrate, titanium is used in many types of prostheses in humans (see Figure 11-17, p. 376). (Howmedica, Inc.)

fect performance, as highly polished surfaces resist corrosion and wear more effectively than nonpolished surfaces.

Titanium is another example of a biomaterial that can be used for joint replacements and dental reconstruction (refer back to Section 12.2). Figure 28-9 shows applications of titanium implants for use in the human body. Titanium and its alloys do not need chromium to develop surface passivity, but a coating process is used to improve bone growth around prosthetic devices and substantially reduce the risk of implant failure. The implant is first coated with a thin layer of ***hydroxyapatite (HA)*** (more about this below). Then an ion implantation beam is focused on the surface, causing the HA to combine with the titanium to strengthen the bond. The implant is then dipped into a HA solution filled with ceramic particles and heated to form a hard coating around the initial implant coating. The bone grows easily into the outer, porous HA layer. Improved HA powders and a new technology increase the adhesion of a new coating called Laserapatite to the substrate for both dental and orthopedic implants. There is also a new polished-metal alloy that includes titanium that provides, for the first time, different size ball joints for shoulder joint replacement. Up until now this procedure, total shoulder replacement, was limited to those patients who could use a replacement ball of only one or two sizes. With this new alloy, different sizes are now available allowing for a more precise or custom fit for each patient. New materials to repair broken and deformed bones that mimic the natural composition of bone have been formulated. Injected as a paste into the joint they bring about speedy recovery without the normal heavy plaster cast. One of these new, bone-simulating biomaterials, called ***Hedrocal,*** can support injured bone areas. It is an open-celled lattice substance that appears similar to coral and cork and is reinforced with tantalum (Ta). Having the properties of living bone, it can be implanted between two vertebra (any of the single bones or segments of the spinal column) where human bone will infiltrate its pores and form a true bone fusion across the joint. The hydroxyapatite coating used on titanium implants works well because human (living) bone consists of 75% hydroxyapatite (HA), which has the chemical formula $[Ca_5(PO_4)_3OH]$. The word *hydroxy* comes from the word *hydroxides,* which chemists use to describe compounds containing the polyatomic ion made up of oxygen and hydrogen, $-OH^{-1}$. For example, the compound NaOH, or lye, is called sodium hydroxide. *Apatite* is the name for a class of minerals that contain essentially calcium phosphate along with either fluorine, chlorine, or ions of hydroxy or carbonate. Human bone and teeth enamel are made principally from this chemical compound. Bacteria in the mouth produce acids from food trapped between teeth. These acids cause tooth enamel to dissolve and eventually produce cavities. The use of fluoride-containing toothpaste to control tooth decay results in the conversion of some hydroxyapatite into *fluorapatite* $[-Ca_5(PO_4)_3F]$. The crystalline structure of fluorapatite is more densely packed (less porous) than hydroxyapatite, which makes it more resistant to penetration by acids in the mouth. The chemical reaction involves the replacement of the OH ion with the

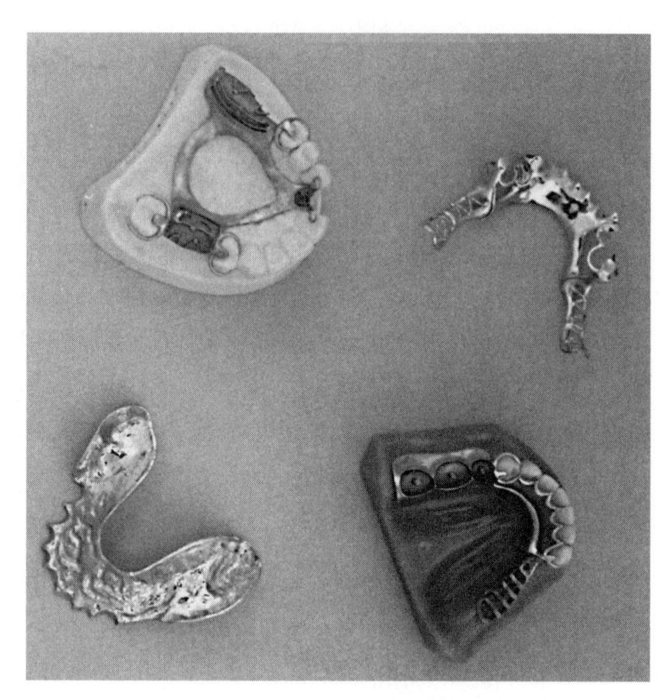

Figure 28-10 Vitallium dental castings. (Howmet)

fluorine ion. This same replacement phenomenon occurs in the fossilization of bone; ancient bones are nearly all fluorapatite. One human-made ceramic, derived from natural coral, called *Pro Osteon*™ is being used to surgically repair bone defects. Provided in small blocks that can be cut and shaped, the surgeon can replace portions of bone with this ceramic, which is sufficiently porous to allow tissue and bone to grow into its pores.

Vitallium is a biomaterial used for human prosthetics. The development of this alloy (cobalt, chromium, molybdenum, and nickel), first as a dental material (Figure 28-10), then as a material for orthopedic use (Figure 28-11), is typical of the synergy of engineering materials technology. Once Vitallium was found to be a suitable biomaterial during the 1920s and 1930s, then new methods of molding the alloy were sought. Because of the high melting temperature of the cobalt-based alloy, ethyl silicates were developed to serve as binders for the ceramic refractory to counter the problems of mold reaction and expansion. The process that evolved is called *investment casting* (see Section 11.5). During the 1930s, the experience gained casting Vitallium dentures and orthopedic prosthetics was refined and used to produce Vitallium and titanium components for aircraft jet engines. That evolution has resulted in directional and single-grained castings. A similar synergy has led to the development of a new ceramic composite. Orthopedic implants such as hip and knee joints last about 10 years, after which patients must undergo another surgery to have them replaced. Factors such as longer human life spans and an aging population have led to great demand for a material whose properties include retention of great strength over much longer periods of time. Two ceramic materials, alumina and zirconia, have been found to produce a composite that has greater resistance to crack propagation resulting from surface cracks or scratches as well as the ability to withstand loads twice as great as either of the two ceramics alone.

These synergies typify how materials science, engineering, and manufacturing have interacted and progressed. Wise technologists, engineers, and technicians always keep informed of new developments. Serendipitous events may occur by chance, but the alert person will recognize their value.*

* Conduct Internal searches for new developments on these issues.

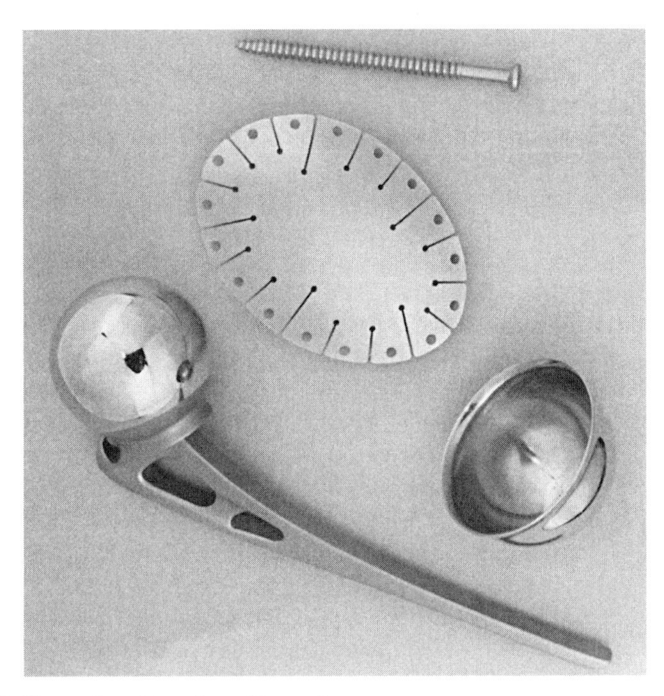

Figure 28-11 Vitallium orthopedic castings. (Howmet)

28.3 MICROTECHNOLOGY/MICROELECTROMECHANICAL SYSTEMS (MEMS)

Much active research in engineering involves building electronic and mechanical systems on ever-smaller scales using microtechnologies. The microscale (Module 3 and Tables 3-1 and 10-6) is, in terms of volume, 10^6 times larger than the nanometer scale, but there is no mechanism available for gaining precise, molecular control of the surface and interior of a complex, three-dimensional structure on this scale. The materials involved can include diamond films, organic films, semiconductors, metal films, dielectrics, ferroelectrics, and piezoelectric films. To control these structures, the technology relies on processing techniques, such as photolithographic pattern definition, plasma and wet etchings, physical and chemical vapor deposition, diffusion, and ion implantation. Refer to Table 6-4 to see how the free-electron laser supports these manufacturing processes. Some examples of the products made by these methods are flat-panel displays, solid-state sensors, and products of the microelectronics and optoelectronics industries. Microtechnology techniques are essentially unrelated to those of molecular manufacturing. Microtechnology uses a top-down approach (starting with large complex and irregular structures); nanotechnology, the chemistry of molecular manufacturing, follows its bottom-up path (starting with small, simple, and exact structures). In essence, microtechnology's goal is to make imprecise structures smaller, while the goal of molecular manufacturing is to make precise structures larger.

Micromachines are made from polysilicon, the same materials used in the manufacture of integrated circuits. As a result, micromachines and integrated circuits can be constructed on one chip. Current production methods for quartz-crystal timing devices and integrated circuits involve separate manufacturing steps for both the devices and the circuit, followed by assembly operations. First, the micromachines are embedded in a trench on a Si wafer, as a starting material for the conventional ***complementary metal-oxide semiconductor (CMOS)*** manufacturing process of integrated circuits. The circuits are built on the surface of the wafer; the MEMS are sealed in the trench. The MEMS function as a clock source, the oscillator generating frequencies of about 1 MHz with very low noise. Because of the low noise, the signals are constant, not disruptive, and are accurate. The frequency provides the constant timing signals necessary for the digital electronics device to operate.

Present-day applications of MEMS are as triggers for air bags, projection lights in digital theaters, and controllers that split colors in inkjet printers. A minuscule combination lock, called the Recordable Locking Device, developed by Sandia National Laboratory, uses MEMS technology to prevent unauthorized use of data on a computer's hard drive. Embedded onto a computer chip, the entire device is about the size of a shirt button and consists of a series of timing gears that move to an unlocked position when the operator enters the correct code. A series of six code wheels each less than 300 μm in diameter (about the size of the period at the end of this sentence) are driven by electrostatic drives that turn electrical impulses into mechanical motion. The lock "owner" sets a combination to any value from 1 to 1 million. The device "locks up" upon a wrong entry. The development phase concluded in 2001. Sandia National Laboratories has also developed a microtransmission for its similar-size microengine. It has a 3 million to 1 gear ratio that theoretically can generate a force capable of moving a 1-pound object (see Figures 28-12 and 28-13). The size of these mechanical devices is comparable to a grain of sand, or about one-third the width of a human hair. Possible applications are aircraft sensors, optical switches for telephones, microsurgical instruments, and more effective air bag sensors for automobiles.

Battery technology has failed to miniaturize batteries sufficiently for use in small electronic and mechanical components. Scientists at MIT's Lincoln Laboratory have turned to MEMS devices to develop a turbine engine the size of a shirt button. Linked to a tiny electric generator, the turbine engine would be capable of producing 10 to 20 times as much power as the best chemical batteries. If successful, inexpensive miniature turbine generators could eventually be stamped out in large quantities, like computer chips. Micromachinery fabricated from silicon would be limited in operation due to the high-temperature restrictions of silicon. Further research points toward the use of silicon carbide—a tougher, more temperature-resistant material. Silicon micromachines, also called MEMS, are projected to find routine applications in industries such as automotive and medical diagnostics. A class of medical cutting tools etched from wafers of silicon has blades said to be 10 times sharper than metal tools. These tools, using ultrasonic vibrations, vibrate up to 200,000 times per second. Such technology could lead to greater precision for highly sensitive procedures such as cataract surgery and neurosurgery and could be the basis for truly painless needles. Typical of products being de-

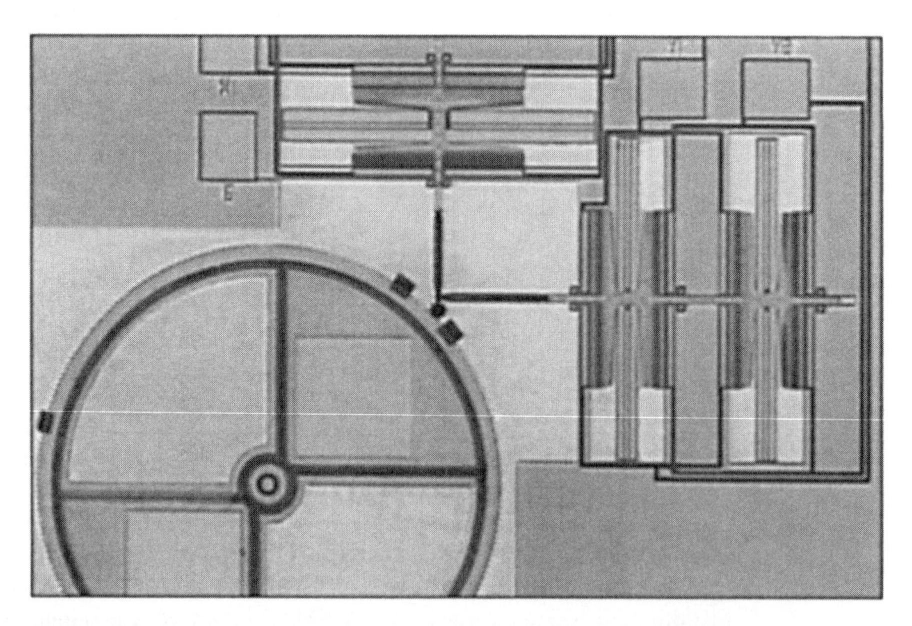

Figure 28-12 Three micromotors (rectangular structures, top and right), when energized by on–off voltages, power a microgear (at center) to drive the larger gear. The motor consists of two silicon combs with a shuttle between them. The edges of the shuttle also have teeth that interdigitate with those of the stationary combs. When energized, the stationary combs alternately pull the shuttle via electrostatic interaction. (Ernest J. Garcia, Sandia National Laboratories)

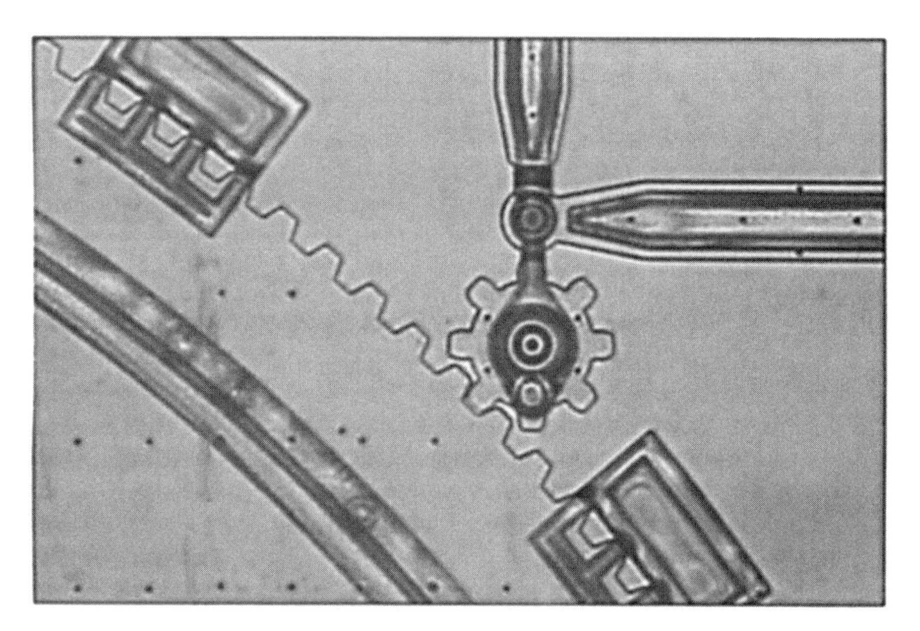

Figure 28-13 A close-up of the microgear (drive gear) meshing with the much larger optical-shutter gear. The attached shaft turns a drive gear in a quarter of a circle during the shaft's power stroke. Another micromotor, at right angles to the first, is timed to turn the microgear on the second quarter of its rotation. The two drives, alternating their force, convert reciprocating motion into rotary motion to drive the gear completely around. (Ernest J. Garcia, Sandia National Laboratories)

veloped using microtechnology (Figures 28-14 and 28-15) are three micromotors driving a microgear smaller in diameter than a human hair reported to be fabricated from silicon by researchers at Sandia National Laboratories. These devices were fabricated using both etching processes and silicon materials common to the microelectronics industry to facilitate their mass production. Each micromotor develops 0.5 μW of power rotating at a speed of 30 revolutions per second—sufficient to drive the larger gear (30 times larger than a microgear) at an angular velocity of 1 revolution per second. Open spaces in the large gear in some positions allow light to pass through, creating an optical shutter.

Our final example of microtechnology is illustrated in Figure 28-16, a photograph of a Africanized, or killer, bee with a chip attached to its thorax. This solar-powered electronic package, with dimensions of 1 × 3 × 5 mm and weighing less than 50 mg, contains a microminiature infrared LED transmitter and tracking system. Developed to allow entomologists to continuously track bees in the field to study their mating and foraging habits, the chip, when mounted on the bee, prevented it from flying. This illustrates how research often results in apparent solutions that do not succeed or that require further modifications, a result that is normally discovered during field testing.

28.4 MOLECULAR NANOTECHNOLOGY, MOLECULAR SYSTEMS ENGINEERING, OR MOLECULAR MANUFACTURING

Much of what has been covered in this module and many of the innovations discussed throughout this book rely on manipulating the very small nanostructures of materials to construct such advanced materials as smart materials and biomaterials. Emerging from the nation's laboratories are electronic components—transistors, capacitors, and diodes—with dimensions of less than 0.25 μm. Units such as the micrometer are becoming less useful as units of measure. Instead, the next smallest unit of measure, the ***nanometer (nm),*** defined as one-billionth of a meter, will be the norm. Compared to a human hair, the diameter of a nanometer is 10 times smaller. The diameter of most atoms is 0.1 to 0.4 nm. This new technology is being investigated for use in manufacturing and we will describe it in more detail in the following paragraphs.

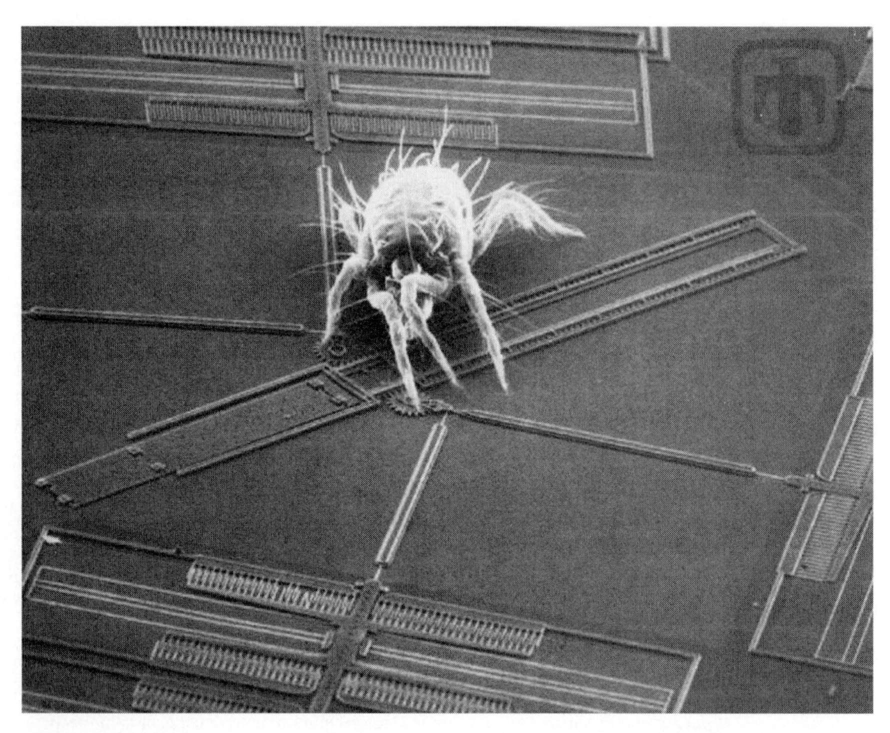

Figure 28-14 Spider mite on mirror assembly. Imagine a machine so small that it is imperceptible to the human eye. Imagine working machines with gears no bigger than a grain of pollen. Imagine these machines being batch fabricated tens of thousands at a time, at a cost of only a few pennies each. Imagine a realm where the world of design is turned upside down and the seemingly impossible suddenly becomes easy—a place where gravity and inertia are no longer important, but the effects of atomic forces and surface science dominate. The microdomain is a world now occupied by an explosive new technology known as MEMS (microelectromechanical systems) or, more simply, micromachines. Sandia National Laboratories, motivated by a guiding vision for MEMS, has become a recognized leader in this emerging field. To learn more, visit the MEMS site. (Sandia National Laboratories Intelligent Machining Institute)

Figure 28-15 No, this is not Godzilla, but rather a spider mite (a miniscule white fleck to the human eye) hanging out on a microlock mechanism. Note the scale key in the lower right corner, which shows 100 μm. (Sandia National Laboratories Intelligent Machining Institute)

Figure 28-16 A microminiature infrared LED transmitter and tracking system is under development that will emit infrared pulses, detectable over a range of 1–2 km. The chip shown is the proper size but it is not the actual transmitter. This chip was used to test the bee's ability to fly with the microchip on its back. (Oak Ridge National Laboratories)

Module 1 stated that one constant of materials technology is change and discussed the difference between using designer materials versus off-the-shelf materials with their limited structures and properties. Module 2 stated that materials will be produced in new ways as materials engineers work with manufacturing engineers and materials scientists to invent new processing systems. The Pause & Ponder for Module 3 pointed out the synthesis and processing capabilities of contemporary materials technology at the various scales of measurement, i.e., macro, micro, nano, and atomic. Module 3 also contains a reminder that in the race to develop new materials both developed and newly developing countries are competing to improve or maintain their positions.

The heading of this section uses present-day terms to refer to the emerging field of engineering nanoscale mechanical systems to guide the placement of reactive molecules while building complex structures atom-by-atom. ***Molecular manufacturing*** is defined as the construction of objects to complex, atomic specifications using sequences of chemical reactions directed by nonbiological molecular machinery. ***Molecular nanotechnology*** describes the field as a whole, comprising molecular manufacturing together with its techniques, its products, and their design and analysis. This new field is related to, yet distinct from, mechanical engineering, microtechnology (to be defined), chemistry, and molecular biology. As physics is related to engineering, so chemistry is related to molecular engineering. Molecular manufacturing/microfabrication applies the principles of mechanical engineering to chemistry. Figure 28-17 is an example of a molecular manufacturing process.

Recent research into ***carbon nanotubes (CNT),*** rolled sheets of carbon just 1/50,000 the thickness of a human hair, has revealed that they are excellent conductors of electricity. Carbon nanotubes are formed by vaporized carbon that condenses into a series of hexagons that naturally curl into hollow tubes. The twist of the tube determines its particular electronic properties. If the carbon sheet forming a nanotube is rolled up evenly, then the carbon molecules conduct electricity as if the sheet were a metallic substance. If the nanotube is rolled up in a twisted manner, then the nanotube acts as a semiconductor. Further, if the nanotubes are joined together end-to-

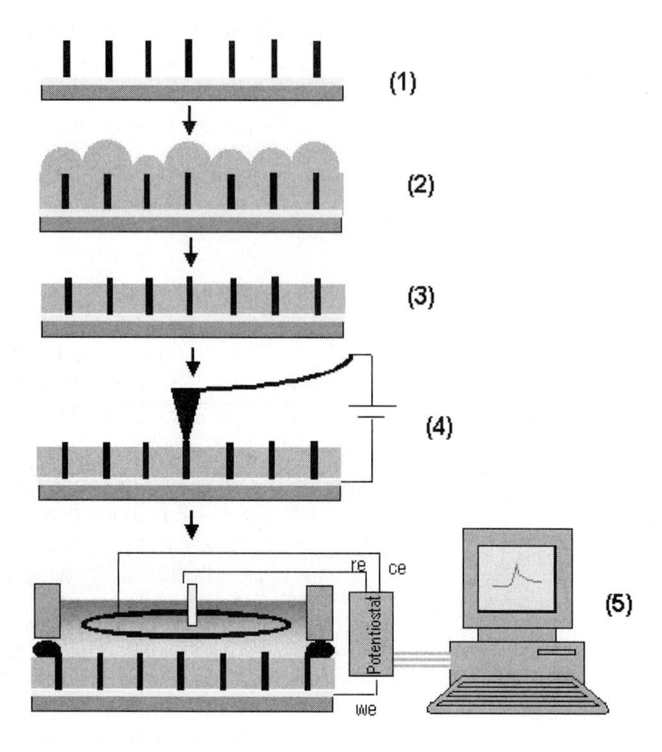

Figure 28-17 Fabrication of CNT nanoelectrode array. (1) Growth of vertically aligned CNT array, (2) dielectric encapsulation, (3) planarization, (4) electrical property characterization by current-sensing AFM (atomic force microscope), and (5) electrochemical characterization. (Ames Research Center)

end, then their junction acts as an electronic diode. A ***diode*** is the basis for devices that convert alternating current into direct current. Research is proceeding in constructing ***nanotransmitters*** that could be built into electrical circuits. The growth of 3-D carbon nanotube structures can now be controlled through a selective process that allows the nanotubes to grow in any direction. When combined with silicon in future computer chips, integrated circuits, and MEMS devices, carbon nanotube structures that can be grown in any predetermined direction under very fine control will provide the power necessary to run these devices.

The National Nanotechnology Initiative (NTI) was created by the U.S. Congress to provide federal funding for nanoscale research and development. Its initial funding level was doubled after only two years and several key objectives have been reached ahead of time. The broad bipartisan support for this research in Congress is due in large part to its long-term societal implications in terms of potential applications. Nanotechnology will have a large impact on warfare and national security. One of the largest contributors of funds necessary for development is the U.S. Department of Defense. Some likely military applications will be the detection of and protection from explosives. ***Nanoengineered fabrics*** for masks and protective clothing and multifunctional adaptable "smart" materials for soldiers' uniforms are under design. Decontamination equipment—high-surface-area adsorbents, nanopowders, and nanoreactors for chemical and biological-agent detection and destruction—are high on the list of priorities. The major impact nanotechnology is already having on the commercial marketplace is evidenced by the following examples. Magnetic resistance molecules are now routinely used in magnetic storage applications. Nanoparticles used in drug delivery systems and as reinforcements in polymer materials are now commonplace, and a new class of polymer blends contains functional nanoparticles. For example, a new Wilson tennis ball holds pressure twice as long because of a butyl rubber nanocomposite coating on the inside of the ball, an example of the new class of ***polymer nanocomposites*** (see Figure 16-13). Using a variety of inorganic materials, including clay minerals, vermiculites, carbon nanotubes, and silver dendrimers, polymer nanocomposites have a long list of improved physical properties such as improved tensile strength and modulus without loss of impact resistance and improved conductivity. Nanotechnology is a major trend in science and engineering both in the United States and in Europe. Materials scientists are now able to engineer materials at the atomic and molecular levels with precision by using proteins, viruses, and other biosystems to assemble molecular-scale nanodevices. Reaching that level of production of 1500 tons a year in 2004 reduced the price many times.

The ability to control structure and arrangement (assembly of the microstructure and nanostructure of materials) has led to the development of many new materials and many new applications. Carbon nanotubes (CNTs) like those shown in Figure 28-18 can now be arranged into specific configurations with distinct electrical and mechanical properties. By using an optimized catalytic chemical vapor deposition (CVD) technique, long, continuous strands of single-walled nanotubes (SWNTs) can be synthesized. The ***SWNTs*** typically form bundles of nanotubes that are 10s of micrometers in length (Figure 28-19). Multiwalled nanotubes (***MWNTs***) aligned normal to the substrate surface can be grown into densely packed cylindrical pillars whose heights can be controlled to within 1–2 μm. CNTs can also be grown in multiple directions simultaneously to produce vertically and horizontally aligned periodic arrays that form repeated patterns. This capability is now leading to new nanoscale applications such as highly conductive microcables made out of long strands of SWNTs and integrated systems and electromechanical devices (see Figure 28-20) manufactured from well-ordered patterns. Improvements in nanoscale manufacturing capabilities, in instruments and tools for measurement, manipulation, and analysis and standards are ongoing. Advances will depend, however, on the education and training of the future technical workforce.

Over the next 20 years, this new manufacturing system for the bulk processing of materials and the fabrication of custom products will rely on numerous small manufacturing systems working with locally available materials. The products will increasingly exhibit order-of-magnitude improvements in mechanical properties and will be of high quality and low cost. When these machines have sufficient general capabilities to, first, manufacture copies of themselves and, second, be reprogrammed to manufacture finished products directly from raw materials, bulk processing will be attained and traditional fabricating techniques will be bypassed (Figure 28-21).

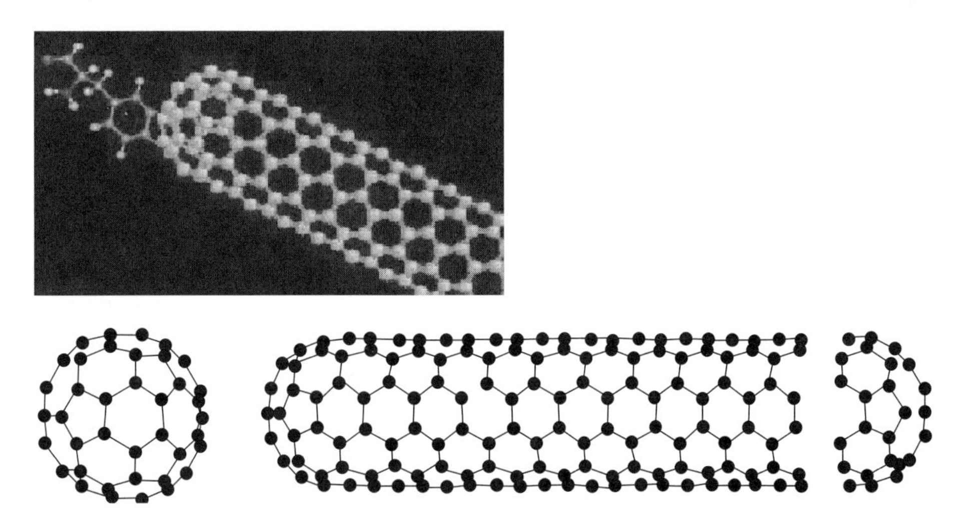

Figure 28-18 (a) Single-walled nanotubes (SWNT), carbon nanotubes composed of all-carbon molecules that resemble rolled-up chicken wire with curved ends. Tubes can be as narrow as 1 nanometer (nm) and as long as 100 micrometers (μm) or longer. Electrical conductivity can be six orders greater than copper. (b) Nanotubes form individually or within one another as multiwalled tubes (MWNTs) or in mixtures. Lightweight and durable nanotubes and composite nanotubes will assist in producing more fuel-efficient vehicles. They are 100 times stronger than steel at a sixth of its weight. Their electronic properties can be tailored through external magnetic fields or mechanical deformation applications. These building blocks can be further assembled into larger structures with designed properties that will revolutionize materials manufacturing. (Meyyapan, Meyya. "Nanotechnology: Opportunities and Challenges," *National Educators' Workshop: Update 2002.* NASA/CP–2003–212403, p. 27.

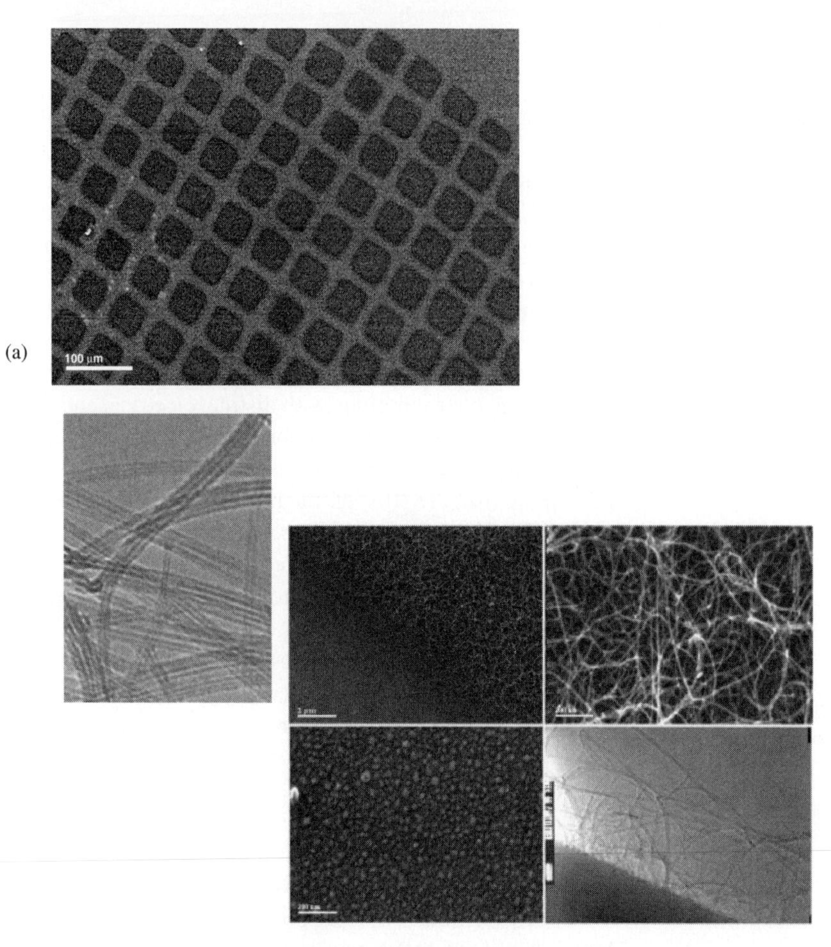

Figure 28-19 (a) Carbon nanotubes viewed with a transmission electron microscope (TEM) with a surface-masked 400-mesh grid. (b) Weaving nanotubes into a membrane with pores only a few atoms across produces nanofilters that are extremely effective. Scientists are hoping to mix nanotubes with other materials to create high-strength plastic and composites. Many applications are foreseen, such as using nanowires that self-assemble into semiconducting circuits. A sheet of aligned nanotubes has the capability to bend in response to an electrical charge acting as an actuator. (O'Brien, Stephen P. "A Chemical Perspective to Strategy and Design of Nanoscale Materials: The Science Behind Technology," *National Educators' Workshop: Update 2002.* NASA/CP–2002–212403, p. 276)

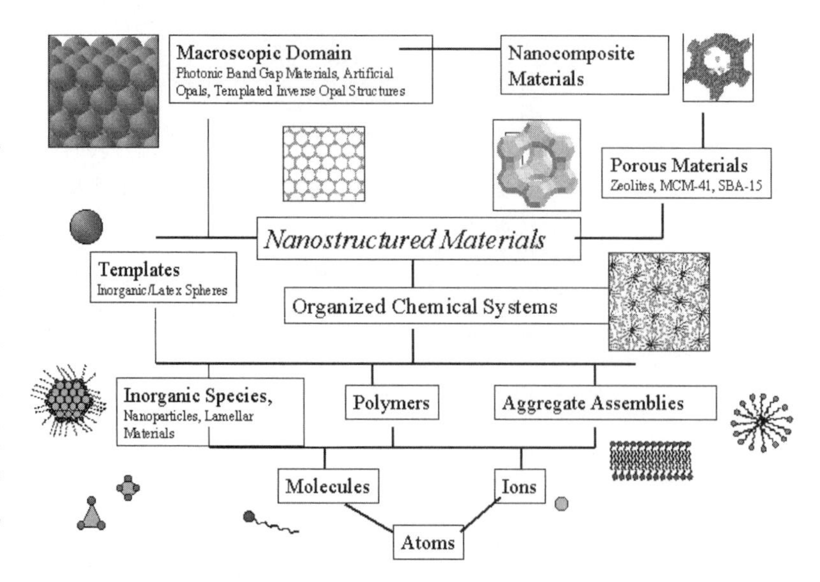

Figure 28-20 A chemical approach to the hierarchical construction of matter. This graphic emphasizes the fact that chemistry is the science behind nanotechnology, forming a firm foundation for the development of materials not only in the past but in the future. Recall that nanostructured materials are only one atom thick, a few atoms across, and hundreds of atoms long.The graphic bears out the definition of nanotechnology as the science and engineering of creating materials, functional structures, and devices on the nanometer scale, i.e., a bottom-up, atom-by-atom manufacturing synthesis for the production of advanced materials. (O'Brien, Stephen P. "A Chemical Perspective to Strategy and Design of Nanoscale Materials: The Science Behind Technology," *National Educators' Workshop: Update 2002.* NASA/CP–2002–212403, p. 276)

Since the introduction of the concept in 1981, molecular nanotechnology has led to the development of the scanning probe microscope (Module 8), new proteins designed by molecular biologists, molecules that can trap other molecules and ions, scientific workstations and molecular modeling software that permit faster testing and construction of new designs, biomimetics, and near-net-shape processes (Modules 3 and 5) that have eliminated the traditional secondary processes accompanying manufacturing (Figure 28-20). The next step is the development of new conceptual and mathematical tools to design and analyze a limited class of molecular manufacturing systems. The total costs of manufacturing, excluding the cost of development and distribution, will then be almost wholly determined by the cost of materials. For the general class of manufactured goods, the relevant materials would cost about \$0.1/kg to \$0.5/kg. Most of today's manufactured products fall in the range between about \$10/kg (i.e., automobiles, appliances) and $\$10^4$/kg (i.e., aerospace vehicles).

Mechatronics is a new term describing the integration of mechanical systems with new electronic components and intelligent software control. Until recently, mechatronics systems were used mostly in aircraft and industrial equipment or in small precision components for products such as cameras and photocopiers. Today, automotive researchers are turning to mechatronics. This new direction has been fueled by the falling prices of microprocessors and sensors and more stringent vehicle emissions standards in Europe and California. Customer demand for additional comfort and performance features in vehicles has also had an effect on the development of this new technology. It is reported that some 80% of the innovations in the development of engines and cars are due to electronics and mechatronics. Some luxury BMWs have more than 70 microprocessors that control more than 120 tiny motors. Software identifies and corrects flaws in real time to make sure the technology functions impeccably.

 Micro World with tutorials—*http://www.lbl.gov/MicroWorlds/*

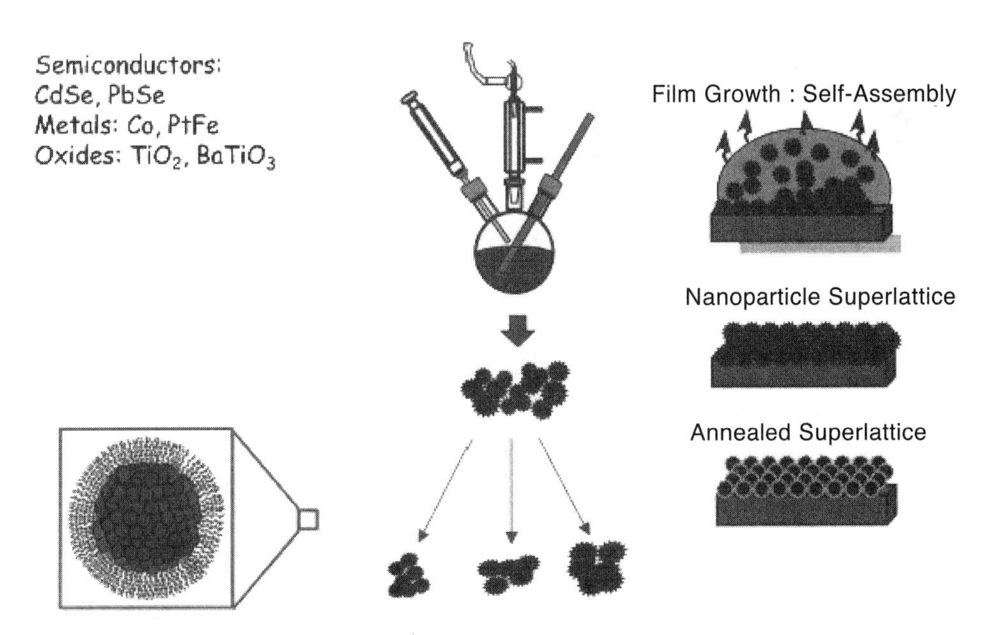

Figure 28-21 Nanoparticles of various materials are synthesized and assembled into thin films. Size distribution is important, with variations in the size of particles being the objective. Particles can be obtained from semiconductors, metals, or oxides. The particles are precipitated with the controlled addition of a flocculent and then centrifuged. These films have potential for biosensor applications as antifogging or antireflection surfaces. The size of the nanoparticles and their concentration can be controlled such that they can change the optical properties of nanocomposite films. (O'Brien, Stephen P. "A Chemical Perspective to Strategy and Design of Nanoscale Materials: The Science Behind Technology," *National Educators' Workshop: Update 2002.* NASA/CP–2002–212403, p. 276)

28.5 CHALLENGES IN MATERIALS SCIENCE AND TECHNOLOGY

Throughout this book we have sought to transport you from the basic concepts of engineering materials technology, dealing with traditional materials, to advanced and future materials and processes technology. You have been exposed to new instruments, unique processes, and innovations in materials. At the same time, we have tried to help you to develop problem-solving skills, to learn to look around your environment and to observe what and how materials are currently being used and abused. We encouraged you to become familiar with the databases outside this book. Now we implore you to join a technical society such as ASM International, the Society of Manufacturing Engineers, the American Society for Testing and Materials, or any of the other groups who can assist you in keeping up with developments in our ever-expanding field. Student membership rates for these organizations are excellent, as are the benefits.

We encouraged you to consider the materials cycle and the consequences of each stage of the cycle. All who venture into technology must carry with them a concern for the environment. Selection of certain technologies and the associated materials to solve environmental problems is sometimes guided by politics and the perceptions of the general public, but these decisions may not be based on the best use of the facts. An example is the mandating of electric-powered vehicles to relieve air pollution caused by internal combustion and gasoline-powered vehicles. Figure 28-22 reflects efforts to make thinner, lighter, and more powerful batteries. The construction of newer batteries and the use of thin materials eliminates the need for liquids and therefore heavy, bulky casings. We discussed electric vehicles and the new batteries in Module 11. As we suggested there, battery technology has a long way to go before electric vehicles become a strong option for most needs. Take, for example, the issue of the weight of a gasoline-powered, internal combustion engine versus a battery-powered motor. One gallon of gasoline weighs about 6 pounds. To produce the same energy as the 6 pounds of gasoline requires nearly 400 pounds of battery. In terms of air pollution, battery-powered vehicles also are responsible for harmful emissions, although not the vehicles themselves. To produce electricity for recharging batteries, generating plants using fossil fuel waste as much as 65% of the energy to transform the fuel into electricity; and then another loss of 5–10% is incurred when the electricity is transmitted to the battery-charging receptacle. So generating

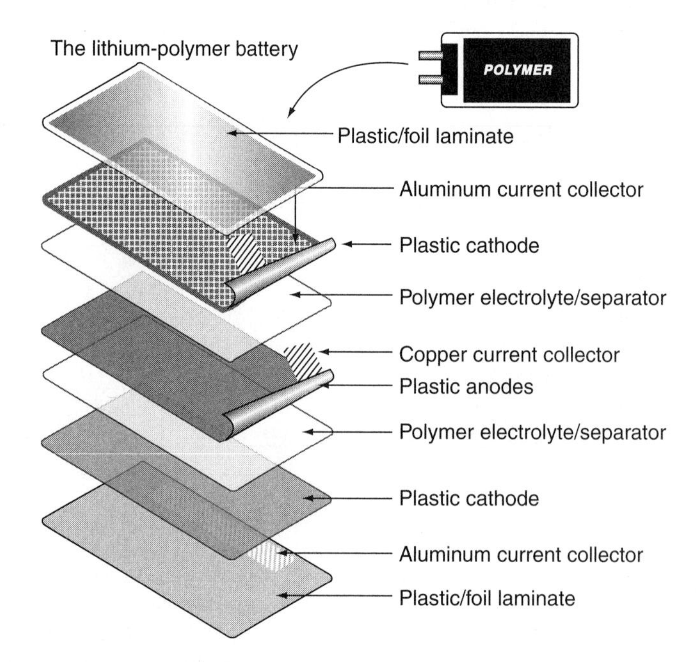

Figure 28-22 Batteries store energy chemically and release it electrically. A typical battery is made up of two terminals connected by an electrically conductive substance (electrolyte). Lithium–polymer batteries are lighter and thinner than other batteries since they contain no liquids requiring a heavy casing. (*USA Today*)

Figure 28-23 "Smart coatings" may buy more time for the auto industry to improve power systems in vehicles that will be easier on our air quality. Smog-eating radiators use PremAir™ coatings to convert ozone (O_3)—the main component of smog—into breathable oxygen (O_2). (Ford Motor Company)

electricity for battery charging not only creates harmful emissions, it is an inefficient use of fossil fuel. Many other issues challenge the viability of battery-powered vehicles, including cost and fire hazards associated with current batteries, and materials science has yet to develop reliable, long-life batteries. Experts feel that we should pursue more viable and cost-effective solutions for dealing with the many environmental issues of the internal combustion engine for vehicles and wait for battery technology to mature before committing to electric vehicles. *Fuel cell technology* is garnering considerable attention.

Research is being conducted aimed at cleaner internal combustion engines. One example involves using a coating on radiators (Figure 28-23) that actually offers the potential to clean rather than pollute the air as the engine runs. Volvo sedans are equipped with the PremAir™ coated radiators. Such a development may buy more time for economical and practical improvements in vehicles powered by electricity, hydrogen, fuel cells, or other alternative fuels.

28.6 ADVANCED PROCESSING TECHNOLOGIES

28.6.1 3DP (Three-Dimensional Printing)

Three-dimensional printing is a solid, free-form technology for fabricating tools and parts from a variety of ***engineered materials.*** The parts are built up in layers from a computer-assisted design (CAD). Figure 28-24 is a four-part diagram of the 3-D dimensional printing process. Each layer begins with a thin distribution of powdered metal spread over the surface of a part-build bed. Using an inkjet printing technology, the binder material is deposited selectively to join the metal particles in the shape of the part. This layer-by-layer process repeats until the part is completed and removed from the powder bed. Following densification using heat-treating processes such as debinding, sintering, and infiltration of a second metal, near full density is achieved (8.10 g/cc). The fabricated parts or tools are made from a variety of materials, including ceramics and tool steels, and can have an array of geometries, such as intricate internal passages, overhangs, undercuts, and other internal volumes. Future developments will concentrate on nozzles that will allow the composition of the printed part on a 100-μm scale as well as production by a machine that can print a full layer in one pass of the print head. This new process, truly a ***rapid (direct) manufacturing*** process, will permit the direct mass production or customized production of many types of products that are near-net-shape and require little, if any, final machining.

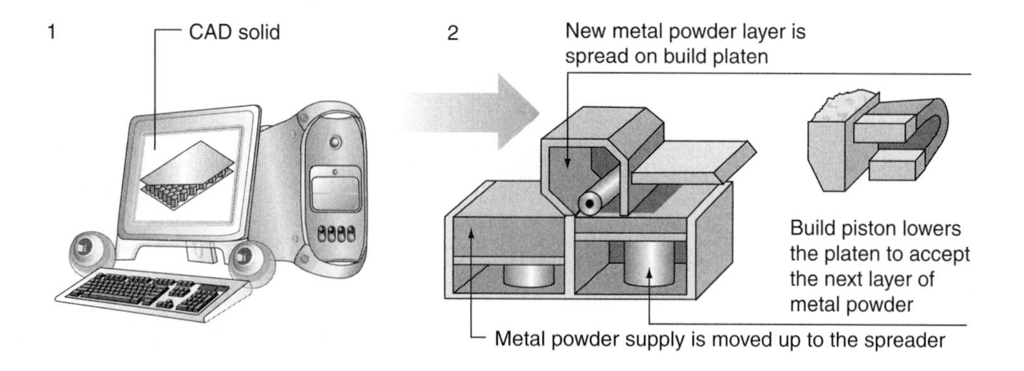

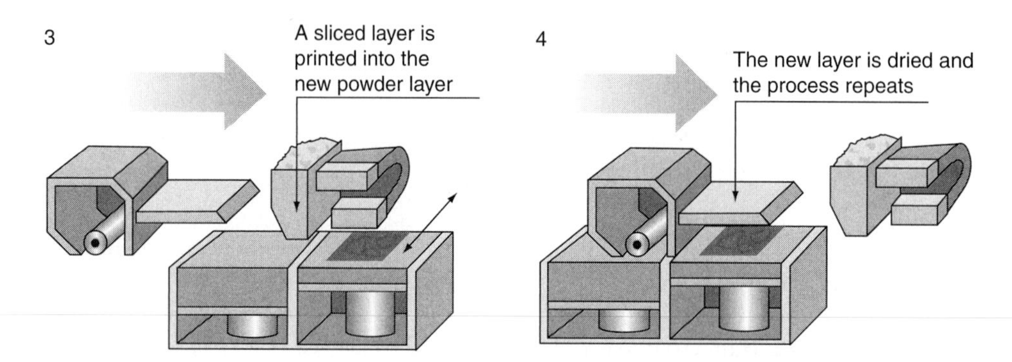

Figure 28-24 Diagram of the three-dimensional printing process used to fabricate tools and parts from a variety of materials. Engineered materials can be produced using powder forms from such materials as ceramics, metals, cermets, and polymers. (*Advanced Materials & Processes* and Extrude Hone Corporation)

28.6.2 Ultrasonic Consolidation (UC)

This newly commercialized rapid tooling process fabricates aluminum tooling in a single operation using a single machine. The tooling can be used in other applications, such as injection molding or extrusion. Figure 28-25 is a schematic of how the UC process works. UC is a microfriction process that produces true metallurgical bonds between layers of metal without melting. As surfaces vibrate against each other at very high frequencies, surface contaminates such as oxides are removed. Modest pressures at temperatures that are typically below half the melting point of the metal are applied to these atomically clean surfaces and strong, featureless (no course remelted zones) bonding results.

Like other rapid prototyping and rapid tooling processes, UC generates objects directly from CAD representations of the component. RPCAM, the software from Solidica, Inc., Ann Arbor, MI, slices the geometry into layers 0.1 mm thick and generates the machine program to build, trim, and finish the part. This solid-state processing has a number of benefits: Less energy is required due to the low temperatures involved, residual stresses and distortion are reduced primarily because there is no liquid–solid transformation, no additional finishing or machining is required, surface finishes are consistent with those of CNC machining, part build can be suspended during operation and resumed at a later date or time, a range of engineering plastics can be successfully molded without excessive tool wear, damaged aluminum parts can be repaired, engineering changes to parts can be accepted at any point in the process, and structural as well as optical fibers can be embedded between the layers. Nanoscale fabrication will involve generalized synthesis and self-assembly into thin films. As seen in Figure 28-21, size-selective precipitation is involved. The particles are precipitated with the controlled addition of a flocculent and then centrifuged. The process can be repeated to focus the size distribution to deviations of ± 10% of the diameter ± 1 nm, or in some cases, down to a ± monolayer of atoms).

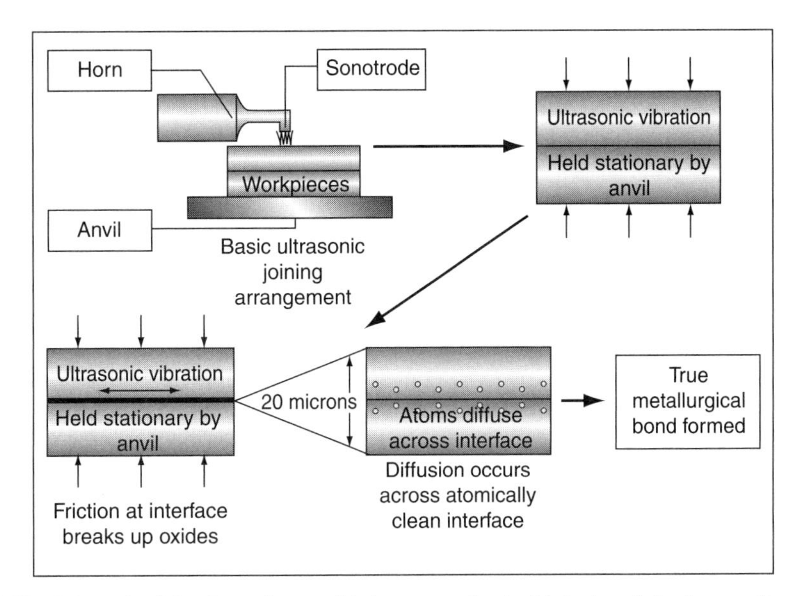

Figure 28-25 Schematic of the ultrasonic consolidation process for the fabrication of aluminum tooling and parts. This is a microfriction process that produces metallurgically strong bonds using CAD representations and computer software. (*Advanced Materials & Processes* and Solidica, Inc.)

28.7 INTO THE FUTURE

The technically literate must enter into debates about the impact of technologies on our world and its inhabitants. Those who fear technology need to hear from those who understand it. To parallel the saying "Those who don't know the lessons of history are doomed to repeat them," we say, "Those who don't understand the fundamentals of technology may suffer from it and not enjoy its potential benefits." The knowledge you have gained through this study should serve you well in your daily life and career. We would love to hear your comments on this book. Good luck in the future!

SELF-ASSESSMENT

28-1. Which smart material might work a robot finger control or an antenna for a cellular phone because of its ability to remember its original shape after being deformed?

a. MMC **b.** Piezoelectric **c.** Titanium **d.** Shape-memory alloy

28-2. Biomimetics can help intelligent materials systems technology by

a. Providing feedback and control
b. Ensuring osseointegration
c. Aiding improved hybrids
d. Prompting biotransfer

28-3. What is a main concern when selecting a material for bioengineering?

a. Ability to resist heat
b. Ability to provide feedback and control
c. Compatibility with human and animal systems
d. Good formability

28-4. Intelligent materials systems or "smart materials" of the future will depend on, or mimic __________ to control and add feedback into their materials structures.

a. Semiconductors **b.** Biological organisms **c.** Microrobots

28-5. Organic films are found everywhere. One example of such a film can be found on

a. Soap

b. Chromium automobile bumpers

c. Ceramic coatings on polymers

28-6. The greatest challenge to research teams of chemists, materials scientists, and surgeons working to find materials for use as implants is to find one that is not harmful to the human body and is

a. Organic **b.** Degradable **c.** Lightweight

28-7. Sensors in aircraft wings are being developed using piezoelectrics. What feature of piezoelectrics is being used in such an application?

a. Ability to sense a temperature change

b. Ability to detect a change in applied force or pressure

c. Ability to respond to a change in electrical current

28-8. The biological nanoscale objects that can be manipulated are

a. Polymer molecules **b.** Cells **c.** Inorganic particles

28-9. The biomaterial Vitallium is composed of chromium, nickel, molybdenum, and

a. Copper **b.** Cadmium **c.** Cobalt

28-10. Silicon micromachines are also referred to as

a. NEMS **b.** Macromachines **c.** MEMS

28-11. As physics is related to engineering so__________is related to molecular engineering.

a. Biology **b.** Chemistry **c.** Metallurgy

28-12. A diode is a device that converts __________ to direct current.

a. Am **b.** Alternating current **c.** dc

28-13. Nanotubes can be arranged into specific structures that exhibit distinct __________ properties.

a. Mechanical **b.** Optical **c.** Thermal

REFERENCES & RELATED READINGS

AEROSPACE VEHICLE SYSTEMS TECHNOLOGY OFFICE, NASA Langley. *Morphing. http://avst.larc.nasa.gov/morphing,* January, 2004.

AKSAY, ILHAN A. "Material World—What Can Biology Teach Us?" Lecture at NASA Langley, May 2003.

ASM INTERNATIONAL. "Smart Materials Alter Shape of Adaptive Aircraft Wings," *Advanced Materials & Processes,* September 1995, p. 9.

Biologically Inspired Materials Institute (BIMat NASA URETI) website—*http://bimat.princeton.edu.* January 2, 2004.

CONSTANTZ, BRENT R. "Mineral Paste Surgically Injected to Repair Bones," *Advanced Materials & Processes,* June 1995, p. 17.

DREXLER, K. ERIC. *Nanosystems—Molecular Machinery, Manufacturing, and Computation.* New York: John Wiley & Sons, Inc., 1992.

FISCHER, ARTHUR. "Superconductivity," *Popular Science,* April 1988, pp. 54–58.

FONASH, STEPHEN J. "Micro and Nanofabrication," Surface Engineering and Manufacturing Technology Center (SEMTC), U.S. Navy Manufacturing Science & Technology Program Report, February 1995, p. 3.

JACOBS, JAMES AND DIANZHA LACLAIRE (EDS.) *National Educators' Workshop: Update 2003, Binder of Proceedings,* October, 2003.

JONES, WALTER. "Air Force Basic Research in Materials and Structures," *National Educators' Workshop,* NASA Langley, October 22, 2003.

KLIENFIELD, SONNY. *A Machine Called Indomitable.* New York: Time Books, 1985, pp. 92–97.

MCWHORTER, PAUL, JEFF SNIEGOWSKI, AND ERNEST GARCIA. "Mighty Micrometers Drive Miniature Gears," *Advanced Materials & Processes,* December 1995, p. 6.

MALLARDI, JOSEPH L. *From Teeth to Jet Engines.* Hampton, VA: Howmet Corporation, 1992.

"Manufacturing Parts Drop by Drop," *Compressed Air Magazine,* March 1995, pp. 38–44.

MIT: *Technology Review* website—*www.techreview.com*, January, 2004

Moving America: New Directions, New Opportunities, report to Congress, June 1990.

"Nanotechnology," *Compressed Air Magazine,* October/November 1994, pp. 16–23.

National MAGLEV Initiative: Annual Report—November 1992. Washington, DC: Federal Railroad Administration, 1992.

NIKON'S SMALL WORLD GALLERY. *http://www.microscopy/u.com,* January, 2004.

"Nimbus Information System." *CD-ROM Replication Guide.* Nimbus Manufacturing, Inc., Charlottesville, VA, 1995.

O'BRIEN, STEPHEN P. "A Chemical Perspective to Strategy and Design of Nanoscale Materials: The Science Behind Technology," *National Educators' Workshop: Update 2002,* NASA/CP–2002–212403, p. 276.

OLIVER, JOYCE. "Superconductors: The Heat Is On," *The Electron,* November/December 1988, pp. 12–17.

SARIKAYA, MEHMET, AND ILHAN A. AKSAY, eds. *Biomimetics: Design and Processing of Materials.* New York: AIP Press, 1995.

SHUART, MARK J. "Structures and Materials at NASA Langley," *National Educators' Workshop,* NASA Langley, October 20, 2003.

SONG, KYO D. *Development of Microwave-Driven Smart Materials Actuators.* NASA project overview, March 1, 2000.

UENOHARA, MICHIYKI. "Electro-Ceramics and Applications in C&C Era," *Journal of Materials Education,* Volume 10, No. 3, 1988, pp. 259–280.

WEBER, WILLIAM J. "9.5 Electroceramics," Pacific Northwest Laboratory, October 1992.

Appendix

TABLE A-1A PERODIC TABLE OF THE ELEMENTS

Group	Atomic number	Symbol	Name	Atomic weight	Oxidation states	Boiling point, K	Melting point, K	Density	Electron configuration
GROUP 1/IA	1	H	Hydrogen	1.00794	1	20.28	13.81	0.0899 †	$1s^{1}$
GROUP 1/IA	3	Li	Lithium	6.941	1	1615	453.7	0.53	$1s^{2}2s^{1}$
GROUP 1/IA	11	Na	Sodium	22.98977	1	1156	371.0	0.97	$[Ne]3s^{1}$
GROUP 1/IA	19	K	Potassium	39.0983	1	1033	336.8	0.86	$[Ar]4s^{1}$
GROUP 1/IA	37	Rb	Rubidium	85.4678	1	961	312.63	1.532	$[Kr]5s^{1}$
GROUP 1/IA	55	Cs	Cesium	132.9054	1	944	301.54	1.87	$[Xe]6s^{1}$
GROUP 1/IA	87	Fr	Francium	(223)	1	950	300	—	$[Rn]7s^{1}$
2/IIA	4	Be	Beryllium	9.01218	2	3243	1560	1.85	$1s^{2}2s^{2}$
2/IIA	12	Mg	Magnesium	24.305	2	1380	922	1.74	$[Ne]3s^{2}$
2/IIA	20	Ca	Calcium	40.078	2	1757	1112	1.55	$[Ar]4s^{2}$
2/IIA	38	Sr	Strontium	87.62	2	1655	1042	2.54	$[Kr]5s^{2}$
2/IIA	56	Ba	Barium	137.33	2	2078	1002	3.59	$[Xe]6s^{2}$
2/IIA	88	Ra	Radium	(226)	2	1413	973	5.0	$[Rn]7s^{2}$
3/IIIA	21	Sc	Scandium	44.9559	3	3109	1814	2.99	$[Ar]3d^{1}4s^{2}$
3/IIIA	39	Y	Yttrium	88.9059	3	3611	1795	4.47	$[Kr]4d^{1}5s^{2}$
3/IIIA	57	La ★	Lanthanum	138.9055	3	3737	1191	6.15	$[Xe]5d^{1}6s^{2}$
3/IIIA	89	Ac ★★	Actinium	(227)	3	3470*	1324	10.07	$[Rn]6d^{1}7s^{2}$
4/IVA	22	Ti	Titanium	47.87	4,3	3560	1935	4.54	$[Ar]3d^{2}4s^{2}$
4/IVA	40	Zr	Zirconium	91.224	4	4682	2128	6.51	$[Kr]4d^{2}5s^{2}$
4/IVA	72	Hf	Hafnium	178.49	4	4875	2504	13.31	$[Xe]4f^{14}5d^{2}6s^{2}$
4/IVA	104	Rf	Rutherfordium	(261)	—	—	—	—	$[Rn]5f^{14}6d^{2}7s^{2}$*
5/VA	23	V	Vanadium	50.9415	5,4,3,2	3650	2163	6.11	$[Ar]3d^{3}4s^{2}$
5/VA	41	Nb	Niobium	92.9064	5,3	5015	2742	8.57	$[Kr]4d^{4}5s^{1}$
5/VA	73	Ta	Tantalum	180.9479	5	5730	3293	16.65	$[Xe]4f^{14}5d^{3}6s^{2}$
5/VA	105	Db	Dubnium	(262)	—	—	—	—	$[Rn]5f^{14}6d^{3}7s^{2}$*
6/VIA	24	Cr	Chromium	51.996	6,3,2	2945	2130	7.19	$[Ar]3d^{5}4s^{1}$
6/VIA	42	Mo	Molybdenum	95.94	6,5,4,3,2	4912	2896	10.22	$[Kr]4d^{5}5s^{1}$
6/VIA	74	W	Tungsten	183.84	6,5,4,3,2	5825	3695	19.3	$[Xe]4f^{14}5d^{4}6s^{2}$
6/VIA	106	Sg	Seaborgium	(263)	—	—	—	—	$[Rn]5f^{14}6d^{4}7s^{2}$*
7/VIIA	25	Mn	Manganese	54.9380	7,6,4,2,3	2235	1518	7.44	$[Ar]3d^{5}4s^{2}$
7/VIIA	43	Tc	Technetium	(98)	7	4538	2477	11.5	$[Kr]4d^{5}5s^{2}$
7/VIIA	75	Re	Rhenium	186.207	7,6,4,2	5870	3455	21.0	$[Xe]4f^{14}5d^{5}6s^{2}$
7/VIIA	107	Bh	Bohrium	(264)	—	—	—	—	$[Rn]5f^{14}6d^{5}7s^{2}$*
8	26	Fe	Iron	55.845	2,3,6	3023	1808	7.874	$[Ar]3d^{6}4s^{2}$
8	44	Ru	Ruthenium	101.07	2,3,4,6,8	4425	2610	12.37	$[Kr]4d^{7}5s^{1}$
8	76	Os	Osmium	190.23	2,3,4,6,8	5300	3300	22.6	$[Xe]4f^{14}5d^{6}6s^{2}$
8	108	Hs	Hassium	(265)	—	—	—	—	$[Rn]5f^{14}6d^{6}7s^{2}$*
9 VIIIA	27	Co	Cobalt	58.9332	2,3	3143	1768	8.90	$[Ar]3d^{7}4s^{2}$
9 VIIIA	45	Rh	Rhodium	102.9055	2,3,4	3970	2236	12.41	$[Kr]4d^{8}5s^{1}$
9 VIIIA	77	Ir	Iridium	192.22	2,3,4,6	4700	2720	22.6	$[Xe]4f^{14}5d^{7}6s^{2}$
9 VIIIA	109	Mt	Meitnerium	(268)	—	—	—	—	$[Rn]5f^{14}6d^{7}7s^{2}$*

* Estimated Values

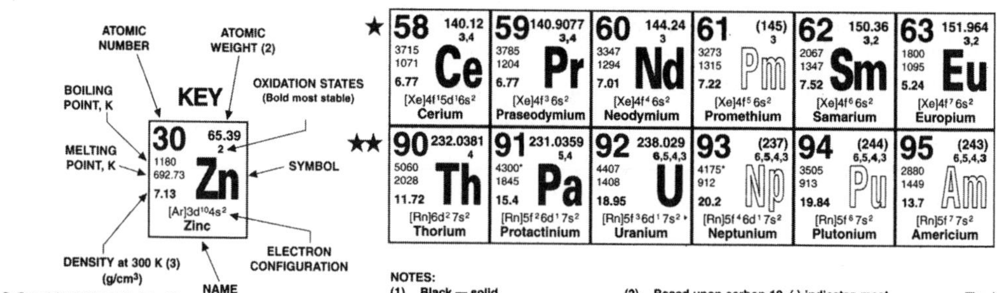

Series	Atomic number	Symbol	Name	Atomic weight	Oxidation states	Boiling point, K	Melting point, K	Density	Electron configuration
★	58	Ce	Cerium	140.12	3,4	3715	1071	6.77	$[Xe]4f^{1}5d^{1}6s^{2}$
★	59	Pr	Praseodymium	140.9077	3,4	3785	1204	6.77	$[Xe]4f^{3}6s^{2}$
★	60	Nd	Neodymium	144.24	3	3347	1294	7.01	$[Xe]4f^{4}6s^{2}$
★	61	Pm	Promethium	(145)	3	3273	1315	7.22	$[Xe]4f^{5}6s^{2}$
★	62	Sm	Samarium	150.36	3,2	2067	1347	7.52	$[Xe]4f^{6}6s^{2}$
★	63	Eu	Europium	151.964	3,2	1800	1095	5.24	$[Xe]4f^{7}6s^{2}$
★★	90	Th	Thorium	232.0381	4	5060	2028	11.72	$[Rn]6d^{2}7s^{2}$
★★	91	Pa	Protactinium	231.0359	5,4	4300*	1845	15.4	$[Rn]5f^{2}6d^{1}7s^{2}$
★★	92	U	Uranium	238.029	6,5,4,3	4407	1408	18.95	$[Rn]5f^{3}6d^{1}7s^{2}$
★★	93	Np	Neptunium	(237)	6,5,4,3	4175*	912	20.2	$[Rn]5f^{4}6d^{1}7s^{2}$
★★	94	Pu	Plutonium	(244)	6,5,4,3	3505	913	19.84	$[Rn]5f^{6}7s^{2}$
★★	95	Am	Americium	(243)	6,5,4,3	2880	1449	13.7	$[Rn]5f^{7}7s^{2}$

KEY

30	65.39
1160	2
692.73	**Zn**
7.13	
$[Ar]3d^{10}4s^{2}$	
Zinc	

ATOMIC NUMBER; ATOMIC WEIGHT (2); OXIDATION STATES (Bold most stable); BOILING POINT, K; MELTING POINT, K; SYMBOL; DENSITY at 300 K (3) (g/cm^3); ELECTRON CONFIGURATION; NAME

NOTES:

(1) Black — solid.
Red — gas.
Blue — liquid.
Outline — synthetically prepared.

(2) Based upon carbon-12. ( ) indicates most stable or best known isotope.

(3) Entries marked with daggers refer to the gaseous state at 273 K and 1 atm and are given in units of g/l.

The /
6 and
Appli

Selected Radioactive Isotopes

Naturally occurring radioactive isotopes are designated by a mass number in blue (although some are also manufactured). Letter m indicates an isomer of another isotope of the same mass number. Half-lives follow in parentheses, where s, min, h, d, and y stand respectively for seconds, minutes, hours, days, and years. The table includes mainly the longer-lived radioactive isotopes; many others have been prepared. Isotopes known to be radioactive but with half-lives exceeding 10^{12} y have not been included. Symbols describing the principal mode (or modes) of decay are as follows (these processes are generally accompanied by gamma radiation):

- α alpha particle emission
- β^- beta particle (electron) emission
- β^+ positron emission
- EC orbital electron capture
- IT isomeric transition from upper to lower isomeric state
- SF spontaneous fission

10	11/IB	12/IIB	13/IIIB	14/IVB	15/VB	16/VIB	17/VIIB	18/VIII
								2 4.00260 He 4.216 0.95 at 26 atm 0.1785† $1s^2$ Helium
			5 10.811 3 B 4275 2365 2.34 $1s^22s^2p^1$ Boron	6 12.011 ±4,2 C 5100* 3825* 2.26 $1s^22s^2p^2$ Carbon	7 14.0067 ±3,5,4,2 N 77.344 63.15 1.251† $1s^22s^2p^3$ Nitrogen	8 15.9994 -2 O 90.188 54.8 1.429† $1s^22s^2p^4$ Oxygen	9 18.99840 -1 F 85.0 53.55 1.696† $1s^22s^2p^5$ Fluorine	10 20.1797 Ne 27.10 24.55 0.900† $1s^22s^2p^6$ Neon
			13 26.98154 3 Al 2740 933.5 2.70 $[Ne]3s^2p^1$ Aluminum	14 28.0855 4,2 Si 2630 1683 2.33 $[Ne]3s^2p^2$ Silicon	15 30.97376 ±3,5,4 P 553 317.3 1.82 $[Ne]3s^2p^3$ Phosphorus	16 32.066 ±2,4,6 S 717.82 392.2 2.07 $[Ne]3s^2p^4$ Sulfur	17 35.4527 ±1,3,5,7 Cl 239.18 172.17 3.214† $[Ne]3s^2p^5$ Chlorine	18 39.948 Ar 87.45 83.95 1.784† $[Ne]3s^2p^6$ Argon
28 58.6934 2,3 Ni 3005 1726 8.90 $[Ar]3d^84s^2$ Nickel	29 63.546 2,1 Cu 2840 1356.6 8.96 $[Ar]3d^{10}4s^1$ Copper	30 65.39 2 Zn 1180 692.73 7.13 $[Ar]3d^{10}4s^2$ Zinc	31 69.723 3 Ga 2478 302.92 5.91 $[Ar]3d^{10}4s^2p^1$ Gallium	32 72.61 4 Ge 3107 1211.5 5.32 $[Ar]3d^{10}4s^2p^2$ Germanium	33 74.9216 ±3,5 As 876 (subl.) 1090 (28 atm.) 5.78 $[Ar]3d^{10}4s^2p^3$ Arsenic	34 78.96 -2,4,6 Se 958 494 4.79 $[Ar]3d^{10}4s^2p^4$ Selenium	35 79.904 ±1,5,7 Br 331.85 265.95 3.12 $[Ar]3d^{10}4s^2p^5$ Bromine	36 83.80 0,2 Kr 120.85 116 3.75† $[Ar]3d^{10}4s^2p^6$ Krypton
46 106.42 2,4 Pd 3240 1825 12.0 $[Kr]4d^{10}$ Palladium	47 107.868 1 Ag 2436 1235.08 10.50 $[Kr]4d^{10}5s^1$ Silver	48 112.41 2 Cd 1040 594.26 8.65 $[Kr]4d^{10}5s^2$ Cadmium	49 114.82 3 In 2350 429.78 7.31 $[Kr]4d^{10}5s^2p^1$ Indium	50 118.710 4,2 Sn 2876 505.12 7.31 $[Kr]4d^{10}5s^2p^2$ Tin	51 121.760 ±3,5 Sb 1860 903.91 6.69 $[Kr]4d^{10}5s^2p^3$ Antimony	52 127.60 -2,4,6 Te 1261 722.72 6.24 $[Kr]4d^{10}5s^2p^4$ Tellurium	53 126.9045 ±1,5,7 I 457.5 386.7 4.93 $[Kr]4d^{10}5s^2p^5$ Iodine	54 131.29 0,2,4,6 Xe 165.1 161.39 5.90† $[Kr]4d^{10}5s^2p^6$ Xenon
78 195.08 2,4 Pt 4100 2042.1 21.45 $[Xe]4f^{14}5d^96s^1$ Platinum	79 196.9665 3,1 Au 3130 1337.58 19.3 $[Xe]4f^{14}5d^{10}6s^1$ Gold	80 200.59 2,1 Hg 629.88 234.31 13.55 $[Xe]4f^{14}5d^{10}6s^2$ Mercury	81 204.383 3,1 Tl 1746 577 11.85 $[Xe]4f^{14}5d^{10}6s^2p^1$ Thallium	82 207.2 4,2 Pb 2023 600.65 11.35 $[Xe]4f^{14}5d^{10}6s^2p^2$ Lead	83 208.9804 3,5 Bi 1837 544.59 9.75 $[Xe]4f^{14}5d^{10}6s^2p^3$ Bismuth	84 (209) 4,2,6 Po — 527 9.3 $[Xe]4f^{14}5d^{10}6s^2p^4$ Polonium	85 (210) ±1,3,5,7 At 610* 575 — $[Xe]4f^{14}5d^{10}6s^2p^5$ Astatine	86 (222) 0,2 Rn 211.4 202 9.73 $[Xe]4f^{14}5d^{10}6s^2p^6$ Radon
110 (269) Uun — — — $[Rn]5f^{14}6d^87s^1$* (Ununnilium)	111 (272) Uuu — — — $[Rn]5f^{14}6d^97s^1$* (Unununium)	112 (277) Uub — — — $[Rn]5f^{14}6d^{10}7s^2$* (Ununbium)	113 Uut (Ununtrium)	114 (285) Uuq — — — $[Rn]5f^{14}6d^{10}7s^27p^2$* (Ununquadium)	115 Uup (Ununpentium)	116 (289) Uuh — — — $[Rn]5f^{14}6d^{10}7s^27p^4$* (Ununhexium)	117 Uus (Ununseptium)	118 (293) Uuo — — — $[Rn]5f^{14}6d^{10}7s^27p^6$* (Ununoctium)

64 157.25 3 Gd 3545 1585 7.90 $[Xe]4f^75d^16s^2$ Gadolinium	65 158.9253 3,4 Tb 3500 1629 8.23 $[Xe]4f^96s^2$ Terbium	66 162.50 3 Dy 2840 1685 8.55 $[Xe]4f^{10}6s^2$ Dysprosium	67 164.9303 3 Ho 2968 1747 8.80 $[Xe]4f^{11}6s^2$ Holmium	68 167.26 3 Er 3140 1802 9.07 $[Xe]4f^{12}6s^2$ Erbium	69 168.9342 3,2 Tm 2223 1818 9.32 $[Xe]4f^{13}6s^2$ Thulium	70 173.04 3,2 Yb 1469 1092 6.97 $[Xe]4f^{14}6s^2$ Ytterbium	71 174.967 3 Lu 3668 1936 9.84 $[Xe]4f^{14}5d^16s^2$ Lutetium
96 (247) 3 Cm — 1620 13.5 $[Rn]5f^76d^17s^2$ Curium	97 (247) 4,3 Bk — — 14* $[Rn]5f^97s^2$ Berkelium	98 (251) 3 Cf — 1170* — $[Rn]5f^{10}7s^2$ Californium	99 (252) 3 Es — 1130* — $[Rn]5f^{11}7s^2$ Einsteinium	100 (257) 3 Fm — 1800* — $[Rn]5f^{12}7s^2$ Fermium	101 (258) 3 Md — 1100* — $[Rn]5f^{13}7s^2$ Mendelevium	102 (259) 2,3 No — 1100* — $[Rn]5f^{14}7s^2$ Nobelium	103 (262) 3 Lr — 1900* — $[Rn]5f^{14}7s^27p^1$* Lawrencium

. & B subgroup designations, applicable to elements in rows 4, 5, 7, are those recommended by the International Union of Pure and d Chemistry.

Catalog Number WLS-18806

P.O. Box 5229, Buffalo Grove, IL 60089-5229
1-800-727-4368 FAX 1-800-676-2540

Side 1

TABLE A-1B PERIODIC PROPERTIES OF THE ELEMENTS

Percent Ionic Character of a Single Che

Difference in electronegativity	0.1	0.2	0.3	0.4	0.5	0.6	0.7	0.8	0.9	1.0	1.1	1.2	1.3	1
Percent ionic character %	0.5	1	2	4	6	9	12	15	19	22	26	30	34	3

DATA CONCERNING THE MORE STABLE ELEMENTARY (SUBATOMIC) PARTICLES

	Neutron	Proton	Electron*	Neutrino*	Photon
Symbol	n	p	e (e-)	ν	γ
Rest mass (kg)	1.67495×10^{-27}	1.67265×10^{-27}	9.1095×10^{-31}	~0	0
Relative atomic mass (^{12}C=12)	1.008665	1.007276	5.48580×10^{-4}	~0	0
Charge (C)	0	1.60219×10^{-19}	-1.60219×10^{-19}	0	0
Radius (m)	8×10^{-16}	8×10^{-16}	$<1 \times 10^{-16}$	~0	0
Spin quantum number	1/2	1/2	1/2	1/2	1
Magnetic Moment†	-1.913 μ_N	2.793 μ_N	1.001 μ_B	0	0

GROUP 1/IA, 2/IIA, 3/IIIA, 4/IVA, 5/VA, 6/VIA, 7/VIIA, 8

Property	H
Covalent radius	0.32
Atomic radius	0.79
Atomic volume	14.10
First ionization potential	13.598
Specific heat capacity	14.304
Electronegativity	2.10
Heat of vaporization	0.4581
Heat of fusion	0.0585
Electrical conductivity	—
Thermal conductivity	0.1815

Property	Li	Be
Covalent radius	1.23	0.90
Atomic radius	2.05	1.40
Atomic volume	13.10	5.0
First ionization potential	5.392	9.322
Specific heat capacity	3.582	1.825
Electronegativity	0.98	1.57
Heat of vaporization	147.1	297
Heat of fusion	3.0	11.71
Electrical conductivity	11.7	25
Thermal conductivity	84.7	200

Property	Na	Mg
Covalent radius	1.54	1.36
Atomic radius	2.23	1.72
Atomic volume	23.70	14.0
First ionization potential	5.139	7.646
Specific heat capacity	1.23	1.02
Electronegativity	0.93	1.31
Heat of vaporization	98.01	127.6
Heat of fusion	2.601	8.95
Electrical conductivity	20.1	22.4
Thermal conductivity	141	156

Property	K	Ca	Sc	Ti	V	Cr	Mn	Fe
Covalent radius	2.03	1.74	1.44	1.32	1.22	1.18	1.17	1.17
Atomic radius	2.77	2.23	2.09	2.00	1.92	1.85	1.79	1.72
Atomic volume	45.30	29.90	15.0	10.60	8.35	7.23	7.39	7.10
First ionization potential	4.341	6.113	6.54	6.82	6.74	6.766	7.435	7.870
Specific heat capacity	0.757	0.647	0.568	0.523	0.489	0.449	0.48	0.449
Electronegativity	0.82	1.00	1.36	1.54	1.63	1.66	1.55	1.83
Heat of vaporization	76.9	154.67	304.80	425.2	446.7	339.5	219.74	349.5
Heat of fusion	2.33	8.53	16.11	18.6	22.8	20	14.64	13.8
Electrical conductivity	16.4	31.3	1.5	2.6	4.0	7.9	0.5	11.2
Thermal conductivity	102.5	200	15.8	21.9	30.7	93.7	7.82	80.2

Property	Rb	Sr	Y	Zr	Nb	Mo	Tc	Ru
Covalent radius	2.16	1.91	1.62	1.45	1.34	1.30	1.27	1.25
Atomic radius	2.98	2.45	2.27	2.16	2.08	2.01	1.95	1.89
Atomic volume	55.9	33.7	19.80	14.10	10.80	9.40	8.5	8.30
First ionization potential	4.177	5.695	6.38	6.84	6.88	7.099	7.28	7.37
Specific heat capacity	0.363	0.30	0.30	0.278	0.265	0.25	0.24	0.238
Electronegativity	0.82	0.95	1.22	1.33	1.6	2.16	1.9	2.2
Heat of vaporization	69.2	136.9	393.3	590.5	690.1	590.4	502.0	567.77
Heat of fusion	2.34	8.2	17.15	21	26.9	36	23	25.52
Electrical conductivity	47.8	5.0	1.8	2.3	6.6	17.3	0.001	14.9
Thermal conductivity	58.2	35.3	17.2	22.7	53.7	138	50.6	117

Property	Cs	Ba	La★	Hf	Ta	W	Re	Os
Covalent radius	2.35	1.98	1.25	1.44	1.34	1.30	1.28	1.26
Atomic radius	3.34	2.78	2.74	2.16	2.09	2.02	1.97	1.92
Atomic volume	70	39.0	22.5	13.6	10.90	9.53	8.85	8.43
First ionization potential	3.894	5.212	5.58	6.65	7.89	7.98	7.88	8.7
Specific heat capacity	0.24	0.204	0.19	0.14	0.14	0.13	0.137	0.13
Electronegativity	0.79	0.89	1.10	1.3	1.5	2.36	1.9	2.2
Heat of vaporization	67.740	140.2	399.57	661.07	737.0	422.58	707.1	627.6
Heat of fusion	2.092	8.01	11.3	21.76	36	35.40	33.05	29.29
Electrical conductivity	5.3	2.8	1.9	3.4	8.1	18.2	5.8	12.3
Thermal conductivity	35.9	18.4	13.5	23.0	57.5	174	47.9	87.6

Property	Fr	Ra	Ac★★	Rf	Db	Sg	Bh	Hs
Covalent radius	—	—	—	—	—	—	—	—
Atomic radius	—	—	—	—	—	—	—	—
Atomic volume	—	45.2	22.5	—	—	—	—	—
First ionization potential	—	5.279	5.17	—	—	—	—	—
Specific heat capacity	—	0.094	0.120	—	—	—	—	—
Electronegativity	0.7	0.89	1.1	—	—	—	—	—
Heat of vaporization	64*	136.82	—	—	—	—	—	—
Heat of fusion	2.1*	8.37	—	—	—	—	—	—
Electrical conductivity	_	1.0	—	—	—	—	—	—
Thermal conductivity	15	18.6	12	—	—	—	—	—

*ESTIMATED VALUES

★

Property	Ce	Pr	Nd	Pm
Covalent radius	1.65	1.65	1.64	1.63
Atomic radius	2.70	2.67	2.64	2.62
Atomic volume	21.0	20.8	20.6	22.4
First ionization potential	5.47	5.42	5.49	5.55
Specific heat capacity	0.19	0.193	0.190	—
Electronegativity	1.12	1.13	1.14	1.1
Heat of vaporization	313.8	332.63	283.68	—
Heat of fusion	9.20	10.04	10.88	—
Electrical conductivity	1.4	1.5	1.6	2
Thermal conductivity	11.4	12.5	16.5	17.

★★

Property	Th	Pa	U	Np
Covalent radius	1.65	—	1.42	—
Atomic radius	—	—	—	—
Atomic volume	19.90	15.0	12.50	21.1
First ionization potential	6.08	5.88	6.05	6.19
Specific heat capacity	0.113	—	0.12	—
Electronegativity	1.3	1.5	1.38	1.3
Heat of vaporization	543.92	—	422.58	—
Heat of fusion	15.65	—	15.48	—
Electrical conductivity	7.1	5.6	3.6	0.8
Thermal conductivity	54	47	27.6	6.3

KEY

Zn	
1.25	1.65
1.53	115.30
9.20	7.38
9.394	16.9
0.388	116

SYMBOL; CRYSTAL STRUCTURE (2); ACID-BASE PROPERTIES (1); ELECTRONEGATIVITY, (Pauling's); COVALENT RADIUS, Å; ATOMIC RADIUS, Å (7); ATOMIC VOLUME, cm^3/mol (8); FIRST IONIZATION POTENTIAL V; SPECIFIC HEAT CAPACITY, $J g^{-1} K^{-1}$ (3); HEAT OF VAPORIZATION kJ / mol (4); HEAT OF FUSION kJ / mol (5); ELECTRICAL CONDUCTIVITY $10^6 \Omega^{-1} m^{-1}$ (6); THERMAL CONDUCTIVITY, $W m^{-1} K^{-1}$ (3)

NOTES: (1) For representative oxides (higher valen
and amphoteric if both colors are show

(2) Cubic, face centered; cub
hexagonal; rhombohedral;

(3) At 300 K (27°C)
(4) At boiling point
(5) At melting point
(6) Generally at 29
for polycrystalli
(7) Quantum mech
for free atom

›mical Bond

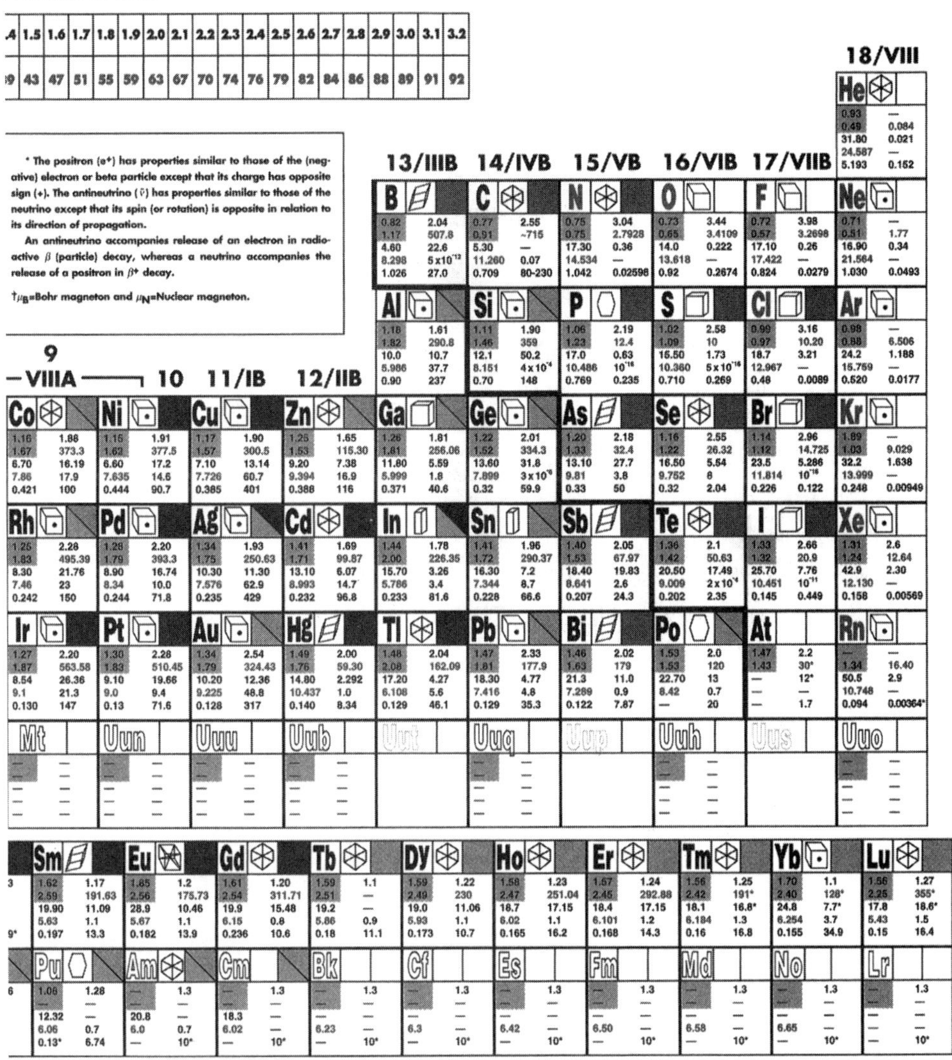

:e) of group. Oxide is acidic if color is red, basic if color is blue
n. Intensity of color indicates relative strength.

ic, body centered; cubic;

tetragonal; orthorhombic; monoclinic.

3 K (20°C) (8) From density at 300 K (27°C) for liquid
ne material and solid elements; values for gaseous elements
ınical value refer to liquid state at boiling point

The A & B subgroup designations, applicable to elements in rows 4, 5, 6 and 7, are those recommended by the International Union of Pure and Applied Chemistry.

P.O. Box 5229, Buffalo Grove, IL 60089-5229
1-800-727-4368 FAX 1-800-676-2540

Catalog Number WLS-18806

SIDE 2

TABLE A-2 SYMBOLS OF THE ELEMENTS AND THEIR ATOMIC NUMBERS AND WEIGHTS[a]

Name	Symbol	Atomic Number	Atomic Weight[b]	Name	Symbol	Atomic Number	Atomic Weight[b]
Actinium	Ac	89	(227)	Lawrencium	Lr	103	(257)
Aluminum	Al	13	27.0	Lead	Pd	82	207.2
Americium	Am	95	(243)	Lithium	Li	3	6.94
Antimony	Sb	51	121.8	Lutetium	Lu	71	175.0
Argon	Ar	18	39.9	Magnesium	Mg	12	24.3
Arsenic	As	33	74.9	Manganese	Mn	25	54.9
Astatine	At	85	(210)	Mendelevium	Md	101	(256)
Barium	Ba	56	137.3	Mercury	Hg	80	200.6
Berkelium	Bk	97	(247)	Molybdenum	Mo	42	95.9
Beryllium	Be	4	9.01	Neodymium	Nd	60	144.2
Bismuth	Bi	83	209.0	Neon	Ne	10	20.2
Boron	B	5	10.8	Neptunium	Np	93	(237)
Bromine	Br	35	79.9	Nickel	Ni	28	58.7
Cadmium	Cd	48	112.4	Niobium	Nb	41	92.9
Calcium	Ca	20	40.1	Nitrogen	N	7	14.01
Californium	Cf	98	(251)	Nobelium	No	102	(254)
Carbon	C	6	12.01	Osmium	Os	76	190.2
Cerium	Ce	58	140.1	Oxygen	O	8	16.00
Cesium	Cs	55	132.9	Palladium	Pd	46	106.4
Chlorine	Cl	17	35.5	Phosphorus	P	15	31.0
Chromium	Cr	24	52.0	Platinum	Pt	78	195.1
Cobalt	Co	27	58.9	Plutonium	Pu	94	(242)
Copper	Cu	29	63.5	Polonium	Po	84	(210)
Curium	Cm	96	(247)	Potassium	K	19	39.1
Dysprosium	Dy	66	162.5	Praseodymium	Pr	59	140.9
Einsteinium	Es	99	(254)	Promethium	Pm	61	(147)
Erbium	Er	68	167.3	Protactinium	Pa	91	(231)
Europium	Eu	63	152.0	Radium	Ra	88	(226)
Fermium	Fm	100	(253)	Radon	Rn	86	(222)
Fluorine	F	9	19.0	Rhenium	Re	75	186.2
Francium	Fr	87	(223)	Rhodium	Rh	45	102.9
Gadolinium	Gd	64	157.3	Rubidium	Rb	37	85.5
Gallium	Ga	31	69.7	Ruthenium	Ru	44	101.1
Germanium	Ge	32	72.6	Samarium	Sm	62	150.4
Gold	Au	79	197.0	Scandium	Sc	21	45.0
Hafnium	Hf	72	178.5	Selenium	Se	34	79.0
Helium	He	2	4.00	Silicon	Si	14	28.1
Holmium	Ho	67	164.9	Silver	Ag	47	107.9
Hydrogen	H	1	1.008	Sodium	Na	11	23.0
Indium	In	49	114.8	Strontium	Sr	38	87.6
Iodine	I	53	126.9	Sulfur	S	16	32.1
Iridium	Ir	77	192.2	Tantalum	Ta	73	180.9
Iron	Fe	26	55.8	Technetium	Te	43	(99)
Krypton	Kr	36	83.8	Tellurium	Te	52	127.6
Lanthanum	La	57	138.9	Terbium	Tb	65	158.9

[a]The newest elements (rutherfordium, Rf-104, and hahnium, Ha-105) are not included.

[b]The values given in parentheses are mass numbers of the principal isotopes of unstable elements.

TABLE A-2 (continued)

Name	Symbol	Atomic Number	Atomic Weight[b]	Name	Symbol	Atomic Number	Atomic weight[b]
Thallium	Tl	81	204.4	Vanadium	V	23	50.9
Thorium	Th	90	232.0	Xenon	Xe	54	131.3
Thulium	Tm	69	168.9	Ytterbium	Yb	70	173.0
Tin	Sn	50	118.7	Yttrium	Y	39	88.9
Titanium	Ti	22	47.9	Zinc	Zn	30	65.4
Tungsten	W	74	183.9	Zirconium	Zr	40	91.2
Uranium	U	92	238.0				

TABLE A-3 GREEK SYMBOLS AND THEIR PRONUNCIATIONS

Name of Letter	Uppercase	Lowercase
Alpha	A	α
Beta	B	β
Gamma	Γ	γ
Delta	Δ	δ
Epsilon	E	ε
Theta	Θ	θ
Kappa	K	κ
Lambda	Λ	λ
Mu	M	μ
Pi	Π	π
Rho	P	ρ
Sigma	Σ	σ
Tau	T	τ
Phi	Φ	φ
Omega	Ω	ω

INTERNATIONAL SYSTEM OF UNITS AND UNIT CONVERSIONS

SI was created in 1960 by international agreement and represents a worldwide measurement system far superior to earlier measurement systems (gravitational) in expressing scientific and technical data. All gravitational systems using force as a fundamental dimension, including the American engineering system with its pound of mass a fundamental unit, are considered obsolete, and the changeover, though not mandatory, is proceeding at a steady pace, with some industries completing the changeover in record time. For most technology users, this means that they will continue to be confronted with problems that arise when two systems of measurement exist. Consequently, they must be very familiar with both systems and demonstrate ability to convert from one to another upon completion of their calculations.

Units. SI units are grouped into three general classes: base or fundamental units, derived units, and supplementary units. Table 10-4 lists the base uits as well as some of the more common derived units used in this book. Special names are given to some of the derived units (see Table 10-5). For example, hertz (Hz) is the special name given to the SI unit for frequency. Units of force, stress, power and energy also have special names and therefore need not be expressed in their base units. The amount of force required to accelerate 1 kilogram of mass 1 meter per second squared is given the special name of newton (N). Thus $1\ N = 1\ kg\ \ m/s^2$.

Prefixes. Prefixes corresponding to pwers of 10 are attached to the units discussed above in order to form larger or smaller units. In technical work the powers of 10 divisible by 3 are preferred. Table 10-6 contains the authorized prefixes.

TABLE A-4 NAMES AND SYMBOLS OF SI UNITS

Quantity	Name of Unit	Symbol	Expressed in Base Units Where Applicable
Base units			
Length	meter	m	
Mass	kilogram	kg	
Time	second	s	
Electric current	ampere	A	
Thermodynamic temperature	kelvin	K	
Luminous intensity	candela	cd	
Amount of substance	mole	mol	
Derived units			
Area	square meter	m^2	
Volume	cubic meter	m^3	
Frequency	hertz, cycle per second	Hz	s^{-1}
Density (mass)	kilogram per cubic meter	kg/m^3	
Velocity (linear)	meter per second	m/s	
Velocity (angular)	radian per second	rad/s	
Acceleration (linear)	meter per second squared	m/s^2	
Acceleration (angular)	radian per second squared	rad/s^2	
Force	newton, kilogram-meter per second squared	N	$kg \cdot m \cdot s^{-2}$
Permeability	henry per meter	H/m	$m \cdot kg \cdot s^{-2} \cdot A^{-2}$
Permittivity	farad per meter	F/m	$m^{-3} \cdot kg^{-1} \cdot s^4 \cdot A^2$
Pressure (mechanical stress)	pascal, newton per square meter	Pa	$N \cdot m^{-2}$
Kinematic viscosity	square meter per second	m^2/s	
Dynamic viscosity	newton-second per square meter	$N \cdot s/m^2$	$m^{-1} \cdot kg \cdot s^{-1}$
Work energy, quantity of heat	joule, newton-meter	J, $N \cdot m$	
Power	watt, joule per second	W, J/s	
Quantity of electricity, electric charge	coulomb	C, $A \cdot s$	
Potential difference, electromotive force	volt	V, W/A	
Electric field strength	volt per meter	V/m	$m \cdot kg \cdot s^{-3} \cdot A^{-1}$
Electric resistance	ohm	Ω, V/A	
Capacitance	farad	F, $A \cdot s/V$	
Magnetic flux	weber	Wb, $V \cdot s$	
Inductance	henry	H, $V \cdot s/A$	
Magnetic flux density	tesla	T, Wb/m^2	
Magnetic field strength	ampere per meter	A/m	
Magnetomotive force	ampere	A	
Luminous flux	lumen	lm	
Luminance	candela per square meter	cd/m^2	
Illuminance	lux	lx	
Wave number	1 per meter	m^{-1}	
Entropy	joule per kelvin	J/K	
Specific heat capacity	joule per kilogram kelvin	$J/(kg \cdot K)$	$m^2 \cdot s^{-2} \cdot K^{-1}$
Thermal conductivity	watt per meter kelvin	$W/(m \cdot K)$	$m \cdot kg \cdot s^{-3} \cdot K^{-1}$
Conductance	siemens	S, A/V	
Torque	newton-meter	N/m	$m^2 \cdot kg \cdot s^{-3} \cdot K^{-1}$
Supplementary units			
Plane angle	radian	rad	

TABLE A-5 SPECIAL NAMES FOR UNITS

Quantity	Unit	Symbol	Formula
Frequency (of a periodic phenomenon)	hertz	Hz	1/s
Force	newton	N	$kg \cdot m/s^2$
Pressure, stress	pascal	Pa	N/m^2
Energy, work, quantity of heat	joule	J	$N \cdot m$
Power, radiant flux	watt	W	J/s
Quantity of electricity, electric charge	coulomb	C	$A \cdot s$
Electric potential, potential difference, electromotive force	volt	V	W/A
Capacitance	farad	F	C/V
Electric resistance	ohm	Ω	V/A
Conductance	siemens	S	A/V
Magnetic flux	weber	Wb	$V \cdot s$
Magnetic flux density	tesla	T	Wb/m^2
Inductance	henry	H	Wb/A
Luminous flux	lumen	lm	$cd \cdot sr$
Illuminance	lux	lx	lm/m^2

TABLE A-6 PREFIXES

Multiplication factor	Prefix	Symbol
1 000 000 000 000 000 000 = 10^{18}	exa	E
1 000 000 000 000 000 = 10^{15}	peta	P
1 000 000 000 000 = 10^{12}	tera	T
1 000 000 000 = 10^{9}	giga	G
1 000 000 = 10^{6}	mega	M
1 000 = 10^{3}	kilo	k
100 = 10^{2}	hecto	h
10 = 10^{1}	deka	da
0.1 = 10^{-1}	deci	d
0.01 = 10^{-2}	centi	c
0.001 = 10^{-3}	milli	m
0.000 001 = 10^{-6}	micro	μ
0.000 000 001 = 10^{-9}	nano	n
0.000 000 000 001 = 10^{-12}	pico	p
0.000 000 000 000 001 = 10^{-15}	femto	f
0.000 000 000 000 000 001 = 10^{-18}	atto	a

Rules for usage. For standardized usage, the following rules should be observed:

1. Uppercase (capitals) and lowercase letters are never interchanged: kg, not KG.
2. The same symbol is used for plurals: N, not Ns; 14 meters or 14 m.
3. No space is left between the prefix and its unit symbol: GHz, not G Hz.
4. To form products, a raised dot is preferred (or a dot on a line): $kN \cdot m$, or kN.m. The dot may be dispensed with if confusion is not created by its absence.
5. To form quotients, one solidus (an oblique line), a fraction line (horizontal), or a negative power is used to express derived units: m/s, $\frac{m}{s}$, or $m \cdot s^{-1}$.

   Note: The solidus must not be repeated on the same line: m/s^2, not m/s/s. Also, $kg/(m \cdot s)$, $\frac{kg}{m \cdot s}$, or $kg \cdot m^{-1} \cdot s^{-1}$, but not kg/m/s. Note also the use of the parentheses to avoid ambiguity.
6. An exponent affixed to a symbol containing a prefix indicates that the multiple or submultiple of the unit is raised to the power expressed by the exponent.

$$1\ mm^3 = 10^{-9}\ m^3, \text{ not } 10^{-3}\ m^3$$
$$1\ cm^3 = 10^{-6}\ m^3, \text{ not } 10^{-2}\ m^3$$
$$1\ cm^{-1} = 10^{2}\ m^{-1}, \text{ not } 10^{-2}\ m^{-1}$$

7. A period is used as a decimal marker. It is not used to separate groups of digits. A space is left for this purpose: 5 279 585 J, and 0.000 34 s.
8. Numbers are preferably expressed between the limits 0.1 and 1000, using the appropriate prefix to change the size of the unit: 5.23 GN.
9. For decimal numbers less than 1, the leading zero is never omitted: 0.625, not .625.
10. When units are written in words, they always start with lowercase letters except at the beginning of a sentence. If the unit is derived from the name of a person, the symbol is capitalized. Plurals of special names are written in the usual manner.

    125 watts or 125 W
    0.25 newton or 0.25 N
    58.6 hertz or 58.6 Hz

11. A space or hyphen may be used to form the product expressed in words: newton - meters or newton meters.
12. For quotients, the word *per* may be used: newton per meter squared, kilogram per cubic meter.
13. The kelvin (symbol K) is the standard unit of temperature. In writing this absolute temperature, the word *degree* or its symbol (°) is not used: 472 K. In addition, K may be used to express an interval or a difference in temperature. Celsius temperature is expressed in degrees Celsius with symbol °C. The unit degrees Celsius is equal to the unit kelvin and may also be used to represent an interval or a difference of Celsius temperature: 25°C. Temperature in K = temperature in °C + 273.15.

Computations. The SI system of units makes computations relatively simple because (1) a single unit is used to represent a particular physical quantity, (2) the system is coherent in that the factor of 1.0 replaces many conversion factors, and (3) SI is based on the decimal system.

Prior to computation, all prefixes should be replaced by their respective powers of 10. The final step is to select a suitable prefix to express the answer once the resulting answer is rounded to the appropriate number of significant digits.

Two examples illustrate the solution of typical problems using SI units and the preceding information.

Problem 1

Given: A metal rod under tensile load of 356 kN is allowed to withstand a unit stress of 110 NM/m^2.
Required: Find the diameter of the rod in millimeters.

Solution

1. Convert SI prefixes to powers of 10:

$$356 \text{ kN} = 356 \times 10^3 \text{ N}$$

$$110 \text{ MN/m}^2 = 110 \times 10^6 \text{ N/m}^2$$

2. Use the direct stress formula ($s = P/A$) and solve for the area (A):

$$A = \frac{P}{s} = \frac{356 \times 10^3 \text{ N}}{110 \times 10^6 \text{ N/m}^2} = 3.24 \times 10^{-3} \text{ m}^2$$

3. Use the formula for circular area $\left(A = \frac{\pi}{4}D^2\right)$ and solve for diameter (D):

$$D = \sqrt{\frac{4}{\pi}(3.24 \times 10^{-3} \text{ m}^2)}$$

or

$$D = \left[\frac{4}{\pi}(3.24 \times 10^{-3} \text{ m}^2)\right]^{1/2}$$

$$= \left[\frac{4}{\pi}(32.4 \times 10^{-4} \text{ m}^2)\right]^{1/2}$$

$$= 6.429 \times 10^{-2} \text{ m}$$

$$= 64.29 \times 10^{-3} \text{ m} = 64.3 \text{ mm (rounding up and using three significant digits)}$$

Problem 2

Given: Modulus of elasticity in tension (E) for steel is 29,120,000 psi.
Required: Express E in GPa.

Solution

1. Express E in terms of powers of 10:

$$E = 29.12 \times 10^6 \text{ psi}$$

2. Locate conversion ratio (see Table 10-7):

$$1 \text{ psi} = 6.895 \times 10^{-3} \text{ MPa}$$

3. Express prefixes in terms of powers of 10 (see Table 10-6):

$$1 \text{ psi} = 6.895 \times 10^{3} \text{ Pa}$$

4. Multiply E by conversion ratio:

$$29.12 \times 10^{6} \text{ psi} \left(\frac{6.895 \times 10^{3} \text{ Pa}}{1 \text{ psi}} \right) = 200.8 \times 10^{9} \text{ Pa}$$

5. Express answer using required SI prefix (see Table 10-6):

   200.8 GPa (using four significant digits)

REFERENCES

ASTM E 380: *Standard for Metric Practice.*
ANSI Z 210.1: *American National Standard for Metric Practice.*
ISO 1000: *SI Units and Recommendations for the Use of Their Multiples and of Certain Other Units.*
NBS Special Publication 330: *The International System of Units (SI).*

TABLE A-7 CONVERSIONS

Quantity	U.S. customary to SI
Acceleration, linear	1 ft/s^2 = 3.048 × 10^{-1} m/s^2
Area	1 in^2 = 6.452 × 10^2 mm^2
	1 ft^2 = 9.290 × 10^{-2} m^2
Density (mass), ρ	1 lb/in^3 = 2.768 × 10^4 kg/m^3
	1 lb/ft^3 = 1.602 × 10^1 kg/m^3
Electric current, I	1 ampere (A) = 1 C/s
Energy, work	1 Btu = 1.055 kJ
	1 in. · lb = 1.129 × 10^{-1} J
	1 ft · lb = 1.356 J
	1 Btu = 0.293 W · h
Force	1 lbf = 4.448 N
	1 kgf = 9.807 N
Impulse	1 lb · s = 4.448 N · s
Length	1 Å = 1 × 10^{-1} nm
	1 microinch = 2.540 × 10^{-2} μm
	1 mil = 2.540 × 10^1 μm
	1 in. = 2.540 × 10^1 mm
	1 ft = 3.048 × 10^{-1} m
Magnetic field strength, H	1 oersted (Oe) = 79.58 ampere turns per meter
Magnetic flux	1 maxwell = 10^{-8} Wb
Magnetic flux density, B	1 Wb/M^2 = 1 T (tesla)
Modulus of elasticity, E	1 lb/in^2 = 6.895 × 10^{-6} GPa
Moment of force, torque	1 lb · in. = 1.130 × 10^{-1} N · m
	1 lb · ft = 1.356 N · m
Moment of inertia, I (of area)	1 in^4 = 4.162 × 10^5 mm^4
Momentum, linear	1 lb · ft/s = 1.383 × 10^{-1} kg · m/s
Power	1 Btu/min = 1.758 × 10^{-2} kW
	1 ft · lb/min = 2.259 × 10^{-2} W
	1 hp = 7.457 × 10^{-1} kW
Stress (pressure)	1 lb/in^2 = 6.895 × 10^{-3} MPa
	1 ksi = 6.895 MPa
Temperature	1°F (difference) = 0.555°C
	1.8 °F = 1°C (difference)
Thermal expansion, linear coefficient, α	in./in./°F = 1.8 K^{-1}, K = °C + 273.15
Thermal conductivity, k	1 Btu/ft · hr · °F = 1.729 W/m · K
Velocity, linear, v	1 in./s = 2.540 × 10^1 mm/s
	1 ft/s = 3.048 × 10^{-1} m/s
	1 in./min = 4.233 × 10^{-1} mm/s
	1 ft/min = 5.080 × 10^{-3} m/s
Velocity, angular, ω	1 rev/min = 1.047 × 10^{-1} rad/s
Viscosity	1 poise (P) = 0.1 Pa · s
Volume	1 in^3 = 1.639 × 10^4 mm^3
	1 ft^3 = 2.832 × 10^{-2} m^3
	1 yd^3 = 7.646 × 10^{-1} m^3

1 joule (J) = 10^7 ergs = 0.625 × 10^{19} eV

1 gauss (G) = 10^{-4} Wb/m^2 = 1 T

1 weber (Wb) = 1 T/m^2

1 Å = 10^{-10} m (obsolete)

1 micrometer (micron) = 10^{-6} m (obsolete)

TABLE A-7A ADDITIONAL USEFUL CONVERSIONS

1 W = 3.413 Btu/hr
1 W = 1 N·m·s^{-1}
1 W = 0.7376 ft·lb·s^{-1}
1 kW = 1.341 Hp
1 J = 0.2390 cal
1 cal = 4.186 J
1 Btu = 252 cal
1 J = 0.737562 lb·ft
1 J = 1 N·m = 1 W·s
1 Pa = 1 N/m^2
1 N/m^2 = 1.450 lb/in^2 (psi)
1 m = 3.28 ft = 39.37 in.
$°C = \frac{5}{9}(F - 32)$
$°F = \frac{9}{5}(C + 32)$
K = °C + 273 (K = kelvin)
°R = °F + 460 (R = Rankine)
1 L = 1.06 qt
1 Pa·s = 10 poise

TABLE A-8 CONSTANTS

Quantity	Symbol	Value
Acceleration of gravity	g	$9.80\ m \cdot s^{-2}$
Atomic mass unit	amu	1.66×10^{-27} kg
Avogadro's number	N_A, N_0	6.022×10^{23} molecules/mole
Bohr radius	a_0	5.292×10^{-11} m
Electron charge	q, $-e$	1.60×10^{-19} C
Electron mass	m	9.11×10^{-31} kg
Electron volt	eV	0.160×10^{-18} J
Magnetic permeability of free space	μ_0	$4\pi \times 10^{-7}$ H/m $= 1.257 \times 10^{-6}$ H/m
Permittivity of free space	ϵ_0	$\frac{10^7}{4\pi c^2} = 8.854 \times 10^{-12}$ C/V·m
Planck's constant	$\hbar$	6.63×10^{-34} J·s
Speed of light (vacuum)	c	$3.00 \times 10^8\ m \cdot s^{-1}$

TABLE A-9 REPRESENTATIVE PLASTICS

Common Name (Chemical Name)	ASTM Abbre-viations	Trade Names	Common Structure	Grouping[a]	Typical Uses
ABS (acrylonitrile–butadiene–styrene)	ABS	Absinol, Abson, Cycolac, Royalite	Amorphous terpolymer 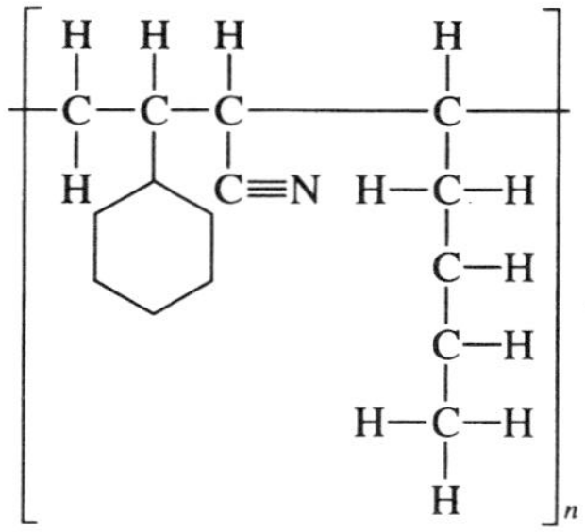	TP, EP	Pipe, toys, luggage, boat hulls, football helmets, gears, chrome-plated plumbing, and auto parts
Acetal (polyoxymethylene) (polymerized formaldehyde)	POM	Delrin, Celcon, Formaldafil	Highly crystalline homopolymer and copolymers 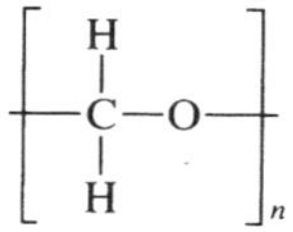	TP, EP	Gears, bearings, fan blades, shower heads, auto parts, and aerosol bottles
Acrylics [poly(methyl methacrylate)]	PMMA	Plexiglas, Lucite, Acrylite	Amorphous 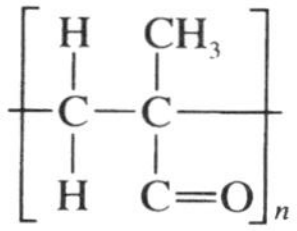	TP, GP	Lenses, windows, signs, sculpture, light pipes, and skylights
(polyacrylonitrile)	PAN	Sayelle			

[a]TP, thermoplastic; TS, thermosetting; EP, engineering plastic; GP, general-purpose or special. ASTM, American Society for Testing and Materials.
[b]Engineering plastics.

TABLE A-9 (continued)

Common Name (Chemical Name)	ASTM Abbreviations	Trade Names	Common Structure	Grouping[a]	Typical Uses
Alkyd plaskon (modified polyester resins)		Premix Dyal Glaskyd	Cross-link network 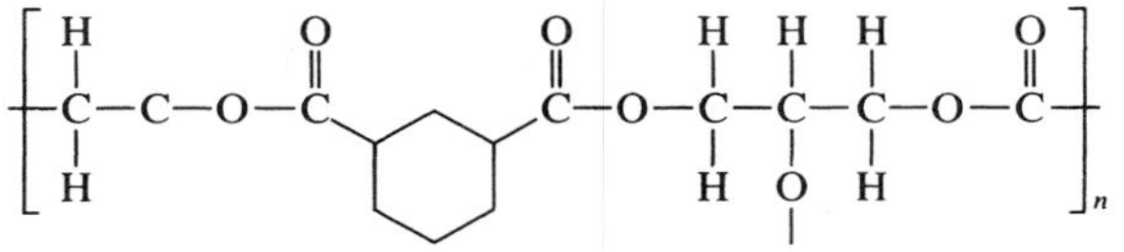	TS, GP	Coatings: enamel, lacquer and paint, molded electrical parts
Allylics (diallyl phthalate) (diallyl isophthalate) (diethylene glycol bisallyl/carbonate)	DAP DAIP CR39-allyl diglycol carbonate	Diall Poly-Dap	Cross-link network $\left[CH_2{=}CH{-}CH_2 \right]_n$	TS, GP TS	Electronic parts, pump impellers, glass-fiber impregnate, dinnerware, watch crystals
Aminos (urea–formaldehyde) (melamine–formaldehyde)	UF MF	Plaskon Cymel	Cross-link network 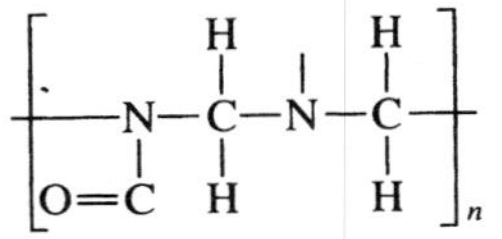	TS, GP	Electrical parts, particle-board binders, coatings, dinnerware, paper impregnate
Cellulosics (cellulose acetate) (cellulose butyrate) (cellulose nitrate) (cellulose propionate) (ethyl cellulose)	CA CAB CN CAP EC	Tenite Uvex Nixonite Forticel Ethocel Methocel	Highly crystalline 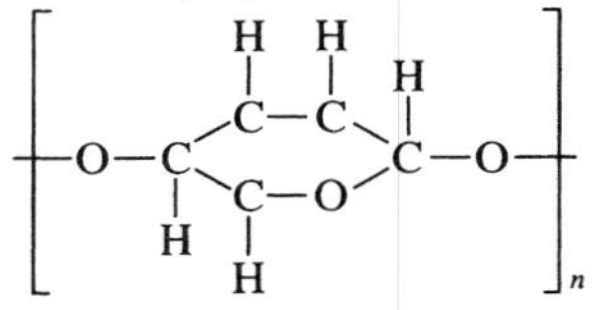	TP, GP	Packaging film, pipe, optical frames, flashbulb shields, helmets, rollers

[a] TP, thermoplastic; TS, thermosetting; EP, engineering plastic; GP, general-purpose or special. ASTM, American Society for Testing and Materials.

[b] Engineering plastics.

TABLE A-9 (continued)

Common Name (Chemical Name)	ASTM Abbreviations	Trade Names	Common Structure	Grouping[a]	Typical Uses
Epoxy	EP	Epi Rez Hysol	Cross-linked network or amorphous 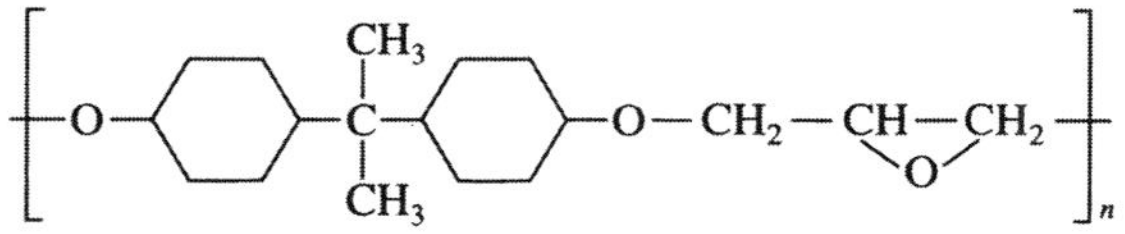 $\left[-O-C_6H_{10}-C(CH_3)_2-C_6H_{10}-O-CH_2-CH-CH_2(-O-)-\right]_n$	TS, TP, GP	Coatings, fiber reinforcement, potting and encapsulating
Fluorocarbons (polytetrafluoroethylene)[b] (polychlorotrifluoroethylene)[b] [poly(vinylidene fluoride)]	FEP PTFE or TFE CTFE PVDF	Teflon Kelf Kynar	Highly crystalline 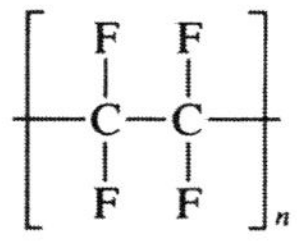 $\left[-CF_2-CF_2-\right]_n$	TP,[b] EP, GP	Seals, cookware, corrosion liquids hardware, chemical-processing equipment
Nylon (polyamide) (nylon 6) (nylon 6/6) (nylon 6/10) (nylon 6/12) (aramid)	PA	Nylon Celanese Zytel Plaskon Zytrel Nomex Kevlar	Crystalline 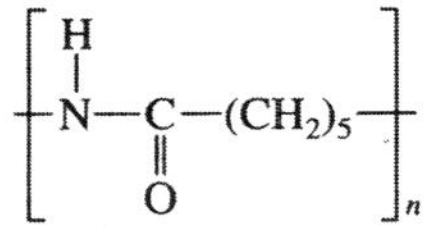 $\left[-NH-C(=O)-(CH_2)_5-\right]_n$ Nylon 6	TP, EP	Bearings, tubing, gears, housings Parts exposed to moisture: nylon 6/10, nylon 6/12 Reinforcing fiber: aramid
Phonelic (phenol–formaldehyde)	PF	Bakelite Durite Resinox	Cross-linked network 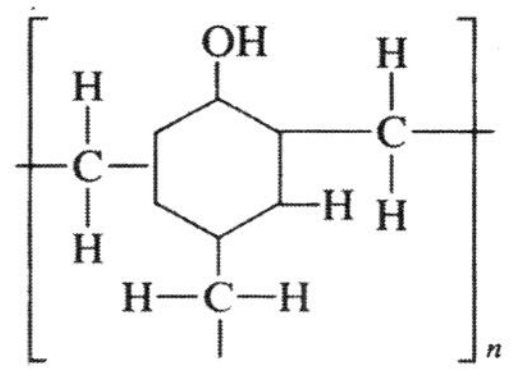 $\left[-CH_2-C_6H(OH)(-CH_2-)(-CH_2-)-\right]_n$	TS, GP	Electrical parts, housings, and binders

[a]TP, thermoplastic; TS, thermosetting; EP, engineering plastic; GP, general-purpose or special. ASTM, American Society for Testing and Materials.
[b]Engineering plastics.

TABLE A-9 (continued)

Common Name (Chemical Name)	ASTM Abbreviations	Trade Names	Common Structure	Grouping[a]	Typical Uses
Phenoxy [poly(hydroxy ethers)]			Amorphous or cross-linked network $\left[-O-C_6H_4-C(CH_3)_2-C_6H_4-O-CH_2-CH(OH)-CH_2-\right]_n$	TP, TS, EP	Gas pipe, sports equipment, electrical housing, adhesives and coatings
Phenylene oxide [poly(phenylene oxide)]	PPO	Noryl	Crystalline $\left[-O-CH_2-CH_2-O-C(=O)-C_6H_4-\right]_n$	TP, EP	Auto trim, panels electrical housing, TV cabinets, pump parts
Polycarbonate	PC	Lexan Xenoy (blend)	Amorphous $\left[-O-C_6H_4-C(CH_3)_2-C_6H_4-O-C(=O)-\right]_n$	TP, EP	Optical lenses, bullet-resistant windows, housings, cookers
Polyester [poly(ethylene terephthalate)] [poly(butylene terephthalate)] (aromatic polyesters)	 PET (TS) PBT (TP)	Mylar Dacron Kodel Fortrel Laminac Selectron Gafite tp Ekonol	Amorphous or cross-linked network $\left[-O-CH_2-CH_2-O-C(=O)-C_6H_4-\right]_n$	TS, TP, GP, EP	Glass-fiber reinforcer, films, fibers

[a] TP, thermoplastic; TS, thermosetting; EP, engineering plastic; GP, general-purpose or special. ASTM, American Society for Testing and Materials.

[b] Engineering plastics.

TABLE A-9 (continued)

Common Name (Chemical Name)	ASTM Abbre-viations	Trade Names	Common Structure	Grouping[a]	Typical Uses
Polyimide	LaRC-ITP1 LaRC-CP1 UPILEX R_1	Kapton	Cross-linked network or amorphous 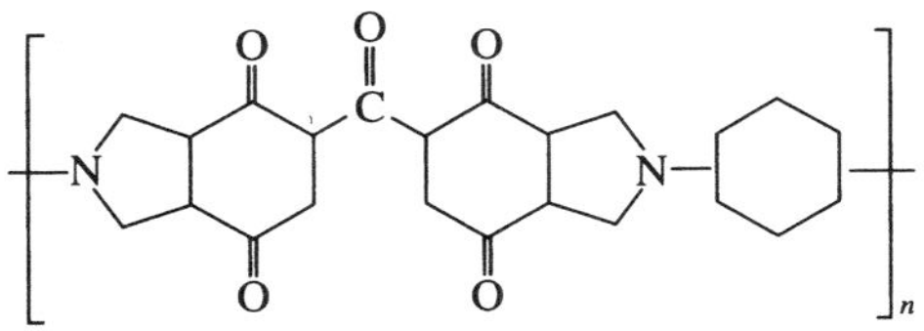	TS, TP, EP	Jet-engine vane bushings, seals, ball-bearing separators, high-temperature film, fiber matrix
Poly(amide/imide)		Torlon Tygon	Amorphous 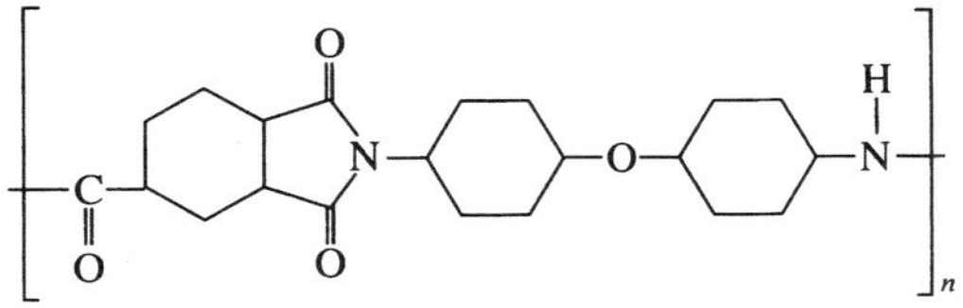	TP, EP	Engineering-plastic gears, structural components, bearings, seals, and valves
Polyolefins			Low to high crystallinity or cross-linked network	TP, TS, GP, EP	PE: packaging, squeeze bottles, electrical insulation and tubing
(polyethylene, low density)	LDPE	Alathon	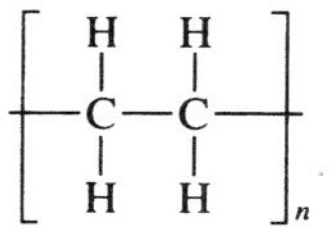 polyethylene		
(polyethylene, high density)[b]	HDPE	Dylan			
(polyethylene, ultra-high molecular weight)[b]	UHMWPE	Marlex			PP: packaging, auto-battery cases, housings, electrical components and fan blades
(polypropylene)	PP	Escon			
(polyallomer)		Tenite			
(ethylene–vinyl acetate)	EVA				
[poly(ethylene terephthalate)]	PET	Arnite Dacron Rynite	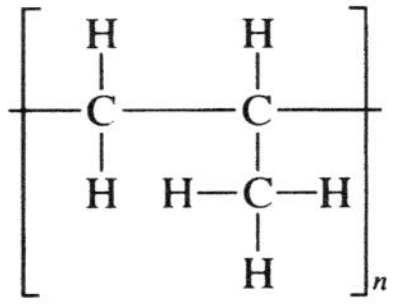		EVA: shoe soles and hypodermic syringes
(polybutylene)			polypropylene		
[poly(methyl pentene)]	TPX	Surlyn			
(ionomer)					

[a] TP, thermoplastic; TS, thermosetting; EP, engineering plastic; GP, general-purpose or special. ASTM, American Society for Testing and Materials.
[b] Engineering plastics.

TABLE A-9 (continued)

Common Name (Chemical Name)	ASTM Abbre-viations	Trade Names	Common Structure	Grouping[a]	Typical Uses
Poly(ether ether ketone)	PEEK	Victrex		EP	Crystalline polymer for high-temperatures and composite matrices
Poly(phenylene sulfide)	PPS	Ryton	Amorphous	TP, EP	Electrical terminal block and connectors, seals, gears
Polystyrene (styrene–acrylonitrile)	PS SAN	Styron Styrofoam Lustrex Dylite Tyril	Amorphous	TP, GP	Packaging, control knobs, TV cabinets, wood substitute, foam insulation
Polysulfone [poly(ether sulfone)] (polyphenylsulfone)		Udel Radel	Amorphous	TP, EP	Electrical insulators, auto-distributor caps, tubing, and aircraft cabin interiors
Polyurethane (isocyanate polyester or polyether)	PUR	Estane Flexane Texin Calspan	Amorphous or cross-linked network	TP, TS, EP	Reaction injection molded (RIM) foamed auto parts, solid tires, auto bumpers, synthetic leather

[a] TP, thermoplastic; TS, thermosetting; EP, engineering plastic; GP, general-purpose or special. ASTM, American Society for Testing and Materials.

[b] Engineering plastics.

TABLE A-9 (continued)

Common Name (Chemical Name)	ASTM Abbre-viations	Trade Names	Common Structure	Grouping[a]	Typical Uses
Silicones	SI	Silastic RTV, Silicone RTV	Amorphous $\left[-Si(CH_3)_2-O-\right]_n$	TS, GP	Room-temperature-vulcanizing (RTV) molds, fiber-matrix, electronic potting and encapsulating, heat seals
Vinyls		Vinylite	Cross-linked network	TS, TP, GP	Plastisol coating, upholstery, pipe, building trim, coatins, film
[poly(vinyl chloride)]	PVC	Naugahyde	$\left[-CH_2-CHCl-\right]_n$ PVC		
[poly(vinyl acetate)]	PVAc	Luxite	$\left[-CH_2-CH(C(=O)-CH_3)-\right]_n$ PVAC		
[poly(vinyl alcohol)]	PVA	Elvanol			
[poly(vinyl butyral)]	PVB	Butrar			
[poly(vinyl fluoride)]	PVF	Tedlar			
[poly(vinylidene chloride)]	PVDC	Saran			

[a]TP, thermoplastic; TS, thermosetting; EP, engineering plastic; GP, general-purpose or special. ASTM, American Society for Testing and Materials.
[b]Engineering plastics.

TABLE A-10 REPRESENTATIVE ELASTOMERS[a]

Abbreviation (ASTM):	NR	IR	SBR	IIR	BR	NBR
Common name:	Natural rubber		GRS or buna S	Butyl		Nitrile or buna N
Chemical name:	Natural polyisoprene	Isoprene Polyisoprene	Styrene Butadiene	Isobutene Isoprene	Polybutadiene Butadiene	Nitrile butadiene
(1) Tensile strength (kPa)	31	27.5	24	20.6	20.6	24
(2) Hardness Shore A	30–100	40–80	40–90	40–75	45–80	40–95
(3) Specific gravity	0.93	0.93	0.94	0.92	0.94	1.00
(4) Abrasion resistance	A	A	A	B	A	A
(5) Tear resistance	A	B	C	B	B	B
(6) Flexibility at low temperature	B	B	C	B	B	C
(7) Impact resistance	A	A	A	B	B	C
(8) Resiliency	A	A	B	C	A	B
(9) Creep	A	B	B	C	A	B
(10) ASTM/SAE type class	AA	AA	AA–BA	AA–BA	AA	BF, BG, PK, CH
(11) Maximum service temp. (°C)	70	70	100	100	70	100–125
(12) Heat-aging resistance	B–C	B–C	B	B–A	C	B
(13) Flame resistance	D	D	D	D	D	D
(14) Oil and gasoline resistance	X	X	X	X	C	A
(15) Oxidation resistance	C	B–C	C	C	C	C
(16) Ozone resistance	C	C–D	C–D	A	C–D	C–D
(17) Ultraviolet resistance	C–D	C–D	B–C	B	B–C	B–C
(18) Acid and base resistance	B–C	B–C	B–C	B	B–C	B
(19) Water absorption resistance	A	A	B	A	A	C
(20) Permeability to gases	C	C	C	A	C	B
(21) Electrical resistivity	A	A	A	A	A	D
(22) Adhesion to metals	A	A	A	B	A	A
(23) Trade names	—	Natsyn, Isoprene, Ameripol SN	K-Resin	Enjay Butyl, Petro-Tex Butyl	Diene, Ameripol CB	Paracril
(24) Typical uses	Tires, seals, bearings, couplings, shoe soles and heels	Same as natural rubber	Shock absorbers, belts, heels, sponges, gaskets, belts	Truck and auto tires, shock absorbers, inner tubes	Pneumatic tires, gaskets, seals, abrasion resistance belts	Gasoline, chemical and oil seals, gaskets and O-rings, belting

[a] A, excellent; B, good; C, fair; D, poor; X, not recommended.

TABLE A-10 (continued)

	ACM Acrylate Polyacrylate	CR Neoprene Chloropene	CSM Hypalom Chlorosulfanyl Polyethylene	FPM Fluorocarbon Fluorinated hydrocarbon	EPDM EPDM Ethylene propylene	SI Silicone rubber Polysilicone	UE Urethane rubber Polyether urethane	PTR Thikol Polysulfide
(1)	17	27.5	27.5	17	20.6	4–10	34–55	4–10
(2)	40–90	40–90	50–95	60–90	30–90	25–80	35–100	20–80
(3)	1.10	1.23	1.12–1.28	1.45	0.86	1.14–2.05	1.06	1.34
(4)	B	B	B	A	B	D	B	D
(5)	C–B	C–B	B	B	C	B	B	D
(6)	D	C	B	D	B	A	C	C
(7)	D	B	C–B	B	B	D–C	B–A	D
(8)	C–B	A	C	C	B	D–A	C–A	C
(9)	C	B	C	B	C–B	C–A	C–A	D
(10)	DF, DH	BC, BE	CE	HK	AA, DA, CA	FC, FE, FK	AD, EC	AK
(11)	125	100	125	250	125	200–225	100	70
(12)	A	A	C	A	A	A	B–A	C–B
(13)	D	B	C	A	D	A	D	D
(14)	A	C	C	A	X	D–C	B	A
(15)	A	A	A	A	A	A	A	B
(16)	B	B	A	A	A	A	A	A
(17)	B	B	B	A	A	A	B	B
(18)	C	B	B	A	A	D	D	X
(19)	A	B	A	A	A	A	A	B
(20)	B	B	B	A	C	D	B	D
(21)	B	C	B	B	B	A	B	C
(22)	B	A	A	C–B	C	A	C	C–B
(23)	Hycar, Acrylon	Neoprene, Perbunanc	Hypalon	Viton, Proflo, Fluorel	Nordel, Epcar, Royalene	Adiprene, RTV, Silastic	Kalrez, Estane, Roylar	Thikol
(24)	Oil hose, colored and white parts, pressure and oil O-rings	Belts, hose, extruded goods, molded sheet, adhesives, chemical-tank liners	Laminated roofing, tarpaulins, reservoir and pond liners, diaphragms, shoe soles and heels, whitewall tires	Brake seals, ducting connectors, carburetors, needle tips, roll coverings	Garden and industrial hoses, belts, bike tires, electrical-wire insulation, paintable auto bumpers	Industrial tires and rolls, mining belts, die pads, gaskets and seals	Chemical O-ring seals, valve seats, gaskets, nuclear, oil, gas, hydraulic, and acid seals	Gasoline hose, printing rolls, caulking, adhesives and binders

TABLE A-11 REPRESENTATIVE METALS

Material	Nominal Composition (Essential Elements) (%)	Form and Condition	Typical Mechanical Properties: Yield Strength (0.2% offset) (1000 psi)	Tensile Strength (1000 psi)	Elongation in 2 in. (%)	Hardness, Brinell or Rockwell
Copper CA 110	Cu 99.90 min	Strip annealed	10[a]	32	45	40 HRF
Sheet—ASTM B152						
Rod—ASTM B124, B133		Spring temper	50[a]	55	4	60 HRB
Wire—ASTM B1, B2, B3						
Commercial bronze CA 220	Cu 90	Strip annealed	10[a]	37	45	53 HRF
Plate, sheet, strip, bar—ASTM B36						
Wire—ASTM B 134	Zn 10	Spring temper	62[a]	72	3	78 HRB
Red brass CA 230	Cu 85	Strip annealed	15[a]	40	50	50
Strip, sheet, plate—ASTM B36						
Wire—ASTM B134	Zn 15	Hard temper	60[a]	75	7	135
Tube—ASTM B135						
Copper–nickel CA 715	Cu bal					
Sheet—ASTM B122	Ni 30	Tube annealed	25[a]	60	45	45 HRB
Plate—ASTM B171	Fe 0.55					
Tube—ASTM B111	Mn 0.5					
Aluminum alloy Alclad 2024	Core: 2024					
Sheet and plate—ASTM B209	Al bal	Sheet annealed	11	26	20	—
	Cu 4.5					
	Mn 0.6	Heat annealed	42	64	19	—
	Mg 1.5					
	Cladding:					
	Al 99.3 min					
Aluminum alloy 3003	Al bal	Sheet annealed	6	16	30	28
Sheet and plate—ASTM B209	Mn 1.2	Cold rolled	27	29	4	55
Aluminum alloy 5052	Al bal	Sheet annealed	13	28	25	47
Sheet and plate—ASTM	Mg 2.5					
B209	Cr 0.25	Cold rolled	37	42	7	77
Aluminum alloy 6061	Al bal					
Sheet and plate—ASTM B209	Si 0.6	Sheet annealed	8	18	25	30
	Cu 0.25					
	Mg 1.0	Heat treated	40	45	12	95
	Cr 0.25					
Aluminum alloy 707S	Al bal					
Bar, rod, wire, and shapes—ASTM B221	Zn 5.6	Bar annealed	15	33	16	60
	Cu 1.6					
	Mg 2.5	Heat treated	73	83	11	150
	Cr 0.3					
Cast alluminum alloy 13	Al bal	Die casting	21	43	2.5	—
Castings—ASTM B85 Grade S12A	Si 12.0	As cast				

Typical physical properties								
Density (lb/cu in)	Specific Gravity	Melting Point (°F)	Specific Heat (32 to 212°F) (Btu/lb/°F)	Thermal-expansion Coefficient (32 to 212°F) (in 10^{-6} in./in./°F)	Thermal Conductivity (32 to 212°F) (Btu/sq ft/hr/°F/in.)	Electrical Resistivity (68°F) (ohms/cir mil ft)	Tensile Modulus of Elasticity (10^6 psi)	Torsional Modulus of Elasticity ($\times$ 10^6 psi)
0.322	8.91	1980	0.092	9.4	2512	10.3	17	6.4
—	—	—	—	9.8	—	—	—	—
0.318	8.80	1910	0.09	10.2	1308	23.6	17	6.4
—	—	—	—	—	—	—	—	—
0.316	8.75	1880	0.09	10.4	1104	28	17	6.4
—	—	—	—	—	—	—	—	—
0.323	8.94	2260	0.09	9.0	204	225	22	8.3
0.100	2.77	1180	0.23	12.6	1340	21	10.6	3.75
—	—	—	—	—	840	35	10.6	4.0
0.099	2.73	1210	0.23	12.9	1340	21	10.0	3.75
—	—	—	—	—	1070	26	10.0	3.75
0.097	2.68	1200	0.23	13.2	960	30	10.2	3.75
—	—	—	—	—	960	30	10.2	3.75
0.098	2.70	1205	0.23	13.0	1190	23	10.0	3.75
—	—	—	—	—	1070	26	10.0	3.75
0.101	2.80	1175	0.23	12.9	—	—	10.4	3.9
—	—	—	—	—	840	35	10.4	3.9
0.096	2.65	1080	0.23	11.5	870	34	10.3	3.85

TABLE A-11 REPRESENTATIVE METALS

Material	Nominal Composition (Essential Elements) (%)		Form and Condition	Typical Mechanical Properties: Yield Strength (0.2% offset) (1000 psi)	Tensile Strength (1000 psi)	Elongation in 2 in. (%)	Hardness, Brinell or Rockwell
Magnesium alloy AZ 31B	Mg	bal	Sheet annealed	22	37	21	56
Plate and sheet—ASTM	Al	3.0					
B90	Zn	1.0	Hard sheet	32	42	15	73
	Mn	0.2 min					
Magnesium alloy AZ 80A	Mg	bal	As forged	33	48	11	69
Forgings—ASTM B91	Al	8.5					
	Zn	0.5	Forged and aged	36	50	6	72
	Mn	0.15 min					
Magnesium alloys AZ 91A and AZ 91B	Mg	bal					
Castings—ASTM B94	Al	9.0	Die cast	22	33	3	63
	Zn	0.7					
	Mn	0.2 min					
Titanium Ti-35A	Ti	bal					
Forgings—ASTM B381	C	0.08 max					
Sheet, strip, and plate—ASTM B265	Fe	0.12 max	Sheet annealed	30	40	30	135
Pipe—ASTM B337	N_2	0.05 max					
Tubes—ASTM B338	H_2	0.015 max					
Bars—ASTM B348							
Ti-6 Al-4 V alloy	Ti	bal					
Sheet, strip, and plate—ASTM B265	Al	6.5	Sheet annealed	130	140	13	39 HRC
Bar—ASTM B348	V	4					
Forgings—ASTM B381	C	0.08 max	Heat treated	165	175	12	—
	Fe	0.25 max					
	N_2	0.05 max					
	H_2	0.015 max					
Nickel 211 ASTM F290	Ni	95.0					
	Mn	4.75	Annealed	35	75	40	140
	C	0.10					
Nickel (cast)	Ni	95.6					
	Cu	0.5					
	Fe	0.5	As cast	25[a]	57	22	110
	Mn	0.8					
	Si	1.5					
	C	0.8					
Duranickel alloy 301	Ni	bal					
	Al	4.5					
	Si	0.55	Hot rolled and aged	132	185	28	330
	Ti	0.5					
	Mn	0.25					
	Fe	0.15					
	C	0.15					

Typical Physical Properties								
Density (lb/cu in)	Specific Gravity	Melting Point (°F)	Specific Heat (32 to 212°F) (Btu/lb/°F)	Thermal-expansion Coefficient (32 to 212°F) (in 10^{-6} in./in./°F)	Thermal Conductivity (32 to 212°F) (Btu/sq ft/hr/°F/in.)	Electrical Resistivity (68°F) (ohms/cir mil ft)	Tensile Modulus of Elasticity (10^6 psi)	Torsional Modulus of Elasticity ($\times 10^6$ psi)
0.064	1.77	1170	0.245	14.5	657	55	6.5	2.4
—	—	—	—	—	—	—	—	—
0.065	1.80	1130	0.25	14.5	522	87	6.5	2.4
—	—	—	—	—	—	—	—	—
0.065	1.80	1105	0.25	14.5	493	102	6.5	2.4
0.163	4.50	3063	0.124	4.8	108	336	14.9	6.5
0.160	4.42	3000	0.135	4.9	50	1026	16.5	6.1
—	—	—	—	—	—	—	—	—
0.315	8.73	2600	0.11	7.4	306	102	30	11
0.301	8.34	2550	0.13	8.85	410	125	21.5	—
0.298	8.75	2620	0.104	7.2	165	255	30	11

TABLE A-11 REPRESENTATIVE METALS

Material	Nominal Composition (Essential Elements) (%)	Form and Condition	Typical Mechanical Properties			
			Yield Strength (0.2% offset) (1000 psi)	Tensile Strength (1000 psi)	Elongation in 2 in. (%)	Hardness, Brinell or Rockwell
Monel alloy 400	Ni bal					
Rod and Bar—ASTM B164	Cu 31.5					
Plate, sheet, and strip—ASTM B127	Fe 1.35	Rod hot-rolled	30	79	48	125
	Mn 0.90	annealed				
Tube—ASTM B165	Si 0.15					
	C 0.12					
Inconel Alloy 600 plate	Ni bal					
Sheet and strip—ASTM B168	Cr 15.8					
Rod and bar—ASTM B166	Fe 7.2	Rod hot-rolled	36	90	47	150
Pipe and tube—ASTM B163 and B167	Mn 0.2	annealed				
	Si 0.2					
	C 0.04					
	Cu 0.10					
Hastelloy Alloy W	Cr 5					
Wire—AMS 5786	Mo 24.5	Sheet annealed	53	123	55	—
Bar and forgings—AMS 5755A	Fe 5.5					
	Ni bal					
Ingot iron		Hot rolled	29	45	26	90
	Fe 99.9 plus					
		Annealed	19	38	45	67
Wrought iron	Fe bal	Hot rolled	30	48	30	100
Forgings—ASTM A73	Slag 2.5					
Carbon steel—SAE 1020	Fe bal	Annealed	38	65	30	130
ASTM A285	C 0.20					
	Mn 0.45	Quenched and				
	Si 0.25	tempered at 1000° F	62	90	25	179
300 M alloy steel	Fe bal					
Bar and forgings—AMS 6416	Mn 0.80					
	Si 1.6					
	Ni 1.85	Hardened	240	290	10	535
	Cr 0.85					
	Mo 0.38					
	V 0.08					
	C 0.43					
Cast gray iron	C 3.4					
ASTM A48 class 30	Si 1.8	As cast	—	32	—	190
	Mn 0.8					
	Fe bal					
Malleable iron	C 2.5					
Castings—ASTM A47	Si 1	Annealed	33	52	12	130
	Mn 0.55 max					
	Fe bal					

Density (lb/cu in)	Specific Gravity	Melting Point (°F)	Specific Heat (32 to 212°F) (Btu/lb/°F)	Thermal-expansion Coefficient (32 to 212°F) (in 10^{-6} in./in./°F)	Thermal Conductivity (32 to 212°F) (Btu/sq ft/hr/°F/in.)	Electrical Resistivity (68°F) (ohms/cir mil ft)	Tensile Modulus of Elasticity (10^6 psi)	Torsional Modulus of Elasticity ($\times 10^6$ psi)
Typical physical properties								
0.319	8.84	2460	0.102	7.7	151	307	26	9.5
0.304	8.43	2600	0.106	7.4	103	620	31	11
0.325	9.03	2400	—	6.3	—	—	—	—
0.284	7.86	2795	0.108	6.8	490	57	30.1	11.8
—	—	—	—	—	—	—	—	—
0.278	7.70	2750	0.11	6.35	418	70	29	—
0.284	7.86	2760	0.107	6.7	360	60	30	—
—	—	—	—	—	—	—	—	—
0.283	7.84	2740	0.107	6.7	360	60	30	11.6
0.260	7.20	2150	—	6.7	310	400	14	—
0.264	7.32	2250	0.122	6.6	—	180	25	—

TABLE A-11 REPRESENTATIVE METALS

Material	Nominal Composition (Essential Elements) (%)		Form and Condition	Typical Mechanical Properties: Yield Strength (0.2% offset) (1000 psi)	Tensile Strength (1000 psi)	Elongation in 2 in. (%)	Hardness, Brinell or Rockwell
Ductile iron	Fe	bal					
(Nickle containing)	C	3.6					
Grade 60-40-18	Si	2.3	Annealed	47	65	24	160
castings—ASTM	Mn	0.5					
A536							
	Ni	0.75					
Ductile iron	Fe	bal					
(Nickel containing)	C	3.6					
Grade 120-90-02	Si	2.3	Oil quenched	120	140	4	325
castings—ASTM	Mn	0.5	and tempered				
A536							
	Ni	0.75					
Type 201 stainless steel	Fe	bal					
(UNS 20100)	Cr	17					
Plate, sheet, and strip—	Ni	4.5	Strip annealed	55	115	60	90 HRB
ASTM A412	Mn	6.5					
Bar—ASTM A429	N_2	0.25 max					
	C	0.15 max					
Type 302 stainless steel	Fe	bal					85 HRB
(UNS 30200)	Cr	18	Sheet annealed	40	90	50	up to
	Ni	9		up to	up to		40 HRC
Plate, sheet, and strip—	C	0.15 max	Cold rolled	165	190	5	
ASTM A167							
and A 240							
Bar—ASTM A276 and							
A314							
Wire—ASTM A313							
Forgings—ASTM A473							
Type 303 and 303 Se stain-	Fe	bal					
less steel (UNS 30323)	Cr	18					
Bar—ASTM A276 and	Ni	9	Bar annealed	35	90	50	160
A314	{ S	0.15 min or					
Forgings—ASTM A473	{ Se	0.15 min					
	C	0.15 max					
Type 314 stainless steel	Fe	bal					
(UNS 31400)	Cr	25					
Bar—ASTM A276 and	Ni	20	Bar annealed	50	100	45	180
A314	Si	2.50					
	C	0.25 max					
Type 405 stainless steel	Fe	bal					
(UNS 40500)	Cr	12.5	Sheet annealed	40	65	25	75 HRB
	C	0.08 max					
Plate, sheet, and strip—	Al	0.20					
ASTM A176 and							
A240							
Tube—ASTM A268							
Bar—ASTM A276 and							
A314							

Source: Courtesy of Inco.

Typical Physical Properties								
Density (lb/cu in)	Specific Gravity	Melting Point (°F)	Specific Heat (32 to 212°F) (Btu/lb/°F)	Thermal-expansion Coefficient (32 to 212°F) (in 10^{-6} in./in./°F)	Thermal Conductivity (32 to 212°F) (Btu/sq ft/hr/°F/in.)	Electrical Resistivity (68°F) (ohms/cir mil ft)	Tensile Modulus of Elasticity (10^6 psi)	Torsional Modulus of Elasticity ($\times 10^6$ psi)
0.250	7.1	2150	—	6.2	276	399	24.5	9.3
0.252	7.2	2150	—	5.9	218	408	24.5	9.3
0.283	7.86	—	—	8.7	—	423	28.6	—
0.29	7.9	2590	0.12	9.6	113	432	28	12.5
—	—	—	—	—	—	—	—	—
0.29	—	2590	0.12	9.6	113	432	28	—
0.279	—	—	0.12	8.4	121	462	29	—
0.28	7.7	2790	0.11	6.0	—	360	29	—

TABLE A-12 REPRESENTATIVE ULTRAHIGH-STRENGTH STEELS[a] (Courtesy Advanced Materials & Processes)

Steel and Composition[a] (%)	Tempering Temperature [°C (°F)]	Tensile Strength [MPa (10^3 psi)]	Yield Strength [MPa (10^3 psi)]	Elongation in 50 mm (2 in.)(%)	Reduction in Area (%)	Hardness[c]	Impact Energy[d] [J (ft · lbf)]
Medium-carbon low-alloy steels							
AISI/SAE 4130 (UNS G41300):[e]	205 (400)	1765 (256)	1520 (220)	10	33	475 HB	18 (13) I
0.28–0.33 C, 0.4–0.6 Mn,	425 (800)	1380 (200)	1170 (170)	16.5	49	375 HB	34 (25) I
0.2–0.35 Si, 0.8–1.1 Cr 0.15–0.25 Mo	650 (1200)	965 (140)	830 (120)	22	63	270 HB	135 (100) I
AISI/SAE 4140 (UNS G41400):[f]	205 (400)	1965 (285)	1740 (252)	11	42	578 HB	15 (11) I
0.38–0.43 C, 0.75–1 Mn,	425 (800)	1450 (210)	1340 (195)	15	50	429 HB	28 (21) I
0.2–0.35 Si, 0.8–1.1 Cr, 0.15–0.25 Mo	650 (1200)	900 (130)	790 (114)	21	61	277 HB	112 (83) I
AISI/SAE 4340 (UNS G43400):[h]	205 (400)	1980 (287)	1860 (270)	11	39	53 HRC	20 (15) I
0.38–0.43 C, 0.6–0.8 Mn,	425 (800)	1500 (217)	1365 (198)	14	48	46 HRC	16 (12) I
0.2–0.35 Si, 0.7–0.9 Cr, 1.65–2 Ni, 0.2–0.3 Mo	650 (1200)	1020 (148)	860 (125)	20	60	31 HRC	100 (74) I
Alloy 300M (UNS K44220):[i]	205 (400)	2140 (310)	1650 (240)	7	27	54.5 HRC	21.7 (16) CVN
0.4–0.46 C, 0.65–0.9 Mn, 1.45–1.8 Si, 0.7–0.95 Cr, 1.65–2 Ni, 0.3–0.45 Mo, 0.05 V min	425 (800)	1790 (260)	1480 (215)	8.5	23	45.5 HRC	13.6 (10) CVN

AISI/SAE 6150 (UNS G61500):[k]	205 (400)	2050 (298)	1810 (263)	1	5	610 HB	—
0.48–0.53 C, 0.7–0.9 Mn,	425 (800)	1585 (230)	1490 (216)	11	42	470 HB	14 (10) I
0.2–0.35 Si, 0.8–1.1 Cr, 0.15–0.25 V	595 (1100)	1150 (167)	1080 (157)	16	47	350 HB	28 (21) I
AISI/SAE 8640 (UNS G86400):[l]	205 (400)	1810 (263)	1670 (242)	8	25.8	55 HRC	11.5 (8.5) I
0.38–0.43 C, 0.75–1 Mn,	425 (800)	1380 (200)	1230 (179)	10.5	46.3	44 HRC	27.8 (20.5) I
0.2–0.35 Si, 0.4–0.6 Cr, 0.4–0.7 Ni, 0.15–0.25 Mo	650 (1200)	870 (126)	760 (110)	20.5	61	28 HRC	96.9 (71.5) I
Medium-alloy air-hardening steels							
H11 Mod (UNS T20811 Mod):[m]	540 (1000)	2010 (291)	1675 (243)	9.6	30.6	56 HRC	21 (15.5) CVN
0.37–0.43 C, 0.2–0.4 Mn, 0.8–1 Si, 4.75–5.25 Cr, 1.2–1.4 Mo, 0.4–0.6 V	650 (1200)	1060 (154)	850 (124)	14.1	41.2	33 HRC	40 (29.5) CVN
AISI/SAE H13 (UNS T20813):[n]	525 (980)	1960 (284)	1570 (228)	13[o]	46.2	52 HRC	16 (12) CVN
0.32–0.45 C, 0.2–0.5 Mn, 0.8–1.2 Si, 4.75–5.5 Cr, 1.1–1.75 Mo, 0.8–1.2 V	605 (1120)	1495 (217)	1290 (187)	15.4[o]	54	44 HRC	30 (22) CVN

TABLE A-12 REPRESENTATIVE ULTRAHIGH-STRENGTH STEELS

Steel and Composition[a] (%)	Tempering Temperature [°C (°F)]	Tensile Strength [MPa (10^3 psi)]	Yield Strength [MPa (10^3 psi)]	Elongation in 50 mm (2 in.)(%)	Reduction in Area (%)	Hardness[c]	Impact Energy[d] [J (ft · lbf)]
						Fracture toughness, [K_{Ic}, MPa · $m^{1/2}$ (ksi · in.$^{1/2}$)]	
High-fracture-toughness steels							
HP 9-4-30 (UNS K91283):[p] 0.29–0.34 C, 0.10–0.35 Mn, 0.2 Si max, 0.9–1.1 Cr, 7–8 Ni,	205 (400)[q]	1650–1790 (240–260)	1380–1450 (200–210)	8–12[o]	25–35	66–99 (60–90)	20–27 (15–20) CVN
0.9–1.1 Mn, 0.06–0.12 V, 4.25–4.75 Co	550 (1025)[r]	1520–1650 (220–240)	1310–1380 (190–200)	12–16[o]	35–50	99–115 (90–105)	24–34 (18–25) CVN
	Quenchant						
AF1410 (UNS K92571):[s]	Air	1680 (244)	1475 (214)	16	69	174 (158)	69 (51) CVN
0.13–0.17 C, 0.1 Mn max,	Oil	1750 (254)	1545 (224)	16	69	154 (140)	65 (48) CVN
0.1 Si max, 1.8–2.2 Cr, 9.5–10.5 Ni, 0.9–1.1 Mo, 13.5–14.5 Co	Water	1710 (248)	1570 (228)	16	70	160 (146)	65 (48) CVN

Source: Bruce A. Becherer and Thomas J. Witheford. "Heat Treating of Ultrahigh-Strength Steels," *ASM Handbook,* 10th ed., Vol. 4, *Heat Treating,* ASM International, Materials Park, Ohio, 1991, pp. 207–218.

[a]The ultrahigh-strength-steel category (1380 MPa, 200 × 10^3 psi, minimum yield strength) also includes the 18% Ni maraging steels and a variety of stainless-type steels.

[b]Phosphorus and sulfur contents may vary with steelmaking practice, but usually do not exceed 0.035% P and 0.04% S.

[c]HB, Brinell; HRC, Rockwell C.

[d]I, Izod; CVN, Charpy V-notch.

[e]Round bars, 25-mm (1 in.) diameter; water quenched from 845 to 870°C (1550 to 1600°F).

[f]Round bars, 13-mm (0.5 in.) diameter; oil quenched from 845°C (1550°F).

[g]EAF-VAR plate, 25-mm (1 in.) thick; oil quenched from 870°C (1600°F), tempered 2 + 2 hr.

[h]Oil quenched from 845°C (1550°F).

[i]Round bars, 25-mm (1 in.) diameter; oil quenched from 860°C (1575°F).

[j]Bars normalized at 900°C (1650°F), oil quenched from 845°C (1550°F), D-6A was developed by Ladish Co. Inc., Cudahy, Wis.

[k]Round bars, 14-mm (0.55 in.) diameter; normalized at 870°C (1600°F), oil quenched from 860°C (1575°F).

[l]Round bars, 13.5-mm (0.53 in.) diameter; oil quenched from 830°C (1525°F).

[m]Longitudinal properties. Heat treatment: air cooled from 1010°C (1850°F), double tempered 2 + 2 hr.

[n]Longitudinal properties of round bars. Heat treatment: oil quenched from 1010°C (1850°F) double tempered 2 + 2 hr.

[o]Elongation in 4D.

[p]Oil quenched from 845°C (1550°F), refrigerated to −73°C (−100°F), double tempered, HP 9-4-30 was developed by Republic Steel Corp. (now LTV Steel Co.), Cleveland, Ohio.

[q]Hardness, 49–53 HRC.

[r]Hardness, 44–48 HRC.

[s]VIM/VAR plate, 50-mm (2 in.) thick. Heat treatment: 675° C (1250°F) for 8 hr, air cooled; 900°C (1650°F) for 1 hr, quenched; 830°C (1525°F) for 1 hr, quenched; refrigerated at −73°C (−100°F) for 1 hr; 510°C (950°F) for 1 hr, air cooled.

TABLE A-13 PROPERTIES OF SELECTED CERAMICS

	Density (lb/in^3) (kg/m^3)	Hardness (M, Mohs) (K, Knoop)	Tensile Strength (psi) (MPa)	Thermal Conductivity ($Btu \cdot in./hr \cdot ft^2 \cdot °F$) ($W/m \cdot K$)		Coefficient of Thermal Expansion (10^{-6} F^{-1}) (10^{-6} K^{-1})
Alumina	0.14	M, 9	25,000	25°C	192–255	77–1830°F 4.3
(Al_2O_3)	3.8	K, 2500	172		27.7–36.7	298–1272 K 8.1
Beryllia	0.11	M, 9	23,000	25°C	1741	68–2550°F 5.28
(BeO)	2.92	K, 2000	159		250	293–1672 K 9.5
Boron carbide	0.087	M, 9	22,500	70°F	104–197	0–2550°F 1.73
(B_4C)	2.41	K, 2800	155		—	255–1672 K 3.1
Boron nitride	0.076	—	3,500	70°F	100–200	70–1800°F 4.17
(BN)	2.10		24.1			294–1255 K 7.5
Cordierite	0.065	M, 6.5	3,500	25°C	12–22	68–212°F 2.08
($2MgO \cdot 2Al_2O_3 \cdot 5SiO_2$)	1.8		24.1		1.7–3.2	293–373 K 3.7
Silicon carbide	0.11	M, 9	24,000	70°F	101	0–2552°F 2.4
(SiC)	3.17	K, 2500	165		—	225–1672 K 4.3
Steatite	0.09	M, 7.5	8,500	25°C	20–41	68–212°F 3.99
($MgO \cdot SiO_2$)	2.7	K, 1500	60		2.9–5.9	293–373 K 7.2
Zircon	0.13	M, 8	12,000	25°C	4.9–6.2	68–212°F 1.84
($ZrO_2 \cdot SiO_2$)	3.7		82.7			293–212 K 3.3

Source: Data adapted from *Machine Design,* '84 Materials Reference Issue 3M, and Ceramics Bulletin No. 757, *Materials Engineering,* '84 Materials Selection Issue.

Dielectric Constant at 10^6 Hz (Except as Noted)	Volume Resistivity	Compressive Strength (psi) (MPa)	Flexural Strength (psi) (MPa) 70°F	Flexural Strength (psi) (MPa) 2,250°F	Impact Strength (in. · lb) (N · m)	Modulus of Elasticity (psi × 10^6) (GPa)	Safe Service Temperature (°F) (°C)
8.0–10.0	$>10^{20}$	340,000	48,000	31,000	6.5	50	3540
		2,344	311	214	0.73	379	1965
6.4–7.0	$>10^{14}$	260,000	33,000	—	—	47	4350
		1,793	228	—	—	324	2414
	—	420,000	44,000	—	—	65	1100
		2,896	303	—	—	448	611
4.1–4.8	—	45,000	—	—	—	7	3000
		310	—	—	—	48	1665
4.02–6.23	$>10^{20}$	40,000	8,000	—	2.5	7	2282
		276	55	—	0.28	48	1250
—	—	100,000	110,000	80,000	—	62	3200
		690	758	552	—	427	1776
5.9–6.3							
6 × 10^1 cps	$>10^{20}$	90,000	19,000	—	5.0	15	1832
(6 × 10^1 Hz)		620	131	—	0.56	103	1016
8.0–10.0	$>10^{20}$	100,000	22,000	—	5.5	23	2012
		690	152	—	0.62	159	1117

TABLE A-14 REPRESENTATIVE GLASSES

U.S. Customary																
				Corrosion Resistance			Thermal Expansion Multiply by 10^{-7} in./in./°F		Upper Working Temperatures (Mechanical Considerations Only)				Thermal Shock Resistance Plates 6 × 6 in.		
									Annealed		Tempered		Annealed		
Type	Color	Principal Use	Class	Weathering	Water	Acid	32 to 572°F	77°F to Setting Point	Normal Service (°F)	Extreme Service (°F)	Normal Service (°F)	Extreme Service (°F)	⅛ in. Thick (°F)	¼ in. Thick (°F)	½ in. Thick (°F)
(1) Soda lime	Clear	Lamp bulbs	I	3	2	2	52	58.3	230	860	428	482	149	122	95
(2) Potash soda lead	Clear	Lamp tubing	I	2	2	2	49.8	53.9	230	716	—	—	—	—	—
(3) Aluminosilicate ..	Clear	Electron tube	I	1	1	3	25.6	30	392	1202	752	842	257	212	158
(4) Borosilicate	Clear	Sealed-beam lamps	I	[3]1	[3]2	[3]2	20.4	21.2	446	860	500	500	320	266	294
(5) Borosilicate	Clear	General	I	2	2	2	18.9	20.6	446	842	482	482	320	266	194
(6) 96% Silica	Clear	High temp.	I	1	1	1	4.2	3.1*	1652	2192	—	—	—	—	—
(7) Fused silica	Clear	Optical	I	1	1	1	3.1	1.9*	1652	2012	—	—	—	—	—
(8) Glass-ceramic ...	White	Missle nose cones	II	—	1	4	31.7	—	1292	—	—	—	572	338	266

Source: Courtesy of Corning Glass Works.

TABLE A-14 (continued)

	Thermal Stress Resistance °F	Viscosity Data				Knoop Hardness, KHN_{100}	Density (lb/ft³)	Young's Modulus (Multiply by 10^6 psi)	Poisson's Ratio	Log_{10} of volume resistivity ($\Omega \cdot cm$)			Dielectric properties at 1 MHz, 68°F			Refractive Index
		Strain Point (°F)	Annealing Point (°F)	Softening Point (°F)	Working Point (°F)					77°F	482°F	662°F	Power Factor (%)	Dielectric Constant	Loss Factor (%)	
(1)	29	883	957	1285	1841	465	154	10.2	0.22	12.4	6.4	5.1	0.9	7.2	6.5	1.512
(2)	36	743	815	1166	1805	382	190.3	8.6	0.22	17+	10.1	8.0	0.12	6.7	0.8	1.560
(3)	47	1229	1310	1666	2134	514	164.7	12.5	0.24	17+	13.5	11.3	0.16	6.3	1.0	1.547
(4)	86	932	1011	1436	2133	—	140.4	9.3	0.19	18	8.1	6.6	0.45	4.85	2.18	1.476
(5)	94	892	973	1436	2188	442	139.8	9.0	0.20	17	9.4	7.7	0.18	4.5	0.79	1.473
(6)	396	1634	1868	2786	—	487	136	9.8	0.19	17+	9.7	8.1	0.04	3.8	0.15	1.458
(7)	515	1753	1983	2876	—	489	137.2	10.5	0.16	17+	11.8	10.2	0.001	3.8	0.0038	1.459
(8)	29	—	—	—	—	657	162.2	17.2	0.24	16.7	10.0	8.7	0.30	5.6	1.7	—

(Corning Glass Works)

TABLE A-14 (continued)

	Thermal Stress Resistance °C	Knoop Hardness, KHN_{100}	Density (g/cm^3)	Young's Modulus (Multiply by 10^3 kg/mm^2)	Poisson's Ratio	Log_{10} of Volume Resistivity $(\Omega \cdot cm)$			Dielectric Properties at 1 MHz, 20°C			Refractive Index
						25°C	250°C	350°F	Power Factor (%)	Dielectric Constant	Loss Factor (%)	
(1)	16	465	2.47	7.1	0.22	12.4	6.4	5.1	0.9	7.2	6.5	1.512
(2)	20	382	3.05	6.0	0.22	17+	10.1	8.0	0.12	6.7	0.8	1.560
(3)	26	514	2.64	8.8	0.24	17+	13.5	11.3	0.16	6.3	1.0	1.547
(4)	21	—	5.42	5.6	0.28	17+	10.6	8.7	0.22	15	3.3	1.86
(5)	52	442	2.24	6.3	0.20	17	9.4	7.7	0.18	4.5	0.79	1.473
(6)	220	487	2.18	6.9	0.19	17+	9.7	8.1	0.04	3.8	0.15	1.458
(7)	286	489	2.20	7.4	0.16	17+	11.8	10.2	0.001	3.8	0.0038	1.459
(8)	16	657	2.6	12	0.24	16.7	10.0	8.7	0.30	5.6	1.7	—

TABLE A-14 (continued)

	Thermal Stress Resistance °C	Knoop Hardness, KHN_{100}	Density (g/cm^3)	Young's Modulus (Multiply by 10^3 kg/mm^2)	Poisson's Ratio	Log_{10} of Volume Resistivity ($\Omega \cdot cm$)			Dielectric Properties at 1 MHz, 20°C			Refractive Index
						25°C	250°C	350°F	Power Factor (%)	Dielectric Constant	Loss Factor (%)	
(1)	16	465	2.47	7.1	0.22	12.4	6.4	5.1	0.9	7.2	6.5	1.512
(2)	20	382	3.05	6.0	0.22	17+	10.1	8.0	0.12	6.7	0.8	1.560
(3)	26	514	2.64	8.8	0.24	17+	13.5	11.3	0.16	6.3	1.0	1.547
(4)	21	—	5.42	5.6	0.28	17+	10.6	8.7	0.22	15	3.3	1.86
(5)	52	442	2.24	6.3	0.20	17	9.4	7.7	0.18	4.5	0.79	1.473
(6)	220	487	2.18	6.9	0.19	17+	9.7	8.1	0.04	3.8	0.15	1.458
(7)	286	489	2.20	7.4	0.16	17+	11.8	10.2	0.001	3.8	0.0038	1.459
(8)	16	657	2.6	12	0.24	16.7	10.0	8.7	0.30	5.6	1.7	—

TABLE A-15
HARDNESS/TENSILE-STRENGTH CONVERSION

Hardness conversion chart for hardenable carbon and alloy steels: approximate relationship between hardnesses and tensile-strength. Conversions from one scale to another are made at the intercepts with the curve crossing the chart. For example, follow the horizontal line representing 400 Diamond Pyramid Hardness to its intersection with the conversion curve. From this point follow vertically downward for equivalent Rockwell C hardness (41), horizontally to the right for Brinell hardness (379) and tensile strength (187,000 psi), and vertically upward for the equivalent Shore hardness (55).

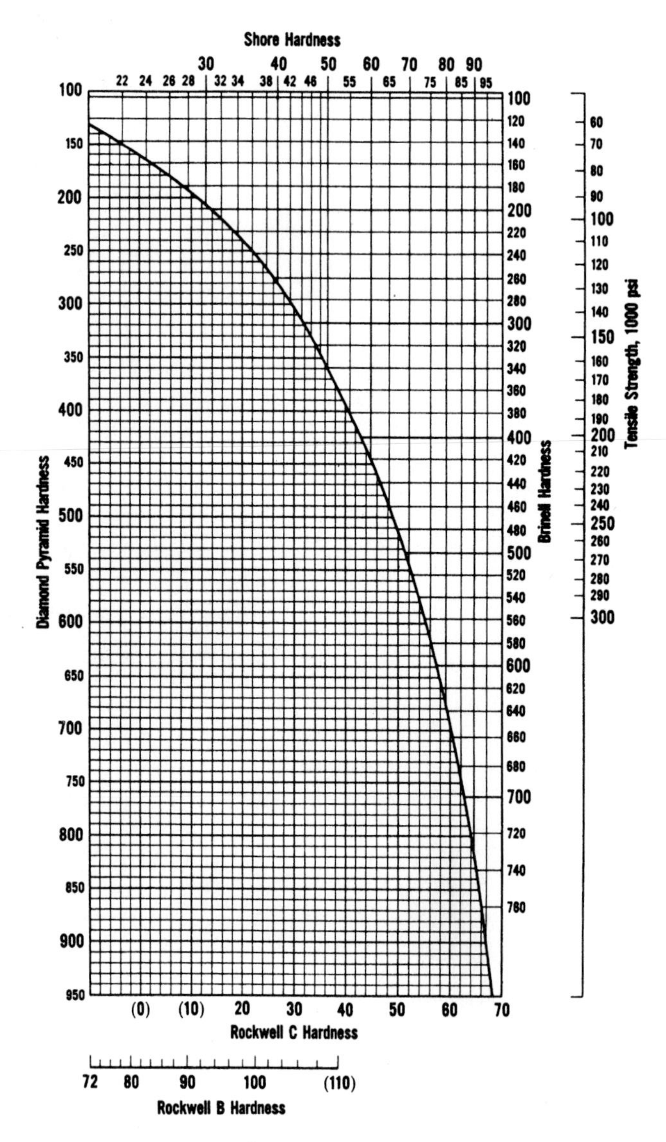

Source: Data from 1966 SAE handbook.

Index

Page numbers on which definitions appear are printed in **boldface.** C denotes page in color section.

Index of Principles/Equations and Illustrative Problems

TABLE A-7 CONVERSIONS

Quantity	U.S. customary to SI
Acceleration, linear	1 ft/s^2 = 3.048 × 10^{-1} m/s^2
Area	1 in^2 = 6.452 × 10^2 mm^2
	1 ft^2 = 9.290 × 10^{-2} m^2
Density (mass), ρ	1 lb/in^3 = 2.768 × 10^4 kg/m^3
	1 lb/ft^3 = 1.602 × 10^1 kg/m^3
Electric current, I	1 ampere (A) = 1 C/s
Energy, work	1 Btu = 1.055 kJ
	1 in. · lb = 1.129 × 10^{-1} J
	1 ft · lb = 1.356 J
	1 Btu = 0.293 W · h
Force	1 lbf = 4.448 N
	1 kgf = 9.807 N
Impulse	1 lb · s = 4.448 N · s
Length	1 Å = 1 × 10^{-1} nm
	1 microinch = 2.540 × 10^{-2} μm
	1 mil = 2.540 × 10^1 μm
	1 in. = 2.540 × 10^1 mm
	1 ft = 3.048 × 10^{-1} m
Magnetic field strength, H	1 oersted (Oe) = 79.58 ampere turns per meter
Magnetic flux	1 maxwell = 10^{-8} Wb
Magnetic flux density, B	1 Wb/M^2 = 1 T (tesla)
Modulus of elasticity, E	1 lb/in^2 = 6.895 × 10^{-6} GPa
Moment of force, torque	1 lb · in. = 1.130 × 10^{-1} N · m
	1 lb · ft = 1.356 N · m
Moment of inertia, I (of area)	1 in^4 = 4.162 × 10^5 mm^4
Momentum, linear	1 lb · ft/s = 1.383 × 10^{-1} kg · m/s
Power	1 Btu/min = 1.758 × 10^{-2} kW
	1 ft · lb/min = 2.259 × 10^{-2} W
	1 hp = 7.457 × 10^{-1} kW
Stress (pressure)	1 lb/in^2 = 6.895 × 10^{-3} MPa
	1 ksi = 6.895 MPa
Temperature	1°F (difference) = 0.555°C
	1.8 °F = 1°C (difference)
Thermal expansion, linear coefficient, α	in./in./°F = 1.8 K^{-1}, K = °C + 273.15
Thermal conductivity, k	1 Btu/ft · hr · °F = 1.729 W/m · K
Velocity, linear, v	1 in./s = 2.540 × 10^1 mm/s
	1 ft/s = 3.048 × 10^{-1} m/s
	1 in./min = 4.233 × 10^{-1} mm/s
	1 ft/min = 5.080 × 10^{-3} m/s
Velocity, angular, ω	1 rev/min = 1.047 × 10^{-1} rad/s
Viscosity	1 poise (P) = 0.1 Pa · s
Volume	1 in^3 = 1.639 × 10^4 mm^3
	1 ft^3 = 2.832 × 10^{-2} m^3
	1 yd^3 = 7.646 × 10^{-1} m^3

1 joule (J) = 10^7 ergs = 0.625 × 10^{19} eV

1 gauss (G) = 10^{-4} Wb/m^2 = 1 T

1 weber (Wb) = 1 T/m^2

1 Å = 10^{-10} m (obsolete)

1 micrometer (micron) = 10^{-6} m (obsolete)